Principles of Instrumental Analysis

SIXTH EDITION

Principles of Instrumental Analysis

Douglas A. Skoog
Stanford University

F. James Holler
University of Kentucky

Stanley R. Crouch
Michigan State University

THOMSON

BROOKS/COLE

Australia • Brazil • Canada • Mexico • Singapore • Spain
United Kingdom • United States

BROOKS/COLE
CENGAGE Learning™

**Principles of Instrumental Analysis,
Sixth Edition**
Douglas A. Skoog, F. James Holler and
Stanley R. Crouch

Publisher: David Harris

Development Editor: Sandra Kiselica

Editorial Assistant: Lauren Oliveira

Technology Project Manager: Lisa Weber

Marketing Manager: Amee Mosley

Project Manager, Editorial Production:
 Belinda Krohmer

Creative Director: Rob Hugel

Art Director: John Walker

Print Buyer: Rebecca Cross

Permissions Editor: Roberta Broyer

Production Service: Newgen–Austin

Text Designer: Ellen Pettengell

Photo Researcher: Terri Wright

Copy Editor: Mary Ann Short

Cover Designer: Irene Morris

Cover Image: Photo Researchers, Inc.

Compositor: Newgen

For product information and technology assistance, contact us at
Cengage Learning Customer & Sales Support, 1-800-354-9706
For permission to use material from this text or product,
submit all requests online at **www.cengage.com/permissions**
Further permissions questions can be e-mailed to
permissionrequest@cengage.com

Library of Congress Control Number: 2006926952

ISBN-13: 978-0-495-12570-9

ISBN-10: 0-495-12570-9

Brooks/Cole Cengage Learning
20 Davis Drive
Belmont, CA 94002-3098
USA

Cengage Learning is a leading provider of customized learning solutions with office locations around the globe, including Singapore, the United Kingdom, Australia, Mexico, Brazil, and Japan. Locate your local office at: **www.cengage.com/global**

Cengage Learning products are represented in Canada by Nelson Education, Ltd.

To learn more about Brooks/Cole, visit **www.cengage.com/brookscole**

Purchase any of our products at your local college store or at our preferred online store **www.cengagebrain.com**

Printed in Canada
9 10 11 12 17 16 15 14

Contents Overview

Contents

Preface

Today, there is a wide and impressive array of powerful and elegant tools for obtaining qualitative and quantitative information about the composition and structure of matter. Students of chemistry, biochemistry, physics, geology, the life sciences, forensic science, and environmental science must develop an understanding of these instrumental tools and their applications to solve important analytical problems in these fields. This book is addressed to meet the needs of these students and other users of analytical instruments.

When instrument users are familiar with the fundamental principles of operation of modern analytical instrumentation, they then will make appropriate choices and efficient use of these measurement tools. There are often a bewildering number of alternative methods for solving any given analytical problem, but by understanding the advantages and limitations of the various tools, users can choose the most appropriate instrumental method and be attuned to its limitations in sensitivity, precision, and accuracy. In addition, knowledge of measurement principles is necessary for calibration, standardization, and validation of instrumental methods. It is therefore our objective to give readers a thorough introduction to the principles of instrumental analysis, including spectroscopic, electrochemical, chromatographic, radiochemical, thermal, and surface analytical methods. By carefully studying this text, readers will discover the types of instruments available and their strengths and limitations.

ORGANIZATION OF THIS EDITION

This text is organized in sections similar to the fifth edition. After the brief introductory chapter, the book is divided into six sections.

- Section 1 contains four chapters on basic electrical circuits, operational amplifiers, digital electronics and computers, signals, noise, and signal-to-noise enhancement.
- Section 2 comprises seven chapters devoted to various atomic spectrometric methods, including an introduction to spectroscopy and spectroscopic instrumentation, atomic absorption, atomic emission, atomic mass spectrometry, and X-ray spectrometry.
- Section 3 treats molecular spectroscopy in nine chapters that describe absorption, emission, luminescence, infrared, Raman, nuclear magnetic resonance, mass spectrometry, and surface analytical methods.
- Section 4 consists of four chapters that treat electroanalytical chemistry, including potentiometry, coulometry, and voltammetry.
- Section 5 contains five chapters that discuss such analytical separation methods as gas and liquid chromatography, supercritical fluid chromatography, electrophoresis, and field-flow fractionation.
- Section 6 consists of four chapters devoted to miscellaneous instrumental methods with emphasis on thermal, radiochemical, and automated methods. A chapter on particle size analysis is also included in this final section.

Since the first edition of this text appeared in 1971, the field of instrumental analysis has grown so large and diverse that it is impossible to treat all of the modern instrumental techniques in a one- or even two-semester course. Also, instructors have differing opinions on which techniques to discuss and which to omit in their courses. Because of this, we have included more material in this text than can be covered in a single instrumental analysis course, and as a result, this comprehensive text will also be a valuable reference for years to come. An important advantage of organizing the material into sections is that instructors have flexibility in picking and choosing topics to be included in reading assignments. Thus, as in the previous edition, the sections on atomic and molecular spectroscopy, electrochemistry, and chromatography begin with introductory chapters that precede the chapters devoted to specific methods of each type. After assigning the introductory chapter in a section, an instructor can select the chapters that follow in any order desired. To assist students in using this book, the answers to most numerical problems are provided at the end of the book.

NEW TO THIS EDITION

- We have included a new chapter on particle size determination (Chapter 34). The physical and chemical properties of many research materials and consumer and industrial products are intimately related to their particle size distributions. As a result, particle size analysis has become an important technique in many research and industrial laboratories.
- Exciting new *Instrumental Analysis in Action* features have been added at the end of each of the six sections. These case studies describe how some of the methods introduced in each section can be applied to a specific analytical problem. These stimulating examples have been selected from the forensic, environmental, and biomedical areas.
- 🗵 Spreadsheet applications have been included throughout to illustrate how these powerful programs can be applied to instrumental methods. Problems accompanied by this icon 🗵 encourage the use of spreadsheets. When a more detailed approach is required or supplemental reading is appropriate, readers are referred to our companion book, *Applications of Microsoft® Excel in Analytical Chemistry* (Belmont, CA: Brooks/Cole, 2004), for assistance in understanding these applications.
- The book is now printed in two colors. This particularly aids in understanding the many figures and diagrams in the text. The second color clarifies graphs; aids in following the data flow in diagrams; provides keys for correlating data that appear in multiple charts, graphs, and diagrams; and makes for a more pleasing overall appearance.
- An open-ended *Challenge Problem* provides a capstone research-oriented experience for each chapter and requires reading the original literature of analytical chemistry, derivations, extensive analysis of real experimental data, and creative problem solving.
- All chapters have been revised and updated with recent references to the literature of analytical chemistry. Among the chapters that have been changed extensively are those on mass spectrometry (Chapters 11 and 20), surface characterization (Chapter 21), voltammetry (Chapter 25), chromatography (Chapters 26 and 27), and thermal analysis (Chapter 31). Throughout the book, new and updated methods and techniques are described, and photos of specific commercial instruments have been added where appropriate. Some of these modern topics include plasma spectrometry, fluorescence quenching and lifetime measurements, tandem mass spectrometry, and biosensors.
- Many new and revised charts, diagrams, and plots contain data, curves, and waveforms calculated from theory or obtained from the original literature to provide an accurate and realistic representation.
- Throughout the text, we have attempted to present material in a student-friendly style that is active and engaging. Examples are sprinkled throughout each chapter to aid in solving relevant and interesting problems. The solutions to the problems in each example are indicated so that students can easily separate the problem setup from the problem solution.

ANCILLARIES

- 🗡 The book's companion website at **www.thomsonedu.com/chemistry/skoog** includes more than 100 interactive tutorials on instrumental methods, simulations of analytical techniques, exercises, and animations to help students visualize important concepts. In addition, Excel files containing data and sample spreadsheets are available for download. Selected papers from the chemical literature are also available as PDF files to engage student interest and to provide background information for study. Throughout the book, this icon 🗡 alerts and encourages students to incorporate the website into their studies.
- An Instructor's Manual containing the solutions to all the text problems and online images from the text can be found at **www.thomsonedu.com/chemistry/skoog**.

ACKNOWLEDGMENTS

We wish to acknowledge the many contributions of reviewers and critics of all or parts of the manuscript. Those who offered numerous helpful suggestions and corrections include:

Larry Bowman, University of Alaska, Fairbanks
John Dorsey, Florida State University
Constantinos E. Efstathiou, University of Athens
Dale Ensor, Tennessee Tech University
Doug Gilman, Louisiana State University
Michael Ketterer, Northern Arizona University
Robert Kiser, University of Kentucky
Michael Koupparis, University of Athens
David Ryan, University of Massachusetts–Lowell
Alexander Scheeline, University of Illinois at Urbana-Champaign
Dana Spence, Wayne State University
Apryll Stalcup, University of Cincinnati
Greg Swain, Michigan State University
Dragic Vukomanovic, University of Massachusetts–Dartmouth
Mark Wightman, University of North Carolina
Charles Wilkins, University of Arkansas
Steven Yates, University of Kentucky

We are most grateful for the expert assistance of Professor David Zellmer, California State University, Fresno, who reviewed most of the chapters and served as the accuracy reviewer for the entire manuscript. His efforts are most heartily appreciated.

We owe special thanks to Ms. Janette Carver, head of the University of Kentucky Chemistry/Physics Library, who, in addition to serving as a superb reference librarian, provided essential library services and technical assistance in the use of the many electronic resources at our disposal.

Numerous manufacturers of analytical instruments and other products and services related to analytical chemistry have contributed by providing diagrams, application notes, and photos of their products. We are particularly grateful to Agilent Technologies, Bioanalytical Systems, Beckman Coulter, Inc., Brinkman Instruments, Caliper Life Sciences, Hach Co., Hamamatsu Photonics, InPhotonics, Inc., Kaiser Optical Systems, Leeman Labs, LifeScan, Inc., Mettler-Toledo, Inc., National Instruments Corp., Ocean Optics, Inc., Perkin-Elmer Corp., Postnova Analytics, Spectro Analytical Instruments, T. A. Instruments, Thermo-Electron Corp., and Varian, Inc. for providing photos.

We are especially indebted to the many members of the staff of Brooks/Cole–Thomson Learning who provided excellent support during the production of this text. Our development editor, Sandra Kiselica, did a wonderful job in organizing the project, in keeping the authors on task, and in making many important comments and suggestions. We also thank the many people involved in the production of the book. We are grateful to Katherine Bishop, who served as the project coordinator, and to Belinda Krohmer, the project manager at Brooks/Cole. Finally, we wish to acknowledge the support and assistance of our publisher David Harris. His patience, understanding, and guidance were of great assistance in the completion of the project.

Douglas A. Skoog
F. James Holler
Stanley R. Crouch

CHAPTER ONE

Introduction

A nalytical chemistry deals with methods for determining the chemical composition of samples of matter. A **qualitative method** yields information about the identity of atomic or molecular species or the functional groups in the sample. A **quantitative method**, in contrast, provides numerical information as to the relative amount of one or more of these components.

Throughout the book, this logo indicates an opportunity for online self-study. Visit the book's companion website at **www.thomsonedu.com/chemistry/skoog** to view interactive tutorials, guided simulations, and exercises.

1A CLASSIFICATION OF ANALYTICAL METHODS

Analytical methods are often classified as being either *classical* or *instrumental*. Classical methods, sometimes called *wet-chemical methods*, preceded instrumental methods by a century or more.

1A-1 Classical Methods

In the early years of chemistry, most analyses were carried out by separating the components of interest (the *analytes*) in a sample by precipitation, extraction, or distillation. For qualitative analyses, the separated components were then treated with reagents that yielded products that could be recognized by their colors, their boiling or melting points, their solubilities in a series of solvents, their odors, their optical activities, or their refractive indexes. For quantitative analyses, the amount of analyte was determined by *gravimetric* or by *volumetric* measurements.

In gravimetric measurements, the mass of the analyte or some compound produced from the analyte was determined. In volumetric, also called *titrimetric*, procedures, the volume or mass of a standard reagent required to react completely with the analyte was measured.

These classical methods for separating and determining analytes are still used in many laboratories. The extent of their general application is, however, decreasing with the passage of time and with the advent of instrumental methods to supplant them.

1A-2 Instrumental Methods

Early in the twentieth century, scientists began to exploit phenomena other than those used for classical methods for solving analytical problems. Thus, measurements of such analyte physical properties as conductivity, electrode potential, light absorption or emission, mass-to-charge ratio, and fluorescence began to be used for quantitative analysis. Furthermore, highly efficient chromatographic and electrophoretic techniques began to replace distillation, extraction, and precipitation for the separation of components of complex mixtures prior to their qualitative or quantitative determination. These newer methods for separating and determining chemical species are known collectively as *instrumental methods of analysis*.

Many of the phenomena underlying instrumental methods have been known for a century or more. Their application by most scientists, however, was delayed by lack of reliable and simple instrumentation. In fact, the growth of modern instrumental methods of analysis has paralleled the development of the electronics and computer industries.

1B   TYPES OF INSTRUMENTAL METHODS

Let us first consider some of the chemical and physical characteristics that are useful for qualitative or quantitative analysis. Table 1-1 lists most of the characteristic properties that are currently used for instrumental analysis. Most of the characteristics listed in the table require a source of energy to stimulate a measurable response from the analyte. For example, in atomic emission an increase in the temperature of the analyte is required first to produce gaseous analyte atoms and then to excite the atoms to higher energy states. The excited-state atoms then emit characteristic electromagnetic radiation, which is the quantity measured by the instrument. Sources of energy may take the form of a rapid thermal change as in the previous example; electromagnetic radiation from a selected region of the spectrum; application of an electrical quantity, such as voltage, current, or charge; or perhaps subtler forms intrinsic to the analyte itself.

Note that the first six entries in Table 1-1 involve interactions of the analyte with electromagnetic radiation. In the first property, radiant energy is produced by the analyte; the next five properties involve changes in electromagnetic radiation brought about by its interaction with the sample. Four electrical properties then follow. Finally, five miscellaneous properties are grouped together: mass, mass-to-charge ratio, reaction rate, thermal characteristics, and radioactivity.

The second column in Table 1-1 lists the instrumental methods that are based on the various physical and chemical properties. Be aware that it is not always easy to select an optimal method from among available instrumental techniques and their classical counterparts. Some instrumental techniques are more sensitive than classical techniques, but others are not. With certain combinations of elements or compounds, an instrumental method may be more selective, but with others, a gravimetric or volumetric approach may suffer less interference. Generalizations on the basis of accuracy, convenience, or expenditure of time are equally difficult to draw. Nor is it necessarily true that instrumental procedures employ more sophisticated or more costly apparatus.

TABLE 1-1 Chemical and Physical Properties Used in Instrumental Methods

Characteristic Properties	Instrumental Methods
Emission of radiation	Emission spectroscopy (X-ray, UV, visible, electron, Auger); fluorescence, phosphorescence, and luminescence (X-ray, UV, and visible)
Absorption of radiation	Spectrophotometry and photometry (X-ray, UV, visible, IR); photoacoustic spectroscopy; nuclear magnetic resonance and electron spin resonance spectroscopy
Scattering of radiation	Turbidimetry; nephelometry; Raman spectroscopy
Refraction of radiation	Refractometry; interferometry
Diffraction of radiation	X-ray and electron diffraction methods
Rotation of radiation	Polarimetry; optical rotary dispersion; circular dichroism
Electrical potential	Potentiometry; chronopotentiometry
Electrical charge	Coulometry
Electrical current	Amperometry; polarography
Electrical resistance	Conductometry
Mass	Gravimetry (quartz crystal microbalance)
Mass-to-charge ratio	Mass spectrometry
Rate of reaction	Kinetic methods
Thermal characteristics	Thermal gravimetry and titrimetry; differential scanning calorimetry; differential thermal analyses; thermal conductometric methods
Radioactivity	Activation and isotope dilution methods

As noted earlier, in addition to the numerous methods listed in the second column of Table 1-1, there is a group of instrumental procedures that are used for separation and resolution of closely related compounds. Most of these procedures are based on chromatography, solvent extraction, or electrophoresis. One of the characteristics listed in Table 1-1 is usually used to complete the analysis following chromatographic separations. Thus, for example, thermal conductivity, ultraviolet and infrared absorption, refractive index, and electrical conductance are used for this purpose.

This text deals with the principles, the applications, and the performance characteristics of the instrumental methods listed in Table 1-1 and of chromatographic and electrophoretic separation procedures as well. No space is devoted to the classical methods because we assume that the reader has previously studied these techniques.

1C   INSTRUMENTS FOR ANALYSIS

An instrument for chemical analysis converts information about the physical or chemical characteristics of the analyte to information that can be manipulated and interpreted by a human. Thus, an analytical instrument can be viewed as a communication device between the system under study and the investigator. To retrieve the desired information from the analyte, it is necessary to provide a stimulus, which is usually in the form of electromagnetic, electrical, mechanical, or nuclear energy, as illustrated in Figure 1-1. The stimulus elicits a response from the system under study whose nature and magnitude are governed by the fundamental laws of chemistry and physics. The resulting information is contained in the phenomena that result from the interaction of the stimulus with the analyte. A familiar example is passing a narrow band of wavelengths of visible light through a sample to measure the extent of its absorption by the analyte. The intensity of

the light is determined before and after its interaction with the sample, and the ratio of these intensities provides a measure of the analyte concentration.

Generally, instruments for chemical analysis comprise just a few basic components, some of which are listed in Table 1-2. To understand the relationships among these instrument components and the flow of information from the characteristics of the analyte through the components to the numerical or graphical output produced by the instrument, it is instructive to explore how the information of interest can be represented and transformed.

1C-1 Data Domains

The measurement process is aided by a wide variety of devices that convert information from one form to another. Before investigating how instruments function, it is important to understand how information can be *encoded* (represented) by physical and chemical characteristics and particularly by *electrical signals*, such as current, voltage, and charge. The various modes of encoding information are called *data domains*. A classification scheme has been developed based on this concept that greatly simplifies the analysis of instrumental systems and promotes understanding of the measurement process.[1] As shown in the data-domain map of Figure 1-2, data domains may be broadly classified into *nonelectrical domains* and *electrical domains*.

1C-2 Nonelectrical Domains

The measurement process begins and ends in nonelectrical domains. The physical and chemical information that is of interest in a particular experiment resides in these data domains. Among these characteristics are length, density, chemical composition, intensity of light, pressure, and others listed in the first column of Table 1-1.

It is possible to make a measurement by having the information reside entirely in nonelectrical domains. For instance, the determination of the mass of an object using a mechanical equal-arm balance involves a comparison of the mass of the object, which is placed on one balance pan, with standard masses placed on a second pan. The information representing the mass of the object in standard units is encoded directly by the experimenter, who provides information processing by

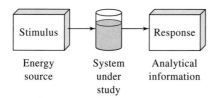

FIGURE 1-1 Block diagram showing the overall process of an instrumental measurement.

Stimulus → System under study → Response

Energy source | System under study | Analytical information

---

[1] C. G. Enke, *Anal. Chem.*, **1971**, *43*, 69A.

TABLE 1-2 Some Examples of Instrument Components

Instrument	Energy Source (stimulus)	Analytical Information	Information Sorter	Input Transducer	Data Domain of Transduced Information	Signal Processor/ Readout
Photometer	Tungsten lamp	Attenuated light beam	Filter	Photodiode	Electrical current	Amplifier, digitizer, LED display
Atomic emission spectrometer	Inductively coupled plasma	UV or visible radiation	Monochromator	Photomultiplier tube	Electrical current	Amplifier, digitizer, digital display
Coulometer	Direct-current source	Charge required to reduce or oxidize analyte	Cell potential	Electrodes	Time	Amplifier, digital timer
pH meter	Sample/ glass electrode	Hydrogen ion activity	Glass electrode	Glass-calomel electrodes	Electrical voltage	Amplifier, digitizer, digital display
Mass spectrometer	Ion source	Mass-to-charge ratio	Mass analyzer	Electron multiplier	Electrical current	Amplifier, digitizer, computer system
Gas chromatograph with flame ionization	Flame	Ion concentration vs. time	Chromatographic column	Biased electrodes	Electrical current	Electrometer, digitizer, computer system

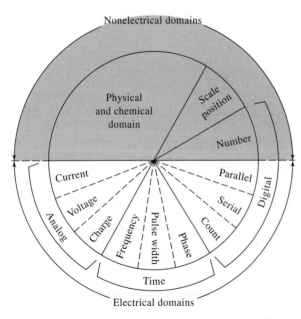

FIGURE 1-2 Data-domain map. The upper (shaded) half of the map consists of nonelectrical domains. The bottom half is made up of electrical domains. Note that the digital domain spans both electrical and nonelectrical domains.

summing the masses to arrive at a number. In certain other mechanical balances, the gravitational force on a mass is amplified mechanically by making one of the balance arms longer than the other, thus increasing the resolution of the measurement.

The determination of the linear dimensions of an object with a ruler and the measurement of the volume of a sample of liquid with a graduated cylinder are other examples of measurements carried out exclusively in nonelectrical domains. Such measurements are often associated with classical analytical methods. The advent of inexpensive electronic signal processors, sensitive transducers, and readout devices has led to the development of a host of electronic instruments, which acquire information from nonelectrical domains, process it in electrical domains, and finally present it in a nonelectrical way. Electronic devices process information and transform it from one domain to another in ways analogous to the multiplication of mass in mechanical balances with unequal arms. Because these devices are available and capable of rapid and so-

 Tutorial: Learn more about **data domains**.

phisticated information processing, instruments that rely exclusively on nonelectrical information transfer are rapidly becoming obsolete. Nonetheless, the information we seek begins in the properties of the analyte and ends in a number, both of which are nonelectrical representations. The ultimate objective in an analytical measurement is to obtain a final numerical result that is in some manner proportional to the sought-for chemical or physical characteristic of the analyte.

1C-3 Electrical Domains

The modes of encoding information as electrical quantities can be subdivided into *analog domains*, *time domains*, and the *digital domain*, as illustrated in the bottom half of the circular map in Figure 1-2. Note that the digital domain is not only composed of electrical signals but also includes one nonelectrical representation, because numbers presented on any type of display convey digital information.

Any measurement process can be represented as a series of *interdomain conversions*. For example, Figure 1-3 illustrates the measurement of the molecular fluorescence intensity of a sample of tonic water containing a trace of quinine and, in a general way, some of the data-domain conversions that are necessary to arrive at a number related to the intensity. The intensity of the fluorescence is significant in this context because it

is proportional to the concentration of the quinine in the tonic water, which is ultimately the information that we desire. The information begins in the solution of tonic water as the concentration of quinine. This information is teased from the sample by applying to it a stimulus in the form of electromagnetic energy from the laser shown in Figure 1-3. The radiation interacts with the quinine molecules in the tonic water to produce fluorescence emission in a region of the spectrum characteristic of quinine and of a magnitude proportional to its concentration. Radiation that is unrelated to the concentration of quinine is removed from the beam of light by an optical filter, as shown in Figure 1-3. The intensity of the fluorescence emission, which is nonelectrical information, is encoded into an electrical signal by a special type of device called an *input transducer*. The particular type of transducer used in this experiment is a phototransducer, of which there are numerous types, some of which are discussed in Chapters 6 and 7. In this example, the input transducer converts the fluorescence from the tonic water to an electrical current I, proportional to the intensity of the radiation. The mathematical relationship between the electrical output and the input radiant power impinging on its surface is called the *transfer function* of the transducer.

The current from the phototransducer is then passed through a resistor R, which according to Ohm's

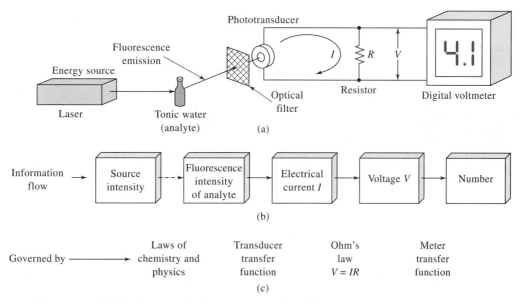

FIGURE 1-3 A block diagram of a fluorometer showing (a) a general diagram of the instrument, (b) a diagrammatic representation of the flow of information through various data domains in the instrument, and (c) the rules governing the data-domain transformations during the measurement process.

law produces a voltage V that is proportional to I, which is in turn proportional to the intensity of the fluorescence. Finally, V is measured by the digital voltmeter to provide a readout proportional to the concentration of the quinine in the sample.

Voltmeters, alphanumeric displays, electric motors, computer screens, and many other devices that serve to convert data from electrical to nonelectrical domains are called *output transducers*. The digital voltmeter of the fluorometer of Figure 1-3 is a rather complex output transducer that converts the voltage V to a number on a liquid-crystal display so that it may be read and interpreted by the user of the instrument. We shall consider the detailed nature of the digital voltmeter and various other electrical circuits and signals in Chapters 2 through 4.

Analog-Domain Signals

Information in the analog domain is encoded as the *magnitude* of one of the electrical quantities — voltage, current, charge, or power. These quantities are contin-

uous in both amplitude and time as shown by the typical analog signals of Figure 1-4. Magnitudes of analog quantities can be measured continuously, or they can be sampled at specific points in time dictated by the needs of a particular experiment or instrumental method as discussed in Chapter 4. Although the data of Figure 1-4 are recorded as a function of time, any variable such as wavelength, magnetic field strength, or temperature may be the independent variable under appropriate circumstances. The correlation of two analog signals that result from corresponding measured physical or chemical properties is important in a wide variety of instrumental techniques, such as nuclear magnetic resonance spectroscopy, infrared spectroscopy, and differential thermal analysis.

Because electrical noise influences the magnitude of electrical signals, analog signals are especially susceptible to electrical noise that results from interactions within measurement circuits or from other electrical devices in the vicinity of the measurement system. Such undesirable noise bears no relationship to

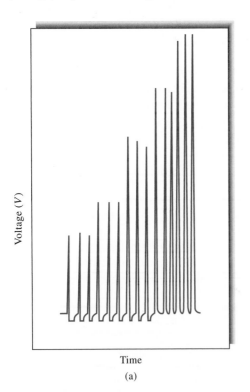
(a)

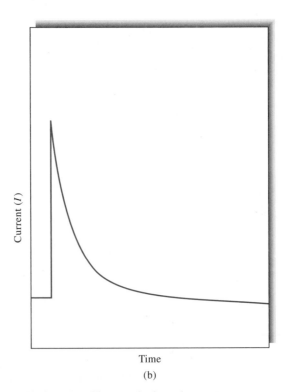
(b)

FIGURE 1-4 Analog signals. (a) Instrument response from the photometric detection system of a flow injection analysis experiment. A stream of reaction mixture containing plugs of red $Fe(SCN)^{2+}$ flows past a monochromatic light source and a phototransducer, which produces a changing voltage as the sample concentration changes. (b) The current response of a photomultiplier tube when the light from a pulsed source falls on the photocathode of the device.

the information of interest, and methods have been developed to minimize the effects of this unwanted information. Signals, noise, and the optimization of instrumental response are discussed in Chapter 5.

Time-Domain Information

Information is stored in the time domain as the time relationship of signal fluctuations, rather than in the amplitudes of the signals. Figure 1-5 illustrates three different time-domain signals recorded as an analog quantity versus time. The horizontal dashed lines represent an arbitrary analog signal threshold that is used to decide whether a signal is HI (above the threshold) or LO (below the threshold). The time relationships between transitions of the signal from HI to LO or from LO to HI contain the information of interest. For instruments that produce periodic signals, the number of cycles of the signal per unit time is the *frequency* and the time required for each cycle is its *period*. Two examples of instrumental systems that produce information encoded in the frequency domain are Raman spectroscopy (Chapter 18) and instrumental neutron activation analysis (Chapter 32). In these methods, the frequency of arrival of photons at a detector is directly related to the intensity of the emission from the analyte, which is proportional to its concentration.

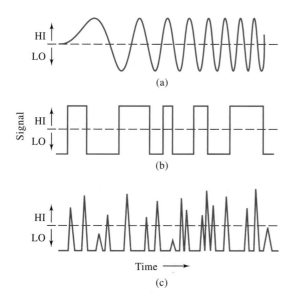

FIGURE 1-5 Time-domain signals. The horizontal dashed lines represent signal thresholds. When each signal is above the threshold, the signal is HI, and when it is below the threshold, the signal is LO.

The time between successive LO to HI transitions is called the *period*, and the time between a LO to HI and a HI to LO transition is called the *pulse width*. Devices such as voltage-to-frequency converters and frequency-to-voltage converters may be used to convert time-domain signals to analog-domain signals and vice versa. These and other such *data-domain converters* are discussed in Chapters 2–4 as a part of our treatment of electronic devices and will be referred to in other contexts throughout this book.

Digital Information

Data are encoded in the digital domain in a two-level scheme. The information can be represented by the state of a lightbulb, a light-emitting diode, a toggle switch, or a logic-level signal, to cite but a few examples. The characteristic that these devices share is that each of them must be in one of only two states. For example, lights and switches may be only ON or OFF and logic-level signals may be only HI or LO. The definition of what constitutes ON and OFF for switches and lights is understood, but in the case of electrical signals, as in the case of time-domain signals, an arbitrary signal level must be defined that distinguishes between HI and LO. Such a definition may depend on the conditions of an experiment, or it may depend on the characteristics of the electronic devices in use. For example, the signal represented in Figure 1-5c is a train of pulses from a nuclear detector. The measurement task is to count the pulses during a fixed period of time to obtain a measure of the intensity of radiation. The dashed line represents a signal level that not only is low enough to ensure that no pulses are lost but also is sufficiently high to reject random fluctuations in the signal that are unrelated to the nuclear phenomena of interest. If the signal crosses the threshold fourteen times, as in the case of the signal in Figure 1-5c, then we may be confident that fourteen nuclear events occurred. After the events have been counted, the data are encoded in the digital domain by HI-LO signals representing the number 14. In Chapter 4 we explore the means for making HI-LO electronic decisions and encoding the information in the digital domain.

As shown by the data-domain map of Figure 1-2, the digital domain spans both electrical and nonelectrical encoding methods. In the example just cited, the nuclear events are accumulated by using an electronic counter and displayed on a digital readout. When the experimenter reads and interprets the display, the number that represents the measured quantity is once again in a nonelectrical domain. Each piece of HI-LO data that represents a nuclear event is a *bit* (binary digit) of information, which is the fundamental unit of information in the digital domain. Bits of information that are transmitted along a single electronic channel or wire may be counted by an observer or by an electronic device that is monitoring the channel; such accumulated data is termed *count digital data*, which appears in the data-domain map of Figure 1-2. For example, the signal in Figure 1-5a might represent the number $n = 8$ because there are eight complete pulses in the signal. Likewise, the signal in Figure 1-5b might correspond to $n = 5$, and that in Figure 1-5c could represent $n = 14$. Although effective, this means of transmitting information is not very efficient.

A far more efficient way to encode information is to use binary numbers to represent numeric and alphabetic data. To see how this type of encoding may be accomplished, let us consider the signals in Figure 1-6. The count digital data of the signal in Figure 1-6a represents the number $n = 5$, as before. We monitor the signal and count the number of complete oscillations. The process requires a period of time that is proportional to the number of cycles of the signal, or in this case, five times the length of a single time interval, as indicated in Figure 1-6. Note that the time intervals are numbered consecutively beginning with zero. In a binary encoding scheme, such as the one shown for the signal in Figure 1-6b, we assign a numerical value to each successive interval of time. For example, the

zeroth time interval represents $2^0 = 1$, the first time interval represents $2^1 = 2$, the second time interval represents $2^2 = 4$, and so forth, as shown in Figure 1-6. During each time interval, we need decide only whether the signal is HI or LO. If the signal is HI during any given time interval, then the value corresponding to that interval is added to the total. All intervals that are LO contribute zero to the total.

In Figure 1-6b, the signal is HI only in interval 0 and interval 2, so the total value represented is $(1 \times 2^0) + (0 \times 2^1) + (1 \times 2^2) = 5$. Thus, in the space of only three time intervals, the number $n = 5$ has been represented. In the count digital example in Figure 1-6a, five time intervals were required to represent the same number. In this limited example, the binary-coded serial data is nearly twice as efficient as the count serial data. A more dramatic example may be seen in the counting of $n = 10$ oscillations, similar to those of the signal in Figure 1-6a. In the same ten time intervals, ten HI-LO bits of information in the serial binary coding scheme enable the representation of the binary numbers from 0 to $2^{10} - 1 = 1024$ numbers, or 0000000000 to 1111111111. The improvement in efficiency is 1024/10, or about 100-fold. In other words, the count serial scheme requires 1024 time intervals to represent the number 1024, but the binary coding scheme requires only ten time intervals. As a result of the efficiency of binary coding schemes, most digital information is encoded, transferred, processed, and decoded in binary form.

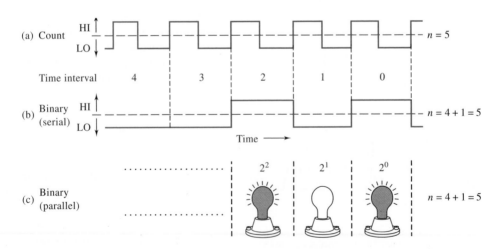

FIGURE 1-6 Diagram illustrating three types of digital data: (a) count serial data, (b) binary-coded serial data, and (c) parallel binary data. In all three cases, the data represent the number $n = 5$.

Data represented by binary coding on a single transmission line are called *serial-coded binary data*, or simply *serial data*. A common example of serial data transmission is the computer modem, which is a device for transmitting data between computers by telephone over a single conductor (and a common connection).

A still more efficient method for encoding data in the digital domain is seen in the signal of Figure 1-6c. Here, we use three lightbulbs to represent the three binary digits: $2^0 = 1, 2^1 = 2$, and $2^2 = 4$. However, we could use switches, wires, light-emitting diodes, or any of a host of electronic devices to encode the information. In this scheme, ON = 1 and OFF = 0, so that our number is encoded as shown in Figure 1-6 with the first and third lights ON and the middle light OFF, which represents $4 + 0 + 1 = 5$. This scheme is highly efficient because all of the desired information is presented to us simultaneously, just as all of the digits on the face of the digital voltmeter in Figure 1-3 appear simultaneously. Data presented in this way are referred to as *parallel digital data*. Data are transmitted within analytical instruments and computers by parallel data transmission. Because data usually travel relatively short distances within these devices, it is economical and efficient to use parallel information transfer. This economy of short distances is in contrast to the situation in which data must be transported over long distances from instrument to instrument or from computer to computer. In such instances, communication is carried out serially by using modems or other more sophisticated or faster serial data-transmission schemes. We will consider these ideas in somewhat more detail in Chapter 4.

1C-4 Detectors, Transducers, and Sensors

The terms *detector*, *transducer*, and *sensor* are often used synonymously, but in fact the terms have somewhat different meanings. The most general of the three terms, *detector*, refers to a mechanical, electrical, or chemical device that identifies, records, or indicates a change in one of the variables in its environment, such as pressure, temperature, electrical charge, electromagnetic radiation, nuclear radiation, particulates, or molecules. This term has become a catchall to the extent that entire instruments are often referred to as *detectors*. In the context of instrumental analysis, we shall use the term *detector* in the general sense in which we have just defined it, and we shall use *detection system* to refer to entire assemblies that indicate or record physical or chemical quantities. An example is the UV (ultraviolet) detector often used to indicate and record the presence of eluted analytes in liquid chromatography.

The term *transducer* refers specifically to those devices that convert information in nonelectrical domains to information in electrical domains and the converse. Examples include photodiodes, photomultipliers, and other electronic photodetectors that produce current or voltage proportional to the radiant power of electromagnetic radiation that falls on their surfaces. Other examples include thermistors, strain gauges, and Hall effect (magnetic-field strength) transducers. As suggested previously, the mathematical relationship between the electrical output and the input radiant power, temperature, force, or magnetic field strength is called the *transfer function* of the transducer.

The term *sensor* also has become rather broad, but in this text we shall reserve the term for the class of analytical devices that are capable of monitoring specific chemical species continuously and reversibly. There are numerous examples of sensors throughout this text, including the glass electrode and other ion-selective electrodes, which are treated in Chapter 23; the Clark oxygen electrode, which is described in Chapter 25; and fiber-optic sensors (optrodes), which appear in Chapter 14. Sensors consist of a transducer coupled with a chemically selective recognition phase, as shown in Figure 1-7. So, for example, optrodes consist of a phototransducer coupled with a fiber optic that is coated on the end opposite the transducer with a substance that responds specifically to a particular physical or chemical characteristic of an analyte.

A sensor that is especially interesting and instructive is made from a *quartz crystal microbalance*, or QCM. This device is based on the *piezoelectric* characteristics of quartz. When quartz is mechanically deformed, an electrical potential difference develops across its surface. Furthermore, when a voltage is impressed across the faces of a quartz crystal, the crystal deforms. A crystal connected in an appropriate electrical circuit oscillates at a frequency that is characteristic of the mass and shape of the crystal and that is amazingly constant as long as the mass of the crystal is constant. This property of some crystalline materials is called the *piezoelectric effect* and forms the basis for the QCM. Moreover, the characteristic constant frequency of the quartz crystal is the basis for modern high-precision clocks, time bases, counters, timers, and frequency meters, which in turn have led to many highly accurate and precise analytical instrumental systems.

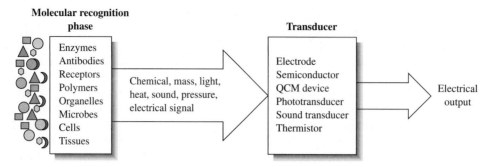

FIGURE 1-7 Chemical sensor. The sensor consists of a molecular recognition element and a transducer. A wide variety of recognition elements are possible. Shown here are some fairly selective recognition elements particularly useful with biosensors. The recognition phase converts the information of interest into a detectable characteristic, such as another chemical, mass, light, or heat. The transducer converts the characteristic into an electrical signal that can be measured.

If a quartz crystal is coated with a polymer that selectively adsorbs certain molecules, the mass of the crystal increases if the molecules are present, thus decreasing the resonant frequency of the quartz crystal. When the molecules are desorbed from the surface, the crystal returns to its original frequency. The relationship between the change in frequency of the crystal Δf and the change in mass of the crystal ΔM is given by

$$\Delta f = \frac{Cf^2 \Delta M}{A}$$

where M is the mass of the crystal, A is its surface area, f is the frequency of oscillation of the crystal, and C is a proportionality constant. This relationship indicates that it is possible to measure very small changes in the mass of the crystal if the frequency of the crystal can be measured precisely. As it turns out, it is possible to measure frequency changes of 1 part in 10^7 quite easily with inexpensive instrumentation. The limit of detection for a piezoelectric sensor of this type is estimated to be about 1 pg, or 10^{-12} g. These sensors have been used to detect a variety of gas-phase analytes, including formaldehyde, hydrogen chloride, hydrogen sulfide, and benzene. They have also been proposed as sensors for chemical warfare agents such as mustard gas and phosgene.

The piezoelectric mass sensor presents an excellent example of a transducer converting a property of the analyte, mass in this case, to a change in an electrical quantity, the resonant frequency of the quartz crystal. This example also illustrates the distinction between a transducer and a sensor. In the QCM, the transducer is the quartz crystal and the selective second phase is the polymeric coating. The combination of the transducer and the selective phase constitute the sensor.

1C-5 Readout Devices

A readout device is a transducer that converts information from an electrical domain to a form that is understandable by a human observer. Usually, the transduced signal takes the form of the alphanumeric or graphical output of a cathode-ray tube, a series of numbers on a digital display, the position of a pointer on a meter scale, or occasionally, the blackening of a photographic plate or a tracing on a recorder paper. In some instances, the readout device may be arranged to give the analyte concentration directly.

1C-6 Computers in Instruments

Most modern analytical instruments contain or are attached to one or more sophisticated electronic devices and data-domain converters, such as operational amplifiers, integrated circuits, analog-to-digital and digital-to-analog converters, counters, microprocessors, and computers. To appreciate the power and limitations of such instruments, investigators need to develop at least a qualitative understanding of how these devices function and what they can do. Chapters 3 and 4 provide a brief treatment of these important topics.

1D CALIBRATION OF INSTRUMENTAL METHODS

A very important part of all analytical procedures is the calibration and standardization process. *Calibration* determines the relationship between the analytical response and the analyte concentration. Usually this is determined by the use of *chemical standards.*

Almost all analytical methods require some type of calibration with chemical standards. Gravimetric methods and some coulometric methods (Chapter 24) are among the few *absolute* methods that do not rely on calibration with chemical standards. Several types of calibration procedures are described in this section.

1D-1 Comparison with Standards

Two types of comparison methods are described here, the direct comparison technique and the titration procedure.

Direct Comparison

Some analytical procedures involve comparing a property of the analyte (or the product of a reaction with the analyte) with standards such that the property being tested matches or nearly matches that of the standard. For example, in early colorimeters, the color produced as the result of a chemical reaction of the analyte was compared with the color produced by reaction of standards. If the concentration of the standard was varied by dilution, for example, it was possible to obtain a fairly exact color match. The concentration of the analyte was then equal to the concentration of the standard after dilution. Such a procedure is called a *null comparison* or *isomation method.*[2]

Titrations

Titrations are among the most accurate of all analytical procedures. In a titration, the analyte reacts with a standardized reagent (the titrant) in a reaction of known stoichiometry. Usually the amount of titrant is varied until chemical equivalence is reached, as indicated by the color change of a chemical indicator or by the change in an instrument response. The amount of the standardized reagent needed to achieve chemical equivalence can then be related to the amount of analyte present. The titration is thus a type of chemical comparison.[3]

1D-2 External-Standard Calibration

An *external standard* is prepared separately from the sample. By contrast, an *internal standard* is added to the sample itself. External standards are used to calibrate instruments and procedures when there are no interference effects from matrix components in the analyte solution. A series of such external standards containing the analyte in known concentrations is prepared. Ideally, three or more such solutions are used in the calibration process. However, in some routine analyses, two-point calibrations can be reliable.

Calibration is accomplished by obtaining the response signal (absorbance, peak height, peak area) as a function of the known analyte concentration. A calibration curve is prepared by plotting the data or by fitting them to a suitable mathematical equation, such as the slope-intercept form used in the method of linear least squares. The next step is the prediction step, where the response signal is obtained for the sample and used to *predict* the unknown analyte concentration, c_x, from the calibration curve or best-fit equation. The concentration of the analyte in the original bulk sample is then calculated from c_x by applying the appropriate dilution factors from the sample preparation steps.

The Least-Squares Method

A typical calibration curve is shown in Figure 1-8 for the determination of isooctane in a hydrocarbon sample. Here, a series of isooctane standards was injected into a gas chromatograph, and the area of the isooctane peak was obtained as a function of concentration. The ordinate is the dependent variable, peak area, and the abscissa is the independent variable, mole percent (mol %) of isooctane. As is typical and usually desirable, the plot approximates a straight line. Note, however, that because of the indeterminate errors in the measurement process, not all the data fall exactly on the line. Thus, the investigator must try to draw the "best" straight line among the data points. *Regression analysis* provides the means for objectively obtaining such a line

 Tutorial: Learn more about **calibration**.

[2]See, for example, H. V. Malmstadt and J. D. Winefordner, *Anal. Chim. Acta,* **1960,** *20,* 283; L. Ramaley and C. G. Enke, *Anal. Chem.,* **1965,** *37,* 1073.

[3]See D. A. Skoog, D. M. West, F. J. Holler, and S. R. Crouch, *Fundamentals of Analytical Chemistry*, 8th ed., Belmont, CA: Brooks/Cole, 2004, Chaps. 13–17.

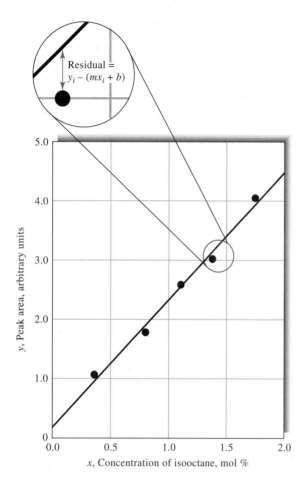

FIGURE 1-8 Calibration curve for the determination of isooctane in a hydrocarbon mixture. The residual is the difference between an experimental data point y_i and that calculated from the regression model, $mx_i + b$, as shown in the insert.

and also for specifying the uncertainties associated with its subsequent use. The uncertainties are related to the *residuals* shown in Figure 1-8, which are a measure of how far away from the best straight line the data points lie. The *method of least squares* (see Appendix 1, Section a1D) is often applied to obtain the equation for the line.[4]

The method of least squares is based on two assumptions. The first is that there is actually a linear relationship between the measured response y and the standard analyte concentration x. The mathematical relationship that describes this assumption is called the *regression model*, which may be represented as

$$y = mx + b$$

where b is the y intercept (the value of y when x is zero) and m is the slope of the line (see Figure 1-8). We also assume that any deviation of the individual points from the straight line arises from error in the *measurement*. That is, we assume there is no error in the x values of the points (concentrations). Both of these assumptions are appropriate for many analytical methods, but bear in mind that whenever there is significant uncertainty in the x data, basic linear least-squares analysis may not give the best straight line. In such a case, a more complex *correlation analysis* may be necessary. In addition, basic least-squares analysis may not be appropriate when the uncertainties in the y values vary significantly with x. In this case, it may be necessary to apply different weighting factors to the points and perform a *weighted least-squares analysis*.[5]

In cases where the data do not fit a linear model, nonlinear regression methods are available.[6] Some of these use polynomial models or multiple regression procedures. There are even computer programs that will find a model that describes a set of experimental data from an internal or user-defined set of equations.[7]

The slope m and intercept b of the linear least-squares line are determined as in Equations a1-34 and a1-35 of Appendix 1. For determining an unknown concentration c_x from the least-squares line, the value of the instrument response y_c is obtained for the unknown, and the slope and intercept are used to calculate the unknown concentration c_x as shown in Equation 1-1.

$$c_x = \frac{y_c - b}{m} \tag{1-1}$$

The standard deviation in concentration s_c can be found from the *standard error of the estimate* s_y, also called the *standard deviation about regression*, as given in Equation 1-2:

$$s_c = \frac{s_y}{m}\sqrt{\frac{1}{M} + \frac{1}{N} + \frac{(\bar{y}_c - \bar{y})^2}{m^2 S_{xx}}} \tag{1-2}$$

[4]For a discussion of using spreadsheets in linear regression analysis, see S. R. Crouch and F. J. Holler, *Applications of Microsoft® Excel in Analytical Chemistry*, Belmont, CA: Brooks/Cole, 2004, Chap. 4.

[5]See P. R. Bevington and D. K. Robinson, *Data Reduction and Error Analysis for the Physical Sciences*, 3rd ed., New York: McGraw-Hill, 2002.
[6]J. L. Devore, *Probability and Statistics for Engineering and the Sciences*, 6th ed., Pacific Grove, CA: Duxbury Press at Brooks/Cole, 2004.
[7]See, for example, *TableCurve*, Systat Software, Point Richmond, CA.

where M is the number of replicate results, N is the number of points in the calibration curve (number of standards), $\bar{y}_c$ is the mean response for the unknown, and $\bar{y}$ is the mean value of y for the calibration results. The quantity S_{xx} is the sum of the squares of the deviations of x values from the mean as given in Equation a1-31 of Appendix 1.

Errors in External-Standard Calibration

When external standards are used, it is assumed that the same responses will be obtained when the same analyte concentration is present in the sample and in the standard. Thus, the calibration functional relationship between the response and the analyte concentration must apply to the sample as well. Usually, in a determination, the raw response from the instrument is not used. Instead, the raw analytical response is corrected by measuring a *blank*. An *ideal blank* is identical to the sample but without the analyte. In practice, with complex samples, it is too time-consuming or impossible to prepare an ideal blank and a compromise must be made. Most often a real blank is either a *solvent blank*, containing the same solvent in which the sample is dissolved, or a *reagent blank*, containing the solvent plus all the reagents used in sample preparation.

Even with blank corrections, several factors can cause the basic assumption of the external-standard method to break down. Matrix effects, due to extraneous species in the sample that are not present in the standards or blank, can cause the same analyte concentrations in the sample and standards to give different responses.[8] Differences in experimental variables at the times at which blank, sample, and standard are measured can also invalidate the established calibration function. Even when the basic assumption is valid, errors can still occur because of contamination during the sampling or sample preparation steps.

Also, systematic errors can occur during the calibration process. For example, if the standards are prepared incorrectly, an error will occur. The accuracy with which the standards are prepared depends on the accuracy of the gravimetric and volumetric techniques and equipment used. The chemical form of the standards must be identical to that of the analyte in the sample; the state of oxidation, isomerization, or complexation of the analyte can alter the response. Once prepared, the concentration of the standards can change because of decomposition, volatilization, or adsorption onto container walls. Contamination of the standards can also result in higher analyte concentrations than expected. A systematic error can occur if there is some bias in the calibration model. For example, errors can occur if the calibration function is obtained without using enough standards to obtain good statistical estimates of the parameters.

Random errors can also influence the accuracy of results obtained from calibration curves, as illustrated in Figure 1-9. The uncertainty in the concentration of analyte s'_c obtained from a calibration curve is lowest when the response is close to the mean value $\bar{y}$. The point $\bar{x}, \bar{y}$ represents the centroid of the regression line. Note that measurements made near the center of the curve will give less uncertainty in analyte concentration than those made at the extremes.

Multivariate Calibration

The least-squares procedure just described is an example of a univariate calibration procedure because only one response is used per sample. The process of relating multiple instrument responses to an analyte or a mixture of analytes is known as *multivariate calibration*. Multivariate calibration methods[9] have become quite popular in recent years as new instruments become available that produce multidimensional responses (absorbance of several samples at multiple wavelengths, mass spectrum of chromatographically separated components, etc.). Multivariate calibration methods are very powerful. They can be used to simultaneously determine multiple components in mixtures and can provide redundancy in measurements to improve precision because repeating a measurement N times provides a $\sqrt{N}$ improvement in the precision of the mean value (see Appendix 1, Section a1B-1). They can also be used to detect the presence of interferences that would not be identified in a univariate calibration.

1D-3 Standard-Addition Methods

Standard-addition methods are particularly useful for analyzing complex samples in which the likelihood of matrix effects is substantial. A standard-addition

---

[8] The *matrix* includes the analyte and other constituents, which are termed *concomitants*.

[9] For a more extensive discussion, see K. R. Beebe, R. J. Pell, and M. B. Seasholtz, *Chemometrics: A Practical Guide*, New York: Wiley, 1998, Chap. 5; H. Martens and T. Naes, *Multivariate Calibration*, New York: Wiley, 1989.

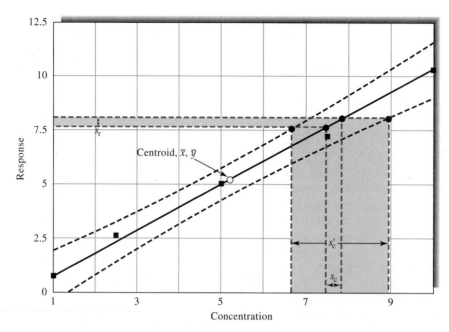

FIGURE 1-9 Effect of calibration curve uncertainty. The dashed lines show confidence limits for concentrations determined from the regression line. Note that uncertainties increase at the extremities of the plot. Usually, we estimate the uncertainty in analyte concentration only from the standard deviation of the response. Calibration curve uncertainty can significantly increase the uncertainty in analyte concentration from s_c to s_c'.

method can take several forms.[10] One of the most common forms involves adding one or more increments of a standard solution to sample aliquots containing identical volumes. This process is often called *spiking* the sample. Each solution is then diluted to a fixed volume before measurement. Note that when the amount of sample is limited, standard additions can be carried out by successive introductions of increments of the standard to a single measured volume of the unknown. Measurements are made on the original sample and on the sample plus the standard after each addition. In most versions of the standard-addition method, the sample matrix is nearly identical after each addition, the only difference being the concentration of the analyte or, in cases involving the addition of an excess of an analytical reagent, the concentration of the reagent. All other constituents of the reaction mixture should be identical because the standards are prepared in aliquots of the sample.

Assume that several aliquots V_x of the unknown solution with a concentration c_x are transferred to volumetric flasks having a volume V_t. To each of these flasks is added a variable volume V_s of a standard solution of the analyte having a known concentration c_s. Suitable reagents are then added, and each solution is diluted to volume. Instrumental measurements are then made on each of these solutions and corrected for any blank response to yield a net instrument response S. If the blank-corrected instrument response is proportional to concentration, as is assumed in the standard-addition method, we may write

$$S = \frac{kV_s c_s}{V_t} + \frac{kV_x c_x}{V_t} \qquad (1\text{-}3)$$

where k is a proportionality constant. A plot of S as a function of V_s is a straight line of the form

$$S = mV_s + b$$

where the slope m and the intercept b are given by

$$m = \frac{kc_s}{V_t}$$

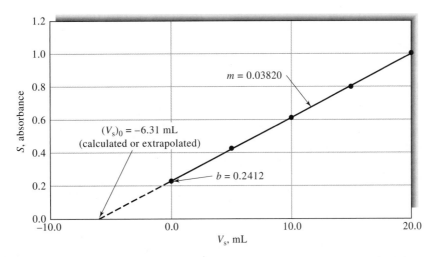

FIGURE 1-10 Linear calibration plot for the method of standard additions. The concentration of the unknown solution may be calculated from the slope m and the intercept b, or it may be determined by extrapolation, as explained in the text.

and

$$b = \frac{kV_x c_x}{V_t}$$

Such a standard-addition plot is shown in Figure 1-10.

A least-squares analysis (Appendix 1, Section a1D) can be used to determine m and b; c_x can then be obtained from the ratio of these two quantities and the known values of c_s, V_x, and V_s. Thus,

$$\frac{b}{m} = \frac{kV_x c_x / V_t}{kc_s / V_t} = \frac{V_x c_x}{c_s}$$

or

$$c_x = \frac{bc_s}{mV_x} \tag{1-4}$$

The standard deviation in concentration can then be obtained by first calculating the standard deviation in volume s_V and then using the relationship between volume and concentration. The standard deviation in volume is found from Equation 1-2 with a few alterations. Because we extrapolate the calibration curve to the x-axis in the standard-addition method, the value of y for the unknown is 0 and the $1/M$ term is absent. Hence, the equation for s_V becomes

$$s_V = \frac{s_y}{m}\sqrt{\frac{1}{N} + \frac{(0 - \overline{y})^2}{m^2 S_{xx}}} \tag{1-5}$$

As shown by the dashed line of Figure 1-10, the difference between the volume of the standard added at the origin (zero) and the value of the volume at the intersection of the straight line with the x-axis, or the x-intercept $(V_s)_0$, is the volume of standard reagent equivalent to the amount of analyte in the sample. In addition, the x-intercept corresponds to zero instrument response, so that we may write

$$S = \frac{kV_s c_s}{V_t} + \frac{kV_x c_x}{V_t} = 0 \tag{1-6}$$

By solving Equation 1-6 for c_x, we obtain

$$c_x = -\frac{(V_s)_0 c_s}{V_x} \tag{1-7}$$

The standard deviation in concentration s_c is then

$$s_c = s_V\left(\frac{c_s}{V_x}\right) \tag{1-8}$$

EXAMPLE 1-1

Ten-millimeter aliquots of a natural water sample were pipetted into 50.00-mL volumetric flasks. Exactly 0.00, 5.00, 10.00, 15.00, and 20.00 mL of a standard solution containing 11.1 ppm of Fe^{3+} were added to each, followed by an excess of thiocyanate ion to give the red complex $Fe(SCN)^{2+}$. After dilution to volume, the instrument response S for each of the five solutions, measured with a colorimeter, was found to be 0.240, 0.437, 0.621, 0.809, and 1.009, respectively. (a) What was the concentration of Fe^{3+} in the water sample? (b) Calculate the standard deviation in the concentration of Fe^{3+}.

▶ *Solution*

(a) In this problem, $c_s = 11.1$ ppm, $V_x = 10.00$ mL, and $V_t = 50.00$ mL. A plot of the data, shown in Figure 1-10, demonstrates that there is a linear relationship between the instrument response and the added amount of iron.

To obtain the equation for the line in Figure 1-10 ($S = mV_s + b$), we follow the procedure illustrated in Example a1-11 in Appendix 1. The result, shown in the spreadsheet in Figure 1-11, is $m = 0.0382$ and $b = 0.2412$ and thus

$$S = 0.0382V_s + 0.2412$$

From Equation 1-4, or from the spreadsheet, we obtain

$$c_x = \frac{0.2412 \times 11.1}{0.0382 \times 10.00} = 7.01 \text{ ppm Fe}^{3+}$$

This value may be determined by graphical extrapolation as the figure also illustrates. The extrapolated value represents the volume of reagent corresponding to zero instrument response, which in this case is −6.31 mL. The unknown concentration of the analyte in the original solution is then calculated from Equation 1-7 as follows:

$$c_x = -\frac{(V_s)_0 c_s}{V_x} = \frac{6.31 \text{ mL} \times 11.1 \text{ ppm}}{10.00 \text{ mL}}$$

$$= 7.01 \text{ ppm Fe}^{3+}$$

(b) The standard deviation in concentration can be obtained from the standard deviation in the volume intercept (Equation 1-5) and Equation 1-8 as shown in the spreadsheet of Figure 1-11. The result is $s_c = 0.11$ ppm. Hence, the concentration of Fe in the unknown is 7.01 ± 0.16 ppm.

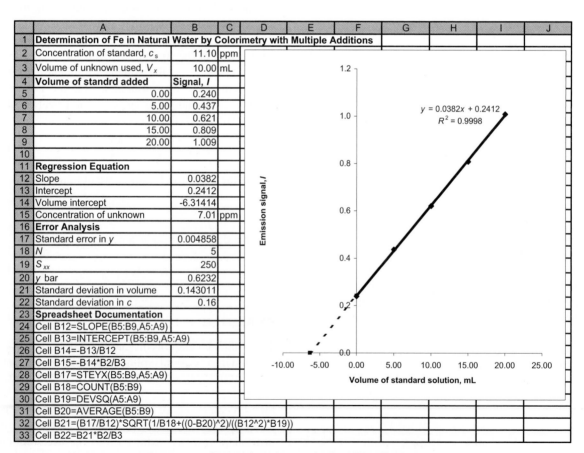

	A	B	C	D	E	F	G	H	I	J
1	Determination of Fe in Natural Water by Colorimetry with Multiple Additions									
2	Concentration of standard, c_s	11.10	ppm							
3	Volume of unknown used, V_x	10.00	mL							
4	Volume of standrd added	Signal, I								
5	0.00	0.240								
6	5.00	0.437								
7	10.00	0.621								
8	15.00	0.809								
9	20.00	1.009								
10										
11	Regression Equation									
12	Slope	0.0382								
13	Intercept	0.2412								
14	Volume intercept	-6.31414								
15	Concentration of unknown	7.01	ppm							
16	Error Analysis									
17	Standard error in y	0.004858								
18	N	5								
19	S_{xx}	250								
20	y bar	0.6232								
21	Standard deviation in volume	0.143011								
22	Standard deviation in c	0.16								
23	Spreadsheet Documentation									
24	Cell B12=SLOPE(B5:B9,A5:A9)									
25	Cell B13=INTERCEPT(B5:B9,A5:A9)									
26	Cell B14=-B13/B12									
27	Cell B15=-B14*B2/B3									
28	Cell B17=STEYX(B5:B9,A5:A9)									
29	Cell B18=COUNT(B5:B9)									
30	Cell B19=DEVSQ(A5:A9)									
31	Cell B20=AVERAGE(B5:B9)									
32	Cell B21=(B17/B12)*SQRT(1/B18+((0-B20)^2)/((B12^2)*B19))									
33	Cell B22=B21*B2/B3									

FIGURE 1-11 Spreadsheet for standard-addition Example 1-1.

In the interest of saving time or sample, it is possible to perform a standard-addition analysis by using only two increments of sample. Here, a single addition of V_s mL of standard would be added to one of the two samples, and we can write

$$S_1 = \frac{kV_xc_x}{V_t}$$

$$S_2 = \frac{kV_xc_x}{V_t} + \frac{kV_sc_s}{V_t}$$

where S_1 and S_2 are the signals resulting from the diluted sample and the diluted sample plus standard, respectively. Dividing the second equation by the first gives, after rearranging,

$$c_x = \frac{S_1c_sV_s}{(S_2 - S_1)V_x}$$

The single-addition method is somewhat dangerous because it presumes a linear relationship and provides no check of this assumption. The multiple-addition method at least gives a check on the linearity supposition.

1D-4 The Internal-Standard Method

An *internal standard* is a substance that is added in a constant amount to all samples, blanks, and calibration standards in an analysis. Alternatively, it may be a major constituent of samples and standards that is present in a large enough amount that its concentration can be assumed to be the same in all cases. Calibration then involves plotting the ratio of the analyte signal to the internal-standard signal as a function of the analyte concentration of the standards. This ratio for the samples is then used to obtain their analyte concentrations from a calibration curve.

An internal standard, if properly chosen and used, can compensate for several types of both random and systematic errors. Thus, if the analyte and internal-standard signals respond proportionally to random instrumental and method fluctuations, the ratio of these signals is independent of such fluctuations. If the two signals are influenced in the same way by matrix effects, compensation of these effects also occurs. In those instances where the internal standard is a major constituent of samples and standards, compensation for errors that arise in sample preparation, solution, and cleanup may also occur.

A major difficulty in applying the internal-standard method is that of finding a suitable substance to serve as the internal standard and of introducing that substance into both samples and standards in a reproducible way. The internal standard should provide a signal that is similar to the analyte signal in most ways but sufficiently different so that the two signals are distinguishable by the instrument. The internal standard must be known to be absent from the sample matrix so that the only source of the standard is the added amount. For example, lithium is a good internal standard for the determination of sodium or potassium in blood serum because the chemical behavior of lithium is similar to both analytes, but it does not occur naturally in blood.

An example of the determination of sodium in blood by flame spectrometry using lithium as an internal standard is shown in Figure 1-12. The upper figure shows the normal calibration curve of sodium intensity versus sodium concentration in ppm. Although a fairly linear plot is obtained, quite a bit of scatter is observed. The lower plot shows the intensity ratio of sodium to lithium plotted against the sodium concentration in ppm. Note the improvement in the calibration curve when the internal standard is used.

In the development of any new internal-standard method, we must verify that changes in concentration of analyte do not affect the signal intensity that results from the internal standard and that the internal standard does not suppress or enhance the analyte signal.

1E SELECTING AN ANALYTICAL METHOD

Column 2 of Table 1-1 shows that today we have an enormous array of tools for carrying out chemical analyses. There are so many, in fact, that the choice among them is often difficult. In this section, we briefly describe how such choices are made.

1E-1 Defining the Problem

To select an analytical method intelligently, it is essential to define clearly the nature of the analytical problem. Such a definition requires answers to the following questions:

1. What accuracy is required?
2. How much sample is available?
3. What is the concentration range of the analyte?
4. What components of the sample might cause interference?

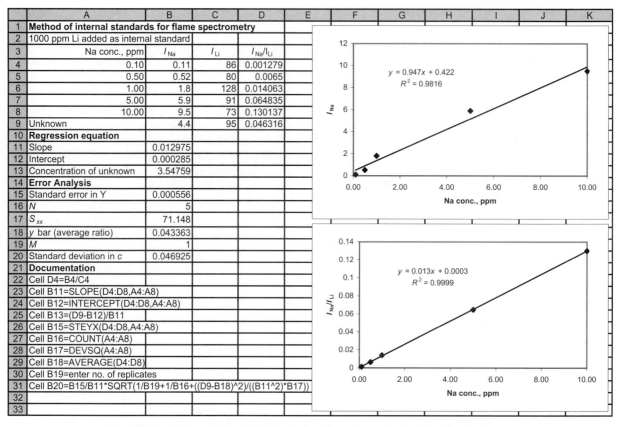

	A	B	C	D	E	F	G	H	I	J	K
1	Method of internal standards for flame spectrometry										
2	1000 ppm Li added as internal standard										
3	Na conc., ppm	I_{Na}	I_{Li}	I_{Na}/I_{Li}							
4	0.10	0.11	86	0.001279							
5	0.50	0.52	80	0.0065							
6	1.00	1.8	128	0.014063							
7	5.00	5.9	91	0.064835							
8	10.00	9.5	73	0.130137							
9	Unknown	4.4	95	0.046316							
10	Regression equation										
11	Slope	0.012975									
12	Intercept	0.000285									
13	Concentration of unknown	3.54759									
14	Error Analysis										
15	Standard error in Y	0.000556									
16	N	5									
17	S_{xx}	71.148									
18	y bar (average ratio)	0.043363									
19	M	1									
20	Standard deviation in c	0.046925									
21	Documentation										
22	Cell D4=B4/C4										
23	Cell B11=SLOPE(D4:D8,A4:A8)										
24	Cell B12=INTERCEPT(D4:D8,A4:A8)										
25	Cell B13=(D9-B12)/B11										
26	Cell B15=STEYX(D4:D8,A4:A8)										
27	Cell B16=COUNT(A4:A8)										
28	Cell B17=DEVSQ(A4:A8)										
29	Cell B18=AVERAGE(D4:D8)										
30	Cell B19=enter no. of replicates										
31	Cell B20=B15/B11*SQRT(1/B19+1/B16+((D9-B18)^2)/((B11^2)*B17))										
32											
33											

FIGURE 1-12 Spreadsheet to illustrate the internal-standard method for the flame spectrometric determination of sodium.

5. What are the physical and chemical properties of the sample matrix?

6. How many samples are to be analyzed?

The answer to question 1 is of vital importance because it determines how much time and care will be needed for the analysis. The answers to questions 2 and 3 determine how sensitive the method must be and how wide a range of concentrations must be accommodated. The answer to question 4 determines the selectivity required of the method. The answers to question 5 are important because some analytical methods in Table 1-1 are applicable to solutions (usually aqueous) of the analyte. Other methods are more easily applied to gaseous samples, and still other methods are suited to the direct analysis of solids.

The number of samples to be analyzed (question 6) is also an important consideration from an economic standpoint. If this number is large, considerable time and money can be spent on instrumentation, method development, and calibration. Furthermore, if the number is large, a method should be chosen that requires the least operator time per sample. On the other hand, if only a few samples are to be analyzed, a simpler but more time-consuming method that requires little or no preliminary work is often the wiser choice.

With answers to these six questions, a method can then be chosen, provided that the performance characteristics of the various instruments shown in Table 1-1 are known.

1E-2 Performance Characteristics of Instruments

Table 1-3 lists quantitative instrument performance criteria that can be used to decide whether a given instrumental method is suitable for attacking an analytical problem. These characteristics are expressed in numerical terms that are called *figures of merit*. Figures of merit permit us to narrow the choice of instruments for a given analytical problem to a relatively few. Selection among these few can then be based on the qualitative performance criteria listed in Table 1-4.

TABLE 1-3 Numerical Criteria
for Selecting Analytical Methods

Criterion	Figure of Merit
1. Precision	Absolute standard deviation, relative standard deviation, coefficient of variation, variance
2. Bias	Absolute systematic error, relative systematic error
3. Sensitivity	Calibration sensitivity, analytical sensitivity
4. Detection limit	Blank plus three times standard deviation of the blank
5. Dynamic range	Concentration limit of quantitation (LOQ) to concentration limit of linearity (LOL)
6. Selectivity	Coefficient of selectivity

TABLE 1-4 Other Characteristics
to Be Considered in Method Choice

1. Speed
2. Ease and convenience
3. Skill required of operator
4. Cost and availability of equipment
5. Per-sample cost

In this section, we define each of the six figures of merit listed in Table 1-3. These figures are then used throughout the remainder of the text in discussing various instruments and instrumental methods.

Precision

As we show in Section a1A-1, Appendix 1, the precision of analytical data is the degree of mutual agreement among data that have been obtained in the same way. Precision provides a measure of the random, or indeterminate, error of an analysis. Figures of merit for precision include *absolute standard deviation, relative standard deviation, standard error of the mean, coefficient of variation*, and *variance*. These terms are defined in Table 1-5.

Bias

As shown in Section a1A-2, Appendix 1, bias provides a measure of the systematic, or determinate, error of an analytical method. Bias Δ is defined by the equation

$$\Delta = \mu - \tau \tag{1-9}$$

TABLE 1-5 Figures of Merit for
Precision of Analytical Methods

Terms	Definition*
Absolute standard deviation, s	$s = \sqrt{\dfrac{\sum_{i=1}^{N}(x_i - \bar{x})^2}{N-1}}$
Relative standard deviation (RSD)	$\text{RSD} = \dfrac{s}{\bar{x}}$
Standard error of the mean, s_m	$s_m = s/\sqrt{N}$
Coefficient of variation (CV)	$\text{CV} = \dfrac{s}{\bar{x}} \times 100\%$
Variance	s^2

*x_i = numerical value of the ith measurement

$\bar{x}$ = mean of N measurements = $\dfrac{\sum_{i=1}^{N} x_i}{N}$

where μ is the population mean for the concentration of an analyte in a sample and τ is the true value.

Determining bias involves analyzing one or more standard reference materials whose analyte concentration is known. Sources of such materials are discussed in Section a1A-2 of Appendix 1. The results from such an analysis will, however, contain both random and systematic errors; but if we repeat the measurements a sufficient number of times, the mean value may be determined with a given level of confidence. As shown in Section a1B-1, Appendix 1, the mean of twenty or thirty replicate analyses can usually be taken as a good estimate of the population mean μ in Equation 1-9. Any difference between this mean and the known analyte concentration of the standard reference material can be attributed to bias.

If performing twenty replicate analyses on a standard is impractical, the probable presence or absence of bias can be evaluated as shown in Example a1-10 in Appendix 1. Usually in developing an analytical method, we attempt to identify the source of bias and eliminate it or correct for it by the use of blanks and by instrument calibration.

Sensitivity

There is general agreement that the sensitivity of an instrument or a method is a measure of its ability to discriminate between small differences in analyte concentration. Two factors limit sensitivity: the slope of the calibration curve and the reproducibility or precision of the measuring device. Of two methods that have equal precision, the one that has the steeper calibration curve will be the more sensitive. A corollary to this statement is that if two methods have calibration curves with equal slopes, the one that exhibits the better precision will be the more sensitive.

The quantitative definition of sensitivity that is accepted by the International Union of Pure and Applied Chemistry (IUPAC) is *calibration sensitivity*, which is the slope of the calibration curve at the concentration of interest. Most calibration curves that are used in analytical chemistry are linear and may be represented by the equation

$$S = mc + S_{bl} \qquad (1\text{-}10)$$

where S is the measured signal, c is the concentration of the analyte, S_{bl} is the instrumental signal for a blank, and m is the slope of the straight line. The quantity S_{bl} is the y-intercept of the straight line. With such curves, the calibration sensitivity is independent of the concentration c and is equal to m. The calibration sensitivity as a figure of merit suffers from its failure to take into account the precision of individual measurements.

Mandel and Stiehler[11] recognized the need to include precision in a meaningful mathematical statement of sensitivity and proposed the following definition for *analytical sensitivity* γ:

$$\gamma = m/s_S \qquad (1\text{-}11)$$

Here, m is again the slope of the calibration curve, and s_S is the standard deviation of the measurement.

The analytical sensitivity offers the advantage of being relatively insensitive to amplification factors. For example, increasing the gain of an instrument by a factor of five will produce a fivefold increase in m. Ordinarily, however, this increase will be accompanied by a corresponding increase in s_S, thus leaving the analytical sensitivity more or less constant. A second advantage of analytical sensitivity is that it is independent of the measurement units for S.

A disadvantage of analytical sensitivity is that it is often concentration dependent because s_S may vary with concentration.

Detection Limit

The most generally accepted qualitative definition of detection limit is that it is the minimum concentration or mass of analyte that can be detected at a known confidence level. This limit depends on the ratio of the magnitude of the analytical signal to the size of the statistical fluctuations in the blank signal. That is, unless the analytical signal is larger than the blank by some multiple k of the variation in the blank due to random errors, it is impossible to detect the analytical signal with certainty. Thus, as the limit of detection is approached, the analytical signal and its standard deviation approach the blank signal S_{bl} and its standard deviation s_{bl}. The minimum distinguishable analytical signal S_m is then taken as the sum of the mean blank signal $\overline{S}_{bl}$ plus a multiple k of the standard deviation of the blank. That is,

$$S_m = \overline{S}_{bl} + ks_{bl} \qquad (1\text{-}12)$$

Experimentally, S_m can be determined by performing twenty to thirty blank measurements, preferably over an extended period of time. The resulting data are then treated statistically to obtain $\overline{S}_{bl}$ and s_{bl}. Finally, the slope from Equation 1-10 is used to convert S_m to c_m, which is defined as the detection limit. The detection limit is then given by

$$c_m = \frac{S_m - \overline{S}_{bl}}{m} \qquad (1\text{-}13)$$

As pointed out by Ingle,[12] numerous alternatives, based correctly or incorrectly on t and z statistics (Section a1B-2, Appendix 1), have been used to determine a value for k in Equation 1-12. Kaiser[13] argues that a reasonable value for the constant is $k = 3$. He points out that it is wrong to assume a strictly normal distribution of results from blank measurements and that when $k = 3$ the confidence level of detection will be 95% in most cases. He further argues that little is to be gained by using a larger value of k and thus a greater confidence level. Long and Winefordner,[14] in a discussion of detection limits, also recommend the use of $k = 3$.

EXAMPLE 1-2

A least-squares analysis of calibration data for the determination of lead based on its flame emission spectrum yielded the equation

$$S = 1.12c_{Pb} + 0.312$$

[11] J. Mandel and R. D. Stiehler, *J. Res. Natl. Bur. Std.*, **1964**, *A53*, 155.

[12] J. D. Ingle Jr., *J. Chem. Educ.*, **1974**, *51*, 100.

[13] H. Kaiser, *Anal. Chem.*, **1987**, *42*, 53A.

[14] G. L. Long and J. D. Winefordner, *Anal. Chem.*, **1983**, *55*, 712A.

where c_{Pb} is the lead concentration in parts per million and S is a measure of the relative intensity of the lead emission line. The following replicate data were then obtained:

Conc., ppm Pb	No. of Replications	Mean Value of S	s
10.0	10	11.62	0.15
1.00	10	1.12	0.025
0.000	24	0.0296	0.0082

Calculate (a) the calibration sensitivity, (b) the analytical sensitivity at 1 and 10 ppm of Pb, and (c) the detection limit.

Solution

(a) By definition, the calibration sensitivity is the slope $m = 1.12$.

(b) At 10 ppm Pb, $\gamma = m/s_S = 1.12/0.15 = 7.5$.
At 1 ppm Pb, $\gamma = 1.12/0.025 = 45$.
Note that the analytical sensitivity is quite concentration dependent. Because of this, it is not reported as often as the calibration sensitivity.

(c) Applying Equation 1-12,

$$S = 0.0296 + 3 \times 0.0082 = 0.054$$

Substituting into Equation 1-13 gives

$$c_m = \frac{0.054 - 0.0296}{1.12} = 0.0022 \text{ ppm Pb}$$

Dynamic Range

Figure 1-13 illustrates the definition of the *dynamic range* of an analytical method, which extends from the lowest concentration at which quantitative measurements can be made (limit of quantitation, or LOQ) to the concentration at which the calibration curve departs from linearity by a specified amount (limit of linearity, or LOL). Usually, a deviation of 5% from linearity is considered the upper limit. Deviations from linearity are common at high concentrations because of nonideal detector responses or chemical effects. The lower limit of quantitative measurements is generally taken to be equal to ten times the standard deviation of repetitive measurements on a blank, or $10s_{bl}$. At this point, the relative standard deviation is about 30% and decreases rapidly as concentrations become larger.

To be very useful, an analytical method should have a dynamic range of at least a few orders of magnitude.

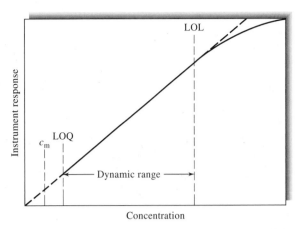

FIGURE 1-13 Useful range of an analytical method. LOQ = limit of quantitative measurement; LOL = limit of linear response.

Some analytical techniques, such as absorption spectrophotometry, are linear over only one to two orders of magnitude. Other methods, such as mass spectrometry and molecular fluorescence, may exhibit linearity over four to five orders of magnitude.

Selectivity

Selectivity of an analytical method refers to the degree to which the method is free from interference by other species contained in the sample matrix. Unfortunately, no analytical method is totally free from interference from other species, and frequently steps must be taken to minimize the effects of these interferences.

Consider, for example, a sample containing an analyte A as well as potential interfering species B and C. If c_A, c_B, and c_C are the concentrations of the three species and m_A, m_B, and m_C are their calibration sensitivities, then the total instrument signal will be given by a modified version of Equation 1-10. That is,

$$S = m_A c_A + m_B c_B + m_C c_C + S_{bl} \qquad (1\text{-}14)$$

Let us now define the selectivity coefficient for A with respect to B, $k_{B,A}$, as

$$k_{B,A} = m_B/m_A \qquad (1\text{-}15)$$

The selectivity coefficient then gives the response of the method to species B relative to A. A similar coefficient for A with respect to C is

$$k_{C,A} = m_C/m_A \qquad (1\text{-}16)$$

Substituting these relationships into Equation 1-14 leads to

$$S = m_A(c_A + k_{B,A}c_B + k_{C,A}c_C) + S_{bl} \quad (1\text{-}17)$$

Selectivity coefficients can range from zero (no interference) to values considerably greater than unity. Note that a coefficient is negative when the interference causes a reduction in the intensity of the output signal of the analyte. For example, if the presence of interferent B causes a reduction in S in Equation 1-14, m_B will carry a negative sign, as will $k_{B,A}$.

Selectivity coefficients are useful figures of merit for describing the selectivity of analytical methods. Unfortunately, they are not widely used except to characterize the performance of ion-selective electrodes (Chapter 23). Example 1-3 illustrates the use of selectivity coefficients when they are available.

EXAMPLE 1-3

The selectivity coefficient for an ion-selective electrode for K^+ with respect to Na^+ is reported to be 0.052. Calculate the relative error in the determination of K^+ in a solution that has a K^+ concentration of 3.00×10^{-3} M if

the Na^+ concentration is (a) 2.00×10^{-2} M; (b) 2.00×10^{-3} M; (c) 2.00×10^{-4} M. Assume that S_{bl} for a series of blanks was approximately zero.

Solution

(a) Substituting into Equation 1-17 yields

$$S = m_{K^+}(c_{K^+} + k_{Na^+,K^+}c_{Na^+}) + 0$$

$$S/m_{K^+} = 3.00 \times 10^{-3} + 0.052 \times 2.00 \times 10^{-2}$$

$$= 4.04 \times 10^{-3}$$

If Na^+ were not present

$$S/m_{K^+} = 3.00 \times 10^{-3}$$

The relative error in c_{K^+} will be identical to the relative error in S/m_{K^+} (see Section a, Appendix 1). Therefore,

$$E_{rel} = \frac{4.04 \times 10^{-3} - 3.00 \times 10^{-3}}{3.00 \times 10^{-3}} \times 100\%$$

$$= 35\%$$

Proceeding in the same way, we find

(b) $E_{rel} = 3.5\%$
(c) $E_{rel} = 0.35\%$

QUESTIONS AND PROBLEMS

*Answers are provided at the end of the book for problems marked with an asterisk.

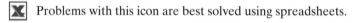 Problems with this icon are best solved using spreadsheets.

1-1 What is a transducer in an analytical instrument?

1-2 What is the information processor in an instrument for measuring the color of a solution visually?

1-3 What is the detector in a spectrograph in which spectral lines are recorded photographically?

1-4 What is the transducer in a smoke detector?

1-5 What is a data domain?

1-6 Name electrical signals that are considered analog. How is the information encoded in an analog signal?

1-7 List four output transducers and describe how they are used.

1-8 What is a figure of merit?

*1-9 A 25.0-mL sample containing Cu^{2+} gave an instrument signal of 23.6 units (corrected for a blank). When exactly 0.500 mL of 0.0287 M $Cu(NO_3)_2$ was added to the solution, the signal increased to 37.9 units. Calculate the molar concentration of Cu^{2+} assuming that the signal was directly proportional to the analyte concentration.

 *1-10 The data in the table below were obtained during a colorimetric determination of glucose in blood serum.

Glucose Concentration, mM	Absorbance, A
0.0	0.002
2.0	0.150
4.0	0.294
6.0	0.434
8.0	0.570
10.0	0.704

  (a) Assuming a linear relationship, find the least-squares estimates of the slope and intercept.
  (b) Use the LINEST function in Excel to find the standard deviations of the slope and intercept.[15] What is the standard error of the estimate?
  (c) Determine the 95% confidence intervals for the slope and intercept.
  (d) A serum sample gave an absorbance of 0.350. Find the glucose concentration and its standard deviation.

1-11 Exactly 5.00-mL aliquots of a solution containing phenobarbital were measured into 50.00-mL volumetric flasks and made basic with KOH. The following volumes of a standard solution of phenobarbital containing 2.000 μg/mL of phenobarbital were then introduced into each flask and the mixture was diluted to volume: 0.000, 0.500, 1.00, 1.50, and 2.00 mL. The fluorescence of each of these solutions was measured with a fluorometer, which gave values of 3.26, 4.80, 6.41, 8.02, and 9.56, respectively.
  (a) Plot the data.
  *(b) Using the plot from (a), calculate the concentration of phenobarbital in the unknown.
  *(c) Derive a least-squares equation for the data.
  *(d) Find the concentration of phenobarbital from the equation in (c).
  (e) Calculate a standard deviation for the concentration obtained in (d).

Challenge Problem

1-12 (a) Use a search engine to find the IUPAC website and locate the *Compendium of Analytical Nomenclature*. What year was the latest edition published? Who were the authors?
  (b) Find the definition of bias recommended in the *Compendium*. Does this definition and the symbology differ from that in this chapter? Explain.

[15]See S. R. Crouch and F. J. Holler, *Applications of Microsoft® Excel in Analytical Chemistry*, Belmont, CA: Brooks/Cole, 2004, Chap. 4.

(c) Find the definition of detection limit recommended in the *Compendium*. Does this definition and the symbology differ from that in this chapter? Explain.

(d) What are the differences between calibration sensitivity and analytical sensitivity?

(e) The following calibration data were obtained by an instrumental method for the determination of species X in aqueous solution.

Conc. X, ppm	No. Replications	Mean Analytical Signal	Standard Deviation
0.00	25	0.031	0.0079
2.00	5	0.173	0.0094
6.00	5	0.422	0.0084
10.00	5	0.702	0.0084
14.00	5	0.956	0.0085
18.00	5	1.248	0.0110

(i) Calculate the calibration sensitivity.

(ii) Find the analytical sensitivity at each analyte concentration.

(iii) Find the coefficient of variation for the mean of each of the replicate sets.

(iv) What is the detection limit for the method?

F. James Holler

Measurement Basics

Our lives have been profoundly influenced by the microelectronics revolution. It is not surprising that the field of instrumental analysis has undergone a similar revolution. The montage of computer and electronic components depicted above is symbolic of both the nature and the pace of the changes that are occurring. By the time you read these lines, one or more of the components in this image will have been supplanted by more sophisticated technology or will have become obsolete.

Chapter 1 laid the foundation for the study of instrumental chemical analysis. In the four chapters of Section I, the fundamental concepts of analog electronics, digital electronics, computers, and data manipulation are presented. These concepts are essential to understanding how instrumental measurements are accomplished. Chapter 2 provides a brief introduction to the components and principles governing basic analog direct-current and alternating-current circuits. Chapter 3 continues the investigation of analog electronics by presenting the principles and applications of operational amplifier circuits. Digital electronics and the boundary between analog and digital domains are explored in Chapter 4, as are the nature of computers and their role in instrumental analysis. In Chapter 5, we complete the inquiry of measurement fundamentals by examining the nature of signals and noise, as well as hardware and software methods for increasing the signal-to-noise ratio. In the Instrumental Analysis in Action feature, we examine a recent development in analytical laboratories, the paperless electronic laboratory.

CHAPTER TWO

Electrical Components and Circuits

I n Chapter 1, we introduced the concept of data domains and pointed out that modern instruments function by converting data from one domain to another. Most of these conversions are between electrical domains. To understand these conversions, and thus how modern electronic instruments work, some knowledge is required of basic direct-current (dc) and alternating-current (ac) circuit components. The purpose of this chapter is to survey these topics in preparation for the two following chapters, which deal with integrated circuits and computers in instruments for chemical analysis. Armed with this knowledge, you will understand and appreciate the functions of the measurement systems and methods discussed elsewhere in this text.

Throughout this chapter, this logo indicates an opportunity for online self-study at **www .thomsonedu.com/chemistry/skoog**, linking you to interactive tutorials, simulations, and exercises.

2A DIRECT-CURRENT CIRCUITS AND MEASUREMENTS

In this section, we consider some basic dc circuits and how they are used in making current, voltage, and resistance measurements. A general definition of a *circuit* is a closed path that may be followed by an electric current. We begin our discussion of circuits with a survey of four important laws of electricity. We adopt here the *positive current convention*. That is, the direction of current in an electrical circuit is from a point of positive potential to a point of negative potential. In many electrical circuits, electrons are the charge carriers, and the positive current convention is opposite to the flow of electrons. However, in semiconductors and ionic solutions, positively charged species can be the major carriers. It is necessary, in any case, to adopt a consistent convention for the direction of current. The positive current convention is nearly universally employed in science and engineering.

2A-1 Laws of Electricity

Ohm's Law

Ohm's law describes the relationship among voltage, resistance, and current in a resistive *series circuit*. In a series circuit, all circuit elements are connected in sequence along a unique path, head to tail, as are the battery and three resistors shown in Figure 2-1. Ohm's law can be written in the form

$$V = IR \qquad (2\text{-}1)$$

where V is the potential difference in volts between two points in a circuit, R is the resistance between the two points in ohms, and I is the resulting current in amperes.[1]

Kirchhoff's Laws

Kirchhoff's current law states that the algebraic sum of currents at any point in a circuit is zero. This means that the sum of currents coming into a point in a circuit has to equal the sum of the currents going out. *Kirchhoff's voltage law* states that the algebraic sum of the voltages around a closed conducting path, or loop, is zero. This means that in a closed loop, the sum of the

---

[1] Throughout most of the text the symbol V denotes the electrical potential difference, or voltage, in circuits. In Chapters 22–25, however, the electrochemical convention will be followed, in which electromotive force is designated as E.

voltage increases (rises) has to equal the sum of the voltage decreases (drops). These laws are a result of the conservation of energy in electrical circuits.

The applications of Kirchhoff's and Ohm's laws to basic dc circuits are considered in Section 2A-2.

Power Law

The power P in watts dissipated in a resistive element is given by the product of the current in amperes and the potential difference across the element in volts:

$$P = IV \tag{2-2}$$

Substituting Ohm's law gives

$$P = I^2R = V^2/R \tag{2-3}$$

2A-2 Direct-Current Circuits

In this section we describe two types of basic dc circuits that find widespread use in electrical devices, namely, *series resistive circuits* and *parallel resistive circuits*, and analyze their properties with the aid of the laws described in the previous section.

Series Circuits

Figure 2-1 shows a basic series circuit, which consists of a battery, a switch, and three resistors in series. Components are in series if they have only one contact in common. When the switch is closed, there is a current

 Simulation: Learn more about **Ohm's law**.

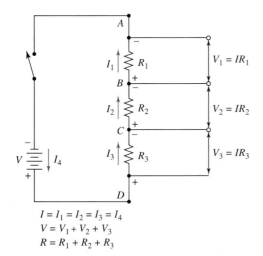

$I = I_1 = I_2 = I_3 = I_4$
$V = V_1 + V_2 + V_3$
$R = R_1 + R_2 + R_3$

FIGURE 2-1 Resistors in series; a voltage divider. Elements are in series if they have only one point in common. The current in a series circuit is everywhere the same. In other words, $I_1 = I_2 = I_3 = I_4$.

in the circuit. Applying Kirchhoff's current law to point D in this circuit gives

$$I_4 - I_3 = 0$$

or

$$I_4 = I_3$$

Note that the current out of point D must be *opposite* in sign to the current into point D. Similarly, application of the law to point C gives

$$I_3 = I_2$$

Thus, *the current is the same at all points in a series circuit*; that is, there is only one current I, given by

$$I = I_1 = I_2 = I_3 = I_4 \tag{2-4}$$

Application of Kirchhoff's voltage law to the circuit in Figure 2-1 yields

$$V - V_3 - V_2 - V_1 = 0$$

or

$$V = V_3 + V_2 + V_1 \tag{2-5}$$

Note that, beginning at the switch and proceeding counterclockwise around the loop, there is one voltage increase (V) and three voltage decreases (V_3, V_2, and V_1). Note also that across a resistive element there is a *voltage drop* in the direction of the current.

Substitution of Ohm's law into Equation 2-5 gives

$$V = I(R_1 + R_2 + R_3) = IR_s \tag{2-6}$$

Equation 2-6 shows that the total resistance R_s of a series circuit is equal to the sum of the resistances of the individual components. That is, for the three resistors of Figure 2-1,

$$R_s = R_1 + R_2 + R_3 \tag{2-7}$$

For n resistors in series, we can extend Equation 2-7 to read

$$R_s = R_1 + R_2 + \cdots + R_n = \sum_{i=1}^{n} R_i \tag{2-8}$$

The Voltage Divider

Resistors in series form a *voltage divider* in that a fraction of the total voltage appears across each resistor. In Figure 2-1 if we apply Ohm's law to the part of the circuit from point B to A, we obtain

$$V_1 = I_1R_1 = IR_1 \tag{2-9}$$

The fraction of the total voltage V that appears across R_1 is V_1/V. By dividing Equation 2-9 by Equation 2-6, we obtain

$$\frac{V_1}{V} = \frac{IR_1}{I(R_1 + R_2 + R_3)} = \frac{R_1}{R_1 + R_2 + R_3} = \frac{R_1}{R_s}$$
$$(2\text{-}10)$$

In a similar way, we may also write

$$\frac{V_2}{V} = \frac{R_2}{R_s}$$

and

$$\frac{V_3}{V} = \frac{R_3}{R_s}$$

Thus, the fraction of the total voltage that appears across a given resistor is the resistance of *that* resistor divided by the total series resistance R_s. This is sometimes referred to as the *voltage divider theorem*.

Voltage dividers are widely used in electrical circuits to provide output voltages that are a fraction of an input voltage. In this mode, they are called *attenuators* and the voltage is said to be *attenuated*. As shown in Figure 2-2a, fixed series resistors can provide voltages in fixed increments. In the switch position shown, the voltage drop across two resistors ($100\,\Omega$ and $100\,\Omega$) is the output voltage V. The fraction of the total voltage V_{AB} selected is $200\,\Omega/R_s$, or $200\,\Omega/500\,\Omega = 0.400$. If V_{AB} were 5 V, for example, the voltage V would be $0.400 \times 5\,\text{V} = 2.00\,\text{V}$. If resistors with accurately known resistances are used, the selected fraction can be very accurate (within 1% or less).

A second type of voltage divider is shown in Figure 2-2b. This type is called a *potentiometer*[2] and provides a voltage that is continuously variable from 0.00 V to the full input voltage V_{AB}. In most potentiometers, the resistance is linear — that is, the resistance between one end, A, and any point, C, is directly proportional to the length, AC, of that portion of the resistor. Then $R_{AC} = kAC$, where AC is expressed in convenient units of length and k is a proportionality constant. Similarly, $R_{AB} = kAB$. Combining these relationships with Equation 2-10 yields

$$\frac{V_{AC}}{V_{AB}} = \frac{AC}{AB}$$

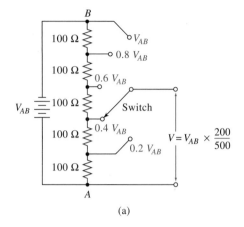

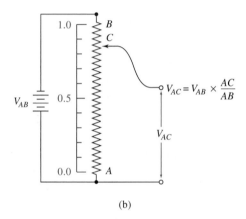

FIGURE 2-2 Voltage dividers: (a) fixed attenuator type and (b) continuously variable type (potentiometer).

or

$$V_{AC} = V_{AB}\frac{AC}{AB} \qquad (2\text{-}11)$$

In commercial potentiometers, R_{AB} is generally a wire-wound resistor formed in a helical coil. A movable contact, called a *wiper*, can be positioned anywhere between one end of the helix and the other, allowing V_{AC} to be varied continuously from zero to the input voltage V_{AB}.

Parallel Circuits

Figure 2-3 depicts a *parallel* dc circuit. Applying Kirchhoff's current law to point A in this figure, we obtain

$$I_1 + I_2 + I_3 - I_t = 0$$

 Tutorial: Learn more about **dc circuits and voltage dividers**.

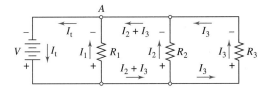

FIGURE 2-3 Resistors in parallel. Parallel circuit elements have two points in common. The voltage across each resistor is equal to V, the battery voltage.

or

$$I_t = I_1 + I_2 + I_3 \qquad (2\text{-}12)$$

Applying Kirchhoff's voltage law to this circuit gives three independent equations. Thus, we may write, for the loop that contains the battery and R_1,

$$V - I_1 R_1 = 0$$
$$V = I_1 R_1$$

For the loop containing V and R_2,

$$V = I_2 R_2$$

For the loop containing V and R_3,

$$V = I_3 R_3$$

Note that the battery voltage V appears across all three resistors.

We could write additional equations for the loop containing R_1 and R_2 as well as the loop containing R_2 and R_3. However, these equations are not independent of the three preceding equations. Substitution of the three independent equations into Equation 2-12 yields

$$I_t = \frac{V}{R_p} = \frac{V}{R_1} + \frac{V}{R_2} + \frac{V}{R_3}$$

Dividing this equation by V, we obtain

$$\frac{1}{R_p} = \frac{1}{R_1} + \frac{1}{R_2} + \frac{1}{R_3} \qquad (2\text{-}13)$$

Because the conductance G of a resistor R is given by $G = 1/R$, we may then write for the three parallel resistors of Figure 2-3

$$G_p = G_1 + G_2 + G_3 \qquad (2\text{-}14)$$

For n resistors in parallel, we can extend Equations 2-13 and 2-14 to read

$$\frac{1}{R_p} = \frac{1}{R_1} + \frac{1}{R_2} + \frac{1}{R_3} + \cdots + \frac{1}{R_n} = \sum_{i=1}^{n} \frac{1}{R_i} \qquad (2\text{-}15)$$

$$G_p = G_1 + G_2 + G_3 = \sum_{i=1}^{n} G_i \qquad (2\text{-}16)$$

Equation 2-16 shows that in a parallel circuit, in contrast to a series circuit, it is the conductances G that are additive rather than the resistances.

For the special case of two resistors in parallel, Equation 2-13 can be solved to give

$$R_p = \frac{R_1 R_2}{R_1 + R_2} \qquad (2\text{-}17)$$

The parallel resistance is just the product of the two resistances divided by the sum.

Current Splitters

Just as series resistances form a voltage divider, parallel resistances create a current divider, or *current splitter*. The fraction of the total current I_t that is present in R_1 in Figure 2-3 is

$$\frac{I_1}{I_t} = \frac{V/R_1}{V/R_p} = \frac{1/R_1}{1/R_p} = \frac{G_1}{G_p}$$

or

$$I_1 = I_t \frac{R_p}{R_1} = I_t \frac{G_1}{G_p} \qquad (2\text{-}18)$$

An interesting special case occurs when two resistances, R_1 and R_2, form a parallel circuit. The fraction of the current in R_1 is given by

$$\frac{I_1}{I_t} = \frac{G_1}{G_p} = \frac{1/R_1}{1/R_p} = \frac{1/R_1}{1/R_1 + 1/R_2} = \frac{R_2}{R_1 + R_2}$$

Similarly,

$$\frac{I_2}{I_t} = \frac{R_1}{R_1 + R_2}$$

In other words, for two parallel resistors, the fraction of the current in one resistor is just the ratio of the resistance of the second resistor to the sum of the resistances of the two resistors. The equations for I_1/I_t and I_2/I_t are often called the *current-splitting equations*.

An illustration of the calculations in series and parallel circuits is given in Example 2-1.

EXAMPLE 2-1

For the accompanying circuit, calculate (a) the total resistance, (b) the current drawn from the battery, (c) the current in each of the resistors, and (d) the potential difference across each of the resistors.

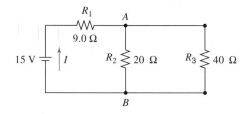

Solution

R_2 and R_3 are parallel resistances. Thus, the resistance $R_{2,3}$ between points A and B will be given by Equation 2-13. That is,

$$\frac{1}{R_{2,3}} = \frac{1}{20\ \Omega} + \frac{1}{40\ \Omega}$$

or

$$R_{2,3} = 13.3\ \Omega$$

We can now reduce the original circuit to the following *equivalent circuit*.

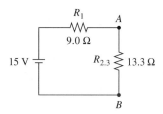

Here, we have the equivalent of two series resistances, and

$$R_s = R_1 + R_{2,3} = 9.0\ \Omega + 13.3\ \Omega = 22.3\ \Omega$$

From Ohm's law, the current I is given by

$$I = 15\ V/22.3\ \Omega = 0.67\ A$$

Employing Equation 2-8, the voltage V_1 across R_1 is

$$V = 15\ V \times 9.0\ \Omega/(9.0\ \Omega + 13.3\ \Omega) = 6.0\ V$$

Similarly, the voltage across resistors R_2 and R_3 is

$$V_2 = V_3 = V_{2,3} = 15\ V \times 13.3\ \Omega/22.3\ \Omega$$
$$= 8.95\ V = 9.0\ V$$

Note that the sum of the two voltages is 15 V, as required by Kirchhoff's voltage law.

The current across R_1 is given by

$$I_1 = I = 0.67\ A$$

The currents through R_2 and R_3 are found from Ohm's law. Thus,

$$I_2 = 9.0\ V/20\ \Omega = 0.45\ A$$
$$I_3 = 9.0\ V/40\ \Omega = 0.22\ A$$

Note that the two currents add to give the net current, as required by Kirchhoff's current law.

2A-3 Direct Current, Voltage, and Resistance Measurements

In this section, we consider (1) how current, voltage, and resistance are measured in dc circuits and (2) the uncertainties associated with such measurements.

Digital Voltmeters and Multimeters

Until about 30 years ago, dc electrical measurements were made with a D'Arsonval moving-coil meter, which was invented more than a century ago. Now such meters are largely obsolete, having been replaced by the ubiquitous digital voltmeter (DVM) and the digital multimeter (DMM).

A DVM usually consists of a single integrated circuit, a power supply that is often a battery, and a liquid-crystal digital display. The heart of the integrated circuit is an *analog-to-digital converter*, which converts the input analog signal to a number that is proportional to the magnitude of the input voltage.[3] A discussion of analog-to-digital converters is given in Section 4C-7. Modern commercial DVMs can be small, are often inexpensive (less than fifty dollars), and generally have input resistances as high as 10^{10} to $10^{12}\ \Omega$.

The DVM is also the heart of a DMM. The DMM can not only measure voltages but has internal circuits that allow it to measure currents and resistances as well. Figure 2-4 illustrates how a DVM can be used to measure dc voltages, currents, and resistances. In each schematic, the reading on the meter display is V_M and the internal resistance of the DVM is R_M. The configuration shown in Figure 2-4a is used to determine the voltage V_x of a voltage source that has an internal resistance of R_s. The voltage displayed by the meter V_M may be somewhat different from the true voltage of the source because of a *loading error*, which is discussed in the section that follows. The input of a DMM

---

[3] An analog signal varies continuously with time and can assume any value within a certain range.

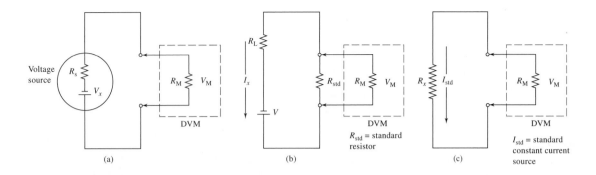

FIGURE 2-4 Uses of a DVM. (a) Measurement of the output V_x of a voltage source. (b) Measurement of current I_x through a load resistor R_L. (c) Measurement of resistance R_x of a resistive circuit element.

is usually a voltage divider, such as that shown in Figure 2-2a, to provide it with several operating ranges.

DMMs can also be used to measure various ranges of current. The unknown current is passed through one of several small standard resistances built into the meter. The voltage drop across this resistance is then measured and is proportional to the current. Figure 2-4b illustrates how an unknown current I_x is measured in a circuit consisting of a dc source and a load resistance R_L. The precision resistors R_{std} in the meter usually range from 0.1 Ω or less to several hundred ohms, thus giving various current ranges. If, for example, $R_{std} = 1.000\ \Omega$ and the DVM reads 0.456 V, then the measured current is 0.456 A (456 mA). By choosing the standard resistors to be even powers of ten and arranging circuitry to move the display decimal point to match the resistor, the DMM reads the current directly.

Figure 2-4c demonstrates how an unknown resistance R_x is determined with a modern DMM. For this application, the meter is equipped with a dc source that produces a constant current I_{std} that is directed through the unknown resistance R_x. The DVM indicates the voltage drop across R_x when the current I_{std} is passed through the resistor. For example, if the standard current is 0.0100 A, then a DVM reading of 0.945 V yields a measured resistance of 0.945 V/0.0100 A = 94.5 Ω. Once again, we just move the decimal point to obtain a direct readout of resistance.

Complete DMMs with current, voltage, and resistance capabilities are available for as little as $30. More sophisticated units with features such as autoranging, semiconductor testing, and ac capabilities can be purchased for less than $100.

Loading Errors in Voltage Measurements

When a meter is used to measure voltage, the presence of the meter tends to perturb the circuit in such a way that a *loading error* is introduced. This situation is not unique to voltage measurements. In fact, it is an example of a fundamental limitation to any physical measurement. That is, the process of measurement inevitably disturbs the system of interest so that the quantity actually measured differs from its value prior to the measurement. Although this type of error can never be completely eliminated, it can often be reduced to insignificant levels.

The magnitude of the loading error in voltage measurements depends on the ratio of the internal resistance of the meter to the resistance of the circuit under consideration. The percentage relative loading error E_r associated with the measured voltage V_M in Figure 2-4a is given by

$$E_r = \frac{V_M - V_x}{V_x} \times 100\%$$

where V_x is the true voltage of the source. Applying Equation 2-11 for a voltage divider, we can write

$$V_M = V_x \left(\frac{R_M}{R_M + R_s} \right)$$

When we substitute this equation into the previous one and rearrange the result, we obtain

$$E_r = -\frac{R_s}{R_M + R_s} \times 100\% \qquad (2\text{-}19)$$

Equation 2-19 shows that the relative loading error becomes smaller as the meter resistance R_M becomes

TABLE 2-1 Effect of Meter Resistance on the Accuracy of Voltage Measurements (see Figure 2-4a)

Meter Resistance R_M, Ω	Resistance of Source R_s, Ω	R_M/R_s	Relative Error %
10	20	0.50	−67
50	20	2.5	−29
500	20	25	−3.8
1.0×10^3	20	50	−2.0
1.0×10^4	20	500	−0.20

TABLE 2-2 Effect of Resistance of Standard Resistor R_{std} on Accuracy of Current Measurement (see Figure 2-4b)

Circuit Resistance R_L, Ω	Standard Resistance R_{std}, Ω	R_{std}/R_L	Relative Error %
1.0	1.0	1.0	−50
10	1.0	0.10	−9.1
100	1.0	0.010	−0.99
1000	1.0	0.0010	−0.10

larger relative to the source resistance R_s. Table 2-1 illustrates this effect. DMMs offer the great advantage of having very high internal resistances of 10^8 to 10^{12} Ω, and thus loading errors are usually avoided except in circuits having resistances of greater than about 10^6 Ω. Often it is the input voltage divider to the DMM that determines the effective input resistance and not the inherent resistance of the meter. An important example of a loading error can occur in the measurement of the voltage of glass pH electrodes, which have resistances of 10^6 to 10^9 Ω or greater. Instruments such as pH meters and pIon meters must have very high resistance inputs to guard against loading errors of this kind.

Loading Errors in Current Measurements

As shown in Figure 2-4b, in making a current measurement, a small, high-precision standard resistor with a resistance R_{std} is introduced into the circuit. In the absence of this resistance, the current in the circuit would be $I = V/R_L$. With resistor R_{std} in place, it would be $I_M = V/(R_L + R_{std})$. Thus, the loading error is given by

$$E_r = \frac{I_M - I_x}{I_x} \times 100\%$$

$$= \frac{\dfrac{V}{(R_L + R_{std})} - \dfrac{V}{R_L}}{\dfrac{V}{R_L}} \times 100\%$$

This equation simplifies to

$$E_r = -\frac{R_{std}}{R_L + R_{std}} \times 100\% \qquad (2\text{-}20)$$

Table 2-2 reveals that the loading error in current measurements becomes smaller as the ratio of R_{std} to R_L becomes smaller.

 Simulation: Learn more about **DMMs and loading**.

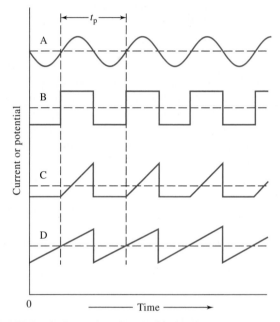

FIGURE 2-5 Examples of periodic signals: (A) sinusoidal, (B) square wave, (C) ramp, and (D) sawtooth.

2B ALTERNATING CURRENT CIRCUITS

The electrical outputs from transducers of analytical signals often fluctuate periodically or can be made to undergo such fluctuations. These changing signals can be represented (as in Figure 2-5) by a plot of the instantaneous current or voltage as a function of time. The *period* t_p for the signal is the time required for the completion of one cycle.

The reciprocal of the period is the *frequency f* of the signal. That is,

$$f = 1/t_p \qquad (2\text{-}21)$$

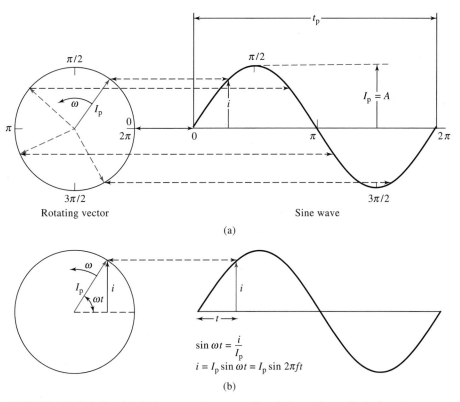

FIGURE 2-6 Relationship between a sine wave of period t_p and amplitude I_p and a corresponding vector of length I_p rotating at an angular velocity $\omega = 2\pi f$ radians/second or a frequency of f Hz.

The unit of frequency is the hertz (Hz), which is defined as one cycle per second.

2B-1 Sinusoidal Signals

The sinusoidal wave (Figure 2-5A) is the most frequently encountered type of periodic electrical signal. A common example is the current produced by rotation of a coil in a magnetic field (as in an electrical generator). Thus, if the instantaneous current or voltage produced by a generator is plotted as a function of time, a sine wave results.

A pure sine wave is conveniently represented as a vector of length I_p (or V_p), which is rotating counterclockwise at a constant angular velocity ω. The relationship between the vector representation and the sine wave plot is shown in Figure 2-6a. The vector rotates at a rate of 2π radians in the period t_p. The angular frequency is thus given by

$$\omega = \frac{2\pi}{t_p} = 2\pi f \qquad (2\text{-}22)$$

If the vector quantity is current or voltage, the instantaneous current i or instantaneous voltage v at time t is given by (see Figure 2-6b)[4]

$$i = I_p \sin \omega t = I_p \sin 2\pi f t \qquad (2\text{-}23)$$

or, alternatively,

$$v = V_p \sin \omega t = V_p \sin 2\pi f t \qquad (2\text{-}24)$$

where I_p and V_p, the maximum, or peak, current and voltage, are called the amplitude, A, of the sine wave.

Figure 2-7 shows two sine waves that have different amplitudes. The two waves are also out of phase by 90°, or $\pi/2$ radians. The phase difference is called the *phase angle* and arises when one vector leads or lags a second

 Simulation: Learn more about **sinusoidal waveforms**.

---

[4]It is useful to symbolize the instantaneous value of time-varying current, voltage, or charge with the lower case letters i, v, and q, respectively. On the other hand, capital letters are used for steady current, voltage, or charge or for a specifically defined variable quantity such as a peak voltage and current, that is, V_p and I_p.

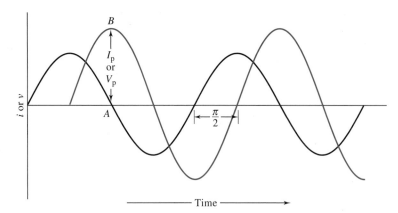

FIGURE 2-7 Sine waves with different amplitudes (I_p or V_p) and with a phase difference of 90°, or $\pi/2$ radians.

by this amount. A more generalized equation for a sine wave is then

$$i = I_p \sin(\omega t + \phi) = I_p \sin(2\pi ft + \phi) \quad (2\text{-}25)$$

where ϕ is the phase angle relative to a reference sine wave. An analogous equation can be written in terms of voltage:

$$v = V_p \sin(\omega t + \phi) = V_p \sin(2\pi ft + \phi) \quad (2\text{-}26)$$

The current or voltage associated with a sinusoidal current can be expressed in several ways. The simplest is the peak amplitude I_p (or V_p), which is the maximum instantaneous current or voltage during a cycle; the peak-to-peak value, which is $2I_p$, or $2V_p$, is also used. An alternating current expressed as the *root-mean-square*, or *rms*, value produces the same heating in a resistor expressed as a direct current of the same magnitude. Thus, the rms current is important in power calculations (Equations 2-2 and 2-3). The rms current is given by

$$I_{rms} = \sqrt{\frac{I_p^2}{2}} = 0.707 I_p$$

$$\quad (2\text{-}27)$$

$$V_{rms} = \sqrt{\frac{V_p^2}{2}} = 0.707 V_p$$

2B-2 Reactance in Electrical Circuits

When the current in an electrical circuit is increased or decreased, energy is required to change the electric and magnetic fields associated with the flow of charge. For example, if the circuit contains a coil of copper wire, or an *inductor*, the coil resists the change in the current as energy is stored in the magnetic field of the inductor. As the current is reversed, the energy is returned to the ac source. As the second half of the cycle is completed, energy is once again stored in a magnetic field of the opposite sense. In a similar way, a capacitor in an ac circuit resists changes in voltage. The resistance of inductors to changes in current and the resistance of capacitors to changes in voltage is called *reactance*. As we shall see, reactances in an ac circuit introduce phase shifts in the ac signal. The two types of reactance that characterize capacitors and inductors are *capacitive reactance* and *inductive reactance*, respectively.

Both capacitive reactance and inductive reactance are frequency-dependent quantities. At low frequency when the rate of change in current is low, the effects of inductive reactance in most of the components of a circuit are sufficiently small to be neglected. With rapid changes, on the other hand, circuit elements such as switches, junctions, and resistors may exhibit inductive reactance. Capacitive reactance, on the other hand, is largest at low frequencies and decreases with increasing frequency. Reactance effects may be undesirable and a result of capacitance and inductance inherent in components. Attempts are made to minimize reactance in such circumstances.

Capacitance and inductance are often deliberately introduced into circuits by using components called *capacitors* and *inductors*. These devices play important roles in such useful functions as converting ac to dc or the converse, discriminating among signals of different frequencies, separating ac and dc signals, and differentiating or integrating signals.

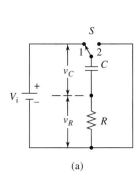
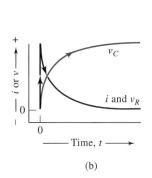
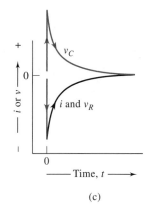

(a)   (b)   (c)

FIGURE 2-8 (a) A series RC circuit. Time response of circuit when switch S is (b) in position 1 and (c) in position 2.

In the sections that follow, only the properties of capacitors are considered because most modern electronic circuits are based on these devices rather than on inductors.

2B-3 Capacitors and Capacitance: Series *RC* Circuits

A typical capacitor consists of a pair of conductors separated by a thin layer of a dielectric substance — that is, by an electrical insulator that contains essentially no mobile, current-carrying, charged species. The simplest capacitor consists of two sheets of metal foil separated by a thin film of a dielectric such as air, oil, plastic, mica, paper, ceramic, or metal oxide. Except for air and mica capacitors, the two layers of foil and the insulator are usually folded or rolled into a compact package and sealed to prevent atmospheric deterioration.

To describe the properties of a capacitor, consider the *series RC circuit* shown in Figure 2-8a, which contains a battery V_i, a resistor R, and a capacitor C, in series. The capacitor is symbolized by a pair of parallel lines of equal length.

When the switch S is moved from position 2 to position 1, electrons flow from the negative terminal of the battery through the resistor R into the lower conductor, or plate, of the capacitor. Simultaneously, electrons are repelled from the upper plate and flow toward the positive terminal of the battery. This movement constitutes a momentary current, which quickly decays to zero as the potential difference builds up across the plates of

 Tutorial: Learn more about **RC circuits**.

the capacitor and eventually reaches the battery voltage V_i. When the current decays to zero, the capacitor is said to be charged.

If the switch then is moved from position 1 to position 2, electrons will flow from the negatively charged lower plate of the capacitor through the resistor R to the positive upper plate. Again, this movement constitutes a current that decays to zero as the potential difference between the two plates disappears; here, the capacitor is said to be discharged.

A useful property of a capacitor is its ability to store an electrical charge for a period of time and then to give up the stored charge when needed. Thus, if S in Figure 2-8a is first held at position 1 until C is charged and is then moved to a position between 1 and 2, the capacitor will maintain its charge for an extended period. When S is moved to position 2, discharge occurs the same way it would if the change from 1 to 2 had been rapid.

The quantity of electricity Q required to fully charge a capacitor depends on the area of the plates, their shape, the spacing between them, and the dielectric constant for the material that separates them. In addition, the charge Q is directly proportional to the applied voltage. That is,

$$Q = CV \tag{2-28}$$

When V is the applied voltage (volts) and Q is the quantity of charge (coulombs), the proportionality constant C is the capacitance of a capacitor in farads (F). A one-farad capacitor, then, stores one coulomb of charge per applied volt. Most of the capacitors used in electronic circuitry have capacitances in the microfarad (10^{-6} F) to picofarad (10^{-12} F) ranges.

Capacitance is important in ac circuits, particularly because a voltage that varies with time gives rise to a time-varying charge — that is, a current. This behavior can be understood by differentiating Equation 2-28 to give

$$\frac{dq}{dt} = C\frac{dv_C}{dt} \qquad (2\text{-}29)$$

By definition, the current i is the time rate of change of charge; that is, $dq/dt = i$. Thus,

$$i = C\frac{dv_C}{dt} \qquad (2\text{-}30)$$

It is important to note that the current in a capacitor is zero when the voltage is time independent — that is, when the voltage across the capacitor is constant. Furthermore, note that a very large current is required to make a rapid change in the voltage across a capacitor. This result imposes a significant limitation on certain electroanalytical methods of analysis, as discussed in Chapter 25.

Rate of Current Change in an RC Circuit

The rate at which a capacitor is charged or discharged is finite. Consider, for example, the circuit shown in Figure 2-8a. From Kirchhoff's voltage law, we know that at any instant after the switch is moved to position 1, the sum of the voltages across $C(v_C)$ and $R(v_R)$ must equal the input voltage V_i. Thus,

$$V_i = v_C + v_R \qquad (2\text{-}31)$$

Because V_i is constant, the increase in v_C that accompanies the charging of the capacitor must be exactly offset by a decrease in v_R.

Substitution of Equations 2-1 and 2-28 into this equation gives, after rearrangement,

$$V_i = \frac{q}{C} + iR \qquad (2\text{-}32)$$

To determine how the current in an RC circuit changes as a function of time, we differentiate Equation 2-32 with respect to time, remembering that V_i is constant. Thus,

$$\frac{dV_i}{dt} = 0 = \frac{dq/dt}{C} + R\frac{di}{dt} \qquad (2\text{-}33)$$

Here, again, we have used lowercase letters to represent instantaneous charge and current.

As noted earlier, $dq/dt = i$. Substituting this expression into Equation 2-33 yields, after rearrangement,

$$\frac{di}{i} = -\frac{dt}{RC}$$

Integration between the limits of the initial current I_{init} and i gives

$$\int_{I_{init}}^{i} \frac{di}{i} = -\int_{0}^{t} \frac{dt}{RC} \qquad (2\text{-}34)$$

and

$$i = I_{init}\, e^{-t/RC} \qquad (2\text{-}35)$$

This equation shows that the current in the RC circuit decays exponentially with time.

Rate of Voltage Change in an RC Circuit

To obtain an expression for the instantaneous voltage across the resistor v_R, we use Ohm's law to substitute $i = v_R/R$ and $I_{init} = V_R/R$ into Equation 2-35 and rearrange to obtain

$$v_R = V_i\, e^{-t/RC} \qquad (2\text{-}36)$$

Substitution of this expression into Equation 2-31 yields, after rearrangement, an expression for the instantaneous voltage across the capacitor v_C:

$$v_C = V_i(1 - e^{-t/RC}) \qquad (2\text{-}37)$$

Note that the product RC that appears in the last three equations has the units of time, because $R = v_R/i$ and $C = q/v_C$,

$$RC = \frac{v_R}{i} \times \frac{q}{v_C}$$

and

$$\frac{\text{volts}}{\text{coulombs/second}} \times \frac{\text{coulombs}}{\text{volt}} = \text{seconds}$$

The product RC is called the *time constant* for the circuit and is a measure of the time required for a capacitor to charge or discharge. This dependence of the charging time on RC can be rationalized from the form of Equation 2-37. Because the ratio $-t/RC$ is the exponent in this equation, RC determines the rate of exponential change of the voltage across the capacitor.

Example 2-2 illustrates the use of the equations that were just derived.

EXAMPLE 2-2

Values for the components in Figure 2-8a are $V_i = 10.0$ V, $R = 1000$ Ω, $C = 1.00$ μF or 1.00×10^{-6} F. Calculate (a) the time constant for the circuit and (b) i, v_C, and v_R after two time constants ($t = 2RC$) have elapsed.

Solution

(a) Time constant $= RC = 1000 \times 1.00 \times 10^{-6} = 1.00 \times 10^{-3}$ s or 1.00 ms.

(b) Substituting Ohm's law, $I_{init} = V_i/R$, and $t = 2.00$ ms into Equation 2-35 reveals

$$i = \frac{V}{R} e^{-t/RC} = \frac{10.0}{1000} e^{-2.00/1.00}$$

$$= 1.35 \times 10^{-3} \text{A} \quad \text{or} \quad 13.5 \text{ mA}$$

We find from Equation 2-36 that

$$v_R = 10.0 \, e^{-2.00/1.00} = 1.35 \text{ V}$$

And by substituting into Equation 2-31, we find that

$$v_C = V_i - v_R = 10.00 - 1.35$$

$$= 10.0(1 - e^{-2.00/1.00}) = 8.65 \text{ V}$$

Phase Relations between Current and Voltage in an RC Circuit

Figure 2-8b shows the changes in i, v_R, and v_C that occur during the charging cycle of an RC circuit. These plots are presented in arbitrary units because the shape of the curves is independent of the time constant of the circuit. Note that v_R and i assume their maximum values the instant the switch in Figure 2-8a is moved to position 1. At the same instant, on the other hand, the voltage across the capacitor increases rapidly from zero and ultimately approaches a constant value. For practical purposes, a capacitor is considered to be fully charged after five time constants ($5RC$) have elapsed. At this point, the current has decayed to less than 1% of its initial value ($e^{-5RC/RC} = e^{-5} = 0.0067 \approx 0.01$).

When the switch in Figure 2-8a is moved to position 2, the battery is removed from the circuit and the capacitor becomes a source of current. The flow of charge, however, will be in the opposite direction from what it was during charging. Thus,

$$dq/dt = -i$$

If it was fully charged, the initial capacitor voltage is that of the battery. That is,

$$V_C = V_i$$

Using these equations and proceeding as in the earlier derivation, we find that for the discharge cycle

$$i = -\frac{V_C}{R} e^{-t/RC} \tag{2-38}$$

$$v_R = -V_C e^{-t/RC} \tag{2-39}$$

and because $V_i = 0 = v_C + v_R$ (Equation 2-31)

$$v_C = V_C e^{-t/RC} \tag{2-40}$$

Figure 2-8c shows how i, v_R, and v_C change with time.

It is important to note that in each cycle the change in voltage across the capacitor is out of phase with and lags behind that of the current and the voltage across the resistor.

2B-4 Response of Series *RC* Circuits to Sinusoidal Inputs

In the sections that follow, we investigate the response of series RC circuits to a sinusoidal ac voltage signal. The input signal v_s is described by Equation 2-24; that is,

$$v_s = V_p \sin \omega t = V_p \sin 2\pi ft \tag{2-41}$$

Phase Changes in Capacitive Circuits

If the switch and battery in the RC circuit shown in Figure 2-8a are replaced with a sinusoidal ac source, the capacitor continuously stores and releases charge, thus causing a current that alternates in direction and changes continuously. A phase difference ϕ between the current and voltage is introduced as a consequence of the finite time required to charge and discharge the capacitor (see Figure 2-8b and c).

We may determine the magnitude of the phase shift by considering a capacitor in an ideal circuit that has no resistance. We first combine Equations 2-23 and 2-30 to give, after rearrangement,

$$C \frac{dv_C}{dt} = I_p \sin 2\pi ft \tag{2-42}$$

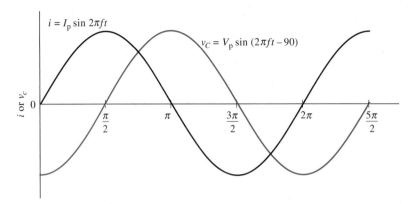

FIGURE 2-9 Sinusoidal current i and voltage v_C signals for a capacitor.

At time $t = 0$, $v_C = 0$. Thus, if we rearrange this equation and integrate between times 0 and t, we obtain

$$v_C = \frac{I_p}{C} \int_0^t \sin 2\pi f t\, dt = \frac{I_p}{2\pi f C}(-\cos 2\pi f t)$$

But from trigonometry, $-\cos x = \sin(x - 90)$. Therefore, we may write

$$v_C = \frac{I_p}{2\pi f C} \sin(2\pi f t - 90) \qquad (2\text{-}43)$$

By comparing Equation 2-43 with Equation 2-26, we see that $I_p/(2\pi f C) = V_p$; therefore, Equation 2-43 can be written in the form

$$v_C = V_p \sin(2\pi f t - 90) \qquad (2\text{-}44)$$

The instantaneous current, however, is given by Equation 2-23. That is,

$$i = I_p \sin 2\pi f t$$

When we compare the last two equations, we find that the voltage across a pure capacitor that results from a sinusoidal input signal is sinusoidal but lags the current by 90° (see Figure 2-9). As we show later, this lag is smaller than 90° in a real circuit that also contains resistance.

Capacitive Reactance

Like a resistor, a capacitor during charging impedes the flow of charge, which results in a continuous decrease in the magnitude of the current. This effect results from the limited capacity of the device to hold charge at a given voltage, as given by the expression $Q = CV$. In contrast to a resistor, however, charging a capacitor does not create a permanent loss of energy

as heat. Here, the energy stored in the charging process is released to the system during discharge.

Ohm's law can be applied to capacitive ac circuits and takes the form

$$V_p = I_p X_C \qquad (2\text{-}45)$$

where X_C is the *capacitive reactance*, a property of a capacitor that is analogous to the resistance of a resistor. A comparison of Equations 2-43 and 2-44 shows, however, that

$$V_p = \frac{I_p}{2\pi f C} = \frac{I_p}{\omega C}$$

Thus, the capacitive reactance is given by

$$X_C = \frac{V_p}{I_p} = \frac{I_p}{I_p 2\pi f C} = \frac{1}{2\pi f C} = \frac{1}{\omega C} \qquad (2\text{-}46)$$

where X_C has the units of ohms.

It should also be noted that, in contrast to resistance, capacitive reactance is *frequency dependent* and becomes smaller at higher frequency. At very high frequencies, capacitive reactance approaches zero, and the capacitor acts like a short circuit. At zero frequency X_C becomes extremely large, so that a capacitor acts as an open circuit (insulator) for a direct current (neglecting the momentary initial charging current). Example 2-3 shows a spreadsheet[5] calculation of capacitive reactance at several different frequencies.

---

[5] For additional information on spreadsheet applications, see S. R. Crouch and F. J. Holler, *Applications of Microsoft® Excel in Analytical Chemistry*, Belmont, CA: Brooks/Cole, 2004.

EXAMPLE 2-3

Use a spreadsheet to calculate the reactance of a 0.0200 μF capacitor at frequencies of 28 Hz, 280 Hz, 2.8 kHz, 28 kHz, 280 kHz, and 2.8 MHz.

▸ *Solution*

	A	B	C
1	Example 2–3 Capacitive Reactance		
2	C	2.00E−08	
3		f, Hz	$X_C = 1/2\pi fC, \Omega$
4		28	2.84E+05
5		280	2.84E+04
6		2800	2.84E+03
7		28000	2.84E+02
8		280000	2.84E+01
9		2.80E+06	2.84E+00
10			
11	Documentation		
12	Cell C4=1/(2*PI()*B4*$B$2)		

We employ Equation 2-46 in cells C4–C9 and note that the reactance varies from 284 kΩ at 28 Hz to only 2.84 Ω at 2.8 MHz.

Impedance in a Series RC Circuit

The impedance Z of an RC circuit is made up of two components: the resistance of the resistor and the reactance of the capacitor. Because of the phase shift with the latter, however, the two cannot be combined directly but must be added vectorially, as shown in Figure 2-10. Here the phase angle for R is chosen as zero. As we have shown, the phase angle for a pure capacitive element is −90°. Thus, the X_C vector is drawn at a right angle to and extends down from the R vector. From the Pythagorean theorem, the quantity Z, called the *impedance*, is given by

$$Z = \sqrt{R^2 + X_C^2} \qquad (2\text{-}47)$$

The phase angle is

$$\phi = \arctan\frac{X_C}{R} \qquad (2\text{-}48)$$

To show the frequency dependence of the impedance and of the phase angle, we can substitute Equation 2-46 into 2-47 and 2-48, giving

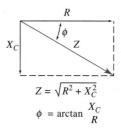

$$Z = \sqrt{R^2 + X_C^2}$$
$$\phi = \arctan\frac{X_C}{R}$$

FIGURE 2-10 Vector diagram for a series RC circuit.

$$Z = \sqrt{R^2 + \left(\frac{1}{2\pi fC}\right)^2} \qquad (2\text{-}49)$$

and

$$\phi = \arctan\frac{1}{2\pi fRC} \qquad (2\text{-}50)$$

Note that the extent to which the voltage lags the current in an RC circuit, ϕ, depends on the frequency f, the resistance R, and the capacitance C, of the circuit.

Ohm's law for a series RC circuit can be written as

$$I_p = \frac{V_p}{Z} = \frac{V_p}{\sqrt{R^2 + \left(\frac{1}{2\pi fC}\right)^2}} \qquad (2\text{-}51)$$

or

$$V_p = I_pZ = I_p\sqrt{R^2 + \left(\frac{1}{2\pi fC}\right)^2}$$

Example 2-4 shows spreadsheet calculations[6] of impedance in a series RC circuit.

EXAMPLE 2-4

A sinusoidal ac source with a peak voltage of 20.0 V was placed in series with a 15 kΩ resistor and a 0.0080 μF capacitor. Use a spreadsheet to calculate the peak current, the phase angle, and the voltage drop across each of the components for frequencies of 75 Hz, 750 Hz, 7.5 kHz, and 75 kHz.

[6] See S. R. Crouch and F. J. Holler, *Applications of Microsoft® Excel in Analytical Chemistry*, Belmont, CA: Brooks/Cole, 2004.

▶ *Solution*

	A	B	C	D	E	F	G	H	I
1	**Example 2–4 Impedance Calculations**								
2	R	1.5E+04	V_P	20.0					
3	C	8.0E−09							
4		f, Hz	X_C, Ω	Z, Ω	ϕ, rad	ϕ, deg	I_P, A	$(V_P)_R$, V	$(V_P)_C$, V
5		75	2.7E+05	2.7E+05	1.5	87	7.5E−05	1.1	20.0
6		750	2.7E+04	3.0E+04	1.1	61	6.6E−04	9.8	17.4
7		7.5E+03	2.7E+03	1.5E+04	0.2	10	1.3E−03	19.7	3.4
8		7.5E+04	2.7E+02	1.5E+04	0.0	1	1.3E−03	20.0	0.354
9									
10	**Spreadsheet Documentation**								
11	Cell C5=1/(2*PI()*B5*$B$3)				Cell G5=$D$2/D5				
12	Cell D5=SQRT($B$2^2+C5^2)				Cell H5=$D$2*$B$2/D5				
13	Cell E5=ATAN(C5/$B$2)				Cell I5=$D$2*C5/D5				
14	Cell F5=DEGREES(E5)								

Note here that in cells C5–C8 we calculate the capacitive reactance from Equation 2-46 as we did in Example 2-3. In cells D5–D8 we obtain the impedance of the circuit from Equation 2-49. In cells E5–E8 we use Equation 2-48 to obtain the phase angle. Note that Excel calculates the arc tangent in radians. In cells F5–F8 we convert the phase angle to degrees by using the Excel function DEGREES(). In cells G5–G8 we obtain the peak current from Equation 2-51. The voltage drops across the resistor and the capacitor are obtained from the voltage divider equations

$$(V_p)_R = V_p\left(\frac{R}{Z}\right) \quad \text{and} \quad (V_p)_C = V_p\left(\frac{X_C}{Z}\right).$$

Several important properties of a series RC circuit are illustrated by the results obtained in Example 2-4. First, the sum of the peak voltages for the resistor and the capacitor are not equal to the peak voltage of the source. At the lower frequency, for example, the sum is 21.1 V compared with 20.0 V for the source. This apparent anomaly is understandable when we realize that the peak voltage occurs in the resistor at an earlier time than in the capacitor because of the voltage lag in the latter. At any time, however, the sum of the instantaneous voltages across the two elements equals the instantaneous voltage of the source.

A second important point shown by the data in Example 2-4 is that the reactance of the capacitor is three orders of magnitude greater at 75 Hz than at the highest frequency. As a result, the impedance at the two highest frequencies is almost entirely due to the resistor, and the current is significantly greater. Associated with the lowered reactance at the higher frequencies are the much smaller voltages across the capacitor compared with those at lower frequencies.

Finally, the magnitude of the voltage lag in the capacitor is of interest. At the lowest frequency, this lag amounts to approximately 87°, but at the highest frequency it is only about 1°.

2B-5 Filters Based on *RC* Circuits

Series RC circuits are often used as filters to attenuate high-frequency signals while passing low-frequency components (a low-pass filter) or, alternatively, to reduce low-frequency components while passing the high frequencies (a high-pass filter). Figure 2-11 shows how a series RC circuit can be arranged to give a high- and a low-pass filter. In each case, the input and output are indicated as the voltages $(V_p)_i$ and $(V_p)_o$.

High-Pass Filters

To use an RC circuit as a high-pass filter, the output voltage is taken across the resistor R (see Figure 2-11a). The peak current in this circuit can be found by substitution into Equation 2-51. Thus,

$$I_p = \frac{(V_p)_i}{Z} = \frac{(V_p)_i}{\sqrt{R^2 + \left(\frac{1}{2\pi fC}\right)^2}} \qquad (2\text{-}52)$$

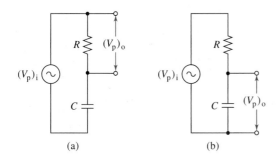

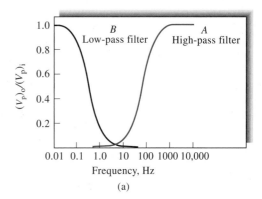

(a)

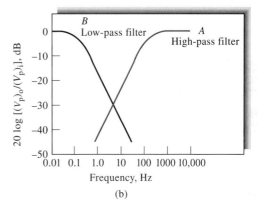

(b)

FIGURE 2-11 Filter circuits: (a) a high-pass filter and (b) a low-pass filter.

Because the voltage across the resistor is in phase with the current,

$$I_p = \frac{(V_p)_o}{R}$$

The ratio of the peak output to the peak input voltage is obtained by dividing the second equation by the first and rearranging. Thus,

$$\frac{(V_p)_o}{(V_p)_i} = \frac{R}{Z} = \frac{R}{\sqrt{R^2 + \left(\dfrac{1}{2\pi f C}\right)^2}} \qquad (2\text{-}53)$$

A plot of this ratio as a function of frequency for a typical high-pass filter is shown as curve A in Figure 2-12a. Note that frequencies below 20 Hz have been largely removed from the input signal.

Low-Pass Filters

For the low-pass filter shown in Figure 2-11b, we can write

$$(V_p)_o = I_p X_C$$

Substituting Equation 2-46 gives, on rearranging,

$$I_p = 2\pi f C (V_p)_o$$

Substituting Equation 2-52 and rearranging yields

$$\frac{(V_p)_o}{(V_p)_i} = \frac{X_C}{Z} = \frac{1/2\pi f C}{\sqrt{R^2 + \left(\dfrac{1}{2\pi f C}\right)^2}} \qquad (2\text{-}54)$$

Curve B in Figure 2-12a shows the frequency response of a typical low-pass filter; the data for the plot were obtained with the aid of Equation 2-54. In this case,

 Simulation: Learn more about **RC filters**.

FIGURE 2-12 (a) Frequency response of high-pass and low-pass filters. (b) Bode diagram for high-pass and low-pass filters. For the high-pass filter, $R = 10$ kΩ and $C = 0.1$ μF. For the low-pass filter, $R = 1$ MΩ and $C = 1$ μF.

direct and low-frequency components of the input signal are transferred to the circuit output, but high-frequency components are effectively removed.

Figure 2-12b shows *Bode diagrams*, or *plots*, for the two filters just described. Bode plots are widely encountered in the electronics literature to show the frequency dependence of input-output ratios for circuits, amplifiers, and filters. The quantity $20 \log[(V_p)_o/(V_p)_i]$ gives the gain (or attenuation) of an amplifier or a filter in *decibels*, dB. Thus, if $[(V_p)_o/(V_p)_i] = 10$, the gain is 20 dB. If $[(V_p)_o/(V_p)_i] = -10$, the attenuation is 20 dB.

Low- and high-pass filters are of great importance in the design of electronic circuits.

2B-6 The Response of *RC* Circuits to Pulsed Inputs

When a pulsed input is applied to an *RC* circuit, the voltage across the capacitor and the voltage across the resistor take various forms, depending on the relation-

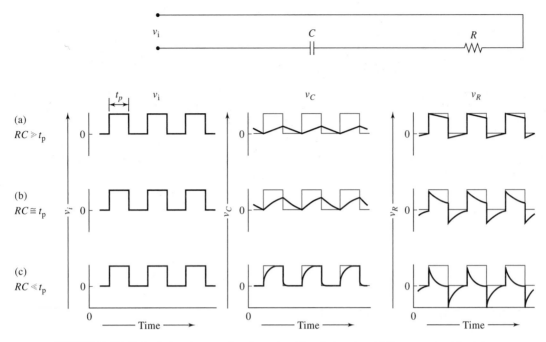

FIGURE 2-13 Output signals v_R and v_C for pulsed input signal v_i. (a) time constant $\gg$ pulse width t_p; (b) time constant $= t_p$; (c) time constant $\ll t_p$.

ship between the width of the pulse and the time constant for the circuit. These effects are illustrated in Figure 2-13, where the input is a square wave with a pulse width of t_p seconds. The second column shows the variation in capacitor voltage as a function of time, and the third column shows the change in resistor voltage at the same times. In the top set of plots (Figure 2-13a), the time constant of the circuit is much greater than the input pulse width. Under these circumstances, the capacitor can become only partially charged during each pulse. The capacitor then discharges as the input voltage returns to zero, and a sawtooth output results. The voltage across the resistor under these circumstances rises instantaneously to a maximum value and then decreases nearly linearly during the pulse lifetime.

The bottom set of curves (Figure 2-13c) illustrates the two outputs when the time constant of the circuit is much shorter than the pulse width. Here, the charge on the capacitor rises rapidly and approaches full charge near the end of the pulse. As a consequence, the voltage across the resistor rapidly decreases to zero after its initial rise. When v_i goes to zero, the capacitor

discharges immediately; the output across the resistor peaks in a negative direction and then quickly approaches zero.

These various output wave forms find applications in electronic circuitry. The sharply peaked voltage output shown in Figure 2-13c is particularly important in timing and trigger circuits.

2B-7 Alternating Current, Voltage, and Impedance Measurements

Alternating current, voltage, and impedance measurements can be carried out with many DMMs. Such multimeters are sophisticated instruments that permit the measurement of both ac and dc voltages and currents as well as resistances or impedances over ranges of many orders of magnitude. As shown in Figure 2-14, the heart of the DMM is the dc DVM as discussed in Section 2A-3. In this type of meter, circuits similar to those shown in Figure 2-4 are used. For ac measurements, the outputs from the various input converter circuits are passed into an ac-to-dc converter before being digitized and displayed.

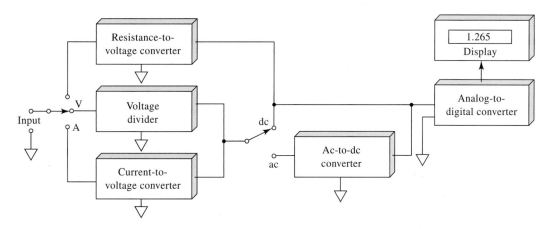

FIGURE 2-14 Block diagram of a DMM. (From H. V. Malmstadt, C. G. Enke, and S. R. Crouch, *Electronics and Instrumentation for Scientists*, Menlo Park, CA: Benjamin-Cummings, 1981, with permission).

2C SEMICONDUCTORS AND SEMICONDUCTOR DEVICES

Electronic circuits usually contain one or more nonlinear devices, such as transistors, semiconductor diodes, and vacuum or gas-filled tubes.[7] In contrast to circuit components such as resistors, capacitors, and inductors, the input and output voltages or currents of nonlinear devices are not linearly proportional to one another. As a consequence, nonlinear components can be made to change an electrical signal from ac to dc (*rectification*) or the reverse, to amplify or to attenuate a voltage or current (*amplitude modulation*), or to alter the frequency of an ac signal (*frequency modulation*).

Historically, the vacuum tube was the predominant nonlinear device used in electronic circuitry. In the 1950s, however, tubes were suddenly and essentially completely displaced by *semiconductor-based diodes* and *transistors*, which have the advantages of low cost, low power consumption, small heat generation, long life, and compactness. The era of the individual, or discrete, transistor was remarkably short, however, and

electronics is now based largely on *integrated circuits*, which contain as many as a million transistors, resistors, capacitors, and conductors on a single tiny semiconductor chip. Integrated circuits permit the scientist or engineer to design and construct relatively sophisticated instruments, armed with only their functional properties and input and output characteristics and without detailed knowledge of the internal electronic circuitry of individual chips.

In this section we examine some of the most common components that make up electronic circuits. We then look in detail at a few devices that are important parts of most electronic instruments.

A semiconductor is a crystalline material with a conductivity between that of a conductor and an insulator. There are many types of semiconducting materials, including elemental silicon and germanium; intermetallic compounds, such as silicon carbide and gallium arsenide; and a variety of organic compounds. Two semiconducting materials that have found widest application for electronic devices are crystalline silicon and germanium. Here, we limit our discussions to these substances.

2C-1 Properties of Silicon and Germanium Semiconductors

Silicon and germanium are Group IV elements and thus have four valence electrons available for bond formation. In a silicon crystal, each of these electrons is

---

[7] For further information about modern electronic circuits and components, see H. V. Malmstadt, C. G. Enke, and S. R. Crouch, *Microcomputers and Electronic Instrumentation: Making the Right Connections*, Washington, DC: American Chemical Society, 1994; A. J. Diefenderfer and B. E. Holton, *Principles of Electronic Instrumentation*, 3rd ed., Philadelphia: Saunders College Publishing, 1994; J. J. Brophy, *Basic Electronics for Scientists*, 5th ed., New York: McGraw-Hill, 1990; P. Horowitz and W. Hill, *The Art of Electronics*, 2nd ed., New York: Cambridge University Press, 1989.

localized by combination with an electron from another silicon atom to form a covalent bond. Thus, in principle, there are no free electrons in crystalline silicon, and the material would be expected to be an insulator. In fact, however, sufficient thermal agitation occurs at room temperature to liberate an occasional electron from its bonded state, leaving it free to move through the crystal lattice and thus to conduct electricity. This thermal *excitation* of an electron leaves a positively charged region, termed a *hole*, associated with the silicon atom. The hole, however, like the electron, is mobile and thus also contributes to the electrical conductance of the crystal. The mechanism of hole movement is stepwise; a bound electron from a neighboring silicon atom jumps to the electron-deficient region and thereby leaves a positive hole in its wake. Thus, conduction by a semiconductor involves motion of thermally excited electrons in one direction and holes in the other.

The conductivity of a silicon or germanium crystal can be greatly enhanced by *doping*, a process whereby a tiny, controlled amount of an impurity is introduced, usually by diffusion, into the heated germanium or silicon crystal. Typically, a silicon or germanium semiconductor is doped with a Group V element, such as arsenic or antimony, or with a Group III element, such as indium or gallium. When an atom of a Group V element replaces a silicon atom in the lattice, one unbound electron is introduced into the structure; only a small amount of thermal energy is then needed to free this electron for conduction. Note that the resulting positive Group V ion does not provide a *mobile* hole because there is little tendency for electrons to move from a covalent silicon bond to this nonbonding position. A semiconductor that has been doped so that it contains nonbonding electrons is termed an *n-type* (negative type) because negatively charged electrons are the *majority carriers* of charge. Holes still exist as in the undoped crystal, which are associated with silicon atoms, but their number is small with respect to the number of electrons; thus, holes represent *minority carriers* in an *n*-type semiconductor.

A *p-type* (positive type) semiconductor is formed when silicon or germanium is doped with a Group III element, which contains only three valence electrons. Here, holes are introduced when electrons from adjoining silicon atoms jump to the vacant orbital associated with the impurity atom. Note that this process imparts a negative charge to the Group III atoms. Movement of the holes from silicon atom to silicon atom, as described earlier, constitutes a current in

which the majority carrier is positively charged. Because holes are less mobile than free electrons, the conductivity of a *p*-type semiconductor is inherently less than that of an *n*-type.

2C-2 Semiconductor Diodes

A *diode* is a nonlinear device that has greater conductance in one direction than in the other. Useful diodes are manufactured by forming adjacent *n*-type and *p*-type regions within a single germanium or silicon crystal; the interface between these regions is termed a *pn junction*.

Properties of a pn Junction

Figure 2-15a is a cross section of one type of *pn* junction, which is formed by diffusing an excess of a *p*-type impurity, such as indium, into a minute silicon chip that has been doped with an *n*-type impurity, such as antimony. A junction of this kind permits movement of holes from the *p* region into the *n* region and movement of electrons in the reverse direction. As holes and electrons diffuse in the opposite direction, a region is created that is depleted of mobile charge carriers and has a very high resistance. This region, referred to as the *depletion region*, is depicted in Figure 2-15d. Because there is separation of charge across the depletion region, a potential difference develops across the region, which causes a migration of holes and electrons in the opposite direction. The current that results from the diffusion of holes and electrons is balanced by the current produced by migration of the carriers in the electric field, and thus there is no net current. The magnitude of the potential difference across the depletion region depends on the composition of the materials used in the *pn* junction. For silicon diodes, the potential difference is about 0.6 V, and for germanium, it is about 0.3 V. When a positive voltage is applied across a *pn* junction, there is little resistance to current in the direction of the *p*-type to the *n*-type material. On the other hand, the *pn* junction offers a high resistance to the flow of holes in the opposite direction and is thus a *current rectifier*.

Figure 2-15b illustrates the symbol for a diode. The arrow points in the direction of low resistance to positive currents. The triangular portion of the diode symbol may be imagined to point in the direction of current in a conducting diode.

Figure 2-15c shows the mechanism of conduction of charge when the *p* region is made positive with respect

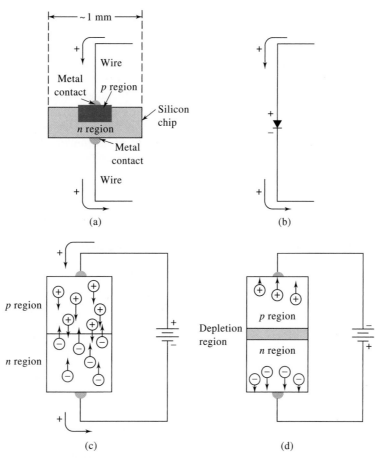

FIGURE 2-15 A *pn*-junction diode. (a) Physical appearance of one type formed by diffusion of a *p*-type impurity into an *n*-type semiconductor, (b) symbol for a diode, (c) current under forward bias, (d) resistance to current under reverse bias.

to the *n* region by application of a voltage; this process is called *forward biasing*. Here, the holes in the *p* region and the excess electrons in the *n* region, which are the majority carriers in the two regions, move under the influence of the electric field toward the junction, where they combine and thus annihilate each other. The negative terminal of the battery injects new electrons into the *n* region, which can then continue the conduction process; the positive terminal, on the other hand, extracts electrons from the *p* region, thus creating new holes that are free to migrate toward the *pn* junction.

When the diode is *reverse biased*, as in Figure 2-15d, the majority carriers in each region drift away from the junction to form the depletion layer, which contains few charges. Only the small concentration of minority

carriers present in each region drifts toward the junction and thus creates a current. Consequently, conductance under reverse bias is typically 10^{-6} to 10^{-8} that of conductance under forward bias.

Current-Voltage Curves for Semiconductor Diodes

Figure 2-16 shows the behavior of a typical semiconductor diode under forward and reverse bias. With forward bias, the current increases nearly exponentially with voltage. For some power diodes, forward biasing can result in currents of several amperes. For a germanium diode under reverse bias, a current on the order of tens of microamperes is observed over a considerable voltage range. Reverse-bias current for a silicon diode is on the order of tens of nanoamperes. In this region of the diode characteristic curve, conduction is by the minority carriers. Ordinarily, this reverse current is of

 Simulation: Learn more about **diodes**.

little consequence. As the reverse-bias voltage is increased, however, the *breakdown voltage* is ultimately reached where the reverse current increases abruptly to very high values. Here, holes and electrons, formed by the rupture of covalent bonds of the semiconductor, are

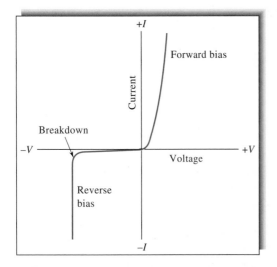

FIGURE 2-16 Current-voltage characteristics of a silicon semiconductor diode. (For clarity the small current under reverse bias before breakdown has been greatly exaggerated.)

accelerated by the field to produce additional electrons and holes by collision. In addition, quantum mechanical tunneling of electrons through the junction layer contributes to the enhanced conductance. This conduction, if sufficiently large, may result in heating and damaging of the diode. The voltage at which the sharp increase in current occurs under reverse bias is called the *Zener breakdown voltage*. By controlling the thickness and type of the junction layer, Zener breakdown voltages that range from a few volts to several hundred volts can be realized. As we shall see, this phenomenon has important practical applications in precision voltage sources.

2C-3 Transistors

The transistor is the basic semiconductor amplifying and switching device. This device performs the same function as the vacuum amplifier tube of yesteryear — that is, it provides an output signal whose magnitude is usually significantly greater than the signal at the input. Several types of transistors are available; two of the most widely used of these, the bipolar junction transistor and the field-effect transistor, are described here.

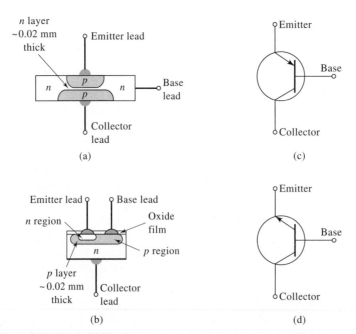

FIGURE 2-17 Two types of BJTs. Construction details are shown in (a) for a *pnp* alloy BJT and in (b) for an *npn* planar transistor. Symbols for a *pnp* and *npn* BJT are shown in (c) and (d), respectively. (Note that alloy junction transistors may also be fabricated as *npn* types and planar transistors as *pnp*.)

Bipolar Junction Transistors

Bipolar junction transistors (BJTs) may be viewed as two back-to-back semiconductor diodes. The *pnp* transistor consists of a very thin *n*-type region sandwiched between two *p*-type regions; the *npn* type has the reverse structure. BJTs are constructed in a variety of ways, two of which are illustrated in Figure 2-17. The symbols for the *pnp* and the *npn* types of transistors are shown on the right in Figure 2-17. In these symbols, the arrow on the emitter lead indicates the direction of positive current. Thus, in the *pnp* type, there is a positive current from the emitter to the base; the reverse is true for the *npn* type.

Electrical Characteristics of a BJT

The discussion that follows focuses on the behavior of a *pnp*-type BJT. It should be appreciated that the *npn* type acts analogously except for the direction of the current, which is opposite to that of the *pnp* transistor.

When a transistor is to be used in an electronic device, one of its terminals is connected to the input and the second serves as the output; the third terminal is connected to both and is the *common* terminal. Three configurations are thus possible: a common-emitter, a common-collector, and a common-base. The common-emitter configuration has the widest application in amplification and is the one we consider in detail.

Figure 2-18 illustrates the current amplification that occurs when a *pnp* transistor is used in the common-emitter mode. Here, a small input current I_B, which is to be amplified, is introduced in the emitter-base circuit; this current is labeled as the base current in the figure. As we show later, an ac current can also be amplified by superimposing it on I_B. After amplification, the dc component can then be removed by a filter.

The emitter-collector circuit is powered by a dc power supply, such as that described in Section 2D. Typically, the power supply provides a voltage between 9 and 30 V.

Note that, as shown by the breadth of the arrows, the collector, or output, current I_C is significantly larger than the base input current I_B. Furthermore, the magnitude of the collector current is directly proportional to the input current. That is,

$$I_C = \beta I_B \qquad (2\text{-}55)$$

where the proportionality constant β is the current gain, which is a measure of the current amplification that has occurred. Values for β for typical transistors range from 20 to 200.

It is important to note that a BJT requires a current into the base or out of the base to initiate conduction between the emitter and the collector. Consequently, circuits built from BJTs draw significant current from their power supplies during operation. We will soon describe another type of transistor, the field-effect transistor, which requires nearly zero current for its operation.

Mechanism of Amplification with a BJT

It should be noted that the emitter-base interface of the transistor shown in Figure 2-18 constitutes a forward-biased *pn* junction similar in behavior to that shown in Figure 2-15c, whereas the base-collector region is a reverse-biased *np* junction similar to the circuit shown in Figure 2-15d. Under forward bias, a significant current I_B develops when an input signal of a few tenths of a volt is applied (see Figure 2-16). In contrast, current through the reverse-biased collector-base junction is inhibited by the migration of majority carriers away from the junction, as shown in Figure 2-15d.

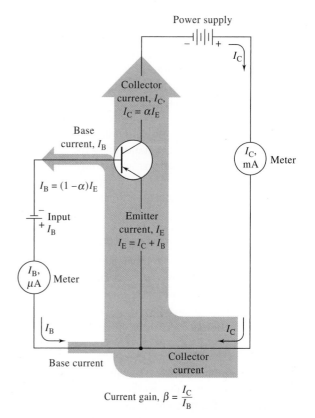

FIGURE 2-18 Currents in a common-emitter circuit with a transistor. Ordinarily, $\alpha = 0.95$ to 0.995 and $\beta = 20$ to 200.

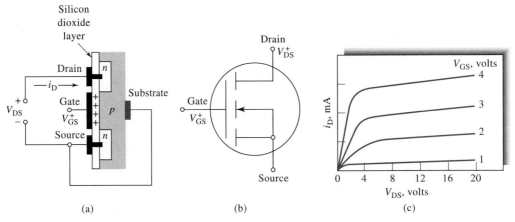

FIGURE 2-19 An *n*-channel enhancement mode MOSFET: (a) structure, (b) symbol, (c) performance characteristics.

In the manufacture of a *pnp* transistor, the *p* region is purposely much more heavily doped than is the *n* region. As a consequence, the concentration of holes in the *p* region is 100 times the concentration of mobile electrons in the *n* layer. Thus, the fraction of the current carried by holes is perhaps 100 times the fraction carried by electrons.

Turning again to Figure 2-18, holes are formed at the *p*-type emitter junction through removal of electrons by the two dc sources, namely, the input signal and the power supply. These holes can then move into the very thin *n*-type base region where some will combine with the electrons from the input source; the base current I_B is the result. The majority of the holes will, however, drift through the narrow base layer and be attracted to the negative collector junction, where they can combine with electrons from the power supply; the collector current I_C is the result.

The magnitude of the collector current is determined by the number of current-carrying holes available in the emitter. This number, however, is a fixed multiple of the number of electrons supplied by the input base current. Thus, when the base current doubles, so does the collector current. This relationship leads to the current amplification exhibited by a BJT.

Field-Effect Transistors

Several types of field-effect transistors (FETs) have been developed and are widely used in integrated circuits. One of these, the insulated-gate field-effect transistor, was the outgrowth of the need to increase the input resistance of amplifiers. Typical insulated-gate

FETs have input impedances that range from 10^9 to 10^{14} Ω. This type of transistor is most commonly referred to as a MOSFET, which is the acronym for *metal oxide semiconductor field-effect transistor*.

Figure 2-19a shows the structural features of an *n-channel* MOSFET. Here, two isolated *n* regions are formed in a *p*-type substrate. Covering both regions is a thin layer of highly insulating silicon dioxide, which may be further covered with a protective layer of silicon nitride. Openings are etched through these layers so that electrical contact can be made to the two *n* regions. Two additional contacts are formed, one to the substrate and the other to the surface of the insulating layer. The contact with the insulating layer is termed the *gate* because the voltage at this contact determines the magnitude of the positive current between the drain and the source. Note that the insulating layer of silicon dioxide between the gate lead and the substrate accounts for the high impedance of a MOSFET.

In the absence of an applied voltage to the gate, essentially no current develops between drain and source because one of the two *pn* junctions is always reverse biased regardless of the sign of the applied voltage V_{DS}. MOFSET devices are designed to operate in either an *enhancement* or a *depletion* mode. The former type is shown in Figure 2-19a, where current enhancement is brought about by application of a positive voltage to the gate. As shown, this positive voltage induces a negative substrate channel immediately below the layer of silicon dioxide that covers the gate electrode. The number of negative charges here, and thus the current, increases as the gate voltage V_{GS}

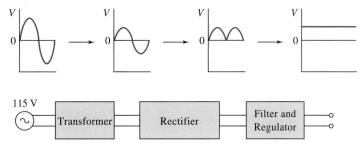

FIGURE 2-20 Diagram showing the components of a power supply and their effects on the 115-V line voltage.

increases. The magnitude of this effect is shown in Figure 2-19c. Also available are *p*-channel enhancement-mode MOSFET devices in which the *p* and *n* regions are reversed from those shown in Figure 2-19a.

Depletion-mode MOSFET devices are designed to conduct in the *absence* of a gate voltage and to become nonconducting as potential is applied to the gate. An *n*-channel MOSFET of this type is similar in construction to the transistor shown in Figure 2-19a except that the two *n* regions are now connected by a narrow channel of *n*-type semiconductor. Application of a negative voltage at V_{DS} repels electrons out of the channel and thus decreases the conduction through the channel. It is important to note that virtually zero current is required at the gate of a MOSFET device to initiate conduction between the source and drain. This tiny power requirement is in contrast to the rather large power requirements of BJTs. The characteristic low power consumption of field-effect devices makes them ideal for portable applications requiring battery power.

2D POWER SUPPLIES AND REGULATORS

Generally, laboratory instruments require dc power to operate amplifiers, computers, transducers, and other components. The most convenient source of electrical power, however, is the nominally 110-V ac, 60-Hz line voltage furnished by public utility companies.[8] As shown in Figure 2-20, laboratory power supply units increase or decrease the voltage from the house supply,

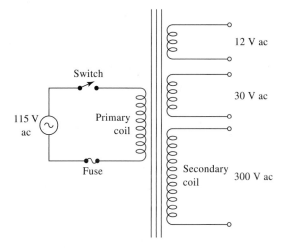

FIGURE 2-21 Schematic of a typical power transformer with multiple secondary windings.

rectify the current so that it has a single polarity, and finally, smooth the output to give a nearly steady dc voltage. Most power supplies also contain a voltage regulator that maintains the output voltage at a constant desired level.

2D-1 Transformers

The voltage from the ac power lines is readily increased or decreased by means of a power transformer such as that shown schematically in Figure 2-21. The varying magnetic field formed around the *primary* coil in this device from the 110-V ac induces alternating currents in the *secondary* coils. The voltage V_x across each is given by

$$V_x = 115 \times N_2/N_1$$

where N_2 and N_1 are the number of turns in the secondary and primary coils, respectively. Power supplies

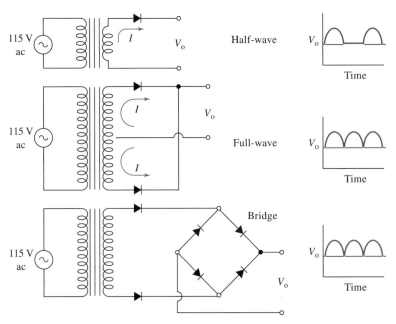

FIGURE 2-22 Three types of rectifiers: half-wave, full-wave, and bridge.

with multiple taps, as in Figure 2-21, are available commercially, and many different voltage combinations may be obtained. Thus, a single transformer can serve as a power source for several components of an instrument with different voltage requirements.

2D-2 Rectifiers and Filters

Figure 2-22 shows three types of rectifiers and their output-signal forms. Each uses semiconductor diodes (see Section 2C-2) to block current in one direction while permitting it in the opposite direction. To minimize the current fluctuations shown in Figure 2-22, the output of a rectifier is usually filtered by placing a capacitor with a large capacitance in parallel with the load R_L, as shown in Figure 2-23. The charge and discharge of the capacitor has the effect of decreasing the variations to a relatively small *ripple*. In some applications, an inductor in series and a capacitor in parallel with the load serve as a filter; this type of filter is known as an *L section*. By suitable choice of capacitance and inductance, the peak-to-peak ripple can be reduced to the millivolt range or lower.

2D-3 Voltage Regulators

Often, instrument components require dc voltages that are constant regardless of the current drawn or of fluctuations in the line voltage. *Voltage regulators* serve this purpose. Figure 2-24 illustrates a simple voltage regulator that uses a *Zener diode*, a *pn* junction that has been designed to operate under breakdown conditions; note the special symbol for this type of diode. Figure 2-16 shows that a semiconductor diode undergoes an abrupt breakdown at a certain reverse bias, whereupon the current changes precipitously. For example, under breakdown conditions, a current change of 20 to 30 mA may result from a voltage change of 0.1 V or less. Zener diodes with a variety of specified breakdown voltages are available commercially.

 Simulation: Learn more about **power supplies**.

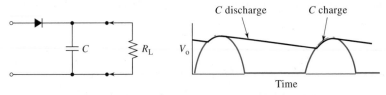

FIGURE 2-23 Filtering the output from a rectifier.

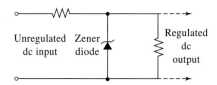

FIGURE 2-24 A Zener-stabilized voltage regulator.

For voltage regulators, a Zener diode is chosen such that it always operates under breakdown conditions; that is, the input voltage to be regulated is greater than the breakdown voltage. For the regulator shown in Figure 2-24, an increase in voltage results in an increase in current through the diode. Because of the steepness of the current-voltage curve in the breakdown region (Figure 2-16), however, the voltage drop across the diode, and thus the load, is virtually constant.

Modern integrated-circuit voltage regulators take advantage of the properties of Zener diodes to provide stable reference voltages. These voltages are used in conjunction with feedback circuitry and power transistors to create power supplies regulated to ± 1 mV or better. Such regulators have three terminals: input, output, and circuit common. The raw output from a rectified and filtered power supply is connected to the three-terminal voltage regulator to produce a supply that is stable with respect to temperature fluctuations and that automatically shuts down when the load current exceeds its maximum rating, which is typically one ampere in the most widely used circuits. Integrated-circuit voltage regulators are found in the power supplies of most electronic devices.

Regulators of this type have the disadvantage of dissipating considerable power, so that with the proliferation of computers and other electronic devices, more efficient regulators have become desirable. The solution to this difficulty has been the advent of *switch-ing regulators*, which provide power to the load only when it is needed while maintaining constant voltage. Most computer power supplies contain switching regulators. The details of operation of switching power supplies are beyond the scope of our discussion, but their principles of operation are discussed in the general references given in Section 2C.

2E   READOUT DEVICES

In this section, three common readout devices are described, namely, the cathode-ray tube, the laboratory recorder, and the alphanumeric display unit.

2E-1 Oscilloscopes

The oscilloscope is a most useful and versatile laboratory instrument that uses a cathode-ray tube (CRT) as a readout device. Both analog and digital oscilloscopes are manufactured. Digital oscilloscopes are used when sophisticated signal processing is required. Analog oscilloscopes are generally simpler than their digital counterparts, are usually portable, are easier to use, and are less expensive, costing as little as a few hundred dollars. We confine our discussion here to simple analog oscilloscopes. The block diagram in Figure 2-25 shows the most important components of such an instrument and the signal pathways for its components. The actual display is provided by a CRT.

Cathode-Ray Tubes

Figure 2-26 is a schematic that shows the main components of a CRT. Here, the display is formed by the interaction of electrons in a focused beam with a phosphorescent coating on the interior of the large curved surface of the evacuated tube. The electron beam is formed at a heated cathode, which is main-

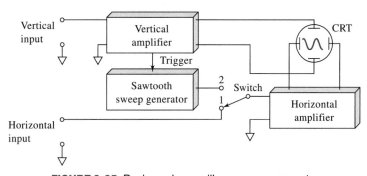

FIGURE 2-25 Basic analog oscilloscope components.

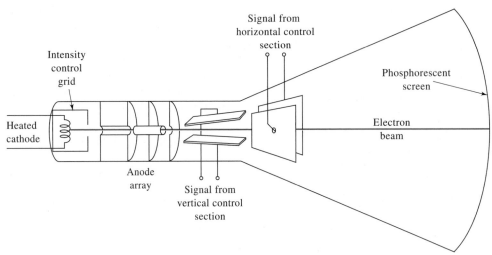

FIGURE 2-26 Schematic of a CRT.

tained at ground potential. A multiple anode-focusing array produces a narrow beam of electrons that has been accelerated through a field of several thousand volts. In the absence of input signals, the beam appears as a small bright dot in the center of the screen.

Horizontal and Vertical Control Plates. Input signals are applied to two sets of plates, one of which deflects the beam horizontally and the other vertically. Thus, it is possible to display an x-y plot of two related signals on the face of the CRT when the switch in Figure 2-25 is in position 1. Because the screen is phosphorescent, the movement of the dot appears as a lighted continuous trace that fades after a brief period.

The most common way of operating the CRT is to cause the dot to sweep periodically at a constant rate across the central horizontal axis of the tube by applying a sawtooth sweep signal to the horizontal deflection plates. The oscilloscope is operated in this way when the switch in Figure 2-25 is moved to position 2. When operated in this way, the horizontal axis of the display corresponds to time. Application of a periodic signal to the vertical plates then provides a display of the waveform of the periodic signal. The fastest sweep rates in most analog oscilloscopes range between 1 μs/cm and 1 ns/cm. Usually the sweep speed can be slowed by factors of 10 until the rate is in the seconds-per-centimeter range.

The horizontal control section of most oscilloscopes can, if desired, be driven by an external signal rather than by the internal sawtooth signal. In this

mode of operation, the oscilloscope becomes an x-y plotter that displays the functional relationship between two input signals.

Trigger Control. To steadily display a repetitive signal, such as a sine wave, on the screen, it is essential that each sweep begin at an identical place on the signal profile — for example, at a maximum, a minimum, a zero crossing, or an abrupt change in the signal. Synchronization is usually accomplished by mixing a portion of the test signal with the sweep signal in such a way as to produce a voltage spike for, say, each maximum or some multiple thereof. This spike then serves to trigger the sweep. Thus, the waveform can be observed as a steady image on the screen.

Oscilloscopes are extremely useful in a variety of display and diagnostic applications. They can be used to view the time profile of signals from transducers, to compare the relationships among repetitive waveforms in analog signal processing circuits, or to reveal high-frequency noise or other signals of interest that cannot be observed by using a DMM or other dc-measuring device. The oscilloscope is an essential diagnostic tool in the instrument laboratory.

2E-2 Recorders

The typical laboratory recorder[9] is an example of a servosystem, a null device that compares two signals and then makes a mechanical adjustment that reduces

Simulation: Learn more about **CRTs**.

---

[9] For a more extensive discussion of laboratory recorders, see G. W. Ewing, *J. Chem. Educ.*, **1976**, *53*, A361, A407.

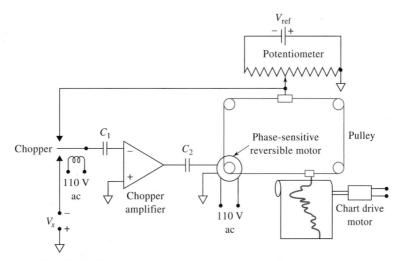

FIGURE 2-27 Schematic of self-balancing recording potentiometer.

their difference to zero; that is, a servosystem continuously seeks the null condition. Laboratory recorders were once very common but today they have been mostly replaced by computer-based systems.

In the laboratory recorder, shown schematically in Figure 2-27, the signal to be recorded, V_x, is continuously compared with the output from a potentiometer powered by a reference signal, V_{ref}. In many recorders, the reference signal is generated by a temperature-compensated Zener diode voltage-reference circuit that provides a stable reference voltage. Any difference in voltage between the potentiometer output and V_x is converted to a 60-cycle ac signal by an electronic chopper; the resulting signal is then amplified sufficiently to activate a small phase-sensitive electric motor that is mechanically geared or linked (by a pulley arrangement in Figure 2-27) to both a recorder pen and the sliding contact of the potentiometer. The direction of rotation of the motor is such that the potential difference between the potentiometer output and V_x is decreased to zero, which causes the motor to stop.

To understand the directional control of the motor, it is important to note that a reversible ac motor has two sets of coils, one of which is fixed (the stator) and the other of which rotates (the rotor). One of these — say, the rotor — is powered from the 110-V power line and thus has a continuously fluctuating magnetic field associated with it. The output from the ac amplifier, on the other hand, is fed to the coils of the stator. The magnetic field induced here interacts with the rotor field and causes the rotor to turn. The direction of motion depends on the *phase* of the stator current with

respect to that of the rotor; the phase of the stator current, however, differs by 180°, depending on whether V_x is greater or smaller than the signal from V_{ref}. Thus, the amplified difference signal can be caused to drive the servomechanism to the null state from either direction.

In most laboratory recorders, the paper is moved at a fixed speed. Thus, a plot of signal intensity as a function of time is obtained. Because the recorder paper is fed from a long roll, or strip, this type of laboratory recorder has come to be called a *strip-chart recorder*. In *x-y recorders*, the paper is fixed as a single sheet mounted on a flat bed. The paper is traversed by an arm that moves along the *x*-axis. The pen travels along the arm in the *y* direction. The arm drive and the pen drive are connected to the *x* and *y* inputs, respectively, thus permitting both to vary continuously. Often recorders of this type are equipped with two pens, thus allowing the simultaneous plotting of two functions on the *y*-axis. An example of an application of this kind is in chromatography, where it is desirable to have a plot of the detector output as a function of time as well as the time integral of this output. Alternatively, a dual-pen recorder might be used to display the outputs of two different detectors that are monitoring the effluent from the same chromatographic column.

A typical laboratory strip-chart recorder has several chart speeds that often range from 0.1 to 20 cm/min. Most provide a choice of several voltage ranges, from 1 mV full scale to several volts. Generally, the precision of these instruments is on the order of a few tenths of a percent of full scale.

Digital recorders and plotters are now widely used. Here, the pen may be driven by a stepper motor, which responds to digitized voltage signals by turning some precise fraction of a rotation for each voltage pulse. Computer-driven x-y plotters often use dc servomotors to move either the pen, the paper, or both, to draw graphs of data from analytical instruments. Graphical output is now commonly obtained from inkjet and laser printers. Software packages such as Excel, SigmaPlot, and Adobe Illustrator© produce the appropriate formats for plotting with these printers.

2E-3 Alphanumeric Displays

The output from digital equipment is most conveniently displayed in terms of decimal numbers and letters, that is, in alphanumeric form. The seven-segment readout device is based on the principle that any alphanumeric character can be represented by lighting an appropriate combination of seven segments, as shown in Figure 2-28. Here, for example, a five is formed when segments a, f, g, c, and d are lighted; the letter C appears when segments a, f, e, and d are lighted. Perhaps the most common method of lighting a seven-segment display is to fashion each segment as a light-emitting diode (LED). A typical LED consists of a pn junction shaped as one of the segments and prepared from gallium arsenide, which is doped with phosphorus. Under forward bias, the junction emits red radiation as a consequence of recombinations of minority carriers in the junction region. Each of the seven segments is connected to a decoder logic circuit so that it is activated at the proper time.

Seven-segment liquid-crystal displays, or LCDs, are also widely encountered. Here, a small amount of a liquid crystal is contained in a thin, flat optical cell, the walls of which are coated with a conducting film.

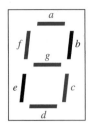

FIGURE 2-28 A seven-segment display.

Application of an electric field to a certain region of the cell causes a change in alignment of the molecules in the liquid crystal and a consequent change in its optical appearance.[10] Both LEDs and LCDs find application in many different types of instruments, and each type of electronic readout has its advantages. LCDs are especially useful for battery-operated instruments because they consume little power, but they can have problems in either very bright or very dim ambient light. On the other hand, LEDs are readable at low ambient light levels as well as in fairly bright light, but they consume considerably more power and thus are not generally used in applications powered by battery.

2E-4 Computers

Many modern instruments use computers and computer monitors as readout devices. Data manipulation, data processing, and formatting for graphical display and printing can all be done with the computer. Computer-based instruments are discussed in detail in Section 4D.

---

[10] For discussions of the properties and applications of liquid crystals, see G. H. Brown and P. P. Crooker, *Chem. Eng. News*, **1983**, *Jan. 31*, 24; G. H. Brown, *J. Chem. Educ.*, **1983**, *60*, 900.

QUESTIONS AND PROBLEMS

*Answers are provided at the end of the book for problems marked with an asterisk.

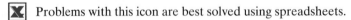 Problems with this icon are best solved using spreadsheets.

*2-1 For assembling the voltage divider shown below, two of each of the following resistors are available: 500 Ω, 1.00 kΩ, and 2.00 kΩ.

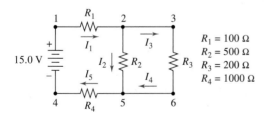

(a) Describe a suitable combination of the resistors that would give the indicated voltages.

(b) What would be the IR drop across R_3?

(c) What current would be drawn from the source?

(d) What power is dissipated by the circuit?

*2-2 Assume that for a circuit similar to that shown in Problem 2-1, $R_1 = 200\ \Omega$, $R_2 = 500\ \Omega$, $R_3 = 1.00\ \text{k}\Omega$, and $V_a = 15\ \text{V}$.

(a) What is the voltage V_2?

(b) What would be the power loss in resistor R_2?

(c) What fraction of the total power lost by the circuit would be dissipated in resistor R_2?

*2-3 For a circuit similar to the one shown in Problem 2-1, $R_1 = 1.00\ \text{k}\Omega$, $R_2 = 2.50\ \text{k}\Omega$, $R_3 = 4.00\ \text{k}\Omega$, and $V_a = 12.0\ \text{V}$. A voltmeter was placed across contacts 2 and 4. Calculate the relative error in the voltage reading if the internal resistance of the voltmeter was (a) 5000 Ω, (b) 50 kΩ, and (c) 500 kΩ.

*2-4 A voltmeter was used to measure the voltage of a cell with an internal resistance of 750 Ω. What must the internal resistance of the meter be if the relative error in the measurement is to be less than (a) -1.0%, (b) -0.10%?

*2-5 For the following circuit, calculate

(a) the potential difference across each of the resistors.

(b) the magnitude of each of the currents shown.

(c) the power dissipated by resistor R_3.

(d) the potential difference between points 3 and 4.

*2-6 For the circuit shown below, calculate

(a) the power dissipated between points 1 and 2.

(b) the current drawn from the source.

(c) the voltage drop across resistor R_A.

(d) the voltage drop across resistor R_D.

(e) the potential difference between points 5 and 4.

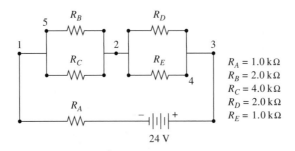

$R_A = 1.0\,k\Omega$
$R_B = 2.0\,k\Omega$
$R_C = 4.0\,k\Omega$
$R_D = 2.0\,k\Omega$
$R_E = 1.0\,k\Omega$

24 V

*2-7 The circuit that follows is for a laboratory potentiometer for measuring unknown voltages V_x.

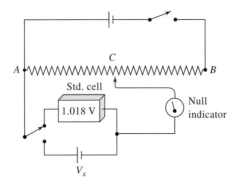

Assume the resistor AB is a slide wire whose resistance is directly proportional to its length. With the standard Weston cell (1.018 V) in the circuit, a null point was observed when contact C was moved to a position 84.3 cm from point A. When the Weston cell was replaced with a cell of an unknown voltage V_x, null was observed at 44.3 cm. Find the value of V_x.

2-8 Show that the data in the last column of Table 2-1 are correct.

2-9 Show that the data in the last column of Table 2-2 are correct.

*2-10 The current in a circuit is to be determined by measuring the voltage drop across a precision resistor in series with the circuit.

(a) What should be the resistance of the resistor in ohms if 1.00 V is to correspond to 50 µA?

(b) What must be the resistance of the voltage-measuring device if the error in the current measurement is to be less than 1.0% relative?

2-11 An electrolysis at a nearly constant current can be performed with the following arrangement:

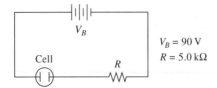

$V_B = 90$ V
$R = 5.0\,k\Omega$

The 90-V source consists of dry cells whose voltage can be assumed to remain constant for short periods. During the electrolysis, the resistance of the cell increases from 20 Ω to 40 Ω because of depletion of ionic species. Calculate the percentage change in the current, assuming that the internal resistance of the batteries is zero.

2-12 Repeat the calculations in Problem 2-11, assuming that $V_B = 9.0$ V and $R = 0.50$ kΩ.

*2-13 A 24-V dc voltage source was connected in series with a resistor and capacitor. Calculate the current after 0.00, 0.010, 0.10, 1.0, and 10 s if the resistance was 10 MΩ and the capacitance 0.20 μF.

*2-14 How long would it take to discharge a 0.015 μF capacitor to 1% of its full charge through a resistance of (a) 10 MΩ, (b) 1 MΩ, (c) 1 kΩ?

*2-15 Calculate time constants for each of the RC circuits described in Problem 2-14.

 2-16 A series RC circuit consists of a 25-V dc source, a 50 kΩ resistor, and a 0.035 μF capacitor.
    (a) Calculate the time constant for the circuit.
    (b) Calculate the current and voltage drops across the capacitor and the resistor during a charging cycle; employ as times 0, 1, 2, 3, 4, 5, and 10 ms.
    (c) Repeat the calculations in (b) for a discharge cycle assuming 10 ms charging time.

 2-17 Repeat the calculations in Problem 2-16, assuming that the voltage source was 15 V, the resistance was 20 MΩ, and the capacitance was 0.050 μF. Use as times 0, 1, 2, 3, 4, 5, and 10 s.

 2-18 Calculate the capacitive reactance, the impedance, and the phase angle ϕ for the following series RC circuits:

Frequency, Hz	R, Ω	C, μF
(a) 1	20,000	0.033
(b) 10^3	20,000	0.033
(c) 10^6	20,000	0.0033
(d) 1	200	0.0033
(e) 10^3	200	0.0033
(f) 10^6	200	0.0033
(g) 1	2,000	0.33
(h) 10^3	2,000	0.33
(i) 10^6	2,000	0.33

 2-19 Derive a frequency response curve for a low-pass RC filter in which $R = 2.5$ kΩ and $C = 0.015$ μF. Cover a range of $(V_p)_o/(V_p)_i$ of 0.01 to 0.9999. Plot $(V_p)_o/(V_p)_i$ versus ln f.

2-20 Derive a frequency response curve for a high-pass RC filter in which $R = 500$ kΩ and $C = 100$ pF (1 pF $= 10^{-12}$ F). Cover a range of $(V_p)_o/(V_p)_i$ of 0.001 to 0.9999. Plot $(V_p)_o/(V_p)_i$ versus ln f.

Challenge Problem

2-21 (a) The circuit shown below is a network of four capacitors connected in parallel. Show that the parallel capacitance C_p is given by $C_p = C_1 + C_2 + C_3 + C_4$.

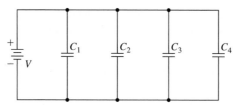

(b) If $V = 5.00$ V, $C_1 = 0.050$ μF, $C_2 = 0.010$ μF, $C_3 = 0.075$ μF, and $C_4 = 0.020$ μF, find the parallel capacitance C_p, the charge on each capacitor, and the total charge Q_p.

(c) A series combination of capacitors is shown in the figure below. Show that the series capacitance C_S is given by

$$\frac{1}{C_S} = \frac{1}{C_1} + \frac{1}{C_2} + \frac{1}{C_3}$$

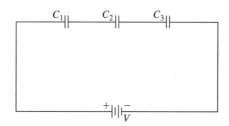

(d) If $V = 3.0$ V, $C_1 = 1.00$ μF, $C_2 = 0.75$ μF, and $C_3 = 0.500$ μF, find the series capacitance C_S and the voltage drops across each capacitor.

(e) For the series circuit of part (d) suppose that there were only two capacitors, C_1 and C_2. Show that the series capacitance in this case is the product of the two capacitances divided by the sum of the two.

(f) The complex capacitive network shown below is wired. Find the capacitance of the network, the voltage across each capacitor, and the charge on each capacitor.

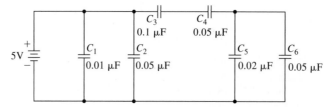

CHAPTER THREE

Operational Amplifiers in Chemical Instrumentation

Most modern analog signal–conditioning circuits owe their success to the class of integrated circuits known as operational amplifiers, which are referred to as **op amps** or **OAs**. Operational amplifiers are ubiquitous. Open any instrument or piece of electronic equipment or scan an instrument schematic, and you will likely find one or more op amps. This fact, coupled with the ease with which relatively complex functions can be accomplished, emphasizes the importance of having a basic understanding of their principles of operation. In this chapter we explore a number of operational amplifier circuits and applications and investigate their properties, advantages, and limitations.

Throughout this chapter, this logo indicates an opportunity for online self-study at **www .thomsonedu.com/chemistry/skoog**, linking you to interactive tutorials, simulations, and exercises.

3A PROPERTIES OF OPERATIONAL AMPLIFIERS

Operational amplifiers derive their name from their original applications in analog computers, where they were used to perform such mathematical operations as summing, multiplying, differentiating, and integrating.[1] Operational amplifiers also find general application in the precise measurement of voltage, current, and resistance, which are measured variables with the transducers that are used in chemical instruments. Operational amplifiers also are widely used as constant-current and constant-voltage sources.[2]

3A-1 Symbols for Operational Amplifiers

Figure 3-1 is an equivalent circuit representation of an operational amplifier. In this figure, the input voltages are represented by v_+ and v_-. The input difference voltage v_s is the difference between these two voltages; that is, $v_s = v_+ - v_-$. The power supply connections are labeled +PS and −PS and usually have values of +15 and −15 V dc or sometimes +5 and −5 V. The so-called *open-loop gain* of the operational amplifier is shown as A, and the output voltage v_o is given by $v_o = Av_s$. Finally, Z_i and Z_o are the input and output impedances of the operational amplifier. The input signal may be either ac or dc, and the output signal will then correspond.[3] Note that all voltages in operational amplifier circuits are measured with respect to the *circuit common* shown in Figure 3-1. Circuit common is often referred to somewhat sloppily as *ground*. These terms are defined carefully and their significance discussed in Section 3A-2.

As shown in Figure 3-2, two alternate versions of Figure 3-1 are commonly used to symbolize operational

[1] For general information on electronics, computers, and instrumentation, including operational amplifiers and their characteristics, see H. V. Malmstadt, C. G. Enke, and S. R. Crouch, *Microcomputers and Electronic Instrumentation: Making the Right Connections*, Washington, DC: American Chemical Society, 1994; A. J. Diefenderfer and B. E. Holton, *Principles of Electronic Instrumentation*, 3rd ed., Philadelphia: Saunders, 1994; J. J. Brophy, *Basic Electronics for Scientists*, 5th ed., New York: McGraw-Hill, 1990; P. Horowitz and W. Hill, *The Art of Electronics*, 2nd ed., New York: Cambridge University Press, 1989.

[2] For more detailed information about operational amplifiers, see R. Kalvoda, *Operational Amplifiers in Chemical Instrumentation*, New York: Halsted Press, 1975. See also the references in note 1.

[3] Throughout this text, we follow the convention of using uppercase I, V, and Q to respectively represent direct current, voltage, and charge and lowercase i, v, and q for the corresponding ac quantities.

amplifiers in circuit diagrams. A symbol as complete as that in Figure 3-2a is seldom encountered, and the simplified diagram in Figure 3-2b is used almost exclusively. Here, the power and common connections are omitted.

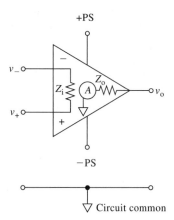

FIGURE 3-1 Equivalent circuit representation of an operational amplifier.

3A-2 General Characteristics of Operational Amplifiers

As shown in Figure 3-3, the typical operational amplifier is an analog device that consists of a high-input-impedance difference amplifier stage, a high-gain voltage amplifier stage, and a low-output-impedance amplifier stage. Most have approximately 20 transistors and resistors that are fabricated on a single integrated-circuit chip. Other components, such as capacitors and diodes, may also be integrated into the device. The physical dimensions of an operational amplifier, excluding power supply, are normally on the order of a centimeter or less. Modern operational amplifiers, in addition to being compact, are remarkably reliable and inexpensive. Their cost ranges from a few cents for general-purpose amplifiers to more than fifty dollars for specialized units. A wide variety of operational amplifiers are available, each differing in gain, input and output impedance, operating voltage, speed, and maximum power. A typical commercially available

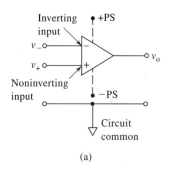

(a)

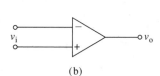

(b)

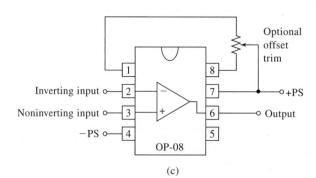

(c)

FIGURE 3-2 Symbols for operational amplifiers. More detail than usual is provided in (a). Note that the two input voltages v_- and v_+ as well as the output voltage v_o are measured with respect to the circuit common, which is usually at or near earth ground potential. (b) The usual way of representing an operational amplifier in circuit diagrams. (c) Representation of typical commercial 8-pin operational amplifier.

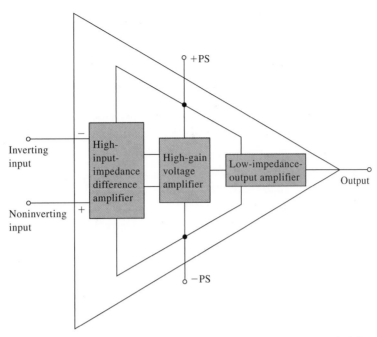

FIGURE 3-3 Circuit design of a typical operational amplifier. The input stage is followed by a high-gain amplification stage. The output stage is capable of supplying current to a load over a voltage range determined by the + and − power supply values.

operational amplifier comes housed in an 8-pin epoxy or ceramic package as shown in Figure 3-2c.

Operational amplifiers have the following properties: (1) large open-loop gains ($A = 10^4$ to $>10^6$); (2) high input impedances ($Z_i = 10^8$ to 10^{15} Ω); (3) low output impedances ($Z_o = 0.001$ to 1 Ω); and nearly zero output voltage for zero input. Most operational amplifiers do in fact exhibit a small output voltage with zero input due to circuit characteristics or component instabilities. The *offset voltage* is the input voltage required to produce zero output voltage. Modern operational amplifiers often have less than 5 mV offset because of laser trimming in the manufacturing process. Some operational amplifiers are provided with an *offset trim* adjustment to reduce the offset to a negligible value (see Figure 3-2c).

Circuit Common and Ground Potential

As shown in Figure 3-2a, each of the two input voltages, as well as the output voltage, of a typical operational amplifier are measured with respect to circuit common, which is symbolized by a downward-pointing open triangle (▽). Circuit common is a conductor that provides a common return for all currents to their sources. As a

consequence, all voltages in the circuit are measured with respect to the circuit common. Ordinarily, electronic equipment is not directly connected to earth ground, which is symbolized by ⏚. Usually, however, the circuit common potential does not differ significantly from the ground potential, but it is important to recognize that circuit common is not necessarily at the same potential as ground. Note that in Figure 3-2b a circuit common is not shown, but you may assume that there is a circuit common and that all voltages are measured with respect to it.

Inverting and Noninverting Inputs

It is important to realize that in Figure 3-2 the *negative and positive signs indicate the inverting and noninverting inputs* of the amplifier and *do not* imply that they are to be connected to positive and negative signals. Either input may be connected to positive or negative signals depending on the application of the circuit. Thus, if a negative voltage is connected to the inverting input, the output of the amplifier is positive with respect to it; if, on the other hand, a positive voltage is connected to the inverting input of the amplifier, a negative output results. An ac signal connected to the inverting input yields an output *that is 180 degrees out of phase* with the signal at the input. The noninverting input of an amplifier, on the other hand, yields an in-phase signal; in the case of a dc signal at the noninverting input, the output will be a dc signal of the same polarity as the input.

3B OPERATIONAL AMPLIFIER CIRCUITS

Operational amplifiers are used in three different modes: the *comparator* mode, the *voltage follower* mode, and the *current follower*, or *operational*, mode.

3B-1 Comparators

In the comparator mode, the operational amplifier is used *open loop*, without any feedback as shown in Figure 3-4a. In this mode, the amplifier is almost always at one of the limits imposed by the + and − power supplies (often ±15-V supplies). Usually, the voltages to be compared are connected directly to the two op amp inputs. The amplifier output is given by $v_o = Av_s = A(v_+ − v_−)$. If $A = 10^6$, for example, and the power supply limits were ±13 V,[4] the amplifier would be at one of the limits except for a small region where $v_s \geq −13$ V/10^6 or $v_s \leq 13$ V/10^6. Thus, unless v_s is in the range −13 µV to +13 µV, the output is at limit as illustrated in Figure 3-4b. Note also that the sign of the output voltage v_o tells us whether $v_+ > v_−$ or $v_+ < v_−$. If $v_+ > v_−$ by more than 13 µV, for example, the output is at positive limit (+13 V in our case). In contrast, if $v_− > v_+$ by more than 13 µV, the output is at negative limit (−13 V). Some applications of comparator circuits are given in Sections 3F and 4C.

3B-2 The Voltage Follower

In Section 2A-3, the problem of loading a voltage source and distorting its output was discussed. To prevent such loading, the input resistance of the measuring device or connected circuit must be much larger than the inherent internal resistance of the voltage source. When the voltage source is a transducer with high internal resistance, such as a glass pH electrode or pIon electrode, it is necessary to introduce a circuit known as a *voltage follower* to prevent the loading error.

---

[4] The limits to which the op amp can be driven are often slightly less than the power supply voltages (±15 V) because of internal voltage drops in the amplifier.

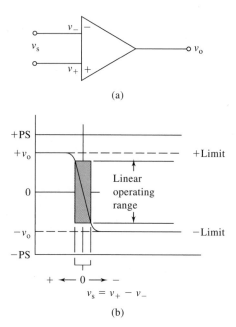

FIGURE 3-4 (a) Comparator mode. Note that the operational amplifier has no feedback and is thus an open-loop amplifier. (b) Output voltage v_o of operational amplifier as a function of input difference voltage v_s. Note that only a very small voltage difference at the two inputs causes the amplifier output to go to one limit or the other.

A typical operational amplifier voltage follower is shown in Figure 3-5. Modern, high-quality operational amplifiers have input impedances of 10^{12} to 10^{15} Ω and output impedances of less than 1 Ω. Therefore, the voltage source connected to the noninverting input is not loaded by circuits or measuring devices connected to the amplifier output v_o. Furthermore, the output is the same voltage as the input, but isolated from it. The voltage follower is a nearly ideal buffer to protect high-impedance sources from being loaded.

When the output signal of an operational amplifier is connected to one of the inputs, the process is called *feedback*. In the case of the voltage follower, the output signal is connected to the inverting input so that it is opposite in sense to the input signal v_i. This type of error-reducing feedback is termed *negative feedback*.

In actuality, v_s is not zero in voltage follower circuits. There must always be a small error, given by v_s, to produce the output voltage of the amplifier. For the follower, we can write, from Kirchhoff's voltage law,

$$v_o = v_i + v_s \qquad (3-1)$$

 Simulation: Learn more about **voltage followers**.

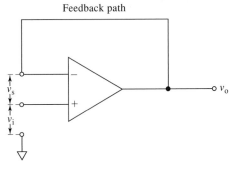

Feedback path

FIGURE 3-5 Voltage follower. The amplifier output is connected directly back to the amplifier inverting input. The input voltage v_i is connected to the noninverting input. The output voltage is the sum of the input voltage and the difference voltage v_s. If the output voltage is not at limit, v_s is very small. Therefore, $v_o \approx v_i$ and the output voltage follows the input voltage.

If the amplifier is not at limit, $v_s = -v_o/A$. Substituting into Equation 3-1,

$$v_o = v_i - \frac{v_o}{A} = v_i\left(\frac{A}{1+A}\right) \qquad (3\text{-}2)$$

When the amplifier is not at limit, v_s must be quite small ($-13\ \mu V$ to $+13\ \mu V$ for an op amp with a gain of 10^6 and ± 13 V limits). Therefore, if $|v_o| \geq 10$ mV, the error $v_s \leq 0.13\%$. For all practical purposes, $v_o = v_i$.

Note that, although this circuit has a unit voltage gain ($v_o/v_i = 1$), it can have a very large power gain because operational amplifiers have high input impedances but low output impedances. To show the effect of this large difference in impedance, let us define the power gain as P_o/P_i, where P_o is the power of the output from the operational amplifier and P_i is the input power. If we then substitute the power law ($P = iv = v^2/R$) and Ohm's law into this definition and recall that v_o and v_i are approximately the same in a voltage follower, we obtain the expression

$$\text{power gain} = \frac{P_o}{P_i} = \frac{i_o v_o}{i_i v_i} = \frac{v_o^2/Z_o}{v_i^2/Z_i} = \frac{Z_i}{Z_o}$$

where Z_i and Z_o are the input and output impedances of the operational amplifier. This result is important because it means that the voltage follower will draw almost no current from an input; the internal circuitry of the operational amplifier and the power supply, however, can supply large currents at the output of the operational amplifier.

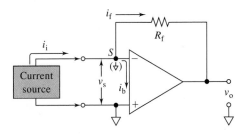

FIGURE 3-6 Operational amplifier current follower. A current source is connected between the inverting input and the noninverting input. The noninverting input is connected to common. A feedback resistor R_f is connected from the output to the inverting input.

3B-3 Current Follower Circuits

Operational amplifiers can be used to measure or process currents by connecting them in the *current follower* mode. This mode provides a nearly zero resistance load to the current source and prevents it from being loaded by a measuring device or circuit. The effect of loading on current measurements was described in Section 2A-3.

The Current Follower

In the operational amplifier current follower mode, the output is connected to the inverting input through a *feedback resistor* R_f as shown in Figure 3-6. If the amplifier is kept within its power supply limits, the difference voltage v_s is very small, as we have seen. Hence, the potential at the inverting input is essentially equal to that at the noninverting input. If the noninverting input is connected to circuit common, the inverting input is kept very nearly at circuit common as long as the amplifier is not at limit. The noninverting input is said to be *virtually at the circuit common* or at *virtual common potential*. From Kirchhoff's current law, the input current i_i is equal to the feedback current i_f plus the amplifier input bias current i_b.

$$i_i = i_f + i_b \qquad (3\text{-}3)$$

With modern amplifiers, typical values of i_b can be $10^{-11}-10^{-15}$ A. Hence, $i_i \approx i_f$.

If point S in Figure 3-6 is at virtual common potential, it follows that the output voltage v_o must equal the iR drop across the feedback resistor R_f and be opposite in sign. We can thus write

$$v_o = -i_f R_f = -i_i R_f \qquad (3\text{-}4)$$

 Simulation: Learn more about **current followers**.

Because point S is at virtual common ($\downarrow$), several current sources can be connected here without interacting with one another. Point S is thus called the *summing point* or *summing junction*.

From Equation 3-4, it can be seen that the operational amplifier generates a feedback current i_f that *follows* the input current i_i and produces an output voltage v_o that is directly proportional to i_i. If, for example, $R_f = 100 \text{ M}\Omega$ ($10^8 \, \Omega$) and $i_i = 5.00 \text{ nA}$ (5.0×10^{-9} A), the output voltage v_o would be 0.50 V, which is readily measured. Because it produces an output voltage proportional to the input *current*, the current follower is often termed a *current-to-voltage converter*.

There is very little loading effect of the current follower circuit. Because the input current source is connected between point S and common and point S is kept at virtual common, the input-signal source senses virtually zero resistance at its output terminals. The effective input resistance R_i is the error voltage v_s divided by the input current, i_i; that is, $R_i = v_s/i_i$. Because $v_s = -v_o/A$ and from Equation 3-4 $i_i = -v_o/R_f$, we can write

$$R_i = \frac{R_f}{A} \qquad (3\text{-}5)$$

The loading effect of R_f on the circuit is reduced by a factor of A. If $R_f = 100 \text{ M}\Omega$, for example, and $A = 10^7$, the effective input resistance is reduced to 10 Ω.

A more exact relationship between v_o and i_i is given in Equation 3-6 and shows the limitations of the current follower:

$$\begin{aligned} v_o &= -R_f(i_i - i_b)\left(\frac{A}{1+A}\right) \\ &= -i_iR_f + i_bR_f\frac{v_o}{A} \end{aligned} \qquad (3\text{-}6)$$

When A is very large and i_b is small, Equation 3-6 reduces to Equation 3-4. Measurement of low currents is limited by the input bias current of the amplifier, typically $10^{-11} \, \Omega$ or less. On the high current end, the output current capability of the amplifier (typically 2 to 100 mA) is a limitation. The amplifier open-loop gain A is a limitation only when the output voltage is small or the bias current is large.

The Inverting Voltage Amplifier

The current follower mode can be used to make an inverting voltage amplifier if the input current i_i comes from a voltage source and series resistor R_i as shown in

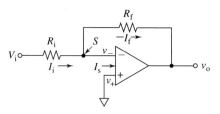

FIGURE 3-7 Inverting voltage amplifier. Here, the input current into the summing point S comes from the input voltage v_i and input resistance R_i.

Figure 3-7. Because the summing point S is at virtual common potential, the input current is

$$i_i = \frac{v_i}{R_i} \qquad (3\text{-}7)$$

If we substitute this result into Equation 3-4, we obtain

$$v_o = -i_iR_f = -v_i\frac{R_f}{R_i} \qquad (3\text{-}8)$$

Thus, the output voltage v_o is the input voltage v_i multiplied by the ratio of two resistors, R_f/R_i, and of opposite polarity. If the two resistors are precision resistors, the amplifier *closed-loop gain*, R_f/R_i, can be made quite accurate. For example, if R_f were 100 kΩ and R_i were 10 kΩ, the gain would be 10 and $v_o = -10 \times v_i$. Note that the accuracy of the gain depends on how accurately the two resistances are known and not on the *open-loop gain*, A, of the operational amplifier.

The amplifier of Figure 3-7 is often called an *inverting amplifier* or an *inverter* when $R_i = R_f$ because in this case the sign of the input voltage is inverted. Note, however, that there is a possibility of loading the input voltage v_i because current given by Equation 3-7 is drawn from the source.

3B-4 Frequency Response of a Negative Feedback Circuit

The gain of a typical operational amplifier decreases rapidly in response to high-frequency input signals. This frequency dependence arises from small capacitances that develop within the transistors inside the operational amplifier. The frequency response for a typical amplifier is usually given in the form of a *Bode diagram*, such as that shown in Figure 3-8 (see also Section 2B-5). Here, the solid curve labeled open-loop gain represents the behavior of the amplifier *in the absence of the feedback resistor R_f in Figure 3-7. Note*

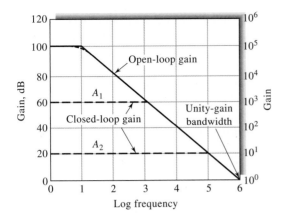

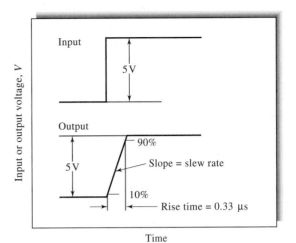

FIGURE 3-8 A Bode diagram showing the frequency response of a typical operational amplifier. Solid line: operational amplifier open-loop gain without negative feedback; dashed line: inverting amplifier configuration as in Figure 3-7 with $A_1 = R_f/R_i = 1000$ and $A_2 = R_f/R_i = 10$.

FIGURE 3-9 Response of an operational amplifier to a rapid step change in input voltage. The slope of the changing portion of the output signal is the slew rate, and the time required for the output to change from 10% to 90% of the total change is the rise time.

that both the ordinate and abscissa are log scales and that the gain is presented in decibels, or dB, where $1\,\text{dB} = 20 \log(v_o/v_i)$.

At low input-signal frequencies, $A = 10^5$, or 100 dB; but as the frequency increases above 10 Hz, the open-loop gain rolls off at -20 dB/decade in the same fashion as in a low-pass filter (Section 2B-5). The frequency range from dc to the frequency at which $A = 1$, or 0 dB, is called the *unity-gain bandwidth*, which for this amplifier is 1 MHz. The point at which $A = 1$ also fixes the *gain bandwidth product* at 1 MHz, which is a constant characteristic of the operational amplifier at all frequencies above 10 Hz. This characteristic permits the calculation of the bandwidth of a given amplifier configuration.

As an example, consider the amplifier of Figure 3-7 with a signal gain $A_1 = R_f/R_i = 1000$, as indicated by the upper dashed line in Figure 3-8. The bandwidth of the amplifier is then 1 MHz/1000 = 1 kHz, which corresponds to the intersection of the upper dashed line with the open-loop gain curve. A similar amplifier with $A_2 = 10$ then has a bandwidth of 100 kHz as indicated by the lower dashed line, and so on for other values of signal gain that might be chosen for a given op amp. Hence, we see that negative feedback increases amplifier bandwidth, which can be calculated from the signal gain and the unity-gain bandwidth of the operational amplifier.

Simulation: Learn more about **op amp frequency response**.

Two other parameters that relate to the speed or bandwidth Δf of an amplifier are illustrated in Figure 3-9. The output response of a voltage follower to a step input is characterized by the *rise time* t_r, which is the time required for the output to change from 10% to 90% of the total change. It can be shown that

$$t_r = \frac{1}{3\Delta f} \tag{3-9}$$

For the voltage follower with closed-loop gain of 1 and unity-gain bandwidth of 10^6, $t_r = 1/(3 \times 1.00 \text{ MHz}) = 0.33$ μs. From the slope of the change in voltage at the output during the transition from 5 V to 10 V, the *slew rate* can be calculated as follows:

$$\text{slew rate} = \frac{\Delta v}{\Delta t} = \frac{5\,\text{V}}{0.33\,\mu s} = 17\,\text{V/μs}$$

The slew rate is the maximum rate of change of the output of an amplifier in response to a step change at the input. Typical values of slew rate are on the order of a few volts per microsecond, but special operational amplifiers may be purchased with slew rates of up to several hundred volts per microsecond.

3C AMPLIFICATION AND MEASUREMENT OF TRANSDUCER SIGNALS

Operational amplifiers find general application in the amplification and measurement of the electrical signals from transducers. Such transducers can produce voltage outputs, current outputs, or charge outputs. The signals from transducers are often related to concentration. This section presents basic applications of operational amplifiers to the measurement of each type of signal.

3C-1 Current Measurement

The accurate measurement of small currents is important to such analytical methods as voltammetry, coulometry, photometry, and chromatography. As pointed out in Chapter 2, an important concern that arises in all physical measurements, including current measurement, is whether the measuring process itself will alter significantly the signal being measured and lead to a loading error. It is inevitable that any measuring process will perturb a system under study in such a way that the quantity actually measured will differ from its original value before the measurement. We must therefore try to ensure that the perturbation can be kept small. For a current measurement, this consideration requires that the internal resistance of the measuring device be minimized so that it does not alter the current significantly. The current follower presented in Section 3B-3 is a nearly ideal current-measuring device.

An example of a current follower being used to measure a small photocurrent is presented in Figure 3-10. The transducer is a phototube that converts light intensity into a related current, I_x. Radiation striking

the photocathode results in ejection of electrons from the cathode surface. If the anode is maintained at a potential that is positive with respect to the cathode (cathode negative with respect to the anode), the ejected photoelectrons are attracted, giving rise to a photocurrent proportional to the power of the incident beam.

From Equation 3-4, the output voltage V_o can be written as

$$V_o = -I_f R_f = -I_x R_f$$

and

$$I_x = -V_o/R_f = kV_o$$

Thus, measuring V_o gives the current I_x provided that R_f is known. Nanoampere currents can be measured with a high degree of accuracy if R_f is a large value.

As shown in Example 3-1, an operational amplifier current follower can give rise to minimal perturbation errors in the measurement of current.

EXAMPLE 3-1

Assume that R_f in Figure 3-10 is 1 MΩ, the internal resistance of the phototube is 5.0×10^4 Ω, and the amplifier open-loop gain is 1.0×10^5. Calculate the relative error in the current measurement that results from the presence of the measuring circuit.

Solution

From Equation 3-5, the input resistance of the current follower R_i is

$$R_i = \frac{R_f}{A} = \frac{1 \times 10^6}{1 \times 10^5} = 10\ \Omega$$

Equation 2-20 shows that the relative loading error (rel error) in a current measurement is given by

$$\text{rel error} = \frac{-R_M}{R_L + R_M}$$

where the meter resistance R_M is the current follower input resistance R_i and R_L is the internal resistance of the phototube. Thus,

$$\text{rel error} = \frac{-10.0\ \Omega}{(5.0 \times 10^4\ \Omega) + 10.0\ \Omega}$$

$$= -2.0 \times 10^{-4}, \text{ or } 0.020\%$$

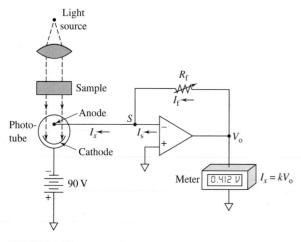

FIGURE 3-10 Application of an operational amplifier current follower to the measurement of a small photocurrent, I_x.

The instrument shown in Figure 3-10 is called a *photometer*. A photometer can be used to measure the attenuation of a light beam by absorption brought about by an analyte in a solution. The absorbance is related to the concentration of the species responsible for the absorption. Photometers are described in detail in Section 13D-3.

In addition to phototubes, other transducers such as oxygen electrodes, flame ionization detectors, photodiodes, and photomultiplier tubes produce output currents related to concentration or to a physical phenomenon of interest. The current follower is an indispensable circuit for measuring the small currents produced.

3C-2 Voltage Measurements

Several transducers produce output voltages related to concentration or to a physical quantity of interest. For example, ion-selective electrodes produce voltage outputs related to pH or the concentration of an ion in solution. Thermocouples produce voltage outputs related to temperature. Similarly, Hall effect transducers produce output voltages proportional to magnetic field strength. Operational amplifier circuits, particularly those based on the voltage follower (see Section 3B-2), are used extensively for such measurements.

Equation 2-19 shows that accurate voltage measurements require that the resistance of the measuring device be large compared to the internal resistance of the voltage source being measured. The need for a high-resistive measuring device becomes particularly acute in the determination of pH with a glass electrode, which typically has an internal resistance on the order of tens to hundreds of megohms. The voltage follower circuit shown in Figure 3-5 presents a very high input resistance to prevent loading of the glass electrode. If amplification is needed, the voltage follower can be combined with the basic inverting amplifier of Figure 3-7 to give a high-impedance voltage-measuring device with amplification as shown in Figure 3-11. Here, the first stage consists of a voltage follower, which typically provides an input impedance in excess of 10^{12} Ω. An inverting amplifier circuit then amplifies the follower output by R_f/R_i, or 20 in this case. An amplifier such as this with a resistance of 100 MΩ or more is often called an *electrometer*. Carefully designed operational amplifier–based electrometers are available commercially.

Occasionally, amplification without inversion of the signal is desired from a voltage-output transducer. In this case, a circuit known as a *voltage follower with gain*

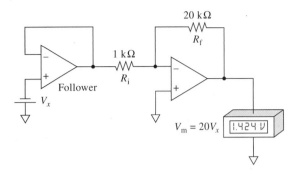

FIGURE 3-11 A high-impedance circuit for voltage amplification and measurement.

can be employed by feeding back only a fraction of the output voltage of the follower of Figure 3-5 to the inverting input.[5]

3C-3 Resistance or Conductance Measurements

Electrolytic cells and temperature-responsive devices, such as thermistors and bolometers, are common examples of transducers whose electrical resistance or conductance varies in response to an analytical signal. These devices are used for conductometric and thermometric titrations, for infrared absorption and emission measurements, and for temperature control in a variety of analytical applications.

The circuit shown in Figure 3-7 provides a convenient means for measurement of the resistance or conductance of a transducer. Here, a constant voltage source is used for V_i and the transducer is substituted for either R_i or R_f in the circuit. The amplified output voltage v_o is then measured with a suitable meter, potentiometer, or a computerized data-acquisition system. Thus, if the transducer is substituted for R_f in Figure 3-7, the output, as can be seen from rearrangement of Equation 3-8, is

$$R_x = \frac{-v_o R_i}{V_i} = k v_o \qquad (3\text{-}10)$$

where R_x is the resistance to be measured and k is a constant that can be calculated if R_i and V_i are known; alternatively, k can be determined from a calibration in which R_x is replaced by a standard resistor.

[5]H. V. Malmstadt, C. G. Enke, and S. R. Crouch, *Microcomputers and Electronic Instrumentation: Making the Right Connections*, Washington, DC: American Chemical Society, 1994, pp. 131, 132.

If conductance rather than resistance is of interest, the transducer conveniently replaces R_i in the circuit. From Equation 3-8, we find that

$$\frac{1}{R_x} = G_x = \frac{-v_o}{V_i R_f} = k' v_o \qquad (3\text{-}11)$$

Figure 3-12 illustrates two basic applications of operational amplifiers for the measurement of conductance or resistance. In (a), the conductance of a cell

for a conductometric titration is of interest. Here, an ac input signal v_i of perhaps 5 to 10 V is provided by an ac power supply. The output signal is then rectified, filtered, and measured as a dc voltage. The variable resistance R_f provides a means for varying the range of conductances that can be measured. Calibration is provided by switching the standard resistor R_s into the circuit in place of the conductivity cell.

Figure 3-12b illustrates how the circuit in Figure 3-7 can be applied to the measurement of a ratio of resis-

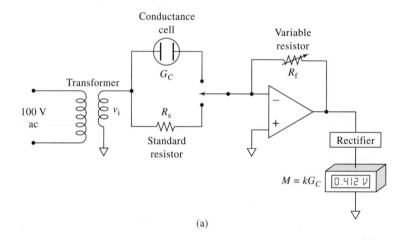

(a)

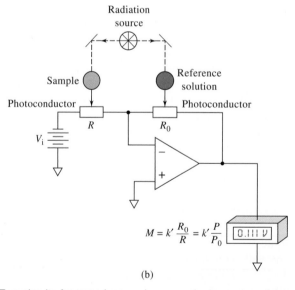

(b)

FIGURE 3-12 Two circuits for transducers whose conductance or resistance is the quantity of interest. (a) The output of the cell is a current proportional to the conductance of the electrolyte. (b) The ratio of the resistances of the photoconductive cells is proportional to the meter reading.

tances or conductances. Here, the absorption of radiant energy by a sample is being compared with that for a reference solution. The two photoconductive transducers, which are devices whose resistances are inversely related to the intensity of light striking their active surfaces, replace R_f and R_i in Figure 3-7. A dc power supply with voltage V_i serves as the source of power, and the output voltage M, as seen from Equation 3-8, is

$$M = V_o = -V_i \frac{R_0}{R}$$

Typically, the resistance of a photoconductive cell is inversely proportional to the radiant power P of the radiation striking it. If R and R_0 are matched photoconductors,

$$R = C \times \frac{1}{P} \quad \text{and} \quad R_0 = C \times \frac{1}{P_0}$$

where C is a constant for both photoconductive cells, giving

$$V_o = M = -V_i \frac{C/P_0}{C/P} = -V_i \frac{P}{P_0} \qquad (3\text{-}12)$$

Thus, the meter reading M is proportional to the ratio of the radiant powers of the two beams (P/P_0).

3C-4 Difference Amplifiers

It is frequently desirable to measure a signal generated by an analyte relative to a reference signal, as in Figure 3-12b. A difference amplifier, such as that shown in Figure 3-13, can also be applied for this purpose. Here, the amplifier is used for a temperature measurement. Note that the two input resistors have equal resistances R_i; similarly, the feedback resistor and the resistor between the noninverting input and common, which are both labeled R_k, are also identical values.

If we apply Ohm's law to the circuit shown in Figure 3-13, we find that

$$I_1 = \frac{V_1 - v_-}{R_i}$$

and

$$I_f = \frac{v_- - V_o}{R_k}$$

Because the operational amplifier has a high input impedance, I_1 and I_f are approximately equal.

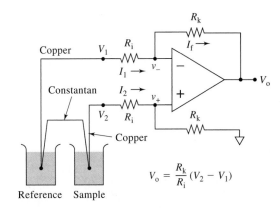

$$V_o = \frac{R_k}{R_i}(V_2 - V_1)$$

FIGURE 3-13 An operational amplifier difference amplifier measuring the output voltage of a pair of thermocouples.

$$I_1 \approx I_f$$

$$\frac{V_1 - v_-}{R_i} = \frac{v_- - V_o}{R_k}$$

Solving this equation for v_- gives

$$v_- = \frac{V_1 R_k + V_o R_i}{R_k + R_i} \qquad (3\text{-}13)$$

The voltage v_+ can be written in terms of V_2 via the voltage divider equation, Equation 2-10:

$$v_+ = V_2 \left(\frac{R_k}{R_i + R_k} \right) \qquad (3\text{-}14)$$

Recall that an operational amplifier with a negative feedback loop will do what is necessary to satisfy the equation $v_+ \approx v_2$. When Equations 3-13 and 3-14 are substituted into this relation we obtain, after rearrangement,

$$V_o = \frac{R_k}{R_i}(V_2 - V_1) \qquad (3\text{-}15)$$

Thus, it is the difference between the two signals that is amplified. Any extraneous voltage *common to the two inputs* shown in Figure 3-13 will be subtracted and will not appear in the output. Thus, any slow drift in the output of the transducers or any 60-cycle currents induced from the laboratory power lines will be eliminated from V_o. This useful property accounts for the widespread use of difference amplifier circuits in the first amplifier stages of many instruments.

An important characteristic of operational amplifier circuits, such as the difference amplifier described here, is the *common mode rejection ratio*, or CMRR. For a difference amplifier, the CMRR is a measure of how well the amplifier rejects signals that are common to both inputs; it is the ratio of the difference gain A_d to the common mode gain A_{cm}; that is,

$$\text{CMRR} = \frac{A_d}{A_{cm}}$$

Suppose that we apply identical signals to the inputs V_1 and V_2, that $R_k = 1000R_i$, and that $V_o = 0.1V_2$. If the difference amplifier were ideal, V_o should be zero. In real difference amplifiers, some fraction of V_2, which is the signal that is to be rejected, appears at the output. In this case, V_2 is the signal to be rejected, or the common mode signal, so that the common mode gain is $A_{cm} = V_o/V_2 = 0.1$. The difference gain A_d is just the gain of the difference amplifier, which is $A_d = R_k/R_i = 1000$. The CMRR for this configuration is then

$$\text{CMRR} = \frac{A_d}{A_{cm}} = 1000/0.1 = 10,000$$

The larger the CMRR of a difference amplifier, the better it is at rejecting common mode signals, that is, signals that are applied to both inputs simultaneously.

The transducers shown in Figure 3-13 are a pair of *thermocouple junctions*; one of the transducers is immersed in the sample, and the other transducer is immersed in a reference solution (often an ice bath) held at constant temperature. A temperature-dependent contact potential develops at each of the two junctions formed from wires made of copper and an alloy called constantan (other metal pairs are also used). The potential difference $v_2 - v_1$ is roughly 5 mV per 100°C temperature difference.

3D APPLICATION OF OPERATIONAL AMPLIFIERS TO VOLTAGE AND CURRENT CONTROL

Operational amplifiers are easily configured to generate constant-voltage or constant-current signals.

3D-1 Constant-Voltage Sources

Several instrumental methods require a dc power source whose voltage is precisely known and from which reasonable currents can be drawn with no change

in voltage. A circuit that meets these qualifications is termed a *potentiostat*.

Two potentiostats are illustrated in Figure 3-14. Both employ a standard voltage source in a feedback circuit. This source is generally an inexpensive, commercially available, Zener-stabilized integrated circuit (see Section 2D-3) that is capable of producing an output voltage that is constant to a few hundredths of a percent. Such a source will not, however, maintain its voltage when it must deliver a large current.

Recall from our earlier discussions that point S in Figure 3-14a is at virtual common potential. For this condition to exist, it is necessary that $V_o = V_{std}$, the standard voltage. That is, the current in the load resistance R_L must be such that $I_L R_L = V_{std}$. It is important to appreciate, however, that this current arises from

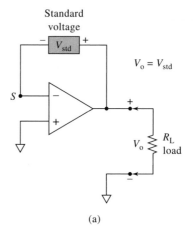

(a)

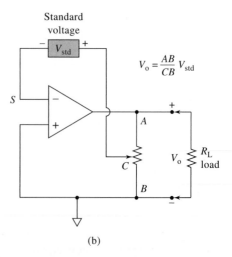

(b)

FIGURE 3-14 Constant-voltage sources.

the power source of the operational amplifier and *not from the standard voltage source*. There is essentially no current in the feedback loop because the impedance of the inverting input is very large. Thus, the standard cell controls V_o but provides essentially none of the current through the load.

Figure 3-14b illustrates a modification of the circuit in (a) that permits the output voltage of the potentiostat to be fixed at a level that is a known multiple of the output voltage of the standard potential source.

3D-2 Constant-Current Sources

Constant-current dc sources, called *amperostats*, find application in several analytical instruments. For example, these devices are usually used to maintain a constant current through an electrochemical cell. An amperostat reacts to a change in input power or a change in internal resistance of the cell by altering its output voltage in such a way as to maintain the current at a predetermined level.

Figure 3-15 shows two amperostats. The first requires a voltage input V_i whose potential is constant while it supplies significant current. Recall from our earlier discussion that

$$I_L = I_i = \frac{V_i}{R_i}$$

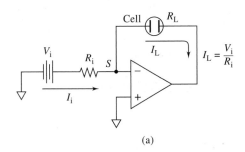

(a)

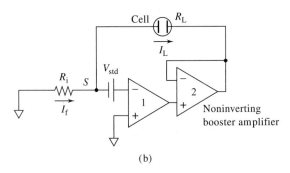

(b)

FIGURE 3-15 Constant-current sources.

Thus, the current will be constant and independent of the resistance of the cell, provided that V_i and R_i remain constant.

Figure 3-15b is an amperostat that employs a standard voltage V_{std} to maintain a constant current. Note that operational amplifier 1 has a negative-feedback loop that contains operational amplifier 2. To satisfy the condition $v_- = v_+$, the voltage at the summing point S must be equal to $-V_{std}$. Furthermore, we may write that at S

$$I_i R_i = I_L R_L = -V_{std}$$

Because R_i and V_{std} in this equation are constant, the operational amplifier functions in such a way as to maintain I_L at a constant level that is determined by R_i.

Operational amplifier 2 in Figure 3-15b is simply a voltage follower, which has been inserted into the feedback loop of operational amplifier 1. A voltage follower used in this configuration is often called a *noninverting booster amplifier* because it can provide the relatively large current that may be required from the amperostat.

3E APPLICATION OF OPERATIONAL AMPLIFIERS TO MATHEMATICAL OPERATIONS

As shown in Figure 3-16, substitution of various circuit elements for R_i and R_f in the circuit shown in Figure 3-7 permits various mathematical operations to be performed on electrical signals as they are generated by an analytical instrument. For example, the output from a chromatographic column usually takes the form of a peak when the electrical signal from a detector is plotted as a function of time. Integration of this peak to find its area is necessary to find the analyte concentration. The operational amplifier shown in Figure 3-16c is capable of performing this integration automatically, giving a signal that is directly proportional to analyte concentration.

3E-1 Multiplication and Division by a Constant

Figure 3-16a shows how an input signal v_i can be multiplied by a constant whose magnitude is $-R_f/R_i$. The equivalent of division by a constant occurs when this ratio is less than unity.

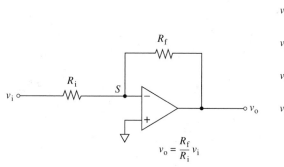

$$v_o = -\frac{R_f}{R_i} v_i$$

(a) Multiplication or Division

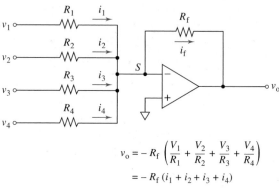

$$v_o = -R_f\left(\frac{V_1}{R_1} + \frac{V_2}{R_2} + \frac{V_3}{R_3} + \frac{V_4}{R_4}\right)$$

$$= -R_f(i_1 + i_2 + i_3 + i_4)$$

(b) Addition or Subtraction

$$v_o = -\frac{1}{R_iC_i}\int_0^t v_i dt$$

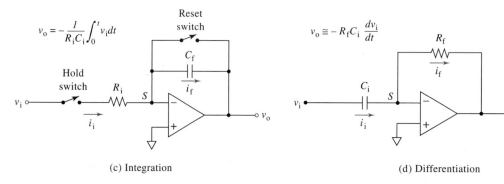

(c) Integration

$$v_o \cong -R_fC_i\frac{dv_i}{dt}$$

(d) Differentiation

FIGURE 3-16 Mathematical operations with operational amplifiers.

3E-2 Addition or Subtraction

Figure 3-16b illustrates how an operational amplifier can produce an output signal that is the sum of several input signals. Because the impedance of the amplifier is large and because the output must furnish a sufficient current i_f to keep the summing point S at virtual common, we may write

$$i_f = i_1 + i_2 + i_3 + i_4 \qquad (3\text{-}16)$$

But $i_f = -v_o/R_f$, and we may thus write

$$v_o = -R_f\left(\frac{v_1}{R_1} + \frac{v_2}{R_2} + \frac{v_3}{R_3} + \frac{v_4}{R_4}\right) \qquad (3\text{-}17)$$

If $R_f = R_1 = R_2 = R_3 = R_4$, the output voltage is the sum of the four input voltages but opposite in sign.

$$v_o = -(v_1 + v_2 + v_3 + v_4)$$

To obtain an average of the four signals, let $R_1 = R_2 = R_3 = R_4 = 4R_f$. Substituting into Equation 3-17 gives

$$v_o = -\frac{R}{4}\left(\frac{v_1}{R} + \frac{v_2}{R} + \frac{v_3}{R} + \frac{v_4}{R}\right)$$

and v_o becomes the average of the four inputs, as shown in Equation 3-18.

$$v_o = \frac{-(v_1 + v_2 + v_3 + v_4)}{4} \qquad (3\text{-}18)$$

In a similar way, a weighted average can be obtained by varying the ratios of the resistances of the input resistors.

Subtraction can be performed by the circuit in Figure 3-16b by introducing an inverter with $R_i = R_f$ in series with one or more of the resistors, thus changing the sign of one or more of the inputs. Weighted subtraction can also be performed by varying the resistance ratios.

3E-3 Integration

Figure 3-16c illustrates a circuit for integrating a variable input signal v_i with respect to time. When the reset switch is open and the hold switch is closed,

$$i_i = i_f$$

and the capacitor C_f begins to charge. The current in the capacitor i_f is given by Equation 2-30 or

$$i_f = -C\frac{dv_o}{dt}$$

From Ohm's law the current i_i is given by $i_i = v_i/R_i$. Thus, we may write

$$\frac{v_i}{R_i} = -C\frac{dv_o}{dt}$$

or

$$dv_o = -\frac{v_i}{R_iC}\,dt \qquad (3\text{-}19)$$

We then integrate Equation 3-19 to obtain an equation for the output voltage v_o

$$\int_{v_{o1}}^{v_{o2}} dv_o = -\frac{1}{R_iC}\int_{t_1}^{t_2} v_i\,dt \qquad (3\text{-}20)$$

or

$$v_{o2} - v_{o1} = -\frac{1}{R_iC}\int_{t_1}^{t_2} v_i\,dt \qquad (3\text{-}21)$$

The integral is ordinarily obtained by first opening the hold switch and closing the reset switch to discharge the capacitor, thus making $v_{o1} = 0$ when $t_1 = 0$. Equation 3-21 then simplifies to

$$v_o = -\frac{1}{R_iC}\int_0^t v_i\,dt \qquad (3\text{-}22)$$

To begin the integration, the reset switch is opened and the hold switch closed. The integration is stopped at time t by opening the hold switch. The integral over the period of 0 to t is v_o.

3E-4 Differentiation

Figure 3-16d is a basic circuit for differentiation, which is useful when the time rate of change of an experimental quantity is the variable of interest. Note that it differs from the integration circuit only in the respect that the positions of C and R have been reversed. Proceeding as in the previous derivation, we may write

$$C\frac{dv_i}{dt} = -\frac{v_o}{R_f}$$

 Simulation: Learn more about **integrators and differentiators**.

or

$$v_o = -R_fC\frac{dv_i}{dt} \qquad (3\text{-}23)$$

The circuit shown in Figure 3-16d is not, in fact, practical for many chemical applications, where the rate of change in the transducer signal is often low. For example, differentiation is a useful way to treat the data from a potentiometric titration; here, the potential change of interest occurs over a period of a second or more ($f \le 1$ Hz). The input signal will, however, contain extraneous 60-, 120-, and 240-Hz components (see Figure 5-3) that are induced by the ac power supply. In addition, signal fluctuations resulting from incomplete mixing of the reagent and analyte solutions are often encountered. These noise components often have a faster rate of change than the desired signal components.

This problem may be overcome to some extent by introducing a small parallel capacitance C_f in the feedback circuit and a small series resistor R_i in the input circuit to filter the high-frequency voltages. These added elements are kept small enough so that significant attenuation of the analytical signal does not occur. In general, differentiators are noise-amplifying circuits, whereas integrators smooth or average the noise. Thus, analog integrators are more widely used than differentiators. If differentiation of a signal is required, it is often accomplished digitally, as is discussed in Chapter 5.

3E-5 Generation of Logarithms and Antilogarithms

The incorporation of an external transistor into an operational amplifier circuit makes it possible to generate output voltages that are either the logarithm or the antilogarithm of the input voltage, depending on the circuit. Operational amplifier circuits of this kind are, however, highly frequency and temperature dependent and accurate to only a few percent; they are, in addition, limited to one or two decades of input voltage. Temperature- and frequency-compensated modules for obtaining logarithms and antilogarithms with accuracies of a few tenths of a percent are available commercially. In the past these circuits were used to produce signals proportional to absorbance in spectrophotometers and to compress data. Currently, logarithms and antilogarithms are computed numerically with computers rather than with operational amplifiers.

3F APPLICATIONS OF OPERATIONAL AMPLIFIERS TO COMPARISON

Another important and widespread application of operational amplifiers is in comparing analog signals. Such circuits are found in a wide variety of applications such as sampling circuits, peak detection circuits, analog timers, circuits designed to produce limited signal levels, and circuits at the interface of the boundary between the digital and analog domains. The comparator mode of operational amplifiers was introduced in Section 3B-1. Figure 3-17a and b show two basic comparator circuits and their output response versus input voltages. In the circuit in (a), the input voltage is compared with the circuit common, and in (b) the comparison is with a reference voltage V_{ref}.

The circuit in Figure 3-17a is often called a *zero-crossing detector* because the sign of the output voltage indicates whether the input voltage is greater than or less than zero (common). If $v_i > 0$ by more than a few microvolts, the output is at negative limit. If $v_i < 0$ by a few microvolts, v_o is at positive limit. Because the amplifier response is very rapid, such a detector can be used to convert a sinusoidal signal into a square-wave signal as shown by the waveforms on the right side. Every time the sine wave crosses zero, the comparator changes state. Such circuits are often used in oscilloscope triggering. A noninverting zero-crossing detector can be made by connecting the sine wave signal to the noninverting input and connecting the inverting input to common.

In Figure 3-17b, the comparison is between v_i and V_{ref}. If $v_i > V_{ref}$, the output is at positive limit, whereas the opposite limit is reached when $v_i < V_{ref}$. This type of comparator is often called a *level detector*. It can be used, for example, to determine whether a transducer output has exceeded a certain level. As shown in Fig-

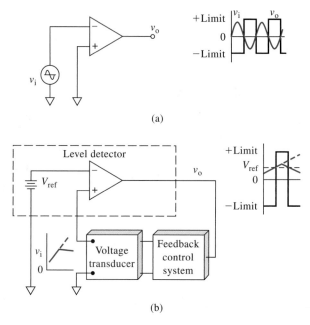

(a)

(b)

FIGURE 3-17 Operational amplifier zero-crossing detector (a) and level detector (b).

ure 3-17b, the level detector can be part of a feedback control system. In a chemical process such a level detector might be used to determine when the temperature has exceeded a critical value and to supply coolant when that occurs or to determine when the pH has fallen below a certain level and to supply base. The level detector is also used in oscilloscope triggering circuits.

Although any operational amplifier can be used in the comparator mode, special amplifiers are available with high gains ($>10^6$) and very fast rise times. These specialized *comparators* are often used in computer interfacing and other switching applications.

 Simulation: Learn more about **comparators**.

QUESTIONS AND PROBLEMS

*Answers are provided at the end of the book for problems marked with an asterisk.

X Problems with this icon are best solved using spreadsheets.

*3-1 An operational amplifier has output voltage limits of $+13$ V and -14 V when used with a ± 15 V power supply. If the amplifier is used as a comparator, by what amount does v_+ have to exceed v_- and v_- have to exceed v_+ for the amplifier to be at limit if the open-loop gain A is
(a) 200,000
(b) 500,000
(c) 1.5×10^6

*3-2 The common mode rejection ratio is important for comparators and other difference amplifiers. If the output voltage changes by 10 V for an input difference voltage v_s of 500 µV, and by 1.0 V for a common mode input voltage of 500 mV, what is the common mode rejection ratio of the amplifier?

*3-3 For a comparator with output voltage limits of ± 13 V, what would the open-loop gain A need to be to keep the absolute value of the difference voltage $|v_s| \leq 5.0$ µV?

*3-4 An operational amplifier with an open-loop gain of 1.0×10^5 and an input resistance of 1.0×10^{12} is used in the voltage follower circuit of Figure 3-5. A voltage source is to be measured with a voltage of 2.0 V and a source resistance of 10.0 kΩ. Find the percentage relative error in the follower output voltage due to (a) the finite gain of the amplifier and (b) the loading of the voltage source.

3-5 By means of a derivation, show that the output voltage v_o and the input voltage v_i for the following circuit are related by $v_o = v_i \left(\dfrac{R_1 + R_2}{R_1} \right)$.

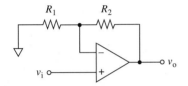

Why is this circuit called a voltage follower with gain?

*3-6 For the circuit shown in Problem 3-5, it is desired that $v_o = 3.5 v_i$. If the total resistance $R_1 + R_2$ is to equal 10.0 kΩ, find suitable values for R_1 and R_2.

3-7 In the following circuit, R is a variable resistor. Derive an equation that describes v_o as a function of v_i and the position x of the movable contact of the voltage divider. Perform the derivation such that x is zero if there is zero resistance in the feedback loop.

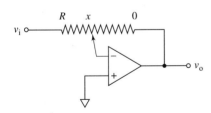

3-8 An operational amplifier to be used in a current follower has an open-loop gain A of 2×10^5 and an input bias current of 2.5 nA.
  (a) Design a current follower that will produce a 1.0 V output for a 10.0 µA input current.
  (b) What is the effective input resistance of the current follower designed in part (a).
  (c) What is the percentage relative error for the circuit designed in part (a) for an input current of 25 µA?

3-9 An operational amplifier to be used in an inverting amplifier configuration has an open-loop gain $A = 1.0 \times 10^5$, an input bias current of 5.0 nA, a linear output voltage range of ± 10 V, and an input resistance of 1.0×10^{12} Ω.

(a) Design an inverting amplifier with a gain of 25 such that the input resistance R_i is 10 kΩ.

(b) Determine the range of usable input voltages for the amplifier in (a).

(c) Find the input resistance of the inverting amplifier designed in (a).

(d) How could you avoid a loading error if the voltage source were loaded by the input resistance found in (c)?

3-10 A low-frequency sine wave voltage is the input to the following circuits. Sketch the anticipated output of each circuit.

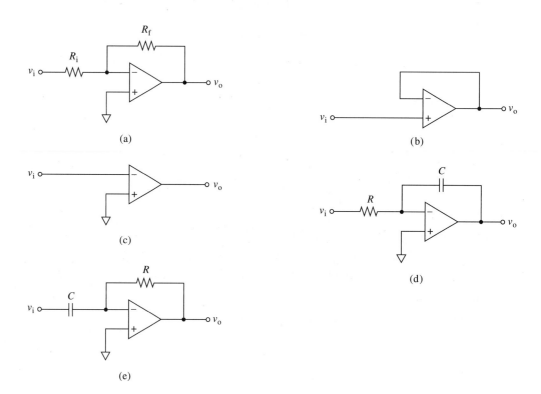

    (a)   (b)

    (c)   (d)

    (e)

*3-11 Calculate the slew rate and the rise time for an operational amplifier with a 50-MHz bandwidth in which the output changes by 10 V.

3-12 Design a circuit having an output given by

$$-V_o = 3V_1 + 5V_2 - 6V_3$$

3-13 Design a circuit for calculating the average value of three input voltages multiplied by 1000.

3-14 Design a circuit to perform the following calculation:

$$-V_o = \frac{1}{10}(5V_1 + 3V_2)$$

3-15 Design a circuit to perform the following function:

$$V_o = -4V_1 - 1000I_1$$

*3-16 For the following circuit

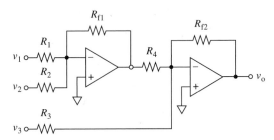

(a) write an expression that gives the output voltage in terms of the three input voltages and the various resistances.

(b) indicate the mathematical operation performed by the circuit when $R_1 = R_{f1} = 200\,\text{k}\Omega$; $R_4 = R_{f2} = 400\,\text{k}\Omega$; $R_2 = 50\,\text{k}\Omega$; $R_3 = 10\,\text{k}\Omega$.

3-17 Show the algebraic relationship between the output voltage and input voltage for the following circuit:

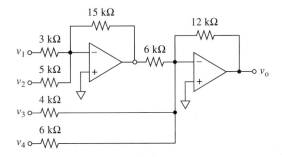

3-18 For the circuit below, sketch the outputs at v_{oA} and v_{oB} if the input is initially zero but is switched to a constant positive voltage at time zero.

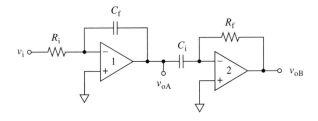

3-19 Derive an expression for the output voltage of the following circuit:

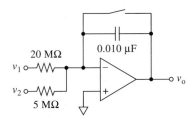

3-20 Show that when the four resistances are equal, the following circuit becomes a subtracting circuit.

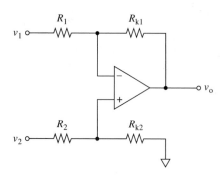

*3-21 The linear slide wire AB in the circuit shown has a length of 100 cm. Where along its length should contact C be placed to provide exactly 3.00 V at V_o? The voltage of the Weston cell is 1.02 V.

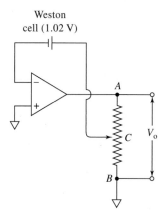

3-22 Design a circuit that will produce the following output:

$$v_o = 4.0 \int_0^t v_1 \, dt + 5.0 \int_0^t v_2 \, dt$$

3-23 Design a circuit that will produce the following output:

$$v_o = 2.0 \int_0^t v_1 \, dt - 6.0(v_2 + v_3)$$

3-24 Plot the output voltage of an integrator 1, 3, 5, and 7 s after the start of integration if the input resistor is 2.0 MΩ, the feedback capacitor is 0.25 μF, and the input voltage is 4.0 mV.

Challenge Problem

3-25 The circuit shown below is an integrating type of differentiator based on a circuit originally described by E. M. Cordos, S. R. Crouch, and H. V. Malmstadt, *Anal. Chem.*, **1968**, *40*, 1812–1818. American Chemical Society.

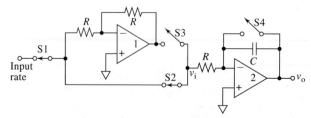

(a) What is the function of operational amplifier 1?

(b) What function does operational amplifier 2 perform?

(c) Assume that the input signal is a linearly increasing voltage and the rate of change of this signal is desired. During the first period Δt_1, switches S1 and S2 are closed and switch S4 opens. Describe and plot the output v_o during this interval Δt_1.

(d) During a second consecutive and identical time period $\Delta t_2 = \Delta t_1 = \Delta t$, switch S2 opens and S3 closes. Now describe and plot the output v_o during this second interval.

(e) At the end of the second interval, switch S1 opens, disconnecting the input signal. Show that the output voltage v_o at the end of the measurement cycle is given by

$$v_o = k \times \text{input rate}$$

(f) What are the advantages and disadvantages of this circuit over the normal operational amplifier differentiator of Figure 3-16d?

(g) What would happen if the input signal were to change slope during the measurement cycle?

(h) What would be the result if the two time intervals were not consecutive but instead were separated by a time delay Δt_3?

(i) What would be the result if the two time intervals were of different duration?

(j) The circuit shown above with consecutive time intervals was the basis of several automatic ratemeters used in instruments for measuring enzyme kinetics. The total measurement time for these instruments is $2\Delta t$. Discuss why it is desirable for $2\Delta t$ to be as long as possible. In measuring enzyme kinetics, what limitations might be imposed if the measurement time is too long?
Hint: Refer to part (g), above.

CHAPTER FOUR

Digital Electronics and Computers

This chapter is a springboard for further study and use of modern instrumental systems. Our goals are (1) to provide a brief overview of how digital information can be encoded, (2) to introduce some of the basic components of digital circuits and microcomputers, (3) to describe some of the most common instrument-computer interactions, and (4) to illustrate how computers and software are used in an analytical laboratory.

Throughout this chapter, this logo indicates an opportunity for online self-study at **www .thomsonedu.com/chemistry/skoog**, linking you to interactive tutorials, simulations, and exercises.

The rate of growth of electronics, instrumentation, and computer technology is nearly incomprehensible.[1] Computers first began to appear in chemical laboratories in the mid-1960s, but they were expensive and difficult to program and use. The advent of the microcomputer in the 1970s gave rise to an increasing number of applications in the chemical laboratory and in chemical instruments. It has been the advent of the inexpensive mass-produced personal computer (PC), however, with its collection of associated peripheral devices, that has brought about revolutionary changes in the way in scientists operate. At present, computers are found in virtually every laboratory instrument. Not only are computers found in the laboratory but today the scientist has a computer on the desktop with a high-speed connection to the Internet and to other computers in the organization. Computers are now used not only for scientific computations (simulations, theoretical calculations, modeling, data acquisition, data analysis, graphical display, and experimental control) but also for manuscript preparation, visualization, document sharing, and communication with other scientists and with funding agencies.

An understanding of the advantages and limitations of modern electronic devices and computers is important for today's scientist. Although it is impossible, and perhaps undesirable, for all chemists to attain knowledge of electronics and computers at the design level, the development of high-function integrated-circuit modules and data-acquisition hardware permits a conceptually straightforward top-down approach to the implementation of electronics and computer technology. In the top-down view, we assume the perspective that an instrument is a collection of functional modules that can be represented as blocks in a schematic diagram such as those depicted in Figures 4-2, 4-4, and 4-5. Using such an approach, it is possible to accomplish very sophisticated physicochemical measurements by connecting several function modules, integrated circuits, or computers in the correct sequence. It is not usually necessary to possess detailed knowledge of the internal design of individual components of an instrumental system. In addition to easing the learning process, the top-down approach aids in diagnosing system malfunctions and in the intelligent application of instrumental systems to the solution of chemical problems.

---

[1] See for example, S. R. Crouch and T. V. Atkinson, "The Amazing Evolution of Computerized Instruments," *Anal. Chem.*, **2000**, *72*, 596A; P. E. Ceruzzi, *A History of Modern Computing*, Cambridge, MA: MIT Press, 1998.

Digital circuits offer some important advantages over their analog counterparts. For example, digital circuits are less susceptible to environmental noise, and digitally encoded signals can usually be transmitted with a higher degree of signal integrity. Second, digital signals can be transmitted directly to digital computers, which means that software can be used to extract the information from signal outputs of chemical instruments.[2]

4A ANALOG AND DIGITAL SIGNALS

As described in Chapter 1, chemical data are encoded in *digital*, *analog*, or *time domains*. An example of a discrete phenomenon in a nonelectrical domain that may be easily converted to the digital domain is the radiant energy produced by the decay of radioactive species. Here, the information consists of a series of pulses of energy that is produced as individual atoms decay. These pulses can be converted to an electrical domain by using an appropriate input transducer, first as analog pulses and then as digital pulses that can be counted. The resulting information can be interpreted and manipulated as an integer number of decays, which is a form of nonelectrical information.

It is important to appreciate that whether a signal resulting from a chemical phenomenon is continuous or discrete may depend on the intensity of the signal and how it is observed. For example, the yellow radiation produced by heating sodium ions in a flame is often measured with a phototransducer that converts the radiant energy into an analog current, which can vary continuously over a considerable range. However, at low radiation intensity, a properly designed transducer can respond to the individual photons, producing a signal that consists of a series of analog pulses that can be converted to digital pulses and then counted.

Often, in modern instruments an analog signal, such as shown in Figure 4-1a, is converted to a digital one (Figure 4-1b) by sampling and recording the analog output at regular time intervals. In a later section, we consider how such a conversion is accomplished with an *analog-to-digital converter*, or ADC.

[2]For further information, see H. V. Malmstadt, C. G. Enke, and S. R. Crouch, *Microcomputers and Electronic Instrumentation: Making the Right Connections*, Washington, DC: American Chemical Society, 1994; A. J. Diefenderfer and B. E. Holton, *Principles of Electronic Instrumentation*, 3rd ed., Philadelphia: Saunders, 1994; K. L. Ratzlaff, *Introduction to Computer Assisted Experimentation*, New York: Wiley, 1987; S. C. Gates and J. Becker, *Laboratory Automation Using the IBM-PC*, New York: Prentice-Hall, 1989.

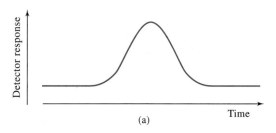

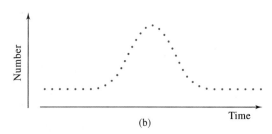

FIGURE 4-1 Detector response versus time plots for the same signal in (a) an analog domain and (b) the digital domain.

4B COUNTING AND ARITHMETIC WITH BINARY NUMBERS

In a typical digital measurement, a high-speed electronic counter is used to count the number of events that occur within a specified set of boundary conditions. Examples of such signals include the number of photons or alpha decay particles emitted by an analyte per second, the number of drops of titrant, or the number of steps of a stepper motor used to deliver reagent from a syringe. Boundary conditions might include a time interval such as 1 second, which provides the frequency of the signal in hertz, or a given change in an experimental variable such as pH, absorbance, current, or voltage.

Counting such signals electronically requires that they first be converted to digital signal levels to provide a series of pulses of equal voltage compatible with the digital circuitry of the counter. Ultimately, these pulses are converted by the counter to a binary number for processing by a computer or to a decimal number for display. Electronic counting is performed in binary-coded-decimal numbers or in binary numbers. In both of these coding schemes, only two digits, 0 and 1, are required to represent any number. In many electronic counters, the 0 is usually represented by a signal of about 0 V and the 1 by a voltage of typically 5 V. It is important to recognize that these voltage levels depend on current technology and are different for different logic families. For example, many state-of-the-art computers internally use about 3 V to represent a 1 and use 0 V to signify a 0.

4B-1 The Binary Number System

Each digit in the decimal numbering system represents the coefficient of some power of 10. Thus, the number 3076 can be written as

3   0   7   6

$6 \times 10^0 = 0006$

$7 \times 10^1 = 0070$

$0 \times 10^2 = 0000$

$3 \times 10^3 = \underline{3000}$

Sum = 3076

Similarly, each digit in the binary system of numbers corresponds to a coefficient of a power of 2.

4B-2 Conversion of Binary and Decimal Numbers

Table 4-1 illustrates the relationship between a few decimal and binary numbers. The examples that follow illustrate methods for conversions between the two systems.

EXAMPLE 4-1

Convert 101011 in the binary system to a decimal number.

Solution

Binary numbers are expressed in terms of base 2. Thus,

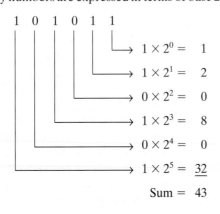

1 0 1 0 1 1

$1 \times 2^0 = 1$

$1 \times 2^1 = 2$

$0 \times 2^2 = 0$

$1 \times 2^3 = 8$

$0 \times 2^4 = 0$

$1 \times 2^5 = \underline{32}$

Sum = 43

TABLE 4-1 Relationship between Some Decimal and Binary Numbers

Decimal Number	Binary Representation
0	0
1	1
2	10
3	11
4	100
5	101
6	110
7	111
8	1000
9	1001
10	1010
12	1100
15	1111
16	10000
32	100000
64	1000000

EXAMPLE 4-2

Convert 710 to a binary number.

Solution

As a first step, we determine the largest power of 2 that is less than 710. Thus, since $2^{10} = 1024$,

$$2^9 = 512 \quad \text{and} \quad 710 - 512 = 198$$

The process is repeated for 198:

$$2^7 = 128 \quad \text{and} \quad 198 - 128 = 70$$

Continuing, we find that

$$2^6 = 64 \quad \text{and} \quad 70 - 64 = 6$$
$$2^2 = 4 \quad \text{and} \quad 6 - 4 = 2$$
$$2^1 = 2 \quad \text{and} \quad 2 - 2 = 0$$

The binary number is then computed as follows:

$$\begin{array}{ccccccccccc} 1 & 0 & 1 & 1 & 0 & 0 & 0 & 1 & 1 & 0 \\ 2^9 & - & 2^7 & 2^6 & - & - & - & 2^2 & 2^1 & - \end{array}$$

It is worthwhile noting that in the binary numbering system, the binary digit, or *bit*, lying farthest to the right

 Tutorial: Learn more about **binary and BCD**.

in a number is termed the *least significant bit*, or *LSB*; the one on the far left is the *most significant bit*, or *MSB*.

4B-3 Binary Arithmetic

Arithmetic with binary numbers is similar to, but simpler than, decimal arithmetic. For addition, only four combinations are possible:

$$
\begin{array}{cccc}
0 & 0 & 1 & 1 \\
+0 & +1 & +0 & +1 \\
\hline
0 & 1 & 1 & 10
\end{array}
$$

Note that in the last sum, a 1 is carried over to the next higher power of 2. Similarly, for multiplication,

$$
\begin{array}{cccc}
0 & 0 & 1 & 1 \\
\times 0 & \times 1 & \times 0 & \times 1 \\
\hline
0 & 0 & 0 & 1
\end{array}
$$

The following example illustrates the use of these operations.

EXAMPLE 4-3

Perform the following calculations with binary arithmetic: (a) 7 + 3, (b) 19 + 6, (c) 7 × 3, and (d) 22 × 5.

Solution

$$
\textbf{(a)} \quad
\begin{array}{r}
7 \\
+3 \\
\hline
10
\end{array}
\qquad
\begin{array}{r}
111 \\
+11 \\
\hline
1010
\end{array}
\qquad
\textbf{(b)} \quad
\begin{array}{r}
19 \\
+6 \\
\hline
25
\end{array}
\qquad
\begin{array}{r}
10011 \\
+110 \\
\hline
11001
\end{array}
$$

$$
\textbf{(c)} \quad
\begin{array}{r}
7 \\
\times 3 \\
\hline
21
\end{array}
\qquad
\begin{array}{r}
111 \\
\times 11 \\
\hline
111 \\
111 \\
\hline
10101
\end{array}
\qquad
\textbf{(d)} \quad
\begin{array}{r}
22 \\
\times 5 \\
\hline
110
\end{array}
\qquad
\begin{array}{r}
10110 \\
\times 101 \\
\hline
10110 \\
00000 \\
10110 \\
\hline
1101110
\end{array}
$$

Note that a carry operation, similar to that in the decimal system, is used. Thus, in (a) the sum of the two 1s in the right column is equal to 0 plus 1 to carry to the next column. Here, the sum of the three 1s is 1 plus 1 to carry to the next column. Finally, this carry combines with the 1 in the next column to give 0 plus 1 as the most significant digit.

4B-4 Binary-Coded-Decimal Scheme

In the binary-coded-decimal (BCD) scheme, binary bits are arranged in groups of four to represent the decimal numbers 0–9. Each group of four represents one decimal digit in a number. The BCD code is the same as normal binary for the numerals 0 through 9. For example, in the number 97, the nine would be represented by the four binary bits 1001 and the seven by the four binary bits 0111 so that 1001 0111 in BCD would represent 97_{10}. Several decimal numbers are presented in both binary coding and BCD coding in Table 4-2 to illustrate the differences.

4C BASIC DIGITAL CIRCUITS

Figure 4-2 is a block diagram of an instrument for counting the number of electrical pulses that are received from a transducer per unit of time. The voltage signal from the transducer first passes into a shaper that removes the small background signals and converts the large signal pulses to rectangular pulses that have the same frequency as the input signal. The resulting signal is then the input to a gate opened by an internal clock that provides a precise time interval t during which input pulses are allowed to accumulate in the counter. Finally, the BCD output of the counter is decoded and presented as a decimal number.

TABLE 4-2 Comparison of Binary and BCD Coding for Various Decimal Numbers

Decimal Number	Binary Equivalent								BCD Equivalent		
	2^7 (128)	2^6 (64)	2^5 (32)	2^4 (16)	2^3 (8)	2^2 (4)	2^1 (2)	2^0 (1)	Hundreds	Tens	Ones
83	0	1	0	1	0	0	1	1	0000	1000	0011
97	0	1	1	0	0	0	0	1	0000	1001	0111
135	1	0	0	0	0	1	1	1	0001	0011	0101
198	1	1	0	0	0	1	1	0	0001	1001	1000
241	1	1	1	1	0	0	0	1	0010	0100	0001

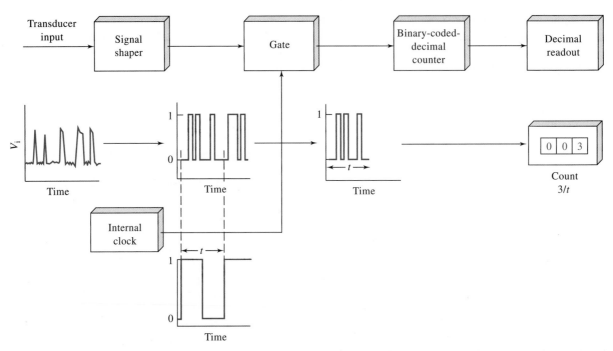

FIGURE 4-2 A counter for determining voltage pulses per second.

4C-1 Signal Shapers

Figure 4-3a shows a circuit of a typical signal shaper. It makes use of a voltage comparator to convert the input signal to the pulsed rectangular waveform shown in Figure 4-3c. As shown in Figure 3-17b, the output of a comparator is either at +limit or −limit. These two levels are often termed *logic levels*. Commercial integrated-circuit comparators are designed so that their outputs have limits of 0 V (LO logic level) or 5 V (HI logic level). Usually, the HI level is chosen to represent the binary 1 and the LO level represents the binary 0, as we show in Figures 4-2 and 4-3. These HI and LO logic levels are compatible with most modern digital integrated circuits. As we can see in Figure 4-3b and c, when the comparator input voltage v_i is greater than the reference voltage V_{ref}, the output is HI (logic level 1). On the other hand, when v_i is less than V_{ref}, the output is LO (logic level 0). Note that the comparator responds only to signals greater than V_{ref} and ignores the fluctuation in the background signal, provided that noise on the signal or any background fluctuations are small enough that the signal remains below V_{ref}. A comparator used for signal shaping is often called a *discriminator*.

4C-2 Binary and BCD Counters

Binary and BCD counters are available in integrated-circuit form for the counting of electrical pulses. In fact, the entire counting system shown in Figure 4-2 is available in a single integrated circuit. Although a BCD counter is shown, binary counting systems are also available. Inside a binary counter chip, the circuitry consists of electronic switches that have only two possible logic states, HI and LO, or 1 and 0. Each circuit can then be used to represent one bit of a binary number (or the coefficient of a power of 2). Two circuits can have four possible outputs: 0/0, 0/1, 1/0, and 1/1. We can readily show that three of these circuits have 8 different combinations and four have 16. Thus, n circuits have 2^n distinguishable output combinations. By using a sufficient number of circuit stages, the number of significant bits in a count can be made as large as desired. Thus, eight stages have 256 states, which would allow counting from 0 to 255. Counting circuits often have an error of ±1 count so that eight binary stages would provide a count that is accurate to 1 part in 256, or better than 0.5% relative.

BCD counters are arranged so that on the 10th count, a pulse is sent to the next decade of a *decade*

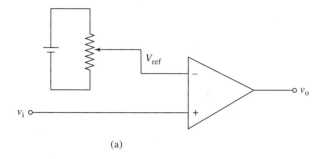

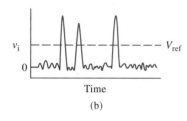

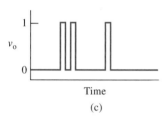

FIGURE 4-3 A signal shaper: (a) circuit, (b) input signal, (c) output signal.

counting unit (DCU), and the first decade is set to zero. Counters with as many as 10 decades of counting are available.

Counters are made from circuits called *flip-flops*. The flip-flops used in many counters change output levels whenever the input signal changes from logic level 1 to 0; *no change in output* is associated with an input change of 0 to 1. Flip-flops are also integrated circuits that are made up of a suitable combination of diodes, resistors, and transistors. In integrated-circuit counters, several flip-flops are integrated on a chip to make the counter, and in integrated-circuit counting systems, a counter, gate, clock, and readout are integrated on a single chip. Integrated-circuit flip-flops are examples of *small-scale integration* (SSI), and complete counting systems are examples of *large-scale integration* (LSI).

 Simulation: Learn more about **counters**.

4C-3 Counting Measurements

Counting is one of the most reliable forms of measurement. As discussed in Section 4B, we always count events that occur within a set of boundary conditions (time per 100 m, apples per bushel, pulses per second, revolutions per minute, etc.). In Figure 4-2, the events to be counted are the pulses from the transducer, and the boundary conditions are used to open the gate. In this case the clock time t is the boundary condition. The counter of Figure 4-2 is shown as a *frequency meter* because it can measure the number of pulses per unit time (frequency). A general frequency meter is shown in Figure 4-4a.

In a frequency meter, the opening and closing of the counting gate is not synchronized with the input pulses. Because of this, there is always an uncertainty of ±1 count in the results. Hence, to obtain frequency results with less than 0.1% uncertainty, at least 1000 counts must be accumulated. For low-frequency signals, we either have to count for a long period of time or rearrange the components of Figure 4-4a. For example, if the input frequency were 10 Hz, we would have to count for 100 s to obtain 1000 counts and an uncertainty of 0.1%. For a 1-Hz signal, we would have to count for 1000 s for a similar counting uncertainty. A *period meter*, shown in Figure 4-4b, achieves less uncertainty for low-frequency signals by using the transducer output after shaping to open and close the gate and counting the number of cycles of the precision clock. If the clock had a time base of 1 ms, for example, and the input signal from the transducer were 1 Hz, 1,000 clock pulses would be counted during 1 cycle of the input signal (1 s). For a 0.1-Hz input signal, 10,000 pulses would be counted during 1 cycle of the input signal (10 s). Thus, the period mode is much better for low-frequency signals.

Figure 4-4c shows one additional mode for a general-purpose digital counting system, the time-interval mode. Here, we might be interested in measuring the time that elapses between two events, such as the firing of a starter's gun and the breaking of the tape in a 100-m dash. As can be seen, one of the events opens the counting gate and the other closes it. The precision clock is again counted during this interval. Time-interval measurements are very important in many areas of science. As one example, the distance from the earth to the moon can be obtained by measuring the time required for a laser beam to travel to the moon, be reflected from a mirror on the surface, and return to earth. Here, the

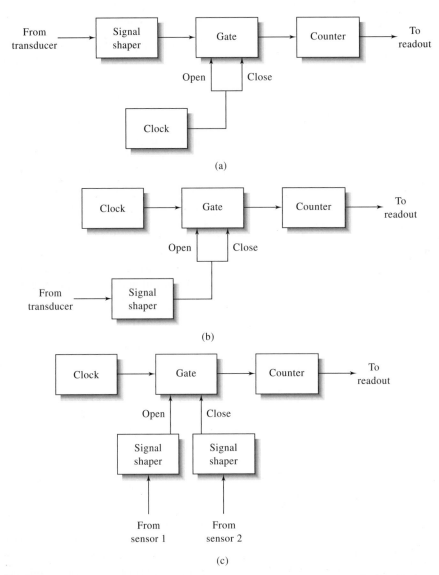

(a)

(b)

(c)

FIGURE 4-4 Counting measurements. In (a) the frequency meter is shown. Shaped pulses from the transducer are the input to a counting gate that is opened and closed by a precision digital clock. In (b), a period meter is shown. Here, the precision clock is counted for one cycle of the input signal. In (c), the time-interval mode is shown. In this mode the gate is opened by a signal from sensor 1 and closed by a signal from sensor 2. The time interval between these two events is measured by counting the number of cycles of the precision clock.

firing of the beam starts the timer (opens the counting gate) and the return pulse stops it (closes the gate). Knowing the speed of light and measuring the time required for a round trip allows calculation of the distance from earth to moon.

4C-4 Scalers

A DCU produces one carry pulse on the 10th count. In counting, this pulse is fed to the next decade counting stage. However, it is important to note that the output frequency of a DCU is the input frequency divided by 10. A 1-MHz input to a DCU will result in a 100-kHz output. The frequency division is exact and the accuracy of the output frequency is the same as that of the input frequency. DCUs can be cascaded to reduce the input frequency by exact multiples of 10. This process is often called *scaling*. Scaling is also used when the input frequency of a signal is greater than a counting device can accommodate. In this situation, a *scaler* is introduced between the signal shaper and the counter of Figure 4-4a. In period measurements (Figure 4-4b), if the frequency of the input signal is too high, a scaler is often introduced between the shaper and the counting gate. This allows multiple periods to be averaged during the measurement.

4C-5 Clocks

Many digital applications require that a highly reproducible and accurately known frequency source be used in conjunction with the measurement of time as shown in Figure 4-4b and c. Generally, electronic frequency sources are based on quartz crystals that exhibit the *piezoelectric effect*, as described in Section 1C-4. The resonant frequency of a quartz crystal depends on the mass and dimensions of the crystal. By varying these parameters, electrical output frequencies that range from 10 kHz to 50 MHz or greater can be obtained. Typically, these frequencies are constant to 100 ppm. With special precautions, such as precise temperature control, crystal oscillators can be constructed for time standards that are accurate to 1 part in 10 million.

The use of a series of decade scalers with a quartz oscillator provides a precise clock, as illustrated in Figure 4-5. The frequency can be selected in decade steps from the original 10 MHz to as low as 0.1 Hz. Because there is no noise or variation in the counting operation, all the outputs are as accurate and precise as the crystal oscillator used. Integrated circuits are available containing several decades of scaling. Such circuits are programmable with the output decade selected by a binary input.

4C-6 Digital-to-Analog Converters

Digital signals are often converted to their analog counterparts for the control of instruments, for display by readout devices such as meters and oscilloscopes, or as part of ADCs. Figure 4-6 illustrates a digital-to-

 Simulation: Learn more about **digital-to-analog converters**.

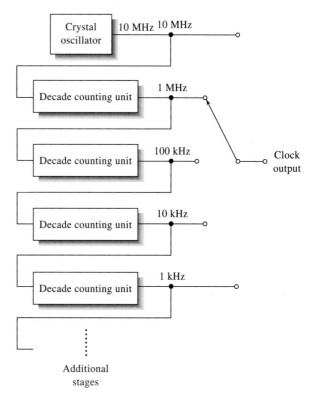

FIGURE 4-5 Precision clock circuit. Each decade counting unit divides the frequency at its input by exactly 10. The accuracy of each output frequency is equal to that of the crystal oscillator.

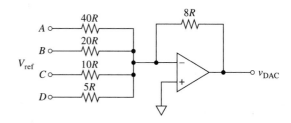

FIGURE 4-6 A 4-bit digital-to-analog converter (DAC). Here, A, B, C, and D are $+5$ V for logic state 1 and 0 V for logic state 0.

analog converter (DAC) and the principle of one of the common ways of accomplishing this conversion, which is based on a *weighted-resistor ladder* network. Note that the circuit is similar to the summing circuit shown in Figure 3-16b, with four resistors weighted in the ratio 8:4:2:1. From the discussion of summing circuits we can show that the output v_{DAC} is given by

$$v_{DAC} = -V_{ref}\left(\frac{D}{1} + \frac{C}{2} + \frac{B}{4} + \frac{A}{8}\right) \quad (4\text{-}1)$$

where V_{ref} is the voltage associated with logic state 1 and D, C, B, and A designate the logic states (0 or 1) for a 4-bit binary number in which A is the least significant bit and D the most significant. Table 4-3 shows the analog output from the weighted-resistor ladder shown in Figure 4-6 when V_{ref} is 5 V.

The resolution of a DAC depends on the number of input bits that the device will accommodate. An n-bit device has a resolution of 1 part in 2^n. Thus, a 10-bit DAC has 2^{10}, or 1024, output voltages and, therefore, a resolution of 1 part in 1024; a 12-bit DAC has a resolution of 1 part in 4096. Note that in our discussion of DACs and ADCs we use the letter n to represent the number of bits of resolution of the device and N to represent the digital output.

TABLE 4-3 Analog Output from the DAC in Figure 4-6

Binary Number DCBA	Decimal Equivalent	v_{DAC}, V
0000	0	0.0
0001	1	−1.0
0010	2	−2.0
0011	3	−3.0
0100	4	−4.0
0101	5	−5.0

4C-7 Analog-to-Digital Converters

The output from most transducers used in analytical instruments is an analog signal. To realize the advantages of digital electronics and computer data processing, it is necessary to convert the analog signal from the analog domain to the digital domain. Figure 4-1 illustrates such a digitization process. Numerous methods are used for this kind of conversion. Two common types of ADC are described here: the staircase ADC and the successive-approximation ADC.

Staircase Analog-to-Digital Converter

Figure 4-7 shows a simplified schematic of a device for converting an unknown analog voltage v_i into a digital number N. Here, an n-bit binary counter, controlled by the signal from a quartz clock, is used to drive an n-bit DAC similar to that described in the previous section. The output of the DAC is the *staircase* voltage output v_{DAC} shown in the lower part of the figure. Each step of this signal corresponds to a voltage increment, such as 1 mV. The output of the DAC is compared with the unknown input v_i by means of the comparator. When the two voltages become identical within the resolution of the DAC, the comparator changes state from HI to

 Simulation: Learn more about **analog-to-digital converters**.

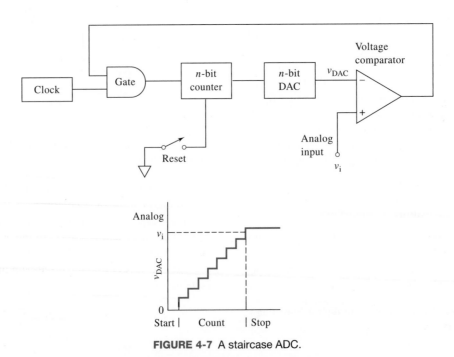

FIGURE 4-7 A staircase ADC.

LO, which in turn stops the counter. The count N then corresponds to the input voltage in units of millivolts. Closing the reset switch sets the counter back to zero in preparation for the conversion of a new voltage, which is begun by opening the reset switch. The input voltage v_i must be held constant during the conversion process to ensure that the digital output corresponds to the desired voltage.

The higher the resolution of the DAC, the more precisely the number will represent v_i. This type of converter clearly illustrates the measurement process. The DAC output serves as the reference standard that is compared to v_i by the difference detector (the comparator). The time of conversion is $t_c = Nt_p$, where N is the counter output, which varies with v_i, and t_p is the period of the clock. This conversion time is advantageous if v_i is known to be relatively small most of the time. If v_i is large most of the time, then t_c will be proportionally longer. This type of ADC is often used in nuclear spectroscopy and related fields in which background signals of low intensity are encountered. The oscillator frequency can be as large as 100 MHz when high-speed counters, DACs, and comparators are used.

The operation of the staircase ADC can be made continuous by replacing the simple counter by an up-down counter controlled by the comparator. If v_i increases, the comparator output goes HI and the counter counts up; and if v_i decreases, the counter counts down. When the DAC output crosses v_i, the counter alternates between N and $N - 1$, a range that is within $\pm\frac{1}{2}$LSB of v_i. This type of ADC works well when v_i varies only slowly relative to the conversion time or when a continuous readout is important.

Successive-Approximation Analog-to-Digital Converter

To understand how a successive-approximation ADC works, consider the following question: What is the minimum number of trials necessary to determine, *with certainty*, a number N that lies between 0 and 15? Assume that following each guess you are told whether your number was high or low. The answer is that no more than four guesses are required. To illustrate, suppose that 10 is the target number. Let us first divide the range in half and guess that the unknown number is 7. The number 7 is less than 10, so the upper half of the range 1–7 is halved and added to 7 to obtain a second trial number, $N = 7 + 4 = 11$. This number is too large, so 4 is dropped, and half of 4 is added to 7, to yield $N = 9$, which is low. Finally, half of 2 is added to 9 to

obtain the target value $N = 10$. The rules for the successive approximation are as follows:

1. Begin with a guess of one half of full range.
2. If too large, drop guess.
3. If too small, retain guess.
4. Add half of previous increment.
5. Repeat steps 2 through 4 until finished.

Note that n guesses are required to determine a number in the range 0 to $2^n - 1$. For example, 12 guesses are required to determine with certainty a number between 0 and 4095.

The successive-approximation ADC uses exactly the same logic to arrive at a binary or BCD number to represent an unknown voltage v_i as depicted in Figure 4-8a. Here, the 4-bit DAC of Figure 4-6 is used to illustrate how the successive-approximation process can be carried out. Assume that $v_i = 5.1$ V and that all

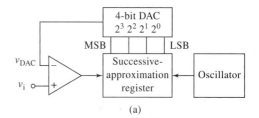

(a)

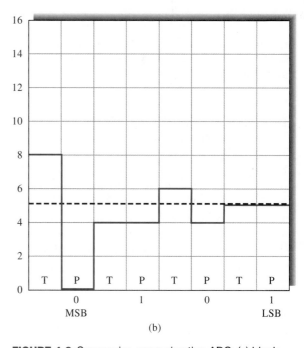

(b)

FIGURE 4-8 Successive-approximation ADC: (a) block diagram of the ADC, (b) output of the DAC during the conversion process.

bits are initially set to 0. The first cycle of the oscillator sets the MSB = 2^3 to 1, which causes the DAC voltage v_{DAC} to change to 8 V as shown in Figure 4-8b. This is called the *test* period and labeled T in the figure. Because $v_{DAC} > v_i$, the successive-approximation register (SAR) clears the 2^3 bit = 1 during the *post* period, labeled P. The next cycle of the oscillator causes the 2^2 bit to be set to logic 1 to give $v_{DAC} = 4$ V. Because $v_{DAC} < v_i$, the comparator output goes to logic 1 (HI), which results in the SAR setting this bit to a logic 1 during the post period. The next cycle then causes bit 2^1 to be set to 1, resulting in $v_{DAC} = 6$ V, which is larger than v_i. The comparator output goes to 0, and thus the SAR clears the 2^1 bit. Finally, the SAR sets $2^0 = 1$, to give $v_{DAC} = 5$ V, which is $< v_i$. This results in retaining bit 2^0 as a logic 1. The process is diagrammed in Figure 4-8b. The resulting binary number, 0101, represents the input voltage 5 V ± 0.5 V. Note the resolution is $\pm\frac{1}{2}$LSB, which in this case is ±0.5 V.

To increase the resolution of the ADC, a DAC of the required resolution must be provided and the SAR must have a correspondingly larger number of bits. Twelve-bit ADCs with input ranges of ±5 V, ±10 V, or 0 to 10 V are typical. Such converters have a fixed conversion time, usually 2 to 8 μs for 12 bits. Successive-approximation converters of this type are widely used for computerized timed data acquisition. Because it is important that the voltage to be measured does not vary during the conversion process, a fast analog memory called a *sample-and-hold* amplifier is almost always used to sample and keep the signal of interest constant during the conversion process.

4D COMPUTERS AND COMPUTERIZED INSTRUMENTS

The first computers to be used in laboratories in the 1960s were constructed from discrete components (transistors, diodes, resistors, capacitors, etc.). These minicomputers were difficult to use and program, and interfacing them to instruments was often a complex and frustrating task. Very few programs were available for experiment control, data acquisition, and data analysis, so scientists had to become computer programmers to develop and use computerized instruments.

In the 1970s the situation changed dramatically with the introduction of the microprocessor and the microprocessor-based microcomputer. A *micropro-*

cessor is a large-scale integrated circuit made up of hundreds of thousands and even millions of transistors, resistors, diodes, and other circuit elements miniaturized to fit on a single silicon chip a few millimeters on a side. A microprocessor often serves as an arithmetic and logic component, called the *central processing unit* (CPU), of a microcomputer. Microprocessors also find widespread use for operating such diverse items as analytical instruments, automobile ignition systems, microwave ovens, cash registers, and electronic games.

Microcomputers consist of one or more microprocessors combined with other circuit components that provide memory, timing, input, and output functions. Microcomputers are finding ever-increasing use for the control of analytical instruments and for processing, storing, and displaying the data produced. There are at least two reasons for connecting a computer to an analytical instrument. The first is that partial or complete automation of measurements becomes possible. Ordinarily, automation leads to more rapid data acquisition, which shortens the time required for an analysis or increases its precision by providing time for additional replicate measurements to be made. Automation, moreover, frequently provides better and faster control over experimental variables than a human operator can achieve; more precise and accurate data are the result.

A second reason for interfacing computers with instruments is to take advantage of their tremendous computational and data-handling capabilities. These capabilities make possible the routine use of techniques that would be impractical because of their excessive computational time requirements. Notable among such applications are Fourier transform calculations, signal averaging, and correlation techniques in spectroscopy to extract small analytical signals from noisy environments.

The interfacing of these devices to instruments is too large a subject to be treated in detail in this text. However, there have been many advances in interface boards and in the software to control them. Today, readily available hardware and software allow sophisticated interfacing tasks to be done by chemists as they perform experiments. The discussion in this chapter is limited to a general summary of computer terminology, the architecture and properties of computers, some useful hardware and software employed in instrumental applications, and the advantages gained by using these remarkable devices.

4D-1 Computer Terminology

One of the problems that faces the newcomer to the field of computers and computer applications is the bewildering array of new terms, acronyms, and initialisms, such as CPU, RAM, ROM, BIOS, FTP, GUI, HTTP, USB, WiFi, LAN, firewall, and TCP/IP. Unfortunately, these terms are not often defined even in elementary presentations. Some of the most important terms and abbreviations are defined here; others will be defined as they appear later in the chapter.

Bits, Bytes, and Words

Bits are represented in a computer by two electrical states (HI/LO, or 1/0) that differ from one another normally by 2 to 5 V. A group of eight bits is called a *byte*. A series of bytes arranged in sequence to represent a piece of data or an instruction is called a *word*. The number of bits (or bytes) per word depends on the computer; some common sizes include 8, 16, 32, and 64 bits, or 1, 2, 4, and 8 bytes.

Registers

The basic building block of digital computers is the *register*, a physical device that can store a complete byte or a word. A 16-bit binary counter, for example, can serve as a register that is capable of holding a 16-bit word.

Data contained in a register can be manipulated in a number of ways. For example, a register can be cleared, a process by which the register is reset to all zeros; the *ones complement* of a register can be taken, that is, every 1 is changed to 0 and every 0 to a 1. The contents of one register can be transferred to another. Moreover, the contents of one register can be added, subtracted, multiplied, or divided by the contents of another. A register in which these processes are performed is often referred to as an *accumulator*. It has been proved that the proper sequence of register operations can solve any computational or informational processing problem no matter how complex provided that there is an *algorithm* for the solution. An algorithm is a detailed statement of the individual steps required to arrive at a solution. One or more algorithms then make up a *computer program*.

Hardware and Software

Computer *hardware* consists of the physical devices that make up a computer. Examples include disk drives, printers, clocks, memory units, data-acquisition modules, and arithmetic logic units. The collection of programs and instructions to the computer, including the disks or tapes for their storage, is the *software*. Hardware and software are equally important to the successful application of computers, and the initial cost of the software may be as great as the cost of the computer. This is especially true of sophisticated software packages designed for special purposes such as data manipulation, curve fitting, or statistical analysis. Over the past few years, the dramatic increase in the availability of fast, high-capacity, low-cost PCs has produced a corresponding demand for useful and user-friendly software packages. The production and sale of tens or perhaps hundreds of millions of computers worldwide virtually guarantees that a wide variety of software is available at reasonable cost. These market forces have produced lowered costs even for special-purpose scientific software, as we shall see in a subsequent section.

4D-2 Operational Modes of Computerized Instruments

Figure 4-9 suggests three ways computers can be used in conjunction with analytical measurements. In the *off-line* method shown in Figure 4-9a, the data are collected by a human operator and subsequently transferred to the computer for data processing. The *on-line* method of Figure 4-9b differs from the off-line procedure in that direct communication between the instrument and the computer is made possible by means of an electronic *interface* where the signal from the instrument is shaped, digitized, and stored. Here, the computer remains a distinct entity with provision for mass storage of data and instructions for processing these data; off-line operation is also possible with this arrangement.

Most modern instruments are configured as shown in Figure 4-9c. In this *in-line* arrangement, a microcomputer or microprocessor is embedded in the instrument. Here, the operator communicates with and directs the instrument operation via the computer. The operator does not, however, necessarily program the computer, although often the option is available for doing so. The primary software suite is usually provided with commercial instruments along with a programming language so that the users may program optional modes of data acquisition and manipulation. Often there are several computers or microprocessors in a given instrument. The user may communicate with

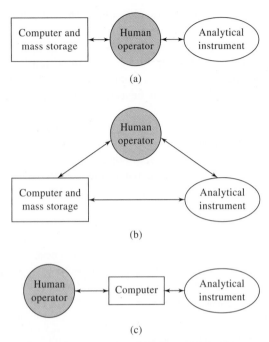

FIGURE 4-9 Three methods of using computers for analytical measurements: (a) off-line, (b) on-line, (c) in-line.

the instrument via one computer, and the instrument may be controlled and data acquired by others.

In in-line and on-line operations, the data are often transferred to the computer in *real time*, that is, as the data are generated by the instrument. Often, the rate at which data are produced by an instrument is low enough that only a small fraction of the computer's time is occupied in data acquisition; under this circumstance the periods between data collection can be used for processing the information in various ways. For example, data processing may involve calculating a concentration, smoothing a curve, combining a data point with previously collected and stored data for subsequent averaging, and plotting or printing out the result. *Real-time processing* involves data treatment performed simultaneously with data acquisition. Real-time processing has two major advantages. First, it may reduce significantly the amount of data storage space required, thus making possible the use of a less sophisticated and less expensive computer. Second, if there is sufficient time between the data acquisitions, the processed signal may be used to adjust instrument parameters to improve the quality of subsequent output signals. As the speed and storage capacity of microcomputers has increased, so has the capability for real-

time processing. Currently, with instruments that contain multiple processors, real-time operations are becoming quite common.

An example of a real-time processing system is a microprocessor-controlled instrument for automatically performing potentiometric titrations. Usually, such instruments have the storage capacity necessary to store a digitized form of the potential versus reagent volume curve, and all other information that might be required in the process of generating a report about the titration. It is also usually the case that such instruments calculate the first derivative of the potential with respect to volume in real time and use this information to control the rate at which the titrant is added by a motor-driven syringe. In the early part of the titration, when the rate of potential change is low, the derivative is small, so the titrant is added rapidly. As the equivalence point is approached, the derivative becomes larger, and the computer slows the rate at which the titrant is added. The reverse process occurs beyond the equivalence point.

4E COMPONENTS OF A COMPUTER

Figure 4-10 is a block diagram showing the major hardware components of a computer and its peripheral devices.

4E-1 Central Processing Unit

The heart of a computer is the CPU, which in the case of a microcomputer is a microprocessor chip. A microprocessor is made up of a control unit and an arithmetic logic unit. The control unit determines the sequence of operations by means of instructions from a program stored in the computer memory. The control unit receives information from the input device, fetches instructions and data from the memory, and transmits instructions to the arithmetic logic unit, output, and memory.

The arithmetic logic unit, or ALU, of a CPU is made up of a series of registers, or accumulators, in which the intermediate results of binary arithmetic and logic operations are accumulated. The Intel Pentium 4 processor contains nearly 50 million transistors and is capable of operating at clock speeds greater than 3.5 GHz. The Intel Itanium processor contains 22 million transistors (the Itanium 2 processor has 410 million transistors). The fastest computers can execute nearly 1 billion instructions per second.

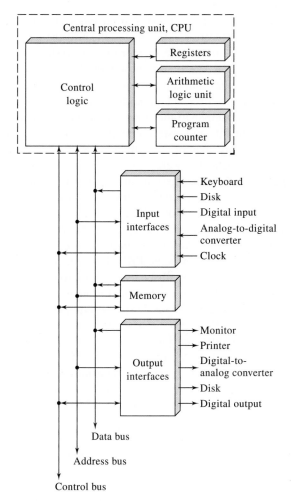

Central processing unit, CPU

Registers

Arithmetic logic unit

Control logic

Program counter

Input interfaces
— Keyboard
— Disk
— Digital input
— Analog-to-digital converter
— Clock

Memory

Output interfaces
→ Monitor
→ Printer
→ Digital-to-analog converter
→ Disk
→ Digital output

Data bus

Address bus

Control bus

FIGURE 4-10 Basic components of the digital computer, including peripheral devices.

4E-2 Buses

The parts of a computer and its memory and peripheral devices are joined by *buses*, each of which is made up of a number of transmission lines. For rapid communication among the various parts of a computer, all of the digital signals making up a word are usually transmitted simultaneously by the parallel lines of the bus. The number of lines in the internal buses of the CPU is then equal to the size of the word processed by the computer. For example, the internal bus for a 32-bit CPU requires 32 parallel transmission lines, each of which transmits one of the 32 bits.

Data are carried into and out of the CPU by a data bus, as depicted in Figure 4-10. Both the origin and destination of the signals in the data bus line are specified by the address bus. An address bus with 32 lines can directly address 2^{32}, or 4,294,967,296, registers or other locations within the computer or 4 gigabytes of memory. The control bus carries control and status information to and from the CPU. These transfers are sequenced by timing signals carried in the bus.

Data must also be transmitted between instrument components or peripheral devices and the CPU; an external bus or communication line is used for this type of data transfer. Table 4-4 summarizes some of the specifications for some popular external communication standards.

4E-3 Memory

In a computer, the memory is a storage area that is directly accessible by the CPU. Because the memory contains both data and program information, the memory must be accessed by the CPU at least once for each program step. The time required to retrieve a piece of information from the memory is called its

TABLE 4-4 Specifications for Common Communication Standards

	RS-232	IEEE-488	Ethernet 10BaseT	Ethernet 100BaseT	USB*	IEEE-1394 (FireWire)
Type	Serial	Parallel	Serial	Serial	Serial	Serial
Distance (m)	30	20	100	100	4.8	4.5
Maximum baud**	19.2 kbps	10 Mbps	10 Mbps	100 Mbps	12 Mbps (1.1) 480 Mbps (2.0)	400 Mbps (1394a) 800 Mbps (1394b)
Cabling	Twisted pair	Shielded bundle	Twisted pair	Twisted pair	Shielded twisted pair	Shielded twisted pair

*USB stands for universal serial bus.

**Baud is a measure of the rate at which information can be transmitted. Baud rate units are bits per second.

access time; access times are usually in the tens-of-nanosecond range.

Memory Chips

The basic unit of a memory chip is a cell, which may have one of two states and is thus capable of storing one bit of information. Typically, up to a few billion of these cells may be contained on a single silicon memory chip. Figure 4-11 illustrates the functions associated with an individual memory cell. With a READ command from the CPU, the logic state (1 or 0) appears as one of two possible states at the output. A WRITE command allows the 1 or 0 state from the input terminal to displace the contents already present in the cell and to store the new value in its place.

Individual cells are produced in arrays on memory chips, which in turn are mounted on printed circuit boards that plug directly into the case of the computer. Typically, PCs have on the order of 128 to 1024 megabytes (MB) of memory, but many different configurations are usually available. Because the actual process of addressing and storing information in memory is either established during manufacture or controlled by the CPU, most chemists have little need for an understanding of the detailed design of memories. However, a little experience with the terminology used when describing memory is often helpful in selecting the proper amount of memory for performing a particular computational task.

Types of Memory

There are two types of memory in most computers: *random access memory* (RAM) and *read-only memory* (ROM). The term *random access* is somewhat misleading, because ROM may also be accessed randomly. Random access means that all locations in the memory are equally accessible and can be reached at about the same speed. Thus, read-write memory is a more descriptive term for RAM. Earlier types of semiconductor RAM were volatile; that is, the information was not retained unless the memory was refreshed regularly. Many RAM boards now have battery-backup power supplies that can prevent the loss of any information if power is lost for 8 hours or more. This sort of memory is similar to that found in pocket calculators that retain data and instructions even when turned off.

Read-only memories contain permanent instructions and data that were placed there at the time of their manufacture. These memories are truly static in the sense that they retain their original states for the life of the computer or calculator. The contents of a ROM cannot be altered by reprogramming. A variant of ROM is the *erasable programmable read-only memory* (EPROM, or erasable PROM) in which the program contents can be erased by exposure to ultraviolet radiation. After this treatment, the memory can be reprogrammed by means of special equipment. Also available are ROMs that can be reprogrammed by relatively straightforward logic signals. These ROMs are designated EAROMs (electrically alterable ROMs). Bootstrap programs that perform the initialization of a computer system when it is turned on are usually stored in some type of ROM. When system information must be stored when power is off but must occasionally be reprogrammed when new devices are installed in the computer, a battery-powered RAM is usually used to store such data.

In some simple digital devices and handheld calculator systems, ROMs are used to store programs needed for performing mathematical operations such as obtaining logarithms, exponentials, and trigonometric functions; calculating statistical quantities such as means, standard deviations, and least-squares parameters; and formatting data in fixed point, scientific notation, or engineering notation. In most computers, however, these operations are carried out in software.

Bulk Storage Devices

In addition to semiconductor memories, computers are usually equipped with bulk storage devices. Magnetic tapes were for years the primary means for bulk storage, but tapes have now been replaced by disks, flash memories that connect via universal serial buses (USBs), CDs, and DVDs. Disk storage capacities are constantly increasing. Floppy disks were once a common means of storage but are now becoming obsolete. Removable disks, such as ZIP disks, are widely available with several-hundred-megabyte capacity. The

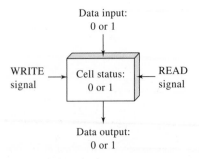

Data input:
0 or 1

WRITE signal → Cell status: 0 or 1 ← READ signal

Data output:
0 or 1

FIGURE 4-11 An individual computer memory cell for the storage of 1 bit.

USB flash drive is a semiconductor memory device that plugs into a USB port and contains from 32 MB to 1 GB or more of nonvolatile memory. These devices are about the size of a ballpoint pen and can be readily carried in a pocket.

The smallest hard disk drives now have capacities in the gigabyte range, but hard disks capable of storing 250 GB or more are now common. The time required to reach a randomly selected location on a disk is the *seek time*, and for most this time is on the order of 10 ms. The CD ROM is a particularly attractive storage device for large databases, encyclopedias, and for backing up data because of its storage capacity of about 750 MB. Originally, CD drives for computers were read-only units. However, read-write CD ROM drives are now quite common. Many computers now have DVD drives. Again, these were originally able to read only prerecorded DVDs. However, drives capable of writing to DVDs are now widespread. Some can store as much as 8.5 GB of data or video. New storage technologies are constantly emerging. DVD drives equipped with short-wavelength lasers capable of writing and reading 25 GB on each of two layers of a disk are now commercially available.

4E-4 Input-Output Systems

Input-output devices provide the means for the user, or attached instruments, to communicate with the computer. Familiar input devices include keyboards, mouses, digital cameras, hard disks, CD ROMs, USB flash memories, and the transduced signals from analytical instruments. Output devices include printers, CD ROMs, USB flash memories, speakers, monitors, and hard disks. It is important to understand that many of these devices provide or use an *analog* signal, although, as we have pointed out, the computer can respond only to digital signals. Thus, an important part of the input-output system is an ADC for providing data in a form the computer can use and a DAC for converting the output from the computer to a usable analog signal.

An important piece of hardware in the computerized acquisition, analysis, and output of analytical information is the data-acquisition module. These devices may be plugged directly into the computer bus and provide a way to acquire data by means of analog-to-digital conversion to feed back data to an experiment by means of an ADC, to provide critical timing in the process, and to transfer digital data directly to the computer. Figure 4-12a shows a block diagram of a typical data-acquisition module. It includes an ADC, a DAC, a programmable-gain amplifier, digital input and output lines, memory for temporarily holding collected data, and a precision real-time clock for critical timing of the data acquisition. The entire data-acquisition process is carried out by a powerful onboard microprocessor, which receives user-selected instructions from the primary computer to which it is attached. A photo of a data-acquisition module is presented in Figure 4-12b.

The main computer carries out data analysis and long-term bulk data storage by using a software package such as those discussed in the next section. The operator controls the overall process through interactions with the main computer. Using data-acquisition modules such as these, virtually any instrument or experiment can be interfaced with a computer for efficient analysis and output of instrumental data.

One current trend in modern analytical instrumentation is to use devices that connect to the computer via a USB port (see Table 4-4). The USB standard allows connection of up to 127 different devices in a simple, straightforward manner. Currently, printers, cameras, disk drives, scanners, webcams, networking components, some data-acquisition systems, and some analytical instruments are available with USB interfaces. Some devices now include a FireWire connection. With these, several devices can be connected in series (daisychained) so that only one FireWire port is required for multiple units. Devices that require very high-speed connections, such as the data-acquisition board shown in Figure 4-12b, still require interface boards that connect directly to the internal computer bus.

4F COMPUTER SOFTWARE

In the past, it was necessary for scientists to write programs in machine code or assembly language to acquire data, process data, and control experiments. The age of PCs has brought about a wide variety of useful programs for scientists, teachers, students, business people, and home users. Currently, commercial software is available for data acquisition, statistical processing, graphical presentation, word processing, spreadsheet applications, database management, and many other applications. Today, scientists write their own programs only for specialized applications for which commercial software is nonexistent. Programs that are written are usually in a higher-level language such as C, PASCAL, or BASIC.

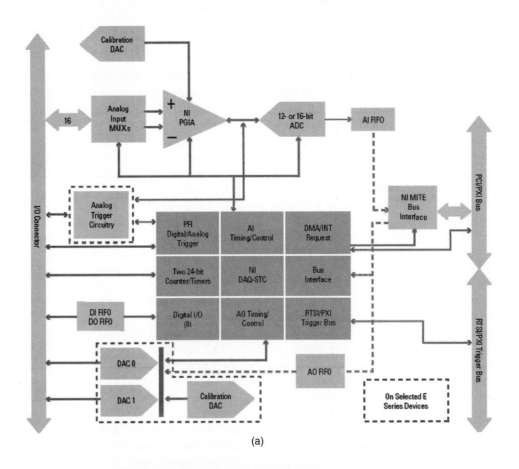

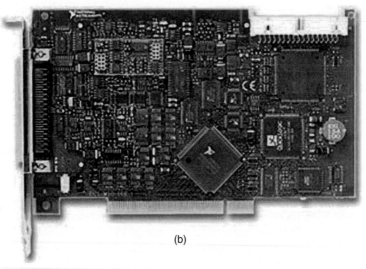

FIGURE 4-12 Computer data-acquisition module: (a) block diagram of the module, AI = analog input, AO = analog output, DAQ = data acquisition, DMA = direct memory access, DI = digital input, DO = digital output, FIFO = first-in first-out memory buffer, I/O = input-output, INT = interrupt, MITE = MXI interface to everything, MUX = multiplexer, MXI = multisystem extension interface, NI = National Instruments, PCI = peripheral compact interface bus, PFI = programmable function input, PGIA = programmable-gain instrumentation amplifier, PXI = PCI extensions for instrumentation bus, RTSI = real-time system integration bus, STC = system timing controller; (b) photo of the printed circuit board containing the module. (Reprinted with permission of National Instruments Corporation.)

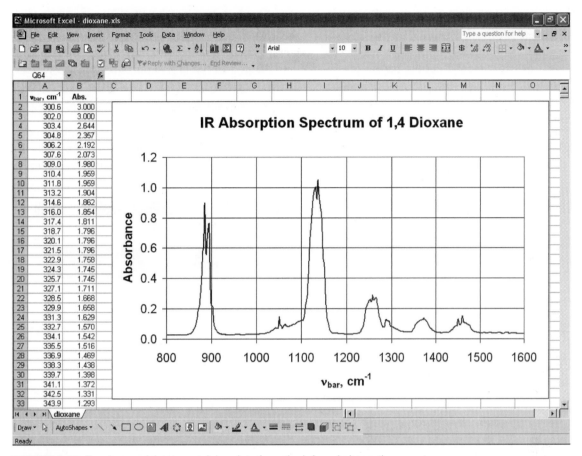

FIGURE 4-13 Excel spreadsheet containing data from the infrared absorption spectrum of 1,4-dioxane.

4F-1 Spreadsheets

Spreadsheets were originally designed as a tool for business, but as they became easier to use, people from all walks of life began to use them to carry out numerical computations in many fields, including science. When software manufacturers realized the diverse nature of their clientele, they began to add highly sophisticated and specialized functions to their spreadsheets. For example, Microsoft® Excel now contains many functions that can be used to save steps when developing complex statistical or engineering analyses.[3] These relationships include basic statistical functions, such as mean, standard deviation, median, mode, and various distribution functions, and advanced statistical functions, such as linear and nonlinear least squares,

t-tests, F tests, random-number generation, and analysis of variance.

Figure 4-13 shows an Excel spreadsheet for analyzing and plotting spectral data files. Columns A and B of the spreadsheet contain data from the infrared spectrum of 1,4-dioxane. The data were collected with a Fourier transform infrared spectrometer. Software is available from several companies to import data directly into spreadsheets such as Excel. The user then manipulates the data by implementing the built-in functions of Excel or writing specialized routines.

4F-2 Statistical Analysis

Many scientists use statistical software packages for doing more complex or more complete analysis than can be done with spreadsheets. Some popular statistics packages include MINITAB, SAS, SYSTAT, Origin, STATISTICA, SigmaStat, and STATGRAPHICS

---

[3]For additional information on spreadsheet applications, see S. R. Crouch and F. J. Holler, *Applications of Microsoft® Excel in Analytical Chemistry*, Belmont, CA: Brooks/Cole, 2004.

Plus. In addition to the normal statistics functions that are included in many spreadsheet programs, these statistical programs can often carry out more advanced functions such as experimental design, partial least-squares regression, principal components analysis, factor analysis, cluster analysis, time-series analysis, control chart generation, and nonparametric statistics.

A typical output from MINITAB is shown in Figure 4-14. Here, the determination of Na by flame spectrometry is shown with and without using a Li internal standard. The open windows show the data (Worksheet window), the regression statistics (Session window), and the two x-y (scatter) plots. The internal standard method is seen to give the better linearity.

4F-3 Mathematics Tools

Several types of mathematics tools are available commercially. A tool of special interest to chemists is the equation solver, a program that quickly solves complex

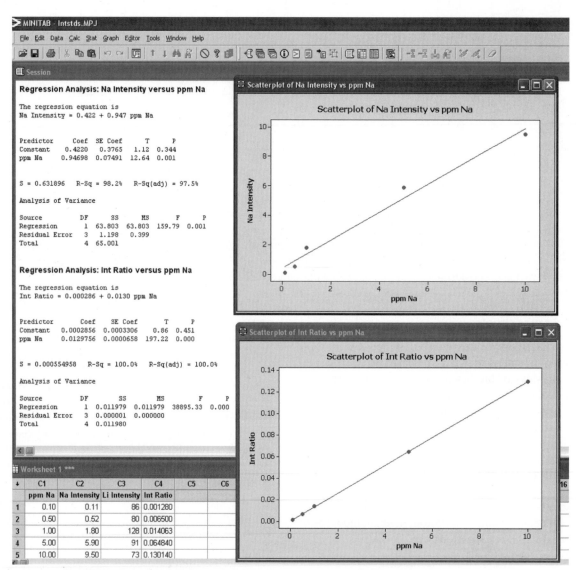

FIGURE 4-14 MINITAB output for internal standard method for determining sodium by flame spectrometry. The data are shown in the Worksheet window at lower left. The upper scatter plot is for the Na intensity versus concentration, and the lower scatter plot is for the ratio of Na intensity to Li intensity versus concentration. The regression statistics in the Session window show that the internal standard method gives better linearity.

mathematical equations such as those found in the study of multiple equilibria. Several equation solvers are available, including TK Solver, Mathematica, and Mathcad. All of these programs have been rated highly, and thus the choice of which to use depends on the tasks at hand and the resources available.

TK Solver has many built-in functions and integrates nicely with Excel. It is a rules-based programming system. Mathematica handles symbolic calculations and readily allows numerical modeling and simulations. Because of its low cost, power, ease of use, and intuitive nature for representing complex mathematical expressions, Mathcad has become popular in the physical sciences and in engineering for solving a tremendous variety of computation problems from basic statistical analysis to eigenvalue-eigenvector problems in quantum chemistry.

Excel has a built-in Solver that is very useful in solving equations and fitting models to data. For chemical problems, Excel's Solver can be used in fitting nonlinear models such as those encountered in kinetics, chromatography, and other areas.[4] Figure 4-15 shows an example of using Excel's Solver to estimate the Michaelis constant K_m and the maximum velocity v_m for an enzyme-catalyzed reaction. The initial estimates gave a very poor fit to the data points, as seen in Figure 4-15a. By using Solver to minimize the sum of the squares of the residuals, SSR, the much better fit in Figure 4-15b was obtained.

Figure 4-16 illustrates the application of MINITAB to the production of a control chart for monitoring a process that produces a weak acid. The molarity of the acid is plotted and the upper and lower control limits are obtained from the data.[5] One of the observations is seen to fall outside the control limits.

Sophisticated programs for dealing with matrix algebra and various applications are also available. MathWorks MATLAB is one of the most popular. In addition to the basic program, a variety of applications modules, called toolboxes, are available for matrix applications. Some of the available toolboxes include statistics, instrument control, chemometrics, image processing, bioinformatics, and signal processing. In addition to those provided by the parent company, many other software developers provide tools for use with MATLAB.

4F-4 Scientific Packages

A number of software packages have been developed specifically for use in chemistry and related sciences. Programs are available for tasks as diverse as drawing organic molecular structures (ChemWindows and ChemDraw), for performing thermodynamic calculations (HSC Chemistry for Windows), for curve fitting (SYSTAT Software's TableCurve and PeakFit), for spectroscopic data analysis and plotting (Thermo Electron's GRAMS software), and for laboratory data acquisition, analysis, and presentation (National Instrument's LabVIEW).

GRAMS

To illustrate the utility of software for data analysis, let us consider GRAMS, which stands for graphic relational-array management system and refers to spectrograms and chromatograms. GRAMS is a suite of integrated software modules centered around the GRAMS/AI spectroscopy data-processing and reporting software. The individual components of the suite can be used alone or in conjunction with GRAMS/AI. The modules include GRAMS/3D for visualizing and plotting multidimensional data sets, SPECTRAL DB for creating databases that can be shared by groups, and SPECTRAL ID for performing qualitative spectral identification. In addition, modules are available for creating chemometric calibration models (PLSplus/ IQ) and for data exchange with Excel. GRAMS is capable of reading, analyzing, and translating data files generated by more than 100 different chemical instruments and other software packages, including spectrometers, chromatographs, and other instruments. Data files may be translated to and from these file formats, which include ASCII format, comma-delimited format, and spreadsheet formats as well as standard spectroscopic formats such as JCAMP. Figure 4-17 illustrates the use of GRAMS/AI for fitting peaks and baselines for a series of overlapped chromatographic peaks. Such software tools are indispensable in finding the individual components of a mixture separated by chromatography.

Plotting Packages

Several packages provide publication-quality graphics designed for use by scientists. Although general-purpose programs such as Excel have fairly extensive plotting capabilities, these more specialized programs

---

[4]For more information on Excel's Solver, see S. R. Crouch and F. J. Holler, *Applications of Microsoft® Excel in Analytical Chemistry*, Belmont, CA: Brooks/Cole, 2004.

[5]For information on control charts, see D. A. Skoog, D. M. West, F. J. Holler, and S. R. Crouch, *Fundamentals of Analytical Chemistry*, 8th ed., Belmont, CA, Brooks/Cole, 2004, pp. 216–19.

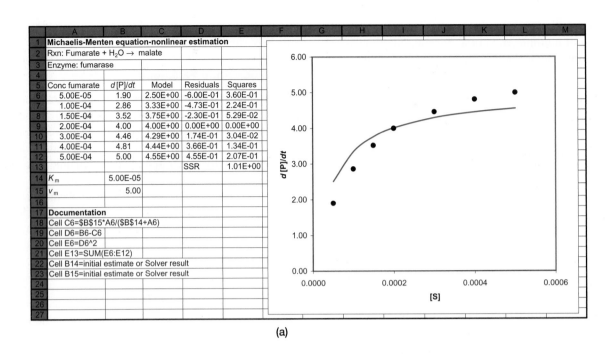

(a)

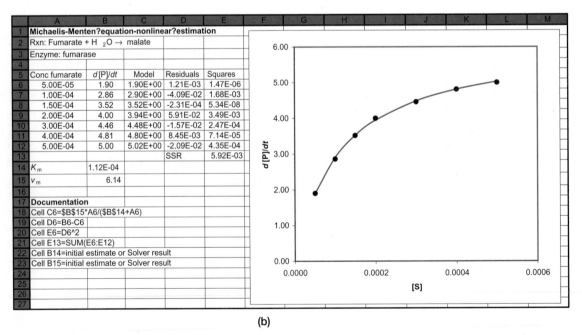

(b)

FIGURE 4-15 Nonlinear estimation by Excel's Solver for enzyme kinetics. In (a), data for the fumarase-catalyzed hydrolysis of fumarate to malate are shown as rate, $d[P]/dt$, versus concentration of fumarate. Initial estimates of the Michaelis constant K_m and the maximum rate v_m result in the line shown in the plot. In (b), Solver has minimized the sum of the squares of the residuals (SSR), and converged on values of K_m and v_m that give a much better fit to the data points. (From S. R. Crouch and F. J. Holler, *Applications of Microsoft® Excel in Analytical Chemistry*, Belmont, CA: Brooks/Cole, **2004**, pp. 264–65.)

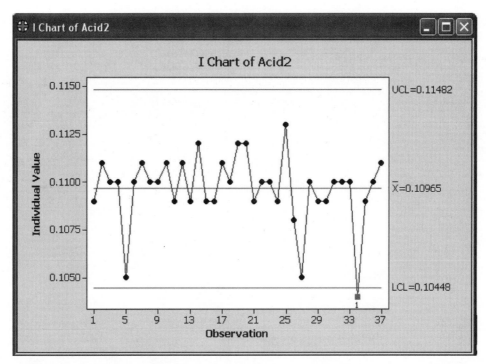

FIGURE 4-16 MINITAB output for control chart for the production of a weak acid. The mean value $\bar{x}$, the upper control limit (UCL) and the lower control limit (LCL) are shown. Observation 34 is beyond the LCL.

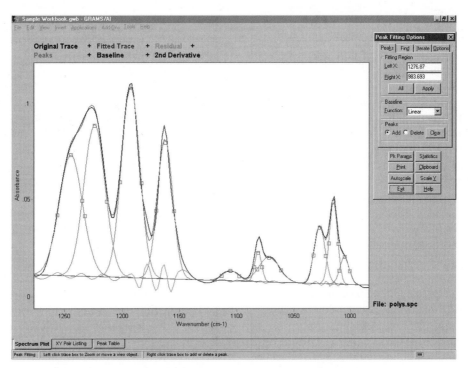

FIGURE 4-17 Computer display of GRAMS/AI being used to fit peaks and baselines for several overlapping peak-shaped signals. (Courtesy of Thermo Electron Corp.)

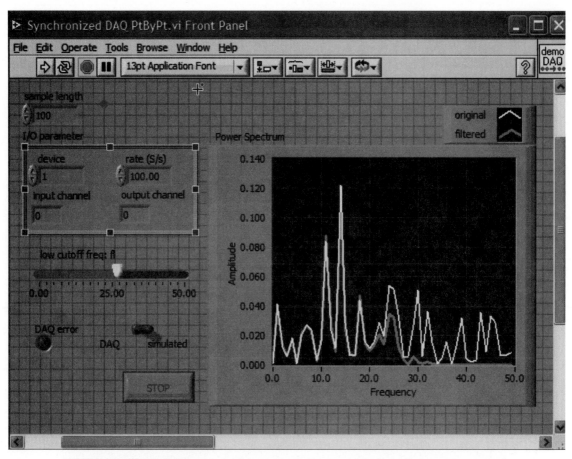

FIGURE 4-18 LabVIEW front panel for a data-acquisition system allows the user to choose parameters such as sampling rate, sample length, and filtering values. (Reprinted with permission of National Instruments Corporation.)

have advanced capabilities, such as two- and three-dimensional plotting, contour plot generation, advanced statistical plotting, multiple-axis plotting, and many other features. Packages such as SYSTAT Software's SigmaPlot, OriginLab's Origin, and GRAMS/AI have features that allow scientists to prepare plots for publication, presentation, and classroom use.

LabVIEW

The LabVIEW program provides a graphical environment for data acquisition from a variety of instruments, for many different types of data analysis, and for sophisticated data presentation. The data-acquisition section of LabVIEW works in conjunction with the National Instruments Data Acquisition boards. Data can be acquired from plug-in boards, from USB devices, and from Ethernet-based systems. LabVIEW allows the user to set up virtual instruments with front panels

that are customized for a particular acquisition and measurement situation as shown in Figure 4-18. Components and processes can be visualized with block diagrams as shown in Figure 4-19 and connectivity changed readily.

The data-analysis part of LabVIEW includes such tools as curve fitting, signal generation, peak analysis, Fourier analysis, deconvolution, smoothing, and various mathematical operations. The program also integrates with standard mathematics software such as Mathcad and MATLAB.

The presentation part of LabVIEW includes a variety of tools for plotting and visualization, such as two- and three-dimensional charting, report-generation tools, and web-publishing tools. Instrument control panels and block diagrams can be published via the web, and virtual instruments can be accessed and controlled remotely on the Internet.

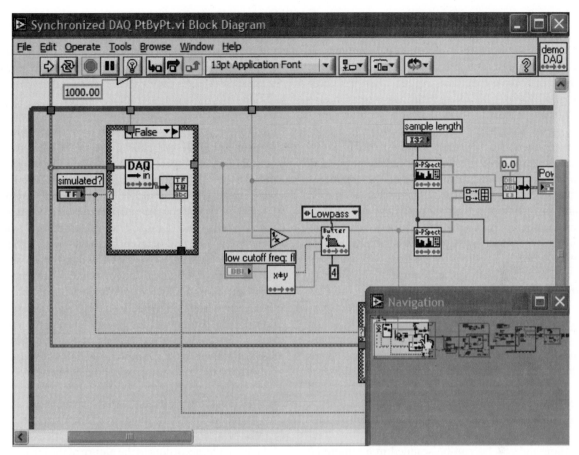

FIGURE 4-19 LabVIEW block diagram of data-acquisition and measurement processes. (Reprinted with permission of National Instruments Corporation.)

The LabVIEW environment is one that allows virtual instruments to be configured and used with relative ease. Many templates and an extensive network of users are available for assistance in developing applications for LabVIEW. The program finds extensive use in industry, in research laboratories, and in teaching laboratories.

4G APPLICATIONS OF COMPUTERS

Computer interactions with analytical instruments are of two types, *passive* and *active*. In passive applications, the computer does not participate in the control of the experiment but is used only for data handling, processing, storing, file searching, or display. In an active interaction, the output from the computer controls the sequence of steps required for operation of the instrument. For example, in a spectroscopic determination,

the computer may choose the proper source, cause this source to be activated and its intensity adjusted to an appropriate level, cause the radiation to pass through the sample and then a blank, control the monochromator so that a proper wavelength is chosen, adjust the detector response, and record the intensity level. In addition, the computer may be programmed to use the data as it is being collected to vary experimental conditions in such a way as to improve the quality of subsequent data. Instruments with computer control are said to be automated.

4G-1 Passive Applications

Data processing by a computer may involve relatively simple mathematical operations such as calculation of concentrations, data averaging, least-squares analysis, statistical analysis, and integration to obtain peak areas. More complex calculations may involve the solution of several simultaneous equations, curve fitting, averaging, and Fourier transformations.

Data storage is another important passive function of computers. For example, a powerful tool for the analysis of complex mixtures results when gas chromatography (GC) is linked with mass spectrometry (MS) (see Chapters 11, 20, and 27). GC separates mixtures on the basis of the time required for the individual components to appear at the end of a suitably packed column. MS permits identification of each component according to the mass of the fragments formed when the compound is bombarded with one of a number of different types of particles such as electrons. Equipment for GC/MS may produce data for as many as 100 spectra in a few minutes, with each spectrum being made up of tens to hundreds of peaks. Conversion of these data to an interpretable form (a graph) in real time is often impossible. Thus, the data are usually stored in digital form for subsequent processing and presentation in graphical form.

Identification of a species from its mass spectrum involves a search of files of spectra for pure compounds until a match is found; done manually, this process is time consuming, but it can be accomplished quickly by using a computer. Here, the spectra of pure compounds, stored on a hard disk, are searched until spectra are found that are similar to the analyte. Several thousand spectra can be scanned in a minute or less. Such a search usually produces several possible compounds. Further comparison of the spectra by the scientist often makes identification possible.

Another important passive application of the power of computers in GC/MS uses the high-speed data fetching and correlating capabilities of the computer. Thus, for example, the computer can be called on to display on a monitor the mass spectrum of any one of the separated components after the component has exited from a gas chromatographic column.

4G-2 Active Applications

In active applications only part of the computer's time is devoted to data collection, and the rest is used for data processing and control. Thus, active applications are real-time operations. Most modern instruments contain one or more microprocessors that perform control functions. Examples include adjustment of (1) the slit width and wavelength settings of a monochromator, (2) the temperature of a chromatographic column, (3) the potential applied to an electrode, (4) the rate of addition of a reagent, and (5) the time at which the integration of a peak is to begin. For the GC/MS instrument considered in the last section, a computer is often used to initiate the collection of mass spectral data each time a compound is sensed at the end of the chromatographic column.

Computer control can be relatively simple, as in the examples just cited, or more complex. For example, the determination of the concentration of elements by atomic emission spectroscopy involves the measurement of the heights of emission lines, which are found at wavelengths characteristic for each element (see Chapter 10). Here, the computer can cause a monochromator to rapidly sweep a range of wavelengths until a peak is detected. The rate of sweep is then slowed to better determine the exact wavelength at which the maximum output signal is obtained. Intensity measurements are repeated at this point until an average value is obtained that gives a suitable signal-to-noise ratio (see Chapter 5). The computer then causes the instrument to repeat this operation for each peak of interest in the spectrum. Finally, the computer calculates and sends to the printer the concentrations of the elements present.

Because of its great speed, a computer can often control variables more efficiently than can a human operator. Furthermore, with some experiments, a computer can be programmed to alter the way the measurement is being made, according to the nature of the initial data. Here, a feedback loop is used in which the signal output is converted to digital data and fed back through the computer, serving to control and optimize how later measurements are performed.

4H COMPUTER NETWORKS

The connection of two or more computers produces a computer network, or simply a *network*. In today's world, computer networks are all around us. We get money from an ATM, access the Internet for information, and watch programs on digital cable television. Each of these examples requires a computer network. Today, networks significantly increase the efficiency with which information can be transmitted and manipulated.[6]

---

[6]See, for example, J. Habraken and M. Hayden, *Sams Teach Yourself Networking in 24 Hours*, Indianapolis, IN: Sams Publishing, 2004; L. L. Peterson and B. S. Davie, *Computer Networks: A Systems Approach*, 3rd ed., New York: Elsevier, 2003; A. S. Tanenbaum, *Computer Networks*, 4th ed., Upper Saddle River, NJ: Pearson/Prentice-Hall PTR, 2002.

4H-1 Network Types

Networks encompass an enormous number of possible interactions between computers, but they can be classified into local area networks, wide area networks, and the Internet. None of the physical networks described here will operate without the appropriate software on all the interconnected machines.

Local Area Networks

A *local area network*, or LAN, is the least complex type of network. A LAN is a group of linked computers all located at a single site. The usual LAN has a high data-transfer rate ranging from a few megabits per second (Mbps) to gigabits per second (Gbps).

Many LANs are physically connected by wires. More recently, wireless networks have become popular, allowing computers to interact through radio waves sent from a transmitter to receivers. In the past, wired LANs employed a bus-type architecture, in which computers were connected to a long cable (the bus) with taps along its length as shown in Figure 4-20a. If any of the links between computers were broken in a bus topology, the entire network went down. Coaxial Ethernet networks (10Base5 and 10Base2) were examples of bus networks. These have been replaced in more modern networks by star topology networks. Twisted-pair Ethernet networks (10BaseT or 100BaseT) use the star topology shown in Figure 4-20b. Star networks are more robust and less prone to interruptions than

bus-type networks. Ring networks use a configuration similar to the star network, but in the ring network information circulates in a ring around the network. The IBM Token Ring network and the fiber-optic distributed-data interface (FDDI) network use ring structures.

The computers in a network where users work are called *workstations*. A computer whose resources are shared with other computers on the network is called a *server*. In addition to these physical devices, hubs, access units, network cards, and the appropriate wiring and cabling are needed along with software to establish a LAN.

Wide Area Networks

A second type of network is the *wide area network*, or WAN. With a WAN, the computers involved are geographically scattered. These networks are usually LANs joined by high-speed interconnections and devices called *routers* that manage data flow. WANs are usually accessed by leased digital phone lines (T-carrier lines) operating in the United States at 1.5 Mbps (T-1 lines) or 45 Mbps (T-3 or DS3 lines). These leased lines can be quite expensive, running thousands of dollars per month for T-1 lines.

The Internet

Finally, there is the Internet, which is capable of rapidly transmitting digital representations of an almost unbelievable variety of textual, graphical, audio, and

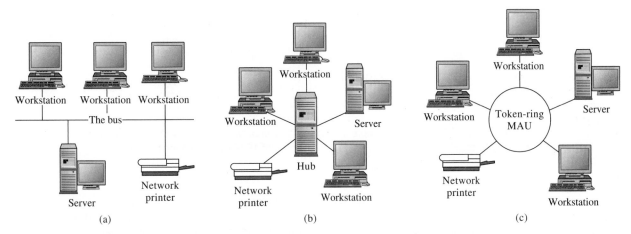

FIGURE 4-20 Network topologies. In (a), the bus topology is shown. Here, computers communicate along a physical bus. Software is necessary on the various devices to allow communication. In (b), a star topology is shown. Here, a junction box or hub connects computers to one another. In (c), a token-ring topology is shown. Information circulates in a ring around the network. MAU = multistation access unit.

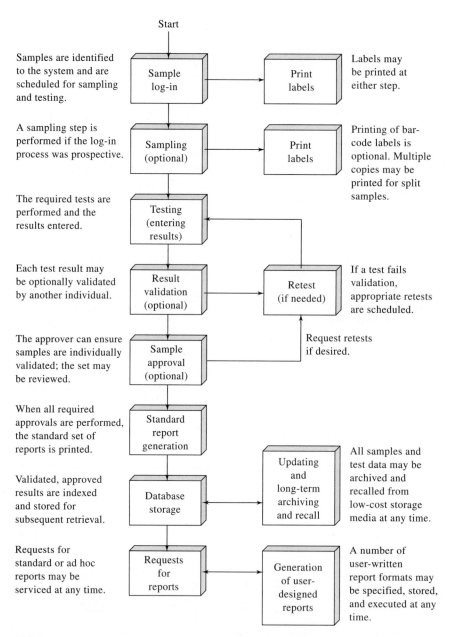

Start

Samples are identified to the system and are scheduled for sampling and testing. — **Sample log-in** → **Print labels** — Labels may be printed at either step.

A sampling step is performed if the log-in process was prospective. — **Sampling (optional)** → **Print labels** — Printing of bar-code labels is optional. Multiple copies may be printed for split samples.

The required tests are performed and the results entered. — **Testing (entering results)**

Each test result may be optionally validated by another individual. — **Result validation (optional)** → **Retest (if needed)** — If a test fails validation, appropriate retests are scheduled.

The approver can ensure samples are individually validated; the set may be reviewed. — **Sample approval (optional)** — Request retests if desired.

When all required approvals are performed, the standard set of reports is printed. — **Standard report generation**

Validated, approved results are indexed and stored for subsequent retrieval. — **Database storage** → **Updating and long-term archiving and recall** — All samples and test data may be archived and recalled from low-cost storage media at any time.

Requests for standard or ad hoc reports may be serviced at any time. — **Requests for reports** → **Generation of user-designed reports** — A number of user-written report formats may be specified, stored, and executed at any time.

FIGURE 4-21 LIMS data and sample management overview. (Reprinted with permission from F. I. Scott, *Amer. Lab.,* **1987**, 19 (11), 50. Copyright 1987 by International Scientific Communications, Inc.)

video information throughout the world. The Internet is really a network of networks. It can be accessed in several different ways: by a standard dial-up telephone line, by a cable modem employing the same coaxial cable lines that provide cable television signals, and by a *digital subscriber line* (DSL), which is a private tele-phone line partitioned for data transmission. The dial-up line is usually limited to a 56 kilobaud transmission rate. The cable modem and DSL connections are usually termed *broadband* connections. A cable modem is much faster than dial-up, with a maximum throughput of 2.8 Mbps. However, because cable communications

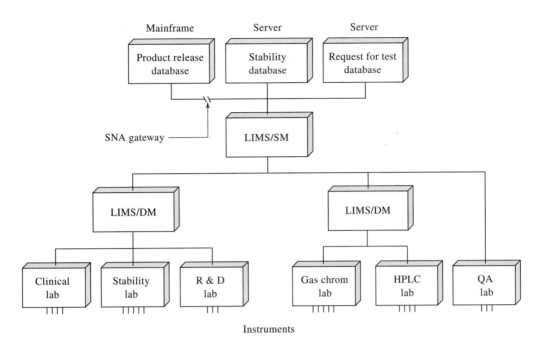

FIGURE 4-22 Block diagram of an entire automated laboratory system. (Reprinted with permission from E. L. Cooper and E. J. Turkel, *Amer. Lab.,* **1988**, 20 (3), 42. Copyright 1988 by International Scientific Communications, Inc.)

are based on a shared network topology, bandwidth is not always available when needed. One type of DSL, the asymmetric DSL, can provide downloading speeds to the subscriber of more than 6 Mbps and uploading speeds of more than 600 kilobits per second (kbps). Because DSL uses a private phone line, there is no degradation of speed as the number of users increases. The speed does, however, depend on the distance of the subscriber from the central telephone office. Security is also less of an issue with DSL than with cable modems.

The Internet will eventually deliver information to virtually every home via high-speed (hundreds of megabits per second) cables or telephone lines. Today, much of the world's information, including scientific data, journals, and other types of reports, is already available on the Internet.

4H-2 Laboratory Information Management Systems

Networking computers in a laboratory environment can result in enormous quantities of data to be handled, manipulated, and stored. In addition, govern-ment regulations, sample validation, and quality control dictate that data be archived and readily recalled at any time. A laboratory information management system (LIMS) can address these concerns.[7]

A well-designed LIMS keeps track of all of the information about all of the samples and projects that have been completed or are in progress. Figure 4-21 summarizes many of the processes that might be controlled by a LIMS in a testing laboratory and provides an overview of some of the options that might be exercised as a sample is processed. Finally, Figure 4-22 is a block diagram of a computer system designed to totally automate an entire laboratory. Note that at the bottom of this figure entire laboratories are designated by boxes; within each of these laboratories a LAN would be used to coordinate activities and communicate with the next level in the hierarchy. In this system we see that two different types of LIMSs are used; those designated DM are standard for data-management LIMSs, and the SM designation stands for system-sample management. Essentially the only

---

[7]For more information, see C. Paszko and E. Crossland, *Laboratory Information Management Systems,* 2nd ed., New York: Dekker, 2001.

difference between these coordinating computers, or servers, is the software that controls the communication and the data handling. The SNA (systems network architecture) gateway represents a means of connecting this laboratory's cluster of computers with the primary server at the corporate headquarters.

QUESTIONS AND PROBLEMS

*Answers are provided at the end of the book for problems marked with an asterisk.

X Problems with this icon are best solved using spreadsheets.

*4-1   Convert each of the following decimal numbers to its binary equivalent.
   (a) 24   (b) 91   (c) 135   (d) 396

*4-2   Convert each of the decimal numbers in Problem 4-1 into binary-coded-decimal (BCD) numbers.

4-3   Based on your results in Problems 4-1 and 4-2, which is more efficient in expressing decimal numbers in the fewest number of bits, binary or BCD? Why is the less efficient coding scheme still very useful?

*4-4   Convert each of the following binary numbers into its decimal equivalent.
   (a) 101   (b) 10101   (c) 1110101   (d) 1101011011

*4-5   Convert each of the following BCD numbers into its decimal equivalent.
   (a) 0100   (b) 1000 1001   (c) 0011 0100 0111   (d) 1001 0110 1000

4-6   Based on your results in Problems 4-4 and 4-5, which of the two coding schemes is easier to convert to decimal, binary or BCD? Why?

*4-7   Perform the following calculations using binary numbers and convert the result back to decimal.
   (a) 9 + 6   (b) 341 + 29   (c) 47 + 16   (d) 3 × 8

*4-8   Three ADCs all have a range of 0 to 10 V and a digitization uncertainty of ±1 LSB. What is the maximum uncertainty in the digitization of a 10-V signal if the converters have
   (a) 8 bits?   (b) 12 bits?   (c) 16 bits?

4-9   Repeat Problem 4-8 if a 1-V signal is being digitized with the same three ADCs and the input signal is (a) not amplified and (b) amplified by 10 to bring it to full scale.

4-10   The maximum percentage error of a voltage processed by an ADC is given by the following equation:

$$\text{max \% error} = (\text{maximum uncertainty/measured voltage}) \times 100\%$$

If the same ADC is used, how do the percentage errors in measured voltages compare if the measured voltages are 10 V and 1 V?

*4-11   ADCs digitize at different rates. What conversion rate is required if a chromatographic peak is to be sampled and digitized 20 times between the first positive deflection from the baseline until the peak returns to the baseline? The total baseline-to-baseline time is (a) 20 s and (b) 1 s.

*4-12 According to the Nyquist sampling criterion (see Section 5C-2), a signal must be digitized at a rate at least twice that of the highest frequency in the signal to avoid a sampling error. If a particular 12-bit ADC has a conversion time of 8 μs, what is the highest frequency that can be accurately digitized while satisfying the Nyquist criterion?

Challenge Problem

4-13 Use a search engine such as Google to find information about Gordon E. Moore and Moore's law, the famous law about technological advances that he proposed.
   (a) What is Moore's law? Give a brief description in your own words.
   (b) Who is Gordon E. Moore? What was his position at the time he first proposed Moore's law? What company did he later cofound? With whom did he cofound this company?
   (c) In what field did Gordon E. Moore obtain his BS degree? At what university did he receive his BS degree? Where did he obtain his PhD degree? In what field was his PhD degree?
   (d) What Nobel Prize–winning physicist gave Gordon E. Moore his first job opportunity?
   (e) What was the number of the first microprocessor developed at Moore's company and how many transistors did it have? When was it introduced?
   (f) One important benchmark of computational progress is the performance-to-price ratio (PPR) of computers.[8] The PPR is the number of bits per word divided by the product of cycle time (1/clock speed) and price. The original IBM PC (1981) with an 8-bit word length, a 4.77 MHz clock, and a price tag of $5000 came in with a PPR of ~7600. Computers based on other important processors are listed in the table below. Calculate the PPR of each of these computers. Does Moore's law hold for the PPR? How did you come to your conclusion?

Processor Type	Year	Clock Speed, MHz	Bits/Word	Computer Price, $
286	1982	6.0	16	5000
486	1989	25	32	4000
Pentium	1993	60	32	3500
Pentium II	1997	266	32	3000
Pentium III	1999	700	32	2500
Pentium 4	2000	3000	64	2000

[8] See S. R. Crouch and T. V. Atkinson, *Anal. Chem.* **2000**, *72*, 596A–603A.

CHAPTER FIVE

Signals and Noise

E*very analytical measurement is made up of two components. One component, the **sig-nal**, carries information about the analyte that is of interest to the scientist. The second, called **noise**, is made up of extraneous information that is unwanted because it degrades the accuracy and precision of an analysis and also places a lower limit on the amount of analyte that can be detected. In this chapter we describe some of the common sources of noise and how their effects can be minimized.*

Throughout this chapter, this logo indicates an opportunity for online self-study at **www .thomsonedu.com/chemistry/skoog**, linking you to interactive tutorials, simulations, and exercises.

5A THE SIGNAL-TO-NOISE RATIO

The effect of noise[1] on a signal is shown in Figure 5-1a, which is a strip-chart recording of a tiny direct current of about 10^{-15} A. Figure 5-1b is a theoretical plot of the same current in the absence of noise.[2] The difference between the two plots corresponds to the noise associated with this experiment. Unfortunately, noise-free data, such as that shown in Figure 5-1b, can never be realized in the laboratory because some types of noise arise from thermodynamic and quantum effects that are impossible to avoid in a measurement.

In most measurements, the average strength of the noise N is constant and independent of the magnitude of the signal S. Thus, the effect of noise on the relative error of a measurement becomes greater and greater as the quantity being measured decreases in magnitude. For this reason, the *signal-to-noise ratio* (S/N) is a much more useful figure of merit than noise alone for describing the quality of an analytical method or the performance of an instrument.

For a dc signal, such as that shown in Figure 5-1a, the magnitude of the noise is conveniently defined as the standard deviation s of numerous measurements of the signal strength, and the signal is given by the mean $\bar{x}$ of the measurements. Thus, S/N is given by

$$\frac{S}{N} = \frac{\text{mean}}{\text{standard deviation}} = \frac{\bar{x}}{s} \qquad (5\text{-}1)$$

Note that the signal-to-noise ratio $\bar{x}/s$ is the reciprocal of the relative standard deviation, RSD (see Section a1B-1, Appendix 1), of the group of measurements. That is,

$$\frac{S}{N} = \frac{1}{\text{RSD}} \qquad (5\text{-}2)$$

For a recorded signal such as that shown in Figure 5-1a, the standard deviation can be estimated easily at a 99% confidence level by dividing the difference

[1]The term *noise* is derived from radio engineering where the presence of an unwanted signal is manifested as audio static, or noise. The term is applied now throughout science and engineering to describe the random fluctuations observed whenever replicate measurements are made on signals that are monitored continuously. Random fluctuations are described and treated by statistical methods (see Section a1B, Appendix 1).

[2]For a more detailed discussion of noise, see T. Coor, *J. Chem. Educ.*, **1968**, *45*, A533, A583; G. M. Hieftje, *Anal. Chem.*, **1972**, *44* (6), 81A; A. Bezegh and J. Janata, *Anal. Chem.*, **1987**, *59*, 494A; M. E. Green, *J. Chem. Educ.*, **1984**, *61*, 600; H. V. Malmstadt, C. G. Enke, and S. R. Crouch, *Microcomputers and Electronic Instrumentation: Making the Right Connections*, Washington, DC: American Chemical Society, 1994.

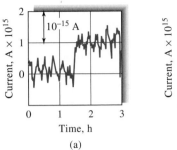
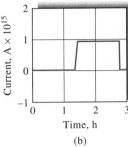

(a)   **(b)**

FIGURE 5-1 Effect of noise on a current measurement: (a) experimental strip-chart recording of a 0.9×10^{-15} A direct current, (b) mean of the fluctuations. (Adapted from T. Coor, *J. Chem. Educ.*, **1968**, *45*, A594. With permission.)

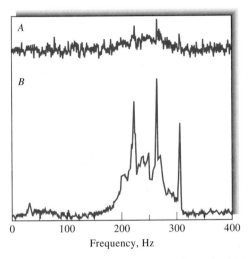

FIGURE 5-2 Effect of signal-to-noise ratio on the NMR spectrum of progesterone: *A, S/N* = 4.3; *B, S/N* = 43. (Adapted from R. R. Ernst and W. A. Anderson, *Rev. Sci. Inst.*, **1966**, *37*, 101. With permission.)

between the maximum and the minimum signal by five. Here, we assume that the excursions from the mean are random and can thus be treated by the methods of statistics. In Figure a1-5 of Appendix 1, it is seen that 99% of the data under the normal error curve lie within $\pm 2.5\ \sigma$ of the mean. Thus, we can say with 99% certainty that the difference between the maximum and minimum encompasses 5σ. One fifth of the difference is then a good estimate of the standard deviation.

As a general rule, it becomes impossible to detect a signal when the signal-to-noise ratio becomes less than about 2 or 3. Figure 5-2 illustrates this rule. The upper plot is a nuclear magnetic resonance (NMR) spectrum for progesterone with a signal-to-noise ratio

of about 4.3. In the lower plot the ratio is 43. At the smaller signal-to-noise ratio, only a few of the several peaks can be recognized with certainty.

5B SOURCES OF NOISE IN INSTRUMENTAL ANALYSES

Chemical analyses are affected by two types of noise: chemical noise and instrumental noise.

5B-1 Chemical Noise

Chemical noise arises from a host of uncontrollable variables that affect the chemistry of the system being analyzed. Examples include undetected variations in temperature or pressure that affect the position of chemical equilibria, fluctuations in relative humidity that cause changes in the moisture content of samples, vibrations that lead to stratification of powdered solids, changes in light intensity that affect photosensitive materials, and laboratory fumes that interact with samples or reagents. Details on the effects of chemical noise appear in later chapters that deal with specific instrumental methods. In this chapter we focus exclusively on instrumental noise.

5B-2 Instrumental Noise

Noise is associated with each component of an instrument — that is, with the source, the input transducer, all signal-processing elements, and the output transducer. Furthermore, the noise from each of these elements may be of several types and may arise from several sources. Thus, the noise that is finally observed is a complex composite that usually cannot be fully characterized. Certain kinds of instrumental noise are recognizable: (1) thermal, or Johnson, noise; (2) shot noise; (3) flicker, or $1/f$, noise; and (4) environmental noise. A consideration of the properties of the four kinds of noise is useful.

Thermal Noise, or Johnson Noise

Thermal noise is caused by the thermal agitation of electrons or other charge carriers in resistors, capacitors, radiation transducers, electrochemical cells, and other resistive elements in an instrument. This agitation of charged particles is random and periodically creates charge inhomogeneities, which in turn create voltage fluctuations that then appear in the readout as noise. It is important to note that thermal noise is present even in the absence of current in a resistive element and disappears only at absolute zero.

The magnitude of thermal noise in a resistive circuit element can be derived from thermodynamic considerations[3] and is given by

$$\bar{v}_{rms} = \sqrt{4kTR\Delta f} \qquad (5\text{-}3)$$

where $\bar{v}_{rms}$ is the root-mean-square noise voltage lying in a frequency bandwidth of Δf Hz, k is Boltzmann's constant $(1.38 \times 10^{-23}$ J/K$)$, T is the temperature in kelvins, and R is the resistance of the resistive element in ohms.

In Section 3B-4 we discussed the relationship between the rise time t_r and the bandwidth Δf of an operational amplifier. These variables are also used to characterize the capability of complete instruments to transduce and transmit information, because

$$\Delta f = \frac{1}{3t_r} \qquad (5\text{-}4)$$

The rise time of an instrument is its response time in seconds to an abrupt change in input and normally is taken as the time required for the output to increase from 10% to 90% of its final value. Thus, if the rise time is 0.01 s, the bandwidth Δf is 33 Hz.

Equation 5-3 suggests that thermal noise can be decreased by narrowing the bandwidth. However, as the bandwidth narrows, the instrument becomes slower to respond to a signal change, and more time is required to make a reliable measurement.

EXAMPLE 5-1

What is the effect on thermal noise of decreasing the response time of an instrument from 1 s to 1 μs?

Solution

If we assume that the response time is approximately equal to the rise time, we find that the bandwidth has been changed from 1 Hz to 10^6 Hz. According to Equation 5-3, such a change will cause an increase in the noise by $(10^6/1)^{1/2}$, or 1000-fold.

As shown by Equation 5-3, thermal noise can also be reduced by lowering the electrical resistance of

instrument circuits and by lowering the temperature of instrument components. The thermal noise in transducers is often reduced by cooling. For example, lowering the temperature of a UV-visible photodiode array from room temperature (298 K) to the temperature of liquid nitrogen (77 K) will halve the thermal noise.

It is important to note that thermal noise, although dependent on the frequency bandwidth, is independent of frequency itself. For this reason, it is sometimes termed *white noise* by analogy to white light, which contains all visible frequencies. Also note that thermal noise in resistive circuit elements is independent of the physical size of the resistor.

Shot Noise

Shot noise is encountered wherever electrons or other charged particles cross a junction. In typical electronic circuits, these junctions are found at *pn* interfaces; in photocells and vacuum tubes, the junction consists of the evacuated space between the anode and cathode. The currents comprise a series of quantized events, the transfer of individual electrons across the junction. These events occur randomly, however, and the rate at which they occur is thus subject to statistical fluctuations, which are described by the equation

$$i_{rms} = \sqrt{2Ie\Delta f} \qquad (5\text{-}5)$$

where i_{rms} is the root-mean-square current fluctuation associated with the average direct current, I; e is the charge on the electron of 1.60×10^{-19} C; and Δf is again the bandwidth of frequencies being considered. Like thermal noise, shot noise is white noise and is thus the same at any frequency.

Equation 5-5 suggests that shot noise in a current measurement can be minimized only by reducing bandwidth.

Flicker Noise

Flicker noise is characterized as having a magnitude that is inversely proportional to the frequency of the signal being observed; it is sometimes termed $1/f$ (*one-over-f*) noise as a consequence. The causes of flicker noise are not totally understood; it is ubiquitous, however, and is recognizable by its frequency dependence. Flicker noise becomes significant at frequencies lower than about 100 Hz. The long-term drift observed in dc amplifiers, light sources, voltmeters, and current meters is an example of flicker noise. Flicker noise can be reduced significantly in some cases by using wire-wound

---

[3]For example, see T. Coor, *J. Chem. Educ.*, **1968**, *45*, A534.

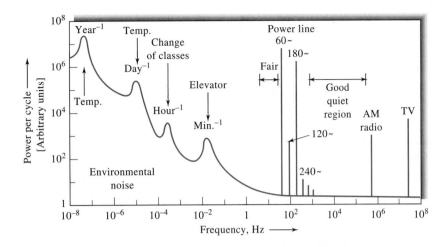

FIGURE 5-3 Some sources of environmental noise in a university laboratory. Note the frequency dependence and regions where various types of interference occur. (From T. Coor, *J. Chem. Educ.*, **1968**, *45*, A540. With permission.)

or metallic-film resistors rather than the more common carbon-composition type.

Environmental Noise

Environmental noise is a composite of different forms of noise that arise from the surroundings. Figure 5-3 suggests typical sources of environmental noise in a university laboratory.

Much environmental noise occurs because each conductor in an instrument is potentially an antenna capable of picking up electromagnetic radiation and converting it to an electrical signal. There are numerous sources of electromagnetic radiation in the environment, including ac power lines, radio and TV stations, gasoline-engine ignition systems, arcing switches, brushes in electric motors, lightning, and ionospheric disturbances. Note that some of these sources, such as power lines and radio stations, cause noise with relatively narrow frequency bandwidths.

Note that the noise spectrum shown in Figure 5-3 contains a large, continuous noise region at low frequencies. This noise has the properties of flicker noise; its sources are not fully known. Superimposed on the flicker noise are noise peaks associated with yearly and daily temperature fluctuations and other periodic phenomena associated with the use of a laboratory building.

Finally, two quiet-frequency regions in which environmental noise is low are indicated in Figure 5-3: the region extending from about 3 Hz to almost 60 Hz and the region from about 1 kHz to about 500 kHz, or a frequency at which AM radio signals are prevalent. Often, signals are converted to frequencies in these regions to reduce noise during signal processing.

5C SIGNAL-TO-NOISE ENHANCEMENT

Many laboratory measurements require only minimal effort to maintain the signal-to-noise ratio at an acceptable level. Examples include the weight determinations made in the course of a chemical synthesis or the color comparison made in determining the chlorine content of the water in a swimming pool. For both examples, the signal is large relative to the noise and the requirements for precision and accuracy are minimal. When the need for sensitivity and accuracy increases, however, the signal-to-noise ratio often becomes the limiting factor in the precision of a measurement.

Both hardware and software methods are available for improving the signal-to-noise ratio of an instrumental method. Hardware noise reduction is accomplished by incorporating into the instrument design components such as filters, choppers, shields, modulators, and synchronous detectors. These devices remove or attenuate the noise without affecting the analytical signal significantly. Software methods are based on various computer algorithms that permit extraction of signals from noisy data. As a minimum, software methods require sufficient hardware to condition the output signal from the instrument and convert it from analog to digital form. Typically, data are collected by using a computer equipped with a data-acquisition module as described in Chapter 4. Signals may then be extracted from noise by using the data-acquisition computer or another that is connected to it via a network.

5C-1 Some Hardware Devices for Noise Reduction

We briefly describe here some hardware devices and techniques used for enhancing the signal-to-noise ratio.

Grounding and Shielding

Noise that arises from environmentally generated electromagnetic radiation can often be substantially reduced by shielding, grounding, and minimizing the lengths of conductors within the instrumental system. Shielding consists of surrounding a circuit, or the most critical conductors in a circuit, with a conducting material that is attached to earth ground. Electromagnetic radiation is then absorbed by the shield rather than by the enclosed conductors. Noise pickup and possibly amplification of the noise by the instrument circuit may thus be minimized. It may be somewhat surprising that the techniques of minimizing noise through grounding and shielding are often more art than science, particularly in instruments that involve both analog and digital circuits. The optimum configuration is often found only after much trial and error.[4]

Shielding becomes particularly important when the output of a high-resistance transducer, such as the glass electrode, is being amplified. Here, even minuscule randomly induced currents produce relatively large voltage fluctuations in the measured signal.

Difference and Instrumentation Amplifiers

Any noise generated in the transducer circuit is particularly critical because it usually appears in an amplified form in the instrument readout. To attenuate this type of noise, most instruments employ a difference amplifier, such as that shown in Figure 3-13, for the first stage of amplification. Common-mode noise in the

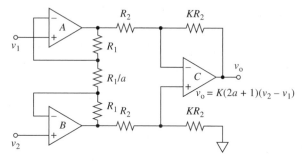

FIGURE 5-4 An instrumentation amplifier for reducing the effects of noise common to both inputs. The gain of the circuit is controlled by resistors R_1/a and KR_2.

transducer circuit generally appears in phase at both the inverting and noninverting inputs of the amplifier and is largely canceled by the circuit so that the noise at its output is diminished substantially. For cases in which a difference amplifier is insufficient to remove the noise, an instrumentation amplifier such as the one shown in Figure 5-4 is used.

Instrumentation amplifiers are composed of three op amps configured as shown in Figure 5-4. Op amp A and op amp B make up the input stage of the instrumentation amplifier in which the two op amps are cross-coupled through three resistors R_1, R_1/a, and R_1. The second stage of the module is the difference amplifier of op amp C. The overall gain of the circuit is given by[5]

$$v_o = K(2a + 1)(v_2 - v_1) \qquad (5\text{-}6)$$

Equation 5-6 highlights two advantages of the instrumentation amplifier: (1) the overall gain of the amplifier may be controlled by varying a single resistor R_1/a, and (2) the second difference stage rejects common-mode signals. In addition, op amps A and B are voltage followers with very high input resistance, so the instrumentation amplifier presents a negligible load to its transducer circuit. The combination of the two stages can provide rejection of common-mode noise by a factor of 10^6 or more while amplifying the signal by as much as 1000.

These devices are used often with low-level signals in noisy environments such as those found in biological organisms in which the organism acts as an antenna.

[4]For an excellent discussion of grounding and shielding, see H. V. Malmstadt, C. G. Enke, and S. R. Crouch, *Microcomputers and Electronic Instrumentation: Making the Right Connections*, pp. 401–9, Washington, DC: American Chemical Society, 1994. A valuable reference is R. Morrison, *Grounding and Shielding Techniques in Instrumentation*, 4th ed., New York: Wiley, 1998.

[5]H. V. Malmstadt, C. G. Enke, and S. R. Crouch, *Microcomputers and Electronic Instrumentation: Making the Right Connections*, pp. 210–11, Washington, DC: American Chemical Society, 1994.

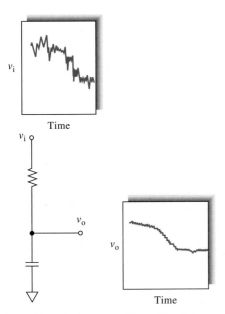

FIGURE 5-5 Use of a low-pass filter with a large time constant to remove noise from a slowly changing dc voltage.

Electrocardiographic instrumentation exploits the advantages of instrumentation amplifiers. Another typical application is in a computer data-acquisition module such as the programmable-gain instrumentation amplifier (PGIA) shown in Figure 4-12. The gain of the instrumentation amplifier is under the control of a computer, which changes resistor R_1/a of Figure 5-4 by using solid-state switches under digital control.

Analog Filtering

One of the most common methods of improving the signal-to-noise ratio in analytical instruments is by use of low-pass analog filters such as that shown in Figure 2-11b. The reason for this widespread application is that many instrument signals are of low frequency with bandwidths extending over a range of only a few hertz. Thus, a low-pass filter characterized by the Bode diagram shown in Figure 2-12b will effectively remove many of the high-frequency components of the signal, including thermal and shot noise. Figure 5-5 illustrates the use of a low-pass RC filter for reducing noise from a slowly varying dc signal.

High-pass analog filters such as that shown in Figure 2-11a also find considerable application in those analytical instruments where the analyte signal is at relatively high frequency. The high-pass filter reduces the effects of drift and other low-frequency flicker noise.

Narrow-band electronic filters are also available to attenuate noise outside an expected band of signal frequencies. We have pointed out that the magnitude of fundamental noise is directly proportional to the square root of the frequency bandwidth of a signal. Thus, significant noise reduction can be achieved by restricting the input signal to a narrow band of frequencies and using an amplifier that is tuned to this band. It is important to note that the bandpass of the filter must be sufficiently wide to include all of the signal frequencies.

Modulation

Direct amplification of a low-frequency or dc signal is particularly troublesome when an instrument exhibits amplifier drift and flicker noise. Often, this $1/f$ noise is several times larger than the types of noise that predominate at higher frequencies, as is illustrated in the noise-power spectrum of Figure 5-3. For this reason, low-frequency or dc signals from transducers are often converted to a higher frequency, where $1/f$ noise is less troublesome. This process is called *modulation*. After amplification, the modulated signal can be freed from amplifier $1/f$ noise by filtering with a high-pass filter; demodulation and filtering with a low-pass filter then produce an amplified dc signal suitable for output.

Figure 5-6 is a schematic showing the flow of information through such a system. The original dc signal shown in power spectrum A is modulated to give a narrow-band 400-Hz signal, which is then amplified by a factor of 10^5. As illustrated in power spectrum B, shown in the center of the figure, amplification introduces $1/f$ and power-line noise. Much of this noise can be removed with the aid of a suitable high-pass filter such as the one shown in Figure 2-11a. Demodulation of the filtered signal produces the amplified dc signal whose power spectrum is shown in C.

Modulation of the analytical signal can be accomplished in several ways. In spectroscopy, radiation sources are often modulated by mechanical devices called *choppers*, as illustrated in Figure 5-7. These devices interrupt the light beam periodically by physically blocking the beam. The radiation passed alternates between full intensity and no intensity. When converted to an electrical signal, a square wave results at the chopping frequency. Alternatively, some light sources can be pulsed electronically to produce the same alternating on-off effect.

Atomic absorption spectroscopy (see Chapter 9) provides an example of the use of a mechanical chopper

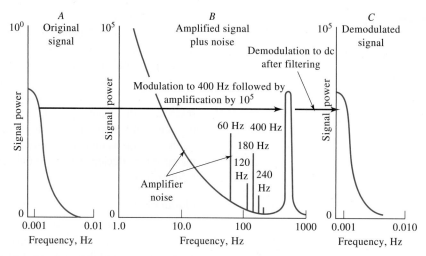

FIGURE 5-6 Amplification of a modulated dc signal. (Adapted from T. Coor, *J. Chem. Educ.*, **1968**, *45*, A540. With permission.)

for signal modulation. Noise is a major concern in detecting and measuring signals from atomic absorption sources of the low-frequency fluctuations inherent in flames and other atomization devices. To minimize these noise problems, light sources in atomic absorption instruments are often chopped by positioning a slotted rotating disk in the beam path, as illustrated in Figure 5-8. The rotation of the chopper produces a radiant signal that fluctuates periodically between zero and some maximum intensity. After interaction with the sample in the flame, the signal is converted by the transducer to a square-wave electrical signal whose frequency depends on the size of the slots and the rate at which the disk rotates. Noise inherent in flames and other atomization devices is usually of low frequency and can be significantly reduced by the use of an ac amplifier that is tuned to the chopping frequency. The ac

 Simulation: Learn more about **modulation and lock-in amplifiers**.

amplifier not only amplifies the signal and discriminates against the noise but also converts the square-wave signal into a sinusoidal signal, as shown by the waveforms in Figure 5-8.

Synchronous Demodulation

The modulation and tuned amplification process shown in Figure 5-8 produces a sine wave signal that is at the chopping frequency and in phase with the chopper. The switch shown in the figure alternates between the low-pass filter at the output and common. If the switch is made to alternate in phase with the chopper, it has the effect of passing the sinusoidal signal during the positive half cycle and blocking it during the negative-going half cycle. As can be seen by the waveform, this is a form of *rectification*, but rectification that is *synchronous* with the chopper. Usually, a reference signal is provided by the chopper to drive the switch. The reference signal is of the same frequency and has a fixed phase relation-

FIGURE 5-7 Mechanical choppers for modulating a light beam: (a) rotating disk chopper, (b) rotating vane chopper, (c) oscillating tuning fork design where rotational oscillation of a vane causes periodic interruptions of a light beam.

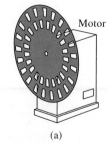

(a)

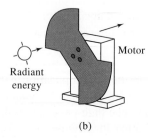

(b)

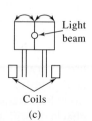

(c)

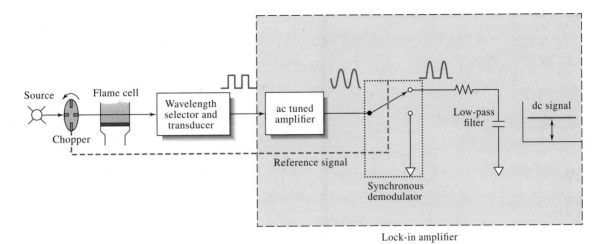

FIGURE 5-8 Lock-in amplifier for atomic absorption spectrometric measurements. The chopper converts the source beam to an on-off signal that passes through the flame cell where absorption occurs. After wavelength selection and transduction to an electrical signal, the ac square-wave input to the lock-in amplifier is amplified and converted to a sinusoidal signal by the tuned amplifier. The synchronous demodulator is phase-locked to the ac signal and provides a half-wave rectification of the signal. The low-pass filter converts the demodulated signal to a dc signal for measurement.

ship with the analytical signal. The reference signal might be provided by another beam passing through the chopper, but not through the flame, or it might be derived from the motor driving the chopper. Synchronous demodulation results in a dc signal that can then be sent through a low-pass filter to provide the final dc output, as shown in Figure 5-8.

Lock-In Amplifiers

The tuned amplifier, synchronous demodulator, reference input, and low-pass filter of Figure 5-8 constitute a *lock-in amplifier*. Such amplifiers allow recovery of signals that are utterly obscured by noise. Usually, the synchronous demodulator of a lock-in amplifier operates in a full-wave rectification mode rather than the half-wave mode shown in Figure 5-8. This is accomplished by directly passing the sinusoidal signal during one half cycle and inverting it during the other half cycle to provide a fluctuating dc signal that is relatively easy to filter with a low-pass filter. Rather than mechanical switches, solid-state devices called *analog multipliers* are usually used in synchronous demodulators.

A lock-in amplifier is generally relatively free of noise because only those signals that are locked in to the reference signal are amplified and demodulated. Extraneous signals of different frequencies and phase are rejected by the system.

5C-2 Software Methods

With the widespread availability of laboratory computers, many of the signal-to-noise enhancement devices described in the previous section are being replaced or supplemented by computer programs. Among these programs are routines for various types of averaging, digital filtering, Fourier transformation, smoothing, and correlation techniques. These procedures are applicable to nonperiodic, or irregular, waveforms, such as an absorption spectrum; to signals having no synchronizing or reference wave; and to periodic signals. Some of these relatively common procedures are discussed briefly here.

Ensemble Averaging

In ensemble averaging, successive sets of data stored in memory as *arrays* are collected and summed point by point for averaging. This process is sometimes called *coaddition*. After the collection and summation are complete, the data are averaged by dividing the sum for each point by the sets of data summed. Figure 5-9 illustrates ensemble averaging of an absorption spectrum.

To understand why ensemble averaging effectively increases the signal-to-noise ratio of digitally

 Simulation: Learn more about **signal averaging**.

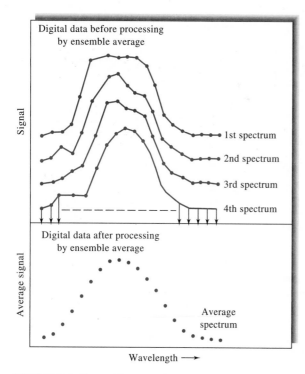

FIGURE 5-9 Ensemble averaging of a spectrum. (From D. Binkley and R. Dessy, *J. Chem. Educ.*, **1979**, *6*, 150. With permission.)

acquired signals, let us suppose that we are attempting to measure the magnitude of a dc signal S. We make n repetitive measurements of S and calculate the mean value of the signal from the equation

$$S_x = \frac{\sum_{i=1}^{n} S_i}{n} \qquad (5\text{-}7)$$

where S_i, $i = 1, 2, 3, \ldots, n$, are the individual measurements of the signal, including noise. The noise for each measurement is then $S_i - S_x$. If we square and sum the deviations of the signal from the mean S_x and divide by the number of measurements n, we obtain the mean-square noise, given by

$$\text{mean-square noise} = \sigma_i^2 = \frac{\sum_{i=1}^{n} (S_i - S_x)^2}{n} \qquad (5\text{-}8)$$

The mean-square noise is the *variance* of the signal σ_i^2 and the *root-mean-square*, or *rms*, noise is its standard deviation (Appendix 1, Section a1B-1):

$$\text{rms noise} = N_i = \sigma_i = \sqrt{\frac{\sum_{i=1}^{n} (S_i - S_x)^2}{n}} \qquad (5\text{-}9)$$

If we sum n measurements to obtain the ensemble average, the signal S_i adds for each repetition. The total signal S_n is given by $S_n = \sum_{i=1}^{n} S_i = nS_i$. For the noise, the variance is additive (see Appendix 1, Table a1-6). The total variance is $\sigma_n^2 = \sum_{i=1}^{n} \sigma_i^2 = n\sigma_i^2$. The standard deviation, or the total rms noise, is $\sigma_n = N_n = \sqrt{n}\sigma_i = \sqrt{n}N_i$. The signal-to-noise ratio after n repetitions $(S/N)_n$ is then given by

$$\left(\frac{S}{N}\right)_n = \frac{nS_i}{\sqrt{n}N_i} \qquad (5\text{-}10)$$

$$\left(\frac{S}{N}\right)_n = \sqrt{n}\left(\frac{S}{N}\right)_i \qquad (5\text{-}11)$$

The last expression shows that the signal-to-noise ratio is proportional to the square root of the number of data points collected to determine the ensemble average. Note that this same signal-to-noise enhancement is realized in boxcar averaging and digital filtering, which we describe in subsequent sections.

The improvement in S/N that is realized by signal averaging is exploited in many areas of science; NMR spectroscopy and Fourier transform infrared spectroscopy are but two of the most prominent examples in chemical instrumentation. We consider signal averaging and other aspects of digital data acquisition in more detail in the chapters on those topics.

To realize the advantage of ensemble averaging and still extract all of the information available in a signal waveform, it is necessary to measure points at a frequency that is at least twice as great as the highest frequency component of the waveform. This assertion is a consequence of the *Nyquist sampling theorem*, which states that for bandwidth-limited signals sampling must occur at a frequency that is at least twice the highest frequency f of the components making up the signal of interest. That is, the data-acquisition frequency must be at least $2f = 1/\Delta t$, where Δt is the time interval between signal samples. For example, if the maximum frequency component in an instrumental signal is 150 Hz, the data must be sampled at a rate of at least 300 samples/s if the signal is to be reproduced accurately.

Sampling frequencies much greater than the Nyquist frequency do not produce significant additional information, and much higher frequencies may actually in-

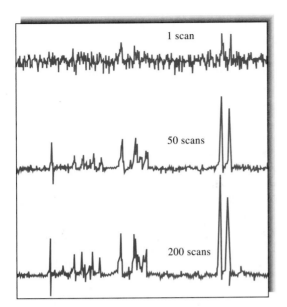

FIGURE 5-10 Effect of signal averaging. Note that the vertical scale is smaller as the number of scans increases. The signal-to-noise ratio is proportional to $\sqrt{n}$. Random fluctuations in the signal tend to cancel as the number of scans increases, but the signal itself accumulates. Thus, the S/N increases with an increasing number of scans.

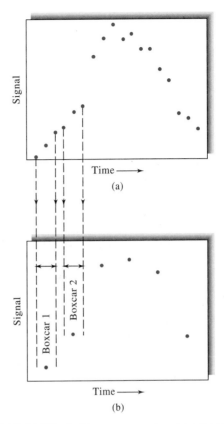

FIGURE 5-11 Effect of boxcar averaging: (a) original data, (b) data after boxcar averaging. (From G. Dulaney, *Anal. Chem.*, **1975**, *47*, 27A. Figure 2, p. 27A. Copyright 1978 American Chemical Society.)

clude undesirable noise. It is customary, however, to sample at a frequency about ten times the Nyquist frequency to ensure signal integrity. Furthermore, it is very important to sample the waveform reproducibly; that is, the waveform must be sampled beginning at precisely the same point on each successive wave. For example, if the waveform is a visible absorption spectrum, each scan of the spectrum must be synchronized to start at exactly the same wavelength, and each successive sample of the waveform must occur at the same wavelength interval. Generally, synchronization is accomplished by means of a synchronizing pulse, which is derived from the waveform itself or the experimental event that initiated the waveform such as a laser pulse or a pulse of radio-frequency radiation. This pulse then initiates the acquisition of data for each scan of the waveform.

Ensemble averaging can produce dramatic improvements in signal-to-noise ratios, as demonstrated by the three NMR spectra in Figure 5-10. Here, only a few of the resonances can be observed in the single scan because their magnitudes are roughly the same as the signal excursions because of random noise. The improvement in the resulting spectrum as a consequence of signal averaging is apparent from the spectra shown

in Figure 5-10. Further discussion of the Nyquist sampling theorem and its consequences may be found in the treatment of Fourier transform NMR spectroscopy in Chapter 19.

Boxcar Averaging

Boxcar averaging is a procedure for smoothing irregularities and enhancing the signal-to-noise ratio in a waveform, the assumption being that these irregularities are the result of noise. That is, it is assumed that the analog analytical signal varies only slowly with time and that the average of a small number of adjacent points is a better measure of the signal than any of the individual points. Figure 5-11b illustrates the effect of the technique on the data plotted in Figure 5-11a. The first point on the boxcar plot is the mean of points 1, 2, and 3 on the original curve; point 2 is the average of points 4, 5, and 6; and so forth. In practice, 2 to 50 points are averaged to generate a final point. Most often this averaging is performed by a computer in real time, that is, as data are being collected (in contrast to ensemble averaging, which requires storage of the data for subsequent processing). As Figure 5-11 shows, detail is lost by boxcar averaging, and its utility is limited for complex signals that change rapidly as a function of time. Boxcar averaging is of considerable importance, however, for square-wave or repetitive-pulsed outputs where only the average amplitude is important.

Boxcar averaging can also be accomplished in the analog domain by using a *boxcar integrator*. This device uses a fast electronic switch to sample a repetitive waveform at a programmed time interval from the origin of the waveform. The sampled waveform is connected to an analog integrator to furnish a low-frequency version of the signal at the selected time interval. The instrument may be programmed to scan a very noisy signal waveform from beginning to end. In this way, a profile of the signal is acquired with a signal-to-noise ratio that is selectable by adjusting the time constant of the integrator, the scan speed of the sampling window, and the time window over which the sampling occurs. This window is called the *aperture time*.

Boxcar integrators are often used to sample and measure instrumental waveforms on the picosecond-to-microsecond time scale. Such integrators are particularly useful in connection with pulsed-laser systems, in which physical and chemical events occur on such very fast time scales. The output of the integrator may be connected to computer data-acquisition systems such as those described in Section 4E-4 for logging of data and subsequent postexperimental analysis and presentation. The advantage of acquisition of signals by means of a boxcar integrator is that the averaging time of the units may be increased to provide *S/N* enhancement. The signal-to-noise ratio is proportional to the square root of the amount of time it takes the integrator to acquire the signal at each time window of the waveform. Such improvement is equivalent to the enhancement realized in digital data acquisition by ensemble averaging.

Digital Filtering

Digital filtering can be accomplished by a number of different well-characterized numerical procedures, including ensemble averaging, which was discussed in the previous section; Fourier transformation; least-squares polynomial smoothing; and correlation. In this section, we briefly discuss the Fourier transform procedure and least-squares polynomial smoothing, which

is one of the most common of all numerical data-enhancement techniques.

In the Fourier transformation, a signal such as the spectrum shown in Figure 5-12a, which is acquired in the time domain, is converted to a frequency-domain signal in which the independent variable is frequency f rather than time as depicted in Figure 5-12b. This transformation, which is discussed in some detail in Section 7I, is accomplished mathematically on a computer by a very fast and efficient algorithm. The frequency-domain signal in (b) is then multiplied by the frequency response of a digital low-pass filter with upper-cutoff frequency f_0 shown in (c), which has the effect of removing all frequency components above f_0 as illustrated in (d). The inverse Fourier transform then recovers the filtered time-domain spectrum of Figure 5-12e. The Fourier transformation is used in most modern infrared and NMR spectrometers, as well as in a host of laboratory test instruments and digital oscilloscopes. The procedure is often built into general-purpose software packages such as Mathcad and Excel and is widely available in subroutine packages in a variety of computer languages.

The last and perhaps most widely used digital data-enhancement technique that we will discuss is least-squares *polynomial data smoothing*. In its simplest form, smoothing is very similar to the boxcar averaging scheme of Figure 5-11. Figure 5-13 illustrates how unweighted data smoothing is accomplished. The eleven data points represented by filled circles in the plot comprise a section of a noisy absorption spectrum. The first five data points encompassed by bracket 1 in the figure are averaged and plotted at the midpoint position on the x-axis, or the point represented by triangle 1. The bracket is then moved one point to the right to position 2, points 2 through 6 are averaged, and the average is plotted as triangle 2. The process is repeated for brackets 3, 4, 5, and so on until all of the points except the last two are averaged to produce a new absorption curve represented by the triangular points and the line connecting them. The new curve is somewhat less noisy than the original data. This procedure is called a five-point *moving average smooth*. In this type of signal-to-noise enhancement, the width of the smoothing function always encompasses an odd number of points and an even number of points remain unsmoothed at each end of the data set. The moving average results in losing $(n - 1)/2$ data points at the beginning and the

 Simulation: Learn more about **FFT**.

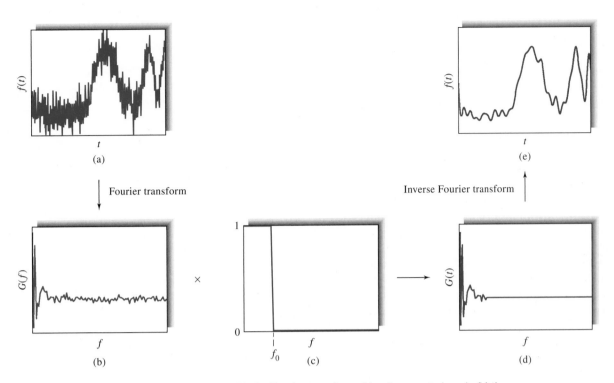

FIGURE 5-12 Digital filtering with the Fourier transform: (a) noisy spectral peak, (b) the frequency-domain spectrum of part (a) resulting from the Fourier transformation, (c) low-pass digital-filter function, (d) product of part (b) and part (c), (e) the inverse Fourier transform of part (d) with most of the high-frequency noise removed.

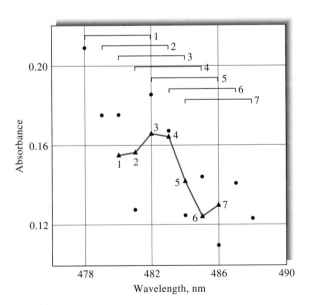

FIGURE 5-13 The operation of an unweighted moving average smoothing function: noisy spectral data (•), smoothed data (▲). See text for a description of the smoothing procedure.

same number at the end of the smooth. For a five-point smooth, a total of four data points are lost. For an absorption spectrum consisting of hundreds or perhaps thousands of data points, the loss of a few points is usually inconsequential.

Further improvements in signal-to-noise ratio can be obtained by increasing the number of points smoothed to, for example, a seven-point smooth, a nine-point smooth, etc. However, the more points in the smooth, the more distortion in the results, and, of course, the more points are lost at the ends of the data set.[6]

Another approach to moving average smoothing is to calculate a weighted moving average. The most applicable weighting scheme in scientific analysis is the exponentially weighted moving average. This type of scheme weights the current data point highest and gives previous data points exponentially decreasing weights. The exponential smooth is similar to the ef-

---

[6]See, for example, S. R. Crouch and F. J. Holler, *Applications of Microsoft® Excel in Analytical Chemistry*, Belmont, CA: Brooks/Cole, 2004, pp. 305–7.

fect of an *RC* filter with the width of the smooth analogous to the size of the time constant of the filter.

An even better procedure than simply averaging points on a data curve is to perform a least-squares fit of a small section of the curve to a polynomial and take the calculated central point of the fitted polynomial curve as the new smoothed data point. This approach is much superior to the unweighted averaging scheme, but it has the disadvantage of being computationally intense and thus requires considerable computer time. Savitzky and Golay[7] showed that a set of integers could be derived and used as weighting coefficients to carry out the smoothing operation. The use of these weighting coefficients, sometimes referred to as *convolution integers*, turns out to be *exactly equivalent* to fitting the data to a polynomial as just described. The convolution integers for a five-point quadratic smoothing function are plotted in Figure 5-14a.[8] The process of smoothing in this manner is often called *least-squares polynomial smoothing*.

The application of the smoothing integers of Figure 5-14a to the data of Figure 5-13 illustrates the smoothing process. We begin by multiplying the leftmost convolution integer, which is −3 in this case, by the absorbance at point 1 in Figure 5-13. The second integer, which is 12, is then multiplied by the second point, and the result is added to the product obtained for the first point. Point 3 is then multiplied by 17, which is the third integer, and the result is summed once again. This process is repeated until each of the five data points has been multiplied by its corresponding integer and the sum of the five results obtained. Finally, the sum of the results is divided by a sixth integer, the so-called normalization integer, which is 35 in this example of a five-point quadratic smooth, and the quotient is taken as the new value of the central point of the smoothing interval. The normalization integer also derives from the Savitzky-Golay treatment, as do other similar sets of integers for smoothing, to generate the first and second derivatives of the data. The first-derivative convolution integers for a five-point cubic smooth are plotted in Figure 5-14b, and the second-derivative integers for a five-point quadratic are shown in Figure 5-14c. These sets of integers may be exploited in exactly the same

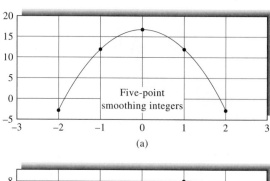

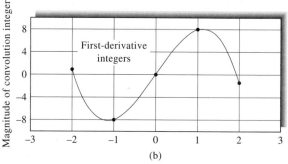

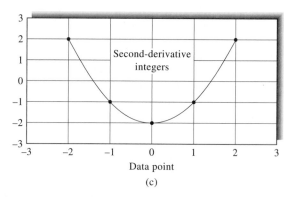

FIGURE 5-14 Least-squares polynomial smoothing convolution integers: (a) quadratic five-point integers, (b) first-derivative cubic five-point integers, (c) second-derivative quadratic five-point integers.

way as the basic smoothing integers to produce the first and second derivatives of the original absorption data. Least-squares polynomial derivative smoothing is used to develop derivative spectra because, as noted in our discussion of analog differentiators in Section 3E-4, differentiation is usually a noise-producing process. The effect of derivative smoothing is to minimize the noise generated in the differentiation.

Because least-squares polynomial smoothing is so widely used to enhance the quality of analytical data, it is important to note the advantages and limitations of

[7] A. Savitzky and M. J. Golay, *Anal. Chem.*, **1964**, *36*, 1627–39.
[8] Savitzky-Golay smoothing is compared to moving average smoothing in S. R. Crouch and F. J. Holler, *Applications of Microsoft® Excel in Analytical Chemistry*, Belmont, CA: Brooks/Cole, 2004, pp. 305–10.

the method. The procedure reduces noise and acts as a low-pass filter on data. As with any filtering process, the signal suffers some distortion because of the bandwidth limitation inherent in the process. Users of smoothing must recognize that noise reduction has to be balanced against possible distortion of the signal. The advantage of the procedure is that variables such as the type of smooth, the smooth width, and the number of times that the data are smoothed may be decided after data collection. Furthermore, the smoothing algorithm is computationally trivial and requires minimal computer time. The *S/N* enhancement resulting from smoothing is relatively modest, generally amounting to about a factor of 4 for spectra that contain peaks with a width of thirty-two data points and with a smooth width twice that value. However, smoothing produces cleaner data than the original does for human interpretation, and it is widely used for this purpose. In fact, it has been stated that "least-squares polynomial smoothing is of cosmetic value only."[9] No additional information is contained in the smoothed data and, indeed, distortion may be introduced. When smoothing is applied to data for quantitative analysis, distortion of the data has a minimal effect on the quantitative results because distortion errors tend to cancel when samples and standards are smoothed in the same manner.

The data of Figure 5-15 illustrate the application of least-squares polynomial smoothing to a rather noisy 501-point absorption spectrum of the dye tartrazine shown at the bottom of the figure in curve *A*. Curve *B* is quadratic 5-point smooth of the data, curve *C* is a fourth-degree 13-point smooth, and curve *D* is a tenth-degree 77-point smooth. Note in curve *D* that 38 points remain unsmoothed at each end of the data set. The effects of the smoothing process are apparent in the progression from curve *A* to curve *D*.

Because of the general utility of smoothing and its wide application, guidelines have been developed for its use, equations are available for calculating the smoothing coefficients, and it has been applied to two-dimensional data such as diode-array spectra.[10]

 Simulation: Learn more about **data smoothing**.

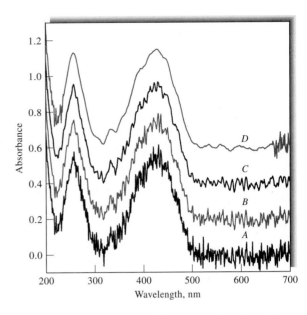

FIGURE 5-15 Effect of smoothing on a noisy absorption spectrum of tartrazine: (*A*) Raw spectrum, (*B*) quadratic 5-point smooth of the data in *A*, (*C*) fourth-degree 13-point smooth of the same data, (*D*) tenth-degree 77-point smooth of the data.

Correlation Methods

Correlation methods are often used for processing data from analytical instruments. These procedures provide powerful tools for performing such tasks as extracting signals that appear to be hopelessly lost in noise, smoothing noisy data, comparing a spectrum of an analyte with stored spectra of pure compounds, and resolving overlapping or unresolved peaks in spectroscopy and chromatography.[11] Correlation methods are based on complex mathematical data manipulations that can be carried out conveniently only by means of a computer or by sophisticated analog instrumentation.

Correlation methods are not discussed further in this text. The interested reader should consult the references given in note 11.

[9]T. A. Nieman and C. G. Enke, *Anal. Chem.*, **1976**, *48*, 705A.

[10]For further details on the nature of the smoothing process and its implementation, see J. Steinier, Y. Termonia, and J. Deltour, *Anal. Chem.*, **1972**, *44*, 1906; T. A. Nieman and C. G. Enke, *Anal. Chem.*, **1976**, *48*, 705A; H. H. Madden, *Anal. Chem.*, **1978**, *50*, 1383; K. L. Ratzlaff, *Introduction to Computer-Assisted Experimentation*, New York: Wiley, 1987.

[11]For a more detailed discussion of correlation methods, see G. Horlick and G. M. Hieftje, in *Contemporary Topics in Analytical and Clinical Chemistry*, D. M. Hercules, et al., eds., Vol. 3, pp. 153–216, New York: Plenum Press, 1978. For a briefer discussion, see G. M. Hieftje and G. Horlick, *American Laboratory*, **1981**, *13* (3), 76.

QUESTIONS AND PROBLEMS

*Answers are provided at the end of the book for problems marked with an asterisk.

[X] Problems with this icon are best solved using spreadsheets.

5-1 What types of noise are frequency dependent? Frequency independent?

5-2 Name the type or types of noise that can be reduced by
(a) decreasing the temperature of a measurement.
(b) decreasing the frequency used for the measurement.
(c) decreasing the bandwidth of the measurement.

5-3 Suggest a frequency range that is well suited for noise minimization. Explain.

5-4 Why is shielding vital in the design of glass electrodes that have an internal resistance of 10^6 ohms or more?

5-5 What type of noise is likely to be reduced by (a) a high-pass filter and (b) a low-pass filter?

5-6 Make a rough estimate of the signal-to-noise ratio for the 0.9×10^{-15} A current shown in Figure 5-1a.

[X] *5-7 The following data were obtained for repetitive weighings of a 1.004-g standard weight on a top-loading balance:

1.003	1.000	1.001
1.004	1.005	1.006
1.001	0.999	1.007

(a) Assuming the noise is random, calculate the signal-to-noise ratio for the balance.
(b) How many measurements would have to be averaged to obtain a S/N of 500?

[X] *5-8 The following data were obtained for a voltage measurement, in mV, on a noisy system: 1.37, 1.84, 1.35, 1.47, 1.10, 1.73, 1.54, 1.08.
(a) Assuming the noise is random, what is the signal-to-noise ratio?
(b) How many measurements would have to be averaged to increase S/N to 10?

*5-9 Calculate the rms thermal noise associated with a 1.0-MΩ load resistor operated at room temperature if an oscilloscope with a 1-MHz bandwidth is used. If the bandwidth is reduced to 100 Hz, by what factor will the noise be reduced?

*5-10 If spectrum A in Figure 5-2 is the result of a single scan and spectrum B is the result of ensemble averaging, how many individual spectra were summed to increase S/N from 4.3 to 43?

*5-11 Calculate the improvement in S/N in progressing from the top spectrum to the bottom in Figure 5-10. By what factor is the S/N of the bottom spectrum improved over the middle spectrum?

*5-12 Calculate the improvement in S/N in progressing from spectrum A to spectrum D in Figure 5-15.

Challenge Problem

5-13 (a) The following Excel spreadsheet represents capillary electrophoresis results monitored by fluorescence detection. Plot the raw data and calculate and plot a five-point, unweighted moving average smooth on the results.

(b) Calculate and plot a five-point Savitzky-Golay smooth using the smoothing coefficients given in the chapter. Compare the improvement in S/N of the Savitzky-Golay smooth to that of the moving average smooth plotted in (a). Also compare the peak distortion of the two smooths.

(c) Use a search engine like Google to find the Savitzky-Golay smoothing coefficients for a seven-point quadratic smooth (order = 2) for the data points shown. Note that in many cases the coefficients have already been normalized by dividing by a normalization integer. You can determine whether the coefficients have been normalized by comparing them to those given in Figure 5-14a, which must be normalized by dividing by 35.

(d) Now perform a seven-point moving average smooth and a seven-point Savitzky-Golay smooth on these data. Compare the improvement in S/N and the peak shape distortion.

(e) Perform a seven-point moving average smooth and a fifteen-point moving average smooth of the data. Which smoothing width gives better improvement in S/N? Which smoothing width gives least distortion of peak shape?

(f) Defend or criticize the conclusion made by Enke and Nieman in their paper on smoothing (*Anal. Chem.*, **1976**, *48*, 705A) that "least-squares smoothing is of cosmetic value only." Begin with a statement of whether you agree or disagree with Enke and Nieman's conclusion. Justify your conclusions by using results and inferences from their paper as well as your own reasoning.

	A	B	C
1	**Smoothing of data**		
2	Capillary electrophoresis peak		
3	detected by fluorescence		
4	time, s	Fl. Intensity	
5	100	0.0956	
6	105	0.1195	
7	110	0.1116	
8	115	0.1275	
9	120	0.0717	
10	125	0.1036	
11	130	0.0319	
12	135	0.0717	
13	140	0.1355	
14	145	0.2231	
15	150	0.1753	
16	155	0.5817	
17	160	1.8646	
18	165	2.6535	
19	170	2.8527	
20	175	2.8846	
21	180	2.8368	
22	185	2.7890	
23	190	2.7093	
24	195	2.5180	
25	200	2.3427	
26	205	2.2312	
27	210	1.9603	
28	215	1.8248	
29	220	1.6017	
30	225	1.4901	
31	230	1.2989	
32	235	1.2590	
33	240	1.1076	
34	245	0.9642	
35	250	0.9164	
36	255	0.8845	
37	260	0.7809	
38	265	0.7172	
39	270	0.6694	
40	275	0.6215	
41	280	0.6454	
42	285	0.6454	
43	290	0.5817	

	A	B	C
44	295	0.5259	
45	300	0.5658	
46	305	0.5259	
47	310	0.5020	
48	315	0.5419	
49	320	0.5419	
50	325	0.3426	
51	330	0.4940	
52	335	0.4383	
53	340	0.4462	
54	345	0.3984	
55	350	0.4064	
56	355	0.4383	
57	360	0.3984	
58	365	0.3905	
59	370	0.4861	
60	375	0.3984	
61	380	0.3267	
62	385	0.3905	
63	390	0.3825	
64	395	0.4303	
65	400	0.3267	
66	405	0.3586	
67	410	0.3905	
68	415	0.3745	
69	420	0.3347	
70	425	0.3426	
71	430	0.2869	
72	435	0.2550	
73	440	0.3506	
74	445	0.3028	
75	450	0.3028	
76	455	0.3108	
77	460	0.2311	
78	465	0.2709	
79	470	0.3028	
80	475	0.2789	
81	480	0.3347	
82	485	0.2311	
83	490	0.1753	
84	495	0.2869	
85	500	0.2311	

The Electronic Analytical Laboratory

In many laboratories, particularly those in the pharmaceutical industry, government regulations, new product development, product testing, and auditing have led to a substantial increase in paperwork. Because of the extensive use of computers and automated instruments in analytical laboratories, there has been discussion for many years that electronic, paperless laboratories could solve the problems caused by this explosion of paperwork and record keeping. In the electronic laboratory, processing, manipulation, reporting, archiving, storage, and data mining would all be handled electronically. In recent years, government regulations have been written to permit and even encourage paperless laboratories when certain criteria are met. The Food and Drug Administration Title 21, Part 11 of the Code of Federal Regulations deals with acceptable electronic record keeping, electronic submissions, and electronic signatures, particularly for the pharmaceutical industry. Compliance with regulations as well as a desire to take full advantage of modern technological advances have spurred developments in implementing such all-electronic laboratories.

Potential Advantages

Electronic laboratories have several possible benefits.[1] Reduction of paperwork is certainly one major advantage. Data acquired from instruments can be manipulated and stored electronically. Data can be entered in electronic notebooks automatically. In paper-based systems, data had to be entered manually into notebooks and chart recordings or photographs had to be stored.

Another positive effect of an electronic laboratory is increased efficiency. With the availability of data electronically, sharing data and information among many coworkers in an organization can be accomplished quite readily. Those workers who need information have nearly immediate access so that decisions can be made much more rapidly than with paper-based laboratories.

A third benefit is to speed the throughput of samples in analytical laboratories. Here, laboratory automation and the electronic manipulation of data help to complete analyses of samples more rapidly so that those who need analytical results can receive them quickly. This can help alleviate production problems, assist in studies of the efficacy of new products, or speed up the release of new materials to the marketplace.

Paperless laboratories can also be of great help in complying with governmental regulations. Systems that are in accord with regulations can automate the compliance process and lead to better detection if laboratory practices do not conform to prescribed regulations. The audit trail is more easily tracked and any potential alterations more readily found with electronic-based systems.

Finally, costs can be reduced with paperless laboratories because fewer workers are required to complete paperwork and produce reports. The more efficient use of available resources can lead to long-range benefits and cost savings.

Components of Electronic Laboratories

What constitutes an electronic analytical laboratory?[2] Figure IA1-1 shows the major components of a paperless laboratory. First, a computerized analytical instrument is required to acquire, manipulate, and process the data. The instrument may be connected to its own data system, as are many chromatography instruments, or directly to an electronic laboratory notebook. The notebook often provides input to a laboratory information management system (LIMS), as discussed in section 4H-2, and to a laboratory data-archiving system. In some cases, information from the data system or notebook flows directly into archival storage. Several of these components are discussed in this case study.

Electronic Laboratory Notebooks

Electronic laboratory notebooks (ELNs) first became available in the mid-1990s. They promised to revolutionize the collection and dissemination of laboratory data.[3] Until recently though, ELNs were not in widespread use in industrial, governmental, or academic laboratories. With new regulations and the desire to enhance efficiency has come the realization that capturing information about new products or analytical samples as soon as possible and making that information available to others in the organization is an idea with a good deal of merit. Hence, the ELN has been making a comeback in the last few years, particularly in the pharmaceutical industry.

The ELN consists of laboratory notebook pages in which a research worker can enter data and graphics or import the data from instruments, data systems, or software like

[1] R. D. McDowall, *Am. Pharm. Rev.*, **2004**, *7* (4), 20.

[2] S. Piper, *Am. Pharm. Rev.* **2005**, *8* (1), 10 (www.americanpharmaceutical review.com).

[3] P. Rees, *Scientific Computing World* **2004**, *79*, 10 (www.scientific-computing.com/scwnovdec04lab_notebooks.html).

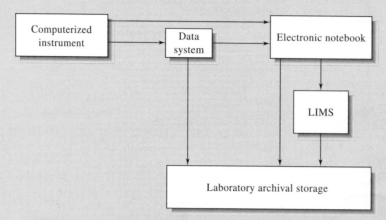

FIGURE IA1-1 Components of an electronic analytical laboratory.

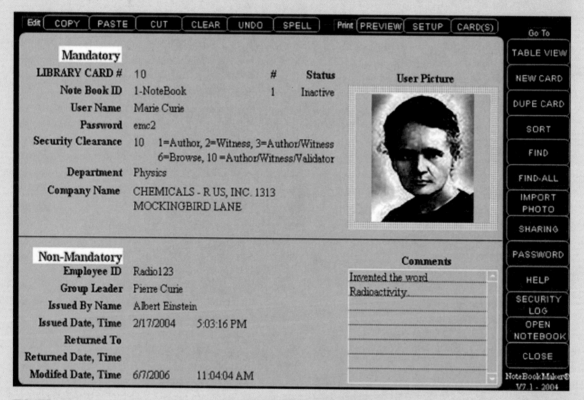

FIGURE IA1-2 Library card for fictitious user. (Courtesy of FileMaker, Inc.)

spreadsheets and scientific packages. In one ELN, a library card is first issued by a librarian. With the type of notebook shown in Figure IA1-1, a librarian gives a library card to each user of the notebook. The library card contains personal information about the user such as that shown in Figure IA1-2.

Once a user has a library card, ID number, and password, the ELN can be opened and pages observed. A typical laboratory notebook page is illustrated in Figure IA1-3. Here, the author of the page is required to sign the page electronically by posting it with a picture and statement. The electronic signature is a unique, nonhandwritten means

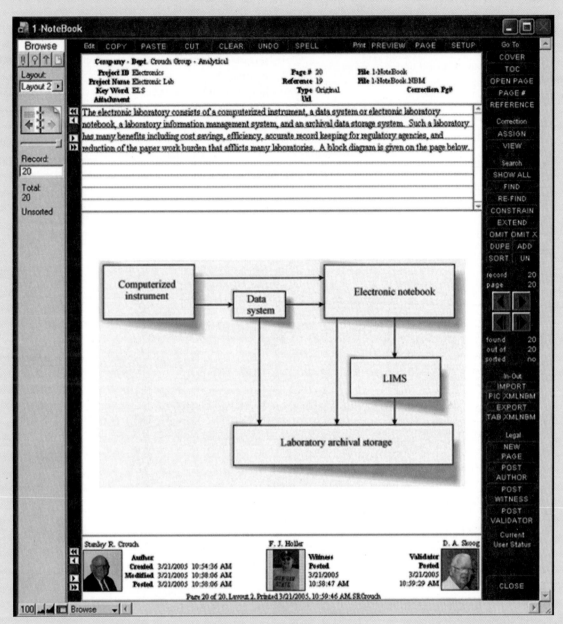

FIGURE IA1-3 Laboratory notebook page showing author, witness, and validator. (Courtesy of FileMaker, Inc.)

of identifying a person that remains with the individual so that other users cannot apply it. These usually include an ID number and password, a bar code, and a personal ID code or a retinal, voice, or fingerprint scan. A witness must then sign in with a valid ID number and password and electronically witness the page. Likewise, a validator signs in and validates the notebook page. Users are assigned various levels of security when the library card is issued.

The ELN is searchable electronically by any registered user. This feature allows a user to quickly find data and to share it with coworkers who have the appropriate security clearance. The ELN does not depend on the penmanship of the user as does a conventional notebook. It can be accessed remotely from home, a meeting in another location, or a worker's office. The ELN can be backed up by normal procedures so that it cannot be lost or destroyed by a researcher. Notebook pages can be printed for use in reports.

Connecting to LIMS and Archival Storage

Many laboratory notebooks are networked to LIMSs or to archival storage systems. The LIMS handles sample tracking and management along with scheduling and data archiving (see Figure 4-21). Some ELNs communicate wirelessly with the LIMS or data-archiving system, and others are part of a wired local area network. Integration of existing instruments, data systems, and LIMSs is often a formidable task. Although it is difficult to fully integrate older laboratories, new or redesigned laboratories can incorporate electronic laboratory principles from the beginning.

Future Developments

It is anticipated that in the future we will see more closely integrated electronic laboratory systems. Companies in the pharmaceutical industry have already begun making the switch. As the benefits become more generally recognized, many other industrial, governmental, and academic laboratories will also become paperless. Scientists often have additional electronic devices such as laptop computers and personal digital assistants to support their work. Increasingly, we should see these devices communicating with ELNs, LIMSs, and laboratory instruments to further increase efficiency and productivity. Although government regulations have been driving the adoption of electronic laboratories in the pharmaceutical industry, the benefits of increased productivity, reduced paper work, and lower costs will eventually be the most important reasons for implementing paperless, electronic laboratories.

© Ted Kinsman/Photo Researchers, Inc.

Atomic Spectroscopy

The image shown contains the optical emission spectra of a number of gaseous substances. Each spectrum provides a unique fingerprint of the corresponding substance that may be used to identify it. From top to bottom, the substances are bromine, deuterium, helium, hydrogen, krypton, mercury, neon, water vapor, and xenon. Note the similarities between hydrogen, deuterium, and water vapor. As we shall see, the intensities of the lines in the spectra yield information regarding the concentrations of the elements. Optical spectroscopy is just one of many spectroscopic methods that we shall explore in Section 2.

Section 2 comprises the fundamental principles and methods of atomic spectroscopy. Chapter 6 investigates the nature of light and its interaction with matter, and Chapter 7 introduces the optical, electronic, and mechanical components of optical instruments. These two chapters also present concepts and instrument components that are useful in our discussion of molecular spectroscopy in Section 3. The general nature of atomic spectra and practical aspects of introducing atomic samples into a spectrometer are addressed in Chapter 8. Chapter 9 explores the practice of atomic absorption and atomic fluorescence spectroscopy, and Chapter 10 provides a similar treatment of atomic emission spectroscopy. Mass spectrometry is introduced in Chapter 11, as are descriptions of various instruments and methods of atomic mass spectrometry. Our presentation of atomic spectrometry is completed by a discussion of X-ray spectrometry in Chapter 12. In the Instrumental Analysis in Action study, we examine analytical methods to monitor and speciate mercury in the environment.

CHAPTER SIX

An Introduction to Spectrometric Methods

This chapter treats in a general way the interactions of electromagnetic waves with atomic and molecular species. After this introduction to spectrometric methods, the next six chapters describe spectrometric methods used by scientists for identifying and determining the elements present in various forms of matter. Chapters 13 through 21 then discuss the uses of spectrometry for structural determination of molecular species and describe how these methods are used for their quantitative determination.

Throughout this chapter, this logo indicates an opportunity for online self-study at **www .thomsonedu.com /chemistry/skoog**, linking you to interactive tutorials, simulations, and exercises.

Spectrometric methods are a large group of analytical methods that are based on atomic and molecular spectroscopy. *Spectroscopy* is a general term for the science that deals with the interactions of various types of radiation with matter. Historically, the interactions of interest were between electromagnetic radiation and matter, but now spectroscopy has been broadened to include interactions between matter and other forms of energy. Examples include acoustic waves and beams of particles such as ions and electrons. *Spectrometry* and *spectrometric methods* refer to the measurement of the intensity of radiation with a photoelectric transducer or other type of electronic device.

The most widely used spectrometric methods are based on electromagnetic radiation, which is a type of energy that takes several forms, the most readily recognizable being light and radiant heat. Less obvious manifestations include gamma rays and X-rays as well as ultraviolet, microwave, and radio-frequency radiation.

6A GENERAL PROPERTIES OF ELECTROMAGNETIC RADIATION

Many of the properties of electromagnetic radiation are conveniently described by means of a classical sinusoidal wave model, which embodies such characteristics as wavelength, frequency, velocity, and amplitude. In contrast to other wave phenomena, such as sound, electromagnetic radiation requires no supporting medium for its transmission and thus passes readily through a vacuum.

The wave model fails to account for phenomena associated with the absorption and emission of radiant energy. To understand these processes, it is necessary to invoke a particle model in which electromagnetic radiation is viewed as a stream of discrete particles, or wave packets, of energy called *photons*. The energy of a photon is proportional to the frequency of the radiation. These dual views of radiation as particles and as waves are not mutually exclusive but, rather, complementary. Indeed, the wave-particle duality is found to apply to the behavior of streams of electrons, protons, and other elementary particles and is completely rationalized by wave mechanics.

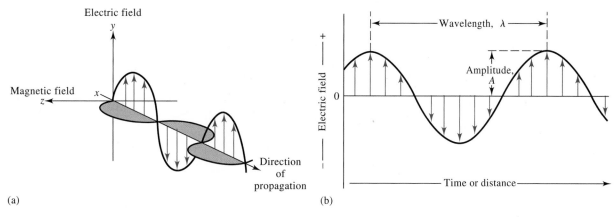

FIGURE 6-1 Wave nature of a beam of single-frequency electromagnetic radiation. In (a), a plane-polarized wave is shown propagating along the x-axis. The electric field oscillates in a plane perpendicular to the magnetic field. If the radiation were unpolarized, a component of the electric field would be seen in all planes. In (b), only the electric field oscillations are shown. The amplitude of the wave is the length of the electric field vector at the wave maximum, while the wavelength is the distance between successive maxima.

6B WAVE PROPERTIES OF ELECTROMAGNETIC RADIATION

For many purposes, electromagnetic radiation is conveniently represented as electric and magnetic fields that undergo in-phase, sinusoidal oscillations at right angles to each other and to the direction of propagation. Figure 6-1a is such a representation of a single ray of plane-polarized electromagnetic radiation. The term *plane polarized* implies that all oscillations of either the electric or the magnetic field lie in a single plane. Figure 6-1b is a two-dimensional representation of the electric component of the ray in Figure 6-1a. The electric field strength in Figure 6-1 is represented as a vector whose length is proportional to its magnitude. The abscissa of this plot is either time as the radiation passes a fixed point in space or distance when time is held constant. Throughout this chapter and most of the remaining text, only the electric component of radiation will be considered because the electric field is responsible for most of the phenomena that are of interest to us, including transmission, reflection, refraction, and absorption. Note, however, that the magnetic component of electromagnetic radiation is responsible for absorption of radio-frequency waves in nuclear magnetic resonance.

6B-1 Wave Characteristics

In Figure 6-1b, the *amplitude A* of the sinusoidal wave is shown as the length of the electric vector at a maximum in the wave. The time in seconds required for the passage of successive maxima or minima through a fixed point in space is called the *period p* of the radiation. The *frequency ν* is the number of oscillations of the field that occur per second[1] and is equal to $1/p$. Another variable of interest is the *wavelength λ*, which is the linear distance between any two equivalent points on successive waves (e.g., successive maxima or minima).[2] Multiplication of the frequency in cycles per second by the wavelength in meters per cycle gives the *velocity of propagation v_i* in meters per second:

$$v_i = \nu \lambda_i \qquad (6\text{-}1)$$

It is important to realize that the frequency of a beam of radiation is determined by the source and

---

[1] The common unit of frequency is the reciprocal second (s^{-1}), or *hertz* (Hz), which corresponds to one cycle per second.

[2] The units commonly used for describing wavelength differ considerably in the various spectral regions. For example, the angstrom unit, Å (10^{-10} m), is convenient for X-ray and short ultraviolet radiation; the nanometer, nm (10^{-9} m), is employed with visible and ultraviolet radiation; the micrometer, μm (10^{-6} m), is useful for the infrared region. (The *micrometer* was called *micron* in the early literature; use of this term is discouraged.)

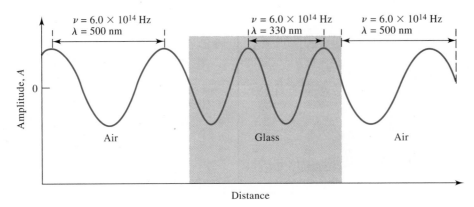

FIGURE 6-2 Change in wavelength as radiation passes from air into a dense glass and back to air. Note that the wavelength shortens by nearly 200 nm, or more than 30%, as it passes into glass; a reverse change occurs as the radiation again enters air.

remains invariant. In contrast, the velocity of radiation depends upon the composition of the medium through which it passes. Thus, Equation 6-1 implies that the wavelength of radiation also depends on the medium. The subscript i in Equation 6-1 indicates these dependencies.

In a vacuum, the velocity of radiation is independent of wavelength and is at its maximum. This velocity, given the symbol c, has been determined to be 2.99792×10^8 m/s. It is significant that the velocity of radiation in air differs only slightly from c (about 0.03% less); thus, for either air or vacuum, Equation 6-1 can be written to three significant figures as

$$c = \nu\lambda = 3.00 \times 10^8 \text{ m/s} = 3.00 \times 10^{10} \text{ cm/s} \quad (6\text{-}2)$$

In any medium containing matter, propagation of radiation is slowed by the interaction between the electromagnetic field of the radiation and the bound electrons in the matter. Since the radiant frequency is invariant and fixed by the source, the wavelength must decrease as radiation passes from a vacuum to another medium (Equation 6-2). This effect is illustrated in Figure 6-2 for a monochromatic beam of visible radiation.[3] Note that the wavelength shortens nearly 200 nm, or more than 30%, as it passes into glass; a reverse change occurs as the radiation again enters air.

The *wavenumber* $\bar{\nu}$, which is defined as the reciprocal of the wavelength in centimeters, is yet another way of describing electromagnetic radiation. The unit for $\bar{\nu}$ is cm^{-1}. Wavenumber is widely used in infrared spec-

troscopy. The wavenumber is a useful unit because, in contrast to wavelength, it is directly proportional to the frequency, and thus the energy, of radiation. Thus, we can write

$$\bar{\nu} = k\nu \quad (6\text{-}3)$$

where the proportionality constant k depends on the medium and is equal to the reciprocal of the velocity (Equation 6-1).

The *power P* of radiation is the energy of the beam that reaches a given area per second, whereas the *intensity I* is the power per unit solid angle. These quantities are related to the square of the amplitude A (see Figure 6-1). Although it is not strictly correct to do so, power and intensity are often used interchangeably.

6B-2 The Electromagnetic Spectrum

As shown in Figure 6-3, the electromagnetic spectrum encompasses an enormous range of wavelengths and frequencies (and thus energies). In fact, the range is so great that a logarithmic scale is required. Figure 6-3 also depicts qualitatively the major spectral regions. The divisions are based on the methods used to generate and detect the various kinds of radiation. Several overlaps are evident. Note that the portion of the spectrum visible to the human eye is tiny when compared with other spectral regions. Also note that spectrochemical methods employing not only visible but also ultraviolet and infrared radiation are often called *optical methods* despite the human eye's inability to sense either of the latter two types of radiation. This somewhat ambiguous terminology arises from the many common features

---

[3] A *monochromatic* beam is a beam of radiation whose rays have identical wavelengths. A *polychromatic* beam is made up of rays of different wavelengths.

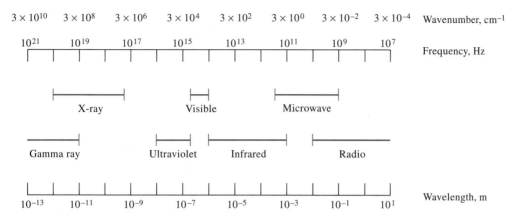

FIGURE 6-3 Regions of the electromagnetic spectrum.

TABLE 6-1 Common Spectroscopic Methods Based on Electromagnetic Radiation

Type of Spectroscopy	Usual Wavelength Range*	Usual Wavenumber Range, cm^{-1}	Type of Quantum Transition
Gamma-ray emission	0.005–1.4 Å	—	Nuclear
X-ray absorption, emission, fluorescence, and diffraction	0.1–100 Å	—	Inner electron
Vacuum ultraviolet absorption	10–180 nm	1×10^6 to 5×10^4	Bonding electrons
Ultraviolet-visible absorption, emission, and fluorescence	180–780 nm	5×10^4 to 1.3×10^4	Bonding electrons
Infrared absorption and Raman scattering	0.78–300 μm	1.3×10^4 to 3.3×10^1	Rotation/vibration of molecules
Microwave absorption	0.75–375 mm	13–0.03	Rotation of molecules
Electron spin resonance	3 cm	0.33	Spin of electrons in a magnetic field
Nuclear magnetic resonance	0.6–10 m	1.7×10^{-2} to 1×10^3	Spin of nuclei in a magnetic field

*1 Å = 10^{-10} m = 10^{-8} cm

1 nm = 10^{-9} m = 10^{-7} cm

1 μm = 10^{-6} m = 10^{-4} cm

of instruments for the three spectral regions and the similarities in how we view the interactions of the three types of radiation with matter.

Table 6-1 lists the wavelength and frequency ranges for the regions of the spectrum that are important for analytical purposes and also gives the names of the various spectroscopic methods associated with each. The last column of the table lists the types of nuclear, atomic, or molecular quantum transitions that serve as the basis for the various spectroscopic techniques.

 Exercise: Learn more about the **electromagnetic spectrum**.

6B-3 Mathematical Description of a Wave

With time as a variable, the wave in Figure 6-1b can be described by the equation for a sine wave. That is,

$$y = A \sin(\omega t + \phi) \tag{6-4}$$

where y is the magnitude of the *electric field* at time t, A is the amplitude or maximum value for y, and ϕ is the *phase angle*, a term defined in Section 2B-1, page 34. The angular velocity of the vector ω is related to the frequency of the radiation ν by the equation

$$\omega = 2\pi\nu$$

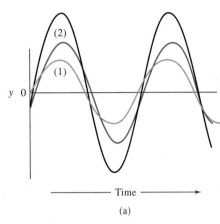

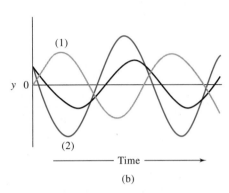

FIGURE 6-4 Superposition of sinusoidal wave: (a) $A_1 < A_2$, $(\phi_1 - \phi_2) = 20°$, $\nu_1 = \nu_2$; (b) $A_1 < A_2$, $(\phi_1 - \phi_2) = 200°$, $\nu_1 = \nu_2$. In each instance, the black curve results from the combination of the two other curves.

Substitution of this relationship into Equation 6-4 yields

$$y = A \sin(2\pi\nu t + \phi) \tag{6-5}$$

6B-4 Superposition of Waves

The *principle of superposition* states that when two or more waves traverse the same space, a disturbance occurs that is the sum of the disturbances caused by the individual waves. This principle applies to electromagnetic waves, in which the disturbances involve an electric field, as well as to several other types of waves, in which atoms or molecules are displaced. When n electromagnetic waves differing in frequency, amplitude, and phase angle pass some point in space simultaneously, the principle of superposition and Equation 6-5 permit us to write

$$y = A_1 \sin(2\pi\nu_1 t + \phi_1) + A_2 \sin(2\pi\nu_2 t + \phi_2)$$
$$+ \cdots + A_n \sin(2\pi\nu_n t + \phi_n) \tag{6-6}$$

where y is the resultant field.

The black curve in Figure 6-4a shows the application of Equation 6-6 to two waves of identical frequency but somewhat different amplitude and phase angle. The resultant is a periodic function with the same frequency but larger amplitude than either of the component waves. Figure 6-4b differs from 6-4a in that the phase difference is greater; here, the resultant amplitude is smaller than the amplitudes of the component waves. A maximum amplitude occurs when the

two waves are completely in phase — a situation that occurs whenever the phase difference between waves $(\phi_1 - \phi_2)$ is 0°, 360°, or an integer multiple of 360°. Under these circumstances, maximum *constructive interference* is said to occur. A maximum *destructive interference* occurs when $(\phi_1 - \phi_2)$ is equal to 180° or 180° plus an integer multiple of 360°. Interference plays an important role in many instrumental methods based on electromagnetic radiation.

Figure 6-5 depicts the superposition of two waves with identical amplitudes but different frequencies. The resulting wave is no longer sinusoidal but does exhibit a periodicity, or *beat*. Note that the period of the beat p_b is the reciprocal of the frequency difference $\Delta\nu$ between the two waves. That is,

$$p_b = \frac{1}{\Delta\nu} = \frac{1}{(\nu_2 - \nu_1)} \tag{6-7}$$

An important aspect of superposition is that a complex waveform can be broken down into simple components by a mathematical operation called the *Fourier transformation*. Jean Fourier, French mathematician (1768–1830), demonstrated that any periodic function, regardless of complexity, can be described by a sum of simple sine or cosine terms. For example, the square waveform widely encountered in electronics can be described by an equation with the form

$$y = A\left(\sin 2\pi\nu t + \frac{1}{3}\sin 6\pi\nu t \right.$$
$$\left. + \frac{1}{5}\sin 10\pi\nu t + \cdots + \frac{1}{n}\sin 2n\pi\nu t \right) \tag{6-8}$$

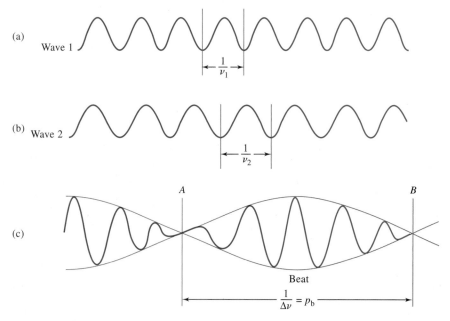

FIGURE 6-5 Superposition of two waves of different frequencies but identical amplitudes: (a) wave 1 with a period of $1/\nu_1$; (b) wave 2 with a period of $1/\nu_2$ ($\nu_2 = 1.25\nu_1$); (c) combined wave pattern. Note that superposition of ν_1 and ν_2 produces a beat pattern with a period of $1/\Delta\nu$ where $\Delta\nu = |\nu_1 - \nu_2|$.

where n takes values of 3, 5, 7, 9, 11, 13, and so forth. A graphical representation of the summation process is shown in Figure 6-6. The blue curve in Figure 6-6a is the sum of three sine waves that differ in amplitude in the ratio of 5:3:1 and in frequency in the ratio of 1:3:5. Note that the resultant approximates the shape of a square wave after including only three terms in Equation 6-8. As shown by the blue curve in Figure 6-6b, the resultant more closely approaches a square wave when nine waves are incorporated.

Decomposing a complex waveform into its sine or cosine components is tedious and time consuming when done by hand. Efficient software, however, makes routine Fourier transformations practical. The application of this technique was mentioned in Section 5C-2 and will be considered in the discussion of several types of spectroscopy.

6B-5 Diffraction of Radiation

All types of electromagnetic radiation exhibit *diffraction*, a process in which a parallel beam of radiation is bent as it passes by a sharp barrier or through a narrow opening. Figure 6-7 illustrates the process. Diffraction is a wave property that can be observed not only for electromagnetic radiation but also for mechanical or acoustical waves. For example, diffraction is easily demonstrated in the laboratory by mechanically generating waves of constant frequency in a tank of water and observing the wave crests before and after they pass through a rectangular opening, or slit. When the slit is wide relative to the wavelength (Figure 6-7a), diffraction is slight and difficult to detect. On the other hand, when the wavelength and the slit opening are of the same order of magnitude, as in Figure 6-7b, diffraction becomes pronounced. Here, the slit behaves as a new source from which waves radiate in a series of nearly 180° arcs. Thus, the direction of the wave front appears to bend as a consequence of passing the two edges of the slit.

Diffraction is a consequence of *interference*. This relationship is most easily understood by considering an experiment, performed first by Thomas Young in 1800, in which the wave nature of light was demonstrated unambiguously. As shown in Figure 6-8a, a parallel beam of light is allowed to pass through a narrow slit A (or in Young's experiment, a pinhole) whereupon it is diffracted and illuminates more or less equally two closely spaced slits or pinholes B and C; the radiation emerging from these slits is then observed on the

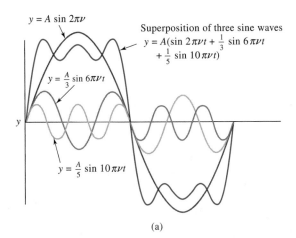

$y = A \sin 2\pi\nu$

Superposition of three sine waves
$y = A(\sin 2\pi\nu t + \frac{1}{3} \sin 6\pi\nu t + \frac{1}{5} \sin 10\pi\nu t)$

$y = \frac{A}{3} \sin 6\pi\nu t$

$y = \frac{A}{5} \sin 10\pi\nu t$

(a)

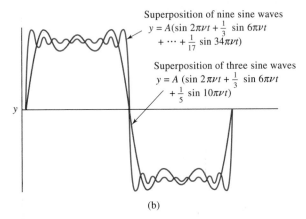

Superposition of nine sine waves
$y = A(\sin 2\pi\nu t + \frac{1}{3} \sin 6\pi\nu t + \cdots + \frac{1}{17} \sin 34\pi\nu t)$

Superposition of three sine waves
$y = A(\sin 2\pi\nu t + \frac{1}{3} \sin 6\pi\nu t + \frac{1}{5} \sin 10\pi\nu t)$

(b)

FIGURE 6-6 Superposition of sine waves to form a square wave: (a) combination of three sine waves; (b) combination of three, as in (a), and nine sine waves.

screen lying in a plane *XY*. If the radiation is monochromatic, a series of dark and light images perpendicular to the plane of the page is observed.

Figure 6-8b is a plot of the intensities of the bands as a function of distance along the length of the screen. If, as in this diagram, the slit widths approach the wavelength of radiation, the band intensities decrease only gradually with increasing distances from the central band. With wider slits, the decrease is much more pronounced.

In Figure 6-8a, the appearance of the central band *E*, which lies in the shadow of the opaque material separating the two slits, is readily explained by noting that the paths from *B* to *E* and *C* to *E* are identical. Thus, constructive interference of the diffracted rays from the two slits occurs, and an intense band is observed. With the aid of Figure 6-8c, the conditions for maximum constructive interference, which result in the other light

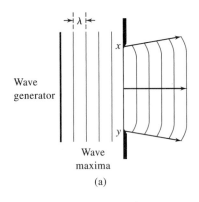

Wave generator

Wave maxima

(a)

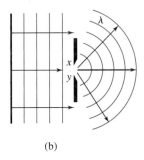

(b)

FIGURE 6-7 Propagation of waves through a slit: (a) $xy \gg \lambda$; (b) $xy = \lambda$.

bands, can be derived. In Figure 6-8c, the angle of diffraction θ is the angle formed by the lines *OE* (the normal) and *OD*, where *D* is the point of maximum intensity. The black lines *BD* and *CD* represent the light paths from the slits *B* and *C* to this point. Ordinarily, the distance $\overline{OE}$ is enormous compared to the distance between the slits $\overline{BC}$. As a consequence, the lines *BD*, *OD*, and *CD* are, for all practical purposes, parallel. Line *BF* is perpendicular to *CD* and forms the triangle *BCF*, which is, to a close approximation, similar to *DOE*; consequently, the angle *CBF* is equal to the angle of diffraction θ. We may then write

$$\overline{CF} = \overline{BC} \sin \theta$$

Because *BC* is so very small compared to $\overline{OE}$, $\overline{FD}$ closely approximates $\overline{BD}$, and the distance $\overline{CF}$ is a good measure of the difference in path lengths of beams *BD* and *CD*. For the two beams to be in phase at *D*, it is necessary that $\overline{CF}$ correspond to the wavelength of the radiation; that is,

$$\lambda = \overline{CF} = \overline{BC} \sin \theta$$

Reinforcement also occurs when the additional path length corresponds to 2λ, 3λ, and so forth. Thus,

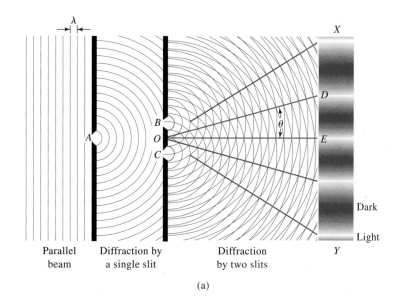

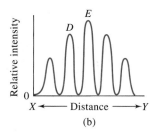

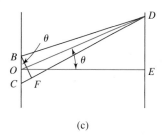

FIGURE 6-8 Diffraction of monochromatic radiation by slits.

a more general expression for the light bands surrounding the central band is

$$\mathbf{n}\lambda = \overline{BC} \sin \theta \qquad (6\text{-}9)$$

where **n** is an integer called the *order* of interference.

The linear displacement $\overline{DE}$ of the diffracted beam along the plane of the screen is a function of the distance $\overline{OE}$ between the screen and the plane of the slits, as well as the spacing between the slits, and is given by

$$\overline{DE} = \overline{OD} \sin \theta$$

Substitution into Equation 6-9 gives

$$\mathbf{n}\lambda = \frac{\overline{BC}\,\overline{DE}}{\overline{OD}} = \frac{\overline{BC}\,\overline{DE}}{\overline{OE}} \qquad (6\text{-}10)$$

Equation 6-10 permits the calculation of the wavelength from the three measurable quantities.

 Simulation: Learn more about **double-slit diffraction**.

EXAMPLE 6-1

Suppose that the screen in Figure 6-8 is 2.00 m from the plane of the slits and that the slit spacing is 0.300 mm. What is the wavelength of radiation if the fourth band is located 15.4 mm from the central band?

Solution

Substituting into Equation 6-10 gives

$$4\lambda = \frac{0.300 \text{ mm} \times 15.4 \text{ mm}}{2.00 \text{ m} \times 1000 \text{ mm/m}} = 0.00231 \text{ mm}$$

$$\lambda = 5.78 \times 10^{-4} \text{ mm} = 578 \text{ nm}$$

6B-6 Coherent Radiation

To produce a diffraction pattern such as that shown in Figure 6-8a, it is necessary that the electromagnetic waves that travel from slits *B* and *C* to any given point on the screen (such as *D* or *E*) have sharply defined phase differences that remain *entirely* constant with time; that is, the radiation from slits *B* and *C* must be *coherent*. The conditions for coherence are that (1) the two sources of radiation must have identical frequencies (or sets of frequencies) and (2) the phase relationships between the two beams must remain constant with time. The necessity for these requirements can be demonstrated by illuminating the two slits in Figure 6-8a with individual tungsten lamps. Under this circumstance, the well-defined light and dark patterns disappear and are replaced by a more or less uniform illumination of the screen. This behavior is a consequence of the *incoherent* character of filament sources (many other sources of electromagnetic radiation are incoherent as well).

With incoherent sources, light is emitted by individual atoms or molecules, and the resulting beam is the summation of countless individual events, each of which lasts on the order of 10^{-8} s. Thus, a beam of radiation from this type of source is not continuous but instead is composed of a series of *wave trains* that are a few meters in length at most. Because the processes that produce trains are random, the phase differences among the trains must also be variable. A wave train from slit *B* may arrive at a point on the screen in phase with a wave train from *C* so that constructive interference occurs; an instant later, the trains may be totally out of phase at the same point, and destructive interference occurs. Thus, the radiation at all points on the screen is governed by the random phase variations among the wave trains; uniform illumination, which represents an average for the trains, is the result.

There are sources that produce electromagnetic radiation in the form of trains with essentially infinite length and constant frequency. Examples include radio-frequency oscillators, microwave sources, and optical lasers. Various mechanical sources, such as a two-pronged vibrating tapper in a water-containing ripple tank, produce a mechanical analog of coherent radiation. When two coherent sources are substituted for slit *A* in the experiment shown in Figure 6-8a, a regular diffraction pattern is observed.

Diffraction patterns can be obtained from random sources, such as tungsten filaments, provided that an arrangement similar to that shown in Figure 6-8a is employed. Here, the very narrow slit *A* assures that the radiation reaching *B* and *C* emanates from the same small region of the source. Under this circumstance, the various wave trains that exit from slits *B* and *C* have a constant set of frequencies and phase relationships to one another and are thus coherent. If the slit

at *A* is widened so that a larger part of the source is sampled, the diffraction pattern becomes less pronounced because the two beams are only partially coherent. If slit *A* is made sufficiently wide, the incoherence may become great enough to produce only a constant illumination across the screen.

6B-7 Transmission of Radiation

Experimental observations show that the rate radiation propagates through a transparent substance is less than its velocity in a vacuum and depends on the kinds and concentrations of atoms, ions, or molecules in the medium. It follows from these observations that the radiation must interact in some way with the matter. Because a frequency change is not observed, however, the interaction *cannot* involve a permanent energy transfer.

The *refractive index* of a medium is one measure of its interaction with radiation and is defined by

$$n_i = \frac{c}{v_i} \qquad (6\text{-}11)$$

where n_i is the refractive index at a specified frequency i, v_i is the velocity of the radiation in the medium, and c is its velocity in a vacuum. The refractive index of most liquids lies between 1.3 and 1.8; it is 1.3 to 2.5 or higher for solids.[4]

The interaction involved in transmission can be ascribed to periodic *polarization* of the atomic and molecular species that make up the medium. Polarization in this context means the temporary deformation of the electron clouds associated with atoms or molecules that is brought about by the alternating electromagnetic field of the radiation. Provided that the radiation is not absorbed, the energy required for polarization is only momentarily retained (10^{-14} to 10^{-15} s) by the species and is reemitted without alteration as the substance returns to its original state. Since there is no net energy change in this process, the frequency of the emitted radiation is unchanged, but the rate of its propagation is slowed by the time that is required for retention and reemission to occur. Thus, transmission through a medium can be viewed as a stepwise process that involves polarized atoms, ions, or molecules as intermediates.

Radiation from polarized particles should be emitted in all directions in a medium. If the particles are small, however, it can be shown that destructive

---

[4]For a more complete discussion of refractive index measurement, see T. M. Niemczyk, in *Physical Methods in Modern Clinical Analysis*, T. Kuwana, Ed., Vol. 2, pp. 337–400. New York: Academic, 1980.

interference prevents the propagation of significant amounts in any direction other than that of the original light path. On the other hand, if the medium contains large particles (such as polymer molecules or colloidal particles), this destructive interference is incomplete, and a portion of the beam is scattered in all directions as a consequence of the interaction step. Scattering is considered in Section 6B-10.

Since the velocity of radiation in matter is wavelength dependent and since c in Equation 6-11 is independent of wavelength, the refractive index of a substance must also change with wavelength. The variation in refractive index of a substance with wavelength or frequency is called its *dispersion*. The dispersion of a typical substance is shown in Figure 6-9. The intricacy of the curve indicates that the relationship is complex; generally, however, dispersion plots exhibit two types of regions. In the *normal dispersion* region, there is a gradual increase in refractive index with increasing frequency (or decreasing wavelength). Regions of *anomalous dispersion* are frequency ranges in which sharp changes in refractive index occur. Anomalous dispersion always occurs at frequencies that correspond to the natural harmonic frequency associated with some part of a molecule, atom, or ion of the substance. At such a frequency, permanent energy transfer from the radiation to the substance occurs and *absorption* of the beam is observed. Absorption is discussed in Section 6C-5.

Dispersion curves are important when choosing materials for the optical components of instruments. A substance that exhibits normal dispersion over the wavelength region of interest is most suitable for the manufacture of lenses, for which a high and relatively constant refractive index is desirable. Chromatic aberrations (formation of colored images) are minimized through the choice of such a material. In contrast, a substance with a refractive index that is not only large but also highly frequency dependent is selected for the fabrication of prisms. The applicable wavelength region for the prism thus approaches the anomalous dispersion region for the material from which it is fabricated.

6B-8 Refraction of Radiation

When radiation passes at an angle through the interface between two transparent media that have different densities, an abrupt change in direction, or *refraction*, of the beam is observed as a consequence of a difference in velocity of the radiation in the two media. When the beam passes from a less dense to a more dense environment, as in Figure 6-10, the bending is toward the normal to the interface. Bending away from the normal occurs when the beam passes from a more dense to a less dense medium.

The extent of refraction is given by Snell's law:

$$\frac{\sin \theta_1}{\sin \theta_2} = \frac{n_2}{n_1} = \frac{v_2}{v_1} \tag{6-12}$$

If M_1 in Figure 6-10 is a *vacuum*, v_1 is equal to c, and n_1 is unity (see Equation 6-11); with rearrangement, Equation 6-12 simplifies to

$$(n_2)_{\text{vac}} = \frac{(\sin \theta_1)_{\text{vac}}}{\sin \theta_2} \tag{6-13}$$

The refractive indexes of substance M_2 can then be computed from measurements of $(\theta_1)_{\text{vac}}$ and θ_2. For convenience, refractive indexes are usually measured

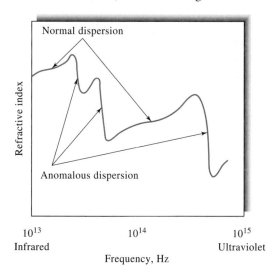

FIGURE 6-9 A typical dispersion curve.

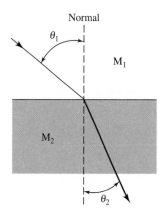

FIGURE 6-10 Refraction of light in passing from a less dense medium M_1 into a more dense medium M_2, where its velocity is lower.

and reported with air, rather than a vacuum, as the reference. The refractive index is then

$$(n_2)_{air} = \frac{(\sin \theta_1)_{air}}{\sin \theta_2} \qquad (6\text{-}14)$$

Most compilations of refractive indexes provide data in terms of Equation 6-14. Such data are readily converted to refractive indexes with vacuum as a reference by multiplying by the refractive index of air relative to a vacuum. That is,

$$n_{vac} = 1.00027 n_{air}$$

This conversion is seldom necessary.

6B-9 Reflection of Radiation

When radiation crosses an interface between media that differ in refractive index, reflection always occurs. The fraction of reflected radiation becomes greater with increasing differences in refractive index. For a beam that enters an interface at right angles, the fraction reflected is given by

$$\frac{I_r}{I_0} = \frac{(n_2 - n_1)^2}{(n_2 + n_1)^2} \qquad (6\text{-}15)$$

where I_0 is the intensity of the incident beam and I_r is the reflected intensity; n_1 and n_2 are the refractive indexes of the two media.

EXAMPLE 6-2

Calculate the percentage loss of intensity due to reflection of a perpendicular beam of yellow light as it passes through a glass cell that contains water. Assume that for yellow radiation the refractive index of glass is 1.50, of water is 1.33, and of air is 1.00.

Solution

The total reflective loss will be the sum of the losses occurring at each of the interfaces. For the first interface (air to glass), we can write

$$\frac{I_{r1}}{I_0} = \frac{(1.50 - 1.00)^2}{(1.50 + 1.00)^2} = 0.040$$

The beam intensity is reduced to $(I_0 - 0.040I_0) = 0.960I_0$. Reflection loss at the glass-to-water interface is then given by

$$\frac{I_{r2}}{0.960I_0} = \frac{(1.50 - 1.33)^2}{(1.50 + 1.33)^2} = 0.0036$$

$$I_{r2} = 0.0035I_0$$

The beam intensity is further reduced to $(0.960I_0 - 0.0035I_0) = 0.957I_0$. At the water-to-glass interface

$$\frac{I_{r3}}{0.957I_0} = \frac{(1.50 - 1.33)^2}{(1.50 + 1.33)^2} = 0.0036$$

$$I_{r3} = 0.0035I_0$$

and the beam intensity becomes $0.953I_0$. Finally, the reflection at the second glass-to-air interface will be

$$\frac{I_{r4}}{0.953I_0} = \frac{(1.50 - 1.00)^2}{(1.50 + 1.00)^2} = 0.0400$$

$$I_{r4} = 0.038I_0$$

The total reflection loss I_{rt} is

$$I_{rt} = 0.040I_0 + 0.0035I_0 + 0.0035I_0 + 0.038I_0$$
$$= 0.085I_0$$

and

$$\frac{I_{rt}}{I_0} = 0.85 \quad \text{or} \quad 8.5\%$$

In later chapters we show that losses such as those shown in Example 6-2 are of considerable significance in various optical instruments.

Reflective losses at a polished glass or quartz surface increase only slightly as the angle of the incident beam increases up to about 60°. Beyond this angle, however, the percentage of radiation that is reflected increases rapidly and approaches 100% at 90°, or grazing incidence.

6B-10 Scattering of Radiation

As noted earlier, the transmission of radiation in matter can be pictured as a momentary retention of the radiant energy by atoms, ions, or molecules followed by reemission of the radiation in all directions as the particles return to their original state. With atomic or molecular particles that are small relative to the wavelength of the radiation, destructive interference removes most but not all of the reemitted radiation except the radiation that travels in the original direction of the beam; the path of the beam appears to be unaltered as a consequence of the interaction. Careful observation, however, reveals that a very small fraction of the radiation is transmitted at all angles from the original path and that the intensity of this *scattered radiation* increases with particle size.

Rayleigh Scattering

Scattering by molecules or aggregates of molecules with dimensions significantly smaller than the wavelength of the radiation is called *Rayleigh scattering*; its intensity is proportional to the inverse fourth power of the wavelength, the dimensions of the scattering particles, and the square of the polarizability of the particles. An everyday manifestation of Rayleigh scattering is the blue color of the sky, which results from greater scattering of the shorter wavelengths of the visible spectrum.

Scattering by Large Molecules

With large particles, scattering can be different in different directions (Mie scattering). Measurements of this type of scattered radiation are used to determine the size and shape of large molecules and colloidal particles (see Chapter 34).

Raman Scattering

The Raman scattering effect differs from ordinary scattering in that part of the scattered radiation suffers quantized frequency changes. These changes are the result of vibrational energy level transitions that occur in the molecules as a consequence of the polarization process. Raman spectroscopy is discussed in Chapter 18.

6B-11 Polarization of Radiation

Ordinary radiation consists of a bundle of electromagnetic waves in which the vibrations are equally distributed among a huge number of planes centered along the path of the beam. Viewed end on, a beam of monochromatic radiation can be visualized as an infinite set of electric vectors that fluctuate in length from zero to a maximum amplitude A. Figure 6-11b depicts an end-on view of these vectors at various times during the passage of one wave of monochromatic radiation through a fixed point in space (Figure 6-11a).

Figure 6-12a shows a few of the vectors depicted in Figure 6-11b at the instant the wave is at its maximum. The vector in any one plane, say XY as depicted in Figure 6-12a, can be resolved into two mutually perpendicular components AB and CD as shown in Figure 6-12b. If the two components for all of the planes shown in Figure 6-12a are combined, the resultant has the appearance shown in Figure 6-12c. Removal of one of the two resultant planes of vibration in Figure 6-12c produces a beam that is *plane polarized*. The resultant electric vector of a plane-polarized beam then occupies

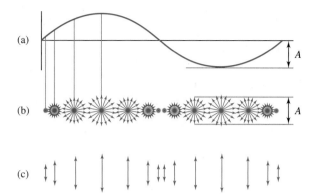

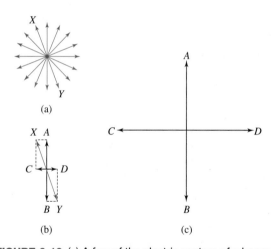

FIGURE 6-11 Unpolarized and plane-polarized radiation: (a) cross-sectional view of a beam of monochromatic radiation, (b) successive end-on view of the radiation in (a) if it is unpolarized, (c) successive end-on views of the radiation of (a) if it is plane polarized on the vertical axis.

FIGURE 6-12 (a) A few of the electric vectors of a beam traveling perpendicular to the page. (b) The resolution of a vector in a plane XY into two mutually perpendicular components. (c) The resultant when all vectors are resolved (not to scale).

a single plane. Figure 6-11c shows an end-on view of a beam of plane-polarized radiation after various time intervals.

Plane-polarized electromagnetic radiation is produced by certain radiant energy sources. For example, the radio waves emanating from an antenna and the microwaves produced by a klystron tube are both plane polarized. Visible and ultraviolet radiation from relaxation of a single excited atom or molecule is also polarized, but the beam from such a source has no net polarization since it is made up of a multitude of individual wave trains produced by an enormous number of individual atomic or molecular events. The plane of polarization of these individual waves is random so that their individual polarizations cancel.

Polarized ultraviolet and visible radiation is produced by passage of radiation through media that selectively absorb, reflect, or refract radiation that vibrates in only one plane.

6C QUANTUM-MECHANICAL PROPERTIES OF RADIATION

When electromagnetic radiation is emitted or absorbed, a permanent transfer of energy from the emitting object or to the absorbing medium occurs. To describe these phenomena, it is necessary to treat electromagnetic radiation not as a collection of waves but rather as a stream of discrete particles called *photons* or *quanta*. The need for a particle model for radiation became apparent as a consequence of the discovery of the photoelectric effect in the late nineteenth century.

6C-1 The Photoelectric Effect

The first observation of the photoelectric effect was made in 1887 by Heinrich Hertz, who reported that a spark jumped more readily between two charged spheres when their surfaces were illuminated with light. Between the time of this observation and the theoretical explanation of the photoelectric effect by Einstein in 1905, several important studies of the photoelectric effect were performed with what is now known as a vacuum phototube. Einstein's explanation of the photoelectric effect was both simple and elegant but was far enough ahead of its time that it was not generally accepted until 1916, when Millikan's systematic studies confirmed the details of Einstein's theoretical conclusions.

Figure 6-13 is a schematic of a vacuum phototube circuit similar to the one used by Millikan to study the photoelectric effect. The surface of the large *photocathode* on the left usually is coated with an alkali metal or one of its compounds, but other metals may be used. When monochromatic radiation impinges on the photocathode, electrons are emitted from its surface with a range of kinetic energies. As long as the voltage V applied between the anode and the cathode is positive, the electrons are drawn from left to right

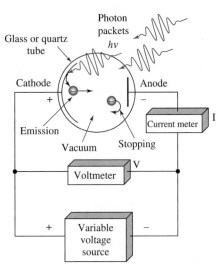

FIGURE 6-13 Apparatus for studying the photoelectric effect. Photons enter the phototube, strike the cathode, and eject electrons. The photoelectrons are attracted to the anode when it is positive with respect to the cathode. When the anode is negative as shown, the electrons are "stopped," and no current passes. The negative voltage between the anode and the cathode when the current is zero is the stopping potential.

through the phototube to produce a current I in the circuit. When the voltage across the phototube is adjusted so that the anode is slightly negative with respect to the cathode, the photoelectrons are repelled by the anode, and the photocurrent decreases as expected. At this point in the experiment, however, some of the electrons have sufficient kinetic energy to overcome the negative potential applied to the anode, and a current is still observed.

This experiment may be repeated for phototubes with different materials coating the photocathode. In each experiment, the photocurrent is measured as a function of the applied voltage, and the voltage V_0 at which the photocurrent becomes precisely zero is noted. The negative voltage at which the photocurrent is zero is called the *stopping voltage*. It corresponds to the potential at which the most energetic electrons from the cathode are just repelled from the anode. If we multiply the stopping voltage by the charge on the electron, $e = 1.60 \times 10^{-19}$ coulombs, we have a measure of the kinetic energy in joules of the *most energetic* of the emitted electrons. When this experiment

 Tutorial: Learn more about the **photoelectric effect**.

is repeated for various frequencies of monochromatic light, the following results are observed:

1. When light of constant frequency is focused on the anode at low applied negative potential, the photocurrent is directly proportional to the intensity of the incident radiation.
2. The magnitude of the stopping voltage depends on the frequency of the radiation impinging on the photocathode.
3. The stopping voltage depends on the chemical composition of the coating on the photocathode.

4. The stopping voltage is *independent of the intensity of the incident radiation.*

These observations suggest that electromagnetic radiation is a form of energy that releases electrons from metallic surfaces and imparts to these electrons sufficient kinetic energy to cause them to travel to a negatively charged electrode. Furthermore, the number of photoelectrons released is proportional to the intensity of the incident beam.

The results of these experiments are shown in the plots of Figure 6-14, in which the maximum kinetic

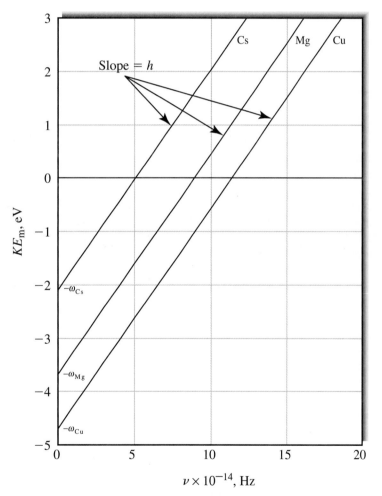

FIGURE 6-14 Maximum kinetic energy of photoelectrons emitted from three metal surfaces as a function of radiation frequency. The y-intercepts $(-\omega)$ are the work functions for each metal. If incident photons do not have energies of at least $h\nu = \omega$, no photoelectrons are emitted from the photocathode.

energy, or stopping energy, $KE_m = eV_0$ of the photoelectrons is plotted against frequency for photocathode surfaces of magnesium, cesium, and copper. Other surfaces give plots with identical slopes, h, but different intercepts, ω. The plots shown in Figure 6-14 are described by the equation

$$KE_m = h\nu - \omega \qquad (6\text{-}16)$$

In this equation, the slope h is Planck's constant, which is equal to 6.6254×10^{-34} joule second, and the intercept $-\omega$ is the *work function*, a constant that is characteristic of the surface material and represents the minimum energy binding electron in the metal. Approximately a decade before Millikan's work that led to Equation 6-16, Einstein had proposed the relationship between frequency ν of light and energy E as embodied by the now famous equation

$$E = h\nu \qquad (6\text{-}17)$$

By substituting this equation into Equation 6-16 and rearranging, we obtain

$$E = h\nu = KE_m + \omega \qquad (6\text{-}18)$$

This equation shows that the energy of an incoming photon is equal to the kinetic energy of the ejected photoelectron plus the energy required to eject the photoelectron from the surface being irradiated.

The photoelectric effect cannot be explained by a classical wave model but requires instead a quantum model, in which radiation is viewed as a stream of discrete bundles of energy, or photons as depicted in Figure 6-13. For example, calculations indicate that no single electron could acquire sufficient energy for ejection if the radiation striking the surface were uniformly distributed over the face of the electrode as it is in the wave model; nor could any electron accumulate enough energy rapidly enough to establish the nearly instantaneous currents that are observed. Thus, it is necessary to assume that the energy is not uniformly distributed over the beam front but rather is concentrated in packets, or bundles of energy.

Equation 6-18 can be recast in terms of wavelength by substitution of Equation 6-2. That is,

$$E = h\frac{c}{\lambda} = KE_m - \omega \qquad (6\text{-}19)$$

Note that although photon energy is directly proportional to frequency, it is a reciprocal function of wavelength.

EXAMPLE 6-3

Calculate the energy of (a) a 5.3-Å X-ray photon and (b) a 530-nm photon of visible radiation.

$$E = h\nu = \frac{hc}{\lambda}$$

Solution

(a) $E = \dfrac{(6.63 \times 10^{-34}\,\text{J}\cdot\text{s}) \times (3.00 \times 10^{8}\,\text{m/s})}{5.30\,\text{Å} \times (10^{-10}\,\text{m/Å})}$

$\qquad = 3.75 \times 10^{-16}\,\text{J}$

The energy of radiation in the X-ray region is commonly expressed in electron volts, the energy acquired by an electron that has been accelerated through a potential of one volt. In the conversion table inside the front cover of this book, we see that $1\,\text{J} = 6.24 \times 10^{18}\,\text{eV}$.

$$E = 3.75 \times 10^{-16}\,\text{J} \times (6.24 \times 10^{18}\,\text{eV/J})$$

$$= 2.34 \times 10^{3}\,\text{eV}$$

(b) $E = \dfrac{(6.63 \times 10^{-34}\,\text{J}\cdot\text{s}) \times (3.00 \times 10^{8}\,\text{m/s})}{530\,\text{nm} \times (10^{-9}\,\text{m/nm})}$

$\qquad = 3.75 \times 10^{-19}\,\text{J}$

Energy of radiation in the visible region is often expressed in kJ/mol rather than kJ/photon to aid in the discussion of the relationships between the energy of absorbed photons and the energy of chemical bonds.

$$E = 3.75 \times 10^{-19}\,\frac{\text{J}}{\text{photon}}$$

$$\times \frac{(6.02 \times 10^{23}\,\text{photons})}{\text{mol}} \times 10^{-3}\,\frac{\text{kJ}}{\text{J}}$$

$$= 226\,\text{kJ/mol}$$

6C-2 Energy States of Chemical Species

The quantum theory was first proposed in 1900 by Max Planck, a German physicist, to explain the properties of radiation emitted by heated bodies. The theory was later extended to rationalize other types of emission and absorption processes. Two important postulates of quantum theory include the following:

1. Atoms, ions, and molecules can exist only in certain discrete states, characterized by definite amounts of energy. When a species changes its state, it absorbs or emits an amount of energy *exactly* equal to the energy difference between the states.

2. When atoms, ions, or molecules absorb or emit radiation in making the transition from one energy state to another, the frequency ν or the wavelength λ of the radiation is related to the energy difference between the states by the equation

$$E_1 - E_0 = h\nu = \frac{hc}{\lambda} \qquad (6\text{-}20)$$

where E_1 is the energy of the higher state and E_0 the energy of the lower state. The terms c and h are the speed of light and the Planck constant, respectively.

For atoms or ions in the elemental state, the energy of any given state arises from the motion of electrons around the positively charged nucleus. As a consequence the various energy states are called *electronic states*. In addition to having electronic states, molecules also have quantized *vibrational states* that are associated with the energy of interatomic vibrations and quantized *rotational states* that arise from the rotation of molecules around their centers of mass.

The lowest energy state of an atom or molecule is its *ground state*. Higher energy states are termed *excited states*. Generally, at room temperature chemical species are in their ground state.

6C-3 Interactions of Radiation and Matter

Spectroscopists use the interactions of radiation with matter to obtain information about a sample. Several of the chemical elements were discovered by spectroscopy. The sample is usually stimulated by applying energy in the form of heat, electrical energy, light, particles, or a chemical reaction. Prior to applying the stimulus, the analyte is predominantly in its lowest energy state, or *ground state*. The stimulus then causes some of the analyte species to undergo a transition to a higher energy, or *excited state*. We acquire information about the analyte by measuring the electromagnetic radiation emitted as it returns to the ground state or by measuring the amount of electromagnetic radiation absorbed or scattered as a result of excitation.

Figure 6-15 illustrates the processes involved in emission and chemiluminescence spectroscopy. Here,

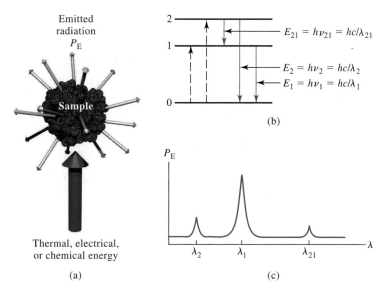

FIGURE 6-15 Emission or chemiluminescence processes. In (a), the sample is excited by the application of thermal, electrical, or chemical energy. These processes do not involve radiant energy and are hence called nonradiative processes. In the energy level diagram (b), the dashed lines with upward-pointing arrows symbolize these nonradiative excitation processes, while the solid lines with downward-pointing arrows indicate that the analyte loses its energy by emission of a photon. In (c), the resulting spectrum is shown as a measurement of the radiant power emitted P_E as a function of wavelength, λ.

the analyte is stimulated by heat or electrical energy or by a chemical reaction. *Emission spectroscopy* usually involves methods in which the stimulus is heat or electrical energy, and *chemiluminescence spectroscopy* refers to excitation of the analyte by a chemical reaction. In both cases, measurement of the radiant power emitted as the analyte returns to the ground state can give information about its identity and concentration. The results of such a measurement are often expressed graphically by a *spectrum*, which is a plot of the emitted radiation as a function of frequency or wavelength.

When the sample is stimulated by application of an external electromagnetic radiation source, several processes are possible. For example, the radiation can be reflected (Section 6B-9), scattered (Section 6B-10), or absorbed (Section 6C-5). When some of the incident radiation is absorbed, it promotes some of the analyte species to an excited state, as shown in Figure 6-16. In *absorption spectroscopy*, we measure the amount of light absorbed as a function of wavelength. This can give both qualitative and quantitative information about the sample. In *photoluminescence spectroscopy* (Figure 6-17), the emission of photons is measured after absorption. The most important forms of photoluminescence for analytical purposes are *fluorescence* and *phosphorescence spectroscopy*.

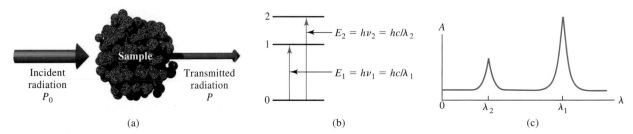

FIGURE 6-16 Absorption methods. Radiation of incident radiant power P_0 can be absorbed by the analyte, resulting in a transmitted beam of lower radiant power P. For absorption to occur, the energy of the incident beam must correspond to one of the energy differences shown in (b). The resulting absorption spectrum is shown in (c).

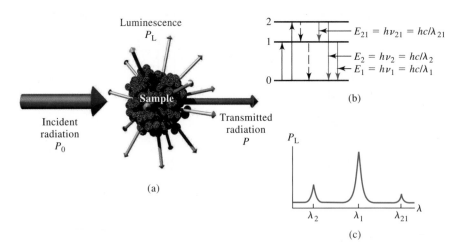

FIGURE 6-17 Photoluminescence methods (fluorescence and phosphorescence). Fluorescence and phosphorescence result from absorption of electromagnetic radiation and then dissipation of the energy emission of radiation (a). In (b), the absorption can cause excitation of the analyte to state 1 or state 2. Once excited, the excess energy can be lost by emission of a photon (luminescence, shown as solid line) or by nonradiative processes (dashed lines). The emission occurs over all angles, and the wavelengths emitted (c) correspond to energy differences between levels. The major distinction between fluorescence and phosphorescence is the time scale of emission, with fluorescence being prompt and phosphorescence being delayed.

When radiation is scattered, the interaction of the incoming radiation with the sample may be elastic or inelastic. In elastic scattering, the wavelength of the scattered radiation is the same as that of the source radiation. The intensity of the elastically scattered radiation is used to make measurements in *nephelometry* and *turbidimetry*, and particle sizing. *Raman spectroscopy*, which is mentioned briefly in Section 6B-10 and is discussed in detail in Chapter 18, uses inelastic scattering to produce a vibrational spectrum of sample molecules, as illustrated in Figure 6-18. In this type of spectroscopic analysis, the intensity of the scattered radiation is recorded as a function of the frequency shift of the incident radiation. The intensity of Raman peaks is related to the concentration of the analyte.

 Simulation: Learn more about the **interaction of radiation with matter**.

6C-4 Emission of Radiation

Electromagnetic radiation is produced when excited particles (atoms, ions, or molecules) relax to lower energy levels by giving up their excess energy as photons. Excitation can be brought about by a variety of means, including (1) bombardment with electrons or other elementary particles, which generally leads to the emission of X-radiation; (2) exposure to an electric current, an ac spark, or an intense heat source (flame, dc arc, or furnace), producing ultraviolet, visible, or infrared radiation; (3) irradiation with a beam of electromagnetic radiation, which produces fluorescence radiation; and (4) an exothermic chemical reaction that produces chemiluminescence.

Radiation from an excited source is conveniently characterized by means of an *emission spectrum*, which usually takes the form of a plot of the relative power of the emitted radiation as a function of wavelength or

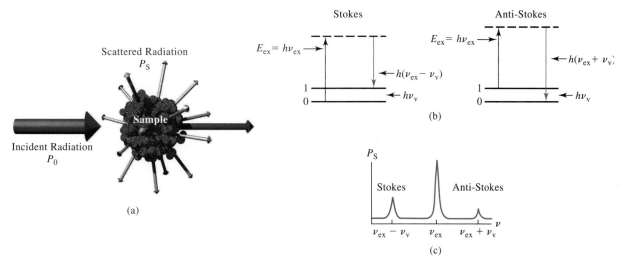

FIGURE 6-18 Inelastic scattering in Raman spectroscopy. (a) As incident radiation of frequency ν_{ex} impinges on the sample, molecules of the sample are excited from one of their ground vibrational states to a higher so-called *virtual state*, indicated by the dashed level in (b). When the molecule relaxes, it may return to the first vibrational state as indicated and emit a photon of energy $E = h(\nu_{ex} - \nu_v)$ where ν_v is the frequency of the vibrational transition. Alternatively, if the molecule is in the first excited vibrational state, it may absorb a quantum of the incident radiation, be excited to the virtual state, and relax back to the ground vibrational state. This process produces an emitted photon of energy $E = h(\nu_{ex} + \nu_v)$. In both cases, the emitted radiation differs in frequency from the incident radiation by the vibrational frequency of the molecule ν_v. (c) The spectrum resulting from the inelastically scattered radiation shows three peaks: one at $\nu_{ex} - \nu_v$ (Stokes), a second intense peak at ν_{ex} for radiation that is scattered without a frequency change, and a third (anti-Stokes) at $\nu_{ex} + \nu_v$. The intensities of the Stokes and anti-Stokes peaks give quantitative information, and the positions of the peaks give qualitative information about the sample molecule.

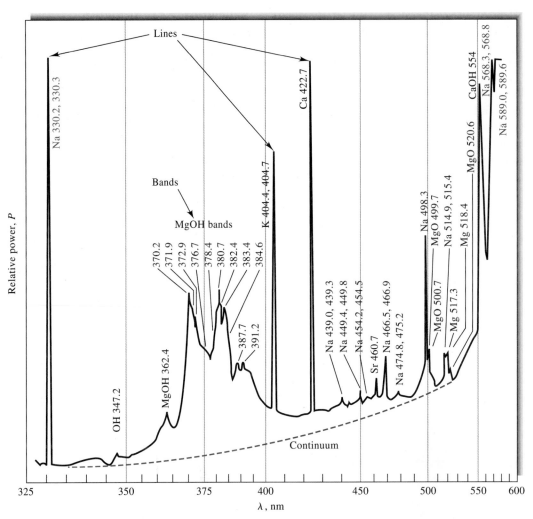

FIGURE 6-19 Emission spectrum of a brine sample obtained with an oxyhydrogen flame. The spectrum consists of the superimposed line, band, and continuum spectra of the constituents of the sample. The characteristic wavelengths of the species contributing to the spectrum are listed beside each feature. (R. Hermann and C. T. J. Alkemade, *Chemical Analysis by Flame Photometry,* 2nd ed., p. 484. New York: Interscience, 1979.)

frequency. Figure 6-19 illustrates a typical emission spectrum, which was obtained by aspirating a brine solution into an oxyhydrogen flame. Three types of spectra appear in the figure: *lines*, *bands*, and a *continuum*. The line spectrum is made up of a series of sharp, well-defined peaks caused by excitation of individual atoms. The band spectrum consists of several groups of lines so closely spaced that they are not completely resolved. The source of the bands consists of small molecules or radicals. Finally, the continuum portion of the spectrum is responsible for the increase in the background that is evident above about 350 nm. The line and band spectra

are superimposed on this continuum. The source of the continuum is described on page 152.

Figure 6-20 is an X-ray emission spectrum produced by bombarding a piece of molybdenum with an energetic stream of electrons. Note the line spectrum superimposed on the continuum. The source of the continuum is described in Section 12A-1.

Line Spectra

Line spectra in the ultraviolet and visible regions are produced when the radiating species are individual atomic particles that are well separated in the gas

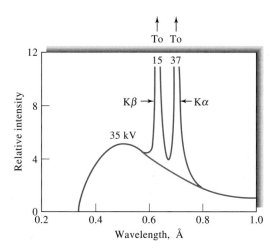

FIGURE 6-20 X-ray emission spectrum of molybdenum metal.

phase. The individual particles in a gas behave independently of one another, and the spectrum consists of a series of sharp lines with widths of about 10^{-5} nm (10^{-4} Å). In Figure 6-19, lines for gas-phase sodium, potassium, and calcium are identified.

The energy-level diagram in Figure 6-21 shows the source of two of the lines in a typical emission spectrum of an element. The horizontal line labeled E_0 corresponds to the lowest, or ground-state, energy of the atom. The horizontal lines labeled E_1 and E_2 are two higher-energy electronic levels of the species. For ex-

ample, the single outer electron in the ground state E_0 for a sodium atom is located in the $3s$ orbital. Energy level E_1 then represents the energy of the atom when this electron has been promoted to the $3p$ state by absorption of thermal, electrical, or radiant energy. The promotion is depicted by the shorter wavy arrow on the left in Figure 6-21a. After perhaps 10^{-8} s, the atom returns to the ground state, emitting a photon whose frequency and wavelength are given by Equation 6-20.

$$\nu_1 = (E_1 - E_0)/h$$
$$\lambda_1 = hc/(E_1 - E_0)$$

This emission process is illustrated by the shorter blue arrow on the right in Figure 6-21a.

For the sodium atom, E_2 in Figure 6-21 corresponds to the more energetic $4p$ state; the resulting emitted radiation λ_2 appears at a shorter wavelength or a higher frequency. The line at about 330 nm in Figure 6-19 results from this transition; the $3p$-to-$3s$ transition provides a line at about 590 nm.

X-ray line spectra are also produced by electronic transitions. In this case, however, the electrons involved are those in the innermost orbitals. Thus, in contrast to ultraviolet and visible emissions, the X-ray spectrum for an element is independent of its environment. For example, the emission spectrum for molybdenum is the same regardless of whether the sample being excited is molybdenum metal, solid molybdenum sulfide, gaseous molybdenum hexafluoride, or an aqueous solution of an anionic complex of the metal.

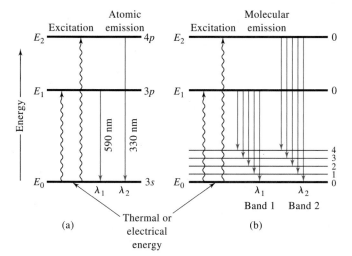

(a)

(b)

FIGURE 6-21 Energy-level diagrams for (a) a sodium atom showing the source of a line spectrum and (b) a simple molecule showing the source of a band spectrum.

Band Spectra

Band spectra are often encountered in spectral sources when gaseous radicals or small molecules are present. For example, in Figure 6-19 bands for OH, MgOH, and MgO are labeled and consist of a series of closely spaced lines that are not fully resolved by the instrument used to obtain the spectrum. Bands arise from numerous quantized vibrational levels that are superimposed on the ground-state electronic energy level of a molecule.

Figure 6-21b is a partial energy-level diagram for a molecule that shows its ground state E_0 and two of its excited electronic states, E_1 and E_2. A few of the many vibrational levels associated with the ground state are also shown. Vibrational levels associated with the two excited states have been omitted because the lifetime of an excited vibrational state is brief compared with that of an electronically excited state (about 10^{-15} s versus 10^{-8} s). A consequence of this tremendous difference in lifetimes is that when an electron is excited to one of the higher vibrational levels of an electronic state, relaxation to the lowest vibrational level of that state occurs before an electronic transition to the ground state can occur. Therefore, the radiation produced by the electrical or thermal excitation of polyatomic species nearly always results from a transition from the *lowest vibrational level of an excited electronic state* to any of the several vibrational levels of the ground state.

The mechanism by which a vibrationally excited species relaxes to the nearest electronic state involves a transfer of its excess energy to other atoms in the system through a series of collisions. As noted, this process takes place at an enormous speed. Relaxation from one electronic state to another can also occur by collisional transfer of energy, but the rate of this process is slow enough that relaxation by photon release is favored.

The energy-level diagram in Figure 6-21b illustrates the mechanism by which two radiation bands that consist of five closely spaced lines are emitted by a molecule excited by thermal or electrical energy. For a real molecule, the number of individual lines is much larger because in addition to the numerous vibrational states, a multitude of rotational states would be superimposed on each. The differences in energy among the rotational levels is perhaps an order of magnitude smaller than that for vibrational states. Thus, a real molecular band would be made up of many more lines than we have shown in Figure 6-21b, and these lines would be much more closely spaced.

Continuum Spectra

As shown in Figure 6-22, truly continuum radiation is produced when solids are heated to incandescence. Thermal radiation of this kind, which is called *blackbody radiation*, is characteristic of the temperature of the emitting surface rather than the material of which that surface is composed. Blackbody radiation is produced by the innumerable atomic and molecular oscillations excited in the condensed solid by the thermal energy. Note that the energy peaks in Figure 6-22 shift to shorter wavelengths with increasing temperature. It is clear that very high temperatures are needed to cause a thermally excited source to emit a substantial fraction of its energy as ultraviolet radiation.

As noted earlier, part of the continuum background radiation exhibited in the flame spectrum shown in Figure 6-19 is probably thermal emission from incandescent particles in the flame. Note that this background decreases rapidly as the ultraviolet region is approached.

Heated solids are important sources of infrared, visible, and longer-wavelength ultraviolet radiation for analytical instruments.

6C-5 Absorption of Radiation

When radiation passes through a layer of solid, liquid, or gas, certain frequencies may be selectively removed by *absorption*, a process in which electromagnetic energy is transferred to the atoms, ions, or molecules composing the sample. Absorption promotes these particles from their normal room temperature state, or ground state, to one or more higher-energy excited states.

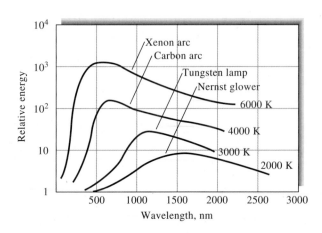

FIGURE 6-22 Blackbody radiation curves.

According to quantum theory, atoms, molecules, and ions have only a limited number of discrete energy levels; for absorption of radiation to occur, the energy of the exciting photon must *exactly* match the energy difference between the ground state and one of the excited states of the absorbing species. Since these energy differences are unique for each species, a study of the frequencies of absorbed radiation provides a means of characterizing the constituents of a sample of matter. For this purpose, a plot of absorbance as a function of wavelength or frequency is experimentally determined (*absorbance*, a measure of the decrease in radiant power, is defined by Equation 6-32 in Section 6D-2). Typical absorption spectra are shown in Figure 6-23.

The four plots in Figure 6-23 reveal that absorption spectra vary widely in appearance; some are made up of numerous sharp peaks, whereas others consist of smooth continuous curves. In general, the nature of a spectrum is influenced by such variables as the complexity, the physical state, and the environment of the absorbing species. More fundamental, however, are

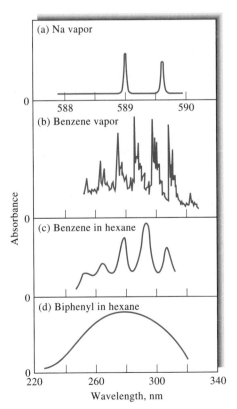

FIGURE 6-23 Some typical ultraviolet absorption spectra.

the differences between absorption spectra for atoms and those for molecules.

Atomic Absorption

The passage of polychromatic ultraviolet or visible radiation through a medium that consists of monoatomic particles, such as gaseous mercury or sodium, results in the absorption of but a few well-defined frequencies (see Figure 6-23a). The relative simplicity of such spectra is due to the small number of possible energy states for the absorbing particles. Excitation can occur only by an electronic process in which one or more of the electrons of the atom are raised to a higher energy level. For example, sodium vapor exhibits two closely spaced, sharp absorption peaks in the yellow region of the visible spectrum (589.0 and 589.6 nm) as a result of excitation of the $3s$ electron to two $3p$ states that differ only slightly in energy. Several other narrow absorption lines, corresponding to other allowed electronic transitions, are also observed. For example, an ultraviolet peak at about 285 nm results from the excitation of the $3s$ electron in sodium to the excited $5p$ state, a process that requires significantly greater energy than does excitation to the $3p$ state (in fact, the peak at 285 nm is also a doublet; the energy difference between the two peaks is so small, however, that most instruments cannot resolve them).

Ultraviolet and visible radiation have enough energy to cause transitions of the outermost, or bonding, electrons only. X-ray frequencies, on the other hand, are several orders of magnitude more energetic (see Example 6-3) and are capable of interacting with electrons that are closest to the nuclei of atoms. Absorption peaks that correspond to electronic transitions of these innermost electrons are thus observed in the X-ray region.

Molecular Absorption

Absorption spectra for polyatomic molecules, particularly in the condensed state, are considerably more complex than atomic spectra because the number of energy states of molecules is generally enormous when compared with the number of energy states for isolated atoms. The energy E associated with the bands of a molecule is made up of three components. That is,

$$E = E_{\text{electronic}} + E_{\text{vibrational}} + E_{\text{rotational}} \quad (6\text{-}21)$$

where $E_{\text{electronic}}$ describes the electronic energy of the molecule that arises from the energy states of its several bonding electrons. The second term on the right refers to the total energy associated with the multitude of interatomic vibrations that are present in molecular species. Generally, a molecule has many more quantized vibrational energy levels than it does electronic levels. Finally, $E_{\text{rotational}}$ is the energy caused by various rotational motions within a molecule; again the number of rotational states is much larger than the number of vibrational states. Thus, for each electronic energy state of a molecule, there are normally several possible vibrational states. For each of these vibrational states, in turn, numerous rotational states are possible. As a consequence, the number of possible energy levels for a molecule is normally orders of magnitude greater than the number of possible energy levels for an atomic particle.

Figure 6-24 is a graphical representation of the energy levels associated with a few of the numerous electronic and vibrational states of a molecule. The heavy line labeled E_0 represents the electronic energy of the molecule in its ground state (its state of lowest elec-

tronic energy); the lines labeled E_1 and E_2 represent the energies of two excited electronic states. Several of the many vibrational energy levels ($e_0, e_1, \ldots, e_n$) are shown for each of these electronic states.

Figure 6-24 shows that the energy difference between the ground state and an electronically excited state is large relative to the energy differences between vibrational levels in a given electronic state (typically, the two differ by a factor of 10 to 100).

The arrows in Figure 6-24a depict some of the transitions that result from absorption of radiation. Visible radiation causes excitation of an electron from E_0 to any of the n vibrational levels associated with E_1 (only five of the n vibrational levels are shown in Figure 6-24). Potential absorption frequencies are then given by n equations, each with the form

$$\nu_i = \frac{1}{h}(E_1 + e_i' - E_0) \tag{6-22}$$

where $i = 1, 2, 3, \ldots, n$.

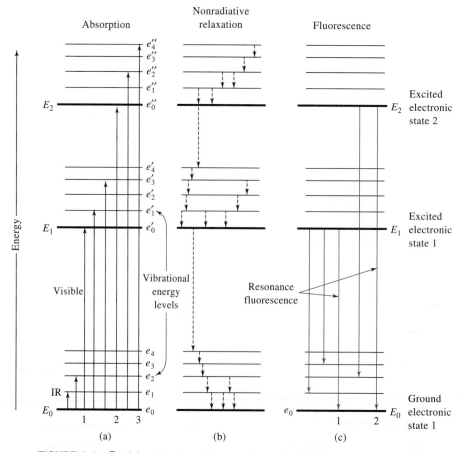

FIGURE 6-24 Partial energy-level diagrams for a fluorescent organic molecule.

Similarly, if the second electronic state has m vibrational levels (four of which are shown), potential absorption frequencies for ultraviolet radiation are given by m equations such as

$$\nu_i = \frac{1}{h}(E_2 + e_i'' - E_0) \qquad (6\text{-}23)$$

where $i = 1, 2, 3, \ldots, m$.

Finally, as shown in Figure 6-24a, the less energetic near- and mid-infrared radiation can bring about transitions only among the k vibrational levels of the ground state. Here, k potential absorption frequencies are given by k equations, which may be formulated as

$$\nu = \frac{1}{h}(e_i - e_0) \qquad (6\text{-}24)$$

where $i = 1, 2, 3, \ldots, k$.

Although they are not shown, several rotational energy levels are associated with each vibrational level in Figure 6-24. The energy difference between the rotational energy levels is small relative to the energy difference between vibrational levels. Transitions between a ground and an excited rotational state are brought about by radiation in the 0.01- to 1-cm-wavelength range, which includes microwave and longer-wavelength infrared radiation.

In contrast to atomic absorption spectra, which consist of a series of sharp, well-defined lines, molecular spectra in the ultraviolet and visible regions are ordinarily characterized by absorption regions that often encompass a substantial wavelength range (see Figure 6-23b, c). Molecular absorption also involves electronic transitions. As shown by Equations 6-23 and 6-24, however, several closely spaced absorption lines will be associated with each electronic transition, because of the existence of numerous vibrational states. Furthermore, as we have mentioned, many rotational energy levels are associated with each vibrational state. As a result, the spectrum for a molecule usually consists of a series of closely spaced absorption lines that constitute an *absorption band*, such as those shown for benzene vapor in Figure 6-23b. Unless a high-resolution instrument is employed, the individual peaks may not be detected, and the spectra will appear as broad smooth peaks such as those shown in Figure 6-23c. Finally, in the condensed state, and in the presence of solvent molecules, the individual lines tend to broaden even further to give nearly *continuous spectra* such as that shown in

Figure 6-23d. Solvent effects are considered in later chapters.

Pure vibrational absorption is observed in the infrared region, where the energy of radiation is insufficient to cause electronic transitions. Such spectra exhibit narrow, closely spaced absorption peaks that result from transitions among the various vibrational quantum levels (see the transition labeled IR at the bottom of Figure 6-24a). Variations in rotational levels may give rise to a series of peaks for each vibrational state; but in liquid and solid samples rotation is often hindered to such an extent that the effects of these small energy differences are not usually detected. Pure rotational spectra for gases can, however, be observed in the microwave region.

Absorption Induced by a Magnetic Field

When electrons of the nuclei of certain elements are subjected to a strong magnetic field, additional quantized energy levels can be observed as a consequence of the magnetic properties of these elementary particles. The differences in energy between the induced states are small, and transitions between the states are brought about only by absorption of long-wavelength (or low-frequency) radiation. With nuclei, radio waves ranging from 30 to 500 MHz ($\lambda = 1000$ to 60 cm) are generally involved; for electrons, microwaves with a frequency of about 9500 MHz ($\lambda = 3$ cm) are absorbed. Absorption by nuclei or by electrons in magnetic fields is studied by *nuclear magnetic resonance* (NMR) and *electron spin resonance* (ESR) techniques, respectively; NMR methods are considered in Chapter 19.

6C-6 Relaxation Processes

Ordinarily, the lifetime of an atom or molecule excited by absorption of radiation is brief because there are several *relaxation processes* that permit its return to the ground state.

Nonradiative Relaxation

As shown in Figure 6-24b, *nonradiative relaxation* involves the loss of energy in a series of small steps, the excitation energy being converted to kinetic energy by collision with other molecules. A minute increase in the temperature of the system results.

As shown by the blue lines in Figure 6-24c, relaxation can also occur by emission of fluorescence radiation. Still other relaxation processes are discussed in Chapters 15, 18, and 19.

Fluorescence and Phosphorescence Relaxation

Fluorescence and phosphorescence are analytically important emission processes in which species are excited by absorption of a beam of electromagnetic radiation; radiant emission then occurs as the excited species return to the ground state. Fluorescence occurs more rapidly than phosphorescence and is generally complete after about 10^{-5} s from the time of excitation. Phosphorescence emission takes place over periods longer than 10^{-5} s and may indeed continue for minutes or even hours after irradiation has ceased. Fluorescence and phosphorescence are most easily observed at a 90° angle to the excitation beam.

Molecular fluorescence is caused by irradiation of molecules in solution or in the gas phase. As shown in Figure 6-24a, absorption of radiation promotes the molecules into any of the several vibrational levels associated with the two excited electronic levels. The lifetimes of these excited vibrational states are, however, only on the order of 10^{-15} s, which is much smaller than the lifetimes of the excited electronic states (10^{-8} s). Therefore, on the average, vibrational relaxation occurs before electronic relaxation. As a consequence, the energy of the emitted radiation is smaller than that of the absorbed by an amount equal to the vibrational excitation energy. For example, for the absorption labeled 3 in Figure 6-24a, the absorbed energy is equal to $(E_2 - E_0 + e_4'' - e_0'')$, whereas the energy of the fluorescence radiation is again given by $(E_2 - E_0)$. Thus, the emitted radiation has a lower frequency, or longer wavelength, than the radiation that excited the fluorescence. This shift in wavelength to lower frequencies is sometimes called the *Stokes shift* as mentioned in connection with Raman scattering in Figure 6-18.

Phosphorescence occurs when an excited molecule relaxes to a metastable excited electronic state (called the *triplet state*), which has an average lifetime of greater than about 10^{-5} s. The nature of this type of excited state is discussed in Chapter 15.

6C-7 The Uncertainty Principle

The *uncertainty principle* was first proposed in 1927 by Werner Heisenberg, who postulated that nature places limits on the precision with which certain pairs of physical measurements can be made. The uncertainty principle, which has important and widespread implications in instrumental analysis, is readily derived from the principle of superposition, which was discussed in

Section 6B-4. Applications of this principle will be found in several later chapters that deal with spectroscopic methods.[5]

Let us suppose that we wish to determine the frequency ν_1 of a monochromatic beam of radiation by comparing it with the output of a standard clock, which is an oscillator that produces a light beam that has a precisely known frequency of ν_2. To detect and measure the difference between the known and unknown frequencies, $\Delta \nu = \nu_1 - \nu_2$, we allow the two beams to interfere as in Figure 6-5 and determine the time interval for a beat (A to B in Figure 6-5). The minimum time Δt required to make this measurement must be equal to or greater than the period of one beat, which as shown in Figure 6-5, is equal to $1/\Delta \nu$. Therefore, the minimum time for a measurement is given by

$$\Delta t \geq 1/\Delta \nu$$

or

$$\Delta t \Delta \nu \geq 1 \tag{6-25}$$

Note that to determine $\Delta \nu$ with negligibly small uncertainty, a huge measurement time is required. If the observation extends over a very short period, the uncertainty will be large.

Let us multiply both sides of Equation 6-25 by Planck's constant to give

$$\Delta t \cdot (h \Delta \nu) = h$$

From Equation 6-17, it is apparent that

$$\Delta E = h \Delta \nu$$

and

$$\Delta t \cdot \Delta E = h \tag{6-26}$$

Equation 6-26 is one of several ways of formulating the Heisenberg uncertainty principle. The meaning in words of this equation is as follows. If the energy E of a particle or system of particles — photons, electrons, neutrons, or protons, for example — is measured for an exactly known period of time Δt, then this energy is uncertain by at least $h/\Delta t$. Therefore, the energy of a particle can be known with zero uncertainty only if it is observed for an infinite period. For finite periods, the energy measurement can never be more precise than $h/\Delta t$. The practical consequences of this limitation will appear in several of the chapters that follow.

[5] A general essay on the uncertainty principle, including applications, is given by L. S. Bartell, *J. Chem. Ed.*, **1985**, *62*, 192.

6D QUANTITATIVE ASPECTS OF SPECTROCHEMICAL MEASUREMENTS

As shown in Table 6-2, spectrochemical methods fall into four major categories. All four require the measurement of radiant *power P*, which is the energy of a beam of radiation that reaches a given area per second. In modern instruments, radiant power is determined with a radiation detector that converts radiant energy into an electrical signal S. Generally S is a voltage or a current that ideally is directly proportional to radiant power. That is,

$$S = kP \qquad (6\text{-}27)$$

where k is a constant.

Many detectors exhibit a small, constant response, known as a *dark current*, in the absence of radiation. In those cases, the response is described by the relationship

$$S = kP + k_d \qquad (6\text{-}28)$$

where k_d is the dark current, which is generally small and constant at least for short periods of time. Spectrochemical instruments are usually equipped with a compensating circuit that reduces k_d to zero whenever measurements are made. With such instruments, Equation 6-27 then applies.

6D-1 Emission, Luminescence, and Scattering Methods

As shown in column 3 of Table 6-2, in emission, luminescence, and scattering methods, the power of the radiation emitted by an analyte after excitation is ordinarily directly proportional to the analyte concentration c ($P_e = kc$). Combining this equation with Equation 6-27 gives

$$S = k'c \qquad (6\text{-}29)$$

where k' is a constant that can be evaluated by exciting analyte radiation in one or more standards and by measuring S. An analogous relationship also applies for luminescence and scattering methods.

6D-2 Absorption Methods

As shown in Table 6-2, quantitative absorption methods require two power measurements: one before a beam has passed through the medium that contains the analyte (P_0) and the other after (P). Two terms, which are widely used in absorption spectrometry and are related to the ratio of P_0 and P, are *transmittance* and *absorbance*.

Transmittance

Figure 6-25 depicts a beam of parallel radiation before and after it has passed through a medium that has a thickness of b cm and a concentration c of an absorbing species. As a consequence of interactions between the photons and absorbing atoms or molecules, the power of the beam is attenuated from P_0 to P. The *transmittance T* of the medium is then the fraction of incident radiation transmitted by the medium:

$$T = \frac{P}{P_0} \qquad (6\text{-}30)$$

Transmittance is often expressed as a percentage or

$$\%T = \frac{P}{P_0} \times 100\% \qquad (6\text{-}31)$$

 Tutorial: Learn more about **transmittance and absorbance**.

TABLE 6-2 Major Classes of Spectrochemical Methods

Class	Radiant Power Measured	Concentration Relationship	Type of Methods
Emission	Emitted, P_e	$P_e = kc$	Atomic emission
Luminescence	Luminescent, P_l	$P_l = kc$	Atomic and molecular fluorescence, phosphorescence, and chemiluminescence
Scattering	Scattered, P_{sc}	$P_{sc} = kc$	Raman scattering, turbidimetry, and particle sizing
Absorption	Incident, P_0, and transmitted, P	$-\log \dfrac{P}{P_0} = kc$	Atomic and molecular absorption

Absorbance

The absorbance A of a medium is defined by the equation

$$A = -\log_{10} T = \log \frac{P_0}{P} \qquad (6\text{-}32)$$

Note that, in contrast to transmittance, the absorbance of a medium increases as attenuation of the beam becomes greater.

Beer's Law

For monochromatic radiation, absorbance is directly proportional to the path length b through the medium and the concentration c of the absorbing species. These relationships are given by

$$A = abc \qquad (6\text{-}33)$$

 Tutorial: Learn more about **Beer's law**.

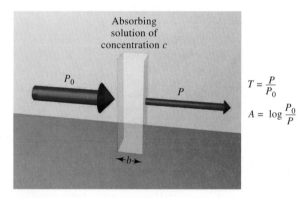

FIGURE 6-25 Attenuation of a beam of radiation by an absorbing solution. The larger arrow on the incident beam signifies a higher radiant power than is transmitted by the solution. The path length of the absorbing solution is b, and the concentration is c.

where a is a proportionality constant called the *absorptivity*. The magnitude of a depends on the units used for b and c. For solutions of an absorbing species, b is often given in centimeters and c in grams per liter. Absorptivity then has units of L g^{-1} cm^{-1}.

When the concentration in Equation 6-33 is expressed in moles per liter and the cell length is in centimeters, the absorptivity is called the *molar absorptivity* and is given the special symbol ϵ. Thus, when b is in centimeters and c is in moles per liter,

$$A = \epsilon bc \qquad (6\text{-}34)$$

where ϵ has the units L mol^{-1} cm^{-1}.

Equations 6-33 and 6-34 are expressions of *Beer's law*, which serves as the basis for quantitative analyses by both atomic and molecular absorption measurements. There are certain limitations to the applicability of Beer's law, and these are discussed in detail in Section 13B-2.

Measurement of Transmittance and Absorbance

Figure 6-26 is a schematic of a simple instrument called a *photometer*, which is used for measuring the transmittance and absorbance of aqueous solutions with a filtered beam of visible radiation. Here, the radiation from a tungsten bulb passes through a colored glass filter that restricts the radiation to a limited band of contiguous wavelengths. The beam then passes through a variable diaphragm that permits adjustment of the power of the radiation that reaches the transparent cell that contains the sample. A shutter can be imposed in front of the diaphragm that completely blocks the beam. With the shutter open, the radiation impinges on a photoelectric transducer that converts the radiant energy of the beam to a direct current that is measured with a microammeter. The output of the

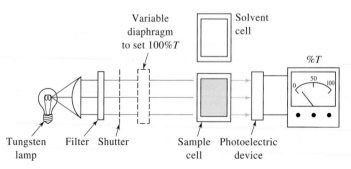

FIGURE 6-26 Single-beam photometer for absorption measurements in the visible region.

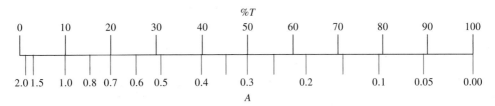

FIGURE 6-27 Readout for an inexpensive photometer. Most modern photometers convert the results directly to absorbance in hardware or software.

meter S is described by Equation 6-28. Note that the meter has a linear scale that extends from 0 to 100.

To make such an instrument direct reading in percent transmittance, two preliminary adjustments are made: the *0% T*, or *dark current, adjustment*, and the *100% T adjustment*. The 0% T adjustment is made with the detector screened from the source by closing the mechanical shutter. Any small dark current in the detector is nulled electrically until the needle of the detector reads zero.

The 100% T adjustment is made with the shutter open and with the solvent cell in the light path. Usually, the solvent is contained in a cell that is as nearly as possible identical to the cell that contains the sample. The 100% T adjustment with this instrument involves varying the power of the beam by means of the vari-

able diaphragm; in some instruments, the same effect is realized by varying the radiant output of the source electrically. The radiant power that reaches the detector is then varied until the meter reads exactly 100. Effectively, this procedure sets P_0 in Equation 6-31 at 100%. When the solvent is replaced by the cell that contains the sample, the scale then indicates the percent transmittance directly, as shown by the equation

$$\% T = \frac{P}{P_0} \times 100\% = \frac{P}{100\%} \times 100\% = P$$

An absorbance scale can also be scribed on the readout device. As shown in Figure 6-27, such a scale will be nonlinear. Modern photometers linearize the readout by conversion to a logarithmic function as discussed in Section 13D.

QUESTIONS AND PROBLEMS

*Answers are provided at the end of the book for problems marked with an asterisk.

| X | Problems with this icon are best solved using spreadsheets.

6-1 Define
  (a) coherent radiation
  (b) dispersion of a transparent substance
  (c) anomalous dispersion
  (d) work function of a substance
  (e) photoelectric effect
  (f) ground state of a molecule
  (g) electronic excitation
  (h) blackbody radiation
  (i) fluorescence
  (j) phosphorescence
  (k) resonance fluorescence
  (l) photon
  (m) absorptivity
  (n) wavenumber
  (o) relaxation
  (p) Stokes shift

*6-2 Calculate the frequency in hertz, the energy in joules, and the energy in electron volts of an X-ray photon with a wavelength of 6.24 Å.

*6-3 Calculate the frequency in hertz, the wavenumber, the energy in joules, and the energy in kJ/mol associated with the 3.517 μm vibrational absorption band of an aliphatic ketone.

*6-4 Calculate the wavelength and the energy in joules associated with an NMR signal at 368 MHz.

*6-5 Calculate the velocity, frequency, and wavelength of the sodium D line (λ = 589 nm) as light from this source passes through a species whose refractive index, n_D, is 1.09.

*6-6 When the D line of sodium light impinges an air-diamond interface at an angle of incidence of 30.0°, the angle of refraction is 11.9°. What is n_D for diamond?

*6-7 What is the wavelength of a photon that has three times as much energy as that of a photon whose wavelength is 779 nm?

*6-8 The silver iodide bond energy is approximately 255 kJ/mol (AgI is one of the possible active components in photogray sunglasses). What is the longest wavelength of light that is capable of breaking the bond in silver iodide?

*6-9 Cesium is used extensively in photocells and in television cameras because it has the lowest ionization energy of all the stable elements.
(a) What is the maximum kinetic energy of a photoelectron ejected from cesium by 555 nm light? Note that if the wavelength of the light used to irradiate the cesium surface becomes longer than 660 nm, no photoelectrons are emitted.
(b) Use the rest mass of the electron to calculate the velocity of the photoelectron in (a).

*6-10 The Wien displacement law for blackbody radiators states that the product of temperature in kelvin and the wavelength of maximum emission is a constant k ($k = T \cdot \lambda_{max}$). Calculate the wavelength of maximum emission for a Globar infrared source operated at 1800 K. Use the data in Figure 6-22 for the Nernst glower for the evaluation of the constant.

*6-11 Calculate the wavelength of
(a) the sodium line at 589 nm in honey, which has a refractive index of 1.50.
(b) the output of a ruby laser at 694.3 when it is passing through a piece of quartz, which has a refractive index of 1.55.

*6-12 Calculate the reflection loss when a beam of radiant energy passes through an empty quartz cell assuming the refractive index of quartz is 1.55.

6-13 Explain why the wave model for radiation cannot account for the photoelectric effect.

*6-14 Convert the following absorbance data into percent transmittance:
(a) 0.278   (b) 1.499   (c) 0.039

*6-15 Convert the following percent transmittance data into absorbance:
(a) 29.9   (b) 86.1   (c) 2.97

*6-16 Calculate the percent transmittance of solutions with half the absorbance of those in Problem 6-14.

*6-17 Calculate the absorbance of solutions with half the percent transmittance of those in Problem 6-15.

*6-18 A solution that was 3.78×10^{-3} M in X had a transmittance of 0.212 when measured in a 2.00-cm cell. What concentration of X would be required for the transmittance to be increased by a factor of 3 when a 1.00-cm cell was used?

*6-19 A compound had a molar absorptivity of 3.03×10^3 L cm^{-1} mol^{-1}. What concentration of the compound would be required to produce a solution that has a transmittance of 9.53% in a 2.50-cm cell?

X *Challenge Problem*

6-20 One of the watershed events in the development of physics and chemistry was the appearance of Einstein's landmark paper explaining the photoelectric effect, establishing the corpuscular nature of light, and leading to the modern view of the wave-particle duality of the microscopic realm.

(a) Look up Millikan's paper on the photoelectric effect, and describe how he characterized Einstein's work.[6]

(b) In Section 6C-1 we described how measurements of the stopping voltage in a phototube as a function of frequency can be used to determine Planck's constant. Describe and discuss three experimental difficulties in the determination of the Planck constant by this method. You may find the paper by Keesing[7] useful in this discussion.

(c) Use the data in Table III of Millikan's paper to determine the stopping potential as a function of wavelength at 433.9, 404.7, 365.0, 312.5, and 253.5 nm.

(d) Enter these data into an Excel spreadsheet, and perform a least-squares analysis of the data to determine Planck's constant and its uncertainty. Compare your results to those of Millikan, and discuss any differences.

(e) One of the difficulties that you discovered in (b) and (c) is related to the determination of the stopping potential. Knudsen[8] has described a method based on the following normalized equations[9]:

$$\phi(\delta) = \frac{4\pi m k^2 T^2}{h^2}\left(e^{\delta} - \frac{e^{2\delta}}{2^2} + \frac{e^{3\delta}}{3^2} - \cdots\right) \quad (\delta \leq 0)$$

$$\phi(\delta) = \frac{4\pi m k^2 T^2}{h^2}\left[\frac{\pi^2}{6} + \frac{1}{2}\delta^2 - \left(e^{-\delta} - \frac{e^{-2\delta}}{2^2} + \frac{e^{-3\delta}}{3^2} - \cdots\right)\right] \quad (\delta \geq 0)$$

where $\phi(\delta)$ is normalized photocurrent and δ is the normalized retarding voltage in units of kT. Create a spreadsheet, and generate a plot of $\Phi(\delta) = \log \phi(\delta)$ versus δ for 56 values in the range $\delta = -5$ to $\delta = 50$. Also plot the following normalized data collected at 365.015 nm.

---

[6] R. A. Millikan, *Phys. Rev.*, **1918**, *7*, 355.

[7] R. G. Keesing, *Eur. J. Phys.*, **1981**, *2*, 139.

[8] A. W. Knudsen, *Am. J. Phys.*, **1983**, *8*, 725.

[9] R. H. Fowler, *Phys. Rev.*, **1931**, *38*, 45.

δ, kT	$\log \phi(\delta)$
33.24	0.17
33.87	0.33
34.72	0.53
35.46	0.79
36.20	0.99
36.93	1.20
37.67	1.39
38.41	1.58
40.43	1.97
42.34	2.27
44.25	2.47
46.16	2.64
47.97	2.77
49.99	2.89
51.90	3.01
53.82	3.10
55.63	3.19
57.65	3.25
59.56	3.33
61.48	3.38
63.29	3.44
65.31	3.51
67.23	3.54
69.04	3.60

Print two copies of the plot in full-page format, and overlay the two copies over a light source. Determine the stopping potential at 365.015 nm as described by Knudsen. Compare your result with his result in the table in (f).

(f) Perform a least-squares analysis of the data in the following table to determine Planck's constant. Compare these results to those of Millikan and your results from (c). Rationalize any differences in the results in terms of experimental differences and other fundamental considerations.

λ, nm	Stopping Potential	
	kT	V
435.834	56.7	1.473
404.656	48.1	1.249
365.015	35.4	0.919
334.148	23.4	0.608
313.170	14.0	0.364
296.728	5.8	0.151
289.36	1.3	0.034

(g) There is an element of circular reasoning in the Knudsen procedure. Describe and discuss critically the use of the curve-matching process to determine the stopping potential.

(h) Planck's constant is no longer determined by measurements on photocells. How is it determined?[10] What other fundamental constants depend on Planck's constant? What are the current values of these constants, and what are their uncertainties?

(i) Least-squares procedures can be used to adjust the values of the fundamental constants so that they are internally consistent.[11] Describe how this procedure might be accomplished. How does an improvement in the quality of the measurement of one constant such as Avogadro's number affect the values of the other constants?

(j) Over the past several decades, much effort has been expended in the determination of the values of the fundamental constants. Why is this effort important in analytical chemistry, or why is it not? What measurable quantities in analytical chemistry depend on the values of the fundamental constants? Why are these efforts important to science and to the world at large? Comment critically on the return on the investment of time and effort that has gone into the determination of the fundamental constants.

---

[10] E. R. Williams, R. L. Steiner, D. B. Newell, and P. T. Olsen, *Phys. Rev. Lett.*, **1998**, *81*, 2404.
[11] J. W. M. DuMond and E. Richard Cohen, *Rev. Modern Phys.*, **1953**, *25*, 691.

CHAPTER SEVEN

Components of Optical Instruments

Instruments for the ultraviolet (UV), visible, and infrared (IR) regions have enough features in common that they are often called optical instruments even though the human eye is not sensitive to ultraviolet or infrared wavelengths. In this chapter, we consider the function, the requirements, and the behavior of the components of instruments for optical spectroscopy for all three types of radiation. Instruments for spectroscopic studies in regions more energetic than the ultraviolet and less energetic than the infrared have characteristics that differ substantially from optical instruments and are considered separately in Chapters 12 and 19.

Throughout this chapter, this logo indicates an opportunity for online self-study at **www.thomsonedu.com/chemistry/skoog**, linking you to interactive tutorials, simulations, and exercises.

7A GENERAL DESIGNS OF OPTICAL INSTRUMENTS

Optical spectroscopic methods are based on six phenomena: (1) absorption, (2) fluorescence, (3) phosphorescence, (4) scattering, (5) emission, and (6) chemiluminescence. Although the instruments for measuring each differ somewhat in configuration, most of their basic components are remarkably similar. Furthermore, the required properties of these components are the same regardless of whether they are applied to the ultraviolet, visible, or infrared portion of the spectrum.[1]

Typical spectroscopic instruments contain five components: (1) a stable source of radiant energy; (2) a transparent container for holding the sample; (3) a device that isolates a restricted region of the spectrum for measurement[2]; (4) a radiation detector, which converts radiant energy to a usable electrical signal; and (5) a signal processor and readout, which displays the transduced signal on a meter scale, a computer screen, a digital meter, or another recording device. Figure 7-1 illustrates the three ways these components are configured to carry out the six types of spectroscopic measurements mentioned earlier. The figure also shows that components (1), (4), and (5) are arranged in the same way for each type of measurement.

The first two instrumental configurations, which are used for the measurement of absorption, fluorescence, and phosphorescence, require an external source of radiant energy. For absorption, the beam from the source passes into the wavelength selector and then through the sample, although in some instruments the positions of the selector and sample are reversed. For fluorescence and phosphorescence, the source induces the sample, held in a container, to emit characteristic radiation, which is usually measured at an angle of 90° with respect to the source.

---

[1] For a more complete discussion of the components of optical instruments, see J. Lindon, G. Tranter, J. Holmes, eds., *Encyclopedia of Spectroscopy and Spectrometry*, Vols. 1–3, San Diego: Academic Press, 2000; J. W. Robinson, ed., *Practical Handbook of Spectroscopy*, Boca Raton, FL: CRC Press, 1991; E. J. Meehan, in *Treatise on Analytical Chemistry*, P. J. Elving, E. J. Meehan, and I. M. Kolthoff, eds., Part I, Vol. 7, Chap. 3, New York: Wiley, 1981; J. D. Ingle Jr. and S. R. Crouch, *Spectrochemical Analysis*, Chaps. 3 and 4, Upper Saddle River, NJ: Prentice Hall, 1988.

[2] Fourier transform instruments, which are discussed in Section 7I-3, require no wavelength selection device but instead use a frequency modulator that provides spectral data in a form that can be interpreted by a mathematical technique called a Fourier transformation.

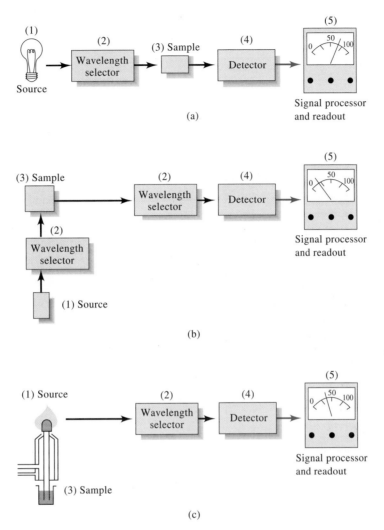

FIGURE 7-1 Components of various types of instruments for optical spectroscopy. In (a), the arrangement for absorption measurements is shown. Note that source radiation of the selected wavelength is sent through the sample, and the transmitted radiation is measured by the detector–signal processing–readout unit. With some instruments, the position of the sample and wavelength selector is reversed. In (b), the configuration for fluorescence measurements is shown. Here, two wavelength selectors are needed to select the excitation and emission wavelengths. The selected source radiation is incident on the sample and the radiation emitted is measured, usually at right angles to avoid scattering. In (c), the configuration for emission spectroscopy is shown. Here, a source of thermal energy, such as a flame or plasma, produces an analyte vapor that emits radiation isolated by the wavelength selector and converted to an electrical signal by the detector.

Emission spectroscopy and chemiluminescence spectroscopy differ from the other types in that no external radiation source is required; the sample itself is the emitter (see Figure 7-1c). In emission spectroscopy, the sample container is a plasma, a spark, or a flame that both contains the sample and causes it to emit characteristic radiation. In chemiluminescence spectroscopy, the radiation source is a solution of the analyte plus reagents held in a transparent sample holder. Emission is brought about by energy released in a chemical reaction in which the analyte takes part directly or indirectly.

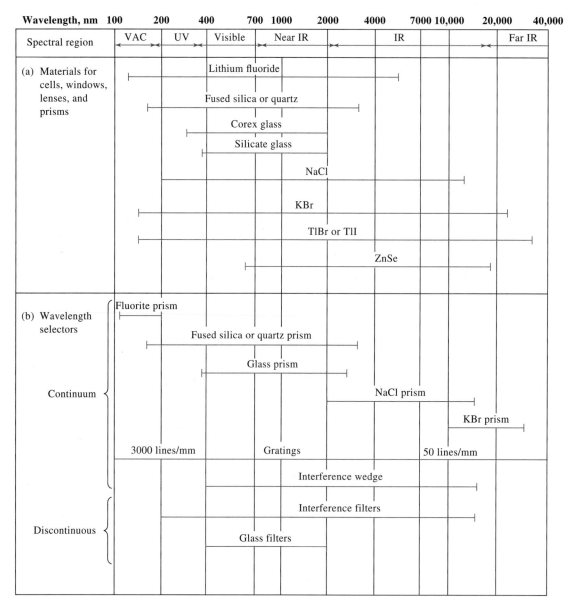

FIGURE 7-2 (a) Construction materials and (b) wavelength selectors for spectroscopic instruments.

Figures 7-2 and 7-3 summarize the optical characteristics of all the components shown in Figure 7-1 with the exception of the signal processor and readout. Note that instrument components differ in detail, depending on the wavelength region within which they are to be used. Their design also depends on whether the instrument is to be used primarily for qualitative or quantitative analysis and on whether it is to be applied to atomic or molecular spectroscopy. Nevertheless, the general function and performance requirements of each type of component are similar, regardless of wavelength region and application.

7B SOURCES OF RADIATION

To be suitable for spectroscopic studies, a source must generate a beam with sufficient radiant power for easy detection and measurement. In addition, its output power should be stable for reasonable periods. Typi-

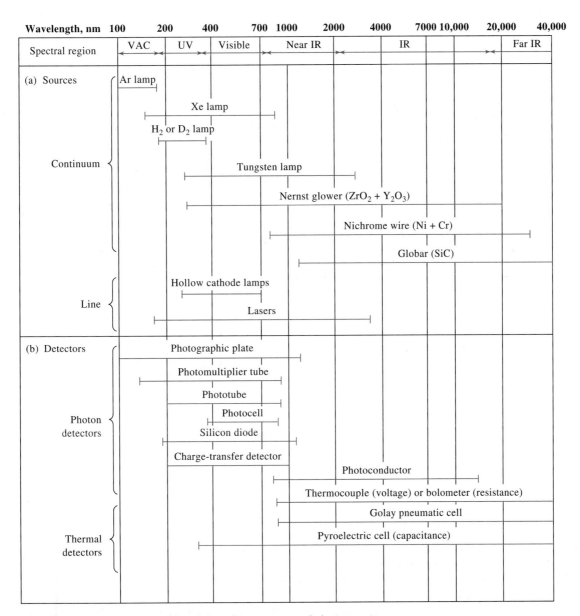

FIGURE 7-3 (a) Sources and (b) detectors for spectroscopic instruments.

cally, the radiant power of a source varies exponentially with the voltage of its power supply. Thus, a regulated power source is almost always needed to provide the required stability. Alternatively, the problem of source stability can sometimes be circumvented by double-beam designs in which the ratio of the signal from the sample to that of the source in the absence of sample serves as the analytical variable. In such designs, the intensities of the two beams are measured simultaneously or nearly simultaneously so that the effect of fluctuations in the source output is largely canceled.

Figure 7-3a lists the most widely used spectroscopic sources. Note that these sources are of two types: *continuum sources*, which emit radiation that changes in intensity only slowly as a function of wavelength, and *line sources*, which emit a limited number of lines, or bands of radiation, each of which spans a limited range of wavelengths.

 Tutorial: Learn more about **optical materials and sources**.

7B-1 Continuum Sources

Continuum sources find widespread use in absorption and fluorescence spectroscopy. For the UV region, the most common source is the deuterium lamp. High-pressure, gas-filled arc lamps that contain argon, xenon, or mercury are used when a particularly intense source is required. For the visible region of the spectrum, a tungsten filament lamp is used almost universally. The common infrared sources are inert solids heated to 1500 to 2000 K, a temperature at which the maximum radiant output occurs at 1.5 to 1.9 μm (see Figure 6-22). Details on the construction and behavior of these various continuum sources appear in the chapters dealing with specific types of spectroscopic methods.

7B-2 Line Sources

Sources that emit a few discrete lines find wide use in atomic absorption spectroscopy, atomic and molecular fluorescence spectroscopy, and Raman spectroscopy (refractometry and polarimetry also use line sources). The familiar mercury and sodium vapor lamps provide a relatively few sharp lines in the ultraviolet and visible regions and are used in several spectroscopic instruments. Hollow-cathode lamps and electrodeless discharge lamps are the most important line sources for atomic absorption and fluorescence methods. Discussion of such sources is deferred to Section 9B-1.

7B-3 Laser Sources

Lasers are highly useful sources in analytical instrumentation because of their high intensities, their narrow bandwidths, and the coherent nature of their outputs.[3] The first laser was described in 1960. Since that time, chemists have found numerous useful applications for these sources in high-resolution spectroscopy, in kinetic studies of processes with lifetimes in the range of 10^{-9} to 10^{-12} s, in the detection and determination of extremely small concentrations of species in the atmo-

sphere, and in the induction of isotopically selective reactions.[4] In addition, laser sources have become important in several routine analytical methods, including Raman spectroscopy, molecular absorption spectroscopy, emission spectroscopy, and as part of instruments for Fourier transform infrared spectroscopy.

The term *laser* is an acronym for **l**ight **a**mplification by **s**timulated **e**mission of **r**adiation. Because of their light-amplifying characteristics, lasers produce spatially narrow (a few hundredths of a micrometer), extremely intense beams of radiation. The process of stimulated emission, which will be described shortly, produces a beam of highly monochromatic (bandwidths of 0.01 nm or less) and remarkably coherent (Section 6B-6) radiation. Because of these unique properties, lasers have become important sources for use in the UV, visible, and IR regions of the spectrum. A limitation of early lasers was that the radiation from a given source was restricted to a relatively few discrete wavelengths or lines. Now, however, dye lasers are available that provide narrow bands of radiation at any chosen wavelength within a somewhat limited range of the source.

Components of Lasers

Figure 7-4 is a schematic representation that shows the components of a typical laser source. The heart of the device is the lasing medium. It may be a solid crystal such as ruby, a semiconductor such as gallium arsenide, a solution of an organic dye, or a gas such as argon or krypton. The lasing material is often activated, or *pumped*, by radiation from an external source so that a few photons of proper energy will trigger the formation of a cascade of photons of the same energy. Pumping can also be accomplished by an electrical current or by an electrical discharge. Thus, gas lasers usually do not have the external radiation source shown in Figure 7-4; instead, the power supply is connected to a pair of electrodes contained in a cell filled with the gas.

A laser normally functions as an oscillator, or a resonator, in the sense that the radiation produced by the lasing action is caused to pass back and forth through the medium numerous times by means of a pair of mirrors as shown in Figure 7-4. Additional photons are

---

[3]For a more complete discussion of lasers, see W. T. Silfvast, *Laser Fundamentals*, 2nd ed., Cambridge: Cambridge Univ. Press, 2004; D. L Andrews and A. A Demidov, eds., *An Introduction to Laser Spectroscopy*, 2nd ed., New York: Plenum, 2002; G. R. Van Hecke and K. K. Karukstis, *A Guide to Lasers in Chemistry*, Boston: Jones and Bartlett, 1998; D. L. Andrews, ed., *Lasers in Chemistry*, 3rd ed., New York: Springer-Verlag, 1997.

[4]For reviews of some of these applications, see J. C. Wright and M. J. Wirth, *Anal. Chem.*, **1980**, *52*, 988A, 1087A; J. K. Steehler, *J. Chem. Educ.*, **1990**, *67*, A37; C. P. Christensen, *Science*, **1984**, *224*, 117; R. N. Zare, *Science*, **1984**, *226*, 1198; E. W. Findsend and M. R. Ondrias, *J. Chem. Educ.*, **1986**, *63*, 479; A. Schawlow, *Science*, **1982**, *217*, 9.

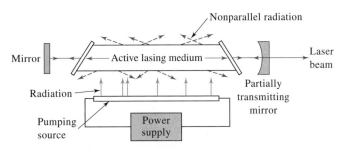

FIGURE 7-4 Schematic representation of a typical laser source.

generated with each passage, thus leading to enormous amplification. The repeated passage also produces a beam that is highly parallel, because nonparallel radiation escapes from the sides of the medium after being reflected a few times (see Figure 7-4). One of the easiest ways to obtain a usable laser beam is to coat one of the mirrors with a sufficiently thin layer of reflecting material so that a fraction of the beam is transmitted rather than reflected.

Mechanism of Laser Action

To illustrate laser action, we will consider a molecular system. Many lasers are, however, atomic or ionic lasers, and although the mechanisms for these lasers are similar to those for molecular lasers, the details are somewhat different. Laser action can be understood by considering the four processes depicted in Figure 7-5: (a) pumping, (b) spontaneous emission (fluorescence), (c) stimulated emission, and (d) absorption. In this figure, we show the behavior of two of the many molecules that make up the lasing medium. Two of the several electronic energy levels of each are shown as having energies E_y and E_x. Note that the higher electronic state for each molecule has several slightly different vibrational energy levels depicted as E_y, E_y', E_y'', and so forth. We have not shown additional levels for the lower electronic state, although such levels usually exist. Note that He-Ne, Ar^+, ruby, Nd-YAG, or other atomic or ionic lasers *do not have vibrational levels*; instead, these lasing media have other electronic states.

Pumping. Pumping, which is necessary for laser action, is a process by which the active species of a laser is excited by means of an electrical discharge, passage of an electrical current, or exposure to an intense radiant source. During pumping in a molecular system, several of the higher electronic and vibrational energy levels

 Simulation: Learn more about **how lasers work**.

of the active species are populated. In diagram (1) of Figure 7-5a, one electron is shown as being promoted to an energy state E_y''; the second is excited to the slightly higher vibrational level E_y'''. The lifetime of an excited vibrational state is brief, so after 10^{-13} to 10^{-15} s, the electron relaxes to the lowest excited vibrational level [E_y in Figure 7-5a(3)] and an undetectable quantity of heat is produced. Some excited electronic states of laser materials have lifetimes considerably longer (often 1 ms or more) than their excited vibrational counterparts; long-lived states are sometimes termed metastable as a consequence.

Spontaneous Emission. As was pointed out in the discussion of fluorescence (Section 6C-5), a species in an excited electronic state may lose all or part of its excess energy by spontaneous emission of radiation. This process is depicted in the three diagrams shown in Figure 7-5b. Note that the wavelength of the fluorescence radiation is given by the relationship $\lambda = hc/(E_y - E_x)$, where h is Planck's constant and c is the speed of light. It is also important to note that the instant at which emission occurs and the path of the resulting photon vary from excited molecule to excited molecule because spontaneous emission is a random process; thus, the fluorescence radiation produced by one of the species in Figure 7-5b(1) differs in direction and phase from that produced by the second species in Figure 7-5b(2). Spontaneous emission, therefore, yields *incoherent* monochromatic radiation.

Stimulated Emission. Stimulated emission, which is the basis of laser behavior, is depicted in Figure 7-5c. Here, the excited laser species are struck by photons that have precisely the same energies ($E_y - E_x$) as the photons produced by spontaneous emission. Collisions of this type cause the excited species to relax immediately to the lower energy state and to simultaneously emit a photon of exactly the same energy as the pho-

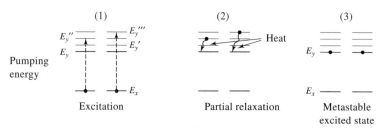

(a) Pumping (excitation by electrical, radiant, or chemical energy)

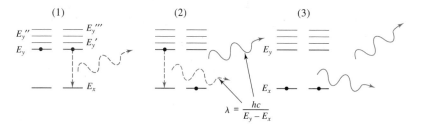

(b) Spontaneous emission

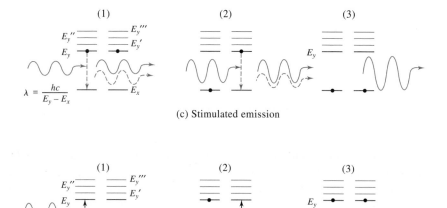

(c) Stimulated emission

(d) Absorption

FIGURE 7-5 Four processes important in laser action: (a) pumping (excitation by electrical, radiant, or chemical energy), (b) spontaneous emission, (c) stimulated emission, and (d) absorption.

ton that stimulated the process. Equally important, the emitted photon *travels* in exactly the same direction and *is precisely in phase* with the photon that caused the emission. Therefore, the stimulated emission is totally *coherent* with the incoming radiation.

Absorption. The absorption process, which competes with stimulated emission, is depicted in Figure 7-5d. Here, two photons with energies exactly equal to $(E_y - E_x)$ are absorbed to produce the metastable excited state shown in Figure 7-5d(3); note that this

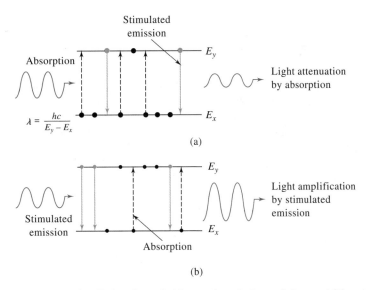

FIGURE 7-6 Passage of radiation through (a) a noninverted population and (b) an inverted population created by excitation of electrons into virtual states by an external energy source (pumping).

state is identical to that attained in Figure 7-5a(3) by pumping.

Population Inversion and Light Amplification

To have light amplification in a laser, the number of photons produced by stimulated emission must exceed the number lost by absorption. This condition prevails only when the number of particles in the higher energy state exceeds the number in the lower; in other words, there must be a *population inversion* from the normal distribution of energy states. Population inversions are created by pumping. Figure 7-6 contrasts the effect of incoming radiation on a noninverted population with that of an inverted one. In each case, nine molecules of the laser medium are in the two states E_x and E_y. In the noninverted system, three molecules are in the excited state and six are in the lower energy level. The medium absorbs three of the incoming photons to produce three additional excited molecules, which subsequently relax very rapidly to the ground state without achieving a steady-state population inversion. The radiation may also stimulate emission of two photons from excited molecules resulting in a net attenuation of the beam by one photon. As shown in Figure 7-6b, pumping two molecules into virtual states E_n followed by relaxation to E_y creates a population inversion between E_y and E_x. Thus, the diagram shows six electrons in state E_y and only three electrons in E_x. In the inverted system, stim-

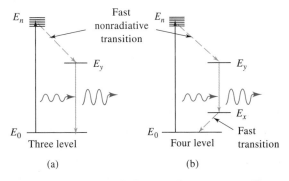

FIGURE 7-7 Energy level diagrams for two types of laser systems.

ulated emission prevails over absorption to produce a net gain in emitted photons. Light amplification, or lasing, then occurs.

Three- and Four-Level Laser Systems

Figure 7-7 shows simplified energy diagrams for the two common types of laser systems. In the three-level system, the transition responsible for laser radiation is between an excited state E_y and the ground state E_0; in a four-level system, on the other hand, radiation is generated by a transition from E_y to a state E_x that has a greater energy than the ground state. Furthermore, it is necessary that transitions between E_x and the ground state be rapid. The advantage of the four-level system is that the population inversions essential for laser action are achieved more easily than in three-level systems. To understand this advantage, note that at room temperature a large majority of the laser species will be in the ground-state energy level E_0 in both systems. Sufficient energy must thus be provided to convert more than 50% of the lasing species to the E_y level of a three-level system. In a four-level system, it is only necessary to pump sufficiently to make the number of particles in the E_y energy level exceed the number in E_x. The lifetime of a particle in the E_x state is brief, however, because the transition to E_0 is fast; thus, the number in the E_x state is generally negligible relative to the number that has energy E_0 and also (with a modest input of pumping energy) with respect to the number in the E_y state. Therefore, the four-level laser usually achieves a population inversion with a small expenditure of pumping energy.

Some Examples of Useful Lasers

Several different types of lasers have been used in analytical chemistry.[5]

Solid State Lasers. The first successful laser, and one that is still used, is a three-level device in which a ruby crystal is the active medium. Ruby is primarily Al_2O_3 but contains approximately 0.05% chromium(III) distributed among the aluminum(III) lattice sites, which accounts for the red coloration. The chromium(III) ions are the active lasing material. In early lasers, the ruby was machined into a rod about 4 cm long and 0.5 cm in diameter. A flash tube (often a low-pressure xenon lamp) was coiled around the cylinder to produce intense flashes of light ($\lambda = 694.3$ nm). Because the flashlamp was pulsed, a pulsed beam was produced. Continuous-wave (CW) ruby sources are now available.

The Nd-YAG laser is one of the most widely used solid-state lasers. The lasing medium consists of neodymium ion in a host crystal of yttrium aluminum garnet. This system offers the advantage of being a four-level laser, which makes it much easier to achieve population inversion than with the ruby laser. The Nd-YAG laser has a very high radiant power output at 1064 nm, which is usually frequency doubled (see page 175)

to give an intense line at 532 nm. This radiation can be used for pumping tunable dye lasers.

Gas Lasers. A variety of gas lasers is available commercially. These devices are of four types: (1) neutral atom lasers such as He-Ne; (2) ion lasers in which the active species is Ar^+ or Kr^+; (3) molecular lasers in which the lasing medium is CO_2 or N_2; and (4) excimer lasers. The helium-neon laser is the most widely encountered of all lasers because of its low initial and maintenance costs, its great reliability, and its low power consumption. The most important of its output lines is at 632.8 nm. It is generally operated in a continuous mode rather than a pulsed mode.

The argon ion laser, which produces intense lines in the green (514.5 nm) and the blue (488.0 nm) regions, is an important example of an ion laser. This laser is a four-level device in which argon ions are formed by an electrical or radio-frequency discharge. The required input energy is high because the argon atoms must first be ionized and then excited from their ground state, with a principal quantum number of 3, to various $4p$ states. Lasing occurs when the excited ions relax to the $4s$ state. The argon ion laser is used as a source in fluorescence and Raman spectroscopy because of the high intensity of its lines.

The nitrogen laser is pumped with a high-voltage spark source that provides a momentary (1 to 5 ns) pulse of current through the gas. The excitation creates a population inversion that decays very quickly by spontaneous emission because the lifetime of the excited state is quite short relative to the lifetime of the lower level. The result is a short (a few nanoseconds) pulse of intense (up to 1 MW) radiation at 337.1 nm. This output is used for exciting fluorescence in a variety of molecules and for pumping dye lasers. The carbon dioxide gas laser is used to produce monochromatic infrared radiation at 10.6 μm.

Excimer lasers contain a gaseous mixture of helium, fluorine, and one of the rare gases argon, krypton, or xenon. The rare gas is electronically excited by a current followed by reaction with fluorine to form excited species such as ArF*, KrF*, or XeF*, which are called *excimers* because they are stable only in the excited state. Because the excimer ground state is unstable, rapid dissociation of the compounds occurs as they relax while giving off a photon. Thus, there is a population inversion as long as pumping is carried on. Excimer lasers produce high-energy pulses in the ultraviolet (351 nm for XeF, 248 nm for KrF, and 193 nm for ArF).

---

[5]For a review of lasers useful in analytical chemistry, see C. Gooijer, *Anal. Chim. Acta*, **1999**, *400*, 281.

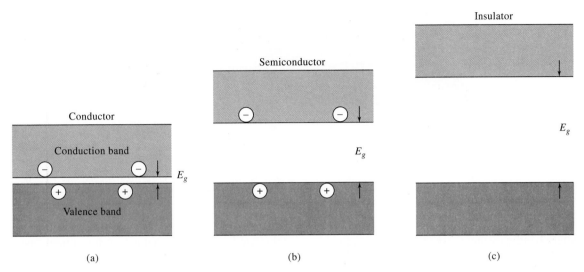

FIGURE 7-8 Conduction bands and valence bands in three types of materials.

Dye Lasers. Dye lasers[6] have become important radiation sources in analytical chemistry because they are continuously tunable over a range of 20 to 50 nm. By changing dyes, the wavelength range of a dye laser can be arranged to be quite broad. The bandwidth of a tunable dye laser is typically a few hundredths of a nanometer or less. The active materials in dye lasers are solutions of organic compounds capable of fluorescing in the ultraviolet, visible, or infrared regions. Dye lasers are four-level systems. In contrast to the other lasers of this type that we have considered, however, the lower energy level for laser action (E_x in Figure 7-6b) is not a single energy but a band of energies that arise from the superposition of a large number of closely spaced vibrational and rotational energy states on the base electronic energy state. Electrons in E_y may then undergo transitions to any of these states, thus producing photons of slightly different energies. Tuning of dye lasers can be readily accomplished by replacing the nontransmitting mirror shown in Figure 7-4 with a reflection grating or a Fabry-Perot etalon (see page 177) that reflects only a narrow bandwidth of radiation into the laser medium; the peak wavelength can be varied by the grating or tilting the etalon. Emission is then stimulated for only part of the fluorescence spectrum, namely, the wavelength selected by the monochromator. Depending on the pump source, dye lasers can be operated in either the pulsed or the CW mode.

Semiconductor Diode Lasers. An increasingly important source of nearly monochromatic radiation is the laser diode.[7] Laser diodes are products of modern semiconductor technology. We can understand their mechanism of operation by considering the electrical conduction characteristics of various materials as illustrated in Figure 7-8. A good conductor, such as a metal, consists of a regular arrangement of atoms immersed in a sea of valence electrons. Orbitals on adjacent atoms overlap to form the *valence band*, which is essentially a molecular orbital over the entire metal containing the valence electrons of all of the atoms. Empty outer orbitals overlap to form the *conduction band*, which lies at a slightly higher energy than the valence band. The difference in energy between the valence band and the conduction band is the band-gap energy E_g. Because the band-gap energy is so small in conductors (see Figure 7-8a), electrons in the valence band easily acquire sufficient thermal energy to be promoted to the conduction band, thus providing mobile charge carriers for conduction.

In contrast, insulators have relatively large band-gap energies, and as a result, electrons in the valence band are unable to acquire enough thermal energy to make the transition to the conduction band. Thus, insulators do not conduct electricity (see Figure 7-8c). Semiconductors, such as silicon or germanium, have

[6]For further information, see M. Stuke, ed., *Dye Lasers: Twenty-Five Years*, New York: Springer, 1992; F. J. Duarte and L. W. Hillman, *Dye Laser Principles with Applications*, San Diego: Elsevier, 1990.

[7]M. G. D. Bauman, J. C. Wright, A. B. Ellis, T. Kuech, and G. C. Lisensky, *J. Chem. Educ.*, **1992**, *69*, 89; T. Imasaka and N. Ishibashi, *Anal. Chem.*, **1990**, *62*, 363A; R. L. Beyer, *Science*, **1989**, *239*, 742; K. Niemax, A. Zybin, C. Schnürer-Patschan, and H. Groll, *Anal. Chem.*, **1996**, *68*, 351A.

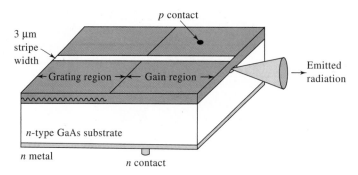

FIGURE 7-9 A distributed Bragg-reflector laser diode. (From D. W. Nam and R. G. Waarts, *Laser Focus World*, *1994*, *30* (8), 52. Reprinted with permission of PennWell Publishing Company.)

intermediate band-gap energies so that their conduction characteristics are intermediate between conductors and insulators (see Figure 7-8b). We should note that whether a material is a semiconductor or an insulator depends not only on the band-gap energy but also on the temperature of operation and the excitation energy of the material, which is related to the voltage applied to the material.

When a voltage is impressed across a semiconductor diode in the forward direction (see Section 2C-2), electrons are excited into the conduction band, hole-electron pairs are created, and the diode conducts. Ultimately, some of these electrons relax and go back into the valence band, and energy is released that corresponds to the band-gap energy $E_g = h\nu$. Some of the energy is released in the form of electromagnetic radiation of frequency $\nu = E_g/h$. Diodes that are fabricated to enhance the production of light are called *light-emitting diodes*, or LEDs. Light-emitting diodes are often made of gallium arsenic phosphide, which has a band-gap energy that corresponds to a wavelength maximum λ_m of 650 nm. Diodes of this type find wide use as indicators and readouts in electronic instruments. Diodes made from gallium aluminum arsenide ($\lambda_m = 900$ nm), gallium phosphide ($\lambda_m = 550$ nm), gallium nitride ($\lambda_m = 465$ nm), and indium gallium nitride ($\lambda_m = 450$ nm) are widely available. LEDs are widely used in simple photometers and other photometric detectors as described in Section 13D-1.

In recent years, semiconductor fabrication techniques have progressed to an extent that permits the construction of highly complex integrated devices such as the distributed Bragg-reflector (DBR) laser diode shown in Figure 7-9. This device contains a gallium arsenide *pn*-junction diode that produces infrared radiation at about 975 nm. In addition, a stripe of material

is fabricated on the chip that acts as a resonant cavity for the radiation so that light amplification can occur within the cavity. An integrated grating provides feedback to the resonant cavity so that the resulting radiation has an extremely narrow bandwidth of about 10^{-5} nm. Laser diodes of this type have achieved continuous power outputs of more than 100 mW with a typical thermal stability of 0.1 nm/°C. Laser diodes may be operated in either a pulsed or CW mode, which increases their versatility in a variety of applications. Rapid development of laser diodes has resulted from their utility as light sources for CD players, CD-ROM drives, DVD players, bar-code scanners, and other familiar optoelectronic devices, and mass production of laser diodes ensures that their cost will continue to decrease.

A major impediment to the use of laser diodes in spectroscopic applications has been that their wavelength range has been limited to the red and infrared regions of the spectrum. This disadvantage may be overcome by operating the laser diode in a pulsed mode to achieve sufficient peak power to use nonlinear optics to provide frequency doubling as shown in Figure 7-10. Here, the output of a laser diode is focused in a doubling crystal to provide output in the blue-green region of the spectrum (~490 nm). With proper external optics, frequency-doubled laser diodes can achieve average output powers of 0.5 to 1.0 W with a tunable spectral range of about 30 nm. The advantages of such light sources include compactness, power efficiency, high reliability, and ruggedness. The addition of external optics to the laser diode increases the cost of the devices substantially, but they are competitive with larger, less efficient, and less reliable gas-based lasers.

Gallium nitride laser diodes produce radiation directly in the blue, green, and yellow region of the

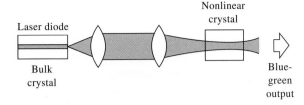

FIGURE 7-10 A frequency-doubling system for converting 975-nm laser output to 490 nm. (From D. W. Nam and R. G. Waarts, *Laser Focus World*, **1994**, *30* (8), 52. Reprinted with permission of PennWell Publishing Company.)

spectrum.[8] These diodes are now being used routinely for spectroscopic studies.

The utility of laser diodes for spectroscopic applications has been demonstrated in molecular absorption spectrometry, molecular fluorescence spectrometry, atomic absorption spectrometry, and as light sources for detectors in various chromatographic methods. Recent advances in laser diode technology fueled by consumer demand for high-speed, high-capacity DVD players have resulted in the availability of blue laser diodes with output powers up to 50 mW at 473 nm. These light sources are appearing routinely in commercial spectrometric systems.

Nonlinear Optical Effects with Lasers

We noted in Section 6B-7 that when an electromagnetic wave is transmitted through a dielectric[9] medium, the electromagnetic field of the radiation causes momentary distortion, or polarization, of the valence electrons of the molecules that make up the medium. For ordinary radiation the extent of polarization P is directly proportional to the magnitude of the electric field E of the radiation. Thus, we may write

$$P = \alpha E$$

where α is the proportionality constant. Optical phenomena that occur when this situation prevails are said to be *linear*.

At the high radiation intensities encountered with lasers, this relationship breaks down, particularly when E approaches the binding energy of the electrons. Under these circumstances, *nonlinear optical effects* are

observed, and the relationship between polarization and electric field is given by

$$P = \alpha E + \beta E^2 + \gamma E^3 + \cdots \qquad (7\text{-}1)$$

where the magnitudes of the three constants are in the order $\alpha > \beta > \gamma$. At ordinary radiation intensities, only the first term on the right is significant, and the relationship between polarization and field strength is linear. With high-intensity lasers, however, the second term and sometimes even the third term are required to describe the degree of polarization. When only two terms are required, Equation 7-1 can be rewritten in terms of the radiation frequency ω and the maximum amplitude of the field strength E_m. Thus,

$$P = \alpha E_m \sin \omega t + \beta E_m^2 \sin^2 \omega t \qquad (7\text{-}2)$$

Substituting the trigonometric identity $\sin^2 \omega t = (1 - \cos 2\omega t)/2$ into Equation 7-2 gives

$$P = \alpha E_m \sin \omega t + \frac{\beta E_m^2}{2}(1 - \cos 2\omega t) \qquad (7\text{-}3)$$

The first term in Equation 7-3 is the normal linear term that predominates at low radiation intensities. At sufficiently high intensity, the second-order term becomes significant and results in radiation that has a frequency 2ω that is *double* that of the incident radiation. This frequency-doubling process is now widely used to produce laser frequencies of shorter wavelengths. For example, the 1064-nm near-infrared radiation from a Nd-YAG laser can be frequency doubled to produce a 30% yield of green radiation at 532 nm by passing the radiation through a crystalline material such as potassium dihydrogen phosphate. The 532-nm radiation can then be doubled again to yield UV radiation at 266 nm by passage through a crystal of ammonium dihydrogen phosphate.

Laser radiation is used in several types of nonlinear spectroscopy, most notably in Raman spectroscopy (see Section 18D-3).

7C WAVELENGTH SELECTORS

Most spectroscopic analyses require radiation that consists of a limited, narrow, continuous group of wavelengths called a *band*.[10] A narrow bandwidth enhances the sensitivity of absorbance measurements, may pro-

---

[8]G. Fasol, *Science*, **1996**, *272*, 1751.

[9]Dielectrics are a class of substances that are nonconductors because they contain no free electrons. Generally, dielectrics are optically transparent in contrast to electrically conducting solids, which either absorb radiation or reflect it strongly.

---

[10]Note that the term *band* in this context has a somewhat different meaning from that used in describing types of spectra in Chapter 6.

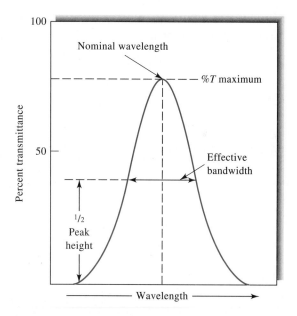

FIGURE 7-11 Output of a typical wavelength selector.

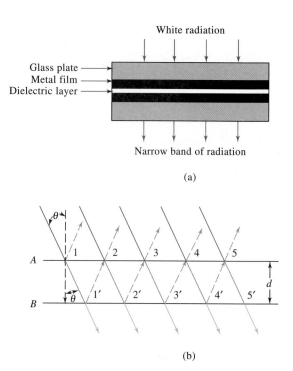

FIGURE 7-12 (a) Schematic cross section of an interference filter. Note that the drawing is not to scale and that the three central bands are much narrower than shown. (b) Schematic to show the conditions for constructive interference.

vide selectivity to both emission and absorption methods, and is frequently required from the standpoint of obtaining a linear relationship between the optical signal and concentration (Equation 6-29). Ideally, the output from a wavelength selector would be radiation of a single wavelength or frequency. No real wavelength selector approaches this ideal; instead, a band, such as that shown in Figure 7-11, is produced. Here, the percentage of incident radiation of a given wavelength that is transmitted by the selector is plotted as a function of wavelength. The *effective bandwidth*, which is defined in Figure 7-11, is an inverse measure of the quality of the device, a narrower bandwidth representing better performance. There are two types of wavelength selectors: filters and monochromators.

7C-1 Filters

Two types of filters are used for wavelength selection: *interference filters*, which are sometimes called *Fabry-Perot filters*, and *absorption filters*. Absorption filters are restricted to the visible region of the spectrum; interference filters, on the other hand, are available for the ultraviolet, visible, and well into the infrared region.

Interference Filters

As the name implies, interference filters rely on optical interference to provide narrow bands of radiation. An interference filter consists of a transparent dielectric

 Tutorial: Learn more about **wavelength selectors**.

(frequently calcium fluoride or magnesium fluoride) that occupies the space between two semitransparent metallic films. This array is sandwiched between two plates of glass or other transparent materials (see Figure 7-12a). The thickness of the dielectric layer is carefully controlled and determines the wavelength of the transmitted radiation. When a perpendicular beam of collimated radiation strikes this array, a fraction passes through the first metallic layer and the remainder is reflected. The portion that is passed undergoes a similar partition when it strikes the second metallic film. If the reflected portion from this second interaction is of the proper wavelength, it is partially reflected from the inner side of the first layer in phase with incoming light of the same wavelength. The result is that this particular wavelength is reinforced, and most other wavelengths, being out of phase, undergo destructive interference.

The relationship between the dielectric layer's thickness d and the transmitted wavelength λ can be found with the aid of Figure 7-12b. For purposes of clarity, the incident beam is shown arriving at an angle θ from the perpendicular to the surface of the dielectric. At point 1, the radiation is partially reflected and partially transmitted to point 1' where partial reflection and transmission again take place. The same process occurs at 2, 2′, and so forth. For reinforcement to occur at point 2, the distance traveled by the beam reflected at 1′ must be some multiple of its wavelength in the medium λ′. Because the path length between surfaces can be expressed as $d/\cos\theta$, the condition for reinforcement is that

$$\mathbf{n}\lambda' = 2d/\cos\theta$$

where $\mathbf{n}$ is a small whole number, λ′ is the wavelength of radiation *in the dielectric*, and d is the thickness of the dielectric.

In ordinary use, θ approaches zero and $\cos\theta$ approaches unity so that the previous equation simplifies to

$$\mathbf{n}\lambda' = 2d \qquad (7\text{-}4)$$

The corresponding wavelength in air is given by

$$\lambda = \lambda'n$$

where n is the refractive index of the dielectric medium. Thus, the wavelengths of radiation transmitted by the filter are

$$\lambda = \frac{2dn}{\mathbf{n}} \qquad (7\text{-}5)$$

The integer $\mathbf{n}$ is the *order* of interference. The glass layers of the filter are often selected to absorb all but one of the reinforced bands; transmission is thus restricted to a single order.

Figure 7-13 illustrates the performance characteristics of typical interference filters. Filters are generally characterized, as shown, by the wavelength of their transmittance peaks, the percentage of incident radiation transmitted at the peak (their *percent transmittance*, Equation 6-31), and their effective bandwidths.

Interference filters are available with transmission peaks throughout the ultraviolet and visible regions and up to about 14 μm in the infrared. Typically, effective bandwidths are about 1.5% of the wavelength at peak transmittance, although this figure is reduced to 0.15% in some narrow-band filters; these have maximum transmittances of 10%.

Fabry-Perot Etalon

Another important device based on interference is the *Fabry-Perot etalon*. The device consists of a plate made of a transparent material with highly parallel faces coated with a nonabsorbing, highly reflective material. Alternatively, a spacer of the chosen thickness made of invar or quartz with highly parallel faces is sandwiched between two mirrors to form the etalon. The bandwidth

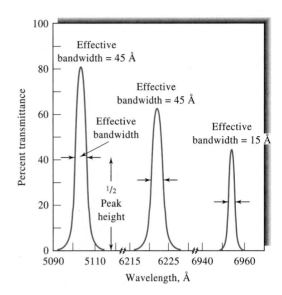

FIGURE 7-13 Transmission characteristics of typical interference filters.

of the device is determined by the reflectivity of the coatings or mirrors, and the separation of the transmitted bands is dictated by the distance between the mirrors. The etalon may be tilted to vary the band that is transmitted. If the reflecting surfaces are configured with an air gap between them so that their separation can be adjusted mechanically, the device is called a *Fabry-Perot interferometer*.[11] Fabry-Perot etalons have many uses in laser experiments, spectroscopy, and in fiber-optic communications, where they are used to separate frequency bands.

Interference Wedges

An interference wedge consists of a pair of mirrored, partially transparent plates separated by a wedge-shape layer of a dielectric material. The length of the plates ranges from about 50 to 200 mm. The radiation transmitted varies continuously in wavelength from one end to the other as the thickness of the wedge varies. By choosing the proper linear position along the wedge, a bandwidth of about 20 nm can be isolated.

Interference wedges are available for the visible region (400 to 700 nm), the near-infrared region (1000 to 2000 nm), and for several parts of the infrared region (2.5 to 14.5 μm). They can be used in place of prisms or gratings in monochromators.

[11] J. D. Ingle Jr. and S. R. Crouch, *Spectrochemical Analysis*, pp. 78–81, Upper Saddle River, NJ: Prentice Hall, 1988.

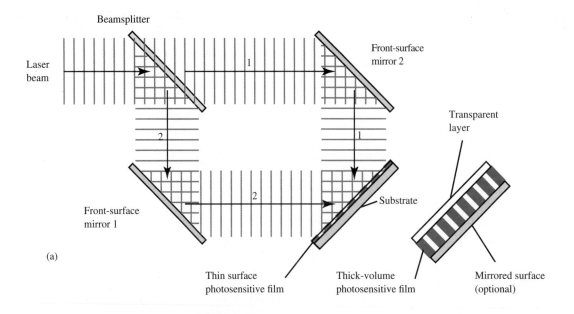

(a)

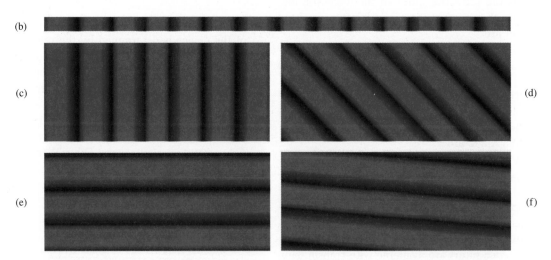

(b)

(c)

(d)

(e)

(f)

FIGURE 7-14 (a) Experimental arrangement for fabricating holograms. A laser beam is split into two beams, which are directed by mirrors to recombine at the surface of a photosensitive film. The apparatus as shown is used to produce holographic reflection gratings. By replacing the surface object with a thick-volume photosensitive film shown on the right, a volume transmission hologram results. (b) Interference pattern at the surface of the photosensitive film, which produces a corresponding refractive-index pattern on the film. (c) Untilted fringe pattern, (d) tilted fringe pattern, (e) conformal fringe pattern produced by making the back surface of the thick-volume film a mirror, (f) nonconformal fringe pattern.

Holographic Filters

Holographic optical devices and, in particular, holographic filters[12] are among a growing repertoire of optical devices and materials that have resulted from the

broad availability of laser technology. Figure 7-14a is a schematic diagram of a typical experimental arrangement for producing holograms. The coherent radiation of a laser beam impinges on a beamsplitter, where it is divided into two beams, 1 and 2, shown in the figure. The two beams are redirected by the two front-surface

[12] J. M. Tedesco et al., *Anal. Chem.*, **1993**, *65*, 441A.

mirrors to be recombined at the surface of the thin (10–50 μm) photosensitive film. Because the two mutually coherent beams have a fixed phase-and-intensity relationship, they produce an extremely uniform interference pattern of light and dark bars, or fringes, at the surface of the film as depicted in Figure 7-14b. These interference fringes sensitize the film, which is a photographic emulsion or photoresistive polymer layer, so that the sensitized areas can be developed or dissolved away, leaving a grooved structure on the surface on the substrate. This structure is then coated with aluminum or other reflective material to produce a *holographic reflection grating*. We describe the characteristics and uses of gratings in detail in Section 7C-2.

A second type of holographic device that can be fabricated in a similar fashion is the *volume transmission hologram*. These devices are formed within a thick layer (>100 μm) of photosensitive material sandwiched between two transparent layers as shown on the right in Figure 7-14a. The thick-volume film assembly is placed at the plane of intersection of the two laser beams as before, and the interference fringes are formed throughout the volume of the film layer rather than just at its surface. When the two beams intersect at equal angles to the surface normal of the film, a sinusoidally modulated fringe pattern such as the one shown in Figure 7-14c is formed within the film, and the entire volume of the film is sensitized with this pattern. The film is then developed chemically to "fix" the

pattern, which produces a corresponding pattern of variation of the refractive index within the film material. By making the angle of incidence of the two beams asymmetric about the film normal, the resulting refractive-index modulation pattern can be tilted as illustrated in Figure 7-14d. By controlling the laser wavelength and angle of incidence of the beam, the frequency of the refractive-index modulation may be tailored to the requirements of the device. These devices are also used as gratings and as filters for removing undesirable radiation such as plasma lines and sidebands from laser radiation.

Perhaps the most useful volume holographic device is formed when the transparent back face of the thick film is replaced by a mirror, and the film is illuminated by a single laser beam. The beam enters the front surface of the film, is reflected from the interface between the back of the film and the mirror, and forms an interference pattern within the film that is parallel to its face as illustrated in Figure 7-14e. These devices are called *conformal reflection holograms*, and their characteristics make them nearly ideal for use as *notch filters*. The transmission characteristics of a typical holographic notch filter are compared to those of a dielectric interference filter in Figure 7-15a. Note that the holographic filter provides extremely flat transmission at all wavelengths except at the notch, where it blocks more than 99.99% of the incident radiation, corresponding to an optical density (absorbance) of 4. The effective band-

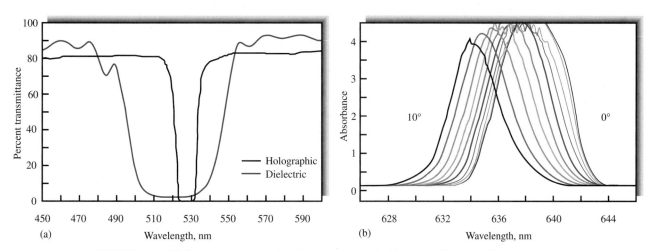

FIGURE 7-15 (a) Comparison of the bandwidths of a typical holographic notch filter and an interference filter. (b) Tilt-tuning of a holographic notch filter by 1° increments from perpendicular to the incident laser beam. The rejection band can be finely adjusted by altering the angle of the notch filter relative to the laser. (Courtesy of Kaiser Optical Systems, Inc., Ann Arbor, MI, with permission.)

width of the filter is less than 10 nm and its edgewidth, or the wavelength range over which its optical density ranges from 0.3 to 4, is less than 4 nm. Figure 7-15b shows how the rejection band of a notch filter may be fine-tuned by tilting it with respect to incoming radiation. These filters are available in a wide variety of sizes and rejection bands tuned to the most common laser lines in the range of 350 to 1400 nm. The commercial availability of holographic notch filters has initiated a revolution in the use of Raman spectroscopy (see Section 18B-3) by virtually eliminating the need for costly high-resolution double monochromators for routine work. The refractive-index modulation characteristics of *nonconformal* holographic optical elements are shown in Figure 7-14f. Because there is some modulation of the refractive index at the surface of the device, it acts in principle both as a notch filter and as a grating.

Absorption Filters

Absorption filters, which are generally less expensive than interference filters, have been widely used for band selection in the visible region. These filters function by absorbing selected portions of the spectrum. The most common type consists of colored glass or of a dye suspended in gelatin and sandwiched between glass plates. Colored glass filters have the advantage of greater thermal stability.

Absorption filters have effective bandwidths that range from 30 to 250 nm (see Figures 7-16 and 7-17). Filters that provide the narrowest bandwidths also absorb a significant fraction of the desired radiation and may have a transmittance of 10% or less at their band peaks.

Glass filters with transmittance maxima throughout the entire visible region are available commercially.

Cutoff filters have transmittances of nearly 100% over a portion of the visible spectrum but then rapidly decrease to zero transmittance over the remainder. A narrow spectral band can be isolated by coupling a cutoff filter with a second filter (see Figure 7-17).

Figure 7-16 shows that the performance characteristics of absorption filters are significantly inferior to those of interference-type filters. Not only are the bandwidths of absorption filters greater than those of interference filters, but for narrow bandwidths the fraction of light transmitted by absorption filters is also smaller. Nevertheless, absorption filters are adequate for some applications.

7C-2 Monochromators

For many spectroscopic methods, it is necessary or desirable to be able to continuously vary the wavelength of radiation over a broad range. This process is called *scanning* a spectrum. Monochromators are designed for spectral scanning. Monochromators for ultraviolet, visible, and infrared radiation are all similar in mechanical construction in the sense that they use slits, lenses, mirrors, windows, and gratings or prisms. The materials from which these components are fabricated depend on the wavelength region of intended use (see Figure 7-2).

Components of Monochromators

Figure 7-18 illustrates the optical elements found in all monochromators, which include the following: (1) an entrance slit that provides a rectangular optical image,

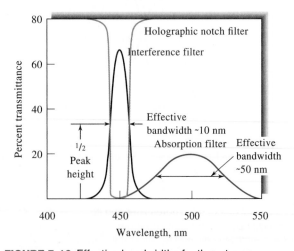

FIGURE 7-16 Effective bandwidths for three types of filters.

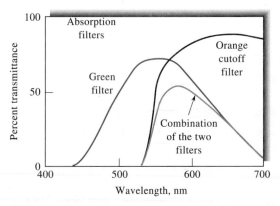

FIGURE 7-17 Comparison of various types of absorption filters for visible radiation.

(2) a collimating lens or mirror that produces a parallel beam of radiation, (3) a prism or a grating that disperses the radiation into its component wavelengths, (4) a focusing element that reforms the image of the entrance slit and focuses it on a planar surface called a *focal plane*, and (5) an exit slit in the focal plane that isolates the desired spectral band. In addition, most monochromators have entrance and exit windows designed to protect the components from dust and corrosive laboratory fumes.

As shown in Figure 7-18, two types of dispersing elements are found in monochromators: reflection gratings and prisms. For purposes of illustration, a beam made up of just two wavelengths, λ_1 and λ_2 ($\lambda_1 > \lambda_2$), is shown. This radiation enters the monochromators via a narrow rectangular opening, or *slit*; is collimated; and then strikes the surface of the dispersing element at an angle. For the grating monochromator, angular dispersion of the wavelengths results from diffraction, which occurs at the reflective surface; for the prism, refraction at the two faces results in angular dispersion of the radiation, as shown. In both designs, the dispersed radi-

ation is focused on the focal plane AB where it appears as two rectangular images of the entrance slit (one for λ_1 and one for λ_2). By rotating the dispersing element, one band or the other can be focused on the exit slit.

In the past, most monochromators were prism instruments. Currently, however, nearly all commercial monochromators are based on reflection gratings because they are cheaper to fabricate, provide better wavelength separation for the same size dispersing element, and disperse radiation linearly along the focal plane. As shown in Figure 7-19a, linear dispersion means that the position of a band along the focal plane for a grating varies linearly with its wavelength. For prism instruments, in contrast, shorter wavelengths are dispersed to a greater degree than are longer ones, which complicates instrument design. The nonlinear dispersion of two types of prism monochromators is illustrated by Figure 7-19b. Because of their more general use, we will focus most of our discussion on grating monochromators.

 Exercise: Learn more about **monochromators**.

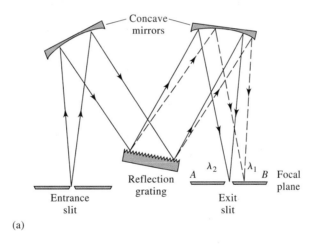

(a)

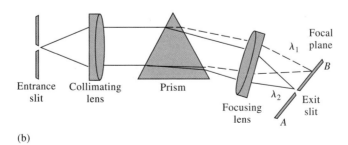

(b)

FIGURE 7-18 Two types of monochromators: (a) Czerney-Turner grating monochromator and (b) Bunsen prism monochromator. (In both instances, $\lambda_1 > \lambda_2$.)

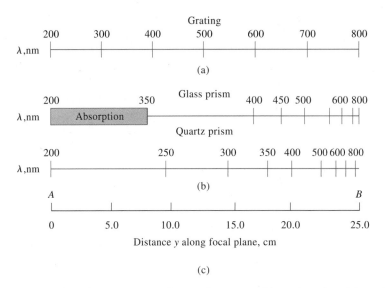

FIGURE 7-19 Dispersion for three types of monochromators. The points A and B on the scale in (c) correspond to the points shown in Figure 7-18.

Prism Monochromators

Prisms can be used to disperse ultraviolet, visible, and infrared radiation. The material used for their construction differs, however, depending on the wavelength region (see Figure 7-2b).

Figure 7-20 shows the two most common types of prism designs. The first is a 60° prism, which is usually fabricated from a single block of material. When crystalline (but not fused) quartz is the construction material, however, the prism is usually formed by cementing two 30° prisms together, as shown in Figure 7-20a; one is fabricated from right-handed quartz and the second from left-handed quartz. In this way, the optically active quartz causes no net polarization of the emitted radiation; this type of prism is called a *Cornu prism*. Figure 7-18b shows a *Bunsen monochromator*, which uses a 60° prism, likewise often made of quartz.

As shown in Figure 7-20b, the *Littrow prism*, which permits more compact monochromator designs, is a 30° prism with a mirrored back. Refraction in this type of prism takes place twice at the same interface so that the performance characteristics are similar to those of a 60° prism in a Bunsen mount.

Grating Monochromators

Dispersion of ultraviolet, visible, and infrared radiation can be brought about by directing a polychromatic beam through a *transmission grating* or onto the surface of a *reflection grating*; the latter is by far the more com-

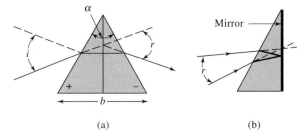

FIGURE 7-20 Dispersion by a prism: (a) quartz Cornu type and (b) Littrow type.

mon type. *Replica gratings*, which are used in most monochromators, are manufactured from a *master grating*.[13] A master grating consists of a hard, optically flat, polished surface that has a large number of parallel and closely spaced grooves, made with a diamond tool. A magnified cross-sectional view of a few typical grooves is shown in Figure 7-21. A grating for the ultraviolet and visible region typically has from 300 to 2000 grooves/mm, with 1200 to 1400 being most common. For the infrared region, gratings typically have 10 to 200 grooves/mm; for spectrophotometers designed for the most widely used infrared range of 5 to 15 μm,

---

[13]For an interesting and informative discussion of the manufacture, testing, and performance characteristics of gratings, see *Diffraction Grating Handbook*, 6th ed., Irvine, CA: Newport Corp., 2005 (www.newport.com). For a historical perspective on the importance of gratings in the advancement of science, see A. G. Ingalls, *Sci. Amer.*, **1952**, *186* (6), 45.

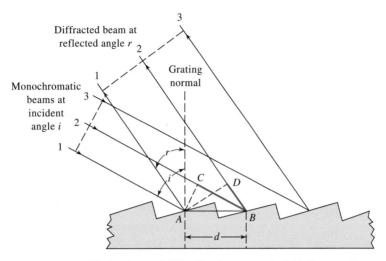

FIGURE 7-21 Mechanisms of diffraction from an echellette-type grating.

a grating with about 100 grooves/mm is suitable. The construction of a good master grating is tedious, time-consuming, and expensive because the grooves must be of identical size, exactly parallel, and equally spaced over the length of the grating (3 to 10 cm).

Replica gratings are formed from a master grating by a liquid-resin casting process that preserves virtually perfectly the optical accuracy of the original master grating on a clear resin surface. This surface is usually made reflective by a coating of aluminum, or sometimes gold or platinum.

The Echellette Grating. Figure 7-21 is a schematic representation of an *echellette-type* grating, which is grooved, or *blazed*, such that it has relatively broad faces from which reflection occurs and narrow unused faces. This geometry provides highly efficient diffraction of radiation, and the reason for blazing is to concentrate the radiation in a preferred direction.[14] Each of the broad faces can be considered to be a line source of radiation perpendicular to the plane of the page; thus interference among the reflected beams 1, 2, and 3 can occur. For the interference to be constructive, it is necessary that the path lengths differ by an integral multiple **n** of the wavelength λ of the incident beam.

In Figure 7-21, parallel beams of monochromatic radiation 1 and 2 are shown striking the grating at an incident angle i to the *grating normal*. Maximum constructive interference is shown as occurring at the re-

flected angle r. Beam 2 travels a greater distance than beam 1 and the difference in the paths is equal to $(\overline{CB} + \overline{BD})$ (shown as a blue line in the figure). For constructive interference to occur, this difference must equal **n**λ. That is,

$$\mathbf{n}\lambda = (\overline{CB} + \overline{BD})$$

where **n**, a small whole number, is called the diffraction *order*. Note, however, that angle *CAB* is equal to angle i and that angle *DAB* is identical to angle r. Therefore, from trigonometry, we can write

$$\overline{CB} = d\sin i$$

where d is the spacing between the reflecting surfaces. It is also seen that

$$\overline{BD} = d\sin r$$

Substitution of the last two expressions into the first gives the condition for constructive interference. Thus,

$$\mathbf{n}\lambda = d(\sin i + \sin r) \qquad (7\text{-}6)$$

Equation 7-6 suggests that there are several values of λ for a given diffraction angle r. Thus, if a first-order line (**n** = 1) of 900 nm is found at r, second-order (450-nm) and third-order (300-nm) lines also appear at this angle. The first-order line is usually the most intense; indeed, it is possible to design gratings with blaze angles and shapes that concentrate as much as 90% of the incident intensity in this order. Filters can generally remove the higher-order lines. For example, glass, which absorbs radiation below 350 nm, eliminates the higher-order spectra associated with first-order radiation in most of

[14] J. D. Ingle Jr. and S. R. Crouch, *Spectrochemical Analysis*, p. 66, Upper Saddle River, NJ: Prentice Hall, 1988.

the visible region. The example that follows illustrates these points.

EXAMPLE 7-1

An echellette grating that contains 1450 blazes/mm was irradiated with a polychromatic beam at an incident angle 48° to the grating normal. Calculate the wavelengths of radiation that would appear at an angle of reflection of +20, +10, and 0° (angle r, Figure 7-21).

Solution

To obtain d in Equation 7-6, we write

$$d = \frac{1 \text{ mm}}{1450 \text{ blazes}} \times 10^6 \frac{\text{nm}}{\text{mm}} = 689.7 \frac{\text{nm}}{\text{blaze}}$$

When r in Figure 7-21 equals +20°,

$$\lambda = \frac{689.7 \text{ nm}}{n} (\sin 48 + \sin 20) = \frac{748.4}{n} \text{ nm}$$

and the wavelengths for the first-, second-, and third-order reflections are 748, 374, and 249 nm, respectively.

Further calculations of a similar kind yield the following data:

		λ, nm	
r,°	$n = 1$	$n = 2$	$n = 3$
20	748	374	249
10	632	316	211
0	513	256	171

Concave Gratings. Gratings can be formed on a concave surface in much the same way as on a plane surface. A concave grating permits the design of a monochromator without auxiliary collimating and focusing mirrors or lenses because the concave surface both disperses the radiation and focuses it on the exit slit. Such an arrangement is advantageous in terms of cost; in addition, the reduction in number of optical surfaces increases the energy throughput of a monochromator that contains a concave grating.

Holographic Gratings. *Holographic gratings*[15] are appearing in ever-increasing numbers in modern optical instruments, even some of the less expensive ones.

Holographic gratings, because of their greater perfection with respect to line shape and dimensions, provide spectra that are relatively free of stray radiation and ghosts (double images).

In the preparation of holographic gratings, the beams from a pair of identical lasers are brought to bear at suitable angles on a glass surface coated with photoresist. The resulting interference fringes from the two beams sensitize the photoresist so that it can be dissolved, leaving a grooved structure that can be coated with aluminum or other reflecting substance to produce a reflection grating. The spacing of the grooves can be altered by changing the angle of the two laser beams with respect to one another.

As described in Section 7C-1 and illustrated in Figure 7-14, holographic gratings are produced by generating an interference pattern on the surface of a thin film of photosensitive material, which is developed to provide the grooved structure of the grating. The grating is then coated with a reflective substance such as aluminum. Nearly perfect, large (~50 cm) gratings with as many as 6000 lines/mm can be manufactured in this way at a relatively low cost. As with ruled gratings, replica gratings can be cast from a master holographic grating. There is apparently no optical test that can distinguish between a master and a replica holographic grating.[16]

Performance Characteristics of Grating Monochromators

The quality of a monochromator depends on the purity of its radiant output, its ability to resolve adjacent wavelengths, its light-gathering power, and its spectral bandwidth. The last property is discussed in Section 7C-3.

Spectral Purity. The exit beam of a monochromator is usually contaminated with small amounts of scattered or stray radiation with wavelengths far different from that of the instrument setting. This unwanted radiation can be traced to several sources. Among these sources are reflections of the beam from various optical parts and the monochromator housing. Reflections from optical parts result from mechanical imperfections, particularly in gratings, introduced during manufacture. Scattering by dust particles in the atmosphere or on the surfaces of optical parts also causes stray radiation to reach the exit slit. Generally, the effects of spurious radiation are minimized by introducing baffles in appro-

---

[15] See J. Flamand, A. Grillo, and G. Hayat, *Amer. Lab.*, **1975**, 7 (5), 47; and J. M. Lerner et al., *Proc. Photo-Opt. Instrum. Eng.*, **1980**, *240*, 72, 82.

[16] I. R. Altelmose, *J. Chem. Educ.*, **1986**, *63*, A221.

priate spots in the monochromator and by coating interior surfaces with flat black paint. In addition, the monochromator is sealed with windows over the slits to prevent entrance of dust and fumes. Despite these precautions, however, some spurious radiation still appears; we shall see that its presence can have serious effects on absorption measurements under certain conditions.[17]

Dispersion of Grating Monochromators. The ability of a monochromator to separate different wavelengths depends on its *dispersion*. The *angular dispersion* is given by $dr/d\lambda$, where dr is the change in the angle of reflection or refraction with a change in wavelength $d\lambda$. The angle r is defined in Figures 7-20 and 7-21.

The angular dispersion of a grating can be obtained by differentiating Equation 7-6 while holding i constant. Thus, at any given angle of incidence,

$$\frac{dr}{d\lambda} = \frac{\mathbf{n}}{d\cos r} \qquad (7\text{-}7)$$

The *linear dispersion D* refers to the variation in wavelength as a function of y, the distance along the line AB of the focal planes as shown in Figure 7-18. If f is the focal length of the monochromator, the linear dispersion can be related to the angular dispersion by the relationship

$$D = \frac{dy}{d\lambda} = \frac{f\,dr}{d\lambda} \qquad (7\text{-}8)$$

A more useful measure of dispersion is the *reciprocal linear dispersion* D^{-1} where

$$D^{-1} = \frac{d\lambda}{dy} = \frac{1}{f}\frac{d\lambda}{dr} \qquad (7\text{-}9)$$

The dimensions of D^{-1} are often nm/mm or Å/mm in the UV-visible region.

By substituting Equation 7-7 into Equation 7-9 we have the reciprocal linear dispersion for a grating monochromator:

$$D^{-1} = \frac{d\lambda}{dy} = \frac{d\cos r}{\mathbf{n}f} \qquad (7\text{-}10)$$

Note that the angular dispersion increases as the distance d between rulings decreases, as the number of lines per millimeter increases, or as the focal length in-

creases. At small angles of diffraction ($>20°$), $\cos r \approx 1$ and Equation 7-10 becomes approximately

$$D^{-1} = \frac{d}{\mathbf{n}f} \qquad (7\text{-}11)$$

Thus, for all practical purposes, if the angle r is small, the *linear dispersion of a grating monochromator is constant*, a property that greatly simplifies monochromator design.

Resolving Power of Monochromators. The *resolving power R* of a monochromator describes the limit of its ability to separate adjacent images that have a slight difference in wavelength. Here, by definition

$$R = \frac{\lambda}{\Delta\lambda} \qquad (7\text{-}12)$$

where λ is the average wavelength of the two images and $\Delta\lambda$ is their difference. The resolving power of typical bench-top UV-visible monochromators ranges from 10^3 to 10^4.

It can be shown[18] that the resolving power of a grating is given by the expression

$$R = \frac{\lambda}{\Delta\lambda} = \mathbf{n}N \qquad (7\text{-}13)$$

when $\mathbf{n}$ is the diffraction order and N is the number of grating blazes illuminated by radiation from the entrance slit. Thus, better resolution is a characteristic of longer gratings, smaller blaze spacings, and higher diffraction orders. This equation applies to both echellette and echelle (see later discussion) gratings.

Light-Gathering Power of Monochromators. To increase the signal-to-noise ratio of a spectrometer, it is necessary that the radiant energy that reaches the detector be as large as possible. The *f-number F*, or *speed*, provides a measure of the ability of a monochromator to collect the radiation that emerges from the entrance slit. The f-number is defined by

$$F = f/d \qquad (7\text{-}14)$$

where f is the focal length of the collimating mirror (or lens) and d is its diameter. The light-gathering power of an optical device increases as the inverse square of the f-number. Thus, an $f/2$ lens gathers four times more light than an $f/4$ lens. The f-numbers for many monochromators lie in the 1 to 10 range.

[17]For discussion of the detection, the measurement, and the effects of stray radiation, see W. Kaye, *Anal. Chem.*, **1981**, *53*, 2201; M. R. Sharpe, *Anal. Chem.*, **1984**, *56*, 339A.

[18]J. D. Ingle Jr. and S. R. Crouch, *Spectrochemical Analysis*, pp. 71–73, Upper Saddle River, NJ: Prentice Hall, 1988.

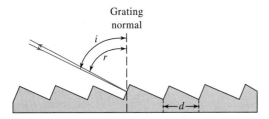

FIGURE 7-22 Echelle grating: i = angle of incidence; r = angle of reflection; d = groove spacing. In usual practice, $i \approx r = \beta = 63°26'$.

Echelle Monochromators. Echelle monochromators contain two dispersing elements arranged in series. The first of these elements is a special type of grating called an *echelle grating*. The second, which follows, is usually a low-dispersion prism, or sometimes a grating. The echelle grating, which was first described by G. R. Harrison in 1949, provides higher dispersion and higher resolution than an echellette of the same size.[19] Figure 7-22 shows a cross section of a typical echelle grating. It differs from the echellette grating shown in Figure 7-21 in several ways. First, to achieve a high angle of incidence, the blaze angle of an echelle grating is significantly greater than the conventional device, and the short side of the blaze is used rather than the long. Furthermore, the grating is relatively coarse, having typically 300 or fewer grooves per millimeter for UV or visible radiation. Note that the angle of reflection r is much higher in the echelle grating than the echellette and approaches the angle of incidence i. That is,

$$r \approx i = \beta$$

Under these circumstances, the grating equation becomes

$$\mathbf{n}\lambda = 2d \sin \beta \qquad (7\text{-}15)$$

With a normal echellette grating, high dispersion, or low reciprocal dispersion, is obtained by making the groove width d small and the focal length f large. A large focal length reduces light gathering and makes the monochromator large and unwieldy. In contrast, the echelle grating achieves high dispersion by making both the angle β and the order of diffraction $\mathbf{n}$ large. The reciprocal linear dispersion for an echelle monochromator of focal length f is

---

[19] For a more detailed discussion of the echelle grating, see P. N. Keliher and C. C. Wohlers, *Anal. Chem.*, **1976**, *48*, 333A; D. L. Anderson, A. R. Forster, and M. L. Parsons, *Anal. Chem.*, **1981**, *53*, 770; A. T. Zander and P. N. Keliher, *Appl. Spectrosc.*, **1979**, *33*, 499.

TABLE 7-1 Comparison of Performance Characteristics of a Conventional and an Echelle Monochromator

	Conventional	Echelle
Focal length	0.5 m	0.5 m
Groove density	1200/mm	79/mm
Diffraction angle, β	10°22'	63°26'
Order $\mathbf{n}$ (at 300 nm)	1	75
Resolution (at 300 nm), $\lambda/\Delta\lambda$	62,400	763,000
Reciprocal linear dispersion, D^{-1}	16 Å/mm	1.5 Å/mm
Light-gathering power, F	$f/9.8$	$f/8.8$

With permission from P. E. Keliher and C. C. Wohlers, *Anal. Chem.*, **1976**, *48*, 334A. Copyright 1976 American Chemical Society.

$$D^{-1} = \frac{d \cos \beta}{\mathbf{n}f} \qquad (7\text{-}16)$$

Table 7-1 lists the performance characteristics for two typical monochromators: one with a conventional echellette grating and the other with an echelle grating. These data demonstrate the advantages of the echelle grating. Note that for the same focal length, the linear dispersion and resolution are an order of magnitude greater for the echelle; the light-gathering power of the echelle is also somewhat superior.

One of the problems encountered with the use of an echelle grating is that the linear dispersion at high orders of refraction is so great that to cover a reasonably broad spectral range it is necessary to use many successive orders. For example, one instrument designed to cover a range of 200 to 800 nm uses diffraction orders 28 to 118 (90 successive orders). Because these orders inevitably overlap, it is essential that a system of cross dispersion, such as that shown in Figure 7-23a, be used with an echelle grating. Here, the dispersed radiation from the grating is passed through a prism (in some systems, a second grating is used) whose axis is 90° to the grating. The effect of this arrangement is to produce a two-dimensional spectrum as shown schematically in Figure 7-23b. In this figure, the locations of 8 of the 70 orders is indicated by short vertical lines. For any given order, the wavelength dispersion is approximately linear, but, as can be seen, the dispersion is relatively small at lower orders or higher wavelengths. An actual two-dimensional spectrum from an echelle monochromator consists of a complex series of short vertical lines lying along 50 to 100 horizontal axes, each axis corresponding to one diffraction order.

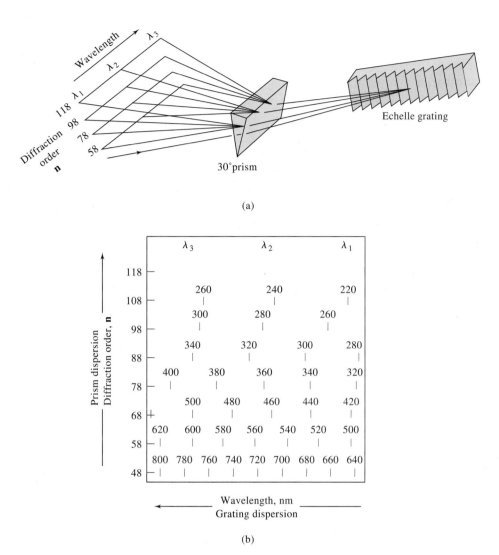

(a)

(b)

FIGURE 7-23 An echelle monochromator: (a) arrangement of dispersing elements, and (b) schematic end-on view of the dispersed radiation from the point of view of the transducer.

To change wavelength with an echelle monochromator, it is necessary to change the angle of both the grating and the prism.

Several instrument manufacturers offer echelle-type spectrometers for simultaneous determination of a multitude of elements by atomic emission spectroscopy. The optical designs of two of these instruments are shown in Figures 10-7, 10-9, and 10-11.

7C-3 Monochromator Slits

The slits of a monochromator play an important role in determining the monochromator's performance characteristics and quality. Slit jaws are formed by carefully machining two pieces of metal to give sharp edges. Care is taken to assure that the edges of the slit are exactly parallel to one another and that they lie on the same plane. In some monochromators, the openings of the two slits are fixed; more often, the spacing can be adjusted with a micrometer mechanism.

The entrance slit (see Figure 7-18) of a monochromator serves as a radiation source; its image is ultimately focused on the focal plane that contains the exit slit. If the radiation source consists of a few discrete wavelengths, a series of rectangular images appears on this surface as bright lines, each corresponding to a given wavelength. A particular line can be brought to focus on the exit slit by rotating the dispersing element.

If the entrance and exit slits are of the same width (as is usually the case), in theory the image of the entrance slit will just fill the exit-slit opening when the setting of the monochromator corresponds to the wavelength of the radiation. Movement of the monochromator setting in one direction or the other produces a continuous decrease in emitted intensity, which reaches zero when the entrance-slit image has moved a distance equal to its full width.

Effect of Slit Width on Resolution

Figure 7-24 shows monochromatic radiation of wavelength λ_2 striking the exit slit. Here, the monochromator is set for λ_2 and the two slits are identical in width. The image of the entrance slit just fills the exit slit. Movement of the monochromator to a setting of λ_1 or λ_3 results in the image being moved completely out of the slit. The lower half of Figure 7-24 shows a plot of the radiant power passing through the slit as a function of monochromator setting. Note that the *bandwidth* is defined as the span of monochromator settings (in units of wavelength, or sometimes in units of cm^{-1}) needed to

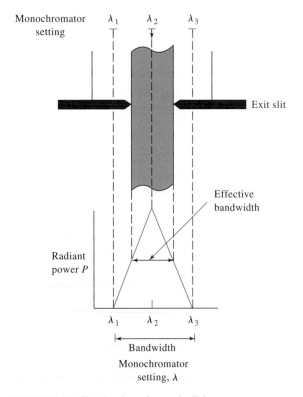

FIGURE 7-24 Illumination of an exit slit by monochromatic radiation λ_2 at various monochromator settings. Exit and entrance slits are identical.

move the image of the entrance slit across the exit slit. If polychromatic radiation were used, it would also represent the span of wavelengths from the exit slit for a given monochromator setting.

The *effective bandwidth* (also called the *spectral bandpass* or *spectral slit width*), which is one half the bandwidth when the two slit widths are identical, is seen to be the range of wavelengths that exit the monochromator at a given wavelength setting. The effective bandwidth can be related to the reciprocal linear dispersion by writing Equation 7-9 in the form

$$D^{-1} = \frac{\Delta\lambda}{\Delta y}$$

where $\Delta\lambda$ and Δy are now finite intervals of wavelength and linear distance along the focal plane, respectively. As shown by Figure 7-24, when Δy is equal to the slit width w, $\Delta\lambda$ is the effective bandwidth. That is,

$$\Delta\lambda_{\text{eff}} = wD^{-1} \qquad (7\text{-}17)$$

Figure 7-25 illustrates the relationship between the effective bandwidth of an instrument and its ability to resolve spectral peaks. Here, the exit slit of a grating monochromator is illuminated with a beam composed of just three equally spaced lines at wavelengths, λ_1, λ_2, and λ_3; each line is assumed to be of the same intensity. In the top figure, the effective bandwidth of the instrument is exactly equal to the difference in wavelength between λ_1 and λ_2 or λ_2 and λ_3. When the monochromator is set at λ_2, radiation of this wavelength just fills the slit. Movement of the monochromator in either direction diminishes the transmitted intensity of λ_2, but increases the intensity of one of the other lines by an equivalent amount. As shown by the solid line in the plot to the right, no spectral resolution of the three wavelengths is achieved.

In the middle drawing of Figure 7-25, the effective bandwidth of the instrument has been reduced by narrowing the openings of the exit and entrance slits to three quarters that of their original dimensions. The solid line in the plot on the right shows that partial resolution of the three lines results. When the effective bandwidth is decreased to one half the difference in wavelengths of the three beams, complete resolution is achieved, as shown in the bottom drawing. Thus, complete resolution of two lines is feasible only if the slit width is adjusted so that the effective bandwidth of the monochromator is equal to one half the wavelength difference of the lines.

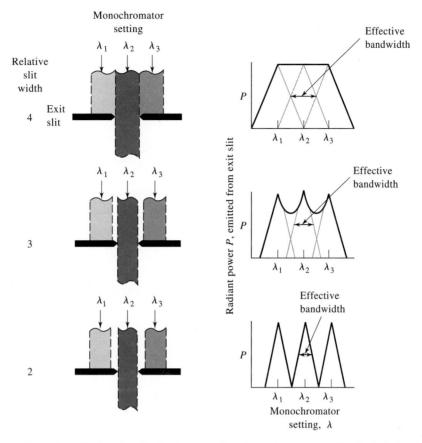

FIGURE 7-25 The effect of the slit width on spectra. The entrance slit is illuminated with λ_1, λ_2, and λ_3 only. Entrance and exit slits are identical. Plots on the right show changes in emitted power as the setting of monochromator is varied.

EXAMPLE 7-2

A grating monochromator with a reciprocal linear dispersion of 1.2 nm/mm is to be used to separate the sodium lines at 588.9950 nm and 589.5924 nm. In theory, what slit width would be required?

Solution

Complete resolution of the two lines requires that

$$\Delta\lambda_{\text{eff}} = \frac{1}{2}(589.5924 - 588.9950) = 0.2987 \text{ nm}$$

Substitution into Equation 7-17 after rearrangement gives

$$w = \frac{\Delta\lambda_{\text{eff}}}{D^{-1}} = \frac{0.2987}{1.2 \text{ nm/mm}} = 0.25 \text{ mm}$$

It is important to note that slit widths calculated as in Example 7-2 are theoretical. Imperfections present in most monochromators are such that slit widths narrower than theoretical are usually required to achieve a desired resolution.

Figure 7-26 shows the effect of bandwidth on experimental spectra for benzene vapor. Note the much greater spectral detail realized with the narrowest slit setting and thus the narrowest bandwidth.

Choice of Slit Widths

The effective bandwidth of a monochromator depends on the dispersion of the grating or prism as well as the width of the entrance and exit slits. Most monochromators are equipped with variable slits so that the effective bandwidth can be changed. The use of the minimal slit width is desirable when narrow absorption or emission bands must be resolved. On the other hand, the available radiant power decreases significantly when the slits

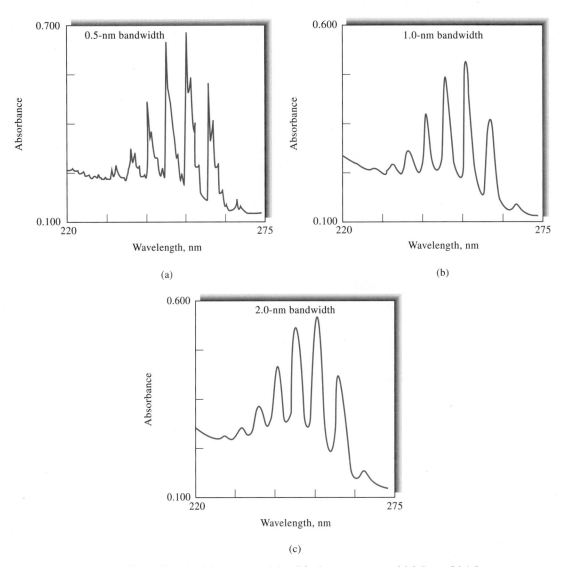

FIGURE 7-26 Effect of bandwidth on spectral detail for benzene vapor: (a) 0.5 nm, (b) 1.0 nm, (c) 2.0 nm. (From V. A. Kohler, *Amer. Lab.*, **1984**, *11*, 132. Copyright 1984 International Scientific Communications Inc. Reprinted with permission.)

are narrowed, and it becomes more difficult to measure the power accurately. Thus, wider slit widths may be used for quantitative analysis rather than for qualitative work, where spectral detail is important.

7D SAMPLE CONTAINERS

Sample containers are required for all spectroscopic studies except emission spectroscopy. Like the optical elements of monochromators, the *cells*, or *cuvettes*, that

hold the samples must be made of material that is transparent to radiation in the spectral region of interest. Thus, as shown in Figure 7-2, quartz or fused silica is required for work in the UV region (below 350 nm). Both of these substances are transparent in the visible region and up to about 3 μm in the infrared region as well. Silicate glasses can be used in the region between 350 and 2000 nm. Plastic containers have also been used in the visible region. Crystalline sodium chloride is the most common substance used for cell windows in the IR region; the other infrared transparent materials listed in Figure 7-2 may also be used for this purpose.

7E RADIATION TRANSDUCERS

7E-1 Introduction

The detectors for early spectroscopic instruments were the human eye or a photographic plate or film. Transducers that convert radiant energy into an electrical signal have almost totally replaced these detection devices. We will confine our discussion to modern transducers.

Properties of the Ideal Transducer

The ideal transducer would have a high sensitivity, a high signal-to-noise ratio, and a constant response over a considerable range of wavelengths. In addition, it would exhibit a fast response time and a zero output signal in the absence of illumination. Finally, the electrical signal produced by the ideal transducer would be directly proportional to the radiant power P. That is,

$$S = kP \qquad (7\text{-}18)$$

where S is the current or voltage output of the transducer, and k is the calibration sensitivity (Section 1E-2).

Many real transducers exhibit a small, constant response, known as a *dark current* in the absence of radiation. For these transducers, the response is described by the relationship

$$S = kP + k_d \qquad (7\text{-}19)$$

where k_d represents the dark current, which is usually constant over short measurement periods. Instruments with transducers that produce a dark current are often equipped with a compensating circuit that reduces k_d to zero; Equation 7-18 then applies.

Types of Radiation Transducers

As indicated in Figure 7-3b, there are two general types of radiation transducers.[20] One type responds to photons, the other to heat. All photon transducers (also called *photoelectric* or *quantum detectors*) have an active surface that absorbs radiation. In some types,

 Tutorial: Learn more about **radiation transducers**.

[20]For a discussion of optical radiation detectors, see J. D. Ingle Jr. and S. R. Crouch, *Spectrochemical Analysis*, pp. 106–17, Upper Saddle River, NJ: Prentice Hall, 1988; E. L. Dereniak and D. G. Crowe, *Optical Radiation Detectors*, New York: Wiley, 1984; F. Grum and R. J. Becherer, *Optical Radiation Measurements*, Vol. 1, New York: Academic Press, 1979.

the absorbed energy causes emission of electrons and the production of a photocurrent. In others, the radiation promotes electrons into conduction bands; detection here is based on the resulting enhanced conductivity (*photoconduction*). Photon transducers are used largely for measurement of UV, visible, and near-infrared radiation. When they are applied to radiation much longer than 3 μm in wavelength, these transducers must be cooled to dry-ice or liquid-nitrogen temperatures to avoid interference from thermal background noise. Photoelectric transducers produce an electrical signal resulting from a series of individual events (absorption of single photons), the probability of which can be described statistically. In contrast, thermal transducers, which are widely used for the detection of infrared radiation, respond to the average power of the incident radiation.

The distinction between photon and heat transducers is important because shot noise often limits the behavior of photon transducers and thermal noise frequently limits thermal transducers. As shown in Section 5B-2, the indeterminate errors associated with the two types of transducers are fundamentally different.

Figure 7-27 shows the relative spectral response of the various kinds of transducers that are useful for UV, visible, and IR spectroscopy. The ordinate function is inversely related to the noise of the transducers and directly related to the square root of its surface area. Note that the relative sensitivity of the thermal transducers (curves H and I) is independent of wavelength but significantly lower than the sensitivity of photoelectric transducers. On the other hand, photon transducers are often far from ideal with respect to constant response versus wavelength.

7E-2 Photon Transducers

Several types of photon transducers are available, including (1) photovoltaic cells, in which the radiant energy generates a current at the interface of a semiconductor layer and a metal; (2) phototubes, in which radiation causes emission of electrons from a photosensitive solid surface; (3) photomultiplier tubes, which contain a photoemissive surface as well as several additional surfaces that emit a cascade of electrons when struck by electrons from the photosensitive area; (4) photoconductivity transducers in which absorption of radiation by a semiconductor produces electrons and holes, thus leading to enhanced conductivity; (5) silicon photodiodes, in which photons cause the formation of

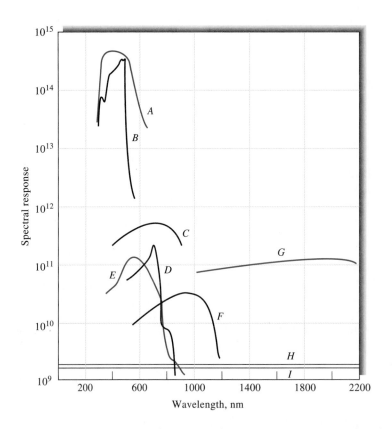

FIGURE 7-27 Relative response of various types of photoelectric transducers (A–G) and heat transducers (H, I): A, photomultiplier tube; B, CdS photoconductivity; C, GaAs photovoltaic cell; D, CdSe photoconductivity cell; E, Se/SeO photovoltaic cell; F, silicon photodiode; G, PbS photoconductivity; H, thermocouple; I, Golay cell. (Adapted from P. W. Druse, L. N. McGlauchlin, and R. B. Quistan, *Elements of Infrared Technology*, pp. 424–25, New York: Wiley, 1962. Reprinted by permission of John Wiley & Sons Inc.)

electron-hole pairs and a current across a reverse-biased *pn* junction; and (6) charge-transfer transducers, in which the charges developed in a silicon crystal as a result of absorption of photons are collected and measured.[21]

Barrier-Layer Photovoltaic Cells

The photovoltaic cell is a simple device that is used for detecting and measuring radiation in the visible region. The typical cell has a maximum sensitivity at about 550 nm; the response falls off to about 10% of

[21] For a comparison of the performance characteristics of the three most sensitive and widely used photon transducers — photomultipliers, silicon diodes, and charge-transfer devices — see W. E. L. Grossman, *J. Chem. Educ.*, **1989**, *66*, 697.

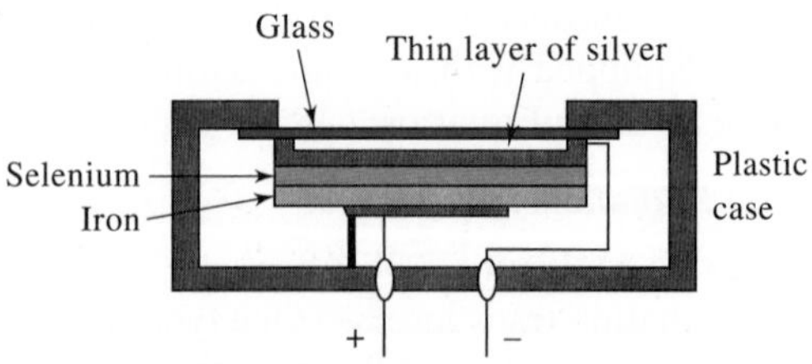

FIGURE 7-28 Schematic of a typical barrier-layer cell.

the maximum at 350 and 750 nm (see curve E in Figure 7-27). Its range is similar to that of the human eye.

The photovoltaic cell consists of a flat copper or iron electrode on which is deposited a layer of semiconducting material, such as selenium (see Figure 7-28). The outer surface of the semiconductor is coated with a thin transparent metallic film of gold or silver, which serves as the second, or collector, electrode; the entire array is protected by a transparent envelope. When radiation of sufficient energy reaches the semiconductor, electrons and holes are formed. The electrons that have been promoted to the conduction band then migrate toward the metallic film and the holes toward the base on which the semiconductor is deposited. The liberated electrons are free to migrate through the external circuit to interact with these holes. The result is an electrical current of a magnitude that is proportional to the number of photons that strike the semiconductor surface. The currents produced by a photovoltaic cell are usually large enough to be measured with a microammeter; if the resistance of the external circuit is kept small ($<400\ \Omega$), the photocurrent is directly proportional to the radiant power that strikes the cell. Currents on the order of 10 to 100 μA are typical.

The barrier-layer cell constitutes a rugged, low-cost means for measuring radiant power. No external source of electrical energy is required. On the other hand, the low internal resistance of the cell makes the amplification of its output less convenient. Consequently, although the barrier-layer cell provides a readily measured response at high levels of illumination, it suffers from lack of sensitivity at low levels. Another disadvantage of the barrier-type cell is that it exhibits *fatigue* in which its current output decreases gradually during continued illumination; proper circuit design and choice of experimental conditions minimize this effect. Barrier-type cells are used in simple, portable instruments where ruggedness and low cost are important. For routine analyses, these instruments often provide perfectly reliable analytical data.

Vacuum Phototubes

A second type of photoelectric device is the vacuum phototube,[22] which consists of a semicylindrical cathode and a wire anode sealed inside an evacuated transparent envelope (see Figure 7-29). The concave surface of the electrode supports a layer of photoemissive material (Section 6C-1) that tends to emit electrons when it is irradiated. When a voltage is applied across the electrodes, the emitted electrons flow to the wire anode generating a photocurrent that is generally about one tenth as great as that associated with a photovoltaic cell for a given radiant intensity. In contrast,

[22]For a very practical, although somewhat dated, discussion of vacuum phototubes and photomultiplier tubes and circuits, see F. E. Lytle, *Anal. Chem.*, **1974**, *46*, 545A.

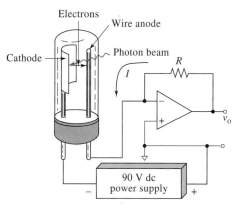

FIGURE 7-29 A phototube and op amp readout. The photocurrent induced by the radiation causes a voltage drop across R, which appears as v_o at the output of the current-to-voltage converter. This voltage may be displayed on a meter or acquired by a data-acquisition system.

however, amplification is easily accomplished because the phototube has a high electrical resistance.

The number of electrons ejected from a photoemissive surface is directly proportional to the radiant power of the beam that strikes that surface. As the voltage applied across the two electrodes of the tube is increased, the fraction of the emitted electrons that reaches the anode rapidly increases; when the saturation voltage is achieved, essentially all of the electrons are collected at the anode. The current then becomes independent of voltage and directly proportional to the radiant power. Phototubes are usually operated at a voltage of about 90 V, which is well within the saturation region.

A variety of photoemissive surfaces are used in commercial phototubes. Typical examples are shown in Figure 7-30. From the user's standpoint, photoemissive surfaces fall into four categories: highly sensitive, red sensitive, ultraviolet sensitive, and flat response. The most sensitive cathodes are bialkali types such as number 117 in Figure 7-30; they are made up of potassium, cesium, and antimony. Red-sensitive materials are multialkali types (for example, Na/K/Cs/Sb) or Ag/O/Cs formulations. The behavior of the Ag/O/Cs surface is shown as S-11 in the figure. Compositions of Ga/In/As extend the red region up to about 1.1 μm. Most formulations are ultraviolet sensitive if the tube is equipped with UV-transparent windows. Relatively flat responses are obtained with Ga/As compositions such as that labeled 128 in Figure 7-30.

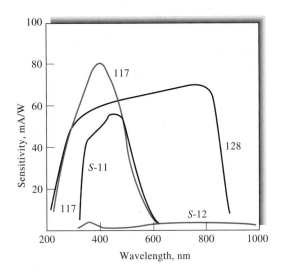

FIGURE 7-30 Spectral response of some typical photoemissive surfaces. (From F. E. Lytle, *Anal. Chem.*, **1974**, *46*, 545A. Figure 1, p. 546A. Copyright 1974 American Chemical Society.)

Phototubes frequently produce a small dark current (see Equation 7-19) that results from thermally induced electron emission and natural radioactivity from ^{40}K in the glass housing of the tube.

Photomultiplier Tubes

For the measurement of low radiant powers, the *photomultiplier tube* (PMT) offers advantages over an ordinary phototube.[23] Figure 7-31 is a schematic of such a device. The photocathode surface is similar in composition to the surfaces of the phototubes described in Figure 7-30, and it emits electrons when exposed to radiation. The tube also contains additional electrodes (nine in Figure 7-31) called *dynodes*. Dynode D1 is maintained at a voltage approximately 90 V more positive than the cathode, and electrons are accelerated toward it as a result. Each photoelectron that strikes the dynode causes emission of several additional electrons. These electrons, in turn, are accelerated toward dynode D2, which is ~90 V more positive than dynode D1. Again, several electrons are emitted for each electron that strikes the surface. By the time this process has been repeated nine times, 10^6 to 10^7 electrons have been formed for each incident photon. This cascade of

electrons is finally collected at the anode and the resulting current is then converted to voltage and measured.

Curve *A* in Figure 7-27 shows that photomultipliers are highly sensitive to ultraviolet and visible radiation. In addition, they have extremely fast response times. Often, the sensitivity of an instrument with a photomultiplier is limited by its dark current. Because thermal emission is the major source of dark-current electrons, the performance of a photomultiplier can be enhanced by cooling. In fact, thermal dark currents can be virtually eliminated by cooling the detector to −30°C. Transducer housings, which can be cooled by circulation of an appropriate coolant, are available commercially.

Photomultiplier tubes are limited to measuring low-power radiation because intense light causes irreversible damage to the photoelectric surface. For this reason, the device is always housed in a light-tight compartment and care is taken to eliminate the possibility of its being exposed even momentarily to daylight or other strong light while powered. With appropriate external circuitry, photomultiplier tubes can be used to detect the arrival of individual photons at the photocathode.

Silicon Photodiode Transducers

A silicon photodiode transducer consists of a reverse-biased *pn* junction formed on a silicon chip. As shown in Figure 7-32, the reverse bias creates a depletion layer that reduces the conductance of the junction to nearly zero. If radiation impinges on the chip, however, holes and electrons are formed in the depletion layer and swept through the device to produce a current that is proportional to radiant power. They require only low-voltage power supplies or can be operated under zero bias and so can be used in portable, battery-powered instruments.

Silicon diodes are more sensitive than vacuum phototubes but less sensitive than photomultiplier tubes (see curve *F* in Figure 7-27). Photodiodes have spectral ranges from about 190 to 1100 nm.

7E-3 Multichannel Photon Transducers

The first multichannel detector used in spectroscopy was a photographic plate or a film strip that was placed along the length of the focal plane of a spectrometer so that all the lines in a spectrum could be recorded simultaneously. Photographic detection is relatively sensitive, with some emulsions that respond to as few

---

[23] For a detailed discussion of the theory and applications of photomultipliers, see R. W. Engstrom, *Photomultiplier Handbook*, Lancaster, PA: RCA Corporation, 1980.

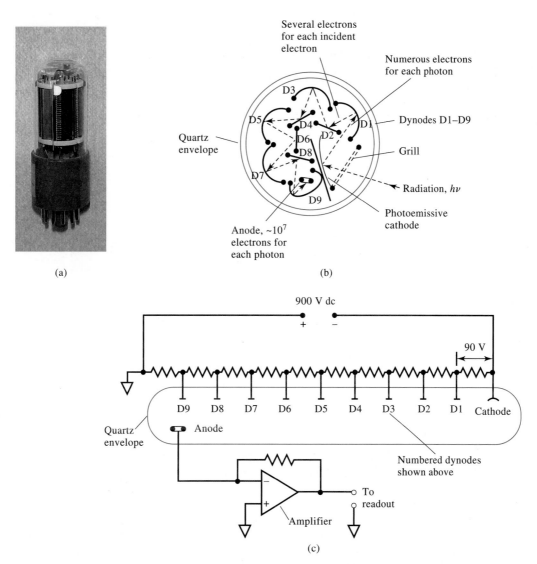

FIGURE 7-31 Photomultiplier tube: (a), photograph of a typical commercial tube; (b), cross-sectional view; (c), electrical diagram illustrating dynode polarization and photocurrent measurement. Radiation striking the photosensitive cathode (b) gives rise to photoelectrons by the photoelectric effect. Dynode D1 is held at a positive voltage with respect to the photocathode. Electrons emitted by the cathode are attracted to the first dynode and accelerated in the field. Each electron striking dynode D1 thus gives rise to two to four secondary electrons. These are attracted to dynode D2, which is again positive with respect to dynode D1. The resulting amplification at the anode can be 10^6 or greater. The exact amplification factor depends on the number of dynodes and the voltage difference between each. This automatic internal amplification is one of the major advantages of photomultiplier tubes. With modern instrumentation, the arrival of individual photocurrent pulses can be detected and counted instead of being measured as an average current. This technique, called photon counting, is advantageous at very low light levels.

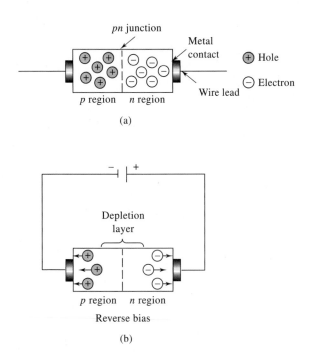

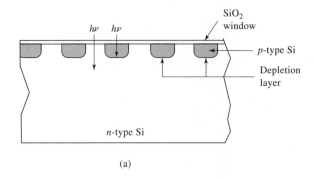

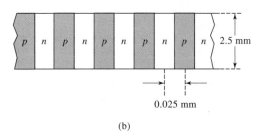

FIGURE 7-32 (a) Schematic of a silicon diode. (b) Formation of depletion layer, which prevents flow of electricity under reverse bias.

FIGURE 7-33 A reverse-biased linear diode-array detector: (a) cross section and (b) top view.

as 10 to 100 photons. The primary limitation of this type of detector, however, is the time required to develop the image of the spectrum and convert the blackening of the emulsion to radiant intensities.

Modern multichannel transducers[24] consist of an array of small photosensitive elements arranged either linearly or in a two-dimensional pattern on a single semiconductor chip. The chip, which is usually silicon and typically has dimensions of a few millimeters on a side, also contains electronic circuitry to provide an output signal from each of the elements either sequentially or simultaneously. For spectroscopic studies, a multichannel transducer is generally placed in the focal plane of a spectrometer so that various elements of the dispersed spectrum can be transduced and measured simultaneously.

Three types of multichannel devices are used in commercial spectroscopic instruments: *photodiode arrays* (PDAs), *charge-injection devices* (CIDs), and *charge-*

coupled devices* (CCDs). PDAs are one-dimensional transducers in which the photosensitive elements are arranged linearly on the transducer face. In contrast, the individual photosensitive elements of CIDs and CCDs are usually formed as two-dimensional arrays. Charge-injection and charge-coupled transducers both function by collecting photogenerated charges in various areas of the transducer surface and then measuring the quantity of charge accumulated in a brief period. In both devices, the measurement is accomplished by transferring the charge from a collection area to a detection area. For this reason, the two types of transducers are sometimes called *charge-transfer devices* (CTDs). These devices have widespread use as image transducers for television applications and in astronomy.

Photodiode Arrays

In a PDA, the individual photosensitive elements are small silicon photodiodes, each of which consists of a reverse-biased *pn* junction (see previous section). The individual photodiodes are part of a large-scale integrated circuit formed on a single silicon chip. Figure 7-33 shows the geometry of the surface region of a few of the transducer elements. Each element consists

[24]For a discussion of multichannel photon detectors, see J. M. Harnly and R. E. Fields, *Appl. Spectros.*, **1997**, *51*, 334A; M. Bonner Denton, R. Fields, and Q. S. Hanley, eds., *Recent Developments in Scientific Optical Imaging*, Cambridge, UK: Royal Soc. Chem., 1996; J. V. Sweedler, K. L. Ratzlaff, and M. B. Denton, eds., *Charge-Transfer Devices in Spectroscopy*, New York: VCH, 1994; J. V. Sweedler, *Crit. Rev. Anal. Chem.*, **1993**, *24*, 59.

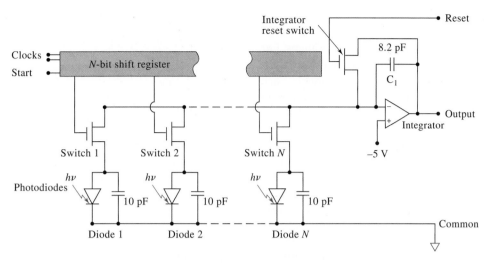

FIGURE 7-34 Schematic of a photodiode-array detector chip.

of a diffused *p*-type bar in an *n*-type silicon substrate to give a surface region that consists of a series of side-by-side elements that have typical dimensions of 2.5 by 0.025 mm (Figure 7-33b). Radiation that is incident on these elements creates charges in both the *p* and *n* regions. The positive charges are collected and stored in the *p*-type bars for subsequent integration (the charges formed in the *n*-regions divide themselves proportionally between the two adjacent *p*-regions). The number of transducer elements in a chip ranges from 64 to 4096, with 1024 being perhaps the most widely used.

The integrated circuit that makes up a diode array also contains a storage capacitor and switch for each diode as well as a circuit for sequentially scanning the individual diode-capacitor circuits. Figure 7-34 is a simplified diagram that shows the arrangement of these components. Note that in parallel with each photodiode is a companion 10-pF storage capacitor. Each diode-capacitor pair is sequentially connected to a common output line by the *N*-bit shift register and the transistor switches. The shift register sequentially closes each of these switches, momentarily causing the capacitor to be charged to -5 V, which then creates a reverse bias across the *pn* junction of the detector. Radiation that impinges on the depletion layer in either the *p* or the *n* region forms charges (electrons and holes) that create a current that partially discharges the capacitor in the circuit. The capacitor charge that is lost in this way is replaced during the next cycle. The resultant charging current is integrated by the preamplifier circuit, which produces a voltage that is proportional to the radiant intensity. After amplification, the analog signal from the preamplifier passes into an analog-to-digital converter and to a computer that controls the readout.

In using a diode-array transducer, the slit width of the spectrometer is usually adjusted so that the image of the entrance slit just fills the surface area of one of the diodes that make up the array. Thus, the information obtained is equivalent to that recorded during scanning with a traditional spectrophotometer. With the array, however, information about the entire spectrum is accumulated essentially simultaneously and in discrete elements rather than in a continuous way.

Some of the photoconductor transducers mentioned in the previous section can also be fabricated into linear arrays for use in the infrared region.

Charge-Transfer Devices

PDAs cannot match the performance of photomultiplier tubes with respect to sensitivity, dynamic range, and signal-to-noise ratio. Thus, they have been used most often for applications in which high sensitivity and large dynamic range is not needed, such as in absorption spectrometry. On the other hand, the performance characteristics of CTDs approach or sometimes surpass those of photomultiplier tubes in addition to having the multichannel advantage. As a result, this type of transducer is now appearing in ever-increasing numbers in modern spectroscopic instruments.[25] A

---
[25]For details on CTDs, see J. V. Sweedler, K. L. Ratzlaff, and M. B. Denton, eds., *Charge-Transfer Devices in Spectroscopy*, New York: VCH, 1994; J. V. Sweedler, *Crit. Rev. Anal. Chem.*, **1993**, *24*, 59; J. V. Sweedler, R. B. Bilhorn, P. M. Epperson, G. R. Sims, and M. B. Denton, *Anal. Chem.*, **1988**, *60*, 282A, 327A.

further advantage of CTDs is that they are usually two dimensional in the sense that individual transducer elements are arranged in rows and columns. For example, one detector that we describe in the next section consists of 244 rows of transducer elements, each row composed of 388 detector elements, resulting in a two-dimensional array of 19,672 individual transducers, or *pixels*, on a silicon chip with dimensions of 6.5 mm by 8.7 mm. With this device, it is possible to simultaneously record an entire two-dimensional spectrum from an echelle spectrometer (Figure 7-23).

CTDs operate like a photographic film in the sense that they integrate signal information as radiation strikes them. Figure 7-35 is a cross-sectional depiction of one of the pixels in a charge-transfer array. In this case, the pixel consists of a pair of conductive electrodes that overlie an insulating layer of silica (note that a pixel in some CTDs is made up of more than two electrodes). This silica layer separates the electrodes from a region of *n*-doped silicon. This assembly constitutes a metal oxide semiconductor capacitor that stores the charges formed when radiation strikes the doped silicon. When, as shown, a negative voltage is applied to the electrodes, a charge inversion region is created under the electrodes that is energetically favorable for the storage of holes. The mobile holes created by the absorption of photons then migrate and collect in this region. Typically, this region, which is called a *potential well*, is capable of holding as many as 10^5 to 10^6 charges before overflowing into an adjacent pixel. In the figure, one electrode is shown as more negative than the other, making the accumulation of charge under this electrode more favorable. The amount of charge generated during exposure to radiation is measured in either of two ways. In a *charge-injection device*, the voltage change that arises from movement of the charge from the region under one electrode to the region under the other is measured. In a *charge-coupled device*, the charge is moved to a charge-sensing amplifier for measurement.

Charge-Injection Devices. Figure 7-36 is a simplified diagram that shows the steps involved in the collection, storage, and measurement of the charge generated when one pixel of a semiconductor is exposed to photons. To monitor the intensity of the radiation that strikes the sensor element, the voltages applied to the capacitors are cycled as shown in steps (a) through (d) in the figure. In step (a), negative voltages are applied to the two electrodes, which leads to formation of po-

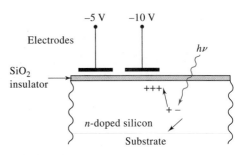

FIGURE 7-35 Cross section of a CTD in the charge integration mode. The positive hole produced by the photon $h\nu$ is collected under the negative electrode.

tential wells that collect and store holes formed in the *n* layer by absorption of photons. Because the electrode on the right is at a more negative voltage, all the holes are initially retained under this electrode. The magnitude of the charge collected in some brief time interval is determined in steps (b) and (c). In (b), the voltage of the capacitor on the left (V_1) is determined after removal of its applied voltage. In step (c), the holes that have accumulated on the right electrode are transferred to the potential well under the left electrode by switching the voltage applied to the right electrode from negative to positive. The new voltage of the electrode V_2 is then measured. The magnitude of the accumulated charge is determined from the difference in voltage ($V_1 - V_2$). In step (d), the detector is returned to its original state by applying positive voltages to both electrodes, which cause the holes to migrate toward the substrate. As an alternative to step (d), however, the detector can be returned to the condition shown in (a) without the loss of charge that has already accumulated. This process is called the *nondestructive readout* (NDRO) *mode*. A major advantage of CIDs over CCDs is that successive measurements can be made while integration is taking place.

As was true for the diode-array detector, the chip that contains the array of charge-injection transducer elements also contains appropriate integrated circuits for performing the cycling and measuring steps.

Charge-Coupled Device. CCDs are marketed by several manufacturers and come in a variety of shapes and forms. Figure 7-37a illustrates the arrangement of individual detectors in a typical array that is made up of 512×320 pixels. Note that in this case the semiconductor is formed from *p*-type silicon, and the capacitor is biased positively so that electrons formed by

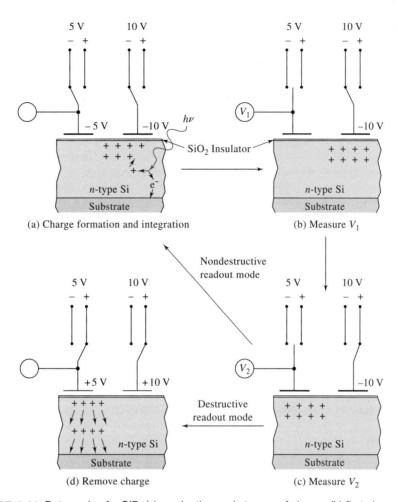

FIGURE 7-36 Duty cycle of a CID: (a) production and storage of charge, (b) first charge measurement, (c) second charge measurement after charge transfer, (d) reinjection of charge into the semiconductor.

the absorption of radiation collect in the potential well below the electrode and holes move away from the *n*-type layer toward the substrate. Note also that each pixel is made up of three electrodes (labeled 1, 2, and 3 in Figure 7-37b) rather than two electrodes as in a CID. To measure the accumulated charge, a three-phase clock circuit is used to shift the charge in a step-wise manner to the right to the high-speed shift register shown in Figure 7-37a. The charges are then transferred downward to a preamplifier and then to the readout. Thus, a row-by-row scan of the detector surface is accomplished. In contrast to a CID, the readout in this case neutralizes the accumulated charge. A CCD offers the advantage of greater sensitivity to low light levels. A disadvantage in some applications, however, is the destructive nature of the readout process.

Linear CCD arrays and cameras are now available with pixel dimensions as great as 12,000. These devices provide very high-resolution detection of spectral information in an increasingly broad range of analytical applications. Cameras with built-in USB ports for data transfer to computers for analysis, storage, and presentation of data can be purchased from a number of manufacturers.

Giles et al.[26] have presented a comprehensive guide for selecting CCD cameras for spectroscopic applications. They provide detailed discussion of array materials, sensitivity, quantum efficiency, noise considerations, dynamic range, resolution, readout modes, and

[26] J. H. Giles, T. D. Ridder, R. H. Williams, D. A. Jones, and M. B. Denton, *Anal. Chem.*, **1998**, 70 (19), 663A.

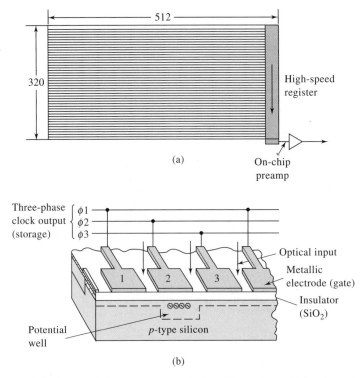

FIGURE 7-37 A CCD array: (a) arrangement of 512 × 320 pixels and (b) schematic showing four of the individual detectors.

hardware and software issues associated with the acquisition and use of these devices.

7E-4 Photoconductivity Transducers

The most sensitive transducers for monitoring radiation in the near-infrared region (0.75 to 3 μm) are semiconductors whose resistances decrease when they absorb radiation within this range. The useful range of photoconductors can be extended into the far-infrared region by cooling to suppress noise that arises from thermally induced transitions among closely lying energy levels. This application of photoconductors is important in infrared Fourier transform instrumentation. Crystalline semiconductors are formed from the sulfides, selenides, and stibnides of such metals as lead, cadmium, gallium, and indium. Absorption of radiation by these materials promotes some of their bound electrons into an energy state in which they are free to conduct electricity. The resulting change in conductivity can then be measured with a circuit such as that shown in Figure 3-10a.

Lead sulfide is the most widely used photoconductive material that can be used at room temperature.

Lead sulfide transducers are sensitive in the region between 0.8 and 3 μm (12,500 to 3300 cm^{-1}). A thin layer of this compound is deposited on glass or quartz plates to form the cell. The entire assembly is then sealed in an evacuated container to protect the semiconductor from reaction with the atmosphere. The sensitivity of cadmium sulfide, cadmium selenide, and lead sulfide transducers is shown by curves *B*, *D*, and *G* in Figure 7-27.

7E-5 Thermal Transducers

The convenient phototransducers just considered are generally not applicable in the infrared because photons in this region lack the energy to cause photoemission of electrons. Thus, thermal transducers or photoconductive transducers (see Section 7E-4) must be used. Neither of these is as satisfactory as photon transducers.

In thermal transducers,[27] the radiation impinges on and is absorbed by a small blackbody; the resultant

---

[27] For a good discussion of optical radiation transducers of all types, including thermal detectors, see E. L. Dereniak and G. D. Growe, *Optical Radiation Detectors*, New York: Wiley, 1984.

temperature rise is measured. The radiant power level from a typical infrared beam is extremely small (10^{-7} to 10^{-9} W), so that the heat capacity of the absorbing element must be as small as possible if a temperature change is to be detected. The size and thickness of the absorbing element is minimized and the entire infrared beam is concentrated on its surface so that, under optimal conditions, temperature changes are limited to a few thousandths of a kelvin.

The problem of measuring infrared radiation by thermal means is compounded by thermal noise from the surroundings. For this reason, thermal detectors are housed in a vacuum and are carefully shielded from thermal radiation emitted by other nearby objects. To further minimize the effects of extraneous heat sources, the beam from the source is generally chopped. In this way, the analyte signal, after transduction, has the frequency of the chopper and can be separated electronically from extraneous noise, which usually varies slowly with time.

Thermocouples

In its simplest form, a thermocouple consists of a pair of junctions formed when two pieces of a metal such as copper are fused to each end of a dissimilar metal such as constantan as shown in Figure 3-13. A voltage develops between the two junctions that varies with the *difference* in their temperatures.

The transducer junction for infrared radiation is formed from very fine wires of bismuth and antimony or alternatively by evaporating the metals onto a nonconducting support. In either case, the junction is usually blackened to improve its heat-absorbing capacity and sealed in an evacuated chamber with a window that is transparent to infrared radiation.

The reference junction, which is usually housed in the same chamber as the active junction, is designed to have a relatively large heat capacity and is carefully shielded from the incident radiation. Because the analyte signal is chopped, only the difference in temperature between the two junctions is important; therefore, the reference junction does not need to be maintained at constant temperature. To enhance sensitivity, several thermocouples may be connected in series to fabricate what is called a *thermopile*.

A well-designed thermocouple transducer is capable of responding to temperature differences of 10^{-6} K. This difference corresponds to a potential difference of about 6 to 8 $\mu V/\mu W$. The thermocouple of an infrared detector is a low-impedance device that is

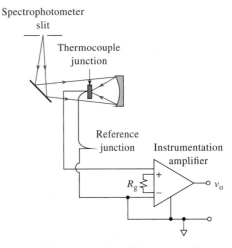

FIGURE 7-38 Thermocouple and instrumentation amplifier. The output voltage v_o is proportional to the thermocouple voltage. The magnitude of v_o is determined by the gain resistor R_g, which performs the same function as R_1/a in Figure 5-4.

usually connected to a high-impedance differential amplifier, such as the instrumentation amplifier shown in Figure 7-38. Instrumentation amplifiers are discussed in Section 5C-1. Difference amplifiers, such as the one shown in Figure 3-13, are also used for signal conditioning in thermocouple detector circuits.

Bolometers

A bolometer is a type of resistance thermometer constructed of strips of metals, such as platinum or nickel, or of a semiconductor. Semiconductor bolometers are often called *thermistors*. These materials exhibit a relatively large change in resistance as a function of temperature. The responsive element is kept small and blackened to absorb radiant heat. Bolometers are not so extensively used as other infrared transducers for the mid-infrared region. However, a germanium bolometer, operated at 1.5 K, is nearly an ideal transducer for radiation in the 5 to 400 cm^{-1} (2000 to 25 μm) range.

Pyroelectric Transducers

Pyroelectric transducers are constructed from single crystalline wafers of *pyroelectric materials*, which are insulators (dielectric materials) with very special thermal and electrical properties. Triglycine sulfate $(NH_2CH_2COOH)_3 \cdot H_2SO_4$ (usually deuterated or with a fraction of the glycines replaced with alanine), is the most important pyroelectric material used in the construction of infrared transducers.

When an electric field is applied across any dielectric material, electric polarization takes place whose magnitude is a function of the dielectric constant of the material. For most dielectrics, this induced polarization rapidly decays to zero when the external field is removed. Pyroelectric substances, in contrast, retain a strong temperature-dependent polarization after removal of the field. Thus, by sandwiching the pyroelectric crystal between two electrodes (one of which is infrared transparent) a temperature-dependent capacitor is produced. Changing its temperature by irradiating it with infrared radiation alters the charge distribution across the crystal, which creates a measurable current in an external electrical circuit that connects the two sides of the capacitor. The magnitude of this current is proportional to the surface area of the crystal and to its rate of change of polarization with temperature. Pyroelectric crystals lose their residual polarization when they are heated to a temperature called the *Curie point*. For triglycine sulfate, the Curie point is 47°C.

Pyroelectric transducers exhibit response times that are fast enough to allow them to track the changes in the time-domain signal from an interferometer. For this reason, most Fourier transform infrared spectrometers use this type of transducer.

7F SIGNAL PROCESSORS AND READOUTS

The signal processor is usually an electronic device that amplifies the electrical signal from the transducer. In addition, it may alter the signal from dc to ac (or the reverse), change the phase of the signal, and filter it to remove unwanted components. Furthermore, the signal processor may perform such mathematical operations on the signal as differentiation, integration, or conversion to a logarithm.

Several types of readout devices are found in modern instruments. Some of these devices include the D'Arsonval meter, digital meters, recorders, cathode-ray tubes, LCD panels, and computer displays.

7F-1 Photon Counting

The output from a photomultiplier tube consists of a pulse of electrons for each photon that reaches the detector surface. This analog signal is often filtered to remove undesirable fluctuations due to the random appearance of photons at the photocathode and measured as a dc voltage or current. If, however, the intensity of the radiation is too low to provide a satisfactory signal-to-noise ratio, it is possible, and often advantageous, to process the signal to a train of digital pulses that may then be counted as discussed in Section 4C. Here, radiant power is proportional to the number of pulses per unit time rather than to an average current or voltage. This type of measurement is called *photon counting*.

Counting techniques have been used for many years for measuring the radiant power of X-ray beams and of radiation produced by the decay of radioactive species (these techniques are considered in detail in Chapters 12 and 32). Photon counting is also used for ultraviolet and visible radiation, but in this application, it is the output of a photomultiplier tube that is counted.[28] In the previous section we indicated that a single photon that strikes the cathode of a photomultiplier ultimately leads to a cascade of 10^6 to 10^7 electrons, which produces a pulse of charge that can be amplified and counted.

Generally, the equipment for photon counting is similar to that shown in Figure 4-2 in which a comparator rejects pulses unless they exceed some predetermined minimum voltage. Comparators are useful for this task because dark-current and instrument noise are often significantly smaller than the signal pulse and are thus not counted; an improved signal-to-noise ratio results.

Photon counting has a number of advantages over analog-signal processing, including improved signal-to-noise ratio, sensitivity to low radiation levels, improved precision for a given measurement time, and lowered sensitivity to photomultiplier tube voltage and temperature fluctuations. The required equipment is, however, more complex and expensive. As a result, the technique has thus not been widely applied for routine molecular absorption measurements in the ultraviolet and visible regions where high sensitivity is not required. However, it has become the detection method of choice in fluorescence, chemiluminescence, and Raman spectrometry, where radiant power levels are low.

7G FIBER OPTICS

In the late 1960s, analytical instruments began to appear on the market that contained fiber optics for transmitting radiation and images from one component of

---

[28]For a review of photon counting, see H. J. Malmstadt, M. L. Franklin, and G. Horlick, *Anal. Chem.*, **1972**, *44* (8), 63A.

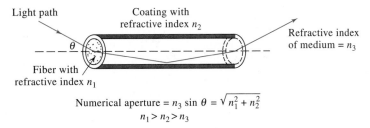

Numerical aperture $= n_3 \sin \theta = \sqrt{n_1^2 + n_2^2}$

$n_1 > n_2 > n_3$

FIGURE 7-39 Schematic of the light path through an optical fiber.

the instrument to another. Fiber optics have added a new dimension of utility to optical instrument designs.[29]

7G-1 Properties of Optical Fibers

Optical fibers are fine strands of glass or plastic that transmit radiation for distances of several hundred feet or more. The diameter of optical fibers ranges from 0.05 μm to as large as 0.6 cm. Where images are to be transmitted, bundles of fibers, fused at the ends, are used. A major application of these fiber bundles has been in medical diagnoses, where their flexibility permits transmission of images of organs through tortuous pathways to the physician. Fiber optics are used not only for observation but also for illumination of objects. In such applications, the ability to illuminate without heating is often very important.

Light transmission in an optical fiber takes place by total internal reflection as shown in Figure 7-39. For total internal reflections to occur, it is necessary that the transmitting fiber be coated with a material that has a refractive index that is somewhat smaller than the refractive index of the fiber material. Thus, a typical glass fiber has a core with a refractive index of about 1.6 and has a glass coating with a refractive index of approximately 1.5. Typical plastic fibers have a polymethylmethacrylate core with a refractive index of 1.5 and have a polymer coating with a refractive index of 1.4.

A fiber (Figure 7-39) will transmit radiation contained in a limited incident cone with a half angle shown as θ in the figure. The *numerical aperture* of the fiber provides a measure of the magnitude of the so-called *acceptance cone*.

By suitable choice of construction materials, fibers that will transmit ultraviolet, visible, or infrared radiation can be manufactured. You will find several examples of their application to conventional analytical instruments in the chapters that follow.

7G-2 Fiber-Optic Sensors

Fiber-optic sensors, which are sometimes called *optrodes*, consist of a reagent phase immobilized on the end of a fiber optic.[30] Interaction of the analyte with the reagent creates a change in absorbance, reflectance, fluorescence, or luminescence, which is then transmitted to a detector via the optical fiber. Fiber-optic sensors are generally simple, inexpensive devices that are easily miniaturized. They are used extensively for sensing biological materials, where they are often called *biosensors*. In fact, such sensors have been miniaturized to the nanometer scale. These devices are called *nanobiosensors*.[31]

7H TYPES OF OPTICAL INSTRUMENTS

In this section, we define the terminology we use to describe various types of optical instruments. It is important to understand that the nomenclature proposed here is not agreed upon and used by all scientists; it is simply a common nomenclature and the one that will be encountered throughout this book.

A *spectroscope* is an optical instrument used for the *visual* identification of atomic emission lines. It consists of a monochromator, such as one of those shown in Figure 7-18, in which the exit slit is replaced by an eyepiece that can be moved along the focal plane. The wavelength of an emission line can then be determined

[29] For a review of applications of fiber optics, see I. Chabay, *Anal. Chem.*, **1982**, *54*, 1071A and J. K. Crum, *Anal. Chem.*, **1969**, *41*, 26A.

[30] M. A. Arnold, *Anal. Chem.*, **1992**, *64*, 1015A; R. E. Dessy, *Anal. Chem.*, **1989**, *61*, 1079A; W. R. Seitz, *Anal. Chem.*, **1984**, *56*, 16A; S. Borman, *Anal. Chem.*, **1987**, *59*, 1161A; Ibid., **1986**, *58*, 766A.

[31] T. Vo-Dinh, in *Encyclopedia of Nanoscience and Nanotechnology*, Stevenson Ranch, CA: American Scientific Publishers, **2004**, *6*, 53–59; T. Vo-Dinh, *J. Cell. Biochem.* Suppl., **2002**, *39*, 154.

from the angle between the incident and dispersed beams when the line is centered on the eyepiece.

We use the term *colorimeter* to designate an instrument for absorption measurements in which the human eye serves as the detector using one or more color-comparison standards. A *photometer* consists of a source, a filter, and a photoelectric transducer as well as a signal processor and readout. Note that some scientists and instrument manufacturers refer to photometers as colorimeters or photoelectric colorimeters. Filter photometers are commercially available for absorption measurements in the ultraviolet, visible, and infrared regions, as well as emission and fluorescence in the first two wavelength regions. Photometers designed for fluorescence measurements are also called *fluorometers*.

A *spectrograph* is similar in construction to the two monochromators shown in Figure 7-18 except that the slit arrangement is replaced with a large aperture that holds a detector or transducer that is continuously exposed to the entire spectrum of dispersed radiation. Historically, the detector was a photographic film or plate. Currently, however, diode arrays or CTDs are often used as transducers in spectrographs.

A *spectrometer* is an instrument that provides information about the intensity of radiation as a function of wavelength or frequency. The dispersing modules in some spectrometers are multichannel so that two or more frequencies can be viewed simultaneously. Such instruments are sometimes called *polychromators*. A *spectrophotometer* is a spectrometer equipped with one or more exit slits and photoelectric transducers that permit the determination of the ratio of the radiant power of two beams as a function of wavelength as in absorption spectroscopy. A spectrophotometer for fluorescence analysis is sometimes called a *spectrofluorometer*.

All of the instruments named in this section thus far use filters or monochromators to isolate a portion of the spectrum for measurement. A *multiplex* instrument, in contrast, obtains spectral information without first dispersing or filtering the radiation to provide wavelengths of interest. The term *multiplex* comes from communication theory, where it is used to describe systems in which many sets of information are transported simultaneously through a single channel. Multiplex analytical instruments are single-channel devices in which all components of an analytical response are collected *simultaneously*. To determine the magnitude of each of these signal components, it is necessary to modulate the analyte signal so that, subsequently, it can be decoded into its components.

Most multiplex analytical instruments depend on the *Fourier transform* (FT) for signal decoding and are thus often called Fourier transform spectrometers. Such instruments are by no means confined to optical spectroscopy. Fourier transform devices have been described for nuclear magnetic resonance spectrometry, mass spectrometry, and microwave spectroscopy. Several of these instruments are discussed in some detail in subsequent chapters. The section that follows describes the principles of operation of Fourier transform optical spectrometers.

7I PRINCIPLES OF FOURIER TRANSFORM OPTICAL MEASUREMENTS

Fourier transform spectroscopy was first developed by astronomers in the early 1950s to study the infrared spectra of distant stars; only by the Fourier technique could the very weak signals from these sources be isolated from environmental noise. The first chemical applications of Fourier transform spectroscopy, which were reported approximately a decade later, were to the energy-starved far-infrared region; by the late 1960s, instruments for chemical studies in both the far-infrared (10 to 400 cm^{-1}) and the ordinary infrared regions were available commercially. Descriptions of Fourier transform instruments for the ultraviolet and visible spectral regions can also be found in the literature, but their adoption has been less widespread.[32]

7I-1 Inherent Advantages of Fourier Transform Spectrometry

The use of Fourier transform instruments has several major advantages. The first is the *throughput*, or *Jaquinot*, *advantage*, which is realized because Fourier transform instruments have few optical elements and

[32]For more complete discussions of optical Fourier transform spectroscopy, consult the following references: A. G. Marshall and F. R. Verdun, *Fourier Transforms in NMR, Optical, and Mass Spectrometry*, New York: Elsevier, 1990; A. G. Marshall, *Fourier, Hadamard, and Hilbert Transforms in Chemistry*, New York: Plenum Press, 1982; *Transform Techniques in Chemistry*, P. R. Griffiths, ed., New York: Plenum Press, 1978. For brief reviews, see P. R. Griffiths, *Science*, **1983**, *222*, 297; W. D. Perkins, *J. Chem. Educ.*, **1986**, 63, A5, A296; L. Glasser, *J. Chem. Educ.*, **1987**, *64*, A228, A260, A306.

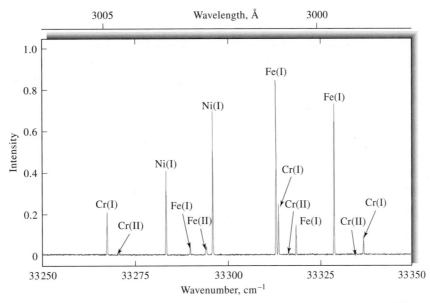

FIGURE 7-40 An iron emission spectrum illustrating the high resolving power of a Fourier transform emission spectrometer. (From A. P. Thorne, *Anal. Chem.*, **1991**, *63*, 57A. Figure 6, p. 63A. Copyright 1991 American Chemical Society.)

no slits to attenuate radiation. As a result, the radiant power that reaches the detector is much greater than in dispersive instruments and much greater signal-to-noise ratios are observed.

A second advantage of Fourier transform instruments is their extremely high resolving power and wavelength reproducibility that make possible the analysis of complex spectra in which the sheer number of lines and spectral overlap make the determination of individual spectral features difficult. Figure 7-40, which is part of an emission spectrum for a steel, illustrates this advantage. The spectrum, which extends from only 299.85 to 300.75 nm, contains thirteen well-separated lines of three elements. The wavelength resolution ($\Delta\lambda/\lambda$) for the closest pair of lines is about 6 ppm.

A third advantage arises because all elements of the source reach the detector simultaneously. This characteristic makes it possible to obtain data for an entire spectrum in 1 second or less. Let us examine the consequences of this last advantage in further detail.

For purposes of this discussion, it is convenient to think of an experimentally derived spectrum as made up of m individual transmittance measurements at equally spaced frequency or wavelength intervals called *resolution elements*. The quality of the spectrum — that is, the amount of spectral detail — increases as the number of resolution elements becomes larger or as the frequency intervals between measure-

ments become smaller.[33] Thus, to increase spectral quality, m must be made larger; clearly, increasing the number of resolution elements must also increase the time required for obtaining a spectrum with a scanning instrument.

Consider, for example, the measurement of an infrared spectrum from 500 to 5000 cm^{-1}. If resolution elements of 3 cm^{-1} were chosen, m would be 1500; if 0.5 s was required for recording the transmittance of each resolution element, 750 s, or 12.5 min, would be needed to obtain the spectrum. Reducing the width of the resolution element to 1.5 cm^{-1} would be expected to provide significantly greater spectral detail; it would also double the number of resolution elements as well as the time required for their measurement.

For most optical instruments, particularly those designed for the infrared region, decreasing the width of the resolution element has the unfortunate effect of decreasing the signal-to-noise ratio because narrower slits must be used, which lead to weaker source signals that reach the transducer. For infrared detectors, the reduction in signal strength is not, however, accompanied by a corresponding decrease in detector noise. Therefore, a degradation in signal-to-noise ratio results.

---

[33] Individual point-by-point measurements are not made with a recording spectrophotometer; nevertheless, the idea of a resolution element is useful, and the ideas generated from it apply to recording instruments as well.

In Chapter 5, we pointed out that signal-to-noise ratios can be improved substantially by signal averaging. We showed (Equation 5-11) that the signal-to-noise ratio S/N for the average of n measurements is given by

$$\left(\frac{S}{N}\right)_n = \sqrt{n}\left(\frac{S}{N}\right)_i \qquad (7\text{-}20)$$

where $(S/N)_i$ is the S/N for one measurement. The application of signal averaging to conventional spectroscopy is, unfortunately, costly in terms of time. Thus, in the example just considered, 750 s is required to obtain a spectrum of 1500 resolution elements. To improve the signal-to-noise ratio by a factor of 2 would require averaging four spectra, which would then require 4×750 s, or 50 min.

Fourier transform spectroscopy differs from conventional spectroscopy in that all of the resolution elements for a spectrum are measured *simultaneously*, thus reducing enormously the time required to obtain a spectrum at any chosen signal-to-noise ratio. An entire spectrum of 1500 resolution elements can then be recorded in about the time required to observe just one element by conventional spectroscopy (0.5 s in our earlier example). This large decrease in observation time is often used to significantly enhance the signal-to-noise ratio of Fourier transform measurements. For example, in the 750 s required to determine the spectrum by scanning, 1500 Fourier transform spectra could be recorded and averaged. According to Equation 7-20, the improvement in signal-to-noise ratio would be $\sqrt{1500}$, or about 39. This inherent advantage of Fourier transform spectroscopy was first recognized by P. Fellgett in 1958 and is termed the *Fellgett*, or *multiplex*, *advantage*. It is worth noting here that, for several reasons, the theoretical $\sqrt{n}$ improvement in S/N is seldom entirely realized. Nonetheless, major improvements in signal-to-noise ratios are generally observed with the Fourier transform technique.

The multiplex advantage is important enough so that nearly all infrared spectrometers are now of the Fourier transform type. Fourier transform instruments are much less common for the ultraviolet, visible, and near-infrared regions, however, because signal-to-noise limitations for spectral measurements with these types of radiation are seldom a result of detector noise but instead are due to shot noise and flicker noise associated with the source. In contrast to detector noise, the magnitudes of both shot and flicker noise increase as the radiant power of the signal increases. Furthermore, the total noise for all of the resolution elements in a Fourier transform measurement tends to be averaged and spread out uniformly over the entire transformed spectrum. Thus, the signal-to-noise ratio for strong peaks in the presence of weak peaks is improved by averaging but degraded for the weaker peaks. For flicker noise, such as is encountered in the background radiation from many spectral sources, degradation of S/N for all peaks is observed. This effect is sometimes termed the *multiplex disadvantage* and is largely responsible for the Fourier transform not being widely applied for UV-visible spectroscopy.[34]

7I-2 Time-Domain Spectroscopy

Conventional spectroscopy can be termed *frequency-domain* spectroscopy in that radiant power data are recorded as a function of frequency or the inversely related wavelength. In contrast, *time-domain* spectroscopy, which can be achieved by the Fourier transform, is concerned with changes in radiant power with *time*. Figure 7-41 illustrates the difference.

The plots in Figure 7-41c and d are conventional spectra of two monochromatic sources with frequencies ν_1 and ν_2 Hz. The curve in Figure 7-41e is the spectrum of a source that contains both frequencies. In each case, the radiant power $P(\nu)$ is plotted with respect to the frequency in hertz. The symbol in parentheses is added to emphasize the frequency dependence of the power; time-domain power is indicated by $P(t)$.

The curves in Figure 7-41a show the time-domain spectra for each of the monochromatic sources. The two have been plotted together to make the small frequency difference between them more obvious. Here, the instantaneous power $P(t)$ is plotted as a function of time. The curve in Figure 7-41b is the time-domain spectrum of the source that contains both frequencies. As is shown by the horizontal arrow, the plot exhibits a periodicity, or *beat*, as the two waves go in and out of phase.

Figure 7-42 is a time-domain signal from a source that contains many wavelengths. It is considerably more complex than those shown in Figure 7-41. Because a large number of wavelengths are involved, a full cycle is not completed in the time period shown. A pattern of beats appears as certain wavelengths pass in and out of phase. In general, the signal power de-

---

[34]For a further description of this *multiplex disadvantage* in atomic spectroscopy, see A. P. Thorne, *Anal. Chem.*, **1991**, *63*, 62A–63A; A. G. Marshall and F. R. Verdun, *Fourier Transforms in NMR, Optical, and Mass Spectrometry*, Chap. 5, New York: Elsevier, 1990.

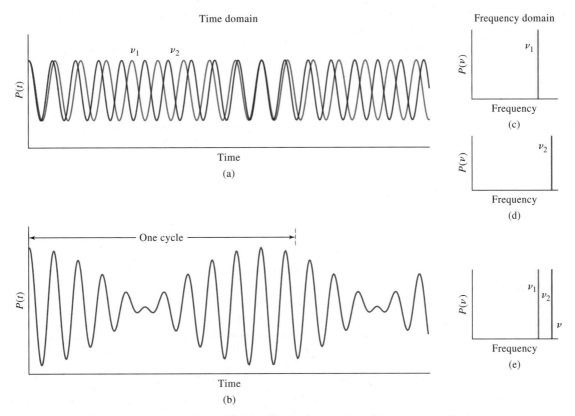

FIGURE 7-41 (a) Time-domain plot of two slightly different frequencies of the same amplitude ν_1 and ν_2. (b) Time-domain plot of the sum of the two waveforms in (a). (c) Frequency-domain plot of ν_1. (d) Frequency-domain plot of ν_2. (e) Frequency-domain plot of the waveform in (b).

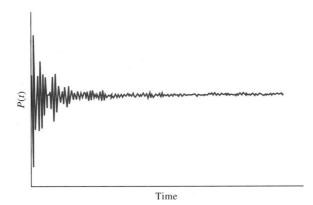

FIGURE 7-42 Time-domain signal of a source made up of many wavelengths.

creases with time because the closely spaced wavelengths become more and more out of phase.

It is important to appreciate that a time-domain signal contains the same information as a spectrum does in the frequency domain, and in fact, one can be converted to the other by numerical computations. Thus, Figure 7-41b was computed from the data of Figure 7-41e using the equation

$$P(t) = k\cos(2\pi\nu_1 t) + k\cos(2\pi\nu_2 t) \quad (7\text{-}21)$$

where k is a constant and t is the time. The difference in frequency between the two lines was approximately 10% of ν_2.

The interconversion of time- and frequency-domain signals is complex and mathematically tedious when more than a few lines are involved; the operation is only practical with a computer. Today fast Fourier transform algorithms allow calculation of frequency-domain spectra from time-domain spectra in seconds or less.

7I-3 Acquiring Time-Domain Spectra with a Michelson Interferometer

Time-domain signals, such as those shown in Figures 7-41 and 7-42, cannot be acquired experimentally with radiation in the frequency range that is associated with optical spectroscopy (10^{12} to 10^{15} Hz) because there are no transducers that will respond to radiant power variations at these high frequencies. Thus, a typical transducer yields a signal that corresponds to the average power of a high-frequency signal and not to its periodic variation. To obtain time-domain signals requires, therefore, a method of converting (or *modulating*) a high-frequency signal to one of measurable frequency without distorting the time relationships carried in the signal; that is, the frequencies in the modulated signal must be directly proportional to those in the original. Different signal-modulation procedures are used for the various wavelength regions of the spectrum. The *Michelson interferometer* is used extensively to modulate radiation in the optical region.

The device used for modulating optical radiation is an interferometer similar in design to one first described by Michelson late in the nineteenth century. The Michelson interferometer is a device that splits a beam of radiation into two beams of nearly equal power and then recombines them in such a way that intensity variations of the combined beam can be measured as a function of differences in the lengths of the paths of the two beams. Figure 7-43 is a schematic of such an interferometer as it is used for optical Fourier transform spectroscopy.

Simulation: Learn more about the **Michelson interferometer and Fourier transform spectrometers**.

As shown in Figure 7-43, a beam of radiation from a source is collimated and impinges on a beamsplitter, which transmits approximately half the radiation and reflects the other half. The resulting twin beams are then reflected from mirrors, one fixed and the other movable. The beams then meet again at the beamsplitter, with half of each beam directed toward the sample and detector and the other two halves directed back toward the source. Only the two halves that pass through the sample to the detector are used for analytical purposes.

Horizontal motion of the movable mirror causes the radiant power that reaches the detector to fluctuate in a reproducible manner. When the two mirrors are equidistant from the splitter (position 0 in Figure 7-43), the two parts of the recombined beam are precisely in phase and the signal power is at a maximum. For a monochromatic source, motion of the movable mirror in either direction by a distance equal to exactly one-quarter wavelength (position B or C in the figure) changes the path length of the corresponding reflected beam by one-half wavelength (one-quarter wavelength for each direction). At this mirror position, destructive interference reduces the radiant power of the recombined beams to zero. When the mirror moves to A or D the two halves of the beams are back in phase so that constructive interference again occurs.

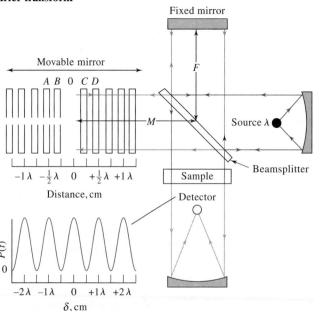

FIGURE 7-43 Schematic of a Michelson interferometer illuminated by a monochromatic source.

The difference in path lengths for the two beams, $2(M - F)$ in the figure, is called the *retardation* δ. A plot of the output power from the detector versus δ is called an *interferogram*. For monochromatic radiation, the interferogram takes the form of a cosine curve such as that shown in the lower left of Figure 7-43 (cosine rather than sine because the power is always at a maximum when δ is zero and the two paths are identical).

The time-varying signal produced by the radiation striking the detector in a Michelson interferometer is much lower in frequency than the source frequency. The relationship between the two frequencies is derived by reference to the $P(t)$ versus δ plot in Figure 7-43. One cycle of the signal occurs when the mirror moves a distance that corresponds to one half a wavelength ($\lambda/2$). If the mirror is moving at a constant velocity of v_M, and we define τ as time required for the mirror to move $\lambda/2$ cm, we may write

$$v_M \tau = \frac{\lambda}{2} \qquad (7\text{-}22)$$

The frequency f of the signal at the detector is simply the reciprocal of τ, or

$$f = \frac{1}{\tau} = \frac{v_M}{\lambda/2} = \frac{2v_M}{\lambda} \qquad (7\text{-}23)$$

We may also relate this frequency to the wavenumber $\bar{v}$ of the radiation. Thus,

$$f = 2v_M \bar{v} \qquad (7\text{-}24)$$

The relationship between the *optical frequency* of the radiation and the frequency of the interferogram is obtained by substitution of $\lambda = c/v$ into Equation 7-23. Thus,

$$f = \frac{2v_M}{c} v \qquad (7\text{-}25)$$

where v is the frequency of the radiation and c is the velocity of light (3×10^{10} cm/s). When v_M is constant, the *interferogram frequency f is directly proportional to the optical frequency v*. Furthermore, the proportionality constant is a very small number. For example, if the mirror is driven at a rate of 1.5 cm/s,

$$\frac{2v_M}{c} = \frac{2 \times 1.5 \text{ cm/s}}{3 \times 10^{10} \text{ cm/s}} = 10^{-10}$$

and

$$f = 10^{-10} v$$

As shown by the following example, the frequency of visible and infrared radiation is easily modulated into the audio range by a Michelson interferometer.

EXAMPLE 7-3

Calculate the frequency range of a modulated signal from a Michelson interferometer with a mirror velocity of 0.20 cm/s, for visible radiation of 700 nm and infrared radiation of 16 μm (4.3×10^{14} to 1.9×10^{13} Hz).

Solution

Using Equation 7-23, we find

$$f_1 = \frac{2 \times 0.20 \text{ cm/s}}{700 \text{ nm} \times 10^{-7} \text{ cm/nm}} = 5700 \text{ Hz}$$

$$f_1 = \frac{2 \times 0.20 \text{ cm/s}}{16 \text{ μm} \times 10^{-4} \text{ cm/μm}} = 250 \text{ Hz}$$

Certain types of visible and infrared transducers are capable of following fluctuations in signal power that fall into the audio-frequency range. Thus, it is possible to record a modulated time-domain signal in the audio frequency range that is an exact translation of the appearance of the very high-frequency time-domain signal from a visible or infrared source. Figure 7-44 shows three examples of such time-domain interferograms. At the top of each column is the image of the interference pattern that appears at the output of the Michelson interferometer. In the middle are the interferogram signals resulting from the patterns at the top, and the corresponding frequency-domain spectra appear at the bottom.

Fourier Transformation of Interferograms

The cosine wave of the interferogram shown in Figure 7-44a (and also in Figure 7-43) can be described by the equation

$$P(\delta) = \frac{1}{2} P(\bar{v}) \cos 2\pi f t \qquad (7\text{-}26)$$

where $P(\bar{v})$ is the radiant power of the beam incident on the interferometer and $P(\delta)$ is the amplitude, or power, of the interferogram signal. The parenthetical symbols emphasize that one power is in the frequency domain and the other is in the time domain. In practice, we modify Equation 7-26 to take into account the fact that the interferometer does not split the source exactly in half and that the detector response and the amplifier behavior are frequency dependent. Thus, it is

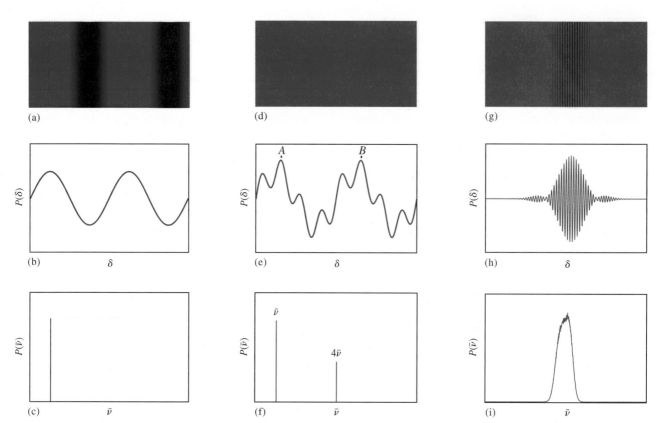

FIGURE 7-44 Formation of interferograms at the output of the Michelson interferometer. (a) Interference pattern at the output of the interferometer resulting from a monochromatic source. (b) Sinusoidally varying signal (interferogram) produced at the detector as the pattern in (a) sweeps across the detector. (c) Frequency spectrum of the monochromatic light source resulting from the Fourier transformation of the signal in (b). (d) Interference pattern at the output of the interferometer resulting from a two-frequency source. (e) Complex signal produced by the interference pattern of (d) as it falls on the detector. Zero retardation is indicated by point *A*. (f) Frequency spectrum of the two-frequency source. (g) Interference pattern resulting from a broad emission band. (h) Interferogram of the source in (f). (i) Frequency spectrum of the emission band.

useful to introduce a new variable, $B(\bar{\nu})$, which depends on $P(\bar{\nu})$ but takes these factors into account. We can then rewrite the equation in the form

$$P(\delta) = B(\bar{\nu}) \cos 2\pi f t \qquad (7\text{-}27)$$

Substitution of Equation 7-24 into Equation 7-27 leads to

$$P(\delta) = B(\bar{\nu}) \cos 4\pi \nu_{\mathrm{M}} \bar{\nu} t \qquad (7\text{-}28)$$

But the mirror velocity can be expressed in terms of retardation or

$$\nu_{\mathrm{M}} = \frac{\delta}{2t}$$

Substitution of this relationship into Equation 7-28 gives

$$P(\delta) = B(\bar{\nu}) \cos 2\pi \delta \bar{\nu}$$

which expresses the magnitude of the interferogram signal as a function of the retardation factor and the wavenumber of the source.

The interferograms shown in Figure 7-44b, e, and h can be described by two terms, one for each wavenumber. Thus,

$$P(\delta) = B_1(\bar{\nu}_1) \cos 2\pi \delta \bar{\nu}_1 + B_2(\bar{\nu}_2) \cos 2\pi \delta \bar{\nu}_2 \quad (7\text{-}29)$$

For a continuum source, as in Figure 7-44i, the interferogram can be represented as a sum of an infinite number of cosine terms. That is,

$$P(\delta) = \int_{-\infty}^{\infty} B(\bar{\nu}) \cos 2\pi\bar{\nu}\delta \, d\bar{\nu} \qquad (7\text{-}30)$$

The Fourier transform of this integral is

$$B(\bar{\nu}) = \int_{-\infty}^{\infty} P(\delta) \cos 2\pi\bar{\nu}\delta \, d\delta \qquad (7\text{-}31)$$

A complete Fourier transformation requires both real (cosine) and imaginary (sine) components; we have presented only the cosine part, which is sufficient for manipulating real and even functions.

Optical Fourier transform spectroscopy consists of recording $P(\delta)$ as a function of δ (Equation 7-30) and then mathematically transforming this relation to one that gives $B(\bar{\nu})$ as a function of $\bar{\nu}$ (the frequency spectrum) as shown by Equation 7-31.

Equations 7-30 and 7-31 cannot be used as written because they assume that the beam contains radiation from zero to infinite wavenumbers and a mirror drive of infinite length. Furthermore, Fourier transformations with a computer require that the detector output be digitized; that is, the output must be sampled periodically and stored in digital form. Equation 7-31 however, demands that the sampling intervals $d\delta$ be infinitely small, that is, $d\delta \rightarrow 0$. From a practical standpoint, only a finite-sized sampling interval can be summed over a finite retardation range of a few centimeters. These constraints have the effect of limiting the resolution of a Fourier transform instrument and restricting its frequency range.

Resolution

The resolution of a Fourier transform spectrometer can be described in terms of the difference in wavenumber between two lines that can be just separated by the instrument. That is,

$$\Delta\bar{\nu} = \bar{\nu}_1 - \bar{\nu}_2 \qquad (7\text{-}32)$$

where $\bar{\nu}_1$ and $\bar{\nu}_2$ are wavenumbers for a pair of barely resolvable lines.

It is possible to show that to resolve two lines, the time-domain signal must be scanned long enough so that one complete cycle, or beat, for the two lines is completed; only then will all of the information contained in the spectrum have been recorded. For example, resolution of the two lines $\bar{\nu}_1$ and $\bar{\nu}_2$ in Figure 7-44f requires recording the interferogram from the maximum A at zero retardation to the maximum B where the two waves are again in phase. The maximum at B occurs, however, when $\delta\bar{\nu}_1$ is larger than $\delta\bar{\nu}_2$ by 1 in Equation 7-29. That is, when

$$\delta\bar{\nu}_2 - \delta\bar{\nu}_1 = 1$$

or

$$\bar{\nu}_2 - \bar{\nu}_1 = \frac{1}{\delta}$$

Substitution into Equation 7-32 gives the resolution

$$\Delta\bar{\nu} = \bar{\nu}_2 - \bar{\nu}_1 = \frac{1}{\delta} \qquad (7\text{-}33)$$

This equation means that resolution in wavenumbers will improve in proportion to the reciprocal of the distance that the mirror travels.

EXAMPLE 7-4

What length of mirror drive will provide a resolution of 0.1 cm^{-1}?

Solution

Substituting into Equation 7-33 gives

$$0.1 \text{ cm}^{-1} = \frac{1}{\delta}$$

$$\delta = 10 \text{ cm}$$

The mirror motion required is one half the retardation, or 5 cm.

Instruments

Details about modern Fourier transform optical spectrometers are found in Section 16B-1. An integral part of these instruments is a state-of-the-art computer for controlling data acquisition, for storing data, for signal averaging, and for computing the Fourier transforms.

QUESTIONS AND PROBLEMS

*Answers are provided at the end of the book for problems marked with an asterisk.

X   Problems with this icon are best solved using spreadsheets.

7-1 Why must the slit width of a prism monochromator be varied to provide constant effective bandwidths but a nearly constant slit width provides constant bandwidth with a grating monochromator?

7-2 Why do quantitative and qualitative analyses often require different monochromator slit widths?

***7-3** The Wien displacement law states that the wavelength maximum in micrometers for blackbody radiation is given by the relationship

$$\lambda_{max}T = 2.90 \times 10^3$$

where T is the temperature in kelvins. Calculate the wavelength maximum for a blackbody that has been heated to (a) 5000 K, (b) 3000 K, and (c) 1500 K.

***7-4** Stefan's law states that the total energy E_t emitted by a blackbody per unit time and per unit area is given by $E_t = \alpha T^4$, where α has a value of 5.69×10^{-8} Wm^{-2} K^{-4}. Calculate the total energy output in W/m^2 for each of the blackbodies described in Problem 7-3.

***7-5** Relationships described in Problems 7-3 and 7-4 may be of help in solving the following.
(a) Calculate the wavelength of maximum emission of a tungsten filament bulb operated at the usual temperature of 2870 K and at a temperature of 3500 K.
(b) Calculate the total energy output of the bulb in terms of W/cm^2.

7-6 Contrast spontaneous and stimulated emission.

7-7 Describe the advantage of a four-level laser system over a three-level type.

7-8 Define the term *effective bandwidth* of a filter.

***7-9** An interference filter is to be constructed for isolation of the nitrobenzene absorption band at 1537 cm^{-1}.
(a) If it is to be based on first-order interference, what should be the thickness of the dielectric layer (refractive index of 1.34)?
(b) What other wavelengths would be transmitted?

7-10 A 10.0-cm interference wedge is to be built that has a linear dispersion from 400 to 700 nm. Describe details of its construction. Assume that a dielectric with a refractive index of 1.32 is to be used.

7-11 Why is glass better than fused silica as a prism construction material for a monochromator to be used in the region of 400 to 800 nm?

***7-12** For a grating, how many lines per millimeter would be required for the first-order diffraction line for $\lambda = 400$ nm to be observed at a reflection angle of 5° when the angle of incidence is 45°?

*7-13 Consider an infrared grating with 84.0 lines per millimeter and 15.0 mm of illuminated area. Calculate the first-order resolution ($\lambda/\Delta\lambda$) of this grating. How far apart (in cm^{-1}) must two lines centered at $1200\ cm^{-1}$ be if they are to be resolved?

7-14 For the grating in Problem 7-13, calculate the wavelengths of first- and second-order diffraction at reflective angles of (a) 25° and (b) 0°. Assume the incident angle is 45°.

7-15 With the aid of Figures 7-2 and 7-3, suggest instrument components and materials for constructing an instrument that would be well suited for
    (a) the investigation of the fine structure of absorption bands in the region of 450 to 750 nm.
    (b) obtaining absorption spectra in the far infrared (20 to 50 μm).
    (c) a portable device for determining the iron content of natural water based on the absorption of radiation by the red $Fe(SCN)^{2+}$ complex.
    (d) the routine determination of nitrobenzene in air samples based on its absorption peak at 11.8 μm.
    (e) determining the wavelengths of flame emission lines for metallic elements in the region from 200 to 780 nm.
    (f) spectroscopic studies in the vacuum ultraviolet region.
    (g) spectroscopic studies in the near infrared.

*7-16 What is the speed (f-number) of a lens with a diameter of 5.4 cm and a focal length of 17.6 cm?

7-17 Compare the light-gathering power of the lens described in Problem 7-16 with one that has a diameter of 37.6 cm and a focal length of 16.8 cm.

*7-18 A monochromator has a focal length of 1.6 m and a collimating mirror with a diameter of 3.5 cm. The dispersing device was a grating with 1500 lines/mm. For first-order diffraction,
    (a) what is the resolving power of the monochromator if a collimated beam illuminated 3.0 cm of the grating?
    (b) what are the first- and second-order reciprocal linear dispersions of the monochromator?

7-19 A monochromator with a focal length of 0.78 m was equipped with an echellette grating of 2500 blazes per millimeter.
    (a) Calculate the reciprocal linear dispersion of the instrument for first-order spectra.
    (b) If 2.0 cm of the grating were illuminated, what is the first-order resolving power of the monochromator?
    (c) At approximately 430 nm, what minimum wavelength difference could in theory be completely resolved by the instrument?

7-20 Describe the basis for radiation detection with a silicon diode transducer.

7-21 Distinguish among (a) a spectroscope, (b) a spectrograph, and (c) a spectrophotometer.

*7-22 A Michelson interferometer had a mirror velocity of 2.75 cm/s. What would be the frequency of the interferogram for (a) UV radiation of 350 nm, (b) visible radiation of 575 nm, (c) infrared radiation of 5.5 μm, and (d) infrared radiation of 25 μm?

*7-23 What length of mirror drive in a Michelson interferometer is required to produce a resolution sufficient to separate
(a) infrared peaks at 500.6 and 500.4 cm^{-1}?
(b) infrared peaks at 4002.1 and 4008.8 cm^{-1}?

☒ Challenge Problem

7-24 The behavior of holographic filters and gratings is described by coupled wave theory.[35] The Bragg wavelength λ_b for a holographic optical element is given by

$$\lambda_b = 2nd\cos\theta$$

where n is the refractive index of the grating material; d is the grating period, or spacing; and θ is the angle of incidence of the beam of radiation.
(a) Create a spreadsheet to calculate the Bragg wavelength for a holographic grating with a grating spacing of 17.1 μm and a refractive index of 1.53 at angles of incidence from 0 to 90°.
(b) At what angle is the Bragg wavelength 462 nm?
(c) The Bragg wavelength is sometimes called the "playback wavelength." Why?
(d) What is the historical significance of the term "Bragg wavelength?"
(e) A tunable volume holographic filter has been developed for communications applications.[36] The filter is said to be tunable over the wavelength range $2nd$ to $2d\sqrt{n^2 - 1}$. If the filter has a Bragg wavelength of 1550 nm and a grating spacing of 535 nm, calculate the angular tuning range of the filter.
(f) Find the Bragg wavelength for a grating with a spacing of 211.5 nm, assuming that the grating is made of the same material. Find the wavelength range over which this grating may be tuned.
(g) Discuss potential spectroscopic applications of tunable holographic filters.
(h) When a volume grating is created in a holographic film, the refractive index of the film material is varied by an amount Δn, which is referred to as the *refractive-index modulation*. In the ideal case, this quantity is given by

$$\Delta n \approx \frac{\lambda}{2t}$$

where λ is the wavelength of the laser forming the interference pattern, and t is the thickness of the film. Calculate the refractive-index modulation in a film 38 μm thick with a laser wavelength of 633 nm.
(i) For real gratings, the refractive-index modulation is

$$\Delta n = \frac{\lambda\sin^{-1}\sqrt{D_e}}{\pi t}$$

where D_e is the grating efficiency. Derive an expression for the ideal grating efficiency, and determine its value.
(j) Calculate the efficiency for a grating 7.5 μm thick if its refractive-index modulation is 0.030 at a wavelength of 633 nm.
(k) Holographic films for fabricating filters and gratings are available commercially, but they can be made in the laboratory using common chemicals. Use a search engine to find a recipe for dichromated gelatin for holographic films. Describe the fabrication of the films, the chemistry of the process, and how the films are recorded.

---

[35] H. Kogelnik, *Bell Syst. Tech. J.* **1969**, *84*, 2909–47.
[36] F. Havermeyer et al., *Optical Engineering*, **2004**, *43*, 2017.

CHAPTER EIGHT

An Introduction to Optical Atomic Spectrometry

I n this chapter, we first present a theoretical discussion of the sources and properties of optical atomic spectra. We then list methods used for producing atoms from samples for elemental analysis. Finally, we describe in some detail the various techniques used for introducing samples into the devices that are used for optical absorption, emission, and fluorescence spectrometry as well as atomic mass spectrometry. Chapter 9 is devoted to atomic absorption methods, the most widely used of all the atomic spectrometric techniques. Chapter 10 deals with several types of atomic emission techniques. Brief chapters on atomic mass spectrometry and atomic X-ray methods follow this discussion.

Throughout this chapter, this logo indicates an opportunity for online self-study at **www .thomsonedu.com/chemistry/skoog**, linking you to interactive tutorials, simulations, and exercises.

Three major types of spectrometric methods are used to identify the elements present in samples of matter and determine their concentrations: (1) optical spectrometry, (2) mass spectrometry, and (3) X-ray spectrometry. In optical spectrometry, discussed in this chapter, the elements present in a sample are converted to gaseous atoms or elementary ions by a process called *atomization*. The ultraviolet-visible absorption, emission, or fluorescence of the atomic species in the vapor is then measured. In atomic mass spectrometry (Chapter 11), samples are also atomized, but in this case, the gaseous atoms are converted to positive ions (usually singly charged) and separated according to their mass-to-charge ratios. Quantitative data are then obtained by counting the separated ions. In X-ray spectrometry (Chapter 12), atomization is not required because X-ray spectra for most elements are largely independent of their chemical composition in a sample. Quantitative results can therefore be based on the direct measurement of the fluorescence, absorption, or emission spectrum of the sample.

8A OPTICAL ATOMIC SPECTRA

In this section, we briefly consider the theoretical basis of optical atomic spectrometry and some of the important characteristics of optical spectra.

8A-1 Energy Level Diagrams

The energy level diagram for the outer electrons of an element is a convenient method to describe the processes behind the various methods of atomic spectroscopy. The diagram for sodium shown in Figure 8-1a is typical. Notice that the energy scale is linear in units of electron volts (eV), with the 3s orbital assigned a value of zero. The scale extends to about 5.14 eV, the energy required to remove the single 3s electron to produce a sodium ion, the *ionization energy*.

Horizontal lines on the diagram indicate the energies of several atomic orbitals. Note that the *p* orbitals are split into two levels that differ only slightly in energy. The classical view rationalizes this difference by invoking the idea that an electron spins on an axis and that the direction of the spin may be in either the same direction as its orbital motion or the opposite direction. Both the spin and the orbital motions create magnetic fields as a result of the rotation of the charge on the electron. The two fields interact in an attractive

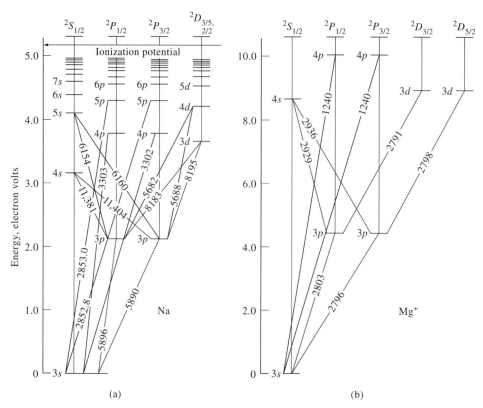

FIGURE 8-1 Energy level diagrams for (a) atomic sodium and (b) magnesium(I) ion. Note the similarity in pattern of lines shown in blue but not in actual wavelengths (Å).

sense if these two motions are in the opposite direction; the fields repel one another when the motions are parallel. As a result, the energy of an electron whose spin opposes its orbital motion is slightly smaller than that of an electron with spin parallel to its orbital motion. There are similar differences in the d and f orbitals, but their magnitudes are usually so small that they are undetectable; thus, only a single energy level is shown for d orbitals in Figure 8-1a.

The splitting of higher energy p, d, and f orbitals into two states is characteristic of all species containing a *single* outer-shell electron. Thus, the energy level diagram for Mg$^+$, shown in Figure 8-1b, has much the same general appearance as that for the uncharged sodium atom. The same is true of the diagrams for Al^{2+} and the remainder of the alkali-metal atoms. Even though all these species are isoelectronic, the energy differences between the 3p and 3s states are different in each case as a result of the different nuclear charges. For example, this difference is about twice as great for Mg$^+$ as for Na.

By comparing Figure 8-1b with Figure 8-2, we see that the energy levels, and thus the spectrum, of an ion

is significantly different from that of its parent atom. For atomic magnesium, with two 1s electrons, there are excited singlet and triplet states with different energies. In the excited singlet state, the spins of the two electrons are opposed and said to be paired; in the triplet states, the spins are unpaired or parallel (Section 15A-1). Using arrows to indicate the direction of spin, the ground state and the two excited states can be represented as in Figure 8-3. As is true of molecules, the triplet excited state is lower in energy than the corresponding singlet state.

The p, d, and f orbitals of the triplet state are split into three levels that differ slightly in energy. We rationalize these splittings by taking into account the interaction between the fields associated with the spins of the two outer electrons and the net field resulting from the orbital motions of all the electrons. In the singlet state, the two spins are paired and their respective magnetic effects cancel; thus, no energy splitting is

 Simulation: Learn more about **atomic spectra**.

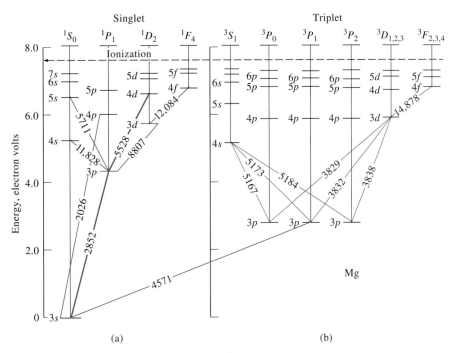

FIGURE 8-2 Energy level diagram for atomic magnesium. The width of the lines between states indicates the relative line intensities. Note that a singlet-to-triplet transition is considerably less probable than a singlet-to-singlet transition. Wavelengths are presented in angstroms.

observed. In the triplet state, however, the two spins are unpaired (that is, their spin moments are in the same direction). The effect of the orbital magnetic moment on the magnetic field of the combined spins produces a splitting of the p level into a triplet. This behavior is characteristic of all of the alkaline-earth atoms as well as B^+, Si^{2+}, and others.

As the number of electrons outside the closed shell becomes larger, the energy level diagrams become increasingly complex. Thus, with three outer electrons, a splitting of energy levels into two and four states occurs; with four outer electrons, there are singlet, triplet, and quintet states.

Although correlating atomic spectra with energy level diagrams for elements such as sodium and magnesium is relatively straightforward and amenable to theoretical interpretation, this is not true for the heavier elements, particularly the transition metals. These species have a larger number of closely spaced energy levels, and as a result, the number of absorption or emission lines can be enormous. For example, a survey[1] of lines observed in the spectra of neutral and

---

[1] Y. Ralchenko, A. E. Kramida, and J. Reader, Developers, *NIST Atomic Spectra Database*, Version 3.0, 2005, http://physics.nist.gov/PhysRefData/ASD/index.html.

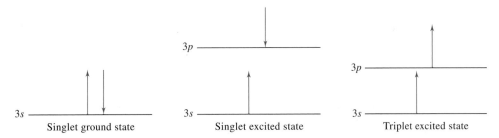

FIGURE 8-3 Spin orientations in singlet ground and excited states and triplet excited state.

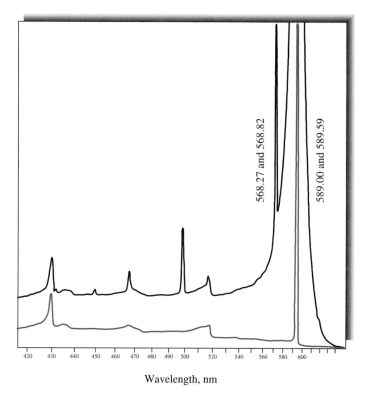

FIGURE 8-4 A portion of the flame emission spectrum for sodium, 800 ppm in naphtha-isopropanol; oxyhydrogen flame; slit 0.02 mm. Note that the scale is expanded in the upper trace and flame conditions were changed to reveal greater detail for Na lines, but not for molecular bands. Note also that the lines at 589.00 and 589.59 nm are off scale in the upper trace. (Adapted from C. T. J. Alkemade and R. Herrmann, *Fundamentals of Analytical Flame Spectroscopy*, p. 229, New York: Wiley, 1979, with permission.)

Wavelength, nm

singly ionized atoms for a variety of elements in the range 300–700 nm (3000–7000 Å) reveals the following numbers of lines. For the alkali metals, this number is 106 for lithium, 170 for sodium, 124 for potassium, and 294 for rubidium; for the alkaline earths, magnesium has 147, calcium 182, and barium 201. Chromium, iron, and scandium with 792, 2340, and 1472 lines, respectively, are typical of transition metals. Although fewer lines are excited in lower-temperature atomizers, such as flames, the flame spectra of the transition metals are still considerably more complex than the spectra of species with low atomic numbers.

Note that radiation-producing transitions shown in Figures 8-1 and 8-2 are observed only between certain of the energy states. For example, transitions from 5s or 4s to 3s states do not occur, nor do transitions among p states or d states. Such transitions are said to be "forbidden," and *quantum mechanical selection rules* permit prediction of which transitions are likely to occur and which are not. These rules are outside the scope of this book.[2]

---

[2]J. D. Ingle Jr. and S. R. Crouch, *Spectrochemical Analysis*, pp. 205–7. Upper Saddle River, NJ: Prentice-Hall, 1988.

Atomic Emission Spectra

At room temperature, essentially all of the atoms of a sample of matter are in the ground state. For example, the single outer electron of a sodium atom occupies the 3s orbital under these circumstances. Excitation of this electron to higher orbitals can be brought about by the heat of a flame, a plasma, or an electric arc or spark. The lifetime of the excited atom is brief, however, and its return to the ground state results in photon emission. The vertical lines in Figure 8-1a indicate some of the common electronic transitions that follow excitation of sodium atoms; the wavelength of the resulting radiation is also shown. The two lines at 589.0 and 589.6 nm (5890 and 5896 Å) are the most intense and are responsible for the yellow color that appears when sodium salts are introduced into a flame.

Figure 8-4 shows a portion of a recorded emission spectrum for sodium. Excitation in this case resulted from spraying a solution of sodium chloride into an oxyhydrogen flame. Note the very large peak at the far right, which is off scale and corresponds to the 3p to 3s transitions at 589.0 and 589.6 nm (5890 and 5896 Å) shown in Figure 8-1a. The resolving power of the monochromator used was insufficient to separate the peaks.

These two lines are *resonance lines*, which result from transitions between an excited electronic state and the ground state. As shown in Figure 8-1, other resonance lines occur at 330.2 and 330.3 nm (3302 and 3303 Å) as well as at 285.30 and 285.28 (2853.0 and 2852.8 Å). The much smaller peak at about 570 nm (5700 Å) in Figure 8-4 is in fact two unresolved nonresonance lines that arise from the two 4d to 3p transitions also shown in the energy level diagram.

Atomic Absorption Spectra

In a hot gaseous medium, sodium atoms are capable of *absorbing* radiation of wavelengths characteristic of electronic transitions from the 3s state to higher excited states. For example, sharp absorption lines at 589.0, 589.6, 330.2, and 330.3 nm (5890, 5896, 3302, and 3303 Å) appear in the experimental spectrum. We see in Figure 8-1a that each adjacent pair of these peaks corresponds to transitions from the 3s level to the 3p and the 4p levels, respectively. Note that nonresonance absorption due to the 3p to 5s transition is so weak that it goes undetected because the number of sodium atoms in the 3p state is generally small at the temperature of a flame. Thus, typically, an atomic absorption spectrum consists predominantly of resonance lines, which are the result of transitions from the ground state to upper levels.

Atomic Fluorescence Spectra

Atoms or ions in a flame fluoresce when they are irradiated with an intense source containing wavelengths that are absorbed by the element. The fluorescence spectrum is most conveniently measured at 90° to the light path. The observed radiation is most often the result of resonance fluorescence involving transitions from excited states returning to the ground state. For example, when magnesium atoms are exposed to an ultraviolet source, radiation of 285.2 nm (2852 Å) is absorbed as electrons are promoted from the 3s to the 3p level (see Figure 8-2); the resonance fluorescence emitted at this same wavelength may then be used for analysis. In contrast, when sodium atoms absorb radiation of wavelength 330.3 nm (3303 Å), electrons are promoted to the 4p state (see Figure 8-1a). A radiationless transition to the two 3p states takes place more rapidly than the fluorescence-producing transition to the ground state. As a result, the observed fluorescence occurs at 589.0 and 589.6 nm (5890 and 5896 Å).

Figure 8-5 illustrates yet a third mechanism for atomic fluorescence that occurs when thallium atoms

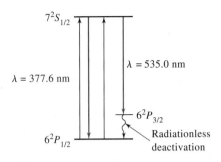

FIGURE 8-5 Energy level diagram for thallium showing the source of two fluorescence lines.

are excited in a flame. Some of the atoms return to the ground state in two steps: a fluorescence emission step producing a line at 535.0 nm (5350 Å) and a radiationless deactivation to the ground state that quickly follows. Resonance fluorescence at 377.6 nm (3776 Å) also occurs.

8A-2 Atomic Line Widths

The widths of atomic lines are quite important in atomic spectroscopy. For example, narrow lines are highly desirable for both absorption and emission spectra because they reduce the possibility of interference due to overlapping lines. Furthermore, as will be shown later, line widths are extremely important in the design of instruments for atomic emission spectroscopy. For these reasons, we now consider some of the variables that influence the width of atomic spectral lines.

As shown in Figure 8-6, atomic absorption and emission lines are generally found to be made up of a symmetric distribution of wavelengths that centers on

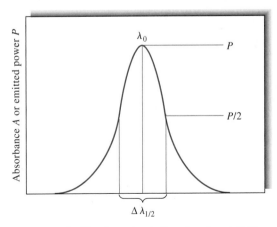

FIGURE 8-6 Profile of an atomic line showing definition of the effective line width $\Delta\lambda_{1/2}$.

a mean wavelength λ_0, which is the wavelength of maximum absorption or maximum intensity for emitted radiation. The energy associated with λ_0 is equal to the exact energy difference between the two quantum states responsible for absorption or emission.

Energy level diagrams, such as that shown in Figure 8-1a, suggest that an atomic line contains only a single-wavelength λ_0—that is, because a line results from a transition of an electron between two discrete, single-valued energy states, the line width will be zero. Several phenomena, however, cause line broadening so that all atomic lines have finite widths, as shown in Figure 8-6. Note that the *line width* or *effective line width* $\Delta\lambda_{1/2}$ of an atomic absorption or emission line is defined as its width in wavelength units when measured at one half the maximum signal. This point is chosen because the measurement can be made more accurately at half-peak intensity than at the base.

There are four sources of line broadening: (1) the uncertainty effect, (2) the Doppler effect, (3) pressure effects due to collisions between atoms of the same kind and with foreign atoms, and (4) electric and magnetic field effects. We consider only the first three of these phenomena here. The magnetic field effect will be discussed in Section 9C-1 in connection with the Zeeman effect.

Line Broadening from the Uncertainty Effect

Spectral lines always have finite widths because the lifetimes of one or both transition states are finite, which leads to uncertainties in the transition times and to line broadening as a consequence of the uncertainty principle (see Section 6C-7). In other words, the breadth of an atomic line resulting from a transition between two states would approach zero only if the lifetimes of two states approached infinity. Although the lifetime of a ground-state electron is long, the lifetimes of excited states are generally short, typically 10^{-7} to 10^{-8} s. Example 8-1 illustrates how we can estimate the width of an atomic emission line from its mean lifetime and the uncertainty principle.

EXAMPLE 8-1

The mean lifetime of the excited state produced by irradiating mercury vapor with a pulse of 253.7 nm radiation is 2×10^{-8} s. Calculate the approximate value for the width of the fluorescence line produced in this way.

 Simulation: Learn more about **line broadening**.

Solution

According to the uncertainty principle (Equation 6-25),

$$\Delta\nu\Delta t \geq 1$$

Substituting 2×10^{-8} s for Δt and rearranging gives the uncertainty $\Delta\nu$ in the frequency of the emitted radiation.

$$\Delta\nu = 1/(2 \times 10^{-8}) = 5 \times 10^7 \text{ s}$$

To evaluate the relationship between this uncertainty in frequency and the uncertainty in wavelength units, we write Equation 6-2 in the form

$$\nu = c\lambda^{-1}$$

Differentiating with respect to frequency gives

$$d\nu = -c\lambda^{-2}\, d\lambda$$

By rearranging and letting $\Delta\nu$ approximate $d\nu$ and $\Delta\lambda_{1/2}$ approximate $d\lambda$, we find

$$|\Delta\lambda_{1/2}| = \frac{\lambda^2\Delta\nu}{c}$$

$$= \frac{(253.7 \times 10^{-9}\text{ m})^2(5 \times 10^7\text{ s}^{-1})}{3 \times 10^8\text{ m/s}}$$

$$= 1.1 \times 10^{-14}\text{ m}$$

$$= 1.1 \times 10^{-14}\text{ m} \times 10^{10}\text{ Å/m} = 1 \times 10^{-4}\text{ Å}$$

Line widths due to uncertainty broadening are sometimes called *natural line widths* and are generally about 10^{-5} nm (10^{-4} Å), as shown in Example 8-1.

Doppler Broadening

The wavelength of radiation emitted or absorbed by a rapidly moving atom decreases if the motion is toward a transducer and increases if the atom is receding from the transducer (see Figure 8-7). This phenomenon is known as the *Doppler shift* and is observed not only with electromagnetic radiation but also sound waves. For example, the Doppler shift occurs when an automobile blows its horn while it passes a pedestrian. As the auto approaches the observer, the horn emits each successive sound vibration from a distance that is increasingly closer to the observer. Thus, each sound wave reaches the pedestrian slightly sooner than would be expected if the auto were standing still. The result is a higher frequency, or pitch, for the horn. When the auto is even with the observer, the waves come directly to the ear of the observer along a line perpendicular to

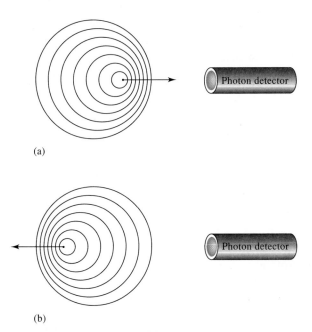

(a)

(b)

FIGURE 8-7 Cause of Doppler broadening. (a) When an atom moves toward a photon detector and emits radiation, the detector sees wave crests more often and detects radiation of higher frequency. (b) When an atom moves away from a photon detector and emits radiation, the detector sees crests less frequently and detects radiation of lower frequency. The result in an energetic medium is a statistical distribution of frequencies and thus a broadening of the spectral lines.

the path of the auto, so there is no shift in the frequency. When the car recedes from the pedestrian, each wave leaves the source at a distance that is larger than that for the previous wave; as a result, the frequency is smaller, and this results in a lower pitch.

The magnitude of the Doppler shift increases with the velocity at which the emitting or absorbing species approaches or recedes from the observer. For relatively low velocities, the relationship between the Doppler shift $\Delta\lambda$ and the velocity v of an approaching or receding atom is

$$\frac{\Delta\lambda}{\lambda_0} = \frac{v}{c}$$

where λ_0 is the wavelength of an unshifted line of a sample of an element at rest relative to the transducer, and c is the velocity of light.

In a collection of atoms in a hot environment, such as a flame, atomic motion occurs in every direction.

 Simulation: Learn more about the **Doppler effect**.

Individual atoms exhibit a Maxwell-Boltzmann velocity distribution, in which the average velocity of a particular atomic species increases as the square root of the absolute temperature. The Doppler shifts of such an ensemble result in broadening of the spectral lines.[3] The maximum Doppler shifts occur for atoms moving with the highest velocities either directly toward or away from the transducer. No shift is associated with atoms moving perpendicular to the path to the transducer. Intermediate shifts occur for the remaining atoms, and these shifts are a function of their speed and direction. Thus, the transducer encounters an approximately symmetrical distribution of wavelengths. In flames, Doppler broadening causes lines to be about two orders of magnitude wider than the natural line width.

Pressure Broadening

Pressure, or collisional, broadening is caused by collisions of the emitting or absorbing species with other atoms or ions in the heated medium. These collisions produce small changes in energy levels and hence a range of absorbed or emitted wavelengths. In a flame, the collisions are largely between the analyte atoms and the various combustion products of the fuel. These collisions produce broadening that is two or three orders of magnitude greater than the natural line widths. Broadening in the hollow cathode lamps and discharge lamps used as sources in atomic absorption spectroscopy results mainly from collisions between the emitting atoms and other atoms of the same kind. In high-pressure mercury and xenon lamps, pressure broadening of this type is so extensive that continuum radiation is produced throughout the ultraviolet and visible region.

8A-3 The Effect of Temperature on Atomic Spectra

Temperature has a profound effect on the ratio between the number of excited and unexcited atomic particles in an atomizer. We calculate the magnitude of this effect from the Boltzmann equation, which takes the form

$$\frac{N_j}{N_0} = \frac{g_j}{g_0} \exp\left(\frac{-E_j}{kT}\right) \tag{8-1}$$

---

[3] For a quantitative treatment of Doppler broadening and pressure broadening, see J. D. Ingle Jr. and S. R. Crouch, *Spectrochemical Analysis*, pp. 210–12, Upper Saddle River, NJ: Prentice-Hall, 1988.

Here, N_j and N_0 are the number of atoms in an excited state and the ground state, respectively, k is Boltzmann's constant (1.38×10^{-23} J/K), T is the absolute temperature, and E_j is the energy difference between the excited state and the ground state. The quantities g_j and g_0 are statistical factors called *statistical weights* determined by the number of states having equal energy at each quantum level. Example 8-2 illustrates a calculation of N_j/N_0.

EXAMPLE 8-2

Calculate the ratio of sodium atoms in the $3p$ excited states to the number in the ground state at 2500 and 2510 K.

Solution

We calculate E_j in Equation 8-1 by using an average wavelength of 589.3 nm (5893 Å) for the two sodium emission lines corresponding to the $3p \rightarrow 3s$ transitions. We compute the energy in joules using the constants found inside the front cover.

$$\bar{\nu} = \frac{1}{589.3 \text{ nm} \times 10^{-7} \text{cm/nm}}$$

$$= 1.697 \times 10^4 \text{ cm}^{-1}$$

$$E_j = 1.697 \times 10^4 \text{ cm}^{-1} \times 1.986 \times 10^{-23} \text{ J cm}$$

$$= 3.37 \times 10^{-19} \text{ J}$$

The statistical weights for the $3s$ and $3p$ quantum states are 2 and 6, respectively, so

$$\frac{g_j}{g_0} = \frac{6}{2} = 3$$

Substituting into Equation 8-1 yields

$$\frac{N_j}{N_0} = 3\exp\left(\frac{-3.37 \times 10^{-19} \text{ J}}{1.38 \times 10^{-23} \text{ J K}^{-1} \times 2500 \text{ K}}\right)$$

$$= 3 \times 5.725 \times 10^{-5} = 1.72 \times 10^{-4}$$

Replacing 2500 with 2510 in the previous equations yields

$$\frac{N_j}{N_0} = 1.79 \times 10^{-4}$$

Example 8-2 demonstrates that a temperature fluctuation of only 10 K results in a 4% increase in the number of excited sodium atoms. A corresponding increase in emitted power by the two lines would result. Thus, an analytical method based on the measurement of emission requires close control of atomization temperature.

Absorption and fluorescence methods are theoretically less dependent on temperature because both measurements are made on initially *unexcited* atoms rather than thermally excited ones. In the example just considered, only about 0.017% of the sodium atoms were thermally excited at 2500 K. Emission measurements are made on this tiny fraction of the analyte. On the other hand, absorption and fluorescence measurements use the 99.98% of the analyte present as unexcited sodium atoms to produce the analytical signals. Note also that although a 10-K temperature change causes a 4% increase in excited atoms, the corresponding *relative* change in fraction of unexcited atoms is negligible.

Temperature fluctuations actually do exert an indirect influence on atomic absorption and fluorescence measurements in several ways. An increase in temperature usually increases the efficiency of the atomization process and hence increases the total number of atoms in the vapor. In addition, line broadening and a decrease in peak height occur because the atomic particles travel at greater rates, which enhances the Doppler effect. Finally, temperature variations influence the degree of ionization of the analyte and thus the concentration of nonionized analyte on which the analysis is usually based (see page 246). Because of these effects, reasonable control of the flame temperature is also required for quantitative absorption and fluorescence measurements.

The large ratio of unexcited to excited atoms in atomization media leads to another interesting comparison of the three atomic methods. Because atomic absorption and atomic fluorescence measurements are made on a much larger population of atoms, these two procedures might be expected to be more sensitive than the emission procedure. This apparent advantage is offset in the absorption method, however, by an absorbance measurement involving evaluation of a ratio ($A = \log P_0/P$). When P and P_0 are nearly equal, we expect larger relative errors in the ratio. Therefore, emission and absorption procedures tend to be complementary in sensitivity, one technique being advantageous for one group of elements and the other for a different group. Based on active atom population,

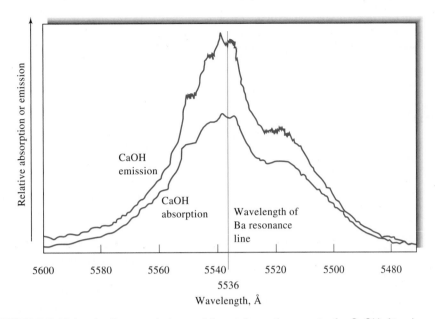

FIGURE 8-8 Molecular flame emission and flame absorption spectra for CaOH. Atomic emission wavelength of barium is also indicated. (Adapted from L. Capacho-Delgado and S. Sprague, *Atomic Absorption Newsletter*, **1965**, *4*, 363. Courtesy of Perkin-Elmer Corporation, Norwalk, CT.)

atomic fluorescence methods should be the most sensitive of the three, at least in principle.

8A-4 Band and Continuum Spectra Associated with Atomic Spectra

Generally, when atomic line spectra are generated, both band and continuum radiation are produced as well. For example, Figure 6-19 (page 150) shows the presence of both molecular bands and a continuum, the latter resulting from the thermal radiation from hot particulate matter in the atomization medium. As we show later, plasmas, arcs, and sparks also produce both bands and continuum radiation.

Band spectra often appear while determining elements by atomic absorption and emission spectrometry. For example, when solutions of calcium ion are atomized in a low-temperature flame, molecular absorption and emission bands for CaOH appear in the region of 554 nm (see Figure 8-8). In this case, the band can be used for the determination of calcium. More often, however, molecular bands and continuum radiation are a potential source of interference that must be minimized by proper choice of wavelength, by background correction, or by a change in atomization conditions.

8B ATOMIZATION METHODS

To obtain both atomic optical and atomic mass spectra, the constituents of a sample must be converted to gaseous atoms or ions, which can then be determined by emission, absorption, fluorescence, or mass spectral measurements. The precision and accuracy of atomic methods depend critically on the atomization step and the method of introduction of the sample into the atomization region. The common types of atomizers are listed in Table 8-1. We describe several of these devices in detail in Chapters 9, 10, and 11.

8C SAMPLE-INTRODUCTION METHODS

Sample introduction has been called the Achilles' heel of atomic spectroscopy because in many cases this step limits the accuracy, the precision, and the detection limits of atomic spectrometric measurements.[4] The primary purpose of the sample-introduction system in atomic spectrometry is to transfer a reproducible and

---

[4]R. F. Browner and A. W. Boorn, *Anal. Chem.*, **1984**, *56*, 786A, 875A; *Sample Introduction in Atomic Spectroscopy*, J. Sneddon, ed., New York: Elsevier, 1990.

TABLE 8-1 Types of Atomizers
Used for Atomic Spectroscopy

Type of Atomizer	Typical Atomization Temperature, °C
Flame	1700–3150
Electrothermal vaporization (ETV)	1200–3000
Inductively coupled argon plasma (ICP)	4000–6000
Direct current argon plasma (DCP)	4000–6000
Microwave-induced argon plasma (MIP)	2000–3000
Glow-discharge plasma (GD)	Nonthermal
Electric arc	4000–5000
Electric spark	40,000 (?)

representative portion of a sample into one of the atomizers listed in Table 8-1 with high efficiency and with no adverse interference effects. Whether it is possible to accomplish this goal easily very much depends on the physical and chemical state of the analyte and the sample matrix. For solid samples of refractory materials, sample introduction is usually a major problem; for solutions and gaseous samples, the introduction step is often trivial. For this reason, most atomic spectroscopic studies are performed on solutions.

For the first five atomization sources listed in Table 8-1, samples are usually introduced in the form of aqueous solutions (occasionally, nonaqueous solutions are used) or less often as slurries (a *slurry* is a suspension of a finely divided powder in a liquid). For samples that are difficult to dissolve, however, several methods have been used to introduce samples into the atomizer in the form of solids or finely dispersed powders. Generally, solid sample-introduction techniques are less reproducible and more subject to various errors and as a result are not nearly as widely applied as aqueous solution techniques. Table 8-2 lists the common sample-introduction methods for atomic spectroscopy and the type of samples to which each method is applicable.

8C-1 Introduction of Solution Samples

Atomization devices fall into two classes: *continuous atomizers* and *discrete atomizers*. With continuous atomizers, such as plasmas and flames, samples are introduced in a steady manner. With discrete atomizers,

 Tutorial: Learn more about **sample introduction**.

samples are introduced in a discontinuous manner with a device such as a syringe or an autosampler. The most common discrete atomizer is the *electrothermal atomizer*.

The general methods for introducing solution samples into plasmas and flames[5] are illustrated in Figure 8-9. Direct *nebulization* is most often used. In this case, the *nebulizer* constantly introduces the sample in the form of a fine spray of droplets, called an *aerosol*. Continuous sample introduction into a flame or plasma produces a steady-state population of atoms, molecules, and ions. When flow injection or liquid chromatography is used, a time-varying plug of sample is nebulized, producing a time-dependent vapor population. The complex processes that must occur to produce free atoms or elementary ions are illustrated in Figure 8-10.

Discrete solution samples are introduced by transferring an aliquot of the sample to the atomizer. The vapor cloud produced with electrothermal atomizers is transient because of the limited amount of sample available. Solid samples can be introduced into plasmas by vaporizing them with an electric spark or with a laser beam. Solutions are generally introduced into the atomizer by one of the first three methods listed in Table 8-2.

Pneumatic Nebulizers

The most common kind of nebulizer is the concentric-tube pneumatic type, shown in Figure 8-11a, in which the liquid sample is drawn through a capillary tube by a

---

[5]For an excellent discussion of liquid introduction methods, see A. G. T. Gustavsson, in *Inductively Coupled Plasmas in Analytical Atomic Spectrometry*, 2nd ed., A. Montaser and D. W. Golightly, eds., Chapter 15. New York: VCH Publishers, Inc., 1992.

TABLE 8-2 Methods of Sample Introduction
in Atomic Spectroscopy

Method	Type of Sample
Pneumatic nebulization	Solution or slurry
Ultrasonic nebulization	Solution
Electrothermal vaporization	Solid, liquid, or solution
Hydride generation	Solution of certain elements
Direct insertion	Solid, powder
Laser ablation	Solid, metal
Spark or arc ablation	Conducting solid
Glow-discharge sputtering	Conducting solid

high-pressure stream of gas flowing around the tip of the tube (the Bernoulli effect). This process of liquid transport is called *aspiration*. The high-velocity gas breaks up the liquid into droplets of various sizes, which are then carried into the atomizer. Cross-flow nebulizers, in which the high-pressure gas flows across a capillary tip at right angles, are illustrated in Figure 8-11b. Figure 8-11c is a schematic of a fritted-disk nebulizer in which the sample solution is pumped onto a fritted surface through which a carrier gas flows. This type of nebulizer produces a much finer aerosol than do the first two. Figure 8-11d shows a Babington nebulizer, which consists of a hollow sphere in which a high-pressure gas is pumped through a small orifice in the sphere's surface. The expanding jet of gas nebulizes the liquid sample flowing in a thin film over the surface of the sphere. This type of nebulizer is less subject to clogging than other devices, and it is therefore useful for samples that have a high salt content or for slurries with a significant particulate content.

Ultrasonic Nebulizers

Several instrument manufacturers also offer ultrasonic nebulizers in which the sample is pumped onto the surface of a piezoelectric crystal that vibrates at a frequency of 20 kHz to several megahertz. Ultrasonic nebulizers produce more dense and more homogeneous aerosols than pneumatic nebulizers do. These devices have low efficiencies with viscous solutions and solutions containing particulates, however.

Electrothermal Vaporizers

An electrothermal vaporizer (ETV) is an evaporator located in a chamber through which an inert gas such as argon flows to carry the vaporized sample into the

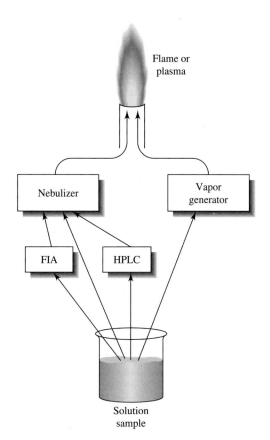

FIGURE 8-9 Continuous sample-introduction methods. Samples are frequently introduced into plasmas or flames by means of a nebulizer, which produces a mist or spray. Samples can be introduced directly to the nebulizer or by means of flow injection analysis (FIA; Chapter 33) or high-performance liquid chromatography (HPLC; Chapter 28). In some cases, samples are separately converted to a vapor by a vapor generator, such as a hydride generator or an electrothermal vaporizer.

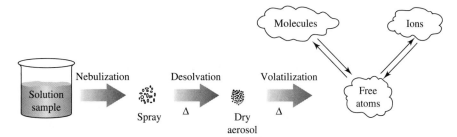

FIGURE 8-10 Processes leading to atoms, molecules, and ions with continuous sample introduction into a plasma or flame. The solution sample is converted into a spray by the nebulizer. The high temperature of the flame or plasma causes the solvent to evaporate, leaving dry aerosol particles. Further heating volatilizes the particles, producing atomic, molecular, and ionic species. These species are often in equilibrium, at least in localized regions.

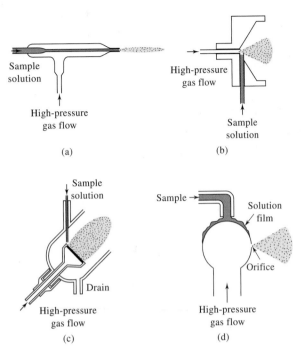

FIGURE 8-11 Types of pneumatic nebulizers: (a) concentric tube, (b) cross-flow, (c) fritted disk, (d) Babington.

atomizer. A small liquid or solid sample is placed on a conductor, such as a carbon tube or tantalum filament. An electric current then evaporates the sample rapidly and completely into the argon flow. In contrast to the nebulizer arrangements we have just considered, an electrothermal system produces a discrete signal rather than a continuous one. That is, the signal from the atomized sample increases to a maximum and then decreases to zero as the sample is swept through the observation region. Peak heights or areas then provide the desired quantitative information.

Hydride Generation Techniques

Hydride generation techniques[6] provide a method for introducing samples containing arsenic, antimony, tin, selenium, bismuth, and lead into an atomizer as a gas. Such a procedure enhances the detection limits for these elements by a factor of 10 to 100. Because several of these species are highly toxic, determining them at

low concentration levels is quite important. This toxicity also dictates that gases from atomization must be disposed of in a safe and efficient manner.

Volatile hydrides can usually be generated by adding an acidified aqueous solution of the sample to a small volume of a 1% aqueous solution of sodium borohydride contained in a glass vessel. A typical reaction is

$$3BH_4^-(aq) + 3H^+(aq) + 4H_3AsO_3(aq) \rightarrow$$
$$3H_3BO_3(aq) + 4AsH_3(g) + 3H_2O(l)$$

The volatile hydride—in this case, arsine (AsH_3)—is swept into the atomization chamber by an inert gas. The chamber is usually a silica tube heated to several hundred degrees in a tube furnace or in a flame where decomposition of the hydride takes place, leading to formation of atoms of the analyte. The concentration of the analyte is then measured by absorption or emission. The signal has a peak shape similar to that obtained with electrothermal atomization.

8C-2 Introduction of Solid Samples

The introduction of solids[7] in the form of powders, metals, or particulates into plasma and flame atomizers has the advantage of avoiding the often tedious and time-consuming step of sample decomposition and dissolution. Such procedures, however, often suffer from severe difficulties with calibration, sample conditioning, precision, and accuracy.

Several techniques have been proposed during the last two decades for the direct introduction of solids into atomizers, thus avoiding the need to dissolve or decompose the sample. These techniques include (1) direct manual insertion of the solid into the atomization device, (2) electrothermal vaporization of the sample and transfer of the vapor into the atomization region, (3) arc, spark, or laser ablation of the solid to produce a vapor that is then swept into the atomizer, (4) slurry nebulization in which the finely divided solid sample is carried into the atomizer as an aerosol consisting of a suspension of the solid in a liquid medium, and (5) sputtering in a glow-discharge device. None of these procedures yields results as satisfactory as those

[6]For a detailed discussion of these methods, see T. Nakahara, in *Sample Introduction in Atomic Spectroscopy*, J. Sneddon, ed., Chap. 10, New York: Elsevier, 1990; J. Didina and D. L. Tsalev, *Hydride Generation Atomic Spectroscopy*, New York: Wiley, 1995.

[7]For a description of solid-introduction techniques, see C. M. McLeod, M. W. Routh, and M. W. Tikkanen, in *Inductively Coupled Plasmas in Analytical Atomic Spectrometry*, 2nd ed., A. Montaser and D. W. Golightly, eds., Chap. 16, New York: VCH, 1992.

obtained by introducing sample solutions by nebulization. Most of these techniques lead to a discrete analytical signal rather than a continuous one.

Direct Sample Insertion

In the direct sample-insertion technique, the sample is physically placed in the atomizer. For solids, the sample may be ground into a powder, which is then placed on or in a probe that is inserted directly into the atomizer. With electric arc and spark atomizers, metal samples are frequently introduced as one or both electrodes that are used to form the arc or spark.

Electrothermal Vaporizers

Electrothermal vaporizers, which were described briefly in the previous section, are also used for various types of solid samples. The sample is heated conductively on or in a graphite or tantalum rod or boat. The vaporized sample is then carried into the atomizer by an inert carrier gas.

Arc and Spark Ablation

Electrical discharges of various types are often used to introduce solid samples into atomizers. The discharge interacts with the surface of a solid sample and creates a plume of a particulate and vaporized sample that is then transported into the atomizer by the flow of an inert gas. This process of sample introduction is called *ablation*.

For arc or spark ablation to be successful, the sample must be electrically conducting or it must be mixed with a conductor. Ablation is normally carried out in an inert atmosphere such as an argon gas stream. Depending on the nature of the sample, the resulting analytical signal may be discrete or continuous. Several instrument manufacturers market accessories for electric arc and spark ablation.

Note that arcs and sparks also atomize samples and excite the resulting atoms to generate emission spectra that are useful for analysis. A spark also produces a significant number of ions that can be separated and analyzed by mass spectrometry (see Section 11D).

Laser Ablation

Laser ablation is a versatile method of introducing solid samples into atomizers. This method is similar to arc and spark ablation; a sufficiently energetic focused laser beam, usually from a Nd-YAG or an excimer laser, impinges on the surface of the solid sample,

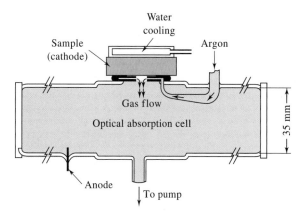

FIGURE 8-12 A glow-discharge atomizer. (From D. S. Gough, P. Hannaford, and R. M. Lowe, *Anal. Chem.*, **1989**, *61*, 1652. Figure 1(a), p. 1653. Copyright 1989 American Chemical Society.)

where ablation takes place to convert the sample into a plume of vapor and particulate matter that is then swept into the atomizer.

Laser ablation is applicable to both conducting and nonconducting solids, inorganic and organic samples, and powder and metallic materials. In addition to bulk analysis, a focused laser beam permits analysis of small areas on the surface of solids. Several instrument makers offer laser samplers.

The Glow-Discharge Technique

A glow-discharge (GD)[8] device is a versatile source that performs both sample introduction and sample atomization simultaneously (see Figure 8-12). A glow discharge takes place in a low-pressure atmosphere (1 to 10 torr) of argon gas between a pair of electrodes maintained at a dc voltage of 250 to 1000 V. The applied voltage causes the argon gas to break down into positively charged argon ions and electrons. The electric field accelerates the argon ions to the cathode surface that contains the sample. Neutral sample atoms are then ejected by a process called *sputtering*. The rate of sputtering may be as high as 100 μg/min.

The atomic vapor produced in a glow discharge consists of a mixture of atoms and ions that can be determined by atomic absorption or fluorescence or by mass spectrometry. In addition, a fraction of the atomized

[8]See R. K. Marcus, T. R. Harville, Y. Mei, and C. R. Shick, *Anal. Chem.*, **1994**, *66*, 902A; W. W. Harrison, C. M. Barshick, J. A. Kingler, P. H. Ratliff, and Y. Mei, *Anal. Chem.*, **1990**, *62*, 943A; R. K. Marcus, *Spectroscopy*, **1992**, 7 (5), 12; *Glow Discharge Spectroscopies*, R. K. Marcus, ed., New York: Plenum Press, 1993.

species present in the vapor is in an excited state. When the excited species relax to their ground states, they produce a low-intensity glow (thus, the name) that can be used for optical emission measurements.

The most important applications of the glow-discharge atomizer have been to the analysis of metals and other conducting samples, although with modification, the device has also been used with liquid samples and nonconducting materials by mixing them with a conductor such as graphite or pure copper powders.

Glow-discharge sources of various kinds are available from several instrument manufacturers.

QUESTIONS AND PROBLEMS

*Answers are provided at the end of the book for problems marked with an asterisk.

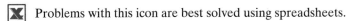 Problems with this icon are best solved using spreadsheets.

8-1 Why is the CaOH spectrum in Figure 8-8 so much broader than the barium emission line?

8-2 What is resonance fluorescence?

8-3 Under what conditions can a Stokes shift (see Section 6C-6) occur in atomic spectroscopy?

8-4 What determines natural line widths for atomic emission and absorption lines? About how broad are these widths, typically?

8-5 In a hot flame, the emission intensities of the sodium lines at 589.0 and 589.6 nm are greater in a sample solution that contains KCl than when this compound is absent. Suggest an explanation.

8-6 The intensity of a line for atomic Cs is much lower in a natural gas flame, which operates at 1800°C, than in a hydrogen-oxygen flame, whose temperature is 2700°C. Explain.

8-7 Name a continuous type and a discrete type of atomizer that are used in atomic spectrometry. How do the output signals from a spectrometer differ for each?

*‍**8-8** The Doppler effect is one of the sources of the line broadening in atomic absorption spectroscopy. Atoms moving toward the light source encounter higher-frequency radiation than atoms moving away from the source. The difference in wavelength $\Delta\lambda$ experienced by an atom moving at speed v (compared to one at rest) is $\Delta\lambda/\lambda = v/c$, where c is the velocity of light. Estimate the line width (in nanometers) of the lithium line at 670.776 (6707.76 Å) when the absorbing atoms are at a temperature of (a) 2100 K and (b) 3150 K. The average speed of an atom is given by $v = \sqrt{8kT/\pi m}$, where k is Boltzmann's constant, T is the absolute temperature, and m is its mass.

*‍**8-9** For Na^+ and Mg^+ ions, compare the ratios of the number of ions in the $3p$ excited state to the number in the ground state in
(a) a natural gas–air flame (1800 K).
(b) a hydrogen-oxygen flame (2950 K).
(c) an inductively coupled plasma source (7250 K).

 8-10 In high-temperature sources, sodium atoms emit a doublet with an average wavelength of 1139 nm. The transition responsible is from the $4s$ to $3p$ state. Set up a spreadsheet to calculate the ratio of the number of excited atoms in the $4s$ state to the number in the ground $3s$ state over the temperature range from an acetylene-oxygen flame (3000°C) to the hottest part of an inductively coupled plasma source (8750°C).

8-11 In the concentration range of 500 to 2000 ppm of U, there is a linear relationship between absorbance at 351.5 nm and concentration. At lower concentrations the relationship is nonlinear unless about 2000 ppm of an alkali metal salt is introduced into the sample. Explain.

X *Challenge Problem*

8-12 In a study of line broadening mechanisms in low-pressure laser-induced plasmas, Gornushkina et al.[9] present the following expression for the half width for Doppler broadening $\Delta\lambda_D$ of an atomic line.

$$\Delta\lambda_D(T) = \lambda_0\sqrt{\frac{8kT\ln 2}{\mathcal{M}c^2}}$$

where λ_0 is the wavelength at the center of the emission line, k is Boltzmann's constant, T is the absolute temperature, $\mathcal{M}$ is the atomic mass, and c is the velocity of light. Ingle and Crouch[10] present a similar equation in terms of frequencies.

$$\Delta\nu_D = 2\left[\frac{2(\ln 2)kT}{\mathcal{M}}\right]^{1/2}\frac{\nu_m}{c}$$

where $\Delta\nu_D$ is the Doppler half width and ν_m is the frequency at the line maximum.

(a) Show that the two expressions are equivalent.

(b) Calculate the half width in nanometers for Doppler broadening of the $4s \rightarrow 4p$ transition for atomic nickel at 361.939 nm (3619.39 Å) at a temperature of 20,000 K in both wavelength and frequency units.

(c) Estimate the natural line width for the transition in (b) assuming that the lifetime of the excited state is 5×10^{-8} s.

(d) The expression for the Doppler shift given in the chapter and in Problem 8-8 is an approximation that works at relatively low speeds. The relativistic expression for the Doppler shift is

$$\frac{\Delta\lambda}{\lambda} = \frac{1}{\sqrt{\dfrac{c-\nu}{c+\nu}}} - 1$$

Show that the relativistic expression is consistent with the equation given in the chapter for low atomic speeds.

(e) Calculate the speed that an iron atom undergoing the $4s \rightarrow 4p$ transition at 385.9911 nm (3859.911 Å) would have if the resulting line appeared at the rest wavelength for the same transition in nickel.

(f) Compute the fraction of a sample of iron atoms at 10,000 K that would have the velocity calculated in (e).

(g) Create a spreadsheet to calculate the Doppler half width $\Delta\lambda_D$ in nanometers for the nickel and iron lines cited in (b) and (e) from 3000–10,000 K.

(h) Consult the paper by Gornushkin et al. (note 9) and list the four sources of pressure broadening that they describe. Explain in detail how two of these sources originate in sample atoms.

[9] I. B. Gornushkin, L. A. King, B. W. Smith, N. Omenetto, and J. D. Winefordner, *Spectrochim. Acta B*, **1999**, *54*, 1207.
[10] J. D. Ingle Jr. and S. R. Crouch, *Spectrochemical Analysis*, p. 212, Upper Saddle River, NJ: Prentice Hall, 1988.

CHAPTER NINE

Atomic Absorption and Atomic Fluorescence Spectrometry

I n this chapter we consider two types of optical atomic spectrometric methods that use similar techniques for sample introduction and atomization. The first is atomic absorption spectrometry (AAS), which for nearly half a century has been the most widely used method for the determination of single elements in analytical samples. The second is atomic fluorescence spectrometry (AFS), which since the mid-1960s has been studied extensively. By contrast to the absorption method, atomic fluorescence has not gained widespread general use for routine elemental analysis, however. Thus, although several instrument makers have in recent years begun to offer special-purpose atomic fluorescence spectrometers, the vast majority of instruments are still of the atomic absorption type. Because of this difference in usage, we devote the bulk of this chapter to AAS and confine our description of AFS to a brief section at the end.

Throughout this chapter, this logo indicates an opportunity for online self-study at **www .thomsonedu.com/chemistry/skoog**, linking you to interactive tutorials, simulations, and exercises.

Prior to discussing AAS[1] in detail, we first present an overview of the types of atomizers used in both AAS and AFS.

9A SAMPLE ATOMIZATION TECHNIQUES

The two most common methods of sample atomization encountered in AAS and AFS, flame atomization and electrothermal atomization, are first described. We then turn to three specialized atomization procedures used in both types of spectrometry.

9A-1 Flame Atomization

In a flame atomizer, a solution of the sample is *nebulized* by a flow of gaseous oxidant, mixed with a gaseous fuel, and carried into a flame where atomization occurs. As shown in Figure 9-1, a complex set of interconnected processes then occur in the flame. The first is *desolvation*, in which the solvent evaporates to produce a finely divided solid molecular aerosol. The aerosol is then *volatilized* to form gaseous molecules. *Dissociation* of most of these molecules produces an atomic gas. Some of the atoms in the gas ionize to form cations and electrons. Other molecules and atoms are produced in the flame as a result of interactions of the fuel with the oxidant and with the various species in the sample. As indicated in Figure 9-1, a fraction of the molecules, atoms, and ions are also excited by the heat of the flame to yield atomic, ionic, and molecular emission spectra. With so many complex processes occurring, it is not surprising that atomization is the most critical step in flame spectroscopy and the one that limits the precision of such methods. Because of the critical nature of the atomization step, it is important to understand the characteristics of flames and the variables that affect these characteristics.

Types of Flames

Table 9-1 lists the common fuels and oxidants used in flame spectroscopy and the approximate range of temperatures realized with each of these mixtures. Note

[1] General references on atomic absorption spectrometry include L. H. J. Lajunen and P. Peramaki, *Spectrochemical Analysis by Atomic Absorption and Emission*, 2nd ed., Cambridge: Royal Society of Chemistry, 2004; J. A. C. Broekaert, *Analytical Atomic Spectrometry with Flames and Plasmas*, Weinheim, Germany: Wiley-VCH, 2002; B. Magyar, *Guide-Lines to Planning Atomic Spectrometric Analysis*, New York: Elsevier, 1982; J. D. Ingle Jr. and S. R. Crouch, *Spectrochemical Analysis*, Chap. 10, Englewood Cliffs, NJ: Prentice Hall, 1988; M. Sperling and B. Welz, *Atomic Absorption Spectrometry*, 3rd ed., New York: VCH Publishers, 1999; N. H. Bings, A. Bogaerts, and J. A. C. Broekaert, *Anal. Chem.*, **2004**, *76*, 3313.

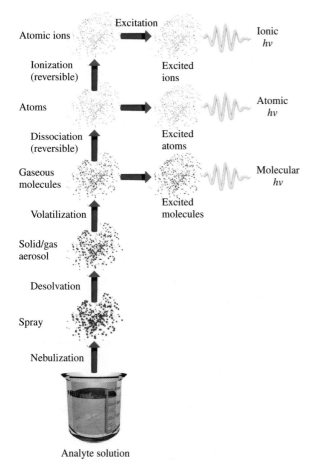

FIGURE 9-1 Processes occurring during atomization.

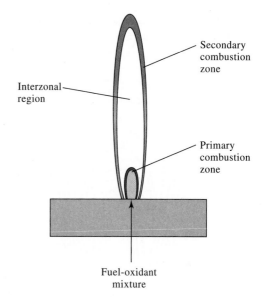

FIGURE 9-2 Regions in a flame.

that temperatures of 1700°C to 2400°C occur with the various fuels when air is the oxidant. At these temperatures, only easily decomposed samples are atomized, so oxygen or nitrous oxide must be used as the oxidant for more refractory samples. These oxidants produce temperatures of 2500°C to 3100°C with the common fuels.

TABLE 9-1 Properties of Flames

Fuel	Oxidant	Temperature, °C	Maximum Burning Velocity, cm s^{-1}
Natural gas	Air	1700–1900	39–43
Natural gas	Oxygen	2700–2800	370–390
Hydrogen	Air	2000–2100	300–440
Hydrogen	Oxygen	2550–2700	900–1400
Acetylene	Air	2100–2400	158–266
Acetylene	Oxygen	3050–3150	1100–2480
Acetylene	Nitrous oxide	2600–2800	285

The burning velocities listed in the fourth column of Table 9-1 are important because flames are stable only in certain ranges of gas flow rates. If the gas flow rate does not exceed the burning velocity, the flame propagates back into the burner, giving *flashback*. As the flow rate increases, the flame rises until it reaches a point above the burner where the flow velocity and the burning velocity are equal. This region is where the flame is stable. At higher flow rates, the flame rises and eventually reaches a point where it blows off the burner. With these facts in mind, it is easy to see why it is so important to control the flow rate of the fuel-oxidant mixture. This flow rate very much depends on the type of fuel and oxidant being used.

Flame Structure

As shown in Figure 9-2, important regions of a flame include the primary combustion zone, the interzonal region, and the secondary combustion zone. The appearance and relative size of these regions vary considerably with the fuel-to-oxidant ratio as well as with the type of fuel and oxidant. The primary combustion zone in a hydrocarbon flame is recognizable by its blue luminescence arising from the band emission of C_2, CH, and other radicals. Thermal equilibrium is usually not achieved in this region, and it is, therefore, rarely used for flame spectroscopy.

The interzonal area, which is relatively narrow in stoichiometric hydrocarbon flames, may reach several centimeters in height in fuel-rich acetylene-oxygen or

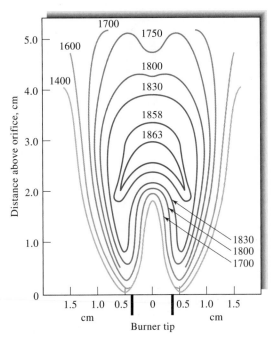

FIGURE 9-3 Temperature profiles in degrees Celsius for a natural gas–air flame. (From B. Lewis and G. van Elbe, *J. Chem. Phys.*, **1943**, *11*, 94. With permission.)

acetylene–nitrous oxide sources. Because free atoms are prevalent in the interzonal region, it is the most widely used part of the flame for spectroscopy. In the secondary reaction zone, the products of the inner core are converted to stable molecular oxides that are then dispersed into the surroundings.

A flame profile provides useful information about the processes that go on in different parts of a flame; it is a contour plot that reveals regions of the flame that have similar values for a variable of interest. Some of these variables include temperature, chemical composition, absorbance, and radiant or fluorescence intensity.

Temperature Profiles. Figure 9-3 shows a temperature profile of a typical flame for atomic spectroscopy. The maximum temperature is located in the flame about 2.5 cm above the primary combustion zone. It is important — particularly for emission methods (Section 10C-1) — to focus the same part of the flame on the entrance slit for all calibrations and analytical measurements.

Flame Absorption Profiles. Figure 9-4 shows typical absorption profiles for three elements. Magnesium exhibits a maximum in absorbance at about the mid-

dle of the flame because of two opposing effects. The initial increase in absorbance as the distance from the base increases results from an increased number of magnesium atoms produced by the longer exposure to the heat of the flame. As the secondary combustion zone is approached, however, appreciable oxidation of the magnesium begins. This process eventually leads to a decrease in absorbance because the oxide particles formed do not absorb at the observation wavelength. To achieve maximum analytical sensitivity, then, the flame must be adjusted up and down with respect to the beam until the region of maximum absorbance is located.

The behavior of silver, which is not easily oxidized, is quite different; as shown in Figure 9-4, a continuous increase in the number of atoms, and thus the absorbance, is observed from the base to the periphery of the flame. By contrast, chromium, which forms very stable oxides, shows a continuous decrease in absorbance beginning close to the burner tip; this observation suggests that oxide formation predominates from the start. These observations suggest that a different portion of the flame should be used for the determination of each of these elements. The more sophisticated instruments for flame spectroscopy are equipped with monochromators that sample the radiation from a relatively small region of the flame, and so a critical step in the optimization of signal output is the adjustment of the position of the flame with respect to the entrance slit.

Flame Atomizers

Flame atomizers are used for atomic absorption, fluorescence, and emission spectroscopy. Figure 9-5 is a diagram of a typical commercial laminar-flow burner

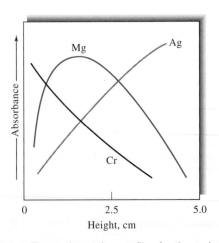

FIGURE 9-4 Flame absorption profiles for three elements.

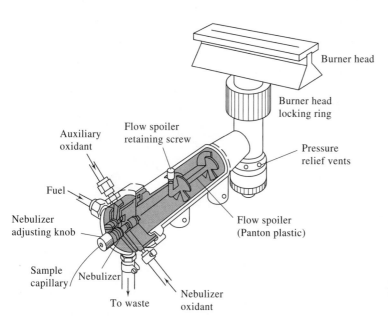

FIGURE 9-5 A laminar-flow burner. (Courtesy of Perkin-Elmer Corporation, Norwalk, CT.)

that uses a concentric-tube nebulizer, such as that shown in Figure 8-11a. The aerosol, formed by the flow of oxidant, is mixed with fuel and passes a series of baffles that remove all but the finest solution droplets. The baffles cause most of the sample to collect in the bottom of the mixing chamber where it drains to a waste container. The aerosol, oxidant, and fuel are then burned in a slotted burner to provide a 5- to 10-cm high flame.

Laminar-flow burners produce a relatively quiet flame and a long path length for maximizing absorption. These properties tend to enhance sensitivity and reproducibility in AAS. The mixing chamber in this type of burner contains a potentially explosive mixture that can flash back if the flow rates are too low. Note that the laminar-flow burner in Figure 9-5 is equipped with pressure relief vents for this reason. Other types of laminar-flow burners and turbulent-flow burners are available for atomic emission spectrometry and AFS.

Fuel and Oxidant Regulators. An important variable that requires close control in flame spectroscopy is the flow rate of both oxidant and fuel. It is desirable to be able to vary each over a broad range so that optimal atomization conditions can be determined experimentally. Fuel and oxidant are usually combined in approximately stoichiometric amounts. For the determination of metals that form stable oxides, however, a flame that

contains an excess of fuel is often desirable. Flow rates are usually controlled by means of double-diaphragm pressure regulators followed by needle valves in the instrument housing. A widely used device for measuring flow rates is the rotameter, which consists of a tapered, graduated, transparent tube that is mounted vertically with the smaller end down. A lightweight conical or spherical float is lifted by the gas flow; its vertical position is determined by the flow rate.

Performance Characteristics of Flame Atomizers. Flame atomization is the most reproducible of all liquid-sample-introduction methods that have been developed for atomic absorption and fluorescence spectrometry to date. The sampling efficiency of other atomization methods and thus the sensitivity, however, are markedly better than in flame atomization. There are two primary reasons for the lower sampling efficiency of the flame. First, a large portion of the sample flows down the drain. Second, the residence time of individual atoms in the optical path in the flame is brief ($\sim 10^{-4}$ s).

9A-2 Electrothermal Atomization

Electrothermal atomizers, which first appeared on the market in the early 1970s, generally provide enhanced sensitivity because the entire sample is atomized in a short period, and the average residence time of the atoms in the optical path is a second or more.[2] Electrothermal atomizers are used for atomic absorption and atomic fluorescence measurements but have not been generally applied for direct production of emission spectra. They are used for vaporizing samples in inductively coupled plasma emission spectroscopy, however.

In electrothermal atomizers, a few microliters of sample is first evaporated at a low temperature and then ashed at a somewhat higher temperature in an electrically heated graphite tube similar to the one in Figure 9-6 or in a graphite cup. After ashing, the current is rapidly increased to several hundred amperes, which causes the temperature to rise to 2000°C to 3000°C; atomization of the sample occurs in a period of a few milliseconds to seconds. The absorption or fluorescence of the atomic vapor is then measured in the region immediately above the heated surface.

Electrothermal Atomizers

Figure 9-6a is a cross-sectional view of a commercial electrothermal atomizer. In this device, atomization occurs in a cylindrical graphite tube that is open at both ends and that has a central hole for introduction of sample by means of a micropipette. The tube is about 5 cm long and has an internal diameter of somewhat less than 1 cm. The interchangeable graphite tube fits snugly into a pair of cylindrical graphite electrical contacts located at the two ends of the tube. These contacts are held in a water-cooled metal housing. Two inert gas streams are provided. The external stream prevents outside air from entering and incinerating the tube. The internal stream flows into the two ends of the tube and out the central sample port. This stream not only excludes air but also serves to carry away vapors generated from the sample matrix during the first two heating stages.

Figure 9-6a illustrates the so-called L'vov platform, which is often used in graphite furnaces such as that shown in the figure. The platform is also made of graphite and is located beneath the sample entrance port. The sample is evaporated and ashed on this platform. When the tube temperature is increased rapidly, however, atomization is delayed because the sample is

[2]For detailed discussions of electrothermal atomizers, see K. W. Jackson, *Electrothermal Atomization for Analytical Atomic Spectrometry*, New York: Wiley, 1999; A. Varma, *CRC Handbook of Furnace Atomic Absorption Spectroscopy*, Boca Raton, FL: CRC Press, 1989; D. J. Butcher and J. Sneddon, *A Practical Guide to Graphite Furnace Atomic Absorption Spectrometry*, New York: Wiley, 1998; K. W. Jackson, *Anal. Chem.*, **2000**, 72, 159, and previous reviews in the series.

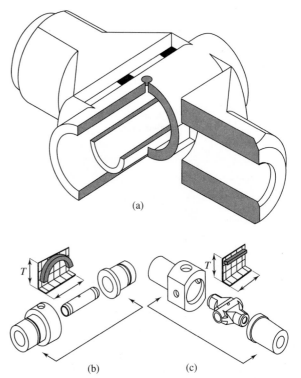

FIGURE 9-6 (a) Cross-sectional view of a graphite furnace with integrated L'vov platform. (b) Longitudinal configuration of the graphite furnace. Note the temperature profile shown in blue along the path of the furnace. In the longitudinal configuration, the temperature varies continuously along the path, reaching a maximum at the center. (c) Transverse configuration of the furnace. The temperature profile is relatively constant along the path. (Courtesy of Perkin-Elmer Life and Analytical Sciences, Shelton, CT.)

no longer directly on the furnace wall. As a result, atomization occurs in an environment in which the temperature is not changing so rapidly, which improves the reproducibility of analytical signals.

Figure 9-6b and c shows the two ways of heating the graphite furnace while it is held in the optical path. Traditionally, the furnace was heated in the longitudinal mode illustrated in Figure 9-6b, which provides a continuously varying temperature profile as shown in the figure. The transverse mode, shown in Figure 9-6c, gives a uniform temperature profile along the entire length of the tube. This arrangement provides optimum conditions for the formation of free atoms throughout the tube. Recombination of atoms to molecules, atom loss, and condensation on the cooler tube ends exhibited in the longitudinal mode are thus minimized in the transverse heating mode.

Experiments show that reducing the natural porosity of the graphite tube minimizes some sample matrix effects and poor reproducibility associated with graphite furnace atomization. During atomization, part of the analyte and matrix apparently diffuse into the surface of the tube, which slows the atomization process, thus giving smaller analyte signals. To overcome this effect, most graphite surfaces are coated with a thin layer of pyrolytic carbon, which seals the pores of the graphite tube. Pyrolytic graphite is a type of graphite that is deposited layer by layer from a highly homogeneous environment. It is formed by passing a mixture of an inert gas and a hydrocarbon such as methane through the tube while it is held at an elevated temperature.

Output Signal

At a wavelength at which absorbance or fluorescence occurs, the transducer output rises to a maximum after a few seconds of ignition followed by a rapid decay back to zero as the atomization products escape into the surroundings. The change is rapid enough (often <1 s) to require a moderately fast data-acquisition system. Quantitative determinations are usually based on peak height, although peak area is also used.

Figure 9-7 shows typical output signals from an atomic absorption spectrophotometer equipped with an electrothermal atomizer. The series of four peaks on the right show the absorbance at the wavelength of a lead peak as a function of time when a 2-μL sample of canned orange juice was atomized. During drying and ashing, three peaks appear that are probably due to molecular evaporation products and particulate ignition products. The three peaks on the left are for lead standards used for calibration. The sample peak on the far right indicates a lead concentration of about 0.05 μg/mL of juice.

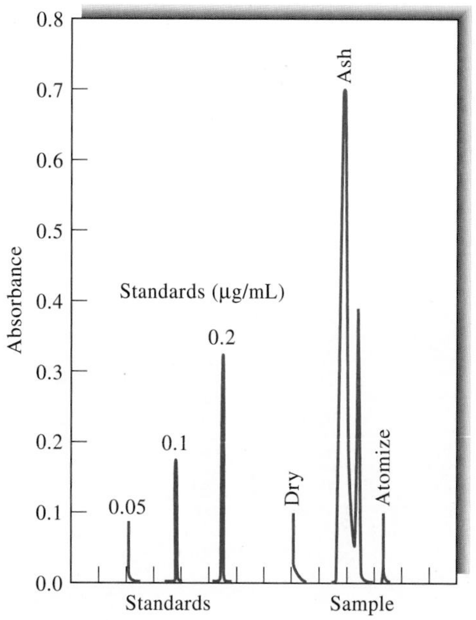

Performance Characteristics of Electrothermal Atomizers

Electrothermal atomizers offer the advantage of unusually high sensitivity for small volumes of sample. Typically, sample volumes between 0.5 and 10 μL are used; under these circumstances, absolute detection limits typically lie in the range of 10^{-10} to 10^{-13} g of analyte.[3]

The relative precision of electrothermal methods is generally in the range of 5% to 10% compared with the

FIGURE 9-7 Typical output for the determination of lead from a spectrophotometer equipped with an electrothermal atomizer. The sample was 2 μL of canned orange juice. The times for drying and ashing are 20 and 60 s, respectively. (Courtesy of Varian Instrument Division, Palo Alto, CA.)

1% or better that can be expected for flame or plasma atomization. Furthermore, because of the heating-cooling cycles, furnace methods are slow — typically requiring several minutes per element. A final disadvantage is that the analytical range is relatively narrow, usually less than two orders of magnitude. As a result, electrothermal atomization is the method of choice when flame or plasma atomization provides inadequate detection limits.

Analysis of Solids with Electrothermal Atomizers

In most methods based on electrothermal atomizers, samples are introduced as solutions. Several reports, however, have described the use of this type of atomizer for the direct analysis of solid samples. One way of performing such measurements is to weigh the finely ground sample into a graphite boat and insert the boat into the furnace manually. A second way is to prepare a slurry of the powdered sample by ultrasonic agitation in an aqueous medium. The slurry is then pipetted into the furnace for atomization.[4]

---

[3]For a comparison of nine commercial furnace spectrometers, see B. E. Erickson, *Anal. Chem.*, **2000**, *72*, 543A.

[4]See, for example, K. Friese and V. Krivan, *Anal. Chem.*, **1995**, *67*, 354.

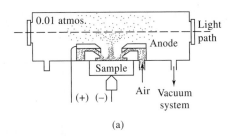

(a)

(b)

FIGURE 9-8 (a) Cross section of a cell for glow-discharge atomization of solid samples. (b) Craters formed on sample surface by six jets of ionized argon. (Teledyne Leeman Labs, Hudson, NH.)

9A-3 Specialized Atomization Techniques

By far, the most common sample-introduction and atomization techniques for atomic absorption analyses are flames or electrothermal vaporizers. Several other atomization methods find occasional use, however. Three of these are described briefly in this section.

Glow-Discharge Atomization

As described in Section 8C-2, a glow-discharge device produces an atomized vapor that can be swept into a cell for absorption measurements. Figure 9-8a shows a glow-discharge cell that can be used as an accessory to most flame atomic absorption spectrometers. It consists of a cylindrical cell about 17 cm long with a circular hole about 2 cm in diameter cut near the middle of the cylinder. An O-ring surrounds the hole. The sample is pressed against this hole with a torque screw so that it seals the tube. Six fine streams of argon gas from tiny nozzles arranged in a circular pattern above the sample impinge on the sample surface in a hexagonal pattern. The argon is ionized by a current between an anode supporting the nozzles and the sample, which acts as a cathode. As a result of sputtering, six craters quickly

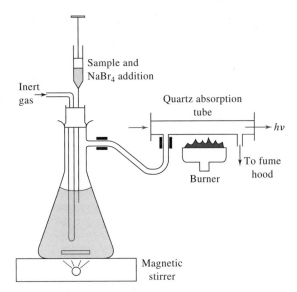

FIGURE 9-9 A hydride generation and atomization system for AAS.

form on the sample surface as shown in Figure 9-8b. The sputtered atoms are drawn by a vacuum to the axis of the cell where they absorb radiation from the spectrometer source.[5]

For this technique to be applicable, the sample must be an electrical conductor or must be made into a pellet with a powdered conductor such as finely ground graphite or copper. Solution samples have also been analyzed by deposition on a graphite, aluminum, or copper cathode. Detection limits with this type of device are reported to be in the low parts-per-million range for solid samples.[6]

Hydride Atomization

In Section 8C-1, we considered methods for introducing solution samples by hydride generation. Atomization of the hydrides requires only that they be heated in a quartz tube, as shown in Figure 9-9.

Cold-Vapor Atomization

The cold-vapor technique is an atomization method applicable only to the determination of mercury because it is the only metallic element that has an appreciable

[5] See E. H. Piepmeier in *Glow Discharge Spectroscopies*, pp. 69–71, R. K. Marcus, ed., New York: Plenum Press, 1993.
[6] For a review of pulsed glow-discharge spectroscopy, see W. W. Harrison, C. Yang, and E. Oxley, *Anal. Chem.*, **2001**, *73*, 480A.

vapor pressure at ambient temperature.[7] The determination of mercury in various types of samples is of vital importance currently because of the toxicity of organic mercury compounds and their widespread distribution in the environment.[8] One popular method for this determination is cold vaporization followed by atomic absorption spectrophotometry. To perform a determination of this type, mercury is converted to Hg^{2+} by treatment of samples with an oxidizing mixture of nitric and sulfuric acids, followed by reduction of the Hg^{2+} to the metal with $SnCl_2$. The elemental mercury is then swept into a long-pass absorption tube similar to the one shown in Figure 9-9 by bubbling a stream of inert gas through the reaction mixture.[9] The determination is completed by measuring the absorbance at 253.7 nm. Detection limits in the parts-per-billion range are achieved. Several manufacturers offer automatic instruments for performing this determination.

9B ATOMIC ABSORPTION INSTRUMENTATION

Instruments for AAS are similar in general design to that shown in Figure 7-1a and consist of a radiation source, a sample holder, a wavelength selector, a detector, and a signal processor and readout.[10] The sample holder in atomic absorption instruments is the atomizer cell that contains the gaseous atomized sample.

9B-1 Radiation Sources

Atomic absorption methods are potentially highly specific because atomic absorption lines are remarkably narrow (0.002 to 0.005 nm) and because electronic transition energies are unique for each element. On the other hand, narrow line widths create a problem that does not normally occur in molecular absorption spectroscopy. In Section 13B-2, we show that a linear relationship between the analytical signal (absorbance) and concentration — that is, for Beer's law as given by Equation 6-34 to be obeyed — requires a narrow source bandwidth relative to the width of an absorption line or band. Even good-quality monochromators, however, have effective bandwidths significantly greater than the width of atomic absorption lines. As a result, nonlinear calibration curves are inevitable when atomic absorbance measurements are made with an ordinary spectrophotometer equipped with a continuum radiation source. Furthermore, the slopes of calibration curves obtained in these experiments are small because only a small fraction of the radiation from the monochromator slit is absorbed by the sample; the result is poor sensitivity. In recent years, the development of high-resolution ($R > 10^5$) continuum-source spectrometers based on the double echelle monochromator coupled with array detection has clouded this issue, and such instruments are beginning to compete with traditional spectrometers equipped with line sources.[11]

The problem created by the limited width of atomic absorption lines has been solved by the use of line sources with bandwidths even narrower than the absorption line width. For example, to use the 589.6-nm line of sodium as the basis for determining the element, a sodium emission line at this same wavelength is isolated to serve as the source. In this instance, a sodium vapor lamp in which sodium atoms are excited by an electrical discharge may be used to produce the line. The other sodium lines emitted from the source are removed with filters or with a relatively inexpensive monochromator. Operating conditions for the source are chosen such that Doppler broadening of the emitted lines is less than the broadening of the absorption line that occurs in the flame or other atomizer. That is, the source temperature and pressure are kept below that of the atomizer. Figure 9-10 illustrates the principle of this procedure. Figure 9-10a shows the emission spectrum of a typical atomic lamp source, which consists of four narrow lines. With a suitable filter or monochromator, all but one of these lines are removed. Figure 9-10b shows the absorption spectrum for the analyte between wavelengths λ_1 and λ_2. Note

[7] See L. H. J. Lajunen and P. Peramaki, *Spectrochemical Analysis by Atomic Absorption and Emission*, 2nd ed., p. 63, Cambridge: Royal Society of Chemistry, 2004.

[8] See, for example, D. A. Skoog, D. M. West, F. J. Holler, and S. R. Crouch, *Fundamentals of Analytical Chemistry*, 8th ed., pp. 865–67, Belmont, CA: Brooks/Cole, 2004.

[9] For a discussion of the importance of determining mercury in the environment, refer to the Instrumental Analysis in Action feature at the end of Section 2.

[10] Reference books on atomic absorption spectroscopy include L. H. J. Lajunen and P. Peramaki, *Spectrochemical Analysis by Atomic Absorption and Emission*, 2nd ed., Cambridge, UK: Royal Society of Chemistry, 2004; M. Sperling and B. Welz, *Atomic Absorption Spectrometry*, 3rd ed., New York: VCH, 1999.

[11] B. Welz, H. Becker-Ross, S. Florek, and U. Heitmann, *High-Resolution Continuum Source AAS*, Hoboken, NJ: Wiley-VCH, 2005; H. Becker-Ross, S. Florek, R. Tischendorf, G. R. Schmecher, *Fresenius J. Anal. Chem.*, **1996**, *355*, 300.

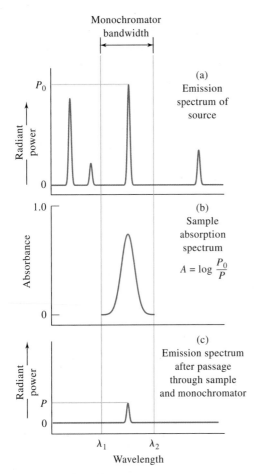

FIGURE 9-10 Absorption of a resonance line by atoms.

that the bandwidth is significantly greater than that of the emission line. As shown in Figure 9-10c, passage of the line from the source through the flame reduces its intensity from P_0 to P; the absorbance is then given by $\log(P_0/P)$, which is linearly related to the concentration of the analyte in the sample.

A disadvantage of the procedure just described is that a separate source lamp is needed for each element (or sometimes group of elements).

Hollow-Cathode Lamps

The most common source for atomic absorption measurements is the hollow-cathode lamp, such as the one shown in Figure 9-11.[12] This type of lamp consists of a tungsten anode and a cylindrical cathode sealed in a

[12] See S. Caroli, *Improved Hollow Cathode Lamps for Atomic Spectroscopy*, New York: Wiley, 1985.

glass tube filled with neon or argon at a pressure of 1 to 5 torr. The cathode is constructed of the metal whose spectrum is desired or serves to support a layer of that metal.

Ionization of the inert gas occurs when a potential difference on the order of 300 V is applied across the electrodes, which generates a current of about 5 to 15 mA as ions and electrons migrate to the electrodes. If the voltage is sufficiently large, the gaseous cations acquire enough kinetic energy to dislodge some of the metal atoms from the cathode surface and produce an atomic cloud in a process called *sputtering*. A portion of the sputtered metal atoms are in excited states and thus emit their characteristic radiation as they return to the ground state. Eventually, the metal atoms diffuse back to the cathode surface or to the glass walls of the tube and are redeposited.

The cylindrical configuration of the cathode tends to concentrate the radiation in a limited region of the metal tube; this design also enhances the probability that redeposition will occur at the cathode rather than on the glass walls.

The efficiency of the hollow-cathode lamp depends on its geometry and the operating voltage. High voltages, and thus high currents, lead to greater intensities. This advantage is offset somewhat by an increase in Doppler broadening of the emission lines from the lamp. Furthermore, the greater currents produce an increased number of unexcited atoms in the cloud. The unexcited atoms, in turn, are capable of absorbing the radiation emitted by the excited ones. This *self-absorption* leads to lowered intensities, particularly at the center of the emission band.

Hollow-cathode lamps are often used as sources in AFS, as discussed in Section 9E-1. In this application, the lamps are pulsed with a duty cycle of 1% to 10%

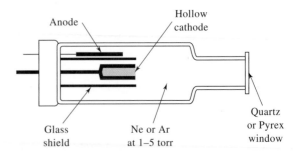

FIGURE 9-11 Schematic cross section of a hollow-cathode lamp.

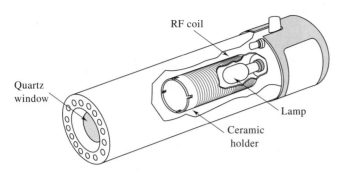

FIGURE 9-12 Cutaway of an EDL. (From W. B. Barnett, J. W. Vollmer, and S. M. DeNuzzo, *At. Absorption Newsletter*, **1976**, *15*, 33. With permission.)

and peak current of 0.1 to 1 A, which increases their peak radiance by a factor of 10 to 100 relative to the steady-state radiance of dc operation.[13]

A variety of hollow-cathode lamps is available commercially. The cathodes of some consist of a mixture of several metals; such lamps permit the determination of more than a single element.

Electrodeless Discharge Lamps

Electrodeless discharge lamps (EDLs) are useful sources of atomic line spectra and provide radiant intensities usually one to two orders of magnitude greater than hollow-cathode lamps.[14] A typical lamp is constructed from a sealed quartz tube containing a few torr of an inert gas such as argon and a small quantity of the metal (or its salt) whose spectrum is of interest. The lamp contains no electrode but instead is energized by an intense field of radio-frequency or microwave radiation. Ionization of the argon occurs to give ions that are accelerated by the high-frequency component of the field until they gain sufficient energy to excite the atoms of the metal whose spectrum is sought.

EDLs are available commercially for fifteen or more elements. Their performance is not as reliable as that of the hollow-cathode lamp, but for elements such as Se, As, Cd, and Sb, EDLs exhibit better detection limits than do hollow-cathode lamps.[15] This occurs because EDLs for these elements are more intense than the corresponding hollow-cathode lamps, and thus, EDLs are quite useful in determining these elements. Figure 9-12 is a schematic of a commercial EDL, which is powered by a 27-MHz radio-frequency source.

---

[13] J. D. Ingle Jr. and S. R. Crouch, *Spectrochemical Analysis*, p. 310, Englewood Cliffs, NJ: Prentice Hall, 1988.
[14] See W. B. Barnett, J. W. Vollmer, and S. M. DeNuzzo, *At. Absorption Newslett.*, **1976**, *15*, 33.
[15] E. Davenport, *Amer. Lab. News*, **1999**, *31* (5), 101S.

Source Modulation

In the typical atomic absorption instrument, it is necessary to eliminate interferences caused by emission of radiation by the flame. Much of this emitted radiation is, of course, removed by the monochromator. Nevertheless, emitted radiation corresponding in wavelength to the monochromator setting is inevitably present in the flame because of excitation and emission of analyte atoms and flame gas species. To eliminate the effects of flame emission, it is necessary to modulate the output of the source so that its intensity fluctuates at a constant frequency. The detector then receives two types of signal, an alternating one from the source and a continuous one from the flame. These signals are converted to the corresponding types of electrical response. A simple high-pass *RC* filter (Section 2B-5) can then be used to remove the unmodulated dc signal and pass the ac signal for amplification.

A simple and entirely satisfactory way of modulating the emission from the source is to interpose a circular metal disk, or chopper, in the beam between the source and the flame. Rotating disk and rotating vane choppers are common (see Figure 5-7a and b). Rotation of the disk or vane at a constant known rate provides a beam that is chopped to the desired frequency. Other types of electromechanical modulators include tuning forks with vanes attached to alternately block and transmit the beam (see Figure 5-7c) and devices that rotate a vane through a fixed arc to perform the same function.[16] As another alternative, the power supply for the source can be designed for intermittent or ac operation so that the source is switched on and off at the desired constant frequency.

---

[16] J. D. Ingle Jr. and S. R. Crouch, *Spectrochemical Analysis*, p. 44, Englewood Cliffs, NJ: Prentice Hall, 1988.

9B-2 Spectrophotometers

Instruments for atomic absorption measurements are offered by numerous manufacturers; both single- and double-beam designs are available. The range of sophistication and cost (upward of a few thousand dollars) is substantial.

In general, the instrument must be capable of providing a sufficiently narrow bandwidth to isolate the line chosen for the measurement from other lines that may interfere with or diminish the sensitivity of the determination. A glass filter suffices for some of the alkali metals, which have only a few widely spaced resonance lines in the visible region. An instrument equipped with easily interchangeable interference filters is available commercially. A separate filter and light source are used for each element. Satisfactory results for the de-

 Exercise: Learn more about **single-beam and double-beam spectrophotometers**.

termination of twenty-two metals are claimed. Most instruments, however, incorporate good-quality ultraviolet-visible monochromators, many of which are capable of achieving a bandwidth on the order of 1 Å.

Most atomic absorption instruments use photomultiplier tubes, which were described in Section 7E-2, as transducers. As pointed out earlier, electronic systems that are capable of discriminating between the modulated signal from the source and the continuous signal from the flame are required. Most instruments currently on the market are equipped with computer systems that are used to control instrument parameters and to control and manipulate data.

Single-Beam Instruments

A typical single-beam instrument, such as that shown in Figure 9-13a, consists of several hollow-cathode sources (only one of which is shown), a chopper or a

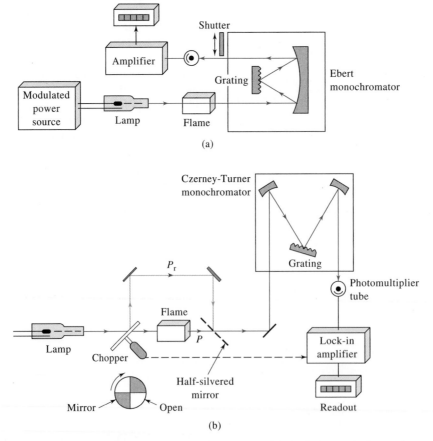

FIGURE 9-13 Typical flame spectrophotometers: (a) single-beam design and (b) double-beam design.

pulsed power supply, an atomizer, and a simple grating spectrophotometer with a photomultiplier transducer. It is used in the manner described on page 159. Thus, the dark current is nulled with a shutter in front of the transducer. The 100% transmittance adjustment is then made while a blank is aspirated into the flame or ignited in a nonflame atomizer. Finally, the transmittance is obtained with the sample replacing the blank.

Double-Beam Instruments

Figure 9-13b is a schematic of a typical double-beam-in-time instrument. The beam from the hollow-cathode source is split by a mirrored chopper, one half passing through the flame and the other half around it. The two beams are then recombined by a half-silvered mirror and passed into a Czerny-Turner grating monochromator; a photomultiplier tube serves as the transducer. The output from the latter is the input to a lock-in amplifier that is synchronized with the chopper drive. The ratio between the reference and sample signal is then amplified and fed to the readout, which may be a digital meter or a computer.

It should be noted that the reference beam in atomic double-beam instruments does not pass through the flame and thus does not correct for loss of radiant power due to absorption or scattering by the flame itself. Methods of correcting for these losses are discussed in the next section.

9C INTERFERENCES IN ATOMIC ABSORPTION SPECTROSCOPY

Interferences of two types are encountered in atomic absorption methods. Spectral interferences arise when the absorption or emission of an interfering species either overlaps or lies so close to the analyte absorption or emission that resolution by the monochromator becomes impossible. Chemical interferences result from various chemical processes occurring during atomization that alter the absorption characteristics of the analyte.

9C-1 Spectral Interferences

Because the emission lines of hollow-cathode sources are so very narrow, interference because of overlapping lines is rare. For such an interference to occur, the separation between the two lines would have to be less than about 0.1 Å. For example, a vanadium line at 3082.11 Å interferes in the determination of aluminum

based on its absorption line at 3082.15 Å. The interference is easily avoided, however, by observing the aluminum line at 3092.7 Å instead.

Spectral interferences also result from the presence of combustion products that exhibit broadband absorption or particulate products that scatter radiation. Both reduce the power of the transmitted beam and lead to positive analytical errors. When the source of these products is the fuel and oxidant mixture alone, the analytical data can be corrected by making absorption measurements while a blank is aspirated into the flame. Note that this correction must be used with both double-beam and single-beam instruments because the reference beam of a double-beam instrument does not pass through the flame (see Figure 9-13b).

A much more troublesome problem occurs when the source of absorption or scattering originates in the sample matrix. In this instance, the power of the transmitted beam P is reduced by the matrix components, but the incident beam power P_0 is not; a positive error in absorbance and thus concentration results. An example of a potential matrix interference because of absorption occurs in the determination of barium in alkaline-earth mixtures. As shown by the solid line in Figure 8-8, the wavelength of the barium line used for atomic absorption analysis appears in the center of a broad absorption band for CaOH. We therefore anticipate that calcium will interfere in barium determinations, but the effect is easily eliminated by substituting nitrous oxide for air as the oxidant. The higher temperature of the nitrous oxide flame decomposes the CaOH and eliminates the absorption band.

Spectral interference because of scattering by products of atomization is most often encountered when concentrated solutions containing elements such as Ti, Zr, and W—which form refractory oxides—are aspirated into the flame. Metal oxide particles with diameters greater than the wavelength of light appear to be formed, and scattering of the incident beam results.

Interference caused by scattering may also be a problem when the sample contains organic species or when organic solvents are used to dissolve the sample. Here, incomplete combustion of the organic matrix leaves carbonaceous particles that are capable of scattering light.

Fortunately, with flame atomization, spectral interferences by matrix products are not widely encountered and often can be avoided by variations in the analytical variables, such as flame temperature and fuel-to-oxidant ratio. Alternatively, if the source of interference is known, an excess of the interfering substance can be added to both sample and standards. Provided the excess added to the standard sample is large with respect to the concentration from the sample matrix, the contribution from the sample matrix will become insignificant. The added substance is sometimes called a *radiation buffer*. The method of standard additions can also be used advantageously in some cases.

In the early days of electrothermal atomization, matrix interference problems were severe. As platform technology, new high-quality graphite materials, fast photometric instrumentation, and Zeeman background correction have developed, matrix interference problems have decreased to the level encountered with flames.[17] Several methods have been developed for correcting for spectral interferences caused by matrix products.[18]

The Two-Line Correction Method

The two-line correction procedure uses a line from the source as a reference. This line should lie as close as possible to the analyte line but must not be absorbed by the analyte. If these conditions are met, it is assumed that any decrease in power of the reference line from that observed during calibration arises from absorption or scattering by the matrix products of the sample. This decrease in power is then used to correct the absorbance of the analyte line.

The reference line may be from an impurity in the hollow cathode, a neon or argon line from the gas contained in the lamp, or a nonresonant emission line of the element that is being determined. Unfortunately, a suitable reference line is often not available.

The Continuum-Source Correction Method

Figure 9-14 illustrates a second method for background corrections that is widely used. In this technique, a deuterium lamp provides a source of continuum radiation throughout the ultraviolet region. The configuration of the chopper is such that radiation from the continuum source and the hollow-cathode lamp are passed alternately through the electrothermal atomizer. The absorbance of the deuterium radiation is then subtracted from that of the analyte beam.

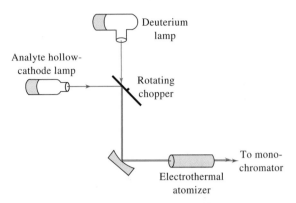

FIGURE 9-14 Schematic of a continuum-source background correction system. Note that the chopper can be eliminated by alternately pulsing each lamp.

The slit width is kept sufficiently wide so that the fraction of the continuum source that is absorbed by the atoms of the sample is negligible. Therefore, the attenuation of the continuum source as it passes through the atomized sample reflects only the broadband absorption or scattering by the sample matrix components. A background correction is thus achieved.

Unfortunately, although most instrument manufacturers offer continuum-source background correction systems, the performance of these devices is often less than ideal, which leads to undercorrection in some systems and overcorrection in others. One of the sources of error is the inevitable degradation of signal-to-noise ratio that accompanies the addition of the correction system. Another is that hot gaseous media are usually highly inhomogeneous both in chemical composition and in particulate distribution; thus if the two lamps are not in perfect alignment, an erroneous correction will result that can cause either a positive or a negative error. Finally, the radiant output of the deuterium lamp in the visible region is low enough to preclude the use of this correction procedure for wavelengths longer than about 350 nm.

Background Correction Based on the Zeeman Effect

When an atomic vapor is exposed to a strong magnetic field (~10 kG), a splitting of electronic energy levels of the atoms takes place that leads to formation of several absorption lines for each electronic transition. These lines are separated from one another by about 0.01 nm, with the sum of the absorbances for the lines being exactly equal to that of the original line from which they

[17] See W. Slavin, *Anal. Chem.*, **1986**, *58*, 590A.
[18] For a comparison of the various methods for background correction, see D. J. Butcher and J. Sneddon, *A Practical Guide to Graphite Furnace Atomic Absorption Spectrometry*, New York: Wiley, 1998, pp. 84–89.

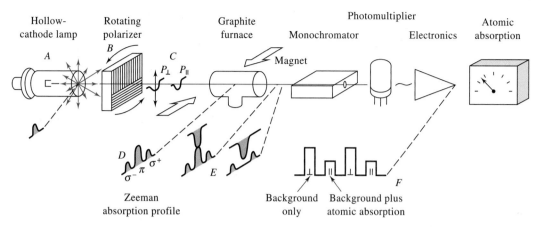

FIGURE 9-15 Schematic of an electrothermal atomic absorption instrument that provides a background correction based on the Zeeman effect. (Courtesy of Hitachi Scientific Instruments, Mountain View, CA.)

were formed. This phenomenon, which is termed the Zeeman effect,[19] is general for all atomic spectra. Several splitting patterns arise depending on the type of electronic transition that is involved in the absorption process. The simplest splitting pattern, which is observed with singlet (Section 8A-1) transitions, leads to a central, or π, line and two equally spaced satellite σ lines. The central line, which is at the original wavelength, has an absorbance that is twice that of each σ line. For more complex transitions, further splitting of the π and σ lines occurs.

Application of the Zeeman effect to atomic absorption instruments is based on the differing response of the two types of absorption lines to polarized radiation. The π line absorbs only that radiation that is plane polarized in a direction parallel to the external magnetic field; the σ lines, in contrast, absorb only radiation polarized at 90° to the field.

Figure 9-15 shows details of an electrothermal atomic absorption instrument, which uses the Zeeman effect for background correction. Unpolarized radiation from an ordinary hollow-cathode source *A* is passed through a rotating polarizer *B*, which separates the beam into two components that are plane-polarized at 90° to one another *C*. These beams pass into a tube-type graphite furnace similar to the one shown in Fig-

ure 9-6a. A permanent 11-kG magnet surrounds the furnace and splits the energy levels into the three absorption peaks shown in *D*. Note that the central peak absorbs only radiation that is plane polarized with the field. During that part of the cycle when the source radiation is polarized similarly, absorption of radiation by the analyte takes place. During the other half cycle, no analyte absorption can occur. Broadband molecular absorption and scattering by the matrix products occur during both half cycles, which leads to the cyclical absorbance pattern shown in *F*. The data-acquisition system is programmed to subtract the absorbance during the perpendicular half cycle from that for the parallel half cycle, thus giving a background corrected value.

A second type of Zeeman effect instrument has been designed in which a magnet surrounds the hollow-cathode source. Here, it is the emission spectrum of the source that is split rather than the absorption spectrum of the sample. This instrument configuration provides an analogous correction. To date, most instruments are of the type illustrated in Figure 9-15.

Zeeman effect instruments provide a more accurate correction for background than the methods described earlier. These instruments are particularly useful for electrothermal atomizers and permit the direct determination of elements in samples such as urine and blood. The decomposition of organic material in these samples leads to large background corrections (background A > 1) and, as a result, susceptibility to significant error.

[19]For a detailed discussion of the application of the Zeeman effect to atomic absorption, see D. J. Butcher and J. Sneddon, *A Practical Guide to Graphite Furnace Atomic Absorption Spectrometry*, New York: Wiley, 1998, pp. 73–84; F. J. Fernandez, S. A. Myers, and W. Slavin, *Anal. Chem.*, **1980**, *52*, 741; S. D. Brown, *Anal. Chem.*, **1977**, *49* (14), 1269A.

 Tutorial: Learn more about the **Zeeman effect**.

Background Correction Based on Source Self-Reversal

A remarkably simple means of background correction appears to offer most of the advantages of a Zeeman effect instrument.[20] This method, which is sometimes called the Smith-Hieftje background correction method, is based on the self-reversal or self-absorption behavior of radiation emitted from hollow-cathode lamps when they are operated at high currents. As was mentioned earlier, high currents produce large concentrations of nonexcited atoms, which are capable of absorbing the radiation produced from the excited species. An additional effect of high currents is to significantly broaden the emission line of the excited species. The net effect is to produce a line that has a minimum in its center, which corresponds exactly in wavelength to that of the absorption peak (see Figure 9-16).

To obtain corrected absorbances, the lamp is programmed to run alternately at low and high currents. The total absorbance is obtained during the low-current operation and the background absorbance is provided by measurements during the second part of the cycle when radiation at the absorption peak is at a minimum. The data-acquisition system then subtracts the background absorbance from the total to give a corrected value. Recovery of the source to its low-current output takes place in milliseconds when the current is reduced. The measurement cycle can be repeated often enough to give satisfactory signal-to-noise ratios. Equipment for this type of correction is available from commercial sources.

9C-2 Chemical Interferences

Chemical interferences are more common than spectral interferences. Their effects can frequently be minimized by a suitable choice of operating conditions.

Both theoretical and experimental evidence suggest that many of the processes occurring in the mantle of a flame are in approximate equilibrium. Therefore, it becomes possible to regard the burned gases as a solvent medium to which thermodynamic calculations can be applied. The equilibria of principal interest include formation of compounds of low volatility, dissociation reactions, and ionization.

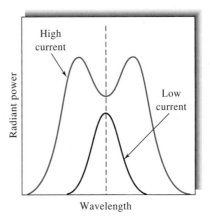

FIGURE 9-16 Emission line profiles for a hollow-cathode lamp operated at high and low currents.

Formation of Compounds of Low Volatility

Perhaps the most common type of interference is by anions that form compounds of low volatility with the analyte and thus reduce the fraction of the analyte that is atomized. Low results are the outcome. An example is the decrease in calcium absorbance observed with increasing concentrations of sulfate or phosphate. These anions form compounds with calcium that are difficult to volatilize. For example, at a fixed calcium concentration, the absorbance falls off nearly linearly with increasing sulfate or phosphate concentrations until the anion-to-calcium ratio is about 0.5; the absorbance then levels off at about 30% to 50% of its original value and becomes independent of anion concentration.

Cation interference has been observed as well. For example, aluminum causes low results in the determination of magnesium, apparently as a result of the formation of a heat-stable aluminum-magnesium compound (perhaps an oxide).

Interferences caused by formation of species of low volatility can often be eliminated or moderated by use of higher temperatures. Alternatively, *releasing agents*, which are cations that react preferentially with the interferant and prevent its interaction with the analyte, can be used. For example, addition of an excess of strontium or lanthanum ion minimizes the interference of phosphate in the determination of calcium. The same two species have also been used as releasing agents for the determination of magnesium in the presence of aluminum. In both instances, the strontium or lanthanum replaces the analyte in the compound formed with the interfering species.

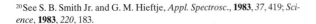

[20]See S. B. Smith Jr. and G. M. Hieftje, *Appl. Spectrosc.*, **1983**, *37*, 419; *Science*, **1983**, *220*, 183.

Protective agents prevent interference by forming stable but volatile species with the analyte. Three common reagents for this purpose are ethylenediaminetetraacetic acid (EDTA), 8-hydroxyquinoline, and APCD, which is the ammonium salt of 1-pyrrolidinecarbodithioic acid. The presence of EDTA has been shown to eliminate the interference of aluminum, silicon, phosphate, and sulfate in the determination of calcium. Similarly, 8-hydroxyquinoline suppresses the interference of aluminum in the determination of calcium and magnesium.

Dissociation Equilibria

In the hot, gaseous environment of a flame or a furnace, numerous dissociation and association reactions lead to conversion of the metallic constituents to the elemental state. It seems probable that at least some of these reactions are reversible and can be treated by the laws of thermodynamics. Thus, it should be possible to formulate equilibria such as

$$MO \rightleftharpoons M + O$$

$$M(OH)_2 \rightleftharpoons M + 2OH$$

where M is the analyte atom and OH is the hydroxyl radical.

In practice, not enough is known about the nature of the chemical reactions in a flame to permit a quantitative treatment such as that for an aqueous solution. Instead, we must rely on empirical observations.

Dissociation reactions involving metal oxides and hydroxides play an important part in determining the nature of the emission or absorption spectra for an element. For example, the alkaline-earth oxides are relatively stable, with dissociation energies in excess of 5 eV. Molecular bands arising from the presence of metal oxides or hydroxides in the flame thus constitute a prominent feature of their spectra (see Figure 8-8). Except at very high temperatures, these bands are more intense than the lines for the atoms or ions. In contrast, the oxides and hydroxides of the alkali metals are much more easily dissociated so that line intensities for these elements are high, even at relatively low temperatures.

Dissociation equilibria that involve anions other than oxygen may also influence flame emission and absorption. For example, the line intensity for sodium is markedly decreased by the presence of HCl. A likely explanation is the mass-action effect on the equilibrium

$$NaCl \rightleftharpoons Na + Cl$$

Chlorine atoms formed from the added HCl decrease the atomic sodium concentration and thereby lower the line intensity.

Another example of this type of interference involves the enhancement of vanadium absorption when aluminum or titanium is present. The interference is significantly more pronounced in fuel-rich flames than in lean flames. These effects can be explained by assuming that the three metals interact with such species as O and OH, which are always present in flames. If the oxygen-bearing species are given the general formula Ox, a series of equilibrium reactions can be postulated. Thus,

$$VOx \rightleftharpoons V + Ox$$

$$AlOx \rightleftharpoons Al + Ox$$

$$TiOx \rightleftharpoons Ti + Ox$$

In fuel-rich combustion mixtures, the concentration of Ox is sufficiently small that its concentration is lowered significantly when aluminum or titanium is present in the sample. This decrease causes the first equilibrium to shift to the right with an accompanying increase in the vanadium concentration and absorbance. In lean mixtures, on the other hand, the concentration of Ox is apparently high relative to the total concentration of metal atoms. In this case, addition of aluminum or titanium scarcely changes the concentration of Ox, and the position of the first equilibrium is relatively undisturbed. Therefore, the position of the first equilibrium is not disturbed significantly.

Ionization Equilibria

Ionization of atoms and molecules is small in combustion mixtures that involve air as the oxidant, and can often be neglected. In higher temperature flames where oxygen or nitrous oxide serves as the oxidant, however, ionization becomes important, and there is a significant concentration of free electrons produced by the equilibrium

$$M \rightleftharpoons M^+ + e^- \tag{9-1}$$

where M represents a neutral atom or molecule and M^+ is its ion. We will focus on equilibria in which M is a metal atom.

The equilibrium constant K for this reaction takes the form

$$K = \frac{[M^+][e^-]}{[M]} \tag{9-2}$$

TABLE 9-2 Degree of Ionization of Metals at Flame Temperatures

| Element | Ionization Potential, eV | Fraction Ionized at the Indicated Pressure and Temperature | | | |
| | | $P = 10^{-4}$ atm | | $P = 10^{-6}$ atm | |
		2000 K	3500 K	2000 K	3500 K
Cs	3.893	0.01	0.86	0.11	>0.99
Rb	4.176	0.004	0.74	0.04	>0.99
K	4.339	0.003	0.66	0.03	0.99
Na	5.138	0.0003	0.26	0.003	0.90
Li	5.390	0.0001	0.18	0.001	0.82
Ba	5.210	0.0006	0.41	0.006	0.95
Sr	5.692	0.0001	0.21	0.001	0.87
Ca	6.111	3×10^{-5}	0.11	0.0003	0.67
Mg	7.644	4×10^{-7}	0.01	4×10^{-6}	0.09

Data from B. L. Vallee and R. E. Thiers, in *Treatise on Analytical Chemistry,* I. M. Kolthoff and P. J. Elving, eds., Part I, Vol. 6, p. 3500, New York: Interscience, 1965. Reprinted with permission of John Wiley & Sons, Inc.

If no other source of electrons is present in the flame, this equation can be written in the form

$$K = \left(\frac{\alpha^2}{1 - \alpha} \right) P$$

where α is the fraction of M that is ionized, and P is the partial pressure of the metal in the gaseous solvent before ionization.

Table 9-2 shows the calculated fraction ionized for several common metals under conditions that approximate those used in flame emission spectroscopy. The temperatures correspond roughly to conditions that exist in air-acetylene and oxygen-acetylene flames, respectively.

It is important to appreciate that treatment of the ionization process as an equilibrium — with free electrons as one of the products — immediately implies that the degree of ionization of a metal will be strongly influenced by the presence of other ionizable metals in the flame. Thus, if the medium contains not only species M but species B as well, and if B ionizes according to the equation

$$B \rightleftharpoons B^+ + e^-$$

then the degree of ionization of M will be decreased by the mass-action effect of the electrons formed from B. Determination of the degree of ionization under these conditions requires a calculation involving the dissociation constant for B and the mass-balance expression

$$[e^-] = [B^+] + [M^+]$$

Atom-ion equilibria in flames create a number of important consequences in flame spectroscopy. For example, intensities of atomic emission or absorption lines for the alkali metals, particularly potassium, rubidium, and cesium, are affected by temperature in a complex way. Increased temperatures cause an increase in the population of excited atoms, according to the Boltzmann relationship (Equation 8-1). Counteracting this effect, however, is a decrease in concentration of atoms resulting from ionization. Thus, under some circumstances a decrease in emission or absorption may be observed in hotter flames. It is for this reason that lower excitation temperatures are usually specified for the determination of alkali metals.

The effects of shifts in ionization equilibria can usually be eliminated by addition of an *ionization suppressor,* which provides a relatively high concentration of electrons to the flame; suppression of ionization of the analyte results. The effect of a suppressor appears in the calibration curves for strontium shown in Figure 9-17. Note the significant increase in slope of these curves as strontium ionization is repressed by the increasing concentration of potassium ions and electrons. Note also the enhanced sensitivity produced by using nitrous oxide instead of air as the oxidant. The higher temperature achieved with nitrous oxide undoubtedly enhances the degree of decomposition and volatilization of the strontium compounds in the plasma.

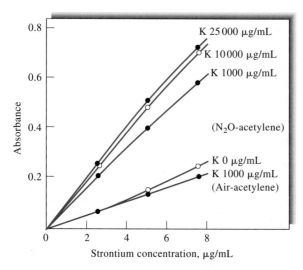

FIGURE 9-17 Effect of potassium concentration on the calibration curve for strontium. (Reprinted with permission from J. A. Bowman and J. B. Willis, *Anal. Chem.*, **1967**, *39*, 1220. Copyright 1967 American Chemical Society.)

9D ATOMIC ABSORPTION ANALYTICAL TECHNIQUES

This section deals with some of the practical details that must be considered in flame or electrothermal atomic absorption analysis.

9D-1 Sample Preparation

A disadvantage of flame spectroscopic methods is the requirement that the sample be introduced into the excitation source in the form of a solution, most commonly an aqueous one. Unfortunately, many materials of interest, such as soils, animal tissues, plants, petroleum products, and minerals are not directly soluble in common solvents, and extensive preliminary treatment is often required to obtain a solution of the analyte in a form ready for atomization. Indeed, the decomposition and solution steps are often more time-consuming and introduce more error than the spectroscopic measurement itself.

Decomposition of materials such as those just cited usually require rigorous treatment of the sample at high temperatures accompanied by a risk of losing the analyte by volatilization or as particulates in smoke. Fur-

 Tutorial: Learn more about **atomic absorption spectroscopy**.

thermore, the reagents used in decomposing a sample often introduce the kinds of chemical and spectral interferences that were discussed earlier. Additionally, the analyte may be present in these reagents as an impurity. In fact, unless considerable care is taken, it is not uncommon in trace analyses to find that reagents are a larger source of the analyte than the samples — a situation that can lead to serious error even with blank corrections.

Some of the common methods used for decomposing and dissolving samples for atomic absorption methods include treatment with hot mineral acids; oxidation with liquid reagents, such as sulfuric, nitric, or perchloric acids (wet ashing); combustion in an oxygen bomb or other closed container to avoid loss of analyte; ashing at a high temperature; and high-temperature fusion with reagents such as boric oxide, sodium carbonate, sodium peroxide, or potassium pyrosulfate.[21]

One of the advantages of electrothermal atomization is that some materials can be atomized directly, thus avoiding the solution step. For example, liquid samples such as blood, petroleum products, and organic solvents can be pipetted directly into the furnace for ashing and atomization. Solid samples, such as plant leaves, animal tissues, and some inorganic substances, can be weighed directly into cup-type atomizers or into tantalum boats for introduction into tube-type furnaces. Calibration is, however, usually difficult and requires standards that approximate the sample in composition.

9D-2 Sample Introduction by Flow Injection

In Section 33B, we describe the methods and instrumentation for flow injection analysis (FIA). FIA methodology serves as an excellent means of introducing samples into a flame atomic absorption spectrometer. Alternatively, we may think of an atomic absorption spectrometer as a useful detector for an FIA system. From any perspective, the peristaltic pump and valve arrangements of FIA described in Chapter 33 are a convenient means to sample analyte solutions reproducibly and efficiently, especially when it is important to conserve sample. The carrier stream of the FIA system consisting of deionized water or dilute electrolyte provides continuous flushing of the flame

---

[21] B. Kebbekus in *Sample Preparation Techniques in Analytical Chemistry*, S. Mitra, ed., New York: Wiley, 2003; R. Bock, *A Handbook of Decomposition Methods in Analytical Chemistry*, New York: Wiley, 1979.

atomizer, which is particularly advantageous for samples containing high levels of salts or suspended solids.

9D-3 Organic Solvents

Early in the development of atomic absorption spectroscopy it was recognized that enhanced absorbances could be obtained if the solutions contained low-molecular-weight alcohols, esters, or ketones. The effect of organic solvents is largely attributable to increased nebulizer efficiency; the lower surface tension of such solutions results in smaller drop sizes and a resulting increase in the amount of sample that reaches the flame. In addition, more rapid solvent evaporation may also contribute to the effect. Leaner fuel-oxidant ratios must be used with organic solvents to offset the presence of the added organic material. Unfortunately, however, the leaner mixture produces lower flame temperatures and an increased potential for chemical interferences.

A most important analytical application of organic solvents to flame spectroscopy is the use of immiscible solvents such as methyl isobutyl ketone to extract chelates of metallic ions. The resulting extract is then nebulized directly into the flame. Here, the sensitivity is increased not only by the enhancement of absorption lines because of the solvent but also because for many systems only small volumes of the organic liquid are required to remove metal ions quantitatively from relatively large volumes of aqueous solution. This procedure has the added advantage that at least part of the matrix components are likely to remain in the aqueous solvent; a reduction in interference often results. Common chelating agents include ammonium pyrrolidinedithiocarbamate, diphenylthiocarbazone (dithizone), 8-hydroxyquinoline, and acetylacetone.

9D-4 Calibration Curves

In theory, atomic absorption should follow Beer's law (Equation 6-34) with absorbance being directly proportional to concentration. Unfortunately, calibration curves are often nonlinear, so it is counterproductive to perform an atomic absorption analysis without experimentally confirming the linearity of the instrument response. Thus, a calibration curve that covers the range of concentrations found in the sample should be prepared periodically. In addition, the number of uncontrolled variables in atomization and absorbance measurements is sufficiently large to warrant measurement of one standard solution each time an analysis is performed. It is even better to use two standards that bracket the analyte concentration. Any deviation of the standard from the original calibration curve can then be used to correct the analytical result.

9D-5 Standard-Addition Method

The standard-addition method, which was described in Section 1D-3, is widely used in atomic absorption spectroscopy to partially or completely compensate for the chemical and spectral interferences introduced by the sample matrix.

9D-6 Applications of AAS

AAS is a sensitive means for the quantitative determination of more than sixty metals or metalloid elements. The resonance lines for the nonmetallic elements are generally located at wavelengths shorter than 200 nm, thus preventing their determination by convenient, nonvacuum spectrophotometers.

Detection Limits

Columns two and three of Table 9-3 present detection limits for a number of common elements by flame and electrothermal atomic absorption. For comparison, detection limits for some of the other atomic procedures are also included. Small differences among the quoted values are not significant. Thus, an order of magnitude is probably meaningful, but a factor of 2 or 3 certainly is not.

For many elements, detection limits for atomic absorption spectroscopy with flame atomization lie in the range of 1 to 20 ng/mL, or 0.001 to 0.020 ppm; for electrothermal atomization, the corresponding figures are 0.002 to 0.01 ng/mL, or 2×10^{-6} to 1×10^{-5} ppm. In a few cases, detection limits well outside these ranges are encountered.

Accuracy

Under the usual conditions, the relative error associated with a flame atomic absorption analysis is of the order of 1% to 2%. With special precautions, this figure can be lowered to a few tenths of a percent. Errors encountered with electrothermal atomization usually exceed those for flame atomization by a factor of 5 to 10.

TABLE 9-3 Detection Limits (ng/mL)[a] for Selected Elements

Element	AAS Flame	AAS Electro-thermal	AES Flame	AES ICP	AFS Flame
Al	30	0.1	5	0.2	5
As	200	0.5	—	2	15
Ca	1	0.25	0.1	0.0001	0.4
Cd	1	0.01	2000	0.07	0.1
Cr	4	0.03	5	0.08	0.6
Cu	2	0.05	10	0.04	0.2
Fe	6	0.25	50	0.09	0.3
Hg	500	5	—	—	5
Mg	0.2	0.002	5	0.003	0.3
Mn	2	0.01	—	0.01	1
Mo	5	0.5	100	0.2	8
Na	0.2	0.02	0.1	0.1	0.3
Ni	3	0.5	600	0.2	0.4
Pb	8	0.1	200	1	5
Sn	15	5	300	—	200
V	25	1	200	0.06	25
Zn	1	0.005	50000	0.1	0.1

From J. D. Ingle Jr. and S. R. Crouch, *Spectrochemical Analysis*, pp. 250–51, 300, 321, Englewood Cliffs, NJ: Prentice Hall, 1988.

Note: Pulsed hollow-cathode-lamp excitation source with ICP atomization.

[a] $1 \text{ ng/mL} = 10^{-3} \text{ μg/mL} = 10^{-3} \text{ ppm}$.

AAS = atomic absorption spectroscopy; AES = atomic emission spectroscopy; AFS = atomic fluorescence spectroscopy; ICP = inductively coupled plasma.

9E ATOMIC FLUORESCENCE SPECTROSCOPY

Over the years, significant research effort has been devoted to the development of analytical methods based on atomic fluorescence.[22] This work has demonstrated clearly that atomic fluorescence spectroscopy provides a useful and convenient means for the quantitative determination of a reasonably large number of elements. To date, however, the procedure has not found widespread use because of the overwhelming successes of atomic emission and especially atomic absorption methods, which were developed prior to atomic fluorescence by more than a decade. As mentioned earlier, these successes have led to the availability of absorption and emission instruments from numerous commercial sources. In recent years, a number of manufacturers have introduced atomic fluorescence spectrometers useful for determining elements that form vapors and hydrides, such as Pb, Hg, Cd, Zn, As, Sb, Bi, Ge, and Se.[23]

The limited use of atomic fluorescence has not arisen so much from any inherent weakness of the procedure but rather because the advantages of atomic fluorescence have been small relative to the well-established absorption and emission methods. Thus, although fluorescence methods, particularly those based on electrothermal atomization, are somewhat more sensitive for several elements, the procedure is also less sensitive and appears to have a smaller useful concentration range for several others. Furthermore, dispersive fluorescence instruments are somewhat more complex and more expensive to purchase and maintain.[24] These disadvantages have been largely overcome in some special-purpose dedicated instruments such as the one described in the Instrumental Analysis in Action feature at the end of Section 2.

9E-1 Instrumentation

The components of instruments for atomic fluorescence measurements are generally arranged as shown in Figure 7-1b. The sample container is most commonly a flame but may also be an electrothermal atomization cell, a glow discharge, or an inductively coupled plasma, as described in Section 10A-1. Flow cells are often used in conjunction with vapor and hydride-based methods.

Sources

A continuum source would be desirable for atomic fluorescence measurements. Unfortunately, however, the output power of most continuum sources over a

---

[22] For further information on atomic fluorescence spectroscopy, see L. H. J. Lajunen and P. Peramaki, *Spectrochemical Analysis by Atomic Absorption and Emission*, 2nd ed., pp. 276–85, Cambridge: Royal Society of Chemistry, 2004; J. A. C. Broekaert, *Analytical Atomic Spectrometry with Flames and Plasmas*, pp. 290–96, Weinheim, Germany: Wiley-VCH, 2002; D. J. Butcher in *Handbook of Instrumental Techniques for Analytical Chemistry*, F. A. Settle, ed., Upper Saddle River, NJ, 1997, pp. 441–58; S. Greenfield, *Trends in Analytical Chemistry*, **1995**, *14*, 435–42; J. C. Van Loon, *Anal. Chem.*, **1981**, *53*, 332A; D. Butcher et al., *J. Anal. Atom. Spectrom.*, **1988**, *3*, 1059.

[23] Examples include Teledyne/Leeman Labs (Hudson, NH) and Aurora Instruments, Ltd. (Vancouver, BC).

[24] See W. B. Barnett and H. L. Kahn, *Anal. Chem.*, **1972**, *44*, 935.

region as narrow as an atomic absorption line is too low to provide sufficient sensitivity for atomic fluorescence.

In the early work on atomic fluorescence, conventional hollow-cathode lamps often served as excitation sources. To enhance the output intensity without destroying the lamp, it was necessary to operate the lamp with short pulses of current that were greater than the lamp could tolerate for continuous operation. The detector was gated to observe the fluorescence signal only during pulses of source radiation.

Perhaps the most widely used sources for atomic fluorescence have been the EDLs (Section 9B-1), which usually produce radiant intensities greater than those of hollow-cathode lamps by an order of magnitude or two. EDLs have been operated in both the continuous and pulsed modes. Unfortunately, this type of lamp is not available for many elements.

Lasers, with their high intensities and narrow bandwidths, would appear to be the ideal source for atomic fluorescence measurements. Their high cost and operational complexities, however, have discouraged their widespread application to routine atomic fluorescence methods.

Dispersive Instruments

A dispersive system for atomic fluorescence measurements consists of a modulated source, an atomizer (flame or nonflame), a monochromator or an interference filter system, a detector, and a signal processor and readout. With the exception of the source, most of these components are similar to those discussed in earlier parts of this chapter.

Nondispersive Instruments

In theory, no monochromator or filter should be necessary for atomic fluorescence measurements when an EDL or hollow-cathode lamp serves as the excitation source because the emitted radiation is, in principle, that of a single element and will thus excite only atoms of that element. A nondispersive system then could be made up of only a source, an atomizer, and a detector. There are several advantages of such a system: (1) simplicity and low-cost instrumentation, (2) adaptability to multielement analysis, (3) high-energy throughput and thus high sensitivity, and (4) simultaneous collection of energy from multiple lines, which also enhances sensitivity.

To realize these important advantages, it is necessary that the output of the source be free of contaminating lines from other elements; in addition, the atomizer should emit no significant background radiation. In some instances with electrothermal atomizers, background radiation is minimal, but certainly, it is not with typical flames. To overcome this problem, filters, located between the source and detector, have often been used to remove most of the background radiation. Alternatively, solar-blind photomultipliers, which respond only to radiation of wavelengths shorter than 320 nm, have been applied. For these devices to be used effectively, analyte emission must be below 320 nm.

9E-2 Interferences

Interferences encountered in atomic fluorescence spectroscopy are generally of the same type and of about the same magnitude as those found in atomic absorption spectroscopy.[25]

9E-3 Applications

Atomic fluorescence methods have been applied to the determination of metals in such materials as lubricating oils, seawater, geological samples, metallurgical samples, clinical samples, environmental samples, and agricultural samples. Table 9-3 lists detection limits for atomic fluorescence procedures.

---

[25] See J. D. Winefordner and R. C. Elser, *Anal. Chem.*, **1971**, *43* (4), 24A.

QUESTIONS AND PROBLEMS

*Answers are provided at the end of the book for problems marked with an asterisk.

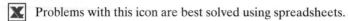 Problems with this icon are best solved using spreadsheets.

9-1 Define the following terms: (a) releasing agent, (b) protective agent, (c) ionization suppressor, (d) atomization, (e) pressure broadening, (f) hollow-cathode lamp,

(g) sputtering, (h) self-absorption, (i) spectral interference, (j) chemical inter-ference, (k) radiation buffer, (l) Doppler broadening.

9-2 Describe the effects that are responsible for the three different absorbance pro-files in Figure 9-4 and select three additional elements you would expect to have similar profiles.

9-3 Why is an electrothermal atomizer more sensitive than a flame atomizer?

9-4 Describe how a deuterium lamp can be used to provide a background correction for an atomic absorption spectrum.

9-5 Why is source modulation used in atomic absorption spectroscopy?

9-6 For the same concentration of nickel, the absorbance at 352.4 nm was found to be about 30% greater for a solution that contained 50% ethanol than for an aqueous solution. Explain.

9-7 The emission spectrum of a hollow-cathode lamp for molybdenum has a sharp line at 313.3 nm as long as the lamp current is less than 50 mA. At higher currents, however, the emission line develops a cuplike crater at its maximum. Explain.

9-8 An analyst attempts to determine strontium with an atomic absorption instru-ment equipped with a nitrous oxide–acetylene burner, but the sensitivity associ-ated with the 460.7-nm atomic resonance line is not satisfactory. Suggest at least three things that might be tried to increase sensitivity.

9-9 Why is atomic emission more sensitive to flame instability than atomic absorption or fluorescence?

9-10 Figure 9-1 summarizes many of the processes that take place in a laminar-flow burner. With specific reference to the analysis of an aqueous $MgCl_2$ solution, describe the processes that are likely to occur.

*__9-11__ Use Equation 7-13 for the resolving power of a grating monochromator to esti-mate the theoretical minimum size of a diffraction grating that would provide a profile of an atomic absorption line at 500 nm having a line width of 0.002 nm. Assume that the grating is to be used in the first order and that it has been ruled at 2400 grooves/mm.

*__9-12__ For the flame shown in Figure 9-3, calculate the relative intensity of the 766.5-nm emission line for potassium at the following heights above the flame (assume no ionization):
(a) 2.0 cm   (b) 3.0 cm   (c) 4.0 cm   (d) 5.0 cm

9-13 In a hydrogen-oxygen flame, the atomic absorption signal for iron was found to decrease in the presence of large concentrations of sulfate ion.
(a) Suggest an explanation for this observation.
(b) Suggest three possible methods for overcoming the potential interference of sulfate in a quantitative determination of iron.

*__9-14__ For Na atoms and Mg^+ ions, compare the ratios of the number of particles in the 3p excited state to the number in the ground state in
(a) a natural gas–air flame (2100 K).
(b) a hydrogen-oxygen flame (2900 K).
(c) an inductively coupled plasma source (6000 K).

*9-15 In higher-temperature sources, sodium atoms emit a doublet with an average wavelength of 1139 nm. The transition responsible is from the $4s$ to $3p$ state. Calculate the ratio of the number of excited atoms in the $4s$ to the number in the ground $3s$ state in
(a) an acetylene-oxygen flame (3000°C).
(b) the hottest part of an inductively coupled plasma source (~9000°C).

*9-16 Assume that the absorption signal shown in Figure 9-7 were obtained for 2-μL aliquots of standards and sample. Calculate the concentration in parts per million of lead in the sample of canned orange juice.

9-17 Suggest sources of the two signals in Figure 9-7 that appear during the drying and ashing processes.

9-18 In the concentration range of 1 to 100 μg/mL P, phosphate suppresses the atomic absorption of Ca in a linear manner. The absorbance levels off, however, between 100 and 300 μg/mL P. Explain. How can this effect be reduced?

9-19 What is the purpose of an internal standard in flame emission methods?

*9-20 A 5.00-mL sample of blood was treated with trichloroacetic acid to precipitate proteins. After centrifugation, the resulting solution was brought to a pH of 3 and was extracted with two 5-mL portions of methyl isobutyl ketone containing the organic lead complexing agent APCD. The extract was aspirated directly into an air-acetylene flame yielding an absorbance of 0.444 at 283.3 nm. Five-milliliter aliquots of standard solutions containing 0.250 and 0.450 ppm Pb were treated in the same way and yielded absorbances of 0.396 and 0.599. Calculate the concentration Pb (ppm) in the sample assuming that Beer's law is followed.

 9-21 The sodium in a series of cement samples was determined by flame emission spectroscopy. The flame photometer was calibrated with a series of NaCl standards that contained sodium equivalent to 0, 20.0, 40.0, 60.0, and 80.0 μg Na_2O per mL. The instrument readings R for these solutions were 3.1, 21.5, 40.9, 57.1, and 77.3.
(a) Plot the data using a spreadsheet.
(b) Obtain a least-squares equation for the data.
(c) Calculate the statistics for the line in (b).
(d) The following data were obtained for replicate 1.000-g samples of cement that were dissolved in HCl and diluted to 100.0 mL after neutralization.

		Emission Reading		
	Blank	Sample A	Sample B	Sample C
Replicate 1	5.1	28.6	40.7	73.1
Replicate 2	4.8	28.2	41.2	72.1
Replicate 3	4.9	28.9	40.2	Spilled

Calculate the percentage of Na_2O in each sample. What are the absolute and relative standard deviations for the average of each determination?

9-22 The chromium in an aqueous sample was determined by pipetting 10.0 mL of the unknown into each of five 50.0-mL volumetric flasks. Various volumes of a standard containing 12.2 ppm Cr were added to the flasks, following which the solutions were diluted to volume.

Unknown, mL	Standard, mL	Absorbance
10.0	0.0	0.201
10.0	10.0	0.292
10.0	20.0	0.378
10.0	30.0	0.467
10.0	40.0	0.554

(a) Plot the data using a spreadsheet.

(b) Determine an equation for the relationship between absorbance and volume of standard.

(c) Calculate the statistics for the least-squares relationship in (b).

(d) Determine the concentration of Cr in ppm in the sample.

(e) Find the standard deviation of the result in (d).

Challenge Problem

9-23 (a) In an investigation of the influence of experimental variables on detection limits in electrothermal AAS, Cabon and Bihan found several factors to be significant in the optimization of the method.[26] List six of these factors, describe in detail the physical basis for each factor, and discuss why each is important.

(b) These workers describe an a priori method for determining the limit of detection (LOD). Compare and contrast this method with the method described in Section 1E-2. How does this method improve on the method as defined by the IUPAC in the "Orange Book"? See http://www.iupac.org/publications/analytical_compendium/. Describe any disadvantages of the method.

(c) The investigations described by Cabon and Bihan treated the data using least-squares polynomial smoothing (see Section 5C-2) prior to determining the LOD. Describe precisely how the data were smoothed. What experimental variable was optimized in the smoothing procedure? How was the width of the smoothing window defined? What effect, if any, did the smoothing procedure have on the LOD as determined by these workers? What effect did smoothing have on the determination of the integration window for the instrumental signal?

(d) These workers compared the determination of the signal magnitude by integration and by measuring peak signals. What was the outcome of this comparison? Explain why these results were obtained by using your understanding of signal-to-noise enhancement procedures.

(e) How were instrument signals integrated? What alternative numerical procedures are available for integrating digital signals? What procedural variable or variables influenced the quality of the integrated signal data? Describe the effect of signal integration on working curves for Pb.

(f) What is *dosing volume*, and what effect did it appear to have on the quality of the results in these procedures?

---

[26] J. Y. Cabon and A. Le Bihan, *Analyst*, **1997**, *122*, 1335.

CHAPTER TEN

Atomic Emission Spectrometry

Plasma, arc, and spark emission spectrometry offer several advantages when compared with the flame and electrothermal absorption methods considered in Chapter 9.[1] Among the advantages is their lower susceptibility to chemical interferences, which is a direct result of their higher temperatures. Second, good emission spectra result for most elements under a single set of excitation conditions; consequently, spectra for dozens of elements can be recorded simultaneously. This property is of particular importance for the multielement analysis of very small samples. Flames are less satisfactory as atomic emission sources because optimum excitation conditions vary widely from element to element; high temperatures are needed for excitation of some elements and low temperatures for others; and finally, the region of the flame that gives rise to optimum line intensities varies from element to element. Another advantage of the more energetic plasma sources is that they permit the determination of low concentrations of elements that tend to form refractory compounds (that is, compounds that are highly resistant to thermal decomposition, such as the oxides of boron, phosphorus, tungsten, uranium, zirconium, and niobium). In addition, plasma sources permit the determination of nonmetals such as chlorine, bromine, iodine, and sulfur. Finally, plasma emission methods usually have concentration ranges of several orders of magnitude, in contrast to a two- or three-decade range for the absorption methods described in the previous chapter.

Emission spectra from plasma, arc, and spark sources are often highly complex and are frequently made up of hundreds, or even thousands, of lines. This large number of lines, although advantageous when seeking qualitative information, increases the probability of spectral interferences in quantitative analysis. Consequently, emission spectroscopy based on plasmas, arcs, and sparks requires higher resolution and more expensive optical equipment than is needed for

This chapter deals with optical atomic emission spectrometry (AES). Generally, the atomizers listed in Table 8-1 not only convert the components of samples to atoms or elementary ions but, in the process, excite a fraction of these species to higher electronic states. As the excited species rapidly relax back to lower states, ultraviolet and visible line spectra arise that are useful for qualitative and quantitative elemental analysis. Plasma sources have become the most important and most widely used sources for AES. These devices, including the popular inductively coupled plasma source, are discussed first in this chapter. Then, emission spectroscopy based on electric arc and electric spark atomization and excitation is described. Historically, arc and spark sources were quite important in emission spectrometry, and they still have important applications for the determination of some metallic elements. Finally several miscellaneous atomic emission sources, including flames, glow discharges, and lasers are presented.

Throughout this chapter, this logo indicates an opportunity for online self-study at **www .thomsonedu.com/chemistry/skoog**, linking you to interactive tutorials, simulations, and exercises.

---

[1]For more extensive treatment of emission spectroscopy, see *Atomic Spectroscopy in Elemental Analysis*, M. Cullen, ed., Chaps. 3–5, Boca Raton, FL: CRC Press, 2004; L. H. J. Lajunen and P. Peramaki, *Spectrochemical Analysis by Atomic Absorption and Emission*, 2nd ed., Chaps. 4–6, Royal Society of Chemistry: Cambridge, 2004; J. A. C. Broekaert, *Analytical Atomic Spectrometry with Flames and Plasmas*, Chap. 5, Hoboken, NJ: Wiley-VCH, 2002; J. D. Ingle Jr. and S. R. Crouch, *Spectrochemical Analysis*, Chaps. 7–9, 11, Upper Saddle River, NJ: Prentice-Hall, 1988.

atomic absorption methods with flame or electrother-mal sources.

Despite their advantages, it is unlikely that emission methods based on high-energy sources will ever completely displace flame and electrothermal atomic absorption procedures. In fact, atomic emission and absorption methods are complementary. Among the advantages of atomic absorption procedures are simpler and less expensive equipment requirements, lower operating costs, somewhat greater precision (presently, at least), and procedures that require less operator skill to yield satisfactory results.[2]

10A EMISSION SPECTROSCOPY BASED ON PLASMA SOURCES

A plasma is an electrically conducting gaseous mixture containing a significant concentration of cations and electrons. (The concentrations of the two are such that the net charge is zero.) In the argon plasma frequently used for emission analyses, argon ions and electrons are the principal conducting species, although cations from the sample are also present in small amounts. Argon ions, once formed in a plasma, can absorb sufficient power from an external source to maintain the temperature at a level where further ionization sustains the plasma indefinitely. Such plasmas achieve temperatures as high as 10,000 K. There are three primary types of high-temperature plasmas: (1) the inductively coupled plasma (ICP), (2) the direct current plasma (DCP), and (3) the microwave induced plasma (MIP).[3] The first two of these sources are marketed by several instrument companies. The microwave induced plasma source is not widely used for general elemental analysis. We shall not consider this source further in this chapter. Note, however, that the microwave plasma source is commercially available as an element detector for gas chromatography. This instrument is described briefly in Section 27B-4.

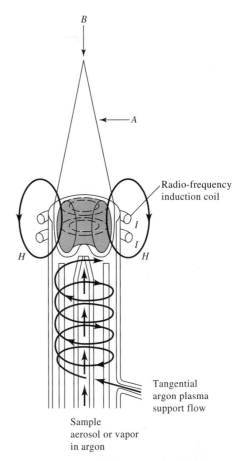

FIGURE 10-1 A typical ICP source. Position *A* shows radial viewing of the torch, and position *B* shows axial viewing. (From V. A. Fassel, *Science*, **1978**, *202*, 185. With permission. Copyright 1978 by the American Association for the Advancement of Science.)

10A-1 The Inductively Coupled Plasma Source

Figure 10-1 is a schematic of a typical ICP source called a *torch*.[4] It consists of three concentric quartz tubes through which streams of argon gas flow. De-

---

[2]For an excellent comparison of the advantages and disadvantages of flames, furnaces, and plasmas as sources for emission spectroscopy, see W. Slavin, *Anal. Chem.*, **1986**, *58*, 589A.

[3]For a comparison of these sources, see R. D. Sacks, in *Treatise on Analytical Chemistry*, 2nd ed., P. J. Elving, E. J. Meehan, and I. M. Kolthoff, eds., Part I, Vol. 7, pp. 516–26, New York: Wiley, 1981; A. T. Zander, *Anal. Chem.*, **1986**, *58*, 1139A.

 Exercise: Learn more about **ICP torches**.

---

[4]For a more complete discussion on ICP sources, see V. A. Fassel, *Science*, **1978**, *202*, 183; V. A. Fassel, *Anal. Chem.*, **1979**, *51*, 1290A; G. A. Meyer, *Anal. Chem.*, **1987**, *59*, 1345A; S. J. Hill, *ICP Spectrometry and Its Applications*, Boca Raton, FL, CRC Press, 1999; A. Varma, *CRC Handbook of Inductively Coupled Plasma Atomic Emission Spectroscopy*, Boca Raton, FL: CRC Press, 1990; *Inductively Coupled Plasma Mass Spectrometry*, A. E. Montaser, ed., Hoboken, NJ: Wiley-VCH, 1998; *Inductively Coupled Plasma in Analytical Atomic Spectroscopy*, 2nd ed., A. Montaser and D. W. Golightly, eds., New York: Wiley-VCH, 1992; *Handbook of Inductively Coupled Plasma Spectroscopy*, M. Thompson and J. N. Walsh, eds., New York: Chapman & Hall, 1989.

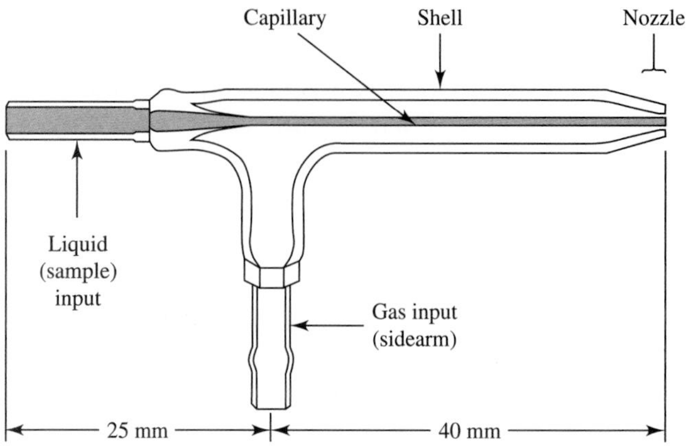

FIGURE 10-2 The Meinhard nebulizer. The nebulizing gas flows through an opening that surrounds the capillary concentrically. This causes a reduced pressure at the tip and aspiration of the sample. The high-velocity gas at the tip breaks up the solution into a mist. (Courtesy of J. Meinhard Associates, Inc.)

pending on the torch design, the total rate of argon consumption is 5 to 20 L/min. The diameter of the largest tube is often about 2.5 cm. Surrounding the top of this tube is a water-cooled induction coil that is powered by a radio-frequency generator, which radiates 0.5 to 2 kW of power at 27.12 MHz or 40.68 MHz.[5] Ionization of the flowing argon is initiated by a spark from a Tesla coil. The resulting ions, and their associated electrons, then interact with the fluctuating magnetic field (labeled H in Figure 10-1) produced by the induction coil. This interaction causes the ions and electrons within the coil to flow in the closed annular paths shown in Figure 10-1. The resistance of the ions and electrons to this flow of charge causes ohmic heating of the plasma.

The temperature of the plasma formed in this way is high enough to require thermal isolation of the outer quartz cylinder. This isolation is achieved by flowing argon tangentially around the walls of the tube as indicated by the arrows in Figure 10-1. The tangential flow cools the inside walls of the center tube and centers the plasma radially.[6]

A design offered by most manufacturers rotates the torch by 90° so that it is aligned axially with the spectrometer system (B in Figure 10-1). The radiation emitted from the center of the plasma is then used for analyses. This axial arrangement is particularly advantageous for ICP mass spectrometry (ICPMS), which we describe in Section 11C-1 (see Figure 10-6 for a

spectrometer with axial viewing geometry and Figure 10-8 for radial geometry).

Note that the argon flow rate through the typical torch is great enough to produce a significant operating cost for an ICP spectrometer (several thousand dollars annually). During the 1980s, low-flow, low-power torches appeared on the market. Typically, these torches require a total argon flow of less than 10 L/min and require less than 800 W of radio-frequency power.

Sample Introduction

Samples can be introduced into the ICP by argon flowing at about 1 L/min through the central quartz tube. The sample can be an aerosol, a thermally generated vapor, or a fine powder. The most common means of sample introduction is the concentric glass nebulizer shown in Figure 10-2. The sample is transported to the tip by the Bernoulli effect (aspiration). The high-velocity gas breaks up the liquid into fine droplets of various sizes, which are then carried into the plasma.

Another popular type of nebulizer has a cross-flow design (Figure 8-11b). Here, a high-velocity gas flows across a capillary tip at right angles, causing the same Bernoulli effect. Often, in this type of nebulizer, the liquid is pumped through the capillary with a peristaltic pump. Many other types of nebulizers are available for higher-efficiency nebulization, for nebulization of samples with high solids content, and for production of ultrafine mists.[7]

Another method of introducing liquid and solid samples into a plasma is by electrothermal vaporization. Here, the sample is vaporized in a furnace similar

---

[5]Most commercial instruments now operate at 40.68 MHz because better coupling efficiency between the coil and the plasma is achieved at the higher frequency and lower background emission is produced.

[6]For a description of commercially available torches, see D. Noble, *Anal. Chem.*, **1994**, *66*, 105A.

[7]See R. F. Browner and A. W. Boorn, *Anal. Chem.*, **1984**, *56*, 787A, 875A.

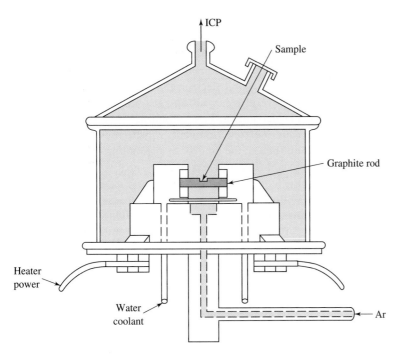

FIGURE 10-3 Device for electrothermal vaporization.

to that described in Section 8C-1 for electrothermal atomization. In plasma applications, however, the furnace is used for *sample introduction only* rather than for *sample atomization*; atomization occurs in the plasma. Figure 10-3 shows an electrothermal vaporizer in which vapor formation takes place on an open graphite rod. The vapor passes into a plasma torch in a stream of argon. The observed signal is a transient peak similar to the peaks obtained in electrothermal atomic absorption. Electrothermal vaporization coupled with a plasma torch offers the microsampling capabilities (~5 μL) and low absolute detection limits (~1 ng) of electrothermal furnaces while maintaining the wide linear working range, acceptable sample-to-sample precision (5%–10%), freedom from interference, and the multielement capabilities of ICP.[8]

The ablation devices for solids described in Section 8C-2 are also available from several makers of ICP instruments. With these types of sample-introduction systems, the plume of vapor and particulate matter produced by interaction of the sample with an electric arc or spark or with a laser beam are transported by a flow of argon into the torch where further atomization and excitation occur.

[8]S. Kim et al., *Microchim. J.*, **2004**, 127.

Plasma Appearance and Spectra

The typical plasma has a very intense, brilliant white, nontransparent core topped by a flamelike tail. The core, which extends a few millimeters above the tube, produces the atomic spectrum of argon superimposed on a continuum spectrum. The continuum is typical of ion-electron recombination reactions and bremsstrahlung, pronounced ′brem(p)-″shträ-lən, which is continuum radiation produced when charged particles are slowed or stopped. In the region 10 to 30 mm above the core, the continuum fades and the plasma is optically transparent. Spectral observations are generally made at a height of 15 to 20 mm above the induction coil, where the temperature is 6000–6500 K. In this region, the background radiation is remarkably free of argon lines and is well suited for analysis. Many of the most sensitive analyte lines in this region of the plasma are from ions, such as Ca^+, Cd^+, Cr^+, and Mn^+.

In ICP spectrometers, the torch may be viewed radially, perpendicular to its axis (*A* in Figure 10-1), or axially (*B* in the figure) or they may have a computer-controlled switching arrangement for both viewing schemes. Advantages of the axial arrangement over the radial configuration include increased radiation intensity resulting from a longer path length and higher precision, which produce lower detection limits (a factor of

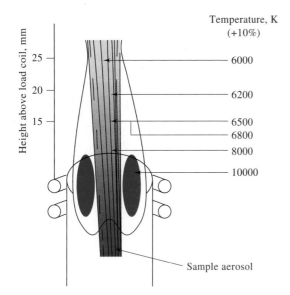

FIGURE 10-4 Temperatures in a typical ICP source. (From V. A. Fassel, *Science*, **1978**, *202*, 186. With permission. Copyright 1978 by the American Association for the Advancement of Science.)

2 to 30 with ultrasonic nebulization). The disadvantages are that the cool plasma tail must be removed from the light path to prevent interference from oxides and that it is more difficult to prevent thermal and contaminant degradation of the spectrometer optics in the axial configuration than in the radial arrangement.[9] The decision as to which viewing arrangement to use depends on the chemical behavior of the analyte in the plasma, the spectral line chosen for the analysis, the quality of the data required, and the detailed nature of the experiment. For example, the axial arrangement is especially useful in ICPMS, which is described in Section 11C.

Analyte Atomization and Ionization

Figure 10-4 shows temperatures in the plasma using isothermal contours. Sample atoms reside in the plasma for about 2 ms before they reach the observation point. During the residence time they experience temperatures ranging from 5500 to 8000 K. The time and temperatures are roughly two to three times greater than those found in the hottest combustion flames (acetylene–nitrous oxide) used in flame spectroscopic methods. As a result, atomization is more complete in

plasmas than in flames, and fewer chemical interferences occur. Surprisingly, ionization interference effects are small, probably because the large electron concentration from ionization of the argon maintains a fairly constant electron concentration in the plasma.

Several other advantages are associated with the plasma source. First, atomization occurs in a chemically inert environment, which tends to enhance the lifetime of the analyte by preventing oxide formation. In addition, and in contrast to arcs, sparks, and flames, the temperature cross section of the plasma is relatively uniform; as a consequence, self-absorption and self-reversal effects do not occur as often. Thus, calibration curves are usually linear over several orders of magnitude of concentration. Finally, the plasma produces significant ionization, which makes it an excellent source for ICPMS.

10A-2 The Direct Current Plasma Source

DCP sources were first described in the 1920s and were systematically investigated as sources for emission spectroscopy for several decades.[10] It was not until the 1970s, however, that the first commercial DCP emission source was introduced. The source became quite popular, particularly among soil scientists and geochemists for multielement analysis.

Figure 10-5 is a schematic of a commercially available DCP source that is well suited for excitation of emission spectra for a wide variety of elements. This plasma jet source consists of three electrodes configured in an inverted Y. A graphite anode is located in each arm of the Y and a tungsten cathode at the inverted base. Argon flows from the two anode blocks toward the cathode. The plasma jet is formed by bringing the cathode into momentary contact with the anodes. Ionization of the argon occurs and a current develops (~14 A) that generates additional ions to sustain the current indefinitely. The temperature at the arc core is more than 8000 K and at the viewing region about 5000 K. The sample is aspirated into the area between the two arms of the Y, where it is atomized, excited, and detected.

Spectra produced by the DCP tend to have fewer lines than those produced by the ICP, and the lines

[9]L. H. J. Lajunen and P. Peramaki, *Spectrochemical Analysis by Atomic Absorption and Emission*, 2nd ed., pp. 253–54, Cambridge: Royal Society of Chemistry, 2004.

[10]For additional details, see G. W. Johnson, H. E. Taylor, and R. K. Skogerboe, *Anal. Chem.*, **1979**, *51*, 2403; *Spectrochim. Acta*, Part B, **1979**, *34*, 197; R. J. Decker, *Spectrochim. Acta*, Part B, **1980**, *35*, 21.

 Exercise: Learn more about the **DCP**.

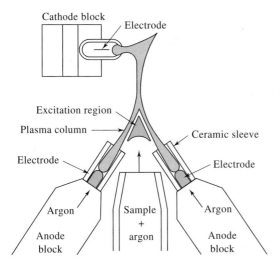

FIGURE 10-5 Diagram of a three-electrode DCP source. Two separate DCPs have a single common cathode. The overall plasma burns in the form of an upside-down Y. Sample can be introduced as an aerosol from the area between the two graphite anodes. Observation of emission in the region beneath the strongly emitting plasma core avoids much of the plasma background emission. (Courtesy of Spectrametrics, Inc. Haverhill, MA.)

TABLE 10-1 Desirable Properties of an Emission Spectrometer

1. High resolution (0.01 nm or $\lambda/\Delta\lambda > 100{,}000$)
2. Rapid signal acquisition and recovery
3. Low stray light
4. Wide dynamic range ($>10^6$)
5. Accurate and precise wavelength identification and selection
6. Precise intensity readings ($<1\%$ relative standard deviation at 500 × the detection limit)
7. High stability with respect to environmental changes
8. Easy background corrections
9. Computerized operation: readout, storage data manipulation, etc.

formed in the DCP are largely from atoms rather than ions. Sensitivities achieved with the DCP range from an order of magnitude lower than those obtainable with the ICP to about the same as the ICP's. The reproducibilities of the two systems are similar. Significantly less argon is required for the DCP, and the auxiliary power supply is simpler and less expensive. The DCP is able to handle organic solutions and aqueous solutions with a high solids content better than the ICP. Sample volatilization is often incomplete with the DCP, however, because of the short residence times in the high-temperature region. Also, the optimum viewing region with the DCP is quite small, so optics have to be carefully aligned to magnify the source image. In addition, the graphite electrodes must be replaced every few hours, whereas the ICP requires little maintenance.

10A-3 Plasma Source Spectrometers

Table 10-1 lists the most important properties of the ideal instrument for plasma emission spectroscopy. The ideal spectrometer is not available today, partly because some of these properties are mutually exclusive. For example, high resolution requires the use of narrow slits, which usually reduces the signal-to-noise ratio and thus the precision of intensity readings. Nevertheless, instruments developed recently approach many of the ideals listed in the table.[11]

A dozen or more instrument manufacturers currently offer plasma emission spectrometers. The designs, performance characteristics, and wavelength ranges of these instruments vary substantially. Most encompass the entire ultraviolet-visible spectrum, from 170 to 800 nm. A few instruments are equipped for vacuum operation, which extends the ultraviolet to 150 to 160 nm. This short-wavelength region is important because elements such as phosphorus, sulfur, and carbon have emission lines in this range.

Instruments for emission spectroscopy are of three basic types: *sequential*, *simultaneous multichannel*, and *Fourier transform*. Fourier transform instruments have not been widely used in emission spectroscopy. Sequential instruments are usually programmed to move from the line for one element to that of a second, pausing long enough (a few seconds) at each to measure line intensities with a satisfactory signal-to-noise ratio. In contrast, multichannel instruments are designed to measure simultaneously, or nearly so, the intensities of emission lines for a large number of elements (sometimes as many as fifty or sixty). When several elements are determined, sequential instruments require significantly greater time for samples to be introduced than is required with the other two types. Thus, these instruments, although simpler, are costly in terms of sample consumption and time.

---

[11]For a review of commercial ICP atomic emission spectrometers, see B. E. Erickson, *Anal. Chem.*, **1998**, *70*, 211A.

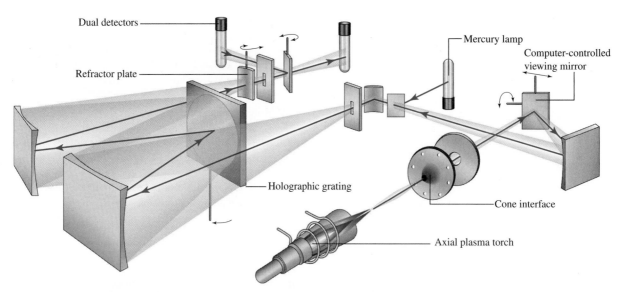

FIGURE 10-6 Optical diagram of a sequential ICP optical emission spectrometer. All moving parts are under computer control, and their modes of motion are indicated by the three-dimensional arrows. Moving parts include the grating, a mirror for transducer selection, a refractor plate for optimizing signal throughput, and a viewing mirror to optimize the plasma viewing position. The spectrometer contains a mercury lamp for automatic wavelength calibration. Notice the axial viewing geometry. (Courtesy of Varian Analytical Instruments.)

Both sequential and multichannel emission spectrometers are of two general types, one using a classical grating spectrometer and the other an echelle spectrometer, such as that shown in Figure 7-23.

Sequential Instruments

Sequential instruments often incorporate a grating monochromator such as that shown in Figure 10-6. Usually, the grating is a holographic type having 2400 or 3600 grooves per millimeter. With some instruments of this type, scanning involves rotating the grating with a digitally controlled stepper motor so that different wavelengths are sequentially and precisely focused on the exit slit. In some designs, however, the grating is fixed and the slit and photomultiplier tube are moved along the focal plane or curve. Instruments such as the one shown in Figure 10-6 have two sets of slits and photomultiplier tubes, one for the ultraviolet region and one for the visible. In such instruments, at an appropriate wavelength, the exit beam is switched from one photomultiplier to the other by movement of the plane mirror located between the two transducers.

Slew-Scan Spectrometers. With complex spectra made up of hundreds of lines, scanning a significant wavelength region takes too long and is thus imprac-

tical. To partially overcome this problem, *slew-scan* spectrometers were developed in which the grating (see Figure 10-6), or the transducer and slit, is driven by a two-speed (or multispeed) motor. In such instruments, the monochromator scans very rapidly, or *slews*, to a wavelength near a line of interest. The scan rate is then quickly reduced so that the instrument scans across the line in a series of small (0.01 to 0.001 nm) steps. With slew scanning, the time spent in wavelength regions containing no useful data is minimized, but sufficient time is spent at analyte lines to obtain satisfactory signal-to-noise ratios. In spectrometers, such as the one illustrated in Figure 10-6, in which the grating movement is under computer control, slewing can be accomplished very efficiently. For example, the spectrometer shown can slew to lines corresponding to fifteen elements and record their intensities in less than 5 minutes. Generally, however, these instruments are slower and consume more sample than multichannel instruments.

Scanning Echelle Spectrometers. Figure 10-7 is a schematic of an echelle spectrometer that can be operated either as a scanning instrument or as a simultaneous multichannel spectrometer. Scanning is accomplished by moving a photomultiplier tube in both *x* and

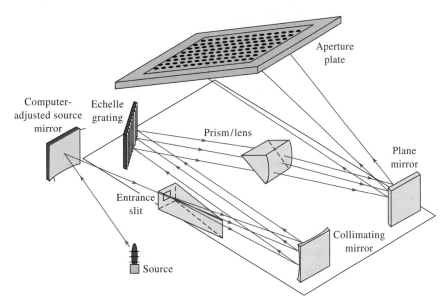

FIGURE 10-7 Schematic of an echelle spectrograph system.

y directions to scan an aperture plate located on the focal plane of the monochromator. The plate contains as many as 300 photo-etched slits. The time required to move from one slit to another is typically 1 s. The instrument can be operated in a slew-scan mode. This instrument can also be converted to a multichannel polychromator by mounting several small photomultiplier tubes behind appropriate slits in the aperture plate.

Multichannel Spectrometers

Simultaneous multichannel instruments are of two general types: *polychromators* and *spectrographs*. Polychromators contain a series of photomultiplier tubes for detection, but spectrographs use two-dimensional charge-injection devices (CIDs) or charge-coupled devices (CCDs) as transducers. Older instruments used photographic emulsions as transducers.

Polychromators. In some multichannel emission spectrometers, photomultipliers are located behind fixed slits along the focal curve of a grating polychromator such as the Paschen-Runge design shown in Figure 10-8. In these instruments, the entrance slit, the exit slits, and the grating surface are located along the circumference of a *Rowland circle*, the curvature of which corresponds to the focal curve of the concave grating. Radiation from each of the fixed slits impinges on the photomultiplier tubes. The slits are factory configured to transmit lines for selected elements. In such instru-

ments, the pattern of lines can be changed relatively inexpensively to accommodate new elements or to delete others. The signals from the photomultiplier tubes are integrated, the output voltages are then digitized and converted to concentrations, and the results are stored and displayed. The entrance slit can be moved tangentially to the Rowland circle by means of a stepper motor. This device permits scanning through peaks and provides information for background corrections.

Polychromator-based spectrometers with photomultipliers as transducers have been used both with plasma and with arc and spark sources. For rapid routine analyses, such instruments can be quite useful. For example, in the production of alloys, quantitative determinations of twenty or more elements can be completed within 5 minutes of receipt of a sample; close control over the composition of a final product is then possible.

In addition to speed, photoelectric multichannel spectrometers often exhibit good analytical precision. Under ideal conditions reproducibilities of the order of 1% relative to the amount present have been demonstrated. Because other instrument components (such as the source) exhibit lower precision than the spectrometer, such high precision is not often achieved in the overall measurement process, however. Multichannel instruments of this type are generally more expensive than the sequential instruments described in the previous section and not as versatile.

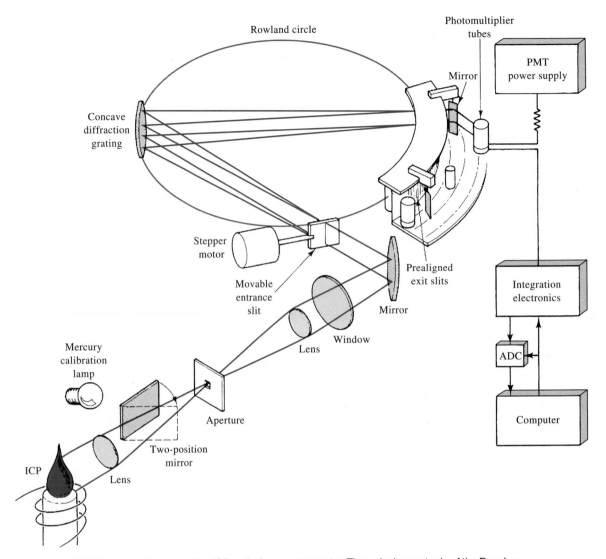

FIGURE 10-8 Direct-reading ICP emission spectrometer. The polychromator is of the Paschen-Runge design. It features a concave grating and produces a spectrum around a Rowland circle. Separate exit slits isolate each spectral line, and a separate photomultiplier tube converts the optical information from each channel into an electrical signal. Notice the radial viewing geometry. PMT = photomultiplier tube. (From J. D. Ingle Jr. and S. R. Crouch, *Spectrochemical Analysis*, p. 241, Upper Saddle River, NJ: Prentice-Hall, 1988, with permission.)

A Charge-Injection Device Instrument. A number of companies offer multichannel simultaneous spectrometers based on echelle spectrometers and two-dimensional array devices. This type of instrument has replaced other types of multichannel emission spectrometers in many applications.

Figure 10-9 is an optical diagram of an echelle spectrometer that has a charge injection device for simultaneous operation.[12] It uses a calcium fluoride prism to sort the spectral orders that are subsequently formed by the echelle grating (see also Figure 7-21). The transducer is a CID (Section 7E-3) 8.7 mm by 6.6 mm and containing 94,672 transducer elements.

---

[12] See M. J. Pilon, M. B. Denton, R. G. Schleicher, P. M. Moran, and S. B. Smith, *Appl. Spectrosc.*, **1990**, *44*, 1613.

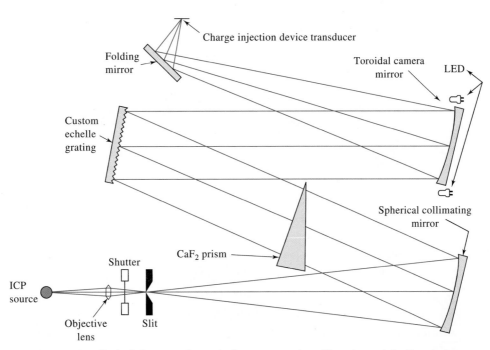

FIGURE 10-9 Optical diagram of an echelle spectrometer with a charge-injection device. (From R. B. Bilhorn and M. B. Denton, *Appl. Spectrosc.*, **1990**, *44*, 1615. With permission.)

A toroidal camera mirror focuses the slit images onto the transducer surface. To eliminate dark currents in the transducer elements, the unit is housed in a liquid nitrogen cryostat that maintains a temperature of 135 K.

A set of 39 transducer elements, called a *read window*, is used to monitor each spectral line as shown in Figure 10-10a. Normally, as shown by the projected image of one of the windows labeled "examination window," the spectral line is focused on the 9 center

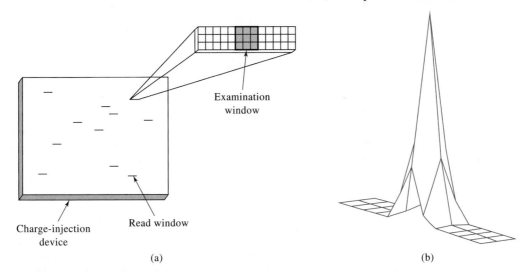

(a)   (b)

FIGURE 10-10 (a) Schematic representing the surface of a CID. The short horizontal lines represent the read windows. A magnified image of one of the read windows is also shown. The nine central elements form the examination window, where a line is positioned. (b) Intensity profile for an iron line. All of the radiation from the line falls on the 3 × 3 examination window. (From R. B. Bilhorn and M. B. Denton, *Appl. Spectrosc.*, **1990**, *44*, 1540. With permission.)

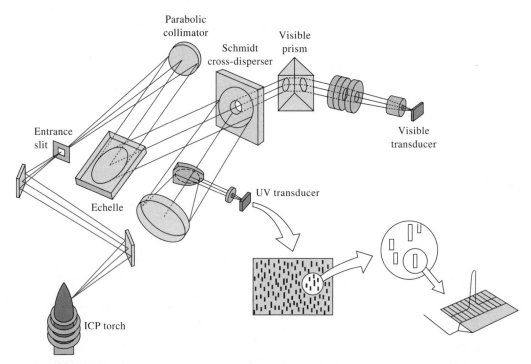

FIGURE 10-11 An echelle spectrometer with segmented array of CCDs. (From T. W. Barnard et al., *Anal. Chem.*, **1993**, *65*, 1231. Figure 1, p. 1232. Copyright 1993 American Chemical Society.)

elements of the window, and the 15 elements on either side of the central set provide background intensity data. Figure 10-10b shows the intensities recorded for the read window for the iron 297.32 nm line. Note that most of the iron radiation falls on the central elements of the window.

One of the useful features of the CID, in contrast to the CCD discussed next, is that the amount of charge accumulated in an element at any instant can be monitored nondestructively; that is, no charge is lost in the measurement process. To make the measurement of line intensity as rapid and efficient as possible, only the charge accumulated in the 9 central elements of the window are read initially to determine when sufficient charge has accumulated to provide a satisfactory signal-to-noise ratio. Only then are the remaining elements in the two 15-element sets read to correct the observed line intensity for the background radiation. This process goes on simultaneously at the read windows for each element. With an intense line, the time required to accumulate the desired charge is brief. With weak lines, the charge accumulated in a brief period is often used to estimate the integration time required to yield a satisfactory signal-to-noise ratio. In some cases integration times of 100 s or more are required.

Periodic wavelength calibration of the spectrometer just described is maintained through reference to the 253.65 nm mercury line from a small mercury lamp. Data files for line positions for more than 40 elements have been developed. The file for each element contains the wavelengths of up to ten lines and the x and y coordinates of each of these spectral lines with respect to the coordinates for the mercury line. Database recalibration is seldom necessary unless something perturbs the optics of the system significantly. Identification of elements in samples is done by visual inspection with the use of a video monitor and interactive markers. With an ICP excitation source, detection limits that range from a few tenths of a nanogram to 10 µg/mL have been reported for most elements. For nonmetals such as phosphorus and arsenic, detection limits are larger by a factor of 100.

A Charge-Coupled Device Instrument. Figure 10-11 is an optical diagram of a commercial spectrometer with two echelle systems and two CCDs, one system for the 160–375-nm region and the other for the 375–782-nm range.[13] Radiation from the plasma enters the

---

[13]For a detailed description of this instrument, see T. W. Barnard et al., *Anal. Chem.*, **1993**, *65*, 1225, 1231.

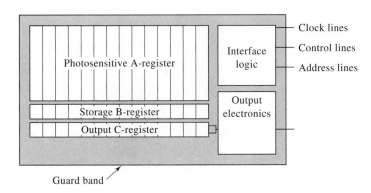

Clock lines
Control lines
Address lines

FIGURE 10-12 Schematic of an array segment showing phototransducers, storage and output registers, and readout circuitry. (From T. W. Barnard et al., *Anal. Chem.*, **1993**, *65*, 1231. Figure 3, p. 1232. Copyright 1993 American Chemical Society.)

spectrometer via a slit and is then dispersed by an echelle grating. The radiation falls on a Schmidt cross-disperser element that separates the orders of the ultraviolet radiation and also separates the ultraviolet and visible optical beams. The Schmidt element consists of a grating ruled on a spherical surface with a hole in the center to pass the visible radiation to a prism where order separation takes place as shown in Figure 7-23. The two dispersed beams are then focused onto the surface of transducer elements as shown. Note that a schematic representation of the surface of the ultraviolet transducer is shown as an insert in Figure 10-11.

These unique detector systems consist of numerous subarrays, or *array segments*, fabricated on silicon chips, with each subarray being custom positioned so that three to four major emission lines for each of 72 elements fall on its surface. Each array segment consists of a *linear* (rather than two-dimensional) CCD that is made up of twenty to eighty pixels. Figure 10-12 is a schematic of one of these array segments, which is made up of the individual photosensitive registers, storage and output registers, and output electronics. Because each array segment can be addressed separately, charge-integration times can be varied over a wide enough range to provide dynamic ranges of 10^5 or greater. Although there are only 224 (235 in the current production version) of these array segments in the system, multiple lines fall on many of the subarrays so that about 6000 lines can be monitored simultaneously.

In the production version of this spectrometer, one of the mirrors that guides the radiation into the optical system is under computer control so that plasma viewing may be axial, radial, or mixed. This arrangement also permits optimization of the spectrometer signal. The entire optical system is contained within a purged temperature-controlled enclosure and is protected from the intense UV radiation of the plasma between samples by a pneumatically operated shutter to extend the life of the input mirror. In addition, a mercury lamp is built into the shutter mechanism to calibrate the spectrometer periodically. Different models of the spectrometer are available that cover the spectral range of 163–782 nm or segments thereof, depending on whether one or both arrays are installed.

A Combination Instrument. An interesting and useful application of both the Paschen-Runge polychromator and modular array detectors is the spectrometer shown in the photo of Figure 10-13a. Fifteen or sixteen linear CCD array modules (eight are visible) are arranged along the circumference of the Rowland circle to provide nearly complete coverage of the range from 140 to 670 nm. Each detector module (see Figure 10-13b) contains a mirror to reflect the radiation to the CCD array, which is arranged parallel to the plane of the Rowland circle. The modules are easily exchanged and positioned in the optical path. Because of its relatively compact design (115 cm wide × 70 cm deep), this spectrometer is particularly well suited to benchtop operation and routine use in industrial and environmental laboratories.

Fourier Transform Spectrometers

Since the early 1980s, several workers have described applications of Fourier transform instruments to the ultraviolet-visible region of the spectrum by using instruments that are similar in design to the infrared instruments discussed in detail in Section 16B-1.[14]

---

[14] A. P. Thorne, *Anal. Chem.*, **1991**, *63*, 57A; L. M. Faires, *Anal. Chem.*, **1986**, *58*, 1023A.

 Tutorial: Learn more about **ICP spectrometers**.

(a)

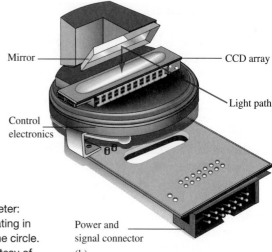

(b)

FIGURE 10-13 Components of a simultaneous CCD-ICP spectrometer: (a) Photo of the optical system. Note the Rowland circle with the grating in the rear, the slit in the front, and the detector array modules along the circle. (b) Array detector module containing a 1024-pixel linear CCD. (Courtesy of Spectro A. I. Inc., Marlborough, MA.)

Much of this work has been devoted to the use of such instruments for multielement analyses with ICP sources. The advantages of Fourier transform instruments include their wide wavelength coverage (170 nm to >1000 nm), speed, high resolution, highly accurate wavelength measurements, large dynamic range, compact size, and large optical throughput. In contrast to Fourier transform infrared instruments, however, ultraviolet-visible instruments of this type often exhibit no multiplex advantage and indeed under some circumstances show a multiplex disadvantage (see Section 7I-1). The reason for this difference is that the performance of infrared instruments is usually limited by transducer noise, whereas ultraviolet-visible spectrometers are limited by shot and flicker noise associated with the source.

An ultraviolet-visible Fourier transform instrument first appeared on the market in the mid-1980s.[15] The mechanical tolerances required to achieve good resolution with this type of instrument are very demanding, however. Consequently, these spectrometers are quite expensive and are generally used for research projects rather than for routine analytical applications.

10A-4 Applications of Plasma Sources

Plasma sources produce spectra rich in characteristic emission lines, which makes them useful for both qual-

itative and quantitative elemental analysis.[16] The ICP and the DCP yield significantly better quantitative analytical data than other emission sources. The quality of these results stems from their high stability, low noise, low background, and freedom from interferences when operated under appropriate experimental conditions. The performance of the ICP source is somewhat better than that of the DCP source in terms of detection limits. The DCP, however, is less expensive to purchase and operate and is entirely satisfactory for many applications.

Sample Preparation

ICP emission spectroscopy is used primarily for the qualitative and quantitative analysis of samples that are dissolved or suspended in aqueous or organic liquids. The techniques for preparation of such solutions are similar to those described in Section 9D-1 for flame absorption methods. With plasma emission, however, it is possible to analyze solid samples directly. These procedures include incorporating electrothermal vaporization, laser and spark ablation, and glow-discharge vaporization, all of which were described in Section 8C-2. Suspensions of solids in solutions can

---

[15] See *Anal. Chem.*, **1985**, *57*, 276A; D. Snook and A. Grillo, *Amer. Lab.*, **1986** (11), 28; F. L. Boudais and J. Buija, *Amer. Lab.*, **1985** (2), 31.

[16] For useful discussions of the applications of plasma emission sources, see J. A. Nolte, *ICP Emission Spectrometry: A Practical Guide*, Hoboken, NJ: Wiley-VCH, 2003; *Inductively Coupled Plasma in Analytical Atomic Spectroscopy*, 2nd ed., A. Montaser and D. W. Golightly, eds., New York: Wiley-VCH, 1992; M. Thompson and J. N. Walsh, *Handbook of Inductively Coupled Plasma Spectrometry*, 2nd ed. Glasgow: Blackie, 1989.

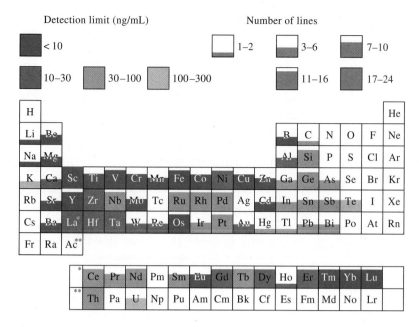

Detection limit (ng/mL)

■	< 10
■	10–30
▨	30–100
▨	100–300

Number of lines

□	1–2
▣	3–6
▨	7–10
▣	11–16
■	17–24

FIGURE 10-14 Periodic table characterizing the detection power and number of useful emission lines of ICP by using a pneumatic nebulizer. The color and degree of shading indicate the range of detection limits for the useful lines. The area of shading indicates the number of useful lines. (Adapted from *Inductively Coupled Plasma Emission Spectroscopy*, Part 1, p. 143, P. W. J. M. Boumans, ed., New York: Wiley, 1987. With permission.)

also be handled with a Babington nebulizer, such as that shown in Figure 8-11d.

Elements Determined

In principle, all metallic elements can be determined by plasma emission spectrometry. A vacuum spectrometer is necessary for the determination of boron, phosphorus, nitrogen, sulfur, and carbon because the emission lines for these elements lie at wavelengths less than 180 nm, where components of the atmosphere absorb radiation. The usefulness for the alkali metals is limited by two difficulties: (1) the compromise operating conditions that can be used to accommodate most other elements are unsuited for the alkalis, and (2) the most prominent lines of Li, K, Rb, and Cs are located at near-infrared wavelengths, which lead to detection problems with many plasma spectrometers that are designed primarily for ultraviolet radiation. Because of problems of this sort, plasma emission spectroscopy is generally limited to the determination of about 60 elements.

The periodic table of Figure 10-14 shows the applicability of ICP emission spectrometry to various elements. The detection limits for the best lines for each of the elements are indicated by the color and degree of shading. The areas of shading indicate the number of lines for each element that yields a detection limit within a factor of 3 of the best line. The more such lines

that are available, the greater the chance that a usable line can be found that is free from interference when the matrix yields a line-rich spectrum.

Line Selection

Figure 10-14 shows that most elements have several prominent lines that can be used for identification and determination purposes. Wavelength data recorded to three decimal places with appropriate intensity information for prominent lines for more than 70 elements can be found in several publications.[17] Thus, a suitable line for the determination of any element can usually be found. Selection depends on a consideration of what elements other than the analyte may be present in the sample and whether there is any likelihood that lines of these elements will overlap analyte lines.

Calibration Curves

Calibration curves for plasma emission spectrometry most often consist of a plot of an electrical signal proportional to line intensity versus analyte concentration.

---

[17] See R. K. Winge, V. A. Fassel, V. J. Peterson, and M. A. Floyd, *Inductively Coupled Plasma Emission Spectroscopy: An Atlas of Spectral Information*, New York: Elsevier, 1985; P. W. J. M. Boumans, *Line Coincidence Tables for Inductively Coupled Plasma Spectrometry*, 2nd ed., Oxford: Pergamon, 1984; C. C. Wohlers, *ICP Information Newslett.*, **1985**, *10*, 601.

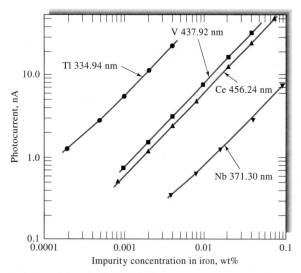

FIGURE 10-15 Typical calibration curves in ICP emission spectrometry. (From V. A. Fassel and R. N. Kniseley, *Anal. Chem.*, **1974**, *46*, 1110A. Figure 1, p. 1117A. Copyright 1974 American Chemical Society.)

When the range of concentrations is large, log-log plots may be used instead. Figure 10-15 shows typical calibration curves for four trace elements present in steel samples. Often, calibration plots are linear, as are the central two curves in the figure. Departures from linearity often occur, however, when large concentration ranges are covered (see the outer two lines in Figure 10-15). A major cause of nonlinearity is *self-absorption*, in which the output signal is reduced because of absorption by ground-state atoms in the medium. Self-absorption becomes evident only at high analyte concentrations and causes the calibration curve to bend toward the horizontal axis. None of the plots in Figure 10-15 show evidence of self-absorption. Nonlinearity also arises from erroneous background corrections, from ionization, and from nonlinear responses of the detection systems. The nonlinearity in the curves for niobium and thallium at low concentrations are probably a result of incorrect background corrections. Note that the departures from linearity are away from the concentration axis.

An internal standard is often used in emission spectrometry. In this case, the vertical axis of the calibration curve is the ratio or the log ratio of the detector signal for the analyte to the detector signal for the internal standard. Figure 10-16 shows calibration curves for several elements. In these experiments a fixed amount of yttrium was incorporated in all of the stan-

dards, and the relative intensity of the analyte line to that of an yttrium line at 242.2 nm served as the analytical variable. Note that all the curves are linear and cover a concentration range of nearly three orders of magnitude. Note also that some of the data were obtained by introducing various amounts of the analyte and internal standard into pure water. Other data are for solutions that contained relatively high concentrations of different salts, thus demonstrating a freedom from interelement interference.

As in atomic absorption spectroscopy, one or more standards should be introduced periodically to correct for the effects of instrument drift. The improvement in precision that results from this procedure is illustrated by the data in Table 10-2. Note also the improved precision when higher concentrations of analyte are measured.[18]

---

[18] For a useful discussion of precision in ICP spectrometry, see R. L. Watters Jr., *Amer. Lab.*, **1983**, *15* (3), 16.

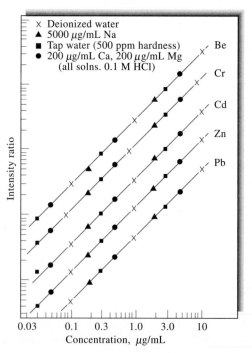

FIGURE 10-16 Internal standard calibration curves with an ICP source. Here, an yttrium line at 242.2 nm served as an internal standard. Notice the lack of interelement interference. (From V. A. Fassel, *Science*, **1978**, *202*, 187. With permission. Copyright 1978 by the American Association for the Advancement of Science.)

TABLE 10-2 Effect of Standardization Frequency on Precision of ICP Data

Frequency of Recalibration, hours	Relative Standard Deviation, %			
	Concentration Multiple above Detection Limit			
	10^1 to 10^2	10^2 to 10^3	10^3 to 10^4	10^4 to 10^5
0.5	3–7	1–3	1–2	1.5–2
2	5–10	2–6	1.5–2.5	2–3
8	8–15	3–10	3–7	4–8

Data from R. M. Barnes, in *Applications of Inductively Coupled Plasmas to Emission Spectroscopy*, R. M. Barnes, ed., p. 16, Philadelphia: The Franklin Institute Press, 1978. With permission.

Interferences

As was noted previously, chemical interferences and matrix effects are significantly lower with plasma sources than with other atomizers. At low analyte concentrations, however, the background emission due to recombination of argon ions with electrons is large enough to require careful corrections. For both single-channel and multichannel instruments, this correction is made by taking background readings on either side of the line of interest. State-of-the-art instruments include software designed to make background corrections automatically or under operator control.

Because ICP spectra for many elements are so rich in lines, spectral interferences are always possible. To avoid this type of error requires knowledge of all of the components likely to be present in the sample and a careful study of the information in the reference works listed in note 16. The software for modern computerized instruments has powerful routines for wavelength and concentration calibration, spectral analysis, and deconvolution of overlapping lines. These features coupled with integrated databases of spectral lines make spotting and correcting for interferences an integral part of the analytical process.

Detection Limits

In general, detection limits with the ICP source are comparable to or better than other atomic spectral procedures. Table 10-3 compares detection limits for several of these methods. Note that more elements can be detected at levels of 10 ppb or less with plasma excitation than with other emission or absorption methods. As we shall see in Chapter 11, the ICP coupled with mass spectrometric detection improves detection limits by two to five orders of magnitude for many elements and is thus strong competition for ICP optical emission spectroscopy.

10B EMISSION SPECTROSCOPY BASED ON ARC AND SPARK SOURCES

Arc and spark source spectroscopies were the first instrumental methods to become widely used for analysis. These techniques, which began to replace the classical gravimetric and volumetric methods for elemental analysis in the 1920s, were based on excitation of emission spectra of elements with electric arcs or high-voltage sparks. These spectra permitted the qualitative

TABLE 10-3 Comparison of Detection Limits for Several Atomic Spectral Methods

Method	Number of Elements Detected at Concentrations of				
	<1 ppb	1–10 ppb	11–100 ppb	101–500 ppb	>500 ppb
ICP emission	9	32	14	6	0
Flame atomic emission	4	12	19	6	19
Flame atomic fluorescence	4	14	16	4	6
Flame atomic absorption	1	14	25	3	14

Data abstracted with permission from V. A. Fassel and R. N. Kniseley, *Anal. Chem.*, **1974**, *46* (13), 1111A. Copyright 1974 American Chemical Society.

Detection limits correspond to a signal that is twice as great as the standard deviation for the background noise.

and quantitative determination of metallic elements in a variety of sample types, including metals and alloys, soils, minerals, and rocks.[19] Arc and spark sources are still used in some situations for qualitative and semi-quantitative analysis, particularly in the metals industries. Arcs and sparks are gradually being displaced by plasma sources, and it appears likely that this trend will continue.

In arc and spark sources, sample excitation occurs in the gap between a pair of electrodes. Passage of electricity from the electrodes through the gap provides the necessary energy to atomize the sample and produce atoms or ions in electronic excited states.

10B-1 Sample Types and Sample Handling

Currently, arc and spark source methods are mostly limited to the elemental analysis of solids because liquid and gaseous samples are handled so much more conveniently by the plasma emission methods that we have discussed in the previous sections.

Metals

If the sample is a metal or an alloy, one or both electrodes can be formed from the sample by milling, by turning, or by casting the molten metal in a mold. Ideally, the electrode will be shaped as a cylindrical rod that is 1/8 to 1/4 in. in diameter and tapered at one end. For some samples, it is more convenient to use the polished, flat surface of a large piece of the metal as one electrode and a tapered cylindrical graphite or metal rod as the other. Regardless of the ultimate shape of the sample, its surface must not become contaminated while it is being formed.

Nonmetallic Solids

For nonmetallic materials, the sample is often supported on an electrode whose emission spectrum will not interfere with the analysis. Carbon is an ideal electrode material for many applications. It can be obtained in a highly pure form, is a good conductor, has good heat resistance, and is easily shaped. Manufacturers offer carbon electrodes in many sizes, shapes, and forms. Frequently, one of the electrodes is a cylinder with a small crater drilled into one end; the finely ground

---

[19]For additional details, see J. D. Ingle Jr. and S. R. Crouch, *Spectrochemical Analysis*, Chap. 9, Upper Saddle River, NJ: Prentice-Hall, 1988; R. D. Sacks, in *Treatise on Analytical Chemistry*, 2nd ed., P. J. Elving, E. J. Meehan, and I. M. Kolthoff, eds., Part I, Vol. 7, Chap. 6, New York: Wiley, 1981; P. W. J. M. Boumans in *Analytical Emission Spectroscopy*, Vol. 1, Part I, E. L. Grove, ed., New York: Marcel Dekker, 1972.

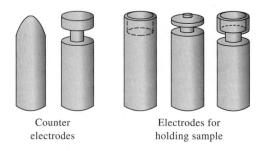

Counter electrodes	Electrodes for holding sample

FIGURE 10-17 Some typical graphite electrode shapes. Narrow necks are to reduce thermal conductivity.

sample is then packed into this cavity. The other electrode is commonly a tapered carbon rod with a slightly rounded tip. This configuration appears to produce the most stable and reproducible arc or spark. Figure 10-17 illustrates some of the common electrode forms.

In another common atomization method, the finely ground sample is mixed with a relatively large amount of powdered graphite, copper, or other conducting and compressible substance. The resulting mixture is then compressed at high pressure into the form of an electrode. This procedure is called *briquetting* or *pelleting*.

10B-2 Instruments for Arc and Spark Source Spectroscopy

Because of their instabilities, it is necessary to integrate the emission signals from arc and spark sources for at least 20 s and often for a minute or more to obtain reproducible analytical data. This requirement makes the use of sequential spectrometers, such as those described in Section 10A-3, impractical for most applications and demands the use of a simultaneous multichannel instrument. Two types of multichannel instruments have been applied to arc and spark spectroscopy: (1) spectrographs, which are considered briefly in the section that follows, and (2) multichannel spectrometers, such as those described in Section 10A-3.

Spectrographs

Beginning in the 1930s, a major portion of industrial laboratories throughout the world adopted spark and arc emission spectroscopy for elemental analysis of raw materials, intermediates, and finished products. In the early years, the only instrument available for this type of analysis was the *spectrograph* with a photographic film or plate located at the focal plane or curve.

A major disadvantage of photographic recording is the time required to obtain and analyze spectra. These operations involve exposing the photographic plate or film while samples and standards are being excited; removing the emulsion to a dark room, where developing, fixing, washing, and drying is carried out; and then examining the exposed plate or film in a comparator that permits line identification. This process often consumes hours of operator time. If a quantitative analysis is required, still more time must be expended to calibrate the film or plates. With the advent of high-speed electronic multichannel detection systems, film- and plate-based spectrographs are now seldom used in chemical analysis except for unique situations. For this reason, we shall not consider them further here.[20]

Multichannel Photoelectric Spectrometers

After the appearance of photoelectric transducers in the 1930s, multichannel photoelectric spectrometers became commercially available.

Multichannel Photomultiplier Instruments. In the metals industry, there has always been a pressing need for a method for determining the amounts of several metals in samples rapidly enough so that the composition of melts could be adjusted before the melt is poured. This need and others where speed is required led directly to the development and widespread use of photoelectric polychromators, similar to the one shown in Figure 10-8. These instruments are large and often not very versatile because changing the position of slits and photomultipliers to accommodate a different set of elements is time-consuming and with some instruments is not possible in the user's laboratory. On the other hand, once an instrument of this type is set up and calibrated, it is capable of determining 20 or more elements with reasonable accuracy in a few minutes. Thus, the use of these instruments is generally limited to situations where there is a demand for a large number of routine analyses of similar samples. In the metals industries, spark sources are used for excitation of samples and can yield precise and accurate results.

Array-Based Multichannel Instruments. Spectrometers that offer the versatility of spectrographs with photographic recording and the speed and precision of spectrometers equipped with multiple photomultipliers are widely available. These instruments are the multichannel instruments equipped with array devices, such as those illustrated in Figures 10-11 and 10-13. Many of these instruments were originally designed for use with plasma sources but now have been applied to spark and arc sources as well.

10B-3 Arc Source Emission Spectroscopy

The usual arc source for a spectrochemical analysis is formed with a pair of graphite or metal electrodes spaced a few millimeters apart. The arc is initially ignited by a low-current spark that causes momentary formation of ions for electrical conduction in the gap; once the arc is struck, thermal ionization maintains the current. Alternatively, the arc can be started by bringing the electrodes together to provide the heat for ionization; the electrodes are then separated to the desired distance.

The typical arc requires currents in the range of 1 to 30 A. A dc arc source usually has an open circuit voltage of about 200 V. There are also ac arc sources that operate in either a high-voltage range of 2200 to 4400 V or a low-voltage range of 100 to 400 V. With either ac type, the arc extinguishes at the end of each half cycle. With the high-voltage type, the arc reignites spontaneously; with the low-voltage arc, reignition is brought about by a low-current spark discharge.

Characteristics of Arc Sources

The current in an arc is caused by the motion of the electrons and ions formed by thermal ionization; the high temperature that develops is a result of the resistance to this motion by the ions in the arc gap. Thus, the arc temperature depends on the composition of the plasma, which in turn depends on the rate of formation of atomic and ionic particles from the sample and the electrodes. Typically, the plasma temperature is 4000 to 5000 K.

The spectra produced in the typical arc are rich in intense lines for atoms and contain a lesser number for ionic species. Note that in collections of spectral line data, wavelengths are often followed by I, II, and III to indicate the source of the line as that of the neutral atom, the singly charged ion, and the doubly charged ion, respectively.

Cyanogen Spectral Bands. Arcs produced from a carbon or graphite electrode in air emit intense bands due to the presence of cyanogen radicals (CN). As a result, most of the region between 350 to 420 nm is useless for

---

[20]For a discussion of quantitative photographic detection, see J. D. Ingle Jr. and S. R. Crouch, *Spectrochemical Analysis*, pp. 267–68, Upper Saddle River, NJ: Prentice-Hall, 1988.

elemental analysis. Unfortunately, several elements have their most sensitive lines in this region. To avoid or minimize cyanogen interference, arc excitation is often performed in a controlled gas atmosphere of carbon dioxide, helium, or argon. The CN band emission is not completely eliminated under these conditions, however, unless the electrodes are heated under vacuum to drive off adsorbed nitrogen.

Rates of Emission. Rates at which various species are volatilized and excited differ widely. The spectra of some species appear early and then disappear as the sample is consumed; the spectra for other species reach their maximum intensities at a later time. Thus, it is necessary to integrate the emission signals for as long as a minute or more. For powder samples, signals for the lines of various elements usually rise rapidly during this period and then decrease to zero as various species vaporize. Because of this type of behavior, powder samples are often "burned" until they are completely consumed; the integrated output signals then serve as analytical variables.

Applications of Arc Sources

Arc sources are particularly useful for qualitative and semiquantitative analysis of nonmetallic samples, such as soils, plant materials, rocks, and minerals. Excitation times and arc currents are usually adjusted so that complete volatilization of the sample occurs; currents of 5 to 30 A for 20 to 200 s are typical. Typically, 2 to 50 mg of sample in the form of a powder, small chips, grindings, or filings, often mixed with a weighed amount of graphite, is packed into the cavity of graphite electrodes. The sample-containing electrode is usually the anode and a second graphite counter electrode is the cathode.

Arc excitation may be used for either qualitative or quantitative analysis. The precision obtainable with an arc is, however, generally poorer than that with a spark and much poorer than that with a plasma or flame. Furthermore, emission intensities of solid samples are highly sample dependent. As a consequence matrix matching of standards and samples is usually required for satisfactory results. The internal-standard method can be used to partially offset this problem.

10B-4 Spark Sources and Spark Spectra

A variety of circuits has been developed to produce high-voltage sparks for emission spectroscopy. It has been found that an intermittent spark that always

propagates in the same direction gives higher precision and lower drift radiant emission. For this reason, the ac line voltage is often rectified before it is stepped up to 10 to 50 kV in a coil. Solid-state circuitry is used to control both the spark frequency and duration. Typically, four spark discharges take place per half cycle of the 60-Hz line current.

The *average* current in a high-voltage spark is usually on the order of a few tenths of an ampere, which is significantly lower than the current of a typical arc. On the other hand, during the initial phase of the discharge the *instantaneous* current may exceed 1000 A. During this initial phase, a narrow streamer that occupies only a fraction of the total space in the spark gap carries the current.[21] The temperature within this streamer is estimated to be as great as 40,000 K. Thus, although the average temperature of a spark source is much lower than that of an arc, the *energy* in the small volume of the streamer may be several times greater. As a result, ionic spectra are more pronounced in a high-voltage spark than in an arc. As a matter of fact, lines emitted by ions are often termed "spark lines" by spectroscopists.

Applications of Spark Source Spectroscopy

Quantitative spark analyses demand precise control of the many variables involved in sample preparation and excitation. In addition, quantitative measurements require a set of carefully prepared standards for calibration; these standards should approximate as closely as possible the composition and physical properties of the samples to be analyzed. Generally, spark source analyses are based on the ratio of intensity of the analyte line to the intensity of an internal-standard line (usually a line of a major constituent of the sample). Under ideal conditions, relative standard deviations of a few percent can be achieved with spark spectral measurements.

Currently, the primary use of spark source emission spectroscopy is for the identification and analysis of metals and other conducting materials. Detection is often carried out with a polychromator equipped with photomultiplier tubes, but a number of vendors offer spectrometers with array detectors as well. In addition, several modern multichannel instruments are now equipped with interchangeable sources that permit excitation by plasmas, arcs, sparks, glow discharge, and lasers. High-voltage sparks have also become

---

[21] For a description of the fundamental nature of analytical spark discharges, see J. P. Walters, *Science*, **1977**, *198*, 787.

important devices for ablating solid samples prior to introduction into plasma excitation sources.

An important use of spark and arc source spectrometers is in foundry and shop floors, scrap yards, and metal-casting facilities. The instruments for these applications are often mobile and may be equipped with a handheld spark or arc gun that the operator can touch to the metal surface to cause excitation. These devices can be used for the rapid identification of alloy types and for the analysis of melts before casting. The smallest spark and arc spectrometers (e.g., 5 lb, 11.5 × 7.5 × 4 in.) are self-contained, battery-powered, CCD-based instruments with programmed calibrations from databases of common alloys and metals. These devices are used to sort metals according to type in scrap yards and recycling centers.

10C MISCELLANEOUS SOURCES FOR OPTICAL EMISSION SPECTROSCOPY

In this section we consider three emission sources other than the plasma and arc and spark sources we have just considered: flame emission sources, glow-discharge sources, and the laser microprobe.

10C-1 Flame Emission Sources

For many years, flames have been used to excite emission spectra for various elements, and most modern atomic absorption spectrometers are readily adapted for flame emission measurements. Flames are not widely used for this purpose, however, because for most single-element determinations, absorption methods provide as good or better accuracy, convenience, and detection limits. For multielement analyses, plasma sources are far superior to flames in most regards. For these reasons flame emission spectrometry is little used except for the determination of the alkali metals and occasionally calcium. These elements are excited at the relatively low temperatures of flames to give spectra that are remarkably simple and free of interference from other metallic species. Alkali-metal spectra generally consist of a relatively few intense lines, many of which are in the visible region and are well suited to quantitative emission measurements.

Because these spectra are simple, basic filter photometers can be quite adequate for routine determinations of the alkali and alkaline-earth metals. A low-temperature flame is used to avoid excitation of most

other metals. As a result, interference filters can isolate the appropriate emission line.

Several instrument manufacturers supply flame photometers designed specifically for the determination of sodium, potassium, lithium, and sometimes calcium in blood serum, urine, and other biological fluids. Single-channel and multichannel (two to four channels) instruments are available for these determinations. In the multichannel instruments, each channel can be used to determine a separate element without an internal standard, or one of the channels can be reserved for an internal standard such as lithium. The ratios of the signals from the other channels to the signal of the lithium channel are then taken to compensate for flame noise and noise from fluctuations in reagent flow rate. Flame photometers such as these have been coupled with flow injection systems to automate the sample-introduction process (see Section 33B-3). Typical precisions for flow-injection-analysis-based flame photometric determinations of lithium, sodium, and potassium in serum are on the order of a few percent or less. Automated flow injection procedures require 1/100 the amount of sample and 1/10 the time of batch procedures.[22]

Although the alkali metals are determined daily in a huge number of samples throughout the world, most are clinical samples that are analyzed potentiometrically (see Chapter 23). Flame photometry is currently used for only a tiny fraction of these samples.

10C-2 Glow-Discharge Sources

The glow discharge, which is described in Section 8C-2 and Section 9A-3, has proven to be a useful source for exciting emission spectra of metals, alloys, and other solid materials.[23] Glow-discharge optical emission spectroscopy (GDOES) has become a mature technique in recent years, and several instrument manufacturers offer glow-discharge sources as well as complete spectrometers configured for this type of analysis. Glow-discharge spectroscopy is a versatile technique in that it is capable of bulk analysis and depth profiling of solids.

Figure 10-18 shows a typical cell for GDOES, and it is similar in many respects to the absorption cells described in Sections 8C-2 and 9A-3. A dc voltage of up to 1 kV applied between the electrodes produces

[22]G. N. Doku and V. P. Y. Gadzekpo, *Talanta*, **1996**, *43*, 735.
[23]*Glow Discharge Plasmas in Analytical Spectroscopy*, R. K. E Marcus and J. A. C. E. Broekaert, eds., Hoboken, NJ: Wiley, 2003.

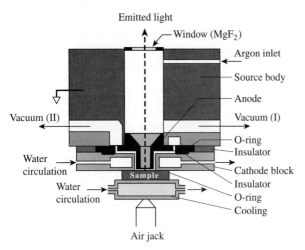

FIGURE 10-18 Diagram of a Grimm-type glow-discharge source. (From M. Boucharcourt and F. Schwoehrer, in *Glow Discharge Optical Emission Spectrometry*, R. Payling, D. G. Jones, and A. Bengtson, eds., p. 54, New York: Wiley, 1997, with permission.)

sputtering of the solid sample at a current of 40–200 mA. In the glow discharge, at the cathode surface, ground-state analyte atoms become excited by collision with high-energy electrons, relax, and emit their characteristic radiation. Radio-frequency (RF) excitation, which permits nonconducting materials to be analyzed by GDOES, is available in some commercial instruments, and pulsed dc and pulsed RF modes have been explored to increase line intensities.[24]

GDOES depth profiling is illustrated in the curves of Figure 10-19 for a sample of brass. The plots show profiles for seven monitored elements as a function of time from the initiation of the glow discharge. During the pre-integration period, surface contaminants volatilize, and over a period of 60 s, the signals reach a relatively constant level corresponding to the composition of the bulk material. The length of the pre-integration period is best determined by the precision of the signal at various times during the sputtering process. For the sample illustrated, the precision was optimal at about 3% relative over the period indicated by I_1. This period was thus chosen for the determination of the composition of similar samples. Depending on the nature of the analysis, the power, the pressure, and the pre-integration and integration periods can be se-

lected to optimize results.[25] The relatively constant signal levels for all of the elements indicate uniform composition throughout the sample. In Figure 10-20, we see a depth profile of an electronic integrated circuit obtained using GDOES with RF excitation. The emergence with time of a peak or a band indicates the appearance of the elements contained in each successive layer of the circuit materials. Labels above each curve indicate the identity and thickness of each layer.[26]

Because of the low background levels of GDOES, detection limits on the order of parts per million are typical using the Grimm source of Figure 10-18. The dynamic range is relatively large compared to arc and spark sources, and relative standard deviations of 1% or lower are common with these devices.[27]

10C-3 Laser-Based Atomic Emission Systems

In recent years, the laser has become very useful in atomic emission spectroscopy. We consider here two laser-based techniques: laser microprobe spectroscopy and laser-induced breakdown spectroscopy (LIBS).

Laser Microprobe Sources

When a powerful laser pulse is focused on a 5- to 50-μm spot on a sample surface, a small amount of the solid is vaporized regardless of whether it is a conductor. The resulting plume is made up of atoms, ions, and molecules. In the microprobe, the contents of the plume are excited by a spark between a pair of small electrodes located immediately above the surface of the sample. The resulting radiation is then focused on a suitable monochromator-detection system. With this type of source, it has been possible to determine the trace element composition of single blood cells and tiny inclusion areas in alloys. The laser can be scanned across a surface to obtain a spatially resolved representation of surface composition.

Laser-Induced Breakdown Spectroscopy

If the pulsed laser beam focused on a sample is powerful enough ($\sim$1 GW cm^{-2} during the pulse), not only can solid samples be ablated but the plume can become superheated and converted into a highly luminous

---

[24]N. Jakubowski, A. Bogaerts, and V. Hoffmann, *Atomic Spectroscopy in Elemental Analysis*, M. Cullen, ed., p. 120, Boca Raton, FL: CRC Press, 2004.

[25]T. A. Nelis and R. A. Payling, *Glow Discharge Optical Emission Spectroscopy*, pp. 23–24, New York: Springer, 2004.
[26]N. Jakubowski, A. Bogaerts, and V. Hoffmann, *Atomic Spectroscopy in Elemental Analysis*, M. Cullen, ed., p. 120, Boca Raton, FL: CRC Press, 2004.
[27]J. A. C. Broekaert, *Analytical Atomic Spectrometry with Flames and Plasmas*, pp. 244–46. Hoboken, NJ: Wiley-VCH, 2002.

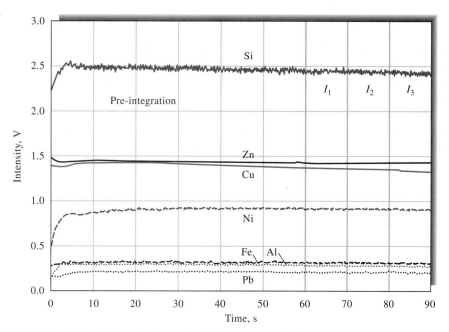

FIGURE 10-19 Qualitative depth profile of a brass sample, showing major and minor elements, indicating the time regions to be used for pre-integration (60 s) and three 10-s integration periods I_1, I_2, and I_3. (From T. A. Nelis and R. A. Payling, *Glow Discharge Optical Emission Spectroscopy*, p. 23, New York: Springer, 2004. With permission.)

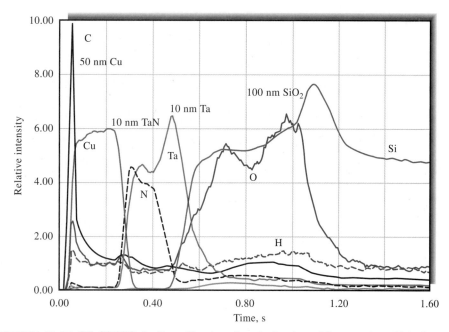

FIGURE 10-20 RF-GDOES depth profile of a microelectronic multilayer system. Note the thickness and composition of each layer shown above each peak or band and the elemental composition indicated by the curves. (From N. Jakubowski, A. Bogaerts, and V. Hoffmann, *Atomic Spectroscopy in Elemental Analysis*, M. Cullen, ed., p. 129, Boca Raton, FL: CRC Press, 2004. With permission.)

plasma. Near the end of the typically 10-ns laser pulse, the plasma cools and the excited ions and atoms emit radiation. By monitoring the plasma spectrometrically at the appropriate time, atom and ion lines of the elements can be observed. Time-gated detection is usually used with LIBS to avoid the intense spectral continuum emitted early during the plasma formation and growth stages and to allow detection of the emission lines later during plasma decay.[28] In addition to single-laser LIBS, two-laser LIBS has also been successful. In two-laser LIBS, much as with the laser microprobe, one laser ablates the sample while the second produces the plasma.

The LIBS technique has found application in several different areas. Metals, semiconductors, ceramics, polymers, and pharmaceuticals have been analyzed by LIBS. In addition to solid samples, gaseous and liquid samples can be used. In fact, the first applications of LIBS were for the remote analyses of hazardous gases in industrial environments. Various process liquids, biological solutions, aqueous environmental solutions, and pharmaceutical preparations have also been analyzed.

---

[28]For more information, see D. A. Cremers and L. J. Radziemski, *Handbook of Laser-Induced Breakdown Spectroscopy*, New York: Wiley, 2006; L. J. Radziemski and D. A. Cremers, in *Laser-Induced Plasmas and Applications*, L. J. Radziemski and D. A. Cremers, eds., New York: Dekker, 1989, Chap. 7.

QUESTIONS AND PROBLEMS

*Answers are provided at the end of the book for problems marked with an asterisk.

 Problems with this icon are best solved using spreadsheets.

10-1 What is an internal standard and why is it used?

10-2 Why are atomic emission methods with an ICP source better suited for multi-element analysis than are flame atomic absorption methods?

10-3 Why do ion lines predominate in spark spectra and atom lines in arc and ICP spectra?

*__10-4__ Calculate the theoretical reciprocal linear dispersion of an echelle grating with a focal length of 0.85 m, a groove density of 120 grooves/mm, and a diffraction angle of 63°26′ when the diffraction order is (a) 30 and (b) 90.

10-5 Why are arc sources often blanketed with a stream of an inert gas?

10-6 Describe three ways of introducing a sample into an ICP torch.

10-7 What are the relative advantages and disadvantages of ICP torches and dc argon torches?

10-8 Why are ionization interferences less severe in ICP than in flame emission spectroscopy?

10-9 What are some of the advantages of plasma sources compared with flame sources for emission spectrometry?

10-10 Why is the internal-standard method often used in plasma emission spectrometry?

 10-11 The chromium in a series of steel samples was determined by ICP emission spectroscopy. The spectrometer was calibrated with a series of standards containing 0, 2.0, 4.0, 6.0, and 8.0 µg $K_2Cr_2O_7$ per milliliter. The instrument readings for these solutions were 3.1, 21.5, 40.9, 57.1, and 77.3, respectively, in arbitrary units.

(a) Plot the data using a spreadsheet.
(b) Find the equation for the regression line.
(c) Calculate standard deviations for the slope and the intercept of the line in (b).
(d) The following data were obtained for replicate 1.00-g samples of cement dissolved in HCl and diluted to 100.0 mL after neutralization.

	Emission Readings			
	Blank	**Sample A**	**Sample B**	**Sample C**
Replicate 1	5.1	28.6	40.7	73.1
Replicate 2	4.8	28.2	41.2	72.1
Replicate 3	4.9	28.9	40.2	spilled

Calculate the percentage of Cr_2O_3 in each sample. What are the absolute and relative standard deviations for the average of each determination?

 10-12 Gold can be determined in solutions containing high concentrations of diverse ions by ICP-AES.[29] Aliquots of 50.0 mL of the sample solution were transferred to each of four 100.0 mL volumetric flasks. A solution was prepared containing 10.0 mg/L Au in 20% H_2SO_4, and quantities of this solution were added to the sample solutions to give 0, 2.5, 5, and 10 mg/L added Au in each of the flasks. The solutions were made up to a total volume of 100.0 mL, mixed, and analyzed by ICP-AES. The resulting data are presented in the following table.

Added Au, mg/L	Emission Intensity, Counts
0.0	12,568
2.5	19,324
5.0	26,622
10.0	40,021

(a) Use a spreadsheet to perform a least-squares analysis to determine the slope, intercept, and regression statistics, including the standard deviation about regression.
(b) Use your results to find the concentration of gold in the sample solution in mg/L.
(c) The known concentration of gold in the sample is 8.51 mg/L. Test the hypothesis that your result is equal to this value at the 95% confidence level.
(d) Compare your result to the result reported by Whitehead et al., and comment on any differences.

10-13 Nakahara and Wasa determined germanium in meteorites by ICP-AES using microwave digestion and hydride generation.[30] Consider the following data table from the referenced paper, which compares results of the work with accepted values for germanium in various meteorites.

[29] J. A. Whitehead, G. A. Lawrance, and A. McCluskey, *Aust. J. Chem.*, **2004**, *57*, 151.
[30] T. Nakahara and T. Wasa, *Microchem. J.*, **1994**, *49*, 202.

Determination of Germanium in Iron Meteorites

Meteorite	Germanium Content, μg/g	
	This Work[a]	Reported Value[b]
Gibeon[c]	0.079 ± 0.01	0.111
Henbury[d]	30.1 ± 1.3	34
Mundrabilla[c]	200.5 ± 11.7	208
Toluca[c]	250.1 ± 15.6	246
Odessa[d]	285.2 ± 11.3	285

[a]The mean ± standard deviation of ten replicate determinations.

[b]J. T. Wasson, *Meteorites, Classification and Properties*. Springer-Verlag, New York, 1974.

[c]Supplied by National Museum of Japanese History.

[d]Supplied by Geological Survey of Japan.

(a) Calculate the 95% confidence intervals for each of the results of Nakahara and Wasa.

(b) Assume that the reported values are the true values for the concentration of germanium in the meteorites, and determine whether the values determined by these workers are equal to or different from the reported values at the 95% confidence level.

(c) Consider the number of significant figures in the standard deviations cited in the table. Is the number of digits justified in each case? Support your answer with statistical calculations.

⊠ Challenge Problem

10-14 Watters et al. have discussed uncertainties associated with calibration curves for ICP-OES and procedures for optimizing the results of least-squares analysis of such data.[31]

(a) The central issue addressed by these workers is whether a weighted or an unweighted least-squares procedure is appropriate for ICP calibration. What is the primary criterion for deciding which procedure to use?

(b) What do the terms *homoscedasticity* and *heteroscedasticity* mean?

(c) The model used by these workers for the ICP curves is represented by the following equation:

$$Y_{ij} = a + bx_i + error_{ij}$$

Define each variable in the equation, and describe the significance of the relationship embodied in it.

(d) Watters et al. chose to model the error in ICP working curves in terms of concentration rather than intensity. What is their rationale for this choice?

(e) One of the suggested models for the error in the calibration curves is

$$\sigma(x) = c + dx + ex^2$$

Describe the significance of each variable, and characterize the nature of the model.

(f) How is $\sigma(x)$ determined in practice?

---

[31]R. L. Watters Jr., R. J. Carroll, and C. H. Spiegelman, *Anal. Chem.*, **1987**, *59*, 1639.

(g) Shot noise and source flicker noise correspond to which variables in the expression shown in (e)?

(h) What experimental noise sources are constant? Which variable in the expression shown in (e) corresponds to these sources (or source)?

(i) What is the relationship of the following alternative model to the model in (e)?

$$\sigma^2(x) = g + hx + kx^2$$

(j) What is the purpose of the models in (e) and (i)?

(k) What is the significance of each of the variables $\hat{a}$ and $\hat{b}$?

(l) The authors conclude that "if heteroscedasticity is ignored, confidence intervals will be too narrow at the high end and too wide at the low end of the ICP calibration curve. The magnitude of these effects will depend on the particular dilution scheme used to make the calibration standard solutions." How does the dilution scheme for standards affect the results of least-squares analysis on ICP calibration curves?

(m) Create an Excel spreadsheet similar to the one shown below containing data from Table I of the paper by Watters et al., and perform both unweighted and weighted least-squares analysis of the data. First use LINEST to perform the unweighted analysis, and then carry out an unweighted analysis using Solver to minimize cell C25 by varying B14 and C14. Compare the results obtained by both methods. Compare and contrast advantages of each method.[32]

	A	B	C	D	E	F	G	H	I
1	Concn,	Intensity	s_i	s	s				
2	µg/µL	(counts)	(obsd)	(10 reps)	(4 reps)			LINEST	
3	0.00	11.33	8.54	7.88	6.97			b	a
4	0.0101	16.60	7.88	7.98	7.03		Value		
5	0.0251	37.92	9.06	8.12	7.12		Uncertainty		
6	0.0503	57.00	8.46	8.36	7.28		R / Sy		
7	0.101	149.88	6.13	8.84	7.58		F / dF		
8	0.251	369.24	11.57	10.25	8.50		SSreg/SSres		
9	0.503	763.36	11.94	12.48	10.02				
10	2.51	3688.46	26.24	25.42	22.25				
11	5.03	7431.08	29.12	29.30	37.55				
12									
13	Unweighted	b	a	Weighted(10)	b	a	Weighted(4)	b	a
14									
15	bx + a	Resid	Resid²	bx + a	Resid	Resid²/s²	bx + a	Resid	Resid²/s²
16									
17	Cell A16=$B$14*A3+$C$14					Cell B14=Initial estimate of slope			
18	Cell B16=B3-A16					Cell C14=Initial estimate of intercept			
19	Cell C16=B16^2								
20	Cell D16=$E$14*A3+$F$14								
21	Cell F16+E16^2*(1/D3^2)								
22	Cell C25=SUM(C16:C24)								
23	Cell C26=SQRT(C25/(COUNT(C16:C24)-2))								
24									
25		SS	0.0000		SS	0.0000		SS	0.000
26		s_r	0.0000		s_r	0.0000		s_r	0.0000

[32] S. R. Crouch and F. J. Holler, *Applications of Microsoft® Excel in Analytical Chemistry*, pp. 68, 88–92, Belmont, CA: Brooks/Cole, 2004.

Repeat the analysis for the ten-replicate data using the formulas provided in the spreadsheet documentation. Formulas entered in row 16 must be copied into rows 17–24. Minimize cell F25 by varying cells E14 and F14 using Solver to obtain estimates of the slope and intercept. Repeat the analysis for the four-replicate data using Solver to minimize cell I25 by varying cells H14 and I14. Compare your procedures and results to those of Watters et al. and comment on any differences. What advantage does weighted least-squares analysis have over unweighted analysis?

(n) Add a section to your Excel spreadsheet to compute the mean and standard deviation of the concentration of an analyte given a number of measurements of sample ICP emission intensity.[33]

(o) There are a number of commercial and noncommercial sources on the Internet for Excel add-ins, which are programs or function packs that supplement the built-in functions of Excel. For example, Solver is actually an add-in that is produced by an independent contractor and that is available from the vendor in an enhanced version. Use a search engine such as Google to locate the websites of programmers and vendors that offer add-ins to perform weighted least-squares analysis. One such vendor is XLSTAT (www.xlstat.com). Download the demonstration version of XLSTAT, install it on your computer, and use it to perform weighted and unweighted least-squares analysis on the data of (n). Note that you must calculate the weighting factors from the standard deviations given in the paper and that you may need to scale them. You should calculate columns in your spreadsheet containing $1/s^2$, and then divide each cell by the largest value in the column. Weighting factors need be only proportional to the variance, so you can scale them in any convenient way. Compare the results from XLSTAT with the results from your spreadsheet, and comment on the ease of use and functionality of XLSTAT. XLSTAT contains many other useful statistics and numerical analysis functions and is available at modest cost to students who desire permanent use of the add-in.

---

[33]D. A. Skoog, D. M. West, F. J. Holler, and S. R. Crouch, *Fundamentals of Analytical Chemistry*, 8th ed., p. 197, Belmont, CA: Brooks/Cole, 2004.

CHAPTER ELEVEN

Atomic Mass Spectrometry

Atomic mass spectrometry is a versatile and widely used tool for identifying the elements present in samples of matter and for determining their concentrations. Nearly all the elements in the periodic table can be determined by mass spectrometry. Atomic mass spectrometry offers a number of advantages over the atomic optical spectrometric methods that we have thus far considered, including (1) detection limits that are, for many elements, as great as three orders of magnitude better than optical methods; (2) remarkably simple spectra that are usually unique and often easily interpretable; and (3) the ability to measure atomic isotopic ratios. Disadvantages include (1) instrument costs that are two to three times that of optical atomic instruments, (2) instrument drift that can be as high as 5% to 10% per hour, and (3) certain types of interference effects that are discussed later.

Throughout this chapter, this logo indicates an opportunity for online self-study at **www.thomsonedu.com/chemistry/skoog**, linking you to interactive tutorials, simulations, and exercises.

11A SOME GENERAL FEATURES OF ATOMIC MASS SPECTROMETRY

An atomic mass spectrometric analysis[1] involves the following steps: (1) atomization, (2) conversion of a substantial fraction of the atoms formed in step 1 to a stream of ions (usually singly charged positive ions), (3) separating the ions formed in step 2 on the basis of their mass-to-charge ratio (m/z), where m is the mass number of the ion in atomic mass units and z is the number of fundamental charges that it bears,[2] and (4) counting the number of ions of each type or measuring the ion current produced when the ions formed from the sample strike a suitable transducer. Because most of the ions formed in step 2 are singly charged, m/z is usually simply the mass number of the ion. Steps 1 and 2 involve the same techniques that were discussed in Section 8C for atomic optical spectroscopy. Steps 3 and 4 are carried out with a mass spectrometer. The data from mass spectrometry are usually presented as a plot of relative intensity versus m/z.

11A-1 Atomic Masses in Mass Spectrometry

At the outset, it is worthwhile to point out that the atomic masses in the literature of mass spectrometry, and in this chapter, differ from those used in most other subdisciplines of analytical chemistry, because mass spectrometers discriminate among the masses of isotopes but other analytical instruments generally do not. We will therefore review briefly some terms related to atomic (and molecular) masses.

Atomic and molecular masses are generally expressed in *atomic mass units* (amu), or *daltons* (Da). The dalton is defined relative to the mass of the carbon isotope $^{12}_{6}C$, and this isotope is assigned a mass of

---

[1] For general treatments of atomic mass spectrometry, see J. R. A. De Laeter, *Applications of Inorganic Mass Spectrometry*, Hoboken, NJ: Wiley-Interscience, 2001; C. M. A. Barshick, D. C. A. Duckworth, and D. H. A. Smith, *Inorganic Mass Spectrometry: Fundamentals and Applications*, Boulder, CO: netLibrary, 2000. For general discussion of mass spectrometry, see E. A. de Hoffmann and V. A. Stroobant, *Mass Spectrometry: Principles and Applications*, 2nd ed., Hoboken, NJ: Wiley, 2002; J. T. Watson, *Introduction to Mass Spectrometry*, 3rd ed., New York: Raven Press, 1997.
[2] Strictly, the *mass number m* is unitless as is z, the number of fundamental charges on an ion. The quantity z, which is an integer, is equal to q/e, where q is the charge on the ion and e is the charge on the electron, both measured in the same units (e.g., coulombs).

exactly 12 amu. Thus, *the amu, or Da, is defined as 1/12 of the mass of one neutral $^{12}_{6}C$ atom*, or

$$\text{mass of 1 atom} \, ^{12}_{6}\text{C} = \frac{12 \text{ g } ^{12}\text{C/mol } ^{12}\text{C}}{6.022142 \times 10^{23} \text{ atoms } ^{12}\text{C/mol } ^{12}\text{C}}$$

$$= 1.992646 \times 10^{-23} \text{ g/atom } ^{12}\text{C}$$

$$= 1.992646 \times 10^{-26} \text{ kg/atom } ^{12}\text{C}$$

The atomic mass unit is then

$$1 \text{ amu} = 1 \text{ Da}$$

$$= \frac{1}{12} (1.992646 \times 10^{-23} \text{ g})$$

$$= 1.6605387 \times 10^{-24} \text{ g}$$

$$= 1.6605387 \times 10^{-27} \text{ kg}$$

The relative atomic mass of an isotope such as $^{35}_{17}\text{Cl}$ is then measured with respect to the mass of the reference $^{12}_{6}\text{C}$ atom. Chlorine-35 has a mass that is 2.914071 times greater than the mass of the carbon isotope. Therefore, the atomic mass of the chlorine isotope is

$$\text{atomic mass } ^{35}_{17}\text{Cl} = 2.914071 \times 12.000000 \text{ Da}$$

$$= 34.968853 \text{ Da}$$

Because 1 mol of $^{12}_{6}\text{C}$ weighs 12.000000 g, the molar mass of $^{35}_{17}\text{Cl}$ is 34.968853 g/mol.[3]

In mass spectrometry, in contrast to most types of chemistry, we are often interested in the exact mass m of particular isotopes of an element or the exact mass of compounds containing a particular set of isotopes. Thus, we may need to distinguish between the masses of compounds such as

$$^{12}\text{C}^1\text{H}_4 \quad m = (12.000000 \times 1) + (1.007825 \times 4)$$

$$= 16.031300 \text{ Da}$$

$$^{13}\text{C}^1\text{H}_4 \quad m = (13.003355 \times 1) + (1.007825 \times 4)$$

$$= 17.034655 \text{ Da}$$

$$^{12}\text{C}^1\text{H}_3{}^2\text{H}_1 \quad m = (12.000000 \times 1) + (1.007825 \times 3)$$

$$+ (2.014102 \times 1)$$

$$= 17.037577 \text{ Da}$$

We have presented the isotopic masses in the calculations above with six digits to the right of the decimal point. Although certain types of state-of-the-art mass spectrometers are capable of such resolution (see Section 20C-4), normally, exact masses are quoted to

---

[3] A listing of the isotopes of all of the elements and their atomic masses can be found in several handbooks, such as *Handbook of Chemistry and Physics*, 85th ed., pp. 11-50 to 11-201, Boca Raton, FL: CRC Press, 2004.

three or four figures to the right of the decimal point because typical high-resolution mass spectrometers make measurements at this level of precision.

In other contexts, we use the term *nominal mass*, which implies a whole-number precision in a mass measurement. Thus, the nominal masses of the three isomers just cited are 16, 17, and 17 Da, respectively.

The *chemical atomic mass*, or the *average atomic mass* (A), of an element in nature is given by the equation

$$A = A_1 p_1 + A_2 p_2 + \cdots + A_n p_n = \sum_{i=1}^{n} A_n p_n$$

where $A_1, A_2, \ldots, A_n$ are the atomic masses in daltons of the n isotopes of the element and $p_1, p_2, \ldots, p_n$ are the fractional abundances of these isotopes in nature. The chemical atomic mass is, of course, the type of mass of interest to chemists for most purposes. The average, or chemical, molecular mass of a compound is then the sum of the chemical atomic masses for the atoms appearing in the formula of the compound. Thus, the chemical molecular mass of CH_4 is 12.01115 + $(4 \times 1.00797) = 16.0434$ Da. The atomic or molecular mass expressed without units is the *mass number*.

11A-2 Mass-to-Charge Ratio

One other term that is used throughout this chapter is the *mass-to-charge ratio* of an atomic or molecular ion. The mass-to-charge ratio of an ion is the unitless ratio of its mass number to the number of fundamental charges z on the ion. Thus, for $^{12}\text{C}^1\text{H}_4{}^+$, $m/z = 16.0313/1 = 16.0313$. For $^{13}\text{C}^1\text{H}_4{}^{2+}$, $m/z = 17.0346/2 = 8.5173$. Because most ions in mass spectrometry are singly charged, the term mass-to-charge ratio is often shortened to the more convenient term *mass*. Strictly speaking, this abbreviation is not correct, but it is widely used in the mass spectrometry literature.

11A-3 Types of Atomic Mass Spectrometry

Table 11-1 lists the most important types of atomic mass spectrometry. Historically, thermal ionization mass spectrometry and spark source mass spectrometry were the first mass spectrometric methods developed for qualitative and quantitative elemental analysis, and these types of procedures still find applications, although they are now overshadowed by some of the other methods listed in Table 11-1, particularly inductively coupled plasma mass spectrometry (ICPMS).

TABLE 11-1 Types of Atomic Mass Spectrometry

Name	Acronym	Atomic Ion Sources	Typical Mass Analyzer
Inductively coupled plasma	ICPMS	High-temperature argon plasma	Quadrupole
Direct current plasma	DCPMS	High-temperature argon plasma	Quadrupole
Microwave-induced plasma	MIPMS	High-temperature argon plasma	Quadrupole
Spark source	SSMS	Radio-frequency electric spark	Double-focusing
Thermal ionization	TIMS	Electrically heated plasma	Double-focusing
Glow discharge	GDMS	Glow-discharge plasma	Double-focusing
Laser microprobe	LMMS	Focused laser beam	Time-of-flight
Secondary ion	SIMS	Accelerated ion bombardment	Double-focusing

Note that the first three entries in the table are *hyphenated methods*, combinations of two instrumental techniques that produce analytical results superior in some way to the results from either of the original individual methods. We shall encounter a number of hyphenated methods elsewhere in this book.

Before examining the various methods for atomization and ionization found in the table, we describe briefly how mass spectrometers are used to separate and measure ionic species.

11B MASS SPECTROMETERS

A mass spectrometer is an instrument that produces ions and separates them according to their mass-to-charge ratios, m/z. Most of the ions we will discuss are singly charged so that the ratio is simply equal to the mass number of the ion. Several types of mass spectrometers are currently available from instrument manufacturers. In this chapter, we describe the three types that are used in atomic mass spectrometry: the *quadrupole mass spectrometer*, the *time-of-flight mass spectrometer*, and the *double-focusing mass spectrometer*. Other types of mass spectrometers are considered in Chapter 20, which is devoted to molecular mass spectrometry. The first column in Table 11-1 indicates the types of atomic mass spectrometry in which each of the three types of mass spectrometer is usually applied.

The block diagram in Figure 11-1 shows the principal components of all types of mass spectrometers. The purpose of the inlet system is to introduce a micro amount of sample into the ion source where the components of the sample are converted into gaseous ions by bombardment with electrons, photons, ions, or molecules. Alternatively, ionization is accomplished by applying thermal or electrical energy. The output of the ion source is a stream of positive (most common) or negative gaseous ions that are then accelerated into the mass analyzer.

The function of the mass analyzer is analogous to the monochromator in an optical spectrometer. In a mass analyzer, however, dispersion depends on the mass-to-charge ratio of analyte ions rather than on the wavelength of photons.

Like an optical spectrometer, a mass spectrometer contains a transducer that converts the beam of ions into an electrical signal that can then be processed, stored in the memory of a computer, and displayed or stored. Unlike most optical spectrometers, mass spectrometers require an elaborate vacuum system to maintain a low pressure in all of the components except the

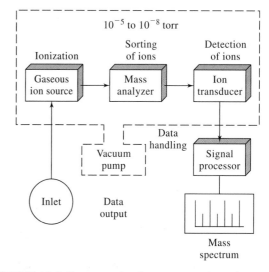

FIGURE 11-1 Components of a mass spectrometer.

signal processor and readout. Low pressure ensures infrequent collisions in the mass spectrometer to produce and maintain free ions and electrons.

In the sections that follow, we first describe the various transducer systems that are used in mass spectrometers. Then we consider the three types of mass analyzers that have been used in atomic mass spectrometers. Sections 11C, D, and E contain material on the nature and operation of common ion sources for atomic mass spectrometers.

11B-1 Transducers for Mass Spectrometry

Several types of transducers are commercially available for mass spectrometers. The *electron multiplier* is the transducer of choice for most routine experiments.

Electron Multipliers

Figure 11-2a is a schematic of a discrete-dynode electron multiplier designed for collecting and converting positive ions into an electrical signal. This device is

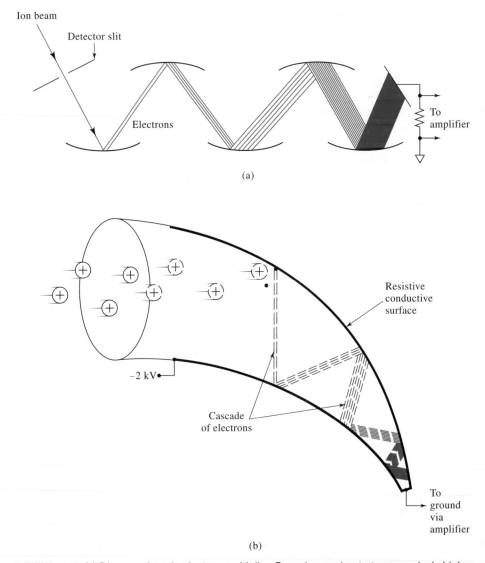

FIGURE 11-2 (a) Discrete-dynode electron multiplier. Dynodes are kept at successively higher voltages via a multistage voltage divider. (b) Continuous-dynode electron multiplier. (Adapted from J. T. Watson, *Introduction to Mass Spectrometry*, 3rd ed., pp. 334–35, New York: Raven Press, 1997. With permission.)

very much like the photomultiplier transducer for ultraviolet-visible radiation, with each dynode held at a successively higher voltage. When energetic ions or electrons strike the Cu-Be surfaces of the cathode and the dynodes, bursts of electrons are emitted. The electrons are then attracted to the next dynode down the chain until, at the last dynode, a huge number of electrons appear for every ion that strikes the cathode. Electron multipliers with up to twenty dynodes are available that typically provide a current gain of 10^7.

Figure 11-2b illustrates a continuous-dynode electron multiplier. The device is shaped like a cornucopia and is made of glass heavily doped with lead to give the material a small conductivity. A voltage of 1.8 to 2 kV applied across the length of the transducer produces a voltage gradient from one end to the other. Ions that strike the surface near the entrance eject electrons that are then attracted to higher-voltage points farther along the tube. These *secondary electrons* skip along the surface, ejecting more electrons with each impact. Transducers of this type typically have gains of 10^5, but in certain applications gains as high as 10^8 can be achieved.

In general, electron multipliers are rugged and reliable and are capable of providing high-current gains and nanosecond response times. These transducers can be placed directly behind the exit slit of a magnetic sector mass spectrometer, because the ions reaching the transducer usually have enough kinetic energy to eject electrons from the first stage of the device. Electron multipliers can also be used with mass analyzers that employ low-kinetic-energy ion beams (that is, quadrupoles), but in these applications the ion beam exiting the analyzer is accelerated to several thousand electron volts prior to striking the first stage.

The Faraday Cup

Figure 11-3 is a schematic of a Faraday cup collector. The transducer is aligned so that ions exiting the analyzer strike the collector electrode. This electrode is surrounded by a cage that prevents the escape of reflected ions and ejected secondary electrons. The collector electrode is inclined with respect to the path of the entering ions so that particles striking or leaving the electrode are reflected from the entrance to the cup. The collector electrode and cage are connected to ground through a large resistor. The charge of the positive ions striking the plate is neutralized by a flow of electrons from ground through the resistor. The resulting voltage drop across the resistor is amplified by a

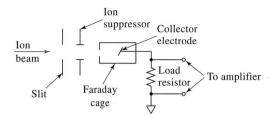

FIGURE 11-3 Faraday cup detector. The voltage on the ion suppressor plates is adjusted to minimize differential response as a function of mass.

high-impedance amplifier. The response of this transducer is independent of the energy, the mass, and the chemical nature of the ion. The Faraday cup is inexpensive and simple mechanically and electrically; its main disadvantage is the need for a high-impedance amplifier, which limits the speed at which a spectrum can be scanned (see Section 3B-4). The Faraday cup transducer is also less sensitive than an electron multiplier, because it provides no internal amplification.

Array Transducers

The analogy between optical spectrometers and mass spectrometers extends to the detection systems. As we have seen, array transducers have revolutionized optical spectrometry for systems in which dispersed electromagnetic radiation is focused at a plane or curved surface. An array transducer positioned in the focal plane of a mass spectrometer (see Figure 11-11) provides several significant advantages over single-channel detection; the most important is the simultaneous detection of multiple resolution elements. Others include greater duty cycle, improved precision using ratio measurements and internal standards, and improved detection of fast transient signals.

Microchannel Plates. One type of array transducer for mass spectrometry is the electrooptical ion detector (EOID) shown in Figure 11-4a.[4] The key element in the EOID is the microchannel electron multiplier, or microchannel plate, which is referred to in some optical applications as an image intensifier. The design and basic principles of operation of the microchannel plate are illustrated in Figure 11-4b. The plate consists of an array of tiny tubes (diameters as small as 6 μm)

[4]C. E. Giffen, H. G. Boettger, and D. D. Norris, *Int. J. Mass Spectrom. Ion. Phys.*, **1974**, *15*, 437.

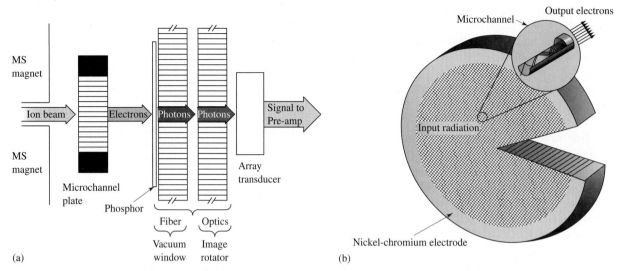

FIGURE 11-4 (a) Electrooptical ion detector. Ions strike the microchannel plate, which produces a cascade of electrons that strike the phosphorescent screen. Radiation from the screen passes to the array transducer via fiber optics, and the transducer then converts the optical signal to an electrical signal for processing. (Reprinted with permission from J. T. Watson, *Introduction to Mass Spectrometry*, 3rd ed., pp. 334–35, New York: Raven Press, 1997. With permission.) (b) Microchannel plate. The tiny channels, which are oriented at a small angle from the normal of the array surface, are individual electron multipliers. Particles striking the inner surface of a channel cascade through the channel and produce about 1000 electrons for every incoming particle. (Diagram courtesy of Photonis, Inc.)

made of lead glass. Metallic electrodes are deposited on both sides of the array, and a voltage of about 1500 V is applied between the ends of the channels. Each tube acts as an electron multiplier so that for every particle that strikes the inside wall of a channel, 1000 electrons emerge from the other end of the channel. If greater gain is required, two or more microchannel plates may be cascaded to provide signals appropriate for the experiment.

In the EOID detector of Figure 11-4a, a phosphorescent screen is placed behind the microchannel plate so that the cascade of electrons produces a flash of light that is directed to an optical array detector via fiber optics. The data collected from the array detector then provide two-dimensional resolution of ions appearing at the focal plane of a mass spectrometer.

Micro-Faraday Array Transducer. Recently, a new type of array transducer for mass spectrometry has been introduced that shows promise for the simultaneous detection of ions at the focal plane of Mattauch-Herzog-type mass spectrometers and other similar applications.[5] An image of the prototype Faraday cup array is shown in Figure 11-5. The device contains gold Faraday electrodes bonded to capacitive transimpedance amplifiers. Each amplifier integrates the charge on the electrode, and the output voltage is switched via a multiplexer to a data system for analog-to-digital conversion. The elements in the image are 145 μm wide, but in principle, they can be as narrow as 10 μm. The detection system exhibits 5% accuracy and 0.007% relative standard deviation (RSD) on isotope-ratio measurements of standards, detection limits of as little as tens to hundreds of parts per quadrillion, and a linear dynamic range of more than seven orders of magnitude. The potential of the detection system has been evaluated in ICP as well as gas chromatography–mass spectrometry (GC/MS) experiments. A recent 128-element version of the transducer exhibited a dynamic range of over

[5] A. K. Knight IV, R. P. Sperline, G. M. Hieftje, E. Young, C. J. Barinaga, D. W. Koppenaal, and M. B. Denton, *Int. J. Mass Spectrom.*, **2002**, *215*, 131; J. H. Barnes IV, G. D. Schilling, G. M. Hieftje, R. P. Sperline, M. B. Denton, C. J. Barinaga, D. W. Koppenaal, *J. Am. Soc. Mass Spectrom.*, **2004**, *15*, 769–76; J. H. Barnes IV, G. D. Schilling, R. P. Sperline, M. B. Denton, E. T. Young, C. J. Barinaga, D. W. Koppenaal, G. M. Hieftje, *Anal. Chem.*, **2004**, *76*, 2531–36; J. H. Barnes IV et al., *J. Anal. At. Spectrom.*, **2004**, *19*, 751–56; G. D. Schilling et al., *Anal. Chem.*, **2006**, *78*, 4319.

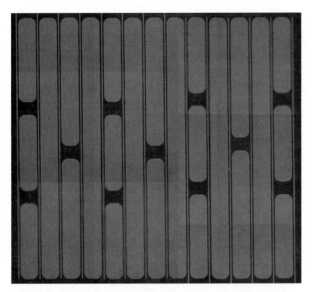

FIGURE 11-5 A micro-Faraday array ion transducer. The transducer elements are 5, 2.45, and 1.60 mm long and 145 μm wide with 10 μm spacing. There is also a 10-μm guard electrode between each element. (Reprinted from A. K. Knight IV, R. P. Sperline, G. M. Hieftje, E. Young, C. J. Barinaga, D. W. Koppenaal, and M. B. Denton, *Int. J. Mass Spectrom.*, **2002**, *215*, 131. Copyright 2002 Wiley.)

7 orders of magnitude, resolving power of up to 600, and isotope-ratio accuracy of 0.2% and precision of 0.018%. The Faraday cup array should be particularly useful for simultaneous detection of ions in experiments involving transient species with performance characteristics comparable to state-of-the-art single-channel transducers.

Other Types of Detection Systems

Scintillation-type transducers consist of a crystalline phosphor dispersed on a thin aluminum sheet that is mounted on the window of a photomultiplier tube. When ions (or electrons produced when the ions strike a cathode) impinge on the phosphor, they produce flashes of light that are detected by the photomultiplier. A specialized version of this type of device is the Daly detector, which consists of an aluminized cathode in the shape of a knob (the Daly knob) held at a very large negative voltage opposite a scintillation transducer. Analyte ions collide with the cathode producing secondary electrons that are then attracted to the sur-

 Tutorial: Learn more about **transducers for mass spectrometry**.

face of a scintillation transducer. The Daly detector has the advantage that all important components of the system except the knob are located outside the vacuum chamber and so can be serviced without disrupting the vacuum. The detector is especially useful for species with high masses.[6]

Just as in optical spectroscopy, where for many years photographic plates and film were used as photon transducers, photographic emulsions are sensitive to energetic ions, and thus they were the original and for a time the only transducers for mass spectrometry. This type of detection is well suited to the simultaneous observation of a wide range of *m/z* values in instruments that focus ions along a plane (Figure 11-11). As electronic detectors have evolved, photographic detection has become rather rare. Undoubtedly, obscure applications still require photographic recording of mass spectra, but with the advent of array transducers for mass spectrometry and the potential elimination of the tedium of developing film or plates, this method of detection is likely to become extinct.

11B-2 Quadrupole Mass Analyzers

The most common type of mass spectrometer used in atomic mass spectroscopy is the quadrupole mass analyzer shown in Figure 11-6. This instrument is more compact, less expensive, and more rugged than most other types of mass spectrometers. It also has the advantage of high scan rates so that an entire mass spectrum can be obtained in less than 100 ms.[7]

The heart of a quadrupole instrument is the four parallel cylindrical (originally hyperbolic) rods that serve as electrodes. Opposite rods are connected electrically, one pair being attached to the positive side of a variable dc source and the other pair to the negative terminal. In addition, variable radio-frequency ac voltages, which are 180° out of phase, are applied to each pair of rods. To obtain a mass spectrum with this device, ions are accelerated into the space between the rods by a potential difference of 5 to 10 V. Meanwhile, the ac and dc voltages on the rods are increased simultaneously while maintaining their ratio constant. At any given moment, all of the ions except those having a certain *m/z* value strike the rods and are converted to neutral molecules. Thus, only ions having a limited range of *m/z* values reach the transducer. Typically,

[6]J. T. Watson, *Introduction to Mass Spectrometry*, 3rd ed., pp. 336–37, New York: Raven Press, 1997.
[7]P. E. Miller and M. B. Denton, *J. Chem. Educ.*, **1986**, *63*, 617.

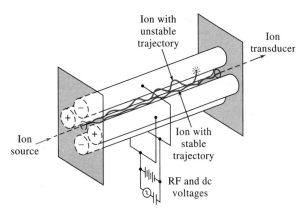

FIGURE 11-6 A quadrupole mass spectrometer.

quadrupole instruments easily resolve ions that differ in mass by one unit.

Note that a quadrupole mass spectrometer is more analogous to an optical variable-band filter photometer than to an optical spectrometer in which a grating simultaneously disperses a spectrum of electromagnetic radiation. For this reason, the device is sometimes referred to as a mass filter rather than a mass analyzer.

Ion Trajectories in a Quadrupole

To understand the filtering capability of a quadrupole, we need to consider the effect of the dc and ac voltages on the trajectory of ions as they pass through the channel between the rods.[8] Let us first focus on the pair of positive rods, which are shown in Figure 11-7 as lying in the xz plane. In the absence of a dc voltage, ions in the channel will tend to converge in the center of the channel during the positive half of the ac cycle and will tend to diverge during the negative half. This behavior is illustrated at points A and B in the figure. If during the negative half cycle an ion strikes the rod, the positive charge will be neutralized, and the resulting molecule will be carried away. Whether a positive ion strikes the rod depends on the rate of movement of the ion along the z-axis, its mass-to-charge ratio, and the frequency and magnitude of the ac signal.

Now let us consider the effect of a positive dc voltage that is superimposed on the ac signal. From Newtonian physics, the momentum of ions of equal kinetic energy is directly proportional to the square root of mass. It is therefore more difficult to deflect a heavier ion than to deflect a lighter one. If an ion in the channel is heavy or the frequency of the ac voltage is large, the ion will not respond significantly to the alternating voltage and will be influenced largely by the dc voltage. Under these circumstances, the ion will tend to remain in the space between the rods. In contrast, if the ion is light or the frequency is low, the ion may collide with the rod and be eliminated during the negative excursion of the ac voltage. Thus, as shown in Figure 11-8a, the pair of positive rods forms a high-pass mass filter for positive ions traveling in the xz plane.

Now let us turn to the pair of rods that are maintained at a negative dc voltage. In the absence of the ac field, all positive ions will tend to be drawn to the rods, where they are annihilated. For the lighter ions, however, this movement may be offset by the excursion of the ac voltage. Thus, as shown in Figure 11-8b, the rods in the yz plane operate as a low-pass filter.

For an ion to travel through the quadrupole to the detector, it must have a stable trajectory in both the xz and yz planes. Thus, the ion must be sufficiently heavy so that it will not be eliminated by the high-mass filter in the xz plane and must be sufficiently light so that it will not be removed by the low-mass filter in the yz plane. Therefore, as is shown in Figure 11-8c, the entire quadrupole transmits a band of ions that have a limited range of m/z values. The center of this band can be varied by adjusting the ac and dc voltages.

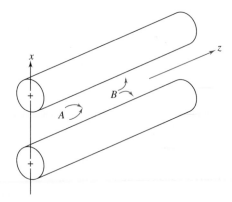

FIGURE 11-7 Operation of a quadrupole in the xz plane. *A*: ions are focused toward the z-axis; *B*: ions are attracted toward the x rods.

---

[8]The discussion of quadrupole mass filters that follows is largely based on P. E. Miller and M. B. Denton, *J. Chem. Educ.*, **1986**, *63*, 617. See also R. E. Marchand and R. J. Hughes, *Quadrupole Storage Mass Spectrometry*, New York: Wiley, 1989; J. J. Leary and R. L. Schmidt, *J. Chem. Educ.*, **1996**, *73*, 1142.

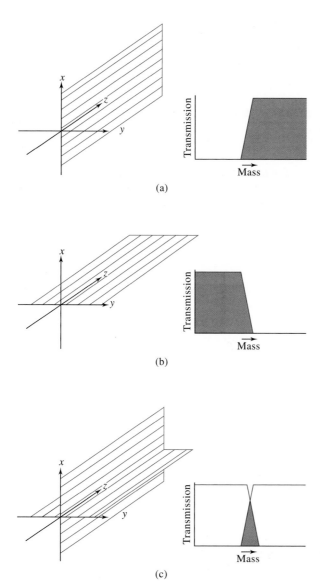

(a)

(b)

(c)

FIGURE 11-8 Quadrupole acts as (a) a high-pass mass filter in the xz plane, (b) a low-pass mass filter in the yz plane, and (c) a narrow-band filter when high-pass and low-pass filters are both in operation. (Reprinted with permission from P. E. Miller and M. B. Denton, *J. Chem. Educ.*, **1986**, *63*, 619. Figures 6, 7, and 8. Copyright 1986 Journal of Chemical Education.)

Scanning with a Quadrupole Filter

The differential equations necessary to describe the behavior of ions of different masses in a quadrupole are complex and difficult to treat analytically and are beyond the scope of this book. These equations reveal,

 Simulation: Learn more about **quadrupole mass analyzers**.

however, that the oscillations of charged particles in a quadrupole fall into two categories: (1) those in which the amplitudes of the oscillations are finite and (2) those in which the oscillations grow exponentially and ultimately approach infinity. The variables in these equations are mass-to-charge ratio, the dc voltage, the frequency and magnitude of the ac voltage, and the distance between the rods. The resolution of a quadrupole is determined by the ratio of the ac to dc voltage and becomes a maximum when this ratio is just slightly less than 6. Thus, quadrupole spectrometers are operated at a constant voltage ratio of this value.

To scan a mass spectrum with a quadrupole instrument, the ac voltage V and the dc voltage U are increased simultaneously from zero to some maximum value while their ratio is maintained at slightly less than 6. The changes in voltage during a typical scan are shown in Figure 11-9. The two diverging straight lines show the variation in the two dc voltages as a function of time. The time for a single sweep is a few milliseconds. While the dc voltages are varied from 0 to about ±250 V, the ac voltages increase linearly from zero to approximately 1500 V. Note that the two ac voltages are 180° out of phase.

Quadrupole mass spectrometers with ranges that extend up to 3000 to 4000 m/z are available from several instrument manufacturers. These instruments easily resolve ions that differ in mass by one unit. Generally, quadrupole instruments are equipped with a circular aperture, rather than a slit (see Figure 11-6), to introduce the sample into the dispersing region. The aperture provides a much greater sample throughput

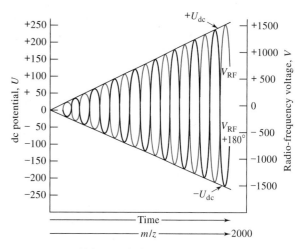

FIGURE 11-9 Voltage relationships during a mass scan with a quadrupole analyzer.

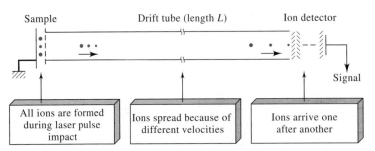

FIGURE 11-10 Principle of a TOF mass spectrometer. A spatially tightly bunched group of ions produced by a laser probe is accelerated into the drift tube where separation occurs. (From A. H. Verbueken, F. J. Bruynseels, R. Van Grieken, and F. Adams, in *Inorganic Mass Spectrometry*, p. 186, F. Adams, R. Gijbels, and R. Van Grieken, eds., New York: Wiley, 1988. With permission.)

than can be tolerated in magnetic sector instruments, in which resolution is inversely related to slit width.

11B-3 Time-of-Flight Mass Analyzers

In time-of-flight (TOF) instruments, positive ions are produced periodically by bombardment of the sample with brief pulses of electrons, secondary ions, or laser-generated photons. These pulses typically have a frequency of 10 to 50 kHz and a lifetime of 0.25 ms. The ions produced in this way are then accelerated by an electric field pulse of 10^3 to 10^4 V that has the same frequency as, but lags behind, the ionization pulse. The accelerated particles pass into a field-free *drift tube* about a meter long (Figure 11-10). Because all ions entering the tube ideally have the same kinetic energy, their velocities in the tube must vary inversely with their masses, with the lighter particles arriving at the detector earlier than the heavier ones. Typical flight times are in the microsecond range for a 1-m flight tube.[9] The equations governing TOF mass spectrometry are given in Section 20C-3.

The transducer in a TOF mass spectrometer is usually an electron multiplier whose output is displayed on the vertical deflection plates of an oscilloscope and the horizontal sweep is synchronized with the accelerator pulses; an essentially instantaneous display of the mass spectrum appears on the oscilloscope screen. Because typical flight times are in the microsecond range, digital data acquisition requires extremely fast elec-

tronics. Variations in ion energies and initial positions cause peak broadening that generally limits attainable resolution.

From the standpoint of resolution and reproducibility, instruments equipped with TOF mass analyzers are not as satisfactory as those with magnetic or quadrupole analyzers. ICP-TOF mass spectrometers, for example, typically are an order of magnitude poorer in sensitivity and in detection limits than comparable quadrupole systems.[10] Several advantages partially offset these limitations, however, including simplicity and ruggedness, ease of accessibility to the ion source, virtually unlimited mass range, and rapid data-acquisition rate. Several instrument manufacturers now offer TOF instruments, but they are less widely used than are quadrupole mass spectrometers.

11B-4 Double-Focusing Analyzers

As shown in Figure 11-11, a double-focusing mass spectrometer contains two devices for focusing a beam of ions: an *electrostatic analyzer* and a *magnetic sector analyzer*. In this instrument, the ions from a source are accelerated through a slit into a curved electrostatic field that focuses a beam of ions having a narrow band of kinetic energies onto a slit that leads to a curved magnetic field. In this field, the lightest ions are deflected the most and the heaviest the least. The dispersed ions then fall on a photographic plate and are thus recorded. A detailed discussion of how the electrostatic and magnetic

---

[9] W. N. Delgass and R. G. Cooks, *Science*, **1987**, *235*, 545.

 Simulation: Learn more about **time-of-flight mass analyzers**.

---

[10] For a review of the principles and applications of mass analyzers for ICP-TOF mass spectrometry, see R. J. Ray and G. M. Hieftje, *J. Anal. At. Spectrom.*, **2001**, *16*, 1206–16.

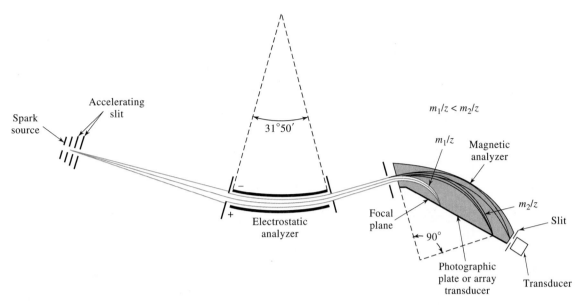

FIGURE 11-11 Mattauch-Herzog-type double-focusing mass spectrometer. Resolution $>10^5$ has been achieved with instruments based on this design.

analyzers function is in Section 20C-3. As shown in Table 11-1, analyzers of this type find use in several types of atomic mass spectrometry.

11C INDUCTIVELY COUPLED PLASMA MASS SPECTROMETRY

Since the early 1980s, ICPMS has grown to be one of the most important techniques for elemental analysis because of its low detection limits for most elements, its high degree of selectivity, and its reasonably good precision and accuracy.[11] In these applications an ICP torch serves as an atomizer and ionizer. For solutions, sample introduction is accomplished by a conventional or an ultrasonic nebulizer. For solids, one of the other sample-introduction techniques discussed in Section 8C-2, such as spark or laser ablation or glow discharge, are used. Commercial versions of instruments for these various techniques have been on the

market since 1983. In these instruments, positive metal ions, produced in a conventional ICP torch, are sampled through a differentially pumped interface linked to a mass analyzer, usually a quadrupole. The spectra produced in this way, which are remarkably simple compared with conventional ICP optical spectra, consist of a simple series of isotope peaks for each element present. These spectra are used to identify the elements present in the sample and for their quantitative determination. Usually, quantitative analyses are based on calibration curves in which the ratio of the ion count for the analyte to the count for an internal standard is plotted as a function of concentration. Analyses can also be performed by the isotope dilution technique, which is described in detail in Section 32D.

11C-1 Instruments for ICPMS

Figure 11-12 shows schematically the components of a commercial ICPMS system.[12] A critical part of the instrument is the interface that couples the ICP torch, which operates at atmospheric pressure with the mass spectrometer that requires a pressure of less than 10^{-4} torr. This coupling is accomplished by a differentially pumped interface coupler that consists of a sampling cone, which is a water-cooled nickel cone with a small

---

[11] For detailed descriptions of this technique, see R. A. Thomas, *Practical Guide to ICP-MS*, Vol. 33, New York: Dekker, 2003; H. E. Taylor, *Inductively Coupled Plasma-Mass Spectrometry: Practices and Techniques*, San Diego, CA: Elsevier, 2001; *Inductively Coupled Plasma Mass Spectrometry*, A. E Montaser, ed. Hoboken, NJ: Wiley-VCH, 1998; K. E. Jarvis, A. L. Gray, and R. S. Houk, *Handbook of Inductively Coupled Plasma Mass Spectrometry*, New York: Chapman & Hall, 1992.

 Simulation: Learn more about **double-focusing mass analyzers**.

[12] For a review of available instruments, see K. Cottingham, *Anal. Chem.*, **2004**, *76*, 35A.

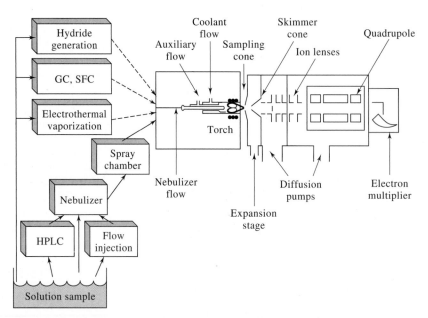

FIGURE 11-12 Schematic of an ICPMS system. Dotted lines show introduction of gaseous samples; solid lines show introduction of liquid samples. HPLC = high-performance liquid chromatography, SFC = supercritical fluid chromatography. (From N. P. Vela, L. K. Olson, and J. A. Caruso, *Anal. Chem.*, **1993**, *65*, 585A. Figure 1, p. 587A. Copyright 1993 American Chemical Society.)

orifice (<1.0 mm) in its center. The hot plasma gas is transmitted through this orifice into a region that is maintained at a pressure of about 1 torr by means of a mechanical pump. In this region, rapid expansion of the gas occurs, which results in its being cooled. A fraction of the gas in this region then passes through a small hole in a second cone called a skimmer and into a chamber maintained at the pressure of the mass spectrometer. Here, the positive ions are separated from electrons and molecular species by a negative voltage, are accelerated, and are focused by a magnetic ion lens onto the entrance orifice of a quadrupole mass analyzer.

Performance specifications for typical commercial atomic mass spectrometers equipped with an ICP torch include a mass range of 3 to 300, the ability to resolve ions differing in m/z by 1, and a dynamic range of six orders of magnitude. More than 90% of the elements in the periodic table have been determined by ICPMS. Measurement times of 10 s per element, with detection limits in the range of 0.1 to 10 ppb for most elements, are achieved. Relative standard deviations of 2% to 4% for concentrations in the middle regions of calibration curves are reported.

In recent years, laser ablation has come into its own as a method of solid-sample introduction with minimal

preparation for ICPMS.[13] In this technique, shown in Figure 11-13, a pulsed laser beam is focused onto a few square micrometers of a solid, giving power densities of as great as 10^{12} W/cm^2. This high-intensity radiation rapidly vaporizes most materials, even refractory ones. A flow of argon then carries the vaporized sample into an ICP torch where atomization and ionization take place. The resulting plasma then passes into the mass spectrometer. Instruments of this type are particularly well suited for semiquantitative analysis (or quantitative analysis if the internal standard can be used) of samples that are difficult to decompose or dissolve, such as geological materials, alloys, glasses, gemstones, agricultural products, urban particulates, and soils. Figure 11-14 is a mass spectrum of a standard rock sample obtained by laser ablation–ICPMS. To obtain this spectrum, the rock was ground into a fine powder and formed into a small disk under pressure. After ablation with the laser beam, the resulting vapor was introduced into an ICP torch by injector gas flow, where ionization took place. The gaseous ions were then analyzed in a quadrupole mass spectrometer.

 Tutorial: Learn more about **ICPMS**.

[13] See B. Hattendorf, C. Latkoczy, D. Gunther, *Anal. Chem.*, **2003**, *75*, 341A; A. Montaser et al., in *Inductively Coupled Plasma Mass Spectrometry*, A. Montaser, ed., Hoboken, NJ: Wiley-VCH, 1998, pp. 194–218.

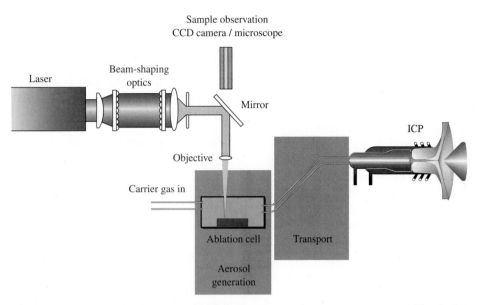

FIGURE 11-13 Laser ablation sample-introduction system for mass spectrometry. The device is similar to most commercially available instruments for spatially resolved elemental analysis. CCD = charge-coupled device. (Reprinted from B. Hattendorf, C. Latkoczy, and D. Gunther, *Anal. Chem.*, **2003**, *75*, 341A. Figure 1, p. 342A. Copyright 2003 American Chemical Society.)

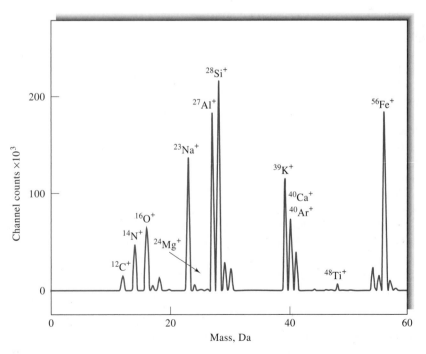

FIGURE 11-14 Spectrum of a standard rock sample obtained by laser ablation–ICPMS. Major components (%): Na, 5.2; Mg, 0.21; Al, 6.1; Si, 26.3; K, 5.3; Cu, 1.4; Ti, 0.18; Fe, 4.6. (From *Inorganic Mass Spectrometry*, F. Adams, R. Gijbek, and R. Van Grieken, eds., p. 297, New York: Wiley, 1988. With permission.)

Although detection limits for laser ablation in ICPMS depend on a broad range of experimental variables, including instrument variables as well as sample type and quality, typical limits are below the microgram-per-gram level; under certain experimental conditions, for certain elements such as thallium and uranium, detection limits are in the nanogram-per-gram range. Standard deviations on the order of a few percent relative are typical. A major advantage of laser ablation is that the beam can be directed at various locations on the sample to provide mass spectra with spatial resolution.

Accessories that permit use of most of the other ion sources listed in Table 11-1 in conjunction with an ICP torch are available commercially.

11C-2 Atomic Mass Spectra and Interferences

One of the advantages of using mass spectrometric detection with ICPs as opposed to optical detection is that mass spectra are usually much simpler and easier to interpret than corresponding optical spectra. This property is especially true for those elements such as the rare earths that may exhibit thousands of emission lines. Figure 11-15 illustrates this advantage for a solution containing cerium. Figure 11-15a is the optical emission spectrum of a solution containing 100 ppm cerium; the lower plot is the mass spectrum of a solution having a concentration of 10 ppm of the element. The optical spectrum contains a dozen or more strong cerium lines and hundreds of weak lines for the element, all of which are superimposed on a complicated background spectrum. The optical background is made up of molecular bands from atmospheric contaminants, such as NH, OH, N_2, and H_2, and a continuum from the recombination of argon and other ions with electrons. In contrast, the mass spectrum for cerium, shown in Figure 11-15b, is simpler, consisting of two isotope peaks for $^{140}Ce^+$ and $^{142}Ce^+$ and a small peak for the doubly charged $^{140}Ce^{2+}$, which is located at $m/z = 70$. Note that the background for the spectrum consists of only a few molecular ion species, and all of these occur at m/z values of 40 or less.

Spectra such as that in Figure 11-15b led early workers in the field of ICPMS to have hopes of an interference-free method. Unfortunately, this hope was not realized in further studies, and serious interference problems are sometimes encountered in atomic mass spectrometry just as in optical atomic spectroscopy.

Interference effects in atomic mass spectroscopy fall into two broad categories: spectroscopic interferences and nonspectroscopic interferences. The latter effects are similar to the matrix effects encountered in optical emission and absorption methods and generally can be managed by means of the various methods described in the two previous chapters.

Spectroscopic Interferences

Spectroscopic interferences occur when an ionic species in the plasma has the same m/z values as an analyte ion. Such interferences fall into four categories: (1) isobaric ions, (2) polyatomic or adduct ions, (3) doubly charged ions, and (4) refractory oxide ions.[14]

Isobaric Interferences. Isobaric species are two elements that have isotopes of essentially the same mass. For atomic mass spectrometry with a quadrupole mass spectrometer, isobaric species are isotopes that differ in mass by less than one unit. With higher-resolution instruments, smaller differences can be tolerated.

Most elements in the periodic table have one, two, or even three isotopes that are free from isobaric overlap. An exception is indium, which has two stable isotopes, $^{113}In^+$ and $^{115}In^+$. The former overlaps with $^{113}Cd^+$ and the latter with $^{115}Sn^+$. More often, an isobaric interference occurs with the most abundant and thus the most sensitive isotope. For example, the very large peak for $^{40}Ar^+$ (see Figure 11-15b) overlaps the peak for the most abundant calcium isotope $^{40}Ca^+$ (97%), making it necessary to use the second-most abundant isotope $^{40}Ca^+$ (2.1%). As another example, the most abundant nickel isotope, $^{58}Ni^+$, suffers from an isobaric overlap by $^{58}Fe^+$. This interference can be corrected by measuring the peak for $^{56}Fe^+$. From the natural-abundance ratio of the $^{56}Fe^+$ isotope to that of $^{58}Fe^+$ isotope, the contribution of Fe to the peak at m/z 58 can be determined and the correction made. Because isobaric overlaps are exactly predictable from abundance tables, corrections for the problem can be carried out with appropriate software. Some current instruments are capable of automatically making such corrections.

[14] For comprehensive discussions of interferences in ICPMS, see H. E. Taylor, *Inductively Coupled Plasma-Mass Spectrometry: Practices and Techniques*, San Diego, CA: Elsevier, 2001; G. Horlick and A. Montaser in *Inductively Coupled Plasma Mass Spectrometry*, A. Montaser, ed., Chap. 7, Hoboken, NJ: Wiley-VCH, 1998; K. E. Jarvis, A. L. Gray, and R. S. Houk, *Handbook of Inductively Coupled Plasma Mass Spectrometry*, Chap. 5, New York: Blackie, 1992.

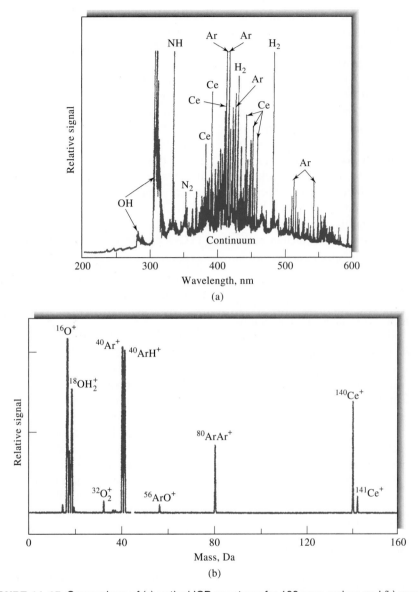

FIGURE 11-15 Comparison of (a) optical ICP spectrum for 100 ppm cerium and (b) mass ICP spectrum for 10 ppm cerium.

Polyatomic Ion Interferences. An interference problem that is usually more serious than that caused by isobaric elements occurs when polyatomic species form from interactions between species in the plasma and species in the matrix or the atmosphere. These species may then produce several molecular ions that interfere. This type of interference is found usually at m/z values of below about 82. The presence of several of these species is shown in Figure 11-15b. Among the potential interferents are $^{40}Ar^+$, $^{40}ArH^+$, $^{16}O_2^+$, $^{16}OH_2^+$, $^{16}OH^+$,

$^{14}N^+$, and several others. Some of these can cause serious problems. Examples of serious polyatomic interferences include $^{14}N_2^+$ with $^{28}Si^+$, NOH^+ with $^{31}P^+$, $^{16}O_2^+$ with $^{32}S^+$, $^{40}ArO^+$ with $^{56}Fe^+$, and $^{40}Ar_2^+$ with $^{80}Se^+$. A blank can correct for some of these. With others, a different analyte isotope may have to be used.

Oxide and Hydroxide Species' Interference. The most serious general class of interferences in ICPMS involves oxides and hydroxides formed from the analyte itself, the matrix components, the solvent, and the plasma gases. Some of these oxides are apparent in Figure 11-15b. The most serious interferences of this type arise from the oxides and hydroxides of the analytes and of the matrix components. Almost all of these species tend to form MO^+ and MOH^+ ions, where M represents the analyte or matrix element. The peaks from these species have the potential to overlap a peak of one of the analyte ions. For example, the singly charged oxide ions of the five naturally occurring titanium isotopes have masses of 62, 63, 64, 65, and 66. The peaks for these ions have the potential to interfere with the analyte peaks for $^{62}Ni^+$, $^{63}Cu^+$, $^{64}Zn^+$, $^{65}Cu^+$, and $^{66}Zn^+$. As a further example, the potential problems created by the oxide and hydroxide isotopes of calcium in the determination of several metallic species are shown in Table 11-2.

Much research has been focused on reducing oxide and hydroxide formation in plasmas. Oxide formation depends on such experimental variables as injector flow rate, radio-frequency power, sampler skimmer spacing, sample orifice size, plasma gas composition, oxygen elimination, and solvent removal efficiencies. All of these variables can be adjusted to address specific oxide and hydroxide overlap problems.

Matrix Effects

In ICPMS, matrix effects become noticeable at concomitant concentrations of greater than about 500 to 1000 mg/mL. Usually these effects cause a reduction

in the analyte signal, although under certain experimental conditions signal enhancement is observed. The matrix effect is quite general in that a high concentration of nearly any concomitant element results in a matrix effect. Generally, matrix effects can be minimized by using more dilute solutions, by altering the sample introduction procedure, or by separating out the offending species. The effects can also be largely eliminated by the use of an appropriate internal standard, that is, by introduction of an internal standard element that has about the same mass and ionization potential as those of the analyte.

11C-3 Applications of ICPMS

ICPMS can be used for qualitative, semiquantitative, and quantitative determination of one or more elements in samples of matter.

Qualitative and Semiquantitative Applications

Because ICPMS is easily adapted to multielement analyses, it is well suited to the rapid characterization and semiquantitative analysis of various types of naturally occurring and manufactured complex materials. Generally, detection limits are better than those for optical emission ICP and compete with detection limits for electrothermal atomic absorption spectroscopy.

Usually, atomic mass spectra are considerably simpler and easier to interpret than optical emission spectra. This property is particularly important for samples that contain rare earth elements and other heavy metals, such as iron, that yield complex emission spectra. Figure 11-15b illustrates this advantage. This spectral simplicity is further illustrated in Figure 11-16, which is the atomic mass spectrum for a mixture of fourteen rare earth elements that range in atomic mass number from 139 to 175. The optical emission spectrum for such a mixture would be so complex that interpretation would be tedious, time-consuming, and perhaps impossible.

Semiquantitative analysis for one or more of the components in a mixture such as the one that yielded Figure 11-16 is possible by measuring the peak ion current or intensity for a solution having a known concentration of the element in question. The analyte concentration in the sample is then computed by assuming that the ion current is proportional to concentration. Concentrations calculated by this crude but simple procedure are generally accurate to $\pm 100\%$ relative.

TABLE 11-2 Calcium Oxide and Hydroxide Species and Other Potential Interferences in the Mass Region for Ni Determination

m/z	Element[a]	Interferences
56	Fe (91.66)	^{40}ArO, ^{40}CaO
57	Fe (2.19)	$^{40}ArOH$, $^{40}CaOH$
58	Ni (67.77), Fe (0.33)	^{42}CaO, $NaCl$
59	Co (100)	^{43}CaO, $^{42}CaOH$
60	Ni (26.16)	$^{43}CaOH$, ^{44}CaO
61	Ni (1.25)	$^{44}CaOH$
62	Ni (3.66)	^{46}CaO, Na_2O, NaK
63	Cu (69.1)	$^{46}CaOH$, $^{40}ArNa$
64	Ni (1.16), Zn (48.89)	$^{32}SO_2$, $^{32}S_2$, ^{48}CaO
65	Cu (30.9)	^{33}S, ^{32}S, $^{33}SO_2$, $^{48}CaOH$

From M. A. Vaughan and D. M. Templeton, *Appl. Spectrosc.*, **1990**, *44*, 1685. With permission.

[a] Percentage natural abundance in parentheses.

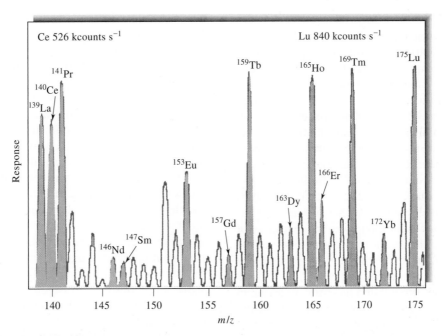

FIGURE 11-16 ICPMS spectrum for the rare earth elements. Solutions contain 1 mg/mL of each element. (From K. E. Jarvis, *J. Anal. Atom. Spectrom.*, **1989**, *4*, 563. With permission.)

Detection Limits

One of the main attractions of ICPMS lies with the lower detection limits attainable with mass spectrometric detection than with optical detection. These limits in many cases equal and sometimes exceed those that can be realized by electrothermal atomic absorption methods. The ICPMS procedure, of course, offers the great advantages of speed and multielement capability.

Figure 11-17 compares detection limits for ICPMS with those for ICP optical emission spectroscopy (ICP-OES) and those for electrothermal atomic absorption spectroscopy (ETAAS) for selected elements. These data are typical for most other elements in the periodic table. Generally, detection limits with mass spectrometric detection range from 0.02 to 0.7 ppb with the majority of elements in the range of 0.02 to 0.1 ppb.

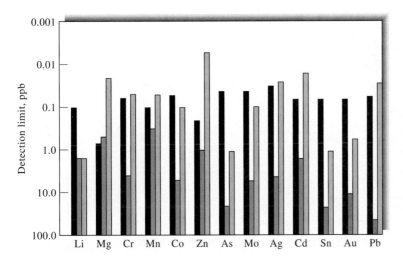

FIGURE 11-17 Detection limits for selected elements by ICPMS (black bars) compared with those for ICP-OES (blue bars) and ETAAS (gray bars), plotted on a logarithmic scale in concentrations of ppb (or mg/L). Because ETAAS detection limits are inherently in mass units (pg), they have been converted to concentration by assuming a 20-μL sample. (From M. Selby and G. M. Hieftje, *Amer. Lab.*, **1987** (8), 20. With permission.)

Quantitative Analyses

The most widely used quantitative method of ICPMS uses a set of calibration standards for preparing a calibration curve. Simple aqueous standards are usually adequate if the unknown solutions are sufficiently dilute — less than 2000 μg/mL of total dissolved solids. With higher concentrations of matrix elements, attempts are often made to match the matrix elements in the samples with those in the standards. To compensate for instrument drift, instabilities, and matrix effects, an internal standard is usually introduced into the standards and the unknowns. The internal standard is an element that is absent from the samples and that has an atomic mass and ionization potential near those of the analytes. Two elements frequently used for internal standards are indium and rhodium. Both produce ions in the central part of the mass range (^{115}In, ^{113}In, and ^{103}Rh) and are not often found naturally occurring in samples. Generally, log-log plots of ion current, ion count, or intensity ratios for sample and internal standards are linear over several orders of magnitude of

concentration. Figure 11-18 shows calibration curves for the determination of several rare earth elements. Note that a linear relationship between peak count and concentration is observed over four orders of magnitude in concentration.

Table 11-3 illustrates typical results obtained for the quantitative analysis of trace elements in a natural water sample from the National Bureau of Standards (NBS), now the National Institute of Standards and Technology (NIST). Note that the concentrations are in parts per billion. The thirteen analyses were completed in an elapsed time of only 15 min. For most of the elements the results show remarkably good agreement with the standard sample data.

For the most accurate quantitative analyses, an isotope dilution method may be used. The basis of this technique is the measurement of the change in ratio of signal intensities for two isotopes of an element that result from the addition of a known quantity, or a spike, of a standard solution that is enriched in one of the isotopes. The principles and mathematical relationships

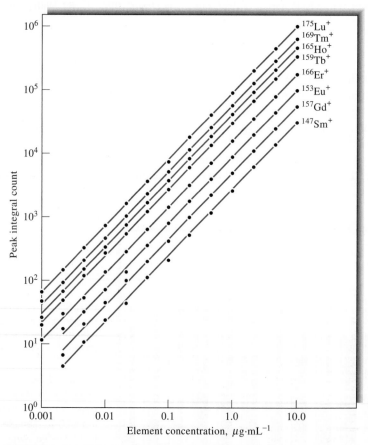

FIGURE 11-18 ICPMS calibration curves for several rare earth elements.

TABLE 11-3 Quantitative Determination of Trace Elements in a Standard Sample of Water

Element	Ion	NBS[a]	ICPMS[a]	
			Mean[b]	RSD (%)[b]
Beryllium	$^9Be^+$	19	21	20
Vanadium	$^{51}V^+$	54	52	6
Chromium	$^{52}Cr^+$	17	18	12
Manganese	$^{55}Mn^+$	32	34	5
Cobalt	$^{59}Co^+$	19	21	7
Zinc	$^{66}Zn^+$	69	57	11
Arsenic	$^{75}As^+$	77	76	5
Strontium	$^{88}Sr^+$	243	297	7
Molybdenum	$^{98}Mo^+$	97	134	9
Silver	$^{107}Ag^+$	2.8	3.5	16
Cadmium	$^{114}Cd^+$	10	13	22
Barium	$^{138}Ba^+$	47	74	17
Lead	$^{208}Pb^+$	27	31	8

[a] Concentration in parts per billion.

[b] Based on ten determinations.

that relate analyte concentration to the quantity of spike added and the change in signal ratio are described in detail in Section 32D, which deals with the use of this technique with radioactive isotopes. The advantages of the isotope dilution method are improved accuracy and immunity to a wide variety of chemical and physical interferences. Its main limitation is the time required to complete this type of analysis.

Isotope Ratio Measurements

The measurement of isotope ratios is of considerable importance in several fields of science and medicine. For example, archeologists and geologists use such data to establish the age of artifacts and various types of deposits. Chemists and clinicians use isotopically enriched materials as tracers in various types of studies. The outcome of these studies is based on isotope ratio measurements. Historically, isotope ratio measurements were based on thermal atomization and ionization in which the sample was decomposed, atomized, and ionized on one or more electrically heated filaments. The ions formed in this way were then directed into a double-focusing mass spectrometer, where isotopic ratios were measured. These measurements were time-consuming but quite precise, with relative standard deviations on the order of 0.01%. Now,

isotope ratios are often determined by ICPMS, with which measurements can often be completed in 5 or 10 minutes. The precision in this case, however, is generally poorer by a factor of 10 to 100. For many purposes, reproducibility of this order is perfectly satisfactory.

11D SPARK SOURCE MASS SPECTROMETRY

Spark source mass spectrometry (SSMS)[15] was first introduced in the 1930s as a general tool for multielement and isotope trace analyses. It was not until 1958, however, that the first commercial spark source atomic mass spectrometer appeared on the market. After a period of rapid development in the 1960s, the use of this technique leveled off and then declined with the appearance of ICPMS and some of the other mass spectrometric methods discussed in this chapter. Currently, SSMS is still applied to samples that are not easily dissolved and analyzed by ICP.

In SSMS, the atomic constituents of a sample are converted by a high-voltage (~30 kV), radio-frequency spark to gaseous ions for mass analysis. The spark is housed in a vacuum chamber located immediately adjacent to the mass analyzer. The chamber is equipped with a separate high-speed pumping system that quickly reduces the internal pressure to about 10^{-8} torr after sample changes. Often, the sample serves as one or both electrodes. Alternatively, it is mixed with graphite and loaded into a cup-shape electrode. The gaseous positive ions formed in the spark plasma are drawn into the analyzer by a dc voltage.

A spark source produces ions with a wide range of kinetic energies. Consequently, expensive double-focusing mass spectrometers are required for mass analysis of the ions. The Mattauch-Herzog type shown in Figure 11-11 is generally used. Modern spark source instruments are designed for both photographic and electrical detection systems. The latter are generally based on electron multipliers, which are described in Section 11B-1. When electron multipliers are used with double-focusing instruments, the spectrum is scanned by varying the magnetic field of the magnetic analyzer. Generally, the resolution of the double-focusing mass spectrometers used with spark sources is

[15] See G. Ramendik, J. Verlinden, and R. Gijbels, in *Inorganic Mass Spectrometry*, F. Adams, R. Gijbels, and R. Van Grieken, eds., Chap. 2, New York: Wiley, 1988.

orders of magnitude greater than that of the quadrupole instruments used with ICP. Thus, the number of isobaric interferences is significantly smaller. For example, in the determination of iron in a silicate matrix, as found in samples of rock or glass, there is potential for interference by silicon. Here, the determination is based on the peak for $^{56}Fe^+$ ($m/z = 55.93494$), but a detectable peak for $^{28}Si_2^+$ ($m/z = 55.95386$) is also observed. The resolving power of most double-focusing mass spectrometers is sufficient to separate these peaks, but with a quadrupole instrument, an isobaric interference would result from the silicon dimer.

11D-1 Spectra

Like ICP mass spectra, spark source mass spectra are much simpler than atomic emission spectra, consisting of one major peak for each isotope of an element as well as a few weaker lines corresponding to multiply charged ions and ionized oxide and hydroxide species. The presence of these additional ions creates the potential for interference just as in ICPMS.

11D-2 Qualitative Applications

SSMS is a powerful tool for qualitative and semiquantitative analysis. All elements in the periodic table from ^7Li through ^{238}U can be identified in a single excitation. By varying data-acquisition parameters, it is possible to determine order of magnitude concentrations for major constituents of a sample as well as for constituents in the parts-per-billion concentration range. Interpretation of spectra does require skill and experience, however, because of the presence of multiply charged species, polymeric species, and molecular ions.

11D-3 Quantitative Applications

A radio-frequency spark is not a very reproducible source over short periods. As a consequence, it is necessary to integrate the output signals from a spark for periods that range from several seconds to hundreds of seconds if good quantitative data are to be obtained. The detection system must be capable of electronic signal integration, either in hardware or software. In addition to integration, it is common practice to improve reproducibility by using the ratio of the analyte signal to that of an internal standard as the analytical variable. Often, one of the major elements of the sample matrix is chosen as the standard; alternatively, a fixed amount of a pure compound is added to each sample and each standard used for calibration. In the latter case, the internal standard substance must be absent from the samples. With these precautions, relative standard deviations that range from a few percent to as much as 20% can be realized.

The advantages of SSMS include its high sensitivity, its applicability to a wide range of sample matrices, the large linear dynamic range of the output signal (often several orders of magnitude), and the wide range of elements that can be detected and measured quantitatively.

11E GLOW DISCHARGE MASS SPECTROMETRY

As shown in Section 8C-2, a glow-discharge source is used as an atomization device for various types of atomic spectroscopy.[16] In addition to atomizing samples, it also produces a cloud of positive analyte ions from solid samples. This device consists of a simple two-electrode closed system containing argon at a pressure of 0.1 to 10 torr. A voltage of 5 to 15 kV from a pulsed dc power supply is applied between the electrodes, causing the formation of positive argon ions, which are then accelerated toward the cathode. The cathode is fabricated from the sample, or alternatively, the sample is deposited on an inert metal cathode. Just as in the hollow-cathode lamp (Section 9B-1), atoms of the sample are sputtered from the cathode into the region between the two electrodes, where they are converted to positive ions by collision with electrons or positive argon ions. Analyte ions are then drawn into the mass spectrometer by differential pumping. The ions are then filtered in a quadrupole analyzer or dispersed with a magnetic sector analyzer for detection and determination.

As mentioned earlier, glow-discharge sources are often used with ICP torches. The glow discharge serves as the atomizer and the ICP torch as the ionizer.

The glow-discharge source appears to be more stable than the spark source and is less expensive to purchase and operate. It is particularly useful for the direct analysis of solid samples. Commercial quadrupole and double-focusing mass spectrometers with glow-discharge sources are now on the market.

---

[16] See R. K. Marcus and J. A. C. Broekaert, *Glow Discharge Plasmas in Analytical Spectroscopy*, New York: Wiley, 2003; W. W. Harrison, in *Inorganic Mass Spectrometry*, Chap. 3, F. Adams, R. Gijbels, and R. Van Grieken, eds., New York: Wiley, 1988.

11F OTHER MASS SPECTROMETRIC METHODS

Over the past quarter century, the science and technology of mass spectrometry has grown explosively. In the last two sections of this chapter we refer briefly to two rather exotic but very important developments in this burgeoning field.

11F-1 Accelerator Mass Spectrometry

In recent years, a variation on the isotope-ratio method called *accelerator mass spectrometry* (AMS) has been developed. In AMS, the target element is first chemically separated from the sample before it is placed in a sample holder in the AMS instrument. The sample element is then bombarded by cesium ions to sputter the analyte element from the sample as negative ions. The analyte ions are then accelerated down a beam tube by a positive potential difference of several million volts, passed through an electron stripper to convert them to positive ions, and accelerated back down the beam tube toward common potential where ion velocities approach a few percent of the speed of light. Using a series of magnetic and electrostatic mass filters, the ion beam containing all isotopes of the analyte element is then separated into separate beams containing the (usually unstable) isotope of interest and other isotopes, and each of the isotopes is counted by a separate detector. As an example, in carbon dating, a single ^{14}C nucleus can be separated and counted in the presence of as many as 10^{15} nuclei of ^{13}C and ^{12}C.

Although there are a relatively small number of AMS facilities in the world, the technique is invaluable for determining the concentration of an unstable isotope in the presence of a massive amount of a second stable isotope, particularly in radiocarbon dating and the determination of nuclides produced from the interaction of cosmic rays with both extraterrestrial and terrestrial sources. These nuclides include ^{10}Be, ^{14}C, ^{26}Al, ^{36}Cl, ^{41}Ca, and ^{129}I.[17]

11F-2 Methods for Elemental Surface Analysis

In certain areas of science and engineering, the elemental composition of a surface layer of solid that is a few angstroms to a few tens of angstroms thick is of much greater interest than the bulk elemental composition. Fields where surface composition is extremely important include heterogeneous catalysis, corrosion and adhesion studies, embrittlement properties, and studies of the behavior and functions of biological membranes.

Two atomic mass spectrometric methods are often used to determine the elemental composition of solid surfaces: secondary-ion mass spectrometry and laser microprobe mass spectrometry. These techniques are discussed in detail in Sections 21D-1 and 21D-3.

[17]For several interesting descriptions of AMS facilities, do a Google search on "accelerator mass spectrometry."

QUESTIONS AND PROBLEMS

*Answers are provided at the end of the book for problems marked with an asterisk.

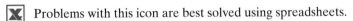 Problems with this icon are best solved using spreadsheets.

11-1 What types of mass spectrometers are used in ICPMS? How do they differ from one another?

11-2 What function does the ICP torch have in ICPMS?

11-3 What are the ordinate and the abscissa of an ordinary atomic mass spectrum?

11-4 Why has ICPMS become an important and widely used analytical method?

11-5 Describe the interface between the ICP torch and the mass spectrometer in an ICPMS instrument.

11-6 How are lasers used as a means of sampling solids for ICPMS?

11-7 What types of interferences are encountered in atomic mass spectrometry?

11-8 Why is an internal standard often used in quantitative analysis by ICPMS?

11-9 Describe how the isotope dilution technique is used in atomic mass spectrometry.

11-10 Describe how glow-discharge mass spectrometry is carried out. What are the advantages of this technique?

11-11 What is secondary-ion mass spectrometry? What type of information does this method provide?

Challenge Problem

11-12 Lezius has described a B-TOF mass spectrometer, presented the theoretical background for the instrument, and provided early experimental results.[18]

(a) What is a B-TOF mass spectrometer?

(b) Explain how the instrument works.

(c) For what types of experiments is the instrument useful?

(d) Define the term *start-up energy*.

(e) The *deflection velocity* $\dot{x}$ of an ion in the flight tube is given by

$$\dot{x} = \frac{qdB}{m}$$

where $q = ze$, and B and m have their usual significance, and d, shown in Figure 1 of the paper, is the width of the magnetic field region. The deflection velocity is the component of the velocity perpendicular to the initial flight path of the ion. Derive this expression. What is curious about the deflection velocity?

(f) Show that the deflection x of the ion perpendicular to the initial flight path of the ion is given by

$$x = (l + d/2)\frac{qdB}{mv \cos \alpha}$$

where l is the distance traveled by an ion in its initial direction along the axis of the spectrometer after it leaves the magnetic field region, and v is the velocity of the ion along the deflected path. Why are the distances x and l important in the B-TOF experiment?

(g) The time of flight t_{MD} of an ion to traverse the distance l from its exit from the magnetic field is given by

$$t_{MD} = \frac{1}{\sqrt{\dfrac{2E_{tot}}{m} - \dot{x}^2}}$$

What is E_{tot}? There is an error in equation 6 of the paper. Find and explain the error.

(h) Using the conservation of energy, derive the expression in (g).

(i) For small deflection angles, $\cos \alpha \approx 1$. What are the consequences of this condition for ions of the same charge and mass?

(j) Look up some typical values for E_{tot}, l, and v, and use Excel to calculate typical values for t_{MD} for a realistic range of values for m. Produce a plot of t_{MD} versus m.

[18] M. Lezius, *J. Mass Spectrom.*, **2002**, *37*, 305.

CHAPTER TWELVE

Atomic X-ray Spectrometry

X-ray spectroscopy, like optical spectroscopy, is based on measurement of emission, absorption, scattering, fluorescence,
and diffraction of electromagnetic radiation. X-ray
fluorescence and X-ray absorption methods are
widely used for the qualitative and quantitative
determination of all elements in the periodic table
having atomic numbers greater than that of sodium.
With special equipment, elements with atomic numbers in the range of 5 to 10 can also be determined.

Throughout this chapter, this logo indicates
an opportunity for online self-study at **www**
.thomsonedu.com/chemistry/skoog, linking you to
interactive tutorials, simulations, and exercises.

12A FUNDAMENTAL PRINCIPLES

X-rays are short-wavelength electromagnetic radiation
produced by the deceleration of high-energy electrons
or by electronic transitions of electrons in the inner
orbitals of atoms.[1] The wavelength range of X-rays is
from about 10^{-5} Å to 100 Å; conventional X-ray spectroscopy is, however, largely confined to the region of
about 0.1 Å to 25 Å (1 Å = 0.1 nm = 10^{-10} m).

12A-1 Emission of X-rays

For analytical purposes, X-rays are generated in four
ways: (1) by bombardment of a metal target with a beam
of high-energy electrons, (2) by exposure of a substance
to a primary beam of X-rays to generate a secondary
beam of X-ray fluorescence, (3) by use of a radioactive
source whose decay process results in X-ray emission,[2]
and (4) from a synchrotron radiation source. Only a
few laboratories in the United States have facilities to
produce X-rays from synchrotron radiation.[3] For this
reason, we will consider only the first three sources.

X-ray sources, like ultraviolet and visible emitters,
often produce both continuum and line spectra; both
types are of importance in analysis. Continuum radiation is also called *white radiation* or *bremsstrahlung*.
Bremsstrahlung means radiation that arises from retardation of particles; such radiation is generally a spectral
continuum.

Continuum Spectra from Electron Beam Sources

In an X-ray tube, electrons produced at a heated
cathode are accelerated toward a metal anode (the *target*) by a potential difference as great as 100 kV; when

---

[1] For a more extensive discussion of the theory and analytical applications
of X-rays, see B. E. Beckhoff, B. E. Kanngießer, N. E. Langhoff, R. E.
Wedell, eds., *Handbook of Practical X-Ray Fluorescence Analysis*, New
York: Springer, 2005; R. E. Van Grieken and A. A. Markowicz, eds., *Handbook of X-ray Spectrometry*, 2nd ed., New York: Marcel Dekker, 2002;
R. Jenkins, *X-Ray Fluorescence Spectrometry*, 2nd ed., New York: Wiley,
1999; R. Jenkins, R. W. Gould, and D. Gedcke, *Quantitative X-Ray Spectrometry*, 2nd ed., New York: Marcel Dekker, 1995; B. Dziunikowski, *Energy-
Dispersive X-Ray Fluorescence Analysis*, New York: Elsevier, 1989.

[2] Electromagnetic radiation produced by radioactive sources is often
called *gamma radiation*. Gamma radiation is indistinguishable from X-ray
radiation.

[3] The high-intensity and collimated nature of beams from a synchrotron
source leads to applications that cannot be accomplished by the other
three X-ray sources. For a review of the applications of synchrotron-
induced X-rays, see K. W. Jones and B. M. Gordon, *Anal. Chem.*, **1989**, *61*,
341A; for a discussion of the use of total-reflectance X-ray fluorescence
induced by synchrotron radiation, see K. Baur, S. Brennan, P. Pianetta,
and R. Opila, *Anal. Chem*, **2002**, *74*, 608A.

electrons collide with the anode, part of the energy of the beam is converted to X-rays. Under some conditions, only a continuum spectrum such as that shown in Figure 12-1 results; under other conditions, a line spectrum is superimposed on the continuum (see Figure 12-2).

The continuum X-ray spectrum shown in the two figures is characterized by a well-defined, short-wavelength limit (λ_0), which depends on the accelerating voltage V but is independent of the target material. Thus, λ_0 (0.35 Å) for the spectrum produced with a molybdenum target at 35 kV (Figure 12-2) is identical to λ_0 for a tungsten target at the same voltage (Figure 12-1).

The continuum radiation from an electron beam source results from collisions between the electrons of the beam and the atoms of the target material. At each collision, the electron is decelerated and a photon of X-ray energy is produced. The energy of the photon is equal to the difference in kinetic energies of the electron before and after the collision. Generally, the electrons in a beam are decelerated in a series of collisions; the resulting loss of kinetic energy differs from collision to collision. Thus, the energies of the emitted X-ray photons vary continuously over a considerable range. The maximum photon energy generated corresponds to the instantaneous deceleration of the electron to zero kinetic energy in a single collision. For such an event, we can write

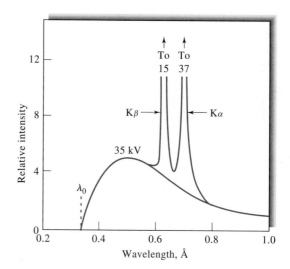

FIGURE 12-2 Line spectrum for an X-ray with a molybdenum target.

$$hv_0 = \frac{hc}{\lambda_0} = Ve \tag{12-1}$$

where Ve, the product of the accelerating voltage and the charge on the electron, is the kinetic energy of all of the electrons in the beam; h is Planck's constant; and c is the velocity of light. The quantity v_0 is the maximum frequency of radiation that can be produced at voltage V, and λ_0 is the low wavelength limit for the radiation. This relationship is known as the *Duane-Hunt law*. It is interesting to note that Equation 12-1 provides a direct means for the highly accurate determination of Planck's constant. When we substitute numerical values for the constants and rearrange, Equation 12-1 becomes

$$\lambda_0 = 12{,}398/V \tag{12-2}$$

where λ_0 and V have units of angstroms and volts, respectively.

Line Spectra from Electron Beam Sources

As shown in Figure 12-2, bombardment of a molybdenum target produces intense emission lines at about 0.63 and 0.71 Å; an additional simple series of lines occurs in the longer-wavelength range of 4 to 6 Å.

The emission behavior of molybdenum is typical of all elements having atomic numbers larger than 23; that is, the X-ray line spectra are remarkably simple when compared with ultraviolet emission and consist of two series of lines. The shorter-wavelength group is called

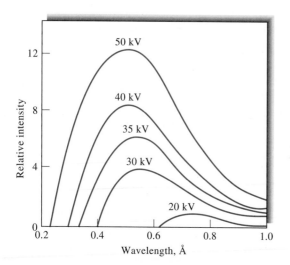

FIGURE 12-1 Distribution of continuum radiation from an X-ray tube with a tungsten target. The numbers above the curves indicate the accelerating voltages.

TABLE 12-1 Wavelengths in Angstroms of the More Intense Emission Lines for Some Typical Elements

Element	Atomic Number	K Series		L Series	
		α_1	β_1	α_1	β_1
Na	11	11.909	11.617	—	—
K	19	3.742	3.454	—	—
Cr	24	2.290	2.085	21.714	21.323
Rb	37	0.926	0.829	7.318	7.075
Cs	55	0.401	0.355	2.892	2.683
W	74	0.209	0.184	1.476	1.282
U	92	0.126	0.111	0.911	0.720

the K series and the other the L series.[4] Elements with atomic numbers smaller than 23 produce only a K series. Table 12-1 presents wavelength data for the emission spectra of a few elements.

A second characteristic of X-ray spectra is that the minimum acceleration voltage required for the excitation of the lines for each element increases with atomic number. Thus, the line spectrum for molybdenum (atomic number = 42) disappears if the excitation voltage drops below 20 kV. As shown in Figure 12-1, bombardment of tungsten (atomic number = 74) produces no lines in the region of 0.1 to 1.0 Å, even at 50 kV. Characteristic K lines appear at 0.18 and 0.21 Å, however, if the voltage increases to 70 kV.

Figure 12-3 illustrates the linear relationship between the square root of the frequency for a given (K or L) line and the atomic number of the element responsible for the radiation. This property was first discovered by English physicist H. G. J. Moseley in 1914.[5]

X-ray line spectra result from electronic transitions that involve the innermost atomic orbitals. The short-wavelength K series is produced when the high-energy electrons from the cathode remove electrons from those orbitals nearest the nucleus of the target atom. The collision results in the formation of excited *ions*, which then emit quanta of X-radiation as electrons from outer orbitals undergo transitions to the vacant orbital. As shown in Figure 12-4, the lines in the K series arise from electronic transitions between higher energy levels and the K shell. The L series of lines results when an electron is lost from the second principal quantum level, either as a result of ejection by an electron from the cathode or from the transition of an L electron to the K level that accompanies the production of a quantum of K radiation. It is important to appreciate that the energy scale in Figure 12-4 is logarithmic. Thus, the energy difference between the L and K levels is significantly larger than that between the M and L levels. The K lines therefore appear at shorter wavelengths. It is also important to note that the energy differences between the transitions labeled α_1 and α_2 as well as those between β_1 and β_2 are so small that only single lines are observed in all but the highest-resolution spectrometers (see Figure 12-2).

The energy level diagram in Figure 12-4 is applicable to any element with sufficient electrons to permit the number of transitions shown. The differences in energies between the levels increase regularly with

[4] For the heavier elements, additional series of lines (M, N, and so forth) are found at longer wavelengths. Their intensities are low, however, and little use is made of them. The designations K and L arise from the German words *kurtz* and *lang* for short and long wavelengths. The additional alphabetical designations were then added for lines occurring at progressively longer wavelengths.

[5] Henry Gwyn Jeffreys Moseley (1887–1915) discovered this important relationship, which is now called Moseley's law, while working with Ernest Rutherford at the University of Manchester. His studies revealed gaps in the sequence of atomic numbers at 43, 61, 72, 75, and 87, which are now filled by the elements technetium, promethium, hafnium, rhenium, and francium—all subsequently discovered or produced artificially. When World War I broke out, Moseley enlisted in the Royal Engineers and was killed by a sniper while serving in Gallipoli. It is thought that his tragic death was the driving force to forbid British scientists from serving in combat. If he had lived, Moseley would likely have received the Nobel Prize for his groundbreaking research. For more details on Moseley's life and work, see R. Porter and M. Ogilvie, eds., *The Biographical Dictionary of Scientists*, 3rd ed., New York: Oxford, 2000.

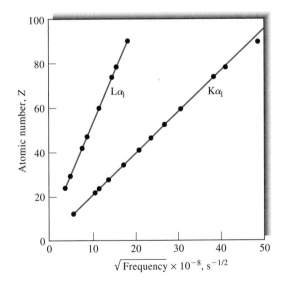

FIGURE 12-3 Relationship between X-ray emission frequency and atomic number for $K\alpha_1$ and $L\alpha_1$ lines.

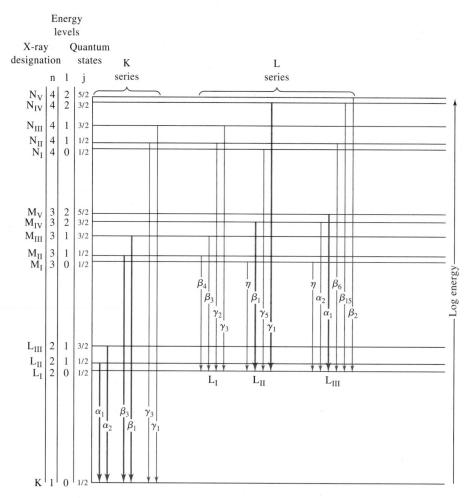

FIGURE 12-4 Partial energy level diagram showing common transitions producing X-rays. The most intense lines are indicated by the wider arrows.

atomic number because of the increasing charge on the nucleus; therefore, the radiation for the K series appears at shorter wavelengths for the heavier elements (see Table 12-1). The effect of nuclear charge is also reflected in the increase in minimum voltage required to excite the spectra of these elements.

It is important to note that for all but the lightest elements, the wavelengths of characteristic X-ray lines are independent of the physical and chemical state of the element because the transitions responsible for these lines involve electrons that take no part in bonding. Thus, the position of the $K\alpha$ lines for molybdenum is the same regardless of whether the target is the pure metal, its sulfide, or its oxide.

Line Spectra from Fluorescent Sources

Another convenient way of producing a line spectrum is to irradiate the element or one of its compounds with the continuum radiation from an X-ray tube. This process is explored further in a later section.

Spectra from Radioactive Sources

X-radiation is often a product of radioactive decay processes. *Gamma rays*, which are indistinguishable from X-rays, are produced in intranuclear reactions. Many α and β emission processes (see Section 32A-2) leave a nucleus in an excited state; the nucleus then releases one or more quanta of gamma rays as it returns to its ground state. *Electron capture* or *K capture*

TABLE 12-2 Common Radioisotopic Sources for X-ray Spectroscopy

Source	Decay Process	Half-Life	Type of Radiation	Energy, keV
$^3_1$H-Ti[a]	β^-	12.3 years	Continuum	3–10
			Ti-K X-rays	4–5
$^{55}_{26}$Fe	EC[b]	2.7 years	Mn-K X-rays	5.9
$^{57}_{27}$Co	EC	270 days	Fe-K X-rays	6.4
			γ rays	14, 122, 136
$^{109}_{48}$Cd	EC	1.3 years	Ag-K X-rays	22
			γ rays	88
$^{125}_{53}$I	EC	60 days	Te-K X-rays	27
			γ rays	35
$^{147}_{61}$Pm-Al	β^-	2.6 years	Continuum	12–45
$^{210}_{82}$Pb	β^-	22 years	Bi-L X-rays	11
			γ rays	47

[a] Tritium adsorbed on nonradioactive titanium metal.

[b] EC = Electron capture.

also produces X-radiation. This process involves capture of a K electron (less commonly, an L or an M electron) by the nucleus and formation of an element of the next lower atomic number. As a result of K capture, electronic transitions to the vacated orbital occur, and we observe the X-ray line spectrum of the newly formed element. The half-lives of K-capture processes range from a few minutes to several thousand years.

Artificially produced radioactive isotopes provide a very simple source of monoenergetic radiation for certain analytical applications. The best-known example is iron-55, which undergoes a K-capture reaction with a half-life of 2.6 years:

$$^{55}_{26}\text{Fe} \rightarrow {}^{55}_{25}\text{Mn} + h\nu$$

The resulting manganese $K\alpha$ line at about 2.1 Å has proved to be a useful source for both fluorescence and absorption methods. Table 12-2 lists some additional common radioisotopic sources for X-ray spectroscopy.

12A-2 Absorption Spectra

When a beam of X-rays is passed through a thin layer of matter, its intensity, or power, is generally diminished as a result of absorption and scattering. The effect of scattering for all but the lightest elements is ordinarily small and can be neglected in those wavelength regions where appreciable absorption occurs. As shown in Figure 12-5, the absorption spectrum of an element, like its emission spectrum, is simple and consists of a few well-

defined absorption peaks. Here again, the wavelengths of the absorption maxima are characteristic of the element and are largely independent of its chemical state.

A peculiarity of X-ray absorption spectra is the appearance of sharp discontinuities, called *absorption edges*, at wavelengths immediately beyond absorption maxima.

The Absorption Process

Absorption of an X-ray quantum causes ejection of one of the innermost electrons from an atom, which results in the production of an excited ion. In this process, the entire energy $h\nu$ of the radiation is partitioned between the kinetic energy of the electron (the *photoelectron*) and the potential energy of the excited ion. The highest probability for absorption occurs when the energy of the quantum is exactly equal to the energy required to remove the electron just to the periphery of the atom (that is, as the kinetic energy of the ejected electron approaches zero).

The absorption spectrum for lead, shown in Figure 12-5, exhibits four peaks, the first occurring at 0.14 Å. The energy of the quantum corresponding to this wavelength exactly matches the energy required to just eject the highest-energy K electron of the element. At wavelengths just larger than this wavelength, the energy of the radiation is insufficient to bring about removal of a K electron, and an abrupt decrease in absorption occurs. At wavelengths shorter than 0.14 Å, the probability of interaction between the electron

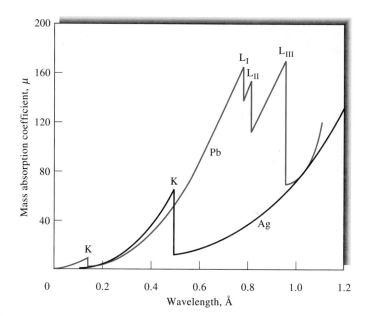

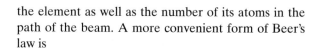

FIGURE 12-5 X-ray absorption spectra for lead and silver.

and the radiation gradually diminishes; this results in a smooth decrease in absorption. In this region, the kinetic energy of the ejected photoelectron increases continuously with the decrease in wavelength.

The additional peaks at longer wavelengths correspond to the removal of an electron from the L energy levels of lead. Three sets of L levels, differing slightly in energy, exist (see Figure 12-4); three maxima are, therefore, observed. Another set of lines, arising from ejections of M electrons, are located at still longer wavelengths.

Figure 12-5 also shows the K absorption edge for silver, which occurs at 0.485 Å. The longer wavelength for the silver peak reflects the lower atomic number of the element compared with lead.

The Mass Absorption Coefficient

Beer's law is as applicable to the absorption of X-radiation as to other types of electromagnetic radiation; thus, we may write

$$\ln\frac{P_0}{P} = \mu x$$

where x is the sample thickness in centimeters and P and P_0 are the powers of the transmitted and incident beams, respectively. The constant μ is called the *linear absorption coefficient* and is characteristic of

Tutorial: Learn more about **X-ray absorption and fluorescence**.

the element as well as the number of its atoms in the path of the beam. A more convenient form of Beer's law is

$$\ln\frac{P_0}{P} = \mu_M\rho x \qquad (12\text{-}3)$$

where ρ is the density of the sample and μ_M is the *mass absorption coefficient*, a quantity that is *independent* of the physical and chemical state of the element. Thus, the mass absorption coefficient for bromine has the same value in gaseous HBr as in solid sodium bromate. Note that the mass absorption coefficient carries units of cm^2/g.

Mass absorption coefficients are additive functions of the weight fractions of elements contained in a sample. Thus,

$$\mu_M = W_A\mu_A + W_B\mu_B + W_C\mu_C + \cdots \qquad (12\text{-}4)$$

where μ_M is the mass absorption coefficient of a sample containing the weight fractions W_A, W_B, and W_C of elements A, B, and C. The terms μ_A, μ_B, and μ_C are the respective mass absorption coefficients for each of the elements. Tables of mass absorption coefficients for the elements at various wavelengths are found in many handbooks, monographs, and research papers and on the web.[6]

[6]For example, J. H. Hubbell, *International Journal of Applied Radiation and Isotopes*, **1982**, *33*, 1269 and http://physics.nist.gov/PhysRefData/XrayMassCoef/cover.html.

12A-3 X-ray Fluorescence

The absorption of X-rays produces electronically excited ions that return to their ground state by transitions involving electrons from higher energy levels. Thus, an excited ion with a vacant K shell is produced when lead absorbs radiation of wavelengths shorter than 0.14 Å (Figure 12-5); after a brief period, the ion returns to its ground state via a series of electronic transitions characterized by the emission of X-radiation (fluorescence) of wavelengths identical to those that result from excitation produced by electron bombardment. The wavelengths of the fluorescence lines are always somewhat greater than the wavelength of the corresponding absorption edge, however, because absorption requires a complete removal of the electron (that is, ionization), whereas emission involves transitions of an electron from a higher energy level within the ion. For example, the K absorption edge for silver occurs at 0.485 Å, but the K emission lines for the element have wavelengths at 0.497 and 0.559 Å. When fluorescence is to be excited by radiation from an X-ray tube, the operating voltage must be sufficiently great so that the cutoff wavelength λ_0 (Equation 12-2) is shorter than the absorption edge of the element whose spectrum is to be excited. Thus, to generate the K lines for silver, the tube voltage would need to be (Equation 12-2)

$$V \geq \frac{12{,}398 \text{ V} \cdot \text{Å}}{0.485 \text{ Å}} = 25{,}560 \text{ V or } 25.6 \text{ kV}$$

12A-4 Diffraction of X-rays

Like other types of electromagnetic radiation, when X-radiation passes through a sample of matter, the electric vector of the radiation interacts with the electrons

 Simulation: Learn more about **X-ray diffraction**.

in the atoms of the matter to produce scattering. When X-rays are scattered by the ordered environment in a crystal, constructive and destructive interference occurs among the scattered rays because the distances between the scattering centers are of the same order of magnitude as the wavelength of the radiation. Diffraction is the result.

Bragg's Law

When an X-ray beam strikes a crystal surface at some angle θ, part of the beam is scattered by the layer of atoms at the surface. The unscattered part of the beam penetrates to the second layer of atoms where again a fraction is scattered, and the remainder passes on to the third layer (Figure 12-6), and so on. The cumulative effect of this scattering from the regularly spaced centers of the crystal is diffraction of the beam in much the same way as visible radiation is diffracted by a reflection grating (Section 7C-2). The requirements for X-ray diffraction are (1) the spacing between layers of atoms must be roughly the same as the wavelength of the radiation and (2) the scattering centers must be spatially distributed in a highly regular way.

In 1912, W. L. Bragg treated the diffraction of X-rays by crystals as shown in Figure 12-6. Here, a narrow beam of radiation strikes the crystal surface at angle θ; scattering occurs as a result of interaction of the radiation with atoms located at O, P, and R. If the distance

$$AP + PC = \mathbf{n}\lambda$$

where $\mathbf{n}$ is an integer, the scattered radiation will be in phase at OCD, and the crystal will appear to reflect the X-radiation. But

$$AP = PC = d \sin \theta \qquad (12\text{-}5)$$

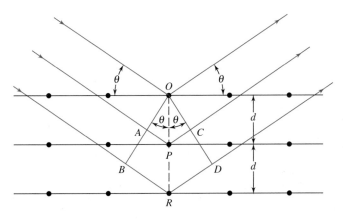

FIGURE 12-6 Diffraction of X-rays by a crystal.

where d is the interplanar distance of the crystal. Thus, the conditions for constructive interference of the beam at angle θ are

$$\mathbf{n}\lambda = 2d \sin \theta \qquad (12\text{-}6)$$

Equation 12-6 is the fundamentally important *Bragg equation*. Note that X-rays appear to be reflected from the crystal only if the angle of incidence satisfies the condition

$$\sin \theta = \frac{\mathbf{n}\lambda}{2d}$$

At all other angles, destructive interference occurs.

12B   INSTRUMENT COMPONENTS

Absorption, emission, fluorescence, and diffraction of X-rays are all applied in analytical chemistry. Instruments for these applications contain components that are analogous in function to the five components of instruments for optical spectroscopic measurement; these components include a source, a device for restricting the wavelength range of incident radiation, a sample holder, a radiation detector or transducer, and a signal processor and readout. These components differ considerably in detail from the corresponding optical components. Their functions, however, are the same, and the ways they combine to form instruments are often similar to those shown in Figure 7-1.

As with optical instruments, both X-ray photometers and spectrophotometers are encountered, the first using filters and the second using monochromators to transmit radiation of the desired wavelength from the source. In addition, however, a third method is available for obtaining information about isolated portions of an X-ray spectrum. Here, isolation is achieved electronically with devices that discriminate among various parts of a spectrum based on the *energy* rather than the *wavelength* of the radiation. Thus, X-ray instruments are often described as *wavelength-dispersive instruments* or *energy-dispersive instruments*, depending on the method by which they resolve spectra.

12B-1 Sources

Three types of sources are used in X-ray instruments: tubes, radioisotopes, and secondary fluorescence sources.

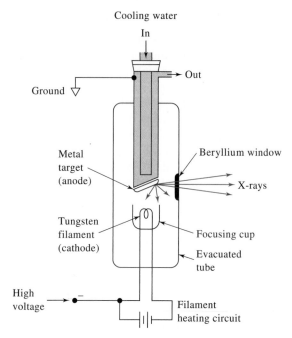

FIGURE 12-7 Schematic of an X-ray tube.

The X-ray Tube

The most common source of X-rays for analytical work is the X-ray tube (sometimes called a *Coolidge tube*), which can take a variety of shapes and forms; one design is shown schematically in Figure 12-7. An X-ray source is a highly evacuated tube in which is mounted a tungsten filament cathode and a massive anode. The anode generally consists of a heavy block of copper with a metal target plated on or embedded in the surface of the copper. Target materials include such metals as tungsten, chromium, copper, molybdenum, rhodium, scandium, silver, iron, and cobalt. Separate circuits are used to heat the filament and to accelerate the electrons to the target. The heater circuit provides the means for controlling the intensity of the emitted X-rays, whereas the accelerating voltage determines their energy, or wavelength. For quantitative work, both circuits must be operated with stabilized power supplies that control the current or the voltage to 0.1% relative.

The production of X-rays by electron bombardment is a highly inefficient process. Less than 1% of the electrical power is converted to radiant power, the remainder being dissipated as heat. As a result, until relatively recently, water cooling of the anodes of X-ray tubes was required. With modern equipment, however, cooling is often unnecessary because tubes can be operated at significantly lower power than formerly. This reduction in power is made possible by the greater sensitivity of modern X-ray transducers.

Radioisotopes

A variety of radioactive substances have been used as sources in X-ray fluorescence and absorption methods (see Table 12-2). Generally, the radioisotope is encapsulated to prevent contamination of the laboratory and shielded to absorb radiation in all but certain directions.

Many of the best radioactive sources provide simple line spectra; others produce a continuum (see Table 12-2). Because of the shape of X-ray absorption curves, a given radioisotope will be suitable for excitation of fluorescence or for absorption studies for a range of elements. For example, a source producing a line in the region between 0.30 and 0.47 Å is suitable for fluorescence or absorption studies involving the K absorption edge for silver (see Figure 12-5). Sensitivity improves as the wavelength of the source line approaches the absorption edge. Iodine-125 with a line at 0.46 Å is ideal for determining silver from this standpoint.

Secondary Fluorescent Sources

In some applications, the fluorescence spectrum of an element that has been excited by radiation from an X-ray tube serves as a source for absorption or fluorescence studies. This arrangement has the advantage of eliminating the continuum emitted by the primary source. For example, an X-ray tube with a tungsten target (Figure 12-1) could be used to excite the Kα and Kβ lines of molybdenum (Figure 12-2). The resulting fluorescence spectrum would then be similar to the spectrum in Figure 12-2 except that the continuum would be absent.

12B-2 Filters for X-rays

In many applications, it is desirable to use an X-ray beam with a narrow wavelength range. As in the visible region, both filters and monochromators are used for this purpose.

Figure 12-8 illustrates a common technique for producing a relatively monochromatic beam by use of a filter. Here, the Kβ line and most of the continuum emitted from a molybdenum target is removed by a zirconium filter having a thickness of about 0.01 cm. The pure Kα line is then available for analytical purposes. Several other target-filter combinations of this type have been developed, each of which serves to

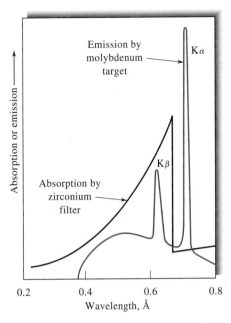

FIGURE 12-8 Use of a filter to produce monochromatic radiation.

isolate one of the intense lines of a target element. Monochromatic radiation produced in this way is widely used in X-ray diffraction studies. The choice of wavelengths available by this technique is limited by the relatively small number of target-filter combinations that are available.

Filtering the continuum from an X-ray tube is also feasible with thin strips of metal. As with glass filters for visible radiation, relatively broad bands are transmitted with a significant attenuation of the desired wavelengths.

12B-3 X-ray Monochromators

Figure 12-9 shows the essential components of an X-ray spectrometer. The monochromator consists of a pair of beam collimators, which serve the same purpose as the slits in an optical instrument, and a dispersing element. The latter is a single crystal mounted on a *goniometer*, or rotatable table, that permits variation and precise determination of the angle θ between the crystal face and the collimated incident beam. From Equation 12-6, it is evident that, for any given angular setting of the goniometer, only a few wavelengths are diffracted (λ, $\lambda/2$, $\lambda/3$, ..., $\lambda/\mathbf{n}$, where $\lambda = 2d \sin \theta$).

To produce a spectrum, it is necessary that the exit beam collimator and the detector be mounted on a

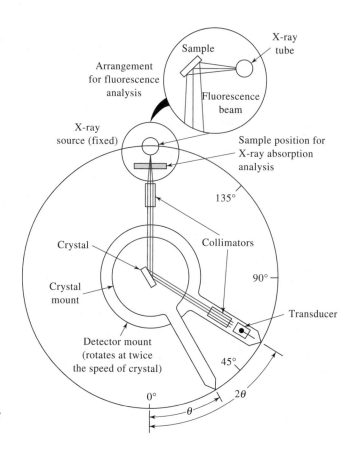

FIGURE 12-9 An X-ray monochromator and detector. Note that the angle of the detector with respect to the beam (2θ) is twice that of the crystal face. For absorption analysis, the source is an X-ray tube and the sample is located in the beam as shown. For emission measurements, the sample becomes a source of X-ray fluorescence as shown in the insert.

second table that rotates at twice the rate of the first; that is, as the crystal rotates through an angle θ, the detector must simultaneously move through an angle 2θ. Clearly, the interplanar spacing d for the crystal must be known precisely (Equation 12-6). Many modern X-ray monochromators have computer-controlled motors to drive the crystal and the detector independently without a gear-based mechanical connection. These units are capable of scanning at very rapid rates (ca. 240°/min). Some instruments feature slew-scan operation, in which the monochromator can be slewed to the wavelength region of interest (ca. 1400°/min) and then slowly scanned over analyte peaks.

The collimators for X-ray monochromators ordinarily consist of a series of closely spaced metal plates or tubes that absorb all but the parallel beams of radiation.

X-radiation longer than about 2 Å is absorbed by constituents of the atmosphere. Therefore, provision is usually made for a continuous flow of helium through the sample compartment and monochromator when longer wavelengths are required. Alternatively, these areas can be evacuated by pumping.

The loss of intensity is high in a monochromator equipped with a flat crystal because as much as 99% of the radiation is sufficiently divergent to be absorbed in the collimators. Increased intensities, by as much as a factor of 10, have been achieved by using a curved crystal surface that acts not only to diffract but also to focus the divergent beam from the source on the exit collimator.

As illustrated in Table 12-1, most analytically important X-ray lines lie in the region between about 0.1 and 10 Å. The data in Table 12-3, however, lead to the conclusion that no single crystal satisfactorily disperses radiation over this entire range. As a result, an X-ray monochromator must be provided with at least two (and preferably more) interchangeable crystals.

The useful wavelength range for a crystal is determined by its lattice spacing d and the problems associated with detection of the radiation when 2θ approaches zero or 180°. When a monochromator is set at angles of 2θ that are much less than 10°, the amount of polychromatic radiation scattered from the crystal surface becomes prohibitively high. Generally, values of

TABLE 12-3 Properties of Typical Diffracting Crystals

Crystal	Lattice Spacing d, Å	Wavelength Range,[a] Å		Dispersion, °/Å	
		λ_{max}	λ_{min}	at λ_{max}	at λ_{min}
Topaz	1.356	2.67	0.24	2.12	0.37
LiF	2.014	3.97	0.35	1.43	0.25
NaCl	2.820	5.55	0.49	1.02	0.18
EDDT[b]	4.404	8.67	0.77	0.65	0.11
ADP[c]	5.325	10.50	0.93	0.54	0.09

[a]Based on the assumption that the measurable range of 2θ is from 160° for λ_{max} to 10° for λ_{min}.

[b]EDDT = Ethylenediamine d-tartrate.

[c]ADP = Ammonium dihydrogen phosphate.

2θ greater than about 160° cannot be measured because the location of the source unit prohibits positioning of the detector at such an angle (see Figure 12-9). The minimum and maximum values for λ in Table 12-3 were determined from these limitations.

Table 12-3 shows that a crystal with a large lattice spacing, such as ammonium dihydrogen phosphate, has a much greater wavelength range than a crystal in which this variable is small. The advantage of large values of d is offset, however, by the resulting lower dispersion. This effect can be seen by differentiation of Equation 12-6, which leads to

$$\frac{d\theta}{d\lambda} = \frac{n}{2d \cos \theta}$$

Here, $d\theta/d\lambda$, a measure of dispersion, is seen to be inversely proportional to d. Table 12-3 provides dispersion data for the various crystals at their maximum and minimum wavelengths. The low dispersion of ammonium dihydrogen phosphate prohibits its use in the region of short wavelengths; here, a crystal such as topaz or lithium fluoride must be substituted.

12B-4 X-ray Transducers and Signal Processors

Early X-ray equipment used photographic emulsions for detection and recording of radiation. For convenience, speed, and accuracy, however, modern instruments are generally equipped with transducers that convert radiant energy into an electrical signal. Three types of transducers are used: *gas-filled transducers*, *scintillation counters*, and *semiconductor transducers*. Before considering the function of each of these de-

vices, it is worthwhile to discuss *photon counting*, a signal-processing method that is often used with X-ray transducers as well as detectors of radiation from radioactive sources (Chapter 32). As was mentioned earlier (Section 7F-1), photon counting is also used in ultraviolet and visible spectroscopy.

Photon Counting

In contrast to the various photoelectric detectors we have thus far considered, X-ray detectors are usually operated as *photon counters*. In this mode of operation, individual pulses of charge are produced as quanta of radiation, are absorbed by the transducer, and are counted; the power of the beam is then recorded digitally as the number of counts per unit of time. Photon counting requires rapid response times for the transducer and signal processor so that the arrival of individual photons may be accurately transduced and recorded. In addition, the technique is applicable only to beams of relatively low intensity. As the beam intensity increases, photon pulses begin to overlap and only a steady-state current, which represents an average number of pulses per second, can be measured. If the response time of the transducer is long, pulse overlap occurs at relatively low photon intensities. As its response time becomes shorter, the transducer is more capable of detecting individual photons without pulse overlap.[7]

For weak sources of radiation, photon counting generally provides more accurate intensity data than are obtainable by measuring average currents. The improvement can be traced to signal pulses being generally substantially larger than the pulses arising from background noise in the source, transducer, and associated electronics; separation of the signal from noise can then be achieved with a *pulse-height discriminator*, an electronic device that is discussed further in Section 12B-5.

Photon counting is used in X-ray work because the power of available sources is often low. In addition, photon counting permits spectra to be acquired without using a monochromator. This property is considered in the section devoted to energy-dispersive systems.

Gas-Filled Transducers

When X-radiation passes through an inert gas such as argon, xenon, or krypton, interactions occur that produce a large number of positive gaseous ions and

 Exercise: Learn more about **X-ray spectrometers.**

---

[7]E. J. Darland, G. E. Leroi, and C. G. Enke, *Anal. Chem.*, **1980**, *52*, 714.

electrons (ion pairs) for each X-ray quantum. Three types of X-ray transducers, *ionization chambers*, *proportional counters*, and *Geiger tubes*, are based on the enhanced conductivity of the gas resulting from this phenomenon.

A typical gas-filled transducer is shown schematically in Figure 12-10. Radiation enters the chamber through a transparent window of mica, beryllium, aluminum, or Mylar. Each photon of X-radiation that interacts with an atom of argon causes it to lose one of its outer electrons. This *photoelectron* has a large kinetic energy, which is equal to the difference between the X-ray photon energy and the binding energy of the electron in the argon atom. The photoelectron then loses this excess kinetic energy by ionizing several hundred additional atoms of the gas. Under the influence of an applied voltage, the mobile electrons migrate toward the central wire anode and the slower-moving cations migrate toward the cylindrical metal cathode.

Figure 12-11 shows the effect of applied voltage on the number of electrons that reach the anode of a gas-filled transducer for each entering X-ray photon. The diagram indicates three characteristic voltage regions. At voltages less than V_1, the accelerating force on the ion pairs is low, and the rate the positive and negative species separate is insufficient to prevent partial recombination. As a result, the number of electrons reaching the anode is smaller than the number produced initially by the incoming radiation.

In the *ionization chamber region* between V_1 and V_2, the number of electrons reaching the anode is reasonably constant and represents the total number formed by a single photon.

In the *proportional counter region* between V_3 and V_4, the number of electrons increases rapidly with applied voltage. This increase is the result of secondary ion-pair production caused by collisions between the accelerated electrons and gas molecules; amplification

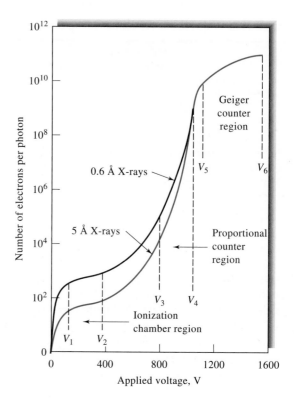

FIGURE 12-11 Gas amplification for various types of gas-filled detectors.

(*gas amplification*) of the ion current occurs under these conditions.

In the *Geiger range* V_5 to V_6, amplification of the electrical pulse is enormous but is limited by the positive space charge created as the faster-moving electrons migrate away from the slower positive ions. Because of this effect, the number of electrons reaching the anode is independent of the type and energy of incoming radiation and is governed instead by the geometry and gas pressure of the tube.

Figure 12-11 also illustrates that a larger number of electrons is produced by the more energetic 0.6 Å radiation than the longer-wavelength 5 Å X-rays. Thus, the size of the pulse (the pulse height) is greater for high-frequency X-rays than for low-frequency X-rays.

The Geiger Tube

The Geiger tube is a gas-filled transducer operated in the voltage region between V_5 and V_6 in Figure 12-11; in this region, gas amplification is greater than 10^9. Each photon produces an avalanche of electrons and cations; the resulting currents are thus large and relatively easy to detect and measure.

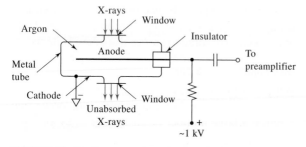

FIGURE 12-10 Cross section of a gas-filled detector.

The conduction of electricity through a chamber operated in the Geiger region, as well as in the proportional region, is not continuous because the space charge mentioned earlier terminates the flow of electrons to the anode. The net effect is a momentary pulse of current followed by an interval during which the tube does not conduct. Before conduction can again occur, this space charge must be dissipated by migration of the cations to the walls of the chamber. During the *dead time*, when the tube is nonconducting, response to radiation is impossible; the dead time thus represents a lower limit in the response time of the tube. Typically, the dead time of a Geiger tube is in the range from 50 to 200 μs.

Geiger tubes are usually filled with argon; a low concentration of an organic *quenching gas*, often alcohol or methane, is also present to minimize the production of secondary electrons when the cations strike the chamber wall. The lifetime of a tube is limited to some 10^8 to 10^9 counts, when the quencher becomes depleted.

With a Geiger tube, radiation intensity is determined by counting the pulses of current. The device is applicable to all types of nuclear and X-radiation. However, it lacks the large counting range of other detectors because of its relatively long dead time; this factor limits its use in X-ray spectrometers. Although quantitative applications of Geiger tube transducers have decreased, transducers of this type are still frequently used in portable instruments.

Proportional Counters

The proportional counter is a gas-filled transducer that is operated in the V_3 to V_4 voltage region in Figure 12-11. In this region, the pulse produced by a photon is amplified by a factor of 500 to 10,000, but the number of positive ions produced is small enough so that the dead time is only about 1 μs. In general, the pulses from a proportional counter tube must be amplified before being counted.

The number of electrons per pulse, which is proportional to the *pulse height*, produced in the proportional region depends directly on the energy, and thus the *frequency*, of the incoming radiation. A proportional counter can be made sensitive to a restricted range of X-ray frequencies with a *pulse-height analyzer*, which counts a pulse only if its amplitude falls within selectable limits. A pulse-height analyzer in effect permits electronic sorting of radiation; its function is analogous to that of a monochromator.

Proportional counters have been widely used as detectors in X-ray spectrometers.

Ionization Chambers

Ionization chambers are operated in the voltage range from V_1 to V_2 in Figure 12-11. In this region, the currents are small (10^{-13} to 10^{-16} A typically) and relatively independent of applied voltage. Ionization chambers are not useful for X-ray spectrometry because of their lack of sensitivity. They are, however, applied in radiochemical measurements, which are discussed in Chapter 32.

Scintillation Counters

The luminescence produced when radiation strikes a phosphor is one of the oldest methods of detecting radioactivity and X-rays, and it is still used widely for certain types of measurements. In its earliest application, the technique involved the manual counting of flashes that resulted when individual photons or radiochemical particles struck a zinc sulfide screen. The tedium of counting individual flashes by eye led Geiger to develop gas-filled transducers, which were not only more convenient and reliable but more responsive to radiation. The advent of the photomultiplier tube (Section 7E-2) and better phosphors, however, has given new life to the technique, and scintillation counting has again become one of the important methods for radiation detection.

The most widely used modern *scintillation detector* consists of a transparent crystal of sodium iodide that has been activated by the introduction of 0.2% thallium iodide. Often, the crystal is shaped as a cylinder that is 3 to 4 in. in each dimension; one of the plane surfaces then faces the cathode of a photomultiplier tube. As the incoming radiation passes through the crystal, its energy is first lost to the *scintillator*; this energy is subsequently released in the form of photons of fluorescence radiation. Several thousand photons with a wavelength of about 400 nm are produced by each primary particle or photon over a period of about 0.25 μs, which is the dead time. The dead time of a scintillation counter is thus significantly smaller than the dead time of a gas-filled detector.

The flashes of light produced in the scintillator crystal are transmitted to the photocathode of a photomultiplier tube and are in turn converted to electrical pulses that can be amplified and counted. An important characteristic of scintillators is that the number of photons produced in each flash is proportional to

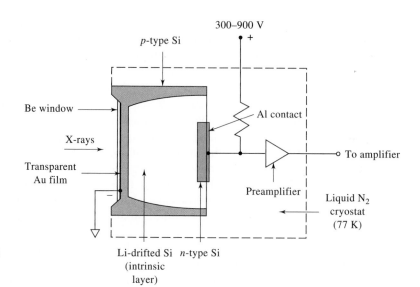

FIGURE 12-12 Vertical cross section of a lithium-drifted silicon detector for X-rays and radiation from radioactive sources.

the energy of the incoming radiation. Thus, incorporation of a pulse-height analyzer to monitor the output of a scintillation counter forms the basis of energy-dispersive photometers, to be discussed later.

In addition to sodium iodide crystals, a number of organic scintillators such as stilbene, anthracene, and terphenyl have been used. In crystalline form, these compounds have decay times of 0.01 and 0.1 μs. Organic liquid scintillators have also been developed and are used to advantage because they exhibit less self-absorption of radiation than do solids. An example of a liquid scintillator is a solution of *p*-terphenyl in toluene.

Semiconductor Transducers

Semiconductor transducers have assumed major importance as detectors of X-radiation. These devices are sometimes called *lithium-drifted silicon detectors*, Si(Li), or *lithium-drifted germanium detectors*, Ge(Li).

Figure 12-12 illustrates one form of a lithium-drifted detector, which is fashioned from a wafer of crystalline silicon. There are three layers in the crystal: a *p*-type semiconducting layer that faces the X-ray source, a central *intrinsic* zone, and an *n*-type layer. The outer surface of the *p*-type layer is coated with a thin layer of gold for electrical contact; often, it is also covered with a thin beryllium window that is transparent to X-rays. The signal output is taken from an aluminum layer that coats the *n*-type silicon; this output is fed into a preamplifier with a gain of about 10. The preamplifier is frequently a field-effect transistor that is fabricated as an integral part of the detector.

A lithium-drifted detector is formed by vapor-depositing lithium on the surface of a *p*-doped silicon crystal. When the crystal is heated to 400°C to 500°C, the lithium diffuses into the crystal. Because lithium easily loses electrons, its presence converts the *p*-type region to an *n*-type region. While still at an elevated temperature, a dc voltage applied across the crystal causes withdrawal of the electrons from the lithium layer and holes from the *p*-type layer. Current across the *pn* junction causes migration, or drifting, of lithium *ions* into the *p* layer and formation of the intrinsic layer, where the lithium ions replace the holes lost by conduction. When the crystal cools, this central layer has a high resistance relative to the other layers because the lithium ions in this medium are less mobile than the holes they displaced.

The intrinsic layer of a silicon detector functions in a way that is analogous to argon in the gas-filled transducer. Initially, absorption of a photon results in formation of a highly energetic photoelectron, which then loses its kinetic energy by elevating several thousand electrons in the silicon to the conduction band; a marked increase in conductivity results. When a voltage is applied across the crystal, a current pulse accompanies the absorption of each photon. As in proportional detectors, the size of the pulse is directly proportional to the energy of the absorbed photons. Unlike proportional detectors, however, secondary amplification of the pulse does not occur.

As shown in Figure 12-12, the detector and preamplifier of a lithium-drifted detector must be thermostatted at the temperature of liquid nitrogen (77 K) to decrease electronic noise to a tolerable level. The original Si(Li) detectors had to be cooled at all times because at room temperature the lithium atoms would diffuse throughout the silicon, thereby degrading the performance of the detector. Modern Si(Li) detectors need to be cooled only during use.

Germanium is used in place of silicon to give lithium-drifted detectors particularly useful for detection of radiation shorter in wavelength than 0.3 Å. These detectors must be cooled at all times. Germanium detectors that do not require lithium drifting have been produced from very pure germanium. These detectors, called *intrinsic germanium detectors*, need to be cooled only during use.

Distribution of Pulse-Heights from X-ray Transducers

To understand the properties of energy-dispersive spectrometers, it is important to appreciate that the size of current pulses resulting from absorption of successive X-ray photons of identical energy by the transducer will not be exactly the same. Variations resulting from the ejection of photoelectrons and their subsequent generation of conduction electrons are random processes governed by the laws of probability. Thus, there is a Gaussian distribution of pulse heights around a mean value. The width of this distribution varies from one type of transducer to another, with semiconductor detectors providing significantly narrower bands of pulses. It is this property that has made lithium-drifted detectors so important for energy-dispersive X-ray spectroscopy.

12B-5 Signal Processors

The signal from the preamplifier of an X-ray spectrometer is the input to a linear fast-response amplifier whose gain can be varied by a factor up to 10,000. Voltage pulses as large as 10 V result.

Pulse-Height Selectors

Most modern X-ray spectrometers (wavelength dispersive as well as energy dispersive) are equipped with *discriminators* that reject pulses of less than about 0.5 V (after amplification). In this way, transducer and amplifier noise is reduced significantly. Some instruments have *pulse-height selectors*, or *window discriminators*, which are electronic circuits that reject not only pulses

 Exercise: Learn more about **X-ray transducers**.

with heights below some predetermined minimum level but also those above a preset maximum level; that is, they remove all pulses except those that lie within a limited *channel* or *window* of pulse heights.[8]

Dispersive instruments are often equipped with pulse-height selectors to reject noise and to supplement the monochromator in separating the analyte line from higher-order, more energetic radiation that is diffracted at the same crystal setting.

Pulse-Height Analyzers

Pulse-height analyzers consist of one or more pulse-height selectors configured in such a way as to provide energy spectra. A single-channel analyzer typically has a voltage range of about 10 V or more with a window of 0.1 to 0.5 V. The window can be manually or automatically adjusted to scan the entire voltage range, thus providing data for an energy-dispersive spectrum. Multichannel analyzers typically contain up to a few thousand separate channels, each of which acts as a single channel that corresponds to a different voltage window. The signal from each channel is then accumulated in a memory location of the analyzer corresponding to the energy of the channel, thus permitting simultaneous counting and recording of an entire spectrum.

Scalers and Counters

To obtain convenient counting rates, the output from an X-ray transducer is sometimes scaled—that is, the number of pulses is reduced by dividing by some multiple of ten or two, depending on whether the circuit is a decade or a binary device. A brief description of electronic scalers is found in Section 4C-4. Counting of the scaled pulses is carried out with electronic counters such as those described in Sections 4C-2 and 4C-3.

12C X-RAY FLUORESCENCE METHODS

Although it is feasible to excite an X-ray emission spectrum by incorporating the sample into the target area of an X-ray tube, this approach is extremely inconvenient for many types of materials. Instead, excitation is usually accomplished by irradiating the sample with a beam of X-rays from an X-ray tube or a radioactive source. In this method, the elements in the sample are excited by absorption of the primary beam and emit their own characteristic fluorescence X-rays.

---

[8]For a discussion of signal shaping, see Section 4C-1.

This procedure is thus properly called an *X-ray fluorescence*, or *X-ray emission* method. X-ray fluorescence (XRF) is a powerful tool for rapid, quantitative determinations of all but the lightest elements. In addition, XRF is used for the qualitative identification of elements having atomic numbers greater that of oxygen (>8) and is often used for semiquantitative or quantitative elemental analyses.[9] A particular advantage of XRF is that it is nondestructive, in contrast to most other elemental analysis techniques.

12C-1 Instruments

Various combinations of the instrument components discussed in the previous section lead to two primary types of spectrometers: *wavelength-dispersive X-ray fluorescence* (WDXRF) and *energy-dispersive X-ray fluorescence* (EDXRF) instruments.[10]

Wavelength-Dispersive Instruments

Wavelength-dispersive instruments always contain tube sources because of the large energy losses that occur when an X-ray beam is collimated and dispersed into its component wavelengths. Radioactive sources produce X-ray photons at a rate less than 10^{-4} that of an X-ray tube; the added attenuation by a monochromator would produce a beam that was difficult or impossible to detect and measure accurately.

Wavelength-dispersive instruments are of two types, *single channel* (or *sequential*) and *multichannel* (or *simultaneous*). The spectrometer shown in Figure 12-9 is a sequential instrument that can be easily used for X-ray fluorescence analysis. In this type of instrument, the X-ray tube and sample are arranged as shown in the circular insert at the top of the figure. Single-channel instruments may be manual or automatic. Manual instruments are quite satisfactory for the quantitative determination of a few elements. In this application, the crystal and transducer are set at the proper angles (θ and 2θ) and counting progresses until sufficient counts have accumulated for precise results. Automatic instruments are much more convenient for qualitative analysis, where an entire spectrum must be scanned. In such a device, the motions of the crystal and detector are synchronized and the detector output is connected to the data-acquisition system.

Most modern single-channel spectrometers are provided with two X-ray sources; typically, one has a chromium target for long wavelengths and the other a tungsten target for short wavelengths. For $\lambda > 2$ Å, it is necessary to remove air between the source and detector by pumping or by displacement with a continuous flow of helium. In this type of instrument, dispersing crystals must also be easily interchangeable. Single-channel instruments cost more than $50,000.

Multichannel dispersive instruments are large, expensive (>$150,000) installations that permit the simultaneous detection and determination of as many as twenty-four elements. In these instruments, individual channels consisting of an appropriate crystal and a detector are arranged radially around an X-ray source and sample holder. The crystals for all or most of the channels are usually fixed at an appropriate angle for a given analyte line; in some instruments, one or more of the crystals can be moved to permit a spectral scan.

Each transducer in a multichannel instrument has its own amplifier, pulse-height selector, scaler, and counter or integrator. These instruments are equipped with a computer for instrument control, data processing, and display of analytical results. A determination of twenty or more elements can be completed in a few seconds to a few minutes.

Multichannel instruments are widely used for the determination of several components in industrial materials such as steel, other alloys, cement, ores, and petroleum products. Both multichannel and single-channel instruments are equipped to handle samples in the form of metals, powdered solids, evaporated films, pure liquids, or solutions. When necessary, the materials are placed in a cell with a Mylar or cellophane window.

Energy-Dispersive Instruments

As shown in Figure 12-13a, an EDXRF spectrometer consists of a polychromatic source — which may be either an X-ray tube or a radioactive material — a sample holder, a semiconductor detector, and the various electronic components required for energy discrimination.[11]

An obvious advantage of energy-dispersive systems is the simplicity and lack of moving parts in the excitation and detection components of the spectrometer. Furthermore, the absence of collimators and a crystal

---

[9] See R. Jenkins, *X-Ray Fluorescence Spectrometry*, 2nd ed., New York: Wiley, 1999.

[10] For a review of recent X-ray fluorescence instruments, see P. J. Potts, A. T. Ellisb, P. Kregsamerc, C. Strelic, C. Vanhoofd, M. Weste, and P. Wobrauschekc, *J. Anal. Atomic Spectrometry*, **2005**, *20*, 1124.

[11] See R. Jenkins, *X-Ray Fluorescence Spectrometry*, 2nd ed., New York: Wiley, 1999.

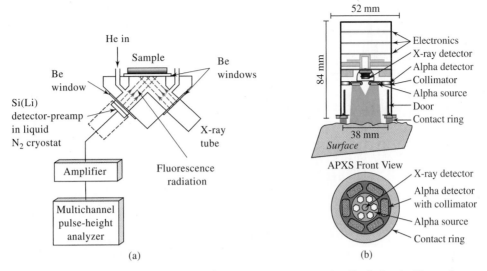

FIGURE 12-13 Energy-dispersive X-ray fluorescence spectrometer. Excitation by X-rays from (a) an X-ray tube and (b) a radioactive substance (curium-244, a 5.81 MeV alpha particle and X-ray source) as shown in the sensor head for the Mars alpha proton X-ray spectrometer. The X-ray detector is a new room-temperature type. (Reprinted with permission from R. Gellert et al., *J. Geophys. Res.*, **2006**, *111*, E02S05.)

diffractor, as well as the closeness of the detector to the sample, result in a 100-fold or more increase in energy reaching the detector. These features permit the use of weaker sources such as radioactive materials or low-power X-ray tubes, which are cheaper and less likely to cause radiation damage to the sample. Generally, energy-dispersive instruments cost about one fourth to one fifth the price of wavelength-dispersive systems.

Figure 12-13b shows the sensor head from the Mars rover missions of 2004. The head contains a curium-244 source that emits X-rays and 5.81 MeV alpha particles. The X-rays cause fluorescence in Martian rock samples, and the alpha particles stimulate X-ray emission as well. X-ray emission stimulated by bombardment by alpha and other subatomic particles such as protons is called *particle induced X-ray emission*, or PIXE. The X-ray detector is a new room-temperature type, which in the low temperature of the Martian night (below −40°C) exhibits low noise and high signal-to-noise ratio for excellent resolution and sensitivity. Note the concentric design of the sensor head with six Cm-244 sources arranged around the central detector. The X-ray spectrum of Figure 12-14 was acquired with the sensor head.

In a multichannel, energy-dispersive instrument, all of the emitted X-ray lines are measured simultane-ously. Increased sensitivity and improved signal-to-noise ratio result from the Fellgett advantage (see Section 7I-1). The principal disadvantage of energy-dispersive systems, when compared with crystal spectrometers, is their lower resolutions at wavelengths longer than about 1 Å. On the other hand, at shorter wavelengths, energy-dispersive systems exhibit superior resolution.

Figure 12-15a is a photo of a basic, commercial, benchtop EDXRF instrument that is used for the routine determination of a broad range of elements from sodium to uranium in samples from many industrial processes. The diagram of Figure 12-15b shows a close-up of the bottom of the sample turntable, which is visible from the top in the photo of Figure 12-15a, and the optical layout of the instrument. Radiation from the X-ray tube passes through an appropriate filter before striking the bottom of the rotating sample. The X-ray fluorescence emitted by the sample passes to the silicon detector, which provides the signal for the multichannel counting system. The system is equipped with a rhodium anode X-ray tube, five programmable filters, a helium purge system, a twelve-position sample changer, and a spinner to rotate each sample during the data-acquisition process. Spinning the sample reduces errors due to sample heterogeneity. We describe a quantitative application of this instrument in Section 12C-3.

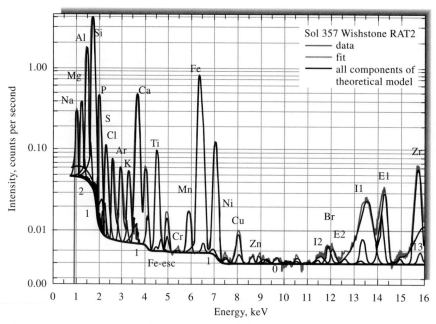

FIGURE 12-14 X-ray spectrum obtained on Mars together with the deconvolution model components. The main elemental characteristic peaks (Kα lines) are labeled and Kβ lines are unlabeled. (Reprinted with permission from R. Gellert et al., *J. Geophys. Res.*, **2006**, *111*, E02S05.)

12C-2 Qualitative and Semiquantitative Analysis

Figure 12-16 is a spectrum obtained with an energy-dispersive instrument. With such equipment, the abscissa is generally calibrated in channel number or energy in keV. Each dot represents the number of counts accumulated in one of the several hundred channels.

Qualitative information, such as that shown in Figures 12-16, can be converted to semiquantitative data by careful measurement of peak heights. To obtain a rough estimate of concentration, the following relationship is used:

$$P_x = P_s W_x \qquad (12\text{-}7)$$

where P_x is the relative line intensity measured in terms of number of counts for a fixed period, and W_x is the weight fraction of the desired element in the sample. The factor P_s is the relative intensity of the line that would be observed under identical counting conditions if W_x were unity. The value of P_s is determined with a sample of the pure element or a standard sample of known composition.

The use of Equation 12-7, as outlined in the previous paragraph, carries with it the assumption that the emission from the species of interest is unaffected by the presence of other elements in the sample. We shall see that this assumption may not be justified. As a result, concentration estimates may be in error by a factor of 2 or more. On the other hand, this uncertainty is significantly smaller than that associated with a semiquantitative analysis by optical emission where order-of-magnitude errors are not uncommon.

12C-3 Quantitative Analysis

Modern X-ray fluorescence instruments are capable of producing quantitative analyses of complex materials with precision greater than or equal to that of the classical wet chemical methods or other instrumental methods.[12] For the accuracy of such analyses to approach this level, however, requires either the availability of calibration standards that closely approach the samples in overall chemical and physical composition or suitable methods for dealing with matrix effects.

[12] See R. E. Van Grieken and A. A. Markowicz, eds., *Handbook of X-ray Spectrometry*, 2nd ed., New York: Marcel Dekker, 2002; R. Jenkins, R. W. Gould, and D. Gedcke, *Quantitative X-Ray Spectrometry*, 2nd ed., New York: Marcel Dekker, 1995.

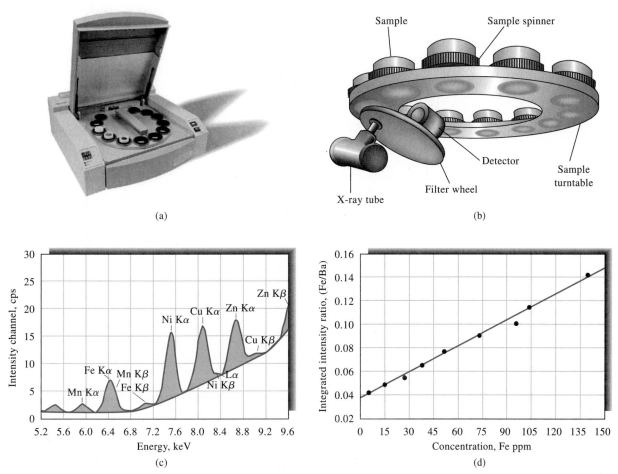

FIGURE 12-15 (a) MiniPal 4 benchtop X-ray fluorescence spectrometer showing removable turntable for up to twelve samples. (b) Diagram showing the X-ray source, filter wheel, detector, and sample turntable from the bottom. (c) X-ray fluorescence spectrum of a rice sample. The shaded areas under the curves are proportional to the amount of each element in the sample. (d) Calibration curve obtained from nine rice samples. Integrated areas from spectra similar to the one in (c) are plotted against the certified concentrations of iron in the samples. (Reprinted with permission from V. Sethi, M. Mizuhira, and Y. Xiao, *G.I.T. Laboratory Journal*, **2005**, *6*, 22. Copyright 2005 by GIT Verlag GmbH & Co.; photo courtesy of PANalytical B.V.)

Matrix Effects

It is important to realize that the X-rays produced in the fluorescence process are generated not only from atoms at the surface of a sample but also from atoms well below the surface. Thus, a part of both the incident radiation and the resulting fluorescence traverse a significant thickness of sample within which absorption and scattering can occur. The extent either beam is attenuated depends on the mass absorption coefficient of the medium, which in turn is determined by the absorption coefficients of *all* of the elements in the sample. There-

fore, although the net intensity of a line reaching the detector in an X-ray fluorescence measurement depends on the concentration of the element producing the line, the concentration and mass absorption coefficients of the matrix elements affect it as well.

Absorption effects by the matrix may cause results calculated by Equation 12-7 to be either high or low. If, for example, the matrix contains a significant amount of an element that absorbs either the incident or the emitted beam more strongly than the element being determined, then W_x will be low, because P_s was evaluated

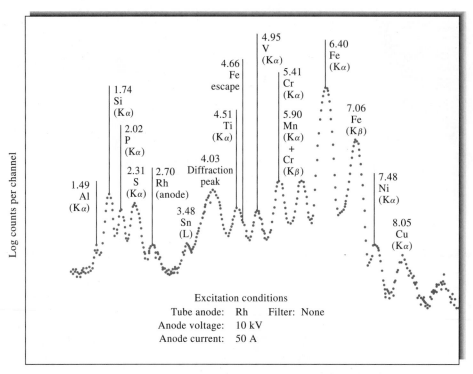

FIGURE 12-16 Spectrum of an iron sample obtained with an energy-dispersive instrument with a Rh anode X-ray tube source. The numbers above the peaks are energies in keV. (Reprinted with permission from J. A. Cooper, *Amer. Lab.*, **1976**, *8* (11), 44. Copyright 1976 by International Scientific Communications, Inc.)

with a standard in which absorption was smaller. On the other hand, if the matrix elements of the sample absorb less than those in the standard, high values for W_x result.

A second matrix effect, called the *enhancement effect*, can also yield results that are greater than expected. This behavior occurs when the sample contains an element whose characteristic emission spectrum is excited by the incident beam, and this spectrum in turn causes a secondary excitation of the analytical line.

Several techniques have been developed to compensate for absorption and enhancement effects in X-ray fluorescence analyses.

External Standard Calibration

The relationship between the analytical line intensity and the concentration must be determined empirically with a set of standards that closely approximate the samples in overall composition. We then assume that absorption and enhancement effects are identical for samples and standards, and the empirical data are used

to convert emission data to concentrations. The degree of compensation achieved in this way depends on the closeness of the match between samples and standards.

Use of Internal Standards

In this procedure, an element is introduced in known and fixed concentration into both the calibration standards and the samples; the added element must be absent in the original sample. The ratio of the intensities between the element being determined and the internal standard is the analytical variable. The assumption here is that absorption and enhancement effects are the same for the two lines and that use of intensity ratios compensates for these effects.

Dilution of Samples and Standards

Both sample and standards are diluted with a substance that absorbs X-rays only weakly, that is, a substance containing elements with low atomic numbers. Examples of such diluents include water; organic solvents only containing carbon, hydrogen, oxygen, and nitrogen; starch; lithium carbonate; alumina; and boric acid or borate glass. By adding a diluent in excess, matrix effects become essentially constant for the diluted standards and samples, and adequate compensation is achieved. This procedure has proved particularly useful for mineral analyses, where both samples and standards are dissolved in molten borax; after cooling, the fused mass is excited in the usual way.

Methods and Data Analysis for Multiple Elements

When it is possible to establish that total sample X-ray absorption does not vary appreciably over an expected range of analyte concentrations and when enhancement effects are absent, we generally find a linear relationship between XRF line intensity and element concentration.[13] When such conditions occur, it is often possible to use *type standardization* techniques. In these techniques, standards are prepared and characterized or, more often, purchased from the National Institute of Standards and Technology (NIST) or other sources. These standards are then used to calibrate spectrometers and establish the working curves for analysis. Hundreds of such standards are available for matching samples with particular types of standards.

When the nature and composition of the sample is not well known, it is necessary to use *influence correction methods*, of which there are three primary types: *fundamental*, *derived*, and *regression*. In the fundamental approach, the intensity of fluorescence can be calculated for each element in a standard sample from variables such as the source spectrum, the fundamental equations for absorption and fluorescence, matrix effects, the crystal reflectivity (in a WDXRF instrument), instrument aperture, the detector efficiency, and so forth. The XRF spectrum of the standard is measured, and in an iterative process the instrument variables are refined and combined with the fundamental variables to obtain a *calibration function* for the analysis. Then the spectrum for an unknown sample is measured, and the iterative process is repeated using initial estimates of the concentrations of the analytes. Iteration continues until the calculated spectrum matches the unknown spectrum according to appropriate statistical criteria. This method gives good results with accuracies on the order of 1%–4% but is generally considered to be less accurate than derived

or regression methods.[14] So-called *standardless analysis* is accomplished using variations on fundamental calibration methods. Instruments are carefully characterized, and instrument and fundamental parameters are stored in the computer. Spectra are then calculated and matched with samples by iterative methods to estimate concentrations of analytes.

Derived methods simplify calculations of the fundamental method by lumping detailed fundamental calculations into generalized parameters to account for instrument functions, but the calculation of fluorescence intensities from analyte concentration is the same as in fundamental methods. Derived methods often show some improvement in accuracy and precision over fundamental methods.

Regression methods rely on models of the form

$$W_i = B_i + K_i I_i \left[1 + \sum_{j=1}^{n} m_{ij} W_{ij} \right]$$

where B_i is the background, K_i is a sensitivity coefficient, I_i is the intensity due to element i, m_{ij} is the *binary influence coefficient* that describes the matrix effect of element j on analyte i, and W_{ij} is the average composition of the sample in i and j. To calibrate an instrument, spectra are acquired for standards and regression analysis is performed to determine K_i, B_i, and m_{ij} for each binary combination of elements in the standards. These parameters are then used in a regression analysis of spectral data from unknown samples to determine the unknown concentrations W_i. A broad range of models and algorithms for regression analysis in XRF are available, and many commercial instruments use combinations and variations on these algorithms in their software suites.[15] Accuracies and precisions attained using these methods are typically 1% or better under optimal conditions. Regression methods were used to analyze the data of Figure 12-14 from the Mars rover to determine by EDXRF the concentrations of many elements in rocks and soil.

In the early days of XRF, influence correction methods were difficult and time-consuming because sufficient computational power to accomplish them was available only on large, expensive mainframe computers. With the advent of powerful, low-cost, dedicated computers, sophisticated data analysis is now

[13] R. Jenkins, *X-Ray Fluorescence Spectrometry*, 2nd ed., New York: Wiley, 1999, pp. 182–86.

[14] J. L. de Vries and B. A. R. Vrebos, in R. E. Van Grieken and A. A. Markowicz, eds., *Handbook of X-ray Spectrometry*, 2nd ed., New York: Marcel Dekker, 2002, p. 378.

[15] Ibid., pp. 341–405.

routine, and all commercial instruments have mature software suites for instrument operation, calibration, data reduction, and analysis.

Some Quantitative Applications of X-ray Fluorescence

With proper correction for matrix effects, X-ray fluorescence spectrometry is one of the most powerful tools available for the rapid quantitative determination of all but the lightest elements in complex samples. For example, Rose, Bornhorst, and Sivonen [16] have demonstrated that twenty-two elements can be determined in powdered rock samples with a commercial EDXRF spectrometer in about 2 hours (1 hour instrument time), including grinding and pellet preparation. Relative standard deviations for the method are better than 1% for major elements and better than 5% for trace elements. Accuracy and detection limits as determined by comparison to results from international standard rock samples were comparable or better than other published procedures. For an excellent overview of XRF analysis of geological materials, see the paper by Anzelmo and Lindsay.[17]

X-ray methods are also applied widely for quality control in the manufacture of metals and alloys. Because of the speed of the analysis in these applications, it is possible to correct the composition of the alloy during its manufacture.

X-ray fluorescence methods are easily adapted to liquid samples. Thus, as mentioned earlier, methods have been devised for the direct quantitative determination of lead and bromine in aviation fuel samples. Similarly, calcium, barium, and zinc have been determined in lubricating oils by exciting fluorescence in the liquid hydrocarbon samples. The method is also convenient for the direct determination of pigments in paint samples.

X-ray fluorescence methods are being widely applied to the analysis of atmospheric pollutants. For example, one procedure for detecting and determining contaminants involves drawing an air sample through a stack consisting of a micropore filter for particulates and three filter paper disks impregnated with orthotolidine, silver nitrate, and sodium hydroxide, respectively. The reagents retain chlorine, sulfides, and sulfur dioxide in that order. The filters containing trapped analytes then serve as samples for X-ray fluorescence analysis.

As an example of the routine application of EDXRF for determining the elemental composition of foodstuffs, consider the data plots of Figure 12-15c and 12-15d.[18] In this analysis, iron, copper, and zinc were determined in rice using the spectrometer shown in Figure 12-15a. The spectrometer was calibrated using nine standard samples of rice. The standards were ground into fine powders, pressed into pellets, and loaded into the sample turntable. Each sample was irradiated with X-rays (Rh anode) through an aluminum filter for 5 minutes, and data were acquired to produce spectra such as the one shown in Figure 12-15c. The instrument computer then analyzed the data and calculated the areas under the various peaks in the spectrum as shown by the shaded areas in the figure, and calibration curves similar to Figure 12-15d were produced for each of the analytes. Determination of the analytes in another well-characterized sample yielded concentrations of 73.49 ppm, 7.46 ppm, and 38.95 ppm for iron, copper, and zinc, respectively. Relative accuracies for the three elements were 0.9% (Fe), 5% (Cu), and 0.3% (Zn) as determined by comparison with values determined by optical spectroscopic methods.

Another indication of the versatility of X-ray fluorescence is its use for the quantitative determination of elements heavier than sodium in rocks and soil encountered near the landing site of the Mars Pathfinder mission.[19] The Pathfinder's microrobot *Sojourner* was equipped with a sensor head that could be placed flat against a material to be analyzed. The head contained curium-244, which emitted α particles and X-rays that bombarded the sample surface as in Figure 12-13b. The X-ray emission from the sample struck the transducer of an energy-dispersive spectrometer in which the radiant power was recorded as a function of energy, was transmitted from Mars to Earth, and ultimately was analyzed on Earth. Lighter elements were determined by backscattering or proton emission. With these three methods of detection, all of the elements in the periodic table but hydrogen can be determined at concentration levels of a few tenths of a percent. In 2004 two new rovers, *Spirit* and *Opportunity*, landed on Mars with X-ray fluorescence spectrometers aboard.[20] As we write this, the rovers have

---

[16] W. I. Rose, T. J. Bornhorst, and S. J. Sivonen, *X-ray Spectrometry*, **1986**, *15*, 55.

[17] J. E. Anzelmo and J. R. Lindsay, *J. Chem. Educ.*, **1987**, *64*, A181 and A200.

[18] V. Sethi, M. Mizuhira, and Y. Xiao, *G.I.T. Laboratory Journal*, **2005**, *6*, 22.

[19] "Mars Pathfinder," 1997, http://mars.jpl.nasa.gov/MPF/ (25 July 1997).

[20] "Spirit and Opportunity," 2004, http://mpfwww.jpl.nasa.gov/missions/present/2003.html (13 Oct 2005).

spent nearly 2 years moving about their respective landing sites, and the spectrometers continue to send data to Earth revealing the elemental composition of Mars rocks and soil. The spectrum shown in Figure 12-14 was transmitted from one of the rovers to a ground station where the data were analyzed using a regression algorithm, and concentrations of the various elements in the sample were extracted from the data. Details of the intricate calibration procedure appear in the paper by Gellert et al.[21]

Advantages and Disadvantages of X-ray Fluorescence Methods

X-ray fluorescence offers a number of impressive advantages. The spectra are relatively simple, so spectral line interference is minimal. Generally, the X-ray method is nondestructive and can be used for the analysis of paintings, archaeological specimens, jewelry, coins, and other valuable objects without harm to the sample. Furthermore, analyses can be performed on samples ranging from a barely visible speck to a massive object. Other advantages include the speed and convenience of the procedure, which permit multielement analyses to be completed in a few minutes. Finally, the accuracy and precision of X-ray fluorescence methods often equal or exceed those of other methods.[22]

X-ray fluorescence methods are generally not as sensitive as the optical methods discussed earlier. In the most favorable cases, concentrations of a few parts per million or less can be measured. More often, however, the concentration range of the method is from about 0.01% to 100%. X-ray fluorescence methods for the lighter elements are inconvenient; difficulties in detection and measurement become progressively worse as atomic numbers become smaller than 23 (vanadium), in part because a competing process, called *Auger emission* (see Section 21C-2), reduces the fluorescence intensity (see Figure 21-7). Today's commercial instruments have a lower atomic number limit of 5 (boron) or 6 (carbon). Another disadvantage of the X-ray emission procedure is the high cost of instruments, which ranges from less than $10,000 for an energy-dispersive system with a radioactive source to well over $500,000 for automated and computerized wavelength-dispersive systems.

12D X-RAY ABSORPTION METHODS

In contrast to optical spectroscopy, where absorption methods are most important, X-ray absorption applications are limited when compared with X-ray emission and fluorescence procedures. Although absorption measurements can be made relatively free of matrix effects, the required techniques are somewhat cumbersome and time-consuming when compared with fluorescence methods. Thus, most applications are confined to samples in which matrix effects are minimal.

Absorption methods are analogous to optical absorption procedures in which the attenuation of a band or line of X-radiation is the analytical variable. Wavelength selection is accomplished with a monochromator such as that shown in Figure 12-9 or by a filter technique similar to that illustrated in Figure 12-8. Alternatively, the monochromatic radiation from a radioactive source may be used.

Because of the width of X-ray absorption peaks, direct absorption methods are generally useful only when a single element with a high atomic number is to be determined in a matrix consisting of only lighter elements. Examples of applications of this type are the determination of lead in gasoline and the determination of sulfur or the halogens in hydrocarbons.

12D-1 X-ray Diffraction Methods

Since its discovery in 1912 by von Laue, X-ray diffraction has provided a wealth of important information to science and industry. For example, much of what is known about the arrangement and the spacing of atoms in crystalline materials has been determined directly from diffraction studies. In addition, such studies have led to a much clearer understanding of the physical properties of metals, polymeric materials, and other solids. X-ray diffraction is one of the most important methods for determining the structures of such complex natural products as steroids, vitamins, and antibiotics. The details of these applications are beyond the scope of this book.

X-ray diffraction also provides a convenient and practical means for the qualitative identification of crystalline compounds. The X-ray powder diffraction method is the only analytical method that is capable of

[21] R. Gellert, R. Rieder, J. Brückner, B. C. Clark, G. Dreibus, G. Klingelhöfer, G. Lugmair, D. W. Ming, H. Wänke, A. Yen, J. Zipfel, and S. W. Squyres, *J. Geophys. Res.*, **2006**, *111*, E02S05.

[22] For a comparison of X-ray fluorescence and inductively coupled plasma for the analysis of environmental samples, see T. H. Nguyen, J. Boman, and M. Leermakers, *X-Ray Spectrometry*, **1998**, *27*, 265.

 Tutorial: Learn more about **applications of X-ray fluorescence**.

providing qualitative and quantitative information about the *compounds* present in a solid sample. For example, the powder method can determine the percentage KBr and NaCl in a solid mixture of these two compounds. Other analytical methods reveal only the percentage K^+, Na^+, Br^-, and Cl^- in the sample.[23]

Because each crystalline substance has a unique X-ray diffraction pattern, X-ray powder methods are well suited for qualitative identification. Thus, if an exact match can be found between the pattern of an unknown and an authentic sample, identification is assured.

12D-2 Identification of Crystalline Compounds

Sample Preparation

For analytical diffraction studies, the crystalline sample is ground to a fine homogeneous powder. In such a form, the enormous number of small crystallites are oriented in every possible direction; thus, when an X-ray beam passes through the material, a significant number of the particles are oriented in such ways as to fulfill the Bragg condition for reflection from every possible interplanar spacing.

Samples are usually placed in a sample holder that uses a depression or cavity to mount the sample. These mounts are commonly made of aluminum, bronze, Bakelite, glass, or Lucite. Cavity mounts are most commonly side loaded or back loaded. A frosted glass surface, ceramic, or cardboard is placed over the front, and the sample is carefully added via the open side or back. Top-loading mounts are also available as are special mounts known as zero background mounts. Alternatively, a specimen may be mixed with a suitable non-crystalline binder and molded into an appropriate shape.

Automatic Diffractometers

Diffraction patterns are generally obtained with automated instruments similar in design to that shown in Figure 12-9. In this instrument, the source is an X-ray tube with suitable filters. The powdered sample, however, replaces the single crystal on its mount. In some instances, the sample holder may be rotated to increase the randomness of the orientation of the crystals. The

diffraction pattern is then recorded by automatic scanning in the same way as for an emission or absorption spectrum. Instruments of this type offer the advantage of high precision for intensity measurements and automated data reduction and report generation.

Photographic Recording

The classical photographic method for recording powder diffraction patterns is still used, particularly when the amount of sample is small. The most common instrument for this purpose is the *Debye-Scherrer* powder camera, which is shown schematically in Figure 12-17a. Here, the beam from an X-ray tube is filtered to produce a nearly monochromatic beam (often the copper or molybdenum $K\alpha$ line), which is collimated by passage through a narrow tube.

Figure 12-8 shows how a filter can be used to produce a relatively monochromatic beam. Note that the $K\beta$ line and most of the continuum from a Mo target are removed by a zirconium filter with a thickness of about 0.01 cm. The pure $K\alpha$ line is then available for diffractometry. Several other target-filter combinations have been developed to isolate one of the intense lines of the target material.

The undiffracted radiation T then passes out of the camera via a narrow exit tube as shown in Figure 12-17a. The camera itself is cylindrical and equipped to hold a strip of film around its inside wall. The inside diameter of the cylinder usually is 5.73 or 11.46 cm, so that each lineal millimeter of film is equivalent to $1.0°$ or $0.5°$ in θ, respectively. The sample is held in the center of the beam by an adjustable mount.

Figure 12-17b shows the appearance of the exposed and developed film; each set of lines (D_1, D_2, and so forth) represents diffraction from one set of crystal planes. The Bragg angle θ for each line is easily evaluated from the geometry of the camera.

12D-3 Interpretation of Diffraction Patterns

The identification of a species from its powder diffraction pattern is based on the position of the lines (in terms of θ or 2θ) and their relative intensities. The diffraction angle 2θ is determined by the spacing between a particular set of planes; with the aid of the Bragg equation, this distance d is calculated from the known wavelength of the source and the measured angle. Line intensities depend on the number and kind of atomic reflection centers in each set of planes.

Crystals are identified empirically. A powder diffraction database is maintained by the International

---

[23]For a more detailed discussion of the X-ray powder diffraction method, see R. Jenkins and R. Snyder, *Introduction to X-Ray Powder Diffractometry*, New York: Wiley, 1996.

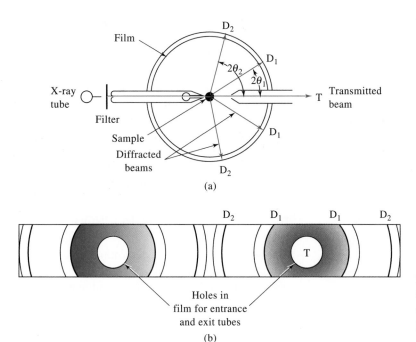

FIGURE 12-17 Schematic of (a) a Debye-Scherrer powder camera; (b) the film strip after development. D_2, D_1, and T indicate positions of the film in the camera.

Centre for Diffraction Data, Newtown Square, Pennsylvania. As of 2005, this file contained powder diffraction patterns for more than 477,000 reference materials. Because the file is so large, the powder data file has been broken down into subfiles that contain listings for inorganics, organics, minerals, metals, alloys, forensic materials, and others. The sub-databases are available on CD-ROM, and software is available for searching the databases using any combination of several search criteria. Each record in the database contains a wealth of information regarding the substances and materials, including name, formula (if appropriate), composition, color, line strengths, melting point, mineral classification, density, and a host of other characteristics of the materials as well as bibliographic information. A variety of presentation modes are available so that graphs and other important images may be viewed and printed.[24]

If the sample contains two or more crystalline compounds, identification becomes more complex. Here, various combinations of the more intense lines are used until a match can be found. Computer searching and matching of data greatly facilitates this task.

One very important application is the determination of the percentage crystallinity of materials. In the analysis of polymeric and fibrous materials, determining the crystalline-to-amorphous ratio has long been of importance, and X-ray powder methods have unique advantages in these determinations. In the pharmaceutical area, the degree of crystallinity can influence the long-term stability of a formulation as well as its bioactivity. X-ray diffraction methods are being increasingly applied to pharmaceuticals.

Crystalline materials produce well-defined diffraction peaks whose widths are related to the crystalline "quality." High-quality materials produce sharp peaks, and poor-quality materials give rise to more diffuse diffraction peaks. Amorphous phases come in different forms depending on how they were formed. A glassy phase produces a diffraction signal that is the radial distribution of nearest neighbor interactions. An amorphous phase derived from a crystalline phase usually corresponds to a poor-quality, or paracrystalline, material. Both glassy and paracrystalline specimens produce a low-frequency halo, which can appear as a broad background.

One approach to determining the crystalline-to-amorphous ratio is to use conventional quantitative analysis methods. Non-overlapped X-ray diffraction peaks are chosen for the phase to be analyzed. Either peak height or peak area is used for quantitative analysis. Standards of known concentration are then used to prepare a calibration curve.

In the Vainshtein approach, the amorphous phase is used as a normalizing factor for the integrated

[24]For further information regarding the powder diffraction database, see http://www.icdd.com/.

intensities of the crystalline peaks.[25] This eliminates the effects of sample preparation and instrument drift. Analysis is based on *Vainshtein's law*, which states that the diffracted intensity from a material is independent of its state of order within identical regions of reciprocal space. To apply the law, a single standard with a known percentage of crystallinity is used to establish the normalization ratio between the integrated crystalline peaks and the amorphous "background." The same measurements are then made on the specimen of unknown crystallinity. The percentage crystallinity of the unknown is then found from

$$\% C_u = \% C_{std} \left[\frac{(C/A)_u}{(C/A)_{std}} \right] \quad (12\text{-}8)$$

where $\% C_u$ and $\% C_{std}$ are the percentage crystallinities of the unknown and standard respectively, and C/A

is the ratio of the integrated intensity of the crystalline phase C to the amorphous background A.

Yet another approach is to use Fourier transform methods to split the power spectrum into low- and high-frequency regions. The amorphous phase is associated with the low-frequency region, and the crystalline phase is associated with the high-frequency region. After filtering the undesired region, the inverse transform gives the amorphous intensity and the crystalline intensity. Standards are used to determine the width and frequencies of the filters.

12E THE ELECTRON MICROPROBE

An important method for the determination of the elemental composition of surfaces is based on the *electron microprobe*. In this technique, X-ray emission from the elements on the surface of a sample is stimulated by a narrowly focused beam of electrons. The resulting X-ray emission is detected and analyzed with either a wavelength or an energy-dispersive spectrometer. This method is discussed in detail in Section 21F-1.

---

[25] B. K. Vainshtein (1921–1996) was a prominent Russian X-ray crystallographer. His monograph *Diffraction of X-rays by Chain Molecules* (Amsterdam: Elsevier, 1966) played an important role in the development of structural studies of polymers.

QUESTIONS AND PROBLEMS

*Answers are provided at the end of the book for problems marked with an asterisk.

[X] Problems with this icon are best solved using spreadsheets.

*12-1 What is the short-wavelength limit of the continuum produced by an X-ray tube having a silver target and operated at 90 kV?

*12-2 What minimum tube voltage would be required to excite the Kβ and Lβ series of lines for (a) U, (b) K, (c) Rb, (d) W?

[X] 12-3 The Kα_1 lines for Ca, Zn, Zr, and Sn have wavelengths of 3.36, 1.44, 0.79, and 0.49 Å, respectively. Calculate an approximate wavelength for the Kα lines of (a) V, (b) Ni, (c) Se, (d) Br, (e) Cd, (f) Sb.

[X] 12-4 The Lα lines for Ca, Zn, Zr, and Sn have wavelengths of 36.3, 11.9, 6.07, and 3.60 Å, respectively. Estimate the wavelengths for the Lα lines for the elements listed in Problem 12-3.

*12-5 The mass absorption coefficient for Ni, measured with the Cu Kα line, is 49.2 cm^2/g. Calculate the thickness of a nickel foil that was found to transmit 47.8% of the incident power of a beam of Cu Kα radiation. The density of Ni is 8.90 g/cm^3.

*12-6 For Mo Kα radiation (0.711 Å), the mass absorption coefficients for K, I, H, and O are 16.7, 39.2, 0.0, and 1.50 cm^2/g, respectively.
   (a) Calculate the mass absorption coefficient for a solution prepared by mixing 11.00 g of KI with 89.00 g of water.
   (b) The density of the solution described in (a) is 1.086 g/cm^3. What fraction of the radiation from a Mo Kα source would be transmitted by a 0.60-cm layer of the solution?

*12-7 Aluminum is to be used as windows for a cell for X-ray absorption measurements with the Ag Kα line. The mass absorption coefficient for aluminum at this wavelength is 2.74; its density is 2.70 g/cm^3. What maximum thickness of aluminum foil could be used to fabricate the windows if no more than 3.5% of the radiation is to be absorbed by them?

*12-8 A solution of I$_2$ in ethanol had a density of 0.794 g/cm^3. A 1.50-cm layer was found to transmit 27.3% of the radiation from a Mo Kα source. Mass absorption coefficients for I, C, H, and O are 39.2, 0.70, 0.00, and 1.50, respectively.
(a) Calculate the percentage of I$_2$ present, neglecting absorption by the alcohol.
(b) Correct the results in part (a) for the presence of alcohol.

*12-9 Calculate the goniometer setting, in terms of 2θ, required to observe the Kα_1 lines for Fe (1.76 Å), Se (0.992 Å), and Ag (0.497 Å) when the diffracting crystal is (a) topaz, (b) LiF, (c) NaCl.

12-10 Calculate the goniometer setting, in terms of 2θ, required to observe the Lβ_1 lines for Br at 8.126 Å when the diffracting crystal is
(a) ethylenediamine *d*-tartrate.
(b) ammonium dihydrogen phosphate.

*12-11 Calculate the minimum tube voltage required to excite the following lines. The numbers in parentheses are the wavelengths in Å for the corresponding absorption edges.
(a) K lines for Ca (3.064)
(b) Lα lines for As (9.370)
(c) Lβ lines for U (0.592)
(d) K lines for Mg (0.496)

 12-12 Manganese was determined in samples of geological interest via X-ray fluorescence using barium as an internal standard. The fluorescence intensity of isolated lines for each element gave the following data:

Wt. % Mn	Ba	Mn
0.00	156	80
0.10	160	106
0.20	159	129
0.30	160	154
0.40	151	167

What is the weight percentage manganese in a sample that had a Mn-to-Ba count ratio of 0.735?

 Challenge Problem

12-13 As discussed in Section 12C-3, the APXS, or alpha proton X-ray spectrometer, has been an important experiment aboard all of the Mars exploration rovers. Journal articles provide details of the APXS experiments on the most recent missions in 2004 and compare the instrumentation and measurement strategies with those on board the Pathfinder mission of 1997.[26]

[26] S. W. Squyres et al., *J. Geophys. Res.*, **2003**, *108*, 8062; R. Rieder et al., *J. Geophys. Res.*, **2003**, *108*, 8066; J. Brückner et al., *J. Geophys. Res.*, **2003**, *108*, 8094; R. Gellert et al., *Science*, **2004**, *305*, 829; H. Y. McSween et al., *Science*, **2004**, *305*, 842; S. W. Squyres et al., *Science*, **2004**, *306*, 1709; L. A. Soderblom et al., *Science*, **2004**, *306*, 1723; G. Klingelhofer et al., *Science*, **2004**, *306*, 1740; R. Rieder et al., *Science*, **2004**, *306*, 1746; A. Banin, *Science*, **2005**, *309*, 888.

(a) Consult the cited articles, and describe the construction and operation of the detector head on the *Spirit* and *Opportunity* rovers. Illustrate your answer with basic diagrams of the instrument components, and describe the function of each.

(b) What determines the selectivity and sensitivity of the APXS system?

(c) What elements cannot be determined by APXS? Why not?

(d) Characterize the general elemental composition of all of the Martian landing sites. What are the similarities and differences among the sites. What explanation do these workers and others give for the similarities?

(e) How were the APXS measurements limited by the time in the Martian day when the experiments were completed? Explain the cause of these limitations. How and why were the APXS experiments terminated?

(f) What effect did temperature have on the results of the X-ray mode of the APXS experiments? How does this effect relate to your discussion in (e)?

(g) The APXS experiments were used to compare the surface composition with the composition of the interiors. How was this accomplished? What differences were discovered? What explanation was given for the differences?

(h) Halogen fractionation was apparent in the Martian rock samples. What explanation was given for this phenomenon?

(i) Characterize the overall precision and accuracy of the APXS experiments. The APXS team states that "accuracy is mainly determined by precision." Explain what is meant by this statement. Under what circumstances is it true?

(j) Explain in some detail the calibration procedure for the APXS experiments. What calibration standards were used on Mars, and where were they located? How were individual element peaks extracted from the X-ray spectra?

(k) Figure 12-18 shows a plot of corrected calibration data for Fe from the APXS experiments. Note that there are uncertainties in both the *x* and *y* data. The

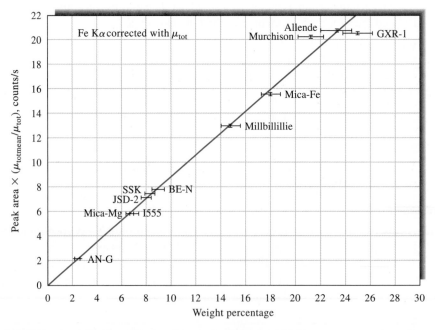

FIGURE 12-18 APXS calibration curve for Fe corrected for attenuation coefficient. (From R. Rieder et al., *J. Geophys. Res.*, **2003**, *108*, 8066, with permission.)

data from the plot are tabulated in the table below. Carry out a normal least-squares analysis of the data using Excel. Determine the slope, intercept, correlation coefficient, and the standard deviation about regression.

Peak Area	σ	Mass %	σ
2.36	0.28	2.12	0.13
7.04	0.54	5.75	0.13
6.68	0.54	5.85	0.13
8.01	0.39	7.14	0.10
8.42	0.44	7.45	0.13
9.04	0.44	7.87	0.10
14.74	0.74	13.20	0.10
17.93	0.90	15.58	0.18
21.27	1.05	20.24	0.21
23.22	1.18	20.71	0.23
25.02	1.26	20.55	0.18

(l) Carry out a weighted least-squares analysis of the data using a procedure similar to the one described in Problem 10-14 that allows for uncertainties in both the x and y variables. Compare your results to those obtained in (k).

Monitoring Mercury

Mercury is an extremely important element in the environment, in foods, and in industrial processes. The health scares concerning mercury have been many since the best-known case, the Minamata Bay disaster in Japan in 1956. Minamata is a small industrial town on the coast of the Shiranui Sea. The major industrial firm in Minamata, the Chisso Corporation, dumped some 37 tons of mercury-containing compounds into Minamata Bay from 1932 to 1968. As a result, thousands of local residents, whose normal diet included fish from the bay, developed symptoms of methylmercury poisoning. The disease became known as *Minamata disease*. Since then, there have been many warnings about eating fish and shellfish known to contain high levels of mercury. The most recent recommendations of the United States Environmental Protection Agency (EPA) and the Food and Drug Administration warn women of child-bearing age and young children not to eat shark, swordfish, king mackerel, or tilefish. These agencies also advise women and young children to eat no more than 12 ounces per week of fish that contain lower levels of mercury, such as shrimp, canned tuna, salmon, pollock, and catfish. More stringent recommendations apply to fish from certain lakes, rivers, and coastal areas.

Sources of Mercury

Mercury is, of course, a naturally occurring element. However, industrial pollution is a major source of environmental mercury. The pollution comes from many sources, such as coal-burning power plants, refineries, runoff from factories, and industrial waste. Mercury also enters the environment from such sources as automobile exhausts, sewage treatment plants, medical and dental facilities, and water runoff from mercury- and gold-mining operations. The Clean Air Act, first enacted in 1970 in the United States, mandated levels of air pollution, including mercury. Likewise, the EPA has set water-quality criteria for levels of mercury in both fresh and saltwater systems.[1] The Clean Water Act requires that individual states achieve safe concentration levels for pollutants like mercury.

Mercury from the air or from water sources accumulates in streams and lakes. Bacteria in the water convert the mercury into methylmercury, which is readily absorbed by insects and other aquatic organisms. The mercury-containing compounds rapidly move up the food chain as small fish eat the small organisms and big fish eat the small fish. The

concentrations of mercury are determined by the life span of the fish and its feeding habits. The highest concentrations are found in large fish such as swordfish and sharks.

Analytical Chemistry Challenges

Because of its importance in the environment, accurate, reliable determinations of mercury are crucial. Although determination of the various forms that mercury takes in the environment is of interest, current regulations focus on measurements of total mercury, which is in itself challenging. The challenge arises because in complex environmental samples mercury is present in very small amounts. The EPA has established a level of 12 ppt as the upper limit for freshwater ecosystems and 25 ppt as the limit for saltwater. The National Toxics rule of the EPA[2] recommends that the safe level for mercury is as low as 1.3 parts per trillion (ppt). Frequently, concentrations lower than this criterion are found for total dissolved mercury (inorganic plus organic forms) in seawater. Some ocean surface waters have been found to have concentrations lower than 0.05 ppt, and intermediate water layers can have levels as high as 2 ppt. These low concentrations can be near or lower than the limits of detection for many analytical techniques. In freshwater systems, much higher concentrations have been observed. In some California lakes, for example, levels of 0.5 to 100 ppt have been determined. In any case, detection of these ultratrace levels can be quite a challenge for analytical chemists.

Traditional Methods for Determining Mercury

The concentrations of interest in satisfying regulations in the United States and the United Kingdom are in the parts-per-trillion range and lower. However, conventional analytical methods such as flame atomic absorption, inductively coupled plasma atomic emission, and inductively coupled plasma mass spectrometry cannot achieve parts-per-trillion detection limits because the use of nebulizers to transfer samples to the atomizer is quite inefficient (often only 2% or less of the sample is transferred).[3]

Cold-vapor atomic absorption (see Section 9A-3) can sometimes achieve sub-parts-per-billion detection limits. In the most sensitive EPA cold-vapor atomic absorption method, the sample is digested with a permanganate-

[1] *Quality Criteria for Water*, EPA 440/5-86-001, US Environmental Protection Agency, Office of Water Regulation and Standards, 1986.

[2] http://www.epa.gov/mercury/report.htm.

[3] "Monitoring the Mercury Menace," P. Stockwell, *Today's Chemist at Work*, p. 27, November 2003; http://pubs.acs.org/subscribe/journals/tcaw/12/i11/pdf/1103instruments.pdf.

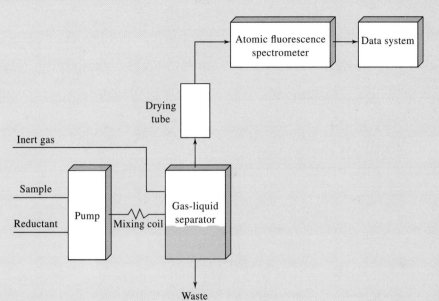

FIGURE IA2-1 Cold-vapor atomic fluorescence system for ultratrace determinations of mercury.

persulfate solution, which oxidizes all the mercury forms to Hg(II). Stannous chloride ($SnCl_2$) is then used to reduce the Hg(II) to elemental mercury, which is swept into the observation cell of the atomic absorption spectrometer by a stream of inert gas. Method 245.1 of the EPA can achieve sub-parts-per-billion sensitivity but gives nonlinear responses. EPA method 7474 uses a microwave digestion procedure and has a range from about 1 ppb to several parts per million.

Atomic Fluorescence Methods

Because regulatory requirements are becoming more stringent, it is desirable to have a more sensitive method for mercury. Atomic fluorescence spectrometry (AFS; see Section 9E) can achieve the required detection limits when combined with newly developed cold-vapor technology. Several instrument companies now market mercury analyzers based on atomic fluorescence spectrometry. Concurrent with development of commercial atomic fluorescence spectrometers for mercury monitoring has been the development of vapor generation techniques based on liquid-gas-separation technology. With these systems, essentially 100% of the sample is introduced into the spectrometer observation cell. Scattering of light from water vapor is also eliminated.

A block diagram of a cold-vapor atomic fluorescence instrument for mercury determination is given in Figure IA2-1. Here, the predigested and oxidized sample is mixed with $SnCl_2$ or another suitable reductant to generate elemental mercury. A gas-liquid separator is then used to separate the

liquid reagents from the mercury vapor, which is transferred to the observation cell by a stream of inert gas. A high-intensity mercury vapor lamp excites atomic fluorescence.

The EPA method 245.7 uses a system similar to Figure IA2-1 for mercury detection.[4] In some commercial systems, the digestion-oxidation step is also automated. The digestion is with a mixture of HCl, Br^-, and BrO_3^-. The method can achieve a detection limit of 0.1 to 0.2 ppt.

EPA method 1631[5] uses oxidation followed by a purge-and-trap method and cold-vapor atomic fluorescence. Mercury is preconcentrated in the gold-coated sand trap by amalgamation. One commercial instrument uses an atomic fluorescence detection system and claims a working range from <0.05 ppt to ~250 ppb. This method is quite useful for a variety of sample types ranging from seawater to sewage.

Conclusions

Measurements of mercury at the ultratrace levels demanded by current regulations are feasible with cold-vapor AFS. However, a good deal of care must be taken to assure high reliability. Getting rid of mercury when high levels are found in the environment is, however, another problem.

[4] *Mercury in Water by Cold Vapor Atomic Fluorescence Spectrometry*, EPA-821-R-01-008, US Environmental Protection Agency, Office of Water, 2001; http://h2o.enr.state.nc.us/lab/qa/epamethods/245_7.pdf.
[5] *Mercury in Water by Oxidation, Purge and Trap, and Cold Vapor Atomic Fluorescence Spectrometry*, Method 1631, Revision E, EPA-821-R-02-019, US Environmental Protection Agency Office of Water, 2002; http://www.epa.gov/waterscience/methods/1631.html.

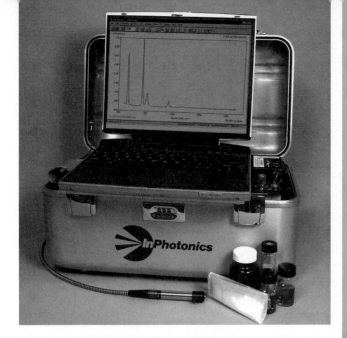

The picture shown above is of a portable Raman spectrometer designed specifically for forensics, civil defense, homeland security, and other applications where suspicious liquids or solids might be encountered. The unit has a Littrow spectrograph, a CCD detector, and a fiber-optic probe for excitation and collection. Raman spectroscopy is the subject of Chapter 18. (InPhotonics, Inc.)

Section 3 begins with an introduction to the theory and practice of ultraviolet, visible, and near-infrared spectrometry in Chapter 13, followed by applications of these regions of the spectrum in Chapter 14. Molecular luminescence spectrometry, including fluorescence, phosphorescence, and chemiluminescence measurement techniques, is described in Chapter 15. Vibrational spectroscopy is discussed in Chapter 16 (theory and instrumentation of infrared spectrometry), Chapter 17 (applications in the near-infrared, mid-infrared, and far-infrared regions of the spectrum), and Chapter 18 (Raman spectroscopy and its applications). Nuclear magnetic resonance spectrometry is investigated in Chapter 19. Molecular mass spectrometry is explored in Chapter 20. Chapter 21 addresses surface analytical techniques such as electron spectroscopy, secondary-ion mass spectrometry, scanning electron microscopy, and scanning probe microscopy. The Instrumental Analysis in Action feature discusses the use of several spectroscopic analytical tools, including surface analytical methods, in the effort to determine the authenticity of the supposedly medieval-era Vinland map.

CHAPTER THIRTEEN

An Introduction to Ultraviolet-Visible Molecular Absorption Spectrometry

Molecular absorption spectroscopy in the ultraviolet and visible spectral regions is widely used for the quantitative determination of a large number of inorganic, organic, and biological species. In this chapter, we introduce the principles of molecular absorption spectroscopy based on electromagnetic radiation in the wavelength region of 190 to 800 nm. Many of these principles, however, are applicable to spectroscopic measurements in other spectral regions such as the infrared region.

Molecular absorption spectroscopy[1] is based on the measurement of the transmittance T or the absorbance A of solutions contained in transparent cells having a path length of b centimeters. Ordinarily, the concentration of an absorbing analyte is linearly related to absorbance as given by Beer's law:

$$A = -\log T = \log \frac{P_0}{P} = \varepsilon bc \qquad (13\text{-}1)$$

All of the variables in this equation are defined in Table 13-1.

13A MEASUREMENT OF TRANSMITTANCE AND ABSORBANCE

Transmittance and absorbance, as defined in Table 13-1, cannot normally be measured in the laboratory because the analyte solution must be held in a transparent container, or cell. As shown in Figure 13-1, reflection occurs at the two air-wall interfaces as well as at the two wall-solution interfaces. The resulting beam attenuation is substantial, as we demonstrated in Example 6-2, where it was shown that about 8.5% of a beam of yellow light is lost by reflection in passing through a glass cell containing water. In addition, attenuation of a beam may occur as a result of scattering by large molecules and sometimes from absorption by the container walls. To compensate for these effects, the power of the beam transmitted by the analyte solution is usually compared with the power of the beam transmitted by an identical cell containing only solvent. An experimental transmittance and absorbance that closely approximate the true transmittance and absorbance are then obtained with the equations

$$T = \frac{P_{\text{solution}}}{P_{\text{solvent}}} \approx \frac{P}{P_0} \qquad (13\text{-}2)$$

$$A = \log \frac{P_{\text{solvent}}}{P_{\text{solution}}} \approx \log \frac{P_0}{P} \qquad (13\text{-}3)$$

The terms P_0 and P, as used in the remainder of this book, refer to the power of radiation after it has passed

---

[1] For further study, see F. Settle, ed., *Handbook of Instrumental Techniques for Analytical Chemistry*, Upper Saddle River, NJ: Prentice-Hall, 1997, Sections III and IV; J. D. Ingle Jr. and S. R. Crouch, *Spectrochemical Analysis*, Upper Saddle River, NJ: Prentice Hall, 1988; E. J. Meehan, in *Treatise on Analytical Chemistry*, 2nd ed., Part I, Vol. 7, Chaps. 1–3, P. J. Elving, E. J. Meehan, and I. M. Kolthoff, eds., New York, NY: Wiley, 1981; J. E. Crooks, *The Spectrum in Chemistry*, New York: Academic Press, 1978.

TABLE 13-1 Important Terms and Symbols for Absorption Measurements

Term and Symbol*	Definition	Alternative Name and Symbol
Incident radiant power, P_0	Radiant power in watts incident on sample	Incident intensity, I_0
Transmitted radiant power, P	Radiant power transmitted by sample	Transmitted intensity, I
Absorbance, A	$\log(P_0/P)$	Optical density, D; extinction, E
Transmittance, T	P/P_0	Transmission, T
Path length of sample, b	Length over which attenuation occurs	l, d
Concentration of absorber, c	Concentration in specified units	
Absorptivity,[†] a	$A/(bc)$	Extinction coefficient, k
Molar absorptivity,[‡] ε	$A/(bc)$	Molar extinction coefficient

*Terminology recommended by the American Chemical Society (*Anal. Chem.*, **1990**, *62*, 91).

[†] c may be expressed in grams per liter or in other specified concentration units; b may be expressed in centimeters or other units of length.

[‡] c is expressed in moles per liter; b is expressed in centimeters.

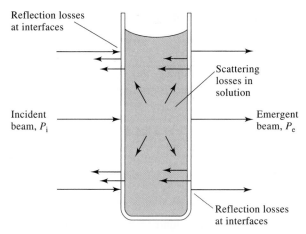

FIGURE 13-1 Reflection and scattering losses with a solution contained in a typical glass cell. Losses by reflection can occur at all the boundaries that separate the different materials. In this example, the light passes through the air-glass, glass-solution, solution-glass, and glass-air interfaces.

through cells containing the solvent and the analyte solutions, respectively.

13B BEER'S LAW

Equation 13-1 represents Beer's law. This relationship can be rationalized as follows.[2] Consider the block of absorbing matter (solid, liquid, or gas) shown in

[2]The discussion that follows is based on a paper by F. C. Strong, *Anal. Chem.*, **1952**, *24*, 338.

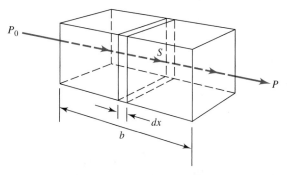

FIGURE 13-2 Radiation of initial radiant power P_0 is attenuated to transmitted power P by a solution containing c moles per liter of absorbing solution with a path length of b centimeters.

Figure 13-2. A beam of parallel monochromatic radiation with power P_0 strikes the block perpendicular to a surface. After passing through a length b of the material, which contains n absorbing atoms, ions, or molecules, its power is decreased to P as a result of absorption. Consider now a cross section of the block having an area S and an infinitesimal thickness dx. Within this section there are dn absorbing particles; we can imagine a surface associated with each particle at which photon capture will occur. That is, if a photon reaches one of these areas by chance, absorption will follow immediately. The total projected area of these capture surfaces within the section is designated as dS; the ratio of the capture area to the total area, then, is dS/S. On a statistical average, this ratio represents the probability for the capture of photons within the section.

The power of the beam entering the section, P_x, is proportional to the number of photons per unit area,

and dP_x represents the power absorbed within the section. The fraction absorbed is then $-dP_x/P_x$, and this ratio also equals the average probability for capture. The term is given a minus sign to indicate that the radiant power P decreases as it passes through the absorbing region. Thus,

$$-\frac{dP_x}{P_x} = \frac{dS}{S} \qquad (13\text{-}4)$$

Recall, now, that dS is the sum of the capture areas for particles within the section; it must therefore be proportional to the number of particles, or

$$dS = a\,dn \qquad (13\text{-}5)$$

where dn is the number of particles and a is a proportionality constant, which can be called the *capture cross section*. Combining Equations 13-4 and 13-5 and integrating over the interval between 0 and n, we obtain

$$-\int_{P_0}^{P}\frac{dP_x}{P_x} = \int_{0}^{n}\frac{a\,dn}{S}$$

When these integrals are evaluated, we find that

$$-\ln\frac{P}{P_0} = \frac{an}{S}$$

After converting to base 10 logarithms and inverting the fraction to change the sign, we obtain

$$\log\frac{P_0}{P} = \frac{an}{2.303S} \qquad (13\text{-}6)$$

where n is the total number of particles within the block shown in Figure 13-2. The cross-sectional area S can be expressed as the ratio of the volume of the block V in cubic centimeters to its length b in centimeters. Thus,

$$S = \frac{V}{b}\ \text{cm}^2$$

Substitution of this quantity into Equation 13-6 yields

$$\log\frac{P_0}{P} = \frac{anb}{2.303V} \qquad (13\text{-}7)$$

Note that n/V is the number of particles per cubic centimeter, which has the units of concentration. We can then convert n/V to moles per liter because the number of moles is given by

$$\text{number mol} = \frac{n\ \cancel{\text{particles}}}{6.02 \times 10^{23}\ \cancel{\text{particles}/mol}}$$

and c in moles per liter is given by

$$c = \frac{n}{6.02 \times 10^{23}}\ \text{mol} \times \frac{1000\ \cancel{\text{cm}^3}/\text{L}}{V\ \cancel{\text{cm}^3}}$$

$$= \frac{1000n}{6.02 \times 10^{23}V}\ \text{mol/L}$$

Combining this relationship with Equation 13-7 yields

$$\log\frac{P_0}{P} = \frac{6.02 \times 10^{23}abc}{2.303 \times 1000}$$

Finally, the constants in this equation can be collected into a single term ε to give

$$\log\frac{P_0}{P} = \varepsilon bc = A \qquad (13\text{-}8)$$

which is Beer's law.

13B-1 Application of Beer's Law to Mixtures

Beer's law also applies to a medium containing more than one kind of absorbing substance. Provided that the species do not interact, the total absorbance for a multicomponent system is given by

$$A_{\text{total}} = A_1 + A_2 + \cdots + A_n$$
$$= \varepsilon_1 bc + \varepsilon_2 bc + \cdots + \varepsilon_n bc \qquad (13\text{-}9)$$

where the subscripts refer to absorbing components $1, 2, \ldots, n$.

13B-2 Limitations to Beer's Law

Few exceptions are found to the generalization that absorbance is linearly related to path length. On the other hand, deviations from the direct proportionality between the measured absorbance and concentration frequently occur when b is constant. Some of these deviations, called *real deviations*, are fundamental and represent real limitations of the law. Others are a result of how the absorbance measurements are made (*instrumental deviations*) or a result of chemical changes that occur when the concentration changes (*chemical deviations*).

Real Limitations to Beer's Law

Beer's law describes the absorption behavior of media containing relatively low analyte concentrations; in this sense, it is a limiting law. At high concentrations (usually >0.01 M), the extent of solute-solvent interactions, solute-solute interactions, or hydrogen bonding can

 Exercise: Learn more about **absorption spectrophotometry**.

affect the analyte environment and its absorptivity. For example, at high concentrations, the average distances between the molecules or ions responsible for absorption are diminished to the point where each particle affects the charge distribution of its neighbors. These solute-solute interactions can alter the ability of the analyte species to absorb a given wavelength of radiation. Because the extent of interaction depends on concentration, deviations from the linear relationship between absorbance and concentration occur. A similar effect is sometimes encountered in media containing low absorber concentrations but high concentrations of other species, particularly electrolytes. The proximity of ions to the absorber alters the molar absorptivity of the latter by electrostatic interactions; the effect is lessened by dilution.

Although the effect of molecular interactions is ordinarily not significant at concentrations below 0.01 M, some exceptions occur among certain large organic ions or molecules. For example, the molar absorptivity at 436 nm for the cation of methylene blue in aqueous solutions is reported to increase by 88% as the dye concentration is increased from 10^{-5} to 10^{-2} M; even below 10^{-6} M, strict adherence to Beer's law is not observed.

Deviations from Beer's law also arise because absorptivity depends on the refractive index of the medium.[3] Thus, if concentration changes cause significant alterations in the refractive index n of a solution, departures from Beer's law are observed. A correction for this effect can be made by substituting the quantity $\varepsilon n/(n^2 + 2)^2$ for ε in Equation 13-8. In general, this correction is never very large and is rarely significant at concentrations less than 0.01 M.

Apparent Chemical Deviations

Apparent deviations from Beer's law arise when an analyte dissociates, associates, or reacts with a solvent to produce a product with a different absorption spectrum than the analyte. A common example of this behavior is found with aqueous solutions of acid-base indicators. For example, the color change associated with a typical indicator HIn arises from shifts in the equilibrium

$$\underset{\text{color 1}}{\text{HIn}} \rightleftharpoons \text{H}^+ + \underset{\text{color 2}}{\text{In}^-}$$

Example 13-1 demonstrates how the shift in this equilibrium with dilution results in deviation from Beer's law.

---

[3] G. Kortum and M. Seiler, *Angew. Chem.*, **1939**, *52*, 87.

EXAMPLE 13-1

Solutions containing various concentrations of the acidic indicator HIn with $K_a = 1.42 \times 10^{-5}$ were prepared in 0.1 M HCl and 0.1 M NaOH. In both media, plots of absorbance at either 430 nm or 570 nm versus the total indicator concentration are nonlinear. However, in both media, Beer's law is obeyed at 430 nm and 570 nm for the individual species HIn and In^-. Hence, if we knew the equilibrium concentrations of HIn and In^-, we could compensate for the dissociation of HIn. Usually, though, the individual concentrations are unknown and only the total concentration $c_{\text{total}} = [\text{HIn}] + [\text{In}^-]$ is known.

We now calculate the absorbance for a solution with $c_{\text{total}} = 2.00 \times 10^{-5}$ M. The magnitude of the acid dissociation constant suggests that, for all practical purposes, the indicator is entirely in the undissociated HIn form in the HCl solution and completely dissociated as In^- in NaOH. The molar absorptivities at 430 and 570 nm of the weak acid HIn and its conjugate base In^- were determined by measurements of strongly acidic and strongly basic solutions of the indicator. The results were

	ε_{430}	ε_{570}
HIn	6.30×10^2	7.12×10^3
In$^-$	2.06×10^4	9.61×10^2

What are the absorbances (1.00-cm cell) of unbuffered solutions of the indicator ranging in concentration from 2.00×10^{-5} M to 16.00×10^{-5} M?

Solution

First, we find the concentration of HIn and In^- in the unbuffered 2×10^{-5} M solution. Here,

$$\text{HIn} \rightleftharpoons \text{H}^+ + \text{In}^-$$

and

$$K_a = 1.42 \times 10^{-5} = \frac{[\text{H}^+][\text{In}^-]}{[\text{HIn}]}$$

From the equation for the dissociation reaction, we know that $[\text{H}^+] = [\text{In}^-]$. Furthermore, the mass balance expression for the indicator tells us that $[\text{In}^-] + [\text{HIn}] = 2.00 \times 10^{-5}$ M. Substitution of these relationships into the K_a expression yields

$$\frac{[\text{In}^-]^2}{2.00 \times 10^{-5} - [\text{In}^-]} = 1.42 \times 10^{-5}$$

Rearrangement yields the quadratic expression

$$[In^-]^2 + (1.42 \times 10^{-5}[In^-]) - (2.84 \times 10^{-10}) = 0$$

The positive solution to this equation is

$$[In^-] = 1.12 \times 10^{-5}\ M$$

and

$$[HIn] = (2.00 \times 10^{-5}) - (1.12 \times 10^{-5})$$
$$= 0.88 \times 10^{-5}\ M$$

We are now able to calculate the absorbance at the two wavelengths. For 430 nm, we can substitute into Equation 13-9 and obtain

$$A = \varepsilon_{In^-}b[In^-] + \varepsilon_{HIn}b[HIn]$$
$$A_{430} = (2.06 \times 10^4 \times 1.00 \times 1.12 \times 10^{-5})$$
$$+ (6.30 \times 10^2 \times 1.00 \times 0.88 \times 10^{-5})$$
$$= 0.236$$

Similarly at 570 nm,

$$A_{570} = (9.61 \times 10^2 \times 1.00 \times 1.12 \times 10^{-5})$$
$$+ (7.12 \times 10^3 \times 1.00 \times 0.88 \times 10^{-5})$$
$$= 0.073$$

Additional data, obtained in the same way, are shown in Table 13-2.

The plots in Figure 13-3, which contain the data from Table 13-2, illustrate the types of departures from Beer's law that occur when the absorbing system undergoes dissociation or association. Note that the direction of curvature is opposite at the two wavelengths.

Instrumental Deviations due to Polychromatic Radiation

Beer's law strictly applies only when measurements are made with monochromatic source radiation. In practice, polychromatic sources that have a continuous distribution of wavelengths are used in conjunction with a grating or with a filter to isolate a nearly symmetric band of wavelengths surrounding the wavelength to be employed (see Figures 7-13, 7-16, and 7-17, for example).

The following derivation shows the effect of polychromatic radiation on Beer's law. Consider a beam of radiation consisting of just two wavelengths λ' and λ''.

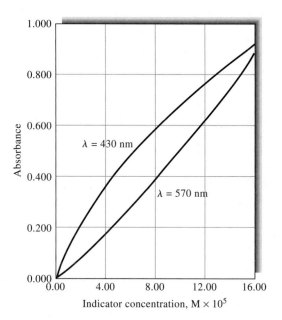

FIGURE 13-3 Chemical deviations from Beer's law for unbuffered solutions of the indicator HIn. For the data, see Example 13-1. Note that there are positive deviations at 430 nm and negative deviations at 570 nm. At 430 nm, the absorbance is primarily due to the ionized In⁻ form of the indicator and is proportional to the fraction ionized, which varies nonlinearly with the total indicator concentration. At 570 nm, the absorbance is due principally to the undissociated acid HIn, which increases nonlinearly with the total concentration.

TABLE 13-2 Calculated Absorbance Data for Various Indicator Concentrations

c_{HIn}, M	[HIn], M	[In⁻], M	A_{430}	A_{570}
2.00×10^{-5}	0.88×10^{-5}	1.12×10^{-5}	0.236	0.073
4.00×10^{-5}	2.22×10^{-5}	1.78×10^{-5}	0.381	0.175
8.00×10^{-5}	5.27×10^{-5}	2.73×10^{-5}	0.596	0.401
12.0×10^{-5}	8.52×10^{-5}	3.48×10^{-5}	0.771	0.640
16.0×10^{-5}	11.9×10^{-5}	4.11×10^{-5}	0.922	0.887

Assuming that Beer's law applies strictly for each wavelength, we may write for λ'

$$A' = \log \frac{P_0'}{P'} = \varepsilon' bc$$

or

$$\frac{P_0'}{P'} = 10^{\varepsilon' bc}$$

and

$$P' = P_0' 10^{-\varepsilon' bc}$$

Similarly, for the second wavelength λ''

$$P'' = P_0'' 10^{-\varepsilon'' bc}$$

When an absorbance measurement is made with radiation composed of both wavelengths, the power of the beam emerging from the solution is the sum of the powers emerging at the two wavelengths, $P' + P''$. Likewise, the total incident power is the sum $P_0' + P_0''$. Therefore, the measured absorbance A_m is

$$A_m = \log \frac{(P_0' + P_0'')}{(P' + P'')}$$

We then substitute for P' and P'' and find that

$$A_m = \log \frac{(P_0' + P_0'')}{(P_0' 10^{-\varepsilon' bc} + P_0'' 10^{-\varepsilon'' bc})}$$

When the molar absorptivities are the same at the two wavelengths ($\varepsilon' = \varepsilon''$), this equation simplifies to

$$A_m = \varepsilon' bc = \varepsilon'' bc$$

and Beer's law is followed. As shown in Figure 13-4, however, the relationship between A_m and concentration is no longer linear when the molar absorptivities differ. In addition, as the difference between ε' and ε'' increases, the deviation from linearity increases. This derivation can be expanded to include additional wavelengths; the effect remains the same.

If the band of wavelengths selected for spectrophotometric measurements corresponds to a region of the absorption spectrum in which the molar absorptivity of the analyte is essentially constant, departures from Beer's law are minimal. Many molecular bands in the UV-visible region fit this description. For these, Beer's law is obeyed as demonstrated by Band A in Figure 13-5. On the other hand, some absorption bands in

 Tutorial: Learn more about the **limitations of Beer's law**.

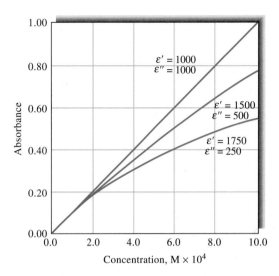

FIGURE 13-4 Deviations from Beer's law with polychromatic radiation. The absorber has the indicated molar absorptivities at the two wavelengths λ' and λ''.

the UV-visible region and many in the IR region are very narrow, and departures from Beer's law are common, as illustrated for Band B in Figure 13-5. Hence, to avoid deviations, it is advisable to select a wavelength band near the wavelength of maximum absorption where the analyte absorptivity changes little with wavelength.

It is also found experimentally that for absorbance measurements at the maximum of narrow bands, departures from Beer's law are not significant if the effective bandwidth of the monochromator or filter $\Delta\lambda_{eff}$ (Equation 7-17) is less than one tenth of the width of the absorption band at half height (full width at half maximum).

Instrumental Deviations in the Presence of Stray Radiation

We showed in Chapter 7 that the radiation exiting from a monochromator is usually contaminated with small amounts of scattered or stray radiation. This radiation, commonly called *stray light*, is defined as radiation from the instrument that is outside the nominal wavelength band chosen for the determination. This stray radiation often is the result of scattering and reflection off the surfaces of gratings, lenses or mirrors, filters, and windows. The wavelength of stray radiation often differs greatly from that of the principal radiation and, in addition, the radiation may not have passed through the sample.

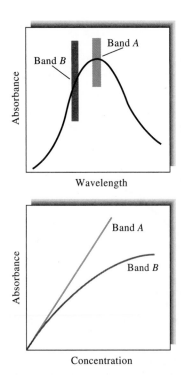

FIGURE 13-5 The effect of polychromatic radiation on Beer's law. In the spectrum at the top, the absorptivity of the analyte is nearly constant over Band *A* from the source. Note in the Beer's law plot at the bottom that using Band *A* gives a linear relationship. In the spectrum, Band *B* corresponds to a region where the absorptivity shows substantial changes. In the lower plot, note the dramatic deviation from Beer's law that results.

When measurements are made in the presence of stray radiation, the observed absorbance A' is given by

$$A' = \log \frac{P_0 + P_s}{P + P_s}$$

where P_s is the power of nonabsorbed stray radiation. Figure 13-6 shows a plot of A' versus concentration for various ratios of P_s to P_0. Note that at high concentrations and at longer path lengths stray radiation can also cause significant deviations from the linear relationship between absorbance and path length.[4]

Note also that the instrumental deviations illustrated in Figures 13-5 and 13-6 result in absorbances that are smaller than theoretical. It can be shown that

---

[4]For a discussion of the effects of stray radiation, see M. R. Sharpe, *Anal. Chem.*, **1984**, *56*, 339A.

instrumental deviations always lead to negative absorbance errors.[5]

Mismatched Cells

Another almost trivial, but important, deviation from adherence to Beer's law is caused by mismatched cells. If the cells holding the analyte and blank solutions are not of equal path length and equivalent in optical characteristics, an intercept k will occur in the calibration curve and $A = \varepsilon bc + k$ will be the actual equation instead of Equation 13-1. This error can be avoided by using either carefully matched cells or a linear regression procedure to calculate both the slope and intercept of the calibration curve. In most cases linear regression is the best strategy because an intercept can also occur if the blank solution does not totally compensate for interferences. Another way to avoid the

---

[5]E. J. Meehan, in *Treatise on Analytical Chemistry*, 2nd ed., P. J. Elving, E. J. Meehan, and I. M. Kolthoff, eds., Part I, Vol. 7, p. 73, New York: Wiley, 1981. For a spreadsheet approach to stray-light calculations, see S. R. Crouch and F. J. Holler, *Applications of Microsoft® Excel in Analytical Chemistry*, pp. 227–29, Belmont, CA: Brooks/Cole, 2004.

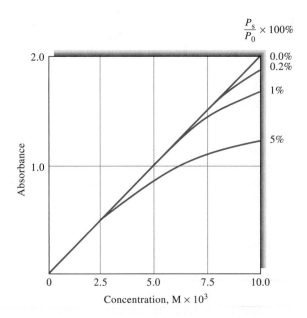

FIGURE 13-6 Apparent deviation from Beer's law brought about by various amounts of stray radiation. Note that the absorbance begins to level off with concentration at high stray-light levels. Stray light always limits the maximum absorbance that can be obtained because when the absorbance is high, the radiant power transmitted through the sample can become comparable to or lower than the stray-light level.

mismatched cell problem with single-beam instruments is to use only one cell and keep it in the same position for both blank and analyte measurements. After obtaining the blank reading, the cell is emptied by aspiration, washed, and filled with analyte solution.

13C THE EFFECTS OF INSTRUMENTAL NOISE ON SPECTROPHOTOMETRIC ANALYSES

The accuracy and precision of spectrophotometric analyses are often limited by the uncertainties or noise associated with the instrument.[6] A general discussion of instrumental noise and signal-to-noise optimization is found in Chapter 5.

13C-1 Instrumental Noise as a Function of Transmittance

As was pointed out earlier, a spectrophotometric measurement entails three steps: a $0\%\,T$ measurement or adjustment, a $100\%\,T$ measurement, and a measurement of $\%\,T$ with the sample in the radiation path. The noise associated with each of these steps combines to give a net uncertainty for the final value obtained for T. The relationship between the noise encountered in the measurement of T and the resulting *concentration uncertainty* can be derived by writing Beer's law in the form

$$c = -\frac{1}{\varepsilon b} \log T = -\frac{0.434}{\varepsilon b} \ln T \qquad (13\text{-}10)$$

To relate the standard deviation in concentration σ_c to the standard deviation in transmittance σ_T we proceed as in Section a1B-3, Appendix 1 by taking the partial derivative of this equation with respect to T, holding b and c constant. That is,

$$\frac{\partial c}{\partial T} = -\frac{0.434}{\varepsilon b T}$$

Application of Equation a1-29 (Appendix 1) gives

$$\sigma_c^2 = \left(\frac{\partial c}{\partial T}\right)^2 \sigma_T^2 = \left(\frac{-0.434}{\varepsilon b T}\right)^2 \sigma_T^2 \qquad (13\text{-}11)$$

---

[6]See L. D. Rothman, S. R. Crouch, and J. D. Ingle Jr., *Anal. Chem.*, **1975**, *47*, 1226; J. D. Ingle Jr. and S. R. Crouch, *Anal. Chem.*, **1972**, *44*, 1375; H. L. Pardue, T. E. Hewitt, and M. J. Milano, *Clin. Chem.*, **1974**, *20*, 1028; J. O. Erickson and T. Surles, *Amer. Lab.*, **1976**, *8* (6), 41; *Optimum Parameters for Spectrophotometry*, Palo Alto, CA: Varian Instruments Division, 1977.

Note that we use the population variance σ^2 instead of the sample variance s^2 when applying Equation a1-29. Dividing Equation 13-11 by the square of Equation 13-10 gives

$$\left(\frac{\sigma_c}{c}\right)^2 = \left(\frac{\sigma_T}{T \ln T}\right)^2$$

$$\frac{\sigma_c}{c} = \frac{\sigma_T}{T \ln T} = \frac{0.434 \sigma_T}{T \log T} \qquad (13\text{-}12)$$

When there is a limited number of measurements and thus a small statistical sample, we replace the population standard deviations σ_c and σ_T with the sample standard deviations s_c and s_T (Section a1B-1, Appendix 1) and obtain

$$\frac{s_c}{c} = \frac{0.434 s_T}{T \log T} \qquad (13\text{-}13)$$

This equation relates the relative standard deviation of c (s_c/c) to the absolute standard deviation of the transmittance measurement (s_T). Experimentally, s_T can be evaluated by making, say, twenty replicate transmittance measurements ($N = 20$) of the transmittance of a solution in exactly the same way and substituting the data into Equation a1-10, Appendix 1.

Equation 13-13 shows that the uncertainty in a photometric concentration measurement varies nonlinearly with the magnitude of the transmittance. The situation is somewhat more complicated than is suggested by Equation 13-13, however, because the uncertainty s_T is also *dependent on T*.

13C-2 Sources of Instrumental Noise

In a detailed theoretical and experimental study, Rothman, Crouch, and Ingle have described several sources of instrumental uncertainties and shown their net effect on the precision of absorbance or transmittance measurements.[7] These uncertainties fall into one of three categories depending on how they are affected by the magnitude of the photocurrent and thus T. For Case I uncertainties, the precision is independent of T; that is, s_T is equal to a constant k_1. For Case II uncertainties, the precision is directly proportional to $\sqrt{T^2 + T}$. Finally, Case III uncertainties are directly proportional to T. Table 13-3 summarizes information about the sources of these three types of uncertainty and the kinds of instruments where each is likely to be encountered.

---

[7]L. D. Rothman, S. R. Crouch, and J. D. Ingle Jr., *Anal. Chem.*, **1975**, *47*, 1226.

TABLE 13-3 Types and Sources of Uncertainties in Transmittance Measurements

Category	Characterized by[a]	Typical Sources	Likely To Be Important In
Case I	$s_T = k_1$	Limited readout resolution	Inexpensive photometers and spectrophotometers having small meters or digital displays
		Heat detector Johnson noise	IR and near-IR spectrophotometers and photometers
		Dark current and amplifier noise	Regions where source intensity and detector sensitivity are low
Case II	$s_T = k_2\sqrt{T^2 + T}$	Photon detector shot noise	High-quality UV-visible spectrophotometers
Case III	$s_T = k_3 T$	Cell positioning uncertainties	High-quality UV-visible and IR spectrophotometers
		Source flicker	Inexpensive photometers and spectrophotometers

[a] k_1, k_2, and k_3 are constants for a given system.

Case I: $s_T = k_1$

Case I uncertainties often appear in less expensive ultraviolet and visible spectrophotometers or photometers equipped with meters or digital readouts with limited resolution. For example, some digital instruments have 3½-digit displays. These can display the result to $0.1\%\,T$. Here, the readout resolution can limit the measurement precision such that the absolute uncertainty in T is the same from $0\%\,T$ to $100\%\,T$. A similar limitation occurs with older analog instruments with limited meter resolution.

Infrared and near-infrared spectrophotometers also exhibit Case I behavior. With these, the limiting random error usually arises from Johnson noise in the thermal detector. Recall (Section 5B-2) that this type of noise is independent of the magnitude of the photocurrent; indeed, fluctuations are observed even in the absence of radiation when there is essentially zero net current.

Dark current and amplifier noise are usually small compared with other sources of noise in photometric and spectrophotometric instruments and become important only under conditions of low photocurrents where the lamp intensity or the transducer sensitivity is low. For example, such conditions are often encountered near the wavelength extremes for an instrument.

The precision of concentration data obtained with an instrument that is limited by Case I noise can be obtained directly by substituting an experimentally determined value for $s_T = k_1$ into Equation 13-13. Here, the precision of a particular concentration determina-

 Simulation: Learn more about the **effects of instrumental noise**.

tion depends on the magnitude of T even though the instrumental precision is independent of T. The third column of Table 13-4 shows data obtained with Equation 13-13 when an absolute standard deviation s_T of ±0.003, or $\pm0.3\%\,T$, was assumed. Curve A in Figure 13-7 shows a plot of the data. Note that a minimum is reached at an absorbance of about 0.5. Note also that the relative concentration error rapidly rises at absorbances lower than about 0.1 and greater than about 1.0.

An uncertainty of $0.3\%\,T$ is typical of many moderately priced spectrophotometers or photometers. With these instruments, concentration errors of 1% to 2% relative are to be expected if the absorbance of the sample lies between about 0.1 and 1.

Case II: $s_T = k_2\sqrt{T^2 + T}$

This type of uncertainty often limits the precision of the highest quality instruments. It has its origin in shot noise (Section 5B-2), which occurs whenever the current involves transfer of charge across a junction, such as the movement of electrons from the cathode to the anode of a photomultiplier tube. Here, an electric current results from a series of discrete events (emission of electrons from a cathode). The average number of these events per unit time is proportional to the photon flux. The frequency of events and thus the current is randomly distributed about the average value. The magnitude of the current fluctuations is proportional to the square root of current (see Equation 5-5). The effect of shot noise on s_c is derived by substituting s_T into Equation 13-13. Rearrangement leads to

TABLE 13-4 Relative Precision of Concentration Measurements as a Function of Transmittance and Absorbance for Three Categories of Instrument Noise

Transmittance, T	Absorbance, A	Relative Standard Deviation in Concentration[a]		
		Case I Noise[b]	Case II Noise[c]	Case III Noise[d]
0.95	0.022	±6.2	±8.4	±25.3
0.90	0.046	±3.2	±4.1	±12.3
0.80	0.097	±1.7	±2.0	±5.8
0.60	0.222	±0.98	±0.96	±2.5
0.40	0.398	±0.82	±0.61	±1.4
0.20	0.699	±0.93	±0.46	±0.81
0.10	1.00	±1.3	±0.43	±0.56
0.032	1.50	±2.7	±0.50	±0.38
0.010	2.00	±6.5	±0.65	±0.2
0.0032	2.50	±16.3	±0.92	±0.23
0.0010	3.00	±43.4	±1.4	±0.19

[a] $(s_c/c) \times 100\%$.

[b] From Equation 13-13 with $s_T = k_1 = \pm 0.0030$.

[c] From Equation 13-14 with $k_2 = \pm 0.0030$.

[d] From Equation 13-15 with $k_3 = \pm 0.013$.

$$\frac{s_c}{c} = \frac{0.434 k_2}{\log T} \sqrt{\frac{1}{T} + 1} \qquad (13\text{-}14)$$

The data in column 4 of Table 13-4 were calculated using Equation 13-14. Curve B in Figure 13-7 is a plot

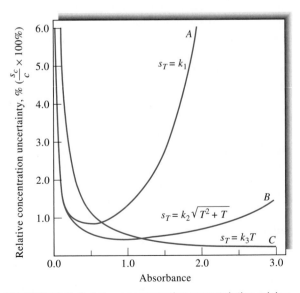

FIGURE 13-7 Relative concentration uncertainties arising from various categories of instrumental noise. A, case I; B, Case II; C, Case III. The data are taken from Table 13-4.

of such data. Note the much broader minimum in the concentration uncertainty. Note also that a large range of absorbances can be measured without the concentration error becoming greater than 1% to 2%. This increased range represents a major advantage of photon-type detectors over thermal types, which are represented by curve A in the figure. As with Johnson-noise-limited instruments, shot-limited instruments do not give very reliable concentration data at transmittances greater than 95% (or $A < 0.02$).

Case III: $s_T = k_3 T$

By substituting $s_T = k_3 T$ into Equation 13-13, we see that the relative standard deviation in concentration from this type of uncertainty is inversely proportional to the logarithm of the transmittance.

$$\frac{s_c}{c} = \frac{0.434 k_3}{\log T} \qquad (13\text{-}15)$$

Column 5 of Table 13-4 contains data obtained from Equation 13-15 when k_3 is assumed to have a value of 0.013, which approximates the value observed in the Rothman, Crouch, and Ingle study. The data are plotted as curve C in Figure 13-7. Note that this type of uncertainty is important at low absorbances (high transmittances) but approaches zero at high absorbances.[8]

One source of noise of this type is the slow drift in the radiant output of the source. This type of noise can be called *source flicker noise* (Section 5B-2). The effects of fluctuations in the intensity of a source can be minimized by the use of a constant-voltage power supply or a feedback system in which the source intensity is maintained at a constant level. Modern double-beam spectrophotometers (Sections 13D-2 and 13D-3) can also help cancel the effect of flicker noise. With many instruments, source flicker noise does not limit performance.

An important and widely encountered noise source, one that is proportional to transmittance, results from failure to position sample and reference cells reproducibly with respect to the beam during replicate transmittance measurements. All cells have minor imperfections. As a consequence, reflection and scattering losses vary as different sections of the cell window are exposed to the beam; small variations in transmittance result. Rothman, Crouch, and Ingle have shown that this uncertainty often is the most common limitation to the accuracy of high-quality ultraviolet-visible spectrophotometers. It is also a serious source of uncertainty in infrared instruments.

One method of reducing the effect of cell positioning with a double-beam instrument is to leave the cells in place during calibration and analysis. New standards and samples are then introduced after washing and rinsing the cell in place with a syringe. Care must be taken to avoid touching or jarring the cells during this process.

13C-3 Effect of Slit Width on Absorbance Measurements

As shown in Section 7C-3, narrow slit widths are required to resolve complex spectra.[9] For example, Figure 13-8 illustrates the loss of detail that occurs when slit widths are increased from small values on the left to larger values in the middle and right. In this example, the absorption spectrum of benzene vapor was obtained at slit settings that provided effective bandwidths of 1.6, 4, and 10 nm. For qualitative studies, the loss of resolution that accompanies the use of wider slits is often important because the details of spectra are useful for identifying species.

Figure 13-9 illustrates a second effect of slit width on spectra made up of narrow peaks. Here, the spectrum of a praseodymium chloride solution was obtained at slit widths of 1.0, 0.5, and 0.1 mm. Note that the peak absorbance values increase significantly (by as much as 70% in one instance) as the slit width decreases. At slit settings less than about 0.14 mm, absorbances were found to become independent of slit width. Careful inspection of Figure 13-8 reveals the same type of effect. In both sets of spectra, the areas under the individual peaks are the same, but wide slit widths result in broader peaks with lower maximum absorbances.

From both of these illustrations, we can conclude that quantitative measurement of narrow absorption bands requires using narrow slit widths or, alternatively, very reproducible slit-width settings. Unfortunately, a decrease in slit width by a factor of 10 reduces the radiant power by a factor of 100 because the radiant power is proportional to the square of the slit width.[10] There is thus a trade-off between resolution and signal-to-noise ratio. Often, a compromise slit width must be chosen. The situation becomes particularly serious in spectral regions where the output of the source or the sensitivity of the detector is low. Under such circumstances, adequate signal-to-noise ratio may require slit widths large enough to result in partial or total loss of spectral fine structure.

In general, it is good practice to narrow slits no more than is necessary for resolution of the spectrum at hand. With a variable-slit spectrophotometer, proper slit adjustment can be determined by acquiring spectra at progressively narrower slits until maximum absorbances become constant. Generally, constant peak heights are observed when the effective bandwidth of the instrument is less than one tenth the full width at half maximum of the absorption band.

[8]For a spreadsheet approach to plotting the relative concentration errors for the cases considered here, see S. R. Crouch and F. J. Holler, *Applications of Microsoft® Excel in Analytical Chemistry*, pp. 229–32. Belmont, CA: Brooks/Cole, 2004.

[9]For a discussion of the effects of slit width on spectra, see *Optimum Parameters for Spectrophotometry*, Palo Alto, CA: Varian Instruments Division, 1977; F. C. Strong III, *Anal. Chem.*, **1976**, *48*, 2155; D. D. Gilbert, *J. Chem. Educ.*, **1991**, *68*, A278.

[10]J. D. Ingle Jr. and S. R. Crouch, *Spectrochemical Analysis*, p. 366, Upper Saddle River, NJ: Prentice-Hall, 1988.

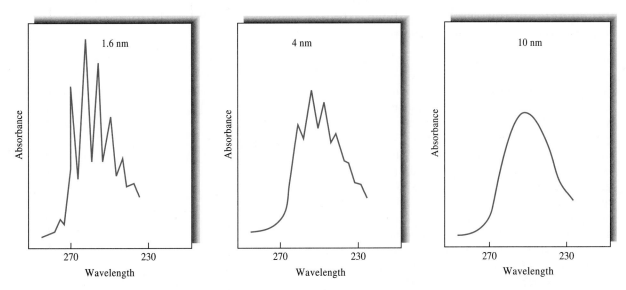

FIGURE 13-8 Effect of bandwidth on spectral detail for a sample of benzene vapor. Note that as the spectral bandwidth increases, the fine structure in the spectrum is lost. At a bandwidth of 10 nm, only a broad absorption band is observed.

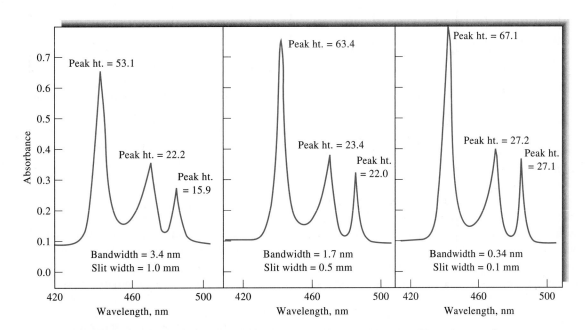

FIGURE 13-9 Effect of slit width (spectral bandwidth) on peak heights. Here, the sample was a solution of praseodymium chloride. Note that as the spectral bandwidth decreases by decreasing the slit width from 1.0 mm to 0.1 mm, the peak heights increase. (From *Optimum Spectrophotometer Parameters*, Application Report, 14-2, Varian Inc., Palo Alto, CA.)

13C-4 Effect of Scattered Radiation at Wavelength Extremes of an Instrument

We have already noted that scattered radiation may cause instrumental deviations from Beer's law. When measurements are made at the wavelength extremes of an instrument, the effects of stray radiation may be even more serious and on occasion may lead to the appearance of false absorption bands. For example, consider an older visible spectrophotometer equipped with glass optics, a tungsten source, and a photovoltaic cell detector. At wavelengths below about 380 nm, the windows, cells, and prism begin to absorb radiation, thus reducing the energy reaching the transducer. The output of the source falls off rapidly in this region as does the sensitivity of the transducer. Thus, the total signal for the 100% T adjustment may be as low as 1% to 2% of that in the region between 500 and 650 nm.

The scattered radiation, however, is often made up of wavelengths to which the instrument is highly sensitive. Thus, the effects of stray radiation can be greatly enhanced. Indeed, in some instances the output signal produced by the stray radiation may exceed that produced by the monochromator output beam. In such cases, the component of the measured transmittance due to the stray radiation may be as large as or exceed the true transmittance.

An example of a false band appearing at the wavelength extremes of a visible-region spectrophotometer is shown in Figure 13-10. The spectrum of a solution of cerium(IV) obtained with an ultraviolet-visible spectrophotometer, sensitive in the region of 200 to 750 nm, is shown by curve B. Curve A is a spectrum of the same solution obtained with a simple visible spectrophotometer. The apparent maximum shown in curve A arises from the instrument responding to stray wavelengths longer than 400 nm. As can be seen in curve B, these longer wavelengths are not absorbed by the cerium(IV) ions. This same effect is sometimes observed with ultraviolet-visible instruments when attempts are made to measure absorbances at wavelengths lower than about 200 nm.

13D INSTRUMENTATION

Instruments for making molecular absorption measurements in the ultraviolet, visible, and near-infrared regions are produced by dozens of companies. There are hundreds of instrument makes and models from which to choose. Some are simple and inexpensive

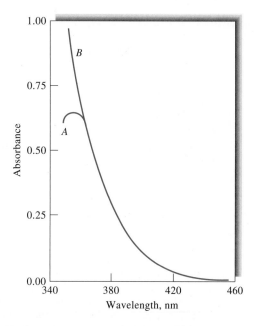

FIGURE 13-10 Spectrum of cerium(IV) obtained with a spectrophotometer having glass optics (A) and quartz optics (B). The false peak in A arises from transmission of stray radiation of longer wavelengths.

(a few hundred dollars); others are complex, computer-controlled, scanning instruments costing $30,000 or more. The simpler instruments are often useful only in the visible region for quantitative measurements at a single wavelength. The more complex instruments can provide spectral scanning at selectable resolution, measurements in the ultraviolet as well as the visible regions, compensation for source intensity fluctuations, and several other features.[11]

13D-1 Instrument Components

Instruments for measuring the absorption of ultraviolet, visible, and near-infrared radiation are made up of one or more (1) sources, (2) wavelength selectors, (3) sample containers, (4) radiation transducers, and (5) signal processors and readout devices. The design and performance of components (2), (4), and (5) were described in considerable detail in Chapter 7 and thus are not discussed further here. We will, however, consider briefly the characteristics of sources and sample containers for the region of 190 to 3000 nm.

---

[11] For an interesting discussion of commercial instruments for UV-visible absorption measurements, see R. Jarnutowski, J. R. Ferraro, and D. C. Lankin, *Spectroscopy*, **1992**, *7* (7), 22.

Sources

For the purposes of molecular absorption measurements, a continuum source is required whose radiant power does not change sharply over a considerable range of wavelengths.

Deuterium and Hydrogen Lamps. A continuum spectrum in the ultraviolet region is produced by electrical excitation of deuterium or hydrogen at low pressure. The mechanism by which a continuum spectrum is produced involves initial formation of an excited molecular species followed by dissociation of the excited molecule to give two atomic species plus an ultraviolet photon. The reactions for deuterium are

$$D_2 + E_e \rightarrow D_2^* \rightarrow D' + D'' + h\nu$$

where E_e is the electrical energy absorbed by the molecule and D_2^* is the excited deuterium molecule. The energetics for the overall process can be represented by the equation

$$E_e = E_{D_2^*} = E_{D'} + E_{D''} + h\nu$$

Here, $E_{D_2^*}$ is the fixed quantized energy of D_2^* whereas $E_{D'}$ and $E_{D''}$ are the kinetic energies of the two deuterium atoms. The sum of $E_{D'}$ and $E_{D''}$ can vary continuously from zero to $E_{D_2^*}$; thus, the energy and the frequency of the photon can also vary continuously. That is, when the two kinetic energies are by chance small, $h\nu$ will be large, and conversely. The consequence is a true continuum spectrum from about 160 nm to the beginning of the visible region, as shown in Figure 13-11b.

Most modern lamps of this type contain deuterium and are of a low-voltage type in which an arc is formed between a heated, oxide-coated filament and a metal electrode (see Figure 13-11a). The heated filament provides electrons to maintain a direct current when about 40 V is applied between the filament and the electrode. A regulated power supply is required for constant intensities.

An important feature of deuterium and hydrogen discharge lamps is the shape of the aperture between the two electrodes, which constricts the discharge to a narrow path. As a consequence, an intense ball of radiation about 1 to 1.5 mm in diameter is produced. Deuterium gives a somewhat larger and brighter ball than hydrogen, which accounts for the widespread use of deuterium.

Both deuterium and hydrogen lamps produce outputs in the range of 160–800 nm. In the ultraviolet region (190–400 nm), a continuum spectrum exists as

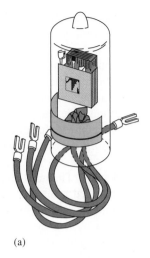

(a)

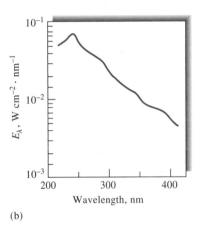

(b)

FIGURE 13-11 (a) A deuterium lamp of the type used in spectrophotometers and (b) its spectrum. The plot is of irradiance E_λ (proportional to radiant power) versus wavelength. Note that the maximum intensity occurs at ~225 nm. Typically, instruments switch from deuterium to tungsten at ~350 nm.

can be seen in Figure 13-11b. At longer wavelengths (>400 nm), the spectra from these lamps are no longer continua, but consist of emission lines and bands superimposed on a weak continuum. For many applications, the line or band emission represents a nuisance. However, some modern array-detector instruments use a deuterium source at wavelengths as long as 800 nm. With these, the array can be exposed for longer times to compensate for the low source intensity in the visible region. Because the entire source spectrum is readily obtained with a solvent blank and then with the sample, the presence of emission lines does not interfere with calculation of the absorption spectrum. The line emission has also been used for wavelength calibration of absorption instruments.

Quartz windows must be employed in deuterium and hydrogen lamps because glass absorbs strongly at wavelengths less than about 350 nm. Although the deuterium lamp continuum spectrum extends to wavelengths as short as 160 nm, the useful lower limit is about 190 nm because of absorption by the quartz windows.

Tungsten Filament Lamps. The most common source of visible and near-infrared radiation is the tungsten filament lamp. The energy distribution of this source approximates that of a blackbody (see Figure 6-22) and is thus temperature dependent. In most absorption instruments, the operating filament temperature is 2870 K; the bulk of the energy is thus emitted in the infrared region. A tungsten filament lamp is useful for the wavelength region between 350 and 2500 nm. Figure 13-12 shows a typical tungsten lamp and its spectrum.

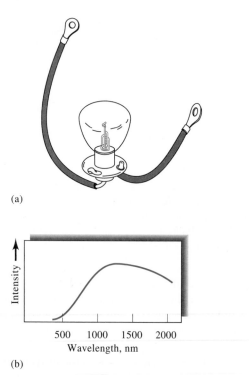

(a)

(b)

FIGURE 13-12 (a) A tungsten lamp of the type used in spectroscopy and its spectrum (b). Intensity of the tungsten source is usually quite low at wavelengths shorter than about 350 nm. Note that the intensity reaches a maximum in the near-IR region of the spectrum (~1200 nm in this case).

Absorption of radiation by the glass envelope that houses the filament imposes the lower wavelength limit.

In the visible region, the energy output of a tungsten lamp varies approximately as the fourth power of the operating voltage. As a consequence, close voltage control is required for a stable radiation source. Electronic voltage regulators or feedback-controlled power supplies are usually employed to obtain the required stability.

Tungsten-halogen lamps, also called quartz-halogen lamps, contain a small quantity of iodine within a quartz envelope that houses the tungsten filament. Quartz allows the filament to be operated at a temperature of about 3500 K, which leads to higher intensities and extends the range of the lamp well into the UV region. The lifetime of a tungsten-halogen lamp is more than double that of the ordinary lamp. This added life results from the reaction of the iodine with gaseous tungsten that forms by sublimation and ordinarily limits the life of the filament; the product is the volatile WI_2. When molecules of this compound strike the filament, decomposition occurs, which redeposits tungsten. Tungsten-halogen lamps are significantly more efficient and extend the output wavelength range well into the ultraviolet. For these reasons, they are found in many modern spectroscopic instruments.

Light-Emitting Diodes. Light-emitting diodes (LEDs) are used as sources in some absorption spectrometers. An LED is a *pn*-junction device that, when forward biased, produces radiant energy. Diodes made from gallium aluminum arsenide ($\lambda_m = 900$ nm), gallium arsenic phosphide ($\lambda_m = 650$ nm), gallium phosphide ($\lambda_m = 550$ nm), gallium nitride ($\lambda_m = 465$ nm), and indium gallium nitride ($\lambda_m = 450$ nm) are widely available. Mixtures of these compounds are used to shift the wavelength maximum to anywhere in the region of 375 nm to 1000 nm or more. LEDs produce a spectral continuum over a narrow wavelength range. Typically, the full width at half maximum of an LED is 20 to 50 nm. As spectroscopic sources, LEDs can be used as "semimonochromatic" sources or in conjunction with interference filters to further narrow the spectral output. They can be operated in a continuous mode or in a pulsed mode.

"White" LEDs are also available in which the light from a blue LED (InGaN) strikes a phosphor, which emits a spectral continuum typically in the range of 400–800 nm. Such LEDs are being used to make lighting products such as flashlights. They have the advantage of long lifetimes and a smaller environmental impact than tungsten filament lamps.

Xenon Arc Lamps. The xenon arc lamp produces intense radiation by the passage of current through an atmosphere of xenon. The spectrum is a continuum over the range between about 200 and 1000 nm, with the peak intensity occurring at about 500 nm (see Figure 6-22). In some instruments, the lamp is operated intermittently by regular discharges from a capacitor; high intensities are obtained.

Sample Containers

In common with the other optical elements of an absorption instrument, the cells, or cuvettes, that hold the sample and solvent must be constructed of a material that passes radiation in the spectral region of interest. Thus, as shown in Figure 7-2a, quartz or fused silica is required for work in the ultraviolet region (below 350 nm). Both of these substances are transparent throughout the visible and near-infrared regions to about 3 μm. Silicate glasses can be employed in the region between 350 and 2000 nm. Plastic containers are also used in the visible region.

The best cells minimize reflection losses by having windows that are perfectly normal to the direction of the beam. The most common path length for studies in the ultraviolet and visible regions is 1 cm. Matched, calibrated cells of this size are available from several commercial sources. Other path lengths, from 0.1 cm (and shorter) to 10 cm, can also be purchased. Transparent spacers for shortening the path length of 1-cm cells to 0.1 cm are also available.

Cylindrical cells are sometimes employed in the ultraviolet and visible regions because they are inexpensive. Special care must be taken to reproduce the position of the cell with respect to the beam; otherwise, variations in path length and reflection losses at the curved surfaces can cause significant errors.

The quality of absorbance data depends critically on the way the cells are used and maintained. Fingerprints, grease, or other deposits on the walls markedly alter the transmission characteristics of a cell. Thus, it is essential to clean cells thoroughly before and after use. The surface of the windows must not be touched during handling. Matched cells should never be dried by heating in an oven or over a flame — such treatment may cause physical damage or a change in path length. The cells should be regularly calibrated against each other with an absorbing solution.

13D-2 Types of Instruments

In this section, we consider four general types of spectroscopic instruments: (1) single beam, (2) double beam in space, (3) double beam in time, and (4) multichannel.

Single-Beam Instruments

Figure 13-13a is a schematic of a single-beam instrument for absorption measurements. It consists of a tungsten or deuterium lamp, a filter or a monochromator for wavelength selection, matched cells that can be placed alternately in the radiation beam, one of the transducers described in Section 7E, an amplifier, and a readout device. Normally, a single-beam instrument requires a stabilized voltage supply to avoid errors resulting from changes in the beam intensity during the time required to make the $100\%\,T$ measurement and determine $\%\,T$ for the analyte.

Single-beam instruments vary widely in their complexity and performance characteristics. The simplest and least expensive consists of a battery-operated tungsten bulb as the source, a set of glass filters for wavelength selection, test tubes for sample holders, a photovoltaic cell as the transducer, and an analog meter as the readout device. At the other extreme are sophisticated, computer-controlled instruments with a range of 200 to 1000 nm or more. These spectrophotometers have interchangeable tungsten and deuterium lamp sources, use rectangular silica cells, and are equipped with a high-resolution grating monochromator with variable slits. Photomultiplier tubes are used as transducers, and the output is often digitized, processed, and stored in a computer so that it can be printed or plotted in several forms.

Double-Beam Instruments

Many modern photometers and spectrophotometers are based on a double-beam design. Figure 13-13b illustrates a double-beam-in-space instrument in which two beams are formed in space by a V-shape mirror called a *beamsplitter*. One beam passes through the reference solution to a photodetector, and the second simultaneously traverses the sample to a second, matched detector. The two outputs are amplified, and their ratio (or the logarithm of their ratio) is determined electronically or by a computer and displayed by the readout device. With manual instruments, the measurement is a two-step operation involving first the zero adjustment with a shutter in place between selector and

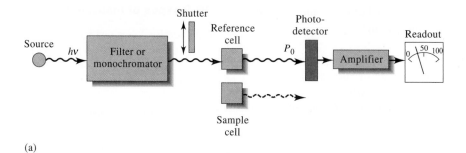

(a)

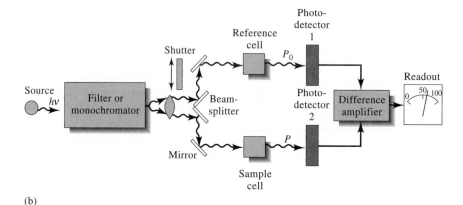

(b)

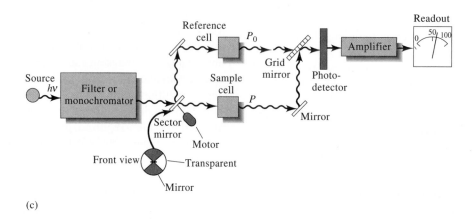

(c)

FIGURE 13-13 Instrumental designs for UV-visible photometers or spectrophotometers. In (a), a single-beam instrument is shown. Radiation from the filter or monochromator passes through either the reference cell or the sample cell before striking the photodetector. In (b), a double-beam-in-space instrument is shown. Here, radiation from the filter or monochromator is split into two beams that simultaneously pass through the reference and sample cells before striking two matched photodetectors. In the double-beam-in-time instrument (c), the beam is alternately sent through reference and sample cells before striking a single photodetector. Only a matter of milliseconds separate the beams as they pass through the two cells.

Exercise: Learn more about **single-beam, double-beam, and multichannel instruments**.

beamsplitter. In the second step, the shutter is opened and the transmittance or absorbance is displayed directly.

The second type of double-beam instrument is illustrated in Figure 13-13c. Here, the beams are separated in time by a rotating sector mirror that directs the entire beam from the monochromator first through the reference cell and then through the sample cell. The pulses of radiation are recombined by another sector mirror, which transmits one pulse and reflects the other to the transducer. As shown by the insert labeled "front view" in Figure 13-13c, the motor-driven sector mirror is made up of pie-shape segments, half of which are mirrored and half of which are transparent. The mirrored sections are held in place by blackened metal frames that periodically interrupt the beam and prevent its reaching the transducer. The detection circuit is programmed to use these periods to perform the dark-current adjustment. The double-beam-in-time approach is generally preferred because of the difficulty in matching the two detectors needed for the double-beam-in-space design.

Double-beam instruments offer the advantage that they compensate for all but the most short-term fluctuations in the radiant output of the source as well as for drift in the transducer and amplifier. They also compensate for wide variations in source intensity with wavelength (see Figures 13-11 and 13-12). Furthermore, the double-beam design lends itself well to the continuous recording of transmittance or absorbance spectra.

Multichannel Instruments

A new type of spectrophotometer appeared on the market in the early 1980s and is based on one of the array detectors (photodiode array or linear charge-coupled-device [CCD] array) described in Section 7E. These instruments are usually of the single-beam design shown in Figure 13-14. With multichannel systems the dispersive system is a grating spectrograph placed after the sample or reference cell. The array detector is placed in the focal plane of the spectrograph, where the dispersed radiation strikes it.

The photodiode array discussed in Section 7E-3, consists of a linear array of several hundred photodiodes (256, 512, 1024, 2048) that have been formed along the length of a silicon chip. Typically, the chips are 1 to 6 cm long, and the widths of the individual diodes are 15 to 50 μm (see Figure 13-15). Linear CCD

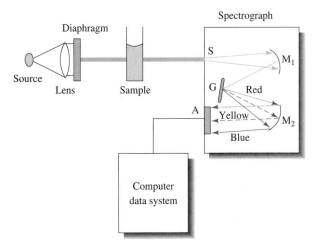

FIGURE 13-14 Diagram of a multichannel spectrometer based on a grating spectrograph with an array detector. Radiation from the tungsten or deuterium source is made parallel and reduced in size by the lens and diaphragm. Radiation transmitted by the sample enters the spectrograph through slit S. Collimating mirror M_1 makes the beam parallel before it strikes the grating G. The grating disperses the radiation into its component wavelengths, which are then focused by focusing mirror M_2 onto the photodiode or CCD array A. The output from the array detector is then processed by the computer data system.

arrays typically consist of 2048 elements, with each element being ~14 μm wide. The linear CCD is substantially more sensitive than the photodiode array and behaves much like a linear array of miniature photomultiplier tubes.

FIGURE 13-15 Diode arrays of various sizes. (Courtesy of Hamamatsu Photonics, Bridgewater, NJ.)

With single-beam designs, the array dark current is measured and stored in computer memory. Next, the spectrum of the source is obtained and stored in memory after dark-current subtraction. Finally, the raw spectrum of the sample is obtained and, after dark-current subtraction, the sample values are divided by the source values at each wavelength to give absorbances. Multichannel instruments can also be configured as double-beam-in-time spectrophotometers.

The spectrograph entrance slit of multichannel instruments is usually variable from about the width of one of the array elements to many times wider. Some spectrometers have no entrance slit but instead use a fiber optic as the entrance aperture. Multichannel spectrometers using both photodiode arrays and CCD arrays are capable of obtaining an entire spectrum in a few milliseconds. With array detectors the light can be integrated on a chip or multiple scans can be averaged in computer memory to enhance the signal-to-noise ratio.

A multichannel instrument is a powerful tool for studies of transient intermediates in moderately fast reactions, for kinetic studies, and for the qualitative and quantitative determination of the components exiting from a liquid chromatographic column or a capillary electrophoresis column. They are also useful for general-purpose scanning experiments. Some have the software necessary to analyze the time dependence at four or more wavelengths for kinetic studies.

Complete general-purpose array-detector-based spectrophotometers are available commercially for $5000 to $10,000 and up. Several instrument companies combine array-detector systems with fiber-optic probes that transport the light to and from the sample.

13D-3 Some Typical Instruments

In the sections that follow, some typical photometers and spectrophotometers are described. The instruments were chosen to illustrate the wide variety of design variables that are encountered.

Photometers

Photometers provide simple, relatively inexpensive tools for performing absorption measurements. Filter photometers are often more convenient and more rugged and are easier to maintain and use than the more sophisticated spectrophotometers. Furthermore, photometers characteristically have high radiant energy throughputs and thus good signal-to-noise ratios even with relatively simple and inexpensive transducers and circuitry. Filter photometers are particularly useful in portable instruments intended for field use or for use in measuring the absorbances of flowing streams. These photometers are also employed for quantitative determinations in clinical laboratories.

Visible Photometers. Figure 13-16 presents schematics for two visible photometers or colorimeters. The upper figure illustrates a single-beam, direct-reading instrument consisting of a tungsten filament lamp or an LED as a source, a lens to provide a parallel beam of light, a filter, and a photodiode transducer. The current produced by the photodiode is processed with electronics or a computer to give a direct readout in absorbance (shown) or, in some cases, transmittance. For most instruments, the dark current ($0\% T$) is obtained by blocking the light beam with a shutter. The $100\% T$ ($0\ A$) is adjusted with solvent or a reagent blank in the light path. With some instruments, the $0\ A$ adjustment is made by changing the voltage applied to the lamp. In others, the aperture size of a diaphragm located in the light path is altered. The sample is then inserted into the light path. Most modern colorimeters store the photodiode signal for the reference (proportional to P_0) and compute the ratio of this signal to the photodiode signal for the sample (proportional to the radiant power P). The absorbance is calculated as the logarithm of the ratio of these signals (Equation 13-3).

A modern LED-based photometer is shown in Figure 13-17. With some instruments, the wavelength is automatically changed by changing the LED or filter. Some instruments are made to operate at only one fixed wavelength. Calibration can be accomplished with two or more standards. The instrument shown is of the fixed-wavelength design with a bandwidth of 15 nm. Wavelengths of 420, 450, 476, 500, 550, 580, 600, and 650 nm are available.

Figure 13-16b is a schematic representation of a double-beam photometer used to measure the absorbance of a sample in a flowing stream. Here, the light beam is split by a two-branched (bifurcated) fiber optic, which transmits about 50% of the radiation striking it in the upper arm and about 50% in the lower arm. One beam passes through the sample, and the other passes through the reference cell. Filters are placed after the cells before the photodiode transducers. Note that this is the double-beam-in-space design, which requires photodiodes with nearly identical response. The electrical outputs from the two photodiodes are converted

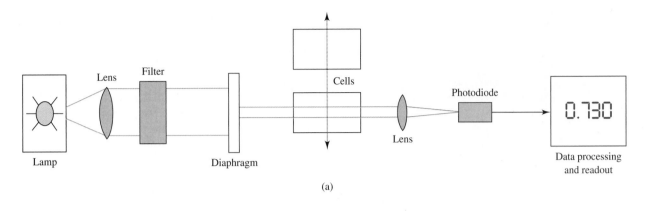

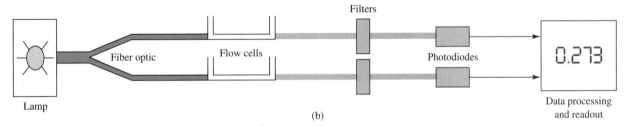

FIGURE 13-16 Single-beam photometer (a) and double-beam photometer for flow analysis (b). In the single-beam system, the reference cell is first placed in the light path and later replaced by the sample cell. In the double-beam system (b), a fiber optic splits the beam into two branches. One passes through the sample cell and the other through the reference cell. Two matched photodiodes are used in this double-beam-in-space arrangement.

FIGURE 13-17 Photograph of a simple LED-based colorimeter. (Hach Company, USA.)

to voltages and the signals are processed by a log ratio amplifier or a computer to give a readout proportional to absorbance.

Probe-Type Photometers. Figure 13-18 shows a photograph and a schematic of an interesting, commercially available, dipping-type photometer, which employs an optical fiber to transmit light from a source to a layer of solution lying between the glass seal at the end of the fiber and a mirror. The reflected radiation from the latter passes to a photodiode detector via a second glass fiber. The photometer uses an amplifier with an electronic chopper synchronized with the light source; as a result, the photometer does not respond to extraneous radiation. Filter options include "drop-in" filters and a filter wheel with six interference filters for selected applications. Custom filters are also available. Probe tips are manufactured from stainless steel, Swagelok® Stainless Steel, Pyrex®, and Acid-resistant Lexan® Plastic. Light path lengths that vary from 1 mm to 10 cm are available.

Absorbance is measured by first dipping the probe into the solvent and then into the solution to be

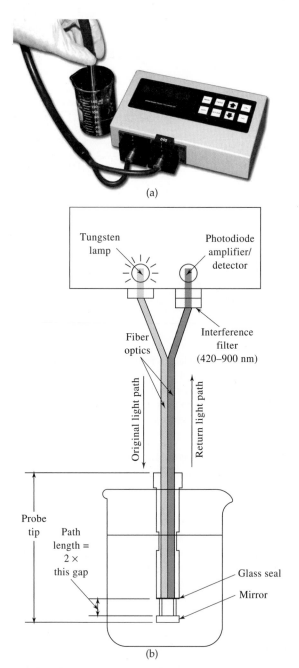

FIGURE 13-18 Photograph (a) and schematic diagram (b) of a probe-type photometer. (Courtesy of Brinkman Instruments, Inc.)

measured. The device is particularly useful for photometric titrations (Section 14E).

Filter Selection. General-purpose photometers are supplied with several filters, each of which transmits a different portion of the visible spectrum. Selection of the proper filter for a given application is important because the sensitivity of the measurement directly depends on choice. Usually for a new analytical method, the absorption spectrum of the solution to be analyzed is taken on a scanning spectrophotometer. The filter that most closely matches the wavelength of maximum absorption is then chosen. In some cases, measurements are made away from the absorption maximum to minimize interferences. Whenever possible, measurements are made near the absorption maximum to minimize Beer's law deviations due to polychromatic radiation (Section 13B-2).

When a spectrophotometer is not available to aid in filter selection, a filter can be chosen by remembering that the color of the light absorbed is the complement of the color of the solution itself. A solution appears red, for example, because it transmits the red portion of the spectrum but absorbs the green. It is the intensity of radiation in the green that varies with concentration. Hence, a green filter should be used. In general, the most suitable filter will be the color complement of the solution being analyzed.

Ultraviolet Absorption Photometers

Ultraviolet photometers often serve as detectors in high-performance liquid chromatography. In this application, a mercury vapor lamp usually serves as a source, and the emission line at 254 nm is isolated by filters. This type of detector is described briefly in Section 28C-6.

Ultraviolet photometers are also available for continuously monitoring the concentration of one or more constituents of gas or liquid streams in industrial plants. The instruments are double beam in space (see Figure 13-16b) and often employ one of the emission lines of mercury, which has been isolated by a filter system. Typical applications include the determination of low concentrations of phenol in wastewater; monitoring the concentration of chlorine, mercury, or aromatics in gases; and the determination of the ratio of hydrogen sulfide to sulfur dioxide in the atmosphere.

Spectrophotometers

Numerous spectrophotometers are available from commercial sources. Some have been designed for the visible region only; others are applicable in the ultraviolet and visible regions. A few have measuring capabilities from the ultraviolet through the near infrared (190 to 3000 nm).

Instruments for the Visible Region. Several spectrophotometers designed to operate within the wavelength range of about 380 to 800 nm are available from commercial sources. These instruments are frequently simple, single-beam grating instruments that are relatively inexpensive (less than $1000 to perhaps $3000), rugged, and readily portable. At least one is battery operated and light and small enough to be handheld. The most common application of these instruments is for quantitative analysis, although several produce good absorption spectra as well.

Figure 13-19 shows a simple and inexpensive spectrophotometer, the Spectronic 20. The original version of this instrument first appeared in the market in the mid-1950s, and the modified version shown in the figure is still being manufactured and widely sold. More of these instruments are currently in use throughout the world than any other single spectrophotometer model. The instrument owes its popularity, particularly as a teaching tool, to its relatively low cost, its ruggedness, and its satisfactory performance characteristics.

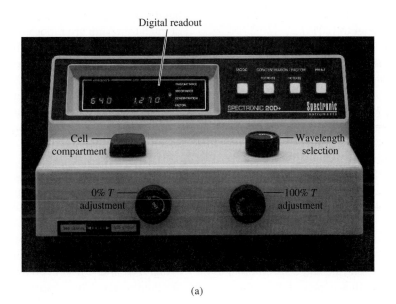

(a)

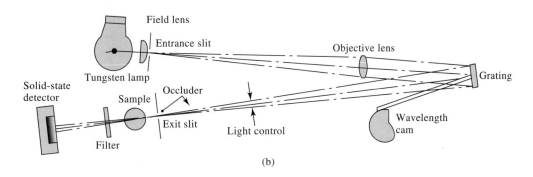

(b)

FIGURE 13-19 (a) The Spectronic 20 spectrophotometer and (b) its optical diagram. Radiation from the tungsten filament source passes through an entrance slit into the monochromator. A reflection grating diffracts the radiation, and the selected wavelength band passes through the exit slit into the sample chamber. A solid-state detector converts the light intensity into a related electrical signal that is amplified and displayed on a digital readout. (Courtesy of Thermo Electron Corp., Madison, WI.)

The Spectronic 20 employs a tungsten filament light source, which is operated by a stabilized power supply that provides radiation of constant intensity. After diffraction by a simple reflection grating, the radiation passes through the sample or reference cuvettes to a solid state detector. The Spectronic 20 reads out in transmittance or in absorbance on an LED display or, in the analog model, in transmittance on a meter.

The instrument is equipped with an occluder, a vane that automatically falls between the beam and the detector whenever the cuvette is removed from its holder; the 0%T adjustment can then be made. As shown in Figure 13-20, the light-control device in the Spectronic 20 consists of a V-shape slot that can be moved in or out of the beam to set the meter to 100%T.

To obtain a percent transmittance reading, the digital readout is first zeroed with the sample compartment empty so that the occluder blocks the beam and no radiation reaches the detector. This process is called the *0%T calibration*, or *adjustment*. A cell containing the blank (often the solvent) is then inserted into the cell holder, and the pointer is brought to the 100%T mark by adjusting the position of the light-control aperture and thus the amount of light reaching the detector. This adjustment is called the *100%T calibration*, or *adjustment*. Finally the sample is placed in the cell compartment, and the percent transmittance or the absorbance is read directly from the LED display.

The spectral range of the Spectronic 20 is 400 to 900 nm. Other specifications include a spectral bandpass of 20 nm, a wavelength accuracy of ±2.5 nm, and a photometric accuracy of ±4%T. The instrument may be interfaced to a computer for data storage and analysis if this option is available.

Single-Beam Instruments for the Ultraviolet-Visible Region. Several instrument manufacturers offer nonrecording single-beam instruments that can be used for both ultraviolet and visible measurements. The lower wavelength extremes for these instruments vary from 190 to 210 nm, and the upper from 800 to 1000 nm. All are equipped with interchangeable tungsten and hydrogen or deuterium lamps. Most employ photomultiplier tubes or photodiodes as transducers and gratings for dispersion. Digital readouts are now standard on almost all these spectrophotometers. The prices for these instruments range from $2000 to $8000.

As might be expected, performance specifications vary considerably among instruments and are related, at least to some degree, to instrument price. Typically, bandwidths vary from 2 to 8 nm; wavelength accuracies of ±0.5 to ±2 nm are reported.

The optical designs for the various grating instruments do not differ greatly from those shown in Figures 13-13a and 13-19. One manufacturer, however, employs a concave rather than a plane grating; a simpler and more compact design results. Holographic gratings are now becoming standard in many spectrophotometers.

Single-Beam Computerized Spectrophotometers. Several manufacturers offer computerized, recording, single-beam spectrophotometers, which operate in the UV-visible range. With these instruments, a wavelength scan is first performed with the reference solution in the beam path. The resulting transducer output is digitized and stored in computer memory. Samples are then scanned and absorbances calculated with the aid of the stored reference solution data. The complete spectrum is displayed within a few seconds of data acquisition. Because the reference and sample spectra are taken at different times, it is necessary that source intensity remain constant. The computer associated with these instruments provides several options with regard to data processing and presentation such as log absorbance, transmittance, derivatives, overlaid spectra, repetitive scans, concentration calculations, peak location and height determinations, and kinetic measurements.

As noted earlier, single-beam instruments have the inherent advantages of greater energy throughput, superior signal-to-noise ratios, and less cluttered sample compartments. On the other hand, the double-beam instruments described next provide better baseline flatness and long-term stability than do the single-beam systems. The highest-quality spectrophotometers still employ the double-beam design.

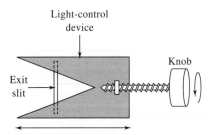

FIGURE 13-20 End view of the exit slit of the Spectronic 20 spectrophotometer pictured in Figure 13-19.

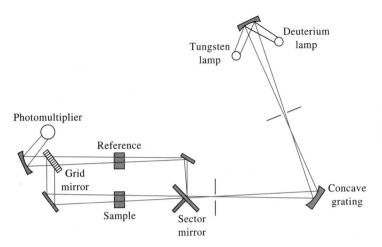

FIGURE 13-21 Schematic of a typical manual double-beam spectrophotometer for the UV-visible region.

Double-Beam Instruments. Numerous double-beam spectrophotometers for the ultraviolet-visible region of the spectrum are now available. Generally, these instruments are more expensive than their single-beam counterparts, with the nonrecording variety ranging from about $5000 to more than $30,000.

Figure 13-21 shows construction details of a typical, relatively inexpensive, manual, double-beam ultraviolet-visible spectrophotometer. In this instrument, the radiation is dispersed by a concave grating, which also focuses the beam on a rotating sector mirror. The instrument design is similar to that shown in Figure 13-13c.

The instrument has a wavelength range of 195 to 850 nm, a bandwidth of 4 nm, a photometric accuracy of 0.5% T, and a reproducibility of 0.2% A; stray radiation is less than 0.1% of P_0 at 240 and 340 nm. This instrument is typical of several spectrophotometers offered by various instrument companies. Such instruments are well suited for quantitative measurements where acquisition of an entire spectrum is not often required.

Figure 13-22 shows the optical design of the Varian Cary 100, a more sophisticated, double-beam-in-time recording spectrophotometer. This instrument employs a 30 × 35 mm plane grating having 1200 lines/mm. Its range is from 190 to 900 nm. Bandwidths of 0.2 to

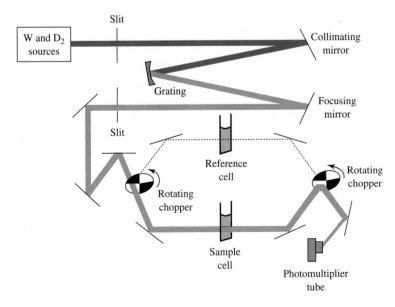

FIGURE 13-22 Schematic of the Varian Cary 100 double-beam spectrophotometer for the UV-visible region. Radiation from one of the sources passes through an entrance slit into the grating monochromator. After exiting the monochromator the radiation is split into two beams by the chopper. The chopper contains a transparent segment and a mirrored segment in addition to the two dark segments. After passing through the cells, the beams are recombined by the second chopper and strike the photomultiplier tube at different times. The photomultiplier tube sees the following sequence: sample beam, dark, reference beam, dark. (Courtesy of Varian Inc., Palo Alto, CA.)

4.00 nm can be chosen in 0.1-nm steps by means of a motor-driven slit control system. The instrument has a photometric accuracy of ± 0.00016 A; its stray radiation is less than 0.0013% of P_0 at 370 nm and 0.0074% at 220 nm. The absorbance range is from 0 to 3.7 absorbance units. The performance of this instrument is significantly better than that of the double-beam instrument of Figure 13-21; its price is correspondingly higher.

Double-Dispersing Instruments. To enhance spectral resolution and achieve a marked reduction in scattered radiation, a number of instruments have been designed with two gratings serially arranged with an intervening slit; in effect, then, these instruments consist of two monochromators in a series configuration.

The Varian Cary 300 shown in Figure 13-23 uses a premonochromator in front of the same double-beam-in-time instrument shown in Figure 13-22. The second monochromator reduces the stray-light levels to 0.000041% at 370 nm and 0.00008% at 220 nm. This extends the absorbance range to 5.0 absorbance units. Most of the other characteristics are identical to that of the Varian Cary 100. Both units have a double-chopper arrangement that ensures nearly identical light paths for both beams. The two beams strike the photomultiplier tube at essentially the same point, which minimizes errors due to photocathode nonuniformity.

Multichannel Instruments. Array detectors began appearing in UV-visible spectrophotometers in the

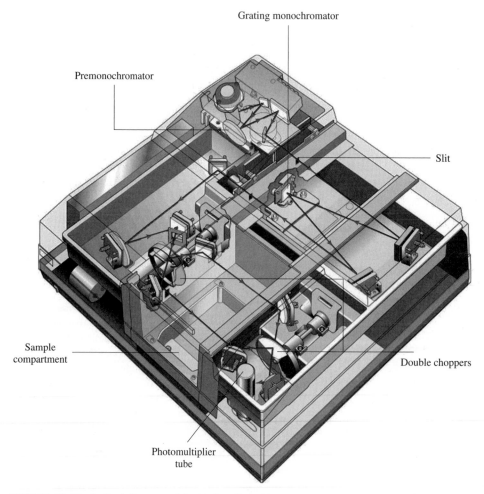

FIGURE 13-23 Optical diagram of the Varian Cary 300 double-dispersing spectrophotometer. The instrument is essentially identical to that shown in Figure 13-22, except that a second monochromator is added immediately after the source. (Varian Inc., Palo Alto, CA.)

1980s. With a diode array, or more recently a CCD array, located at the focal plane of a spectrograph, a spectrum can be obtained by electronic rather than mechanical scanning. All of the data points needed to define a spectrum can thus be gathered essentially simultaneously. The concept of multichannel instruments is attractive because of the potential speed at which spectra can be acquired as well as their applicability to simultaneous multicomponent determinations. By now, several instrument companies offer such instruments as either stand-alone spectrophotometers or as miniature versions.

Figure 13-24 shows a photograph of a miniature fiber-optic spectrometer using a linear CCD array. The optical diagram is similar to that shown in Figure 13-14, except that fiber optics are used to transport the radiation to and from the sample cell. In the version shown, the spectrometer is external to the computer. The output of the array connects to an analog-to-digital converter board in the computer. In other models, the spectrometer contains the converter and interfaces to the computer via a USB port. Such spectrometers are available from about $1800 to about $5000.

Both tungsten and deuterium sources are available for the spectrophotometer shown in Figure 13-24. With deuterium, the wavelength range is 200–400 nm. With the tungsten source, the range is 360–850 nm. A combined deuterium and tungsten-halogen source is also available with a wavelength range of 200–1100 nm. Some fourteen different gratings are available along with six different entrance slits. No entrance slit is needed with fiber-optic coupling. The spectrometer resolution depends on the grating dispersion and the entrance aperture.

A stand-alone diode-array-based spectrophotometer is illustrated in Figure 13-25. This instrument uses a deuterium source for the UV-visible region (190 to 800 nm) and a tungsten lamp for the visible near-infrared region (370 to 1100 nm). A shutter blocks the collimated source radiation or allows it to pass through to the sample cell. A filter can be inserted for stray-light correction. The spectrograph consists of the entrance slit, the grating, and the diode-array detector. A 1024-element photodiode array is used. The nominal bandwidth of the spectrophotometer is 1 nm over the range from 190 to 1100 nm. Stray light is less than 0.05% at 220 nm and less than 0.03% at 340 nm.

With the spectrophotometer of Figure 13-25, scan times can be as short as 0.1 s, but more typically the array is exposed for 1 to 1.5 s. With such short exposure

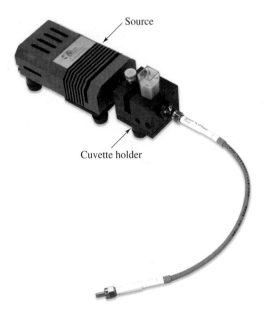

FIGURE 13-24 A multichannel miniature fiber-optic spectrometer. A fiber-optic cable transports the light beam from the cell holder on the left to the spectrograph and detector on the right. In some models, the spectrograph and detector are mounted on a circuit board inserted into the computer. (Courtesy of Ocean Optics, Inc., Dunedin, FL.)

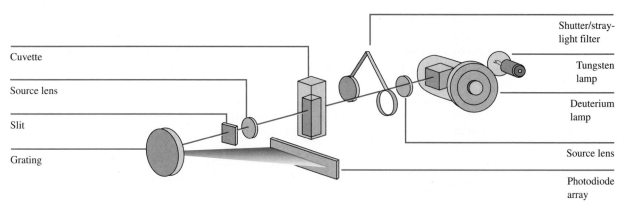

FIGURE 13-25 A multichannel diode-array spectrophotometer, the Agilent Technologies 8453. (Courtesy of Agilent Technologies, Palo Alto, CA.)

times, photodecomposition of samples is minimized despite the location of the sample between the source and the monochromator. The stability of the source and the electronic system is such that the solvent signal needs to be observed and stored only every 5 to 10 min.

The spectrophotometer shown in Figure 13-25 is designed to interface with most personal computer systems. The instrument (without the computer) sells for about $10,000; the exact price depends on accessories and options.

QUESTIONS AND PROBLEMS

*Answers are provided at the end of the book for problems marked with an asterisk.

[X] Problems with this icon are best solved using spreadsheets.

*13-1 Express the following absorbances in terms of percent transmittance:
(a) 0.038 (d) 0.241
(b) 0.958 (e) 0.435
(c) 0.399 (f) 0.692

*13-2 Convert the following transmittance data to absorbances:
(a) 15.8% (d) 23.8%
(b) 0.492 (e) 0.085
(c) 39.4% (f) 5.38%

*13-3 Calculate the percent transmittance of solutions having half the absorbance of the solutions in Problem 13-1.

*13-4 Calculate the absorbance of solutions having twice the percent transmittance of those in Problem 13-2.

*13-5 A solution containing 6.23 ppm $KMnO_4$ had a transmittance of 0.195 in a 1.00-cm cell at 520 nm. Calculate the molar absorptivity of $KMnO_4$ at 520 nm.

13-6 A solution containing 5.24 mg/100 mL of A (335 g/mol) has a transmittance of 55.2% in a 1.50-cm cell at 425 nm. Calculate the molar absorptivity of A at this wavelength.

*13-7 A solution containing the complex formed between Bi(III) and thiourea has a molar absorptivity of 9.32×10^3 L mol^{-1} cm^{-1} at 470 nm.

(a) What is the absorbance of a 3.79×10^{-5} M solution of the complex at 470 nm in a 1.00-cm cell?
(b) What is the percent transmittance of the solution described in (a)?
(c) What is the molar concentration of the complex in a solution that has the absorbance described in (a) when measured at 470 nm in a 2.50-cm cell?

*13-8 At 580 nm, which is the wavelength of its maximum absorption, the complex $Fe(SCN)^{2+}$ has a molar absorptivity of 7.00×10^3 L cm^{-1} mol^{-1}. Calculate
(a) the absorbance of a 3.49×10^{-5} M solution of the complex at 580 nm in a 1.00-cm cell.
(b) the absorbance of a solution in a 2.50-cm cell in which the concentration of the complex is one half that in (a).
(c) the percent transmittance of the solutions described in (a) and (b).
(d) the absorbance of a solution that has half the transmittance of that described in (a).

*13-9 A 2.50-mL aliquot of a solution that contains 7.9 ppm iron(III) is treated with an appropriate excess of KSCN to form the $Fe(SCN)^{2+}$ complex and diluted to 50.0 mL. What is the absorbance of the resulting solution at 580 nm in a 2.50-cm cell? See Problem 13-8 for absorptivity data.

13-10 Zinc(II) and the ligand L form a 1:1 complex that absorbs strongly at 600 nm. As long as the molar concentration of L exceeds that of zinc(II) by a factor of 5, the absorbance depends only on the cation concentration. Neither zinc(II) nor L absorbs at 600 nm. A solution that is 1.59×10^{-4} M in zinc(II) and 1.00×10^{-3} M in L has an absorbance of 0.352 in a 1.00-cm cell at 600 nm. Calculate
(a) the percent transmittance of this solution.
(b) the percent transmittance of this solution in a 2.50-cm cell.
(c) the molar absorptivity of the complex.

 13-11 The equilibrium constant for the conjugate acid-base pair

$$HIn + H_2O \rightleftharpoons H_3O^+ + In^-$$

is 8.00×10^{-5}. From the additional information in the following table,
(a) calculate the absorbance at 430 nm and 600 nm for the following indicator concentrations: 3.00×10^{-4} M, 2.00×10^{-4} M, 1.00×10^{-4} M, 0.500×10^{-4} M, and 0.250×10^{-4} M.
(b) plot absorbance as a function of indicator concentration.

| | | Molar Absorptivity | |
Species	Absorption Maximum, nm	430 nm	600 nm
HIn	430	8.04×10^3	1.23×10^3
In$^-$	600	0.775×10^3	6.96×10^3

 13-12 The equilibrium constant for the reaction

$$2CrO_4^{2-} + 2H^+ \rightleftharpoons Cr_2O_7^{2-} + H_2O$$

is 4.2×10^{14}. The molar absorptivities for the two principal species in a solution of $K_2Cr_2O_7$ are

λ	$\varepsilon_1(CrO_4^{2-})$	$\varepsilon_2(Cr_2O_7^{2-})$
345	1.84×10^3	10.7×10^2
370	4.81×10^3	7.28×10^2
400	1.88×10^3	1.89×10^2

Four solutions were prepared by dissolving 4.00×10^{-4}, 3.00×10^{-4}, 2.00×10^{-4}, and 1.00×10^{-4} moles of $K_2Cr_2O_7$ in water and diluting to 1.00 L with a pH 5.60 buffer. Derive theoretical absorbance values (1.00-cm cells) for each solution and plot the data for (a) 345 nm, (b) 370 nm, and (c) 400 nm.

13-13 Describe the differences between the following and list any particular advantages possessed by one over the other.
(a) hydrogen and deuterium discharge lamps as sources for ultraviolet radiation.
(b) filters and monochromators as wavelength selectors.
(c) photovoltaic cells and phototubes as detectors for electromagnetic radiation.
(d) photodiodes and photomultiplier tubes.
(e) double-beam-in-space and double-beam-in-time spectrophotometers.
(f) spectrophotometers and photometers.
(g) single-beam and double-beam instruments for absorbance measurements.
(h) conventional and multichannel spectrophotometers.

13-14 A portable photometer with a linear response to radiation registered 63.8 µA with the solvent in the light path. The photometer was set to zero with no light striking the detector. Replacement of the solvent with an absorbing solution yielded a response of 41.6 µA. Calculate
(a) the percent transmittance of the sample solution.
(b) the absorbance of the sample solution.
(c) the transmittance to be expected for a solution in which the concentration of the absorber is one third that of the original sample solution.
(d) the transmittance to be expected for a solution that has twice the concentration of the sample solution.

*13-15 A photometer with a linear response to radiation gave a reading of 498 mV with the solvent in the light path and 256 mV when the solvent was replaced by an absorbing solution. The photometer was set to zero with no light striking the detector. Calculate
(a) the percent transmittance and absorbance of the absorbing solution.
(b) the expected transmittance if the concentration of absorber is one half that of the original solution.
(c) the transmittance to be expected if the light path through the original solution is doubled.

13-16 Why does a deuterium lamp produce a continuum rather than a line spectrum in the ultraviolet?

13-17 Why can photomultiplier tubes not be used with infrared radiation?

13-18 Why is iodine sometimes introduced into a tungsten lamp?

13-19 Describe the origin of shot noise in a spectrophotometer. How does the relative uncertainty vary with concentration if shot noise is the major noise source?

13-20 Define
(a) dark current.
(b) transducer.

(c) scattered radiation (in a monochromator).

(d) source flicker noise.

(e) cell positioning uncertainty.

(f) beamsplitter.

13-21 Describe how a monochromator, a spectrograph, and a spectrophotometer differ from each other.

 13-22 The following data were taken from a diode-array spectrophotometer in an experiment to measure the spectrum of the Co(II)-EDTA complex. The column labeled $P_{solution}$ is the relative signal obtained with sample solution in the cell after subtraction of the dark signal. The column labeled $P_{solvent}$ is the reference signal obtained with only solvent in the cell after subtraction of the dark signal. Find the transmittance at each wavelength, and the absorbance at each wavelength. Plot the spectrum of the compound.

Wavelength, nm	$P_{solvent}$	$P_{solution}$
350	0.002689	0.002560
375	0.006326	0.005995
400	0.016975	0.015143
425	0.035517	0.031648
450	0.062425	0.024978
475	0.095374	0.019073
500	0.140567	0.023275
525	0.188984	0.037448
550	0.263103	0.088537
575	0.318361	0.200872
600	0.394600	0.278072
625	0.477018	0.363525
650	0.564295	0.468281
675	0.655066	0.611062
700	0.739180	0.704126
725	0.813694	0.777466
750	0.885979	0.863224
775	0.945083	0.921446
800	1.000000	0.977237

13-23 Why do quantitative and qualitative analyses often require different monochromator slit widths?

 13-24 The absorbances of solutions containing K_2CrO_4 in 0.05 M KOH were measured in a 1.0-cm cell at 375 nm. The following results were obtained:

Conc. of K_2CrO_4, g/L	A at 375 nm
0.0050	0.123
0.0100	0.247
0.0200	0.494
0.0300	0.742
0.0400	0.991

Find the absorptivity of the chromate ion, CrO_4^{2-} in $L\ g^{-1}\ cm^{-1}$ and the molar absorptivity of chromate in $L\ mol^{-1}\ cm^{-1}$ at 375 nm.

 13-25 The absorbances of solutions containing Cr as dichromate $Cr_2O_7^{2-}$ in 1.0 M H_2SO_4 were measured at 440 nm in a 1.0-cm cell. The following results were obtained:

Conc. of Cr, μg/mL	A at 440 nm
10.00	0.034
25.00	0.085
50.00	0.168
75.00	0.252
100.00	0.335
200.00	0.669

Find the absorptivity of dichromate $(L\ g^{-1}\ cm^{-1})$ and the molar absorptivity $(L\ mol^{-1}\ cm^{-1})$ at 440 nm.

 13-26 A compound X is to be determined by UV-visible spectrophotometry. A calibration curve is constructed from standard solutions of X with the following results: 0.50 ppm, $A = 0.24$; 1.5 ppm, $A = 0.36$; 2.5 ppm, $A = 0.44$; 3.5 ppm, $A = 0.59$; 4.5 ppm, $A = 0.70$. A solution of unknown X concentration had an absorbance of $A = 0.50$. Find the slope and intercept of the calibration curve, the standard error in Y, the concentration of the solution of unknown X concentration, and the standard deviation in the concentration of X. Construct a plot of the calibration curve and determine the unknown concentration by hand from the plot. Compare it to that obtained from the regression line.

Challenge Problem

13-27 The following questions concern the relative concentration uncertainty in spectrophotometry.

(a) If the relative concentration uncertainty is given by Equation 13-13, use calculus to show that the minimum uncertainty occurs at 36.8% T. What is the absorbance that minimizes the concentration uncertainty? Assume that s_T is independent of concentration.

(b) Under shot-noise-limited conditions, the relative concentration uncertainty is given by Equation 13-14. Another form of the equation for the shot-noise-limited case is [12]

$$\frac{s_c}{c} = \frac{-kT^{-1/2}}{\ln T}$$

where k is a constant. Use calculus and derive the transmittance and absorbance that minimize the concentration uncertainty.

(c) Describe how you could experimentally determine whether a spectrophotometer was operating under Case I, Case II, or Case III conditions.

---

[12] J. D. Ingle Jr. and S. R. Crouch, *Spectrochemical Analysis*, Upper Saddle River, NJ: Prentice Hall, 1988, Chap. 13.

CHAPTER FOURTEEN

Applications of Ultraviolet-Visible Molecular Absorption Spectrometry

U ltraviolet and visible absorption measurements are widely used for the identification and determination of many different inorganic and organic species. In fact, UV-visible molecular absorption methods are probably the most widely used of all quantitative analysis techniques in chemical, environmental, forensic, and clinical laboratories throughout the world.

Throughout this chapter, this logo indicates an opportunity for online self-study at **www .thomsonedu.com/chemistry/skoog**, linking you to interactive tutorials, simulations, and exercises.

14A THE MAGNITUDE OF MOLAR ABSORPTIVITIES

Empirically, molar absorptivities (ε values) that range from zero up to a maximum on the order of 10^5 L mol^{-1} cm^{-1} are observed in UV-visible molecular absorption spectrometry.[1] For any particular absorption maximum, the magnitude of ε depends on the capture cross section (Section 13B, Equation 13-5) of the species and the probability for an energy-absorbing transition to occur. The relationship between ε and these variables has been shown to be

$$\varepsilon = 8.7 \times 10^{19} PA$$

where P is the transition probability and A is the cross-section target area in square centimeters per molecule.[2] The area for typical organic molecules has been estimated from electron diffraction and X-ray studies to be about 10^{-15} cm^2/molecule; transition probabilities vary from zero to one. For quantum mechanically allowed transitions, values of P range from 0.1 to 1, which leads to strong absorption bands ($\varepsilon_{max} = 10^4$ to 10^5 L mol^{-1} cm^{-1}). Absorption maxima having molar absorptivities less than about 10^3 are classified as being of low intensity. They result from forbidden transitions, which have probabilities of occurrence that are less than 0.01.

14B ABSORBING SPECIES

The absorption of ultraviolet or visible radiation by an atomic or molecular species M can be considered to be a two-step process. The first step involves electronic excitation as shown by the equation

$$M + h\nu \rightarrow M^*$$

The product of the absorption of the photon $h\nu$ by species M is an electronically excited species symbolized by M*. The lifetime of the excited species is brief (10^{-8}

---

[1] Some useful references on absorption methods include E. J. Meehan, in *Treatise on Analytical Chemistry*, 2nd ed., P. J. Elving, E. J. Meehan, and I. M. Kolthoff, eds., Part I, Vol. 7, Chaps. 1–3, New York: Wiley, 1981; R. P. Bauman, *Absorption Spectroscopy*, New York: Wiley, 1962; F. Grum, in *Physical Methods of Chemistry*, A. Weissberger and B. W. Rossiter, eds., Vol. I, Part III B, Chap. 3, New York: Wiley-Interscience, 1972; H. H. Jaffé and M. Orchin, *Theory and Applications of Ultraviolet Spectroscopy*, New York: Wiley, 1962; G. F. Lothian, *Absorption Spectrophotometry*, 3rd ed., London: Adam Hilger, 1969; J. D. Ingle Jr. and S. R. Crouch, *Spectrochemical Analysis*, Chap. 13, Upper Saddle River, NJ: Prentice-Hall, 1988.
[2] E. A. Braude, *J. Chem. Soc.*, **1950**, 379.

to 10^{-9} s). Any of several relaxation processes can lead to deexcitation of M*. The most common type of relaxation involves conversion of the excitation energy to heat as shown by

$$M^* \rightarrow M + heat$$

Relaxation may also occur by a photochemical process such as decomposition of M* to form new species. Alternatively, relaxation may involve reemission of fluorescence or phosphorescence. It is important to note that the lifetime of M* is usually so very short that its concentration at any instant is ordinarily negligible. Furthermore, the amount of thermal energy evolved by relaxation is quite small. Thus, absorption measurements create a minimal disturbance of the system under study except when photochemical decomposition occurs.

The absorption of ultraviolet or visible radiation generally results from excitation of bonding electrons. Because of this, the wavelengths of absorption bands can be correlated with the types of bonds in the species under study. Molecular absorption spectroscopy is, therefore, valuable for identifying functional groups in a molecule. More important, however, are the applications of ultraviolet and visible absorption spectroscopy to the quantitative determination of compounds containing absorbing groups.

As noted in Section 6C, absorption of ultraviolet and visible radiation by molecules generally occurs in one or more electronic absorption bands, each of which is made up of many closely packed but discrete lines. Each line arises from the transition of an electron from the ground state to one of the many vibrational and rotational energy states associated with each excited electronic energy state. Because there are so many of these vibrational and rotational states and because their energies differ only slightly, many closely spaced lines are contained in the typical band.

As can be seen in Figure 14-1a, the visible absorption spectrum for 1,2,4,5-tetrazine vapor shows the fine structure that is due to the numerous rotational and vibrational levels associated with the excited electronic states of this aromatic molecule. In the gaseous state, the individual tetrazine molecules are sufficiently separated from one another to vibrate and rotate freely, and the many individual absorption lines appear as a result of the large number of vibrational and rotational energy states. In the condensed state or in solution, however, the tetrazine molecules have little freedom to rotate, so lines due to differences in rotational energy

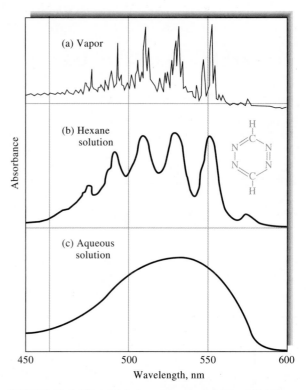

FIGURE 14-1 Ultraviolet absorption spectra for 1,2,4,5-tetrazine. In (a), the spectrum is shown in the gas phase, where many lines due to electronic, vibrational, and rotational transitions can be seen. In a nonpolar solvent (b), the electronic transitions can be observed, but the vibrational and rotational structure has been lost. In a polar solvent (c), the strong intermolecular forces cause the electronic peaks to blend, giving only a single smooth absorption band. (From S. F. Mason, *J. Chem. Soc.*, **1959**, 1265.)

levels disappear. Furthermore, when solvent molecules surround the tetrazine molecules, energies of the various vibrational levels are modified in a nonuniform way, and the energy of a given state in a sample of solute molecules appears as a single, broad peak. This effect is more pronounced in polar solvents, such as water, than in nonpolar hydrocarbon media. This solvent effect is illustrated in Figure 14-1b and c.

14B-1 Absorption by Organic Compounds

All organic compounds are capable of absorbing electromagnetic radiation because all contain valence electrons that can be excited to higher energy levels. The excitation energies associated with electrons forming

TABLE 14-1 Absorption Characteristics of Some Common Chromophores

Chromophore	Example	Solvent	λ_{max}, nm	ε_{max}	Transition Type
Alkene	$C_6H_{13}CH=CH_2$	*n*-Heptane	177	13,000	$\pi \rightarrow \pi^*$
Alkyne	$C_5H_{11}C\equiv C-CH_3$	*n*-Heptane	178	10,000	$\pi \rightarrow \pi^*$
			196	2000	—
			225	160	—
Carbonyl	CH_3CCH_3 $\underset{O}{\parallel}$	*n*-Hexane	186	1000	$n \rightarrow \sigma^*$
			280	16	$n \rightarrow \pi^*$
	CH_3CH $\underset{O}{\parallel}$	*n*-Hexane	180	large	$n \rightarrow \sigma^*$
			293	12	$n \rightarrow \pi^*$
Carboxyl	CH_3COOH	Ethanol	204	41	$n \rightarrow \pi^*$
Amido	CH_3CNH_2 $\underset{O}{\parallel}$	Water	214	60	$n \rightarrow \pi^*$
Azo	$CH_3N=NCH_3$	Ethanol	339	5	$n \rightarrow \pi^*$
Nitro	CH_3NO_2	Isooctane	280	22	$n \rightarrow \pi^*$
Nitroso	C_4H_9NO	Ethyl ether	300	100	—
			665	20	$n \rightarrow \pi^*$
Nitrate	$C_2H_5ONO_2$	Dioxane	270	12	$n \rightarrow \pi^*$

most single bonds are sufficiently high that absorption occurs in the so-called vacuum ultraviolet region ($\lambda <$ 185 nm), where components of the atmosphere also absorb radiation strongly. Such transitions involve the excitation of nonbonding *n* electrons to σ^* orbitals. The molar absorptivities of $n \rightarrow \sigma^*$ transitions are low to intermediate and usually range between 100 and 3000 L mol^{-1} cm^{-1}. Because of experimental difficulties associated with the vacuum ultraviolet region, most spectrophotometric investigations of organic compounds have involved longer wavelengths than 185 nm.

Most applications of absorption spectroscopy to organic compounds are based on transitions for *n* or π electrons to the π^* excited state because the energies required for these processes bring the absorption bands into the ultraviolet-visible region (200 to 700 nm). Both $n \rightarrow \pi^*$ and $\pi \rightarrow \pi^*$ transitions require the presence of an unsaturated functional group to provide the π orbitals. Molecules containing such functional groups and capable of absorbing ultraviolet-visible radiation are called *chromophores*.

The electronic spectra of organic molecules containing chromophores are usually complex, because the superposition of vibrational transitions on the electronic transitions leads to an intricate combination of overlapping lines. The result is a broad band of absorption that often appears to be continuous. The complex na-

ture of the spectra makes detailed theoretical analysis difficult or impossible. Nevertheless, qualitative or semiquantitative statements concerning the types of electronic transitions responsible for a given absorption spectrum can be deduced from molecular orbital considerations.

Table 14-1 lists the common organic chromophores and the approximate wavelengths at which they absorb. The data for position (λ_{max}) and peak intensity (ε_{max}) can serve as only a rough guide for identification purposes, because both are influenced by solvent effects as well as other structural details of the molecule. In addition, conjugation between two or more chromophores tends to cause shifts in absorption maxima to longer wavelengths. Finally, vibrational effects broaden absorption peaks in the ultraviolet and visible regions, which often makes precise determination of an absorption maximum difficult. The molar absorptivities for $n \rightarrow \pi^*$ transitions are normally low and usually range from 10 to 100 L mol^{-1} cm^{-1}. On the other hand, values for $\pi \rightarrow \pi^*$ transitions generally range between 1000 and 15,000 L mol^{-1} cm^{-1}. Typical absorption spectra are shown in Figure 14-2.

Saturated organic compounds containing such heteroatoms as oxygen, nitrogen, sulfur, or halogens have nonbonding electrons that can be excited by radiation in the range of 170 to 250 nm. Table 14-2 lists a

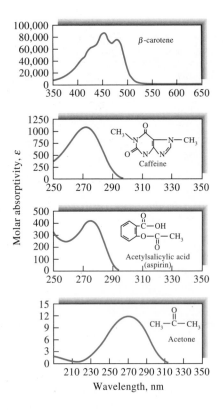

FIGURE 14-2 Absorption spectra for typical organic compounds.

TABLE 14-2 Absorption by Organic Compounds Containing Heteroatoms with Nonbonding Electrons

Compound	λ_{max}, nm	ε_{max}
CH_3OH	167	1480
$(CH_3)_2O$	184	2520
CH_3Cl	173	200
CH_3I	258	365
$(CH_3)_2S$	229	140
CH_3NH_2	215	600
$(CH_3)_3N$	227	900

depend on the ligands bonded to the metal ions. The energy differences between these d-orbitals (and thus the position of the corresponding absorption maximum) depend on the position of the element in the periodic table, its oxidation state, and the nature of the ligand bonded to it.

Absorption spectra of ions of the lanthanide and actinide transitions series differ substantially from those shown in Figure 14-3. The electrons responsible for absorption by these elements ($4f$ and $5f$, respectively) are shielded from external influences by electrons that occupy orbitals with larger principal quantum numbers. As a result, the bands tend to be narrow and relatively

few examples of such compounds. Some of these compounds, such as alcohols and ethers, are common solvents, so their absorption in this region prevents measuring absorption of analytes dissolved in these compounds at wavelengths shorter than 180 to 200 nm. Occasionally, absorption in this region is used for determining halogen and sulfur-bearing compounds.

14B-2 Absorption by Inorganic Species

A number of inorganic anions exhibit ultraviolet absorption bands that are a result of exciting nonbonding electrons. Examples include nitrate (313 nm), carbonate (217 nm), nitrite (360 and 280 nm), azido (230 nm), and trithiocarbonate (500 nm) ions.

In general, the ions and complexes of elements in the first two transition series absorb broad bands of visible radiation in at least one of their oxidation states and are, as a result, colored (see, for example, Figure 14-3). Here, absorption involves transitions between filled and unfilled d-orbitals with energies that

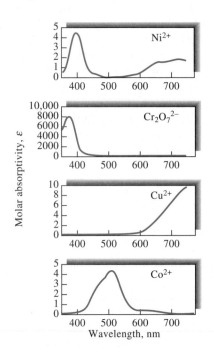

FIGURE 14-3 Absorption spectra of aqueous solutions of transition metal ions.

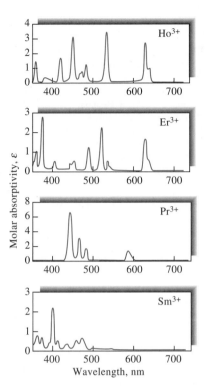

FIGURE 14-4 Absorption spectra of aqueous solutions of rare earth ions.

unaffected by the species bonded by the outer electrons (see Figure 14-4).

14B-3 Charge-Transfer Absorption

For quantitative purposes, charge-transfer absorption is particularly important because molar absorptivities are unusually large ($\varepsilon > 10,000$), which leads to high sensitivity. Many inorganic and organic complexes exhibit this type of absorption and are therefore called charge-transfer complexes.

A charge-transfer complex consists of an electron-donor group bonded to an electron acceptor. When this product absorbs radiation, an electron from the donor is transferred to an orbital that is largely associated with the acceptor. The excited state is thus the product of a kind of internal oxidation-reduction process. This behavior differs from that of an organic chromophore in which the excited electron is in a molecular orbital shared by two or more atoms.

Familiar examples of charge-transfer complexes include the phenolic complex of iron(III), the 1,10-

 Simulation: Learn more about **absorption spectra**.

phenanthroline complex of iron(II), the iodide complex of molecular iodine, and the hexacyanoferrate(II)-hexacyanoferrate(III) complex responsible for the color of Prussian blue. The red color of the iron(III)-thiocyanate complex is a further example of charge-transfer absorption. Absorption of a photon results in the transfer of an electron from the thiocyanate ion to an orbital that is largely associated with the iron(III) ion. The product is an excited species involving predominantly iron(II) and the thiocyanate radical SCN. As with other types of electronic excitation, the electron in this complex ordinarily returns to its original state after a brief period. Occasionally, however, an excited complex may dissociate and produce photochemical oxidation-reduction products. Three spectra of charge-transfer complexes are shown in Figure 14-5.

In most charge-transfer complexes involving a metal ion, the metal serves as the electron acceptor. Exceptions are the 1,10-phenanthroline complexes of iron(II) and copper(I), where the ligand is the acceptor and the metal ion the donor. A few other examples of this type of complex are known.

Organic compounds form many interesting charge-transfer complexes. An example is quinhydrone (a 1:1 complex of quinone and hydroquinone), which exhibits strong absorption in the visible region. Other examples include iodine complexes with amines, aromatics, and sulfides.

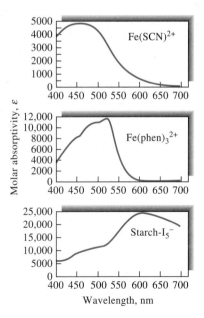

FIGURE 14-5 Absorption spectra of aqueous charge-transfer complexes.

14C QUALITATIVE APPLICATIONS OF ULTRAVIOLET-VISIBLE ABSORPTION SPECTROSCOPY

Spectrophotometric measurements with ultraviolet radiation are useful for detecting chromophoric groups, such as those shown in Table 14-1.[3] Because large parts of even the most complex organic molecules are transparent to radiation longer than 180 nm, the appearance of one or more peaks in the region from 200 to 400 nm is clear indication of the presence of unsaturated groups or of atoms such as sulfur or halogens. Often, the identity of the absorbing groups can be determined by comparing the spectrum of an analyte with those of simple molecules containing various chromophoric groups.[4] Usually, however, ultraviolet spectra do not have enough fine structure to permit an analyte to be identified unambiguously. Thus, ultraviolet qualitative data must be supplemented with other physical or chemical evidence such as infrared, nuclear magnetic resonance, and mass spectra as well as solubility and melting- and boiling-point information.

14C-1 Solvents

Ultraviolet spectra for qualitative analysis are usually measured using dilute solutions of the analyte. For volatile compounds, however, gas-phase spectra are often more useful than liquid-phase or solution spectra (for example, compare Figure 14-1a and b). Gas-phase spectra can often be obtained by allowing a drop or two of the pure liquid to evaporate and equilibrate with the atmosphere in a stoppered cuvette.

In choosing a solvent, consideration must be given not only to its transparency, but also to its possible effects on the absorbing system. Quite generally, polar solvents such as water, alcohols, esters, and ketones tend to obliterate spectral fine structure arising from vibrational effects. Spectra similar to gas-phase spectra (see Figure 14-6) are more likely to be observed in

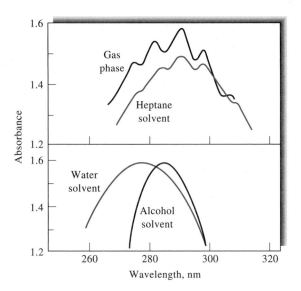

FIGURE 14-6 Effect of solvent on the absorption spectrum of acetaldehyde.

nonpolar solvents such as hydrocarbons. In addition, the positions of absorption maxima are influenced by the nature of the solvent. As a rule, the same solvent must be used when comparing absorption spectra for identification purposes.

Table 14-3 lists some common solvents and the approximate wavelength below which they cannot be used because of absorption. These wavelengths, called the *cutoff wavelengths*, depend strongly on the purity of the solvent.[5] Common solvents for ultraviolet spec-

---

[3]For a detailed discussion of ultraviolet absorption spectroscopy in the identification of organic functional groups, see R. M. Silverstein, G. C. Bassler, and T. C. Morrill, *Spectrometric Identification of Organic Compounds*, 5th ed., Chap. 6, New York: Wiley, 1991.

[4]H. H. Perkampus, *UV-VIS Atlas of Organic Compounds*, 2nd ed., Hoboken, NJ: Wiley-VCH, 1992. In addition, in the past, several organizations have published catalogs of spectra that may still be useful, including American Petroleum Institute, Ultraviolet Spectral Data, *A.P.I. Research Project 44*. Pittsburgh: Carnegie Institute of Technology; *Sadtler Handbook of Ultraviolet Spectra*, Philadelphia: Sadtler Research Laboratories; American Society for Testing Materials, Committee E-13, Philadelphia.

[5]Most major suppliers of reagent chemicals in the United States offer spectrochemical grades of solvents. Spectral-grade solvents have been treated to remove absorbing impurities and meet or exceed the requirements set forth in *Reagent Chemicals*, 9th ed., Washington, DC: American Chemical Society, 2000. Supplements and updates are available at http://pubs.acs.org/reagents/index.html.

TABLE 14-3 Solvents for the Ultraviolet and Visible Regions

Solvent	Lower Wavelength Limit, nm	Solvent	Lower Wavelength Limit, nm
Water	180	Diethyl ether	210
Ethanol	220	Acetone	330
Hexane	200	Dioxane	320
Cyclohexane	200	Cellosolve	320
Carbon tetrachloride	260		

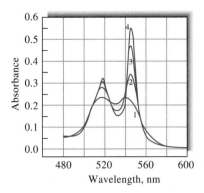

FIGURE 14-7 Spectra for reduced cytochrome c at four spectral bandwidths. (1) 20 nm, (2) 10 nm, (3) 5 nm, and (4) 1 nm. (Courtesy of Varian, Inc., Palo Alto, CA.)

trophotometry include water, 95% ethanol, cyclohexane, and 1,4-dioxane. For the visible region, any colorless solvent is suitable.

14C-2 The Effect of Slit Width

The effect of variation in slit width, and hence effective bandwidth, was shown previously for gas-phase spectra in Figure 13-8. The effect on solution spectra is illustrated for reduced cytochrome c in Figure 14-7. Clearly, peak heights and separation are distorted at wider bandwidths. Because of this, spectra for qualitative applications should be measured with minimum slit widths.

14C-3 Detection of Functional Groups

Even though it may not provide the unambiguous identification of an organic compound, an absorption spectrum in the visible and the ultraviolet regions is nevertheless useful for detecting the presence of certain functional groups that act as chromophores. For example, a weak absorption band in the region of 280 to 290 nm, which is displaced toward shorter wavelengths with increased solvent polarity, strongly indicates the presence of a carbonyl group. Such a shift is termed a *hypsochromic*, or *blue*, *shift*. A weak absorption band at about 260 nm with indications of vibrational fine structure constitutes evidence for the existence of an aromatic ring. Confirmation of the presence of an aromatic amine or a phenolic structure may be obtained by comparing the effects of pH on the spectra of solutions containing the sample with those shown in Table 14-4 for phenol and aniline.

The ultraviolet spectra of aromatic hydrocarbons are characterized by three sets of bands that originate from $\pi \rightarrow \pi^*$ transitions. For example, benzene has a strong absorption peak at 184 nm ($\varepsilon_{max} \approx 60,000$); a weaker band, called the E_2 band, at 204 nm ($\varepsilon_{max} = 7900$); and a still weaker peak, termed the B band, at 256 ($\varepsilon_{max} = 200$). The long-wavelength bands of benzene vapor, 1,2,4,5-tetrazine (see Figure 14-1a), and many other aromatics contain a series of sharp peaks due to the superposition of vibrational transitions on the basic electronic transitions. As shown in Figure 14-1, solvents tend to reduce (or sometimes eliminate) this fine structure as do certain types of substitution.

TABLE 14-4 Absorption Characteristics of Aromatic Compounds

Compound		E_2 Band		B Band	
		λ_{max}, nm	ε_{max}	λ_{max}, nm	ε_{max}
Benzene	C_6H_6	204	7900	256	200
Toluene	$C_6H_5CH_3$	207	7000	261	300
m-Xylene	$C_6H_4(CH_3)_2$	—	—	263	300
Chlorobenzene	C_6H_5Cl	210	7600	265	240
Phenol	C_6H_5OH	211	6200	270	1450
Phenolate ion	$C_6H_5O^-$	235	9400	287	2600
Aniline	$C_6H_5NH_2$	230	8600	280	1430
Anilinium ion	$C_6H_5NH_3^+$	203	7500	254	160
Thiophenol	C_6H_5SH	236	10,000	269	700
Naphthalene	$C_{10}H_8$	286	9300	312	289
Styrene	$C_6H_5CH{=}CH_2$	244	12,000	282	450

All three of the characteristic bands for benzene are strongly affected by ring substitution; the effects on the two longer-wavelength bands are of particular interest because they can be studied with ordinary spectrophotometric equipment. Table 14-4 illustrates the effects of some common ring substituents.

By definition, an *auxochrome* is a functional group that does not itself absorb in the ultraviolet region but has the effect of shifting chromophore peaks to longer wavelengths as well as increasing their intensities. Such a shift to longer wavelengths is called a *bathochromic*, or *red*, *shift*. Note in Table 14-4 that $-OH$ and $-NH_2$ have an auxochromic effect on the benzene chromophore, particularly with respect to the B band. Auxochromic substituents have at least one pair of *n* electrons capable of interacting with the π electrons of the ring. This interaction apparently has the effect of stabilizing the π^* state, thereby lowering its energy, and increasing the wavelength of the corresponding band. Note that the auxochromic effect is more pronounced for the phenolate anion than for phenol itself, probably because the anion has an extra pair of unshared electrons to contribute to the interaction. With aniline, on the other hand, the nonbonding electrons are lost by formation of the anilinium cation, and the auxochromic effect disappears.

14D QUANTITATIVE ANALYSIS BY ABSORPTION MEASUREMENTS

Absorption spectroscopy based on ultraviolet and visible radiation is one of the most useful tools available to the scientist for quantitative analysis.[6] Important characteristics of spectrophotometric and photometric methods include (1) wide applicability to both organic and inorganic systems, (2) typical detection limits of 10^{-4} to 10^{-5} M (in some cases, certain modifications can lead to lower limits of detection),[7] (3) moderate to high selectivity, (4) good accuracy (typically, relative uncertainties are 1% to 3%, although with special precautions, errors can be reduced to a few tenths of a percent), and (5) ease and convenience of data acquisition.

14D-1 Scope

The applications of quantitative, ultraviolet-visible absorption methods not only are numerous but also touch on every field that requires quantitative chemical information. The reader can obtain a notion of the scope of spectrophotometry by consulting the series of review articles that were published in *Analytical Chemistry*[8] as well as monographs on the subject.[9]

Applications to Absorbing Species

Tables 14-1, 14-2, and 14-4 list many common organic chromophoric groups. Spectrophotometric determination of any organic compound containing one or more of these groups is potentially feasible. Many examples of this type of determination are found in the literature.

A number of inorganic species also absorb UV-visible radiation and are thus susceptible to direct determination. We have noted that many ions of the transition metals are colored in solution and can thus be determined by spectrophotometric measurement. In addition, a number of other species show characteristic absorption bands, including nitrite, nitrate, and chromate ions, the oxides of nitrogen, the elemental halogens, and ozone.

Applications to Nonabsorbing Species

Numerous reagents react selectively with nonabsorbing species to yield products that absorb strongly in the ultraviolet or visible regions. The successful application of such reagents to quantitative analysis usually requires that the color-forming reaction be forced to near completion. If the amount of product is limited by the analyte, the absorbance of the product is proportional to the analyte concentration. Color-forming reagents are frequently employed as well for the determination of absorbing species, such as transition-metal ions. The molar absorptivity of the product is frequently orders of magnitude greater than that of the species before reaction.

A host of complexing agents are used to determine inorganic species. Typical inorganic reagents include thiocyanate ion for iron, cobalt, and molybdenum;

[6]For a wealth of detailed, practical information on spectrophotometric practices, see *Techniques in Visible and Ultraviolet Spectrometry*, Vol. I, *Standards in Absorption Spectroscopy*, C. Burgess and A. Knowles, eds., London: Chapman & Hall, 1981; J. R. Edisbury, *Practical Hints on Absorption Spectrometry*, New York: Plenum Press, 1968.
[7]See, for example, T. D. Harris, *Anal. Chem.*, **1982**, *54*, 741A.

[8]L. G. Hargis, J. A. Howell, and R. E. Sutton, *Anal. Chem.* (Review), **1996**, *68*, 169; J. A. Howell and R. E. Sutton, *Anal. Chem.*, (Review), **1998**, *70*, 107.
[9]H. Onishi, *Photometric Determination of Traces of Metals*, Part IIA, Part IIB, 4th ed., New York: Wiley, 1986, 1989; *Colorimetric Determination of Nonmetals*, 2nd ed., D. F. Boltz, ed., New York: Interscience, 1978; E. B. Sandell and H. Onishi, *Photometric Determination of Traces of Metals*, 4th ed., New York: Wiley, 1978; F. D. Snell, *Photometric and Fluorometric Methods of Analysis*, New York: Wiley, 1978.

hydrogen peroxide for titanium, vanadium, and chromium; and iodide ion for bismuth, palladium, and tellurium. Of even more importance are organic chelating agents that form stable, colored complexes with cations. Common examples include diethyldithiocarbamate for the determination of copper, diphenylthiocarbazone for lead, 1,10-phenanthrolene for iron, and dimethylglyoxime for nickel. In the application of the last reaction to the photometric determination of nickel, an aqueous solution of the cation is extracted with a solution of the chelating agent in an immiscible organic liquid. The absorbance of the resulting bright red organic layer serves as a measure of the concentration of the metal.

14D-2 Procedural Details

A first step in any photometric or spectrophotometric analysis is the development of conditions that yield a reproducible relationship (preferably linear) between absorbance and analyte concentration.

Selection of Wavelength

For highest sensitivity, spectrophotometric absorbance measurements are ordinarily made at a wavelength corresponding to an absorption maximum because the change in absorbance per unit of concentration is greatest at this point. In addition, the absorbance is nearly constant with wavelength at an absorption maximum, which leads to close adherence to Beer's law (see Figure 13-5). Finally, small uncertainties that arise from failing to reproduce precisely the wavelength setting of the instrument have less influence at an absorption maximum.

Variables That Influence Absorbance

Common variables that influence the absorption spectrum of a substance include the nature of the solvent, the pH of the solution, the temperature, high electrolyte concentrations, and the presence of interfering substances. The effects of these variables must be known and conditions for the analysis must be chosen such that the absorbance will not be materially influenced by small, uncontrolled variations in their magnitudes.

Cleaning and Handling of Cells

Accurate spectrophotometric analysis requires the use of good-quality, matched cells. These should be regularly calibrated against one another to detect differences that can result from scratches, etching, and wear. It is equally important to use proper cell-cleaning and drying techniques. Erickson and Surles[10] recommend the following cleaning sequence for the outside windows of cells. Prior to measurement, the cell surfaces should be cleaned with a lens paper soaked in spectrograde methanol. While wiping, it is best to hold the paper with a hemostat. The methanol is then allowed to evaporate, leaving the cell surfaces free of contaminants. Erickson and Surles showed that this method was superior to the usual procedure of wiping the cell surfaces with a dry lens paper, which can leave lint and a film on the surface.

Determining the Relationship between Absorbance and Concentration

The method of external standards (see Section 1D-2) is most often used to establish the absorbance versus concentration relationship. After deciding on the conditions for the analysis, the calibration curve is prepared from a series of standard solutions that bracket the concentration range expected for the samples. Seldom, if ever, is it safe to assume adherence to Beer's law and use only a single standard to determine the molar absorptivity. It is never a good idea to base the results of an analysis on a literature value for the molar absorptivity.

Ideally, calibration standards should approximate the composition of the samples to be analyzed not only with respect to the analyte concentration but also with regard to the concentrations of the other species in the sample matrix. This can minimize the effects of various components of the sample on the measured absorbance. For example, the absorbance of many colored complexes of metal ions is decreased to a varying degree in the presence of sulfate and phosphate ions because these anions can form colorless complexes with metal ions. The desired reaction is often less complete as a consequence, and lowered absorbances are the result. The matrix effect of sulfate and phosphate can often be counteracted by introducing into the standards amounts of the two species that approximate the amounts found in the samples. Unfortunately, matrix matching is often impossible or quite difficult when complex materials such as soils, minerals, and tissues are being analyzed. When this is the case, the *standard-addition method* is often helpful in counteracting

---

[10] J. O. Erickson and T. Surles, *Amer. Lab.*, **1976**, *8* (6), 50.

matrix effects that affect the slope of the calibration curve. However, the standard-addition method does not compensate for extraneous absorbing species unless they are present at the same concentration in the blank solution.

The Standard-Addition Method

The standard-addition method can take several forms.[11] The one most often chosen for photometric or spectrophotometric analyses, and the one that was discussed in some detail in Section 1D-3, involves adding one or more increments of a standard solution to sample aliquots. Each solution is then diluted to a fixed volume before measuring its absorbance. Example 14-1 illustrates a spreadsheet approach to the multiple-additions method for the photometric determination of nitrite.

EXAMPLE 14-1

Nitrite is commonly determined by a spectrophotometric procedure using the Griess reaction. The sample containing nitrite is reacted with sulfanilimide and N-(1-naphthyl)ethylenediamine to form a colored species that absorbs radiation at 550 nm. Five-milliliter aliquots of the sample were pipetted into five 50.00-mL volumetric flasks. Then, 0.00, 2.00, 4.00, 6.00, and 8.00 mL of a standard solution containing 10.00 μM nitrite were pipetted into each flask, and the color-forming reagents added. After dilution to volume, the absorbance for each of the five solutions was measured at 550 nm. The absorbances were 0.139, 0.299, 0.486, 0.689, and 0.865, respectively. Devise a spreadsheet to calculate the nitrite concentration in the original sample and its standard deviation.

Solution

The spreadsheet is shown in Figure 14-8. Note that the final result indicates the concentration of nitrite in the original sample is 2.8 ± 0.3 μM. The standard deviation is found from the regression line[12] by using an extrapolated x value of -1.385 mL and a y value of 0.000 as illustrated in Example 1-1 of Chapter 1.

In the interest of saving time or sample, it is possible to perform a standard-addition analysis using only two increments of sample. Here, a single addition of V_s mL of standard would be added to one of the two samples. This approach is based on Equation 14-1 (see Section 1D-3).

$$c_x = \frac{A_1 c_s V_s}{(A_2 - A_1)V_x}$$ (14-1)

The single-addition method is illustrated in Example 14-2.

EXAMPLE 14-2

A 2.00-mL urine specimen was treated with reagent to generate a color with phosphate, following which the sample was diluted to 100 mL. To a second 2.00-mL sample was added exactly 5.00 mL of a phosphate solution containing 0.0300 mg phosphate/mL, which was treated in the same way as the original sample. The absorbance of the first solution was 0.428, and that of the second was 0.538. Calculate the concentration of phosphate in milligrams per millimeter of the specimen.

Solution

Here we substitute into Equation 14-1 and obtain

$$c_x = \frac{(0.428)(0.0300 \text{ mg PO}_4^{3-}/\text{mL})(5.00 \text{ mL})}{(0.538 - 0.428)(2.00 \text{ mL sample})}$$
$$= 0.292 \text{ mg PO}_4^{3-}/\text{mL sample}$$

Analysis of Mixtures of Absorbing Substances

The total absorbance of a solution at any given wavelength is equal to the sum of the absorbances of the individual components in the solution (Equation 13-9). This relationship makes it possible in principle to determine the concentrations of the individual components of a mixture even if their spectra overlap completely. For example, Figure 14-9 shows the spectrum of a solution containing a mixture of species M and species N as well as absorption spectra for the individual components. It is apparent that there is no wavelength where the absorbance is due to just one of these components. To analyze the mixture, molar absorptivities for M and N are first determined at wavelengths λ_1 and λ_2 with sufficient concentrations of the two standard solutions to be sure that Beer's law is obeyed over an absorbance range that encompasses the absorbance of the sample. Note that the wavelengths selected are

[11] See M. Bader, *J. Chem. Educ.*, **1980**, 57, 703.

[12] For more information on spreadsheet approaches to standard addition methods, see S. R. Crouch and F. J. Holler, *Applications of Microsoft® Excel in Analytical Chemistry*, Chaps. 4 and 12, Belmont, CA: Brooks/Cole, 2004.

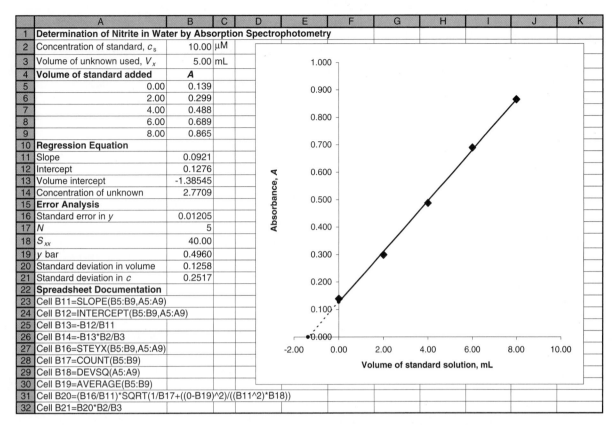

	A	B	C	D	E	F	G	H	I	J	K
1	Determination of Nitrite in Water by Absorption Spectrophotometry										
2	Concentration of standard, c_s	10.00	μM								
3	Volume of unknown used, V_x	5.00	mL								
4	**Volume of standard added**	*A*									
5	0.00	0.139									
6	2.00	0.299									
7	4.00	0.488									
8	6.00	0.689									
9	8.00	0.865									
10	**Regression Equation**										
11	Slope	0.0921									
12	Intercept	0.1276									
13	Volume intercept	-1.38545									
14	Concentration of unknown	2.7709									
15	**Error Analysis**										
16	Standard error in y	0.01205									
17	N	5									
18	S_{xx}	40.00									
19	y bar	0.4960									
20	Standard deviation in volume	0.1258									
21	Standard deviation in c	0.2517									
22	**Spreadsheet Documentation**										
23	Cell B11=SLOPE(B5:B9,A5:A9)										
24	Cell B12=INTERCEPT(B5:B9,A5:A9)										
25	Cell B13=-B12/B11										
26	Cell B14=-B13*B2/B3										
27	Cell B16=STEYX(B5:B9,A5:A9)										
28	Cell B17=COUNT(B5:B9)										
29	Cell B18=DEVSQ(A5:A9)										
30	Cell B19=AVERAGE(B5:B9)										
31	Cell B20=(B16/B11)*SQRT(1/B17+((0-B19)^2)/((B11^2)*B18))										
32	Cell B21=B20*B2/B3										

FIGURE 14-8 Spreadsheet to determine concentration of nitrite by multiple standard additions.

ones at which the molar absorptivities of the two components differ significantly. Thus, at λ_1, the molar absorptivity of component M is much larger than that for component N. The reverse is true for λ_2. To complete the analysis, the absorbance of the mixture is determined at the same two wavelengths. From the known molar absorptivities and path length, the following equations hold:

$$A_1 = \varepsilon_{M_1} b c_M + \varepsilon_{N_1} b c_N \qquad (14\text{-}2)$$

$$A_2 = \varepsilon_{M_2} b c_M + \varepsilon_{N_2} b c_N \qquad (14\text{-}3)$$

where the subscript 1 indicates measurement at wavelength λ_1, and the subscript 2 indicates measurement at wavelength λ_2. With the known values of ε and b, Equations 14-2 and 14-3 represent two equations in two unknowns (c_M and c_N) that can be solved. The relationships are valid only if Beer's law holds at both wavelengths and the two components behave independently of one another. The greatest accuracy is obtained by choosing wavelengths at which the differences in molar absorptivities are large.

 Tutorial: Learn more about **calibration and analysis of mixtures**.

Mixtures containing more than two absorbing species can be analyzed, in principle at least, if a further absorbance measurement is made for each added component. The uncertainties in the resulting data become

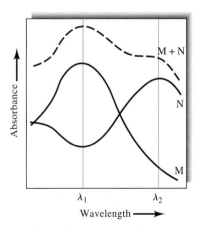

FIGURE 14-9 Absorption spectrum of a two-component mixture (M + N) with spectra of the individual components. Vertical dashed lines indicate optimal wavelengths for determination of the two components.

greater, however, as the number of measurements increases. Some array-detector spectrophotometers are capable of reducing these uncertainties by overdetermining the system. That is, these instruments use many more data points than unknowns and effectively match the entire spectrum of the unknown as closely as possible by least-squares techniques using the methods of matrix algebra. The spectra for standard solutions of each component are required for the analysis.

Computer data-processing methods based on factor analysis or principal components analysis have been developed to determine the number of components and their concentrations or absorptivities in mixtures.[13] These methods are usually applied to data obtained from array-detector-based spectrometers.

14D-3 Derivative and Dual-Wavelength Spectrophotometry

In derivative spectrophotometry, spectra are obtained by plotting the first- or a higher-order derivative of absorbance with respect to wavelength as a function of wavelength.[14] Often, these plots reveal spectral detail that is lost in an ordinary spectrum. In addition, concentration measurements of an analyte in the presence of an interference or of two or more analytes in a mixture can sometimes be made more easily or more accurately using derivative methods. Unfortunately, the advantages of derivative spectra are at least partially offset by the degradation in signal-to-noise ratio that accompanies obtaining derivatives. In some parts of the ultraviolet and visible regions, however, signal-to-noise ratio is not a limiting factor. Even if signal-to-noise ratio is degraded by differentiation, smoothing methods can be applied to help improve precision.

Several different methods have been used to obtain derivative spectra. For modern computer-controlled digital spectrophotometers, the differentiation can be performed numerically using procedures such as derivative least-squares polynomial smoothing, which is discussed in Section 5C-2. With older analog instruments, derivatives of spectral data could be obtained electronically with a suitable operational amplifier circuit (see

Section 3E-4). Another technique uses mechanical oscillation of a refractor plate to sweep a wavelength interval of a few nanometers repetitively across the exit slit of a monochromator while the spectrum is scanned, a technique known as *wavelength modulation*. Alternatively, the spectrum can be scanned using two wavelengths offset by a few nanometers, which is called *dual-wavelength spectrophotometry*.

Applications of Derivative Spectra

Many of the most important applications of derivative spectroscopy in the ultraviolet and visible regions have been for qualitative identification of species. The enhanced detail of a derivative spectrum makes it possible to distinguish among compounds having overlapping spectra, a technique often called *feature enhancement*.[15] Figure 14-10 illustrates how a derivative plot can reveal details of a spectrum consisting of three overlapping absorption peaks. It should be noted that taking a derivative enhances noise, and so high-quality spectra are a must for using this technique. If high-quality spectra are not used, derivative spectra can sometimes create features that are not actually present.

Derivative spectrophotometry is also proving quite useful for the simultaneous determination of two or more components in mixtures. With mixtures, several methods have been proposed for quantitative analysis. The peak-to-peak height has been used as has the peak height at the zero crossing wavelengths for the individual components. More recently, multivariate statistical techniques, such as partial least squares and principal components analysis, have been used to determine concentrations. Derivative methods have been used to determine trace metals in mixtures. For example, trace amounts of Mn and Zn can be determined in mixtures by forming complexes with 5,8-dihydroxy-1,4-naphthoquinone.[16] Derivative methods have also been widely applied to pharmaceutical preparations and to vitamin mixtures.[17]

Derivative and dual-wavelength spectrophotometry have also proved particularly useful for extracting ultraviolet-visible absorption spectra of analytes present in turbid solutions, where light scattering obliterates the details of an absorption spectrum. For example,

[13] E. R. Malinowski, *Factor Analysis in Chemistry*, 3rd ed., Chap. 9, New York: Wiley, 2002.

[14] For additional information see G. Talsky, *Derivative Spectrophotometry Low and High Order*, New York: VCH, 1994; T. C. O'Haver, *Anal. Chem.*, **1979**, *51*, 91A; F. Sanchez Rojas, C. Bosch Ojeda, and J. M. Cano Pavon, *Talanta*, **1988**, *35*, 753; C. Bosch Ojeda, F. Sanchez Rojas, and J. M. Cano Pavon, *Talanta*, **1995**, *42*, 1195.

[15] Spreadsheet applications involving derivative techniques for feature enhancement are given in S. R. Crouch and F. J. Holler, *Applications of Microsoft® Excel in Analytical Chemistry*, pp. 312–15, Belmont, CA: Brooks/Cole, 2004.

[16] H. Sedaira, *Talanta*, **2000**, *51*, 39.

[17] F. Aberastuuri, A. I. Jimenez, F. Jimenez, and J. J. Arias, *J. Chem. Educ.*, **2001**, *78*, 793.

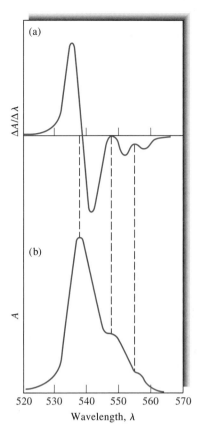

FIGURE 14-10 Comparison of a derivative spectrum (a) with a standard absorption spectrum (b).

14E PHOTOMETRIC AND SPECTROPHOTOMETRIC TITRATIONS

Photometric or spectrophotometric measurements are useful for locating the equivalence point of a titration, provided the analyte, the reagent, or the titration product absorbs radiation.[18] Alternatively, an absorbing indicator can provide the absorbance change necessary for location of the equivalence point.

14E-1 Titration Curves

A photometric titration curve is a plot of absorbance, corrected for volume changes, as a function of the volume of titrant. For many titrations, the curve consists of two linear regions with differing slopes, one occurring early in the titration and the other located well beyond the equivalence-point region. The end point is

---

[18] For further information concerning this technique, see J. B. Headridge, *Photometric Titrations*, New York: Pergamon, 1961; M. A. Leonard, in *Comprehensive Analytical Chemistry*, G. Svehla, ed., Vol. 8, Chap. 3, New York: Elsevier, 1977.

three amino acids, tryptophan, tyrosine, and phenylalanine, contain aromatic side chains, which exhibit sharp absorption bands in the 240 to 300-nm region. These sharp peaks are not, however, apparent in spectra of typical protein preparations, such as bovine or egg albumin, because the large protein molecules scatter radiation severely, yielding only a smooth absorption peak such as that shown in Figure 14-11a. As shown in curves (b) and (c), the aromatic fine structure is revealed in first- and second-derivative spectra.

Dual-wavelength spectrophotometry has also proved useful for determination of an analyte in the presence of a spectral interference. Here, the instrument is operated in the nonscanning mode with absorbances being measured at two wavelengths at which the interference has identical molar absorptivities. In contrast, the analyte must absorb radiation more strongly at one of these wavelengths than the other. The differential absorbance is then directly proportional to the analyte concentration.

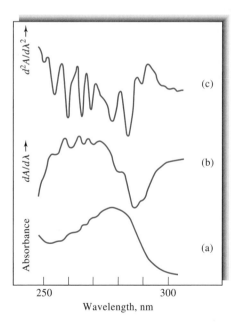

FIGURE 14-11 Absorption spectra of bovine albumin: (a) ordinary spectrum, (b) first-derivative spectrum, (c) second-derivative spectrum. (Reprinted with permission from J. E. Cahill and F. G. Padera, *Amer. Lab.*, **1980**, *12* (4), 109. Copyright 1980 by International Scientific Communications, Inc.)

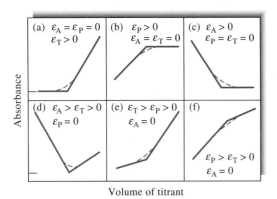

FIGURE 14-12 Typical photometric titration curves. Molar absorptivities of the analyte, the product, and the titrant are given by ε_A, ε_P, ε_T, respectively.

the intersection of the extrapolated linear portions of the curve. End points can also be determined automatically by titration to a fixed absorbance or by taking the derivative to convert the linear-segment curve to a sigmoid-shape curve.

Figure 14-12 shows typical photometric titration curves. Figure 14-12a is the curve for the titration of a nonabsorbing species with an absorbing titrant that reacts with the analyte to form a nonabsorbing product. An example is the titration of thiosulfate ion with triiodide ion. The titration curve for the formation of an absorbing product from nonabsorbing reactants is shown in Figure 14-12b; an example is the titration of iodide ion with a standard solution of iodate ion to form triiodide. The remaining figures illustrate the curves obtained with various combinations of absorbing analytes, titrants, and products.

To obtain titration curves with linear portions that can be extrapolated, the absorbing systems must obey Beer's law. In addition, absorbances must be corrected for volume changes by multiplying the observed absorbance by $(V + v)/V$, where V is the original volume of the solution and v is the volume of added titrant. Many methods, however, use only changes in absorbance to locate the endpoints by various techniques. With these, strict adherence to Beer's law is not a necessity.

14E-2 Instrumentation

Photometric titrations are ordinarily performed with a spectrophotometer or a photometer that has been modified so that the titration vessel is held stationary in the light path. Alternatively, a probe-type cell, such as that shown in Figure 13-18, can be employed. After the instrument is set to a suitable wavelength or an appropriate filter is inserted, the 0% T adjustment is made in the usual way. With radiation passing through the analyte solution to the transducer, the instrument is then adjusted to a convenient absorbance reading by varying the source intensity or the transducer sensitivity. It is not usually necessary to measure the true absorbance because relative values are perfectly adequate for end-point detection. Titration data are then collected without changing the instrument settings. The power of the radiation source and the response of the transducer must remain constant during a photometric titration. Cylindrical containers are often used in photometric titrations, and it is important to avoid moving the cell, so that the path length remains constant.

Both filter photometers and spectrophotometers have been used for photometric titrations. Several instrument companies currently produce photometric titration equipment.

14E-3 Applications of Photometric Titrations

Photometric titrations often provide more accurate results than a direct photometric analysis because the data from several measurements are used to determine the end point. Furthermore, the presence of other absorbing species may not interfere, because only a change in absorbance is being measured.

An advantage of end points determined from linear-segment photometric titration curves is that the experimental data are collected well away from the equivalence-point region where the absorbance changes gradually. Consequently, the equilibrium constant for the reaction need not be as large as that required for a sigmoid titration curve that depends on observations near the equivalence point (for example, potentiometric or indicator end points). For the same reason, more dilute solutions may be titrated using photometric detection.

The photometric end point has been applied to many types of reactions. For example, most standard oxidizing agents have characteristic absorption spectra and thus produce photometrically detectable end points. Although standard acids or bases do not absorb, the introduction of acid-base indicators permits photometric neutralization titrations. The photometric end point has also been used to great advantage in titrations with EDTA (ethylenediaminetetraacetic

 Simulation: Learn more about **spectrophotometric titrations**.

acid) and other complexing agents.[19] Figure 14-13a illustrates the application of this technique to the determination of total hardness in tap water using Eriochrome Black T as the indicator. The absorbance of the indicator is monitored at 610 nm.[20]

The photometric end point has also been adapted to precipitation titrations. In *turbidimetric titrations*, the suspended solid product causes a decrease in the radiant power of the light source by scattering from the particles of the precipitate. The end point is observed when the precipitate stops forming and the amount of light reaching the detector becomes constant. The end point in some precipitation titrations can also be detected as shown in Figure 14-13b by addition of an indicator. Here, the Ba^{2+} titrant reacts with SO_4^{2-} to form insoluble $BaSO_4$. Once the end point has been reached, the excess Ba^{2+} ions react with an indicator, Thorin, to form a colored complex that absorbs light at 523 nm.[21]

14F SPECTROPHOTOMETRIC KINETIC METHODS

Kinetic methods of analysis[22] differ in a fundamental way from the equilibrium, or stoichiometric, methods we have been considering. In kinetic methods, measurements are made under *dynamic* conditions in which the concentrations of reactants and products are changing as a function of time. In contrast, titrations or procedures using complexing agents to form absorbing products are performed on systems that have come to equilibrium or steady state so that concentrations are *static*. The majority of kinetic methods use spectrophotometry as the reaction monitoring technique.

The distinction between the two types of methods is illustrated in Figure 14-14, which shows the progress over time of the reaction

$$A + R \rightleftharpoons P \qquad (14\text{-}4)$$

where A represents the analyte, R the reagent, and P the product. Equilibrium methods operate in the re-

[19] For a discussion of EDTA titrations, see D. A. Skoog, D. M. West, F. J. Holler, and S. R. Crouch, *Fundamentals of Analytical Chemistry*, 8th ed., Belmont, CA: Brooks/Cole, 2004, Chap. 17.

[20] For information on this method, see Metrohm, Application Bulletin, 33/4e, Metrohm AG, Herisau, Switzerland.

[21] See Metrohm, Application Bulletin, 140/3e, Metrohm AG, Herisau, Switzerland.

[22] For additional information, see D. A. Skoog, D. M. West, F. J. Holler, and S. R. Crouch, *Fundamentals of Analytical Chemistry*, 8th ed., Chap. 29, Belmont, CA: Brooks/Cole, 2004; H. O. Mottola, *Kinetic Aspects of Analytical Chemistry*, New York: Wiley, 1988.

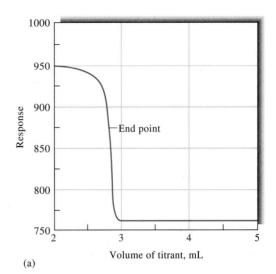

(a)

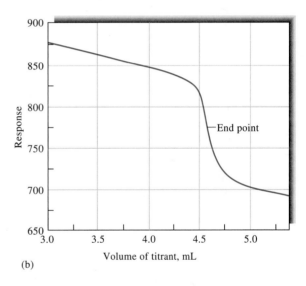

(b)

FIGURE 14-13 Photometric titration curves: (a) total hardness of water, (b) determination of sulfate. In (a), total water hardness is obtained by titration with 0.10 M EDTA at 610 nm for 100 mL of a solution that contained 2.82 mmol/L total hardness. Eriochrome Black T was the indicator. In (b), 10.0 mL of a solution containing sulfate was titrated with 0.050 M $BaCl_2$ using Thorin as an indicator and a wavelength of 523 nm. The response shown is proportional to transmittance. (From A. L. Underwood, *Anal. Chem.*, **1954**, 26, 1322. Figure 1, p. 1323. Copyright 1954 American Chemical Society.)

gion beyond time t_e, when the bulk concentrations of reactants and product have become constant and the chemical system is at equilibrium. In contrast, kinetic methods are carried out during the time interval from

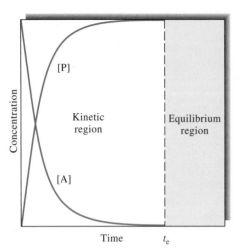

FIGURE 14-14 Change in concentration of analyte [A] and product [P] as a function of time. Until time t_e the analyte and product concentrations are continuously changing. This is the kinetic regime. In the equilibrium region, after t_e, the analyte and product concentrations are static.

0 to t_e, when analyte and product concentrations are changing continuously.

Kinetic methods can be more selective than equilibrium methods if reagents and conditions are chosen to maximize differences in the rates at which the analyte and potential interferents react. In equilibrium-based methods, selectivity is realized by maximizing differences in equilibrium constants.

14F-1 Types of Reactions

Kinetic methods can employ several different types of reactions. Catalyzed reactions are among the most popular. With these, a catalyst is determined by its influence on the reaction rate or one of the reactants is determined. For example, iodide is a catalyst in the reaction of Ce(IV) with As(III). Trace quantities of I$^-$ can be determined by measuring the rate of this reaction as a function of the I$^-$ concentration. Normally, the method of external standards is used to prepare a calibration curve of rate versus iodide concentration. More than forty inorganic cations and more than fifteen anions have been determined based on their catalytic effect.

Organic catalysts have also been determined by kinetic methods. The most important applications of catalyzed reactions to organic analyses involve the use of enzymes as catalysts. The behavior of a large number

 Simulation: Learn more about **kinetic methods**.

of enzymes is consistent with the general mechanism

$$E + S \underset{k_{-1}}{\overset{k_1}{\rightleftharpoons}} ES \overset{k_2}{\longrightarrow} P + E \qquad (14\text{-}5)$$

In this so-called *Michaelis-Menten mechanism*, the enzyme E reacts reversibly with the substrate S to form an enzyme-substrate complex ES. This complex then decomposes irreversibly to form the products and the regenerated enzyme. The rate of this reaction often follows the rate law

$$\frac{d[P]}{dt} = \frac{k_2[E]_0[S]}{\dfrac{k_{-1} + k_2}{k_1} + [S]} = \frac{k_2[E]_0[S]}{K_m + [S]} \qquad (14\text{-}6)$$

where K_m is the Michaelis constant $(k_{-1} + k_2)/k_1$. Under conditions where the enzyme is saturated with substrate, $[S] \gg K_m$, the rate $d[P]/dt$ is directly proportional to the initial enzyme concentration $[E]_0$:

$$\frac{d[P]}{dt} = k_2[E]_0$$

Hence, measurements of the rate can be used to obtain the enzyme activity (concentration), $[E]_0$.

Substrates can also be determined by kinetic methods. Under conditions where $[S] \ll K_m$, Equation 14-6 reduces to

$$\frac{d[P]}{dt} = \frac{k_2}{K_m}[E]_0[S] = k'[S]$$

where $k' = k_2/K_m$. Here, the reaction rate is directly proportional to the substrate concentration, [S]. If measurements are made near the beginning of the reaction (<5% reaction), $[S] \approx [S]_0$ and the rate is directly proportional to the initial substrate concentration.[23]

The regions where enzyme and substrate can be determined by kinetic methods are illustrated in Figure 14-15, which shows a plot of initial rate versus substrate concentration. We can see that the initial rate is proportional to substrate concentration at very low concentrations, but the rate is proportional to the enzyme concentration when the substrate concentration is very high.

In addition to catalyzed reactions, kinetic methods of analysis also can employ uncatalyzed reactions. As an example, phosphate can be determined by measuring the rate of its reaction with molybdate to form a heteropoly species, 12-molybdophosphate. More sensitivity can be achieved by reducing the 12-molyb-

[23] For the conditions necessary for an enzyme reaction to be in its initial stages, see J. D. Ingle Jr. and S. R. Crouch, *Anal. Chem.*, **1971**, *43*, 697.

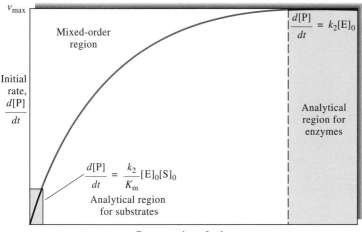

FIGURE 14-15 Plot of initial rate of product formation as a function of substrate concentration, showing the parts of the curve useful for the determination of substrate and enzyme.

dophosphate with ascorbic acid or another reducing agent to form phosphomolybdenum blue, an intensely colored species. In either case, the rate of formation of the product is directly proportional to the phosphate concentration.

14F-2 Instrumentation

Kinetic methods based on reactions that, with half-lives greater than about 10 s, can be performed in an ordinary spectrophotometer equipped with a thermostatted cell compartment and provision to introduce and mix samples and reagents. Rates of reaction are highly dependent on temperature, and so temperature control to about $0.1°C$ is necessary for good reproducibility. Many commercial spectrophotometers have attachments that allow rates to be obtained. For very slow reactions, sample introduction and mixing can be accomplished prior to placing the reaction mixture in the cell compartment. Usually, however, a stationary cell is used, and all reagents except one are placed in the cell. The reagent needed to start the reaction is then introduced by syringe or pipette and the ensuing reaction is monitored while the mixture is stirred. With single-channel spectrophotometers, the reaction is monitored at a single wavelength by measuring the absorbance as a function of time. Array-detector-based instruments allow entire spectra to be taken at different time intervals for later analysis.

Continuous flow methods, such as flow injection analysis (see Chapter 33), are also used for sample introduction and reaction monitoring. A popular method with flow injection is to introduce sample and reagents

in flowing streams and then to stop the flow with the reaction mixture in the spectrophotometric flow cell. For reactions with half-lives of less than 10 s, the *stopped-flow mixing* technique is popular. In this technique, streams of reagent and sample are mixed rapidly, and the flow of mixed solution is stopped suddenly. The reaction progress is then monitored at a position slightly downstream from the mixing point. The apparatus shown in Figure 14-16 is designed to perform stopped-flow mixing and to allow measurements on the millisecond time scale.

14F-3 Types of Kinetic Methods

Kinetic methods can be classified according to how the measurement is made.[24] *Differential methods* compute the rate of reaction and relate it to the analyte concentration. Rates are determined from the slope of the absorbance versus time curve. *Integral methods* use an integrated form of the rate equation and determine the concentration of analyte from the absorbance changes that occur over various time intervals. Curve-fitting methods fit a mathematical model to the absorbance versus time curve and compute the parameters of the model, including the analyte concentration. The most sophisticated of these methods use the parameters of the model to estimate the value of the equilibrium or steady-state response. These methods can provide error compensation because the equilibrium position is

[24] For spreadsheet approaches to kinetic methods, see S. R. Crouch and F. J. Holler, *Applications of Microsoft® Excel in Analytical Chemistry*, Chap. 13, Belmont, CA: Brooks/Cole, 2004.

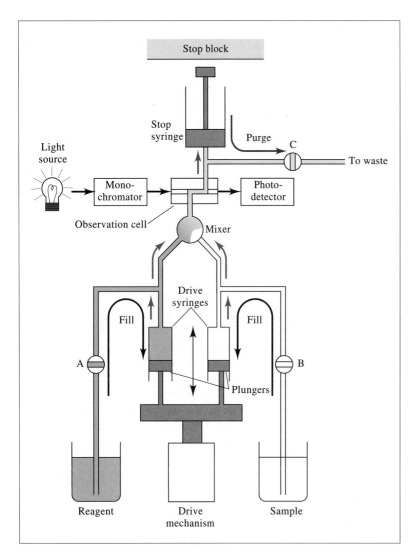

FIGURE 14-16 Stopped-flow mixing apparatus. To begin the experiment, the drive syringes are filled with reagent and sample and valves A, B, and C are closed. The drive mechanism is then activated to move the drive syringe plungers forward rapidly. The reagent and sample are mixed in the mixer and pass immediately into the observation cell and stop syringe. When the stop syringe fills, the plunger strikes the stop block and the flow ceases almost instantly with a recently mixed plug of solution in the spectrophotometric observation cell. For well-designed systems, the time between mixing and observation can be on the order of 2 to 4 ms.

less sensitive to such experimental variables as temperature, pH, and reagent concentrations. Figure 14-17 illustrates the use of this approach to predict the equilibrium absorbance from data obtained during the kinetic regime of the absorbance versus time curve. The equilibrium absorbance is then related to the analyte concentration in the usual way.

14G SPECTROPHOTOMETRIC STUDIES OF COMPLEX IONS

Spectrophotometry is a valuable tool for discovering the composition of complex ions in solution and for determining their formation constants (K_f values).

Quantitative absorption measurements are very useful for studying complexation because they can be made without disturbing the equilibria under consideration. Although many spectrophotometric studies of complexes involve systems in which a reactant or a product absorbs radiation, nonabsorbing systems can also be investigated successfully. For example, the composition and formation constant for a complex of iron(II) and a nonabsorbing ligand may often be determined by measuring the absorbance decreases that occur when solutions of the absorbing iron(II) complex of 1,10-phenanthroline are mixed with various amounts of the nonabsorbing ligand. The success of this approach depends on the well-known values of the *formation constant* ($K_f = 2 \times 10^{21}$) and the

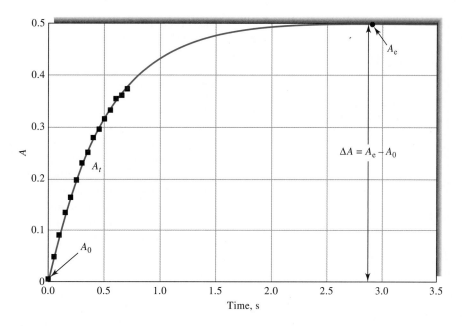

FIGURE 14-17 The predictive approach in kinetic methods. A mathematical model, shown as the solid line, is used to fit the response, shown as the squares, during the kinetic regime of a reaction. The model is then used to predict the equilibrium absorbance, A_e, which is related to the analyte concentration. In the example shown, the absorbance is plotted versus time and the early time data used to predict A_e, the equilibrium value, shown as the circle. (From G. E. Mieling and H. L. Pardue, *Anal. Chem.*, **1978**, *50*, 1611–1618. American Chemical Society.)

composition of the 1,10-phenanthroline (3:1) complex of iron(II).

The most common techniques used for complex-ion studies are (1) the method of continuous variations, (2) the mole-ratio method, (3) the slope-ratio method, and (4) computer-based curve-fitting methods.

14G-1 The Method of Continuous Variations

In the method of continuous variations, cation and ligand solutions with identical analytical concentrations are mixed in such a way that the total volume and the total number of moles of reactants in each mixture are constant but the mole ratio of reactants varies systematically (for example, 1:9, 8:2, 7:3, and so forth). The absorbance of each solution is then measured at a suitable wavelength and corrected for any absorbance the mixture might exhibit if no reaction had occurred. For example, if only the ligand absorbs UV or visible radiation, the corrected absorbance would be the absorbance of the reaction mixture minus the absorbance of the ligand had it not reacted. The corrected absor-

bance is plotted against the volume fraction of one reactant, that is, $V_M/(V_M + V_L)$, where V_M is the volume of the cation solution and V_L is the volume of the ligand solution. A typical continuous-variations plot is shown in Figure 14-18. A maximum (or minimum if the complex is less absorbing than the reactants) occurs at a volume ratio V_M/V_L corresponding to the combining ratio of cation and ligand in the complex. In Figure 14-18, $V_M/(V_M + V_L)$ is 0.33 and $V_L/(V_M + V_L)$ is 0.66; thus, V_M/V_L is 0.33/0.66, which suggests that the complex has the formula ML_2.[25]

The curvature of the experimental lines in Figure 14-18 is the result of incompleteness of the complex-formation reaction. The formation constant for the complex can be evaluated from measurements of the deviations from the theoretical straight lines, which represent the curve that would result if the reac-

[25] A spreadsheet approach for the method of continuous variation is given in S. R. Crouch and F. J. Holler, *Applications of Microsoft® Excel in Analytical Chemistry*, Chap. 12, Belmont, CA: Brooks/Cole Publishing Co., 2004.

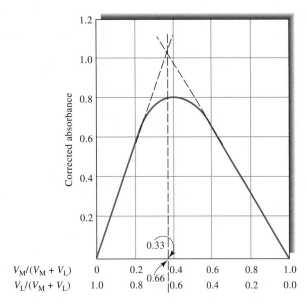

FIGURE 14-18 Continuous-variation plot for the 1:2 complex ML_2.

tion between the ligand and the metal proceeded to completion. Mathematical models can be derived to allow calculation of the K_f value or computer curve-fitting methods can be used (see section 14G-4).

14G-2 The Mole-Ratio Method

In the mole-ratio method, a series of solutions is prepared in which the analytical concentration of one reactant (usually the cation) is held constant while that of the other is varied. A plot of absorbance versus mole ratio of the reactants is then prepared. If the formation constant is reasonably favorable, two straight lines of different slopes that intersect at a mole ratio that corresponds to the combining ratio in the complex are obtained. Typical mole-ratio plots are shown in Figure 14-19. Notice that the ligand of the 1:2 complex absorbs at the wavelength selected so that the slope beyond the equivalence point is greater than zero. We deduce that the uncomplexed cation involved in the 1:1 complex absorbs radiation, because the initial point has an absorbance greater than zero.

Formation constants can be evaluated from the data in the curved portion of mole-ratio plots where the reaction is least complete. If two or more complexes form, successive slope changes in the mole-ratio plot

 Simulation: Learn more about **determining the composition of complexes.**

may occur provided the complexes have different molar absorptivities and different formation constants.

14G-3 The Slope-Ratio Method

This approach is particularly useful for weak complexes but is applicable only to systems in which a single complex is formed. The method assumes (1) that the complex-formation reaction can be forced to completion by a large excess of either reactant and (2) that Beer's law is followed under these circumstances.

Let us consider the reaction in which the complex M_xL_y is formed by the reaction of x moles of the cation M with y moles of a ligand L:

$$x\text{M} + y\text{L} \rightleftharpoons \text{M}_x\text{L}_y$$

Mass-balance expressions for this system are

$$c_\text{M} = [\text{M}] + x[\text{M}_x\text{L}_y]$$
$$c_\text{L} = [\text{L}] + y[\text{M}_x\text{L}_y]$$

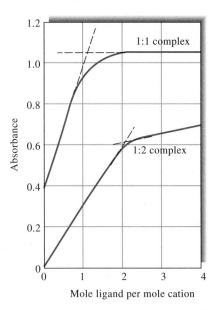

FIGURE 14-19 Mole-ratio plots for a 1:1 and a 1:2 complex. The 1:2 complex is the more stable of the two complexes as indicated by closeness of the experimental curve to the extrapolated lines. The closer the curve is to the extrapolated lines, the larger the formation constant of the complex; the larger the deviation from the straight lines, the smaller the formation constant of the complex.

where c_M and c_L are the molar analytical concentrations of the two reactants. We now assume that at very high analytical concentrations of L, the equilibrium is shifted far to the right and $[M] \ll x[M_xL_y]$. Under this condition, the first mass-balance expression simplifies to

$$c_M = x[M_xL_y]$$

If the system obeys Beer's law,

$$A_1 = \varepsilon b[M_xL_y] = \varepsilon b c_M/x$$

where ε is the molar absorptivity of M_xL_y and b is the path length. A plot of absorbance as a function of c_M is linear when there is sufficient L present to justify the assumption that $[M] \ll x[M_xL_y]$. The slope of this plot is $\varepsilon b/x$.

When c_M is made very large, we assume that $[L] \ll y[M_xL_y]$, and the second mass-balance equation reduces to

$$c_L = y[M_xL_y]$$

and

$$A_2 = \varepsilon b[M_xL_y] = \varepsilon b c_L/y$$

Again, if our assumptions are valid, we find that a plot of A versus c_L is linear at high concentrations of M. The slope of this line is $\varepsilon b/y$.

The ratio of the slopes of the two straight lines gives the combining ratio between M and L:

$$\frac{\varepsilon b/x}{\varepsilon b/y} = \frac{y}{x}$$

14G-4 Computer-Based Methods for Determining Formation Constants of Complexes

Several different methods depend on computer curve fitting to determine formation constants of complexes. We illustrate one approach here, but many others are used.[26]

---

[26]See for example, K. A. Connors, *Binding Constants: The Measurement of Molecular Complex Stability*, New York: Wiley, 1988; F. J. C. Rossotti, and H. Rossotti, *The Determination of Stability Constants*, New York: McGraw-Hill, 1961.

Principles

Let us consider the formation of a single 1:1 complex ML from metal ion M and ligand L. Once again, we leave the charges off for generality:

$$M + L \rightleftharpoons ML \qquad K_f = \frac{[ML]}{[M][L]}$$

If the uncomplexed metal ion and the complex both absorb radiation at the analysis wavelength, we can write

$$A = \varepsilon_{ML}b[ML] + \varepsilon_M b[M] \qquad (14\text{-}7)$$

The mass-balance expression for the metal ion is

$$c_M = [M] + [ML]$$

If we solve for $[M]$ and substitute into Equation 14-7, we get

$$A = \varepsilon_{ML}b[ML] + \varepsilon_M b(c_M - [ML])$$
$$= \varepsilon_{ML}b[ML] + \varepsilon_M b c_M - \varepsilon_M b[ML] \qquad (14\text{-}8)$$

When the ligand concentration is zero, $[ML] = 0$ and the absorbance $A_{L=0}$ is given by

$$A_{L=0} = \varepsilon_M b c_M$$

If we substitute this expression into Equation 14-8 and rearrange, we get

$$\Delta A = A - A_{L=0} = \varepsilon_{ML}b[ML] - \varepsilon_M b[ML]$$
$$= \Delta\varepsilon b[ML] \qquad (14\text{-}9)$$

where ΔA is the difference in absorbance with and without the ligand present, and $\Delta\varepsilon$ is the difference in the molar absorptivities of ML and M.

From the formation constant expression, we can write $[ML] = K_f \times [M][L]$. Furthermore, if our experiments are carried out in the presence of excess ligand, $c_L \approx [L]$. Substituting these expressions and the mass balance expression for $[M]$ into Equation 14-9, we obtain

$$\frac{\Delta A}{b} = \Delta\varepsilon K_f c_L[M] = \Delta\varepsilon K_f c_L\{c_M - [ML]\}$$

This expression can be manipulated to obtain

$$\frac{\Delta A}{b} = \frac{\Delta\varepsilon K_f c_L c_M}{1 + K_f c_L} \qquad (14\text{-}10)$$

Data Analysis

Equation 14-10 is the basis for several computer-based methods for determining the formation constant K_f. In the usual experiment, a constant concentration of metal is used and the total ligand concentration c_L is varied. The change in absorbance ΔA is then measured as a function of total ligand concentration and the results statistically analyzed to obtain K_f. Unfortunately, the relationship shown in Equation 14-10 is nonlinear, and thus nonlinear regression must be used unless the equation is transformed to a linear form.[27] We can linearize the equation by taking the reciprocal of both sides to obtain

$$\frac{b}{\Delta A} = \frac{1 + K_f c_L}{\Delta \varepsilon K_f c_L c_M} = \frac{1}{\Delta \varepsilon K_f c_L c_M} + \frac{1}{\Delta \varepsilon c_M} \quad (14\text{-}11)$$

A double reciprocal plot of $b/\Delta A$ versus $1/c_L$ should be a straight line with a slope of $1/\Delta \varepsilon K_f c_M$ and an intercept of $1/\Delta \varepsilon c_M$. This equation is sometimes called the Benesi-Hildebrand equation.[28]

Linear regression can thus be used to obtain the formation constant as well as $\Delta \varepsilon$ if c_M is known. The least-squares parameters obtained are optimal only for the linearized equation and may not be the optimum values for the nonlinear equation. Hence, nonlinear regression is usually preferred for determining the parameters K_f and $\Delta \varepsilon$. Many computer programs are available for nonlinear regression. Spreadsheet programs such as Excel can be employed, but they do not give statistical estimates of the goodness of fit or standard errors for the parameters. Example 14-3 illustrates the use of Excel for these calculations. Other programs, including Origin, Minitab, GraphPad Prism, and TableCurve, give more complete statistics.

EXAMPLE 14-3

To obtain the formation constant of a 1:1 complex, the absorbance data in the table below were obtained at various ligand concentrations and a metal concentration of 1.00×10^{-3} M. Both the uncomplexed metal ion and the complex absorb radiation at the

analysis wavelength. The cell path length was 10 cm. Use a spreadsheet to determine the formation constant.

[L], M	A
0.0500	1.305
0.0400	1.215
0.0300	1.158
0.0200	1.034
0.0100	0.787
0.0050	0.525
0.0000	0.035

Solution

We will first use Equation 14-11, the linearized form of Equation 14-10. Here, we need to calculate $1/[L]$ and $b/\Delta A$. The resulting spreadsheet is shown in Figure 14-20. The value of K_f determined by this method is 96 and $\Delta \varepsilon$ is 152. Next we will use Excel's Solver to obtain the results for the nonlinear Equation 14-10.[29] We start out with initial estimates of the parameters $K_f = 10$ and $\Delta \varepsilon = 50$. As can be seen in Figure 14-21a, the fit is not good with these values. Using the initial estimates we can calculate the values predicted by the model and obtain the differences between the model and the data values (residuals). Solver then minimizes the sum of the squares of the residuals to obtain the best-fit values shown in Figure 14-21b. The fit is now quite good, as seen in the plot. The nonlinear regression parameters are $K_f = 97$ and $\Delta \varepsilon = 151$.

Extensions to Complicated Equilibria

Equation 14-10 can be modified to fit many other cases. It can be extended to account for the formation of polynuclear complexes and for multiple equilibria. Ultraviolet and visible absorption spectroscopy is not, however, particularly well suited for multiple equilibria because of its lack of specificity and because each additional equilibrium adds two unknowns, a formation constant term and a molar absorptivity term.

The introduction of array-detector spectrometric systems has brought about several new data analysis

---

[27] See D. A. Skoog, D. M. West, F. J. Holler, and S. R. Crouch, *Fundamentals of Analytical Chemistry*, 8th ed., Chap. 8, Belmont, CA: Brooks/Cole, 2004.

[28] H. Benesi and J. H. Hildebrand, *J. Am. Chem. Soc.*, **1949**, *71*, 2703.

[29] For more information on nonlinear regression using Excel, see S. R. Crouch and F. J. Holler, *Applications of Microsoft® Excel in Analytical Chemistry*, Chap. 13, Belmont, CA: Brooks/Cole, 2004.

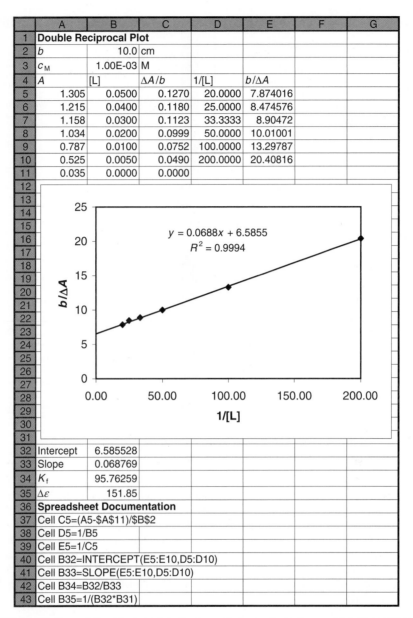

	A	B	C	D	E	F	G
1	**Double Reciprocal Plot**						
2	b	10.0	cm				
3	c_M	1.00E-03	M				
4	A	[L]	$\Delta A/b$	1/[L]	$b/\Delta A$		
5	1.305	0.0500	0.1270	20.0000	7.874016		
6	1.215	0.0400	0.1180	25.0000	8.474576		
7	1.158	0.0300	0.1123	33.3333	8.90472		
8	1.034	0.0200	0.0999	50.0000	10.01001		
9	0.787	0.0100	0.0752	100.0000	13.29787		
10	0.525	0.0050	0.0490	200.0000	20.40816		
11	0.035	0.0000	0.0000				
12							
32	Intercept	6.585528					
33	Slope	0.068769					
34	K_f	95.76259					
35	$\Delta\varepsilon$	151.85					
36	**Spreadsheet Documentation**						
37	Cell C5=(A5-$A$11)/$B$2						
38	Cell D5=1/B5						
39	Cell E5=1/C5						
40	Cell B32=INTERCEPT(E5:E10,D5:D10)						
41	Cell B33=SLOPE(E5:E10,D5:D10)						
42	Cell B34=B32/B33						
43	Cell B35=1/(B32*B31)						

In the chart region:

$$y = 0.0688x + 6.5855$$
$$R^2 = 0.9994$$

FIGURE 14-20 Spreadsheet to calculate the formation constant of a complex from absorbance data using a double reciprocal plot. Equation 14-10 is linearized by calculating the reciprocals in columns D and E. The least-squares slope and intercept are given in the plot and in the cells below the plot. The parameters K_f and $\Delta\varepsilon$ are calculated in cells B34 and B35.

approaches. Instead of using data at only a single wavelength, these approaches can simultaneously use data at multiple wavelengths. With modern computers, equations similar to Equation 14-10 can be fit nearly simultaneously at multiple wavelengths using curve-fitting programs such as TableCurve (Systat Software).

If formation constants are known or determined by the above methods, several programs are available for determining the species composition at various concentrations. The programs HALTAFALL[30] and

[30] N. Ingri, W. Kalolowicz, L. G. Sillén, and B. Warnqvist, *Talanta*, **1967**, *14*, 1261.

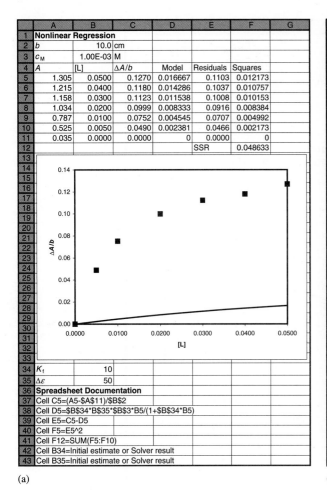

(a)

(b)

FIGURE 14-21 Spreadsheets to calculate formation constants using nonlinear regression. (a) The results from the model (Equation 14-10) are calculated in column D using the initial estimates of $K_f = 10$ and $\Delta\varepsilon = 50$. The plot shows the model values (solid line) and the data values (points). The difference between the data values (column C) and the model values (column D) are the residuals shown in column E. The squares of these are calculated in column F and summed in cell F12 (SSR). In (b), Excel's Solver has minimized the value in cell F12 to obtain the best-fit values shown in cells B34 and B35. The model values are again shown as the solid line in the plot.

COMICS[31] have been popular for many years for speciation in complex systems. These programs use the formation constants and mass-balance expressions to calculate the species composition as a function of initial concentrations. Several modern versions, including some that operate in the Windows environment, are available.

[31] D. D. Perrin and I. G. Sayce, *Talanta*, **1967**, *14*, 833.

QUESTIONS AND PROBLEMS

*Answers are provided at the end of the book for problems marked with an asterisk.

☒ Problems with this icon are best solved using spreadsheets.

*14-1 A 25.0-mL aliquot of an aqueous quinine solution was diluted to 50.0 mL and found to have an absorbance of 0.656 at 348 nm when measured in a 2.50-cm cell.

A second 25.0-mL aliquot was mixed with 10.00 mL of a solution containing 25.7 ppm of quinine; after dilution to 50.0 mL, this solution had an absorbance of 0.976 (2.50-cm cell). Calculate the concentration of quinine in parts per million in the sample.

*14-2 A 0.5990-g pesticide sample was decomposed by wet ashing and then diluted to 200.0 mL in a volumetric flask. The analysis was completed by treating aliquots of this solution as indicated.

Volume of Sample Taken, mL	Reagent Volumes Used, mL			Absorbance, A, 545 nm (1.00-cm cells)
	2.75 ppm Cu^{2+}	Ligand	H_2O	
5.00	0.00	20.0	25.0	0.723
5.00	1.00	20.0	24.0	0.917

Calculate the percentage of copper in the sample.

14-3 Sketch a photometric titration curve for the titration of Sn^{2+} with MnO_4^-. What color radiation should be used for this titration? Explain.

14-4 Iron(III) reacts with thiocyanate ion to form the red complex $Fe(SCN)^{2+}$. Sketch a photometric titration curve for Fe(III) with thiocyanate ion when a photometer with a green filter is used to collect data. Why is a green filter used?

14-5 Ethylenediaminetetraacetic acid abstracts bismuth(III) from its thiourea complex:

$$Bi(tu)_6^{3+} + H_2Y^{2-} \rightarrow BiY^- + 6tu + 2H^+$$

where tu is the thiourea molecule $(NH_2)_2CS$. Predict the shape of a photometric titration curve based on this process, given that the Bi(III) or thiourea complex is the only species in the system that absorbs light at 465 nm, the wavelength selected for the analysis.

*14-6 The accompanying data (1.00-cm cells) were obtained for the spectrophotometric titration of 10.00 mL of Pd(II) with 2.44×10^{-4} M Nitroso R (O. W. Rollins and M. M. Oldham, *Anal. Chem.*, **1971**, *43*, 262).

Volume of Nitroso R, mL	A_{500}
0	0
1.00	0.147
2.00	0.271
3.00	0.375

Calculate the concentration of the Pd(II) solution, given that the ligand-to-cation ratio in the colored product is $2:1$

*14-7 A 3.65-g petroleum specimen was decomposed by wet ashing and subsequently diluted to 500 mL in a volumetric flask. Cobalt was determined by treating 25.00-mL aliquots of this diluted solution as follows:

Reagent Volume, mL			
Co(II), 4.25 ppm	Ligand	H_2O	Absorbance (1.00-cm cell)
0.00	20.00	5.00	0.276
5.00	20.00	0.00	0.491

Assume that the Co(II)-ligand chelate obeys Beer's law, and calculate the percentage of cobalt in the original sample.

*14-8 A simultaneous determination for cobalt and nickel can be based on absorption by their respective 8-hydroxyquinolinol complexes. Molar absorptivities corresponding to their absorption maxima are as follows:

	Molar Absorptivity, ε	
	365 nm	**700 nm**
Co	3529	428.9
Ni	3228	10.2

Calculate the molar concentration of nickel and cobalt in each of the following solutions using the following data:

	Absorbance, A (1.00-cm cells)	
Solution	**365 nm**	**700 nm**
(a)	0.426	0.026
(b)	0.792	0.081

*14-9 When measured with a 1.00-cm cell, a 7.50×10^{-5} M solution of species A exhibited absorbances of 0.155 and 0.755 at 475 and 700 nm, respectively. A 4.25×10^{-5} M solution of species B gave absorbances of 0.702 and 0.091 under the same circumstances. Calculate the concentrations of A and B in solutions that yielded the following absorbance data in a 2.50-cm cell: (a) 0.439 at 475 nm and 1.025 at 700 nm; (b) 0.662 at 475 nm and 0.815 at 700 nm.

*14-10 The acid-base indicator HIn undergoes the following reaction in dilute aqueous solution:

$$\underset{\text{color 1}}{\text{HIn}} \rightleftharpoons \text{H}^+ + \underset{\text{color 2}}{\text{In}^-}$$

The following absorbance data were obtained for a 5.00×10^{-4} M solution of HIn in 0.1 M NaOH and 0.1 M HCl. Measurements were made at wavelengths of 485 nm and 625 nm with 1.00-cm cells.

$$\text{0.1 M NaOH} \qquad A_{485} = 0.075 \qquad A_{625} = 0.904$$
$$\text{0.1 M HCl} \qquad A_{485} = 0.487 \qquad A_{625} = 0.181$$

In the NaOH solution, essentially all of the indicator is present as In^-; in the acidic solution, it is essentially all in the form of HIn.

(a) Calculate molar absorptivities for In^- and HIn at 485 and 625 nm.

(b) Calculate the acid dissociation constant for the indicator if a pH 5.00 buffer containing a small amount of the indicator exhibits an absorbance of 0.567 at 485 nm and 0.395 at 625 nm (1.00-cm cells).

(c) What is the pH of a solution containing a small amount of the indicator that exhibits an absorbance of 0.492 at 485 nm and 0.245 at 635 nm (1.00-cm cells)?

(d) A 25.00-mL aliquot of a solution of purified weak organic acid HX required exactly 24.20 mL of a standard solution of a strong base to reach a phenolphthalein end point. When exactly 12.10 mL of the base was added to a second 25.00-mL aliquot of the acid, which contained a small amount of the indicator under consideration, the absorbance was found to be 0.333 at 485 nm and 0.655 at 625 nm (1.00-cm cells). Calculate the pH of the solution and K_a for the weak acid.

(e) What would be the absorbance of a solution at 485 and 625 nm (1.50-cm cells) that was 2.00×10^{-4} M in the indicator and was buffered to a pH of 6.000?

14-11 A standard solution was put through appropriate dilutions to give the concentrations of iron shown below. The iron(II)-1,10,phenanthroline complex was then formed in 25.0-mL aliquots of these solutions, following which each was diluted to 50.0 mL. The following absorbances (1.00-cm cells) were recorded at 510 nm:

Fe(II) Concentration in Original Solutions, ppm	A_{510}
4.00	0.160
10.0	0.390
16.0	0.630
24.0	0.950
32.0	1.260
40.0	1.580

(a) Plot a calibration curve from these data.
(b) Use the method of least squares to find an equation relating absorbance and the concentration of iron(II).
(c) Calculate the standard deviation of the slope and intercept.

14-12 The method developed in Problem 14-11 was used for the routine determination of iron in 25.0-mL aliquots of groundwater. Express the concentration (as ppm Fe) in samples that yielded the accompanying absorbance data (1.00-cm cell). Calculate the relative standard deviation of the result. Repeat the calculation assuming the absorbance data are means of three measurements.
(a) 0.143   (c) 0.068   (e) 1.512
(b) 0.675   (d) 1.009   (f) 0.546

14-13 Copper(II) forms a 1:1 complex with the organic complexing agent R in acidic medium. The formation of the complex can be monitored by spectrophotometry at 480 nm. Use the following data collected under pseudo-first-order conditions to construct a calibration curve of rate versus concentration of R. Find the concentration of copper(II) in an unknown whose rate under the same conditions was $6.2 \times 10^{-3} A \text{ s}^{-1}$.

$c_{Cu^{2+}}$, ppm	Rate, $A \text{ s}^{-1}$
3.0	3.6×10^{-3}
5.0	5.4×10^{-3}
7.0	7.9×10^{-3}
9.0	1.03×10^{-2}

*14-14 Aluminum forms a 1:1 complex with 2-hydroxy-1-naphthaldehyde *p*-methoxybenzoylhydraxonal, which absorbs UV radiation at 285 nm. Under pseudo-first-order conditions, a plot of the initial rate of the reaction (absorbance units per second) versus the concentration of aluminum (in μM) yields a straight line described by the equation

$$\text{rate} = 1.74c_{Al} - 0.225$$

Find the concentration of aluminum in a solution that exhibits a rate of 0.76 absorbance units per second under the same experimental conditions.

*14-15 The enzyme monoamine oxidase catalyzes the oxidation of amines to aldehydes. For tryptamine, K_m for the enzyme is 4.0×10^{-4} M and $v_{max} = k_2[E]_0 = 1.6 \times 10^{-3}$ μM/min at pH 8. Find the concentration of a solution of tryptamine that reacts at a rate of 0.18 μm/min in the presence of monoamine oxidase under the above conditions. Assume that $[\text{tryptamine}] \ll K_m$.

14-16 The sodium salt of 2-quinizarinsulfonic acid (NaQ) forms a complex with Al^{3+} that absorbs radiation strongly at 560 nm.[32] (a) Use the data from Owens and Yoe's paper to find the formula of the complex. In all solutions, $c_{Al} = 3.7 \times 10^{-5}$ M, and all measurements were made in 1.00-cm cells. (b) Find the molar absorptivity of the complex.

c_Q, M	A_{560}
1.00×10^{-5}	0.131
2.00×10^{-5}	0.265
3.00×10^{-5}	0.396
4.00×10^{-5}	0.468
5.00×10^{-5}	0.487
6.00×10^{-5}	0.498
8.00×10^{-5}	0.499
1.00×10^{-4}	0.500

⊠ 14-17 The accompanying data were obtained in a slope-ratio investigation of the complex formed between Ni^{2+} and 1-cyclopentene-1-dithiocarboxylic acid (CDA). The measurements were made at 530 nm in 1.00-cm cells.

$c_{CDA} = 1.00 \times 10^{-3}$ M		$c_{Ni} = 1.00 \times 10^{-3}$ M	
c_{Ni}, M	A_{530}	c_{CDA}, M	A_{530}
5.00×10^{-6}	0.051	9.00×10^{-6}	0.031
1.20×10^{-5}	0.123	1.50×10^{-5}	0.051
3.50×10^{-5}	0.359	2.70×10^{-5}	0.092
5.00×10^{-5}	0.514	4.00×10^{-5}	0.137
6.00×10^{-5}	0.616	6.00×10^{-5}	0.205
7.00×10^{-5}	0.719	7.00×10^{-5}	0.240

(a) Determine the formula of the complex. Use linear least-squares to analyze the data.
(b) Find the molar absorptivity of the complex and its uncertainty.

⊠ 14-18 The accompanying absorption data were recorded at 390 nm in 1.00-cm cells for a continuous-variation study of the colored product formed between Cd^{2+} and the complexing reagent R.

---

[32] E. G. Owens and J. H. Yoe, *Anal. Chem.*, **1959**, *31*, 385.

Solution	Reagent Volumes, mL		A_{390}
	$c_{Cd} = 1.25 \times 10^{-4}\,M$	$c_R = 1.25 \times 10^{-4}\,M$	
0	10.00	0.00	0.000
1	9.00	1.00	0.174
2	8.00	2.00	0.353
3	7.00	3.00	0.530
4	6.00	4.00	0.672
5	5.00	5.00	0.723
6	4.00	6.00	0.673
7	3.00	7.00	0.537
8	2.00	8.00	0.358
9	1.00	9.00	0.180
10	0.00	10.00	0.000

(a) Find the ligand-to-metal ratio in the product.

(b) Calculate an average value for the molar absorptivity of the complex and its uncertainty. Assume that in the linear portions of the plot the metal is completely complexed.

(c) Calculate K_f for the complex using the stoichiometric ratio determined in (a) and the absorption data at the point of intersection of the two extrapolated lines.

 14-19 Palladium(II) forms an intensely colored complex at pH 3.5 with arsenazo III at 660 nm.[33] A meteorite was pulverized in a ball mill, and the resulting powder was digested with various strong mineral acids. The resulting solution was evaporated to dryness, dissolved in dilute hydrochloric acid, and separated from interferents by ion-exchange chromatography. The resulting solution containing an unknown amount of Pd(II) was then diluted to 50.00 mL with pH 3.5 buffer. Ten-milliliter aliquots of this analyte solution were then transferred to six 50-mL volumetric flasks. A standard solution was then prepared that was $1.00 \times 10^{-5}\,M$ in Pd(II). Volumes of the standard solution shown in the table were then pipetted into the volumetric flasks along with 10.00 mL of 0.01 M arsenazo III. Each solution was then diluted to 50.00 mL, and the absorbance of each solution was measured at 660 nm in 1.00-cm cells.

Volume Standard Solution, mL	A_{660}
0.00	0.216
5.00	0.338
10.00	0.471
15.00	0.596
20.00	0.764
25.00	0.850

(a) Enter the data into a spreadsheet, and construct a standard-additions plot of the data.

(b) Determine the slope and intercept of the line.

---

[33] J. G. Sen Gupta, *Anal. Chem.*, **1967**, *39*, 18.

    (c) Determine the standard deviation of the slope and of the intercept.

    (d) Calculate the concentration of Pd(II) in the analyte solution.

    (e) Find the standard deviation of the measured concentration.

14-20 Given the information that

$$Fe^{3+} + Y^{4-} \rightleftharpoons FeY^- \qquad K_f = 1.0 \times 10^{25}$$
$$Cu^{2+} + Y^{4-} \rightleftharpoons CuY^{2-} \qquad K_f = 6.3 \times 10^{18}$$

and the further information that, among the several reactants and products, only CuY^{2-} absorbs radiation at 750 nm, describe how Cu(II) could be used as an indicator for the photometric titration of Fe(III) with H_2Y^{2-}. Reaction: $Fe^{3+} + H_2Y^{2-} \rightarrow FeY^- + 2H^+$.

*__14-21__ The chelate CuA_2^{2-} exhibits maximum absorption at 480 nm. When the chelating reagent is present in at least a tenfold excess, the absorbance depends only on the analytical concentration of Cu(II) and conforms to Beer's law over a wide range. A solution in which the analytical concentration of Cu^{2+} is 2.15×10^{-4} M and that for A^{2-} is 9.00×10^{-3} M has an absorbance of 0.759 when measured in a 1.00-cm cell at 480 nm. A solution in which the analytical concentrations of Cu^{2+} and A^{2-} are 2.15×10^{-4} M and 4.00×10^{-4} M, respectively, has an absorbance of 0.654 when measured under the same conditions. Use this information to calculate the formation constant K_f for the process

$$Cu^{2+} + 2A^{2-} \rightleftharpoons CuA_2^{2-}$$

*__14-22__ Mixing the chelating reagent B with Ni(II) forms the highly colored NiB_2^{2+}, whose solutions obey Beer's law at 395 nm over a wide range. Provided the analytical concentration of the chelating reagent exceeds that of Ni(II) by a factor of 5 (or more), the cation exists, within the limits of observation, entirely in the form of the complex. Use the accompanying data to evaluate the formation constant K_f for the process

$$Ni^{2+} + 2B \rightleftharpoons NiB_2^{2+}$$

Analytical Concentration, M		
Ni^{2+}	**B**	**A_{395} (1.00-cm cells)**
2.00×10^{-4}	2.20×10^{-1}	0.844
2.00×10^{-4}	1.50×10^{-3}	0.316

14-23 To determine the formation constant of a 1:1 complex, the absorbances below were measured at 470 nm in a 2.50-cm cell for the ligand concentrations shown. The total metal concentration was $c_M = 7.50 \times 10^{-4}$ M.

[L], M	**A**
0.0750	0.679
0.0500	0.664
0.0300	0.635
0.0200	0.603
0.0100	0.524
0.0050	0.421
0.0000	0.056

(a) Use linear regression and the Benesi-Hildebrand equation (Equation 14-11) to determine the formation constant and the difference in molar absorptivities at 470 nm.

(b) Use nonlinear regression and Equation 14-10 to find the values of K_f and $\Delta\varepsilon$. Start with initial estimates of $K_f = 50$ and $\Delta\varepsilon = 50$.

Challenge Problem

14-24 (a) Prove mathematically that the peak in a continuous-variations plot occurs at a combining ratio that gives the complex composition.

(b) Show that the overall formation constant for the complex ML_n is

$$K_f = \frac{\left(\dfrac{A}{A_{extr}}\right)c}{\left[c_M - \left(\dfrac{A}{A_{extr}}\right)c\right]\left[c_L - n\left(\dfrac{A}{A_{extr}}\right)c\right]^n}$$

where A is the experimental absorbance at a given value on the x-axis in a continuous-variations plot, A_{extr} is the absorbance determined from the extrapolated lines corresponding to the same point on the x-axis, c_M is the molar analytical concentration of the metal, c_L is the molar analytical concentration of the ligand, and n is the ligand-to-metal ratio in the complex.[34]

(c) Under what assumptions is the equation valid?

(d) What is c?

(e) Discuss the implications of the occurrence of the maximum in a continuous-variations plot at a value of less than 0.5.

(f) Calabrese and Khan[35] characterized the complex formed between I_2 and I^- using the method of continuous variations. They combined 2.60×10^{-4} M solutions of I_2 and I^- in the usual way to obtain the following data set. Use the data to find the composition of the I_2/I^- complex.

$V(I_2$ soln), mL	A_{350}
0.00	0.002
1.00	0.121
2.00	0.214
3.00	0.279
4.00	0.312
5.00	0.325
6.00	0.301
7.00	0.258
8.00	0.188
9.00	0.100
10.00	0.001

(g) The continuous-variations plot appears to be asymmetrical. Consult the paper by Calabrese and Khan and explain this asymmetry.

(h) Use the equation in part (a) to determine the formation constant of the complex for each of the three central points on the continuous-variations plot.

---

[34] J. Inczédy, *Analytical Applications of Complex Equilibria*, New York: Wiley, 1976.
[35] V. T. Calabrese and A. Khan, *J. Phys. Chem. A*, **2000**, *104*, 1287.

(i) Explain any trend in the three values of the formation constant in terms of the asymmetry of the plot.

(j) Find the uncertainty in the formation constant determined by this method.

(k) What effect, if any, does the formation constant have on the ability to determine the composition of the complex using the method of continuous variations?

(l) Discuss the various advantages and potential pitfalls of using the method of continuous variations as a general method for determining the composition and formation constant of a complex compound.

CHAPTER FIFTEEN

Molecular Luminescence Spectrometry

Three related types of optical methods are considered in this chapter: *molecular fluorescence, phosphorescence,* and *chemiluminescence. In each of these methods, molecules of the analyte are excited to produce a species whose emission spectrum provides information for qualitative or quantitative analysis. The methods are known collectively as **molecular luminescence** procedures.*

Throughout this chapter, this logo indicates an opportunity for online self-study at **www.thomsonedu.com/chemistry/skoog**, linking you to interactive tutorials, simulations, and exercises.

Fluorescence and phosphorescence are alike in that excitation is brought about by absorption of photons. As a consequence, the two phenomena are often referred to by the more general term *photoluminescence.* As will be shown in this chapter, fluorescence differs from phosphorescence in that the electronic energy transitions responsible for fluorescence do not involve a change in electron spin. Because of this, the excited states involved in fluorescence are short-lived ($<10^{-5}$ s). In contrast, a change in electron spin accompanies phosphorescence, and the lifetimes of the excited states are much longer, often on the order of seconds or even minutes. In most instances, photoluminescence, be it fluorescence or phosphorescence, occurs at wavelengths longer than that of the excitation radiation.

The third type of luminescence, chemiluminescence, is based on the emission of radiation by an excited species formed during a chemical reaction. In some instances, the excited species is the product of a reaction between the analyte and a suitable reagent (usually a strong oxidant such as ozone or hydrogen peroxide). The result is an emission spectrum characteristic of the oxidation product of the analyte or the reagent rather than the analyte itself. In other instances, the analyte is not directly involved in the chemiluminescence reaction. Instead, the analyte inhibits or has a catalytic effect on a chemiluminescence reaction.

Measurement of the intensity of photoluminescence or chemiluminescence permits the quantitative determination of a variety of important inorganic and organic species in trace amounts. Currently, the number of fluorometric methods is far greater than the number of applications of phosphorescence and chemiluminescence procedures.

One of the most attractive features of luminescence methods is their inherent sensitivity, with detection limits that are often one to three orders of magnitude lower than those encountered in absorption spectroscopy. In fact, for selected species under controlled conditions, single molecules have been detected by fluorescence spectroscopy. Another advantage of photoluminescence methods is their large linear concentration ranges, which also are often significantly greater than those encountered in absorption methods. Because excited states are quite susceptible to being deactivated by collisions and other processes, many molecules do not fluoresce or phosphoresce at all. Because of such deactivation processes, quantitative luminescence methods are often subject to serious interference effects.

For this reason luminescence measurements are often combined with such separation techniques as chromatography and electrophoresis. Fluorescence detectors are particularly valuable as detectors for liquid chromatography (Chapter 28) and capillary electrophoresis (Chapter 30).

Generally, luminescence methods are less widely applicable for quantitative analyses than are absorption methods because many more species absorb ultraviolet and visible radiation than exhibit photoluminescence when radiation is absorbed in this region of the spectrum.[1]

15A THEORY OF FLUORESCENCE AND PHOSPHORESCENCE

Fluorescence occurs in simple as well as in complex gaseous, liquid, and solid chemical systems. The simplest kind of fluorescence is that exhibited by dilute atomic vapors, which was described in Chapter 9. For example, the $3s$ electrons of vaporized sodium atoms can be excited to the $3p$ state by absorption of radiation of wavelengths 589.6 and 589.0 nm. After about 10^{-8} s, the electrons return to the ground state and in so doing emit radiation of the same two wavelengths in all directions. This type of fluorescence, in which the absorbed radiation is reemitted without a change in frequency, is known as *resonance radiation* or *resonance fluorescence*.

Many molecular species also exhibit resonance fluorescence. Much more often, however, molecular fluorescence (or phosphorescence) bands center at wavelengths longer than the resonance line. This shift toward longer wavelengths is termed the *Stokes shift* (see Section 6C-6).

15A-1 Excited States Producing Fluorescence and Phosphorescence

The characteristics of fluorescence and phosphorescence spectra can be rationalized by means of the molecular orbital considerations described in Sec-

---

[1]For further discussion of the theory and applications of fluorescence, phosphorescence, and luminescence, see J. R. Lakowicz, *Principles of Fluorescence Spectroscopy*, 2nd ed., New York: Kluwer Academic Publishing/Plenum Press, 1999; *Molecular Luminescence Spectroscopy*, S. Schulman, ed., New York: Wiley, Part 1, 1985; Part 2, 1988; Part 3, 1993; E. L. Wehry, in *Physical Methods of Chemistry*, 2nd ed., Volume VIII, Chap. 3, B. W. Rossiter and R. C. Baetzold, eds., New York: Wiley, 1993; G. G. Guilbault, *Practical Fluorescence*, 2nd ed., New York: Marcel Dekker, 1990.

tion 14B-1. However, an understanding of the difference between the two photoluminescence phenomena requires a review of electron spin and the differences between singlet and triplet excited states.

Electron Spin

The Pauli exclusion principle states that no two electrons in an atom can have the same set of four quantum numbers. This restriction requires that no more than two electrons occupy an orbital and furthermore the two have opposed spin states. Under this circumstance, the spins are said to be paired. Because of spin pairing, most molecules exhibit no net magnetic field and are thus said to be diamagnetic — that is, they are neither attracted nor repelled by static magnetic fields. In contrast, free radicals, which contain unpaired electrons, have a magnetic moment and consequently are attracted by a magnetic field. Free radicals are thus said to be *paramagnetic*.

Singlet and Triplet Excited States

A molecular electronic state in which all electron spins are paired is called a *singlet state*, and no splitting of electronic energy levels occurs when the molecule is exposed to a magnetic field. The ground state for a free radical, on the other hand, is a *doublet state* because there are two possible orientations for the odd electron in a magnetic field, and each imparts slightly different energies to the system.

When one of a pair of electrons of a molecule is excited to a higher energy level, a singlet or a *triplet state* is formed. In the excited singlet state, the spin of the promoted electron is still paired with the ground-state electron. In the triplet state, however, the spins of the two electrons have become unpaired and are thus parallel. These states can be represented as in Figure 15-1, where the arrows represent the direction of spin. The nomenclature of singlet, doublet, and triplet derives from spectroscopic *multiplicity* considerations, which need not concern us here. Note that the excited triplet state is less energetic than the corresponding excited singlet state.

The properties of a molecule in the excited triplet state differ significantly from those of the excited singlet state. For example, a molecule is paramagnetic in the triplet state and diamagnetic in the singlet. More important, however, is that a singlet-to-triplet transition (or the reverse), which also involves a change in electronic state, is a significantly less probable event than the corresponding singlet-to-singlet transition. As

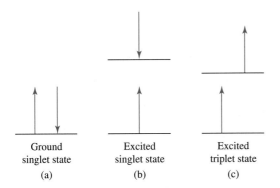

FIGURE 15-1 Electronic spin states of molecules. In (a) the ground electronic state is shown. In the lowest energy, or ground, state, the spins are always paired, and the state is said to be a singlet state. In (b) and (c), excited electronic states are shown. If the spins remain paired in the excited state, the molecule is in an excited singlet state (b). If the spins become unpaired, the molecule is in an excited triplet state (c).

a consequence, the average lifetime of an excited triplet state may range from 10^{-4} to several seconds, compared with an average lifetime of $\sim 10^{-8}$ s for an excited singlet state. Furthermore, radiation-induced

excitation of a ground-state molecule to an excited triplet state has a low probability of occurring, and absorption bands due to this process are several orders of magnitude less intense than the analogous singlet-to-singlet absorption. We shall see, however, that an excited triplet state can be populated from an excited singlet state of certain molecules. Phosphorescence emission is often a result of such a process.

Energy-Level Diagrams for Photoluminescent Molecules

Figure 15-2 is a partial energy-level diagram, called a *Jablonski diagram*, for a typical photoluminescent molecule. The lowest heavy horizontal line represents the ground-state energy of the molecule, which is normally a singlet state, and is labeled S_0. At room temperature, this state represents the energies of most of the molecules in a solution.

The upper heavy lines are energy levels for the ground vibrational states of three excited electronic states. The two lines on the left represent the first (S_1) and second (S_2) electronic singlet states. The one on the right (T_1) represents the energy of the first electronic triplet state. As is normally the case, the energy

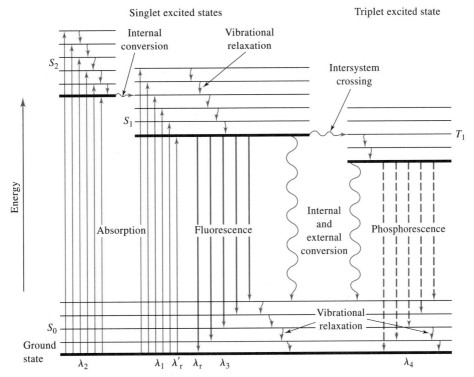

FIGURE 15-2 Partial energy-level diagram for a photoluminescent system.

of the first excited triplet state is lower than the energy of the corresponding singlet state.

Numerous vibrational energy levels are associated with each of the four electronic states, as suggested by the lighter horizontal lines. As shown in Figure 15-2, absorption transitions can occur from the ground singlet electronic state (S_0) to various vibrational levels of the excited singlet electronic states (S_1 and S_2). Note that direct excitation to the triplet state is not shown. Because this transition involves a change in multiplicity, it has a very low probability of occurrence. A low-probability transition of this type is called a *forbidden transition*.

Molecules excited to electronic states S_1 and S_2 rapidly lose any excess vibrational energy and relax to the ground vibrational level of that electronic state. This nonradiational process is termed *vibrational relaxation*.

15A-2 Rates of Absorption and Emission

The rate of photon absorption is very rapid and takes place in about 10^{-14} to 10^{-15} s. Fluorescence emission, on the other hand, occurs at a significantly slower rate. Typically, fluorescence occurs in 10^{-5} to 10^{-10} s. As we have noted, the average rate of a triplet-to-singlet transition is less than that of a corresponding singlet-to-singlet transition. Thus, phosphorescence emission requires 10^{-4} to 10 s or more to occur.

15A-3 Deactivation Processes

An excited molecule can return to its ground state by a combination of several mechanistic steps. As shown by the straight, downward pointing, vertical arrows in Figure 15-2, two of these steps, fluorescence and phosphorescence, involve the emission of a photon of radiation. The other deactivation steps, indicated by wavy arrows, are radiationless processes. The favored route to the ground state is the one that minimizes the lifetime of the excited state. Thus, if deactivation by fluorescence is rapid with respect to the radiationless processes, such emission is observed. On the other hand, if a radiationless path has a more favorable rate constant, fluorescence is either absent or less intense.

Photoluminescence is limited to systems incorporating structural and environmental features that cause the rate of radiationless relaxation or deactivation processes to be slowed to a point where the emission process can compete kinetically. We have a good quantitative understanding of luminescence processes.

However, we are only beginning to understand deactivation processes in any detail.

Vibrational Relaxation

As shown in Figure 15-2, a molecule may be promoted to any of several vibrational levels during the electronic excitation process. Collisions between molecules of the excited species and those of the solvent lead to rapid energy transfer with a minuscule increase in temperature of the solvent. Vibrational relaxation is so efficient that the average lifetime of a *vibrationally* excited molecule is 10^{-12} s or less, a period significantly shorter than the average lifetime of an *electronically* excited state. As a consequence, fluorescence from solution always involves a transition from the *lowest vibrational level of an excited electronic state*. Several closely spaced emission lines are produced, however, and the transition can terminate in any of the vibrational levels of the ground state (see Figure 15-2).

A consequence of the efficiency of vibrational relaxation is that the fluorescence band for a given electronic transition is displaced toward lower frequencies or longer wavelengths from the absorption band (the Stokes shift). Overlap occurs only for the resonance peak involving transitions between the lowest vibrational level of the ground state and the corresponding level of an excited state. In Figure 15-2, the wavelength of absorbed radiation that produces the resonance peak λ_r is labeled λ_r'.

Internal Conversion

The term *internal conversion* describes intermolecular processes by which a molecule passes to a lower-energy electronic state without emission of radiation. These processes are neither well defined nor well understood, but they are often highly efficient.

Internal conversion is a crossover between two states of the same multiplicity (singlet-singlet or triplet-triplet). It is particularly efficient when two electronic energy levels are sufficiently close for there to be an overlap in vibrational energy levels. This situation is illustrated in Figure 15-2 for the two excited singlet states S_2 and S_1. At the overlaps shown, the potential energies of the two excited states are essentially equal, which permits an efficient crossover from S_2 to S_1. Internal conversion can also occur between state S_1 and the ground electronic state S_0. Internal conversion through overlapping vibrational levels is usually more probable than the loss of energy by fluorescence from a higher excited state. Thus, referring again to Figure 15-2, excitation by the band of radiation labeled λ_2 usu-

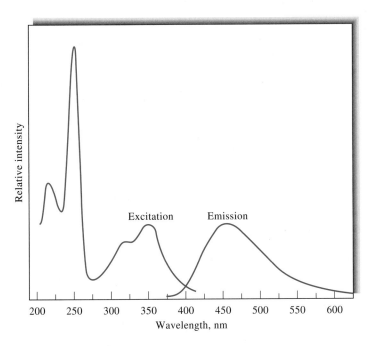

FIGURE 15-3 Fluorescence excitation and emission spectra for a solution of quinine.

ally produces a fluorescence band centered at wavelength λ_3 to the exclusion of a band that would result from a transition between S_2 and S_0. Here, the excited molecule proceeds from the higher electronic state to the lowest vibrational state of the lower electronic excited state via a series of vibrational relaxations, an internal conversion, and then further relaxations. Under these circumstances, the fluorescence occurs at λ_3 only, regardless of whether radiation of wavelength λ_1 or λ_2 was responsible for the excitation. Quinine provides a classical example of this type of behavior (see Problem 15-13). This naturally occurring substance possesses two analytically useful excitation bands, one centered at 250 nm and the other at 350 nm. Regardless of which wavelength is used to excite the molecule, however, the wavelength of maximum emission is 450 nm (see Figure 15-3).

The mechanisms of the internal conversion process $S_1 \rightarrow S_0$ shown in Figure 15-2 are not totally understood. For some molecules, the vibrational levels of the ground state overlap those of the first excited electronic state, and deactivation occurs rapidly by the mechanism just described. This situation prevails with aliphatic compounds, for example, and accounts for these species seldom fluorescing. With such compounds, deactivation by energy transfer through overlapping vibrational levels occurs so rapidly that fluorescence does not have time to occur.

Internal conversion may also result in the phenomenon of *predissociation*. Here, the electron moves from a higher electronic state to an upper vibrational level of a lower electronic state in which the vibrational energy is great enough to cause rupture of a bond. Large molecules have an appreciable probability for the existence of bonds with strengths less than the electronic excitation energy of the chromophores. Rupture of these bonds can occur as a consequence of absorption by the chromophore followed by internal conversion of the electronic energy to vibrational energy associated with the weak bond.

Predissociation should be differentiated from *dissociation*, in which the absorbed radiation directly excites the electron of a chromophore to a sufficiently high vibrational level to cause rupture of the chromophoric bond. In this case, no internal conversion is involved. Dissociation processes also compete with the fluorescence process.

External Conversion

Deactivation of an excited electronic state may involve interaction and energy transfer between the excited molecule and the solvent or other solutes. This process is called *external conversion*. Evidence for external conversion includes a marked solvent effect on the fluorescence intensity of most species. Furthermore, those conditions that tend to reduce the number of collisions between particles (low temperature and high viscosity) generally lead to enhanced fluorescence. The details of external conversion processes are not well understood.

Radiationless transitions to the ground state from the lowest excited singlet and triplet states (Figure 15-2) probably involve external conversion, as well as internal conversion.

Intersystem Crossing

Intersystem crossing is a process in which there is a crossover between electronic states of different multiplicity. The most common process is from the singlet state to the triplet state $(S_1 \rightarrow T_1)$ as shown in Figure 15-2. As with internal conversion, the probability of intersystem crossing is enhanced if the vibrational levels of the two states overlap. Note that in the singlet-triplet crossover shown in Figure 15-2 the lowest singlet vibrational level overlaps one of the upper triplet vibrational levels and a change in spin state is thus more probable.

Intersystem crossing is most common in molecules that contain heavy atoms, such as iodine or bromine (the *heavy-atom effect*). Spin and orbital interactions increase in the presence of such atoms, and a change in spin is thus more favorable. The presence of paramagnetic species such as molecular oxygen in solution also enhances intersystem crossing with a consequent decrease in fluorescence.

Phosphorescence

Deactivation of electronic excited states may also involve phosphorescence. After intersystem crossing to the triplet state, further deactivation can occur either by internal or external conversion or by phosphorescence. A triplet $\rightarrow$ singlet transition is much less probable than a singlet-singlet conversion. Transition probability and excited-state lifetime are inversely related. Thus, the average lifetime of the excited triplet state with respect to emission is large and ranges from 10^{-4} to 10 s or more. Emission from such a transition may persist for some time after irradiation has ceased.

External and internal conversion compete so successfully with phosphorescence that this kind of emission is ordinarily observed only at low temperatures in highly viscous media or by using special techniques to protect the triplet state.

15A-4 Variables Affecting Fluorescence and Phosphorescence

Both molecular structure and chemical environment influence whether a substance will or will not luminesce. These factors also determine the intensity of emission when luminescence does occur. The effects of some of these variables are considered briefly in this section.

Quantum Yield

The *quantum yield*, or *quantum efficiency*, for fluorescence or phosphorescence is simply the ratio of the number of molecules that luminesce to the total number of excited molecules. For a highly fluorescent molecule such as fluorescein, the quantum efficiency approaches unity under some conditions. Chemical species that do not fluoresce appreciably have efficiencies that approach zero.

Figure 15-2 and our discussion of deactivation processes suggest that the fluorescence quantum yield ϕ for a compound is determined by the relative rate constants k_x for the processes by which the lowest excited singlet state is deactivated. These processes are fluorescence (k_f), intersystem crossing (k_i), external conversion (k_{ec}), internal conversion (k_{ic}), predissociation (k_{pd}), and dissociation (k_d). We can express these relationships by the equation

$$\phi = \frac{k_f}{k_f + k_i + k_{ec} + k_{ic} + k_{pd} + k_d} \quad (15\text{-}1)$$

where the k terms are the respective rate constants for the various deactivation processes.

Equation 15-1 permits a qualitative interpretation of many of the structural and environmental factors that influence fluorescence intensity. Those variables that lead to high values for the fluorescence rate constant k_f and low values for the other k_x terms enhance fluorescence. The magnitude of k_f, the predissociation rate constant k_{pd}, and the dissociation rate constant k_d mainly depend on chemical structure. Environment and to a somewhat lesser extent structure strongly influence the remaining rate constants.

Transition Types in Fluorescence

It is important to note that fluorescence seldom results from absorption of ultraviolet radiation of wavelengths shorter than 250 nm because such radiation is sufficiently energetic to cause deactivation of the excited states by predissociation or dissociation. For example, 200-nm radiation corresponds to about 140 kcal/mol. Most organic molecules have at least some bonds that can be ruptured by energies of this magnitude. As a consequence, fluorescence due to $\sigma^* \rightarrow \sigma$ transitions is seldom observed. Instead, such emission is confined to the less energetic $\pi^* \rightarrow \pi$ and $\pi^* \rightarrow n$ processes.

As we have noted, an electronically excited molecule ordinarily returns to its *lowest excited state* by a series of rapid vibrational relaxations and internal conversions that produce no emission of radiation. Thus, fluorescence most commonly arises from a transition from the lowest vibrational level of the first excited electronic state to one of the vibrational levels of the electronic ground state. For most fluorescent compounds, then, radiation is produced by either a $\pi^* \rightarrow n$ or a $\pi^* \rightarrow \pi$ transition, depending on which of these is the less energetic.

Quantum Efficiency and Transition Type

It is observed empirically that fluorescence is more commonly found in compounds in which the lowest energy transition is of a $\pi \rightarrow \pi^*$ type (π, π^* excited singlet state) than in compounds in which the lowest energy transition is of the $n \rightarrow \pi^*$ type (n, π^* excited state); that is, the quantum efficiency is greater for $\pi^* \rightarrow \pi$ transitions.

The greater quantum efficiency associated with the π, π^* state can be rationalized in two ways. First, the molar absorptivity of a $\pi \rightarrow \pi^*$ transition is ordinarily 100- to 1000-fold greater than for an $n \rightarrow \pi^*$ process, and this quantity represents a measure of the transition probability. Thus, the inherent lifetime associated with the π, π^* state is shorter (10^{-7} to 10^{-9} s compared with 10^{-5} to 10^{-7} s for the n, π^* state) and k_f in Equation 15-1 is larger.

The most efficient phosphorescence often occurs from the n, π^* excited state, which tends to be shorter lived and thus less susceptible to deactivation than a π, π^* triplet state. Also, intersystem crossing is less probable for π, π^* excited states than for n, π^* states because the energy difference between the singlet and triplet states is larger and spin-orbit coupling is less likely.

In summary, then, fluorescence is more commonly associated with the π, π^* state because such excited states exhibit relatively short average lifetimes (k_f is larger) and because the deactivation processes that compete with fluorescence are less likely to occur.

Fluorescence and Structure

The most intense and the most useful fluorescence is found in compounds containing aromatic functional groups with low-energy $\pi \rightarrow \pi^*$ transitions. Compounds containing aliphatic and alicyclic carbonyl structures or highly conjugated double-bond structures may also exhibit fluorescence, but the number of these is smaller than the number in the aromatic systems.

Most unsubstituted aromatic hydrocarbons fluoresce in solution, the quantum efficiency usually increasing with the number of rings and their degree of condensation. The simple heterocyclics, such as pyridine, furan, thiophene, and pyrrole,

pyridine   furan

thiophene   pyrrole

do not exhibit fluorescence. On the other hand, fused-ring structures ordinarily do fluoresce. With nitrogen heterocyclics, the lowest-energy electronic transition is believed to involve an $n \rightarrow \pi^*$ system that rapidly converts to the triplet state and prevents fluorescence. Fusion of benzene rings to a heterocyclic nucleus, however, results in an increase in the molar absorptivity of the absorption band. The lifetime of an excited state is shorter in such structures, and fluorescence is observed for compounds such as quinoline, isoquinoline, and indole.

quinoline   isoquinoline   indole

Substitution on the benzene ring causes shifts in the wavelength of absorption maxima and corresponding changes in the fluorescence emission. In addition, substitution frequently affects the quantum efficiency. Some of these effects are illustrated by the data for benzene derivatives in Table 15-1.

The influence of halogen substitution is striking; the decrease in fluorescence with increasing molar mass of the halogen is an example of the heavy-atom effect (page 404), which increases the probability for intersystem crossing to the triplet state. Predissociation is thought to play an important role in iodobenzene and in nitro derivatives as well, because these compounds

TABLE 15-1 Effect of Substitution on the Fluorescence of Benzene

Compound	Formula	Wavelength of Fluorescence, nm	Relative Intensity of Fluorescence
Benzene	C_6H_6	270–310	10
Toluene	$C_6H_5CH_3$	270–320	17
Propylbenzene	$C_6H_5C_3H_7$	270–320	17
Fluorobenzene	C_6H_5F	270–320	10
Chlorobenzene	C_6H_5Cl	275–345	7
Bromobenzene	C_6H_5Br	290–380	5
Iodobenzene	C_6H_5I	—	0
Phenol	C_6H_5OH	285–365	18
Phenolate ion	$C_6H_5O^-$	310–400	10
Anisole	$C_6H_5OCH_3$	285–345	20
Aniline	$C_6H_5NH_2$	310–405	20
Anilinium ion	$C_6H_5NH_3^+$	—	0
Benzoic acid	C_6H_5COOH	310–390	3
Benzonitrile	C_6H_5CN	280–360	20
Nitrobenzene	$C_6H_5NO_2$	—	0

have easily ruptured bonds that can absorb the excitation energy following internal conversion.

Substitution of a carboxylic acid or carbonyl group on an aromatic ring generally inhibits fluorescence. In these compounds, the energy of the $n \rightarrow \pi^*$ transition is less than in the $\pi \rightarrow \pi^*$ transition, and as discussed previously, the fluorescence yield from the $n \rightarrow \pi^*$ systems is ordinarily low.

Effect of Structural Rigidity

It is found empirically that fluorescence is particularly favored in molecules with rigid structures. For example, the quantum efficiencies for fluorene and biphenyl are nearly 1.0 and 0.2, respectively, under similar conditions of measurement. The difference in behavior is largely a result of the increased rigidity furnished by the bridging methylene group in fluorene. Many similar examples can be cited.

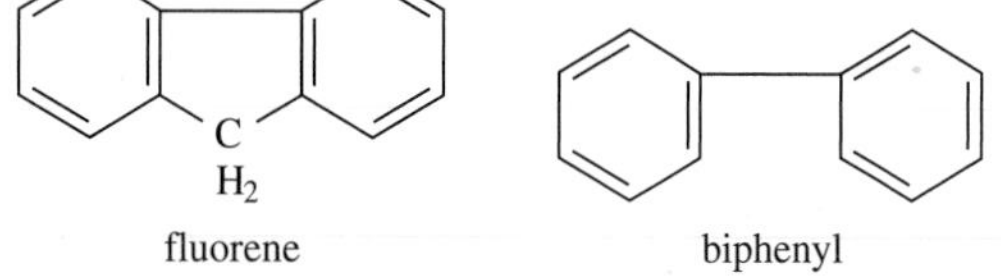

fluorene   biphenyl

The influence of rigidity has also been invoked to account for the increase in fluorescence of certain organic chelating agents when they are complexed with a metal ion. For example, the fluorescence intensity of 8-hydroxyquinoline is much less than that of its zinc complex:

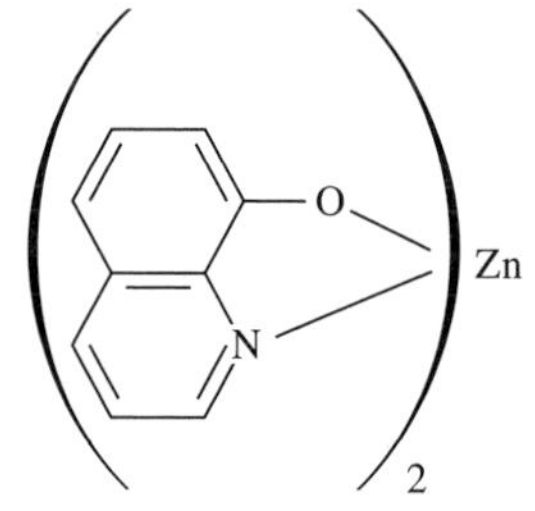

Lack of rigidity in a molecule probably causes an enhanced internal conversion rate (k_{ic} in Equation 15-1) and a consequent increase in the likelihood for radiationless deactivation. One part of a nonrigid molecule can undergo low-frequency vibrations with respect to its other parts; such motions undoubtedly account for some energy loss.

Temperature and Solvent Effects

The quantum efficiency of fluorescence in most molecules decreases with increasing temperature because the increased frequency of collisions at elevated temperatures improves the probability for deactivation by external conversion. A decrease in solvent viscosity also increases the likelihood of external conversion and leads to the same result.

The fluorescence of a molecule is decreased by solvents containing heavy atoms or other solutes with such atoms in their structure; carbon tetrabromide and ethyl iodide are examples. The effect is similar to that which occurs when heavy atoms are substituted into fluorescing compounds; orbital spin interactions result in an increase in the rate of triplet formation and a corresponding decrease in fluorescence. Compounds containing heavy atoms are frequently incorporated into solvents when enhanced phosphorescence is desired.

Effect of pH on Fluorescence

The fluorescence of an aromatic compound with acidic or basic ring substituents is usually pH dependent. Both the wavelength and the emission intensity are likely to be different for the protonated and unprotonated forms of the compound. The data for phenol and aniline shown in Table 15-1 illustrate this effect. The changes in emission of compounds of this type arise from the differing number of resonance species that are associated with the acidic and basic forms of the molecules. For example, aniline has several resonance forms but anilinium has only one. That is,

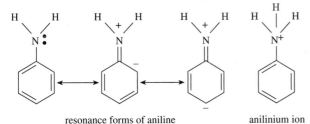

resonance forms of aniline   anilinium ion

The additional resonance forms lead to a more stable first excited state; fluorescence in the ultraviolet region is the consequence.

The fluorescence of certain compounds as a function of pH has been used for the detection of end points in acid-base titrations. For example, fluorescence of the phenolic form of 1-naphthol-4-sulfonic acid is not detectable by the eye because it occurs in the ultraviolet region. When the compound is converted to the phenolate ion by the addition of base, however, the emission band shifts to visible wavelengths, where it can readily be seen. It is significant that this change occurs at a different pH than would be predicted from the acid dissociation constant for the compound. The explanation of this discrepancy is that the acid dissociation constant for the excited molecule differs from that for the same species in its ground state. Changes in acid or base dis-

 Simulation: Learn more about **luminescence spectroscopy**.

sociation constants with excitation are common and are occasionally as large as four or five orders of magnitude. These observations suggest that analytical procedures based on fluorescence frequently require close control of pH.

The presence of dissolved oxygen often reduces the intensity of fluorescence in a solution. This effect may be the result of a photochemically induced oxidation of the fluorescing species. More commonly, however, the quenching takes place as a consequence of the paramagnetic properties of molecular oxygen, which promotes intersystem crossing and conversion of excited molecules to the triplet state. Other paramagnetic species also tend to quench fluorescence.

Effect of Concentration on Fluorescence Intensity

The power of fluorescence emission F is proportional to the radiant power of the excitation beam that is absorbed by the system. That is,

$$F = \phi_f K''(P_0 - P) = K'(P_0 - P) \quad (15\text{-}2)$$

where P_0 is the power of the beam incident on the solution, P is its power after traversing a length b of the medium, ϕ_f is the quantum efficiency of the fluorescence process, and K'' is a constant dependent on geometry and other factors. The quantum efficiency of fluorescence is a constant for a given system, and so the product $\phi_f K''$ is lumped into a new constant K' on the right side of Equation 15-2. To relate F to the concentration c of the fluorescing species, we write Beer's law in the form

$$\frac{P}{P_0} = 10^{-\varepsilon bc} \quad (15\text{-}3)$$

where ε is the molar absorptivity of the fluorescing molecules and εbc is the absorbance A. By substitution of Equation 15-3 into Equation 15-2, we obtain

$$F = K'P_0(1 - 10^{-\varepsilon bc}) \quad (15\text{-}4)$$

The exponential term in Equation 15-4 can be expanded as a Maclaurin series to

$$F = K'P_0\left[2.303\varepsilon bc - \frac{(2.303\varepsilon bc)^2}{2!} + \frac{(2.303\varepsilon bc)^3}{3!} + \cdots\right] \quad (15\text{-}5)$$

Provided that $2.303\varepsilon bc = A < 0.05$, all of the subsequent terms in the brackets are small with respect to the first term. Under these conditions, the maximum relative error caused by dropping all but the first term is 0.13%. Thus, we may write

$$F = 2.303 \, K' \varepsilon b c P_0 \qquad (15\text{-}6)$$

or, at constant P_0

$$F = Kc = 2.303 K' \varepsilon b c P_0 = 2.303 \phi_f K'' \varepsilon b c P_0 \qquad (15\text{-}7)$$

Thus, a plot of the fluorescence radiant power of a solution versus concentration of the emitting species should be linear at low concentrations. When c becomes great enough that the absorbance is larger than about 0.05, the higher-order terms in Equation 15-5 become important and linearity is lost; F then lies below an extrapolation of the straight-line plot. This excessive absorption is known as *primary absorption*.

Another factor responsible for negative departures from linearity at high concentration is *secondary absorption*. Secondary absorption occurs when the wavelength of emission overlaps an absorption band. Fluorescence is then decreased as the emission traverses the solution and is reabsorbed by other molecules in solution. Secondary absorption can be absorption by the analyte species itself or absorption by other species in the solution. The effects of these phenomena are such that a plot of fluorescence versus concentration may exhibit a maximum. Absorption effects are often termed *inner filter effects*.

Dynamic Quenching

The term *quenching* usually refers to nonradiative energy transfer from an excited species to other molecules. *Dynamic quenching*, also called *collisional quenching*, requires contact between the excited species and the quenching agent (Q). Dynamic quenching occurs as rapidly as the collision partners can diffuse together. The rate is temperature and viscosity dependent. The quencher concentration must be high enough that there is a high probability of a collision between the excited species and the quencher during the lifetime of the excited state.

For external conversion controlled by dynamic quenching with a single quencher, the external conversion rate constant can be written as

$$k_{ec} = k_q[Q] \qquad (15\text{-}8)$$

where k_q is the rate constant for quenching and [Q] is the concentration of the quencher. In the absence of quenching ($k_{ec} = 0$) and where predissociation and dissociation are absent ($k_{pd} = k_d = 0$), the fluorescence quantum efficiency ϕ_f^0 can be written from Equation 15-1 as

$$\phi_f^0 = \frac{k_f}{k_f + k_i + k_{ic}} \qquad (15\text{-}9)$$

With quenching, we can write

$$\phi_f = \frac{k_f}{k_f + k_i + k_{ic} + k_q[Q]} \qquad (15\text{-}10)$$

By taking the ratio of Equation 15-9 to Equation 15-10, we obtain the Stern-Volmer equation

$$\frac{\phi_f^0}{\phi_f} = 1 + K_q[Q] \qquad (15\text{-}11)$$

where K_q is the Stern-Volmer quenching constant defined as $K_q = k_q/(k_f + k_i + k_{ic})$. Rearrangement of this equation yields,

$$\frac{1}{\phi_f} = \frac{1}{\phi_f^0} + \frac{K_q[Q]}{\phi_f^0}$$

A plot of $1/\phi_f$ versus [Q] should be a straight line with a slope of K_q/ϕ_f^0 and an intercept of $1/\phi_f^0$. The Stern-Volmer constant can be obtained from the ratio of the intercept and the slope. Dynamic quenching reduces both the fluorescence quantum yield and the fluorescence lifetime.

Because the fluorescence emission F is directly proportional to the quantum efficiency ϕ_f (Equation 15-2), we can also write the Stern-Volmer equation in terms of easily measured quantities

$$\frac{F_0}{F} = 1 + K_q[Q] \qquad (15\text{-}12)$$

where F_0 and F are the fluorescence signals in the absence and in the presence of quencher, respectively. The Stern-Volmer constant is then just the slope of a plot of F_0/F versus [Q], and the intercept of the plot is unity. Example 15-1 illustrates the use of Equation 15-12 in determining K_q values.

EXAMPLE 15-1

The fluorescence of quinine sulfate (see Figure 15-3) is quenched by high concentrations of chloride ion. The following fluorescence signals F were obtained as a function of the concentration of chloride ion.

Fluorescence Intensity, F	[Cl$^-$], M
180.0	0.00
87.5	0.005
58.0	0.010
43.2	0.015
35.0	0.020
28.5	0.025
14.5	0.030
19.1	0.040
15.7	0.050

Set up a spreadsheet and determine the quenching constant K_q.

Solution

The spreadsheet and Stern-Volmer plot are shown in Figure 15-4. The data are entered into columns B and C, and F_0/F is calculated in column D. The slope and its standard deviation are $K_q = 210 \pm 0.9$ M^{-1}, as obtained from the spreadsheet statistics. Note that the intercept is very nearly unity.

The presence of dissolved oxygen often reduces the intensity of fluorescence in a solution. This effect is sometimes the result of a photochemically induced oxidation of the fluorescing species. More commonly,

however, the quenching takes place as a consequence of the paramagnetic properties of molecular oxygen, which promotes intersystem crossing and conversion of excited molecules to the triplet state. Other paramagnetic species also tend to quench fluorescence. Dissolved oxygen is also an efficient quencher of the triplet state. For this reason, solutions are usually deoxygenated before phosphorescence measurements are made.

Other Types of Quenching

In *static quenching*, the quencher and the ground-state fluorophore form a complex called the *dark complex*. Fluorescence is usually observed only from the unbound fluorophore. The decrease in fluorescence intensity can also be described by the Stern-Volmer equation in the case of static quenching. However, K_q in Equation 15-11 is now the formation constant of the fluorophore-quencher complex. With static quenching, the lifetime is not affected. Hence, lifetime measurements can allow dynamic quenching to be distinguished from static quenching.

In *long-range*, or *Förster, quenching*, energy transfer occurs without collisions between molecules. Dipole-dipole coupling between the excited fluorophore and the quencher account for the transfer. The dependence of the degree of quenching on the quencher

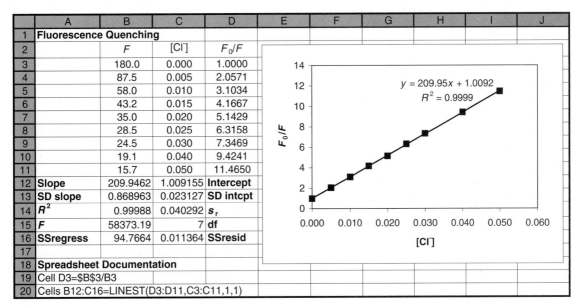

FIGURE 15-4 Spreadsheet and plot to determine quenching constant for chloride ion quenching of quinine.

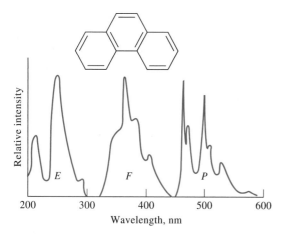

FIGURE 15-5 Spectra for phenanthrene: E, excitation; F, fluorescence; P, phosphorescence. (From W. R. Seitz, in Treatise on Analytical Chemistry, 2nd ed., P. J. Elving, E. J. Meehan, and I. M. Kolthoff, eds., Part I, Vol. 7, p. 169, New York: Wiley, 1981. Reprinted by permission of John Wiley & Sons, Inc.)

concentration is complicated and does not follow Stern-Volmer behavior.

15A-5 Emission and Excitation Spectra

Figure 15-5 shows three types of photoluminescence spectra for phenanthrene. An excitation spectrum labeled E in the figure is obtained by measuring luminescence intensity at a fixed wavelength while the excitation wavelength is varied. Because the first step in generating fluorescence emission is absorption of radiation to create excited states, an excitation spectrum is essentially identical to an absorption spectrum taken under the same conditions. Fluorescence and phosphorescence spectra (F and P, respectively), on the other hand, involve excitation at a fixed wavelength while recording the emission intensity as a function of wavelength.

Another type of luminescence spectrum is shown in Figure 15-6. The *total luminescence spectrum* is either a three-dimensional representation or a contour plot. Both simultaneously show the luminescence signal as a function of excitation and emission wavelengths. Such data are often called an *excitation-emission matrix*. Although the total luminescence spectrum can be obtained on a normal computerized instrument, it can be acquired more rapidly with array-detector-based systems (see next section).

Some luminescence instruments allow simultaneously scanning both the excitation and the emission wavelengths with a small wavelength difference between them. The spectrum that results is known as a *synchronous spectrum*. A luminescence signal is ob-

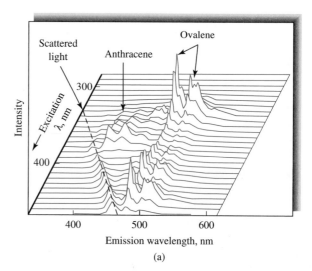

(a)

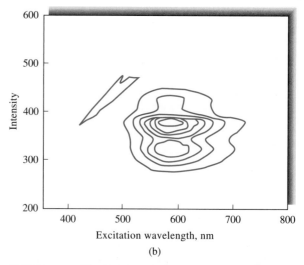

(b)

FIGURE 15-6 Total luminescence spectra. In (a), the total fluorescence spectrum of a mixture of anthracene and ovalene is shown as a three-dimensional plot. In (b), the total fluorescence spectrum of 8-hydroxybenzo[a]pyrene is shown as a contour plot. Each line represents a particular fluorescence intensity. (Part a from Y. Talmi et al., *Anal. Chem.*, **1978**, *50*, 936A. Figure 11, p. 948A. Part b adapted from J. H. Rho and J. L. Stewart, *Anal. Chem.*, **1978**, *50*, 620. Figure 2, p. 622. Copyright 1978 American Chemical Society.)

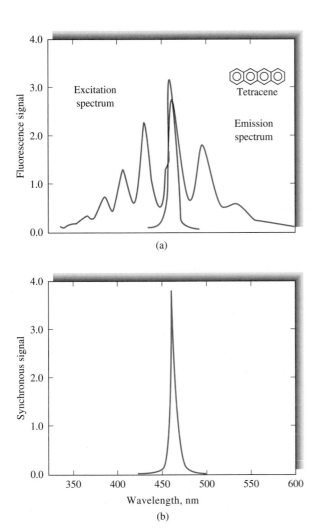

(a)

(b)

FIGURE 15-7 Synchronous fluorescence spectra. In (a), the excitation and emission spectra of tetracene are shown. In (b), the synchronous spectrum is shown for a fixed-wavelength difference of 3 nm. (From T. Vo-Dinh, *Anal. Chem.*, **1978**, *50*, 396. Figure 1, p. 397. Copyright 1978 American Chemical Society.)

tained only at wavelengths where both excitation and emission occur for the wavelength difference chosen as shown in Figure 15-7. The synchronous spectrum can also be generated from the total luminescence spectrum by appropriate software.

As has been pointed out earlier, photoluminescence usually occurs at wavelengths that are longer than the excitation wavelength. Furthermore, phosphorescence bands are generally found at longer wavelengths than fluorescence bands because the excited

triplet state is, in most instances, lower in energy than the corresponding singlet state. In fact, the wavelength difference between the two provides a convenient measure of energy difference between triplet and singlet states.

15B INSTRUMENTS FOR MEASURING FLUORESCENCE AND PHOSPHORESCENCE

The components of instruments for measuring photoluminescence are similar to those found in ultraviolet-visible photometers or spectrophotometers. Figure 15-8 shows a typical configuration for these components in *fluorometers* and *spectrofluorometers*. Nearly all fluorescence instruments employ double-beam optics as shown to compensate for fluctuations in radiant power. The upper sample beam first passes through an excitation wavelength selector (filter or monochromator), which transmits radiation that excites fluorescence but excludes or limits radiation of the fluorescence emission wavelength. Fluorescence is emitted from the sample in all directions, but is most conveniently observed at right angles to the excitation beam. The right-angle geometry minimizes the contributions from scattering and from the intense source radiation. The emitted radiation then passes through an emission wavelength selector (filter or monochromator) that isolates the fluorescence emission. The isolated radiation then strikes a phototransducer, where it is converted into an electrical signal for measurement.

The lower reference beam passes through an attenuator that reduces its power to approximately that of the fluorescence radiation (the power reduction is usually by a factor of 100 or more). The attenuated reference beam then strikes a second transducer and is converted to an electrical signal. Electronics and a computer data system then process the signals to compute the ratio of the fluorescence emission intensity to the excitation source intensity and produce the resulting spectrum or single-wavelength data.

The sophistication, performance characteristics, and costs of fluorometers and spectrofluorometers differ as widely as do those of the corresponding instruments for absorption measurements. If only filters are employed for wavelength selection, the instrument is

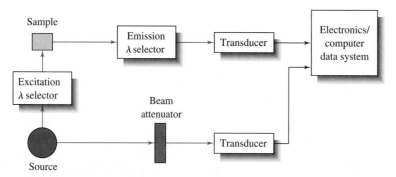

FIGURE 15-8 Components of a fluorometer or spectrofluorometer. Source radiation is split into two beams. The sample beam passes through the excitation wavelength selector to the sample. The emitted fluorescence is isolated by the emission wavelength selector before striking the transducer. The reference beam is attenuated before striking the transducer. The electronics and computer system compute the ratio of the fluorescence intensity to the reference beam intensity, which cancels the effect of source intensity fluctuations.

called a *fluorometer*. True *spectrofluorometers* employ two monochromators for wavelength isolation. Some instruments are hybrids in that they use a filter for selecting the excitation wavelength and a monochromator for choosing the emission wavelength. These are, often, still called spectrofluorometers. Several commercial spectrophotometers can be purchased with adapters that permit their use as spectrofluorometers.

True spectrofluorometers allow production of a fluorescence excitation spectrum or a fluorescence emission spectrum. Figure 15-9a shows an excitation spectrum for anthracene in which the fluorescence emission was measured at a fixed wavelength while the excitation wavelength was scanned. With suitable corrections for variations in source output intensity and detector response as a function of wavelength, an absolute excitation spectrum is obtained that closely resembles an absorption spectrum.

Figure 15-9b is the fluorescence emission spectrum for anthracene. The spectrum was obtained by holding the excitation wavelength constant while the emission wavelengths were scanned. These two spectra are approximately mirror images of one another because the vibrational energy differences for the ground and excited electronic states are roughly the same (see Figure 15-2).

The selectivity of spectrofluorometers is important in electronic and structural characterization of molecules and is valuable in both qualitative and quantitative analytical work. For concentration measurements, however, relatively inexpensive fluorometers often suf-

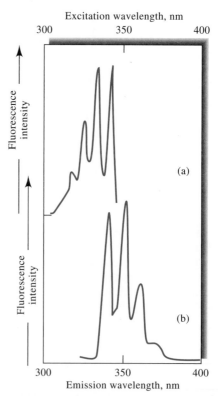

FIGURE 15-9 Fluorescence spectra for 1 ppm anthracene in alcohol: (a) excitation spectrum; (b) emission spectrum.

fice. These are usually designed specifically to solve the measurement problems peculiar to fluorescence methods and are frequently as specific and selective as modified absorption spectrophotometers.

15B-1 Components of Fluorometers and Spectrofluorometers

The components of fluorometers and spectrofluorometers are quite similar to those we have discussed for absorption photometers and spectrophotometers. However, several differences are discussed here.[2]

Sources

As shown by Equation 15-6, the magnitude of the output signal in luminescence measurements, and thus the sensitivity, is directly proportional to the source radiant power P_0. For this reason, more intense sources are used in luminescence methods than the tungsten or deuterium lamps used in absorption measurements.

Lamps. The most common source for filter fluorometers is a low-pressure mercury vapor lamp equipped with a fused silica window. This source produces useful lines for exciting fluorescence at 254, 302, 313, 546, 578, 691, and 773 nm. Individual lines can be isolated with suitable absorption or interference filters. Because fluorescence can be induced in most fluorescing compounds by a variety of wavelengths, at least one of the mercury lines ordinarily proves suitable.

For spectrofluorometers, where a source of continuum radiation is required, a 75- to 450-W high-pressure xenon arc lamp is commonly employed. Such lamps require a power supply capable of producing direct currents of 5 to 20 A at 15 to 30 V. The spectrum from a xenon arc lamp is a continuum from about 300 to 1300 nm. The spectrum approximates that of a blackbody (see Figure 6-22). In some instruments, a capacitor is discharged through the lamp at a constant frequency to provide flashes that are regularly spaced in time; higher peak intensities are obtained by pulsing. In addition, the outputs of the transducers are then ac signals that can be readily amplified and processed.

Blue light-emitting diodes (LEDs) have also been used in fluorescence instruments. These lamps emit radiation at 450–475 nm and are suitable for exciting some fluorophores. Mixtures of phosphors in some LEDs can provide wavelengths in the UV region to about 375 nm (see Section 13D-1).

Lasers. Since the 1970s, various types of lasers have also been used as excitation sources for photoluminescence measurements. Of particular interest are tunable dye lasers pumped by a pulsed nitrogen laser or a Nd-YAG laser. Fixed-wavelength lasers are also used, particularly in detectors for chromatography and electrophoresis.

Most commercial spectrofluorometers use lamp sources because they are less expensive and more applicable to determining multiple analytes with different excitation wavelengths. Laser sources, however, offer significant advantages in certain instances: for example, (1) when samples are very small, as in microbore chromatography and capillary electrophoresis where the amount of sample is a microliter or less; (2) in remote sensing, as in fluorometric detection of hydroxyl radicals in the atmosphere or of chlorophyll in bodies of water, where the collimated nature of laser beams is vital; or (3) when highly monochromatic excitation is needed to minimize the effects of fluorescing interferences.

Filters and Monochromators

Interference and absorption filters have been used in fluorometers for wavelength selection of both the excitation beam and the resulting fluorescence radiation. Spectrofluorometers are equipped with at least one and often two grating monochromators.

Transducers

Luminescence emission signals are typically quite low in intensity. Therefore, sensitive transducers are required. Photomultiplier tubes are the most common transducers in sensitive fluorescence instruments. Often, these are operated in the photon-counting mode to give improved signal-to-noise ratios (see page 202). Transducers are sometimes cooled to improve signal-to-noise ratios. Charge-transfer devices, such as charge-coupled devices (CCDs), are also used for spectrofluorometry.[3] This type of transducer permits the rapid recording of both excitation and emission spectra and is particularly useful in chromatography and electrophoresis.

Cells and Cell Compartments

Both cylindrical and rectangular cells fabricated of glass or silica are employed for fluorescence measurements. Care must be taken in the design of the cell

---

[2]For a review of commercial fluorescence instruments, see J. Kling, *Anal. Chem.*, **2000**, *72*, 219A.

[3]R. S. Pomeroy in *Charge Transfer Devices in Spectroscopy*, J. V. Sweedler, K. L. Ratzlaff, and M. B. Denton, eds., New York: VCH, 1994, pp. 281–314; P. M. Epperson, R. D. Jalkian, and M. B. Denton, *Anal. Chem.*, **1989**, *61*, 282.

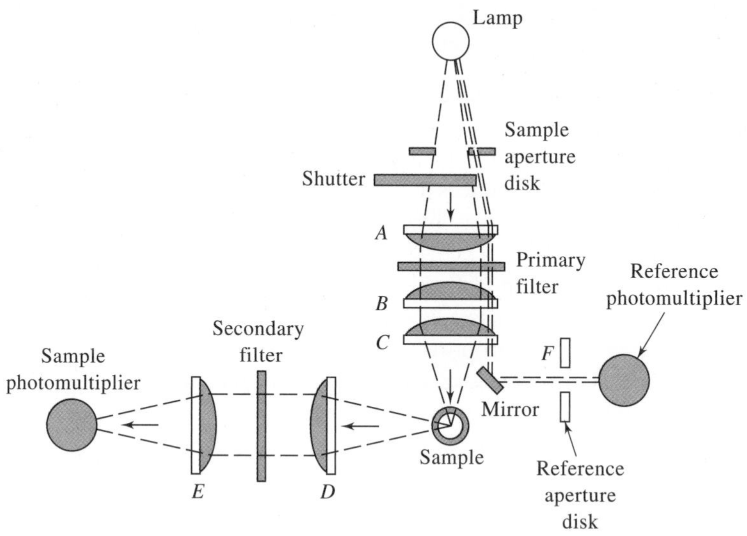

FIGURE 15-10 A typical fluorometer. (Courtesy of Farrand Optical Components and Instruments Division of Ruhle Companies, Inc.)

compartment to reduce the amount of scattered radiation reaching the detector. Baffles are often introduced into the compartment for this purpose. Even more than in absorbance measurements, it is important to avoid fingerprints on cells because skin oils often fluoresce.

Low-volume micro cells are available for situations in which sample volumes are limited. Several companies make flow cells for fluorescence detection in chromatography and in continuous flow analysis. Sample-handling accessories include micro-plate readers, microscope attachments, and fiber-optic probes. Low-volume cells are often used for room-temperature phosphorescence and for chemiluminescence. Special cells and sample handling are needed for low-temperature phosphorescence measurements.

Although the right-angle geometry shown in Figure 15-8 is most common, two other cell configurations are used. Front-surface geometry allows measurements on solutions with high absorbances or on opaque solids. The 180°, or in-line, geometry is rarely used because the weak luminescence signal must be isolated from the intense excitation beam by the emission wavelength selector.

Data Manipulation

Modern computer-based luminescence instruments have many different data-manipulation schemes available in software. Common data-manipulation and display options include blank signal subtraction, production of corrected excitation and emission spectra, calculation and display of difference and derivative

spectra, peak detection and processing, deconvolution, production of three-dimensional total luminescence plots, fitting of calibration data, calculation of statistical parameters, and smoothing of spectra by various methods. Specialized software is available for kinetics, for high-performance liquid chromatography (HPLC) detection, for analysis of mixtures, and for time-resolved measurements.

15B-2 Instrument Designs

Fluorometers

Filter fluorometers provide a relatively simple, low-cost way of performing quantitative fluorescence analyses. As noted earlier, either absorption or interference filters are used to limit the wavelengths of the excitation and emitted radiation. Generally, fluorometers are compact, rugged, and easy to use.

Figure 15-10 is a schematic of a typical filter fluorometer that uses a mercury lamp for fluorescence excitation and a pair of photomultiplier tubes as transducers. The source beam is split near the source into a reference beam and a sample beam. The reference beam is attenuated by the aperture disk so that its intensity is roughly the same as the fluorescence intensity. Both beams pass through the primary filter, with the reference beam then being reflected to the reference photomultiplier tube. The sample beam is focused on the sample by a pair of lenses and causes fluorescence emission. The emitted radiation passes through a second filter and

 Tutorial: Learn more about **fluorescence instrumentation**.

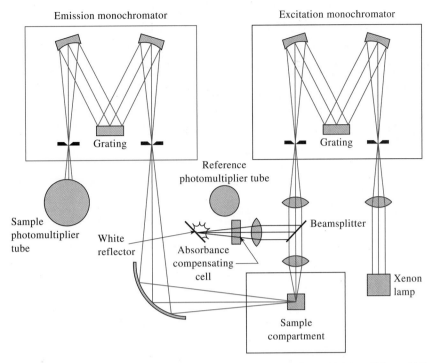

FIGURE 15-11 A spectrofluorometer. (Courtesy of Jobin Yvon, Spex Division, Edison, NJ.)

then is focused on the second photomultiplier tube. The electrical outputs from the two transducers are then processed to compute the ratio of the sample to reference intensities, which serves as the analytical variable.

The instrument just described is representative of the dozen or more fluorometers available commercially. Some of these are simpler single-beam instruments. Some are portable instruments for field work. The cost of such fluorometers ranges from several hundred dollars to more than $10,000.

Spectrofluorometers

Several instrument manufacturers offer spectrofluorometers capable of providing both excitation and emission spectra. The optical design of one of these, which employs two grating monochromators, is shown in Figure 15-11. Radiation from the excitation monochromator is split, part passing to a reference photomultiplier and part to the sample. The resulting fluorescence radiation, after dispersion by the emission monochromator, is detected by a second photomultiplier.

An instrument such as that shown in Figure 15-11 provides perfectly satisfactory spectra for quantitative analysis. The emission spectra obtained will not, however, necessarily compare well with spectra from other instruments because the output depends not only on the intensity of fluorescence but also on the characteristics of the lamp, transducer, and monochromators. All of these instrument characteristics vary with wavelength and differ from instrument to instrument. A number of methods have been developed for obtaining a corrected spectrum, which is the true fluorescence spectrum freed from instrumental effects (see next section).

Spectrofluorometers Based on Array Detectors

A number of spectrofluorometers based on diode-array and charge-transfer devices have been described that permit fluorescence spectra to be obtained in fractions of a second.[4] Several commercial instruments are now available with CCD detectors.[5]

---

[4]See, for example, R. S. Pomeroy in *Charge Transfer Devices in Spectroscopy*, J. V. Sweedler, K. L. Ratzlaff, and M. B. Denton, eds., New York: VCH, 1994, pp. 281–314; P. M. Epperson, R. D. Jalkian, and M. B. Denton, *Anal. Chem.*, **1989**, *61*, 282.

[5]For example, see instruments made by Ocean Optics, Inc., Dunedin, Florida, and Spex Division, Jobin Yvon, Edison, New Jersey.

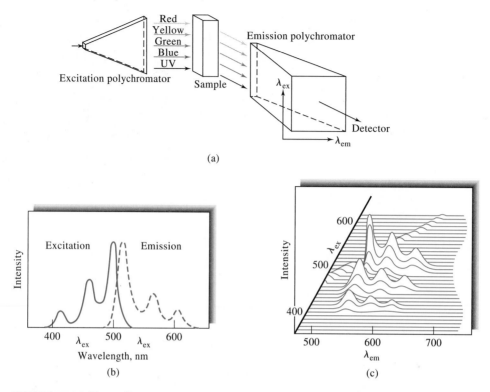

FIGURE 15-12 Three-dimensional spectrofluorometer. (a) Schematic of an optical system for obtaining total luminescence spectra with a CCD detector. (With permission from G. W. Suter, A. J. Kallir, and U. P. Wild, *Chimia*, **1983**, *37*, 413.) (b) Excitation and emission spectra of a hypothetical compound. (c) Total luminescence spectrum of compound in (b). (With permission of D. W. Johnson, J. P. Callis, and G. C. Christian, *Anal. Chem.*, **1977**, *49*, 747A. Figure 3, p. 749A. Copyright 1977 American Chemical Society.)

One of the unique uses of an array-detector spectrofluorometer is the production of total luminescence spectra. Such a spectrum is a plot of the emission spectrum at every excitation wavelength usually presented as a three-dimensional plot. Total luminescence spectra can be obtained in a conventional manner with a standard photomultiplier transducer, although collection of the spectra involved are quite time-consuming. Here, the emission spectrum is obtained at one excitation wavelength. Then, the excitation monochromator is moved to another wavelength and the emission spectrum scanned again. A computer stores the various spectra and presents the total three-dimensional luminescence display.

The principle of a sophisticated array-detector-based instrument to produce total luminescence spectra is illustrated in Figure 15-12a. Here, the length of a sample cell is irradiated with an excitation beam that has been dispersed along the *xy* plane by a monochromator that has been rotated 90° with respect to its exit slit. The transducer is a two-dimensional charge-coupled device that sees the dispersed excitation radiation in the *xy* plane and the dispersed radiation from the emission monochromator in the *yz* plane. Figure 15-12b shows traditional excitation and emission spectra for a hypothetical molecular species. Figure 15-12c shows the total luminescence spectrum for this compound, which is an isometric projection, sometimes called a *stack plot*, of the complete excitation and emission spectra of the compound obtained with the arrangement shown in Figure 15-12a. Total spectra of this type can be obtained in a few seconds or less and are of particular use for analyzing mixtures of fluorescing species. A commercial instrument of this type is now available.[6]

Fiber-Optic Fluorescence Sensors

Fiber-optic probes have been used to demonstrate that several fluorescence determinations can be carried out at various locations well away from a source and a

---

[6]Spex Division, Jobin Yvon, Edison, New Jersey.

detector. Here, radiation from a laser source travels through an optical fiber and excites fluorescence in sample solutions. Fluorescence emission then travels back through the same fiber to a detector for measurement. The applicability of this type of device has been extended to nonfluorescing analytes by fixing a fluorescing indicator material to the end of the fiber.[7] A discussion of the properties of fiber optics and applications to chemical instrumentation is found in Section 7G.

Many fluorescence sensors are based, not on direct fluorescence, but on the quenching of fluorescence.[8] Molecular oxygen, for example, is one of the best collisional quenchers. Oxygen can quench the fluorescence from polycyclic aromatic hydrocarbons; complexes of ruthenium, osmium, iridium, and platinum; and a number of surface-adsorbed heterocyclic molecules. An oxygen sensor can be made by immobilizing the fluorophore in a thin layer of silicone on the end of a fiber-optic bundle.[9] Sensors for SO_2, halides, H_2O_2, and several other molecules have been based on fluorescence quenching.

Phosphorimeters

Instruments used for studying phosphorescence are similar in design to the fluorometers and spectrofluorometers just considered except that two additional components are required.[10] The first is a device that alternately irradiates the sample and, after a suitable time delay, measures the intensity of phosphorescence. The time delay is required to differentiate between long-lived phosphorescence emission and short-lived fluorescence emission, both of which would originate from the same sample. Both mechanical and electronic devices are used, and many commercial fluorescence instruments have accessories for phosphorescence measurements. Many of the current instruments use a gated scheme for the delay. A pulsed xenon arc lamp is often used to excite the sample. After a delay time, specified by the user, the data-acquisition system is activated to obtain the phosphorescence signal. Often, the signal is integrated during this period when

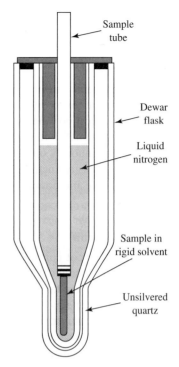

FIGURE 15-13 Dewar flask and cell for low-temperature phosphorescence measurements. The optical path traverses the unsilvered part of the flask.

the lamp is off and fluorescence has decayed to a very small value.

A second new component is needed because phosphorescence measurements are usually performed at liquid nitrogen temperature in a rigid medium to minimize collisional deactivation of the long-lived triplet state. Usually, a Dewar flask with quartz windows, as shown in Figure 15-13, is a part of a phosphorimeter. At the temperature used, the analyte exists as a solute in a glass or solid solvent. A common solvent for this purpose is a mixture of diethylether, pentane, and ethanol.

15B-3 Correction and Compensation Schemes

Several different schemes are used to correct luminescence spectra for some of the many variables that influence them. Source stability, source spectral distribution, inner filter effects, efficiencies of optical components, and spectral responses of instrument components are among the variables that can influence luminescence intensities and spectra. Many instru-

[7]For a discussion of fiber-optic fluorescence sensors, see O. S. Wolfbeis, in *Molecular Luminescence Spectroscopy*, S. G. Schulman, ed., Part 2, Chap. 3, New York: Wiley, 1988.

[8]See W. Trettnak, in *Fluorescence Spectroscopy: New Methods and Applications*, O. S. Wolfbeis, ed., New York: Springer-Verlag, 1993, Chap. 7.

[9]J. N. Demas, B. A. DeGraff, and P. B. Coleman, *Anal. Chem.*, **1999**, 71, 793A.

[10]See R. J. Hurtubise, *Anal. Chem.*, **1983**, 55, 669A; R. J. Hurtubise, *Phosphorimetry: Theory, Instrumentation, and Applications*, Chap. 3, New York: VCH, 1990.

transition-metal complexes are characterized by many closely spaced energy levels, which enhance the likelihood of deactivation by internal conversion. The principal inorganic applications of fluorimetry are thus to nontransition-metal ions, which are less susceptible to these deactivation processes. Note that such cations are generally colorless and tend to form colorless chelates. Thus, fluorometry often complements spectrophotometry.

Fluorometric Reagents

The most successful fluorometric reagents for cation analyses have aromatic structures with two or more donor functional groups that permit chelate formation with the metal ion.[12] The structures of four common reagents are shown in Figure 15-15.

Selected fluorometric reagents and their applications are presented in Table 15-2. As can be seen, limits of detection (LODs) are quite low by most of these methods.

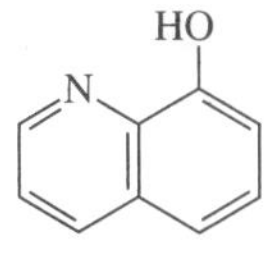

15C-2 Methods for Organic and Biochemical Species

The number of applications of fluorometric methods to organic chemistry is impressive. Dean has summarized the most important of these in a table.[13] The heading *Fluorescence Spectroscopy of Some Organic Compounds* has more than 200 entries, including such diverse compounds as adenine, anthranilic acid, aromatic polycyclic hydrocarbons, cysteine, guanine, isoniazid, naphthols, nerve gases sarin and tabun, proteins, salicylic acid, skatole, tryptophan, uric acid, and warfarin (coumadin). Many medicinal agents that can be determined fluorometrically are listed, including adrenaline, morphine, penicillin, phenobarbital, procaine, reserpine, and lysergic acid diethylamide (LSD). Without question, the most important application of fluorometry is in the analysis of food products, pharmaceuticals, clinical samples, and natural products. The sensitivity and selectivity of the method make it a particularly valuable tool in these fields. Numerous physiologically important compounds fluoresce.

8-hydroxyquinoline
(reagent for Al, Be, and other metal ions)

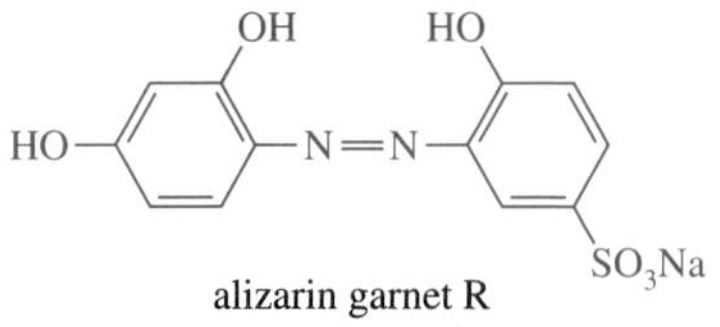

alizarin garnet R
(reagent for Al, F^-)

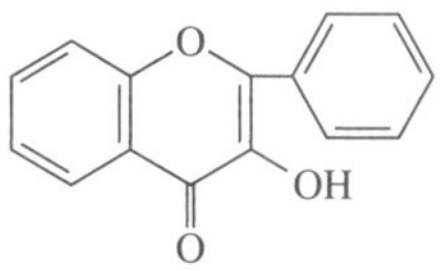

flavanol
(reagent for Zr and Sn)

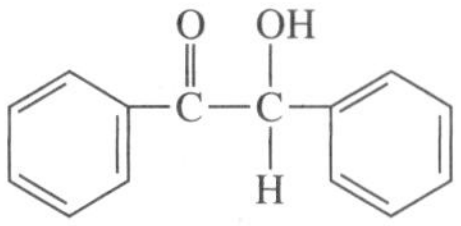

benzoin
(reagent for B, Zn, Ge, and Si)

FIGURE 15-15 Some fluorometric chelating agents for metal cations. Alizarin garnet R can detect Al^{3+} at levels as low as 0.007 μg/mL. Detection of F^- with alizarin garnet R is based on fluorescence quenching of the Al^{3+} complex. Flavanol can detect Sn^{4+} at the 0.1-μg/mL level.

15C-3 Phosphorimetric Methods

Phosphorescence and fluorescence methods tend to be complementary because strongly fluorescing compounds exhibit weak phosphorescence and vice versa.[14] For example, among condensed-ring aromatic hydrocarbons, those containing heavier atoms such as halogens or sulfur often phosphoresce strongly. However, the same compounds in the absence of the heavy atom tend to exhibit fluorescence rather than phosphorescence.

Phosphorimetry has been used for determination of a variety of organic and biochemical species, including such substances as nucleic acids, amino acids, pyrine and pyrimidine, enzymes, petroleum hydro-

[12] For a more detailed discussion of fluorometric reagents, see pp. 384–426 of *Molecular Luminescence Spectroscopy* (note 11), and G. Guilbault, in *Comprehensive Analytical Chemistry*, G. Svehla, ed., Vol. VIII, Chap. 2, pp. 167–78, New York: Elsevier, 1977.

[13] J. A. Dean, *Analytical Chemistry Handbook*, pp. 5.63–5.69, New York: McGraw-Hill, 1995.

[14] See R. J. Hurtubise, *Phosphorimetry: Theory, Instrumentation, and Applications*, Chap. 3, New York: VCH, 1990.

TABLE 15-2 Selected Fluorometric Methods for Inorganic Species

| Ion | Reagent | Wavelength, nm | | LOD, µg/mL | Interferences |
		Absorption	Fluorescence		
Al^{3+}	Alizarin garnet R	470	500	0.007	Be, Co, Cr, Cu, F^-, NO_3^-, Ni, PO_4^{3-}, Th, Zr
F^-	Quenching of Al^{3+} complex of alizarin garnet R	470	500	0.001	Be, Co, Cr, Cu, Fe, Ni, PO_4^{3-}, Th, Zr
$B_4O_7^{2-}$	Benzoin	370	450	0.04	Be, Sb
Cd^{2+}	2-(o-Hydroxyphenyl)-benzoxazole	365	Blue	2	NH_3
Li^+	8-Hydroxyquinoline	370	580	0.2	Mg
Sn^{4+}	Flavanol	400	470	0.1	F^-, PO_4^{3-}, Zr
Zn^{2+}	Benzoin	—	Green	10	B, Be, Sb, colored ions

From J. A. Dean, *Analytical Chemistry Handbook*, New York: McGraw-Hill, 1995, pp. 5.60–5.62

carbons, and pesticides. The method has not, however, found as widespread use as fluorometry, perhaps because of the need for low temperatures and the generally poorer precision of phosphorescence measurements. On the other hand, the potentially greater selectivity of phosphorescence procedures is attractive. The reason for this difference in behavior is that efficient phosphorescence requires rapid intersystem crossing to populate the excited triplet state, which in turn reduces the excited singlet concentration and thus the phosphorescence intensity.

During the past two decades, considerable effort has been expended in the development of phosphorimetric methods that can be carried out at room temperature.[15] The first observations of room-temperature phosphorescence were made with the analyte bound to a solid support, such as filter paper or silica gel. In these applications, a solution of the analyte is dispersed on the solid, and the solvent is evaporated. The phosphorescence of the surface is then measured. The rigid matrix minimizes deactivation of the triplet state by collisional quenching. Collisional quenching has much more of an effect on phosphorescence than on fluorescence because of the much longer lifetime of the triplet state.

Room-temperature phosphorescence in solution has been observed in organized media containing micelles. With micelles, the analyte is incorporated into the core of the micelle, which serves to protect the triplet state. Cyclodextrin molecules, which are

doughnut-shape polymers, have also been used. In most room-temperature experiments, heavy atoms, such as Tl(I), Pb(II), Ag(I), and halide ions, are used to promote intersystem crossing.

15C-4 Fluorescence Detection in Liquid Chromatography

Photoluminescence measurements provide an important method for detecting and determining components of a sample as they elute from a chromatographic or capillary electrophoresis column. Laser-excited fluorescence is particularly important for these applications because the beam can be readily focused to a size on the order of the column diameter. Applications in liquid chromatography and capillary electrophoresis are discussed in more detail in Chapters 28 and 30.

15C-5 Lifetime Measurements

Fluorescence lifetime measurements can give information about collisional deactivation processes, about energy transfer rates, and about excited-state reactions. Lifetime measurements can also be used analytically to provide additional selectivity in the analysis of mixtures containing luminescent species. The measurement of luminescence lifetimes was initially restricted to phosphorescent systems, where decay times were long enough to permit the easy measurement of emitted intensity as a function of time. In recent years, however, it has become relatively routine to measure rates of luminescence decay on the fluorescence time scale (10^{-5} to $<10^{-9}$ s).

[15] T. Vo-Dinh, *Room Temperature Phosphorimetry for Chemical Analysis*, New York: Wiley, 1984.

Two widely used approaches are used for lifetime measurements, the *time-domain* approach and the *frequency-domain* approach. In time-domain measurements, a pulsed source is employed and the time-dependent decay of fluorescence is measured. In the frequency-domain method, a sinusoidally modulated source is used to excite the sample. The phase shift and demodulation of the fluorescence emission relative to the excitation waveform provide the lifetime information. Commercial instrumentation is available to implement both techniques.[16]

15C-6 Fluorescence Imaging Methods

In recent years, it has become possible to combine fluorescence spectroscopy with optical microscopy to produce localized images of fluorophores in complex matrices such as single cells.[17] In some cases the intrinsic (native) fluorescence of biomolecules can be used in conjunction with microscopy to monitor dynamics in cells.[18] In the absence of a native fluorophore, fluorescent indicators can be used to probe biological events. A particularly interesting probe is the so-called ion probe that changes its excitation or emission spectrum on binding to specific ions such as Ca^{2+} or Na^+. These indicators can be used to record events that take place in different parts of individual neurons or to monitor simultaneously the activity of a collection of neurons.

In neurobiology, for example, the dye Fura-2 has been used to monitor the free intracellular calcium concentration following pharmacological or electrical stimulation. By following the fluorescence changes as a function of time at specific sites in the neuron, researchers can determine when and where a calcium-dependent electrical event took place. One cell that has been studied is the Purkinje neuron in the cerebellum, which is one of the largest in the central nervous system. When this cell is loaded with the Fura-2 fluorescent indicator, sharp changes in fluorescence can be measured that correspond to individual calcium action potentials. The changes are correlated to specific sites in the cell by means of fluorescence imaging techniques. Figure 15-16 shows the fluorescence image

on the right along with fluorescence transients, recorded as the change in fluorescence relative to the steady fluorescence $\Delta F/F$, correlated with sodium action potential spikes. The interpretation of these kinds of patterns can have important implications in understanding the details of synaptic activity. Commercial fluorescence microscopes and microscopy attachments are available from several sources.

Fluorescence microscopy and fluorescence lifetime methods have been combined in the technique known as *fluorescence lifetime imaging*. Here, molecular lifetimes can be used to create contrast in two-dimensional fluorescence images.[19]

15D CHEMILUMINESCENCE

The application of chemiluminescence to analytical chemistry is a relatively recent development. The number of chemical reactions that produce chemiluminescence is small, which limits the procedure to a relatively small number of species. Nevertheless, some of the compounds that do react to give chemiluminescence are important components of the environment. For these, the high selectivity, the simplicity, and the extreme sensitivity of the method account for its recent growth in usage.[20]

15D-1 The Chemiluminescence Phenomenon

Chemiluminescence is produced when a chemical reaction yields an electronically excited species that emits light as it returns to its ground state. Chemiluminescence reactions are encountered in a number of biological systems, where the process is often termed *bioluminescence*. Examples of species that exhibit bioluminescence include the firefly, the sea pansy, and certain jellyfish, bacteria, protozoa, and crustacea. The chemistry of the various natural bioluminescence processes is incompletely understood.

---

[16] For references dealing with lifetime measurements, see F. V. Bright and C. A. Munson, *Anal. Chim. Acta*, **2003**, *500*, 72; J. R. Lakowicz, *Principles of Fluorescence Spectroscopy*, 2nd ed., Chaps. 4 and 5, New York: Kluwer Academic Publishers/Plenum Press, 1999.

[17] X. F. Wang, B. Herman, eds., *Fluorescence Imaging Spectroscopy and Microscopy*, New York: Wiley, 1996.

[18] See, for example, E. S. Yeung, *Anal. Chem.*, **1999**, *71*, 522A.

[19] See, for example, J. R. Lakowicz, *Principles of Fluorescence Spectroscopy*, 2nd ed., New York: Kluwer Academic Publishing/Plenum Press, 1999; B. Herman, *Fluorescence Microscopy*, 2nd ed., New York: Springer-Verlag, 1998.

[20] For some reviews of chemiluminescence and its analytical applications, see K. A. Fletcher et al., *Anal. Chem.*, **2006**, *78*, 4047; A. M. Powe et al., *Anal. Chem.*, **2004**, *76*, 4614; L. J. Kricka, *Anal. Chim. Acta*, **2003**, *500*, 279; R. A. Agbaria et al., *Anal. Chem.*, **2002**, *74*, 39, 52; T. A. Nieman, in *Handbook of Instrumental Techniques for Analytical Chemistry*, F. A. Settle, ed., Chap. 27, Upper Saddle River, NJ: Prentice-Hall, 1997.

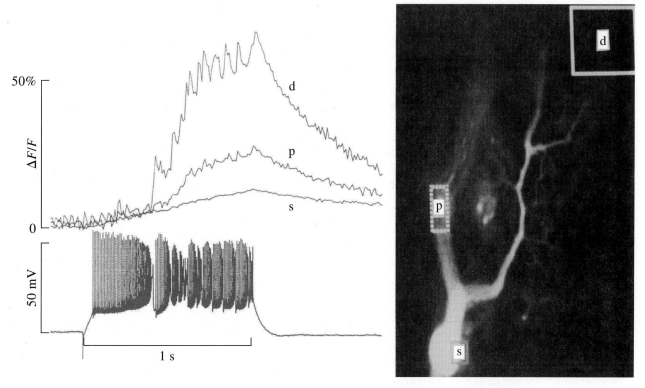

FIGURE 15-16 Calcium transients in a cerebellar Purkinje cell. The image on the right is of the cell filled with a fluorescent dye that responds to the calcium concentration. Fluorescent transients are shown on the top left recorded at areas d, p, and s in the cell. The transients in region d correspond to the dendrite region of the cell. Specific calcium signals can be correlated to the action potentials shown on the bottom left. (From V. Lev-Ram, H. Mikayawa, N. Lasser-Ross, W. N. Ross, *J. Neurophysiol.*, **1992**, *68*, 1170.)

Over a century ago, it was discovered that several relatively simple organic compounds also are capable of exhibiting chemiluminescence. The simplest type of reaction of such compounds to produce chemiluminescence can be formulated as

$$A + B \rightarrow C^* + D$$
$$C^* \rightarrow C + h\nu$$

where C* represents the excited state of the species C. Here, the luminescence spectrum is that of the reaction product C. Most chemiluminescence processes are considerably more complicated than is suggested by these simple reactions.

For chemiluminescence, the radiant intensity I_{CL} (photons emitted per second) depends on the rate of the chemical reaction ($d[C]/dt$) and the chemiluminescence quantum yield ϕ_{CL} (photons per molecule reacted). The latter term is equal to the product of the excitation quantum yield ϕ_{EX} (excited states per molecule reacted) and the emission quantum yield ϕ_{EM} (photons per excited state). These relationships are described by the equation

$$I_{CL} = \phi_{CL} \frac{d[C]}{dt} = \phi_{EX}\phi_{EM} \frac{d[C]}{dt} \quad (15\text{-}13)$$

Chemiluminescence systems useful in analytical chemistry generally have values of ϕ_{CL} of 0.01 to 0.2.

15D-2 Measurement of Chemiluminescence

The instrumentation for chemiluminescence measurements is remarkably simple and may consist of only a suitable reaction vessel and a photomultiplier tube. Generally, no wavelength selection device is necessary because the only source of radiation is the chemical reaction between the analyte and reagent. Several instrument manufacturers offer chemiluminescence photometers.

The typical signal from a chemiluminescence experiment is a time-dependent signal that rises rapidly to a maximum as mixing of reagent and analyte completes. Then, a more or less exponential decay of signal follows (see Figure 15-17). Usually for quantitative analysis, the signal is integrated for a fixed time and compared with standards treated in an identical way. Alternatively, peak heights are used for quantitation. Often a linear relationship between signal and concentration is observed over a concentration range of several orders of magnitude.

15D-3 Analytical Applications of Chemiluminescence

Chemiluminescence methods[21] are generally highly sensitive because low light levels are readily monitored in the absence of noise. Furthermore, radiation attenuation by a filter or a monochromator is usually unnecessary. In fact, detection limits are usually determined not by transducer sensitivity but rather by reagent purity. Typical detection limits lie in the parts-per-billion (or sometimes less) to parts-per-million range. The precision of determinations varies depending on the instrumentation and care that is used.

Analysis of Gases

Chemiluminescence methods for determining components of gases originated with the need for highly sensitive methods for determining atmospheric pollutants such as ozone, oxides of nitrogen, and sulfur compounds. One of the most widely used of these methods is for the determination of nitrogen monoxide with the reactions

$$NO + O_3 \rightarrow NO_2^* + O_2$$
$$NO_2^* \rightarrow NO_2 + h\nu (\lambda = 600 \text{ to } 2800 \text{ nm})$$

Ozone from an electrogenerator and the atmospheric sample are drawn continuously into a reaction vessel, where the luminescence radiation is monitored by a photomultiplier tube. A linear response is reported for nitrogen monoxide concentrations of 1 ppb to 10,000 ppm. Chemiluminescence has become the predominant method for monitoring the concentration of

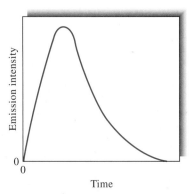

FIGURE 15-17 Chemiluminescence emission intensity as a function of time after mixing reagents.

this important atmospheric constituent from ground level to altitudes as high as 20 km.

The reaction of nitric oxide with ozone has also been applied to the determination of the higher oxides of nitrogen. For example, the nitrogen dioxide content of automobile exhaust gas has been determined by thermal decomposition of the gas at 700°C in a steel tube. The reaction is

$$NO_2 \rightleftharpoons NO + O$$

At least two manufacturers now offer an instrument for determination of nitrogen in solid or liquid materials containing 0.1% to 30% nitrogen. The samples are pyrolyzed in an oxygen atmosphere under conditions whereby the nitrogen is converted quantitatively to nitrogen monoxide. The NO concentration is then measured by the method just described.

Another important chemiluminescence method is used for monitoring atmospheric ozone. In this instance, the determination is based on the luminescence produced when the analyte reacts with the dye Rhodamine B adsorbed on an activated silica gel surface. This procedure is sensitive to less than 1 ppb ozone. The response is linear up to 400 ppb ozone. Ozone can also be determined in the gas phase based on the chemiluminescence produced when the analyte reacts with ethylene. Both reagents are reported to be specific for ozone.

Still another important gas-phase chemiluminescence method is used for the determination of atmospheric sulfur compounds such as hydrogen sulfide, sulfur dioxide, and mercaptans. Here, the sample is combusted in a hydrogen flame to give a sulfur dimer, which then decomposes with the emission of light. For example, with sulfur dioxide the reactions are

[21]For some recent applications, see A. Roda, M. Guardigli, E. Michelini, M. Mirasoli, and P. Pasini, *Anal. Chem.*, **2003**, *75*, 462A.

$$4H_2 + 2SO_2 \rightleftharpoons S_2^* + 4H_2O$$

$$S_2^* \longrightarrow S_2 + h\nu$$

Here, the emission occurs in the blue with maxima at 384 and 394 nm. The chemiluminescent intensity is proportional to the concentration of the excited sulfur dimer. Similarly, combustion of phosphorus compounds in a hydrogen flame gives emission due to HPO* at 526 nm. Linear working curves over four decades of concentration are reported. Both of these flame chemiluminescence techniques have been employed for detection of sulfur and phosphorus species in the effluent from gas chromatographic columns.

Analysis for Inorganic Species in the Liquid Phase

Many of the analyses carried out in the liquid phase make use of organic chemiluminescing substances containing the functional group

$$\underset{\substack{\| \\ O}}{-C}-NH-NHR$$

These reagents react with oxygen, hydrogen peroxide, and many other strong oxidizing agents to produce a chemiluminescing oxidation product. Luminol is the most common example of these compounds. Its reaction with strong oxidants, such as oxygen, hydrogen peroxide, hypochlorite ion, and permanganate ion, in the presence of strong base is given in the following. Often a catalyst is required for this reaction to proceed at a useful rate. The emission produced matches the fluorescence spectrum of the product, 3-aminophthalate anion. The chemiluminescence appears blue and is centered around 425 nm.

Within certain limits, the chemiluminescence intensity of luminol is directly proportional to the concentration of either the oxidant, the catalyst, or the luminol. Consequently, the reaction provides a sensitive method for determining any one of these species. For example, using hydrogen peroxide as the oxidant, the

catalyst Co^{2+} can be estimated at concentrations down to 0.01 nmol/L, Cr^{3+} down to 0.5 nmol/L, and Cu^{2+} down to 1 nmol/L. With a few cations, inhibition of luminescence occurs. For these, the decrease in intensity permits determination of concentrations.

Determinations of Organic Species

To increase the selectivity of chemiluminescence reactions and to extend chemiluminescence to analytes not directly involved in such reactions, it is common practice to precede a chemiluminescence step by an enzyme reaction for which the desired analyte is the substrate and one of the products is detected by chemiluminescence. This is most commonly done in flow systems with reactors containing immobilized enzyme. Recently, however, attention has been directed toward biosensor designs using enzymes attached to optical fibers.

Oxidase enzymes that generate H_2O_2 are commonly used in the predetection step. Not only can H_2O_2 be determined with several different chemiluminescence systems, but the necessary oxidant (O_2) is already present in most samples. Assuming quantitative conversion by the enzyme, substrates can be determined down to 10 to 100 nM, just as can H_2O_2. Substrates detected this way include glucose, cholesterol, choline, uric acid, amino acids, aldehydes, and lactate. For example:

$$\text{uric acid} + O_2 \xrightarrow{\text{uricase}} \text{allantoin} + H_2O_2$$

The approach can be extended by using sequential enzyme steps to ultimately convert the analyte to an equivalent amount of chemiluminescent reactant. In this way, sugars other than glucose, glucosides, cholesterol esters, creatinine, and acetylcholine have been determined.[22] For example,

$$\text{sucrose} + H_2O \xrightarrow{\text{invertase}} \alpha\text{-D-glucose} + \text{fructose}$$

$$\alpha\text{-D-glucose} \xrightarrow{\text{mutarotase}} \beta\text{-D-glucose}$$

$$\beta\text{-D-glucose} + O_2 \xrightarrow{\text{glucose oxidase}} \text{gluconic acid} + H_2O_2$$

Luminol plus a peroxidase catalyst appears to be an excellent reaction medium for determining H_2O_2. Peak chemiluminescence intensity is reached in about 100 ms. The solvent is water and is compatible with some organic components. The detection limit is about 0.1 pM, with linearity for three to four decades of concentration.

---

[22] C. A. K. Swindlehurst and T. A. Nieman, *Anal. Chim. Acta*, **1988**, *205*, 195.

QUESTIONS AND PROBLEMS

*Answers are provided at the end of the book for problems marked with an asterisk.

[X] Problems with this icon are best solved using spreadsheets.

15-1 Explain the difference between a fluorescence emission spectrum and a fluorescence excitation spectrum. Which more closely resembles an absorption spectrum?

15-2 Define the following terms: (a) fluorescence, (b) phosphorescence, (c) resonance fluorescence, (d) singlet state, (e) triplet state, (f) vibrational relaxation, (g) internal conversion, (h) external conversion, (i) intersystem crossing, (j) predissociation, (k) dissociation, (l) quantum yield, (m) chemiluminescence.

15-3 Why is spectrofluorometry potentially more sensitive than spectrophotometry?

15-4 Which compound in each of the pairs below would you expect to have a greater fluorescence quantum yield? Explain.

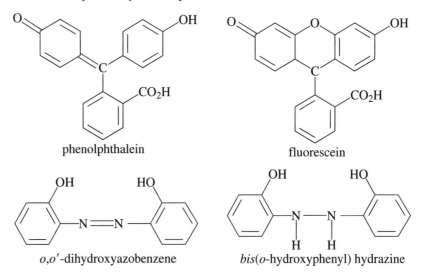

phenolphthalein

fluorescein

o,o'-dihydroxyazobenzene

bis(o-hydroxyphenyl) hydrazine

15-5 Why do some absorbing compounds fluoresce but others do not?

15-6 Discuss the major reasons why molecular phosphorescence spectrometry has not been as widely used as molecular fluorescence spectrometry.

[X] **15-7** The reduced form of nicotinamide adenine dinucleotide (NADH) is an important and highly fluorescent coenzyme. It has an absorption maximum of 340 nm and an emission maximum at 465 nm. Standard solutions of NADH gave the following fluorescence intensities:

Conc. NADH, μmol/L	Relative Intensity
0.100	2.24
0.200	4.52
0.300	6.63
0.400	9.01
0.500	10.94
0.600	13.71
0.700	15.49
0.800	17.91

(a) Construct a spreadsheet and use it to draw a calibration curve for NADH.

*(b) Find the least-squares slope and intercept for the plot in (a).

(c) Calculate the standard deviation of the slope and the standard deviation about regression for the curve.

*(d) An unknown exhibits a relative fluorescence intensity of 12.16. Use the spreadsheet to calculate the concentration of NADH.

*(e) Calculate the relative standard deviation for the result in part (d).

*(f) Calculate the relative standard deviation for the result in part (d) if a result of 7.95 was the mean of three measurements.

 15-8 The volumes of a solution containing 1.10 ppm of Zn^{2+} shown in the table were pipetted into separatory funnels, each containing 5.00 mL of an unknown zinc solution. Each was extracted with three 5-mL aliquots of CCl_4 containing an excess of 8-hydroxyquinoline. The extracts were then diluted to 25.0 mL and their fluorescence measured with a fluorometer. The results were the following:

Volume Std. Zn^{2+}, mL	Fluorometer Reading
0.000	6.12
4.00	11.16
8.00	15.68
12.00	20.64

(a) Construct a working curve from the data.

(b) Calculate a linear least-squares equation for the data.

(c) Calculate the standard deviation of the slope and intercept and the standard deviation about regression.

(d) Calculate the concentration of zinc in the sample.

(e) Calculate a standard deviation for the result in part (d).

*15-9 Quinine in a 1.664-g antimalarial tablet was dissolved in sufficient 0.10 M HCl to give 500 mL of solution. A 20.00-mL aliquot was then diluted to 100.0 mL with the acid. The fluorescence intensity for the diluted sample at 347.5 nm provided a reading of 245 on an arbitrary scale. A standard 100-ppm quinine solution registered 125 when measured under conditions identical to those for the diluted sample. Calculate the mass in milligrams of quinine in the tablet.

*15-10 The determination in Problem 15-9 was modified to use the standard-addition method. In this case, a 4.236-g tablet was dissolved in sufficient 0.10 M HCl to give 1.000 L. Dilution of a 20.00-mL aliquot to 100 mL yielded a solution that gave a reading of 448 at 347.5 nm. A second 20.00-mL aliquot was mixed with 10.0 mL of 50-ppm quinine solution before dilution to 100 mL. The fluorescence intensity of this solution was 525. Calculate the percentage of quinine in the tablet.

*15-11 Iron(II) ions catalyze the oxidation of luminol by H_2O_2. The intensity of the resulting chemiluminescence has been shown to increase linearly with iron(II) concentration from 10^{-10} to 10^{-8} M.

Exactly 1.00 mL of water was added to a 2.00-mL aliquot of an unknown Fe(II) solution, followed by 2.00 mL of a dilute H_2O_2 solution and 1.00 mL of an alkaline solution of luminol. The chemiluminescence from the mixture was integrated over a 10.0-s period and found to be 14.3.

To a second 2.00-mL aliquot of the sample was added 1.00 mL of a 3.58×10^{-5} M Fe(II) solution followed by the same volume of H_2O_2 and luminol. The integrated intensity was 33.3. Find the concentration of Fe(II) in the sample.

15-12 Equations for the chemiluminescence determination of SO_2 are given on page 425. Derive an expression for the relationship between the concentration of SO_2 in a sample, the luminescence intensity, and the equilibrium constant for the first reaction.

15-13 Quinine is one of the best-known fluorescent molecules, and the sensitivities of fluorometers are often specified in terms of the detection limit for this molecule. The structure of quinine is given below. Predict the part of the molecule that is most likely to behave as the chromophore and fluorescent center.

15-14 The quantum efficiency of fluorescence ϕ_f can be written as

$$\phi_f = \frac{\tau}{\tau_0}$$

where τ is the observed lifetime of the excited state in the presence of a quenching agent and τ_0 is the natural lifetime in the absence of a quencher. The fluorescence radiant power F is given by Equation 15-7. This quantity is affected by collisional quenching because the lifetime τ is influenced by collisional quenching. Derive an equation to show that the F-τ ratio is independent of collisional quenching and directly related to concentration. (From G. M. Hieftje and G. R. Haugen, *Anal. Chim. Acta*, **1981**, *123*, 255.)

15-15 The following lifetimes were measured for the chloride quenching of quinine sulfate given in Example 15-1. The fluorescence intensities are given in the example.

Fluorescence Lifetime τ, ns	$[Cl^-]$, M
18.1	0.000
8.9	0.005
5.7	0.010
4.5	0.015
3.6	0.020
2.8	0.025
2.5	0.030
1.9	0.040
1.6	0.050

(a) Plot fluorescence intensity versus $[Cl^-]$.
(b) Plot the ratio of intensity to lifetime, F-τ versus $[Cl^-]$.

(c) Develop a normalization factor to correct the measured fluorescence intensity to that of the solution without quencher.

(d) Plot on the same graph F versus $[Cl^-]$ and F_{corr} versus $[Cl^-]$.

 Challenge Problem

15-16 The following volumes of a standard 10.0 ppb F^- solution were added to four 10.00-mL aliquots of a water sample: 0.00, 1.00, 2.00, and 3.00 mL. Precisely 5.00 mL of a solution containing an excess of the strongly absorbing Al-acid alizarin garnet R complex was added to each of the four solutions, and they were each diluted to 50.0 mL. The fluorescence intensities of the four solutions were as follows:

V_s, mL	Meter Reading
0.00	68.2
1.00	55.3
2.00	41.3
3.00	28.8

(a) Explain the chemistry of the analytical method.

(b) Construct a plot of the data.

(c) Use the fact that the fluorescence decreases with increasing amounts of the F^- standard to derive a relationship like Equation 1-3 for multiple standard additions. Use that relationship further to obtain an equation for the unknown concentration c_x in terms of the slope and intercept of the standard-additions plot, similar to Equation 1-4.

(d) Use linear least squares to find the equation for the line representing the decrease in fluorescence relative to the volume of standard fluoride V_s.

(e) Calculate the standard deviation of the slope and intercept.

(f) Calculate the concentration of F^- in the sample in parts per billion.

(g) Calculate the standard deviation of the result in (e).

CHAPTER SIXTEEN

An Introduction to Infrared Spectrometry

The infrared (IR) region of the spectrum encompasses radiation with wavenumbers ranging from about 12,800 to 10 cm^{-1} or wavelengths from 0.78 to 1000 μm. Because of similar applications and instrumentation, the IR spectrum is usually subdivided into three regions, the near-IR, the mid-IR, and the far-IR. The techniques and the applications of methods based on the three IR spectral regions differ considerably as discussed in this chapter.

Throughout this chapter, this logo indicates an opportunity for online self-study at **www.thomsonedu.com/chemistry/skoog**, linking you to interactive tutorials, simulations, and exercises.

Table 16-1 gives the rough limits of each of the three regions. Measurements in the near-IR region are often made with photometers and spectrophotometers similar in design and components to the instruments described in earlier chapters for ultraviolet-visible spectrometry. The most important applications of this spectral region have been to the quantitative analysis of industrial and agricultural materials and for process control. Applications of near-IR spectrometry are discussed in Section 17D.

Until the early 1980s, instruments for the mid-IR region were largely of the dispersive type and used diffraction gratings. Since that time, however, mid-IR instrumentation has dramatically changed so that now the majority of new instruments are of the Fourier transform type. Photometers based on interference filters also find use for measuring the composition of gases and atmospheric contaminants.

The appearance of relatively inexpensive Fourier transform spectrometers in the last decade has markedly increased the number and type of applications of mid-IR radiation. This increase has come about because interferometric instruments can produce improvements of an order of magnitude, or more, in signal-to-noise ratios and detection limits over dispersive instruments. Before the appearance of these instruments, the mid-IR spectral region was used largely for qualitative organic analysis and structure determinations based on absorption spectra. Now, in contrast, mid-IR spectrometry is beginning to be used in addition for quantitative analysis of complex samples by both absorption and emission spectrometry. Applications of this spectral region are also beginning to appear for microscopic studies of surfaces, analysis of solids by attenuated total reflectance and diffuse reflectance, photoacoustic measurements, and other uses. Several of these applications are described in Chapter 17.

In the past the far-IR region of the spectrum, although potentially quite useful, had limited use because of experimental difficulties. The few sources of this type of radiation are notoriously weak and are further attenuated by the need for order-sorting filters that prevent radiation of higher grating orders from reaching the detector. Fourier transform spectrometers, with their much higher throughput, largely alleviate this problem and make the far-IR spectral region much more accessible to chemists. Several applications of far-IR spectroscopy are described in Section 17E.

In this chapter, we first deal with the mechanisms of the absorption, emission, and reflection of IR radia-

TABLE 16-1 IR Spectral Regions

Region	Wavelengths (λ), μm	Wavenumbers ($\bar{\nu}$), cm^{-1}	Frequencies (ν), Hz
Near	0.78 to 2.5	12800 to 4000	3.8×10^{14} to 1.2×10^{14}
Middle	2.5 to 50	4000 to 200	1.2×10^{14} to 6.0×10^{12}
Far	50 to 1000	200 to 10	6.0×10^{12} to 3.0×10^{11}
Most used	2.5 to 15	4000 to 670	1.2×10^{14} to 2.0×10^{13}

tion using absorption spectroscopy as the basis for this discussion. We follow this discussion with a description of the components of IR instruments and how these are arranged in dispersive and nondispersive instruments as well as in Fourier transform spectrometers.[1]

16A THEORY OF IR ABSORPTION SPECTROMETRY

IR absorption, emission, and reflection spectra for molecular species can be rationalized by assuming that all arise from various changes in energy brought about by transitions of molecules from one vibrational or rotational energy state to another. In this section we use molecular absorption to illustrate the nature of these transitions.

16A-1 Introduction

Figure 16-1 shows a typical output from a commercial IR spectrophotometer. Although the *y*-axis is shown as linear in transmittance, modern computer-based spectrophotometers can also produce spectra that are linear in absorbance. The abscissa in this spectrum is linear in wavenumbers with units of reciprocal centimeters. A wavelength scale is also shown at the top of the plot. Computer-based spectrophotometers can also produce a variety of other spectral formats such as linear in wavelength, baseline corrected, and derivative and smoothed spectra.

A linear wavenumber scale is usually preferred in IR spectroscopy because of the direct proportionality between this quantity and both energy and frequency. The frequency of the absorbed radiation is, in turn, the

molecular vibrational frequency actually responsible for the absorption process. Frequency, however, is seldom if ever used as the abscissa because of the inconvenient size of the unit; that is, a frequency scale of the plot in Figure 16-1 would extend from 1.2×10^{14} to 2.0×10^{13} Hz. Although the axis in terms of wavenumbers is often referred to as a frequency axis, keep in mind that this terminology is not strictly correct because the wavenumber $\bar{\nu}$ is only proportional to frequency ν. The relationships are given in Equation 16-1.

$$\bar{\nu}\,(\text{cm}^{-1}) = \frac{1}{\lambda\,(\mu\text{m})} \times 10^4\,(\mu\text{m/cm}) = \frac{\nu\,(\text{Hz})}{c\,(\text{cm/s})} \quad (16\text{-}1)$$

Finally, note that the horizontal scale of Figure 16-1 changes at 2000 cm^{-1}, with the units at higher wavenumbers being represented by half the linear distance of those at lower wavenumbers. The expanded scale in the region from 2000 to 650 cm^{-1} permits easier identification of spectral features. Numerous IR bands usually appear in this region.

Dipole Moment Changes during Vibrations and Rotations

IR radiation is not energetic enough to bring about the kinds of electronic transitions that we have encountered in our discussions of ultraviolet and visible radiation. Absorption of IR radiation is thus confined largely to molecular species that have small energy differences between various vibrational and rotational states.

To absorb IR radiation, a molecule must undergo a net change in dipole moment as it vibrates or rotates. Only under these circumstances can the alternating electric field of the radiation interact with the molecule and cause changes in the amplitude of one of its motions. For example, the charge distribution around a molecule such as hydrogen chloride is not symmetric because the chlorine has a higher electron density than the hydrogen. Thus, hydrogen chloride has a significant dipole moment and is said to be polar. The dipole moment is determined by the magnitude of the charge

---

[1] For detailed treatments of IR spectrometry, see N. B. Colthup, L. H. Daly, and S. E. Wiberley, *Introduction to Infrared and Raman Spectroscopy*, 3rd ed., San Diego: Academic Press, 1990; B. Schrader, *Infrared and Raman Spectroscopy*, New York: VCH, 1995; C.-P. S. Hsu, in *Handbook of Instrumental Techniques for Analytical Chemistry*, F. Settle, ed., Upper Saddle River, NJ: Prentice Hall, 1997, Chap. 15.

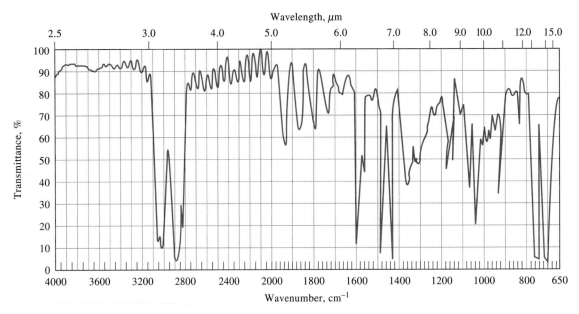

FIGURE 16-1 IR absorption spectrum of a thin polystyrene film. Note the scale change on the *x*-axis at 2000 cm^{-1}.

difference and the distance between the two centers of charge. As a hydrogen chloride molecule vibrates, a regular fluctuation in its dipole moment occurs, and a field is established that can interact with the electric field associated with radiation. If the frequency of the radiation exactly matches a natural vibrational frequency of the molecule, absorption of the radiation takes place that produces a change in the amplitude of the molecular vibration. Similarly, the rotation of asymmetric molecules around their centers of mass results in periodic dipole moment fluctuations that allow interaction with the radiation field.

No net change in dipole moment occurs during the vibration or rotation of homonuclear species such as O_2, N_2, or Cl_2. As a result, such compounds cannot absorb IR radiation. With the exception of a few compounds of this type, all other molecular species absorb IR radiation.

Rotational Transitions

The energy required to cause a change in rotational level is quite small and corresponds to radiation of $\bar{\nu} \leq 100$ cm^{-1} ($\lambda > 100$ μm). Because rotational levels are quantized, absorption by gases in this far-IR region is characterized by discrete, well-defined lines. In liquids or solids, intramolecular collisions and interactions cause broadening of the lines into a continuum.

 Tutorial: Learn more about **IR absorption**.

Vibrational-Rotational Transitions

Vibrational energy levels are also quantized, and for most molecules the energy differences between quantum states correspond to the mid-IR region. The IR spectrum of a gas usually consists of a series of closely spaced lines, because there are several rotational energy levels for each vibrational level. On the other hand, rotation is highly restricted in liquids and solids; in such samples, discrete vibrational-rotational lines disappear, leaving only somewhat broadened vibrational bands.

Types of Molecular Vibrations

The relative positions of atoms in a molecule are not fixed but instead fluctuate continuously as a consequence of a multitude of different types of vibrations and rotations about the bonds in the molecule. For a simple diatomic or triatomic molecule, it is easy to define the number and nature of such vibrations and relate these to energies of absorption. An analysis of this kind becomes difficult if not impossible for molecules made up of many atoms. Not only do large molecules have a large number of vibrating centers, but also interactions among several centers can occur and must be taken into account for a complete analysis.

Vibrations fall into the basic categories of stretching and bending. A stretching vibration involves a continu-

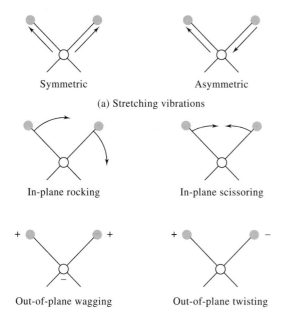

Symmetric Asymmetric

(a) Stretching vibrations

In-plane rocking In-plane scissoring

+ + +

Out-of-plane wagging Out-of-plane twisting

(b) Bending vibrations

FIGURE 16-2 Types of molecular vibrations. Note that + indicates motion from the page toward the reader and − indicates motion away from the reader.

ous change in the interatomic distance along the axis of the bond between two atoms. Bending vibrations are characterized by a change in the angle between two bonds and are of four types: scissoring, rocking,

wagging, and twisting. These are shown schematically in Figure 16-2.

All of the vibration types shown in Figure 16-2 may be possible in a molecule containing more than two atoms. In addition, interaction or coupling of vibrations can occur if the vibrations involve bonds to a single central atom. The result of coupling is a change in the characteristics of the vibrations involved.

In the treatment that follows, we first consider isolated vibrations represented by a simple harmonic oscillator model. Modifications to the theory of the harmonic oscillator, which are needed to describe a molecular system, are taken up next. Finally, the effects of vibrational interactions in molecular systems are discussed.

16A-2 Mechanical Model of a Stretching Vibration in a Diatomic Molecule

The characteristics of an atomic stretching vibration can be approximated by a mechanical model consisting of two masses connected by a spring. A disturbance of one of these masses along the axis of the spring results in a vibration called a *simple harmonic motion*.

Let us first consider the vibration of a single mass attached to a spring that is hung from an immovable object (see Figure 16-3a). If the mass is displaced a distance y from its equilibrium position by application of a force along the axis of the spring, the restoring force

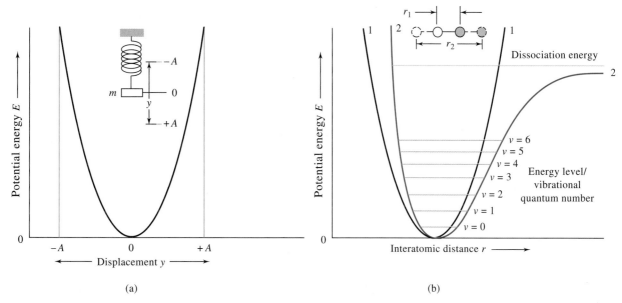

FIGURE 16-3 Potential-energy diagrams. (a) harmonic oscillator. (b) Curve 1, harmonic oscillator; curve 2, anharmonic motion.

F is proportional to the displacement (Hooke's law). That is,

$$F = -ky \qquad (16\text{-}2)$$

where k is the force constant, which depends on the stiffness of the spring. The negative sign indicates that F is a restoring force. This means that the direction of the force is opposite the direction of the displacement. Thus, the force tends to restore the mass to its original position.

Potential Energy of a Harmonic Oscillator

The potential energy E of the mass and spring can be arbitrarily assigned a value of zero when the mass is in its rest, or equilibrium, position. As the spring is compressed or stretched, however, the potential energy of this system increases by an amount equal to the work required to displace the mass. If, for example, the mass is moved from some position y to $y + dy$, the work and hence the change in potential energy dE is equal to the force F times the distance dy. Thus,

$$dE = -F\,dy \qquad (16\text{-}3)$$

Combining Equations 16-3 and 16-2 yields

$$dE = ky\,dy$$

Integrating between the equilibrium position $y = 0$ and y gives

$$\int_0^E dE = \int_0^y y\,dy$$

$$E = \frac{1}{2}ky^2 \qquad (16\text{-}4)$$

The potential-energy curve for a simple harmonic oscillation, derived from Equation 16-4, is a parabola, as depicted in Figure 16-3a. Notice that the potential energy is a maximum when the spring is stretched or compressed to its maximum amplitude A, and it decreases to zero at the equilibrium position.

Vibrational Frequency

The motion of the mass as a function of time t can be deduced from classical mechanics as follows. Newton's second law states that

$$F = ma$$

where m is the mass and a is its acceleration. But acceleration is the second derivative of distance with respect to time. Thus,

$$a = \frac{d^2y}{dt^2}$$

Substituting these expressions into Equation 16-2 gives

$$m\frac{d^2y}{dt^2} = -ky \qquad (16\text{-}5)$$

A solution to this equation must be a periodic function such that its second derivative is equal to the original function times $-k/m$. A suitable cosine relationship meets this requirement. Thus, the instantaneous displacement of the mass at time t can be written as

$$y = A\cos 2\pi \nu_m t \qquad (16\text{-}6)$$

where ν_m is the natural vibrational frequency and A is the maximum amplitude of the motion. The second derivative of Equation 16-6 is

$$\frac{d^2y}{dt^2} = -4\pi^2\nu_m^2 A\cos 2\pi \nu_m t \qquad (16\text{-}7)$$

Substitution of Equations 16-6 and 16-7 into Equation 16-5 gives

$$A\cos 2\pi\nu_m t = \frac{4\pi^2\nu_m^2 m}{k}A\cos 2\pi\nu_m t$$

The natural frequency of the oscillation is then

$$\nu_m = \frac{1}{2\pi}\sqrt{\frac{k}{m}} \qquad (16\text{-}8)$$

where ν_m is the natural frequency of the mechanical oscillator. Although it depends on the force constant of the spring and the mass of the attached body, the natural frequency is independent of the energy imparted to the system; changes in energy merely result in a change in the amplitude A of the vibration.

The equation just developed may be modified to describe the behavior of a system consisting of two masses m_1 and m_2 connected by a spring. Here, it is only necessary to substitute the reduced mass μ for the single mass m where

$$\mu = \frac{m_1 m_2}{m_1 + m_2} \qquad (16\text{-}9)$$

Thus, the vibrational frequency for such a system is given by

$$\nu_m = \frac{1}{2\pi}\sqrt{\frac{k}{\mu}} = \frac{1}{2\pi}\sqrt{\frac{k(m_1 + m_2)}{m_1 m_2}} \qquad (16\text{-}10)$$

Molecular Vibrations

The approximation is ordinarily made that the behavior of a molecular vibration is analogous to the mechanical model just described. Thus, the frequency of the molecular vibration is calculated from Equation 16-10 after substituting the masses of the two atoms for m_1 and m_2. The quantity k is the force constant of the chemical bond, which is a measure of its stiffness.

16A-3 Quantum Treatment of Vibrations

The equations of ordinary mechanics that we have used thus far do not completely describe the behavior of particles of atomic dimensions. For example, the quantized nature of molecular vibrational energies, and of other atomic and molecular energies as well, does not appear in these equations. We may, however, invoke the concept of the simple harmonic oscillator to develop the wave equations of quantum mechanics. Solutions of these equations for potential energies have the form

$$E = \left(v + \frac{1}{2} \right) \frac{h}{2\pi} \sqrt{\frac{k}{\mu}} \qquad (16\text{-}11)$$

where h is Planck's constant, and v is the *vibrational quantum number*, which can take only positive integer values (including zero). Thus, in contrast to ordinary mechanics where vibrators can assume any potential energy, quantum mechanical vibrators can take on only certain discrete energies.

It is interesting to note that the factor $\sqrt{k/\mu}/2\pi$ appears in both the classical and the quantum equations; by substituting Equation 16-10 into 16-11, we find

$$E = \left(v + \frac{1}{2} \right) h\nu_m \qquad (16\text{-}12)$$

where ν_m is the vibrational frequency of the classical model.[2]

We now assume that transitions in vibrational energy levels can be brought about by absorption of radiation, provided the energy of the radiation exactly matches the difference in energy levels ΔE between the vibrational quantum states and provided also the vibration causes a change in dipole moment. This difference is identical between any pair of adjacent levels,

because v in Equations 16-11 and 16-12 can assume only whole numbers; that is,

$$\Delta E = h\nu_m = \frac{h}{2\pi} \sqrt{\frac{k}{\mu}} \qquad (16\text{-}13)$$

At room temperature, the majority of molecules are in the ground state $v = 0$; thus, from Equation 16-12,

$$E_0 = \frac{1}{2} h\nu_m$$

Promotion to the first excited state $v = 1$ with energy

$$E_1 = \frac{3}{2} h\nu_m$$

requires radiation of energy

$$\left(\frac{3}{2} h\nu_m - \frac{1}{2} h\nu_m \right) = h\nu_m$$

The frequency of radiation ν that will bring about this change is identical to the classical vibrational frequency of the bond ν_m. That is,

$$E_{\text{radiation}} = h\nu = \Delta E = h\nu_m = \frac{h}{2\pi} \sqrt{\frac{k}{\mu}}$$

or

$$\nu = \nu_m = \frac{1}{2\pi} \sqrt{\frac{k}{\mu}} \qquad (16\text{-}14)$$

If we wish to express the radiation in wavenumbers, we substitute Equation 6-3 and rearrange:

$$\bar{\nu} = \frac{1}{2\pi c} \sqrt{\frac{k}{\mu}} = 5.3 \times 10^{-12} \sqrt{\frac{k}{\mu}} \qquad (16\text{-}15)$$

where $\bar{\nu}$ is the wavenumber of an absorption maximum (cm^{-1}), k is the force constant for the bond in newtons per meter (N/m), c is the velocity of light (cm s^{-1}), and μ is the reduced mass (kg) defined by Equation 16-9.[3]

IR measurements in conjunction with Equation 16-14 or 16-15 permit the evaluation of the force constants for various types of chemical bonds. Generally, k has been found to lie in the range between 3×10^2 and 8×10^2 N/m for most single bonds, with 5×10^2 serving as a reasonable average value. Double and triple bonds are found by this same means to have force constants of about two and three times this value $(1 \times 10^3$ and 1.5×10^3 N/m, respectively). With these

---

[2]Unfortunately, the generally accepted symbol for the vibrational quantum number v is similar in appearance to the Greek nu, ν, which symbolizes frequency. Thus, constant care must be exercised to avoid confusing the two in equations such as Equation 16-12.

[3]By definition, the newton has the units of N = kg m/s^2. Thus, $\sqrt{k/\mu}$ has units of s^{-1}.

average experimental values, Equation 16-15 can be used to estimate the wavenumber of the fundamental absorption band, or the absorption due to the transition from the ground state to the first excited state, for a variety of bond types. The following example demonstrates such a calculation.

EXAMPLE 16-1

Calculate the approximate wavenumber and wavelength of the fundamental absorption due to the stretching vibration of a carbonyl group $C{=}O$.

▶ *Solution*

The mass of the carbon atom in kilograms is given by

$$m_1 = \frac{12 \times 10^{-3}\ \text{kg/mol}}{6.0 \times 10^{23}\ \text{atom/mol}} \times 1\ \text{atom}$$

$$= 2.0 \times 10^{-26}\ \text{kg}$$

Similarly, for oxygen,

$$m_2 = (16 \times 10^{-3})/(6.0 \times 10^{23}) = 2.7 \times 10^{-26}\ \text{kg}$$

and the reduced mass μ is given by (Equation 16-9)

$$\mu = \frac{2.0 \times 10^{-26}\ \text{kg} \times 2.7 \times 10^{-26}\ \text{kg}}{(2.0 + 2.7) \times 10^{-26}\ \text{kg}}$$

$$= 1.1 \times 10^{-26}\ \text{kg}$$

As noted earlier, the force constant for the typical double bond is about 1×10^3 N/m. Substituting this value and μ into Equation 16-15 gives

$$\bar{\nu} = 5.3 \times 10^{-12}\ \text{s/cm} \sqrt{\frac{1 \times 10^3\ \text{N/m}}{1.1 \times 10^{-26}\ \text{kg}}}$$

$$= 1.6 \times 10^3\ \text{cm}^{-1}$$

The carbonyl stretching band is found experimentally to be in the region of 1600 to 1800 cm^{-1} (6.3 to 5.6 μm).

Selection Rules

As given by Equations 16-12 and 16-13, the energy for a transition from energy level 1 to 2 or from level 2 to 3 should be identical to that for the 0 to 1 transition. Furthermore, quantum theory indicates that the only transitions that can take place are those in which the vibrational quantum number changes by unity; that is, the *selection rule* states that $\Delta v = \pm 1$. Because the vibrational levels are equally spaced for a harmonic oscillator, only a single absorption peak should be observed for a given molecular vibration. In addition to the $\Delta v = \pm 1$ selection rule, there must be a change in dipole moment during the vibration.

Anharmonic Oscillator

Thus far, we have considered the classical and quantum mechanical treatments of the harmonic oscillator. The potential energy of such a vibrator changes periodically as the distance between the masses fluctuates (Figure 16-3a). From qualitative considerations, however, this description of a molecular vibration appears imperfect. For example, as the two atoms approach one another, coulombic repulsion between the two nuclei produces a force that acts in the same direction as the restoring force of the bond. For this reason, the potential energy can be expected to rise more rapidly than the harmonic approximation predicts. At the other extreme of oscillation, a decrease in the restoring force, and thus the potential energy, occurs as the interatomic distance approaches that at which dissociation of the atoms takes place.

In theory, the wave equations of quantum mechanics permit the derivation of more nearly correct potential-energy curves for molecular vibrations. Unfortunately, however, the mathematical complexity of these equations prevents their quantitative application to all but simple diatomic systems. Qualitatively, the curves take the anharmonic form shown as curve 2 in Figure 16-3b. Such curves depart from harmonic behavior by varying degrees, depending on the nature of the bond and the atoms involved. Note, however, that the harmonic and anharmonic curves are nearly alike at low potential energies.

Anharmonicity leads to deviations of two kinds. At higher quantum numbers, ΔE becomes smaller (see curve 2 in Figure 16-3b), and the selection rule is not rigorously followed. As a result, weaker transitions called *overtones* are sometimes observed. These transitions correspond to $\Delta v = \pm 2$ or ± 3. The frequencies of such overtone transitions are approximately two or three times that of the fundamental frequency, and the intensities are lower than that of the fundamental.

Vibrational spectra are further complicated by two different vibrations in a molecule interacting to give absorptions at frequencies that are approximately the sums or differences of their fundamental frequencies.

Again, the intensities of these sum and difference bands are generally low.

16A-4 Vibrational Modes

It is ordinarily possible to deduce the number and kinds of vibrations in simple diatomic and triatomic molecules and whether these vibrations will lead to absorption. Complex molecules may contain several types of atoms as well as bonds. For these molecules, the many types of possible vibrations give rise to IR spectra that are much more difficult to analyze.

The number of possible vibrations in a polyatomic molecule can be calculated as follows. Three coordinates are needed to locate a point in space. To fix N points requires a set of three coordinates for each point, for a total of $3N$. Each coordinate corresponds to one degree of freedom for one of the atoms in a polyatomic molecule. For this reason, a molecule containing N atoms is said to have $3N$ *degrees of freedom*.

In defining the motion of a molecule, we need to consider (1) the motion of the entire molecule through space (that is, the translational motion of its center of gravity); (2) the rotational motion of the entire molecule around its center of gravity; and (3) the motion of each of its atoms relative to the other atoms, or in other words, its individual vibrations. Because all atoms in the molecule move in concert through space, definition of translational motion requires three coordinates and thus this common motion requires three of the $3N$ degrees of freedom. Another three degrees of freedom are needed to describe the rotation of the molecule as a whole. The remaining $3N - 6$ degrees of freedom involve interatomic motion and hence represent the number of possible vibrations within the molecule. A linear molecule is a special case because, by definition, all of the atoms lie on a single, straight line. Rotation about the bond axis is not possible, and two degrees of freedom suffice to describe rotational motion. Thus, the number of vibrations for a linear molecule is given by $3N - 5$. Each of the $3N - 6$ or $3N - 5$ vibrations is called a *normal mode*.

For each normal mode of vibration, there is a potential-energy relationship such as that shown by the solid lines in Figure 16-3b. The same selection rules discussed earlier apply for each of these relationships. In addition, to the extent that a vibration approximates harmonic behavior, the differences between the energy levels of a given vibration are the same; that is,

a single absorption peak appears for each vibration having a change in dipole moment.

Four factors tend to produce fewer experimental bands than would be expected from the theoretical number of normal modes. Fewer absorption bands are found when (1) the symmetry of the molecules is such that no change in dipole moment results from a particular vibration; (2) the energies of two or more vibrations are identical or nearly identical; (3) the absorption intensity is so low as to be undetectable by ordinary means; or (4) the vibrational energy is in a wavelength region beyond the range of the instrument.

Occasionally, more absorption bands are found than are expected based on the number of normal modes. We have already mentioned the overtone bands that occur at two or three times the frequency of the fundamental. In addition, *combination bands* are sometimes encountered when a photon excites two vibrational modes simultaneously. The frequency of the combination band is approximately the sum or difference of the two fundamental frequencies. This phenomenon occurs when a quantum of energy is absorbed by two bonds rather than one.

16A-5 Vibrational Coupling

The energy of a vibration, and thus the wavelength of the corresponding absorption maximum, may be influenced by (or coupled with) other vibrators in the molecule. A number of factors influence the extent of such coupling.

1. Strong coupling between stretching vibrations occurs only when there is an atom common to the two vibrations.
2. Interaction between bending vibrations requires a common bond between the vibrating groups.
3. Coupling between a stretching and a bending vibration can occur if the stretching bond forms one side of the angle that varies in the bending vibration.
4. Interaction is greatest when the coupled groups have individual energies that are nearly equal.
5. Little or no interaction is observed between groups separated by two or more bonds.
6. Coupling requires that the vibrations be of the same symmetry species.[4]

---

[4]For a discussion of symmetry operations and symmetry species, see F. A. Cotton, *Chemical Applications of Group Theory*, 3rd ed., New York: Wiley 1990; R. S. Drago, *Physical Methods for Chemists*, 2nd ed., Philadelphia: Saunders, 1992.

As an example of coupling effects, let us consider the IR spectrum of carbon dioxide. If no coupling occurred between the two C=O bonds, an absorption band would be expected at the same wavenumber as that for the C=O stretching vibration in an aliphatic ketone (about 1700 cm^{-1}, or 6 μm; see Example 16-1). Experimentally, carbon dioxide exhibits two absorption maxima, one at 2350 cm^{-1} (4.3 μm) and the other at 667 cm^{-1} (15 μm).

Carbon dioxide is a linear molecule and thus has $(3 \times 3) - 5 = 4$ normal modes. Two stretching vibrations are possible; furthermore, interaction between the two can occur because the bonds involved are associated with a common carbon atom. As can be seen, one of the coupled vibrations is symmetric and the other is asymmetric.

Symmetric   Asymmetric

The symmetric vibration causes no change in dipole moment, because the two oxygen atoms simultaneously move away from or toward the central carbon atom. Thus, the symmetric vibration is IR inactive. In the asymmetric vibration, one oxygen moves away from the carbon atom as the carbon atom moves toward the other oxygen. As a consequence, a net change in charge distribution occurs periodically, producing a change in dipole moment, so absorption at 2350 cm^{-1} results.

The remaining two vibrational modes of carbon dioxide involve scissoring, as shown here.

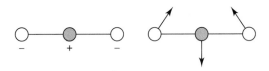

These two bending vibrations are the resolved components at 90° to one another of the bending motion in all possible planes around the bond axis. The two vibrations are identical in energy and thus produce a single absorption band at 667 cm^{-1}. Quantum states that are identical, as these are, are said to be *degenerate*.

It is of interest to compare the spectrum of carbon dioxide with that of a nonlinear, triatomic molecule such as water, sulfur dioxide, or nitrogen dioxide. These molecules have $(3 \times 3) - 6 = 3$ vibrational modes that take the following forms:

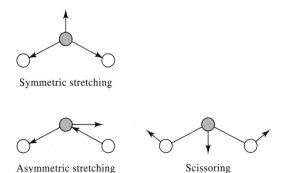

Symmetric stretching

Asymmetric stretching   Scissoring

Because the central atom is not in line with the other two, the symmetric stretching vibration produces a change in dipole moment and is thus IR active. For example, stretching peaks at 3657 and 3766 cm^{-1} (2.74 and 2.66 μm) appear in the IR spectrum for the symmetric and asymmetric stretching vibrations of the water molecule. There is only one component to the scissoring vibration for this nonlinear molecule because motion in the plane of the molecule constitutes a rotational degree of freedom. For water, the bending vibration causes absorption at 1595 cm^{-1} (6.27 μm). The difference in behavior of linear and nonlinear triatomic molecules with two and three absorption bands, respectively, illustrates how IR absorption spectroscopy can sometimes be used to deduce molecular shapes.

Coupling of vibrations is a common phenomenon. As a result, the position of an absorption band corresponding to a given organic functional group cannot be specified exactly. For example, the C—O stretching frequency in methanol is 1034 cm^{-1} (9.67 μm), in ethanol it is 1053 cm^{-1} (9.50 μm), and in 2-butanol it is 1105 cm^{-1} (9.05 μm). These variations result from a coupling of the C—O stretching with adjacent C—C stretching or C—H vibrations.

Although interaction effects may lead to uncertainties in the identification of functional groups contained in a compound, it is this very effect that provides the unique features of an IR absorption spectrum that are so important for the positive identification of a specific compound.

16B IR INSTRUMENTATION

Three types of instruments for IR absorption measurements are commonly available: (1) dispersive spectrophotometers with a grating monochromator; (2) Fourier transform spectrometers employing an interferometer (Section 7I); and (3) nondispersive photometers using a filter or an absorbing gas that are used for analysis of atmospheric gases at specific wavelengths.

Until the 1980s, the most widely used instruments for IR measurements were dispersive spectrophotometers. Now, however, this type of instrument has been largely displaced for mid- and far-IR measurements by Fourier transform spectrometers because of their speed, reliability, signal-to-noise advantage, and convenience. Dispersive spectrometers are still used in the near-IR where they are often extensions of UV-visible instruments, but many dedicated near-IR instruments are of the Fourier transform–IR (FTIR) type.

16B-1 Fourier Transform Spectrometers

The theoretical basis and the inherent advantages of Fourier transform (FT) and other multiplex instruments were discussed in some detail in Section 7I, and the reader may find it worthwhile to review that section before proceeding further. Two types of multiplex instruments have been described for the IR region. In the FT spectrometer, coding is accomplished by splitting the source into two beams whose path lengths can be varied periodically to give interference patterns. The FT is then used for data processing.[5] The second is the Hadamard transform spectrometer, which is a dispersive instrument that employs a moving mask at the focal plane of a monochromator for encoding the spectral data. Hadamard transform IR instruments have not been widely adopted and will, therefore, not be discussed further in this book.[6]

When FTIR spectrometers first appeared in the marketplace, they were bulky, expensive (>$100,000), and required frequent mechanical adjustments. For these reasons, their use was limited to special applications where their unique characteristics (speed, high resolution, sensitivity, and unparalleled wavelength precision and accuracy) were essential. FT instruments have now been reduced to benchtop size and have become reliable and easy to maintain. Furthermore, the price of simpler models has been reduced to the point where they are competitive with all but the simplest dispersive instruments (~$15,000 and more). For these reasons, FT instruments have largely displaced dispersive instruments in the laboratory.[7]

Components of FT Instruments

The majority of commercially available FTIR instruments are based on the Michelson interferometer, although other types of optical systems are also encountered. We shall consider the Michelson design only, which is illustrated in Figure 7-43.[8]

Drive Mechanism. Requirements for satisfactory interferograms (and thus satisfactory spectra) are that the moving mirror have constant speed and a position exactly known at any instant. The planarity of the mirror must also remain constant during its entire sweep of 10 cm or more.

In the far-IR region, where wavelengths range from 50 to 1000 μm (200 to 10 cm^{-1}), displacement of the mirror by a fraction of a wavelength, and accurate measurement of its position, can be accomplished by means of a motor-driven micrometer screw. A more precise and sophisticated mechanism is required for the mid- and near-IR regions, however. Here, the mirror mount is generally floated on an air bearing held within close-fitting stainless steel sleeves (see Figure 16-4). The mount is driven by a linear drive motor and an electromagnetic coil similar to the voice coil in a loudspeaker; an increasing current in the coil drives the mirror at constant velocity. After reaching its terminus, the mirror is returned rapidly to the starting point for the next sweep by a rapid reversal of the current. The length of travel varies from 1 to about 20 cm; the scan rates range from 0.01 to 10 cm/s.

Two additional features of the mirror system are necessary for successful operation. The first is a means of sampling the interferogram at precisely spaced retardation intervals. The second is a method for determining exactly the zero retardation point to permit signal averaging. If this point is not known precisely,

[5]For detailed discussions of Fourier transform IR spectroscopy, see S. Davis, M. Abrams, J. Brault, *Fourier Transform Spectrometry*, San Diego: Academic Press, 2001; B. C. Smith, *Fundamentals of Fourier Transform Infrared Spectroscopy*, Boca Raton, FL: CRC Press, 1996; P. R. Griffiths and J. A. deHaseth, *Fourier Transform Infrared Spectroscopy*, New York: Wiley, 1986.
[6]For a description of the Hadamard transform and Hadamard transform spectroscopy, see D. K. Graff, *J. Chem. Educ.*, **1995**, *72*, 304; *Fourier, Hadamard, and Hilbert Transforms in Chemistry*, A. G. Marshall, ed., New York: Plenum Press, 1982.

[7]For a review of commercial FTIR spectrometers, see J. P. Smith and V. Hinson-Smith, *Anal. Chem.*, **2003**, *75*, 37A.
[8]The Michelson interferometer was designed and built in 1891 by A. A. Michelson. He was awarded the 1907 Nobel Prize in physics for the invention of interferometry.

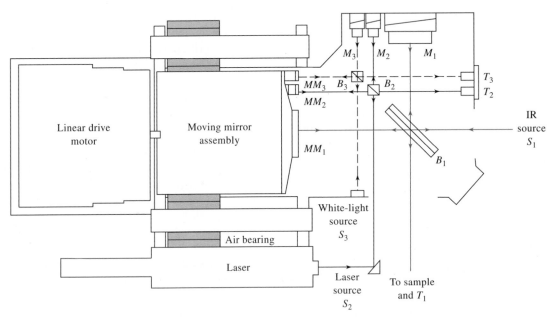

FIGURE 16-4 Interferometers in an FTIR spectrometer. Subscript 1 defines the radiation path in the IR interferometer. Subscripts 2 and 3 refer to the laser and white-light interferometers, respectively. (Courtesy of Thermo Electron Corp., Franklin, MA.)

the signals from repetitive sweeps would not be in phase; averaging would then tend to degrade rather than improve the signal.

The problem of precise signal sampling and signal averaging can be accomplished by using two or three interferometers rather than one, with a single mirror mount holding the movable mirrors. Figure 16-4 is a schematic showing such an arrangement. The components and radiation paths for each of the three interferometer systems are indicated by the subscripts 1, 2, and 3, respectively. System 1 is the IR system that ultimately provides an interferogram similar to that shown as curve A in Figure 16-5. System 2 is a so-called laser-fringe reference system, which provides sampling-interval information. It consists of a He-Ne laser S_2, an interferometric system including mirrors MM_2 and M_2, a beamsplitter B_2, and a transducer T_2. The output from this system is a cosine wave, as shown in C of Figure 16-5. This signal is converted electronically to the square-wave form shown in D; sampling begins or terminates at each successive zero crossing. The laser-fringe reference system gives a highly reproducible and regularly spaced sampling interval. In most instruments, the laser signal is also used to control the speed of the mirror-drive system at a constant level.

The third interferometer system, sometimes called the white-light system, employs a tungsten source S_3 and transducer T_3 sensitive to visible radiation. Its mirror system is fixed to give a zero retardation displaced to the left from that for the analytical signal (see interferogram B, Figure 16-5). Because the source is polychromatic, its power at zero retardation is much larger than any signal before and after that point. Thus, this maximum is used to trigger the start of data sampling for each sweep at a highly reproducible point.

The triple interferometer system just described leads to remarkable precision in determining spectral frequencies, which significantly exceeds that realizable with conventional grating instruments. This high reproducibility is particularly important when many scans are to be averaged. Contemporary instruments, such as the benchtop unit pictured in Figure 16-6, are able to achieve excellent frequency precision with one or two interferometers. In the instrument diagrammed in Figure 16-7, the interferometer is actually two parallel interferometers, one to modulate the IR radiation from the source before it passes through the sample and the second to modulate the red light from the He-Ne laser to provide the reference signal for acquiring data from the IR detector. No white-light

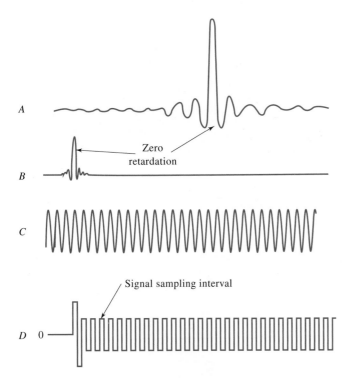

Zero retardation

Signal sampling interval

D 0

FIGURE 16-5 Time-domain signals for the three interferometers contained in many FTIR instruments. Curve A, IR signal; curve B, white-light signal; curve C, laser-fringe reference signal; curve D, square-wave electrical signal formed from the laser signal. (From P. R. Griffiths, *Chemical Infrared Fourier Transform Spectroscopy*, p. 102, New York: Wiley, 1975. Reprinted with permission.)

source is employed, and the IR interferogram is used to establish zero retardation. The maximum in the IR interferogram is an excellent reference because this is the only point at which all wavelengths interfere constructively.

The benchtop system shown in Figure 16-6 is capable of providing spectra with a resolution of approxi-

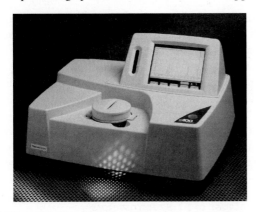

FIGURE 16-6 Photo of a basic, benchtop FTIR spectrometer suitable for student use. Spectra are recorded in a few seconds and displayed on the LCD panel for viewing and interpretation. The spectra may be stored in a memory card for later retrieval and analysis, or they may be printed. (Courtesy of Thermo Electron Corp., Madison, WI.)

mately 4 cm^{-1}. Higher resolution, down to ~0.01 cm^{-1}, can be achieved with a more sophisticated system for maintaining the alignment of the moving mirror. One mirror-alignment system uses three laser-fringe reference systems that are directed at different points on the moving mirror instead of one. Because three points are adequate to define a plane, the use of three lasers significantly increases the accuracy with which the position and orientation of the mirror can be known at any instant.

Instrument Designs

FTIR spectrometers can be single-beam or double-beam instruments. Figure 16-8 shows the optics of a basic single-beam spectrometer, which sells in the range of $15,000 to $20,000. A typical procedure for determining transmittance or absorbance with this type of instrument is to first obtain a reference interferogram by scanning a reference (usually air) twenty or thirty times, coadding the data, and storing the results in the memory of the instrument computer (usually after transforming it to the spectrum). A sample is then inserted in the radiation path and the process repeated. The ratio of sample and reference spectral data is then computed

 Simulation: Learn more about **IR instrumentation**.

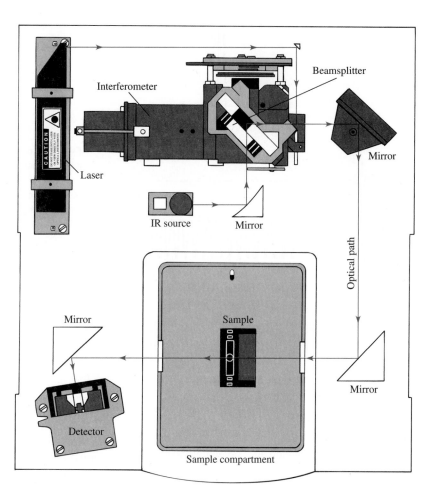

FIGURE 16-7 Diagram of a basic FTIR spectrometer. Radiation of all frequencies from the IR source is reflected into the interferometer where it is modulated by the moving mirror on the left. The modulated radiation is then reflected from the two mirrors on the right through the sample in the compartment at the bottom. After passing through the sample, the radiation falls on the transducer. A data-acquisition system attached to the transducer records the signal and stores it in the memory of a computer as an interferogram. (Courtesy of Thermo Electron Corp., Franklin, MA.)

to give the transmittance at various frequencies. From this ratio the absorbance is calculated as a function of wavenumber. Ordinarily, modern IR sources and detectors are sufficiently stable so that reference spectra need to be obtained only occasionally.

A double-beam spectrometer is illustrated in Figure 16-9. The mirrors directing the interferometer beam through the sample and reference cells are oscillated rapidly compared to the movement of the interferometer mirror so that sample and reference information can be obtained at each mirror position. The double-beam design compensates for source and detector drifts.

Performance Characteristics of Commercial Instruments

A number of instrument manufacturers offer several models of FTIR instruments. The least expensive has a range of 7800 to 350 cm^{-1} (1.3 to 29 μm) with a resolution of 4 cm^{-1}. This performance can be obtained with a scan time as brief as 1 second. More expensive instruments with interchangeable beamsplitters, sources, and transducers offer expanded frequency ranges and higher resolutions. For example, one instrument is reported to produce spectra from the far-IR (10 cm^{-1} or 1000 μm) through the visible region to 50,000 cm^{-1}, or 200 nm. Resolutions for commercial instruments vary from 8 to less than 0.01 cm^{-1}. Several minutes are required to obtain a complete spectrum at the highest resolution.[9]

Advantages of FT Spectrometers

Over most of the mid-IR spectral range, FT instruments have signal-to-noise ratios that are better than those of a good-quality dispersive instrument, usually by more than an order of magnitude. The enhanced

---

[9]D. Noble, *Anal. Chem.*, **1995**, *67*, 381A.

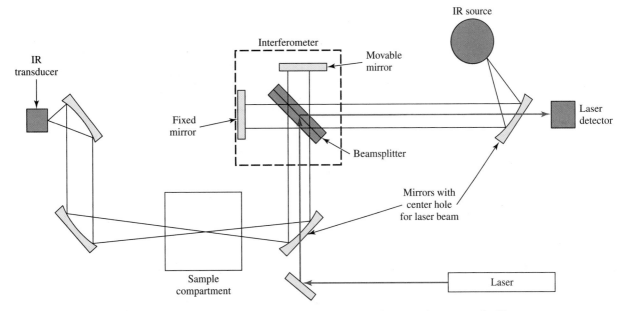

FIGURE 16-8 Single-beam FTIR spectrometer. In one arm of the interferometer, the IR source radiation travels through the beamsplitter to the fixed mirror, back to the beamsplitter, and through the sample to the IR transducer. In the other arm, the IR source radiation travels to the beamsplitter, is reflected to the movable mirror, and travels back through the beamsplitter to the sample and to the transducer. When the two beams meet again at the beamsplitter, they can interfere with each other if the phase difference (path difference) is appropriate. A plot of the signal versus mirror displacement is the interferogram. The interferogram contains information about all the frequencies present. The spectrum, intensity versus wavenumber, is the FT of the interferogram. It can be calculated with a computer from the signal versus mirror displacement. An empty sample compartment allows the reference spectrum to be calculated. Next, the sample is placed in the sample compartment and the sample spectrum is obtained. The absorbance is then calculated at each wavenumber from the ratio of the sample intensity to the reference intensity.

signal-to-noise ratio can, of course, be traded for rapid scanning, with good spectra being attainable in a few seconds in most cases. Interferometric instruments are also characterized by high resolutions (<0.1 cm^{-1}) and highly accurate and reproducible frequency determinations. The latter property is particularly helpful when spectra are to be subtracted for background correction.

A theoretical advantage of FT instruments is that their optics provide a much larger energy throughput (one to two orders of magnitude) than do dispersive instruments, which are limited in throughput by the need for narrow slit widths. The potential gain here, however, may be partially offset by the lower sensitivity of the fast-response detector required for the interferometric measurements. Finally, it should be noted that

the interferometer is free from the problem of stray radiation because each IR frequency is, in effect, chopped at a different frequency.

The areas of chemistry where the extra performance of interferometric instruments has been particularly useful include (1) very high-resolution work that is encountered with gaseous mixtures having complex spectra resulting from the superposition of vibrational and rotational bands; (2) the study of samples with high absorbances; (3) the study of substances with weak absorption bands (for example, the study of compounds that are chemisorbed on catalyst surfaces); (4) investigations requiring fast scanning such as kinetic studies or detection of chromatographic effluents; (5) collecting IR data from very small samples; (6) obtaining reflection spectra; and (7) IR emission studies.

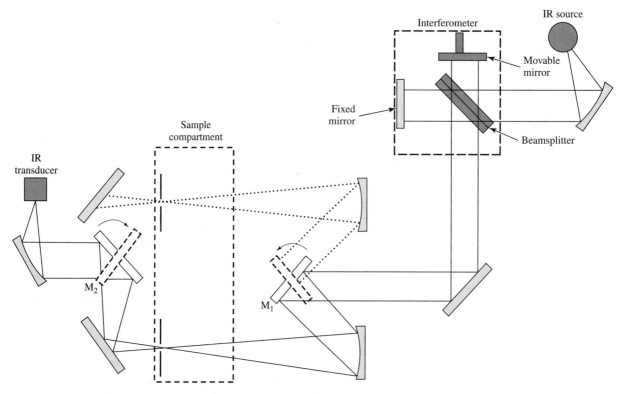

FIGURE 16-9 Double-beam FTIR spectrometer. The beam emerging from the interferometer strikes mirror M_1, which in one position directs the beam through the reference cell and in the other position directs it through the sample cell. Mirror M_2, which is synchronized to M_1, alternately directs the reference beam and the sample beam to the transducer.

16B-2 Dispersive Instruments

Although most instruments produced today are FT systems, many dispersive spectrophotometers are still found in laboratories. Dispersive IR spectrophotometers are generally double-beam, recording instruments, which use reflection gratings for dispersing radiation. As was pointed out in Section 13D-2, the double-beam design is less demanding with respect to the performance of sources and detectors — an important characteristic because of the relatively low intensity of IR sources, the low sensitivity of IR transducers, and the consequent need for large signal amplifications (see Section 16C).

An additional reason for the general use of double-beam instruments in the IR region is shown in Figure 16-10. The lower curve reveals that atmospheric water and carbon dioxide absorb radiation in some important spectral regions and can cause serious interference problems. The upper curve shows that the reference beam compensates nearly perfectly for absorption by both compounds. A stable $100\% T$ baseline results.

Generally, dispersive IR spectrophotometers incorporate a low-frequency chopper (five to thirty cycles per second) that permits the detector to discriminate between the signal from the source and signals from extraneous radiation, such as IR emission from various bodies surrounding the transducer. Low chopping rates are demanded by the slow response times of the IR transducers used in most dispersive instruments. In general, the optical designs of dispersive instruments do not differ greatly from the double-beam UV-visible spectrophotometers discussed in the previous chapter except that the sample and reference compartments are always located between the source and the monochromator in IR instruments. This arrangement is possible because IR radiation, in contrast to UV-visible, is not sufficiently energetic to cause photochemical decomposition of the sample. Placing the sample and

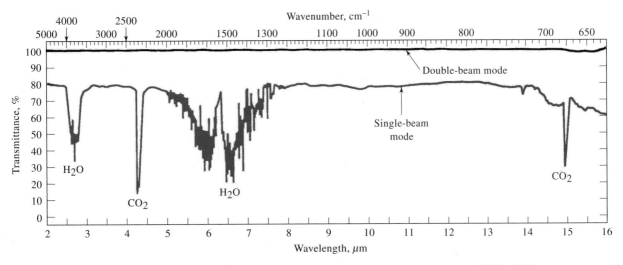

FIGURE 16-10 Single- and double-beam spectra of atmospheric water vapor and CO_2. In the lower, single-beam trace, the absorption of atmospheric gases is apparent. The top, double-beam spectrum shows that the reference beam compensates very well for this absorption and allows for a stable $100\%T$ baseline to be obtained. (From J. D. Ingle Jr. and S. R. Crouch, *Spectrochemical Analysis*, p. 409, Upper Saddle River, NJ: Prentice-Hall, 1988. With permission.)

reference before the monochromator, however, has the advantage that most scattered radiation and IR emission, generated within the cell compartment, is effectively removed by the monochromator and thus does not reach the transducer.

Figure 16-11 shows schematically the arrangement of components in a typical IR spectrophotometer. Like many inexpensive dispersive IR instruments, it is an optical null type, in which the radiant power of the reference beam is reduced, or attenuated, to match

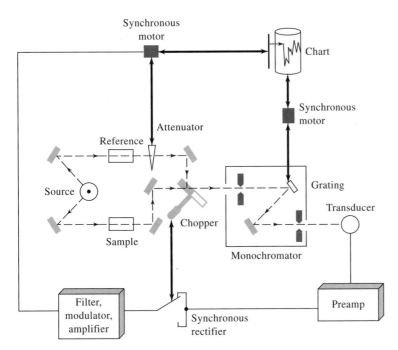

FIGURE 16-11 Schematic diagram of a double-beam, dispersive IR spectrophotometer. The heavy black lines indicate mechanical linkages, and the light lines indicate electrical connections. The radiation path is designated by dashed lines.

that of the beam passing through the sample. Attenuation is accomplished by imposing a device that removes a continuously variable fraction of the reference beam. The attenuator commonly takes the form of a comb, the teeth of which are tapered so that there is a linear relationship between the lateral movement of the comb and the decrease in power of the beam. Movement of the comb occurs when a difference in power of the two beams is sensed by the detection system. Most dispersive IR instruments are older instruments that use mechanical recorders instead of computers. For such instruments, the comb movement is synchronized with the recorder pen so that its position gives a measure of the relative power of the two beams and thus the transmittance of the sample.

Note that three types of systems link the components of the instrument in Figure 16-11: (1) a radiation linkage indicated by dashed lines, (2) a mechanical linkage shown by thick dark lines, (3) an electrical connection shown by narrow solid lines.

Radiation from the source is split into two beams, half passing into the sample-cell compartment and the other half into the reference area. The reference beam then passes through the attenuator and on to a chopper. The chopper consists of a motor-driven disk that alternately reflects the reference or transmits the sample beam into the monochromator. After dispersion by a prism or grating, the alternating beams fall on the transducer and are converted to an electrical signal. The signal is amplified and passed to the synchronous rectifier,

a device that is mechanically or electrically coupled to the chopper to cause the rectifier switch and the beam leaving the chopper to change simultaneously. If the two beams are identical in power, the signal from the rectifier is constant direct current. If, on the other hand, the two beams differ in power, a fluctuating, or ac, current is produced, the phase of which is determined by which beam is the more intense. The current from the rectifier is filtered and further amplified to drive a synchronous motor in one direction or the other, depending on the phase of the input current. The synchronous motor is mechanically linked to both the attenuator and the pen drive of the recorder and causes both to move until a null balance is achieved. A second synchronous motor simultaneously drives the chart and varies the wavelength. Frequently, a mechanical linkage between the wavelength and slit drives varies the slit width to keep the radiant power reaching the detector approximately constant.

The reference-beam attenuator system, such as the one just described, creates three limitations to the performance of dispersive IR instruments. First, the response of the attenuator system always lags behind the transmittance changes, particularly in scanning regions where the signal is changing most rapidly. Second, the momentum associated with both the mechanical attenuator and the recorder system may result in the pen drive overshooting the true transmittance. Third, in regions where the transmittance approaches zero, almost no radiation reaches the transducer, and the exact null

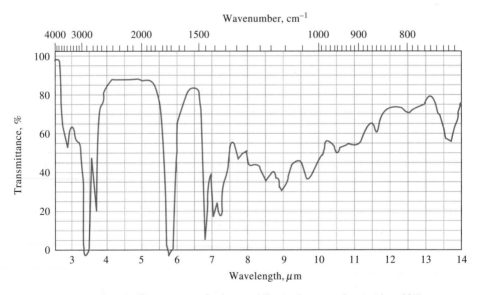

FIGURE 16-12 IR spectrum of *n*-hexanal illustrating overshoot at low %*T*.

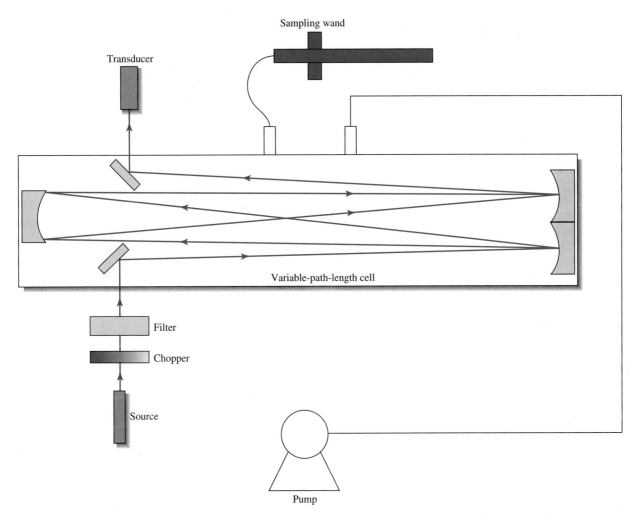

FIGURE 16-13 A portable IR photometer designed for gas analysis. The sample is introduced into the cell by means of a pump. The path length can be changed by altering the number of mirror reflections. (Courtesy of Thermo Electron Corp., Franklin, MA.)

position cannot be established accurately. The result is sluggish transducer response and rounded peaks. Figure 16-12 illustrates transmittance overshoot and rounded peaks in regions of low transmittance (1700 and 3000 cm^{-1}).

16B-3 Nondispersive Instruments

Several simple, rugged instruments have been designed for quantitative IR analysis. Some are simple filter photometers, whereas others employ filter wedges in lieu of a dispersing element to provide entire spectra. Finally, some gas analyzers use no wavelength-selection device at all. Generally, these instruments are less complex, more rugged, easier to maintain, and less expensive than the instruments we have previously described in this chapter.

Filter Photometers

Figure 16-13 shows a schematic diagram of a portable, IR filter photometer designed for quantitative analysis of various substances in the atmosphere.[10] Different models are available that are factory calibrated for 1, 5, 30, or more than 100 gases. The instrument is computer controlled. For many compounds, fixed bandpass filters of 1.8, 3.3, 3.6, 4.0, 4.2, 4.5, 4.7, 8, 11, and 14 µm can be employed. A continuously variable filter,

---

[10]See P. A. Wilks, *Amer. Lab.*, **1994** (12), 44; *Proc. Control Qual.*, **1992**, *3*, 283.

which transmits in the range between about 7.7 and 14.1 μm (1300 to 710 cm^{-1}), can also be used for selecting alternative wavelengths or for spectral scanning applications. The source is a nichrome-wire filament, and the transducer is a pyroelectric device (see Section 16C for descriptions of sources and transducers).

The gaseous sample is brought into the cell by means of a battery-operated pump at a rate of 20 L/min. In the cell, three gold-plated mirrors are used in a folded-path-length design. Path lengths of 0.5 m and 12.5 m may be selected. Detection of many gases at sub-parts-per-million levels, particularly with the long-path-length setting, have been reported with this photometer.

Photometers without Filters

Photometers, which have no wavelength-restricting device, are widely employed to monitor gas streams for a single component.[11] Figure 16-14 shows a typical nondispersive instrument designed to determine carbon monoxide in a gaseous mixture. The reference cell is a sealed container filled with a nonabsorbing gas; as shown in the figure, the sample flows through a second cell that is of similar length. The chopper blade is so arranged that the beams from identical sources are chopped simultaneously at the rate of about five times per second. Selectivity is obtained by filling both compartments of the sensor cell with the gas being analyzed, carbon monoxide in this example. The two chambers of the detector are separated by a thin, flexible, metal diaphragm that serves as one plate of a capacitor; the second plate is contained in the sensor compartment on the left.

In the absence of carbon monoxide in the sample cell, the two sensor chambers are heated equally by IR radiation from the two sources. If the sample contains carbon monoxide, however, the right-hand beam is attenuated somewhat and the corresponding sensor chamber becomes cooler with respect to its reference counterpart. As a result, the diaphragm moves to the right and the capacitance of the capacitor changes. This change in capacitance is sensed by the amplifier system. The amplifier output drives a servomotor that moves the beam attenuator into the reference beam until the two compartments are again at the same temperature. The instrument thus operates as a null balance device.

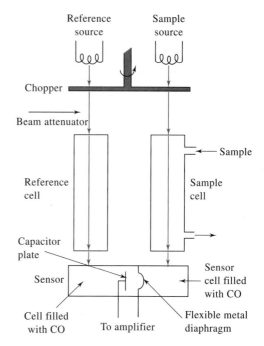

FIGURE 16-14 A nondispersive IR photometer for monitoring carbon monoxide.

The chopper serves to provide an ac signal, which is less sensitive to drift and 1/f noise.

This type of instrument is highly selective because heating of the sensor gas occurs only from that narrow portion of the spectrum of radiation absorbed by the carbon monoxide in the sample. The device can be adapted to the determination of any IR-absorbing gas.

Filter Correlation Analyzers

As shown in Figure 16-15, filter correlation analyzers use a rotating gas filter through which the IR beam passes. The filter has two compartments, one for the gas of interest and the other for a nonabsorbing gas such as nitrogen. When the gas of interest is in the beam, it selectively attenuates the IR source to produce a reference beam. The sample beam is produced by the transparent gas. Typically, the IR source is chopped at a fairly high frequency (360 Hz) whereas the filter rotates at a fairly low frequency (30 Hz). A modulated signal is produced that is related to the concentration of the analyte gas.

Filter correlation analyzers are available for such gases as CO_2 and CO. They can be set up to detect trace levels (<0.1 ppm) or higher amounts. The analyzers are calibrated by constructing a working curve by dilution of a gas standard.

---

[11] For a description of process IR measurements, see W. V. Daily, *Proc. Control Qual.*, **1992**, *3*, 99.

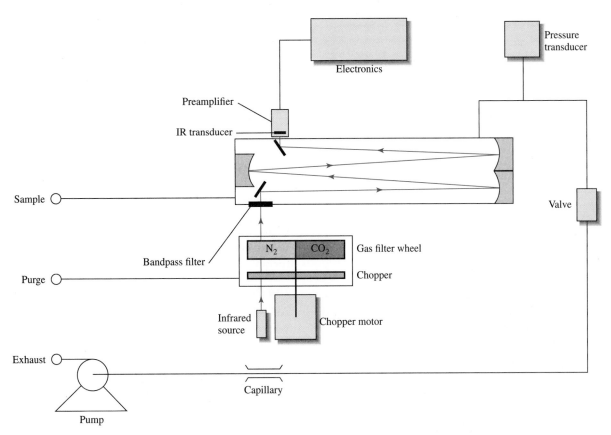

FIGURE 16-15 An IR filter correlation analyzer for determining CO_2. The sample is drawn into the sample cell by a pump. The chopped IR source radiation (360 Hz) alternates between the N_2 and CO_2 sides of the filter wheel, which rotates at 30 Hz. The CO_2 side provides a reference beam that cannot be further attenuated by CO_2 in the sample cell. The N_2 side produces the sample beam by allowing the IR radiation to pass through to the cell where it can be attenuated by CO_2 in the sample. The modulated amplitude of the chopped detector signal is related to the CO_2 concentration in the sample. Other gases do not modulate the detector signal because they absorb the reference and sample beams equally. (Courtesy of Thermo Electron Corp., Franklin, MA.)

16C IR SOURCES AND TRANSDUCERS

Instruments for measuring IR absorption all require a source of continuous IR radiation and an IR transducer. The desirable characteristics of these instrument components were listed in Sections 7B and 7E. In this section we describe sources and transducers that are found in modern IR instruments.

16C-1 Sources

IR sources consist of an inert solid that is heated electrically to a temperature between 1500 and 2200 K. These sources produce continuum radiation approxi-

mating that of a blackbody (see Figure 6-22). The maximum radiant intensity at these temperatures occurs between 5000 and 5900 cm^{-1} (2 and 1.7 μm). At longer wavelengths, the intensity falls off smoothly until it is about 1% of the maximum at 670 cm^{-1} (15 μm). On the short wavelength side, the decrease is much more rapid, and a similar reduction in intensity is observed at about 10,000 cm^{-1} (1 μm).

The Nernst Glower

The Nernst glower is composed of rare earth oxides formed into a cylinder having a diameter of 1 to 3 mm and a length of 2 to 5 cm. Platinum leads are sealed to the ends of the cylinder to permit electrical connections

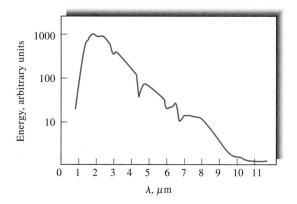

FIGURE 16-16 Spectral distribution of energy from a Nernst glower operated at approximately 2200 K.

to what amounts to a resistive heating element. As current passes through the device, its temperature increases to between 1200 K and 2200 K. The Nernst glower has a large negative temperature coefficient of electrical resistance, and it must be heated externally to a dull red heat before the current is large enough to maintain the desired temperature. Because the resistance decreases with increasing temperature, the source circuit must be designed to limit the current or the glower rapidly becomes so hot that it is destroyed.

Figure 16-16 shows the spectral output of a Nernst glower operated at approximately 2200 K. Note that the overall curve shape is similar to that of a blackbody. The small peaks and depressions are a function of the chemical composition of the device.

The Globar

A Globar is a silicon carbide rod, usually about 5 cm long and 5 mm in diameter. It also is electrically heated (1300 to 1500 K) and has the advantage of a positive coefficient of resistance. On the other hand, water cooling of the electrical contacts is required to prevent arcing. Spectral energies of the Globar and the Nernst glower are comparable except in the region below 5 μm, where the Globar provides a significantly greater output.

Incandescent Wire Source

A source with somewhat lower intensity but with longer life than the Globar or Nernst glower is a tightly wound spiral of nichrome wire heated to about 1100 K by an electrical current. A rhodium wire heater sealed in a ceramic cylinder has similar properties, although it is more expensive. Nichrome wire sources are less intense than many IR sources. However, the incandescent wire source requires no cooling and is nearly maintenance free. For this reason, the nichrome wire source is often used where reliability is paramount, such as in process analyzers.

The Mercury Arc

For the far-IR region of the spectrum ($\lambda > 50$ μm), none of the thermal sources just described provides sufficient radiant power for convenient detection. Here, a high-pressure mercury arc is used. This device consists of a quartz-jacketed tube containing mercury vapor at a pressure greater than 1 atmosphere. Passage of electricity through the vapor forms an internal plasma source that provides continuum radiation in the far-IR region.

The Tungsten Filament Lamp

An ordinary tungsten filament lamp is a convenient source for the near-IR region of 4000 to 12,800 cm^{-1} (2.5 to 0.78 μm).

The Carbon Dioxide Laser Source

A tunable carbon dioxide laser is used as an IR source for monitoring the concentrations of certain atmospheric pollutants and for determining absorbing species in aqueous solutions.[12] A carbon dioxide laser produces a band of radiation in the range of 900 to 1100 cm^{-1} (11 to 9 μm), which consists of about 100 closely spaced discrete lines. As described in Section 7B-3, any one of these lines can be chosen by tuning the laser. Although the range of wavelengths available is limited, the region from 900 to 1100 cm^{-1} is one particularly rich in absorption bands arising from the interactive stretching modes of CO_2. Thus, this source is useful for quantitative determination of a number of important species such as ammonia, butadiene, benzene, ethanol, nitrogen dioxide, and trichloroethylene. An important property of the laser source is the radiant power available in each line, which is several orders of magnitude greater than that of blackbody sources.

Carbon dioxide lasers are widely used in remote-sensing applications such as light detection and ranging (lidar). The operating principle of lidar is similar to that of radar. The lidar system transmits radiation out to a target where it interacts with and is altered by the

[12]See A. A. Demidov, in *Introduction to Laser Spectroscopy*, 2nd ed., D. L. Andrews and A. A. Demidov, eds., New York: Kluwer Academic/Plenum Press, 2002; Z. Zelinger, M. Strizik, P. Kubat, and S. Civis, *Anal. Chim. Acta*, **2000**, *422*, 179; P. L. Meyer, M. W. Sigrist, *Rev. Sci. Instrum.*, **1990**, *61*, 1779.

target. Some of the radiation is then reflected back to the lidar instrument where it is analyzed and used to obtain information about the target. By means of lidar, distance, speed, rotation, chemical composition, and concentration of remote targets can be obtained.

16C-2 IR Transducers

IR transducers are of three general types: (1) pyroelectric transducers, (2) photoconducting transducers, and (3) thermal transducers. The first is found in photometers, some FTIR spectrometers, and dispersive spectrophotometers. Photoconducting transducers are found in many FTIR instruments. Thermal detectors are found in older dispersive instruments but are too slow to be used in FTIR spectrometers.

Pyroelectric Transducers

Pyroelectric transducers are constructed from single crystalline wafers of pyroelectric materials, which are insulators (dielectric materials) with very special thermal and electrical properties. Triglycine sulfate $(NH_2CH_2COOH)_3 \cdot H_2SO_4$ (usually deuterated or with a fraction of the glycines replaced with alanine), is the most important pyroelectric material used for IR-detection systems.

When an electric field is applied across any dielectric material, polarization takes place, with the magnitude of the polarization being a function of the dielectric constant of the material. For most dielectrics, this induced polarization rapidly decays to zero when the external field is removed. Pyroelectric substances, in contrast, retain a strong temperature-dependent polarization after removal of the field. Thus, by sandwiching the pyroelectric crystal between two electrodes, one of which is IR transparent, a temperature-dependent capacitor is produced. Changing its temperature by irradiating it with IR radiation alters the charge distribution across the crystal, which can be detected as a current in an external electrical circuit connecting the two sides of the capacitor. The magnitude of this current is proportional to the surface area of the crystal and the rate of change of polarization with temperature. Pyroelectric crystals lose their residual polarization when they are heated to a temperature called the *Curie point*. For triglycine sulfate, the Curie point is 47°C.

Pyroelectric transducers exhibit response times that are fast enough to allow them to track the changes in the time-domain signal from an interferometer. For this reason, many FTIR spectrometers for the mid-IR region employ this type of transducer.

Photoconducting Transducers

IR photoconducting transducers consist of a thin film of a semiconductor material, such as lead sulfide, mercury telluride–cadmium telluride (MCT), or indium antimonide, deposited on a nonconducting glass surface and sealed in an evacuated envelope to protect the semiconductor from the atmosphere. Absorption of radiation by these materials promotes nonconducting valence electrons to a higher energy-conducting state, thus decreasing the electrical resistance of the semiconductor. Typically, a photoconductor is placed in series with a voltage source and load resistor, and the voltage drop across the load resistor serves as a measure of the power of the beam of radiation.

A lead sulfide photoconductor is the most widely used transducer for the near-IR region of the spectrum from 10,000 to 333 cm^{-1} (1 to 3 μm). It can be operated at room temperature. For mid- and far-IR radiation, MCT photoconductor transducers are used. They must be cooled with liquid nitrogen (77 K) to minimize thermal noise. The long-wavelength cutoff, and many of the other properties of these transducers, depend on the ratio of the mercury telluride to cadmium telluride, which can be varied continuously.

The MCT transducer is faster and more sensitive than the deuterated triglycine sulfate transducer discussed in the previous section. For this reason, the MCT transducer also finds widespread use in FTIR spectrometers, particularly those requiring fast response times, such as spectrometers interfaced to gas chromatographs.

Thermal Transducers

Thermal transducers, whose responses depend on the heating effect of radiation, are found in older dispersive spectrometers for detection of all but the shortest IR wavelengths. With these devices, the radiation is absorbed by a small blackbody and the resultant temperature rise is measured. The radiant power level from a spectrophotometer beam is minute (10^{-7} to 10^{-9} W), so that the heat capacity of the absorbing element must be as small as possible if a detectable temperature change is to be produced. Under the best of circumstances, temperature changes are confined to a few thousandths of a kelvin.

The problem of measuring IR radiation by thermal means is compounded by thermal noise from the surroundings. For this reason, thermal transducers are housed in a vacuum and are carefully shielded from thermal radiation emitted by other nearby objects. To further minimize the effects of extraneous heat sources, the beam from the source is always chopped. In this way, the analyte signal, after transduction, has the frequency of the chopper and can be separated electronically from extraneous noise signals, which are ordinarily broad band or vary only slowly with time.

Thermocouples. In its simplest form, a thermocouple consists of a pair of junctions formed when two pieces of a metal such as bismuth are fused to each end of a dissimilar metal such as antimony. A potential difference between the two junctions varies with their difference in temperature.

The transducer junction for IR radiation is formed from very fine wires or alternatively by evaporating the metals onto a nonconducting support. In either case, the junction is usually blackened (to improve its heat-absorbing capacity) and sealed in an evacuated chamber with a window that is transparent to IR radiation. The response time is typically about 30 ms.

The reference junction, which is usually housed in the same chamber as the active junction, is designed to have a relatively large heat capacity and is carefully shielded from the incident radiation. Because the analyte signal is chopped, only the difference in temperature between the two junctions is important; therefore, the reference junction does not need to be maintained at constant temperature. To enhance sensitivity, several thermocouples may be connected in series to give what is called a *thermopile*.

Bolometers. A bolometer is a type of resistance thermometer constructed of strips of metals, such as platinum or nickel, or from a semiconductor. This type of semiconductor device is sometimes called a *thermistor*. Semiconductor materials exhibit a relatively large change in resistance as a function of temperature. The responsive element in a bolometer is kept small and blackened to absorb the radiant heat. Bolometers are not so extensively used as other IR transducers for the mid-IR region. However, a germanium bolometer, operated at 1.5 K, is an excellent transducer for radiation in the range of 5 to 400 cm^{-1} (2000 to 25 μm). The response time is a few milliseconds.

QUESTIONS AND PROBLEMS

*Answers are provided at the end of the book for problems marked with an asterisk.

[X] Problems with this icon are best solved using spreadsheets.

*16-1 The IR spectrum of CO shows a vibrational absorption band centered at 2170 cm^{-1}.
(a) What is the force constant for the CO bond?
(b) At what wavenumber would the corresponding peak for ^{14}CO occur?

*16-2 Gaseous HCl exhibits an IR absorption at 2890 cm^{-1} due to the hydrogen-chlorine stretching vibration.
(a) Calculate the force constant for the bond.
(b) Calculate the wavenumber of the absorption band for HCl assuming the force constant is the same as that calculated in part (a).

16-3 Calculate the absorption frequency corresponding to the —C—H stretching vibration treating the group as a simple diatomic C—H molecule with a force constant of $k = 5.0 \times 10^2$ N/m. Compare the calculated value with the range found in correlation charts (such as the one shown in Figure 17-6). Repeat the calculation for the deuterated bond.

*16-4 The wavelength of the fundamental O—H stretching vibration is about 1.4 μm. What is the approximate wavenumber and wavelength of the first overtone band for the O—H stretch?

*16-5 The wavelength of the fundamental N—H stretching vibration is about 1.5 μm. What is the approximate wavenumber and wavelength of the first overtone band for the N—H stretch?

*16-6 Sulfur dioxide is a nonlinear molecule. How many vibrational modes will this compound have? How many IR absorption bands would sulfur dioxide be expected to have?

*16-7 Indicate whether the following vibrations are active or inactive in the IR spectrum.

Molecule	Motion
(a) $CH_3—CH_3$	C—C stretching
(b) $CH_3—CCl_3$	C—C stretching
(c) SO_2	Symmetric stretching
(d) $CH_2=CH_2$	C—H stretching:
(e) $CH_2=CH_2$	C—H stretching:
(f) $CH_2=CH_2$	CH_2 wag:
(g) $CH_2=CH_2$	CH_2 twist:

16-8 What are the advantages of an FTIR spectrometer compared with a dispersive instrument?

 16-9 What length of mirror drive in an FTIR spectrometer would be required to provide a resolution of (a) 0.050 cm^{-1}, (b) 0.40 cm^{-1}, and (c) 4.0 cm^{-1}?

*16-10 It was stated that at room temperature (25°C) the majority of molecules are in the ground vibrational energy level ($v = 0$).

    (a) Use the Boltzmann equation (Equation 8-1) to calculate the excited-state and ground-state population ratios for HCl: $N(v = 1)/N(v = 0)$. The fundamental vibrational frequency of HCl occurs at 2885 cm^{-1}.

    (b) Use the results of part (a) to find $N(v = 2)/N(v = 0)$.

16-11 Why are nondispersive IR instruments often used for the determination of gases rather than dispersive IR spectrometers?

16-12 The first FTIR instruments used three different interferometer systems. Briefly, describe how it has been possible to simplify the optical systems in more contemporary instruments.

*16-13 In a particular trace analysis via FTIR, a set of sixteen interferograms were collected. The signal-to-noise ratio (S/N) associated with a particular spectral peak was approximately 5:1. How many interferograms would have to be collected and averaged if the goal is to obtain a $S/N = 20:1$?

🅇 *16-14 If a Michelson interferometer has a mirror velocity of 1.00 cm/s, what will be the frequency at the transducer due to light leaving the source at frequencies of (a) 4.8×10^{13} Hz? (b) 4.9×10^{13} Hz, and (c) 5.0×10^{13} Hz? What are the corresponding wavenumbers of these frequencies?

Challenge Problem

16-15 (a) The IR spectrum of gaseous N_2O shows three strong absorption bands at 2224 cm^{-1}, 1285 cm^{-1}, and 2089 cm^{-1}. In addition two quite weak bands are observed at 2563 cm^{-1} and 2798 cm^{-1}. It is known that N_2O is a linear molecule, but assume it is not known whether the structure is N—N—O or N—O—N. Use the IR data to decide between the two structures. What vibrations can be assigned to the strong absorption bands? What are possible causes of the weak absorptions?

    (b) The IR spectrum of HCN shows three strong absorption bands at 3312 cm^{-1}, 2089 cm^{-1}, and 712 cm^{-1}. From this information alone, can you deduce whether HCN is linear or nonlinear? Assuming that HCN is linear, assign vibrations to the three absorption bands.

    (c) How many fundamental vibrational modes are expected for BF_3? Which of these are expected to be IR active? Why? Sketch the vibrations.

    (d) How many fundamental vibrational modes would you predict for (1) methane, (2) benzene, (3) toluene, (4) ethylene, and (5) carbon tetrachloride?

CHAPTER SEVENTEEN

Applications of Infrared Spectrometry

Modern IR spectrometry is a versatile tool that is applied to the qualitative and quantitative determination of molecular species of all types. In this chapter we first focus on the uses of mid-IR absorption and reflection spectrometry for structural investigations of molecular compounds, particularly organic compounds and species of interest in biochemistry. We then examine in less detail several of the other applications of IR spectroscopy.

Throughout this chapter, this logo indicates an opportunity for online self-study at **www .thomsonedu.com/chemistry/skoog**, linking you to interactive tutorials, simulations, and exercises.

As shown in Table 17-1, the applications of IR spectrometry fall into three major categories based on the three IR spectral regions. The most widely used region is the mid-IR, which extends from about 670 to 4000 cm^{-1} (2.5 to 14.9 μm). Here, absorption, reflection, and emission spectra are employed for both qualitative and quantitative analysis. The near-IR region, from 4000 to $14,000 \text{ cm}^{-1}$ (0.75 to 2.5 μm), also finds considerable use for the routine quantitative determination of certain species, such as water, carbon dioxide, sulfur, low-molecular-weight hydrocarbons, amine nitrogen, and many other simple compounds of interest in agriculture and in industry. These determinations are often based on diffuse-reflectance measurements of untreated solid or liquid samples or absorption studies of gases. The primary use of the far-IR region (15 to 1000 μm) has been for the determination of the structures of inorganic and metal-organic species based on absorption measurements.

17A MID-IR ABSORPTION SPECTROMETRY

Mid-IR absorption and reflection spectrometry are major tools for determining the structure of organic and biochemical species. In this section we examine mid-IR absorption applications. Section 17B is devoted to mid-IR reflectance measurements.[1]

17A-1 Sample Handling

As we have seen in earlier chapters, ultraviolet and visible molecular spectra are obtained most conveniently from dilute solutions of the analyte. Absorbance measurements in the optimal range are obtained by suitably adjusting either the concentration or the cell length. Unfortunately, this approach is often not applicable for IR spectroscopy because no good solvents are transparent throughout the region of interest. As a consequence, sample handling is frequently the most difficult and time-consuming part of an IR spectrometric analysis.[2] In this section we outline some of the common

---

[1] For further reading see R. M. Silverstein, F. X. Webster, and D. Kiemle, *Spectrometric Identification of Organic Compounds*, 7th ed., Chap. 2, New York: Wiley, 2005; B. Schrader, *Infrared and Raman Spectroscopy*, New York: VCH, 1995; N. B. Colthup, L. H. Daly, and S. E. Wiberley, *Introduction to Infrared and Raman Spectroscopy*, 3rd ed., San Diego: Academic Press, 1990.

[2] See *Practical Sampling Techniques for Infrared Analysis*, P. B. Coleman, ed., Boca Raton, FL: CRC Press, 1993; T. J. Porro and S. C. Pattacini, *Spectroscopy*, **1993**, *8* (7), 40; Ibid., *8* (8), 39; A. L. Smith, *Applied Infrared Spectroscopy*, New York: Wiley, 1979, Chap. 4.

TABLE 17-1 Major Applications of IR Spectrometry

Spectral Regions	Measurement Type	Kind of Analysis	Applicable Samples
Near-IR	Diffuse reflectance	Quantitative	Solid or liquid commercial materials
	Absorption	Quantitative	Gaseous mixtures
Mid-IR	Absorption	Qualitative	Pure solid, liquid, or gases
		Quantitative	Complex liquid, solid, or gaseous mixtures
		Chromatographic	Complex liquid, solid, or gaseous mixtures
	Reflectance	Qualitative	Pure solids or liquids°
	Emission	Quantitative	Atmospheric samples
Far-IR	Absorption	Qualitative	Pure inorganic or organometallic species

techniques for preparation of samples for IR absorption measurements.

Gases

The spectrum of a low-boiling-point liquid or gas can be obtained by permitting the sample to expand into an evacuated cylindrical cell equipped with suitable windows. For this purpose, a variety of cylindrical cells are available with path lengths that range from a few centimeters to 10 m or more. The longer path lengths are obtained in compact cells by providing reflecting internal surfaces, so that the beam makes numerous passes through the sample before exiting the cell (see Figure 16-13).

Solutions

When feasible, a convenient way of obtaining IR spectra is on solutions prepared to contain a known concentration of sample, as is generally done in ultraviolet-visible spectrometry. This technique is somewhat limited in its applications, however, by the availability of solvents transparent over significant regions in the IR.

Solvents. Figure 17-1 lists several common solvents employed for IR studies of organic compounds. This figure illustrates that no single solvent is transparent throughout the entire mid-IR region.

Water and the alcohols are difficult to use as solvents in IR spectrometry. Water shows several strong absorption bands in the IR region, as can be seen in Figure 17-2. Here, the spectrum of water is shown along with the spectrum of an aqueous solution of aspirin. The computer-calculated difference spectrum reveals the spectrum of the water-soluble aspirin. Water and alcohols also attack alkali-metal halides, the most common materials used for cell windows. Hence, water-insoluble window materials, such as barium fluoride, must be used with such solvents. Care must also be taken to dry the solvents shown in Figure 17-1 before use with typical cells.

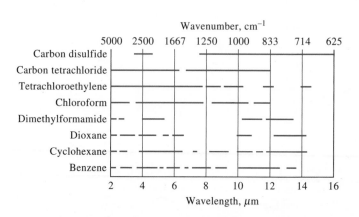

FIGURE 17-1 Solvents for IR spectroscopy. Horizontal lines indicate useful regions.

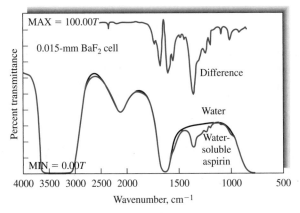

FIGURE 17-2 IR spectra of water and of an aqueous solution containing water-soluble aspirin. The top spectrum shows the computer-generated difference spectrum obtained by subtracting the water spectrum from that of the solution. (From *Practical Sampling Techniques for Infrared Analysis*, P. B. Coleman, ed., Boca Raton, FL: CRC Press, 1993. With permission.)

Cells. Because of the tendency for solvents to absorb IR radiation, IR liquid cells are ordinarily much narrower (0.01 to 1 mm) than those employed in the ultraviolet and visible regions. Often, relatively high sample concentrations (from 0.1% to 10%) are required because of the short path lengths and the low molar absorptivities of IR bands. Liquid cells are frequently designed for easy disassembly and use Teflon spacers to allow variation in path length (see Figure 17-3). Fixed-path-length cells can be filled or emptied with a hypodermic syringe.

A variety of window materials are available as listed in Table 17-2. Window materials are often chosen on

the basis of cost, range of transparency, solubility in the solvent, and reactivity with the sample or solvent. Sodium chloride and potassium bromide windows are most commonly employed and are the least expensive. Even with care, however, their surfaces eventually become fogged because of absorption of moisture. Polishing with a buffing powder returns them to their original condition.

For samples that are wet or aqueous, calcium and barium fluoride are suitable, although neither transmits well throughout the entire mid-IR region. Silver bromide is often used, although it is more expensive, is

TABLE 17-2 Common IR Window Materials

Window Material	Applicable Range, cm^{-1}	Water Solubility, $g/100\ g\ H_2O$, 20°C
Sodium chloride	40,000–625	36.0
Potassium bromide	40,000–385	65.2
Potassium chloride	40,000–500	34.7
Cesium iodide	40,000–200	160.0
Fused silica	50,000–2,500	Insoluble
Calcium fluoride	50,000–1,100	1.51×10^{-3}
Barium fluoride	50,000–770	0.12 (25°C)
Thallium bromide-iodide, KRS-5	16,600–250	<0.0476
Silver bromide	20,000–285	1.2×10^{-7}
Zinc sulfide, Irtran-2	10,000–715	Insoluble
Zinc selenide, Irtran-4	10,000–515	Insoluble
Polyethylene	625–30	Insoluble

From *Practical Sampling Techniques for Infrared Analysis*, P. B. Coleman, ed., Boca Raton, FL: CRC Press, 1993.

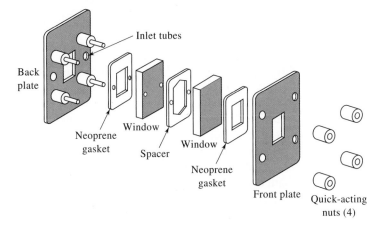

FIGURE 17-3 Expanded view of a demountable IR cell for liquid samples. Teflon spacers ranging from 0.015 to 1 mm thick are available. (Courtesy of Perkin-Elmer Corp., Norwalk, CT.)

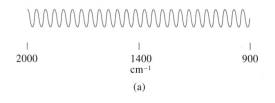

2000	1400	900
	cm^{-1}	

(a)

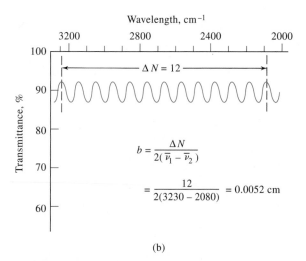

(b)

FIGURE 17-4 Determination of the path length of IR cells from interference fringes. Fringes are observed with an empty cell in the light path (a). In (b) an illustration of the calculation of the path length is given.

photosensitive, and is easily scratched. The Irtran materials are also resistant to chemical attack and are water insoluble. They are somewhat expensive and have high indexes of refraction (2.2 to 2.4), which can lead to interference fringes because of mismatches with the sample index of refraction (often around 1.4 to 1.6).

Figure 17-4 shows how the thickness b of very narrow IR cells can be determined by the interference fringe method. The interference pattern of Figure 17-4a is obtained by inserting an empty cell in the light path of the sample beam and recording the transmittance against air in the reference beam. The reflected radiation from the cell walls interferes with the transmitted beam to produce the interference maxima and minima. The maxima occur when the radiation reflected from the two internal surfaces of the cell has traveled a distance that is an integral multiple N of the wavelength of the radiation transmitted without reflection. Constructive interference then occurs whenever the wavelength is equal to $2b/N$. That is,

$$\frac{2b}{N} = \lambda \qquad (17\text{-}1)$$

As shown in Figure 17-4b, the number of interference fringes ΔN between two known wavelengths λ_1 and λ_2 are counted and introduced into the equation

$$\Delta N = \frac{2b}{\lambda_1} - \frac{2b}{\lambda_2} = 2b\bar{\nu}_1 - 2b\bar{\nu}_2$$

or

$$b = \frac{\Delta N}{2(\bar{\nu}_1 - \bar{\nu}_2)} \qquad (17\text{-}2)$$

It should be noted that interference fringes are ordinarily not seen when a cell is filled with liquid because the refractive index of most liquids approaches that of the window material and reflection is minimized (Equation 6-15). On the other hand, interference can be observed between 2800 and 2000 cm^{-1} in Figure 16-1. Here, the sample is a sheet of polystyrene, which has a refractive index considerably different from that of air. For this reason, significant reflection occurs at the two interfaces of the sheet. Equation 17-2 is often used to calculate the thickness of thin polymer films.

Liquids

When the amount of a liquid sample is small or when a suitable solvent is unavailable, it is common practice to obtain spectra on the pure (neat) liquid. Here, only a very thin film has a sufficiently short path length to produce satisfactory spectra. Commonly, a drop of the neat liquid is squeezed between two rock-salt plates to give a layer 0.015 mm thick or less. The two plates, held together by capillary action, are then placed in the beam path. This technique does not give particularly reproducible transmittance data, but the resulting spectra are usually satisfactory for qualitative investigations.

Solids

Most organic compounds exhibit numerous absorption bands throughout the mid-IR region, and finding a solvent that does not have overlapping peaks is often impossible. Because of this, spectra are often obtained on dispersions of the solid in a liquid or solid matrix. Generally, in these techniques, the solid sample must be ground until its particle size is less than the wavelength of the radiation to avoid the effects of scattered radiation.

Pelleting. One of the most popular techniques for handling solid samples has been KBr pelleting (other alkali-metal halides have also been used). Halide salts have the property of cold flow, in which they have glasslike transparent or translucent properties when sufficient pressure is applied to the finely powdered materials. In using this technique, a milligram or less of the finely ground sample is intimately mixed with about 100 mg of dried potassium bromide powder. Mixing can be carried out with a mortar and pestle or, better, in a small ball mill. The mixture is then pressed in a special die at 10,000 to 15,000 pounds per square inch to yield a transparent disk. Best results are obtained if the disk is formed in a vacuum to eliminate occluded air. The disk is then held in the instrument beam for spectroscopic examination. The resulting spectra frequently exhibit bands at 3450 and 1640 cm^{-1} (2.9 and 6.1 μm) due to absorbed moisture.

With many compounds, KBr pelleting produces excellent spectra that appear in many spectral libraries. Being ionic, KBr transmits throughout most of the IR region with a lower cutoff of about 400 cm^{-1}. Ion exchange can occur with some samples such as amine hydrochlorides or inorganic salts. With the former, bands of the amine hydrobromide are often found. Polymorphism can also occur because of the forces involved in grinding and pressing the pellet. These can convert one polymorph into another. Although KBr is the most frequently used pelleting salt, materials such as CsI and CsBr are sometimes used. Cesium iodide has greater transparency at low frequencies than KBr.

Mulls. IR spectra of solids that are not soluble in an IR-transparent solvent or are not conveniently pelleted in KBr are often obtained by dispersing the analyte in a mineral oil or a fluorinated hydrocarbon *mull*. Mulls are formed by grinding 2 to 5 mg of the finely powdered sample (particle size <2 μm) in the presence of one or two drops of a heavy hydrocarbon oil (Nujol). If hydrocarbon bands are likely to interfere, Fluorolube, a halogenated polymer, can be used instead. In either case, the resulting mull is then examined as a film between flat salt plates.

Other Methods for Solids. The IR behavior of solids can also be obtained by reflectance techniques and by the photoacoustic method. These spectra are often similar to absorption spectra and provide the same kinds of information. These techniques are discussed in Sections 17B and 17C.

17A-2 Qualitative Analysis

The general use of mid-IR spectroscopy by chemists for identifying organic compounds began in the late 1950s with the appearance of inexpensive and easy-to-use double-beam recording spectrophotometers that produce spectra in the region of 5000 to 670 cm^{-1} (2 to 15 μm). The appearance of this type of instrument (as well as nuclear magnetic resonance and mass spectrometers) revolutionized the way chemists identify organic, inorganic, and biological species. Suddenly, the time required to perform a qualitative analysis or structural determination was substantially reduced.

Figure 17-5 shows four typical spectra obtained with an inexpensive double-beam instrument. Identification of an organic compound from a spectrum of this kind is a two-step process. The first step involves determining what functional groups are most likely present by examining the *group frequency region*, which encompasses radiation from about 3600 cm^{-1} to approximately 1250 cm^{-1} (see Figure 17-5). The second step involves a detailed comparison of the spectrum of the unknown with the spectra of pure compounds that contain all of the functional groups found in the first step. Here, the fingerprint region, from 1200 to 600 cm^{-1} (Figure 17-5) is particularly useful because small differences in the structure and constitution of a molecule result in significant changes in the appearance and distribution of absorption bands in this region. Consequently, a close match between two spectra in the fingerprint region (as well as others) constitutes almost certain evidence that the two compounds are identical.

Group Frequencies

We have noted that the approximate frequency (or wavenumber) at which an organic functional group, such as C$=$O, C$=$C, C$-$H, C$\equiv$C, or O$-$H, absorbs IR radiation can be calculated from the masses of the atoms and the force constant of the bond between them (Equations 16-14 and 16-15). These frequencies, called *group frequencies*, are seldom totally invariant because of interactions with other vibrations associated with one or both of the atoms composing the group. On the other hand, such interaction effects are ordinarily small; as a result, a range of frequencies can be assigned within which it is highly probable that the absorption maximum for a given functional group will be found. Table 17-3 lists group frequencies for several common functional groups. A more detailed presentation of group frequencies is found in the *correlation chart* shown in Figure 17-6.[3] Note that, although most group frequencies fall in the range of 3600 to 1250 cm^{-1}, a few fall in the fingerprint region. These include the C—O—C stretching vibration at about 1200 cm^{-1} and the C—Cl stretching vibration at 700 to 800 cm^{-1}.

Several group frequencies are identified in the four spectra in Figure 17-5. All four spectra contain an

absorption band at 2900 to 3000 cm^{-1}, which corresponds to a C—H stretching vibration and generally indicates the presence of one or more alkane groups (see Table 17-3). The two peaks at about 1375 cm^{-1} and 1450 cm^{-1} are also characteristic group frequencies for C—H groups and result from bending vibrations in the molecule. The spectrum in Figure 17-5c illustrates the group frequency for an O—H stretching vibration at about 3200 cm^{-1} as well as the alkane group frequencies (these bands are also present in the spectrum for *n*-hexanal shown in Figure 16-12).

Finally, the characteristic group frequency for a C—Cl stretching vibration is shown at about 800 cm^{-1} in Figure 17-5d.

[3]Other correlation charts can be found in R. M. Silverstein, F. X. Webster, and D. Kienle, *Spectrometric Identification of Organic Compounds*, 7th ed., Chap. 2, New York: Wiley, 2005.

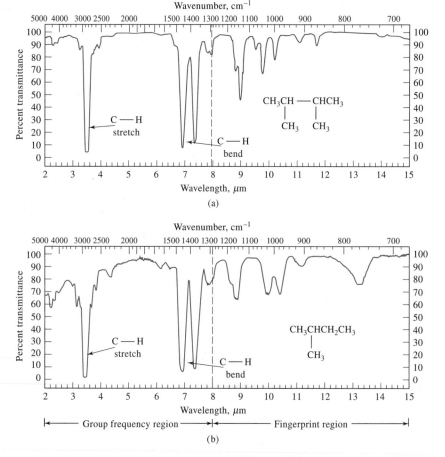

FIGURE 17-5 The group frequency region of the mid-IR (~3600–1250 cm^{-1}) is used to identify common functional groups. The fingerprint region (~1200–600 cm^{-1}) is used to identify compounds. (From R. M. Roberts, J. C. Gilbert, L. B. Rodewald, and A. S. Wingrove, *Modern Experimental Organic Chemistry*, 4th ed., Philadelphia: Saunders, 1985. With permission.)

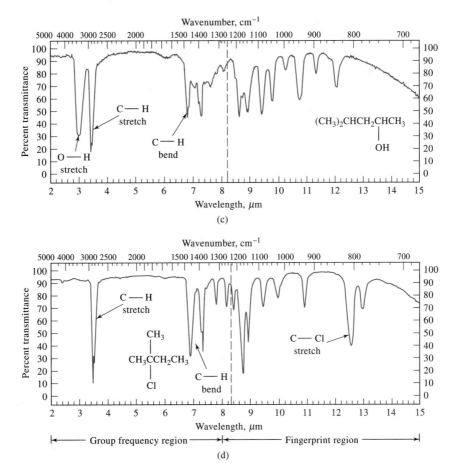

FIGURE 17-5 (*continued*)

TABLE 17-3 Abbreviated Table of Group Frequencies for Organic Functional Groups

Bond	Type of Compound	Frequency Range, cm⁻¹	Intensity
C—H	Alkanes	2850–2970	Strong
		1340–1470	Strong
C—H	Alkenes ($>C=C<^H$)	3010–3095	Medium
		675–995	Strong
C—H	Alkynes (—C≡C—H)	3300	Strong
C—H	Aromatic rings	3010–3100	Medium
		690–900	Strong
O—H	Monomeric alcohols, phenols	3590–3650	Variable
	Hydrogen-bonded alcohols, phenols	3200–3600	Variable, sometimes broad
	Monomeric carboxylic acids	3500–3650	Medium
	Hydrogen-bonded carboxylic acids	2500–2700	Broad
N—H	Amines, amides	3300–3500	Medium
C=C	Alkenes	1610–1680	Variable
C=C	Aromatic rings	1500–1600	Variable
C≡C	Alkynes	2100–2260	Variable
C—N	Amines, amides	1180–1360	Strong
C≡N	Nitriles	2210–2280	Strong
C—O	Alcohols, ethers, carboxylic acids, esters	1050–1300	Strong
C=O	Aldehydes, ketones, carboxylic acids, esters	1690–1760	Strong
NO₂	Nitro compounds	1500–1570	Strong
		1300–1370	Strong

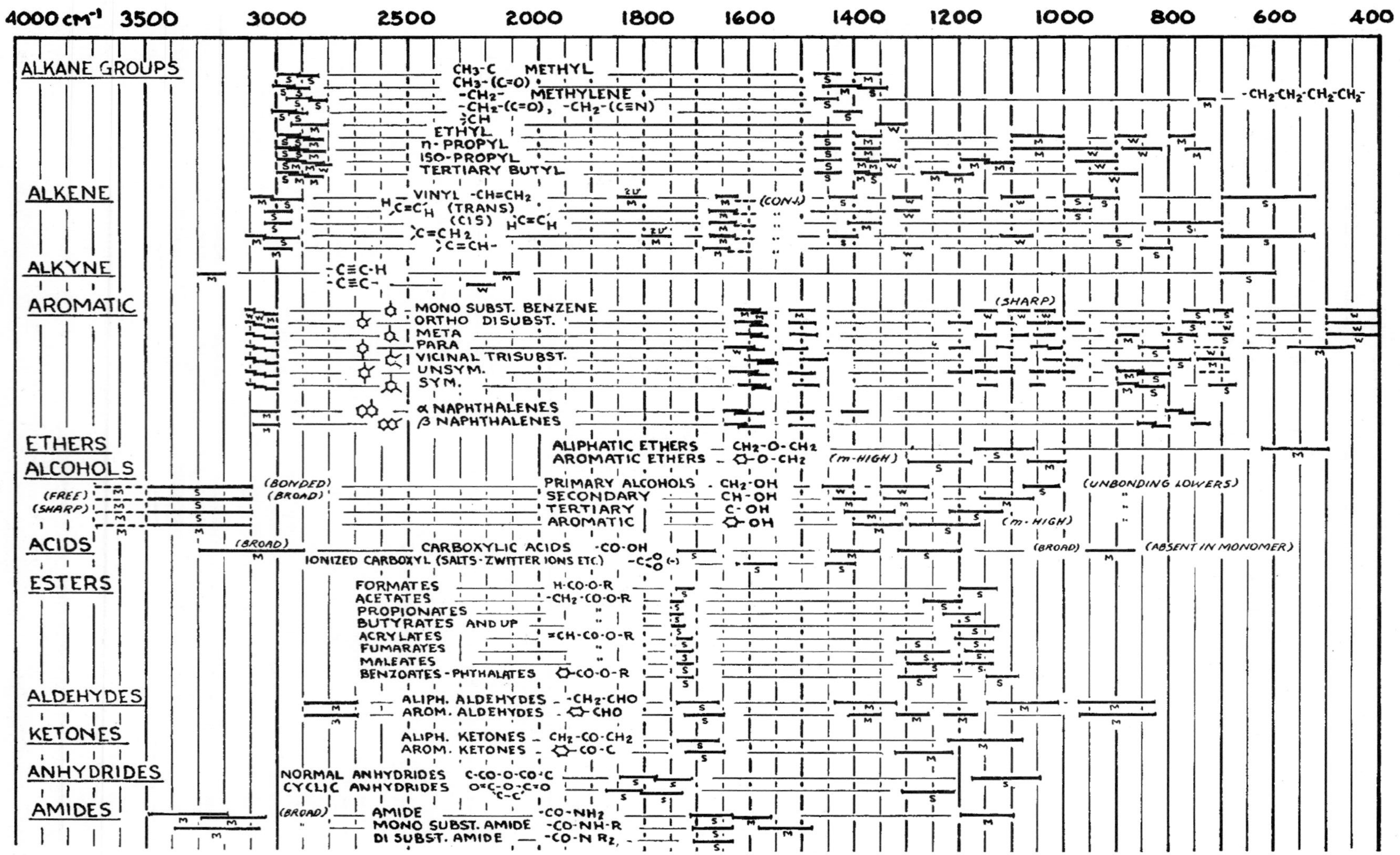
4000 cm⁻¹ 3500 3000 2500 2000 1800 1600 1400 1200 1000 800 600 400
ALKANE GROUPS
CH3-C METHYL
CH3-(C=O)
-CH2- METHYLENE
-CH2-(C=O), -CH2-(C≡N)
≥CH
ETHYL
n-PROPYL
ISO-PROPYL
TERTIARY BUTYL
-CH2-CH2-CH2-CH2-
ALKENE
VINYL -CH=CH2
(TRANS)
(CIS)
>C=CH2
>C=CH-
(CONJ)
ALKYNE
-C≡C-H
-C≡C-
AROMATIC
MONO SUBST. BENZENE
ORTHO DISUBST.
META
PARA
VICINAL TRISUBST.
UNSYM.
SYM.
α NAPHTHALENES
β NAPHTHALENES
(SHARP)
ETHERS
ALIPHATIC ETHERS - CH2-O-CH2
AROMATIC ETHERS - ⌬-O-CH2
(m-HIGH)
ALCOHOLS
(FREE)
(SHARP)
(BONDED)
(BROAD)
PRIMARY ALCOHOLS - CH2-OH
SECONDARY CH-OH
TERTIARY C-OH
AROMATIC ⌬-OH
(UNBONDING LOWERS)
(m-HIGH)
ACIDS
(BROAD)
CARBOXYLIC ACIDS -CO-OH
IONIZED CARBOXYL (SALTS-ZWITTER IONS ETC.)
(BROAD)
(ABSENT IN MONOMER)
ESTERS
FORMATES H-CO-O-R
ACETATES -CH2-CO-O-R
PROPIONATES
BUTYRATES AND UP
ACRYLATES =CH-CO-O-R
FUMARATES
MALEATES
BENZOATES-PHTHALATES ⌬-CO-O-R
ALDEHYDES
ALIPH. ALDEHYDES - -CH2-CHO
AROM. ALDEHYDES - ⌬-CHO
KETONES
ALIPH. KETONES - CH2-CO-CH2
AROM. KETONES - ⌬-CO-C
ANHYDRIDES
NORMAL ANHYDRIDES C-CO-O-CO-C
CYCLIC ANHYDRIDES
AMIDES
(BROAD)
AMIDE -CO-NH2
MONO SUBST. AMIDE -CO-NH-R
DI SUBST. AMIDE -CO-N R2

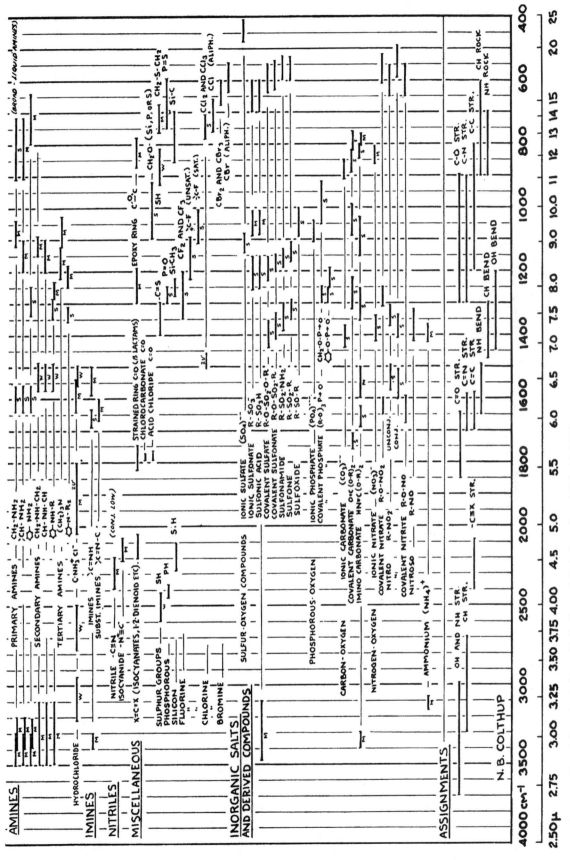

FIGURE 17-6 Correlation chart. (From N. Colthup, *J. Opt. Soc. Am.*, 1950, 40, 397.)

Group frequencies and correlation charts permit intelligent guesses as to what functional groups are likely to be present or absent in a molecule. Ordinarily, it is impossible to identify unambiguously either the sources of all of the bands in a given spectrum or the exact identity of the molecule. Instead, group frequencies and correlation charts serve as a starting point in the identification process.

The Fingerprint Region

Small differences in the structure and constitution of a molecule result in significant changes in the distribution of absorption maxima in this region of the spectrum that extends from about 1200 to 600 cm^{-1} (8 to 14 μm). The fingerprint region is thus well suited for identifying compounds based on spectral comparisons. Most single bonds give rise to absorption bands at these frequencies; because their energies are about the same, strong interaction occurs between neighboring bonds. The absorption bands are thus composites of these various interactions and depend on the overall skeletal structure of the molecule. Exact interpretation of spectra in this region is seldom possible because of the complexity of the spectra. However, for identification purposes, the richness of spectral features is a distinct advantage. Many compounds show unique absorption bands in this region, which is very useful for final identification.

Figure 17-5a and b illustrates the unique character of IR spectra, particularly in the fingerprint region. The two molecules shown differ by just one methyl group, yet the two spectra differ dramatically in the fingerprint region.

As shown in Figure 17-6, a number of inorganic groups such as sulfate, phosphate, nitrate, and carbonate absorb IR radiation in the fingerprint region (<1200 cm^{-1} or >8.3 μm).

Limitations to the Use of Correlation Charts

The unambiguous establishment of the identity or the structure of a compound is seldom possible from correlation charts alone. Uncertainties frequently arise from overlapping group frequencies, spectral variations as a function of the physical state of the sample (that is, whether it is a solution, a mull, in a pelleted form, and so forth), and instrumental limitations.

In using group frequencies, it is essential that the entire spectrum, rather than a small isolated portion, be considered and interrelated. Interpretation based on one part of the spectrum should be confirmed or rejected by studying other regions.

To summarize, correlation charts serve only as a guide for further and more careful study. Several excellent monographs describe the absorption characteristics of functional groups in detail.[4] A study of these characteristics, as well as the other physical properties of the sample, may permit unambiguous identification. IR spectroscopy, when used in conjunction with mass spectrometry, nuclear magnetic resonance, and elemental analysis, usually leads to positive identification of a species.

Collections of Spectra

As just noted, correlation charts seldom suffice for the positive identification of an organic compound from its IR spectrum. However, several catalogs of IR spectra are available that assist in qualitative identification by providing comparison spectra for a large number of pure compounds.[5] Manually searching large catalogs of spectra is slow and tedious. For this reason, computer-based search systems are widely used.

Computer Search Systems

Virtually all IR instrument manufacturers now offer computer search systems to aid in identifying compounds from stored IR spectral data.[6] The position and relative magnitudes of peaks in the spectrum of the analyte are determined and stored in memory to give a peak profile, which can be compared with profiles of pure compounds stored on disk or CD-ROM. The computer then matches profiles and prints a list of compounds having spectra similar to that of the analyte. Usually, the spectrum of the analyte and that of each potential match can be shown simultaneously on the computer display for comparison, as shown in Figure 17-7.

In 1980 the Sadtler Standard IR Collection and the Sadtler Commercial IR Collection became available as

---

[4]R. M. Silverstein, F. X. Webster, and D. Kienle, *Spectrometric Identification of Organic Compounds*, 7th ed., New York: Wiley, 2005; B. Schrader, *Infrared and Raman Spectroscopy*, New York: VCH, 1995; N. B. Colthup, L. H. Daly, and S. E. Wiberley, *Introduction to Infrared and Raman Spectroscopy*, 3rd ed., San Diego: Academic Press, 1990.
[5]See Informatics/Sadtler Group, Bio-Rad Laboratories, Inc., Philadelphia, PA; C. J. Pouchert, *The Aldrich Library of Infrared Spectra*, 3rd ed., Milwaukee, WI: Aldrich Chemical Co., 1981; Thermo Galactic, *Spectra Online* (http://spectra.galactic.com).
[6]See E. Pretsch, G. Toth, M. E. Monk, and M. Badertscher, *Computer-Aided Structure Elucidation: Spectra Interpretation and Structure Generation*, New York: VCH-Wiley, 2003; B. C. Smith, *Infrared Spectral Interpretation: A Systematic Approach*, Boca Raton, FL: CRC Press, 1998, Chap. 9; W. O. George and H. Willis, *Computer Methods in UV, Visible and IR Spectroscopy*, New York: Springer-Verlag, 1990.

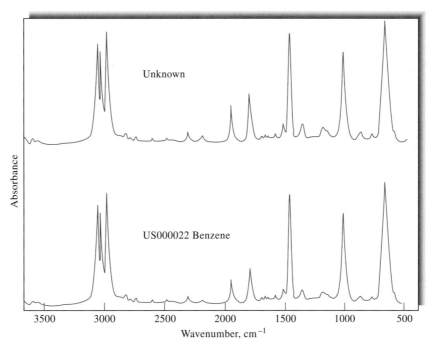

FIGURE 17-7 Plot of an unknown spectrum and the best match from a computer search report. (Courtesy of Informatics/Sadtler Group, Bio-Rad Laboratories, Inc., Philadelphia, PA.)

software packages. Currently, this library contains more than 220,000 spectra.[7] Included are vapor-phase spectra and condensed-phase spectra of pure compounds. Individual databases in application areas such as polymers, industrial products, pure organic compounds, forensic sciences, and the environment are available. Several manufacturers of Fourier transform IR (FTIR) instruments have now incorporated these packages into their instrument computers, thus creating instantly available IR libraries of hundreds of thousands of compounds.

The Sadtler algorithm consists of a search system in which the spectrum of the unknown compound is first coded according to the location of its strongest absorption peak; then each additional strong band ($\%T < 60\%$) in ten regions 200 cm^{-1} wide from 4000 to 2000 cm^{-1} are coded by their location. Finally, the strong bands in seventeen regions 100 cm^{-1} wide from 2100 to 400 cm^{-1} are coded in a similar way. The compounds in the library are coded in this same way. The data are organized by the location of the strongest band with only those compounds having the same strongest

band being considered in any sample identification. This procedure is rapid and produces a list of potential matches within a short period. For example, only a few seconds are required to search a base library of 50,000 compounds by this procedure.

In addition to search systems, artificial intelligence-based computer programs seek to determine the structure or substructure from spectral profiles. Some of these programs use data from different types of instruments (nuclear magnetic resonance, mass, and IR spectrometers, etc.) to determine structures.[8]

17A-3 Quantitative Applications

Quantitative IR absorption methods differ somewhat from ultraviolet-visible molecular spectroscopic methods because of the greater complexity of the spectra, the narrowness of the absorption bands, and the instrumental limitations of IR instruments. Quantitative data obtained with older dispersive IR instruments were generally significantly inferior in quality to data obtained with UV-visible spectrophotometers. The precision and accuracy of measurements with modern FTIR instruments, however, is distinctly better than those

---

[7]See note 5.

 Tutorial: Learn more about **IR spectral interpretation and identification**.

---

[8]See, for example, M. E. Munk, *J. Chem. Inf. Comput. Sci.*, **1998**, *38*, 997.

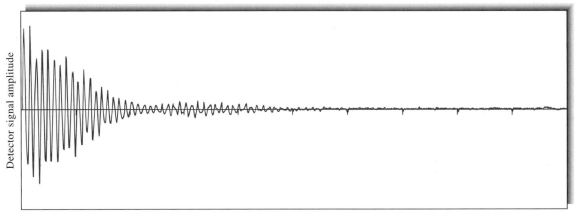

(a)

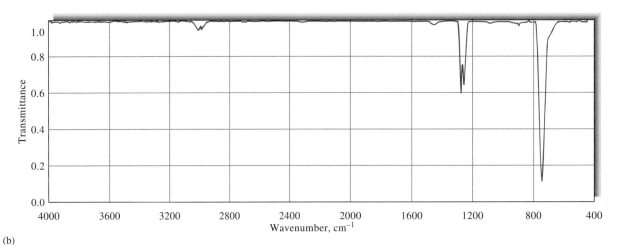

(b)

FIGURE 17-8 (a) Interferogram obtained from a typical FTIR spectrometer for methylene chloride. The plot shows detector signal output as a function of time or displacement of the moving mirror of the interferometer. (b) IR spectrum of methylene chloride produced by the Fourier transformation of the data in (a). Note that the Fourier transform takes signal intensity collected as a function of time and produces transmittance as a function of frequency after subtraction of a background interferogram and proper scaling.

with dispersive instruments. Meticulous attention to detail is, however, essential for obtaining good-quality results.[9]

Deviations from Beer's Law

With IR radiation, instrumental deviations from Beer's law are more common than with ultraviolet and visible wavelengths because IR absorption bands are

relatively narrow. Furthermore, with dispersive instruments, the low intensity of sources and low sensitivities of transducers in this region require the use of relatively wide monochromator slit widths; thus, the bandwidths employed are frequently of the same order of magnitude as the widths of absorption peaks. We have pointed out (Section 13B-2) that this combination of circumstances usually leads to a nonlinear relationship between absorbance and concentration. As discussed in Section 16B-1, FTIR instruments have better performance characteristics. Hence, Beer's law deviations are not quite as serious as with dispersive instru-

---

[9]For a discussion of quantitative FTIR spectroscopy, see B. C. Smith, *Fundamentals of Fourier Transform Infrared Spectroscopy*, Boca Raton, FL: CRC Press, 1995; P. R. Griffiths and J. A. deHasech, *Fourier Transform Infrared Spectrometry*, New York: Wiley, 1986.

ments. In either case, quantitative analysis usually requires that empirical calibration curves be used.

Absorbance Measurement

Matched absorption cells for solvent and solution are ordinarily employed in the ultraviolet and visible regions, and the measured absorbance is then found from the relation

$$A = \log \frac{P_{solvent}}{P_{solution}}$$

The use of the solvent in a matched cell as a reference absorber has the advantage of largely canceling out the effects of radiation losses due to reflection at the various interfaces, scattering and absorption by the solvent, and absorption by the container windows. This technique is seldom practical for measurements in the IR region because of the difficulty in obtaining cells whose transmission characteristics are identical. Most IR cells have very short path lengths, which are difficult to duplicate exactly. In addition, the cell windows are readily attacked by contaminants in the atmosphere and the solvent. Because of this, their transmission characteristics change continually with use. For these reasons, a reference absorber is often dispensed with entirely in IR work, and the intensity of the radiation passing through the sample is simply compared with that of the unobstructed beam. In either case, the resulting transmittance is ordinarily less than 100%, even in regions of the spectrum where no absorption by the sample occurs (see Figure 17-5).

For quantitative work, it is necessary to correct for the scattering and absorption by the solvent and the cell. Two methods are employed. In the so-called cell-in-cell-out procedure, spectra of the pure solvent and the analyte solution are obtained successively with respect to the unobstructed reference beam. The same cell is used for both measurements. The transmittance of each solution versus the reference beam is then determined at an absorption maximum of the analyte. These transmittances can be written as

$$T_0 = P_0/P_r$$

and

$$T_s = P/P_r$$

where P_r is the power of the unobstructed beam and T_0 and T_s are the transmittances of the solvent and analyte solution, respectively, against this reference. If P_r remains constant during the two measurements, then the transmittance of the sample with respect to the solvent can be obtained by division of the two equations. That is,

$$T = T_s/T_0 = P/P_0$$

With modern FTIR spectrometers, the reference interferogram is obtained with no sample in the sample cell. Then, the sample is placed in the cell, and a second interferogram obtained. Figure 17-8a shows an interferogram collected using an FTIR spectrometer with methylene chloride, CH_2Cl_2, in the sample cell. The Fourier transform is then applied to the two interferograms to compute the IR spectra of the reference and the sample. The ratio of the two spectra can then be computed to produce an IR spectrum of the analyte such as the one illustrated in Figure 17-8b.

An alternative way of obtaining P_0 and T for a single absorption band is the baseline method. For an instrument that displays transmittance, the solvent transmittance is assumed to be constant or at least to change linearly between the shoulders of the absorption peak, as illustrated in Figure 17-9. The quantities T_0 and T_s are then obtained as shown in the figure. For direct absorbance readout, as is shown in Figure 17-7, the absorbance is assumed to be constant or to change linearly under the absorption band. The peak absorbance is then obtained by subtracting the baseline absorbance.

Typical Applications

With the exception of homonuclear molecules, all organic and inorganic molecular species absorb radiation in the IR region. IR spectrophotometry thus offers the potential for determining an unusually large number of substances. Moreover, the uniqueness of an IR spectrum leads to a degree of specificity that is

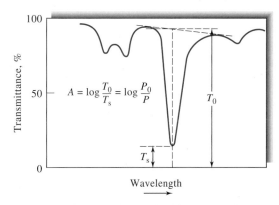

FIGURE 17-9 The baseline method for determining the absorbance of an absorption maximum.

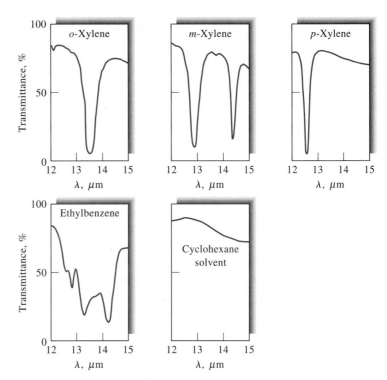

FIGURE 17-10 Spectra of C_8H_{10} isomers in cyclohexane.

matched or exceeded by relatively few other analytical methods. This specificity has found particular application to analysis of mixtures of closely related organic compounds. Two examples that typify these applications follow.

Analysis of a Mixture of Aromatic Hydrocarbons. A typical application of quantitative IR spectroscopy involves the resolution of C_8H_{10} isomers in a mixture that includes *o*-xylene, *m*-xylene, *p*-xylene, and ethylbenzene. The IR absorption spectra of the individual components in the range of 12 to 15 μm in cyclohexane solvent is shown in Figure 17-10. Useful absorption bands for determination of the individual compounds occur at 13.47, 13.01, 12.58, and 14.36 μm, respectively. Unfortunately, however, the absorbance of a mixture at any one of these wavelengths is not entirely determined by the concentration of just one component because of overlapping absorption bands. Thus, molar absorptivities for each of the four compounds must be determined at the four wavelengths. Then, four simultaneous equations can be written that permit the calculation of the concentration of each species from four absorbance measurements (see Section 14D-2). Alternatively, chemometric techniques, such as factor analytical

methods,[10] can use entire spectral regions to determine the individual components. Such methods can be used even when the relationship between absorbance and concentration is nonlinear (as frequently occurs in the IR region).

Determination of Air Contaminants. The recent proliferation of government regulations with respect to atmospheric contaminants has demanded the development of sensitive, rapid, and highly specific methods

[10] E. R. Malinowski, *Factor Analysis in Chemistry*, 3rd ed., New York: Wiley, 2002.

TABLE 17-4 An Example of IR Determinations of Air Contaminants

Contaminant	Concn, ppm	Found, ppm	Relative Error, %
Carbon monoxide	50.0	49.1	1.8
Methyl ethyl ketone	100.0	98.3	1.7
Methanol	100.0	99.0	1.0
Ethylene oxide	50.0	49.9	0.2
Chloroform	100.0	99.5	0.5

Courtesy of Thermo Electron Corp.

TABLE 17-5 IR Determination of Gases for OSHA Compliance

Compound	Allowable Exposure, ppm*	Wavelength, μm	Minimum Detectable Concentration, ppm[†]
Carbon disulfide	4	4.54	0.5
Chloroprene	10	11.4	4
Diborane	0.1	3.9	0.05
Ethylenediamine	10	13.0	0.4
Hydrogen cyanide	4.7[‡]	3.04	0.4
Methyl mercaptan	0.5	3.38	0.4
Nitrobenzene	1	11.8	0.2
Pyridine	5	14.2	0.2
Sulfur dioxide	2	8.6	0.5
Vinyl chloride	1	10.9	0.3

Courtesy of Thermo Electron Corp.

*1992–1993 OSHA exposure limits for 8-h weighted average.

[†]For a 20.25-m cell.

[‡]Short-term exposure limit: 15-min time-weighted average that shall not be exceeded at any time during the work day.

for a variety of chemical compounds. IR absorption procedures appear to meet this need better than any other single analytical tool.

Table 17-4 demonstrates the potential of IR spectroscopy for the analysis of mixtures of gases. The standard sample of air containing five species in known concentration was analyzed with a computerized version of the instrument shown in Figure 16-13; a 20-m gas cell was employed. The data were printed out within a minute or two after sample introduction.

Table 17-5 shows potential applications of IR filter photometers (such as that shown in Figure 16-13) for the quantitative determination of various chemicals in the atmosphere for the purpose of assuring compliance with Occupational Safety and Health Administration (OSHA) regulations.

Of the more than 400 chemicals for which maximum tolerable limits have been set by OSHA, more than half appear to have absorption characteristics suitable for determination by means of IR filter photometers or spectrophotometers. With such a large number of absorbing compounds, we expect overlapping absorption bands. However, the method can provide a moderately high degree of selectivity.

Disadvantages and Limitations to Quantitative IR Methods

There are several disadvantages to quantitative analysis by IR spectrometry. Among these are the frequent nonadherence to Beer's law and the complexity of spectra.

The richness of spectral features enhances the probability of overlapping absorption bands. In addition, with older dispersive instruments, the narrowness of bands and the effects of stray radiation make absorbance measurements critically dependent on the slit width and the wavelength setting. Finally, the narrow-path-length cells required for many analyses are inconvenient to use and may lead to significant analytical uncertainties. For these reasons, the analytical errors associated with a quantitative IR analysis often cannot be reduced to the level associated with ultraviolet and visible methods, even with considerable care and effort.

17B MID-IR REFLECTION SPECTROMETRY

IR reflection spectrometry has found a number of applications, particularly for dealing with solid samples that are difficult to handle, such as polymer films and fibers, foods, rubbers, agriculture products, and many others.[11] Mid-IR reflection spectra, although not identical to the corresponding absorption spectra, are similar in general appearance and provide the same information as do their absorption counterparts. Reflectance spectra can be used for both qualitative and

[11]See F. M. Mirabella, ed., *Modern Techniques in Applied Molecular Spectroscopy*, New York: Wiley, 1998; G. Kortum, *Reflectance Spectroscopy*, New York: Springer, 1969; N. J. Harrick, *Internal Reflection Spectroscopy*, New York: Wiley, 1967.

quantitative analysis. Most instrument manufacturers now offer adapters that fit into the cell compartments of IR absorption instruments and make it possible to obtain reflection spectra readily.

17B-1 Types of Reflection

Reflection of radiation is of four types: *specular reflection*, *diffuse reflection*, *internal reflection*, and *attenuated total reflectance* (ATR). Specular reflection occurs when the reflecting medium is a smooth polished surface. Here, the angle of reflection is identical to the incident angle of the radiation. If the surface is made up of an IR absorber, the relative intensity of reflection is less for wavelengths that are absorbed than for wavelengths that are not. Thus, a plot of reflectance R, which is the fraction of the incident radiant energy reflected, versus wavelength or wavenumber provides a spectrum for a compound similar in general appearance to a transmission spectrum for the species. Specular reflection spectra find some use for examining and characterizing the smooth surfaces of solids and coated solids but are not as widely used as diffuse- and total-reflection spectra. We will, therefore, focus on the latter two types of spectra.

17B-2 Diffuse-Reflectance Spectrometry

Diffuse-reflectance IR Fourier transform spectrometry (DRIFTS) is an effective way of directly obtaining IR spectra on powdered samples with a minimum of sample preparation.[12] In addition to saving time in sample preparation, it permits conventional IR spectral data to be gathered on samples not appreciably altered from their original state. The widespread use of diffuse-reflectance measurements had to await the general availability of FTIR instruments in the mid-1970s because the intensity of radiation reflected from powders is too low to be measured at medium resolution and adequate signal-to-noise ratios with dispersive instruments.

Diffuse reflection is a complex process that occurs when a beam of radiation strikes the surface of a finely divided powder. With this type of sample, specular reflection occurs at each plane surface. However, because there are many of these surfaces and they are randomly oriented, radiation is reflected in all directions. Typically, the intensity of the reflected radiation is roughly independent of the viewing angle.

A number of models have been developed to describe in quantitative terms the intensity of diffuse reflected radiation. The most widely used of these models was developed by Kubelka and Munk.[13] Fuller and Griffiths in their discussion of this model show that the relative reflectance intensity for a powder $f(R'_\infty)$ is given by[14]

$$f(R'_\infty) = \frac{(1 - R'_\infty)^2}{2R'_\infty} = \frac{k}{s}$$

where R'_∞ is the ratio of the reflected intensity of the sample to that of a nonabsorbing standard, such as finely ground potassium chloride. The quantity k is the molar absorption coefficient of the analyte, and s is a scattering coefficient. For a diluted sample, k is related to the molar absorptivity ε and the molar concentration of the analyte c by the relationship

$$k = 2.303\varepsilon c$$

Reflectance spectra then consist of a plot of $f(R'_\infty)$ versus wavenumber (see Figure 17-12b).

Instrumentation

Currently, most manufacturers of FTIR instruments offer adapters that fit in cell compartments and permit diffuse-reflectance measurements. Figure 17-11 illustrates one type of adapter. The collimated beam from the interferometer is directed to an ellipsoidal mirror and then to the sample. The sample is usually ground and mixed with KBr or KCl as a diluent. The mixture is then placed in a sample cup 3–4 mm deep and about 10–15 mm in diameter. A complex combination of reflection, absorption, and scattering occurs before the beam is directed to the detector.

To obtain a spectrum with a single-beam instrument, the signal for the sample is first stored. A reference signal with a good reflector, such as finely ground KBr or KCl, is then recorded in place of the sample. The ratio of these signals is then taken to give the reflectance.

Comparison of Absorption and Reflection Spectra

Figure 17-12 compares the conventional IR absorption spectrum for carbazole obtained by means of a KBr pellet with the diffuse-reflectance spectrum of a 5%

[12] For additional information, see M. Milosevic, S. L. Berets, *Appl. Spectrosc. Rev.*, **2002**, *37*, 347; P. R. Griffiths and M. P. Fuller in *Advances in Infrared and Raman Spectroscopy*, R. J. H. Clark and R. E. Hester, eds., Vol. 9, Chap. 2, London: Heydon and Sons, 1982.

[13] P. Kubelka and E. Munk, *Tech. Phys.*, **1931**, *12*, 593; P. Kubelka, *J. Opt. Soc. Am.*, **1948**, *38*, 448.

[14] M. P. Fuller and P. R. Griffiths, *Anal. Chem.*, **1978**, *50*, 1906.

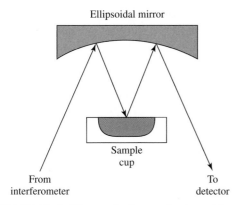

FIGURE 17-11 A diffuse-reflectance attachment for an FTIR spectrometer.

finely ground mixture of carbazole in potassium chloride. Note that the peak locations are the same in the two spectra but that the relative peak heights differ considerably. The differences are typical, with minor peaks generally appearing larger in reflection spectra.

17B-3 ATR Spectrometry

Internal-reflection spectroscopy is a technique for obtaining IR spectra of samples that are difficult to deal with, such as solids of limited solubility, films, threads, pastes, adhesives, and powders.[15]

Principles

When a beam of radiation passes from a more dense to a less dense medium, reflection occurs. The fraction of the incident beam reflected increases as the angle of incidence becomes larger; beyond a certain critical angle, reflection is complete. It has been shown both theoretically and experimentally that during the reflection process the beam penetrates a small distance into the less dense medium before reflection occurs. The depth of penetration, which varies from a fraction of a wavelength up to several wavelengths, depends on the wavelength, the index of refraction of the two materials, and the angle of the beam with respect to the interface. The penetrating radiation is called the *evanescent wave*. At wavelengths where the less dense medium absorbs the evanescent radiation, attenuation of the beam occurs, which is known as *attenuated total reflectance*, or

ATR. The resulting ATR spectrum resembles that of a conventional IR spectrum with some differences.

Instrumentation

Figure 17-13 shows an apparatus for ATR measurements. As can be seen from the upper figure, the sample (here, a solid) is placed on opposite sides of a transparent crystalline material of high refractive index. By proper adjustment of the incident angle, the radiation undergoes multiple internal reflections before passing from the crystal to the detector. Absorption and attenuation take place at each of these reflections.

Figure 17-13b is an optical diagram of an adapter that fits into the cell area of most IR spectrometers and permits ATR measurements. Cells for liquid samples are also available.

ATR Spectra

ATR spectra are similar but not identical to ordinary absorption spectra. In general, although the same bands are observed, their relative intensities differ.

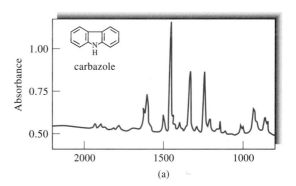

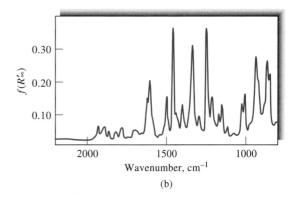

FIGURE 17-12 Comparison of the absorption spectrum (a) for carbazole with its diffuse-reflectance spectrum (b).

---

[15] See F. M. Mirabella, ed., *Modern Techniques in Applied Molecular Spectroscopy*, New York: Wiley, 1998; G. Kortum, *Reflectance Spectroscopy*, New York: Springer, 1969; N. J. Harrick, *Internal Reflection Spectroscopy*, New York: Wiley, 1967.

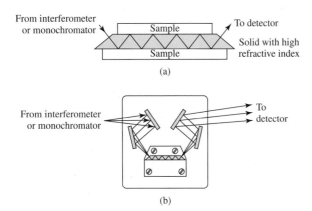

FIGURE 17-13 ATR apparatus. In (a) a solid sample is shown mounted on an internal reflection crystal of high refractive index. The materials used as ATR crystals include KRS-5, AgCl, Ge, Si, and the Irtran materials. Solid samples can be pressed against the crystal to obtain optical contact. In (b), a typical attachment for ATR is shown. With many attachments, the internal reflection plate can be positioned in the holder to provide several incident angles.

With ATR spectra, the absorbances, although dependent on the angle of incidence, are independent of sample thickness, because the radiation penetrates only a few micrometers into the sample. The *effective penetration depth d_p* depends on the wavelength of the beam, the refractive indexes of the crystal and the sample, and the beam angle. The penetration depth can be calculated from

$$d_p = \frac{\lambda_c}{2\pi \left[\sin^2\theta - (n_s/n_c)^2 \right]^{1/2}}$$

where λ_c is the wavelength in the crystal (λ/n_c), θ is the angle of incidence, and n_s and n_c are the refractive indexes of the sample and crystal, respectively. Note that the effective penetration depth can be changed by changing the crystal material, the angle of incidence, or both. It is possible to obtain a depth profile of a surface using ATR spectroscopy. In practice, a multi-reflection crystal with a 45° angle can accommodate most routine samples.

One of the major advantages of ATR spectroscopy is that absorption spectra are readily obtainable on a wide variety of sample types with a minimum of preparation. Threads, yarns, fabrics, and fibers can be stud-

ied by pressing the samples against the dense crystal. Pastes, powders, or suspensions can be handled in a similar way. Aqueous solutions can also be accommodated provided the crystal is not water soluble. There are even ATR flow cells available. ATR spectroscopy has been applied to many substances, such as polymers, rubbers, and other solids. It is of interest that the resulting spectra are free from the interference fringes mentioned previously.

The spectra obtained with ATR methods can differ from IR absorption spectra because of distortions that occur near strong absorption bands where the sample refractive index may change rapidly. Also, the orientation of the sample on the ATR crystal can influence band shapes and relative intensities. However, the ATR band intensity is usually proportional to concentration so that quantitative measurements can be made.

17C PHOTOACOUSTIC IR SPECTROSCOPY

Photoacoustic spectroscopy (PAS) provides a way to obtain ultraviolet, visible, and IR absorption spectra of solids, semisolids, or turbid liquids.[16] Acquisition of spectra for these materials by ordinary methods is usually difficult at best and often impossible because of light scattering and reflection.

17C-1 The Photoacoustic Effect

PAS is based on a light absorption effect that was first investigated in the 1880s by Alexander Graham Bell and others. This effect is observed when a gas in a closed cell is irradiated with a chopped beam of radiation of a wavelength that is absorbed by the gas. The absorbed radiation causes periodic heating of the gas, which in turn results in regular pressure fluctuations within the chamber. If the chopping rate lies in the acoustical frequency range, these pulses of pressure can be detected by a sensitive microphone. The photoacoustic effect has been used since the turn of the century for the analysis of absorbing gases and has recently taken on new importance for this purpose with the advent of tunable IR lasers as sources. Of greater importance, however, has been the use of the photo-

---

[16] See K. H. Michaelian, *Photoacoustic Infrared Spectroscopy*, Hoboken, NJ: Wiley, 2003.

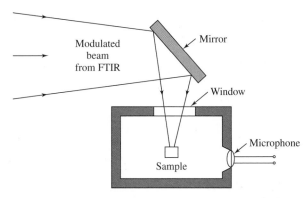

FIGURE 17-14 Diagram of a photoacoustic attachment for an FTIR spectrometer.

acoustic effect for obtaining absorption spectra of solids and turbid liquids.

17C-2 Photoacoustic IR Spectra

Photoacoustic IR spectroscopy saw limited application before the advent of FTIR instruments. Now, several manufacturers make photoacoustic accessories for FTIR instruments. In photoacoustic measurements, the sample is placed in a small sample cup within the photoacoustic attachment as illustrated in Figure 17-14. The photoacoustic spectrometer chamber is filled with a high-thermal-conductivity gas such as helium or nitrogen and placed in the FTIR sample compartment. The mirror shown deflects the modulated beam onto the sample. Absorption of the IR beam by the sample can result in nonradiative decay of the excited vibrational states of the sample molecules. This can transfer heat to the surface of the sample and result in the generation of a modulated acoustic wave in the gas inside the chamber. A very sensitive microphone then detects the acoustic wave.

Photoacoustic spectra are normally plotted in a format similar to absorption spectra as shown in Figure 17-15. The spectra are usually plotted as a ratio to a background scan of a totally absorbing material such as carbon black. Although the frequencies of transitions are the same as in absorption spectra, relative intensities depend on wavelength and modulation frequency.

Photoacoustic spectra can be obtained on samples with essentially no sample preparation. The only requirement is that the sample must fit within the sample cup. In addition to solids and turbid liquids, photoacoustic methods have been used for detecting the components of mixtures separated by thin-layer and high-performance liquid chromatography and for monitoring the concentrations of gaseous pollutants in the atmosphere.

17D NEAR-IR SPECTROSCOPY

The near-IR (NIR) region of the spectrum extends from the upper wavelength end of the visible region at about 770 nm to 2500 nm (13,000 to 4000 cm^{-1}).[17] Absorption bands in this region are overtones or combinations (Section 16A-4) of fundamental stretching vibrations that occur in the region of 3000 to 1700 cm^{-1}. The bonds involved are usually C—H, N—H, and O—H. Because the bands are overtones or combinations, their molar absorptivities are low and detection limits are on the order of 0.1%.

In contrast to mid-IR spectrometry, the most important uses of NIR radiation are for the routine *quantitative* determination of species, such as water, proteins, low-molecular-weight hydrocarbons, and fats, in products of the agricultural, food, petroleum, and chemical industries. Both diffuse-reflection and transmission measurements are used, although diffuse reflectance is by far the more widely used.

---

[17]For a general reference on NIR, see *Handbook of Near-Infrared Analysis*, 2nd ed., D. A. Burns and E. W. Ciurczak, eds., New York: Marcel Dekker, 2001. For a review of commercial instrumentation, see C. M. Henry, *Anal. Chem.*, **1999**, *71*, 625A. In the literature of NIR spectrometry, the abscissa is usually *wavelength* in nanometers or micrometers in contrast to mid-IR spectra in which the abscissa is *wavenumber* in units of (cm)$^{-1}$.

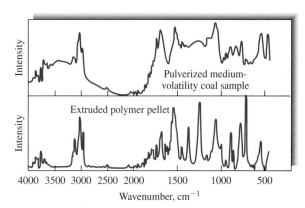

FIGURE 17-15 Photoacoustic IR spectra of (a) a pulverized coal sample and (b) an extruded polymer pellet. (From *Practical Sampling Techniques for Infrared Analysis*, P. B. Coleman, ed., Boca Raton, FL: CRC Press, 1993.)

17D-1 Instrumentation

Four different types of instruments are available for the NIR region. Grating instruments are similar to those used for UV-visible absorption spectroscopy. There are also discrete filter instruments usually containing filter wheels for selecting different wavelengths. These are less flexible than other instruments but useful for fixed, well-characterized samples. In addition, acoustooptic tunable filter (AOTF) instruments are available. The acoustooptic filter is a solid-state device that diffracts radiation at wavelengths determined by a radio frequency signal applied to the crystal. Speed and ruggedness are the main advantages of AOTF devices. Fourier transform spectrometers are also available commercially for NIR spectrometry. The advantages normally associated with FT instruments in the mid-IR region, such as high throughput and resolution, are less applicable in the NIR region. However, the wavelength reproducibility and signal-to-noise-ratio characteristics are major advantages of FT systems.

Most spectrometers use tungsten-halogen lamps with quartz windows. Cells for absorption measurements are usually quartz or fused silica transparent up to about 3000 nm. Cell lengths vary from 0.1 to 10 cm. Detectors range from PbS and PbSe photoconductors to InSb and InAs photodiodes. Array detectors, such as InGaAs detectors, have also become available for the region. Several commercial UV-visible spectrophotometers are designed to operate from 180 to 2500 nm and can thus be used to obtain NIR spectra.

Several solvents are used for NIR studies. Some of these are listed in Figure 17-16. Note that only carbon tetrachloride and carbon disulfide are transparent throughout the entire NIR region.

17D-2 Data Processing in NIR spectrometry

NIR spectral bands are normally broad and often overlapping. There are rarely clean spectral bands that allow simple correlation with analyte concentration. Instead, multivariate calibration techniques are used.[18] Most commonly, partial least squares, principal components regression, and artificial neural networks are em-

ployed. Such calibration involves development of a calibration model through obtaining results on a "training set" that includes as many of the conditions encountered in the samples as possible. Problems and pitfalls in developing the models have been discussed.[19]

The manufacturers of NIR instruments include software packages for developing calibration models. In addition, third-party software for multivariate calibration is readily available.

17D-3 Applications of NIR Absorption Spectrometry

In contrast to mid-IR spectroscopy, NIR absorption spectra are less useful for identification and more useful for quantitative analysis of compounds containing functional groups made up of hydrogen bonded to carbon, nitrogen, and oxygen. Such compounds can often be determined with accuracies and precisions equivalent to ultraviolet-visible spectroscopy, rather than mid-IR spectroscopy. Some applications include the determination of water in a variety of samples, including glycerol, hydrazine, organic films, and fuming nitric acid; the quantitative determination of phenols, alcohols, organic acids, and hydroperoxides based on the first overtone of the O—H stretching vibration that absorbs radiation at about 7100 cm^{-1} (1.4 µm); and the determination of esters, ketones, and carboxylic acids based on their absorption in the region of 3300 to 3600 cm^{-1} (2.8 to 3.0 µm). The absorption in this last case is the first overtone of the carbonyl stretching vibration.

NIR spectrophotometry is also a valuable tool for identification and determination of primary and secondary amines in the presence of tertiary amines in mixtures. The analyses are generally carried out in carbon tetrachloride solutions in 10-cm cells. Primary amines are determined directly by measurement of the absorbance of a combination N—H stretching band at about 5000 cm^{-1} (2.0 µm); neither secondary nor tertiary amines absorb radiation in this region. Primary and secondary amines have several overlapping absorption bands in the 3300 to 10,000 cm^{-1} (1 to 3 µm) region due to various N—H stretching vibrations and their overtones, whereas tertiary amines can have no such bands. Thus, one of these bands gives the second-

---

[18]For a discussion of multivariate techniques, see K. R. Beebe, R. J. Pell, and M. B. Seasholtz, *Chemometrics: A Practical Guide*, Chap. 5, New York: Wiley, 1998; H. Martens and T. Naes, *Multivariate Calibration*, New York: Wiley, 1989.

[19]For example, see M. A. Arnold, J. J. Burmeister, and G. W. Small, *Anal. Chem.*, **1998**, *70*, 1773; L. Zhang, G. W. Small, and M. A. Arnold, *Anal. Chem.*, **2003**, *75*, 5905.

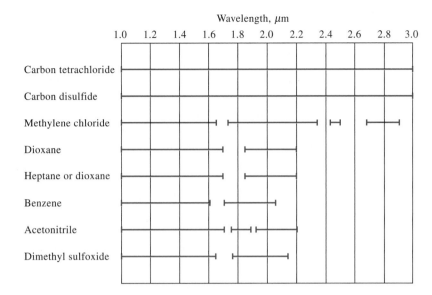

Wavelength, μm

FIGURE 17-16 Some useful solvents for NIR spectroscopy. Solid lines indicate satisfactory transparency for use with 1-cm cells.

ary amine concentration after correction for the absorption by the primary amine.

17D-4 Applications of NIR Reflectance Spectrometry

Near-IR reflectance spectroscopy has become a most important tool for the routine quantitative determination of constituents in finely ground solids. The most widespread use of this technique has been for the determination of protein, moisture, starch, oil, lipids, and cellulose in agricultural products such as grains and oilseeds. For example, Canada sells all its wheat based on guaranteed protein content, and the Canadian Grain Commission formerly ran more than 600,000 Kjeldahl protein determinations annually as a consequence. It was estimated in 1984 that 80% to 90% of all Canadian grain was analyzed for protein by NIR reflectance spectroscopy at a savings of more than $500,000 a year in analytical costs.[20]

In NIR reflectance spectroscopy, the finely ground solid sample is irradiated with one or more narrow bands of radiation ranging in wavelength from 1 to 2.5 μm or 10,000 to 4000 cm^{-1}. Diffuse reflectance occurs in which the radiation penetrates the surface layer of the particles, excites vibrational modes of the analyte molecule, and is then scattered in all directions. A reflectance spectrum is thus produced that is depen-

dent on the composition of the sample. A typical reflectance spectrum for a sample of wheat is shown in Figure 17-17. The ordinate in this case is the logarithm of the reciprocal of reflectance R, where R is the ratio of the intensity of radiation reflected from the sample to the reflectance from a standard reflector, such as finely ground barium sulfate or magnesium oxide. The reflectance band at 1940 nm is a water band used for moisture determinations. The band at about 2100 nm is, in fact, two overlapping bands, one for starch and the other for protein. By making measurements at several wavelengths in this region, the concentrations of each of these components can be determined.

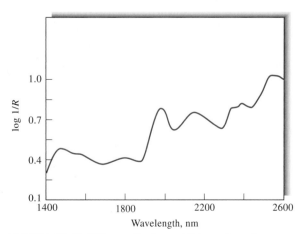

FIGURE 17-17 Diffuse-reflectance NIR spectrum for a sample of wheat.

[20] S. A. Borman, *Anal. Chem.*, **1984**, *56*, 933A.

Instruments for diffuse-reflectance measurements are available from commercial sources. Some of these employ several interference filters to provide narrow bands of radiation. Others are equipped with grating monochromators. Ordinarily, reflectance measurements are made at two or more wavelengths for each analyte species being determined.

The great advantage of NIR reflectance methods is their speed and the simplicity of sample preparation. Once method development has been completed, analysis of solid samples for several species can be completed in a few minutes. Accuracies and precisions of 1% to 2% relative are regularly reported.

17E FAR-IR SPECTROSCOPY

The far-IR region is particularly useful for inorganic studies because absorption due to stretching and bending vibrations of bonds between metal atoms and both inorganic and organic ligands generally occurs at frequencies lower than 650 cm^{-1} ($>15 \text{ μm}$). For example, heavy-metal iodides generally absorb radiation in the region below 100 cm^{-1}, and the bromides and chlorides have bands at higher frequencies. Absorption frequencies for metal-organic bonds ordinarily depend on both the metal atom and the organic portion of the species.

Far-IR studies of inorganic solids have also provided useful information about lattice energies of crystals and transition energies of semiconducting materials.

Molecules composed only of light atoms absorb far-IR radiation if they have skeletal bending modes that involve more than two atoms other than hydrogen. Important examples are substituted benzene derivatives, which generally show several absorption bands. The spectra are frequently quite specific and useful for identifying a particular compound. There are also characteristic group frequencies in the far-IR region.

Pure rotational absorption by gases is observed in the far-IR region, provided the molecules have permanent dipole moments. Examples include H_2O, O_3, HCl, and AsH_3. Absorption by water is troublesome; elimination of its interference requires evacuation or at least purging of the spectrometer.

Prior to the advent of Fourier transform spectrometers, experimental difficulties made it difficult to obtain good far-IR spectra. The sources available in this region are quite low in intensities. Order-sorting filters are needed with grating instruments in this region to minimize radiation diffracted from higher grating orders. These reduced the already low throughputs of dispersive spectrometers. It is no surprise that Fourier transform spectrometry was first applied in this spectral region. The high throughput of interferometers and the relatively low mechanical tolerances required made it fairly simple to obtain FTIR spectra in the far-IR region with good signal-to-noise ratios.

17F IR EMISSION SPECTROSCOPY

When they are heated, molecules that absorb IR radiation are also capable of emitting characteristic IR wavelengths. The principal deterrent to the analytical application of this phenomenon has been the poor signal-to-noise characteristics of the IR emission signal, particularly when the sample is at a temperature only slightly higher than its surroundings. With interferometry, however, interesting and useful applications have appeared in the literature.

An early example of the application of IR emission spectroscopy is found in a paper that describes the use of a Fourier transform spectrometer for the identification of microgram quantities of pesticides.[21] Samples were prepared by dissolving them in a suitable solvent followed by evaporation on a NaCl or KBr plate. The plate was then heated electrically near the spectrometer entrance. Pesticides such as DDT, malathion, and dieldrin were identified in amounts as low as 1 to 10 μg.

Equally interesting has been the use of interferometry for the remote detection of components emitted from industrial smoke stacks. In one of these applications, an interferometer was mounted on an 8-inch reflecting telescope.[22] With the telescope focused on the plume from an industrial plant, CO_2 and SO_2 were readily detected at a distance of several hundred feet.

The Mars Global Surveyor launched from Cape Canaveral in November 1996. It reached Mars orbit in 1997. One of five instruments on board was an IR emission spectrometer, called a thermal emission spectrometer. The spacecraft completed its mapping mission in 2001, providing measurements of the Martian surface and atmosphere. The Mars rover *Spirit* has a

[21] J. Coleman and M. J. D. Low, *Spectrochim. Acta*, **1966**, *22*, 1293.
[22] M. J. D. Low and F. K. Clancy, *Env. Sci. Technol.*, **1967**, 1, 73.

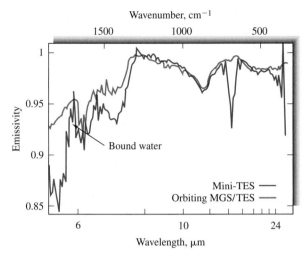

FIGURE 17-18 Spectra from Mars rover *Spirit* showing an unidentified mineral containing bound water in its crystal structure. Minerals such as zeolites and gypsum are possible candidates. Mini-TES = mini thermal-emission spectrometer. (Courtesy of NASA/JPL/Arizona State University.)

mini thermal-emission spectrometer on board capable of indicating the composition of nearby soils and rocks. Figure 17-18 shows an IR emission spectrum taken in early 2004 that showed evidence of an unidentified water-containing mineral. Thermal emission spectra have also indicated the presence of carbonates and other hydrated minerals. These may have been produced by long-standing bodies of water.

17G IR MICROSCOPY

IR microspectrometry is used for obtaining absorption or reflection spectra of species in samples having physical dimensions in the range of 10 to 500 μm.[23] IR microscopes, which were first introduced by several instrument manufacturers in the 1980s, generally consist of two microscopes, one an ordinary optical microscope and the other an IR device with reflection optics that reduce the size of the IR beam to about that of the sample. The optical microscope is used to visually locate the sample particle or spot to be studied with the IR beam. The IR source is an ordinary Fourier transform spectrometer and not a grating instrument because of the much greater sensitivity of FTIR instruments. The detector is usually a liquid-nitrogen-cooled mercury telluride–cadmium telluride photoconductive device that is more sensitive than other types of IR detectors.

The use of IR microscopy is still fairly new but appears to be a technique that will find increasing applications in the future. Some of its current applications include identification of polymer contaminants, imperfections in polymer films, and individual layers of laminated polymer sheets; identification of tiny samples of fibers, paint, and explosives in criminalistics; characterization of single fibers in the textile industry; and identification of contaminants on electronic components.

---

[23]See *Practical Guide to Infrared Microspectroscopy*, H. J. Humecki, ed., New York: Marcel Dekker, 1995; J. Katon, *Infrared Microspectroscopy in Modern Techniques in Molecular Spectroscopy*, F. Mirabella, ed., New York: Wiley, 1998.

QUESTIONS AND PROBLEMS

*Answers are provided at the end of the book for problems marked with an asterisk.

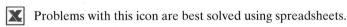 Problems with this icon are best solved using spreadsheets.

17-1 Cyclohexanone exhibits its strongest IR absorption band at 5.86 μm, and at this wavelength there is a linear relationship between absorbance and concentration.
  (a) Identify the part of the molecule responsible for the absorbance at this wavelength.
  (b) Suggest a solvent that would be suitable for a quantitative analysis of cyclohexanone at this wavelength.
  (c) A solution of cyclohexanone (4.0 mg/mL) in the solvent selected in part (b) exhibits a blank-corrected absorbance of 0.800 in a cell with a path length of 0.025 mm. What is the detection limit for this compound under these conditions if the noise associated with the spectrum of the solvent is 0.001 absorbance units?

*17-2 The spectrum in Figure 17-19 was obtained for a liquid with the empirical formula of C_3H_6O. Identify the compound.

17-3 The spectrum in Figure 17-20 is that of a high-boiling-point liquid having the empirical formula $C_9H_{10}O$. Identify the compound as closely as possible.

*17-4 The spectrum in Figure 17-21 is for an acrid-smelling liquid that boils at 50°C and has a molecular weight of about 56. What is the compound? What impurity is clearly present?

*17-5 The spectrum in Figure 17-22 is that of a nitrogen-containing substance that boils at 97°C. What is the compound?

17-6 Why are quantitative analytical methods based on NIR radiation often more precise and accurate than methods based on mid-IR radiation?

17-7 Discuss the origins of photoacoustic IR spectra. What are the usual types of samples for photoacoustic measurements?

17-8 Why do IR spectra seldom show regions at which the transmittance is 100%?

*17-9 An empty cell showed 15 interference fringes in the wavelength range of 6.0 to 12.2 µm. Calculate the path length of the cell.

*17-10 An empty cell exhibited 11.5 interference fringes in the region of 1050 to 1580 cm^{-1}. What was the path length of the cell?

*17-11 Estimate the thickness of the polystyrene film that yielded the spectrum shown in Figure 16-1.

X *Challenge Problem*

17-12 (a) Consider an ATR experiment with a sample having a refractive index of 1.03 at 2000 cm^{-1}. The ATR crystal was AgCl with a refractive index of 2.00 at this wavelength. For an angle of incidence of 45°, what is the effective penetration depth of the evanescent wave? How would the penetration depth change if the angle were changed to 60°?

(b) For the same experiment in part (a) and a 60° incidence angle, find the penetration depths for sample refractive indexes varying from 1.00 to 1.70 in steps of 0.10. Plot penetration depth as a function of refractive index. Determine the refractive index for which the penetration depth becomes zero. What happens at this point?

(c) For a sample with a refractive index of 1.37 at 2000 cm^{-1} and an incidence angle of 45°, plot the penetration depth versus the ATR crystal refractive index. Vary the crystal refractive index over the range of 2.00 to 4.00 in steps of 0.25. Which crystal, AgCl ($n_c = 2.00$) or Ge ($n_c = 4.00$), gives the smaller penetration depth? Why?

(d) An aqueous solution with a refractive index of 1.003 is measured with an ATR crystal with a refractive index of 2.8. The incidence angle is 45°. What is the effective penetration depth at 3000 cm^{-1}, 2000 cm^{-1}, and 1000 cm^{-1}? Is absorption by the aqueous solvent as much of a problem in ATR as in normal IR absorption measurements? Why or why not?

(e) Work by S. Ekgasit and H. Ishida (S. Ekgasit and H. Ishida, *Appl. Spectrosc.*, **1996**, *50*, 1187; *Appl. Spectrosc.*, **1997**, *51*, 461) describes a new method to obtain a depth profile of a sample surface using ATR spectroscopy. Describe the principles of this new approach. Why are spectra taken with different degrees of polarization? What is the complex index of refraction?

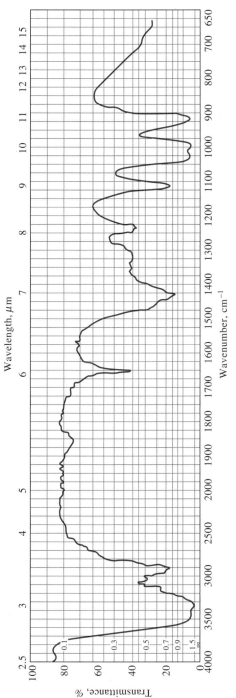

FIGURE 17-19 See Problem 17-2. (Spectrum courtesy of Thermodynamics Research Center Data Project, Texas A&M University, College Station, Texas.)

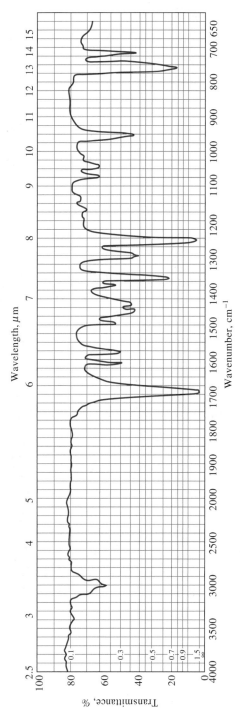

FIGURE 17-20 See Problem 17-3. (Spectrum courtesy of Thermodynamics Research Center Data Project, Texas A&M University, College Station, Texas.)

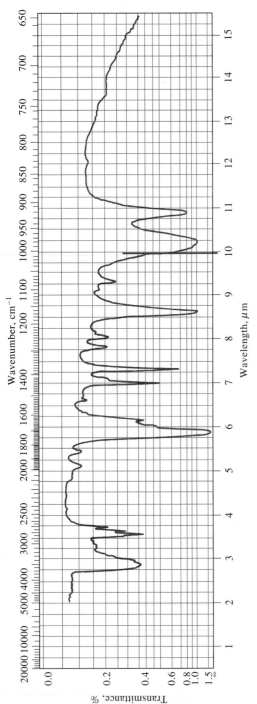

FIGURE 17-21 See Problem 17-4. (Spectrum courtesy of Thermodynamics Research Center Data Project, Texas A&M University, College Station, Texas.)

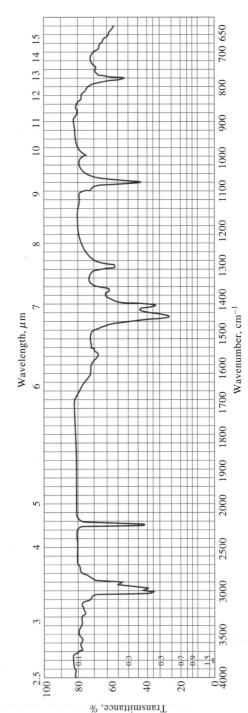

FIGURE 17-22 See Problem 17-5. (Spectrum courtesy of Thermodynamics Research Center Data Project, Texas A&M University, College Station, Texas.)

CHAPTER EIGHTEEN

Raman Spectroscopy

When radiation passes through a transparent medium, the species present scatter a fraction of the beam in all directions (Section 6B-10). Raman scattering results from the same type of quantized vibrational changes associated with IR absorption. Thus, the difference in wavelength between the incident and scattered visible radiation corresponds to wavelengths in the mid-IR region. Indeed, the Raman scattering spectrum and IR absorption spectrum for a given species often resemble one another quite closely. As discussed in this chapter, for some problems, IR spectroscopy is the superior tool, but for others, Raman spectroscopy can provide more useful information.

Throughout this chapter, this logo indicates an opportunity for online self-study at **www .thomsonedu.com/chemistry/skoog**, linking you to interactive tutorials, simulations, and exercises.

In 1928 the Indian physicist C. V. Raman discovered that the visible wavelength of a small fraction of the radiation scattered by certain molecules differs from that of the incident beam and furthermore that the shifts in wavelength depend on the chemical structure of the molecules responsible for the scattering. Raman was awarded the 1931 Nobel Prize in physics for this discovery and for his systematic exploration of it.[1]

Although there can be striking similarities between Raman spectra and IR spectra, enough differences remain between the kinds of groups that are IR active and Raman active to make the techniques complementary rather than competitive. An important advantage of Raman spectroscopy over IR lies in water being a quite useful solvent. In addition, because signals are usually in the visible or near-IR region, glass or quartz cells can be employed, avoiding the inconvenience of working with sodium chloride or other atmospherically unstable window materials. Despite these advantages, Raman spectroscopy was not widely used by chemists until lasers became available in the 1960s, which made spectra a good deal easier to obtain. In recent years, Raman spectroscopy has become a routine tool thanks to the laser, the array detector, and the availability of commercial instrumentation at moderate cost.

18A THEORY OF RAMAN SPECTROSCOPY

Raman spectra are acquired by irradiating a sample with a powerful laser source of visible or near-IR monochromatic radiation. During irradiation, the spectrum of the scattered radiation is measured at some angle (often 90°) with a suitable spectrometer. To avoid fluorescence, the excitation wavelengths are usually well removed from an absorption band of the analyte. The Raman experiment was illustrated previously in Figure 6-18. At the very most, the intensities of Raman lines are 0.001% of the intensity of the source. Because of this, it might seem more difficult to detect and measure Raman bands than IR vibrational bands. However, the Raman scattered radiation is in the visible

[1] For more complete discussions of the theory and practice of Raman spectroscopy, see J. R. Ferraro, K. Nakamoto, and C. W. Brown, *Introductory Raman Spectroscopy*, 2nd ed., San Diego: Academic Press, 2003; R. L. McCreery, *Raman Spectroscopy for Chemical Analysis*, New York: Wiley, 2000; D. P. Strommen in *Handbook of Instrumental Analysis*, F. A. Settle, ed., Chap. 16, Upper Saddle River, NJ: Prentice Hall, 1997; *Analytical Raman Spectroscopy*, J. G. Grasseli and B. J. Bulkin, eds., New York: Wiley, 1991.

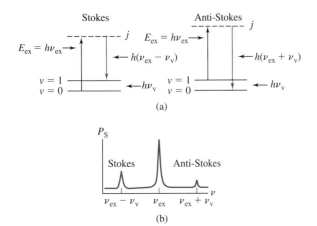

(a)

(b)

FIGURE 18-1 The origin of Raman spectra. In (a) radiation from a source that is incident on the sample produces scattering at all angles. The incident radiation causes excitation (a) to a virtual level j and subsequent reemission of a photon of lower (left) or higher (right) energy. The Raman spectrum (b) consists of lower-frequency emissions called Stokes scattering and higher-frequency emissions termed anti-Stokes scattering. Usually, the ground vibrational level ($v = 0$) is more highly populated than the excited vibration levels so that the Stokes lines are more intense than the anti-Stokes lines. Elastically scattered radiation is of the same frequency as the excitation beam and is called Rayleigh scattering.

or near-IR regions for which more sensitive detectors are available. Hence, today, measurement of Raman spectra is nearly as easy as measurement of IR spectra.

18A-1 Excitation of Raman Spectra

In Figure 18-1, the sample is irradiated by a monochromatic beam of energy $h\nu_{ex}$. Because the excitation wavelength is well away from an absorption band, excitation can be considered to involve a *virtual state* of energy level j, indicated by the dashed line in Figure 18-1a. A molecule in the ground vibrational level ($v = 0$) can absorb a photon of energy $h\nu_{ex}$ and reemit a photon of energy $h(\nu_{ex} - \nu_v)$, as shown on the left side of Figure 18-1a. When the scattered radiation is of a lower frequency than the excitation radiation, it is called *Stokes scattering*. Molecules in a vibrationally excited state ($v = 1$) can also scatter radiation inelastically and produce a Raman signal of energy $h(\nu_{ex} + \nu_v)$. Scattered radiation of a higher frequency than the source radiation is called *anti-Stokes scattering*. Elastic scattering can also occur with emission of a photon of the same

energy as the excitation photon, $h\nu_{ex}$. Scattered radiation of the same frequency as the source is termed *Rayleigh scattering*. Note that the frequency shifts of the inelastically scattered radiation $(\nu_{ex} + \nu_v) - \nu_{ex} = \nu_v$ and $(\nu_{ex} - \nu_v) - \nu_{ex} = -\nu_v$ correspond to the vibrational frequency, ν_v. The simplified Raman spectrum corresponding to the transitions shown is given in Figure 18-1b.

Figure 18-2 depicts a portion of the Raman spectrum of carbon tetrachloride that was obtained by using an argon-ion laser having a wavelength of 488.0 nm as the source. As is usually the case for Raman spectra, the abscissa of Figure 18-2 is the wavenumber shift $\Delta\bar{\nu}$, which is defined as the difference in wavenumbers (cm^{-1}) between the observed radiation and that of the source. Note that three Raman lines are found on both sides of the Rayleigh lines and that the pattern of shifts on each side is identical. That is, Stokes lines are found at wavenumbers that are 218, 314, and 459 cm^{-1} smaller than the Rayleigh lines, and anti-Stokes lines occur at 218, 314, and 459 cm^{-1} greater than the wavenumber of the source. It should also be noted that additional lines can be found at ±762 and $\pm790\ cm^{-1}$. Because the anti-Stokes lines are appreciably less intense than the corresponding Stokes lines, only the Stokes part of a spec-

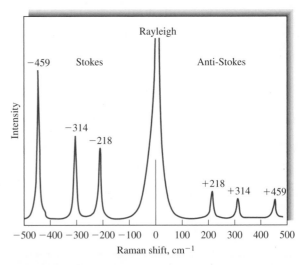

FIGURE 18-2 Raman spectrum of CCl_4, excited by laser radiation of $\lambda_{ex} = 488$ nm ($\bar{\nu}_{ex} = 20{,}492\ cm^{-1}$). The number above the Raman lines is the Raman shift, $\Delta\nu = \nu_{ex} \pm \nu_v$, in cm^{-1}. Stokes-shifted lines are often given positive values rather than negative values as shown. (From J. R. Ferraro, K. Nakamoto, and C. W. Brown, *Introductory Raman Spectroscopy*, 2nd ed., San Diego: Academic Press, 2003. Reprinted with permission.)

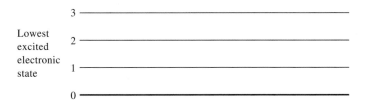

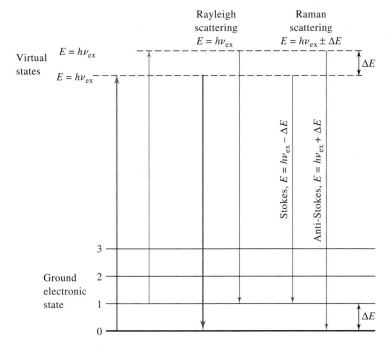

FIGURE 18-3 Origins of Rayleigh and Raman scattering.

trum is generally used. Furthermore, the abscissa of the plot is often labeled simply "wavenumber $\bar{\nu}$, cm^{-1}" rather than "wavenumber shift $\Delta\bar{\nu}$." It is noteworthy that fluorescence may interfere seriously with the observation of Stokes shifts but not with anti-Stokes. With fluorescing samples, anti-Stokes signals may, therefore, be more useful despite their lower intensities.

It is important to appreciate that the magnitude of Raman shifts is *independent of the wavelength of excitation*. Thus, Raman shifts identical to those shown in Figure 18-2 are observed for carbon tetrachloride regardless of whether excitation was carried out with an argon-ion laser (488.0), or a helium-neon laser (632.8 nm).

Superficially, the Raman Stokes shifts to lower energies (longer wavelengths) are analogous to the Stokes shifts found in molecular fluorescence (see Section 6C-6). We shall see, however, that Raman and

fluorescence spectra arise from fundamentally different processes.

18A-2 Mechanisms of Raman and Rayleigh Scattering

In Raman spectroscopy, spectral excitation is normally carried out by radiation having a wavelength that is well away from any absorption bands of the analyte. The energy-level diagram of Figure 18-1a is expanded in Figure 18-3 and provides a qualitative picture of the sources of Raman and Rayleigh scattering. The heavy arrow on the far left depicts the energy change in the molecule when it interacts with a photon from the source. The increase in energy is equal to the energy of the photon $h\nu_{ex}$. It is important to appreciate that the process shown is *not quantized*; thus, depending on the frequency of the radiation from the source, the energy of the molecule can assume any of an infinite number of values, or *virtual states*, between the ground state and the lowest (first) electronic excited state shown in the upper part of the diagram. The second and narrower arrow on the left shows the type of change that would occur if the molecule encountered by the photon happened to be in the first vibrational level of the electronic ground state. At room temperature, the fraction of the molecules in this state is small. Thus, as indicated by the width of the arrows, the probability of this process occurring is much smaller.

The middle set of arrows depicts the changes that produce Rayleigh scattering. Again the more probable change is shown by the wider arrow. Note that no energy is lost in Rayleigh scattering. As a consequence, the collisions between the photon and the molecule are said to be *elastic*.

Finally, the energy changes that produce Stokes and anti-Stokes emission are depicted on the right. The two differ from the Rayleigh radiation by frequencies corresponding to $\pm \Delta E$, the energy of the first vibrational level of the ground state, $h\nu_\mathrm{v}$. Note that if the bond were IR active, the energy of its absorption would also be ΔE. Thus, the Raman *frequency shift* and the *IR absorption frequency* are identical.

Note also that the relative populations of the two upper energy states are such that Stokes emission is much favored over anti-Stokes. In addition, Rayleigh scattering has a considerably higher probability of occurring than Raman scattering because the most probable event is the energy transfer to molecules in the ground state and reemission by the return of these molecules to the ground state. Finally, it should be noted that the ratio of anti-Stokes to Stokes intensities increases with temperature because a larger fraction of the molecules is in the first vibrationally excited state under these circumstances.

18A-3 Wave Model of Raman and Rayleigh Scattering

Let us assume that a beam of radiation having a frequency ν_ex is incident on a solution of an analyte. The electric field E of this radiation can be described by the equation

$$E = E_0 \cos(2\pi\nu_\mathrm{ex}t) \qquad (18\text{-}1)$$

where E_0 is the amplitude of the wave. When the electric field of the radiation interacts with an electron cloud of an analyte bond, it induces a dipole moment m in the bond that is given by

$$m = \alpha E = \alpha E_0 \cos(2\pi\nu_\mathrm{ex}t) \qquad (18\text{-}2)$$

where α is a proportionality constant called the *polarizability* of the bond. This constant is a measure of the deformability of the bond in an electric field.

The polarizability α varies as a function of the distance between nuclei according to the equation

$$\alpha = \alpha_0 + (r - r_\mathrm{eq})\left(\frac{\partial\alpha}{\partial r}\right) \qquad (18\text{-}3)$$

where α_0 is the polarizability of the bond at the equilibrium internuclear distance r_eq and r is the internuclear separation at any instant. The change in internuclear separation varies with the frequency of the vibration ν_v as given by

$$r - r_\mathrm{eq} = r_\mathrm{m} \cos(2\pi\nu_\mathrm{v}t) \qquad (18\text{-}4)$$

where r_m is the maximum internuclear separation relative to the equilibrium position.

Substituting Equation 18-4 into 18-3 gives

$$\alpha = \alpha_0 + \left(\frac{\partial\alpha}{\partial r}\right)r_\mathrm{m}\cos(2\pi\nu_\mathrm{v}t) \qquad (18\text{-}5)$$

We can then obtain an expression for the induced dipole moment m by substituting Equation 18-5 into Equation 18-2. Thus,

$$m = \alpha_0 E_0 \cos(2\pi\nu_\mathrm{ex}t)$$
$$+ E_0 r_\mathrm{m}\left(\frac{\partial\alpha}{\partial r}\right)\cos(2\pi\nu_\mathrm{v}t)\cos(2\pi\nu_\mathrm{ex}t) \quad (18\text{-}6)$$

If we use the trigonometric identity for the product of two cosines

$$\cos x \cos y = [\cos(x + y) + \cos(x - y)]/2$$

we obtain from Equation 18-6

$$m = \alpha_0 E_0 \cos(2\pi\nu_\mathrm{ex}t)$$
$$+ \frac{E_0}{2}r_\mathrm{m}\left(\frac{\partial\alpha}{\partial r}\right)\cos[2\pi(\nu_\mathrm{ex} - \nu_\mathrm{v})t]$$
$$+ \frac{E_0}{2}r_\mathrm{m}\left(\frac{\partial\alpha}{\partial r}\right)\cos[2\pi(\nu_\mathrm{ex} + \nu_\mathrm{v})t] \quad (18\text{-}7)$$

The first term in this equation represents Rayleigh scattering, which occurs at the excitation frequency ν_ex. The second and third terms in Equation 18-7 correspond respectively to the Stokes and anti-Stokes frequencies of $\nu_\mathrm{ex} - \nu_\mathrm{v}$ and $\nu_\mathrm{ex} + \nu_\mathrm{v}$. Here, the excitation frequency has been modulated by the vibrational frequency of the bond. It is important to note that the selection rules for Raman scattering require that there be a change in polarizability during the vibration — that is, $\partial\alpha/\partial r$ in Equation 18-7 must be greater than zero for Raman lines to appear. The selection rules also predict that Raman lines corresponding to fundamental modes of vibration occur with $\Delta\nu = \pm 1$. Just as with IR spectroscopy, much weaker overtone transitions appear at $\Delta\nu = \pm 2$.

We have noted that, for a given bond, the energy *shifts* observed in a Raman experiment should be identical to the *energies* of its IR absorption bands, provided the vibrational modes involved are both IR and Raman active. Figure 18-4 illustrates the similarity of the two types of spectra; it is seen that there are several bands with identical $\bar{\nu}$ and $\Delta\bar{\nu}$ values for the two compounds. We should also note, however, that the relative intensities of the corresponding bands are frequently quite different. Moreover, certain peaks that occur in one spectrum are absent in the other.

The differences between a Raman spectrum and an IR spectrum are not surprising when it is considered

 Tutorial: Learn more about **Raman and IR spectra**.

that the basic mechanisms, although dependent on the same vibrational modes, arise from processes that are mechanistically different. Infrared absorption requires that there be a change in dipole moment or charge distribution during the vibration. Only then can radiation of the same frequency interact with the molecule and promote it to an excited vibrational state. In contrast, scattering involves a momentary distortion of the electrons distributed around a bond in a molecule, followed by reemission of the radiation as the bond returns to its normal state. In its distorted form, the molecule is temporarily polarized; that is, it develops momentarily an induced dipole that disappears on relaxation and reemission. Because of this fundamental difference in mechanism, the Raman activity of a given vibrational mode may differ markedly from its IR activity. For example, a homonuclear molecule such as nitrogen, chlorine, or hydrogen has no dipole moment either in its equilibrium position or when a stretching vibration causes a change in the distance between the two nuclei. Thus, absorption of radiation (IR) of the vibrational frequency cannot occur. On the other hand, the polarizability of the bond between the two atoms of such a molecule varies periodically in phase with the stretching vibrations, reaching a maximum at the greatest

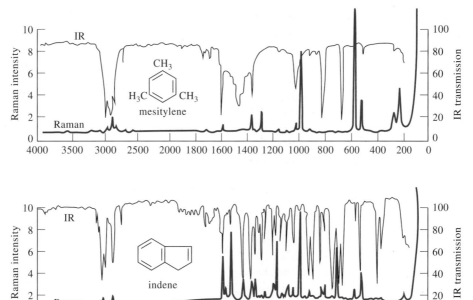

FIGURE 18-4 Comparison of Raman and IR spectra for mesitylene and indene. (Courtesy of Perkin-Elmer Corp., Norwalk, CT.)

separation and a minimum at the closest approach. A Raman shift corresponding in frequency to that of the vibrational mode results.

It is of interest to compare the IR and the Raman activities of coupled vibrational modes such as those described earlier (page 438) for the carbon dioxide molecule. In the symmetric mode, no change in the dipole moment occurs as the two oxygen atoms move away from or toward the central carbon atom; thus, this mode is IR inactive. The polarizability, however, fluctuates in phase with the vibration because distortion of bonds becomes easier as they lengthen and more difficult as they shorten. Raman activity is associated with this mode.

In contrast, the dipole moment of carbon dioxide fluctuates in phase with the asymmetric vibrational mode. Thus, an IR absorption band arises from this mode. On the other hand, as the polarizability of one of the bonds increases as it lengthens, the polarizability of the other decreases, resulting in no net change in the molecular polarizability. Thus, the asymmetric stretching vibration is Raman inactive. For molecules with a center of symmetry, such as CO_2, no IR active transitions are in common with Raman active transitions. This is often called the *mutual exclusion principle*.

Often, as in the foregoing examples, parts of Raman and IR spectra are complementary, each being associated with a different set of vibrational modes within a molecule. For noncentrosymmetric molecules, many vibrational modes may be both Raman and IR active. For example, all of the vibrational modes of sulfur dioxide yield both Raman and IR bands. The intensities of the bands differ, however, because the probabilities for the transitions are different for the two mechanisms. Raman spectra are often simpler than IR spectra because the occurrence of overtone and combination bands is rare in Raman spectra.

18A-4 Intensity of Normal Raman Bands

The intensity or radiant power of a normal Raman band depends in a complex way on the polarizability of the molecule, the intensity of the source, and the concentration of the active group, as well as other factors. In the absence of absorption, the power of Raman emission increases with the fourth power of the frequency of the source. However, advantage can seldom be taken of this relationship because of the likelihood that ultraviolet irradiation will cause photodecomposition of the analyte.

Raman intensities are usually directly proportional to the concentration of the active species. In this regard, Raman spectroscopy more closely resembles fluorescence than absorption, in which the concentration-intensity relationship is logarithmic.

18A-5 Raman Depolarization Ratios

Raman measurements provide, in addition to intensity and frequency information, one additional variable that is sometimes useful in determining the structure of molecules, namely, the *depolarization ratio*.[2] Here, it is important to distinguish carefully between the terms *polarizability* and *polarization*. The former term describes a *molecular* property having to do with the deformability of a bond. Polarization, in contrast, is a property of a beam of radiation and describes the plane in which the radiation vibrates.

When Raman spectra are excited by plane-polarized radiation, as they are when a laser source is used, the scattered radiation is found to be polarized to various degrees depending on the type of vibration responsible for the scattering. The nature of this effect is illustrated in Figure 18-5, where radiation from a laser source is shown as being polarized in the yz plane. Part of the resulting scattered radiation is shown as being polarized parallel to the original beam, that is, in the xz plane; the intensity of this radiation is symbolized by the subscript $\parallel$. The remainder of the scattered beam is polarized in the xy plane, which is perpendicular to the polarization of the original beam; the intensity of this perpendicularly polarized radiation is shown by the subscript $\perp$. The depolarization ratio p is defined as

$$p = \frac{I_\perp}{I_\parallel} \qquad (18\text{-}8)$$

Experimentally, the depolarization ratio may be obtained by inserting a Polaroid sheet or other polarizer between the sample and the monochromator. Spectra are then obtained with the axis of the sheet oriented parallel with first the xz and then the xy plane shown in Figure 18-5.

The depolarization ratio depends on the symmetry of the vibrations responsible for the scattering. For example, the band for carbon tetrachloride at 459 cm^{-1} (Figure 18-2) arises from a totally symmetric "breathing" vibration involving the simultaneous movement of the four tetrahedrally arranged chlorine atoms toward

---

[2]D. P. Strommen, *J. Chem. Educ.*, **1992**, *69*, 803.

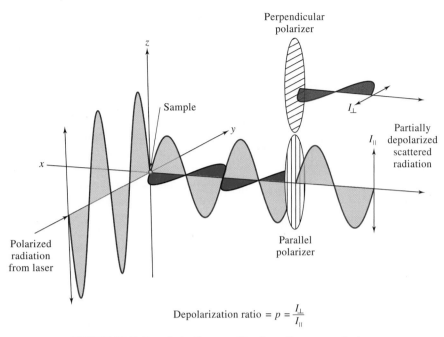

$$\text{Depolarization ratio} = p = \frac{I_\perp}{I_\parallel}$$

FIGURE 18-5 Depolarization resulting from Raman scattering.

and away from the central carbon atom. The depolarization ratio is 0.005, indicating minimal depolarization; the 459-cm^{-1} line is thus said to be polarized. In contrast, the carbon tetrachloride bands at 218 and 314 cm^{-1}, which arise from nonsymmetrical vibrations, have depolarization ratios of about 0.75. From scattering theory it is possible to demonstrate that the maximum depolarization for nonsymmetric vibrations is 6/7, and for symmetric vibrations the ratio is always less than this number. The depolarization ratio is thus useful in correlating Raman lines with modes of vibration.

18B   INSTRUMENTATION

Instrumentation for modern Raman spectroscopy consists of a laser source, a sample illumination system, and a suitable spectrometer as illustrated in Figure 18-6.[3] The performance requirements for these components are more stringent than for the molecular spectrometers we have already described, however, because of the inherent weakness of the Raman scattering signal compared with the signal produced by the Rayleigh scattering.

[3]For reviews of Raman instrumentation, see C. M. Harris, *Anal. Chem.*, **2003**, *75*, 75A and **2002**, *74*, 433A.

18B-1 Sources

The sources used in modern Raman spectrometry are nearly always lasers because their high intensity is necessary to produce Raman scattering of sufficient intensity to be measured with a reasonable signal-to-noise ratio. Five of the most common lasers used for Raman spectroscopy are listed in Table 18-1. Because

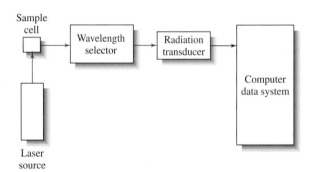

FIGURE 18-6 Block diagram of a Raman spectrometer. The laser radiation is directed into a sample cell. The Raman scattering is usually measured at right angles to avoid viewing the source radiation. A wavelength selector isolates the desired spectral region. The transducer converts the Raman signal into a proportional electrical signal that is processed by the computer data system.

TABLE 18-1 Some Common Laser Sources for Raman Spectroscopy

Laser Type	Wavelength, nm
Argon ion	488.0 or 514.5
Krypton ion	530.9 or 647.1
Helium-neon	632.8
Diode	785 or 830
Nd-YAG	1064

the intensity of Raman scattering varies as the fourth power of the frequency, argon and krypton ion sources that emit in the blue and green region of the spectrum have an advantage over the other sources shown in the table. For example, the argon ion line at 488 nm provides Raman lines that are nearly three times as intense as those excited by the He-Ne source, given the same input power. However, these short-wavelength sources can produce significant fluorescence and cause photodecomposition of the sample.

The last two sources in the table, which emit near-IR radiation, are finding more and more use as excitation sources. Near-IR sources have two major advantages over shorter-wavelength lasers. The first is that they can be operated at much higher power (up to 50 W) without causing photodecomposition of the sample. The second is that they are not energetic enough to populate a significant number of fluorescence-producing excited electronic energy states in most molecules. Consequently, fluorescence is generally much less intense or nonexistent with these lasers. The Nd-YAG laser, used in Fourier transform Raman spectrometers is particularly effective in eliminating fluorescence. The two lines of the diode laser at 785 and 830 nm also markedly reduce fluorescence in most cases.

Figure 18-7 illustrates an example where the Nd-YAG source completely eliminates background fluorescence. The upper curve was obtained with conventional Raman equipment using the 514.5-nm line from an argon-ion laser for excitation. The sample was anthracene, and most of the recorded signal arises from the fluorescence of that compound. The lower curve in blue is for the same sample recorded with a Fourier transform spectrometer equipped with a Nd-YAG laser that emitted at 1064 nm. Note the total absence of fluorescence background signal.

The excitation wavelength in Raman spectrometry must be carefully chosen. Not only is photodecomposition and fluorescence a problem but colored samples

and some solvents can absorb the incident radiation or the Raman-scattered radiation. Thus, there is a need for more than one source or for multiple wavelength sources.

18B-2 Sample-Illumination System

Sample handling for Raman spectroscopic measurements is simpler than for IR spectroscopy because glass can be used for windows, lenses, and other optical components instead of the more fragile and atmospherically less stable crystalline halides. In addition, the laser source is easily focused on a small sample area and the emitted radiation efficiently focused on a slit or entrance aperture of a spectrometer. As a result, very small samples can be investigated. In fact, a common sample holder for nonabsorbing liquid samples is an ordinary glass-melting-point capillary.

Gas Samples

Gases are normally contained in glass tubes 1–2 cm in diameter and about 1 mm thick. Gases can also be sealed in small capillary tubes. For weak scatterers, an external multiple-pass setup with mirrors can be used

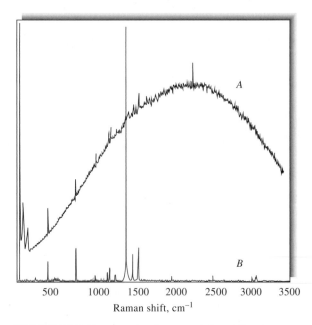

FIGURE 18-7 Spectra of anthracene taken with a conventional Raman instrument with an argon-ion laser source at 514.5 nm (*A*) and with an FT-Raman instrument with a Nd-YAG source at 1064 nm (*B*). (From B. Chase, *Anal. Chem.*, **1987**, *59*, 881A. Figure 5, p. 888A. Copyright 1987 American Chemical Society.)

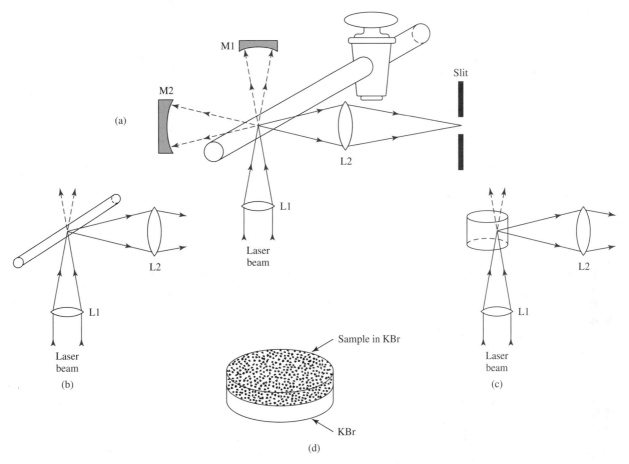

FIGURE 18-8 Sample illumination systems for Raman spectrometry. In (a), a gas cell is shown with external mirrors for passing the laser beam through the sample multiple times. Liquid cells can be capillaries (b) or cylindrical cells (c). Solids can be determined as powders packed in capillaries or as KBr pellets (d).

as shown in Figure 18-8a. The resulting Raman scattering perpendicular to the sample tube and to the excitation laser beam is then focused on the entrance slit of the spectrometer by a large lens (L2 in the figure).

Liquid Samples

Liquids can be sealed in ampoules, glass tubes, or capillaries. Figure 18-8b and c show two of many systems for illuminating liquids. In Figure 18-8b a capillary cell is shown. Capillaries can be as small as 0.5–0.1 mm bore and 1 mm long. The spectra of nanoliter volumes of sample can be obtained with capillary cells. A large cylindrical cell, such as that illustrated in Figure 18-8c, can be used to reduce local heating, particularly for absorbing samples. The laser beam is focused to an area near the wall to minimize absorption of the incident

beam. Further reduction of localized heating is often achieved by rotating the cell.

A major advantage of sample handling in Raman spectroscopy compared with IR arises because water is a weak Raman scatterer but a strong absorber of IR radiation. Thus, aqueous solutions can be studied by Raman spectroscopy but only with difficulty by IR. This advantage is particularly important for biological and inorganic systems and in studies dealing with water pollution.

Solid Samples

Raman spectra of solid samples are often acquired by filling a small cavity or capillary with the sample after it has been ground to a fine powder. Polymers can usually be examined directly with no sample pretreatment.

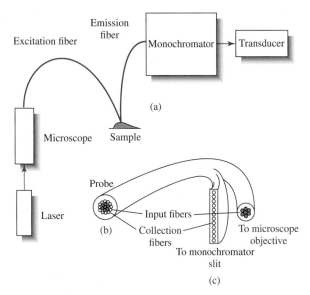

(a)

(b)

(c)

FIGURE 18-9 Raman spectrometer with fiber-optic probe. In (a) a microscope objective focuses the laser radiation onto excitation fibers that transport the beam to the sample. The Raman scattering is collected by emission fibers and carried to the entrance slit of a monochromator or to the entrance of an interferometer. A radiation transducer, such as a photomultiplier tube, converts the scattered light intensity to a proportional current or pulse rate; (b) end view of the probe; (c) end view of collection fibers at entrance slit of monochromator. The colored circles represent the input fiber and the uncolored circles the collection fibers. (Adapted from R. L. McCreery, M. Fleischmann, and P. Hendra, *Anal. Chem.*, **1983**, *55*, 146. Figure 1, p. 147. Copyright 1983 American Chemical Society.)

In some cases, KBr pellets similar to those used in IR spectroscopy are employed as shown in Figure 18-8d. Dilution with KBr can reduce decomposition of the sample produced by local heating.

Fiber-Optic Sampling

One of the significant advantages of Raman spectrometry is that it is based on visible or near-IR radiation that can be transmitted for a considerable distance (as much as 100 m or more) through optical fibers. Figure 18-9 shows the arrangement of a typical Raman instrument that uses a fiber-optic probe. Here, a microscope objective lens is used to focus the laser excitation beam on one end of an excitation fiber of a fiber bundle. These fibers bring the excitation radiation to the sample. Fibers can be immersed in liquid samples or used to illuminate solids. A second fiber or fiber bundle collects the Raman scattering and transports it to

the entrance slit of the spectrometer. Several commercial instruments are now available with such probes.

The Raman spectrum shown in Figure 18-10 illustrates how a fiber-optic probe can be used to monitor chemical processes. In this case a fiber-optic probe was used to monitor the hanging drop crystallization of aprotinin (a serine protease inhibitor) and $(NH_4)_2SO_4$ in aqueous solution. Raman bands were attributed to both the protein and the salt. By using chemometric techniques, changes in the spectrum during crystallization were correlated with depletion of both the protein and the salt. The authors were able to determine accurately supersaturation of aprotinin using this technique.

Fiber-optic probes are proving very useful for obtaining Raman spectra in locations remote from the sample site. Examples include hostile environments, such as hazardous reactors or molten salts; biological samples, such as tissues and arterial walls; and environmental samples, such as groundwater and seawater.

Raman Microprobe

A popular accessory for Raman spectrometers is the Raman microprobe. The first developments in Raman microscopy occurred in the 1970s. Today, several instrument companies make microprobe attachments. With these, the sample is placed on the stage of a microscope where it is illuminated by visible light. After selecting the area to be viewed and adjusting the focus, the illumination lamp is turned off and the exciting laser beam is directed to the sample. With modern optics, the Raman microprobe can obtain high-quality Raman spectra without sample preparation on picogram amounts of sample with 1-μm spatial resolution.

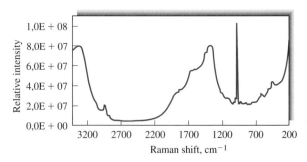

FIGURE 18-10 Raman spectrum of an aqueous solution containing aprotinin (100 mg/mL) and $(NH_4)_2SO_4$ (1.0 M) in 50 mM sodium acetate buffer at pH 4.5 and 24°C. A diode laser source at 785 nm was used with a CCD detector. (From R. E. Tamagawa, E. A. Miranda, and K. A. Berglund, *Cryst. Growth Des.*, **2002**, *2*, 511. Figure 1, p. 512. Copyright 2002 American Chemical Society.)

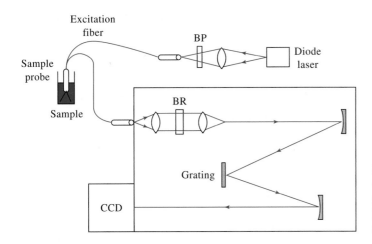

FIGURE 18-11 Fiber-optic Raman spectrometer with spectrograph and CCD detector. The bandpass filter (BP) is used to isolate a single laser line. The band-rejection filter (BR) minimizes the Rayleigh-scattered radiation.

18B-3 Raman Spectrometers

Until the early 1980s, Raman spectrometers were similar in design and used the same type of components as the classical UV-visible dispersing instruments described in Section 13D-3. Most spectrometers employed double-grating systems to minimize the amount of stray and Rayleigh-scattered radiation reaching the transducer. Photomultipliers served as transducers. Now, however, most Raman spectrometers being marketed are either Fourier transform instruments equipped with cooled germanium transducers or multichannel instruments based on charge-coupled devices.

Wavelength-Selection Devices and Transducers

A high-quality wavelength-selection device is required in Raman spectroscopy to separate the relatively weak Raman lines from the intense Rayleigh-scattered radiation. Traditional dispersive Raman spectrometers used double- or even triple-grating monochromators for this purpose. In recent years, holographic interference filters, called *notch filters*, and holographic gratings have improved to the extent that they have virtually eliminated the need for multiple-grating monochromators. In fact, the combination of a notch filter and a high-quality grating monochromator is now found in most commercial dispersive instruments.

Instruments with monochromators invariably use photomultiplier tubes as transducers because of the weak signals being measured. Most spectrometers also employ photon-counting systems to measure the Raman intensity. Because photon counting is inherently a

![icon] *Tutorial*: Learn more about **Raman instrumentation**.

digital technique, such systems are readily interfaced to modern computer data systems.

Many newer Raman instruments have replaced the single-wavelength output monochromator with a spectrograph and an array detector. The photodiode array was the first array detector to be used. It allows the simultaneous collection of entire Raman spectra. Photodiode arrays are typically used in conjunction with an image intensifier to amplify the weak Raman signal.

More recently, charge-transfer devices, such as charge-coupled devices (CCDs) and charge-injection devices (CIDs), have been employed in Raman spectrometers. Figure 18-11 shows a fiber-optic Raman spectrometer that uses a CCD as a multichannel detector. Here, high-quality bandpass and band-rejection (notch) filters provide good stray light rejection. The CCD array can be a two-dimensional array or in some cases a linear array.

Fourier Transform Raman Spectrometers

The Fourier transform Raman (FT-Raman) instrument uses a Michelson interferometer, similar to that used in FTIR spectrometers, and a continuous-wave (CW) Nd-YAG laser as shown in Figure 18-12. The use of a 1064-nm (1.064-μm) source virtually eliminates fluorescence and photodecomposition of samples. Hence, dyes and other fluorescing compounds can be investigated with FT-Raman instruments. The FT-Raman instrument also provides superior frequency precision relative to conventional instruments, which enable spectral subtractions and high-resolution measurements.

One disadvantage of the FT-Raman spectrometer is that water absorbs in the 1000-nm region, which can cancel the Raman advantage of being able to use aque-

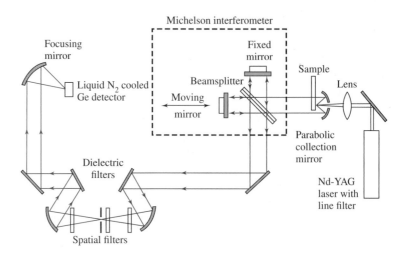

FIGURE 18-12 Optical diagram of an FT-Raman instrument. The laser radiation passes through the sample and then into the interferometer, consisting of the beamsplitter and the fixed and movable mirrors. The output of the interferometer is then extensively filtered to remove stray laser radiation and Rayleigh scattering. After passing through the filters, the radiation is focused onto a cooled Ge detector.

ous solutions. Also, optical filtering, as shown in Figure 18-12, is a necessity. The stray light from the exciting laser must be eliminated because it can saturate many transducers. The Rayleigh-scattered line is often six orders of magnitude greater than the Stokes-shifted Raman lines, and the intensity of this line must be minimized before striking the transducer. Holographic notch filters and other filter types are used for this purpose. Because the Raman scattering from a Nd-YAG laser can occur at wavelengths as long as 1700 nm, photomultipliers and many array detectors are not used. Most FT-Raman instruments instead use InGaAs, Ge, and other photoconductive devices as transducers. These devices are usually operated at cryogenic temperatures.

The FT-Raman spectrometer has a number of unique advantages for Raman spectrometry. However, the limitations previously noted mean that dispersive Raman instruments will be widely used for some time.

18C APPLICATIONS OF RAMAN SPECTROSCOPY

Raman spectroscopy has been applied to the qualitative and quantitative analysis of inorganic, organic, and biological systems.[4]

18C-1 Raman Spectra of Inorganic Species

The Raman technique is often superior to IR spectroscopy for investigating inorganic systems because aqueous solutions can usually be employed.[5] In addition, the vibrational energies of metal-ligand bonds are generally in the range of 100 to 700 cm^{-1}, a region of the IR that is experimentally difficult to study. These vibrations are frequently Raman active, however, and lines with $\Delta\bar{\nu}$ values in this range are readily observed. Raman studies are potentially useful sources of information concerning the composition, structure, and stability of coordination compounds. For example, numerous halogen and halogenoid complexes produce Raman spectra and thus are amenable to investigation by this means. Metal-oxygen bonds are also Raman active. Spectra for such species as VO_3^{4-}, $Al(OH)_4^-$, $Si(OH)_6^{2-}$, and $Sn(OH)_6^{2-}$ have been obtained. Raman studies have been useful in determining the probable structures of these and similar species. For example, in perchloric acid solutions, vanadium(IV) appears to be present as $VO^{2+}(aq)$ rather than as $V(OH)_2^{2+}(aq)$. Studies of boric acid solutions show that the anion formed by acid dissociation is the tetrahedral $B(OH)_4^-$ rather than $H_2BO_3^-$. Dissociation constants for strong acids such as H_2SO_4, HNO_3, H_2SeO_4, and H_5IO_6 have been calculated from Raman measurements. It seems probable that the future will see even wider use of

[4] See *Analytical Raman Spectroscopy*, J. G. Grasselli and B. J. Bulkin, eds., New York: Wiley, 1991.

[5] See K. Nakamoto, *Infrared and Raman Spectra of Inorganic and Coordination Compounds*, 5th ed., New York: Wiley, 1996.

Raman spectroscopy for theoretical verification and structural studies of inorganic systems.

18C-2 Raman Spectra of Organic Species

Raman spectra are similar to IR spectra in that they have regions useful for functional group detection and fingerprint regions that permit the identification of specific compounds. Daimay et al. have published a comprehensive treatment of Raman functional group frequencies.[6]

Raman spectra yield more information about certain types of organic compounds than do their IR counterparts. For example, the double-bond stretching vibration for olefins results in weak and sometimes undetected IR absorption. On the other hand, the Raman band (which like the IR band, occurs at about 1600 cm^{-1}) is intense, and its position is sensitive to the nature of substituents as well as to their geometry. Thus, Raman studies are likely to yield useful information about the olefinic functional group that may not be revealed by IR spectra. This statement applies to cycloparaffin derivatives as well; these compounds have a characteristic Raman band in the region of 700 to 1200 cm^{-1}. This band has been attributed to a breathing vibration in which the nuclei move in and out symmetrically with respect to the center of the ring. The position of the band decreases continuously from 1190 cm^{-1} for cyclopropane to 700 cm^{-1} for cyclooctane; Raman spectroscopy thus appears to be an excellent diagnostic tool for the estimation of ring size in paraffins. The IR band associated with this vibration is weak or nonexistent.

18C-3 Biological Applications of Raman Spectroscopy

Raman spectroscopy has been applied widely for the study of biological systems.[7] The advantages of this technique include the small sample requirement, the minimal sensitivity to water, the spectral detail, and the conformational and environmental sensitivity.

18C-4 Quantitative Applications

Raman spectra tend to be less cluttered with bands than IR spectra. As a consequence, peak overlap in mixtures is less likely, and quantitative measurements are simpler. In addition, Raman sampling devices are not subject to attack by moisture, and small amounts of water in a sample do not interfere. Despite these advantages, Raman spectroscopy has only recently been exploited widely for quantitative analysis. The increasing use of Raman spectroscopy is due to the availability of inexpensive, routine Raman instrumentation.

Because laser beams can be precisely focused, it becomes possible to perform quantitative analyses on very small samples. The Raman microprobe has been used to determine analytes in single bacterial cells, components in individual particles of smoke and fly ash, and species in microscopic inclusions in minerals. Surfaces have been examined by tuning the instrument to a given vibrational mode. This results in an image of regions on a surface where a particular bond or functional group is present.

The Raman microprobe has played a critical role in the authentication of some presumed ancient documents such as the Vinland map (see the Instrumental Analysis in Action feature at the end of Section 3). In the case of the map, the presence of TiO_2 in the ink was shown conclusively by Raman microscopy.

18D OTHER TYPES OF RAMAN SPECTROSCOPY

Advancements in tunable lasers led to several new Raman spectroscopic methods in the early 1970s. A brief discussion of the applications of some of these techniques follows.

18D-1 Resonance Raman Spectroscopy

Resonance Raman scattering refers to a phenomenon in which Raman line intensities are greatly enhanced by excitation with wavelengths that closely approach that of an electronic absorption band of an analyte.[8] Under this circumstance, the magnitudes of Raman

---

[6]L. Daimay, N. B. Colthup, W. G. Fately, and J. G. Grasselli, *The Handbook of Infrared and Raman Characteristic Frequencies of Organic Molecules,* New York: Academic Press, 1991.

[7]See J. R. Ferraro, K. Nakamoto, and C. W. Brown, *Introductory Raman Spectroscopy,* 2nd ed., San Diego: Academic Press, 2003, Chap. 6; *Infrared and Raman Spectroscopy of Biological Materials,* H. U. Gremlich and B. Yan, eds., New York: Marcel Dekker, 2001; *Biological Applications of Raman Spectroscopy,* T. G. Spiro, ed., Vols. 1–3, New York: Wiley, 1987–88.

[8]For brief reviews, see T. G. Spiro and R. S. Czernuszewicz in *Physical Methods in Bioinorganic Chemistry,* L. Que, ed., Sausalito, CA: University Science Books, 2000; S. A. Asher, *Anal. Chem.,* **1993,** *65,* 59A.

lines associated with the most symmetric vibrations are enhanced by a factor of 10^2 to 10^6. As a consequence, resonance Raman spectra have been obtained at analyte concentrations as low as 10^{-8} M. This level of sensitivity is in contrast to normal Raman studies, which are ordinarily limited to concentrations greater than 0.1 M. Furthermore, because resonance enhancement is restricted to the Raman bands associated with the chromophore, resonance Raman spectra are usually quite selective.

Figure 18-13a illustrates the energy changes responsible for resonance Raman scattering. This figure differs from the energy diagram for normal Raman scattering (Figure 18-3) in that the electron is promoted into an excited electronic state followed by an immediate relaxation to a vibrational level of the electronic ground state. As shown in the figure, resonance Raman scattering differs from fluorescence (Figure 18-13b) in that relaxation to the ground state is not preceded by prior relaxation to the lowest vibrational level of the excited electronic state. The time scales for the two phenomena are also quite different, with Raman relaxation occurring in less than 10^{-14} s compared with the 10^{-6} to 10^{-10} s for fluorescence emission.

Line intensities in a resonance Raman experiment increase rapidly as the excitation wavelength approaches the wavelength of the electronic absorption band. Thus, to achieve the greatest signal enhancement for a broad range of absorption maxima, a tunable laser is required. With intense laser radiation, sample decomposition can become a major problem because electronic absorption bands often occur in the ultraviolet region. To circumvent this problem, it is common practice to circulate the sample past the focused beam of the laser. Circulation is normally accomplished in one of two ways: by pumping a solution or liquid through a capillary mounted in the sample position or by rotating a cylindrical cell containing the sample through the laser beam. Thus, only a small fraction of the sample is irradiated at any instant, and heating and sample decomposition are minimized.

Perhaps the most important application of resonance Raman spectroscopy has been to the study of biological molecules under physiologically significant conditions; that is, in the presence of water and at low to moderate concentration levels. As an example, the technique has been used to determine the oxidation state and spin of iron atoms in hemoglobin and cytochrome c. In these molecules, the resonance Raman bands are due solely to vibrational modes of the tetra-

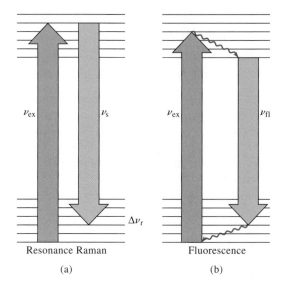

FIGURE 18-13 Energy diagram for (a) resonance Raman scattering and (b) fluorescence emission. Radiationless relaxation is shown as wavy arrows. In the resonance Raman case, the excited electron immediately relaxes into a vibrational level of the ground electronic state giving up a Stokes photon ν_s. In fluorescence, relaxation to the lowest vibrational level of the excited electronic state occurs prior to emission. Resonance Raman scattering is nearly instantaneous, and the spectral bands are very narrow. Fluorescence emission usually takes place on the nanosecond time scale. Fluorescence spectra are usually broad because of the many vibrational states.

pyrrole chromophore. None of the other bands associated with the protein is enhanced, and at the concentrations normally used these bands do not interfere as a consequence.

Time-resolved resonance Raman spectrometry is a technique that allows collection of Raman spectra of excited state molecules. It has been used to study intermediates in enzyme reactions, the spectra of carotenoid excited states, ultrafast electron transfer steps, and a variety of other biological and bioinorganic processes.[9] Time-discrimination methods have been used to overcome a major limitation of resonance Raman spectroscopy, namely, fluorescence interference either by the analyte itself or by other species present in the sample.

[9]J. R. Kincaid and K. Czarnecki, in *Comprehensive Coordination Chemistry II*, J. A. McCleverty and T. J. Meyer, eds., Oxford: Elsevier, 2004.

18D-2 Surface-Enhanced Raman Spectroscopy

Surface-enhanced Raman spectroscopy (SERS)[10] involves obtaining Raman spectra in the usual way on samples that are adsorbed on the surface of colloidal metal particles (usually silver, gold, or copper) or on roughened surfaces of pieces of these metals. For reasons that are finally becoming understood, at least semiquantitatively, the Raman lines of the adsorbed molecule are often enhanced by a factor of 10^3 to 10^6. When surface enhancement is combined with the resonance enhancement technique discussed in the previous section, the net increase in signal intensity is roughly the product of the intensity produced by each of the techniques. Consequently, detection limits in the range of 10^{-9} to 10^{-12} M have been observed.

Several sample-handling techniques are employed for SERS. In one technique, colloidal silver or gold particles are suspended in a dilute solution (usually aqueous) of the sample. The solution is then held or flowed through a narrow glass tube while it is excited by a laser beam. In another method, a thin film of colloidal metal particles is deposited on a glass slide and a drop or two of the sample solution spotted on the film. The Raman spectrum is then obtained in the usual manner. Alternatively, the sample may be deposited electrolytically on a roughened metal electrode, which is then removed from the solution and exposed to the laser excitation source.

18D-3 Nonlinear Raman Spectroscopy

In Section 7B-3, we pointed out that many lasers are intense enough to produce significant amounts of nonlinear radiation. Throughout the 1970s and 1980s, many Raman techniques were developed that depend on polarization induced by second-order and higher field strengths. These techniques are termed *nonlinear Raman methods*.[11] Included in these methods are *stimulated Raman scattering, the hyper-Raman effect, stimulated Raman gain, inverse Raman spectroscopy, coherent anti-Stokes Raman spectroscopy*, and *coherent Stokes Raman spectroscopy*. The most widely used of these methods is coherent anti-Stokes Raman spectroscopy, or CARS.

Nonlinear techniques have been used to overcome some of the drawbacks of conventional Raman spectroscopy, particularly its low efficiency, its limitation to the visible and near-ultraviolet regions, and its susceptibility to interference from fluorescence. A major disadvantage of nonlinear methods is that they tend to be analyte specific and often require several different tunable lasers to be applicable to diverse species. To date, none of the nonlinear methods has found widespread application among nonspecialists. However, many of these methods have shown considerable promise. As less expensive and more routinely useful lasers become available, nonlinear Raman methods, particularly CARS, should become more widely used.

[10]See M. J. Weaver, S. Zou, and H. Y. Chan, *Anal. Chem.*, **2000**, *72*, 38A–47A. American Chemical Society.

[11]See J. R. Ferraro, K. Nakamoto, and C. W. Brown, *Introductory Raman Spectroscopy*, 2nd ed., San Diego: Academic Press, 2003, pp. 194–202.

QUESTIONS AND PROBLEMS

*Answers are provided at the end of the book for problems marked with an asterisk.

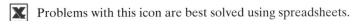 Problems with this icon are best solved using spreadsheets.

18-1 What is a virtual state?

18-2 Why does the ratio of anti-Stokes to Stokes intensities increase with sample temperature?

***18-3** At what wavelengths in nanometers would the Stokes and anti-Stokes Raman lines for carbon tetrachloride ($\Delta \bar{\nu} = 218, 314, 459, 762$, and 790 cm^{-1}) appear if the source were
(a) a helium-neon laser (632.8 nm)?
(b) an argon-ion laser (488.0 nm)?

***18-4** Assume the excitation sources in Problem 18-3 have the same power. (a) Compare the relative intensities of the CCl_4 Raman lines when each of the two excitation sources is used. (b) If the intensities were recorded with a typical monochromator-photomultiplier system, why would the measured intensity ratios differ from the ratio calculated in part (a)?

*18-5 For vibrational states, the Boltzmann equation can be written as

$$\frac{N_1}{N_0} = \exp(-\Delta E / kT)$$

where N_0 and N_1 are the populations of the lower and higher energy states, respectively, ΔE is the energy difference between the states, k is Boltzmann's constant, and T is the temperature in kelvins.

For temperatures of 20°C and 40°C, calculate the ratios of the intensities of the anti-Stokes and Stokes lines for CCl_4 at (a) 218 cm^{-1}; (b) 459 cm^{-1}; (c) 790 cm^{-1}.

18-6 The following questions deal with laser sources in Raman spectroscopy.
   (a) Under what circumstances would a helium-neon laser be preferable to an argon-ion laser?
   (b) Under what circumstances would a diode laser be preferable to an argon-ion or helium-neon laser?
   (c) Why are ultraviolet emitting sources avoided?

*18-7 The following Raman data were obtained for $CHCl_3$ with the polarizer of the spectrometer set (1) parallel to the plane of polarization of the laser and (2) at 90° to the plane of the source.

Relative Intensities

	$\Delta \bar{\nu}$, cm^{-1}	(1) $I_{\parallel}$	(2) $I_{\perp}$
(a)	760	0.60	0.46
(b)	660	8.4	0.1
(c)	357	7.9	0.6
(d)	258	4.2	3.2

Calculate the depolarization ratio and indicate which Raman lines are polarized.

18-8 Discuss the advantages and disadvantages of FT-Raman spectrometers compared to conventional dispersive Raman instruments.

Challenge Problem

18-9 The following questions all deal with the similarities and differences between IR spectrometry and Raman spectrometry.
   (a) What are the requirements for a vibrational mode in a molecule to show IR absorption? What are the requirements for a vibrational mode to be Raman active? Why do these requirements differ? Under what circumstances will vibrational modes be both Raman and IR active? Under what circumstances will vibrational modes be Raman active but not IR active and vice versa?
   (b) Consider the molecule chloroacetonitrile ($ClCH_2CN$). How many vibrational modes should this molecule have? Why might one observe fewer Raman bands than expected?
   (c) Chloroacetonitrile shows a strong Raman band at 2200 cm^{-1} due to the C—N stretching mode. The corresponding IR absorption is very weak or absent. By

comparing spectra in the 2200 cm^{-1} region, what can you conclude about the C—N stretching mode in chloroacetonitrile?

(d) Compare and contrast IR and Raman spectrometry with respect to optics, cell materials, sample handling, solvent compatibility, and applicability to various sample types.

(e) Compare and contrast the sources and transducers used in Raman spectrometers to those used in FTIR instruments. Consider both FT-Raman and dispersive Raman spectrometers in your comparison.

(f) Compare and contrast IR and Raman spectrometry with respect to qualitative usefulness, detection limits, quantitative analysis, and instrumental complexity.

CHAPTER NINETEEN

Nuclear Magnetic Resonance Spectroscopy

Nuclear magnetic resonance (NMR) spectroscopy is based on the measurement of absorption of electromagnetic radiation in the radio-frequency region of roughly 4 to 900 MHz. In contrast to UV, visible, and IR absorption, nuclei of atoms rather than outer electrons are involved in the absorption process. Furthermore, to cause nuclei to develop the energy states required for absorption to occur, it is necessary to place the analyte in an intense magnetic field. In this chapter we describe the theory, instrumentation, and applications of NMR spectroscopy.

NMR spectroscopy is one of the most powerful tools available to chemists and biochemists for elucidating the structure of chemical species. The technique is also useful for the quantitative determination of absorbing species.

Throughout this chapter, this logo indicates an opportunity for online self-study at **www.thomsonedu.com/chemistry/skoog**, linking you to interactive tutorials, simulations, and exercises.

The theoretical basis for NMR spectroscopy[1] was proposed by W. Pauli in 1924. He suggested that certain atomic nuclei have the properties of spin and magnetic moment and that, as a consequence, exposure to a magnetic field would lead to splitting of their energy levels. During the next decade, these postulates were verified experimentally. It was not until 1946, however, that Felix Bloch at Stanford and Edward Purcell at Harvard, working independently, demonstrated that nuclei absorb electromagnetic radiation in a strong magnetic field as a result of the energy level splitting that is induced by the magnetic field. The two physicists shared the 1952 Nobel Prize in Physics for their work.

In the first few years following the discovery of NMR, chemists became aware that the molecular environment influences the absorption of radio-frequency (RF) radiation by a nucleus in a magnetic field and that this effect can be correlated with molecular structure. In 1953 the first high-resolution NMR spectrometer designed for chemical structural studies was marketed by Varian Associates. Since then, the growth of NMR spectroscopy has been explosive, and the technique has had profound effects on the development of organic and inorganic chemistry and biochemistry. It is unlikely that there has ever been as short a delay between a scientific discovery and its widespread acceptance and application.[2]

Two general types of NMR spectrometers are currently in use, *continuous-wave* (CW) and *pulsed*, or *Fourier transform* (FT-NMR), spectrometers. All early studies were carried out with CW instruments. In about 1970, however, FT-NMR spectrometers became available commercially, and now this type of instrument dominates the market. In both types of instruments, the sample is positioned in a powerful magnetic field that has a strength of several tesla.[3] CW spectrometers are

---

[1] The following references are recommended for further study: J. B. Lambert, E. P. Mazzola, *Nuclear Magnetic Resonance Spectroscopy*, Upper Saddle River, NJ: Pearson/Prentice-Hall, 2004; R. M. Silverstein, F. X. Webster, and D. Kiemle, *Spectrometric Identification of Organic Compounds*, 7th ed., New York: Wiley, 2004; M. H. Levitt, *Spin Dynamics: Basics of Nuclear Magnetic Resonance*, New York: Wiley, 2001; E. D. Becker, *High Resolution NMR*, 3rd ed., New York: Academic Press, 2000; H. Günther, *NMR Spectroscopy: Basic Principles, Concepts and Applications in Chemistry*, 2nd ed., Chichester, UK: Wiley, 1995; L. D. Field and B. Sternhell, eds., *Analytical NMR*, Chichester, UK: Wiley, 1989.

[2] For interesting discussions of the history of NMR, see D. L Rabenstein, *Anal. Chem.*, **2001**, *73*, 214A; E. D. Becker, *Anal. Chem.*, **1993**, *65*, 295A; D. C. Lankin, R. R. Ferraro, and R. Jarnutowski, *Spectroscopy*, **1992**, *7* (8), 18.

[3] The SI symbol for magnetic fields is B; an older convention, however, which is still widely used, employed the symbol H instead. The derived unit for describing the field strength is the tesla (T), which is defined as $1\ T = 1\ kg\ s^{-2}\ A^{-1}$. Another unit that was popular in the past and still is frequently encountered is the gauss (G). The relationship between the two units is $10^4\ G = 1\ T$. Also, $1\ T = 1\ Vs/m^2$, where V = volts.

similar in principle to optical absorption instruments in that an absorption signal is monitored as the frequency of the source is slowly scanned. In some instruments, the frequency of the source is held constant while the strength of the field is scanned. In pulsed instruments, the sample is irradiated with periodic pulses of RF energy that are directed through the sample at right angles to the magnetic field. These excitation pulses elicit a time-domain signal that decays in the interval between pulses. This signal is then converted to a frequency-domain signal by using a Fourier transformation to give a spectrum similar to that obtained by using a CW instrument.

Nearly all NMR instruments produced today are of the FT type, and the use of CW instruments is largely limited to special routine applications, such as the determination of the extent of hydrogenation in petroleum process streams and the determination of water in oils, food products, and agricultural materials. Despite this predominance of pulsed instruments in the marketplace, we find it convenient to base our initial development of NMR theory on CW experiments and move from there to a discussion of pulsed NMR measurements.

19A THEORY OF NMR

In common with optical spectroscopy, both classical mechanics and quantum mechanics are useful in explaining the NMR phenomenon. The two treatments yield identical relationships. Quantum mechanics, however, provides a useful relationship between absorption frequencies and nuclear energy states, whereas classical mechanics yields a clear physical picture of the absorption process and how it is measured.

In this section, we first provide a quantum description of NMR applicable to both CW and pulsed NMR measurements. Then, we take a classical approach to NMR and show how it provides a useful picture of CW-NMR. Finally, we complete this section with a discussion of Fourier transform measurements based again on a classical picture.

19A-1 Quantum Description of NMR

To account for the properties of certain nuclei, we must assume that they rotate about an axis and thus have the property of *spin*. Nuclei with spin have angular momentum p. Furthermore, the maximum observable component of this angular momentum is quan-

tized and must be an integral or a half-integral multiple of $h/2\pi$, where h is Planck's constant. The maximum number of spin components or values for p for a particular nucleus is its spin quantum number I. The nucleus will then have $2I + 1$ discrete states. The component of angular momentum for these states in any chosen direction will have values of $I, I - 1, I - 2, \ldots, -I$. In the absence of an external field, the various states have identical energies.

The four nuclei that have been of greatest use to organic chemists and biochemists are ^1H, ^{13}C, ^{19}F, and ^{31}P, and they are the only four we will discuss. The spin quantum number for these nuclei is 1/2. Thus, each nucleus has two spin states corresponding to $I = +1/2$ and $I = -1/2$. Heavier nuclei have spin numbers that range from zero, which implies that they have no net spin component, to at least 9/2.

A spinning, charged nucleus creates a magnetic field analogous to the field produced when electricity flows through a coil of wire. The resulting magnetic moment μ is oriented along the axis of spin and is proportional to the angular momentum p. Thus

$$\mu = \gamma p \qquad (19\text{-}1)$$

where the proportionality constant γ is the *magnetogyric*, or *gyromagnetic*, *ratio*, which has a different value for each type of nucleus. As we shall see, the magnetogyric ratio is also a factor in the proportionality constant in the relationship between the frequency of the absorbed energy and the magnetic field strength (see Equation 19-5). Magnetogyric ratios for the four elements we will be dealing with are found in the second column of Table 19-1.

The relationship between nuclear spin and magnetic moment leads to a set of observable magnetic quantum states m given by

$$m = I, I - 1, I - 2, \ldots, -I \qquad (19\text{-}2)$$

Thus, the nuclei that we will consider have two magnetic quantum numbers, $m = +1/2$ and $m = -1/2$. Note that the rules for determining nuclear quantum numbers are similar to those for electronic quantum numbers.

Energy Levels in a Magnetic Field

As shown in Figure 19-1, when a nucleus with a spin quantum number of 1/2 is brought into an external magnetic field B_0, its magnetic moment becomes oriented in one of two directions with respect to the field, depending on its magnetic quantum state. The poten-

TABLE 19-1 Magnetic Properties of Important Nuclei with Spin Quantum Numbers of 1/2

Nucleus	Magnetogyric Ratio, radian $T^{-1}\,s^{-1}$	Isotopic Abundance, %	Relative Sensitivity[a]	Absorption Frequency, MHz[b]
^1H	2.6752×10^8	99.98	1.00	200.00
^{13}C	6.7283×10^7	1.11	0.016	50.30
^{19}F	2.5181×10^8	100.00	0.83	188.25
^{31}P	1.0841×10^8	100.00	0.066	81.05

[a] At constant field for equal number of nuclei.

[b] At a field strength of 4.69 T.

tial energy E of a nucleus in these two orientations, or quantum states, is given by

$$E = -\frac{\gamma m h}{2\pi} B_0 \qquad (19\text{-}3)$$

The energy for the lower energy $m = +1/2$ state (see Figure 19-1) is given by

$$E_{+1/2} = -\frac{\gamma h}{4\pi} B_0$$

For the $m = -1/2$ state the energy is

$$E_{-1/2} = \frac{\gamma h}{4\pi} B_0$$

Thus, the difference in energy ΔE between the two states is given by

$$\Delta E = \frac{\gamma h}{4\pi} B_0 - \left(-\frac{\gamma h}{4\pi} B_0\right) = \frac{\gamma h}{2\pi} B_0 \qquad (19\text{-}4)$$

As in other types of spectroscopy, transitions between energy states can be brought about by absorption or emission of electromagnetic radiation of a frequency ν_0 that corresponds in energy to ΔE. Thus, by substituting the Planck relationship $\Delta E = h\nu_0$ into Equation 19-4, we obtain the frequency of the radiation required to bring about the transition

$$\nu_0 = \frac{\gamma B_0}{2\pi} \qquad (19\text{-}5)$$

As we suggested previously, the frequency of a magnetic transition is proportional to the applied field strength B_0 with a proportionality constant of $\gamma/2\pi$.

Example 19-1 reveals that RF radiation of approximately 200 MHz is required to bring about a change in alignment of the magnetic moment of the proton from a direction that parallels the field to one that opposes it.

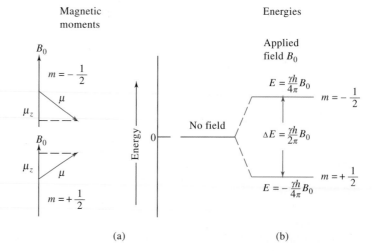

FIGURE 19-1 Magnetic moments and energy levels for a nucleus with a spin quantum number of 1/2.

EXAMPLE 19-1

Many proton NMR instruments employ a magnet that provides a field strength of 4.69 T. At what frequency would the hydrogen nucleus absorb in such a field?

Solution

Substituting the magnetogyric ratio for the proton (Table 19-1) into Equation 19-5, we find

$$\nu_0 = \frac{(2.68 \times 10^8 \text{ T}^{-1}\text{s}^{-1})(4.69 \text{ T})}{2\pi}$$

$$= 2.00 \times 10^8 \text{ s}^{-1} = 200 \text{ MHz}$$

Distribution of Particles between Magnetic Quantum States

In the absence of a magnetic field, the energies of the magnetic quantum states of a nucleus are identical. Consequently, a large collection of protons contains an identical number of nuclei with magnetic quantum numbers $m = +1/2$ and $m = -1/2$. When placed in a magnetic field, however, the nuclei tend to orient themselves so that the lower energy state ($m = +1/2$) predominates. It is instructive to calculate the extent of this predominance in a typical NMR experiment. For this purpose, the Boltzmann equation (Equation 8-1) can be written in the form

$$\frac{N_j}{N_0} = \exp\left(\frac{-\Delta E}{kT}\right) \qquad (19\text{-}6)$$

where N_j is the number of protons in the higher energy state ($m = -1/2$), N_0 is the number in the lower state ($m = +1/2$), k is Boltzmann's constant (1.38×10^{-23} J K^{-1}), T is the absolute temperature, and ΔE is defined by Equation 19-4.

Substituting Equation 19-4 into 19-6 gives

$$\frac{N_j}{N_0} = \exp\left(\frac{-\gamma h B_0}{2\pi kT}\right) \qquad (19\text{-}7)$$

Example 19-2 illustrates that the success of the NMR measurement depends on a remarkably small, ~33 ppm, excess of lower-energy protons. If the numbers of nuclei in the two states were identical, however, we would observe no net absorption because the number of particles excited by the radiation would exactly equal the number producing *induced* emission.

EXAMPLE 19-2

Calculate the relative number of protons in the higher and lower magnetic states when a sample is placed in a 4.69 T field at 20°C.

Solution

Substituting numerical values into Equation 19-7 gives

$$\frac{N_j}{N_0} = \exp\left(\frac{-(2.68 \times 10^8 \text{ T}^{-1}\text{ s}^{-1})(6.63 \times 10^{-34} \text{ J}\cdot\text{s})(4.69 \text{ T})}{2\pi(1.38 \times 10^{-23}\text{J K}^{-1})(293 \text{ K})}\right)$$

$$= e^{-3.28 \times 10^{-5}} = 0.999967$$

or

$$\frac{N_0}{N_j} = 1.000033$$

Thus, for exactly 10^6 protons in higher energy states there will be

$$N_0 = 10^6/0.999967 = 1,000,033$$

in the lower energy state. This figure corresponds to a 33-ppm excess.

If we expand the right side of Equation 19-7 as a Maclaurin series and truncate the series after the second term, we obtain the important result that

$$\frac{N_j}{N_0} = 1 - \frac{\nu h B_0}{2\pi kT} \qquad (19\text{-}8)$$

Equation 19-8 demonstrates that the relative number of excess low-energy nuclei is linearly related to the magnetic field strength. Thus, the intensity of an NMR signal increases linearly as the field strength increases. This dependence of signal sensitivity on magnetic field strength has led manufacturers to produce magnets with field strengths as large as 14 T.

19A-2 Classical Description of NMR

To understand the absorption process, and in particular the measurement of absorption, a classical picture of the behavior of a charged particle in a magnetic field is helpful.

Precession of Nuclei in a Field

Let us first consider the behavior of a nonrotating magnetic body, such as a compass needle, in an external magnetic field. If momentarily displaced from

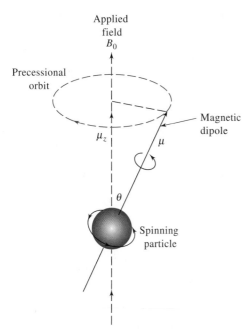

FIGURE 19-2 Precession of a rotating particle in a magnetic field.

alignment with the field, the needle will swing in a plane about its pivot as a consequence of the force exerted by the field on its two ends; in the absence of friction, the ends of the needle will fluctuate back and forth indefinitely about the axis of the field. A quite different motion occurs, however, if the magnet is spinning rapidly around its north-south axis. Because of the gyroscopic effect, the force applied by the field to the axis of rotation causes movement not in the plane of the force but perpendicular to this plane; the axis of the rotating particle, therefore, moves in a circular path. That is, the rotational axis of the rotating particle precesses around the vector representing the applied magnetic field. This motion, illustrated in Figure 19-2, is similar to the motion of a gyroscope when it is displaced from the vertical by application of a lateral force. The angular velocity of this motion ω_0, in radians per second, is given by

$$\omega_0 = \gamma B_0 \tag{19-9}$$

The angular velocity can be converted to the frequency of precession ν_0, known as the *Larmor frequency*, by dividing by 2π. Thus,

$$\nu_0 = \frac{\gamma B_0}{2\pi} \tag{19-10}$$

A comparison of Equation 19-10 with Equation 19-5 reveals that the Larmor frequency is identical to the

frequency of absorbed radiation derived from quantum mechanical considerations.

Absorption in CW Experiments

The potential energy E of the precessing charged particle shown in Figure 19-2 is given by

$$E = -\mu_z B_0 = -\mu B_0 \cos \theta \tag{19-11}$$

where θ is the angle between the magnetic field vector and the spin axis of the particle, μ is the magnetic moment of the particle, and μ_z is the component of μ in the direction of the magnetic field. When RF energy is absorbed by a nucleus, its angle of precession θ must change. Hence, we imagine for a nucleus having a spin quantum number of 1/2 that absorption involves a flipping of the magnetic moment oriented in the field direction to the opposite direction. The process is pictured in Figure 19-3. For the magnetic dipole to flip, there must be a magnetic force at right angles to the fixed field that moves in a circular path in phase with the precessing dipole. The magnetic moment of *circularly polarized radiation* of a suitable frequency has these necessary properties; that is, the magnetic vector of such radiation has a circular component, as represented by the dashed circle in Figure 19-3.[4] If the rotational frequency of the magnetic vector of the radiation is the same as the precessional frequency of a nucleus, absorption and flipping can occur. As discussed in the next paragraph, circularly polarized radiation of suitable frequency can be produced by an RF oscillator coil.

The radiation produced by the coil of an RF oscillator, which serves as the source in NMR instruments, is plane polarized. Plane-polarized radiation, however, consists of d and l circularly polarized radiation. As shown in Figure 19-4b, the vector of the d component rotates clockwise as the radiation approaches the observer; the vector of the l component rotates in the opposite sense. Addition of the two vectors leads to a vector sum that vibrates in a single plane (Figure 19-4a).

Thus, electromagnetic radiation from an oscillator coil oriented at 90° to the direction of the fixed magnetic field introduces circularly polarized radiation into the sample volume in the proper plane for absorption by sample nuclei. Only the magnetic compo-

---

[4] It is important to note here that in contrast to optical spectroscopy, where it is the electric field of electromagnetic radiation that interacts with absorbing species, in NMR spectroscopy it is the *magnetic field* of the radiation that excites absorbing species.

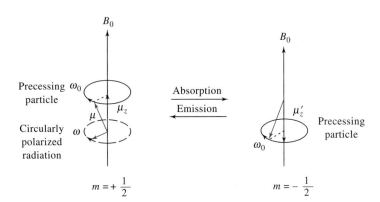

$m = +\frac{1}{2}$ $\qquad$ $m = -\frac{1}{2}$

FIGURE 19-3 Model for the absorption of radiation by a precessing particle.

nent of the excitation radiation that rotates in the precessional direction is absorbed.

Relaxation Processes in NMR

When a nucleus is exposed to radiation of a suitable frequency, absorption occurs because of the slight excess of lower-energy-state nuclei present in the strong magnetic field. This excess is small, as indicated by the result of Example 19-2, so there is always danger that the absorption process will equalize the number of nuclei in the two states and cause the absorption signal to decrease and to approach zero. When this occurs, the spin system is said to be *saturated*. To avoid saturation, the rate of relaxation of excited nuclei to their lower energy state must be as great or greater than the rate at which they absorb the RF energy. One apparent relaxation path is the emission of radiation of

 Tutorial: Learn more about **NMR theory**.

a frequency corresponding to the energy difference between the states, which results in fluorescence. Radiation theory, however, shows that the probability of spontaneous reemission of photons varies as the cube of the frequency and that at radio frequencies this process does not occur to a significant extent. In NMR studies, then, nonradiative relaxation processes are of prime importance.

To reduce saturation and produce a readily detectable absorption signal, relaxation should occur as rapidly as possible; that is, the lifetime of the excited state should be small. A second factor — the inverse relationship between the lifetime of an excited state and the width of its absorption line — negates the advantage of very short lifetimes. Thus, when relaxation rates are high, or the lifetimes low, line broadening prevents high-resolution measurements. These two opposing factors cause the optimal half-life for an excited species to range from about 0.1 to 10 s.

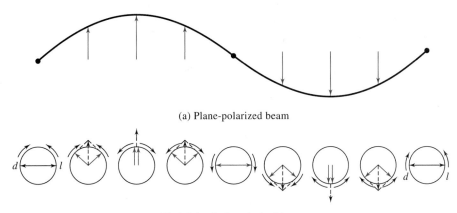

(a) Plane-polarized beam

(b) *d, l* circularly polarized beams

FIGURE 19-4 Equivalency of a plane-polarized beam to two (*d, l*) circularly polarized beams of radiation.

Two types of relaxation processes are important in NMR spectroscopy: (1) *spin-lattice*, or *longitudinal*, *relaxation* and (2) *spin-spin*, or *transverse*, *relaxation*.

Spin-Lattice Relaxation. The absorbing nuclei in an NMR experiment are part of the larger collection of atoms that constitute the sample. The entire collection is termed the *lattice*, regardless of whether the sample is a solid, a liquid, or a gas. In the latter two states, particularly, the various nuclei comprising the lattice are in violent vibrational and rotational motion, which creates a complex field about each magnetic nucleus. As a result, the lattice field contains a continuum of magnetic components, at least some of which must correspond in frequency and phase with the precessional frequency of the magnetic nucleus of interest. These vibrationally and rotationally developed components interact with and convert nuclei from a higher to a lower spin state; the absorbed energy then simply increases the amplitude of the thermal vibrations or rotations. This change produces a minuscule temperature rise in the sample.

Spin-lattice relaxation is a *first-order exponential decay* characterized by a relaxation time T_1, which is a measure of the average lifetime of the nuclei in the higher-energy state. In addition to depending on the magnetogyric ratio of the absorbing nuclei, T_1 is strongly influenced by the mobility of the lattice. In crystalline solids and viscous liquids, where mobilities are low, T_1 is large. As the mobility increases (at higher temperatures, for example), the vibrational and rotational frequencies increase, enhancing the probability of a magnetic fluctuation of the proper magnitude for a relaxation transition. As a consequence, T_1 becomes shorter. At very high mobilities, on the other hand, the fluctuation frequencies are further increased and spread over such a broad range that the probability of a suitable frequency for a spin-lattice transition again decreases. The result is a minimum in the relationship between T_1 and lattice mobility.

Spin-Spin Relaxation. Several other effects tend to diminish relaxation times and thereby broaden NMR lines. These effects are normally lumped together and described by a transverse, or spin-spin, relaxation time T_2. Values for T_2 are generally so small for crystalline solids or viscous liquids (as low as 10^{-4} s) as to preclude the use of samples of these kinds for high-resolution spectra unless special techniques are employed. These

techniques are described briefly in a later section dealing with ^{13}C NMR studies of solids.

When two neighboring nuclei of the same kind have identical precession rates, but are in different magnetic quantum states, the magnetic fields of each can interact to cause an interchange of states. That is, a nucleus in the lower spin state is excited, and the excited nucleus relaxes to the lower energy state. No net change in the relative spin-state population, and thus no decrease in saturation, results, but the average lifetime of a particular excited nucleus is shortened. Line broadening is the result.

Two other causes of line broadening should be noted. Both arise if B_0 in Equation 19-10 differs slightly from nucleus to nucleus. Under these circumstances, a band of frequencies, rather than a single frequency, is absorbed. One cause for such a variation in the static field is the presence in the sample of other magnetic nuclei whose spins create local fields that may enhance or diminish the external field acting on the nucleus of interest. In a mobile lattice, these local fields tend to cancel because the nuclei causing them are in rapid and random motion. In a solid or a viscous liquid, however, the local fields may persist long enough to produce a range of field strengths and thus a range of absorption frequencies. Variations in the static field also result from small inhomogeneities in the source field itself. This effect can be largely offset by rapidly spinning the entire sample in the magnetic field.

19A-3 Fourier Transform NMR

In pulsed NMR measurements,[5] nuclei in a strong magnetic field are subjected periodically to very brief pulses of intense RF radiation as shown in Figure 19-5. The waveform in part (a) of the figure illustrates the pulse train, pulse width, and time interval between pulses. The expanded view of one of the pulses shows that each pulse is actually a packet of RF radiation. The waveforms are intended to be illustrative and are not drawn to scale. The packet of radiation consists of many more cycles than are depicted. The length of the pulses τ is usually less than 10 µs, and the frequency of the radiation is on the order of 10^2 to 10^3 MHz. The

---

[5] For more extensive discussions see J. B. Lambert, E. P. Mazzola, *Nuclear Magnetic Resonance Spectroscopy*, Upper Saddle River, NJ: Pearson/Prentice-Hall, 2004; S. Berger and S. Braun, *200 and More NMR Experiments*, New York: Wiley-VCH, 2004; D. L. Rabenstein, *Anal. Chem.*, **2001**, *73*, 214A; A. E. Derome, *Modern NMR Techniques for Chemistry Research*, New York: Pergamon Press, 1987.

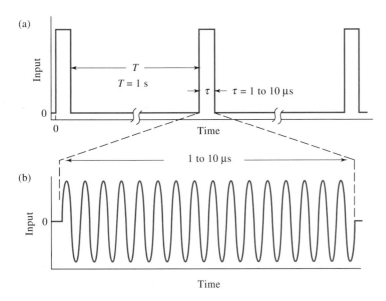

(a)

(b)

FIGURE 19-5 Typical input signal for pulsed NMR: (a) pulse sequence; (b) expanded view of RF pulse, typically at a frequency of several hundred MHz. The time axis is not drawn to scale.

interval between pulses T is typically one to several seconds. During T, a time-domain RF signal, called the *free-induction decay* (FID) signal, is emitted by the excited nuclei as they relax. The FID signal can be detected with a radio receiver coil perpendicular to the static magnetic field. As a matter of fact, a single coil is frequently used to both irradiate the sample with RF pulses and detect the decay signal. The FID signal is digitized and stored in a computer for data processing. Ordinarily, the time-domain decay signals from numerous successive pulses are added to improve the signal-to-noise ratio as described in Section 5C-2. The resulting summed data are then converted to a frequency-domain signal by a Fourier transformation, and finally, digital filtering may be applied to the data to further increase the signal-to-noise ratio. The resulting frequency-domain output is similar to the spectrum produced by a scanning CW experiment.

To describe the events that occur in a pulsed NMR experiment, it is helpful to use a set of Cartesian coordinates with the magnetic field pointing along the z-axis as shown in Figure 19-6a. The narrow arrows are the magnetic moment vectors of a few of the nuclei in the lower energy ($m = +1/2$) state. The orientations of these vectors around the z-axis are random, and they are rotating at the Larmor frequency ν_0. These excess nuclei impart a stationary net magnetic moment M aligned along the z-axis as shown by the blue arrow.

It is helpful in the discussion that follows to imagine that the coordinates in Figure 19-6 are rotating around

the z-axis at exactly the Larmor frequency. With such a rotating frame of reference, the individual magnetic moment vectors in Figure 19-6a become fixed in space at the orientation shown in the figure. Unless otherwise noted, the remaining parts of this figure are discussed in terms of this *rotating frame of reference* rather than a static, or *laboratory, frame of reference*.

Pulsed Excitation

Figure 19-6b shows the position of the net magnetic moment at the instant the pulse of RF radiation, traveling along the x-axis, strikes the sample. The magnetic field of the incident electromagnetic radiation is given the symbol B_1. In the rotating frame, B_1 and the sample magnetization vector M are both static, one along the x-axis and the other at right angles to it. Basic physics tells us that, with each pulse, M experiences a torque that tips it off the z-axis. As shown in Figure 19-6c and d, this torque rotates the sample magnetic moment M around the x-axis in the yz plane.[6] The extent of rotation depends on the length of the pulse τ as given by the equation

$$\alpha = \gamma B_1 \tau \tag{19-12}$$

---

[6] An insight into why NMR is a "resonance" technique can be gained from Figure 19-6. Resonance is a condition in which energy is transferred in such a way that a small periodic perturbation produces a large change in some parameter of the system being perturbed. NMR is a resonance technique because the small periodic perturbation B_1 produces a large change in the orientation of the sample magnetization vector M (Figure 19-6d). In most experiments, B_1 is two or more orders of magnitude smaller than B_0.

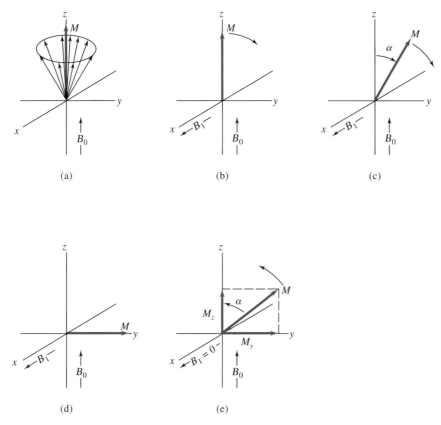

FIGURE 19-6 Behavior of magnetic moments of nuclei in a rotating field of reference, 90° pulse experiment: (a) magnetic vectors of excess lower-energy nuclei just before pulse; (b), (c), (d) rotation of the sample magnetization vector M during lifetime of the pulse; (e) relaxation after termination of the pulse.

where α is the angle of rotation in radians. For many Fourier transform experiments, a pulse length is chosen so that α is 90°, or $\pi/2$ radians, as shown in Figure 19-6d. Typically, the time required to achieve this angle is 1 to 10 μs. Once the pulse is terminated, nuclei begin to relax and return to their equilibrium position as shown in Figure 19-6e. As discussed in the previous section, relaxation takes place by two independent mechanisms: spin-lattice and spin-spin interactions. After several seconds, as a result of these interactions, the nuclei return to their original state as depicted in Figure 19-6a.

When a nucleus returns to its equilibrium state after being tipped by a pulse of RF radiation as shown in Figure 19-6e, the magnetic moment M_y along the y-axis decreases and the magnetic moment M_z along the z-axis increases. Figure 19-7 provides a more detailed picture of the mechanisms of the two relaxation processes as viewed now in the stationary frame of reference. In spin-lattice relaxation, the magnetization along the z-axis increases until it returns to its original value as shown in Figure 19-6a. In spin-spin relaxation, nuclei exchange spin energy with one another so that some now precess faster than the Larmor frequency and others proceed more slowly. The result is that the spins begin to fan out in the xy plane as shown on the right-hand side of Figure 19-7. Ultimately, this leads to a decrease to zero of the magnetic moment along the y-axis. No residual magnetic component can remain in the xy plane by the time relaxation is complete along the z-axis, which means that $T_2 \leq T_1$.

Free-Induction Decay

Let us turn again to Figure 19-6d and consider the situation when the signal B_1 goes to zero at the end of the pulse. Now, however, it is useful to picture what is happening in the static frame of reference rather than the

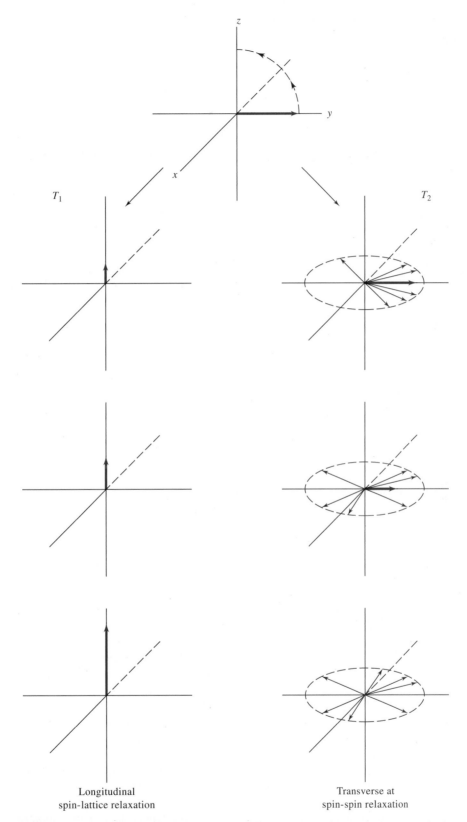

FIGURE 19-7 Two nuclear relaxation processes. Longitudinal relaxation takes place in the *xy* plane; transverse relaxation in the *xy* plane. (Courtesy of Professor Stanford L. Smith, University of Kentucky, Lexington, KY.)

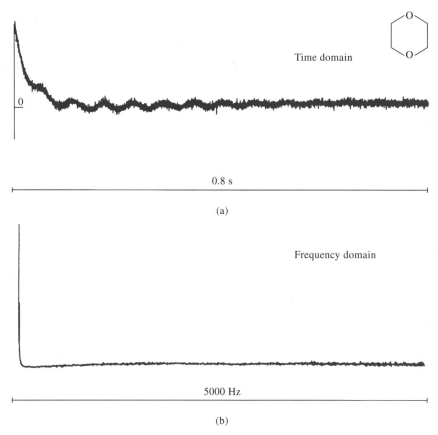

FIGURE 19-8 (a) ^{13}C FID signal for dioxane when pulse frequency is identical to Larmor frequency; (b) Fourier transform of (a). (From R. J. Abraham, J. Fisher, and P. Loftus, *Introduction to NMR Spectroscopy*, p. 89, New York: Wiley, 1988. Reprinted by permission of John Wiley & Sons, Inc.)

rotating frame. If the coordinates are fixed, the magnetic moment M must now rotate clockwise around the z-axis at the Larmor frequency. This motion gives rise to an RF signal that can be detected by a coil along the x-axis. As mentioned earlier, it can be detected with the same coil used to produce the original pulse. As relaxation proceeds, this signal decreases exponentially and approaches zero as the magnetic vector approaches the z-axis. This time-domain signal is the FID mentioned earlier; it is ultimately converted to a frequency-domain signal by the Fourier transformation.

Figure 19-8 illustrates the FID that is observed for ^{13}C nuclei when they are excited by an RF pulse having a frequency that is *exactly* the same as the Larmor frequency of the nuclei. The signal is produced by the four ^{13}C nuclei in dioxane, which behave identically in the magnetic field. The FID in Figure 19-8a is an ex-

ponential curve that approaches zero after a few tenths of a second. The apparent noise superimposed on the decay pattern is an experimental artifact caused by spinning sidebands and for the purposes of our discussion can be disregarded. Figure 19-8b is the Fourier transform of the curve in Figure 19-8a, which shows on the left the single ^{13}C absorption peak for dioxane. When the irradiation frequency ν differs from the Larmor frequency $\omega_0/2\pi$ by a small amount, as it usually will, the exponential decay is modulated by a sine wave of frequency $|\nu - (\omega_0/2\pi)|$. This effect is shown in Figure 19-9, in which the difference in the two frequencies is 50 Hz.

When magnetically different nuclei are present, the FID develops a distinct beat pattern such as that in Figure 19-10a, which is the spectrum for the ^{13}C nuclei in cyclohexene. This compound contains three pairs of

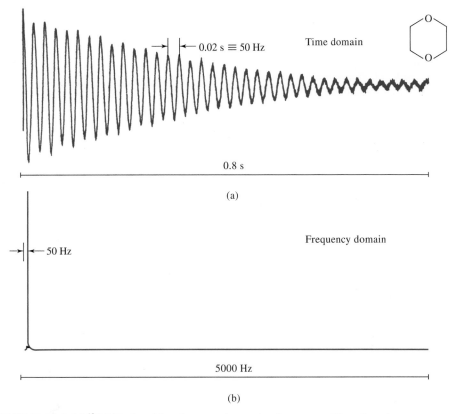

FIGURE 19-9 (a) ^{13}C FID signal for dioxane when pulse frequency differs from Larmor frequency by 50 Hz; (b) Fourier transform of (a). (From R. J. Abraham, J. Fisher, and P. Loftus, *Introduction to NMR Spectroscopy*, p. 90, New York: Wiley, 1988. Reprinted by permission of John Wiley & Sons, Inc.)

magnetically different carbon atoms: the pair of olefinic carbons, the pair of aliphatic carbons adjacent to the olefinic pair, and the pair directly opposite the olefinic group. The lines in Figure 19-10b that differ by 62 Hz arise from the two pairs of aliphatic carbon atoms. The pair of olefinic carbons is responsible for the single resonance on the left. With compounds having several absorption lines, the FID becomes very complex. In every case, however, the time-domain decay signal contains all of the information required to produce an absorption spectrum in the frequency domain using Fourier transformation.

19A-4 Types of NMR Spectra

There are several types of NMR spectra, depending on the kind of instrument used, the type of nucleus involved, the physical state of the sample, the environment of the analyte nucleus, and the purpose of the data collection. Most NMR spectra can, however, be categorized as either *wide line* or *high resolution*.

Wide-Line Spectra

Wide-line spectra are those in which the bandwidth of the source of the lines is large enough that the fine structure due to chemical environment is obscured. Figure 19-11 is a wide-line spectrum for a mixture of several isotopes. A single resonance is associated with each species. Wide-line spectra are useful for the quantitative determination of isotopes and for studies of the physical environment of the absorbing species. Wide-line spectra are usually obtained at relatively low magnetic field strength.

High-Resolution Spectra

Most NMR spectra are *high resolution* and are collected by instruments capable of differentiating between very small frequency differences of 0.01 ppm or

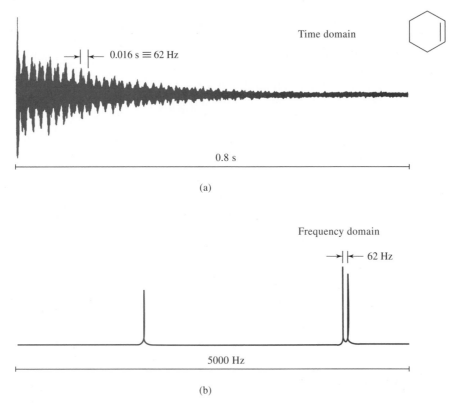

(a)

(b)

FIGURE 19-10 (a) ^{13}C FID signal for cyclohexene; (b) Fourier transform of (a). (From R. J. Abraham, J. Fisher, and P. Loftus, *Introduction to NMR Spectroscopy*, p. 91, New York: Wiley, 1988. Reprinted by permission of John Wiley & Sons, Inc.)

less. For a given isotope such spectra usually exhibit several resonances that result from differences in their chemical environment. Figure 19-12 illustrates two high-resolution spectra for the protons in ethanol. In the upper spectrum, three peaks arise from absorption by the CH$_3$, CH$_2$, and OH protons. As shown in the higher-resolution spectrum of Figure 19-12b, two of the three peaks can be resolved into additional resonances. The discussions that follow deal exclusively with high-resolution spectra.

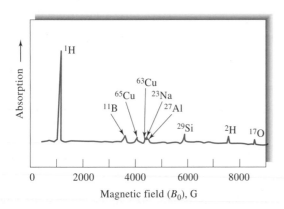

FIGURE 19-11 A low-resolution NMR spectrum of water in a glass container. Frequency = 5 MHz.

19B ENVIRONMENTAL EFFECTS ON NMR SPECTRA

The frequency of RF radiation that is absorbed by a given nucleus is strongly affected by its chemical environment — that is, by nearby electrons and nuclei. As a consequence, even simple molecules provide a wealth of spectral information that can be used to elucidate their chemical structure. The discussion that follows emphasizes proton spectra because ^{1}H is the isotope that has been studied most widely. Most of the concepts of this discussion also apply to the spectra of other isotopes as well.

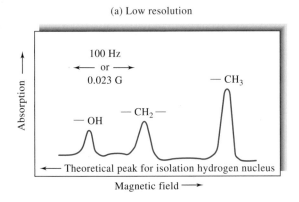

(a) Low resolution

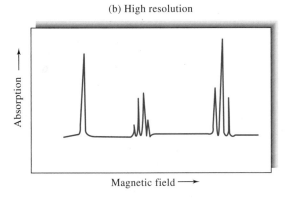

(b) High resolution

FIGURE 19-12 NMR spectra of ethanol at a frequency of 60 MHz. Resolution: (a) ~1/10^6; (b) ~1/10^7.

19B-1 Types of Environmental Effects

The spectra for ethyl alcohol, shown in Figure 19-12, illustrate two types of environmental effects. The spectrum in Figure 19-12a, obtained with a lower-resolution instrument, shows three proton peaks with areas in the ratio 1:2:3 (left to right). On the basis of this ratio, it is logical to attribute the peaks to the hydroxyl, the methylene, and the methyl protons, respectively. Other evidence confirms this conclusion; for example, if the hydrogen atom of the hydroxyl group is replaced by deuterium, the first peak disappears from this part of the spectrum. Thus, small differences occur in the absorption frequency of the proton; such differences depend on the group to which the hydrogen atom is bonded. This effect is called the *chemical shift*.

The higher-resolution spectrum of ethanol, shown in Figure 19-12b, reveals that two of the three proton resonances are split into additional peaks. This secondary environmental effect, which is superimposed

on the chemical shift, has a different cause and is termed *spin-spin splitting*.

Both the chemical shift and spin-spin splitting are important in structural analysis. Experimentally, the two are easily distinguished, because the peak separations resulting from a chemical shift are directly proportional to the field strength or to the oscillator frequency. Thus, if the spectrum in Figure 19-12a is acquired at 100 MHz rather than at 60 MHz, the horizontal distance between any set of resonances is increased by 100/60 as illustrated in Figure 19-13. In contrast, the distance between the fine-structure peaks within a group, caused by spin-spin coupling, is not altered by this frequency change.

Origin of the Chemical Shift

The chemical shift is caused by small magnetic fields generated by electrons as they circulate around nuclei. These fields usually oppose the applied field. As a consequence, the nuclei are exposed to an effective field that is usually somewhat smaller than the external field. The magnitude of the field developed internally is directly proportional to the applied external field, so that we may write

$$B_0 = B_{appl} - \sigma B_{appl} = B_{appl}(1 - \sigma) \quad (19\text{-}13)$$

where B_{appl} is the magnitude of the applied field and B_0 *is the magnitude of the resultant field, which determines the resonance behavior of the nucleus*. The quantity σ is the screening constant, which is determined by the electron density and its spatial distribution around the nucleus. The electron density depends on the structure of the compound containing the nucleus. Substituting Equation 19-5 into Equation 19-13 gives the resonance condition in terms of frequency. That is,

$$\nu_0 = \frac{\gamma}{2\pi} B_0 (1 - \sigma) = k(1 - \sigma) \quad (19\text{-}14)$$

where $k = \gamma B_0 / 2\pi$.

The screening constant for protons in a methyl group is larger than the corresponding constant for methylene protons, and it is even smaller for the proton in an —OH group. For an isolated hydrogen nucleus, the screening constant is zero. Thus, to bring any of the protons in ethanol into resonance at a given excitation oscillator frequency ν, it is necessary to apply a field B_{appl} that is greater than B_0 (Equation 19-13), the resonance value for the isolated proton. Alternatively, if the applied field is held constant, the oscillator frequency must be increased to bring about the reso-

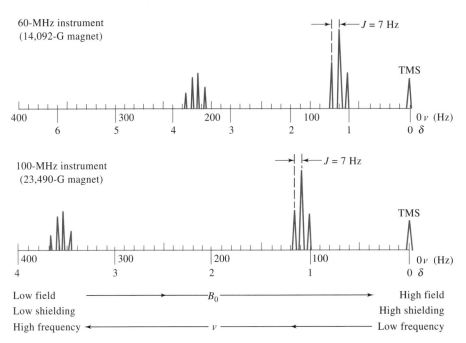

FIGURE 19-13 Abscissa scales for NMR spectra.

nance condition. Because σ differs for protons in various functional groups, the required applied field differs from group to group. This effect is shown in the ethanol spectrum of Figure 19-12a, in which the hydroxyl proton appears at the lowest applied field, the methylene protons next, and finally the methyl protons. Notice that all of these resonances occur at an applied field greater than the theoretical one for the isolated hydrogen nucleus, which lies far to the left of the spectra of Figure 19-12a. Note also that if the strength of the applied magnetic field is held constant at a level necessary to excite the methyl proton, the frequency must be increased to bring the methylene protons into resonance.

Origin of Spin-Spin Splitting

The splitting of chemical-shift resonances occurs as the magnetic moment of a nucleus interacts with the magnetic moments of immediately adjacent nuclei. The magnetic field created by a spinning nucleus affects the distribution of electrons in its bonds to other nuclei. This change in electron distribution then produces changes in the magnetic fields of adjacent nuclei and causes splitting of energy levels and hence multiple transitions. This magnetic coupling of nuclei that is transmitted by bonding electrons is often referred to as a *polarization interaction*. Thus, the fine structure of

the methylene peak shown in Figure 19-12b can be attributed to the effect of the spins of the adjacent methyl protons. Conversely, the splitting of the methyl peak into three smaller peaks is caused by the adjacent methylene protons. These effects are independent of the applied field and are superimposed on the effects of the chemical shift. Spin-spin splitting is discussed in greater detail in Section 19B-3.

Abscissa Scales for NMR Spectra

The determination of the absolute magnetic field strength with the accuracy required for high-resolution NMR measurements is difficult or impossible. On the other hand, as will be shown in Section 19C, it is entirely feasible to determine the magnitude of a *change* in field strength to within a tenth of a milligauss or better. Thus, it is expedient to report the position of resonances relative to the resonance for an internal standard substance that can be measured during the experiment. The use of an internal standard is also advantageous in that chemical shifts can be reported in terms that are independent of the oscillator frequency.

The internal standard used depends on the nucleus being studied and the solvent system. The compound most generally used for proton studies is tetramethylsilane (TMS), $(CH_3)_4Si$. All of the protons in this compound are identical; and for reasons to be considered later, the screening constant for TMS is larger than for most other protons. Thus, the compound provides a single sharp line at a high applied field that is nearly always isolated from the resonances of interest in a spectrum. In addition, TMS is inert, readily soluble in most organic liquids, and easily removed from samples by distillation (boiling point = 27°C). Unfortunately, TMS is not water soluble; in aqueous media, the sodium salt of 2,2-dimethyl-2-silapentane-5-sulfonic acid (DSS), $(CH_3)_3SiCH_2CH_2CH_2SO_3Na$, is normally used in its place. The methyl protons of this compound produce a line at virtually the same place in the spectrum as that of TMS. The methylene protons of DSS give a series of small resonances that may interfere, however. For this reason, most DSS now on the market has the methylene groups deuterated, which eliminates these undesirable absorptions.

To express the chemical shift for a sample nucleus relative to TMS in quantitative terms when measurements are made at a constant field strength B_0, we apply Equation 19-14 to the sample and the TMS resonances to obtain

$$\nu_s = k(1 - \sigma_s) \tag{19-15}$$

$$\nu_r = k(1 - \sigma_r) \tag{19-16}$$

where the subscripts r and s refer to the TMS reference and the analyte sample, respectively. Subtracting the first equation from the second gives

$$\nu_r - \nu_s = k(\sigma_s - \sigma_r) \tag{19-17}$$

Dividing this equation by Equation 19-15 to eliminate k gives

$$\frac{\nu_r - \nu_s}{\nu_r} = \frac{\sigma_r - \sigma_s}{1 - \sigma_r}$$

Generally, σ_r is much less than 1, so that this equation simplifies to

$$\frac{\nu_r - \nu_s}{\nu_r} = \sigma_r - \sigma_s \tag{19-18}$$

We then define the *chemical-shift parameter* δ as

$$\delta = (\sigma_r - \sigma_s) \times 10^6 \tag{19-19}$$

The quantity δ is dimensionless and expresses the relative shift in parts per million. A distinct advantage of this approach is that for a given resonance, δ is the same regardless of whether a 200- or a 800-MHz instrument is used. Most proton resonances lie in the δ range of 1 to 13. For other nuclei, the range of chemical shifts is greater because of the associated $2p$ elec-

trons. For example, the chemical shift for ^{13}C in various functional groups typically lies in the range 0 to 220 ppm but may be as large as 400 ppm or more. For ^{19}F, the range of chemical shifts may be as large as 800 ppm, whereas for ^{31}P it is 300 ppm or more.

Generally, NMR spectra are plotted with the abscissa linear in δ, and historically the data were plotted with the field increasing from left to right (see Figure 19-13). Thus, if TMS is used as the reference, its resonance appears on the far right-hand side of the plot, because σ for TMS is quite large. As shown, the zero value for the δ scale corresponds to the TMS peak, and the value of δ increases from right to left. Refer again to Figure 19-13 and note that the various peaks appear at the same δ value despite the two spectra having been obtained with instruments having markedly different fixed fields.

Spin-spin splitting is generally reported in units of hertz. It can be seen in Figure 19-13 that the spin-spin splitting in frequency units (J) is the same for the 60-MHz and the 100-MHz instruments. Note, however, that the chemical shift *in frequency units* is enhanced with the higher-frequency instrument.

19B-2 Theory of the Chemical Shift

As noted earlier, chemical shifts arise from the secondary magnetic fields produced by the circulation of electrons in the molecule. These so-called local *diamagnetic currents*[7] are induced by the fixed magnetic field and result in secondary fields that may either decrease or enhance the field to which a given proton responds. The effects are complex, and we consider only the major aspects of the phenomenon here. More complete treatments can be found in several reference works.[8]

Under the influence of the magnetic field, electrons bonding the proton tend to precess around the nucleus in a plane perpendicular to the magnetic field (see Figure 19-14). A consequence of this motion is the development of a secondary field, which opposes the primary field, analogous to what happens when electrons

---

[7]The intensity of magnetization induced in a diamagnetic substance is smaller than that produced in a vacuum with the same field. Diamagnetism is the result of motion induced in bonding electrons by the applied field; this motion, termed a diamagnetic current, creates a secondary field that opposes the applied field. Paramagnetism and the resulting *paramagnetic currents* operate in just the opposite sense.

[8]See, for example, J. B. Lambert, E. P. Mazzola, *Nuclear Magnetic Resonance Spectroscopy*, Upper Saddle River, NJ: Pearson/Prentice-Hall, 2004; E. D. Becker, *High Resolution NMR*, 3rd ed., New York: Academic Press, 2000.

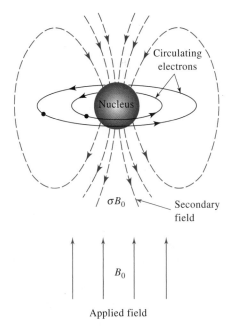

FIGURE 19-14 Diamagnetic shielding of a nucleus.

pass through a wire loop. The nucleus then experiences a resultant field that is smaller, so the nucleus is said to be *shielded* from the full effect of the primary field. As a consequence, the external field must be increased to cause nuclear resonance.

The shielding experienced by a given nucleus is directly related to the electron density surrounding it. Thus, in the absence of the other influences, shielding would be expected to decrease with increasing electronegativity of adjacent groups. This effect is illustrated by the δ values for the protons in the methyl halides, CH_3X, which lie in the order I (2.16), Br (2.68), Cl (3.05), and F (4.26). In this example, the least electronegative halogen, iodine, is also the least effective at withdrawing electrons from the methyl protons. Thus, the electrons of iodine provide the smallest shielding effect. The position of the proton peaks in TMS is also explained by this model, because silicon is relatively electropositive.

Effect of Magnetic Anisotropy

An examination of the spectra of compounds containing double or triple bonds reveals that local diamagnetic effects are not sufficient to explain the position of certain proton resonances. Consider, for example, the irregular change in δ values for protons in the following hydrocarbons, arranged in order of increasing acidity, or increased electronegativity, of the groups to

which the protons are bonded: CH_3—CH_3 ($\delta = 0.9$), CH_2=CH_2 ($\delta = 5.8$), and HC≡CH ($\delta = 2.9$). Furthermore, the aldehydic proton RCHO ($\delta \approx 10$) and the protons in benzene ($\delta \approx 7.3$) appear considerably farther downfield than is expected on the basis of the electronegativity of the attached groups.

The effects of multiple bonds on the chemical shift can be explained by taking into account the anisotropic magnetic properties of these compounds. For example, the magnetic susceptibilities[9] of crystalline aromatic compounds have been found to differ appreciably from one another, depending on the orientation of the ring with respect to the applied field. This anisotropy can be understood from the model shown in Figure 19-15. In this model, the plane of the ring is perpendicular to the magnetic field. In this position, the field can induce a flow of the π electrons around the ring to create a so-called ring current. A ring current is similar to a current in a wire loop; that is, a secondary field is induced that acts in opposition to the applied field. This induced field, however, exerts a magnetic effect on the protons attached to the ring in the direction of the field as shown in Figure 19-15. Thus, the aromatic protons require a lower external field to bring them into resonance. This effect is either absent or self-canceling in other orientations of the ring.

An analogous model operates for ethylenic or carbonyl double bonds. In such cases, we may imagine π electrons circulating in a plane along the bond axis when the molecule is oriented with the field, as pre-

---

[9]The magnetic susceptibility of a substance can be thought of as the extent to which it is susceptible to induced magnetization by an external field.

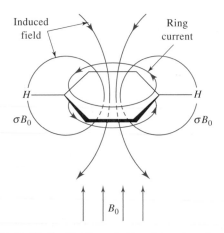

FIGURE 19-15 Deshielding of aromatic protons brought about by ring current.

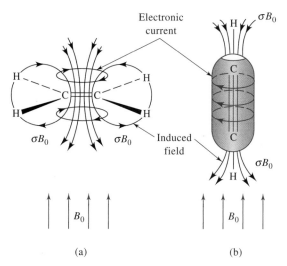

FIGURE 19-16 Deshielding of ethylene and shielding of acetylene brought about by electronic currents.

sented in Figure 19-16a. Again, the secondary field produced acts on the proton to reinforce the applied field. Thus, deshielding shifts the peak to larger values of δ. With an aldehyde, this effect combines with the deshielding brought about by the electronegative nature of the carbonyl group, and a very large value of δ results.

In an acetylenic bond, the symmetric distribution of π electrons about the bond axis permits electrons to circulate around the bond. In contrast, such circulation is prohibited by the nodal plane in the electron distribution of a double bond. From Figure 19-16b, we see that in this orientation the protons are shielded. This effect is apparently large enough to offset the deshielding resulting from the acidity of the protons and from the electronic currents at perpendicular orientations of the bond.

Correlation of Chemical Shift with Structure

The chemical shift is used to identify functional groups and to aid in determining structural arrangements of groups. These applications are based on empirical correlations between structure and shift. A number of correlation charts and tables[10] have been published, two of which are shown in Figure 19-17 and Table 19-2. Keep in mind that the exact values for δ may depend on the nature of the solvent as well as on the concen-

[10] R. M. Silverstein, F. X. Webster, and D. Kiemle, *Spectrometric Identification of Organic Compounds*, 7th ed., Chap. 3, New York: Wiley, 2004; B. Lambert, E. P. Mazzola, *Nuclear Magnetic Resonance Spectroscopy*, Chap. 3, Upper Saddle River, NJ: Pearson/Prentice-Hall, 2004.

tration of solute. These effects are particularly pronounced for protons involved in hydrogen bonding. An excellent example of this effect is the proton of the alcoholic functional group.

19B-3 Spin-Spin Splitting

As can be seen in Figure 19-12, the absorption bands for the methyl and methylene protons in ethanol consist of several narrow resonances that can be routinely separated with a high-resolution instrument. Careful examination shows that the spacing for the three components of the methylene band is identical to that for the four peaks of the methylene band. This spacing in hertz is called the *coupling constant* for the interaction and is given the symbol J. Moreover, the areas of the lines in a multiplet approximate an integer ratio to one another. Thus, for the methyl triplet, the ratio of areas is 1:2:1; for the quartet of methylene peaks, it is 1:3:3:1.

Origin

The foregoing observations are based on the effect that the spins of one set of nuclei exert on the resonance behavior of another. In other words, there is a small interaction or coupling between the two adjacent groups of protons. The results of detailed theoretical calculations are consistent with the concept that coupling takes place via interactions between the nuclei and the bonding electrons rather than through free space. For our purposes only a brief discussion of the mechanism is necessary.

Let us first consider the effect of the methylene protons in ethanol on the resonance of the methyl protons. Recall that the ratio of protons in the two possible spin states is very nearly unity, even in a strong magnetic field. We then imagine that the two methylene protons in the molecule can have four possible combinations of spin states and that in an entire sample the number of each of these combinations will be approximately equal. If we represent the spin orientation of each nucleus with a small arrow, the four states are as follows:

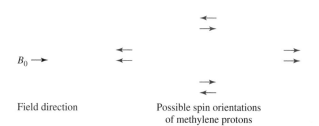

Field direction

Possible spin orientations of methylene protons

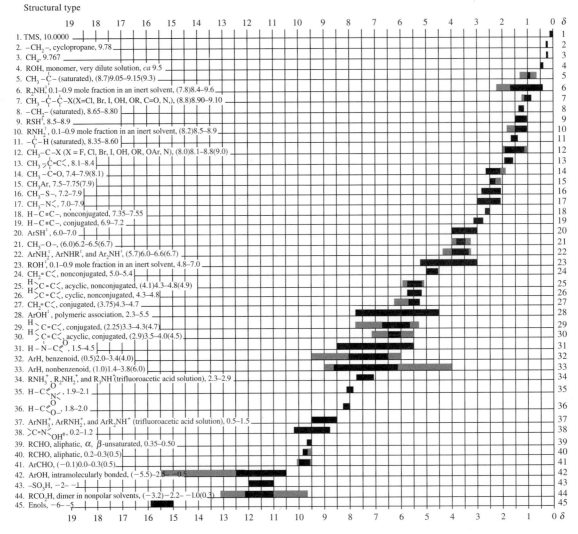

FIGURE 19-17 Absorption positions of protons in various structural environments. (Table taken from J. R. Dyer, *Applications of Absorption Spectroscopy by Organic Compounds*, p. 85, Englewood Cliffs, NJ: Prentice-Hall, 1965. With permission.)

In the first combination on the left, spins of the two methylene protons are paired and aligned against the field, whereas in the second combination on the right, the paired spins are reversed. In the two combinations shown in the center, the spins are opposed to one another. The magnetic effect transmitted to the methyl protons on the adjacent carbon atom is determined by the instantaneous spin combinations in the methylene group. If the spins are paired and opposed to the external field, the effective applied field on the methyl protons is slightly decreased. Thus, a somewhat higher field is needed to bring them into resonance, and an upfield shift results. Spins paired and aligned with the field result in a downfield shift. Neither of the combinations of opposed spin has an effect on the resonance of the methyl protons. Thus, splitting into a triplet re-

TABLE 19-2 Approximate Chemical Shifts for Certain Methyl, Methylene, and Methine Protons

	δ, ppm		
Structure	**M = CH$_3$**	**M = CH$_2$**	**M = CH**
Aliphatic α substituents			
M—Cl	3.0	3.5	4.0
M—Br	2.7	3.4	4.1
M—NO$_2$	4.3	4.4	4.6
M—OH (or OR)	3.2	3.4	3.6
M—O—C$_6$H$_5$	3.8	4.0	4.6
M—OC(=O)R	3.6	4.1	5.0
M—C=C	1.6	1.9	—
M—C≡C	1.7	2.2	2.8
M—C(=O)H	2.2	2.4	—
M—C(=O)R	2.1	2.4	2.6
M—C(=O)C$_6$H$_5$	2.4	2.7	3.4
M—C(=O)OR	2.2	2.2	2.5
M—C$_6$H$_5$	2.2	2.6	2.8
Aliphatic β substituents			
M—C—Cl	1.5	1.8	2.0
M—C—Br	1.8	1.8	1.9
M—C—NO$_2$	1.6	2.1	2.5
M—C—OH (or OR)	1.2	1.5	1.8
M—C—OC(=O)R	1.3	1.6	1.8
M—C—C(=O)H	1.1	1.7	—
M—C—C(=O)R	1.1	1.6	2.0
M—C—C(=O)OR	1.1	1.7	1.9
M—C—C$_6$H$_5$	1.1	1.6	1.8

sults (see Figure 19-12b). The area under the middle peak of the triplet is twice that of either of the other two, because two spin combinations are involved.

Let us now consider the effect of the three methyl protons on the methylene peak (middle peak in Figure 19-12a). Possible spin combinations for the methyl protons are as follows:

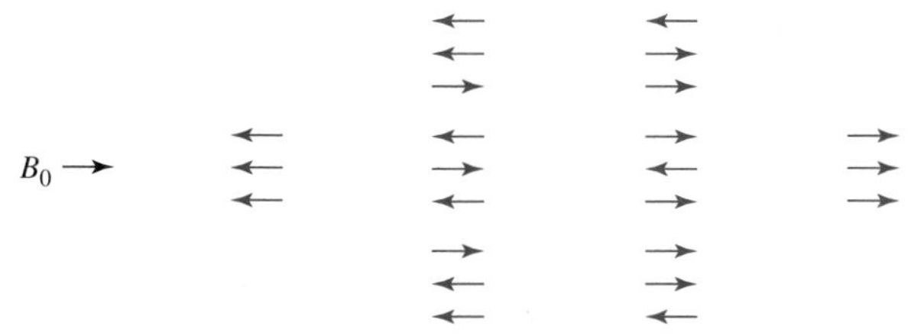

$B_0 \longrightarrow$

In this instance, we have eight possible spin combinations; however, among these are two groups containing three combinations that have equivalent magnetic effects. The methylene peak is thus split into a quartet

having areas in the ratio 1:3:3:1 (Figure 19-12b). These two examples of the adjacent methyl and methylene groups in ethanol suggest the general rule that the number of peaks in a split band in a first-order spectrum is equal to the number n of magnetically equivalent protons [11] on adjacent atoms plus one. The number of such peaks is referred to as the *multiplicity*.

The interpretation of spin-spin splitting patterns is relatively straightforward for *first-order* spectra. First-order spectra are those in which the chemical shift between interacting groups of nuclei is large with respect to their coupling constant J. Rigorous first-order behavior requires that J/δ be smaller than 0.05. Frequently, however, analysis of spectra by first-order techniques can be accomplished down to a value of $\Delta\nu/J$ of somewhat greater than 0.1. The ethanol spectrum

---

[11] Magnetically equivalent protons are those that have identical chemical shifts and identical coupling constants.

TABLE 19-3 Relative Intensities of First-Order Multiplets ($I = 1/2$)

Number of Equivalent Protons, n	Multiplicity, $(n + 1)$	Relative Peak Areas														
0	1								1							
1	2							1		1						
2	3						1		2		1					
3	4					1		3		3		1				
4	5				1		4		6		4		1			
5	6			1		5		10		10		5		1		
6	7		1		6		15		20		15		6		1	
7	8	1		7		21		35		35		21		7		1

shown in Figure 19-13 is an example of a pure first-order spectrum, with J for the methyl and methylene peaks being 7 Hz and the separation between the centers of the two multiplets being about 140 Hz.

Interpretation of second-order NMR spectra is relatively complex and will not be considered here. Note, however, that because δ increases with increases in the magnetic field but J does not, spectra obtained with an instrument having a high magnetic field are much more readily interpreted than those produced by a spectrometer with a weaker magnet.

Rules Governing the Interpretation of First-Order Spectra

The following rules govern the appearance of first-order spectra:

1. Equivalent nuclei do not interact with one another to give multiple absorption peaks. The three protons in the methyl groups in ethanol give rise to splitting of only the adjacent methylene protons and not to splitting among themselves.
2. Coupling constants decrease significantly with separation of groups, and coupling is seldom observed at distances greater than four bond lengths.
3. The multiplicity of a band is determined by the number n of magnetically equivalent protons on the neighboring atoms and is given by the quantity $n + 1$. Thus, the multiplicity for the methylene band in ethanol is determined by the number of protons in the adjacent methyl groups and is equal to $3 + 1 = 4$.
4. If the protons on atom B are affected by protons on atoms A and C that are nonequivalent, the multiplicity of B is equal to $(n_A + 1)(n_C + 1)$, where n_A and n_C are the number of equivalent protons on A and C, respectively.

5. The approximate relative areas of a multiplet are symmetric around the midpoint of the band and are proportional to the coefficients of the terms in the expansion $(x + 1)^n$. The application of this rule is demonstrated in Table 19-3 and in the examples that follow.
6. The coupling constant is independent of the applied field; thus, multiplets are readily distinguished from closely spaced chemical-shift peaks by running spectra at two different field strengths.

EXAMPLE 19-3

For each of the following compounds, calculate the number of multiplets for each band and their relative areas: (a) $Cl(CH_2)_3Cl$; (b) $CH_3CHBrCH_3$; (c) $CH_3CH_2OCH_3$.

Solution

(a) The multiplicity associated with the four equivalent protons on the two ends of the molecule is determined by the number of protons on the central carbon; thus, the multiplicity is $2 + 1 = 3$ and the areas have the ratio 1:2:1. The multiplicity of the central methylene protons is determined by the four equivalent protons at the ends and is $4 + 1 = 5$. Expansion of $(x + 1)^4$ gives the coefficients (Table 19-3), which are proportional to the areas of the peaks 1:4:6:4:1.

(b) The band for the six methyl protons is made up of $1 + 1 = 2$ peaks having relative areas of 1:1; the proton on the central carbon atom has a multiplicity of $6 + 1 = 7$. These peaks have areas (Table 19-3) in the ratio of 1:6:15:20:15:6:1.

(c) The right methyl protons are separated from the others by more than three bonds so that only a single peak is observed. The protons of the central methylene group have a multiplicity of $3 + 1 = 4$ and a ratio of 1:3:3:1. The left methyl protons have a multiplicity of $2 + 1 = 3$ and an area ratio of 1:2:1.

The preceding examples are relatively straightforward because all of the protons influencing the multiplicity of any single band are magnetically equivalent. A more complex splitting pattern results when a set of protons is influenced by two or more nonequivalent protons. As an example, consider the spectrum of 1-iodopropane, $CH_3CH_2CH_2I$. If we label the three carbon atoms (a), (b), and (c) from left to right, the chemical-shift bands are found at $\delta_{(a)} = 1.02$, $\delta_{(b)} = 1.86$, and $\delta_{(c)} = 3.17$. The band at $\delta_{(a)} = 1.02$ will be split by the two methylene protons on (b) into $2 + 1 = 3$ lines having relative areas of 1:2:1. A similar splitting of the band $\delta_{(c)} = 3.17$ will also be observed. The experimental coupling constants for the two shifts are $J_{(ab)} = 7.3$ and $J_{(bc)} = 6.8$. The band for the methylene protons (b) is influenced by two groups of protons that are not magnetically equivalent, as is evident from the difference between $J_{(ab)}$ and $J_{(bc)}$. Thus, invoking rule 4, the number of peaks is $(3 + 1)(2 + 1) = 12$. In cases such as this, derivation of a splitting pattern, as shown in Figure 19-18, is helpful. In this example, the effect of the (a) proton is first shown and leads to four peaks of relative areas 1:3:3:1 spaced at 7.3 Hz. Each of these is then split into three new peaks spaced at 6.8 Hz, having relative areas of 1:2:1. The same final pattern is produced if the original band is first split into a triplet. At a very high resolution, the spectrum for 1-iodopropane exhibits a series of lines that approximates the series shown at the bottom of Figure 19-18. If the resolution is sufficiently low that the instrument does not detect the difference between $J_{(ab)}$ and $J_{(bc)}$, only six peaks are observed with relative areas of 1:5:10:10:5:1.

Modern spectrometers contain software that can calculate spectra fairly accurately for systems with three or four spins. Usually, a trial-and-error approach is first used to estimate chemical shifts and coupling constants to produce a simulated spectrum to match the experimental spectrum. Then, chemical shifts are varied until the widths and locations of the multiplets approximately agree. Finally, coupling constants or their sums and differences are varied until a suitable agreement between observed and simulated spectra is obtained.

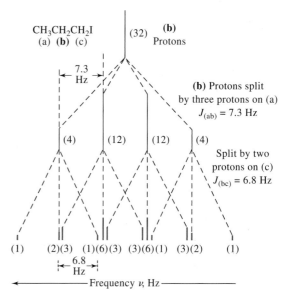

FIGURE 19-18 Splitting pattern for methylene (b) protons in $CH_3CH_2CH_2I$. Figures in parentheses are relative areas under peaks.

Second-Order Spectra

Coupling constants are usually smaller than 20 Hz, although chemical shifts may be as high as several thousand Hz. Therefore, the splitting behavior described by the rules in the previous section is common. However, when J/δ becomes greater than about 0.1 to 0.15, these rules no longer apply. Generally, as δ approaches J, the lines on the inner side of two multiplets tend to be enhanced at the expense of the lines on the outer side, and the symmetry of each multiplet is thus destroyed, as noted earlier. In addition, more, and sometimes many more, lines appear, so that the spacing between lines no longer has anything to do with the magnitude of the coupling constants. Analysis of a spectrum under these circumstances is difficult.

Effect of Chemical Exchange on Spectra

We now turn again to the NMR spectrum of ethanol (Figure 19-12) and consider why the OH proton resonance appears as a singlet rather than a triplet. The methylene protons and the OH proton are separated by only three bonds, so coupling should increase the multiplicity of both OH and the methylene bands. As shown in Figure 19-19, we actually observe the expected multiplicity in the NMR spectrum of a highly purified sample of the alcohol. Examine the triplet OH peaks and the eight methylene peaks in this spectrum.

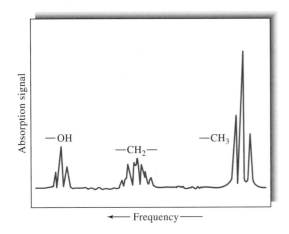

FIGURE 19-19 Spectrum of highly purified ethanol showing additional splitting of OH and CH$_2$ peaks (compare with Figure 19-12).

If we add a trace of acid or base to the pure sample, the spectrum reverts to the form shown in Figure 19-12.

The exchange of OH protons among alcohol molecules is known to be catalyzed by both acids and bases, as well as by the impurities that commonly occur in alcohol. It is thus plausible to associate the decoupling observed in the presence of these catalysts to an exchange process. If exchange is rapid, each OH group will have several protons associated with it during any brief period; within this interval, all of the OH protons will experience the effects of the three spin arrangements of the methylene protons. Thus, the magnetic effects on the alcoholic proton are averaged, and a single sharp resonance is observed. Spin decoupling always occurs when the exchange frequency is greater than the separation frequency between the interacting components.

Chemical exchange can affect not only spin-spin coupling but also chemical shifts. Purified alcohol-water mixtures have two well-defined and easily separated OH proton peaks. When acid or base is added to the mixture, however, the two peaks coalesce to form a single sharp line. In this instance, the catalyst enhances the rate of proton exchange between the alcohol and the water and thus averages the shielding effect. A single sharp line is obtained when the exchange rate is significantly greater than the separation frequency between the alcohol and water lines. On the other hand, if the exchange frequency is about the same as this frequency difference, shielding is only partially averaged and a broad line results. The correlation of line

breadth with exchange rates has provided a direct means for investigating the kinetics of such processes and represents an important application of the NMR experiment. Such studies are often accomplished by acquiring spectra at several different temperatures and determining the temperature at which multiplets coalesce to a simple band. Armed with this information, the activation energy and other kinetic parameters for the exchange process may be computed, and the mechanisms of the exchange process may be investigated.[12]

19B-4 Double Resonance Techniques

Double resonance experiments include a group of techniques in which a sample is simultaneously irradiated with two or more signals of different radio frequencies. Among these methods are *spin decoupling*, the *nuclear Overhauser effect*, *spin tickling*, and *internuclear double resonance*. These procedures aid in the interpretation of complex NMR spectra and enhance the information that can be obtained from them.[13] Only the first of these techniques, spin decoupling, will be described here. The nuclear Overhauser effect is discussed briefly in Section 19E-1.

Figure 19-20 illustrates the spectral simplification that accompanies homonuclear spin decoupling, which is decoupling carried out between similar nuclei. Spectrum *B* shows the absorption associated with the four protons on the pyridine ring of nicotine. Spectrum *C* was obtained by examining the same portion of the spectrum and simultaneously irradiating the sample with a second RF signal having a frequency of about 8.6 ppm, which corresponds to a chemical shift, centered on the absorption peaks of protons (c) and (d). The strength of the second signal is sufficient to cause saturation of the signal for these protons. The consequence is a decoupling of the interaction between the (c) and (d) protons and protons (a) and (b). Here, the complex absorption spectrum for (a) and (b) collapses to two doublet peaks that arise from coupling between these protons. Similarly, the spectrum for (c) and (d) is simplified by decoupling with a signal of frequency corresponding to the proton resonances (a) or (b).

[12]R. S. Drago, *Physical Methods for Chemists*, 2nd ed., Chap. 8, p. 290, Philadelphia: Saunders, 1992.
[13]For a more detailed discussion of double resonance methods, see J. B. Lambert and E. P. Mazzola, *Nuclear Magnetic Resonance Spectroscopy*, Upper Saddle River, NJ: Pearson/Prentice-Hall, 2004; M. H. Levitt, *Spin Dynamics: Basics of Nuclear Magnetic Resonance*, New York, Wiley, 2001; E. D. Becker, *High Resolution NMR*, 3rd ed., New York: Academic Press, 2000.

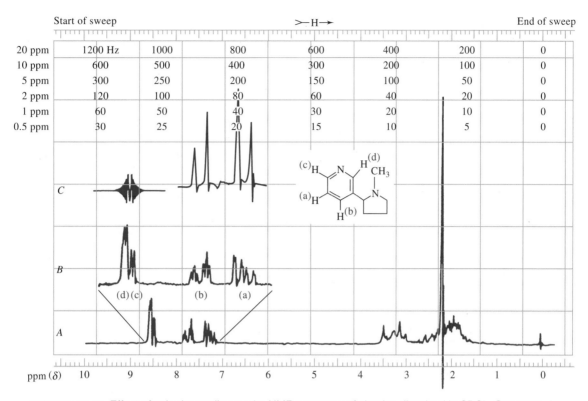

FIGURE 19-20 Effect of spin decoupling on the NMR spectrum of nicotine dissolved in $CDCl_3$. Spectrum *A*, the entire spectrum. Spectrum *B*, expanded spectrum for the four protons on the pyridine ring. Spectrum *C*, spectrum for protons (a) and (b) when decoupled from (d) and (c) by irradiation with a second beam that has a frequency corresponding to about 8.6 ppm. (Courtesy of Varian Inc., Palo Alto, CA.)

Heteronuclear decoupling, which is decoupling of the interaction between dissimilar nuclei, is easily accomplished with modern NMR instrumentation. The most important example is encountered in ^{13}C NMR, where the technique is used to simplify spectra by decoupling protons (Section 19E-1).

19C NMR SPECTROMETERS

Instrument manufacturers market two general types of NMR spectrometers: *wide-line spectrometers* and *high-resolution spectrometers*. Wide-line instruments have magnets with strengths of a few tenths of a tesla and are considerably simpler and less expensive than are high-resolution instruments. High-resolution instruments are equipped with magnets with strengths that range from 1.4 to 23 T, which correspond to proton frequencies of 60 to 1000 MHz (1 GHz). Before

 Tutorial: Learn more about **NMR instrumentation**.

about 1970, high-resolution NMR spectrometers were all of the CW type that used permanent magnets or electromagnets to supply the magnetic field. This type of instrument has been largely replaced by Fourier transform spectrometers equipped with superconducting magnets to provide the magnetic field. Computers are an integral part of modern NMR instruments; they digitize and store the signal, perform Fourier transformation of the FID to provide the frequency-domain signal, and deliver many other data-treatment and instrument-control functions. A major reason why Fourier transform instruments have become so popular is that they permit efficient signal averaging and thus yield greatly enhanced sensitivity (see Sections 5C-2 and 16B-1). Because of this greater sensitivity, routine applications of NMR to naturally occurring carbon-13, to protons in microgram quantities, and to other nuclei, such as fluorine, phosphorus, and silicon, have become widespread.

High-resolution NMR spectrometers are not inexpensive, costing $100,000 to $1,000,000 or more.

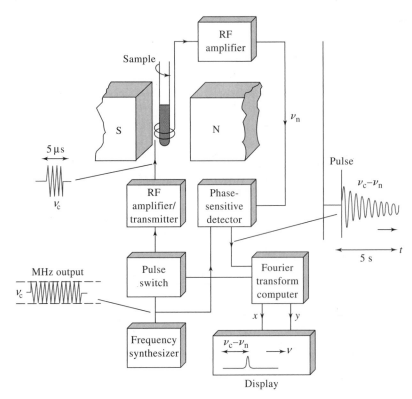

FIGURE 19-21 Block diagram of an FT-NMR spectrometer. (Adapted from J. W. Akitt, *NMR and Chemistry*, 2nd ed., p. 14, London: Chapman & Hall, 1983. With permission.)

Because almost all commercial NMR spectrometers today operate in the pulsed, or FT, mode, we confine our discussion here to FT-NMR instruments.

19C-1 Components of Fourier Transform Spectrometers

Figure 19-21 is a simplified block diagram showing the instrument components of a typical FT-NMR spectrometer. The central component of the instrument is a highly stable magnet in which the sample is placed. The sample is surrounded by a transmitter-receiver coil.

RF radiation is produced by a crystal-controlled frequency synthesizer having an output carrier frequency of ν_c. This signal passes into a pulse switch and power amplifier, which creates an intense and reproducible pulse of RF current in the transmitter coil. The resulting RF radiation impinges on the sample, contained inside the coil. The length, amplitude, shape, and phase of the pulse are selected by the operator; entered into the console; and controlled by the computer. In Figure 19-21, a 5-μs pulse is shown. The resulting FID signal is picked up by the same coil, which now serves as a receiver. The signal is then amplified and transmit-

ted to a phase-sensitive detector. The detector circuitry produces the difference between the nuclear signals ν_n and the crystal oscillator output ν_c, which leads to the low-frequency, time-domain signal shown on the right of the figure. This signal is digitized and collected in the memory of the computer for analysis by a Fourier transform program and other data-analysis software. The output from this program is plotted on a display device, giving a frequency-domain spectrum.

19C-2 Magnets

The heart of all NMR instruments is the magnet. The sensitivity and resolution of spectrometers are both critically dependent on the strength and quality of their magnets (see Example 19-2 and Figure 19-13). Because both sensitivity and resolution increase with increasing field strength, it is advantageous to operate at the highest possible field strength. In addition, the field must be highly homogeneous and reproducible. These requirements ensure that the magnet is by far the most expensive component of an NMR spectrometer.

Spectrometer magnets of three types have been used in NMR spectrometers: permanent magnets, conventional electromagnets, and superconducting solenoids. At present, conventional electromagnets are seldom incorporated into NMR instruments. Permanent magnets with field strengths of 0.7, 1.4, and 2.1 T have been used in commercial CW instruments; corresponding oscillator frequencies for proton studies are 30, 60, and 90 MHz. Permanent magnets are highly temperature sensitive and require extensive thermostatting and shielding as a consequence. Because of field-drift problems, permanent magnets are not ideal for extended periods of data accumulation, such as are often employed in Fourier transform experiments.

Superconducting magnets are used in most modern high-resolution instruments. These magnets attain fields as large as 23 T, corresponding to a proton frequency of 1 GHz. To superconduct, the solenoid, wound from superconducting niobium-tin or niobium-titanium wire, is bathed in liquid helium at a temperature of 4 K. The helium Dewar flask is held in an outer liquid-nitrogen Dewar. Most superconducting magnet systems must be filled with liquid nitrogen about once a week and with liquid helium every few months. The advantages of superconducting solenoids, in addition to their high field strengths, are their high stability, low operating cost, and simplicity and small size compared with an electromagnet.

The performance specifications for a spectrometer magnet are stringent. The field produced must be homogeneous to a few parts per billion within the sample area and must be stable to a similar degree for the length of time required to acquire the sample data. Unfortunately, the inherent stability of most magnets is considerably lower than this figure, with variations as large as one part in 10^7 being observed over a period of 1 hour. Several measures are employed in modern NMR instruments to compensate for both drift and field inhomogeneity.

Locking the Magnetic Field

To offset the effect of field fluctuations, a *field-frequency lock system* is employed in commercial NMR instruments. In these systems, a reference nucleus is continuously irradiated and monitored at a frequency corresponding to its resonance maximum at the rated field strength of the magnet. Changes in the intensity of the reference signal control a feedback circuit, whose output is fed into coils in the magnetic gap in such a way as to correct for the drift. Recall that for a given type of nucleus, the ratio between the field strength and resonance frequencies is a constant *regardless of the nucleus*

involved (Equation 19-5). Thus, the drift correction for the reference signal is applicable to the signals for all nuclei in the sample area. In modern spectrometers, the reference signal is provided by the deuterium in the solvent, and a second transmitter coil tuned to the frequency for deuterium monitors the reference. Most modern superconducting magnets are sufficiently stable that spectra can be obtained in an unlocked mode for a period of 1 to 20 min.

Shimming

Shim coils are pairs of wire loops through which carefully controlled currents are passed, producing small magnetic fields that compensate for inhomogeneities in the primary magnetic field. In contemporary instruments, the shim controls are under computer control with several algorithms available for optimizing field homogeneity. Ordinarily, shimming must be carried out each time a new sample is introduced into the spectrometer. Shim coils are not shown in the simplified diagram shown in Figure 19-21.

Sample Spinning

The effects of field inhomogeneities are also counteracted by spinning the sample along its longitudinal axis. Spinning is accomplished by means of a small plastic turbine that slips over the sample tube. A stream of air drives the turbine at a rate of 20 to 50 revolutions per second. If this frequency is much greater than the frequency spread caused by magnetic inhomogeneities, the nuclei experience an averaged environment that causes apparent frequency dispersions to collapse toward zero. A minor disadvantage of spinning is that the magnetic field is modulated at the spinning frequency, which may lead to *sidebands*, or *spinning sidebands*, on each side of absorption bands.

19C-3 The Sample Probe

A key component of an NMR spectrometer is the sample probe, which serves multiple functions. It holds the sample in a fixed position in the magnetic field, contains an air turbine to spin the sample, and houses the coil or coils that permit excitation and detection of the NMR signal. In addition, the probe ordinarily contains two other transmitter coils, one for locking and the other for decoupling experiments that are discussed in Section 19E-1. Finally, most probes have variable temperature capability. The usual NMR sample cell consists of a 5-mm (outside diameter) glass tube containing 500–650 µL of liquid. Microtubes for smaller sample volumes and larger tubes for special needs are also available.

Transmitter-Receiver Coils

The adjustable probes in modern NMR spectrometers contain transmitter and receiver coils. The receiver coils are arranged according to the experiment purpose. For observing protons, the probe usually has an inner coil for ^{1}H detection and an outer coil, called the X-nucleus coil, for detecting nuclei such as ^{13}C or ^{15}N. The inner coil is generally the closest to the sample to maximize sensitivity. For experiments to detect nuclei other than protons, the X-nucleus coil is placed on the inside and the ^{1}H coil on the outside. Usually, proton spectra have such good signal-to-noise ratios that probe tuning is not extremely critical. When the X-nucleus coil is the primary observation coil, however, tuning can be very important.

The Pulse Generator. RF generators and frequency synthesizers produce a signal of essentially a single frequency. To generate Fourier transform spectra, however, the sample must be irradiated with a range of frequencies sufficiently great to excite nuclei with different resonance frequencies. Fortunately, a sufficiently short pulse of radiation, such as that shown in Figure 19-5, provides a relatively broad band of frequencies centered on the frequency of the oscillator. The frequency range of this band is about $1/(4\tau)$ Hz, where τ is the length in seconds of each pulse. Thus, a 1-μs pulse from a 100-MHz oscillator produces a frequency range of 100 MHz $\pm$ 125 kHz. This production of a band of frequencies from a narrow pulse can be understood by reference to Figure 6-6, where a rectangular waveform is synthesized from a series of sine or cosine waves differing from one another by small frequency increments. Conversely, Fourier analysis of a square waveform reveals that it consists of a broad range of frequency components. The narrower the square wave, the broader the range of component frequencies. Thus, a narrow pulse generated by rapidly switching an RF oscillator consists of a band of frequencies capable of exciting all nuclei whose resonances occur in the vicinity of the frequency of the oscillator.

As shown in Figure 19-21, the typical pulse generator consists of three parts: a frequency synthesizer, a gate to switch the pulse on and off at appropriate times, and a power amplifier to amplify the pulse to about 50 to 100 W.

The Receiver System. Voltages generated by the current in the detector coil are in the nanovolt-to-microvolt range and thus must be amplified to a range of about 0 to 10 V before the signal can be further processed and digitized. The first stage of amplification generally takes place in a preamplifier, which is mounted in the probe so that it is as close to the receiver coil as possible to minimize the effects of noise from other parts of the instrument. Further amplification then is carried out in an external RF amplifier as shown in Figure 19-21.

19C-4 The Detector and Data-Processing System

In the detector system shown in Figure 19-21, the high-frequency radio signal is first converted to an audio-frequency signal, which is much easier to digitize. The signal from the RF amplifier can be thought of as being made up of two components: a *carrier signal*, which has the frequency of the oscillator producing it, and a superimposed NMR signal from the analyte. The analyte signal differs in frequency from that of the carrier by a few parts per million. For example, the chemical shifts in a proton spectrum typically encompass a range of 10 ppm. Thus, the proton NMR data generated by a 500-MHz spectrometer would lie in the frequency range of 500,000,000 to 500,005,000 Hz. To digitize such high-frequency signals and extract the small differences attributed to the sample in the digital domain is not practical; therefore, the carrier frequency ν_c is subtracted electronically from the analyte signal frequency ν_n in the analog domain. In our example, this process yields a difference signal $(\nu_n - \nu_c)$ that lies in the audio-frequency range of 0 to 5000 Hz. This process is identical to the separation of audio signals from an RF carrier signal in home radios.

Sampling the Audio Signal

The sinusoidal audio signal obtained after subtracting the carrier frequency is then digitized by sampling the signal voltage, periodically converting it to a digital form with an analog-to-digital converter. To accurately represent a sine or a cosine wave digitally, it is necessary, according to the *Nyquist sampling theorem* (see Section 5C-2), to sample the signal at least twice during each cycle. If sampling is done at a frequency that is less than twice the signal frequency, *folding*, or *aliasing*, of the signal occurs. The effect of folding is illus-

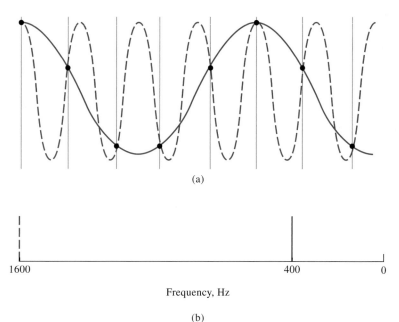

(a)

Frequency, Hz

(b)

FIGURE 19-22 Folding of a spectral line brought about by sampling at a frequency that is less than the Nyquist frequency of 1600 Hz and that is sampled at a frequency of 2000 samples per second as shown by dots; solid line is a cosine wave having a frequency of 400 Hz. (b) Frequency-domain spectrum of dashed signal in (a) showing the folded line at 400 Hz. (Adapted from D. Shaw, *Fourier Transform NMR Spectroscopy*, 2nd ed., p. 159, New York: Elsevier, 1987. With permission, Elsevier Science Publishers.)

trated in Figure 19-22a, which shows, as a colored curve, a 1600-Hz cosine signal that is being sampled at a rate of 2000 data per second. The solid dots represent the times at which the computer sampled and digitized the data. This sampling rate is less than the Nyquist frequency, which is 2×1600 Hz = 3200 Hz. The effect of this inadequate sampling rate is demonstrated by the solid curve in Figure 19-22a, which is a 400-Hz cosine wave. Thus, as shown in Figure 19-22b, the line at 1600 Hz is absent and a folded line appears at 400 Hz in the frequency-domain spectrum.

In modern NMR spectrometers, quadrature phase-sensitive detectors make possible the determination of positive and negative differences between the carrier frequency and NMR frequencies. Because these detectors are able to sense the sign of the frequency difference, folding is avoided.

Signal Integration

The areas of proton NMR signals are almost always directly proportional to the number of protons involved. Hence, modern NMR spectrometers provide not only the NMR signal itself, but for ^{1}H spectra, the integral of the signal. For ^{13}C spectra, the signals are in general *not* proportional to the number of nuclei present. Usually, the integral data appear as step functions superimposed on the NMR spectrum as illustrated in Figure 19-23. Generally, the area data are reproducible to a few percent relative.

19C-5 Sample Handling

Until recently, high-resolution NMR studies have been restricted to samples that could be converted to a nonviscous liquid state. Most often, a 2% to 15% solution of the sample is used, although pure liquid samples can also be examined if their viscosities are sufficiently low.

Solvents must have no resonances of their own in the spectral region of interest. Usually, NMR solvents are deuterated to provide the field-frequency lock signal (see Section 19C-2). The most commonly used

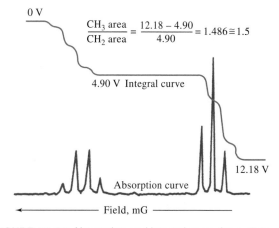

$$\frac{CH_3\ area}{CH_2\ area} = \frac{12.18 - 4.90}{4.90} = 1.486 \cong 1.5$$

FIGURE 19-23 Absorption and integral curve for a dilute ethylbenzene solution (aliphatic region). (Courtesy of Varian Inc., Palo Alto, CA.)

solvent for organic compounds is deuterated chloroform ($CDCl_3$). Polar compounds are often not sufficiently soluble in $CDCl_3$ to obtain good spectra. For these types of compounds, CD_3OD, acetone-d_6, and DMSO-d_6 are often used. For highly polar and ionic compounds, D_2O can be used along with a water-suppression method to remove the intense HOD solvent signal.

High-quality NMR tubes are necessary with modern high-field spectrometers. If tubes have been previously used, they must be carefully cleaned and thoroughly dried prior to use. Sample-tube cleaners are available commercially for this purpose. Cleaning solutions, such as chromic acid–sulfuric acid, should never be used because paramagnetic chromium ions can adsorb to the walls and broaden proton NMR spectra.

It is now possible to routinely collect high-resolution spectra for solid samples. Techniques have been developed and are being applied in increasing numbers to obtain ^{13}C spectra of polymers, fossil fuels, and other high-molecular-mass substances. A brief discussion of the modifications necessary to produce useful NMR spectra of solids is in Section 19E.

19D APPLICATIONS OF PROTON NMR

The most important chemical applications of proton NMR spectroscopy have been to the identification and structural elucidation of organic, metal-organic, and biochemical molecules. In addition, however, the method is often useful for quantitative determination of absorbing species.

19D-1 Identification of Compounds

An NMR spectrum, like an infrared spectrum, seldom suffices by itself for the identification of an organic compound. However, in conjunction with mass, infrared, and ultraviolet spectra, as well as elemental analysis, NMR is a powerful and indispensable tool for the characterization of pure compounds. The examples that follow give some idea of the kinds of information that can be extracted from NMR spectra.

EXAMPLE 19-4

The proton NMR spectrum shown in Figure 19-24 is for an organic compound having the empirical formula $C_5H_{10}O_2$. Identify the compound.

▸ *Solution*

The spectrum suggests the presence of four types of protons. From the peak integrals and the empirical formula, we deduce that these four types are populated by 3, 2, 2, and 3 protons, respectively. The single peak at $\delta = 3.6$ must be due to an isolated methyl group. Inspection of Figure 19-17 and Table 19-3 suggests that the peak may result from the functional group

$$\begin{array}{c} O \\ \parallel \\ CH_3OC- \end{array}$$

The empirical formula and the 2:2:3 distribution of the remaining protons indicate the presence of an *n*-propyl group as well. The structure

$$\begin{array}{c} O \\ \parallel \\ CH_3OC-CH_2CH_2CH_3 \end{array}$$

is consistent with all of these observations. In addition, the positions and the splitting patterns of the three remaining peaks are entirely compatible with this hypothesis. The triplet at $\delta = 0.9$ is typical of a methyl group adjacent to a methylene. From Table 19-3, the two protons of the methylene adjacent to the carboxylate peak should yield the observed triplet peak at about $\delta = 2.2$. The other methylene group would be expected to produce a pattern of $3 \times 4 = 12$ peaks at about $\delta = 1.7$. Only 6 are observed, presumably because the resolution of the instrument is insufficient to produce the fine structure of the band.

EXAMPLE 19-5

The proton NMR spectra shown in Figure 19-25 are for colorless, isomeric liquids containing only carbon and hydrogen. Identify the two compounds.

▸ *Solution*

The single peak at about $\delta = 7.2$ in the upper figure suggests an aromatic structure; the relative area of this peak corresponds to five protons. This information suggests that we may have a monosubstituted derivative of benzene. The seven peaks for the single proton appearing at $\delta = 2.9$ and the six-proton doublet at $\delta = 1.2$ can only be explained by the structure

$$\begin{array}{c} CH_3 \\ | \\ -C-CH_3 \\ | \\ H \end{array}$$

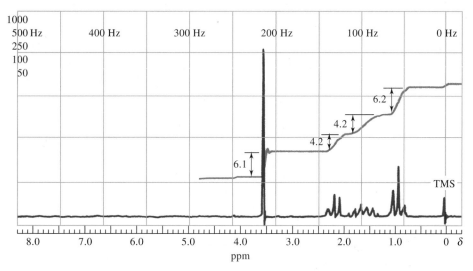

FIGURE 19-24 NMR spectrum and peak integral curve for the organic compound $C_5H_{10}O_2$ in CCl_4. (From R. M. Silverstein, G. C. Bassler, and T. C. Morrill, *Spectrometric Identification of Organic Compounds*, 3rd ed., p. 296, New York: Wiley, 1974. With permission.)

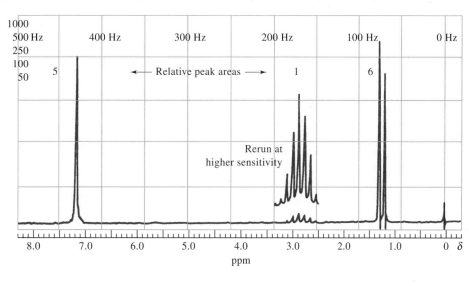

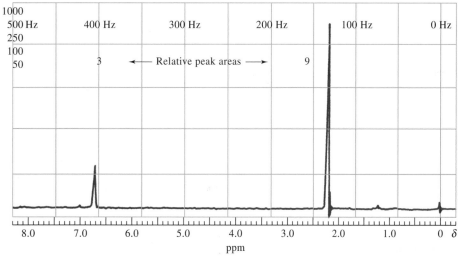

FIGURE 19-25 NMR spectra for two organic isomers in $CDCl_3$. (Courtesy of Varian Inc., Palo Alto, CA.)

Thus, we conclude that this compound is cumene.

H
|
◯—C—CH₃
|
CH₃

The isomeric compound has an aromatic peak at $\delta = 6.8$. Its relative area suggests a trisubstituted benzene, which can only mean that the compound is $C_6H_3(CH_3)_3$. The relative peak areas confirm this deduction.

EXAMPLE 19-6

The proton spectrum shown in Figure 19-26 is for an organic compound having a molecular mass of 72 and containing carbon, hydrogen, and oxygen only. Identify the compound.

The triplet peak at $\delta \approx 9.8$ appears to be that of an aliphatic aldehyde, RCHO (see Figure 19-17). If this hypothesis is valid, R has a molecular mass of 43, which corresponds to a C_3H_7 fragment. The triplet at $\delta \approx 9.8$ means that there is a methylene group adjacent to the carbonyl. Thus, the compound appears to be *n*-butyraldehyde, $CH_3CH_2CH_2CHO$.

 Simulation: Learn more about **NMR spectral interpretation**.

The triplet peak at $\delta = 0.97$ appears to be that of the terminal methyl. We expect the protons on the adjacent methylene to show a complicated splitting pattern of $4 \times 3 = 12$ peaks, and the grouping of peaks around $\delta = 1.7$ is compatible with this prediction.

Finally, the peak for the protons on the methylene group adjacent to the carbonyl should appear as a sextet downfield from the other methylene proton resonances. The group at $\delta = 2.4$ is consistent with our prediction.

19D-2 Application of NMR to Quantitative Analysis

A unique aspect of proton NMR spectra is the direct proportionality between peak areas and the number of nuclei responsible for the peak. As a result, a quantitative determination of a specific compound does not require pure samples for calibration. Thus, if an identifiable resonance for one of the constituents of a sample does not overlap resonances of the other constituents, the area of this peak can be used to establish the concentration of the species directly, provided only that the signal area per proton is known. This latter quantity can be obtained conveniently from a known concentration of an internal standard. For example, if the solvent present in a known amount were benzene, cyclohexane, or water, the areas of the single-proton peak for these compounds could be used to give the desired information. Of course, the resonance of the internal standard should not overlap with any of the sample resonances.

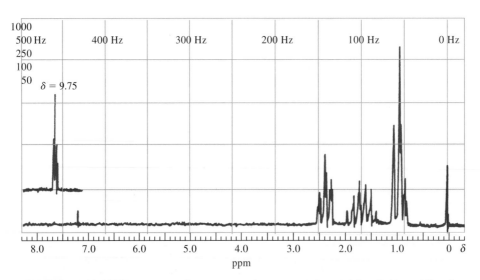

FIGURE 19-26 NMR spectrum of a pure organic compound containing C, H, and O only.

Organic silicon derivatives are uniquely attractive for calibration purposes because of the high upfield location of their proton resonances.

Methods for the analysis of many multicomponent mixtures have been reported. For example, aspirin, phenacetin, and caffeine have been determined in commercial analgesic preparations. Benzene, heptane, ethylene, glycol, and water have been determined in a wide range of mixtures.

One of the useful applications of NMR has been in the determination of functional groups, such as hydroxyl groups in alcohols and phenols, aldehydes, carboxylic acids, olefins, acetylenic hydrogens, amines, and amides.[14] Relative errors in the range of 1% to 5% are reported.

NMR spectroscopy has also been used for elemental analysis. For example, it is possible to make accurate quantitative determinations of total hydrogen in organic mixtures. Likewise, the resonance of fluorine-19 can be used for the quantitative analysis of that element in organic compounds — an analysis that is difficult to carry out by classical methods. For quantitative work, a low-resolution or wide-line spectrometer may be used.

Despite the preceding examples, the widespread use of NMR spectroscopy for quantitative work has been inhibited by the cost of the instruments. In addition, the probability of overlapping resonances becomes greater as the complexity of the sample increases. Also, NMR is often neither as sensitive nor as convenient as competing techniques.

19E   CARBON-13 NMR

Carbon-13 NMR was first studied in 1957, but its widespread use did not begin until the early 1970s. The reason for this delay was the time required for the development of instruments sensitive enough to detect the weak NMR signals from the ^{13}C nucleus. These weak signals result from the low natural abundance of the isotope (1.1%) and the small magnetogyric ratio, which is about 0.25 that of the proton. These factors combine to make ^{13}C NMR about 6000 times less sensitive than proton NMR.

The most important developments in NMR signal enhancement that have led directly to the explosive growth of ^{13}C magnetic resonance spectroscopy include higher-field-strength magnets and Fourier transform instruments. Without these developments, ^{13}C NMR would be restricted to the study of highly soluble low-molecular-mass solids, neat liquids, and isotopically enriched compounds.[15]

Carbon-13 NMR has several advantages over proton NMR in terms of its power to elucidate organic and biochemical structures. First, ^{13}C NMR provides information about the backbone of molecules rather than about the periphery. In addition, the chemical-shift range for ^{13}C for most organic compounds is about 200 ppm, compared with 10 to 15 ppm for the proton. As a result, there is less spectral overlap in ^{13}C spectra than in proton spectra. For example, it is often possible to observe individual resonances for each carbon atom in compounds ranging in molecular mass from 200 to 400. In addition, *homonuclear* spin-spin coupling between carbon atoms is not observed because in natural-abundance samples the probability of two ^{13}C atoms occurring adjacent to each other is small. Furthermore, *heteronuclear* spin coupling between ^{13}C and ^{12}C does not occur because the spin quantum number of ^{12}C is zero. Finally, a number of excellent methods can be used to decouple the interaction between ^{13}C atoms and protons. With decoupling, the spectrum for a particular type of carbon generally exhibits only a single line.

19E-1 Proton Decoupling

Three primary types of proton decoupling experiments are used in ^{13}C NMR, *broadband decoupling*; *off-resonance decoupling*; and *pulsed*, or *gated, decoupling*.

Broadband Decoupling

Broadband decoupling is a type of heteronuclear decoupling in which spin-spin splitting of ^{13}C lines by ^1H nuclei is avoided by irradiating the sample with a broadband RF signal that encompasses the entire proton spectral region, whereas the ^{13}C spectrum is obtained in the usual way. Ordinarily, the proton signal is produced by a second coil located in the sample probe.

---

[14]See R. H. Cox and D. E. Leyden in *Treatise on Analytical Chemistry*, 2nd ed., P. J. Elving, M. M. Bursey, and I. M. Kolthoff, eds., Part I, Vol. 10, pp. 127–36, New York: Wiley, 1983.

[15]For additional information, see K. Pihlaja and E. Kleinpeter, *Carbon-13 NMR Chemical Shifts in Structural and Stereochemical Analysis*, New York: Wiley-VCH, 1994; N. Beckmann, *Carbon-13 NMR Spectroscopy of Biological Systems*, New York: Academic Press, 1995; E. Breitmaier and W. Voelters, *Carbon-13 NMR Spectroscopy*, 3rd ed., New York: VCH, 1987.

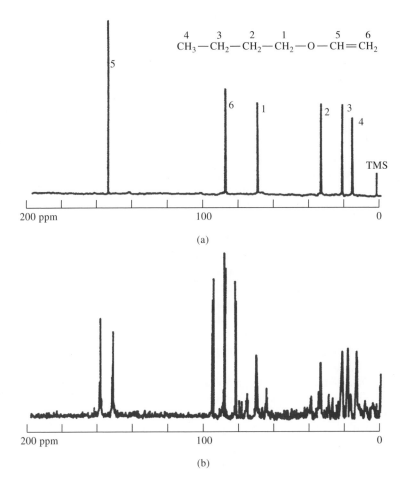

FIGURE 19-27 Carbon-13 NMR spectra for *n*-butylvinylether obtained at 25.2 MHz: (a) proton decoupled spectrum; (b) spectrum showing effect of coupling between ^{13}C atom and attached protons. (From R. J. Abraham and P. Loftus, *Proton and Carbon-13 NMR Spectroscopy*, p. 103, Philadelphia: Heyden, 1978. With permission.)

The effect of broadband decoupling is demonstrated in Figure 19-27.

Off-Resonance Decoupling

Although broadband decoupling considerably simplifies most ^{13}C spectra, it also removes spin-spin splitting information that may be of importance in structural assignments. In the past, this limitation was sometimes rectified by substituting off-resonance decoupling.

In this technique, the decoupling frequency is set at 1000 to 2000 Hz above the proton spectral region, which leads to a partially decoupled spectrum in which all but the largest spin-spin shifts are absent. Under this circumstance, primary carbon nuclei (bearing three protons) yield a quartet, secondary carbons give triplets, tertiary carbon nuclei appear as doublets, and quaternary carbons exhibit a single line. Figure 19-28 demonstrates the utility of this technique for identifying the source of the resonances in a ^{13}C spectrum.

Pulsed Decoupling

Modern NMR spectrometers provide decoupling information through the application of complex pulsing schemes, which yield higher signal-to-noise ratios more rapidly than off-resonance decoupling. These techniques are beyond the scope of this chapter.[16]

---

[16] See J. K. M. Sanders and B. K. Hunter, *Modern NMR Spectroscopy: A Guide for Chemists*, 2nd ed., New York: Oxford, 1993.

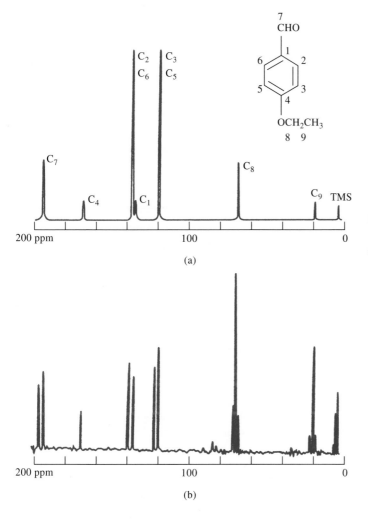

FIGURE 19-28 Comparison of (a) broadband and (b) off-resonance decoupling in ^{13}C spectra of *p*-ethoxybenzaldehyde. (From R. J. Abraham, J. Fisher, and P. Loftus, *Introduction to NMR Spectroscopy*, p. 106, New York: Wiley, 1988. With permission.)

Nuclear Overhauser Effect

Under conditions of broadband decoupling, it is found that the areas of ^{13}C peaks are enhanced by a factor that is significantly greater than would be expected from the collapse of the multiple structures into single lines. This phenomenon is a manifestation of the *nuclear Overhauser effect* (NOE), which is a general effect encountered in decoupling experiments. The enhancement arises from direct magnetic coupling between a decoupled proton and a neighboring ^{13}C nucleus that results in an increase in the population of the lower energy state of the ^{13}C nucleus over that predicted from the Boltzmann relation. The ^{13}C signal is enhanced by as much as a factor of 3 as a result. Although the NOE does increase the sensitivity of ^{13}C measurements, it has the disadvantage that the proportionality between peak

areas and number of nuclei may be lost. The theory of the NOE, which is based on dipole-dipole interactions, is beyond the scope of this text.[17]

19E-2 Application of ^{13}C NMR to Structure Determination

As with proton NMR, the most important and widespread applications of ^{13}C NMR are for the determination of structures of organic and biochemical species. Such determinations are based largely on chemical shifts, with spin-spin data playing a lesser role

---

[17] See D. Neuhaus and M. P. Williamson, *The Nuclear Overhauser Effect in Structural and Conformational Analysis*, 2nd ed., New York: Wiley-VCH, 2000; D. Shaw, *Fourier Transform NMR Spectroscopy*, 2nd ed., p. 233, New York: Elsevier, 1984.

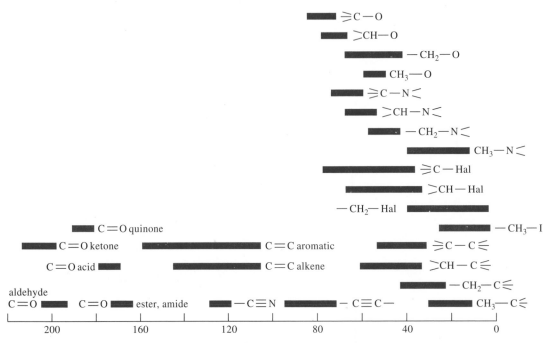

FIGURE 19-29 Chemical shifts for ^{13}C. (From D. E. Leyden and R. H. Cox, *Analytical Applications of NMR*, p. 196, New York: Wiley, 1977. With permission.)

than in proton NMR. Figure 19-29 shows some of the chemical shifts that are observed for ^{13}C in various chemical environments. As with proton spectra, these shifts are relative to tetramethylsilane, with δ values ranging from 0 to 200 ppm. In general, environmental effects are analogous to those for the proton, which were described in Section 19B-2. In contrast to proton spectra, however, the effect of substituents on ^{13}C shifts is not limited to the nearest atom. For example, substitution of chlorine on the C1 carbon in *n*-pentane results in a chemical shift for that carbon of 31 ppm. When chlorine substitution is on the C2 carbon, the shift for the C1 carbon is 10 ppm; similarly, substitutions on the C3, C4, and C5 carbons result in shifts of -5.3, -0.5, and -0.1 ppm, respectively.

Application of ^{13}C NMR to Solid Samples

As was noted earlier, NMR spectra for solids[18] have in the past not been very useful for structural studies because of line broadening, which eliminates or obscures the characteristic sharp individual peaks of NMR. Much of this broadening is attributable to two causes:

---

[18] See M. J. Duer, *Solid-State NMR Spectroscopy*, Oxford: Blackwell Science, 2002; J. P. Smith, *Anal. Chem.*, **2002**, *74*, 45A.

static dipolar interactions between ^{13}C and ^{1}H and from anisotropy in ^{13}C-shielding tensors. In isotropic liquids, these effects are averaged to zero because of the rapid and random motion of molecules. In solids, heteronuclear dipolar interactions between magnetic nuclei, such as ^{13}C and protons, result in characteristic dipolar line splittings, which depend on the angle between C—H bonds and the external field. In an amorphous solid, there is a large number of fixed orientations of these bonds and therefore a large number of splittings can occur. The broad absorption bands in this instance are made up of the numerous lines arising from these individual dipolar interactions. It is possible to remove dipolar splitting from a ^{13}C spectrum by irradiating the sample at proton frequencies while the spectrum is being obtained. This procedure, called *dipolar decoupling*, is similar to spin decoupling, which was described earlier for liquids, except that a much higher power level is required.

A second type of line broadening for solids is caused by chemical-shift anisotropy, which was discussed in Section 19B-2. The broadening produced here results from changes in the chemical shift with the orientation of the molecule or part of the molecule with respect to an external magnetic field. From NMR theory, it is known that the chemical shift $\Delta\delta$ brought about by magnetic anisotropy is given by the equation

$$\Delta\delta = \Delta\chi(3\cos^2\theta - 1)/R^3 \qquad (19\text{-}20)$$

where θ is the angle between the double bond and the applied field, $\Delta\chi$ is the difference in magnetic susceptibilities between the parallel and perpendicular orientation of the double bond, and R is the distance between the anisotropic functional group and the nucleus. When θ is exactly 54.7°, $\Delta\delta$, as defined by Equation 19-20, is zero. Experimentally, line broadening due to chemical-shift anisotropy is eliminated by *magic angle spinning*, which involves rotating solid samples rapidly at a frequency greater than 2 kHz in a special sample holder that is maintained at an angle of 54.7° with respect to the applied field. In effect, the solid then acts like a liquid being rotated in the field.

One further limitation in ^{13}C FT-NMR of solids is the long spin-lattice relaxation time for excited ^{13}C nuclei. The rate at which the sample can be pulsed depends on the relaxation rate. That is, after each excitation pulse, enough time must elapse for the nuclei to return to the equilibrium ground state. Unfortunately, spin-lattice relaxation times for ^{13}C nuclei in solids are often several minutes, which means that several hours or even days of signal averaging would be required to give a good spectrum.

The problem caused by slow spin-lattice relaxation times is overcome by *cross polarization*, a complicated pulsed technique that causes the Larmor frequencies of proton nuclei and ^{13}C nuclei to become identical — that is, $\gamma_C B_{1C} = \gamma_H B_{1H}$. Under these conditions, the magnetic fields of the precessing proton nuclei interact with the fields of the ^{13}C nuclei, causing the latter to relax.

Instruments are now available commercially that incorporate dipolar decoupling, magic angle spinning, and cross polarization, thus making possible the acquisition of high-resolution ^{13}C spectra from solids. Figure 19-30, which shows spectra for crystalline adamantane collected under various conditions, illustrates the power of these instruments.

adamantane

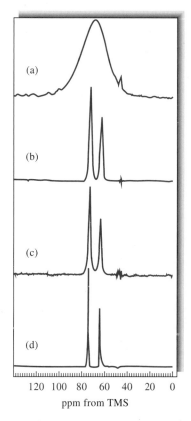

FIGURE 19-30 Carbon-13 spectra of crystalline adamantane: (a) nonspinning and with no proton decoupling; (b) nonspinning but with dipolar decoupling and cross polarization; (c) with magic angle spinning but without dipolar decoupling or cross polarization; (d) with spinning, decoupling, and cross polarization. (From F. A. Bovey, *Nuclear Magnetic Resonance Spectroscopy*, 2nd ed., p. 415, New York: Academic Press, 1988. With permission.)

19F APPLICATION OF NMR TO OTHER NUCLEI

More than 200 isotopes have magnetic moments and thus, in principle, can be studied by NMR. Among the most widely studied nuclei are ^{31}P, ^{15}N, ^{19}F, ^{2}D, ^{11}B, ^{23}Na, ^{14}N, ^{29}Si, ^{109}Ag, ^{199}Hg, ^{113}Cd, and ^{207}Pb. The first three of these are particularly important in the fields of organic chemistry, biochemistry, and biology.

19F-1 Phosphorus-31

Phosphorus-31, with spin number 1/2, exhibits sharp NMR peaks with chemical shifts extending over a range of 700 ppm. The resonance frequency of ^{31}P at

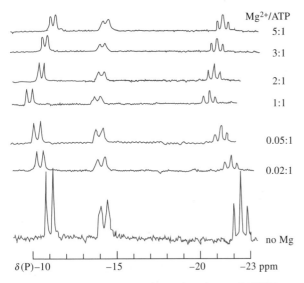

FIGURE 19-31 Fourier transform phosphorus-31 NMR spectra for ATP solution containing magnesium ions. The ratios on the right are moles of Mg^{2+} to moles of ATP. (From J. W. Akitt, *NMR and Chemistry*, 2nd ed., p. 245, London: Chapman & Hall, 1983. With permission.)

4.7 T is 81.0 MHz. Numerous investigations, particularly in the biochemical field, have been based on ^{31}P resonance. An example is shown in Figure 19-31. The species under study is adenosine triphosphate (ATP), a triply charged anion that plays a vital role in carbohydrate metabolism and in energy storage and release in the body.

adenosine-5′-triphosphate (ATP)

The bottom spectrum, which is for ATP in an aqueous environment, is made up of three sets of peaks corresponding to the three phosphorus atoms. The triplet undoubtedly arises from the central phosphorus, which is coupled to the other two phosphorus atoms. The doublet at about 14 ppm shows some poorly defined indications

of proton coupling and thus probably arises from the phosphorus that is adjacent to the methylene group.

Magnesium ions are known to play a part in the metabolic role of ATP, and the upper six spectra in Figure 19-31 suggest that complex formation between the anionic phosphorus and the cation takes place to cause the phosphorus chemical shifts to move downfield as the magnesium ion concentration is increased.

19F-2 Fluorine-19

Fluorine-19 has a spin quantum number of 1/2 and a magnetogyric ratio close to that of ^1H. Thus, the resonance frequency of fluorine at 188 MHz is only slightly lower than that of the proton at 200 MHz when both are measured at 4.69 T.

Fluorine absorption is also sensitive to the environment, and the resulting chemical shifts extend over a range of about 300 ppm. In addition, the solvent plays a much more important role in determining fluorine peak positions than in determining those of protons.

Empirical correlations of the fluorine shift with structure are relatively sparse when compared with information concerning proton behavior. It seems probable, however, that the future will see further developments in this field, particularly for structural investigation of organic fluorine compounds.

Proton, ^{19}F, and ^{31}P spectra for the inorganic species PHF_2 are shown in Figure 19-32. In each spectrum, spin-spin splitting assignments can easily be made based on the discussion in Section 19B-3. Using a modern multinuclear instrument, three of these spectra could be obtained at a single magnetic field strength.

19G MULTIPLE PULSE AND MULTIDIMENSIONAL NMR

Multiple pulse and multidimensional NMR have brought about entirely new experiments that allow us to obtain new levels of information about organic and biological molecules.

19G-1 Multiple-Pulse NMR

In the late 1960s NMR spectroscopists discovered that a vast amount of chemical information could be obtained from experiments based on multiple-pulse sequences. Here, techniques such as *inversion-recovery* and *spin-echo NMR* enabled the measurement of spin-

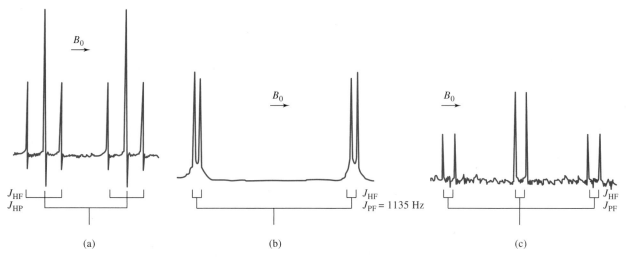

FIGURE 19-32 Spectra of liquid PHF$_2$ at −20°C: (a) ^{1}H spectrum at 60 MHz; (b) ^{19}F spectrum at 94.1 MHz; (c) ^{31}P spectrum at 40.4 MHz. (From R. J. Myers, *Molecular Magnetism and Molecular Resonance Spectroscopy*, Englewood Cliffs, NJ: Prentice-Hall, 1973. With permission.)

lattice T_1 and spin-spin T_2 relaxation times. In fact, these relaxation times began to be used as additional resolution parameters. For example, in ^{1}H NMR, differences between the T_1 values of solute protons and water protons were used to reduce the contribution of the intense solvent resonances in aqueous solutions. In mixtures containing proteins and small biological molecules, T_2 differences were exploited to essentially eliminate the proton resonances of macromolecules and allow enhancement of the small molecule signals. Most significant, however, was the recognition that multiple pulse sequences could be used to add a second frequency dimension to NMR experiments.

19G-2 Two-Dimensional NMR

Two-dimensional NMR, or 2D NMR, comprises a relatively new set of multipulse techniques that make it possible to unravel complex spectra.[19] The two-dimensional methods can identify resonances connected by through-bond coupling, by through-space interactions, or by chemical exchange. In two-dimensional meth-

ods, data are acquired as a function of time t_2 just as in ordinary FT-NMR. Prior to obtaining this FID signal, however, the system is perturbed by a pulse for a period t_1. Fourier transformation of the FID as a function of t_2 for a fixed t_1 yields a spectrum similar to that obtained in an ordinary pulse experiment. This process is then repeated for various values of t_1, thus giving a two-dimensional spectrum in terms of two frequency variables ν_1 and ν_2 or sometimes the chemical-shift parameters δ_1 and δ_2. The nature and timing of the pulses that have been used in 2D NMR vary widely, and in some cases more than two repetitive pulses are used. Thus, the number of types of two-dimensional experiments that have appeared in the literature is large. Some of the most popular methods based on coherence transfer include *homonuclear correlation spectroscopy* (COSY), *total correlation spectroscopy* (TOCSY), the *incredible natural-abundance double-quantum transfer experiment* (INADEQUATE), *heteronuclear correlation* (HETCOR) *spectroscopy*, and *heteronuclear multiple quantum coherence* (HMQC) *spectroscopy*.

Figure 19-33a is a ^{13}C 2D NMR spectrum for a COSY experiment with 1,3-butanediol. Figure 19-33a is the ordinary one-dimensional spectrum for the compound. The two-dimensional spectrum is obtained as follows: With the proton broadband decoupler turned off, a 90° pulse is applied to the sample. After a time t_1, the decoupler is turned on, another pulse is applied, and the resulting FID is digitized and transformed.

---

[19] For a brief review of 2D NMR, see D. L. Rabenstein, *Anal. Chem.*, **2001**, *73*, 214A. For a more detailed treatment, see J. B. Lambert, E. P. Mazzola, *Nuclear Magnetic Resonance Spectroscopy*, Chap. 6, Upper Saddle River, NJ: Pearson/Prentice-Hall, 2004; F. J. M. van de Ven, *Multidimensional NMR in Liquids*, New York: Wiley-VCH, 1995; R. R. Ernst, G. Bodenhausen, and A. Wokaun, *Principles of Nuclear Magnetic Resonance in One and Two Dimensions*, Oxford, UK: Oxford University Press, 1987.

$$CH_3 - CHOH - CH_2 - CH_2OH$$
$$1 \quad\quad 2 \quad\quad\quad 3 \quad\quad\quad 4$$

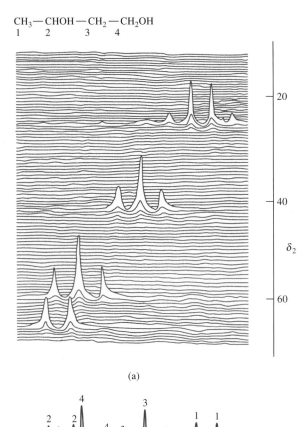

(a)

(b)

FIGURE 19-33 Illustration of the use of the two-dimensional spectrum (a) to identify the ^{13}C resonances in a one-dimensional spectrum (b). Note that the ordinary one-dimensional spectrum is obtained from the peaks along the diagonal. The presence of off-diagonal cross peaks can identify resonances linked by spin-spin coupling. (Adapted from S. Borman, *Anal. Chem.*, **1982**, *54*, 1129A. Figure 1, p. 1129A. Copyright 1982 American Chemical Society.)

After equilibrium has been reestablished, this process is repeated for other values of t_1, which leads to a series of spectra that are plotted horizontally in the figure. That is, the projections along the δ_1-axis are the spectra that would be obtained without decoupling. The projection along the δ_2-axis is the same as the completely decoupled carbon-13 spectrum. It is obvious that the spectrum is made up of a quartet, two triplets,

and a doublet, whose source is apparent from a consideration of the number of protons bonded to each of the four ^{13}C atoms in the molecule. This information is not obvious in the one-dimensional spectrum.

The COSY, TOCSY, HETCOR, and HMQC experiments are great aids in interpreting proton and ^{13}C spectra. However, the INADEQUATE experiment is perhaps the definitive 2D NMR technique. This experiment is based on spin-spin coupling between directly bonded pairs of nuclei. It can trace out the carbon backbone of an organic compound one carbon at a time. The low sensitivity of INADEQUATE has, however, limited its practical utility.

Another class of two-dimensional experiments is based on the incoherent transfer of magnetization by the NOE or by chemical exchange. In *NOE spectroscopy* (NOESY), *rotating-frame Overhauser effect spectroscopy* (ROESY), and *exchange spectroscopy* (EXSY), cross peaks are observed between resonances linked by dipole-dipole interactions or chemical exchange. As an example, the one-dimensional and ROESY spectra for a nineteen-amino-acid peptide are shown in Figure 19-34. Not only could the one-dimensional proton NMR spectrum be completely assigned but the ROESY spectrum also indicated that the peptide is an α-helix in solution.

19G-3 Multidimensional NMR

Overlapping resonances in 2D NMR have limited protein-structure elucidation to fairly small proteins. However, three- and four-dimensional methods have been developed that enable NMR spectroscopy to be further extended to larger and larger protein structures. A third dimension can be added, for example, to spread apart a ^1H-^1H two-dimensional spectrum on the basis of the chemical shift of another nucleus, such as ^{15}N or ^{13}C. In most three-dimensional experiments, the most effective methods for large molecules are used. Thus, COSY is not often employed, but experiments like NOESY-TOCSY and TOCSY-HMQC are quite effective. In some cases, the three dimensions all represent different nuclei such as ^1H-^{13}C-^{15}N. These are considered variants of the HETCOR experiment. Multidimensional NMR is now capable of providing complete solution-phase structures to complement crystal structures from X-ray crystallography. Hence, NMR spectroscopy is now an important technique for determining structures and orientations of complex molecules in solution.

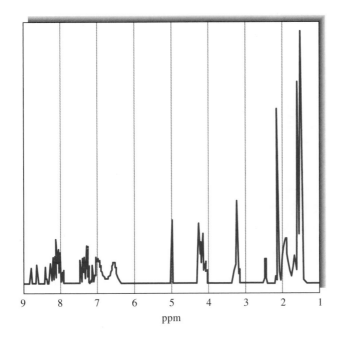

(a)

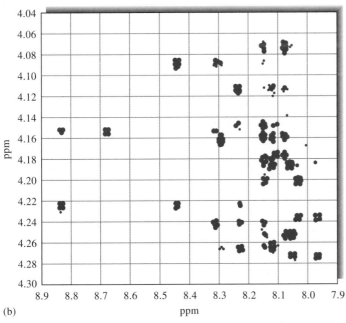

(b)

FIGURE 19-34 The 500-MHz one-dimensional ^{1}H NMR spectrum (a) and a portion of the two-dimensional ROESY spectrum (b) of a nineteen-amino-acid protein. The pulse sequence used collapses multiplets due to ^{1}H-^{1}H spin-spin coupling into singlets. The cross peaks of the dipolar interactions make it possible to completely assign the proton NMR spectrum. (Adapted from A. Kaerner and D. L. Rabenstein, *Magn. Reson. Chem.*, **1998**, *36*, 601. Copyright 1998 Interscience/Wiley.)

19H MAGNETIC RESONANCE IMAGING

Since the 1970s, NMR technology has been applied increasingly to fields other than chemistry, such as biology, engineering, industrial quality control, and medicine. One of the most prominent NMR applications is that of magnetic resonance imaging, or MRI. In MRI, data from pulsed RF excitation of solid or semisolid objects are subjected to Fourier transformation and converted to three-dimensional images of the interior of the objects. The primary advantage of MRI is that images of objects are formed noninvasively. This means that there is little or no potential of radiation injury or other damage to human or animal subjects as might be encountered with X-ray computerized axial tomography (CAT) or other similar

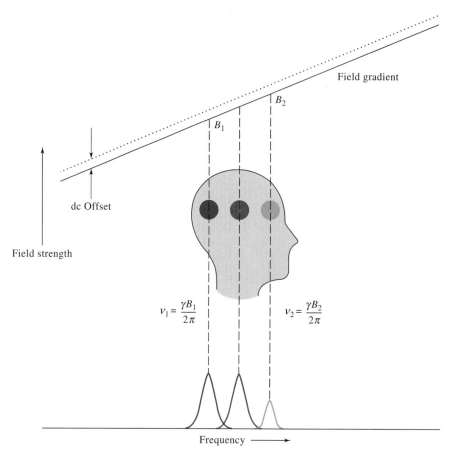

FIGURE 19-35 Fundamental concept of MRI.

methods.[20] In 2003 Dr. Paul Lauterbur of the University of Illinois and Sir Peter Mansfield of the University of Nottingham in the United Kingdom shared the Nobel Prize in Physiology or Medicine for their discoveries concerning MRI.

The fundamental concept of MRI is depicted in Figure 19-35. In MRI, the magnetic field strength is deliberately varied throughout the subject under study to give a profile such as that shown at the top of the figure. This linear variation, or field gradient, in B is created by auxiliary coils in the magnet bore that are under the control of the computer of the magnetic resonance instrument. Protons in different locations in the subject experience different magnetic field strengths B_1 and B_2 for two of the regions shown in the figure. Equa-

[20] For detailed discussions of MRI theory and applications, consult J. P. Hornak, *The Basics of MRI* (http://www.cis.rit.edu/htbooks/mri/); H. Witjes, A. W. Simonetti, L. Buydens, *Anal. Chem.*, **2001**, *73*, 548A; S. A. Huettel, A. W. Song, and G. McCarthy, *Functional Magnetic Resonance Imaging*, Sunderland, MA: Sinauer Associates, 2004; R. B. Buxton, *Introduction to Functional Magnetic Resonance Imaging*, Cambridge, UK: Cambridge University Press, 2002.

tion 19-10, the Larmor equation, suggests that these nuclei exhibit different resonance frequencies, ν_1 and ν_2 in this instance. For example, if a magnetic field gradient of 1×10^{-5} T/cm is applied along the bore, or the z-axis, of an MRI magnet, a resonance frequency range of $(2.68 \times 10^8$ radians s^{-1} T$^{-1})(1 \times 10^{-5}$ T/cm)/$(2\pi$ radians$) = 425$ Hz results. In other words, protons 1 cm apart along the field gradient in the subject have resonance frequencies that differ by 425 Hz. Thus, by changing the center frequency of the NMR probe pulse in increments of 425 Hz, it is possible to probe successive 1-cm positions in the direction of the magnetic field gradient. Each consecutive RF pulse produces an FID signal that encodes the concentration of protons at each 1-cm position along the direction of the field gradient. When the FIDs are subjected to Fourier transformation, concentration information is produced as indicated by the heights of the peaks at the bottom of Figure 19-35. In practice, the position of the slice along the z-axis may be changed by adding a dc offset to the auxiliary coils as shown by the dashed

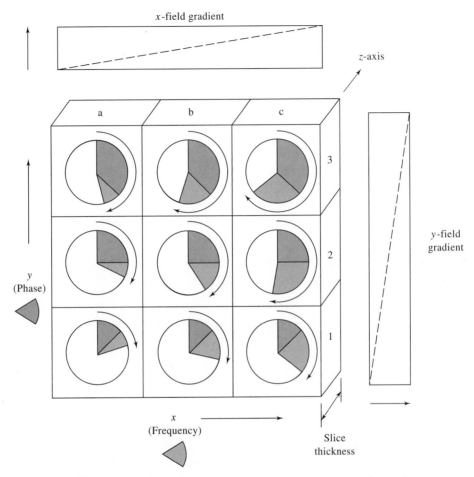

FIGURE 19-36 Acquisition of information within slices along the z-axis.

field gradient in Figure 19-35. Each slice is probed by changing the pulse width of the RF pulse to tune its frequency to the precession frequency of the protons in the slice. The middle peak in Figure 19-35 corresponds to the center circle representing the signal from protons in the center position of the subject's head. If the applied field gradient is made larger, the thickness of the examined slice is made smaller.

The position-selection process described in the previous paragraph provides spatial information in one dimension along the z-axis by flipping the spins of the protons within the selected slice. Information within each of the slices along the z-axis is acquired in a slightly different way, as shown schematically in Figure 19-36. The chosen slice is represented in the figure as a two-dimensional grid lying in the x-y slice with rows 1, 2, and 3 and columns a, b, and c. For simplicity, the grid is shown as a 3×3 array of elements, but in practice the slice might have 128×128, 256×256, or 512×512

elements, with each element, or pixel, having an area of 1 mm $\times$ 1 mm in the sample. In such three-dimensional imaging experiments, the z-axis field gradient is selected to give a slice thickness of 1 mm, then it is turned off. At this point in the measurement sequence, all protons in the slice have their spins flipped. A second gradient is applied to the subject perpendicular to the z-axis by using a second set of y-axis coils oriented at a right angle to the axis of the magnet bore. Nuclei at different positions along the y-axis precess at different frequencies depending on their position in the field gradient. When this second y-axis field gradient is turned off, all nuclei at a distance corresponding to row 1 on the y-axis of Figure 19-36 have precessed through a fixed phase angle whose magnitude is proportional to the size of the gray, pie-shape angular sections of the circles in row 1. Thus, the distance along the y-axis *is encoded in the phase angle* of the precessing protons represented by the gray pie slices.

As soon as the y-axis gradient is terminated, a third set of magnet coils perpendicular to both the z-axis and the y-axis coils are then activated. This action produces a field gradient in the third dimension along the x-axis. Because the nuclei in column b have experienced a stronger magnetic field than those in column a, they precess at a higher frequency whose magnitude is represented by the blue pie slices. Similarly, the nuclei in column c have the largest frequency in the 3×3 pixel array of Figure 19-36. The result is that the positions of the nuclei in the x dimension *are encoded by their precession frequencies*. The total precession angle, which is indicated by the circular arrows in the figure, is unique for each element of the grid, and thus the po-

sition of each is encoded in the total angle. The receiver coils are turned on during the application of this last frequency-encoding phase, and an FID is recorded. The complete cycle of (1) slice selection, (2) phase encoding in the y dimension, (3) frequency encoding in the x dimension, and (4) recording the FID requires only a few hundred milliseconds at most. The next step in collecting the data is to turn on the z-axis coils to once again flip the spins of the nuclei in the slice and repeat the sequence with a somewhat larger y-axis gradient to phase-encode row 2. This step is followed by the application of the x-axis gradient to frequency-encode columns a, b, and c and the acquisition of the FID for row 2. Finally, the entire measurement se-

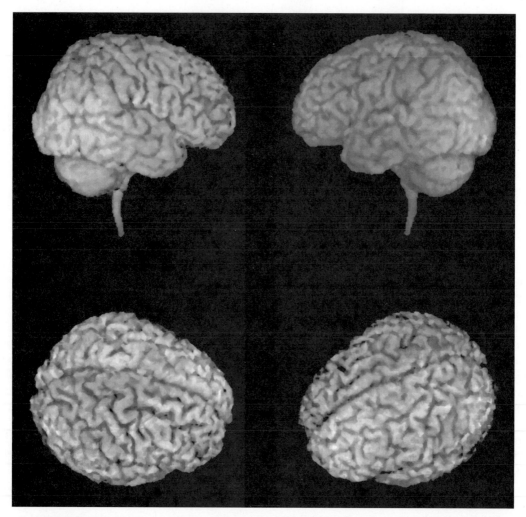

FIGURE 19-37 Structures inside subjects may be reconstructed from the three-dimensional data arrays. (Image courtesy of C. D. Smith Memory Disorders Clinic, Sanders-Brown Research Center on Aging, University of Kentucky Medical Center. With permission.)

quence is repeated for row 3, and a corresponding FID is acquired. The result of the process for the entire 3×3 slice is a set of three FIDs representing the time-domain resonance signal S, each of which is a function of the x and y field gradients G_x and G_y, respectively. The time-domain signal $S(G_x, G_y)$, which contains information related to the total precession angle, is subjected to a two-dimensional Fourier transformation. The result is a frequency-domain signal $S(\omega_x, \omega_y)$ whose frequencies in the x and y dimensions are directly proportional to the distances d_x and d_y. That is,

$$S(G_x, G_y) \rightarrow S(\omega_x, \omega_y) \rightarrow S(d_x, d_y)$$

The two-dimensional distance information in $S(d_x, d_y)$ is at last combined for all of the slices in the z dimension to provide a three-dimensional array of data. Each element of the array contains an intensity that is proportional to the concentration of protons in a volume element, or *voxel*, corresponding to each set of coordinates x, y, and z. Note that with the exception of the magnet, which must have a rather large bore and a large static field strength of 0.5 to 4.7 T, instruments for MRI and high-resolution NMR are identical in

function. Precisely the same pulse-sequencing techniques and data-enhancement procedures are used in both types of instruments. Although the foregoing discussion is considerably simplified, the basic sequencing and encoding schemes are realistic. Clever and timely application of various RF pulse sequences and magnetic field gradients, appropriate Fourier transformations, and data analysis and reconstruction software routines produce three-dimensional images. Structures inside subjects may be reconstructed from the three-dimensional data arrays as illustrated in the group of four MRI images of Figure 19-37. The images are true three-dimensional brain reconstructions calculated by using a single set of MRI data from a young female patient with Rasmussen's encephalitis. These images show the dramatic shrinkage in the front right portion of the brain characteristic of this rare illness. The ability to generate high-resolution (millimeter-scale) images from true three-dimensional data is a unique strength of the MRI technique.

A second magnetic resonance image presented in Figure 19-38 demonstrates functional MRI (fMRI) mapping of visual-confrontation naming in the left

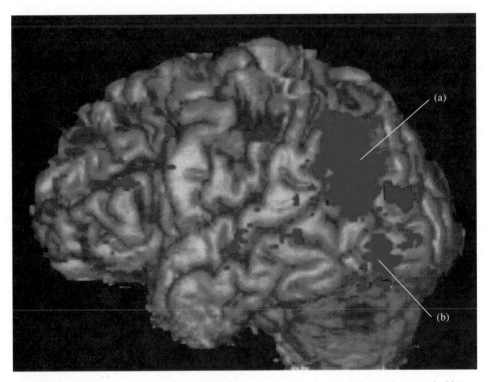

FIGURE 19-38 Brain activity in the left hemisphere resulting from naming tasks revealed by fMRI. (From C. D. Smith, A. H. Andersen, Z. Chen, L. X. Blonder, J. E. Kirsch, and M. J. Avison, *Neuro. Report*, **1996**, *7*, 2. With permission.)

hemisphere of the brain. The dark portion of the map demonstrates statistical differences in signal intensity due to changes in cerebral blood oxygenation from activation of brain regions engaged in a naming task compared to a control task. The task undertaken by the subject in this case was the naming of standardized line drawings of a series of seven subjects. The map shows activation of the *angular gyrus*, the large dark patch in the upper-right region of the brain, and the so-called *area 37*, the smaller light gray region in the lower-back region. Both regions of the brain are known to be involved in language processing. Mapping of brain regions involved in specific types of human brain processing and execution is a current intensive area of research.

Advantages of fMRI compared with other techniques are the noninvasive nature of the technique, reproducibility, speed of data acquisition, and high intrinsic spatial resolution.

MRI has become a mainstay in the arsenal of medical diagnostic tools. The application of MRI in the food industry and in other areas of science and commerce has been hampered by the high cost of MRI installations. However, as the sophistication and ease of application of data-enhancement procedures, exotic pulse sequences, and data-acquisition protocols continue to evolve, especially for nuclei other than ^1H, MRI will doubtless become an indispensable tool for the noninvasive investigation of materials.

QUESTIONS AND PROBLEMS

*Answers are provided at the end of the book for problems marked with an asterisk.

X Problems with this icon are best solved using spreadsheets.

19-1 Explain the difference in the way a CW and an FT-NMR experiment is performed.

19-2 What are the advantages of an FT-NMR measurement over a CW measurement? What are the disadvantages?

19-3 In NMR spectroscopy, what are the advantages of using a magnet with as great a field strength as possible?

19-4 How can spin-spin splitting lines be differentiated from chemical-shift lines?

19-5 Define
(a) magnetic anisotropy
(b) the screening constant
(c) the chemical-shift parameter
(d) CW-NMR measurements
(e) Larmor frequency
(f) coupling constants
(g) first-order NMR spectra

19-6 A nucleus has a spin quantum number of 5/2. How many magnetic energy states does this nucleus have? What is the magnetic quantum number of each?

*19-7 What is the absorption frequency in a 7.05-T magnetic field of (a) ^1H, (b) ^{13}C, (c) ^{19}F, (d) ^{31}P?

*19-8 What is the Larmor frequency for protons in magnetic fields of (a) 1.41 T, (b) 4.69 T, (c) 7.05 T, (d) 11.7 T, (e) 18.8 T, and (f) 21.2 T?

*19-9 A resonance is displaced 90 Hz from TMS at a magnetic field strength of 1.41 T. What will be the frequency difference at (a) 4.69 T, (b) 7.05 T, (c) and 18.8 T? What will be the chemical shifts δ at these same magnetic field strengths?

19-10 Why is ^{13}C-^{13}C spin-spin splitting not observed in ordinary organic compounds?

*19-11 Calculate the relative number of ^{13}C nuclei in the higher and lower magnetic states at 25°C in magnetic fields of (a) 2.4 T, (b) 4.69 T, and (c) 7.05 T.

19-12 What is the difference between longitudinal and transverse relaxation?

19-13 Explain the source of an FID signal in FT-NMR.

19-14 What is a rotating frame of reference?

19-15 How will ΔE for an isolated ^{13}C nucleus compare with that of a ^1H nucleus?

19-16 Calculate the resonance frequency of each of the following nuclei in a 7.05-T magnetic field: (a) ^{19}F and (b) ^{31}P.

*__19-17__ What is the ratio of the number of nuclei in the upper magnetic energy state to the number in the lower energy state of ^{13}C in a 500-MHz instrument if the temperature is 300 K?

19-18 Briefly compare the ^1H and ^{31}P NMR spectra of methyl phosphorous acid $P(OCH_3)_3$ at 4.69 T. There is a weak spin-spin coupling between phosphorus and hydrogen nuclei in the compound.

19-19 In the room-temperature ^1H spectrum of methanol, no spin-spin coupling is observed, but when a methanol sample is cooled to $-40°C$, the exchange rate of the hydroxyl proton slows sufficiently so that splitting is observed. Sketch spectra for methanol at the two temperatures.

19-20 Use the following coupling constant data to predict the ^1H and the ^{19}F spectra of the following:

Species	J, Hz
(a) F—C≡C—H	21
(b) CF_3—CH_3	12.8
(c) $(CH_3)_3$—CF	20.4

19-21 Predict the appearance of the high-resolution ^{13}C spectrum of (proton decoupled)
(a) methyl formate.
(b) acetaldehyde.
(c) acetone.

19-22 Repeat Question 19-21 for the case when the protons are not decoupled. Note that ^{13}C—^1H coupling constants are generally in the range of 100 to 200 Hz.

19-23 What is a frequency lock system in an NMR spectrometer? Describe the two types of lock systems.

19-24 What are shims in an NMR spectrometer, and what are their purpose?

19-25 Why are liquid samples spun while being examined in an NMR spectrometer?

19-26 Predict the appearance of the high-resolution proton NMR spectrum of propionic acid.

19-27 Predict the appearance of the high-resolution proton NMR spectrum of
(a) acetone.
(b) acetaldehyde.
(c) methyl ethyl ketone.

19-28 Predict the appearance of the high-resolution proton NMR spectrum of
(a) ethyl nitrite.
(b) acetic acid.
(c) methyl-*i*-propyl ketone.

19-29 Predict the appearance of the high-resolution proton NMR spectrum of
(a) cyclohexane.
(b) diethyl ether.
(c) 1,2-dimethoxyethane, $CH_3OCH_2CH_2OCH_3$.

19-30 Predict the appearance of the high-resolution proton NMR spectrum of
(a) toluene.
(b) ethyl benzene.
(c) *i*-butane.

19-31 The proton NMR spectrum in Figure 19-39 is for an organic compound containing a single atom of bromine. Identify the compound.

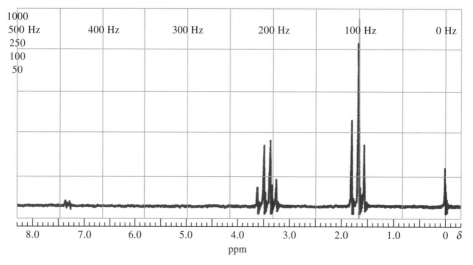

FIGURE 19-39 Proton NMR spectrum. (Courtesy of Varian, Inc., Palo Alto, CA.)

19-32 The proton NMR spectrum in Figure 19-40 is for a compound having an empirical formula $C_4H_7BrO_2$. Identify the compound.

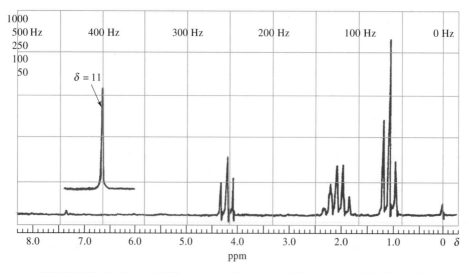

FIGURE 19-40 Proton NMR spectrum. (Courtesy of Varian, Inc., Palo Alto, CA.)

19-33 The proton NMR spectrum in Figure 19-41 is for a compound of empirical formula C_4H_8O. Identify the compound.

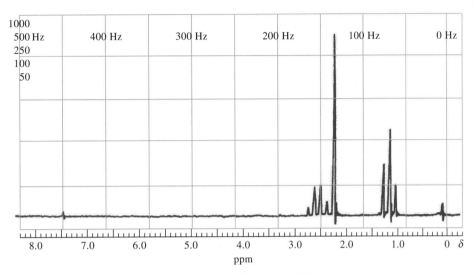

FIGURE 19-41 Proton NMR spectrum. (Courtesy of Varian, Inc., Palo Alto, CA.)

19-34 The proton NMR spectrum in Figure 19-42 is for a compound having an empirical formula $C_4H_8O_2$. Identify the compound.

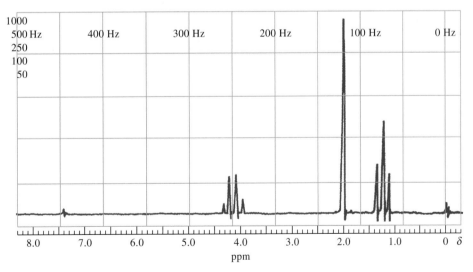

FIGURE 19-42 Proton NMR spectrum. (Courtesy of Varian, Inc., Palo Alto, CA.)

19-35 The proton spectra in Figure 19-43a and b are for compounds with empirical formula C_8H_{10}. Identify the compounds.

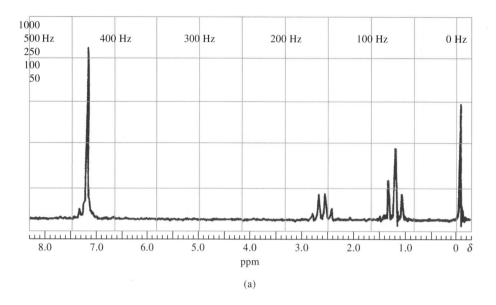

(a)

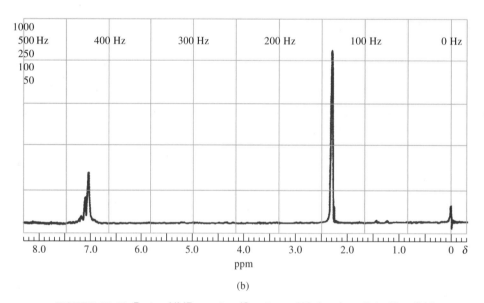

(b)

FIGURE 19-43 Proton NMR spectra. (Courtesy of Varian, Inc., Palo Alto, CA.)

19-36 From the proton NMR spectrum in Figure 19-44, deduce the structure of this hydrocarbon.

19-37 From the proton spectrum given in Figure 19-45, determine the structure of this compound, a commonly used pain killer; its empirical formula is $C_{10}H_{13}NO_2$.

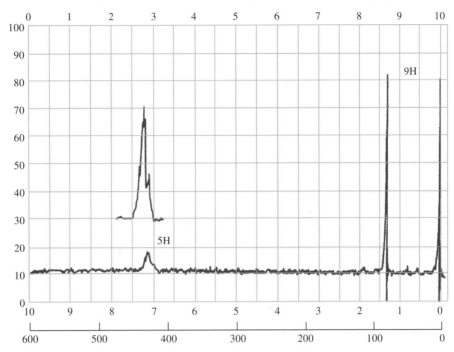

FIGURE 19-44 Proton NMR spectrum. (From C. J. Pouchert, *The Aldrich Library of NMR Spectra*, 2nd ed., Milwaukee, WI: The Aldrich Chemical Company. With permission.)

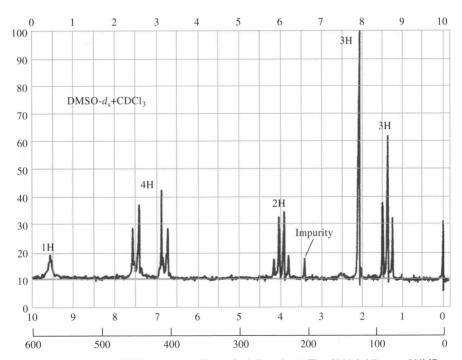

FIGURE 19-45 Proton NMR spectrum. (From C. J. Pouchert, *The Aldrich Library of NMR Spectra*, 2nd ed., Milwaukee, WI: The Aldrich Chemical Company. With permission.)

19-38 Explain how a band of frequencies is obtained from an oscillator, which is essentially a monochromatic source of RF radiation. How could a band broad enough to cover the entire ^{13}C spectrum (200 ppm) be obtained?

19-39 Describe sources of folded spectral lines.

19-40 What is the nuclear Overhauser effect and its source?

19-41 What are the sources of band broadening in ^{13}C spectra of solids? How are lines narrowed so that high-resolution spectra can be obtained?

Challenge Problem

19-42 (a) What are the major advantages and disadvantages of two-dimensional NMR methods over conventional one-dimensional techniques?
   (b) Describe in detail the pulse sequences that are used in the COSY and HETCOR experiments. For both, describe the behavior of the magnetization vector M as a result of the applied pulses. Consult note 19 for assistance in learning more about the details of two-dimensional NMR methods.
   (c) From the 300-MHz ^1H NMR spectrum of phenanthro[3,4-b]thiophene reproduced in Figure 19-46, identify the resonances of H_1 and H_{11} as a pair. Can you assign any of the other resonances?

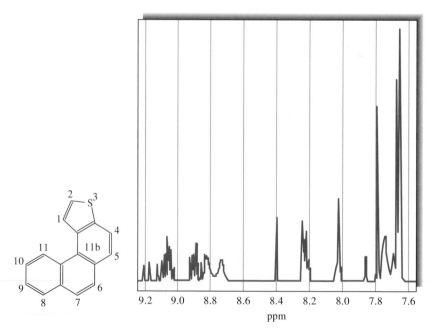

FIGURE 19-46 Proton NMR spectrum of phenanthro[3,4-b]thiophene at 300 MHz. (From G. E. Martin and A. S. Zektzer, *Two-Dimensional NMR Methods for Establishing Connectivity*, New York: Wiley-VCH, 1988. With permission.)

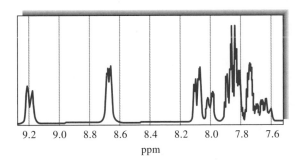

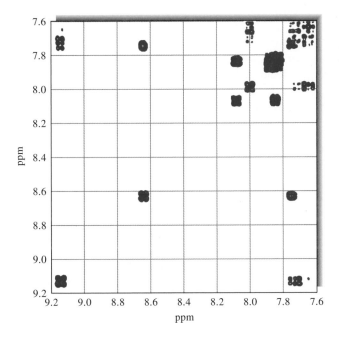

FIGURE 19-47 300-MHz COSY spectrum of phenan-thro[3,4-b]thiophene. (From J. B. Lambert and E. P. Mazzola, *Nuclear Magnetic Resonance Spectroscopy*, Upper Saddle River, NJ: Pearson/Prentice-Hall, 2004. With permission.)

(d) From the 300-MHz COSY spectrum in Figure 19-47, describe how you would distinguish the resonances of H_1 and H_{11}.

(e) In the COSY spectrum of Figure 19-47, assign resonances to H_2, H_8, H_9, and H_{10}. Give the rationale for your assignments.

CHAPTER TWENTY

Molecular Mass Spectrometry

Mass spectrometry is perhaps the most widely applicable of all the analytical tools available in the sense that the technique is capable of providing information about (1) the elemental composition of samples of matter; (2) the structures of inorganic, organic, and biological molecules; (3) the qualitative and quantitative composition of complex mixtures; (4) the structure and composition of solid surfaces; and (5) isotopic ratios of atoms in samples.

We have already discussed in Chapter 11 how mass spectrometry is used by chemists for the identification and quantitative determination of one or more elements in a sample of matter. This chapter is devoted to describing how mass spectrometry is used to obtain the type of information listed in items (2) and (3) in the previous paragraph. Chapter 21 describes how mass spectrometry is employed for elucidating the structure and composition of surfaces. Finally, in Section 32D, the use of isotopic ratios determined by mass spectrometry is discussed.

Throughout this chapter, this logo indicates an opportunity for online self-study at **www.thomsonedu.com/chemistry/skoog**, linking you to interactive tutorials, simulations, and exercises.

The first general application of molecular mass spectrometry for routine chemical analysis occurred in the early 1940s, when the techniques began to be adopted by the petroleum industry for the quantitative analysis of hydrocarbon mixtures produced in catalytic crackers. Before this time, analyses of mixtures of this type, which often contained as many as nine hydrocarbon components, were carried out by fractional distillation followed by refractive-index measurements of the separated components. Typically, 200 hours or more of operator time was required to complete an analysis. It was found that similar information could be obtained in a few hours or less with a mass spectrometer. This improved efficiency led to the appearance and rapid improvement of commercial mass spectrometers. Beginning in the 1950s, these commercial instruments began to be adapted by chemists for the identification and structural elucidation of a wide variety of organic compounds. This use of the mass spectrometer combined with the invention of nuclear magnetic resonance and the development of infrared spectrometry revolutionized the way organic chemists identify and determine the structure of molecules. This application of mass spectrometry is still extremely important.

Applications of molecular mass spectrometry dramatically changed in the decade of the 1980s, as a result of the development of new methods for producing ions from nonvolatile or thermally labile molecules, such as those frequently encountered by biochemists and biologists. Since about 1990, explosive growth in the area of biological mass spectrometry has occurred as a consequence of these new ionization methods. Now, mass spectrometry is being applied to the determination of the structure of polypeptides, proteins, and other high-molecular-mass biopolymers.

In this chapter, we first describe the nature of molecular mass spectra and define some terms used in molecular mass spectrometry. We next consider the various techniques used to form ions from analyte molecules in mass spectrometers and the types of spectra produced by these techniques. We then describe in some detail the various types of mass spectrometers used in molecular mass spectrometry (other than the quadrupole and time-of-flight instruments, which received detailed treatment in Section 11B). Finally, we

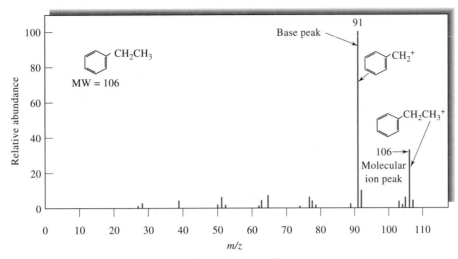

FIGURE 20-1 Mass spectrum of ethyl benzene.

describe several of the current applications of molecular mass spectrometry.[1]

20A MOLECULAR MASS SPECTRA

Figure 20-1 illustrates how mass spectral data are usually presented. The analyte was ethyl benzene, which has a nominal molecular mass of 106 daltons (Da). To obtain this spectrum, ethyl benzene vapor was bombarded with a stream of electrons that led to the loss of an electron by the analyte and formation of the molecular ion M^+ as shown by the reaction

$$C_6H_5CH_2CH_3 + e^- \rightarrow C_6H_5CH_2CH_3^{\cdot+} + 2e^- \quad (20\text{-}1)$$

The charged species $C_6H_5CH_2CH_3^{\cdot+}$ is the *molecular ion*. As indicated by the dot, the molecular ion is a radical ion that has the same molecular mass as the molecule.

The collision between energetic electrons and analyte molecules usually imparts enough energy to the molecules to leave them in an excited state. Relaxation then often occurs by fragmentation of part of the molecular ions to produce ions of lower masses. For example, a major product in the case of ethyl benzene is $C_6H_5CH_2^+$, which results from the loss of a CH_3 group. Other smaller positively charged fragments are also formed in lesser amounts.

The positive ions produced on electron impact are attracted through the slit of a mass spectrometer where they are sorted according to their mass-to-charge ratios and displayed in the form of a mass spectrum. Note that the plot shown in Figure 20-1 is in the form of a bar graph that relates the relative intensity of mass peaks to their mass-to-charge ratio. Note also that in each spectrum the largest peak, termed the *base peak*, has been arbitrarily assigned a value of 100. The heights of the remaining peaks have been computed as a percentage of the base-peak height. Modern mass spectrometers are programmed to automatically recognize the base peak. They then normalize the remaining peaks in the spectrum relative to the base peak.

20B ION SOURCES

The starting point for a mass spectrometric analysis is the formation of gaseous analyte ions, and the scope and the utility of a mass spectrometric method is dictated by the ionization process. The appearance of mass spectra for a given molecular species strongly depends on the method used for ion formation. Table 20-1 lists many of the ion sources that have been employed in

[1]For detailed discussions of mass spectrometry, see R. M. Smith, *Understanding Mass Spectra: A Basic Approach*, 2nd ed., New York: Wiley, 2004; E. de Hoffman and V. Stroobant, *Mass Spectrometry: Principles and Applications*, 2nd ed., Chichester, UK: Wiley, 2002; J. T. Watson, *Introduction to Mass Spectrometry*, 3rd ed., Philadelphia: Lippincott-Raven, 1997.

TABLE 20-1 Ion Sources for Molecular Mass Spectrometry

Basic Type	Name and Acronym	Ionizing Agent
Gas phase	Electron impact (EI)	Energetic electrons
	Chemical ionization (CI)	Reagent gaseous ions
	Field ionization (FI)	High-potential electrode
Desorption	Field desorption (FD)	High-potential electrode
	Electrospray ionization (ESI)	High electrical field
	Matrix-assisted desorption-ionization (MALDI)	Laser beam
	Plasma desorption (PD)	Fission fragments from ^{252}Cf
	Fast atom bombardment (FAB)	Energetic atomic beam
	Secondary-ion mass spectrometry (SIMS)	Energetic beam of ions
	Thermospray ionization (TS)	High temperature

molecular mass spectrometry.[2] Note that these methods fall into two major categories: *gas-phase sources* and *desorption sources*. With a gas-phase source, which includes the first three sources in the table, the sample is first vaporized and then ionized. With a desorption source, however, the solid- or liquid-state sample is converted directly into gaseous ions. An advantage of desorption sources is that they are applicable to nonvolatile and thermally unstable samples. Currently, commercial mass spectrometers are equipped with accessories that permit interchangeable use of several of these sources.

Gas-phase sources are generally restricted to the ionization of thermally stable compounds that have boiling points less than about 500°C. In most cases, this requirement limits gaseous sources to compounds with molecular masses less than roughly 10^3 Da. Desorption sources, which do not require volatilization of analyte molecules, are applicable to analytes having molecular masses as large as 10^5 Da.

Ion sources are also classified as being *hard sources* or *soft sources*. Hard ionization sources impart enough energy to analyte molecules to leave them in a highly excited energy state. Relaxation then involves rupture of bonds, producing fragment ions that have mass-to-charge ratios less than that of the molecular ion. Soft ionization sources cause little fragmentation. Thus, the

mass spectrum from a soft ionization source often consists of the molecular ion peak and only a few, if any, other peaks. Figure 20-2 illustrates the difference in spectra obtained from a hard ionization source and a soft ionization source.

Both hard- and soft-source spectra are useful for analysis. The many peaks in a hard-source spectrum provide useful information about the kinds of functional groups and thus structural information about analytes. Soft-source spectra are useful because they supply accurate information about the molecular mass of the analyte molecule or molecules.

20B-1 The Electron-Impact Source

Historically, ions for mass analysis were produced by *electron impact*. In this process, the sample is brought to a temperature high enough to produce a molecular vapor, which is then ionized by bombarding the resulting molecules with a beam of energetic electrons. Despite certain disadvantages, this technique is still of major importance and is the one on which many libraries of mass spectral data are based.

Figure 20-3 is a schematic of a basic electron-impact ion source. Electrons are emitted from a heated tungsten or rhenium filament and accelerated by applying approximately 70 V between the filament and the anode. As shown in the figure, the paths of the electrons and molecules are at right angles and intersect near the center of the source, where collision and ionization occur. The primary product is singly charged positive ions formed when the energetic electrons approach molecules closely enough to cause

---

[2] For more information about modern ion sources, see http://www.jeol.com/ms/docs/ionize.html#FDFI; E. de Hoffman and V. Stroobant, *Mass Spectrometry: Principles and Applications*, 2nd ed., Chichester, UK: Wiley, 2002; J. T. Watson, *Introduction to Mass Spectrometry*, 3rd ed., Philadelphia: Lippincott-Raven, 1997; E. R. Grant and R. G. Cooks, *Science*, **1990**, *250*, 61.

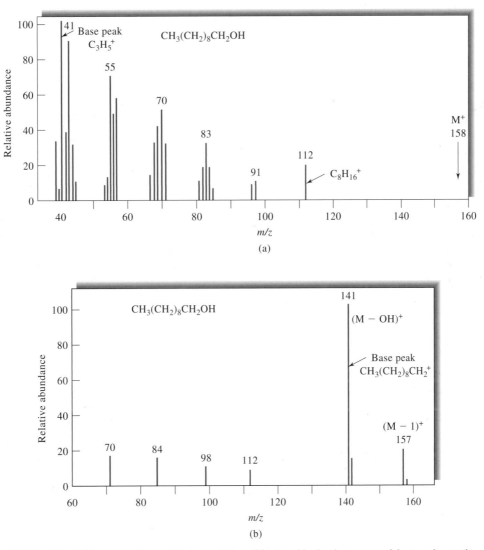

FIGURE 20-2 Mass spectrum of 1-decanol from (a) a hard ionization source (electron impact) and (b) a soft ionization source (chemical ionization).

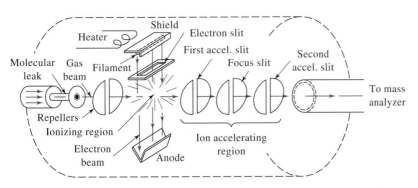

FIGURE 20-3 An electron-impact ion source. (Adapted from R. M. Silverstein and F. X. Webster, *Spectrometric Identification of Organic Compounds*, 6th ed., p. 4, New York: Wiley, 1998. Reprinted with permission of John Wiley & Sons, Inc.)

them to lose electrons by electrostatic repulsion. Electron-impact ionization is not very efficient, and only about one molecule in a million undergoes the primary reaction

$$M + e^- \rightarrow M^{\cdot+} + 2e^- \qquad (20\text{-}2)$$

Here, M represents the analyte molecule, and $M^{\cdot+}$ is its molecular ion. The positive ions produced by electron impact are attracted through the slit in the first accelerating plate by a small potential difference (typically 5 V) that is applied between this plate and the repellers shown in Figure 20-3. With magnetic sector instruments, high voltages (10^3 to 10^4 V) are applied to the accelerator plates, which give the ions their final velocities before they enter the mass analyzer. Commercial electron-impact sources are more complex than that shown in Figure 20-3 and may use additional electrostatic or magnetic fields to manipulate the electron or ion beam. A typical kinetic energy produced in an electron-impact source is calculated in Example 20-1.

EXAMPLE 20-1

(a) Calculate the kinetic energy that a singly charged ion ($z = 1$) will acquire if it is accelerated through a potential of 10^3 V in an electron-impact source. (b) Does the kinetic energy of the ion depend on its mass? (c) Does the velocity of the ion depend on its mass?

▸ *Solution*

(a) The kinetic energy (KE) added to the ion is due to the accelerating potential V and is given by the equation

$$KE = qV = zeV$$

where e is the electronic charge $(1.6 \times 10^{-19}$ coulombs). Thus, for $z = 1$

$$KE = 1 \times 1.6 \times 10^{-19}\,C \times 10^3\,V = 1.6 \times 10^{-16}\,J$$

(b) The kinetic energy that an ion acquires in the source is independent of its mass and depends only on its charge and the accelerating potential.

(c) The translational component of the kinetic energy of an ion is a function of the ion mass m and its velocity v as given by the equation

$$KE = (1/2)mv^2 \quad \text{or} \quad v = (2\,KE/m)^{1/2}$$

Thus, if all ions acquire the same amount of kinetic energy, those ions with largest mass must have the smallest velocity.

Electron-Impact Spectra

To form a significant number of gaseous ions at a reproducible rate, electrons from the filament in the source must be accelerated by a voltage of greater than about 50 V. The low mass and high kinetic energy of the resulting electrons cause little increase in the translational energy of impacted molecules. Instead, the molecules are left in highly excited vibrational and rotational states. Subsequent relaxation then usually takes place by extensive fragmentation, giving a large number of positive ions of various masses that are less than (and occasionally, because of collisions, greater than) that of the molecular ion. These lower-mass ions are called *daughter ions*. Table 20-2 shows some typical fragmentation reactions that follow electron-impact formation of a parent ion from a hypothetical molecule ABCD. In Example 20-2, the kinetic energy of electrons accelerated through a potential difference of 70 V is calculated.

EXAMPLE 20-2

(a) Calculate the energy (in J/mol) that electrons acquire as a result of being accelerated through a potential of 70 V. (b) How does this energy compare to that of a typical chemical bond?

▸ *Solution*

(a) The kinetic energy KE of an individual electron is equal to the product of the charge on the electron e times the potential V through which it has been accelerated. Multiplying the kinetic energy of a single electron by Avogadro's number, N, gives the energy per mole:

$$
\begin{aligned}
KE &= eVN \\
&= (1.60 \times 10^{-19}\,C/e^-)(70\,V)N \\
&= (1.12 \times 10^{-17}\,CV/e^-)(6.02 \times 10^{23}\,e^-/mol) \\
&= 6.7 \times 10^6\,J/mol \quad \text{or} \quad 6.7 \times 10^3\,kJ/mol
\end{aligned}
$$

(b) Typical bond energies fall in the 200- to 600-kJ/mol range. Therefore, an electron that has been accelerated through 70 V generally has considerably more energy than that required to break a chemical bond.

TABLE 20-2 Some Typical Reactions in an Electron-Impact Source

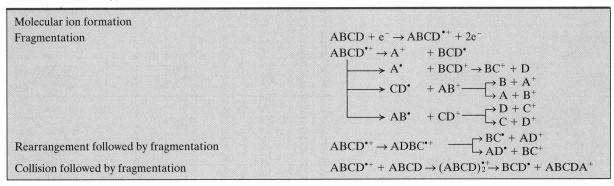

The complex mass spectra that result from electron-impact ionization are useful for compound identification. On the other hand, with certain types of molecules, fragmentation is so complete that no molecular ion persists. Without a molecular ion, important information for determining the molecular mass of the analyte is lost. Figure 20-4 shows typical electron-impact spectra for two simple organic molecules: methylene chloride and 1-pentanol. Note that in each of the spectra, the base peak corresponds to a fragment of the molecule, which has a mass significantly less than the molecular mass of the original compound. For methylene chloride, the base peak occurs at a mass-to-charge ratio m/z of 49, which corresponds to the loss of one Cl atom. For 1-pentanol, the base peak is found at an m/z of 44, which is that of the daughter CH_2CHOH^+. More often than not, the base peaks in electron-impact spectra correspond to fragments such as these rather than the molecular ion.

The molecular ion peak occurs at a mass corresponding to the molecular mass of the analyte. Thus, molecular ion peaks appear at $m/z = 84$ for methylene chloride and at $m/z = 88$ for 1-pentanol. The molecular ion peak is, of course, very important in structural determinations, because its m/z value provides the molecular mass of the unknown. Unfortunately, it is not always possible to identify the molecular ion peak. Indeed, electron-impact ionization of certain molecules yields no molecular ion peak at all (see Figure 20-2a).

Isotope Peaks

It is interesting to note in the spectra shown in Figures 20-1 and 20-4 that peaks occur at mass-to-charge

ratios greater than that of the molecular ion. These peaks are attributable to ions having the same chemical formula, but different isotopic compositions. For example, for methylene chloride, the more important isotopic species are $^{12}C^1H_2^{35}Cl_2$ ($\mathcal{M} = 84$), $^{13}C^1H_2^{35}Cl_2$ ($\mathcal{M} = 85$), $^{12}C^1H_2^{35}Cl^{37}Cl$ ($\mathcal{M} = 86$), $^{13}C^1H_2^{35}Cl^{37}Cl$ ($\mathcal{M} = 87$), and $^{12}C^1H_2^{37}Cl_2$ ($\mathcal{M} = 88$), where $\mathcal{M}$ is molecular mass. Peaks for each of these species can be seen in Figure 20-4a. The size of the various peaks depends on the relative natural abundance of the isotopes. Table 20-3 lists the most common isotopes for atoms that occur widely in organic compounds. Note that fluorine, phosphorus, iodine, and sodium occur only as single isotopes.

The small peak for ethyl benzene at m/z 107 in Figure 20-1 is due to the presence of ^{13}C in the molecules. The intensities of peaks due to incorporation of two or more ^{13}C atoms in ethyl benzene can be predicted with good precision but are normally so small as to be undetectable because of the low probability of there being more than one ^{13}C atom in a small molecule. As will be shown in Section 20D-1, isotope peaks sometimes provide a useful means for determining the formula for a compound.

Collision Product Peaks

Ion-molecule collisions, such as that shown by the last equation in Table 20-2, can produce peaks at higher mass numbers than that of the molecular ion. At ordinary sample pressures, however, the only important reaction of this type is one in which the collision transfers a hydrogen atom to the ion to give a protonated molecular ion. The result is an enhanced $(M + 1)^+$ peak. This transfer is a second-order reaction, and the amount of product depends strongly on the reactant concentration. Consequently, the height of an $(M + 1)^+$

 Tutorial: Learn more about **sources for MS**.

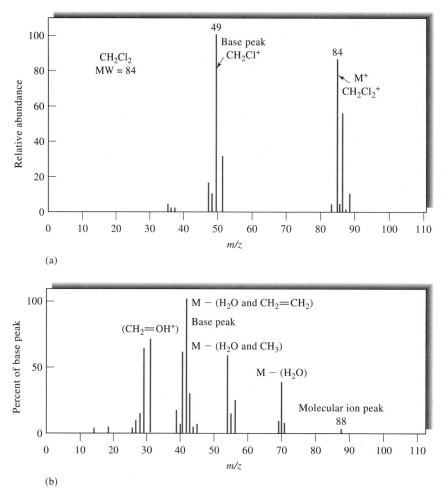

FIGURE 20-4 Electron-impact mass spectra of (a) methylene chloride and (b) 1-pentanol.

TABLE 20-3 Natural Abundance of Isotopes of Some Common Elements

Element[a]	Most Abundant Isotope	Abundance of Other Isotopes Relative to 100 Parts of the Most Abundant[b]	
Hydrogen	^{1}H	^{2}H	0.015
Carbon	^{12}C	^{13}C	1.08
Nitrogen	^{14}N	^{15}N	0.37
Oxygen	^{16}O	^{17}O	0.04
		^{18}O	0.20
Sulfur	^{32}S	^{33}S	0.80
		^{34}S	4.40
Chlorine	^{35}Cl	^{37}Cl	32.5
Bromine	^{79}Br	^{81}Br	98.0
Silicon	^{28}Si	^{29}Si	5.1
		^{30}Si	3.4

[a]Fluorine (^{19}F), phosphorus (^{31}P), sodium (^{23}Na), and iodine (^{127}I) have no additional naturally occurring isotopes.

[b]The numerical entries indicate the average number of isotopic atoms present for each 100 atoms of the most abundant isotope; thus, for every 100 ^{12}C atoms there will be an average of 1.08 ^{13}C atoms.

peak due to this reaction increases much more rapidly with increases in sample pressure than do the heights of other peaks. This phenomenon usually makes it possible to detect this atom-transfer reaction.

Advantages and Disadvantages of Electron-Impact Sources

Electron-impact sources are convenient to use and produce high ion currents, thus giving good sensitivities. The extensive fragmentation and resulting large number of peaks is also an advantage because it often makes unambiguous identification of analytes possible. Extensive fragmentation can also be a disadvantage, however, when it results in the disappearance of the molecular ion peak so that the molecular mass of analytes cannot be easily established. Another limitation of the electron-impact source is the need to volatilize the sample, which may result in thermal degradation of some analytes before ionization can occur. The effects of thermal decomposition can sometimes be minimized by carrying out the volatilization from a heated probe located close to the entrance slit of the spectrometer. At the lower pressure of the source area, volatilization occurs at a lower temperature. Furthermore, less time is allowed for thermal decomposition to take place. As mentioned earlier, electron-impact sources are applicable only to analytes having molecular masses smaller than about 10^3 Da.

20B-2 Chemical Ionization Sources and Spectra

Most modern mass spectrometers are designed so that electron-impact ionization and chemical ionization can be carried out interchangeably. Such sources are called EI-CI sources. In chemical ionization, gaseous atoms of the sample (from either a batch inlet or a heated probe) are ionized by collision with ions produced by electron bombardment of an excess of a reagent gas. Usually, positive ions are used, but negative ion chemical ionization is occasionally used with analytes that contain very electronegative atoms. Chemical ionization is probably the second-most common procedure for producing ions for mass spectrometry.[3]

To carry out chemical ionization experiments, it is necessary to modify the electron beam ionization area

shown in Figure 20-3 by adding vacuum pump capacity and by reducing the width of the slit to the mass analyzer. These measures allow a reagent pressure of about 1 torr to be maintained in the ionization area while maintaining the pressure in the analyzer below 10^{-5} torr. With these changes, a gaseous reagent is introduced into the ionization region in an amount such that the concentration ratio of reagent to sample is 10^3 to 10^4. Because of this large concentration difference, the electron beam reacts nearly exclusively with reagent molecules.

One of the most common reagents is methane, which reacts with high-energy electrons to give several ions such as CH_4^+, CH_3^+, and CH_2^+. The first two predominate and represent about 90% of the reaction products. These ions react rapidly with additional methane molecules as follows:

$$CH_4^+ + CH_4 \rightarrow CH_5^+ + CH_3$$
$$CH_3^+ + CH_4 \rightarrow C_2H_5^+ + H_2$$

Generally, collisions between the analyte molecule MH and CH_5^+ or $C_2H_5^+$ are highly reactive and involve proton or hydride transfer. For example,

$$CH_5^+ + MH \rightarrow MH_2^+ + CH_4 \quad \text{proton transfer}$$
$$C_2H_5^+ + MH \rightarrow MH_2^+ + C_2H_4 \quad \text{proton transfer}$$
$$C_2H_5^+ + MH \rightarrow M^+ + C_2H_6 \quad \text{hydride transfer}$$

Note that proton-transfer reactions give the $(MH + 1)^+$ ion whereas the hydride transfer produces an ion with a mass one less than the analyte, or the $(MH - 1)^+$ ion. With some compounds, an $(MH + 29)^+$ peak is also produced from transfer of a $C_2H_5^+$ ion to the analyte. A variety of other reagents, including propane, isobutane, and ammonia, are used for chemical ionization. Each produces a somewhat different spectrum with a given analyte.

Figure 20-2 contrasts the chemical ionization and electron-impact spectra for 1-decanol. The electron-impact spectrum (Figure 20-2a) shows evidence for rapid and extensive fragmentation of the molecular ion. Thus, no detectable peaks are observed above mass 112, which corresponds to the ion $C_8H_{16}^+$.

The base peak is provided by the ion $C_3H_5^+$ at mass 41. Other peaks for C_3 species are grouped around the base peak. A similar series of peaks, found at 14, 28, and 42 mass units greater, correspond to ions with one, two, and three additional CH_2 groups.

---

[3] For a more detailed discussion of chemical ionization, see A. Harrison, *Chemical Ionization Mass Spectrometry*, Boca Raton, FL: CRC Press, 1983.

Relative to the electron-impact spectrum, the chemical ionization spectrum shown in Figure 20-2b is simple indeed, consisting of the $(M - 1)^+$ peak, a base peak corresponding to a molecular ion that has lost an OH group, and a series of peaks differing from one another by 14 mass units. As in the electron-impact spectrum, these peaks arise from ions formed by cleavage of adjacent carbon-carbon bonds. As we have just noted, chemical ionization spectra generally contain well-defined $(M + 1)^+$ or $(M - 1)^+$ peaks resulting from the addition or abstraction of a proton in the presence of the reagent ion.

20B-3 Field Ionization Sources and Spectra

In *field ionization* sources, ions are formed under the influence of a large electric field (10^8 V/cm). Such fields are produced by applying high voltages (10 to 20 kV) to specially formed emitters consisting of numerous fine tips having diameters of less than 1 μm. The emitter often takes the form of a fine tungsten wire (~10 μm

FIGURE 20-5 Photomicrograph of a carbon microneedle emitter. (Courtesy of R. P. Lattimer, BF Goodrich Research and Development Center.)

diameter) on which microscopic carbon dendrites, or whiskers, have been grown by the pyrolysis of benzonitrile in a high electric field. The result of this treatment is a growth of many hundreds of carbon microtips projecting from the surface of the wire as can be seen in the photomicrograph of Figure 20-5.

Field ionization emitters are mounted 0.5 to 2 mm from the cathode, which often also serves as a slit. The gaseous sample from a batch inlet system is allowed to diffuse into the high-field area around the microtips of the anode. The electric field is concentrated at the emitter tips, and ionization occurs via a quantum mechanical tunneling mechanism in which electrons from the analyte are extracted by the microtips of the anode. Little vibrational or rotational energy is imparted to the analyte, and little fragmentation occurs.

Figure 20-6 shows spectra for glutamic acid obtained by (a) electron-impact ionization and (b) field ionization. In the electron-impact spectrum, the parent ion peak at $m/z = 147$ is not detectable. The highest observable peak ($m/z = 129$) is due to the loss of water by the molecular ion. The base peak at $m/z = 84$ arises from a loss of water and a —COOH group. Numerous other fragments are also found at lower masses. In contrast, the field ionization spectrum is relatively simple, with an easily distinguished $(M + 1)^+$ peak at $m/z = 148$.

A limitation to field ionization is its sensitivity, which is at least an order of magnitude less than that of electron-impact sources; maximum currents are on the order of 10^{-11} A.

20B-4 Desorption Sources

The ionization methods discussed so far require that the ionizing agents act on gaseous samples. Such methods are not applicable to nonvolatile or thermally unstable samples. A number of *desorption ionization* methods have been developed for dealing with this type of sample (see Table 20-1). These methods have enabled mass spectra to be obtained for thermally delicate biochemical species and species having molecular masses of greater than 100,000 Da.

Desorption methods dispense with volatilization followed by ionization of the gaseous analyte molecules. Instead, energy in various forms is introduced into the solid or liquid sample in such a way as to cause direct formation of gaseous ions. As a consequence, spectra are greatly simplified and often consist of only

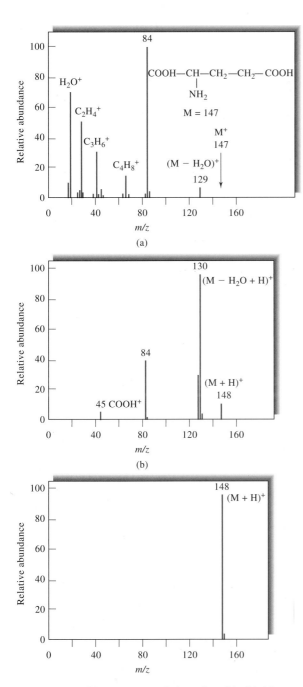

FIGURE 20-6 Mass spectra of glutamic acid with (a) electron-impact ionization, (b) field ionization, and (c) field desorption. (From H. D. Beckey, A. Heindrich, and H. U. Winkler, *Int. J. Mass Spec. Ion Phys.*, **1970**, *3*, App. 11. With permission.)

the molecular ion or the protonated molecular ion. In most cases the exact mechanism of how ions are formed without fragmentation is not well understood.

Field Desorption Methods

In field desorption, a multitipped emitter similar to that used in field ionization sources is employed.[4] In this case, the electrode is mounted on a probe that can be removed from the sample compartment and coated with a solution of the sample. After the probe is reinserted into the sample compartment, ionization again takes place by applying a high voltage to this electrode. With some samples it is necessary to heat the emitter by passing a current through the wire. As a result, thermal degradation may occur before ionization is complete.

Figure 20-6c is a field desorption spectrum for glutamic acid. It is even simpler than the spectrum from field ionization and consists of only the protonated molecular ion peak at $m/z = 148$ and an isotope peak at $m/z = 149$.

Matrix-Assisted Laser Desorption-Ionization

Matrix-assisted laser desorption-ionization (MALDI) spectrometry[5] is an ionization method that can be used to obtain accurate molecular mass information about polar biopolymers ranging in molecular mass from a few thousand to several hundred thousand Da. The method was first described nearly simultaneously in 1988 by two research groups, one German and the other Japanese.[6] Commercial instrumentation is available for MALDI.[7]

In the MALDI technique, a low concentration of the analyte is uniformly dispersed in a solid or liquid matrix deposited on the end of a stainless steel probe or placed on a metal plate. The plate is then placed in a vacuum chamber and a laser beam is focused onto the sample. In addition to the usual vacuum-chamber MALDI, atmospheric-pressure MALDI has also been described.[8] The MALDI matrix must strongly absorb the laser radiation. The matrix and analyte are then de-

---

[4]See L. Prakai, *Field Desorption Mass Spectrometry*, New York: Marcel Dekker, 1990; R. P. Lattimer and H. R. Schulten, *Anal. Chem.*, **1989**, *61*, 1201A.

[5]For additional information, see C. Dass, *Principles and Practice of Biological Mass Spectrometry*, New York: Wiley, 2001; J. R. Chapman, *Mass Spectrometry of Proteins and Peptides*, Totowa, NJ: Humana Press, 2000.

[6]See M. Karas and F. Hillenkamp, *Anal. Chem.*, **1988**, *60*, 2299; K. Tanaka, H. Waki, Y. Ido, S. Akita, Y. Yoshidda, and T. Yoshidda, *Rapid Commun. Mass Spectrosc.*, **1988**, *2*, 151; F. Hillenkamp, M. Karas, and B. T. Chait, *Anal. Chem.*, **1991**, *63*, 1193A.

[7]See D. Noble, *Anal. Chem.*, **1995**, *67*, 497A.

[8]See S. G. Moyer and R. J. Cotter, *Anal. Chem.*, **2002**, *74*, 489A.

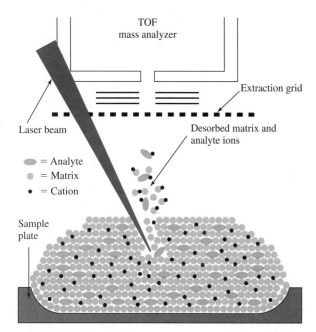

FIGURE 20-7 Diagram of the MALDI process. The analyte is uniformly dispersed in a matrix and placed on a metal sample plate. A pulsed laser beam strikes the sample causing desorption of a plume of matrix, analyte, and other ions. The analyte can be protonated, be deprotonated, or form adducts before entering the TOF analyzer.

sorbed and ionized, creating an ion plume. The overall process is illustrated in Figure 20-7. The most common type of mass analyzer used with MALDI is the time-of-flight (TOF) analyzer. A mass spectrum from a MALDI-TOF instrument is shown in Figure 20-8. Here, the matrix material was nicotinic acid, and the analyte was a monoclonal antibody from a mouse having a molecular mass of approximately 150,000 Da. Note that the spectrum is characterized by very low background noise and a complete absence of fragmentation of the large analyte ion. Multiply charged ions are present as well as peaks for dimer and trimer species.

Although the mechanism of the formation of the MALDI ion plume is not completely understood, it is thought to involve absorption of the laser beam by the matrix, followed by transfer of the energy from the matrix to the analyte. Desorption of the analyte and the matrix then occurs. The analyte is thought to desorb as neutral molecules and then to be ionized by proton-transfer reactions with protonated matrix ions in a dense phase over the surface containing the matrix. A series of photochemical reactions may produce the protonated matrix ions.

Some of the matrix materials used for biomolecules are listed in Table 20-4 along with the laser wavelengths that have been employed. Lasers used include nitrogen (337 nm), Nd-YAG (266 and 355 nm), excimer (308 nm), Er-YAG (2.94 μm), and CO_2 (10.6 μm).

The most common sample-preparation method is the dried-droplet technique in which a droplet of the matrix containing the analyte is deposited on the metal plate and then dried. Typically, the ratio of analyte to matrix is $1:10^3$ to $1:10^5$. Analyte concentrations are usually in the micromolar range.

Electrospray Ionization

Electrospray ionization–mass spectrometry (ESI/MS), which was first described in 1984, has now become one of the most important techniques for analyzing biomolecules, such as polypeptides, proteins, and oligonucleotides, having molecular weights of 100,000 Da or more.[9] In addition, this method is finding more and more application to the characterization of inorganic species and synthetic polymers. For their development of soft desorption ionization methods, such as electrospray ionization, John B. Fenn of Virginia Commonwealth University and Koichi Tanaka of Shimadzu shared the 2002 Nobel Prize in Chemistry.

Electrospray ionization takes place under atmospheric pressures and temperatures in an apparatus such as that shown in Figure 20-9. A solution of the sample is pumped through a stainless steel capillary needle at a rate of a few microliters per minute. The needle is maintained at several kilovolts with respect to a cylindrical electrode that surrounds the needle. The resulting charged spray of fine droplets then passes through a desolvating capillary, where evaporation of the solvent and attachment of charge to the analyte molecules take place. As the droplets become smaller as a consequence of evaporation of the solvent, their charge density becomes greater until, at a point called the *Rayleigh limit*, the surface tension can no longer support the charge. Here a so-called *Coulombic explosion* occurs and the droplet is torn apart into smaller droplets. These small droplets can repeat the process

---

[9]For additional information, see B. N. Pramanik, A. K. Ganguly, and M. L. Gross, *Applied Electrospray Mass Spectrometry*, New York: Marcel Dekker, 2002; *Electrospray Ionization: Fundamentals, Instrumentation, and Applications*, R. B. Cole, ed., New York: Wiley, 1997; J. T. Watson, *Introduction to Mass Spectrometry*, 3rd ed., Philadelphia: Lippincott-Raven, 1997; J. B. Fenn, M. Mann, C. K. Meng, S. F. Wong, and C. M. Whitehouse, *Science*, **1989**, *246*, 64.

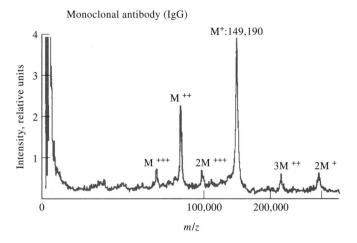

Monoclonal antibody (IgG)

M^+:149,190

M^{++}

M^{+++} $2M^{+++}$

$3M^{++}$ $2M^+$

Intensity, relative units

m/z

0   100,000   200,000

FIGURE 20-8 MALDI-TOF spectrum from nicotinic acid matrix irradiated with a 266-nm laser beam. (From M. Karas and U. Bahr, *Trends Anal. Chem.* (TRAC), **1990**, *9*, 323.)

TABLE 20-4 Common Matrices for MALDI and Usable Wavelengths

Matrix	Analytes	Wavelength, nm
Nitropyridines:		
2-Amino-4-methyl-5-nitropyridine	Proteins, oligonucleotides	355
2-Amino-5-nitropyridine	Oligonucleotides	355
Nicotinic acid	Proteins, glycoproteins, oligonucleotides	266, 220–290
Benzoic acid derivatives:		
2,5-Dihydroxybenzoic acid	Proteins	266, 337, 355, 2940
Vanillic acid	Proteins	266
2-Aminobenzoic acid	Proteins	266, 337, 355
2-(4-Hydroxyphenylazo) benzoic acid	Proteins, gangliosides, polymers	266, 377
2-Pyrazinecarboxylic acid	Proteins	266
3-Aminopyrazine-2-carboxylic acid	Proteins	337
Cinnamic acid derivatives:		
Ferulic acid	Proteins, oligonucleotides	266, 337, 355, 488
Sinapinic acid	Proteins, industrial polymers	337, 355
Caffeic acid	Proteins, oligonucleotides	266, 337, 355, 10600
α-Cyano-4-hydroxy cinnamic acid	Proteins, oligosaccharides	337
3-Nitrobenzyl alcohol	Proteins	266
3-Nitrobenzyl alcohol with rhodamine 6G	Proteins	532
3-Nitrobenzyl alcohol with 1,4-diphenyl-1,3-butadiene	Proteins	337
3-Hydroxypicolinic acid	Oligonucleotides, glycoproteins	266, 308, 355
Succinic acid	Proteins	2940, 10600

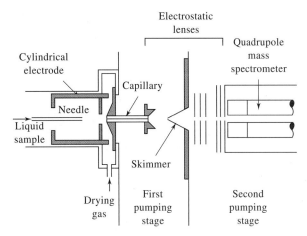

FIGURE 20-9 Apparatus for electrospray ionization. (From J. B. Fenn et al., *Science*, **1989**, *65*, 246. Reprinted with permission.)

until all the solvent is removed from the analyte, leaving a multiply charged analyte molecule.

An interesting and useful feature of the electrospray process is that little fragmentation of large and thermally fragile biomolecules occurs because there is little extra energy retained by the analyte upon ionization. Furthermore, the ions formed are multiply charged so that their m/z values are small enough to make them detectable with a quadrupole instrument with a range of 1500 or less.[10] This important property is demonstrated by the mass spectra of four proteins of varying molecular mass ($\mathcal{M}$) shown in Figure 20-10. In these spectra, adjacent peaks are for analyte ions that differ by one charge. A striking feature of the spectra for proteins, such as those in the figure, is that the average charge state increases in approximately linear fashion with molecular mass. The charge state corresponding to each peak can be determined from peak distribution, thus making it possible to determine the molecular mass of a protein from spectra such as those shown in Figure 20-10.

An important characteristic of electrospray ionization is that it is readily adapted to direct sample introduction from high-performance liquid chromatography and capillary electrophoresis columns. These applications are described in Chapters 28 and 30. That there is little fragmentation of the analyte makes

structural elucidation a difficult task. Usually, tandem mass spectrometry (see Section 20C-5) is employed for this purpose. Here, the ions from the original ionization process are separated and the ion of interest is subjected to a fragmentation step before being mass analyzed.

Fast Atom Bombardment Sources

Fast atom bombardment (FAB) sources, also called *liquid secondary-ion* sources, have assumed a major role in the production of ions for mass spectrometric studies of polar high-molecular-mass species.[11] With this type of source, samples in a condensed state, usually in a viscous solution matrix, are ionized by bombardment with energetic (several keV) xenon or argon *atoms*. Both positive and negative analyte ions are sputtered from the surface of the sample in a desorption process. This treatment provides very rapid sample heating, which reduces sample fragmentation. The liquid matrix helps to reduce the lattice energy, which must be overcome to desorb an ion from a condensed phase, and provides a means of "healing" the damage induced by bombardment. Successful matrices include glycerol, thioglycerol, *m*-nitrobenzyl alcohol, crown ethers (18-crown-6), sulfolane, 2-nitrophenyloctyl ether, diethanolamine, and triethanolamine.

A beam of fast atoms is obtained by passing accelerated argon or xenon ions from an ion source, or gun, through a chamber containing argon or xenon atoms at a pressure of about 10^{-5} torr. The high-velocity ions undergo a resonant electron-exchange reaction with the atoms without substantial loss of translational energy. Thus, a beam of energetic *atoms* is formed. The lower-energy ions from the exchange are readily removed by an electrostatic deflector. Fast atom guns are available from several commercial sources, and modern spectrometers offer FAB sources as accessories.

When FAB is applied to organic or biochemical compounds, significant amounts of molecular ions (as well as ion fragments) are usually produced even for high-molecular-mass and thermally unstable samples. For example, with FAB, molecular masses of more than 10,000 Da have been determined, and detailed structural information has been obtained for compounds with molecular masses on the order of 3000 Da. Among the drawbacks of FAB are the lim-

---

[10] For a description of commercially available spectrometers designed specifically for electrospray ionization, see L. Vorees, *Anal. Chem.*, **1994**, *66*, 481A. For spectrometers combining electrospray with TOF mass analyzers, see C. M. Henry, *Anal. Chem.*, **1999**, *71*, 197A.

[11] For more information, see J. T. Watson, *Introduction to Mass Spectrometry*, 3rd ed., Philadelphia: Lippincott-Raven, 1997, Chap. 9; K. Biemann, *Anal. Chem.*, **1986**, *58*, 1288A.

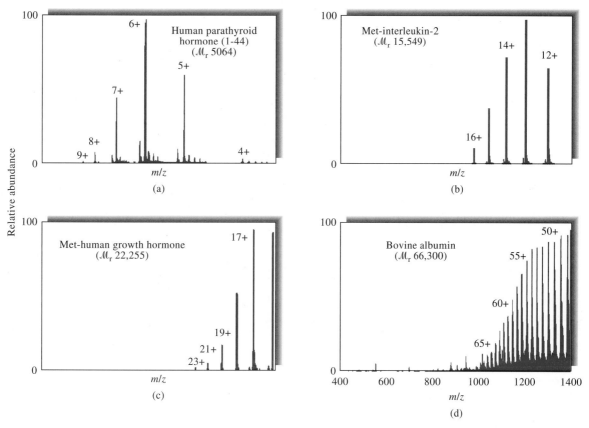

FIGURE 20-10 Typical electrospray mass spectra of proteins and peptides. The numbers above the peaks represent the molecular charge associated with each peak. (From R. D. Smith et al., *Anal. Chem.*, **1990**, *62*, 882. Copyright 1990 American Chemical Society.)

ited molecular mass range, the need for larger sample quantities than electron-impact ionization, the necessity for finding an appropriate matrix in which the analyte is soluble, and baseline noise resulting from formation of ion clusters from the matrix.

Other Desorption Methods

An examination of Table 20-1 reveals that several other desorption methods are available. Generally these produce spectra similar in character to the desorption methods we have just described and have more limited applicability.

20C MASS SPECTROMETERS

Several types of instruments are currently used for molecular mass spectrometric measurements. Two of these, the quadrupole spectrometer and the TOF

spectrometer, have been described in some detail in Sections 11B-2 and 11B-3. In this section, we consider other common types of mass spectrometers.

20C-1 General Description of Instrument Components

The block diagram in Figure 20-11 shows the major components of mass spectrometers. The purpose of the inlet system is to introduce a very small amount of sample (a micromole or less) into the mass spectrometer, where its components are converted to gaseous ions. Often the inlet system contains a means for volatilizing solid or liquid samples.

Ion sources of mass spectrometers, which were discussed in some detail in the previous section and in Chapter 11, convert the components of a sample into ions. In many cases the inlet system and the ion source are combined into a single component. In either case,

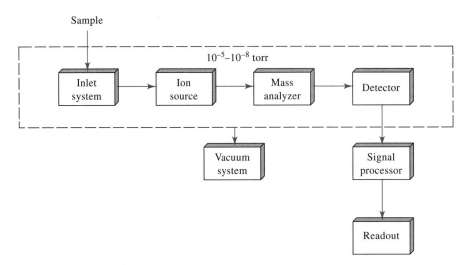

FIGURE 20-11 Components of a mass spectrometer.

the output is a stream of positive or negative ions (more commonly positive) that are then accelerated into the mass analyzer.

The function of the mass analyzer, as discussed in Chapter 11, is analogous to that of the grating in an optical spectrometer. In the mass analyzer, however, ions are dispersed based on the mass-to-charge ratios of the analyte ions. Mass spectrometers fall into several categories, depending on the nature of the mass analyzer.

Like an optical spectrometer, a mass spectrometer contains a transducer (for ions) that converts the beam of ions into an electrical signal that can then be processed, stored in the memory of a computer, and displayed or recorded in a variety of ways. The characteristics of several types of transducers employed in mass spectrometers were described in Section 11B-1.

A characteristic feature of mass spectrometers, which is not shared by optical instruments, is the requirement of an elaborate vacuum system to create low pressures (10^{-4} to 10^{-8} torr) in all of the instrument components except the signal processor and readout. The need for a high vacuum arises because such conditions lead to infrequent collisions with atmospheric components and allow the production and manipulation of free electrons and ions.

In the sections that follow, we first describe inlet systems that are common to all types of mass spectrometers. We then define *resolution*, which describes the ability of a mass spectrometer to differentiate between ions with differing masses. Finally, several types of mass analyzers that lead to different categories of mass spectrometers are described.

20C-2 Sample Inlet Systems

The purpose of the inlet system is to permit introduction of a representative sample into the ion source with minimal loss of vacuum. Most modern mass spectrometers are equipped with several types of inlets to accommodate various kinds of samples; these include batch inlets, direct probe inlets, chromatographic inlets, and capillary electrophoretic inlets.

Batch Inlet Systems

The conventional (and simplest) inlet system is the batch type, in which the sample is volatilized externally and then allowed to leak into the evacuated ionization region. Figure 20-12a is a schematic of a typical system that is applicable to gaseous and liquid samples having boiling points up to about 500°C. For gaseous samples, a small measured volume of gas is trapped between the two valves enclosing the metering area and is then expanded into the reservoir flask. For liquids, a small quantity of sample is introduced into a reservoir, usually with a microliter syringe. In either case, the vacuum system is used to achieve a sample pressure of 10^{-4} to 10^{-5} torr. For samples with boiling points greater than 150°C, the reservoir and tubing must be maintained at an elevated temperature by means of an oven and heating tapes. The maximum temperature of the oven is about 350°C. This maximum limits the system to liquids with boiling points below about 500°C. The sample, which is now in the gas phase, is leaked into the ionization area of the spectrometer via a metal or glass diaphragm containing one or more pinholes.

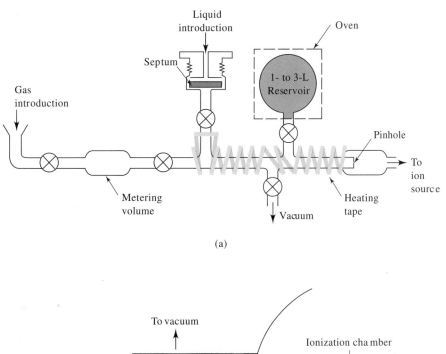

(a)

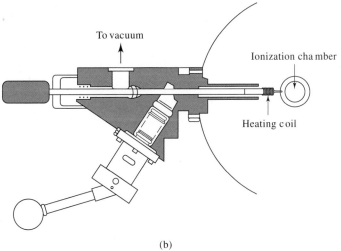

(b)

FIGURE 20-12 Schematic of (a) an external sample-introduction system—note that the various parts are not to scale—and (b) a sample probe for inserting a sample directly into the ion source. (From G. A. Eadon, in *Treatise on Analytical Chemistry*, 2nd ed., J. D. Winefordner, M. M. Bursey, and I. M. Kolthoff, eds., Part I, Vol. 11, p. 9, New York: Wiley, 1989. Reprinted by permission of John Wiley & Sons, Inc.)

The inlet system is often lined with glass to avoid losses of polar analytes by adsorption.

The Direct Probe Inlet

Solids and nonvolatile liquids can be introduced into the ionization region by means of a sample holder, or probe, which is inserted through a vacuum lock (see Figure 20-12b). The lock system is designed to limit the volume of air that must be pumped from the system after insertion of the probe into the ionization region.

Probes are also used when the quantity of sample is limited, because much less sample is wasted than with the batch system. Thus, mass spectra can often be obtained with as little as a few nanograms of sample.

With a probe, the sample is generally held on the surface of a glass or aluminum capillary tube, a fine wire, or a small cup. The probe is positioned within a few millimeters of the ionization source and the slit leading to the spectrometer. Usually, provision is made for both cooling and heating the sample on the probe.

The low pressure in the ionization area and the proximity of the sample to the ionization source often make it possible to obtain spectra of thermally unstable compounds before major decomposition has time to occur. The low pressure also leads to elevated concentrations of relatively nonvolatile compounds in the ionization area. Thus, the probe permits the study of such nonvolatile materials as carbohydrates, steroids, metal-organic species, and low-molecular-mass polymeric substances. The principal sample requirement is attainment of an analyte partial pressure of at least 10^{-8} torr before the onset of decomposition.

Chromatographic and Capillary Electrophoretic Inlet Systems

Mass spectrometers are often coupled with gas or high-performance liquid chromatographic systems or with capillary electrophoresis units to permit the separation and determination of the components of complex mixtures. Linking a chromatographic or electrophoretic column to a mass spectrometer requires the use of specialized inlet systems, some of which are described in Sections 27B-4, 28C-6, and 30B-4.

20C-3 Mass Analyzers

Several devices are available for separating ions with different mass-to-charge ratios. Ideally, the mass analyzer should be capable of distinguishing minute mass differences. In addition, the analyzer should allow passage of a sufficient number of ions to yield readily measurable ion currents. As with an optical monochromator, to which the analyzer is analogous, these two properties are not entirely compatible, and design compromises must be made.

Resolution of Mass Spectrometers

The capability of a mass spectrometer to differentiate between masses is usually stated in terms of its *resolution R*, which is defined as

$$R = \frac{m}{\Delta m} \tag{20-3}$$

where Δm is the mass difference between two adjacent peaks that are just resolved and m is the nominal mass of the first peak (the mean mass of the two peaks is sometimes used instead). Two peaks are considered to be separated if the height of the valley between them is no more than a given fraction of their height (often 10%). Thus, a spectrometer with a resolution of 4000

would resolve peaks occurring at m/z values of 400.0 and 400.1 (or 40.00 and 40.01).

The resolution required in a mass spectrometer depends greatly on its application. For example, discrimination among ions of the same nominal mass such as $C_2H_4^+$, CH_2N^+, N_2^+, and CO^+ (all ions of nominal mass 28 Da but exact masses of 28.0313, 28.0187, 28.0061, and 27.9949 Da, respectively) requires an instrument with a resolution of several thousand. On the other hand, low-molecular-mass ions differing by a unit of mass or more such as NH_3^+ ($m = 17$) and CH_4^+ ($m = 16$), for example, can be distinguished with an instrument having a resolution smaller than 50. Commercial spectrometers are available with resolutions ranging from about 500 to 500,000. Example 20-3 gives a calculation of the resolution needed to separate two species.

EXAMPLE 20-3

What resolution is needed to separate the ions $C_2H_4^+$ and CH_2N^+, with masses of 28.0313 and 28.0187, respectively?

Solution

Here,

$$\Delta m = 28.0313 - 28.0187 = 0.0126$$

The average of the two masses is $(28.0313 + 28.0187)/2 = 28.0250$. Substituting into Equation 20-3 gives

$$R = m/\Delta m = 28.0250/0.0126 = 2.22 \times 10^3$$

Magnetic Sector Analyzers

Magnetic sector analyzers employ a permanent magnet or an electromagnet to cause the beam from the ion source to travel in a circular path, most commonly of 180°, 90°, or 60°. Figure 20-13 shows a 90° sector instrument in which ions formed by electron impact are accelerated through slit B into the metal analyzer tube, which is maintained at an internal pressure of about 10^{-7} torr. Ions of different mass can be scanned across the exit slit by varying the field strength of the magnet or the accelerating potential between slits A and B. The ions passing through the exit slit fall on a collector electrode, resulting in an ion current that is amplified and recorded.

The translational, or kinetic, energy KE of an ion of mass m bearing a charge z on exiting slit B is given by

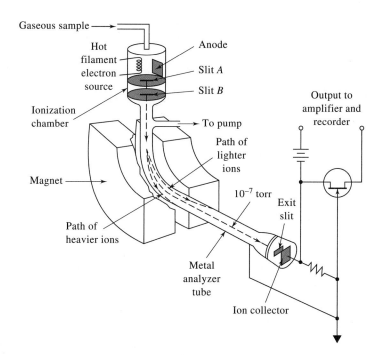

Gaseous sample

Hot filament electron source

Anode

Slit *A*

Slit *B*

Ionization chamber

To pump

Output to amplifier and recorder

Path of lighter ions

Magnet

10^{-7} torr

Exit slit

Path of heavier ions

Metal analyzer tube

Ion collector

FIGURE 20-13 Schematic of a magnetic sector spectrometer.

$$\text{KE} = zeV = \frac{1}{2}mv^2 \qquad (20\text{-}4)$$

where V is the voltage difference between A and B, v is the velocity of the ion after acceleration, and e is the electronic charge ($e = 1.60 \times 10^{-19}$ C). Note that all ions having the same number of charges z are assumed to have the same kinetic energy after acceleration regardless of their mass. This assumption is only approximately true, because before acceleration, the ions possess a statistical distribution of velocities (speeds and directions), which will be reflected in a similar distribution for the accelerated ion. The limitations of this assumption are discussed in the next section when double-focusing instruments are described. Because all ions leaving the slit have approximately the same kinetic energy, the heavier ions must travel through the magnetic sector at lower velocities.

The path in the sector described by ions of a given mass and charge represents a balance between two forces acting on them. The magnetic force F_M is given by the relationship

$$F_M = Bzev \qquad (20\text{-}5)$$

where B is the magnetic field strength. The balancing centripetal force F_c is given by

$$F_c = \frac{mv^2}{r} \qquad (20\text{-}6)$$

where r is the radius of curvature of the magnetic sector. For an ion to traverse the circular path to the collector, F_M and F_c must be equal. Thus, equating Equations 20-5 and 20-6 leads to

$$Bzev = \frac{mv^2}{r} \qquad (20\text{-}7)$$

which rearranges to

$$v = \frac{Bzer}{m} \qquad (20\text{-}8)$$

Substituting Equation 20-8 into Equation 20-4 gives, after rearranging,

$$\frac{m}{z} = \frac{B^2r^2e}{2V} \qquad (20\text{-}9)$$

Equation 20-9 reveals that mass spectra can be acquired by varying one of three variables (B, V, or r) while holding the other two constant. Most modern sector mass spectrometers contain an electromagnet in which ions are sorted by holding V and r constant while varying the current in the magnet and thus B. In older sector spectrometers that used photographic recording, B and V were constant, and r was the variable (see Figure 11-11). In Example 20-4, the use of Equation 20-9 to calculate an appropriate accelerating voltage is illustrated.

EXAMPLE 20-4

What accelerating voltage is required to direct a singly charged water molecule through the exit slit of a magnetic sector mass spectrometer if the magnet has a field strength of 0.240 T (tesla) and the radius of curvature of the ion through the magnetic field is 12.7 cm?

Solution

First, we convert all experimental variables into SI units. Thus, charge per ion $ez = 1.60 \times 10^{-19}$ C $\times 1$

$$\text{radius } r = 0.127 \text{ m}$$

$$\text{mass } m = \frac{18.02 \text{ g H}_2\text{O}^+/\text{mol}}{6.02 \times 10^{23} \text{ g/mol}} \times 10^{-3} \frac{\text{kg}}{\text{g}}$$

$$= 2.99 \times 10^{-26} \text{ kg H}_2\text{O}^+$$

$$\text{magnetic field } B = 0.240 \text{ T} = 0.240 \text{ Vs/m}^2$$

We then substitute into Equation 20-9 and solve for the accelerating voltage V:

$$V = \frac{B^2 r^2 ez}{2m}$$

$$= \frac{[0.240 \text{ Vs/m}^2]^2 [0.127 \text{ m}]^2 [1.60 \times 10^{-19} \text{ C}]}{2 \times 2.99 \times 10^{-26} \text{ kg}}$$

$$= 2.49 \times 10^3 \frac{(\text{Vs})^2\text{C}}{\text{m}^2\text{kg}}$$

$$= 2.49 \times 10^3 \text{ V} \qquad (1 \text{ volt} = 1 \text{ kg m}^2/\text{s}^2 \text{ C})$$

Double-Focusing Spectrometers

The magnetic sector instruments discussed in the previous section are sometimes called *single-focusing* spectrometers. This terminology is used because a collection of ions exiting the source with the same mass-to-charge ratio but with small diverging directional distribution will be acted on by the magnetic field in such a way that a converging directional distribution is produced as the ions leave the field. The ability of a magnetic field to bring ions with different directional orientations to focus means that the distribution of translational energies of ions leaving the source is the factor most responsible for limiting the resolution of magnetic sector instruments ($R \leq 2000$).

The translational energy distribution of ions leaving a source arises from the Boltzmann distribution of energies of the molecules from which the ions are formed and from field inhomogeneities in the source. The spread of kinetic energies causes a broadening of the beam reaching the transducer and thus a loss of resolution. To measure atomic and molecular masses with a precision of a few parts per million, it is necessary to design instruments that correct for both the directional distribution and energy distribution of ions leaving the source. The term *double focusing* is applied to mass spectrometers in which the directional aberrations and the energy aberrations of a population of ions are simultaneously minimized. Double focusing is usually achieved by the use of carefully selected combinations of electrostatic and magnetic fields. In the double-focusing instrument, shown schematically in Figure 20-14, the ion beam is first passed through an electrostatic analyzer (ESA) consisting of two smooth curved metallic plates across which a dc voltage is applied. This voltage has the effect of limiting the kinetic energy of the ions reaching the magnetic sector to a closely defined range. Ions with energies greater than average strike the upper side of the ESA slit and are lost to ground. Ions with energies less than average strike the lower side of the ESA slit and are thus removed.

Directional focusing in the magnetic sector occurs along the focal plane labeled *d* in Figure 20-14; energy focusing takes place along the plane labeled *e*. Thus, only ions of one *m/z* are double focused at the intersection of *d* and *e* for any given accelerating voltage and magnetic field strength. Therefore, the collector slit is located at this locus of the double focus.

Many different types of double-focusing mass spectrometers are available commercially. The most sophisticated of these are capable of resolution in the 10^5 range. More compact double-focusing instruments can also be purchased (for considerably less money). A typical instrument of this type will have a 6-in. electrostatic sector and a 4-in. 90° magnetic deflector. Resolutions of about 2500 are common with such instruments. Often they are employed as detection systems for gas or liquid chromatography.

The spectrometer shown in Figure 20-14 is based on the so-called *Nier-Johnson* design. Another double-focusing design, which employs *Mattauch-Herzog* geometry, is shown in Figure 11-11. The geometry of this type of instrument is unique in that the energy and direction focal planes coincide; for this reason, the Mattauch-Herzog design can use a photographic plate or an array detector for recording the spectrum. The transducer is located along the focal plane, where all of the ions are in focus, regardless of mass-to-charge ratio.

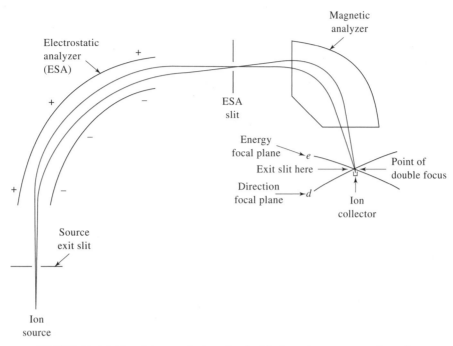

FIGURE 20-14 Nier-Johnson design of a double-focusing mass spectrometer.

Quadrupole Mass Spectrometers

Quadrupole mass spectrometers are usually less expensive and more rugged than their magnetic sector counterparts. They are generally considerably more compact than magnetic sector instruments and are commonly found in commercial benchtop mass spectrometers. They also offer the advantage of low scan times (that is, <100 ms), which is particularly useful for real-time scanning of chromatographic peaks. Quadrupole analyzers are the most common mass analyzers in use today. A detailed discussion of quadrupole mass spectrometers is found in Section 11B-2.

TOF Mass Analyzers

As shown in Section 11B-3, in TOF instruments, positive ions are produced periodically by bombardment of the sample with brief pulses of electrons, secondary ions, or laser-generated photons. The ions produced in this way are then accelerated into a field-free drift tube by an electric field pulse of 10^3 to 10^4 V (see Figure 20-15). Separation of ions by mass occurs during the transit of the ions to the detector located at the end of the tube. Because all ions entering the tube have the same kinetic energy, their velocities in the tube vary inversely with their masses (Equation 20-4), with the lighter particles arriving at the detector earlier than the heavier ones. The flight time t_F is given by

$$t_F = \frac{L}{v} = L\sqrt{\frac{m}{2zeV}} \qquad (20\text{-}10)$$

where L is the distance from the source to the detector. Typical flight times are 1 to 50 μs.

TOF instruments offer several advantages over other types of mass spectrometers, including simplicity and ruggedness, ease of accessibility of the ion source, and virtually unlimited mass range. They suffer, however, from limited resolution and sensitivity. TOF instruments also require fast electronics because ions often arrive at the transducer only fractions of microseconds apart. Several instrument manufacturers offer TOF instruments, but they are less widely used than are magnetic sector and quadrupole mass spectrometers.

Ion-Trap Analyzers

An ion trap is a device in which gaseous anions or cations can be formed and confined for extended periods by electric and magnetic fields. The quadrupole ion trap was first introduced by Paul in 1953.[12] Since that

 Tutorial: Learn more about **mass analyzers**.

[12] W. Paul and H. Steinwedel, *Z. Naturforsch,* **1953**, *8A*, 448.

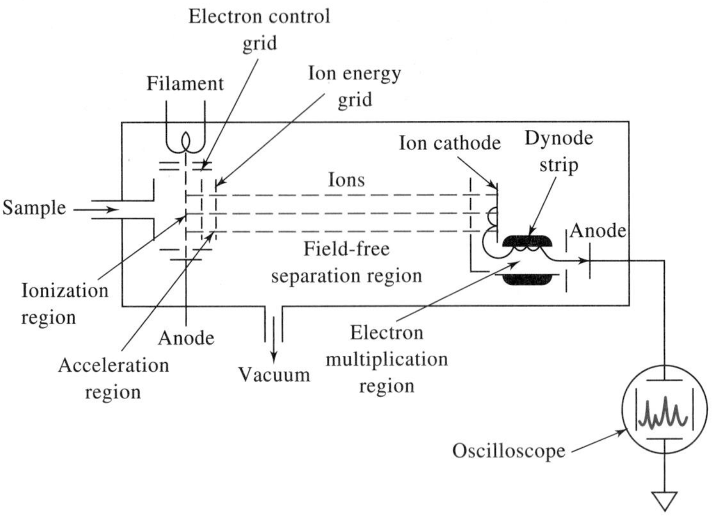

FIGURE 20-15 Schematic of a TOF mass spectrometer.

time several types of ion traps have been developed.[13] Ion-trap mass spectrometers are now used as chromatography detectors and to obtain mass spectra of a variety of analytes. For their work in developing the ion-trap technique, Wolfgang Paul and Hans Dehmelt were awarded the Nobel Prize in Physics in 1989.

Figure 20-16 is a cross-sectional view of a simple, commercially available ion trap.[14] It consists of a central doughnut-shape ring electrode and a pair of end-cap electrodes. A variable radio-frequency voltage is applied to the ring electrode while the two end-cap electrodes are grounded. Ions with an appropriate m/z value circulate in a stable orbit within the cavity surrounded by the ring. As the radio-frequency voltage is increased, the orbits of heavier ions become stabilized, and those for lighter ions become destabilized, causing them to collide with the wall of the ring electrode.

When the device is operated as a mass spectrometer, ions produced by an electron-impact or chemical ionization source are admitted through a grid in the upper end cap. The ionization source is pulsed so as to create a burst of ions. The ions over a large mass range of interest are trapped simultaneously. The ion trap can store ions for relatively long times, up to 15 minutes for some stable ions. A technique called *mass-selective ejection* is then used to sequentially eject the trapped ions in order of mass by increasing the radio-frequency

voltage applied to the ring electrode in a linear ramp. As trapped ions become destabilized, they leave the ring electrode cavity via openings in the lower end cap. The emitted ions then pass into a transducer such as the electron multiplier shown in Figure 20-16.

Ion-trap mass analyzers have the advantage of being rugged, compact, and less costly than other mass analyzers.[15] They have the potential for achieving low detection limits. In addition to electron-impact sources, such analyzers have been interfaced to electrospray ionization sources as well as MALDI sources.

---

[15]For a review of commercially available ion traps, see Z. Ziegler, *Anal. Chem.*, **2002**, *74*, 489A.

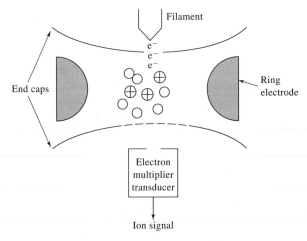

FIGURE 20-16 Ion-trap mass spectrometer. (Adapted from Watson, *Introduction to Mass Spectrometry*, p. 89, Philadelphia: Lippincott-Raven Press, 1997.)

---

[13]R. E. March and J. F. J. Todd, eds., *Practical Aspects of Ion Trap Mass Spectrometry*, Boca Raton, FL: CRC Press, 1995; R. E. March, *Int. J. Mass Spectrom.*, **2000**, *200*, 285.

[14]For a perspective on the development of ion-trap mass spectrometry, see G. Stafford, *J. Am. Soc. Mass Spectrom.*, **2002**, *13*, 589.

20C-4 Fourier Transform Spectrometers

As was true with infrared and nuclear magnetic resonance instruments, Fourier transform mass spectrometers provide improved signal-to-noise ratios, greater speed, and higher sensitivity and resolution.[16] Commercial Fourier transform mass spectrometers appeared on the market in the early 1980s and are now offered by several manufacturers.

The heart of a Fourier transform instrument is an ion trap within which ions can circulate in well-defined orbits for extended periods. Such cavities are constructed to take advantage of a phenomenon known as *ion cyclotron resonance*.

Ion Cyclotron Resonance

When a gaseous ion drifts into or is formed in a strong magnetic field, its motion becomes circular in a plane perpendicular to the direction of the field. The angular frequency of this motion is called the cyclotron frequency, ω_c. Equation 20-8 can be rearranged and solved for v/r, which is the cyclotron frequency in radians per second.

$$\omega_c = \frac{v}{r} = \frac{zeB}{m} \qquad (20\text{-}11)$$

Note that in a fixed field, the cyclotron frequency depends only on the inverse of the m/z value. Increases in the velocity of an ion will be accompanied by a corresponding increase in the radius of rotation of the ion. A measurement of ω_c can provide an accurate indication of z/m and thus the mass-to-charge ratio of the ion.

An ion trapped in a circular path in a magnetic field is capable of absorbing energy from an ac electric field, provided the frequency of the field matches the cyclotron frequency. The absorbed energy then increases the velocity of the ion (and thus the radius of its path) without disturbing ω_c. This effect is illustrated in Figure 20-17. Here, the original path of an ion trapped in a magnetic field is depicted by the inner solid circle. Brief application of an ac voltage creates a fluctuating field between the plates that interacts with the ion, provided the frequency of the source is resonant with the cyclotron frequency of the ion. Under this circumstance, the velocity of the ion increases continuously as does the radius of its path (see dashed line). When the

[16]For reviews of Fourier transform mass spectrometry, see R. M. A. Heeren, A. J. Kleinnijenhuis, L. A. McDonnell, and T. H. Mize, *Anal. Bioanal. Chem.*, **2004**, *378*, 1048; A. G. Marshall, *Int. J. Mass Spectrom.*, **2000**, *200*, 331; A. G. Marshall and P. B. Grosshans, *Anal. Chem.*, **1991**, *63*, 215A.

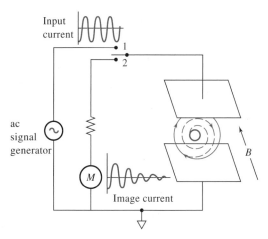

FIGURE 20-17 Path of an ion in a strong magnetic field. Inner solid line represents the original circular path of the ion. Dashed line shows spiral path when switch is moved briefly to position 1. Outer solid line is new circular path when switch is again opened.

ac electrical signal is terminated, the radius of the path of the ion again becomes constant, as is suggested by the outer solid circle in the figure.

When the region between the plates in Figure 20-17 contains an ensemble of ions of the same mass-to-charge ratio, application of the ac signal having the cyclotron resonance frequency sets all of the particles into coherent motion in phase with the field. Ions of different cyclotron frequency, that is, those with different mass-to-charge ratios, are unaffected by the ac field.

Measurement of the ICR Signal

The coherent circular motion of resonant ions creates a so-called *image current* that can be conveniently observed after termination of the frequency sweep signal. Thus, if the switch in Figure 20-17 is moved from position 1 to position 2, a current is observed that decreases exponentially with time. This image current is a capacitor current induced by the circular movement of a packet of ions with the same mass-to-charge ratios. For example, as a packet of positive ions approaches the upper plate in Figure 20-17, electrons are attracted from circuit common to this plate, causing a momentary current. As the packet continues around toward the other plate, the direction of external electron flow is reversed. The magnitude of the resulting alternating current depends on the number of ions in the packet. The frequency of the current (the cyclotron resonance frequency) is characteristic of the

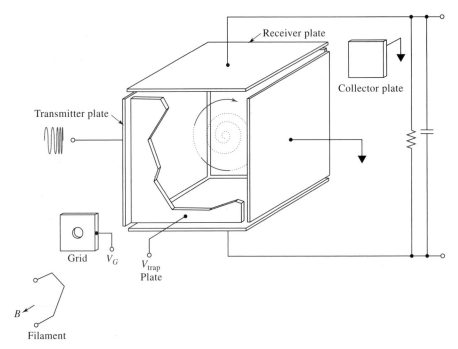

FIGURE 20-18 A trapped-ion analyzer cell. (Reprinted from E. B. Ledford Jr., R. L. White, S. Ghaderi, and C. L. Wilkins, *Anal. Chem.*, **1980**, *52*, 1090. Copyright 1980 American Chemical Society.)

mass-to-charge value of the ions in the packet. This current is employed in ion cyclotron spectrometers to measure the concentration of ions brought into resonance at various applied signal frequencies.

The induced image current just described decays over a period of a few tenths of a second to several seconds as the coherent character of the circulating packet of ions is lost. Collisions between ions provide the mechanism by which the coherently circulating ions lose energy and the ions return to a condition of thermal equilibrium. This decay of the image current provides a time-domain signal that is similar to the free-induction decay signal encountered in Fourier transform–nuclear magnetic resonance experiments (see Section 19A-3).

Fourier Transform Spectrometers

Fourier transform mass spectrometers are generally equipped with a trapped-ion analyzer cell such as that shown in Figure 20-18. Gaseous sample molecules are ionized in the center of the cell by electrons that are accelerated from the filament through the cell to a collector plate. A pulsed voltage applied at the grid serves as a gate to periodically switch the electron beam on and off. The ions are held in the cell by a 1 to 5 V po-

tential applied to the trap plate. The ions are accelerated by a radio-frequency signal applied to the transmitter plate as shown. The receiver plate is connected to a preamplifier that amplifies the image current. This approach for confining ions is highly efficient, and storage times of up to several minutes have been observed. The dimensions of the cell are not critical but are usually a few centimeters on a side.

The basis of the Fourier transform measurement is illustrated in Figure 20-19. Ions are first generated by a brief electron beam pulse (not shown) and stored in the trapped ion cell. After a brief delay, the trapped ions are subjected to a short radio-frequency pulse that increases linearly in frequency during its lifetime. Figure 20-19a shows a pulse of 5 ms, during which time the frequency increases linearly from 0.070 to 3.6 MHz. After the frequency sweep is discontinued, the image current, induced by the various ion packets, is amplified, digitized, and stored in memory. The time-domain decay signal, shown in Figure 20-19b, is then transformed to yield a frequency-domain signal that can be converted to the mass domain via Equation 20-11. Figure 20-20 illustrates the relationship between a time-domain spectrum, its frequency-domain counterpart, and the resulting mass spectrum.

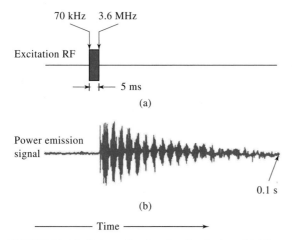

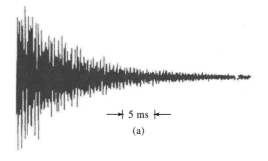

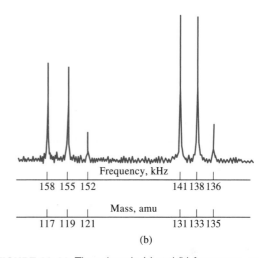

FIGURE 20-19 Schematic showing the timing of (a) the radio-frequency signal and (b) the transient image signal (lower). (Reprinted with permission from R. T. McIver Jr., *Amer. Lab.*, **1980**, *12* (11), 26. Copyright 1980 by International Scientific Communications, Inc.)

Fourier transform spectrometers can be interfaced to a variety of ionization sources, including MALDI, electrospray, FAB, and EI-CI. Resolution in Fourier transform mass spectrometry is limited by the precision of the frequency measurement rather than slits or field measurements. The resolution and mass range also depend on the magnitude and stability of the magnetic field. Because frequency measurements can be made with high precision, extremely high resolution is possible (in excess of 10^6). The accuracy with which mass measurements can be made with Fourier transform instruments is also superb. Fourier transform spectrometers are expensive instruments (>$400,000). Commercial models are available with superconducting magnets with fields varying from 1.2 to 12 T. The high-field models are quite useful in biological applications, including proteomics.

20C-5 Tandem Mass Spectrometry

Tandem mass spectrometry, sometimes called *mass spectrometry–mass spectrometry* (MS/MS), is a method that allows the mass spectrum of preselected and fragmented ions to be obtained.[17] The basic idea is illustrated in Figure 20-21. Here, an ionization source,

---

[17]For additional information, see K. L. Busch, G. L. Glish, and S. A. McLuckey, *Mass Spectrometry/Mass Spectrometry: Techniques and Applications of Tandem Mass Spectrometry*, New York: VCH, 1988; *Tandem Mass Spectrometry*, F. W. McLafferty, ed., New York: Wiley, 1983.

FIGURE 20-20 Time-domain (a) and (b) frequency- or mass-domain spectrum for 1,1,1,2-tetrachloroethane. (Reprinted with permission from E. B. Ledford Jr. et al., *Anal. Chem.*, **1980**, *52*, 463. Copyright 1980 American Chemical Society.)

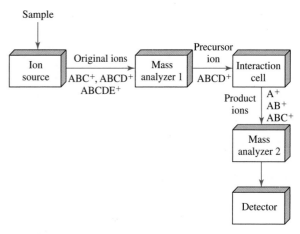

FIGURE 20-21 Block diagram of a tandem mass spectrometer.

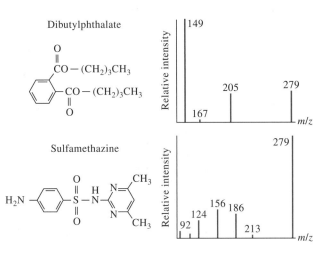

FIGURE 20-22 Product-ion spectra for dibutylphthalate and sulfamethazine obtained after the protonated precursor ion peaks at 279 Da were isolated by the first mass analyzer of an MS/MS instrument. (Reprinted from K. L. Busch and G. C. DiDonato, *Amer. Lab.*, **1986**, *18* (8), 17. Copyright 1986 by International Scientific Communications, Inc.)

often a soft ionization source, produces ions and some fragments. These are then the input to the first mass analyzer, which selects a particular ion called the *precursor ion* and sends it to the interaction cell. In the interaction cell, the precursor ion can decompose spontaneously, react with a collision gas, or interact with an intense laser beam to produce fragments, called *product ions*. These ions are then mass analyzed by the second mass analyzer and detected by the ion detector.

Types of Tandem Mass Spectra

Several different types of spectra can be obtained from the MS/MS experiment. First, the *product-ion spectrum* can be obtained by scanning mass analyzer 2, while mass analyzer 1 is held constant, acting as a mass selector to select one precursor ion. Figure 20-22 shows product-ion spectra for dibutylphthalate and sulfamethazine. Both compounds produce molecular ions with *m/z* values of 279. However, the product-ion spectra of the two compounds are very different.

A *precursor ion spectrum* can be obtained, in addition to product-ion spectra, by scanning the first mass analyzer while holding the second mass analyzer constant to detect a given product ion. In a mixture of compounds, those that give the same products are readily identified by precursor ion spectra. Closely related compounds often give several of the same product ions, so that this method of operation provides a measure of identity and concentration of the *members of a class* of closely related compounds. Consider, for example, a mixture of ABCD, BCDA, IJKL, and IJMN in the sample. To identify species containing the IJ group, the second analyzer is set to the mass corresponding to

the IJ$^+$ ion and the molecular ions ABCD$^+$, BCDA$^+$, IJKL$^+$, and IJMN$^+$ are sequentially selected by the first analyzer. Ion signals in the detector would be observed only when IJKL$^+$ and IJMN$^+$ are selected by the first analyzer, indicating the presence of the IJ group.

By scanning both analyzers simultaneously with an offset in mass between them, a *neutral loss spectrum* can be obtained. This gives the identity of those precursor ions that undergo the same loss such as the loss of a H$_2$O or CO neutral. Finally, by scanning mass analyzer 1 and obtaining the product-ion spectrum for each selected precursor ion, a complete *three-dimensional MS/MS spectrum* can be obtained.

Dissociative Interactions in the Interaction Cell

Several types of interactions can be used to produce fragmentation in the interaction cell. In some cases, the ions selected by mass analyzer 1 in Figure 20-21 are themselves *metastable* and decompose into fragments after a certain time. In general, however, the kinetics of the decomposition process can greatly limit the applicability and sensitivity of the process. In such cases, fragmentation can be induced by adding a collision gas to the interaction cell so that interaction with precursor ions occurs, leading to decomposition into product ions. In this case, the cell is called a *collision cell*, and the interactions are termed *collisionally activated dissociation* (CAD) or alternatively *collision-induced dissociation* (CID).

Another type of interaction is *surface-induced dissociation* (SID), in which precursor ions interact with a surface to induce dissociation. Ions have been reflected off cell walls or trapping plates to increase their internal

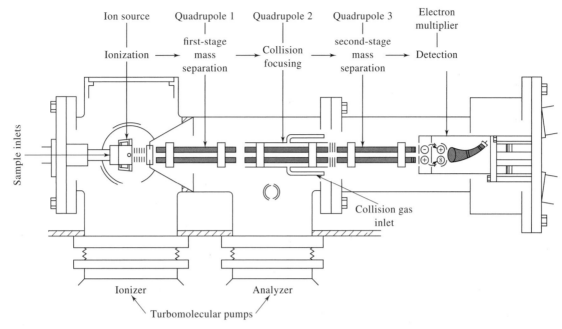

FIGURE 20-23 Schematic of a triple quadrupole mass spectrometer. (Courtesy of Thermo-Finnigan Corp.)

energy and promote dissociation. Chemically modified surfaces such as thin films have also been employed.

Another dissociation technique that has been applied to large multiply charged ions is *electron-capture dissociation* (ECD), in which precursor ions capture a low-energy electron to produce an intermediate that rapidly dissociates. In some cases, a background gas is added to aid in the dissociation process.

Photo-induced dissociation (PID) is another process to stimulate decomposition of precursor ions. In nearly all cases, an intense laser beam is used in the interaction cell to promote the dissociation. A difficulty with PID is that the ion beam and photon beam must overlap in the interaction region for a time long enough for absorption and bond rupture to occur. In some cases ion-trap cells have been used to allow for long periods of overlap.

Instrumentation for Tandem Mass Spectrometry

Tandem mass spectrometry has been implemented in a number of ways.[18] These can be classified as *tandem in space* and *tandem in time*.[19]

---

[18]For a description of commercial tandem mass spectrometers, see D. Noble, *Anal. Chem.*, **1995**, *67*, 265A.
[19]See J. W. Hager, *Anal. Bioanal. Chem.*, **2004**, *378*, 845; S. A. McLuckey and J. M. Wells, *Chem. Rev.*, **2001**, *101*, 571.

Tandem-in-Space Spectrometers. In tandem-in-space instruments, two independent mass analyzers are used in two different regions in space. The triple quadrupole mass spectrometer is the most common of these instruments. In commercial triple quadrupole instruments, such as the instrument illustrated in Figure 20-23, the sample is introduced into a soft ionization source, such as a CI or FAB source. The ions are then accelerated into quadrupole 1 (Q), which is an ordinary quadrupole mass filter. The selected fast-moving ions pass into quadrupole 2 (q), which is a collision chamber where dissociation of the ions selected by quadrupole 1 occurs. This quadrupole is operated in a radio-frequency-only mode in which no dc voltage is applied across the rods. This mode basically traps the precursor and product ions in a relatively high concentration of collision gas so that CAD can occur. Quadrupole 3 (Q) then allows mass analysis of the product ions formed in the collision cell. The configuration is known as the QqQ configuration.

Sector instruments and hybrid quadrupole-sector instruments have also been used in a tandem manner. The first tandem mass spectrometers were sector instruments that combined an electric sector spectrometer with a magnetic sector spectrometer, either in forward geometry (electric sector followed by magnetic sector, or EB) or in reverse geometry (magnetic sector followed by electric sector, or BE). The reverse geometry is sometimes called a *mass-analyzed ion kinetic energy spectrometer* (MIKES). Such instruments were not very efficient, but they allowed the principles of tandem mass spectrometry to be demonstrated.

Hybrid instruments include the BEqQ spectrometer (magnetic sector, B; electric sector, E; RF-only quadrupole, q; quadrupole mass analyzer, Q) and the BTOF (magnetic sector, B; TOF analyzer, TOF) spectrometer. The QqTOF spectrometer is similar to the triple quadrupole (QqQ) instrument except that the final quadrupole mass analyzer is replaced with a TOF analyzer. In another variant, the Qq section can be replaced by a quadrupole ion trap to yield an ion-trap-TOF instrument.

A final type of tandem-in-space spectrometer is the TOF-TOF spectrometer, in which a TOF instrument followed by a timed ion selector separates the precursor ions. A collision cell then induces fragmentation, and the product ions are mass analyzed in the final TOF stage.[20] Mass resolution of several thousand was reported.

Tandem-in-Time Spectrometers. Tandem-in-time instruments form the ions in a certain spatial region and then at a later time expel the unwanted ions and leave the selected ions to be dissociated and mass analyzed in the same spatial region. This process can be repeated many times over to perform not only MS/MS experiments, but also MS/MS/MS and MS^n experiments. Fourier transform ICR and quadrupole ion-trap instruments are well suited for performing MS^n experiments. In principle, tandem-in-time spectrometers can perform MS/MS experiments much more simply than tandem-in-space instruments because of the difficulty in providing different ion focal positions in the latter. Although tandem-in-time spectrometers can readily provide product-ion scans, other scans, such as precursor ion scans and neutral loss scans, are much more difficult to perform than they are with tandem in space instruments.

20C-6 Computerized Mass Spectrometers

Computers are an integral part of modern mass spectrometers. A characteristic of a mass spectrum is the wealth of structural data that it provides. For example, a molecule with a molecular mass of 500 may be fragmented by an EI source into 100 or more different

ions, each of which leads to a discrete spectral peak. For a structural determination, the heights and mass-to-charge ratios of each peak must be determined, stored, and ultimately displayed. Because the amount of information is so large, it is essential that acquisition and processing be rapid; computers are ideally suited for these tasks. Moreover, for mass spectral data to be useful, several instrumental variables must be closely controlled or monitored during data collection. Computers are much more efficient than a human operator in exercising such controls.

Figure 20-24 is a block diagram of the computerized control and data-acquisition system of a triple quadrupole mass spectrometer. This figure shows two features encountered in any modern instrument. The first is a computer that serves as the main instrument controller. The operator communicates via a keyboard with the spectrometer by selecting operating parameters and conditions via easy-to-use interactive software. The computer also controls the programs responsible for data manipulations and output. The second feature common to almost all instruments is a set of microprocessors (often as many as six) that are responsible for specific aspects of instrument control and the transmission of information between the computer and spectrometer.

The interface between a mass spectrometer and a computer usually has provisions for digitizing the amplified ion-current signal plus several other signals that are used for control of instrumental variables. Examples of the latter are source temperature, accelerating voltage, scan rate, and magnetic field strength or quadrupole voltages.

The digitized ion-current signal ordinarily requires considerable processing before it is ready for display. First, the peaks must be normalized, a process by which the height of each peak relative to some reference peak is calculated. Most often the *base peak*, which is the largest peak in a spectrum, serves as the reference and is arbitrarily assigned a peak height of 100 (sometimes 1000). The m/z value for each peak must also be determined. This assignment is frequently based on the time of the peak's appearance and the scan rate. Data are acquired as intensity versus time during a carefully controlled scan of the magnetic or electric fields. Conversion from time to m/z requires careful periodic calibration; for this purpose, perfluorotri-*n*-butylamine (PFTBA) or perfluorokerosene is often used as a standard. For high-resolution work, the standard may be admitted with the sample. The computer is programmed to then recognize and use the peaks of the

[20]K. F. Medzihradszky, J. M. Campbell, M. A. Baldwin, A. M. Falick, P. Juhasz, M. L. Vestal, and A. L. Burlingame, *Anal. Chem.*, **2000**, *72*, 552.

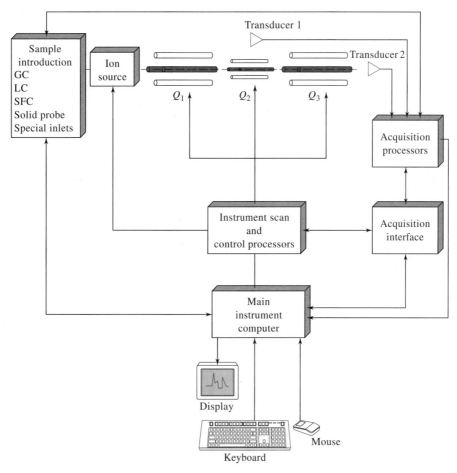

FIGURE 20-24 Instrument control and data processing for a triple quadrupole mass spectrometer.

standard as references for mass assignments. For low-resolution instruments, the calibration must generally be obtained separately from the sample, because of the likelihood of peak overlaps.

With most systems the computer stores all spectra and related information on a disk. In routine applications, bar graphs of the normalized spectra can be sent directly to a printer or display. However, in many cases, the user will employ data-reduction software to extract specific information prior to producing a printed copy of a spectrum. Figure 20-25 is an example of the printout from a computerized mass spectrometer. The first and next-to-last columns in the table list m/z values in increasing order. The second and last columns contain the corresponding ion currents normalized to the largest peak found at mass 156. The current for this ion is assigned the number of 100, and all other peaks are expressed relative to this one. Thus, the height of the peak at mass 141 is 53% of the base peak.

As is true with infrared and nuclear magnetic resonance spectroscopy, large libraries of mass spectra (>150,000 entries) are available in computer-compatible formats.[21] Most commercial mass spectrometer computer systems have the ability to rapidly search all or part of such files for spectra that match or closely match the spectrum of an analyte.

20D APPLICATIONS OF MOLECULAR MASS SPECTROMETRY

The applications of molecular mass spectrometry are so numerous and widespread that describing them adequately in a brief space is not possible. Table 20-5 lists

---

[21] For example, see F. W. McLafferty, *Wiley Registry of Mass Spectral Data*, 7th ed., New York: Wiley, 2000. It is also available combined with *NIST Mass Spectral Library*, on CD-ROM.

Sample: Unknown compound
Date: 11/15/2004

m/z	Rel. Int.			m/z	Rel. Int.
41	9			141	53
43	14			142	4
55	5			155	6
69	4			156	100
71	5			157	21
98	5			197	5

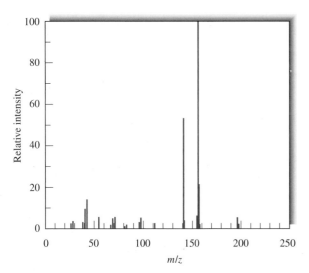

FIGURE 20-25 A computer display of mass-spectral data. The compound was isolated from a blood serum extract by chromatography. The spectrum showed it to be the barbiturate pentobarbital.

some of these applications to provide some idea of the capabilities of mass spectrometry. In this section we describe a few of the most widely used and important of these applications.

20D-1 Identification of Pure Compounds

The mass spectrum of a pure compound provides several kinds of data that are useful for its identification.[22] The first is the molecular mass of the compound, and the second is its molecular formula. In addition, study of fragmentation patterns revealed by the mass spectrum often provides information about the presence or absence of various functional groups. Finally, the actual identity of a compound can often be established by comparing its mass spectrum with those of known compounds until a close match is realized.

[22] R. M. Silverstein, F. X. Webster, and D. Kiemle, *Spectrometric Identification of Organic Compounds*, 7th ed., New York: Wiley, 2004; F. W. McLafferty and F. Turecek, *Interpretation of Mass Spectra*, 4th ed., Mill Valley, CA: University Science Books, 1993.

Molecular Masses from Mass Spectra

For compounds that can be ionized to give a molecular ion or a protonated or a deprotonated molecular ion by one of the methods described earlier, the mass spectrometer is an unsurpassed tool for the determination of molecular mass. This determination, of course, requires the identification of the molecular ion peak, or in some cases, the $(M + 1)^+$ or the $(M - 1)^+$ peak. The location of the peak on the abscissa then gives the molecular mass with an accuracy that cannot be realized easily by any other method.

To determine a molecular mass by mass spectrometry, the identity of the molecular ion peak must be known. Caution is therefore always advisable, particularly with electron-impact sources, when the molecular ion peak is absent or small relative to impurity peaks. When there is doubt, additional spectra by chemical, field, and desorption ionization are particularly useful.

Molecular Formulas from
Exact Molecular Masses

Molecular formulas can be determined from the mass spectrum of a compound, provided the molecular ion peak can be identified and its *exact* mass determined. This application, however, requires a high-resolution instrument capable of detecting mass differences of a few thousandths of a dalton. Consider, for example, the mass-to-charge ratios of the molecular ions of the following compounds: purine, $C_5H_4N_4$ ($\mathcal{M} = 120.044$); benzamidine, $C_7H_8N_2$ ($\mathcal{M} = 120.069$); ethyltoluene, C_9H_{12} ($\mathcal{M} = 120.096$); and acetophenone, C_8H_8O ($\mathcal{M} = 120.058$). If the measured mass of the molecular ion peak is 120.070 (± 0.005), then all but $C_7H_8N_2$ are excluded as possible formulas. Note that the precision in this example is about 40 ppm. Uncertainties on the order of a few parts per million are routinely achievable with high-resolution, double-focusing instruments. Tables that list all reasonable combinations of C, H, N, and O by molecular mass to the third or fourth decimal place have been compiled.[23] A small portion of such a compilation is shown in the fifth column of Table 20-6.

Molecular Formulas from Isotopic Ratios

The data from a low-resolution instrument that can discriminate only between ions differing in mass by whole mass numbers can also yield useful information

[23] J. H. Beynon and A. E. Williams, *Mass and Abundance Tables for Use in Mass Spectrometry*, New York: Elsevier, 1963.

TABLE 20-5 Applications of Molecular Mass Spectrometry

1. Elucidation of the structure of organic and biological molecules
2. Determination of the molecular mass of peptides, proteins, and oligonucleotides
3. Identification of components in thin-layer and paper chromatograms
4. Determination of amino acid sequences in sample of polypeptides and proteins
5. Detection and identification of species separated by chromatography and capillary electrophoresis
6. Identification of drugs of abuse and metabolites of drugs of abuse in blood, urine, and saliva
7. Monitoring gases in patient's breath during surgery
8. Testing for the presence of drugs in blood in race horses and in Olympic athletes
9. Dating archaeological specimens
10. Analyses of aerosol particles
11. Determination of pesticide residues in food
12. Monitoring volatile organic species in water supplies

TABLE 20-6 Isotopic Abundance Percentages and Molecular
Masses for Various Combinations of C, H, O, and N

		Abundance, % M Peak Height		
	Formula	$M + 1$	$M + 2$	Molecular Mass
$M = 83$	C_2HN_3O	3.36	0.24	83.0120
	$C_2H_3N_4$	3.74	0.06	83.0359
	C_3HNO_2	3.72	0.45	83.0007
	$C_3H_3N_2O$	4.09	0.27	83.0246
	$C_3H_5N_3$	4.47	0.08	83.0484
	$C_4H_3O_2$	4.45	0.48	83.0133
	C_4H_5NO	4.82	0.29	83.0371
	$C_4H_7N_2$	5.20	0.11	83.0610
	C_5H_7O	5.55	0.33	83.0497
	C_5H_9N	5.93	0.15	83.0736
	C_6H_{11}	6.66	0.19	83.0861
$M = 84$	CN_4O	2.65	0.23	84.0073
	$C_2N_2O_2$	3.00	0.43	83.9960
	$C_2H_2N_3O$	3.38	0.24	84.0198
	$C_2H_4N_4$	3.75	0.06	84.0437
	C_3O_3	3.36	0.64	83.9847
	$C_3H_2NO_2$	3.73	0.45	84.0085
	$C_3H_4N_2O$	4.11	0.27	84.0324
	$C_3H_6N_3$	4.48	0.08	84.0563
	$C_4H_4O_2$	4.46	0.48	84.0211
	C_4H_6NO	4.84	0.29	84.0449
	$C_4H_8N_2$	5.21	0.11	84.0688
	C_5H_8O	5.57	0.33	84.0575
	$C_5H_{10}N$	5.94	0.15	84.0814
	C_6H_{12}	6.68	0.19	84.0939
	C_7	7.56	0.25	84.0000

Taken from R. M. Silverstein, G. C. Bassler, and T. C. Morrill, *Spectrometric Identification of Organic Compounds*, 4th ed., p. 49, New York: Wiley, 1981.

about the formula of a compound, provided that the molecular ion peak is sufficiently intense that its height and the heights of the $(M + 1)^+$ and $(M + 2)^+$ isotope peaks can be determined accurately. Example 20-5 illustrates this type of analysis.

The use of relative isotope peak heights for the determination of molecular formulas is greatly expedited by the tables referred to in note 21 and by compilations available for computer analysis. In Table 20-6, a listing of all reasonable combinations of C, H, O, and N is given for mass numbers 83 and 84 (the original tables extend to mass number 500). The heights of the $(M + 1)^+$ and $(M + 2)^+$ peaks reported as percentages of the height of the M^+ peak are tabulated. If a reasonably accurate experimental determination of these percentages can be made, a likely formula can be deduced. For example, a molecular ion peak at mass 84 with $(M + 1)^+$ and $(M + 2)^+$ values of 5.6 and 0.3% of M^+ suggests a compound having the formula C_5H_8O.

EXAMPLE 20-5

Calculate the ratios of the $(M + 1)^+$ to M^+ peak heights for the following two compounds: dinitrobenzene, $C_6H_4N_2O_4$ ($\mathcal{M} = 168$), and an olefin, $C_{12}H_{24}$ ($\mathcal{M} = 168$).

Solution

From Table 20-3, we see that for every 100 ^{12}C atoms there are 1.08 ^{13}C atoms. Because nitrobenzene contains six carbon atoms, we would expect each of 6.48 = 6×1.08 molecules of nitrobenzene to have one ^{13}C atom for every 100 molecules having none. Thus, from this effect alone the $(M + 1)^+$ peak will be 6.48% of the M^+ peak. The isotopes of the other elements also contribute to this peak; we may tabulate their effects as follows:

$C_6H_4N_2O_4$	
^{13}C	$6 \times 1.08 = 6.48\%$
^{2}H	$4 \times 0.015 = 0.060\%$
^{15}N	$2 \times 0.37 = 0.74\%$
^{17}O	$4 \times 0.04 = \underline{0.16\%}$
	$(M + 1)^+/M^+ = 7.44\%$

$C_{12}H_{24}$	
^{13}C	$12 \times 1.08 = 12.96\%$
^{2}H	$24 \times 0.015 = \underline{0.36\%}$
	$(M + 1)^+/M^+ = 13.32\%$

Thus, if the heights of the M^+ and $(M + 1)^+$ peaks can be measured, it is possible to discriminate between these two compounds that have identical whole-number molecular masses.

The isotopic ratio is particularly useful for the detection and estimation of the number of sulfur, chlorine, and bromine atoms in a molecule because of the large contribution they make to the $(M + 2)^+$ (see Table 20-3). For example, an $(M + 2)^+$ that is about 65% of the M^+ peak is strong evidence for a molecule containing two chlorine atoms; an $(M + 2)^+$ peak of 4%, on the other hand, suggests the presence of one atom of sulfur.

Structural Information from Fragmentation Patterns

Systematic studies of fragmentation patterns for pure substances have led to rational guidelines to predict fragmentation mechanisms and a series of general rules helpful in interpreting spectra.[24] It is seldom possible (or desirable) to account for all of the peaks in the spectrum. Instead, characteristic patterns of fragmentation are sought. For example, the spectrum in Figure 20-26 is characterized by clusters of peaks differing in mass by 14. Such a pattern is typical of straight-chain paraffins, in which cleavage of adjacent carbon-carbon bonds results in the loss of successive CH_2 groups having this mass. Quite generally, the most stable hydrocarbon fragments contain three or four carbon atoms, and the corresponding peaks are thus the largest.

Alcohols usually have a very weak or nonexistent molecular ion peak but often lose water to give a strong peak at $(M - 18)^+$. Cleavage of the C—C bond next to an oxygen is also common, and primary alcohols always have a strong peak at mass 31 due to the CH_2OH^+. Extensive compilations of generalizations concerning the use of mass spectral data for the identification of organic compounds are available, and the interested reader should consult the references in note 22.

Compound Identification from Comparison Spectra

Generally, after determining the molecular mass of the analyte and studying its isotopic distribution and fragmentation patterns, the experienced mass spectroscopist is able to narrow the possible structures down

---

[24]For example, see R. M. Silverstein, F. X. Webster, and D. Kiemle, *Spectrometric Identification of Organic Compounds*, 7th ed., New York: Wiley, 2004.

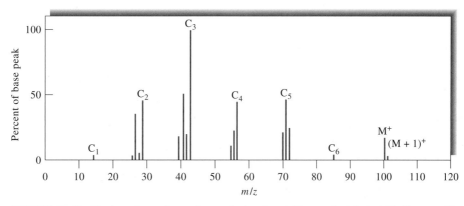

FIGURE 20-26 Electron-impact spectrum of *n*-heptanal. The peaks labeled C_6, C_5, . . . , C_1 correspond to the successive losses of a CH_2 group.

to a handful. When reference compounds are available, final identification is then based on a comparison of the mass spectrum of the unknown with spectra for authentic samples of the suspected compounds. The procedure is based on the assumptions that (1) mass fragmentation patterns are unique and (2) experimental conditions can be sufficiently controlled to produce reproducible spectra. The first assumption often is not valid for spectra of stereo- and geometric isomers and occasionally is not valid for certain types of closely related compounds. The probability that different compounds will yield the same spectrum becomes markedly smaller as the number of spectral peaks increases. Assumption 2 can also be problematic. Electron-impact spectra are fairly reproducible from laboratory to laboratory. However, spectra from other sources can vary significantly For this reason, electron-impact ionization is the method of choice for spectral comparison and for building spectral libraries.

Unfortunately, heights of mass spectral peaks strongly depend on such variables as the energy of the electron beam, the location of the sample with respect to the beam, the sample pressure and temperature, and the general geometry of the mass spectrometer. As a consequence, significant variations in relative abundances are observed for spectra acquired in different laboratories and from different instruments. Nevertheless, it has proven possible in a remarkably large number of cases to identify unknowns from library spectra obtained with a variety of instruments and operating conditions. Generally, however, it is desirable to confirm the identity of a compound by comparing its spectrum to the spectrum of an authentic compound obtained with the same instrument under identical conditions.

Computerized Library Search Systems. Although libraries of mass spectral data are available in text form,[25] most modern mass spectrometers are equipped with highly efficient computerized library search systems. There are two basic types of mass spectral libraries: large comprehensive ones and small specific ones. The largest commercially available mass spectral library (>300,000 spectra) is marketed by John Wiley and Sons.[26] A unique feature of this compilation is that it is available on CD-ROM and can be searched on a personal computer. Small libraries usually contain a few hundred to a few thousand spectra for application to a limited area, such as pesticide residues, drugs, or forensics. Small libraries are often part of the equipment packages offered by instrument manufacturers, and it is almost always possible for the instrument user to generate a library or to add to an existing library. Mass spectra for some 15,000 compounds are available from the National Institute of Standards and Technology (NIST) on the Internet.[27]

For large numbers of spectra, such as are obtained when a mass spectrometer is coupled with a chromatograph for identifying components of a mixture, the instrument's computer system can be used to perform a library search on all, or any subset, of the mass spectra

 Simulation: Learn more about **MS spectral interpretation**.

[25] F. W. McLafferty and D. A. Stauffer, *The Wiley/NBS Registry of Mass Spectral Data*, 7 vols., New York: Wiley, 1989.

[26] F. W. McLafferty, *Wiley Registry of Mass Spectral Data*, 7th ed., New York: Wiley, 2000. It is also available combined with *NIST Mass Spectral Library*, on CD-ROM.

[27] http://webbook.nist.gov/.

associated with a particular sample. The results are reported to the user, and if desired, the reference spectra can be displayed on a monitor or printed for visual comparison.

20D-2 Analysis of Mixtures by Hyphenated Mass Spectral Methods

Although ordinary mass spectrometry is a powerful tool for the identification of pure compounds, its usefulness for analysis of all but the simplest mixtures is limited because of the immense number of fragments of differing m/z values produced. It is often impossible to interpret the resulting complex spectrum. For this reason, chemists have developed methods in which mass spectrometers are coupled with various efficient separation devices in so-called *hyphenated methods*.

Chromatography–Mass Spectrometry

Gas chromatography–mass spectrometry (GC/MS) has become one of the most powerful tools available for the analysis of complex organic and biochemical mixtures. In this application, spectra are collected for compounds as they exit from a chromatographic column. These spectra are then stored in a computer for subsequent processing. Mass spectrometry has also been coupled with liquid chromatography (LC/MS) for the analysis of samples that contain nonvolatile constituents. A major problem that had to be overcome in the development of both of these hyphenated methods is that the sample in the chromatographic column is highly diluted by the gas or liquid carrying it through the column. Thus, methods had to be developed for removing the diluent before introducing the sample into the mass spectrometer. Instruments and applications of GC/MS and LC/MS are described in Sections 27B-4 and 28C-6, respectively.

Capillary Electrophoresis–Mass Spectrometry

The first report on coupling capillary electrophoresis with mass spectrometry was published in 1987.[28] Since then, it has become obvious that this hyphenated method will become a powerful and important tool in the analysis of large biopolymers, such as proteins, polypeptides, and DNA species. In most of the applications reported to date, the capillary effluent is passed directly into an electrospray ionization device, and the products then enter a quadrupole mass filter for analysis. Continuous flow FAB has also been used for ionization in some applications. Capillary electrophoresis–mass spectrometry is discussed in more detail in Section 30B-4.

Applications of Tandem Mass Spectrometry

Dramatic progress in the analysis of complex organic and biological mixtures began when the mass spectrometer was first combined with gas chromatography and subsequently with liquid chromatography. Tandem mass spectrometry offers some of the same advantages as GC/MS and LC/MS but is significantly faster. Separations on a chromatographic column are achieved in a time scale of a few minutes to hours, but equally satisfactory separations in tandem mass spectrometers are complete in milliseconds. In addition, the chromatographic techniques require dilution of the sample with large excesses of a mobile phase and subsequent removal of the mobile phase, which greatly enhances the probability of introduction of interferences. Consequently, tandem mass spectrometry is potentially more sensitive than either of the hyphenated chromatographic techniques because the chemical noise associated with its use is generally smaller. A current disadvantage of tandem mass spectrometry with respect to the other two chromatographic procedures is the greater cost of the required equipment; this gap appears to be narrowing as tandem mass spectrometers gain wider use.

For some complex mixtures the combination of GC or LC and MS does not provide enough resolution. In recent years, it has become feasible to couple chromatographic methods with tandem mass spectrometers to form GC/MS/MS and LC/MS/MS systems.[29] There have also been reports of LC/MS^n instruments.[30]

To date, tandem mass spectrometry has been applied to the qualitative and quantitative determination of the components of a wide variety of complex materials encountered in nature and industry. Some examples include the identification and determination of drug metabolites, insect pheromones, alkaloids in plants, trace contaminants in air, polymer sequences,

[28] J. A. Olivares, N. T. Nguyen, N. T. Yonker, and R. D. Smith, *Anal. Chem.*, **1987**, *59*, 1230. See also R. D. Smith, J. A. Olivares, N. T. Nguyen, and H. R. Hudseth, *Anal. Chem.*, **1988**, *60*, 436.

[29] For recent developments in LC/MS/MS, see R. Thomas, *Spectroscopy*, **2001**, *16*, 28.
[30] See, for example, J. C. A. Wuilloud, S. R. Gratz, B. M. Gamble, and K. A. Wolnik, *Analyst*, **2004**, *129*, 150; E. W. Taylor, W. Jia, M. Bush, and G. D. Dollinger, *Anal. Chem.*, **2002**, *74*, 3232; L. Howells and M. J. Sauer, *Analyst*, **2001**, *126*, 155.

petrochemicals, polychlorinated biphenyls, prostaglandins, diesel exhausts, and odors in air. One of the most promising areas of applications is that of *proteomics*, the study of proteins produced by a cell or by a species.[31]

20E QUANTITATIVE APPLICATIONS OF MASS SPECTROMETRY

Applications of mass spectrometry for quantitative analyses fall into two categories. The first involves the quantitative determination of molecular species or types of molecular species in organic, biological, and occasionally inorganic samples. The second involves the determination of the concentration of elements in inorganic and, less commonly, organic and biological samples. In the first type of analysis, all of the ionization sources listed in Table 20-1 are used. Mass spectroscopic elemental analyses, which are discussed in detail in Chapter 11, are currently based largely on inductively coupled plasma sources, although glow discharge, radio-frequency spark, laser, thermal, and secondary ion sources have also found use.

20E-1 Quantitative Determination of Molecular Species

Mass spectrometry has been widely applied to the quantitative determination of one or more components of complex organic (and sometimes inorganic) systems such as those encountered in the petroleum and pharmaceutical industries and in studies of environmental problems. Currently, such analyses are usually performed by passing the sample through a chromatographic or capillary electrophoretic column and into the spectrometer. With the spectrometer set at a suitable *m/z* value, the ion current is then recorded as a function of time. This technique is termed *selected ion monitoring*. In some instances, currents at three or four *m/z* values are monitored in a cyclic manner by rapid switching from one peak to another. The plot of the data consists of a series of peaks, with each appearing at a time that is characteristic of one of the several components of the sample that yields ions of the chosen value or values for *m/z*. Generally, the areas under the peaks are directly proportional to the component concentrations and are used for determinations. In this type of

procedure, the mass spectrometer simply serves as a sophisticated selective detector for quantitative *chromatographic* or *electrophoretic analyses*. Further details on quantitative gas and liquid chromatography are given in Sections 27B-4 and 28C-6. The use of a mass spectrometer as a detector in capillary electrophoresis is described in Section 30B-4.

In the second type of quantitative mass spectrometry for molecular species, analyte concentrations are obtained directly from the heights of the mass spectral peaks. For simple mixtures, it is sometimes possible to find peaks at unique *m/z* values for each component. Under these circumstances, calibration curves of peak heights versus concentration can be prepared and used for analysis of unknowns. More accurate results can ordinarily be realized, however, by incorporating a fixed amount of an internal standard substance in both samples and calibration standards. The ratio of the peak intensity of the analyte species to that of the internal standard is then plotted as a function of analyte concentration. The internal standard tends to reduce uncertainties arising in sample preparation and introduction. These uncertainties are often a major source of indeterminate error with the small samples needed for mass spectrometry. Internal standards are also used in GC/MS and LC/MS. For these techniques, the ratio of peak areas serves as the analytical variable.

A convenient type of internal standard is a stable, isotopically labeled analog of the analyte. Usually, labeling involves preparation of samples of the analyte in which one or more atoms of deuterium, carbon-13, or nitrogen-15 have been incorporated. It is then assumed that during the analysis the labeled molecules behave in the same way as do the unlabeled ones. The mass spectrometer easily distinguishes between the two. Another type of internal standard is a homolog of the analyte that yields a reasonably intense ion peak for a fragment that is chemically similar to the analyte fragment being measured.

With low-resolution instruments, it is seldom possible to locate peaks that are unique to each component of a mixture. In this situation, it is still possible to complete an analysis by collecting intensity data at a number of *m/z* values that equal or exceed the number of sample components. Simultaneous equations are then developed that relate the intensity of each *m/z* value to the contribution made by each component to

[31] See N. L. Kelleher, *Anal. Chem.*, **2004**, *76*, 196A; F. W. McLafferty, *Int. J. Mass Spectrom.*, **2001**, *212*, 81.

 Tutorial: Learn more about **quantitative applications of MS**.

this intensity. Solving these equations then provides the desired quantitative information. Alternatively, chemometric methods such as partial least squares or principal component analysis are used.

Precision and Accuracy

The precision of quantitative mass spectral measurements by the procedure just described usually ranges between 2% and 10% relative. The analytical accuracy varies considerably depending on the complexity of the mixture being analyzed and the nature of its components. For gaseous hydrocarbon mixtures containing five to ten components, absolute errors of 0.2 to 0.8 mole percent are typical.

Applications

The early quantitative applications of mass spectrometry tended to focus on petroleum products and on industrial materials characterization. In recent years, quantitative mass spectrometry has been applied to many diverse areas, including industrial polymers, environmental and forensic samples, and increasingly to biological materials.

Mass spectrometry is widely used for the characterization and analysis of high-molecular-mass polymeric materials. In recent applications, MALDI mass spectrometry has been the method of choice.[32] Hyphenated techniques such as the combination of size exclusion and liquid chromatography with MALDI MS are very powerful for characterizing complex polymeric substances. Pyrolysis methods combined with GC/MS are also quite popular for polymer characterization. In this application, the sample is first pyrolyzed, and the volatile products are admitted into GC/MS for analysis. Alternatively, heating can be performed on the probe of a direct inlet system. Some polymers yield essentially a single fragment: for example, isoprene from natural rubber, styrene from polystyrene, ethylene from polyethylene, and CF_2=CFCl from Kel-F. Other polymers yield two or more products, which depend in amount and kind on the pyrolysis temperature. Studies of temperature effects can provide information regarding the stabilities of the various bonds, as well as the approximate molecular mass distribution.

In environmental analysis, there has been increasing use of TOF, quadrupole, ion-trap, and Fourier transform mass spectrometers in addition to desorption ionization methods such as MALDI.[33] Detection and quantitative determination of such widely diverse contaminants as perfluoroorganics, polybrominated diphenyl ethers, pharmaceuticals, byproducts of water disinfection, pesticides, algal toxins, surfactants, methyl-*t*-butyl ether, arsenic, and various microorganisms are now carried out using mass spectrometric methods. Mass spectrometry is also used to determine compounds of interest in homeland security.[34]

In forensic science, mass spectrometry and GC/MS are widely used in detecting explosives and materials used by arsonists in setting fires, in analyzing body fluids and hair, in testing athletes for drugs, in testing horses for drugs at equine events, and in examining evidentiary materials such as paints and fibers.[35] Mass spectrometers are now indispensable tools in the forensic laboratory.

In the clinical laboratory, GC/MS, LC/MS, and tandem mass spectrometry are finding increasing application.[36] The GC/MS technique is widely used to analyze urinary profiles of patients suspected of having metabolic disorders. Tandem MS methods are now becoming the standard for screening newborn babies for metabolic disease.[37] The LC/MS and tandem MS methods are replacing some traditional immunological and fluorometric methods in quantitative determinations involving drug monitoring and toxicology.

Many other biological applications of mass spectrometry are appearing. Mass spectrometry has always been important in protein identification, particularly in the analysis of peptides derived from digestions with proteolytic enzymes such as trypsin. Recently, however, intact proteins and large protein fragments have been directly analyzed by tandem mass spectrometry.[38] Electrospray ionization in conjunction with high magnetic field Fourier transform MS/MS appears to be particularly useful in such determinations. Mass spectrometry is now playing a major role in the field of proteomics.

[32]See P. M. Peacock and C. N. McEwen, *Anal. Chem.*, **2004**, *76*, 3417; H. Pasch and W. Schrepp, *MALDI-TOF Mass Spectrometry of Synthetic Polymers*, Berlin: Springer-Verlag, 2003.

[33]S. D. Richardson, *Anal. Chem.*, **2004**, *76*, 3337.

[34]W. D. Smith, *Anal. Chem.*, **2002**, *74*, 462A.

[35]See J. Yinon, *Forensic Applications of Mass Spectrometry*, Boca Raton, FL: CRC Press, 1995.

[36]See D. H. Chace, *Chem. Rev.*, **2001**, *101*, 445.

[37]K. C. Kooley, *Clin. Biochem.*, **2003**, *36*, 471.

[38]G. E. Reid and S. A. McLuckey, *J. Mass Spectrom.*, **2002**, *37*, 663.

QUESTIONS AND PROBLEMS

*Answers are provided at the end of the book for problems marked with an asterisk.

X Problems with this icon are best solved using spreadsheets.

20-1 How do gaseous and desorption sources differ? What are the advantages of each?

20-2 How do the spectra for electron-impact, field ionization, and chemical ionization sources differ from one another?

20-3 Describe the difference between gaseous field ionization sources and field desorption sources.

20-4 The following figure is a simplified diagram of a commercially available electron-impact source.

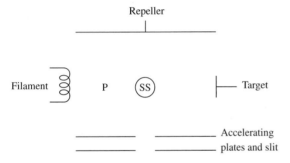

(a) What voltage must be applied between the filament and target so that electrons interacting with molecules at the point marked SS (sample source) will have 70 eV of kinetic energy?

(b) What will happen to a molecule that diffuses toward the filament and is ionized at point P?

*20-5 When a magnetic sector instrument was operated with an accelerating voltage of 3.00×10^3 V, a field of 0.126 T was required to focus the CH_4^+ on the detector.

(a) What range of field strengths would be required to scan the mass range between 16 and 250, for singly charged ions, if the accelerating voltage is held constant?

(b) What range of accelerating voltages would be required to scan the mass range between 16 and 250, for singly charged ions, if the field strength is held constant?

*20-6 Calculate the accelerating voltage that would be required to direct singly charged ions of mass 7,500 through an instrument that is identical to the one described in Example 20-4.

*20-7 The ion-accelerating voltage in a particular quadrupole mass spectrometer is 5.00 V. How long will it take a singly charged cyclohexane ion to travel the 15.0 cm length of the rod assembly? Assume that the initial velocity of the ion in the z direction is zero.

20-8 On page 288 a qualitative discussion described how a positive ion would behave in the xz plane (positive dc potential plane) of a quadrupole mass filter. Construct a similar argument for the behavior of positive ions in the yz plane (negative dc potential plane).

20-9 Why do double-focusing mass spectrometers give narrower peaks and higher resolutions than single-focusing instruments?

20-10 Discuss the differences between quadrupole ion-trap mass spectrometers and Fourier transform ICR mass spectrometers.

*$**20-11**$ Calculate the resolution required to resolve peaks for
(a) CH_2N ($M = 28.0187$) and N_2^+ ($M = 28.0061$).
(b) $C_2H_4^+$ ($M = 28.0313$) and CO^+ ($M = 27.9949$).
(c) $C_3H_7N_3^+$ ($M = 85.0641$) and $C_5H_9O^+$ ($M = 85.0653$).
(d) androst-4-en-3,17,-dione (M^+) at $m/z = 286.1930$ and an impurity at 286.1240.

 20-12 What mass differences can just be resolved at m values of 100, 500, 1500, 3000, and 5000 if the mass spectrometer has a resolution of
(a) 500,   (b) 1000,   (c) 3000,   (d) 5000?

*$**20-13**$ Calculate the ratio of the $(M + 2)^+$ to M^+ and the $(M + 4)^+$ to M^+ peak heights for
(a) $C_{10}H_6Br_2$,   (b) C_3H_7ClBr,   (c) $C_6H_4Cl_2$.

20-14 In a magnetic sector (single-focusing) mass spectrometer, it might be reasonable under some circumstances to monitor one m/z value, to then monitor a second m/z, and to repeat this pattern in a cyclic manner. Rapidly switching between two accelerating voltages while keeping all other conditions constant is called *peak matching*.
(a) Derive a general expression that relates the ratio of the accelerating voltages to the ratio of the corresponding m/z values.
(b) Use this equation to calculate m/z of an unknown peak if m/z of the ion used as a standard, CF_3^+, is 69.00 and the ratio of $V_{unknown}/V_{standard}$ is 0.965035.
(c) Based on your answer in part (b), and the assumption that the unknown is an organic compound that has a mass of 143, draw some conclusions about your answer in part (b), and about the compound.

20-15 Measuring the approximate mass of an ion without using a standard can be accomplished via the following variant of the peak-matching technique described in Problem 20-14. The peak-matching technique is used to alternately cause the P^+ ion and the $(P + 1)^+$ ions to reach the detector. It is assumed that the difference in mass between P^+ and $(P + 1)^+$ is due to a single ^{13}C replacing a ^{12}C atom.
(a) If the accelerating voltage for $(P + 1)^+$ is labeled V_2 and that for P^+ is V_1, derive a relationship that relates the ratio V_2/V_1 to the mass of P^+.
(b) If $V_2/V_1 = 0.987753$, calculate the mass of the P^+ ion.

20-16 Discuss the major differences between a tandem-in-space mass spectrometer and a tandem-in-time mass spectrometer. Include the advantages and disadvantages of each type.

20-17 Identify the ions responsible for the peaks in the mass spectrum shown in Figure 20-20b.

20-18 Identify the ions responsible for the four peaks having greater mass-to-charge ratios than the M^+ peak in Figure 20-4a.

Challenge Problem

20-19 Figure 20-27 shows the mass spectrum of the same compound from an electron-impact ionization source and an ionization source.

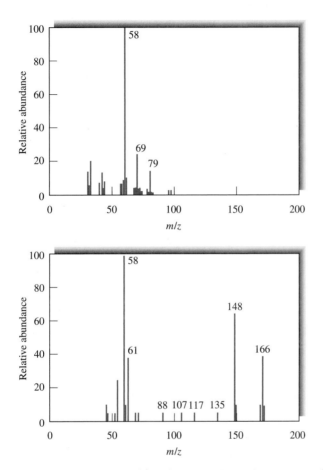

FIGURE 20-27 Electron-impact spectrum (a) and chemical ionization spectrum (b) of the same biologically important compound. (From H. M. Fales, H. A. Lloyd, and G. A. W. Milne, *J. Amer. Chem. Soc.*, **1970**, *92*, 1590–1597. American Chemical Society.)

(a) Which mass spectrum would be best for determining the molecular mass of the compound? Why?

(b) Which mass spectrum would be best for determining the chemical structure? Why?

(c) The electron-impact source was a pulsed source used with a TOF mass analyzer. If the flight tube were 1.0 m long and the accelerating voltage were 3000 V, what would the flight time be for the ion at $m/z = 58$?

(d) For two ions of m/z values m_1/z and m_2/z, derive an equation for the difference in flight times Δt_F as a function of the two masses, the charges, and the accelerating voltage.

(e) For the same TOF analyzer as in part (c), calculate the difference in flight times between ions of $m/z = 59$ and $m/z = 58$.

(f) To get more structural information, the compound of Figure 20-27 was subjected to tandem mass spectrometry. Which ionization source, electron-impact or chemical, would be most suitable for this purpose? Why?

(g) Using the ionization source chosen in part (f), describe the types of mass spectra that could be obtained from an MS/MS experiment by:

(1) holding the first mass analyzer constant and scanning the second analyzer.

(2) scanning both analyzers with a small m/z offset between them.

(3) scanning the first analyzer while holding the second analyzer constant.

(4) scanning the second mass analyzer for every mass selected by the first analyzer.

In your answer, use features of the mass spectrum of Figure 20-27 to illustrate your description.

CHAPTER TWENTY-ONE

Surface Characterization by Spectroscopy and Microscopy

The surface of a solid in contact with a liquid or gaseous phase usually differs substantially from the interior of the solid both in chemical composition and physical properties. Characterization of these surface properties is often of vital importance in a number of fields, including heterogeneous catalysis, sensor development and applications, and semiconductor thin-film technology. Such characterization also aids in understanding corrosion and adhesion mechanisms, activity of metal surfaces, embrittlement properties, and behavior and functions of biological membranes. This chapter deals with the investigation of solid surfaces by spectroscopic and microscopic methods. Although the emphasis is on solid surfaces, some of the techniques are also applicable to other interfaces, such as liquid-liquid and liquid-gas interfaces.

Throughout this chapter, this logo indicates an opportunity for online self-study at **www .thomsonedu.com/chemistry/skoog**, linking you to interactive tutorials, simulations, and exercises.

21A INTRODUCTION TO THE STUDY OF SURFACES

Before considering how surfaces are characterized, we first need to define what constitutes the surface of a solid that is in contact with a gaseous or liquid second phase.

21A-1 Definition of a Solid Surface

We will consider a *surface* to be the boundary layer between a solid and a vacuum, a gas, or a liquid. Generally, we think of a surface as a part of the solid that differs in composition from the average composition of the bulk of the solid. By this definition, the surface comprises not only the top layer of atoms or molecules of a solid but also a transition layer with a nonuniform composition that varies continuously from that of the outer layer to that of the bulk. Thus, a surface may be several or even several tens of atomic layers deep. Ordinarily, however, the difference in composition of the surface layer does not significantly affect the measured overall average composition of the bulk because the surface layer is generally only a tiny fraction of the total solid. From a practical standpoint, it appears best to adopt as an operational definition of a surface that volume of the solid that is sampled by a specific measurement technique. This definition recognizes that if we use several surface techniques, we may in fact be sampling different surfaces and may obtain different, albeit useful, results.

21A-2 Types of Surface Measurements

During the last century, a wide variety of methods have been developed for characterizing surfaces. The classical methods, which are still important, provide much useful information about the *physical* nature of surfaces but less about their chemical nature. These methods involve obtaining optical and electron microscopic images of surfaces as well as measurements of adsorption isotherms, surface areas, surface roughness, pore sizes, and reflectivity. Beginning in the 1950s, spectroscopic surface methods began to appear that provided information about the chemical nature of surfaces.

This chapter is divided into several major parts. After an introduction to surface methods in Section 21B, we then discuss electron spectroscopic techniques, ion spectroscopic techniques, and photon spectroscopic techniques to identify the chemical species making up the surface of solids and to determine their concentrations. Sections 21F and 21G describe modern microscopic methods for imaging surfaces and determining their morphology and their physical features.

21B SPECTROSCOPIC SURFACE METHODS

Generally, the chemical composition of the surface of a solid differs, often significantly, from the interior or bulk of the solid. Thus far in this text, we have focused on analytical methods that provide information about bulk composition of solids only. In certain areas of science and engineering, however, the chemical composition of a surface layer of a solid is much more important than is the bulk composition of the material.

Spectroscopic surface methods provide both qualitative and quantitative *chemical* information about the composition of a surface layer of a solid that is a few tenths of nanometers (a few angstroms) to a few nanometers (tens of angstroms) thick. In this section we describe some of the most widely used of these spectroscopic techniques.[1]

21B-1 Spectroscopic Surface Experiments

Figure 21-1 illustrates the general way spectroscopic examinations of surfaces are performed. Here, the solid sample is irradiated with a *primary beam* made up of photons, electrons, ions, or neutral molecules. Impact of this beam on a surface results in formation of a *secondary beam* also consisting of photons, electrons, molecules, or ions from the solid surface. The secondary beam is detected by the spectrometer. Note that the type of particle making up the primary beam is not necessarily the same as that making up the secondary beam. The secondary beam, which results from scattering, sputtering, or emission, is then studied by a variety of spectroscopic methods.

The most effective surface methods are those in which the primary beam, the secondary beam, or both is made up of either electrons, ions, or molecules and not photons because this limitation assures that the measurements are restricted to the surface of a sample and not to its bulk. For example, the maximum penetration depth of a beam of 1-keV electrons or ions is

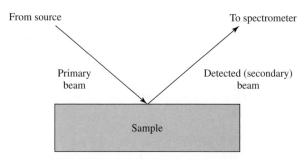

FIGURE 21-1 General scheme for surface spectroscopy. Beams may be photons, electrons, ions, or neutral molecules.

approximately 2.5 nm (25 Å), whereas the penetration depth of a photon beam of the same energy is about 1000 nm (10^4 Å). Thus, for many methods that involve two beams of photons, such as X-ray fluorescence (see Chapter 12), infrared reflection spectroscopy (see Chapter 17), ellipsometry, or resonance Raman spectroscopy (see Chapter 18), precautions must be taken to limit the measurements to a surface layer. The techniques involving primary and detected (secondary) beams of photons discussed in this section are surface plasmon resonance, nonlinear optical spectroscopy, and ellipsometry.

There are several ways to classify surface techniques. Many of these are based on the nature of the primary and detected beams. Table 21-1 lists the most widely used spectroscopic techniques. These will be discussed further in this section.

21B-2 Sampling Surfaces

Regardless of the type of spectroscopic surface method being used, three types of sampling methods are employed. The first involves focusing the primary beam on a single small area of the sample and observing the secondary beam. Often, the spot is chosen visually with an optical microscope. The second method involves mapping the surface, in which a region of the surface is scanned by moving the primary beam across the surface in a *raster pattern* of measured increments and observing changes in the secondary beam that result. The mapping may be linear or two dimensional. The third technique is known as depth profiling. Here, a beam of ions from an ion gun etches a hole in the surface by sputtering. During this process a finer primary beam produces a secondary beam from the center of the hole, which provides the analytical data on the surface composition as a function of depth.

---

[1] For a description of surface spectroscopic techniques, see *Surface Analysis — The Principal Techniques*, J. C. Vickerman, ed., Chichester, UK: Wiley, 1997; *Spectroscopy of Surfaces*, R. G. H. Clark and R. E. Hester, eds., New York: Wiley, 1988.

TABLE 21-1 Some Common Spectroscopic Techniques for Analysis of Surfaces

Method and Acronym	Primary Beam	Detected Beam	Information
X-ray photoelectron spectroscopy (XPS), or electron spectroscopy for chemical analysis (ESCA)	X-ray photons	Electrons	Chemical composition Chemical structure
Auger electron spectroscopy (AES)	Electrons or X-ray photons	Electrons	Chemical composition
Electron energy-loss spectroscopy (EELS)	Electrons	Electrons	Chemical structure Adsorbate binding
Electron microprobe (EM)	Electrons	X-ray photons	Chemical composition
Secondary-ion mass spectrometry (SIMS)	Ions	Ions	Chemical composition Chemical structure
Ion-scattering spectroscopy (ISS) and Rutherford backscattering	Ions	Ions	Chemical composition Atomic structure
Laser-microprobe mass spectrometry (LMMS)	Photons	Ions	Chemical composition Chemical structure
Surface plasmon resonance (SPR)	Photons	Photons	Composition and concentration of thin films
Sum frequency generation (SFG)	Photons	Photons	Interface structure, adsorbate binding
Ellipsometry	Photons	Photons	Thin-film thickness

21B-3 Surface Environment

Most of the surface spectroscopic techniques require a "vacuum" environment. High vacuum conditions ensure that the particles used have long mean free paths to interact with the surface of interest. The vacuum environment also keeps the surface free from adsorbed gases during the surface analysis experiment. The exceptions to the high vacuum requirement are the photon-photon techniques given in the last three rows of Table 21-1. These allow examination of surfaces under conditions more akin to those used in applications such as catalysis, sensing, and corrosion studies.

A problem frequently encountered in surface analyses is contamination of the surface by adsorption of components of the atmosphere, such as oxygen, water, or carbon dioxide. Even in a vacuum, this type of contamination occurs in a relatively short time. For example, at a pressure of 10^{-6} torr (1 torr = 133 Pa), a monolayer of gas molecules will cover a clean surface in just 3 s. At 10^{-8} torr, coverage occurs in about 1 h. At 10^{-10} torr, 10 h is required.[2] Because of adsorption

problems, provisions must often be made to clean the sample surface, usually in the chamber used for irradiating the sample. Cleaning may involve baking the sample at a high temperature; sputtering the sample with a beam of inert gas ions from an electron gun; mechanical scraping or polishing of the sample surface with an abrasive; ultrasonic washing of the sample in various solvents; and bathing the sample in a reducing atmosphere to remove oxides.

In addition to atmospheric contamination, the primary beam itself can alter the surface as a measurement progresses. Damage caused by the primary beam depends on the momentum of the primary beam particles. Thus, of the beams listed in Table 21-1, ions are the most damaging and photons the least.

21C ELECTRON SPECTROSCOPY

The first three methods listed in Table 21-1 are based on detection of emitted electrons produced by incident beams. Here, the signal from the analyte is encoded in a beam of electrons rather than photons. The spectrometric measurements then consist of the determination of the power of this beam as a function of the energy $h\nu$ or frequency ν of the electrons. This type of spectroscopy is termed *electron spectroscopy*.

---

[2]D. M. Hercules and S. H. Hercules, *J. Chem. Educ.*, **1984**, *61*, 403.

 Tutorial: Learn more about **surface methods**.

Although the basic principles of electron spectroscopy were well understood a century ago, the widespread application of this technique to chemical problems did not occur until relatively recently. Studies in the field were inhibited by the lack of technology necessary for performing high-resolution spectral measurements of electrons having energies varying from a few tenths to several thousand electron volts. By the late 1960s, this technology had developed, and commercial electron spectrometers began to appear in the marketplace. With their appearance, an explosive growth in the number of publications devoted to electron spectroscopy occurred.[3]

There are three types of electron spectroscopy for the study of surfaces. The most common type, which is based on irradiation of the sample surface with monochromatic X-radiation, is called *X-ray photoelectron spectroscopy* (XPS). It is also termed *electron spectroscopy for chemical analysis*. Much of the material in this chapter is devoted to XPS. The primary beam for photoelectron spectroscopy can also consist of ultraviolet photons, in which case the technique is called *ultraviolet photoelectron spectroscopy* (UPS). Here, a monochromatic beam of ultraviolet radiation causes ejection of electrons from the analyte. This type of electron spectroscopy is not as common as the other two, and we shall not discuss it further. The second type of electron spectroscopy is called *Auger* (pronounced oh-ZHAY) *electron spectroscopy* (AES). Most commonly, Auger spectra are excited by a beam of electrons, although X-rays are also used. Auger spectroscopy is discussed in Section 21C-2. The third type of electron spectroscopy is electron energy-loss spectroscopy (EELS), in which a low-energy beam of electrons strikes the surface and excites vibrations. The resultant energy loss is then detected and related to the vibrations excited. We briefly describe EELS in Section 21C-3.

Electron spectroscopy is a powerful tool for the identification of all the elements in the periodic table with the exception of hydrogen and helium. More important, the method permits determination of the oxidation state of an element and the type of species to which it is bonded. Finally, the technique provides useful information about the electronic structure of molecules.

Electron spectroscopy has been successfully applied to gases and solids and more recently to solutions and liquids. Because of the poor penetrating power of electrons, however, these methods provide information about solids that is restricted largely to a surface layer a few atomic layers thick (2 to 5 nm). Usually, the composition of such surface layers is significantly different from the average composition of the entire sample. Indeed, the most important and valuable current applications of electron spectroscopy are to the qualitative analysis of the surfaces of solids, such as metals, alloys, semiconductors, and heterogeneous catalysts. Quantitative analysis by electron spectroscopy finds somewhat limited applications.

21C-1 X-ray Photoelectron Spectroscopy

It is important to emphasize the fundamental difference between electron spectroscopy (both XPS and AES) and the other types of spectroscopy we have thus far encountered. In electron spectroscopy, the kinetic energy of emitted electrons is recorded. The spectrum thus consists of a plot of the number of emitted electrons, or the power of the electron beam, as a function of the energy (or the frequency or wavelength) of the emitted electrons (see Figure 21-2).

Principles of XPS

The use of XPS was pioneered by the Swedish physicist K. Siegbahn, who subsequently received the 1981 Nobel Prize in Physics for his work.[4] Siegbahn chose to call the technique electron spectroscopy for chemical analysis (ESCA) because, in contrast to the other two electron spectroscopies, XPS provides information about not only the atomic composition of a sample but also the structure and oxidation state of the compounds being examined. Figure 21-3 is a schematic representation of the physical process involved in XPS. Here, the three lower lines labeled E_b, E_b', and E_b'' represent energies of the inner-shell K and L electrons of an atom. The upper three lines represent some of the energy levels of the outer shell, or valence, electrons. As shown in the illustration, one of the photons of a monochromatic X-ray beam of known energy $h\nu$ displaces an electron

[3]For additional information, see J. F. Watts and J. Wolstenholme, *An Introduction to Surface Analysis by XPS and AES*, Chichester, UK: Wiley, 2003; D. Briggs and M. P. Seah, *Practical Surface Analysis by Auger and X-ray Photoelectron Spectroscopy*, 2nd ed., Chichester, UK: Wiley, 1990.

[4]For a brief description of the history of XPS, see K. Siegbahn, *Science*, **1981**, *217*, 111; D. M Hercules, *J. Chem. Educ.*, **2004**, *81*, 1751. For monographs, see S. Hüfner, *Photoelectron Spectroscopy: Principles and Applications*, Berlin: Springer-Verlag, 1995; T. L. Barr, *Modern ESCA: The Principles and Practice of X-Ray Photoelectron Spectroscopy*, Boca Raton, FL: CRC Press, 1994.

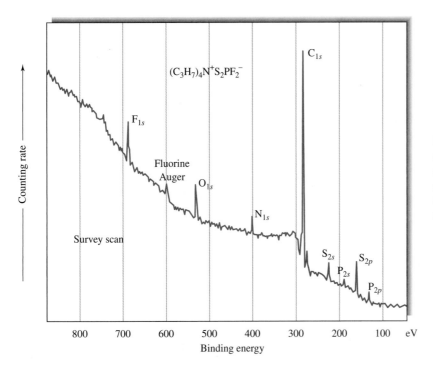

FIGURE 21-2 X-ray photoelectron spectrum of tetrapropylammonium-difluoridethiophosphate. The peaks are labeled according to the element and orbital from which the emitted electrons originate.

e^- from a K orbital of energy E_b. The process can be represented as

$$A + h\nu \rightarrow A^{+*} + e^- \qquad (21\text{-}1)$$

where A can be an atom, a molecule, or an ion and A^{+*} is an electronically excited ion with a positive charge one greater than that of A.

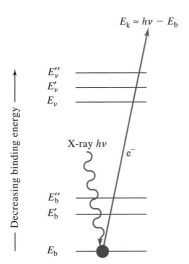

FIGURE 21-3 Schematic representation of the ESCA process. The incident beam consists of monoenergetic X-rays. The emitted beam is made up of electrons.

The kinetic energy of the emitted electron E_k is measured in an electron spectrometer. The *binding energy* of the electron E_b can then be calculated by means of the equation

$$E_b = h\nu - E_k - w \qquad (21\text{-}2)$$

In this equation, w is the *work function* of the spectrometer, a factor that corrects for the electrostatic environment in which the electron is formed and measured. Various methods are available to determine the value of w. The binding energy of an electron is characteristic of the atom and orbital that emit the electron.

Figure 21-2 shows a low-resolution, or survey, XPS spectrum consisting of a plot of electron-counting rate as a function of binding energy E_b. The analyte consisted of an organic compound made up of six elements. With the exception of hydrogen, well-separated peaks for each of the elements can be observed. In addition, a peak for oxygen is present, suggesting that some surface oxidation of the compound had occurred. Note that, as expected, the binding energies for $1s$ electrons increase with atomic number because of the increased positive charge of the nucleus. Note also that more than one peak for a given element can be observed; thus peaks for both $2s$ and $2p$ electrons for sulfur and phosphorus can be seen. The large background count arises because associated with each characteristic peak is a tail of ejected electrons that have lost part of their energy by inelastic collisions within the solid sample. These electrons have less kinetic energy than their nonscattered counterparts and will thus appear at lower kinetic energies or higher binding energies (Equation 21-2). It is evident from Figure 21-2 that XPS provides a means of qualitative identification of the elements present on the surface of solids.

Instrumentation

Instruments for electron spectroscopy are offered by several instrument manufacturers. These products differ considerably in types of components, configurations, and costs. Some are designed for a single type of application, such as XPS, and others can be adapted to AES and UPS by purchase of suitable accessories. All are expensive ($300,000 to >$10^6$).

Electron spectrometers are made up of components whose functions are analogous to those encountered in optical spectroscopic instruments. These components include (1) a source; (2) a sample holder; (3) an analyzer, which has the same function as a monochromator; (4) a detector; and (5) a signal processor and readout. Figure 21-4 shows a typical arrangement of these components. Electron spectrometers generally require elaborate vacuum systems to reduce the pressure in all of the components to as low as 10^{-8} to 10^{-10} torr.[5]

Sources. The simplest X-ray sources for XPS spectrometers are X-ray tubes equipped with magnesium or aluminum targets and suitable filters. The $K\alpha$ lines for these two elements have considerably narrower bandwidths (0.8 to 0.9 eV) than those encountered with higher atomic number targets; narrow bands are desirable because they lead to enhanced resolution. Nonmonochromatic sources typically illuminate a spot a few centimeters in diameter.

Relatively sophisticated XPS instruments, such as that shown in Figure 21-4, employ a crystal monochromator (Section 12B-3) to provide an X-ray beam having a bandwidth of about 0.3 eV. Monochromators eliminate bremsstrahlung background, thus improving signal-to-noise ratios. They also allow much smaller spots on a surface to be examined (spot sizes ~50 μm).

The increased availability of synchrotron radiation in recent years has given XPS experimenters another useful source. The synchrotron produces broadband radiation that is highly collimated and polarized. Such sources when used with a monochromator can provide a source of X-rays that is tunable for photoelectron experiments.

Sample Holders. Solid samples are mounted in a fixed position as close as possible to the photon or electron source and the entrance slit of the spectrometer (see Figure 21-4). To avoid attenuation of the electron beam, the sample compartment must be evacuated to a pressure of 10^{-5} torr or less. Often, however, much better vacuums (10^{-9} to 10^{-10} torr) are required to avoid contamination of the sample surface by substances such as oxygen or water that react with or are adsorbed on the surface.

Gas samples are leaked into the sample area through a slit of such a size as to provide a pressure of perhaps 10^{-2} torr. Higher pressures lead to excessive attenuation of the electron beam, which is due to inelastic collisions; on the other hand, if the sample pressure is too low, weakened signals are obtained.

Analyzers. The analyzer consists of the collection lens or lenses and the electron energy analyzer, which disperses the emitted electrons according to their kinetic energy. The lens system usually allows a wide collection angle (~30°) for high efficiency. In some angle-resolved experiments, an aperture reduces the angles collected. Such experiments are used in depth-profiling studies.

Typically, photoelectron experiments are carried out in constant analyzer energy mode, in which electrons are accelerated or retarded by the lens system to some user-defined energy as they pass through the analyzer (the pass energy, E in Figure 21-4). Often, pass energies of 5–25 eV will give high-resolution spectra, and 100–200 eV pass energies are used for survey scans. The signal intensity decreases as the pass energy decreases.

Most energy analyzers are of the type illustrated in Figure 21-4, in which the electron beam is deflected by the electrostatic field of a hemispherical capacitor. The electrons thus travel in a curved path from the lens to the multichannel transducer. The radius of curvature depends on the kinetic energy of the electrons and the magnitude of the electrostatic field. An entire spec-

[5]Specifications for several representative commercial instruments are given in D. Noble, *Anal. Chem.*, **1995**, *67*, 675A. For a perspective on commercial XPS instrumentation, see M. A. Kelly, *J. Chem. Educ.*, **2004**, *81*, 1726.

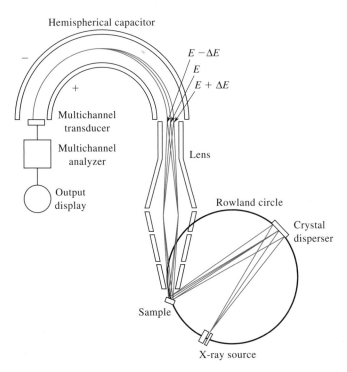

FIGURE 21-4 Principle of a modern ESCA instrument using a monochromatic X-ray source and a hemispherical field spectrometer.

trum is obtained by varying the field so as to focus electrons of various kinetic energies on the transducer.

Transducers. Most modern electron spectrometers are based on solid-state, channel electron multipliers, which consist of tubes of glass that have been doped with lead or vanadium. When a potential difference of several kilovolts is applied across these materials, a cascade or pulse of 10^6 to 10^8 electrons is produced for each incident electron. The pulses are then counted electronically (see Section 4C). Several manufacturers are now offering two-dimensional multichannel electron detectors that are analogous in construction and application to the multichannel photon detectors described in Section 7E-3. Here, all of the resolution elements of an electron spectrum are monitored simultaneously and the data stored in a computer for subsequent display. The advantages of such a system are similar to those realized with multichannel photon detectors.

Data Systems. Modern XPS instruments have nearly all components under computer control. Thus, electron guns, ion guns, valves, lens voltages, sample position, and analyzer parameters are all selected by the computer. Current software on XPS instruments al-

lows many data-analysis options, including peak finding, peak identification, and peak intensity measurement. Many packages also include chemometric data analysis such as multivariate statistical processing and pattern recognition.

Applications of XPS

XPS provides qualitative and quantitative information about the elemental composition of matter, particularly of solid surfaces. It also often provides useful structural information.[6]

Qualitative Analysis. Figure 21-2 shows a low-resolution, wide-scan XPS spectrum, called a *survey spectrum*, which serves as the basis for the determination of the elemental composition of samples. With a magnesium or aluminum $K\alpha$ source, all elements except hydrogen and helium emit core electrons having characteristic binding energies. Typically, a survey spectrum encompasses a kinetic energy range of 250 to 1500 eV, which corresponds to binding energies of about 0 to

---

[6]For reviews of applications of XPS (and AES as well), see J. F. Watts and J. Wolstenholme, *An Introduction to Surface Analysis by XPS and AES*, Chichester, UK: Wiley, 2003; N. H. Turner and J. A. Schreifels, *Anal. Chem.*, **2000**, *72*, 99R; **1998**, *70*, 229R; **1996**, *68*, 309R. See also D. M. Hercules, *J. Chem. Educ.*, **2004**, *81*, 1751.

TABLE 21-2 Chemical Shifts as a Function of Oxidation State[a]

Element[b]	Oxidation State									
	−2	−1	0	+1	+2	+3	+4	+5	+6	+7
Nitrogen (1s)	—	*0[c]	—	+4.5[d]	—	+5.1	—	+8.0	—	—
Sulfur (1s)	−2.0	—	*0	—	—	—	+4.5	—	+5.8	—
Chlorine (2p)	—	*0	—	—	—	+3.8	—	+7.1	—	+9.5
Copper (1s)	—	—	*0	+0.7	+4.4	—	—	—	—	—
Iodine (4s)	—	*0	—	—	—	—	—	+5.3	—	+6.5
Europium (3d)	—	—	—	—	*0	+9.6	—	—	—	—

[a] All shifts are in electron volts measured relative to the oxidation states indicated by (*). (Reprinted with permission from D. M. Hercules, *Anal. Chem.,* **1970**, *42*, 28A. Copyright 1970 American Chemical Society.)

[b] Type of electrons given in parentheses.

[c] Arbitrary zero for measurement, end nitrogen in NaN_3.

[d] Middle nitrogen in NaN_3.

1250 eV. Every element in the periodic table has one or more energy levels that will result in the appearance of peaks in this region. In most instances, the peaks are well resolved and lead to unambiguous identification provided the element is present in concentrations greater than about 0.1%. Occasionally, peak overlap is encountered such as O(1s) with Sb(3d) or Al(2s, 2p) with Cu(3s, 3p). Usually, problems due to spectral overlap can be resolved by investigating other spectral regions for additional peaks. Often, peaks resulting from Auger electrons are found in XPS spectra (see, for example, the peak at about 610 eV in Figure 21-2). Auger lines are readily identified by comparing spectra produced by two X-ray sources (usually magnesium and aluminum $K\alpha$). Auger lines remain unchanged on the kinetic energy scale but photoelectron peaks are displaced. The reason for the behavior of Auger electrons will become apparent in the next section.

Chemical Shifts and Oxidation States. When one of the peaks of a survey spectrum is examined under conditions of higher energy resolution, the position of the maximum depends to a small degree on the chemical environment of the atom responsible for the peak. That is, variations in the number of valence electrons, and the type of bonds they form, influence the binding energies of core electrons. The effect of the number of valence electrons and thus the oxidation state is demonstrated by the data for several elements shown in Table 21-2. Note that in each case, binding energies increase as the oxidation state becomes more positive. This *chemical shift* can be explained by assuming that the attraction of the nucleus for a core electron is di-

minished by the presence of outer electrons. When one of these electrons is removed, the effective charge sensed for the core electron is increased, and an increase in binding energy results.

One of the most important applications of XPS has been the identification of oxidation states of elements in inorganic compounds.

Chemical Shifts and Structure. Figure 21-5 illustrates the effect of structure on the position of peaks for an element. Each peak corresponds to the 1s electron of the carbon atom indicated by dashes above it in the structural formula. Here, the shift in binding energies can be rationalized by taking into account the influence of the various functional groups on the effective nuclear charge experienced by the 1s core electron. For example, of all of the attached groups, fluorine atoms have the greatest ability to withdraw electron density from the carbon atom. The effective nuclear charge felt by the carbon 1s electron is therefore a maximum, as is the binding energy.

Figure 21-6 indicates the position of peaks for sulfur in its several oxidation states and in various types of organic compounds. The data in the top row clearly demonstrate the effect of oxidation state. Note also in the last four rows of the chart that XPS discriminates between two sulfur atoms contained in a single ion or molecule. Thus, two peaks are observed for thiosulfate ion ($S_2O_3^{2-}$), suggesting different oxidation states for the two sulfur atoms.

XPS spectra provide not only qualitative information about types of atoms present in a compound but also the relative number of each type. Thus, the nitro-

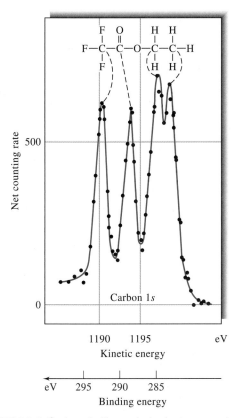

FIGURE 21-5 Carbon 1s X-ray photoelectron spectrum for ethyl trifluoroacetate. (From K. Siegbahn et al., *ESCA: Atomic, Molecular, and Solid-State Studies by Means of Electron Spectroscopy*, p. 21, Upsala: Almquist and Wiksells, 1967. With permission.)

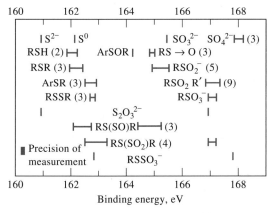

FIGURE 21-6 Correlation chart for sulfur 2s electron binding energies. The numbers in parentheses indicate the number of compounds examined. (Reprinted with permission from D. M. Hercules, *Anal. Chem.*, **1970**, *42*, 35A. Copyright 1970, American Chemical Society.)

gen 1s spectrum for sodium azide ($Na^+N_3^-$) is made up of two peaks having relative areas in the ratio of 2:1 corresponding to the two end nitrogens and the center nitrogen, respectively.

It is worthwhile pointing out again that the photoelectrons produced in XPS are incapable of passing through more than perhaps 1 to 5 nm of a solid. Thus, the most important applications of electron spectroscopy, like X-ray microprobe spectroscopy, are for the accumulation of information about surfaces. Examples of some of its uses include identification of active sites and poisons on catalytic surfaces, determination of surface contaminants on semiconductors, analysis of the composition of human skin, and study of oxide surface layers on metals and alloys.

It is also evident that the method has a substantial potential in the elucidation of chemical structure (see Figures 21-5 and 21-6). Information from XPS spectra is comparable to that from nuclear magnetic resonance (NMR) or IR spectroscopy. The ability of XPS to distinguish among oxidation states of an element is noteworthy.

Note that the information obtained by XPS must also be present in the absorption edge of an X-ray absorption spectrum for a compound. Most X-ray spectrometers, however, do not have sufficient resolution to permit ready extraction of this structural information.

Quantitative Applications. Once, XPS was not considered to be a very useful quantitative technique. However, there has been increasing use of XPS for determining the chemical composition of the surface region of solids.[7] If the solid is homogeneous to a depth of several electron mean free paths, we can express the number of photoelectrons detected each second I as

$$I = n\phi\sigma\varepsilon\eta ATl \qquad (21\text{-}3)$$

where n is the number density of atoms (atoms cm^{-3}) of the sample, ϕ is the flux of the incident X-ray beam (photons $cm^{-2}\,s^{-1}$), σ is the photoelectric cross section for the transition (cm^2/atom), ε is the angular efficiency factor for the instrument, η is the efficiency of producing photoelectrons (photoelectrons/photon), A is the area of the sample from which photoelectrons are detected (cm^2), T is the efficiency of detection of

[7]For a review of quantitative applications of XPS and AES, see K. W. Nebesny, B. L. Maschhoff, and N. R. Armstrong, *Anal. Chem.*, **1989**, *61*, 469A. For a discussion of the reliability of XPS, see C. J. Powell, *J. Chem. Educ.*, **2004**, *81*, 1734.

the photoelectrons, and l is the mean free path of the photoelectrons in the sample (cm).

For a given transition, the last six terms are constant, and we can write the atomic sensitivity factor S as

$$S = \sigma \varepsilon \eta A T l \qquad (2\text{-}4)$$

For a given spectrometer, a set of relative values of S can be developed for the elements of interest. Note that the ratio I/S is directly proportional to the concentration n on the surface. The quantity I is usually taken as the peak area, although peak heights are also used. Often, for quantitative work, internal standards are used. Relative precisions of about 5% are typical. For the analysis of solids and liquids, it is necessary to assume that the surface composition of the sample is the same as its bulk composition. For many applications this assumption can lead to significant errors. Detection of an element by XPS requires that it be present at a level of at least 0.1%. Quantitative analysis can usually be performed if 5% of the element is present.

21C-2 Auger Electron Spectroscopy

In contrast to XPS, AES[8] is based on a two-step process in which the first step involves formation of an electronically excited ion A^{+*} by exposing the analyte to a beam of electrons or sometimes X-rays. With X-rays, the process shown in Equation 21-1 occurs. For an electron beam, the excitation process can be written

$$A + e_i^- \rightarrow A^{+*} + e_i'^- + e_A^- \qquad (21\text{-}5)$$

where e_i^- represents an incident electron from the source, $e_i'^-$ represents the same electron after it has interacted with A and has thus lost some of its energy, and e_A^- represents an electron ejected from one of the inner orbitals of A.

As shown in Figure 21-7a and b, relaxation of the excited ion A^{+*} can occur in two ways:

$$A^{+*} \rightarrow A^{++} + e_A^- \qquad (21\text{-}6)$$

or

$$A^{+*} \rightarrow A^+ + h\nu_f \qquad (21\text{-}7)$$

Here, e_A^- corresponds to an Auger electron and $h\nu_f$ represents a fluorescence photon.

[8]See J. F. Watts and J. Wolstenholme, *An Introduction to Surface Analysis by XPS and AES*, Chichester, UK: Wiley, 2003; M. Thompson, M. D. Baker, A. Christie, and J. F. Tyson, *Auger Electron Spectroscopy*, New York: Wiley, 1985; *Auger Electron Spectrometry Theory Tutorial*, Evans Analytical Group, http://www.eaglabs.com/en-US/references/tutorial/augtheo/caiatheo.html.

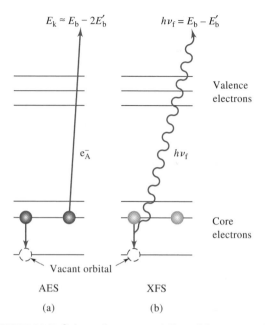

FIGURE 21-7 Schematic representation of the source of (a) Auger electron emission and (b) X-ray fluorescence that competes with Auger emission.

The relaxation process described by Equation 21-7 is X-ray fluorescence, which was described in Chapter 12. Note that the energy of the fluorescence radiation $h\nu_f$ is independent of the excitation energy. Thus, polychromatic radiation can be used for the excitation step. In Auger emission, shown by Equation 21-6, the energy given up in relaxation results in the ejection of an Auger electron e_A^- with kinetic energy E_k. Note that the energy of the Auger electron is *independent* of the energy of the photon or electron that originally created the vacancy in energy level E_b. Thus, as is true in fluorescence spectroscopy, a monoenergetic source is not required for excitation. Because the Auger lines are independent of the input energy, it is possible to differentiate between Auger lines in a spectrum and the XPS peaks.

The kinetic energy of the Auger electron is the difference between the energy released in relaxation of the excited ion ($E_b - E_b'$) and the energy required to remove the second electron from its orbital (E_b'). Thus,

$$E_k = (E_b - E_b') - E_b' = E_b - 2E_b' \qquad (21\text{-}8)$$

Auger emissions are described in terms of the type of orbital transitions involved in the production of the electron. For example, a KLL Auger transition involves an initial removal of a K electron followed by a transition of an L electron to the K orbital with the

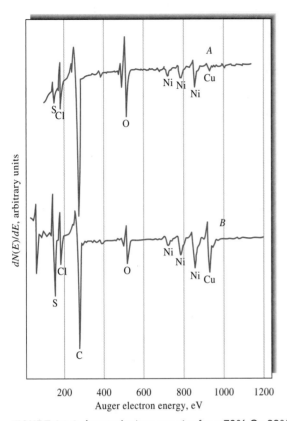

FIGURE 21-8 Auger electron spectra for a 70% Cu:30% Ni alloy. *A*, passivated by anodic oxidation; *B*, not passivated. (Adapted from G. E. McGuire et al., *J. Electrochem. Soc.*, **1978**, *125*, 1802. Reprinted by permission of the publisher, the Electrochemical Society, Inc.)

simultaneous ejection of a second L electron. Other common transitions are LMM and MNN.

Like XPS spectra, Auger spectra consist of a few characteristic peaks lying in the region of 20 to 1000 eV. Figure 21-8 shows typical Auger spectra obtained for two samples of a 70:30 copper-nickel alloy. Note that the derivative of the counting rate as a function of the kinetic energy of the electron $dN(E)/dE$ is the ordinate. Derivative spectra are standard for Auger spectroscopy to enhance the small peaks and to repress the effect of the large, but slowly changing, scattered electron background radiation. Also note that the peaks are well separated, making qualitative identification fairly straightforward.

Auger electron emission and X-ray fluorescence (Figure 21-7) are competitive processes, and their rel-

ative rates depend on the atomic number of the element involved. High atomic numbers favor fluorescence, and Auger emission predominates with atoms of low atomic numbers. As a consequence, X-ray fluorescence is not very sensitive for detecting elements with atomic numbers smaller than about 10.

Auger and XPS provide similar information about the composition of matter. The methods tend to be complementary rather than competitive, however, with Auger spectroscopy being more reliable and efficient for certain applications and XPS for others. As mentioned earlier, most instrument manufacturers recognize the complementary nature of Auger and XPS by making provisions for both kinds of measurements with a single instrument.

The particular strengths of Auger spectroscopy are its sensitivity for atoms of low atomic number, its minimal matrix effects, and above all its high spatial resolution, which permits detailed examination of solid surfaces. The high spatial resolution arises because the primary beam is made up of electrons, which can be more tightly focused on a surface than can X-rays. To date, Auger spectroscopy has not been used extensively to provide the kind of structural and oxidation state information that was described for XPS. Quantitative analysis with AES is not as straightforward as with XPS because of derivative mode data presentation and the presence of fine structure peaks. Often, the integrated area is the preferred parameter for quantitative work. Semiquantitative methods are available and widely used.

Instrumentation

The instrumentation for AES is similar to that for XPS except that the source is usually an *electron gun* rather than an X-ray tube.[9] A schematic of a common type of electron gun is given in Figure 21-9. This source consists of a heated tungsten filament, which is usually about 0.1 mm in diameter and bent into the shape of a hairpin with a V-shape tip. The cathodic filament is maintained at a potential of 1 to 50 kV with respect to the anode contained in the gun. Surrounding the filament is a grid cap, or *Wehnelt cylinder*, which is biased negatively with respect to the filament. The effect of the electric field in the gun is to cause the emitted electrons to converge on a tiny spot called the *crossover* that has a diameter d_0.

---

[9]For a review of Auger spectrometers, see M. J. Felton, *Anal. Chem.*, **2003**, *75*, 269A. For a tutorial, see *Auger Electron Spectrometry Instrumentation Tutorial*, Evans Analytical Group, http://www.eaglabs.com/en-US/references/tutorial/auginst/caiainst.html.

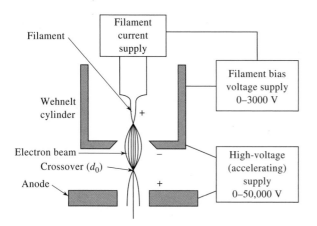

FIGURE 21-9 Block diagram of a tungsten filament source.

Cathodes constructed in the form of lanthanum hexaboride (LaB_6) rods are also used in electron guns when a source of greater brightness is desired. This type of source is expensive and requires a better vacuum system to prevent oxide formation, which causes the efficiency of the source to deteriorate rapidly. The most significant new type of source that has been introduced in recent years is based on *field emission*. Here, the source is a sharp-tip (100 nm or less) tungsten or carbon cathode. When this type of cathode is held at a high voltage, the electric field at the tip is so intense ($>10^7$ V/cm) that electrons are produced by a *quantum mechanical tunneling process*[10] in which no thermal energy is required to free the electrons from the potential barrier that normally prevents their emission. Field emission sources provide a beam of electrons that have a crossover diameter of only 10 nm compared with 10 μm for LaB_6 rods and 50 μm for tungsten hairpins. The disadvantages of this type of source are its fragility and the fact that it also requires a better vacuum than does an ordinary filament source.

Electron guns produce a beam of electrons with energies of 1 to 10 keV, which can be focused on the surface of a sample for Auger electron studies. One of the special advantages of Auger spectroscopy is its capability for very high spatial-resolution scanning of solid surfaces. Normally, electron beams with diameters ranging from 5 to 500 μm are used for this purpose.

Guns producing beams of approximately 5 μm are called Auger microprobes and are employed for scanning solid surfaces to detect and determine the elemental composition of inhomogeneities.

Applications of AES

Qualitative Analysis of Solid Surfaces. Typically, an Auger spectrum is obtained by bombarding a small area (5 to 500 μm diameter) of the surface with a beam of electrons from a gun. A derivative electron spectrum, such as that shown in Figure 21-8, is then obtained with an analyzer. An advantage of Auger spectroscopy for surface studies is that the low-energy Auger electrons (20 to 1000 eV) are able to penetrate only a few atomic layers, 0.3 to 2 nm (3 to 20 Å) of solid. Thus, whereas the electrons from the electron guns penetrate to a considerably greater depth below the sample surface, only those Auger electrons from the first four or five atomic layers escape to reach the analyzer. Consequently, an Auger spectrum is likely to reflect the true surface composition of solids.

The two Auger spectra in Figure 21-8 are for samples of a 70% Cu to 30% Ni alloy, which is often used for structures where saltwater corrosion resistance is required. Corrosion resistance of this alloy is markedly enhanced by preliminary anodic oxidation in a strong solution of chloride. Figure 21-8A is the spectrum of an alloy surface that has been *passivated* in this way. Spectrum B is for another sample of the alloy in which the anodic oxidation potential was not great enough to cause significant passivation. The two spectra clearly reveal the chemical differences between the two samples that account for the greater corrosion resistance of the former. First, the copper-to-nickel ratio in the surface layer of the nonpassivated sample is approximately that for the bulk, whereas in the passivated material the nickel peaks completely overshadow the copper peak. Furthermore, the oxygen-to-nickel ratio in the passivated sample approaches that for pure anodized nickel, which also has a high corrosion resistance. Thus, the resistance toward corrosion of the alloy appears to result from the creation of a surface that is largely nickel oxide. The advantage of the alloy over pure nickel is its significantly lower cost.

Depth Profiling of Surfaces. Depth profiling involves the determination of the elemental composition of a surface as it is being etched away (sputtered) by a beam of argon ions. Either XPS or Auger spectroscopy can be used for elemental detection, although the lat-

---

[10]In quantum mechanics, there is a finite probability that a particle can pass through a potential energy barrier and appear in a region forbidden by classical mechanics. This process is called tunneling. It can be an important process for light particles, such as protons and electrons.

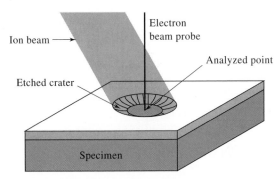

FIGURE 21-10 Schematic representation of the simultaneous use of ion sputter etching and Auger spectroscopy for determining depth profiles. (Courtesy of Physical Electronics, USA, Chanhassen, MN.)

ter is the more common. Figure 21-10 shows schematically how the process is carried out with a highly focused Auger microprobe with a beam diameter of about 5 μm. The microprobe and etching beams are operated simultaneously, with the intensity of one or more of the resulting Auger peaks being recorded as a function of time. Because the etching rate is related to time, a depth profile of elemental composition is obtained. Such information is of vital importance in a variety of studies such as corrosion chemistry, catalyst behavior, and properties of semiconductor junctions.

Figure 21-11 gives a depth profile for the copper-nickel alloy described in the previous section (Figure 21-8). Here, the ratios of the peak intensities for copper versus nickel are recorded as a function of sputtering time. Curve *A* is the profile for the sample that had been passivated by anodic oxidation. With this sample, the copper-to-nickel ratio is essentially zero for the first 10 minutes of sputtering, which corresponds to a depth of about 50 nm. The ratio then rises and approaches that for a sample of alloy that had been chemically etched so that its surface is approximately that of the bulk sample (curve *C*). The profile for the nonpassivated sample (curve *B*) resembles that of the chemically etched sample, although some evidence is seen for a thin nickel oxide coating.

Line Scanning. Line scans are used to characterize the surface composition of solids as a function of distance along a straight line of 100 μm or more. For this purpose, an Auger microprobe is used that produces a beam that can be moved across a surface in a reproducible way. Figure 21-12 shows Auger line scans along the surface of a semiconductor device. In the upper figure, the relative peak amplitude of an oxygen peak is recorded as a function of distance along a line; the lower figure is the same scan produced when the analyzer was set to a peak for gold.

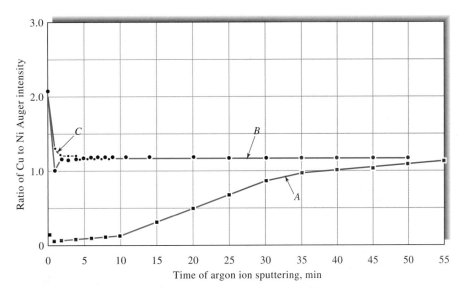

FIGURE 21-11 Auger sputtering profiles for the copper-nickel alloys shown in Figure 21-8: *A*, passivated sample; *B*, nonpassivated sample; *C*, chemically etched sample representing the bulk material. (Adapted from G. E. McGuire et al., *J. Electrochem. Soc.*, **1978**, *125*, 1802. Reprinted by permission of the publisher, the Electrochemical Society, Inc.)

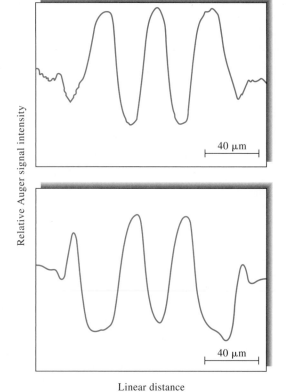

FIGURE 21-12 Auger line scans for oxygen (top) and gold (bottom) obtained for the surface of a semiconductor device. (Courtesy of Physical Electronics, USA, Chanhassen, MN.)

21C-3 Electron Energy-Loss Spectrometry

In EELS,[11] a low-energy (1 to 10 eV) beam of electrons is focused on the surface of a sample and the scattered electrons are analyzed according to scattering energy and scattering angle. Some of the scattered electrons will suffer losses in energy because of vibrational excitation of surface molecules. With high-resolution EELS, a vibrational spectrum can be obtained by counting the number of electrons with a given energy loss relative to the elastically scattered electrons and reporting this count as a function of energy. Such spectra have been used to identify functional groups in the first layer of a surface including adsorbates and provide information on chemical bonding, such as oxidation states and coordination numbers.

In many cases EELS spectra are obtained in conjunction with electron microscopy experiments (see

Section 21G). However, dedicated instruments and instruments combined with other electron spectroscopy techniques (XPS, AES) are available commercially. Typically, the resolution of EELS instruments is $>10\ \text{cm}^{-1}$, which is low compared to IR and Raman instruments but quite suitable for identifying and characterizing surface species.

21D ION SPECTROSCOPIC TECHNIQUES

The techniques named in rows 5–7 of Table 21-1 detect a secondary beam of ions and are classified as ion spectroscopic techniques. These include *secondary-ion mass spectrometry, ion-scattering spectroscopy, Rutherford backscattering spectroscopy,* and *laser-microprobe mass spectrometry.*

21D-1 Secondary-Ion Mass Spectrometry

Secondary-ion mass spectrometry (SIMS) is the most highly developed of the mass spectrometric surface methods, with several manufacturers offering instruments for this technique. SIMS has proven useful for determining both the atomic and the molecular composition of solid surfaces.[12]

There are basically three variations of the SIMS experiment. *Static SIMS* is used for elemental analysis of sub-monolayers on surfaces. Although SIMS is basically a destructive technique, *static* conditions maintain the surface integrity during the time scale of the experiment. *Dynamic SIMS* is used to obtain compositional information as a function of depth below the surface. The dynamic SIMS experiment takes advantage of the destructive nature of SIMS to obtain information on various layers of materials. *Imaging SIMS,* also called *scanning SIMS,* is used to provide spatial images of surfaces.

There are two types of SIMS instruments: *secondary-ion mass analyzers* are used for static and dynamic SIMS, and *microprobe analyzers* are used for imaging SIMS. Both are based on bombarding the surface of the sample with a beam of 5- to 20-keV ions. Usually, Ar^+

---

[11] See R. F. Egerton, *Electron Energy Loss Spectroscopy in the Electron Microscope*, New York: Plenum Press, 1986.

[12] J. C. Vickerman and A. J. Swift, in *Surface Analysis — The Principal Techniques*, J. C. Vickerman, ed., Chichester, UK: Wiley, 1997, Chap. 5; R. G. Wilson, F. A. Stevie, and C. W. Magee, *Secondary Ion Mass Spectrometry: A Practical Handbook for Depth Profiling and Bulk Impurity Analysis*, New York: Wiley, 1989; A. Benninghoven, F. G. Rudenauer, and H. W. Werner, *Secondary Ion Mass Spectrometry: Basic Concepts, Instrumental Aspects, and Applications and Trends*, New York: Wiley, 1987; *Secondary Ion Mass Spectrometry Theory Tutorial*, Evans Analytical Group, http://www.eaglabs.com/en-US/references/tutorial/simstheo/caistheo.html.

ions are used, although Cs^+, N_2^+, or O_2^+ are also common. The ion beam is formed in an ion gun in which the gaseous atoms or molecules are ionized by an electron-impact source. The positive ions are then accelerated by applying a high dc voltage. The impact of these primary ions causes the surface layer of atoms of the sample to be stripped (sputtered) off, largely as neutral atoms. A small fraction, however, forms as positive (or negative) secondary ions that are drawn into a spectrometer for mass analysis. In static SIMS, the sputtering is so slow that consumption of the sample is essentially negligible.

In secondary-ion mass analyzers, which serve for general surface analysis and for depth profiling, the primary ion-beam diameter ranges from 0.3 to 5 mm. Double-focusing, single-focusing, time-of-flight, and quadrupole spectrometers are used for mass determination. Typical transducers for SIMS are electron multipliers, Faraday cups, and imaging detectors. These spectrometers yield qualitative and quantitative information about all of the isotopes (hydrogen through uranium) present on a surface. Relative sensitivity factors vary considerably from ion to ion. Detection limits for most trace elements vary from 1×10^{12} atoms/cm^3 to 1×10^{16} atoms/cm^3. By monitoring peaks for one or a few isotopes, as a function of time, concentration profiles can be obtained with a depth resolution of 5 nm to 10 nm (50 to 100 Å).

Ion microprobe analyzers are more sophisticated (and more expensive) instruments based on a beam of primary ions focused to a diameter of 200 nm to 1 μm. This beam can be moved across a surface (rastered) for about 300 μm in both the x and y directions. A microscope is provided to permit visual adjustment of the beam position. Mass analysis is performed with a double-focusing spectrometer. In some instruments, the primary ion beam passes through an additional low-resolution mass spectrometer so that only a single type of primary ion bombards the sample. The ion microprobe version of SIMS permits detailed studies of solid surfaces.

21D-2 Ion-Scattering and Rutherford Backscattering Spectroscopy

Ion-scattering spectroscopy (ISS) and Rutherford backscattering spectroscopy (RBS) are similar techniques in that a primary ion beam is used to probe a surface.[13] In both techniques, the energy distribution

[13] For more information, see E. Taglauer, in *Surface Analysis — The Principal Techniques*, J. C. Vickerman, ed., Chichester, UK: Wiley, 1997, Chap. 6.

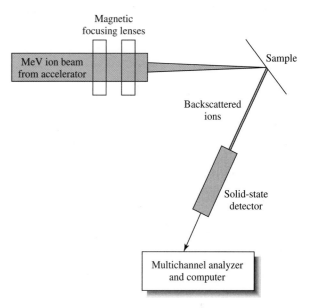

FIGURE 21-13 Rutherford backscattering apparatus. An ion beam from an accelerator is focused on the sample. Backscattered ions are detected with a solid-state particle detector.

of ions backscattered from the sample is measured. Ions striking a solid surface can give rise to a variety of collisional processes as well as electronic excitations. Analysis of the energy spectra of the scattered ions can give information about the atomic masses, their concentrations, and their geometric arrangement on the surface.

The major difference between the two techniques lies in the energies of the incoming ion beam. In ISS, also called low-energy ion scattering, the primary ion energies are in the range of 0.5–5 keV. Noble gas ions, such as He^+, Ar^+, and Ne^+, and alkali ions, such as Li^+, Na^+, or K^+, are used. An electron-impact ion source is most often used with an electrostatic or time-of-flight analyzer. In the ISS technique, information is obtained from the topmost atomic layer or, in some cases, from one or two layers directly below.

In Rutherford backscattering, the primary ion beam ranges in energy from 100 keV for H^+ to several MeV for He^+, He^{2+} (alpha particles), and heavier ions. Sources of energetic ions are often of the Van de Graaff generator type. As shown in Figure 21-13, RBS instruments often use solid-state particle detectors and multichannel analyzers for energy resolution. Magnetic and electrostatic analyzers are also employed for high-resolution studies. Information in RBS arises from a thickness of about 100 nm, although with special techniques surface analysis is also possible. With RBS, it is possible to determine atomic masses and elemental concentrations as a function of depth below the surface.

Because of its sensitivity to the top layer of the surface, ISS has proven very useful in providing surface compositional analysis of materials such as catalysts and alloys. ISS can also provide structural information on metal, semiconductor, metal oxide, and adsorbate surfaces. The major limitations of ISS are associated with difficulties in providing quantitative results, because of neutralization reactions and other interactions. Collisional processes and inelastic energy losses also make absolute mass determinations difficult. In compositional analysis, ISS is complementary to AES although not as generally applicable.

RBS can provide absolute quantitative analysis of elemental composition with an accuracy of about 5%. It can provide depth-profile information from surface layers and thin films to a thickness of about 1 μm. In some cases, however, the high-energy beam can damage the surface. This is particularly a problem with insulating materials, such as polymers, alkali halides, and oxides. The Mars Pathfinder mission in 1997 contained an alpha proton X-ray spectrometer (APXS). In its RBS mode, the spectrometer bombarded samples with alpha particles and determined elemental composition via energy analysis of the backscattered particles. In addition to RBS, the APXS instrument was designed to carry out proton emission and particle-induced X-ray emission (PIXE) experiments. Soil and rock compositions were measured and compared to those from the earlier Viking mission.

21D-3 Laser-Microprobe Mass Spectrometry

Laser-microprobe mass spectrometers are used for the study of solid surfaces. Ablation of the surface is accomplished with a high-power, pulsed laser, usually a Nd-YAG laser. After frequency quadrupling, the Nd-YAG laser can produce 266-nm radiation focused to a spot as small as 0.5 μm. The power density of the radiation within this spot can be as high as 10^{10} to 10^{11} W/cm^2. On ablation of the surface a small fraction of the atoms are ionized. The ions produced are accelerated and then analyzed, usually by time-of-flight mass spectrometry. In some cases laser microprobes have been combined with quadrupole ion traps and with Fourier transform mass spectrometers. Laser-microprobe tandem mass spectrometry is also receiv-

ing current research attention. In addition, the laser microprobe has been used to vaporize samples prior to introduction into inductively-coupled plasmas for either emission or mass spectral analysis.[14]

Laser-microprobe mass spectrometry has an unusually high sensitivity (down to 10^{-20} g), is applicable to both inorganic and organic (including biological) samples, has a spatial resolution of about 1 μm, and produces data at a rapid rate. Some typical applications of laser-microprobe mass spectrometry include determination of Na/K concentration ratios in frog nerve fiber, determination of the calcium distribution in retinas, classification of asbestos and coal mine dusts, determination of fluorine distributions in dental hard tissue, analysis of amino acids, and study of polymer surfaces.[15]

21E SURFACE PHOTON SPECTROSCOPIC METHODS

In this section we discuss methods in which photons provide both the primary beam and the detected beam. The techniques discussed are listed in Table 21-1; namely, *surface plasmon resonance*, *sum frequency generation*, and *ellipsometry*. The electron and ion spectroscopic surface techniques described previously all suffer from one disadvantage: they require an ultra-high vacuum environment and provide no access to buried interfaces. The photon spectroscopic methods described here can all deal with surfaces in contact with liquids and, in some cases, surfaces that are buried under transparent layers.

21E-1 Surface Plasmon Resonance

Surface plasmon waves are surface electromagnetic waves that propagate in the *xy* plane of a metal film when the free electrons interact with photons. An easy way to obtain the resonance condition is to arrange for total internal reflection at an interface, as shown in Figure 21-14. Here, a monochromatic light beam from a laser propagates in a medium of higher refractive index, such as glass. If the radiation strikes an interface to a medium of lower refractive index such as air or water, total internal reflection can occur if the angle of

[14]See, for example, R. E. Russo, X. Mao, and S. S. Mao, *Anal. Chem.*, **2002**, *74*, 70A; B. Hattendorf, C. Latkoczy, and D. Gunther, *Anal. Chem.*, **2003**, *75*, 341A.

[15]For further details, see L. Van Vaeck and R. Gijbels, *Fresenius J. Anal. Chem.*, **1990**, *336*, 743, 755.

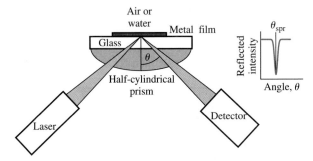

FIGURE 21-14 Surface plasmon resonance. Laser radiation is coupled into the glass substrate coated with a thin metal film by a half-cylindrical prism. If total internal reflection occurs, an evanescent wave is generated in the medium of lower refractive index. This wave can excite surface plasmon waves. When the angle is suitable for surface plasmon resonance, a sharp decrease in the reflected intensity is observed at the detector.

incidence is greater than the critical angle. The radiation is focused and coupled into the interface by a prism coupler or a diffraction grating. With total internal reflection, an *evanescent wave* (see Section 17B-3) is generated in the medium of lower refractive index that decays exponentially with distance from the interface. As we have seen in Chapter 17, the evanescent wave can be absorbed in the less dense medium and the beam attenuated by *attenuated total reflection.*

If the internally reflecting interface is coated with a conducting material, such as a thin metal film, the *p*-polarized component of the evanescent wave may penetrate the metallic layer and excite surface plasmon waves. If the metal is nonmagnetic, such as a gold film, the surface plasmon wave is also *p*-polarized, which creates an enhanced evanescent wave. Because of the penetration of the electric field into the lower-refractive-index medium, the interaction is quite sensitive to the refractive index at the metal film surface. When the angle is suitable for surface plasmon resonance, a sharp decrease in the reflected intensity is observed, as can be seen in Figure 21-14. The resonance condition can be related to the refractive index of the metal film and can be used to measure this quantity and other properties of the surface.

The most interesting aspect of surface plasmon resonance (SPR) is its sensitivity to materials adsorbed onto the metal film and the interactions of these materials, particularly biomolecules.[16] A linear relationship

is often found between the resonant energy and the concentration of biologically relevant materials such as sugars, DNA molecules, and proteins. Because of this sensitivity, SPR has become an important technique for biosensors. Here, biomolecular interactions, such as antibody-antigen binding or enzyme-substrate binding, occur at the sensor surface. Such interactions alter the refractive index and change the SPR angle needed to achieve resonance. The SPR angle, θ_{SPR}, can be monitored as a function of time to give information on the kinetics of binding reactions at the surface. Commercial instrumentation for SPR is available.[17]

Many other potential applications of surface plasmon waves have been envisioned, including miniaturized devices, such as filters, polarizers, and light sources. The field has been termed *molecular plasmonics.*[18]

21E-2 Sum-Frequency Generation

Sum-frequency generation (SFG) is a nonlinear optical technique based on the interaction of two photons at a surface.[19] The result of the wave-mixing interaction is the production of a single photon whose frequency is the sum of the incident frequencies. If the two incident photons are of the same frequency, the technique is called *second-harmonic generation* because the exiting photon has a frequency twice that of the incident photons. Because this is a weak second-order process, intense lasers must be used.

SFG can be applied to solid-liquid, solid-gas, liquid-gas, or liquid-liquid interfaces. The process becomes most efficient when either the incident frequency or the outgoing frequency correspond to an allowed electronic or vibrational transition. Usually, one of the sources is a tunable laser to allow the incident frequency to be varied.

One of the most useful of the possible sum-frequency techniques is vibrational sum-frequency spectroscopy in which one of the incident beams is in the infrared spectral region and the other is in the visible region.[20] Figure 21-15 shows two configurations that can be used. In Figure 21-15a, the arrangement for studying the interface between two immiscible liquids

[16] J. Homola, *Anal. Bioanal. Chem.,* **2003,** *377,* 528.

[17] See R. Mukhopadhyay, *Anal. Chem.,* **2005,** *77,* 313A.

[18] R. P. Van Duyne, *Science,* **2004,** *306,* 985.

[19] See M. E. Pemble, in *Surface Analysis — The Principal Techniques,* J. C. Vickerman, ed., Chichester, UK: Wiley, 1997, Chap. 7.

[20] M. R. Watry, M. G. Brown, and G. L. Richmond, *Appl. Spectrosc.,* **2001,** *55,* 321.

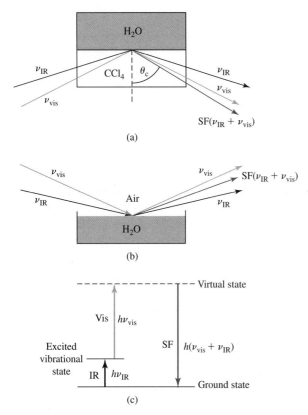

(a)

(b)

(c)

FIGURE 21-15 Sum-frequency generation. In (a) a total internal reflection geometry is shown for studying a liquid-liquid interface. In (b) an external reflection geometry is shown for studying a liquid-gas interface. In both cases an IR photon of frequency ν_{IR} is coincident with a visible photon of frequency ν_{vis}. The sum frequency ($\nu_{IR} + \nu_{vis}$) is generated as seen in (c).

such as water and CCl_4 is shown. Here, a total internal reflection geometry is used. In Figure 21-15b, an external reflection geometry is shown for studying an air-water interface. The sum frequency could also be collected in transmission mode rather than in the reflection modes shown. In both cases, pulsed, high-intensity lasers are used. The most common lasers for vibrational SFG have been Nd-YAG and Ti-sapphire. Various combinations have been used to provide tunable IR radiation and fixed-frequency visible radiation. The pulses must overlap temporally and spatially at the interface. An energy level diagram illustrating the process is shown in Figure 21-15c.

SFG has been applied to several analytical problems. It has been used to study the structure and behavior of surfactants at liquid-liquid interfaces as well as the structure of water at liquid-liquid and liquid-gas

interfaces. The technique has also been used to investigate biological surfactants, such as phospholipid monolayers, at liquid interfaces and to study the adsorption of molecules from the atmosphere on liquid surfaces. The ability to probe buried interfaces and to study the dynamics of processes occurring at these interfaces should make SFG a very useful technique in the future.

21E-3 Ellipsometry

Ellipsometry is a technique that uses polarized light to probe the dielectric properties of samples.[21] It is most commonly applied to the analysis of very thin films on surfaces. In ellipsometry, a polarized incident beam, often from a laser, is reflected from the film, and the reflected light is analyzed to determine a change in the state of polarization. The change in the amplitude and the phase of the reflected light are then related to properties such as film refractive index, absorptivity, optical anisotropy, and thickness.

The basic measurements in ellipsometry involve measuring the reflection coefficients for parallel $R_{\parallel}$ and perpendicularly polarized light $R_{\perp}$ (sometimes called s- and p-polarized light, respectively). The ratio of these values, which is a complex number, gives the elliptical angle Ψ and the phase shift Δ according to

$$\frac{R_{\parallel}}{R_{\perp}} = \tan(\Psi)e^{i\Delta} \qquad (21\text{-}9)$$

The parameters Ψ and Δ can reveal the thickness of the reflecting layer and its optical properties.

Several different types of ellipsometers are available commercially. The earliest type was the null-type ellipsometer in which a circularly polarized incident beam was reflected off the sample surface onto an analyzer. The incident-beam polarization state was chosen by a polarizer and compensator so that linearly polarized light was obtained after reflection. The analyzer was then rotated until it was perpendicular to the polarization axis of the light coming from the sample as indicated by a minimum in the light intensity. Some instruments today still use the null principle, but they are computer controlled and have charge-coupled-device (CCD) cameras as detectors.

Another type of ellipsometer uses a phase modulation technique. In this technique, a rotating quarter-

[21]For more information, see H. G. Tompkins and E. A. Irene, *Handbook of Ellipsometry*, Norwich, NY: William Andrew Publishing, 2005; H. G. Tompkins and W. A. McGahan, *Spectroscopic Ellipsometry and Reflectometry: A User's Guide*, New York: Wiley, 1999.

wave plate or acoustooptic modulator rapidly changes the state of polarization of the incident beam. Characteristics of the reflected light are obtained from an analysis of the modulated detector signal and used to calculate film thickness and other quantities.

Spectroscopic ellipsometry acquires the ellipsometric parameters as a function of wavelength and often angle of incidence. Modern spectroscopic ellipsometers use CCD cameras to collect variable-wavelength data. Many of the more recent instruments also use Fourier transform techniques because of high signal-to-noise ratios and other inherent advantages.

21F ELECTRON-STIMULATED MICROANALYSIS METHODS

Several microanalysis techniques detect the particles emitted after a finely focused beam of electrons strikes the surface of a sample. We discuss here *electron microprobe analysis* and *scanning electron microscopy*.

21F-1 The Electron Microprobe

With the electron microprobe, X-ray emission is stimulated on the surface of the sample by a narrow, focused beam of electrons. The resulting X-ray emission is detected and analyzed with either a wavelength or an energy-dispersive spectrometer.[22]

Instruments

Figure 21-16 is a schematic of an electron microprobe system. The instrument employs three integrated sources of radiation: an electron beam, a visible light beam, and an X-ray beam. In addition, a vacuum system is required that provides a pressure of less than 10^{-5} torr as is a wavelength- or an energy-dispersive X-ray spectrometer (a wavelength-dispersive system is shown in Figure 21-16). The electron beam is produced by a heated tungsten cathode and an accelerating anode (not shown). Two electromagnetic lenses focus the beam on the specimen; the diameter of the beam is between 0.1 and 1 μm. An associated optical microscope is used to locate the area to be bombarded. Finally, the

---

[22]For a detailed discussion of this method, see S. J. B. Reed, *Electron Microprobe Analysis and Scanning Electron Microscopy in Geology*, 2nd ed., Cambridge, UK: Cambridge University Press, 2005; S. J. B. Reed, *Electron Microprobe Analysis*, 2nd ed., Cambridge, UK: Cambridge University Press, 1993; K. F. J. Heinrich, *Electron Beam X-Ray Microanalysis*, New York: Van Nostrand, 1981; L. S. Birks, *Electron Probe Microanalysis*, 2nd ed., New York: Wiley-Interscience, 1971. For a review of the use of the scanning electron microprobe for elemental analysis of surfaces, see D. E. Newbury et al., *Anal. Chem.*, **1990**, *62*, 1159A, 1245A.

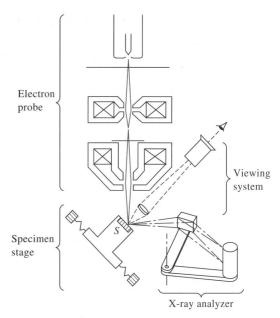

FIGURE 21-16 Schematic view of an electron microprobe. (From Wittr, *Treatise on Analytical Chemistry*, Kolthoff and Elving, Eds., Vol. 5, Part 1, p. 317B, New York: Interscience/Wiley, 1964.)

X-ray fluorescence photons produced by the electron beam are collimated, dispersed by a single crystal, and detected by a gas-filled transducer. Considerable design effort is required to arrange the three systems spatially so that they do not interfere with one another.

As shown in Figure 21-16, the specimen stage is provided with a mechanism to move the sample in two mutually perpendicular directions and rotate it as well, which permits the surface to be scanned.

Applications

The electron microprobe provides a wealth of information about the physical and chemical nature of surfaces. It has had important applications to phase studies in metallurgy and ceramics, the investigation of grain boundaries in alloys, the measurement of diffusion rates of impurities in semiconductors, the determination of occluded species in crystals, and the study of the active sites of heterogeneous catalysts. In all of these applications, both qualitative and quantitative information about surfaces is obtained.

Figure 21-17 illustrates the use of the electron microprobe for the analysis of an α-cohenite (Fe_3C) particle in a lunar rock. The data were obtained by a linear scan of the particle observed visually on the surface and by measurement of the intensity of the characteristic emission line for each of four elements.

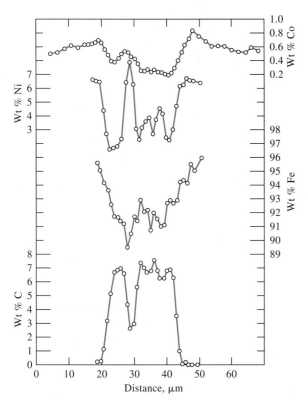

FIGURE 21-17 Scanning electron microprobe output across the surface of an α-cohenite particle in a lunar rock.

21F-2 Scanning Electron Microscopy

The classical method for obtaining detailed information about the physical nature of surfaces was optical microscopy, which is still an important technique. The resolution of optical microscopy is limited by diffraction effects, however, to about the wavelength of light. Much higher-resolution information is obtained by using one of the electron microscopic methods. The two most important methods are *scanning electron microscopy* and *transmission electron microscopy*.[23] The so-called *scanning probe microscopy* methods, featuring *scanning tunneling microscopy* and *atomic force microscopy*, have also become important surface characterization methods and are discussed in Section 21G.

Although scanning and transmission electron microscopy methods have many similarities, scanning electron microscopy can be thought of as providing images of external morphology, similar to those accessed by the human eye. In contrast, transmission electron microscopy probes the internal structure of solids and gives us information about microstructural detail that is not familiar to the eye. We will restrict our discussion here to scanning electron microscope (SEM) methods.

In obtaining an SEM image, a finely focused beam of electrons impinges on the surface of the solid sample. In analog instruments, the beam of electrons is scanned across the sample in a *raster scan* by scan coils. The resulting raster scanning pattern is similar to that used in the cathode-ray tube (CRT) of a television set in which the electron beam is (1) swept across the surface linearly in the x direction, (2) returned to its starting position, and (3) shifted downward in the y direction by a standard increment. This process is repeated until a desired area of the surface has been scanned. In more recent instruments, the same effect is achieved by digital control over the beam position on the sample. In either the analog scanning case or in digital systems, a signal is received above the surface (the z direction) and stored in a computer where it is ultimately converted to an image. Several types of signals are produced from a surface in this process, including backscattered, secondary, and Auger electrons; X-ray fluorescence photons; and other photons of various energies. All of these processes have been used for surface studies. In SEM instruments, backscattered and secondary electrons are detected and used to construct the image. For chemical analysis purposes, many modern SEMs also have X-ray detectors that allow qualitative and quantitative determinations to be made by means of X-ray fluorescence. As discussed in the preceding section, electron microprobe analyzers are instruments specifically made for X-ray analysis.

Instrumentation

Figure 21-18 shows a schematic diagram of an SEM with a microprobe attachment.[24] Both an electron detector and an X-ray detector are present. For simplicity, an analog scanning system is illustrated.

Electron Gun and Optics. The electron source is usually a tungsten filament source, although field emission guns are also employed for high-resolution work. The electrons are accelerated to an energy between 1 and

[23] For additional information, see S. J. B. Reed, *Electron Microprobe Analysis and Scanning Electron Microscopy in Geology*, 2nd ed., Cambridge, UK: Cambridge University Press, 2005; P. J. Goodhew, J. Humphreys, and R. Beanland, *Electron Microscopy and Analysis*, 3rd ed., London: Taylor & Francis, 2001; L. Reimer, *Scanning Electron Microscopy*, Berlin: Springer-Verlag, 1998.

[24] For a review of SEMs in the personal computer era, see E. Zubritsky, *Anal. Chem.*, **2002**, *74*, 215A.

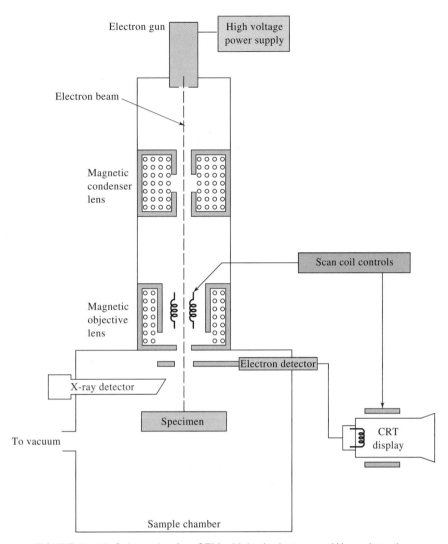

FIGURE 21-18 Schematic of an SEM with both electron and X-ray detection.

30 keV. The magnetic condenser and objective lens systems reduce the spot size to a diameter of 2–10 nm when it reaches the specimen. The condenser lens system, which consists of one or more lenses, is responsible for the throughput of the electron beam reaching the objective lens; the objective lens determines the size of the beam hitting the sample surface.

Scanning with an SEM is accomplished by the two pairs of electromagnetic coils located within the objective lens (see Figure 21-18); one pair deflects the beam in the x direction across the sample, and the other pair deflects it in the y direction. Scanning is controlled by applying an electrical signal to one pair of scan coils, such that the electron beam strikes the sample to one side of the center axis of the lens system. By varying

the electrical signal to this pair of coils (that is, the x coils) as a function of time, the electron beam is moved in a straight line across the sample and then returned to its original position. After completion of the line scan, the other set of coils (y coils in this case) is used to deflect the beam slightly, and the scanning of the beam using the x coils is repeated. Thus, by rapidly moving the beam, the entire sample surface can be irradiated with the electron beam. The signals to the scan coils can be either analog or digital. Digital scanning has the advantage of very reproducible movement and location of the electron beam. The signal from the sample can be encoded and stored in digital form along with digital representations of the x and y positions of the beam.

In analog SEMs, the signals that drive the electron beam in the x and y directions also drive the horizontal and vertical scans of a CRT. The image of the sample is produced by using the output of a detector to control the intensity of the spot on the CRT. Thus, this method of scanning produces a map of the sample in which there is a one-to-one correlation between the signal produced at a particular location on the sample surface and a corresponding point on the CRT display. The magnification (M) achievable in the SEM image is given by

$$M = W/w \qquad (21\text{-}10)$$

where W is the width of the CRT display and w is the width of a single line scan across the sample. Because W is a constant, increased magnification is achieved by decreasing w. For example, if the electron beam is made to scan a raster 10 µm × 10 µm on the sample and the image is displayed on a CRT screen 100 mm × 100 mm, the linear magnification will be 10,000×. The inverse relationship between magnification and the width of the scan across the sample implies that a beam of electrons focused to an infinitely small point could provide infinite magnification. A variety of other factors, however, limit the magnification achievable to a range from about 10× to 100,000×.

Samples and Sample Holders. Sample chambers are designed for rapid changing of samples. Large-capacity vacuum pumps are used to hasten the switch from ambient pressure to ~10^{-6} torr or less for conventional SEMs. The sample holder, or stage, in most instruments is capable of holding samples many centimeters on an edge. Furthermore, the stage can be moved in the x, y, and z directions, and it can be rotated about each axis. As a consequence, the surfaces of most samples can be viewed from almost any perspective. Environmental SEMs, discussed later, use much higher pressures in the sample chamber and allow for variations in temperature and gas composition.

Samples that conduct electricity are easiest to study, because the unimpeded flow of electrons to ground minimizes artifacts associated with the buildup of charge. In addition, samples that are good conductors of electricity are usually also good conductors of heat, which minimizes the likelihood of their thermal degradation. Unfortunately, most biological specimens and most mineral samples do not conduct. A variety of techniques have been developed for obtaining SEM images of nonconducting samples, but the most common approaches involve coating the surface of the sample with

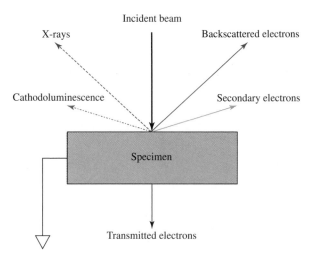

FIGURE 21-19 Pictorial of some of the signals generated with an SEM.

a thin (~10 nm) metallic film produced by sputtering or by vacuum evaporation. Regardless of the method of producing a conductive coating, a delicate balance must be struck between the thinnest uniform coating achievable and an excessively thick coating that obscures surface details. Coating may also interfere with other detection modes (e.g., X-ray emission). The environmental SEM discussed later can be directly used with nonconducting specimens.

Examining nonconducting materials, particularly polymers and biological materials, may present other difficulties, such as thermal degradation, radiation damage, and sample volatility in the high vacuum.

Electron-Beam Interactions. The versatility of the SEM and electron microprobe for the study of solids arises from the wide variety of signals generated when the electron beam interacts with the solid. Figure 21-19 illustrates the signals that can result. We consider just three of these signals: backscattered electrons, secondary electrons, and X-ray emission. The interactions of a solid with an electron beam can be divided into two categories: *elastic* interactions that affect the trajectories of the electrons in the beam without altering their energies significantly and *inelastic* interactions, which result in transfer of part or all of the energy of the electrons to the solid. The excited solid then emits secondary electrons, Auger electrons, X-rays, and sometimes longer-wavelength photons.

When an electron collides elastically with an atom, the direction of the electron changes, but the speed of

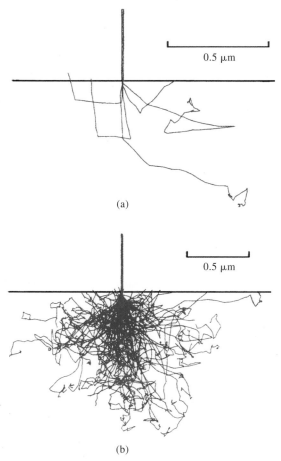

FIGURE 21-20 Simulation of electron trajectories showing the scattering volume of 20-keV electrons in an iron sample. (a) 5 electrons; (b) 100 electrons. (From J. I. Goldstein et al., *Scanning Electron Microscopy and X-Ray Microanalysis*, New York: Plenum Press, 1981, p. 62. With permission.)

the electron is virtually unaffected, so that the kinetic energy remains essentially constant. The angle of deflection for any given collision is random and can vary from 0° to 180°. Figure 21-20 is a computer simulation of the random behavior of 5 electrons and 100 electrons when they enter a solid normal to the surface. The energy of the beam is assumed to be 20 keV, which is typical. Note that such a beam penetrates to a depth of 1.5 μm or more. Some of the electrons eventually lose energy by inelastic collisions and remain in the solid; the majority, however, undergo numerous collisions and, as a result, eventually exit from the surface as backscattered electrons. It is important to note that the beam of backscattered electrons has a much larger diameter than the incident beam — that is, for a

5-nm incident beam, the backscattered beam may have a diameter of several micrometers. The diameter of the backscattered beam is one of the factors limiting the resolution of an electron microscope. Backscattered electrons have a broad energy spread, ranging from 50 eV up to the energy of the incident beam.

When the surface of a solid is bombarded with an electron beam having an energy of several keV, electrons having energies of 50 eV or less are emitted from the surface along with the backscattered electrons. The number of these *secondary electrons* is generally one half to one fifth or less the number of backscattered electrons. Secondary electrons are produced as a result of interactions between the energetic beam electrons and weakly bound conduction electrons in the solid, which leads to ejection of the conduction band electrons with a few electron volts of energy. Secondary electrons are produced from a depth of only 50 to 500 Å and exit in a beam that is slightly larger in diameter than the incident beam. Secondary electrons can be prevented from reaching the detector by applying a small negative bias to the transducer housing.

X-ray photons are yet a third product of electron bombardment of a solid. Both characteristic line spectra and an X-ray continuum are produced. This radiation serves as the basis for the electron microprobe for X-ray fluorescence analysis of SEM images.

The region the electrons penetrate is known as the *interaction volume*. Even though radiation is generated within this volume, it will not be detected unless it escapes from the specimen. X-rays are not easily absorbed and so most escape. For X-rays, the *sampling volume*, which is the volume of material contributing to the X-ray signal, is on the order of the interaction volume as shown in Figure 21-21. The backscattered electrons will not escape if they have penetrated more than a fraction of a micrometer. Thus, the backscattered signal originates from a much smaller volume. The secondary electron signal arises from a region that is on the order of the diameter of the incident electron beam. For this reason, secondary electron signals are capable of giving much higher spatial resolution than the other signals, and they are the most widely used signals in the SEM system.

Transducers. Secondary electrons are most often detected by a scintillator-photomultiplier system, called the Everhart-Thornley detector and illustrated in Figure 21-22. The secondary electrons strike a scintillator that then emits light. The emitted radiation is carried

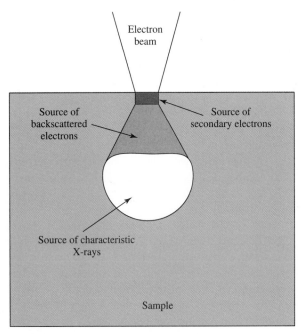

FIGURE 21-21 The interaction volume and the volumes from which each type of SEM signal arises.

by a light pipe to a photomultiplier tube where it is converted into pulses of electrons. These pulses are then used to control the brightness of the electron beam in a CRT.

Because the energy of the secondary electrons is too low (<50 eV) to excite the scintillator directly, the electrons are first accelerated. Acceleration is accomplished by applying a bias voltage of approximately +10 keV to a thin film of aluminum covering the scintillator. A positively biased metal collector grid sur-

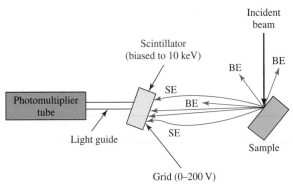

FIGURE 21-22 Diagram of Everhart-Thornley secondary electron detector. Paths of secondary electrons (SE) and backscattered electrons (BE) are shown. The scintillator is a phosphor that emits light when struck by energetic particles such as electrons, gamma rays, or radioactive particles.

rounds the scintillator and prevents the high voltage from affecting the incident electron beam. The grid also improves the collection efficiency by attracting secondary electrons, including those not initially moving toward the detector.

A large-area modification of the scintillation detector is used to detect backscattered electrons. This design maximizes the solid angle of collection. No biased grid is needed because of the high energies of the backscattered electrons. Semiconductor detectors are also widely employed for backscattered electrons. When a high-energy electron strikes the detector, electron-hole pairs are produced that create a photocurrent. A semiconductor detector is small enough to be placed adjacent to the sample, which leads to high collection efficiency. The main disadvantage compared to scintillators is the relatively slow response time.

The X-ray analysis in most SEMs is the energy-dispersive analyzer using a semiconductor detector, such as a *lithium-drifted silicon*, Si(Li), or *lithium-drifted germanium*, Ge(Li), detector, which were discussed in Section 12B-4. Wavelength-dispersive systems have also been used in electron microprobe analyses.

Data Handling and Processing. Data handling and processing depend on whether the SEM is an analog type or a newer digital microscope. With an analog microscope, two image monitors are used. With the first, the operator views the scanned image to identify features of interest and to focus and optimize the system. Observation is usually done at scan rates of fifty frames per second so that the image appears as a television image. The second monitor is a high-resolution CRT for photographic recording. The scan rate is usually slowed to obtain a high-quality image. One frame may take as long as 50 to 100 s. Hence, the requirements for viewing the image and obtaining a photographic record often conflict.

With digital SEMs, the beam stays on each point on the specimen for a predetermined period instead of the raster scanning of the analog system. The image is then constructed by recording each pixel on a *framestore element*.[25] The framestore image is persistent, in contrast to a CRT image. Hence, the operator can view the image without using TV scan rates. With a framestore

---

[25]Some CCD cameras are subdivided into two areas, one to capture the image and one to store frames of the image. The framestore region is protected against photons and only stores information transferred from the image region. Alternatively, separate storage devices can be used as framestore systems.

system, it is possible to employ *frame averaging*, in which the data from successive scans are averaged for each pixel. A permanent record of the image is obtained by saving the framestore contents as a bitmap image. Images can be treated in a variety of ways to provide contrast enhancement, inversion, mixing, subtraction, and color coding.

The Environmental SEM

The conventional SEM operates with a vacuum of 10^{-6} torr or less. Samples must be clean, dry, and electrically conductive. Samples that have volatile components must be pretreated. In the mid-1980s, the so-called *environmental SEM* (ESEM) was developed. The ESEM maintains a high vacuum in the electron gun and microscope column but allows the sample to be placed in a higher-pressure region (1–50 torr). The sample environment can be varied by varying the pressure, temperature, and gas composition. The ESEM typically has three chambers: the gun, the microscope column, and the sample chamber. These regions are separated by small apertures, often corresponding to the electron beam apertures. Each region has its own pumping system.

Both secondary and backscattered electrons may be detected with an ESEM. The Everhart-Thornley detector shown in Figure 21-22 cannot be used because the high bias voltage of the scintillator would cause electrical breakdown at high pressures. Instead, gas-phase secondary electron detectors, which make use of cascade amplification, are employed. These not only enhance the secondary electron signal but also produce positive ions, which are attracted to the insulated specimen surface and suppress charging artifacts. Large-area scintillation detectors can be used to detect backscattered electrons as with conventional SEM systems.

The ESEM allows samples to be observed in their natural states without the extensive modifications or preparations associated with a conventional SEM. Wet, dirty, oily, and nonconductive samples can be examined. Conductive coatings that mask valuable information are not needed. The environmental sample chamber becomes an additional tool that allows interactions between the sample and the environment to be studied by electron microscopy.[26]

The only drawback to ESEM systems compared to conventional SEM instruments is a small loss of resolution due to elastic collisions between the electrons and gas molecules at the higher pressures. However, these collisions help to dissipate any accumulated electrical charges, which aids in the examination of nonconductive samples.

Applications

Scanning electron microscopy provides morphological and topographic information about a wide variety of solid surfaces. Several representative examples shown in Figure 21-23 illustrate the kind of information obtained by this technique.

21G SCANNING PROBE MICROSCOPES

Scanning probe microscopes (SPMs) are capable of resolving details of surfaces down to the atomic level. The first example of this type of microscope, the scanning tunneling microscope, was described in 1982. Only 4 years later, in 1986, its inventors, G. Binnig and H. Roher, were awarded the Nobel Prize in Physics for their work. Currently, the primary use of SPMs is for measuring surface topography of samples.

Unlike optical and electron microscopes, SPMs reveal details not only on the lateral x- and y-axes of a sample but also on the z-axis, which is perpendicular to the surface. Typically, the resolution of SPMs is about 2 nm (20 Å) in the x and y directions, but with ideal samples and the best instruments it can be as low as 0.1 nm (1 Å). Resolution in the z dimension is generally better than 0.1 nm. For comparison, the resolution of a typical electron microscope is about 5 nm.

We shall discuss here two types of scanning probe microscopes that are the most widely used and available from several commercial sources: the *scanning tunneling microscope* (STM) and the *atomic force microscope* (AFM). Both are based on scanning the surface of the sample in an xy raster pattern with a very sharp tip that moves up and down along the z-axis as the surface topography changes. This movement is measured and translated by a computer into an image of the surface topography. This image often shows details on an atomic-size scale.[27] A third type of SPM, the *scanning electrochemical microscope*, is described in Chapter 25 in our discussion of electrochemistry.

---

[26]For a discussion of SEM sample chambers for wet samples, see R. Mukhopadhyay, *Anal. Chem.*, **2004**, *76*, 293A.

[27]For references on scanning probe techniques, see K. S. Birdi, *Scanning Probe Microscopes: Applications in Science and Technology*, Boca Raton, FL: CRC Press, 2003; *Scanning Probe Microscopy and Spectroscopy*, D. Bonnell, ed., New York: Wiley-VCH, 2001; G. J. Leggett, in *Surface Analysis — The Principal Techniques*, J. C. Vickerman, ed., Chichester, UK: Wiley, 1997, Chap. 9; R. Weisendanger, *Scanning Probe Microanalysis and Spectroscopy*, New York: Cambridge University Press, 1994.

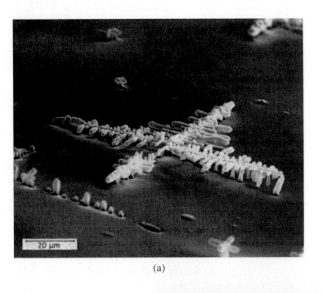

(a)

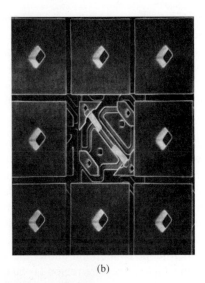

(b)

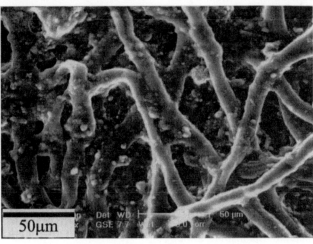

(c)

FIGURE 21-23 Representative examples of SEM photomicrographs. (a) A nickel alloy surface. Note dendritic particles of silica on surface. (From P. J. Goodhew, J. Humphreys, and R. Beanland, *Electron Microscopy and Analysis*, 3rd ed., London: Taylor and Francis, 2001, p. 123.) (b) A nine-micromirror array used to develop digital micromirror spectrometer for atomic spectroscopy. Mirrors are 16 × 16 μm on 17-μm centers. Removing the center mirror reveals underlying components. (From J. D. Batchelor and B. T. Jones, *Anal. Chem.*, **1998**, *70*, 4907. Copyright 1998 American Chemical Society.) (c) An algae mat taken in fully wet condition with an ESEM instrument with a gas-phase secondary-electron detector with cascade amplification. No sample preparation was needed. (From P. J. Goodhew, J. Humphreys, and R. Beanland, *Electron Microscopy and Analysis*, 3rd ed., London: Taylor and Francis, 2001, p. 167.)

21G-1 The Scanning Tunneling Microscope

The Binnig-Roher microscope, which earned its inventors the Nobel Prize, was a scanning tunneling microscope. This device was found to be capable of resolving features on an atomic scale on the surface of a conducting solid surface.[28] These instruments are now available from several instrument manufacturers and are used on a routine basis in hundreds of laboratories

---

[28]G. Binnig, H. Roher, C. Gerber, and E. Weibel, *Phys. Rev. Lett.*, **1982**, *49*, 57.

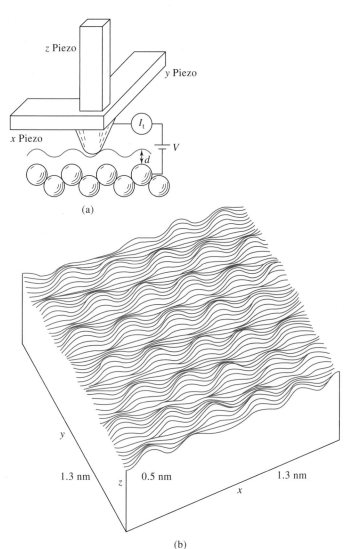

(a)

(b)

FIGURE 21-24 (a) Schematic view of STM tip scanning a sample in the x direction. Blue line is the path of the tip over the individual carbon atoms shown as regularly arranged circles. (b) Contour map of sample surface. (From P. K. Hansma et al., *Science*, **1988**, *242*, 209. With permission.)

throughout the world. Their main disadvantage is the requirement that the surface being examined conduct electricity. The atomic force microscope, discussed in the next section, does not suffer from this limitation.

Principle of STM

In a scanning tunneling microscope, the surface of a sample is scanned in a raster pattern by a very fine metallic tip. The STM is based on the principle of quantum mechanical tunneling, which occurs when a small bias voltage (mV to 3 V) is applied between a sharp tip and a conducting sample and the tip is within a few nanometers of the surface. The magnitude of the tun-

 Simulation: Learn more about **scanning probe techniques**.

neling current I_t is given by

$$I_t = Ve^{-Cd} \qquad (21\text{-}11)$$

where V is the bias voltage, C is a constant, and d is the distance between the probe and the surface. The basic STM is illustrated in Figure 21-24a. The STM can, in principle, be operated in two modes. In the *constant-height mode*, the tip position is kept constant in the z direction while the tunneling current I_t is monitored. In the *constant-current mode*, the tunneling current I_t is kept constant while the z position of the tip changes to keep d constant, as shown in Figure 21-24a. In this case the z position of the tip is monitored. Most modern STMs operate in the constant-current mode. The up-and-down motion of the tip then reflects the topography of the surface. The round balls in the figure represent individual carbon atoms in a sample of pyrolytic graphite. For clarity, the tip is shown as a rounded cone, and the path of the tip during a scan in the x direction is shown by the blue line.

Equation 21-11 reveals that the tunneling current decreases exponentially with the separation between the tip and the sample. This rapid decrease in current with distance causes the tunneling current to be significant only for very small tip-sample separation and is responsible for the high resolution achieved in the z direction.

Sample Scanners

In early scanning tunneling microscopes, three piezo-electric transducers arranged orthogonally controlled the three-dimensional motion of the tip, as shown in Figure 21-24a.[29] Application of a dc voltage along its length varies the length of each transducer, thus making it possible to move the tip in a three-dimensional pattern (see Section 1C-4 for a discussion of the composition and properties of piezoelectric transducers). Depending on the composition of the piezoelectric ceramic material and the dimensions of the transducer, the degree of expansion or contraction can be made as small as 1 nm for every volt applied, which provides remarkably sensitive control of the tip position.

Modern scanning microscopes no longer use the tripod design shown in Figure 21-24a but instead are based on a hollow-tube piezoelectric device, shown in Figure 21-25. The outer surface of the tube, which is typically between 12 and 24 mm long and 6 and 12 mm in diameter, is coated with a thin layer of metal. This conducting layer is divided into four equal segments by vertical strips containing no metal coating. Application of voltages to opposite strips of metal bends the tube in the x and y directions as indicated. Likewise, application of a voltage along the interior axis of the cylinder lengthens or shortens the tube in the z direction. A tip in the center of one end of the cylinder can then be made to move in three dimensions by application of suitable voltages.

Scanning microscopes typically employ scanners with lateral scan ranges of a few nanometers to more than 100 μm. Height differences from less than 0.1 nm to perhaps 10 μm are encountered. The maximum scan

[29]For descriptions of sample scanners, see R. S. Howland, in *Atomic Force/Scanning Tunneling Microscopy*, S. H. Cohen, M. T. Bray, and M. L. Lightbody, eds., pp. 347–58, New York: Plenum Press, 1994; S. Park and R. C. Barrett, in *Scanning Tunneling Microscopy*, J. A. Stroscio and W. J. Kaiser, eds., pp. 51–58, New York: Academic Press, 1993.

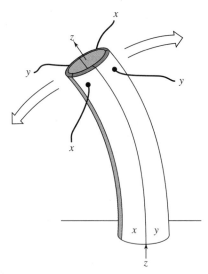

FIGURE 21-25 Piezoelectric scanner of the segmented tube design.

size is determined by the length, diameter, and wall thickness of the cylinder as well as the strain coefficient of its ceramic material.

Computer Interface

Computer control is an essential part of all scanning tunneling microscopes. Most commercial STMs use software and digital-to-analog converters to generate the xy raster scan. The computer processes the voltages applied to the x, y, and z piezoelectric elements and converts them into contour maps such as that shown in Figure 21-24b. With more sophisticated instruments the images may take the form of gray-level images or pseudocolored elevation maps.

The blue line in Figure 21-24a shows the tip path as it scans in the x direction over the surface of a sample of highly oriented pyrolytic graphite. After a scan in the x direction is completed, the tip is returned to its original position and then moved down one line by application of an appropriate voltage to the y piezoelectric transducer. This process is repeated until a plot of the entire sample is obtained as shown in Figure 21-24b. Here, a series of contour lines show clearly the position of the electron cloud of each carbon atom on the surface of the sample.

Because the output signal from the detector is so sensitive to the distance between the sample and the tip, differences in distances along a given contour in the figure are revealed to 1/100 of an atomic dimension. The lateral resolution along a given contour depends on the radius of curvature of the tip. When this radius is that of a single atom, as it usually is, atomic resolution is observed.

Tips

The tunneling tip is a crucial component of the STM. The best images are obtained when tunneling is limited to a single metal atom at the tip end. Fortunately, with a little care, it is possible to construct this type of tip by cutting platinum-iridium wires or by electrochemical etching of tungsten metal. The reason a single-atom tip is not as difficult to prepare as might be expected is because of the exponential increase in tunneling current with decreasing gap (Equation 21-11). Thus, typically, the tunneling current increases by a factor of 10 when the gap distance decreases by 0.1 nm (1 Å). That is, if there is one atom at the tip apex closer to the sample surface by 0.1 nm than any other, nearly all of the current will flow through that atom to the sample, and atomic resolution is achieved.

21G-2 The Atomic Force Microscope

The atomic force microscope, which was invented in 1986,[30] permits resolution of individual atoms on both conducting and *insulating* surfaces. In this procedure, a flexible force-sensing cantilever stylus is scanned in a raster pattern over the surface of the sample. The force acting between the cantilever and the sample surface causes minute deflections of the cantilever, which are detected by optical means. As in the STM, the motion of the tip, or sometimes the sample, is achieved with a piezoelectric tube. During a scan, the force on the tip is held constant by the up-and-down motion of the tip, which then provides the topographic information. The advantage of the AFM is that it is applicable to non-conducting samples.[31]

Figure 21-26 shows schematically the most common method of detecting the deflection of the cantilever holding the tip. A laser beam is reflected off a spot on the cantilever to a segmented photodiode that detects the motion of the probe. The output from the photodiode then controls the force applied to the tip so that

[30] G. Binnig, C. F. Quate, and C. Gerber, *Phys. Rev. Lett.*, **1986**, *56*, 930.
[31] For books treating atomic force microscopy, see note 27. For reviews on atomic force microscopy, see F. J. Giessibl, *Rev. Mod. Phys.*, **2003**, *75*, 949; H. Takano, J. R. Kenseth, S. Wong, J. C. O'Brien, and M. D. Porter, *Chem. Rev.*, **1999**, *99*, 2845; D. R. Louder and B. A. Parkinson, *Anal. Chem.*, **1995**, *67*, 297A. For biological applications, see *Atomic Force Microscopy: Biomedical Methods and Applications*, P. C. Braga, D. Ricci, eds., Totowa, NJ: Humana Press, 2004.

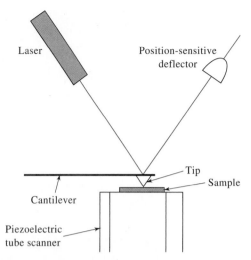

FIGURE 21-26 Side view of an optical beam deflection detector. Typically, deflections of 0.1 nm or less can be detected as the tip scans the sample surface.

it remains constant. In other words, the optical control system is analogous to the tunneling-current control system in the STM.

Figure 21-27 shows a common design of an AFM.[32] The movement system is a tubular piezoelectric device that moves the sample in x, y, and z directions under the tip. The signal from the laser beam detector is then fed

[32] For reviews of AFM instruments, see C. M. Harris, *Anal. Chem.*, **2001**, *73*, 627A; A. Newman, *Anal. Chem.*, **1996**, *68*, 267A.

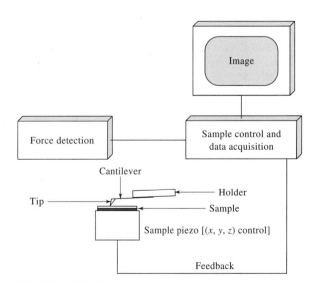

FIGURE 21-27 Typical design of an atomic force microscope. (From D. R. Louder and B. A. Parkinson, *Anal. Chem.*, **1995**, *67*, 298A. With permission.)

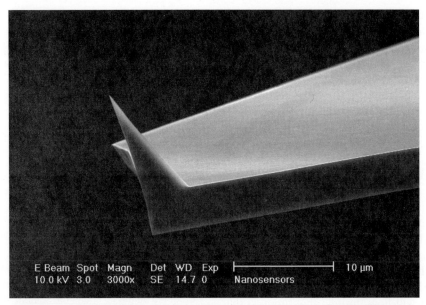

FIGURE 21-28 Micrograph of micromachined silicon cantilever and tip. (Courtesy of NanoWorld AG, Neuchatel, Switzerland.)

back into the sample piezoelectric transducer, which causes the sample to move up and down, maintaining a constant force between the tip and the sample.

The Tip and Cantilever

The performance of an atomic force microscope critically depends on the physical characteristics of the cantilever and tip. In early AFMs, cantilevers were cut from metal foil and tips were made from crushed diamond particles. The tips were painstakingly manually glued to the cantilevers. Currently, this crude method has been replaced by semiconductor mass production methods in which integral cantilever-tip assemblies are produced by etching single chips of silicon, silicon oxide, or silicon nitride. The most common cantilever-tip assemblies in use today are micromachined from monolithic silicon as shown in Figure 21-28. As can be seen, the cantilevers and tips are remarkably small (ideally, a single atom at the tip apex).

AFM Modes

Three modes are commonly used in AFMs: *contact mode*, *noncontact mode*, and *tapping mode*. Contact mode is the most common. Here, the tip is in constant contact with the surface of the sample. The majority of AFM measurements are made under ambient pressure conditions or in liquids, and surface tension forces from adsorbed gases or from the liquid layer may pull the tip downward. These forces, although quite small, may be large enough to damage the sample surface and distort the image. This problem is particularly bothersome with softer materials, such as biological samples, polymers, and even some seemingly hard materials, such as silicon wafers. In addition, many samples can trap electrostatic charges, which can contribute to an attractive force between the probe and the sample. This can lead to additional frictional forces as the tip moves over the sample, which dull the tip and damage the sample.

The problem of surface damage can be largely overcome by a process in which the tip periodically contacts the surface for only a brief time and then is removed from the surface. In this tapping-mode operation, the cantilever oscillates at a frequency of a few hundred kilohertz. The oscillation is driven by a constant driving force and the amplitude is monitored continuously. The cantilever is positioned so that the tip touches the surface at only the bottom of each oscilla-

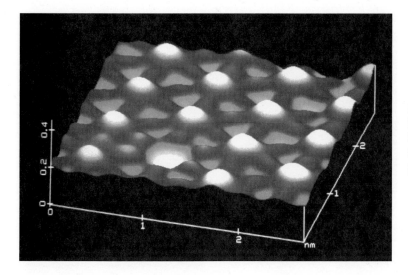

FIGURE 21-29 STM scan of iodine atoms in a 3-nm × 3-nm array adsorbed on platinum. Note the missing iodine atom in the bottom center of the image. (Image provided by Veeco Instruments.)

tion cycle. This technique has been used successfully to image a wide variety of materials that have been difficult or impossible to image by the ordinary contact mode.

The least common mode of operation is the noncontact mode, in which the tip hovers a few nanometers above the sample surface. Attractive van der Waals forces between the tip and the sample are detected as the tip scans over the surface. These forces are substantially weaker than those detected in contact-mode AFMs. Hence, the tip is oscillated and ac detection techniques are used to recover the small signals.

Multimode Scanning Probe Microscopes

Many commercial SPMs are capable of multimode operation. They can be used as AFMs in contact, tapping, and noncontact modes and as STMs. Other operational modes, such as lateral force mode, torsional mode, and magnetic force mode may also be possible, depending on the manufacturer and model.

Some Typical Applications of Scanning Probe Microscopes

SPMs have allowed scientists and engineers to see surface structures with unprecedented resolution. As a consequence, SPMs have found widespread use in a number of fields. For example, in the semiconductor field they have been used for the characterization of sil-

icon surfaces and defects on these surfaces, as well as imaging of magnetic domains on magnetic materials; in biotechnology, imaging of such materials as DNA, chromatin, protein-enzyme interactions, membrane viruses, and so on. An advantage of AFM is that it permits underwater imaging of biological samples under conditions that cause less distortion of the image. For softer samples, distortion often arises because a microdrop of water forms at the tip-surface interface. Capillary forces from this drop exceed the normal force between the tip and sample and obscure surface details. If the sample is in water, water is above the tip as well as below and capillary forces in the up and down directions cancel.

An interesting example of the potential of STM measurements is illustrated in Figure 21-29. The image was produced from the surface of a sample of iodine atoms absorbed on platinum. The hexagonal pattern of absorbed iodine atoms is interrupted by a defect where an iodine atom is absent, which appears as a divot in the bottom center of the image. The scan that is shown represents a 3 nm × 3 nm area of the platinum surface. This example clearly shows how an STM may be used to reveal structures of solid surfaces at the atomic level.

Figure 21-30 is an image of two intertwined DNA molecules on a mica surface obtained by a tapping-mode AFM scan. Such measurements allow biochemists to study the structure of DNA and other biomolecules with relative ease.

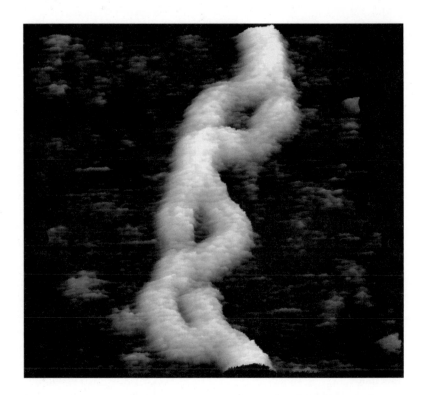

FIGURE 21-30 Two double-strand DNA molecules imaged on mica, showing the ability of an AFM to resolve overlapping molecules. (Courtesy of W. B. Stine, University of California, San Diego. With permission.)

In recent years it has become possible to chemically modify probes to increase selectivity and applicability to chemical problems. Figure 21-31 shows an interesting application of an AFM. In this experiment, a gold-coated Si_3N_4 tip is immersed for a time in a solution containing an organosulfur compound such as 11-thioundecanoic acid. The result is that the ends of the acid molecules opposite the carboxylic acid group become covalently attached to the probe tip, so that the tip is then effectively coated with carboxylic acid

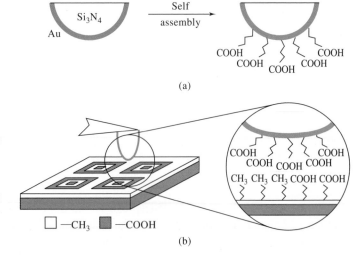

FIGURE 21-31 Detection of a functional group by atomic force microscopy. (a) Carboxylic acid groups are chemically attached to a gold-coated AFM tip. (b) Schematic views of the experiment. (Inset) Interaction between the gold tip coated with —COOH groups and the sample coated with both —CH$_3$ and —COOH groups. (Adapted from C. D. Frisbie, L. F. Rozsnyai, A. Noy, M. S. Wrighton, and C. M. Lieber, *Science*, **1994**, *265*, 2072. With permission.)

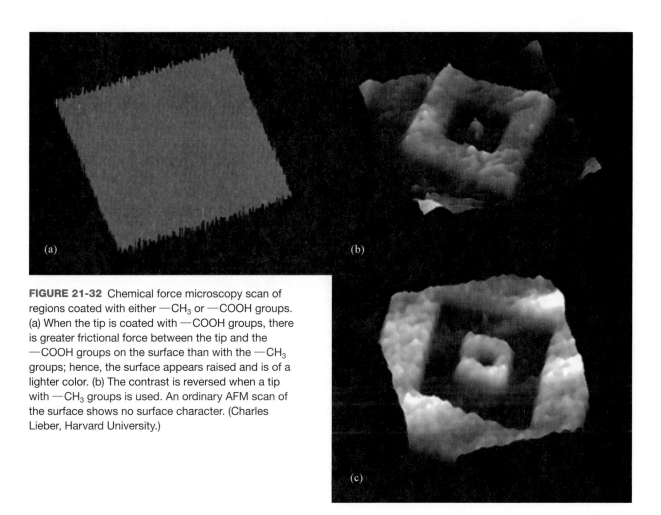

FIGURE 21-32 Chemical force microscopy scan of regions coated with either $-CH_3$ or $-COOH$ groups. (a) When the tip is coated with $-COOH$ groups, there is greater frictional force between the tip and the $-COOH$ groups on the surface than with the $-CH_3$ groups; hence, the surface appears raised and is of a lighter color. (b) The contrast is reversed when a tip with $-CH_3$ groups is used. An ordinary AFM scan of the surface shows no surface character. (Charles Lieber, Harvard University.)

groups as shown in Figure 21-31a. When the probe tip is then scanned across a surface having various attached organic functional groups, as shown in Figure 21-31b, the differences in frictional forces between the acid groups on the tip and other functional groups on the surface of the sample result in an image that is a map of the positions of surface functional groups. This technique is termed *chemical force microscopy* (Figure 21-32). Studies such as these illustrate that scanning probe microscopy provides quite

specific qualitative analytical information as well as information on the spatial arrangement of analytes on surfaces. Because of this, atomic force microscopy and its various modifications, including the inverted design where the sample is placed on the end of the cantilever, have recently been proposed as general analytical instruments for solving a variety of problems.[33]

---

[33] J.-B. D. Green, *Anal. Chim. Acta*, **2003**, *496*, 267.

QUESTIONS AND PROBLEMS

*Answers are provided at the end of the book for problems marked with an asterisk.

☒ Problems with this icon are best solved using spreadsheets.

21-1 Describe the mechanism of the production of an MNN Auger electron.

21-2 Describe how it is possible to distinguish between XPS peaks and Auger electron peaks.

21-3 Explain why the information from an XPS chemical shift must also be contained in an X-ray absorption edge.

*21-4 An XPS electron was found to have a kinetic energy of 1073.5 eV when a Mg Kα source was employed ($\lambda = 0.98900$ nm). The electron spectrometer had a work function of 14.7 eV.
(a) Calculate the binding energy for the emitted electron.
(b) If the signal was from a S(2s) electron, was the analyte S^{2-}, S^0, SO_3^{2-}, or SO_4^{2-}?
(c) What would the kinetic energy have been if an Al Kα source had been used ($\lambda = 0.83393$ nm)?
(d) If the ejected electron with the Mg Kα source had been an Auger electron, what would its kinetic energy be with the Al Kα source?

*21-5 An XPS electron was found to have a kinetic energy of 1052.6 eV when ejected with an Al Kα source ($\lambda = 0.83393$ nm) and measured in a spectrometer with a work function of 27.8 eV. The electron is believed to be a N(1s) electron in $NaNO_3$.
(a) What was the binding energy for the electron?
(b) What would be the kinetic energy of the electron if a Mg Kα ($\lambda = 0.98900$ nm) source were used?
(c) How could one be sure that a peak was an XPS and not an Auger electron peak?
(d) At what binding and kinetic energies would a peak for $NaNO_2$ be expected when the Al Kα source was used with the same spectrometer?

21-6 Compare EELS to conventional infrared and Raman spectroscopy. Focus on what is used to excite and detect vibrations. What are the advantages and limitations of EELS?

21-7 Compare ISS to RBS. For both cases, draw diagrams of the instrumental setup. Describe any differences in the information obtained by the two techniques.

21-8 How does static SIMS instrumentally differ from dynamic SIMS? How does the information obtained from static SIMS differ from that obtained from dynamic SIMS? What is imaging SIMS? What type of information is obtained with imaging SIMS?

21-9 What are the main advantages of surface photon techniques when compared with electron and ion spectroscopic methods? What are the major disadvantages?

21-10 What is a buried interface and what techniques are available to study buried interfacial phenomena?

21-11 Name three possible sources of signals with the SEM. Differentiate between elastic and inelastic scattering of electrons.

21-12 Name the two most common types of scanning probe microscopes.
(a) How do they differ?
(b) What are the advantages of each type?
(c) What are the major limitations of each type?

21-13 If the tunneling current is 10.0 pA when an STM probe is 0.40 nm from a surface and 18.0 pa when the probe is 0.50 nm from the surface, calculate the current on moving the tip in 0.10-nm steps from 0.40 nm to 1.50 nm.

Challenge Problem

21-14 Quantitative X-ray photoelectron spectrometry has become more popular in recent years. The factors relating the intensity of emission to atomic concentration (density) are given in Equation 21-3.
(a) Of the factors influencing the emission intensity, identify those related to the analyte.
(b) Identify the factors influencing the emission intensity related to the spectrometer.
(c) The measured quantity in quantitative XPS is usually I/S, where I is the peak area and S is the sensitivity factor for the element of interest. If this quantity is measured for the analyte $(I/S)_a$ and the corresponding value measured for an internal standard with a different transition $(I/S)_s$, show by means of an equation how the ratio $(I/S)_a/(I/S)_s$ is related to the atomic concentration ratio of the analyte to the internal standard.
(d) If all the elements on a surface are measured by XPS, show that the fractional atomic concentration f_A of element A is given by

$$f_A = \frac{I_A/S_A}{\Sigma(I_n/S_n)}$$

where I_n is the measured peak area for element n, S_n is the atomic sensitivity factor for that peak, and the summation is carried out over all n.
(e) For a polyurethane sample in which carbon, nitrogen, and oxygen were detected by XPS, the sensitivity factor for C was 0.25 on the spectrometer used. The sensitivity factor for N was 0.42 and that for O was 0.66 on the same spectrometer. If the peak areas were $C(1s) = 26,550$, $N(1s) = 4475$, and $O(1s) = 13,222$, what were the atomic concentrations of the three elements?
(f) What are the limitations of quantitative analysis with XPS? Why might the atomic concentrations measured not correspond to the bulk composition?
(g) For polyurethane, the stoichiometric atomic concentrations are $C = 76.0\%$, $N = 8.0\%$, and $O = 16.0\%$. Calculate the percentage error in the values obtained in part (e) for each element.

Assessing the Authenticity of the Vinland Map: Surface Analysis in the Service of History, Art, and Forensics

The Vinland Map

The controversial Vinland map came to light in the 1950s when it was put up for sale by a private library in Europe and eventually purchased by a rare-books dealer in New Haven, Connecticut. It was donated to Yale University in 1965 and is now owned by the Yale University Library. As can be seen in Figure IA3-1, the Vinland map is a medieval-style map of the New World showing a large island in the Western Atlantic called Vinlandia Insula, which generally resembles the northeastern part of North America (Vinland). The map was first revealed to the world in 1965 together with an unknown document dating to the 1440s titled "Tartar Relation." If genuine, it would imply that a large portion of North America was known to western Europeans prior to Columbus.

The case for the authenticity of the map was enhanced when it was discovered that wormholes in the map lined up exactly with wormholes in the "Tartar Relation" and in another authentic medieval document, the "Speculum Historiale." The map was thought to have been bound with these two documents at one time.

The map was largely believed to be authentic until 1974, when the Yale Library hired a surface analysis group (McCrone Associates) to make a detailed examination. Since that time, scientists have used a wide variety of surface analytical techniques to look at the parchment and the ink used in the map. As discussed in the following, many of the measurements indicate that the map is a forgery. However, not all scientists agree, and several nagging questions remain.

Instrumental Techniques

In their comprehensive investigation of the map, McCrone Associates[1] made a careful preliminary examination using optical microscopy. These observations revealed that the drawing consisted of an apparently hand-drawn map with a black ink outline on vellum. The outline is bordered with an underlying pale yellow layer, which is typical of ancient manuscripts. A key early observation was that the black outline was not in perfect registration with the yellow underlayer.

The map was then examined using polarized light microscopy to reveal the likely presence of particles of calcite and titanium dioxide (TiO_2), probably anatase. These observations also suggested that the anatase particles were

similar to those of commercial white pigments produced since the early twentieth century. Following microscopic examination, particles from various parts of the map were subjected to several other ultramicroanalytical methods.

Powder X-ray diffraction (Section 12D-1) measurements were performed on sub-nanogram samples of the yellow pigment layer, which confirmed the presence of both calcite and anatase. Particles of the pigments were examined in a scanning electron microscope (Section 21F-2) equipped with an X-ray fluorescence detection system (Section 12A-3). The yellow pigment particles showed relatively high concentrations of titanium, and the black particles showed high concentrations of iron and chromium. The transmission electron microscope (Section 21F-2) was used to compare the particle shape and size distribution of anatase from the Vinland map to those of the commercial product. These results also suggested that the anatase particles were of modern origin.

The electron microprobe (Section 21F-1), which is capable of elemental analysis on femtogram samples with a beam diameter of 1 μm, was used to show that the yellow pigment contained substantial amounts of titanium, but neither the parchment nor the black pigment contained titanium. Ion microprobe analysis (Section 21D-1) of the pigment samples provided results consistent with those of the electron microprobe. The ion microprobe was also used to compare the pigments of the Vinland map with the inks in the "Tartar Relation" and the "Speculum Historiale." The results from the Vinland map did not match those from the Tartar Relation or the Speculum Historiale, and they did not match results from any known ink. McCrone concluded that the Vinland map was a clever forgery produced by drawing the map with the yellow, anatase-containing pigment and then redrawing the map using a black, carbon-based ink. This conclusion appears to have been triggered by the lack of registration of the yellow and black layers and was consistent with all of the ultramicroanalytical results.

An interdisciplinary group led by Cahill used particle-induced X-ray emission (PIXE) techniques (Section 12C-1) to determine that although titanium is indeed often present in the yellow pigment of the Vinland map, it is found only in minute quantities, far less than is consistent with the impression given by the early results from the McCrone group.[2] The Cahill group, which had previously tested hundreds of other early manuscripts, had detected comparable

[1] W. C. McCrone, *Anal. Chem.*, **1988**, *60*, 1009.

[2] T. A. Cahill et al., *Anal. Chem.*, **1987**, *59*, 829.

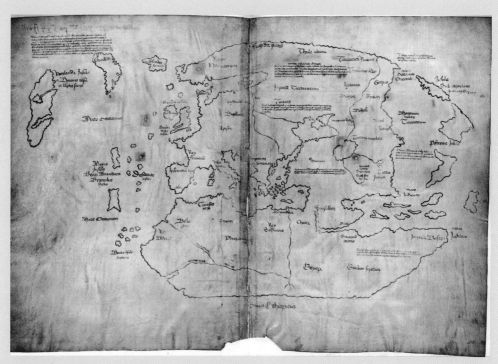

FIGURE IA3-1 The Vinland map. (Medieval and Renaissance Manuscripts, General Collection of Rare Books and Manuscripts, Beinecke Rare Book and Manuscript Library, Yale University.)

levels of titanium in several other undisputedly medieval parchments. In his summary report, McCrone[3] disputed the Cahill study by stating that PIXE, although a trace analysis technique, is not an appropriate tool for ultramicroanalysis. In other words, the sampling area of PIXE is considerably larger than the microanalytical techniques used by McCrone, which leads to apparently lower concentrations of titanium in significant areas of the map.

Raman microprobe measurements (Section 18B-2) on the map by Brown and Clark confirm that the yellow pigment contains anatase, that the black pigment is carbon based, and that these pigments are entirely different than the inks of the "Tartar Relation."[4] These results are consistent with McCrone's conclusion that the map is a forgery. Finally, the age of the parchment of the Vinland map has been determined by radiocarbon dating using accelerator mass spectrometry (Section 11F-1), and these results suggest that the parchment was created sometime between AD 1411 and AD 1468 (95% CL).[5] Thus, if the parchment

is a modern forgery, the forger used an authentic piece of parchment to produce the map.

The Analytical Perspective

The controversy regarding the Vinland map points out several important aspects of any analytical study. The first involves defining exactly the analytical sample and keeping it in mind during the determination. In our case specifically, several different samples could be defined: the parchment, a broad area containing the map's ink and parchment, or small particles of the ink itself. In many analytical studies, we seek to homogenize a sample or average our results over a certain area. In others, however, it is important to look at the individual local heterogeneities in a sample. Our case study shows that scientists looking at small particles may come to conclusions about concentrations that are different from those of others looking at a broader spatial region. The spatial resolution of analytical techniques is also an important aspect of this study. We cannot hope to determine local effects or local concentrations with techniques that observe or average over large spatial regions. In our example, PIXE, although a sensitive trace analysis method, does not

[3]See note 1.
[4]K. L. Brown and R. J. H. Clark, *Anal. Chem.*, **2002**, *74*, 3658.
[5]D. J. Donahue, J. S. Olin, and G. Harbottle, *Radiocarbon*, **2002**, *44*, 45.

have the spatial resolution of many of the surface techniques used to study the map. Another aspect that our case study brings out is the difference between a technique that determines elemental composition and one that determines molecular composition. In our case, X-ray diffraction and Raman microscopy were used to determine conclusively the presence of TiO_2 in the ink. The other techniques employed can establish the presence of specific elements and determine their quantities in the sampled volume. Finally, we should note that the Vinland map controversy is undoubtedly not over. If the map is a forgery, questions remain about how such an old parchment could be obtained. Likewise, if the map is real, the presence of TiO_2 in the particle size and particle distribution must be explained. New and more powerful analytical techniques will certainly be employed in the future to help put to rest these and other remaining questions.[6]

[6]"The Viking Deception," Narr. Gene Galusha, *NOVA*, Granite Productions, PBS, WGBH, Boston, February 8, 2005.

SECTION FOUR

Electro-analytical Chemistry

Electrochemical sensors are used in amazing places. The photo shows the deep-sea vehicle *Alvin* placing an electrochemical sensor array on hydrothermal vents at the bottom of the Pacific Ocean. These sensors measure pH, dissolved hydrogen, and dissolved hydrogen sulfide, species critical to understanding geochemical reactions deep in the ocean crust, as well as biogeochemical processes in the near-vent environment. The sensors make continuous accurate measurements under the extreme environmental conditions of the ocean floor. The temperature measured as the array was placed in position was 365°C and the pressure 250 bar. The operating principles of sensors like these are described in Chapters 23 and 25. (Photo by Woods Hole Oceanographic Institution.)

I n this section the theory and methodology of electro-analytical chemistry are explored. Chapter 22 provides a general foundation for the study of subsequent chapters in this section. Terminology and conventions of electrochemistry as well as theoretical and practical aspects of the measurement of electrochemical potentials and currents are presented. Chapter 23 comprises the many methods and applications of potentiometry, and constant-potential coulometry and constant-current coulometry are discussed in Chapter 24. The many facets of the important and widely used technique of voltammetry are presented in Chapter 25, which concludes the section.

CHAPTER TWENTY-TWO

An Introduction to Electroanalytical Chemistry

Electroanalytical chemistry encompasses a group of qualitative and quantitative analytical methods based on the electrical properties of a solution of the analyte when it is made part of an electrochemical cell. Electroanalytical techniques are capable of producing low detection limits and a wealth of characterization information describing electrochemically accessible systems. Such information includes the stoichiometry and rate of interfacial charge transfer, the rate of mass transfer, the extent of adsorption or chemisorption, and the rates and equilibrium constants for chemical reactions.

Throughout this chapter, this logo indicates an opportunity for online self-study at **www .thomsonedu.com/chemistry/skoog**, linking you to interactive tutorials, simulations, and exercises.

Electroanalytical methods[1] have certain general advantages over other types of procedures discussed in this book. First, electrochemical measurements are often specific for a particular oxidation state of an element. For example, in electrochemical methods it is possible to determine the concentration of each of the species in a mixture of cerium(III) and cerium(IV), whereas most other analytical methods can reveal only the total cerium concentration. A second important advantage of electrochemical methods is that the instrumentation is relatively inexpensive. Most electrochemical instruments cost less than $30,000, and the price for a typical multipurpose commercial instrument is in the range of $8000 to $10,000. In contrast, many spectroscopic instruments cost $50,000 to $250,000 or more. A third feature of certain electrochemical methods, which may be an advantage or a disadvantage, is that they provide information about activities rather than concentrations of chemical species. Generally, in physiological studies, for example, activities of ions such as calcium and potassium are more important than concentrations.

Thoughtful application of the various electroanalytical methods described in Chapters 23–25 requires an understanding of the basic theory and the practical aspects of the operation of electrochemical cells. This chapter is devoted largely to these topics.

22A ELECTROCHEMICAL CELLS

A dc electrochemical cell consists of two electrical conductors called *electrodes*, each immersed in a suitable electrolyte solution. For a current to develop in a cell, it is necessary (1) that the electrodes be connected externally with a metal conductor, (2) that the two electrolyte solutions be in contact to permit movement of ions from one to the other, and (3) that an electron-transfer reaction can occur at each of the two electrodes. Figure 22-1a shows an example of a simple electrochemical cell. It consists of a silver electrode

[1] Some reference works on electrochemistry and its applications include A. J. Bard and L. R. Faulkner, *Electrochemical Methods*, 2nd ed., New York: Wiley, 2001; V. S. Bagotsky and K. Mueller, *Fundamentals of Electrochemistry*, 2nd ed., New York: Wiley, 2005; D. T. Sawyer, A. Sobkowiak, and J. L. Roberts, *Electrochemistry for Chemists*, 2nd ed., New York: Wiley, 1995; J. Koryta and J. Dvorak, *Principles of Electrochemistry*, 2nd ed., New York: Wiley, 1993; *Laboratory Techniques in Electroanalytical Chemistry*, 2nd ed., P. T. Kissinger and W. R. Heineman, eds., New York: Dekker, 1996. The classical, and still useful, monograph dealing with electroanalytical chemistry is J. J. Lingane, *Electroanalytical Chemistry*, 2nd ed., New York: Interscience, 1958.

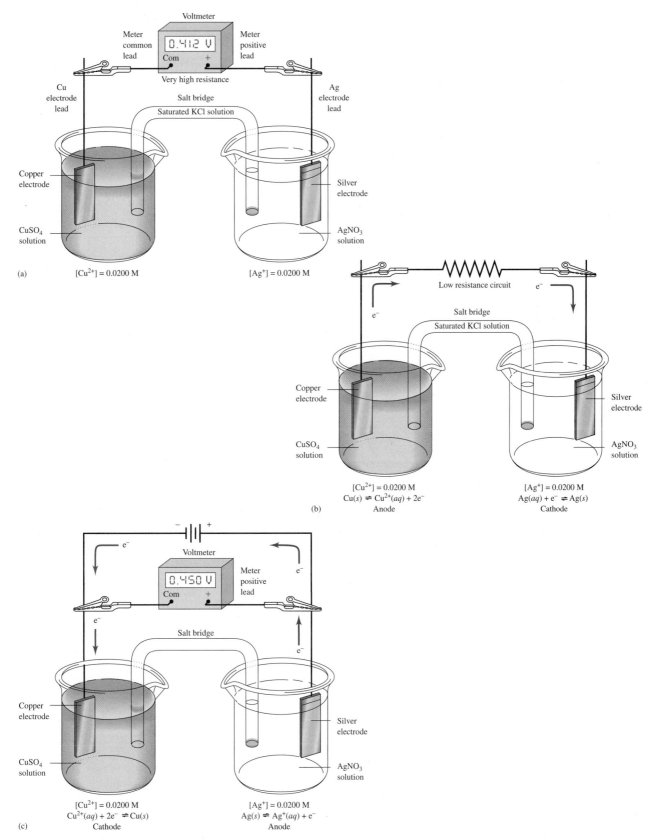

(a) [Cu²⁺] = 0.0200 M

$[Cu^{2+}] = 0.0200\ M$

$[Ag^+] = 0.0200\ M$

(b) $[Cu^{2+}] = 0.0200\ M$
$Cu(s) \rightleftharpoons Cu^{2+}(aq) + 2e^-$
Anode

$[Ag^+] = 0.0200\ M$
$Ag(aq) + e^- \rightleftharpoons Ag(s)$
Cathode

(c) $[Cu^{2+}] = 0.0200\ M$
$Cu^{2+}(aq) + 2e^- \rightleftharpoons Cu(s)$
Cathode

$[Ag^+] = 0.0200\ M$
$Ag(s) \rightleftharpoons Ag^+(aq) + e^-$
Anode

FIGURE 22-1 (a) A galvanic electrochemical cell at open circuit; (b) a galvanic cell doing work; (c) an electrolytic cell.

immersed in a solution of silver nitrate and a copper electrode in a solution of copper sulfate. The two solutions are joined by a *salt bridge*, which consists of a tube filled with a solution that is saturated with potassium chloride or, sometimes, a different electrolyte. The two ends of the tube are fitted with porous plugs or disks that permit the movement of ions across them but prevent siphoning of liquid from one electrolyte solution to the other. The purpose of the bridge is to isolate the contents of the two halves of the cell while maintaining electrical contact between them. Isolation is necessary to prevent direct reaction between silver ions and the copper electrode. The cells in Figure 22-1 contain two so-called *liquid junctions*, one being the interface between the silver nitrate solution and the salt bridge; the second is at the other end of the salt bridge where the electrolyte solution of the bridge contacts the copper sulfate solution. As we show in Section 22B-2, a small *junction potential* at each of these interfaces may influence significantly the accuracy of the analysis.

22A-1 Conduction in a Cell

If we replace the high-impedance meter in Figure 22-1a with a low-resistance wire or load as shown in Figure 22-1b, the circuit is completed, and charge flows. Charge conduction results from three distinct processes in various parts of the cell shown in Figure 22-1b:

1. In the silver and copper electrodes, as well as in the external conductor, electrons are the charge carriers, moving from the copper electrode through the external conductor to the silver electrode.
2. Within the solutions the flow of charge is the result of migration of both cations and anions. In the half-cell on the left, copper ions migrate away from the electrode into the solution, and sulfate and hydrogen sulfate ions move toward it; in the half-cell on the right, silver ions move toward the electrode and anions move away from it. Inside the salt bridge, charge is carried by migration of potassium ions to the right and chloride ions to the left. Thus, all of the ions in the three solutions participate in the flow of charge.
3. A third process occurs at the two electrode surfaces. At these interfaces, an oxidation or a reduction reaction couples the ionic conduction of the solution with the electron conduction of the electrode to provide a complete circuit for the flow of charge.

 Simulation: Learn more about **electrochemical cells**.

The two electrode processes are described by the reactions

$$Cu(s) \rightleftharpoons Cu^{2+}(aq) + 2e^- \qquad (22\text{-}1)$$

$$Ag^+(aq) + e^- \rightleftharpoons Ag(s) \qquad (22\text{-}2)$$

22A-2 Galvanic and Electrolytic Cells

The net cell reaction that occurs in the cell shown in Figure 22-1b is the sum of the two half-cell reactions shown as Equations 22-1 and 22-2.[2] That is,

$$Cu(s) + 2Ag^+ \rightleftharpoons Cu^{2+} + Ag(s)$$

The voltage of this cell is a measure of the tendency for this reaction to proceed toward equilibrium. Thus, as shown in Figure 22-1a, when the copper and silver ion concentrations (actually, activities) are 0.0200 M, the cell voltage is 0.412 V, which shows that the reaction is far from equilibrium. If we connect a resistor or other load in the external circuit as shown in Figure 22-1b, a measurable current in the circuit results and the cell reaction occurs. As the reaction proceeds, the voltage becomes smaller and smaller, and it ultimately reaches 0.000 V when the system achieves equilibrium.

Cells, such as the one shown in Figure 22-1a and 22-1b, that produce electrical energy are called *galvanic cells*. In contrast, *electrolytic cells* consume electrical energy. For example, the Cu-Ag cell can operate in an electrolytic mode by connecting the negative terminal of a battery or dc power supply to the copper electrode and the positive terminal to the silver electrode, as illustrated in Figure 22-1c. If the output of this supply is adjusted to be somewhat greater than 0.412 V, as shown, the two electrode reactions are reversed and the net cell reaction is

$$2Ag(s) + Cu^{2+} \rightleftharpoons Ag^+ + Cu(s)$$

A cell in which reversing the direction of the current simply reverses the reactions at the two electrodes is termed a *chemically reversible cell*.

22A-3 Anodes and Cathodes

By definition, the *cathode* of an electrochemical cell is the electrode at which reduction occurs, and the *anode* is the electrode where oxidation takes place. These definitions apply to both galvanic cells under discharge and to electrolytic cells. For the galvanic cell shown in

---

[2]Note that the silver half-reaction must be multiplied by 2 to maintain charge and mass balance.

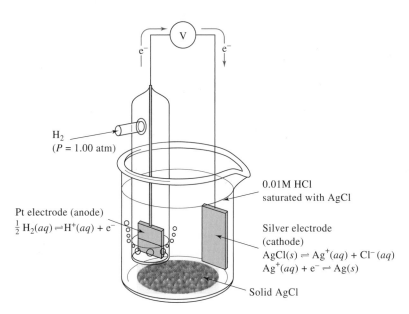

FIGURE 22-2 A galvanic cell without a liquid junction.

H$_2$
(P = 1.00 atm)

Pt electrode (anode)
$\frac{1}{2}$ H$_2$(aq) $\rightleftharpoons$ H$^+$(aq) + e$^-$

0.01M HCl
saturated with AgCl

Silver electrode
(cathode)
AgCl(s) $\rightleftharpoons$ Ag$^+$(aq) + Cl$^-$(aq)
Ag$^+$(aq) + e$^-$ $\rightleftharpoons$ Ag(s)

Solid AgCl

Figure 22-1b, the silver electrode is the cathode and the copper electrode is the anode. On the other hand, in the electrolytic cell of Figure 22-1c, the silver electrode is the anode and the copper electrode the cathode because the half-cell reactions are reversed.

22A-4 Cells without Liquid Junctions

As mentioned previously, the cell shown in Figure 22-1 has two liquid junctions, one between the silver nitrate solution and one end of the salt bridge, the other between the copper sulfate solution and the salt bridge. Sometimes it is possible and advantageous to prepare cells in which the electrodes share a common electrolyte and thus eliminate the effect of junction potentials. An example of a cell of this type is shown in Figure 22-2. If the voltmeter were removed and replaced by a wire, silver would behave as the cathode. The reaction at the cathode would be

$$AgCl(s) + e^- \rightleftharpoons Ag^+(aq) + Cl^-(aq)$$

Under discharge, hydrogen is consumed at the platinum anode:

$$H_2(g) \rightleftharpoons 2H^+(aq) + 2e^-$$

The overall cell reaction is then obtained by multiplying each term in the first equation by 2 and adding the two equations. That is,

$$2AgCl(s) + H_2(g) \rightleftharpoons 2Ag(s) + 2H^+(aq) + 2Cl^-(aq)$$

The direct reaction between hydrogen and solid silver chloride is so slow that the same electrolyte can be used for both electrodes without significant loss of cell efficiency because of direct reaction between cell components.

The cathode reaction in this cell is interesting because it can be considered the result of a two-step process described by the equations

$$AgCl(s) \rightleftharpoons Ag^+ + Cl^-$$
$$Ag^+ + e^- \rightleftharpoons Ag(s)$$

The slightly soluble silver chloride dissolves in the first step to provide an essentially constant concentration of silver ions that are then reduced in the second step.

The anodic reaction in this cell is also a two-step process that can be formulated as

$$H_2(g) \rightleftharpoons H_2(aq)$$
$$H_2(aq) \rightleftharpoons 2H^+ + 2e^-$$

Hydrogen gas is bubbled across the surface of a platinum electrode so that the concentration of the gas at the surface is constant at constant temperature and constant partial pressure of hydrogen. Note that in this case the inert platinum electrode plays no direct role in the reaction but serves only as a surface where electron transfer can occur.

The cell in Figure 22-2 is a galvanic cell with a potential of about 0.46 V. This cell is also chemically reversible and can be operated as an electrolytic cell by applying an external potential of somewhat greater

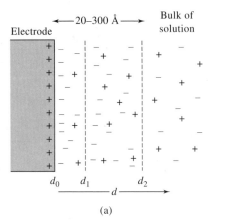

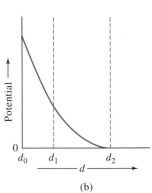

FIGURE 22-3 Electrical double layer formed at electrode surface as a result of an applied potential.

(a)

(b)

than 0.46 V. Note that you cannot tell whether a given electrode will be a cathode or an anode unless you know whether the cell is galvanic under discharge or electrolytic.

22A-5 Solution Structure: The Electrical Double Layer

It is important to realize that electrochemical measurements are made on heterogeneous systems and that an electrode can donate or accept electrons only from a species that is present in a layer of solution that is immediately adjacent to the electrode. Thus, as a result of the chemical and physical changes that occur at the electrode-solution interface, this layer may have a composition that differs significantly from that of the bulk of the solution.

For example, let us consider the structure of the solution immediately adjacent to an electrode in an electrolytic cell when a positive voltage is first applied to the electrode (e.g., the Ag electrode in Figure 22-1c). Immediately after applying the voltage, there is a momentary surge of current, which rapidly decays to zero if no reactive species is present at the surface of the electrode. This current is a charging current that creates an excess (or a deficiency) of negative charge at the surface of the two electrodes. Because of ionic mobility, however, the layers of solution immediately adjacent to the electrodes acquire a charge of the opposite sign. This effect is illustrated in Figure 22-3a. The surface of the metal electrode has an excess of positive charge because of an applied positive voltage. The charged solution layer consists of two parts: (1) a compact inner layer (d_0 to d_1), in which the potential decreases linearly with distance from the electrode surface and (2) a diffuse layer (d_1 to d_2), within which the decrease is approxi-

mately exponential (see Figure 22-3b). This entire array of charged species and oriented dipoles (such as water molecules) at the electrode-solution interface is called the *electrical double layer*.

22A-6 Faradaic and Nonfaradaic Currents

Two types of processes can conduct currents across an electrode-solution interface. One type involves a direct transfer of electrons via an oxidation reaction at one electrode and a reduction reaction at the other. Processes of this type are called *faradaic processes* because they are governed by Faraday's law, which states that the amount of chemical reaction that occurs at an electrode is proportional to the current, called a *faradaic current*.

Under certain conditions a range of voltages may be applied to a cell that do not produce *faradaic* processes at one or both of the electrodes. Faradaic processes may be prevented either because electrons do not have sufficient energy to pass over the potential energy barrier at the electrode-solution interface (thermodynamic reasons) or because the electron-transfer reaction is not fast enough on the time scale of the experiment (kinetic reasons). Under these circumstances, conduction of continuous alternating currents can still take place. With such currents, reversal of the charge relationship occurs with each half-cycle as first negative and then positive ions are attracted alternately to the electrode surface. Electrical energy from the external voltage source is consumed and converted to heat by friction associated with the motion of ions. Another way to look at this consumption of energy is that when the voltage changes, the ions in the double layer have to rearrange and adjust to the new potential, and this rearrangement requires energy. Thus, each electrode surface behaves as one plate of a capacitor, whose capacitance is large (several hundred to several thousand microfarads per square centimeter). The capacitive current increases with frequency and with electrode area (see Section 2B-3 and Equation 2-28); by controlling these variables, it is possible to adjust conditions so that essentially all of the alternating current in a cell is carried across the electrode interface by this *nonfaradaic process*. See Section 25B-4 for a description of a circuit model for this process.

To understand the basic difference between a faradaic and a nonfaradaic current, imagine an electron traveling down the external circuit to an electrode surface. When the electron reaches the solution interface, it can do one of only two things. It can remain at the electrode surface and increase the charge on the double layer, which constitutes a nonfaradaic current. Alternatively, it can leave the electrode surface and transfer to a species in the solution, thus becoming a part of the faradaic current.

22A-7 Mass Transfer in Cells with the Passage of Current

Because an electrode can probe only a very thin layer of solution at the electrode surface (d_0 to d_1 in Figure 22-3a), a faradaic current requires continuous mass transfer of reactive species from the bulk of the solution to the electrode surface. Three mechanisms bring about this mass transfer: *convection*, *migration*, and *diffusion*. Convection results from mechanical motion of the solution as a result of stirring or the flow of the solution past the surface of the electrode. Migration is the movement of ions through the solution brought about by electrostatic attraction between the ions and the electrode of the opposite charge. Diffusion is the motion of species caused by a concentration gradient. There is a detailed discussion of these mechanisms of mass transfer in Section 22E-3.

22A-8 Schematic Representation of Cells

We often use a shorthand notation to simplify the description of cells. For example, the cells shown in Figures 22-1 and 22-2 can be described by

$$Cu \,|\, CuSO_4(a_{Cu^{2+}} = 0.0200) \,\|\, AgNO_3(a_{Ag^+} = 0.0200) \,|\, Ag$$

$$Pt, H_2(p = 1 \text{ atm}) \,|\, H^+(0.01 \text{ M}),$$
$$Cl^-(0.01 \text{ M}), AgCl(sat'd) \,|\, Ag$$

By convention, a single vertical line indicates a phase boundary, or interface, where a potential develops. For example, the first vertical line in this schematic indicates there is a potential difference across the phase boundary between the zinc electrode and the zinc sulfate solution. The double vertical line represents two phase boundaries, one at each end of the salt bridge. There is a *liquid-junction potential* across each of these interfaces. Because the potential of a cell depends on the activities of the cell components, it is common practice to provide the activity or concentration of the cell constituents in parentheses.

In the second cell, there are only two phase boundaries because the electrolyte is common to both electrodes. A shorthand representation of this cell is

$$Pt \,|\, H_2(sat'd), HCl(0.01 \text{ M}), Ag^+(1.8 \times 10^{-5} \text{ M}) \,|\, Ag$$

In this cell, the molecular hydrogen concentration is that of a saturated solution; if the partial pressure of hydrogen is not given, we assume that it is 1.00 atm. The indicated molar silver ion concentration was computed from the solubility-product constant for silver chloride.

22B POTENTIALS IN ELECTROANALYTICAL CELLS

In most electroanalytical methods, we measure either (1) the current in an electrochemical cell at a fixed potential or (2) the potential of a cell while the current is fixed at some constant level or is zero. In general, however, in an electrochemical experiment we can control only the potential of the cell at a desired level and measure the current that results, or vice versa. If we choose to control one variable, it is impossible to control the other independently.

In this section we first consider the thermodynamics of electrochemical cells and the relationship between the activities of the participants in typical cell reactions and the observed potential of the cell. We then describe the source of the junction potentials that occur in most electrochemical cells. In Section 22C we consider the potentials of individual electrodes making up cells.

22B-1 The Thermodynamics of Cell Potentials

The potential of an electrochemical cell is directly related to the *activities* of the reactants and products of the cell reaction and indirectly to their molar concentrations. Although we often make the approximation that these activities are equal to molar concentrations, always remember that this assumption may produce errors in calculated potentials. We review the relationship between the activity of a chemical species and its concentration in Appendix 2.

Recall that the activity a_X of the species X is given by

$$a_X = \gamma_X[X] \tag{22-3}$$

In this equation, γ_X is the activity coefficient of solute X, and the bracketed term is the molar concentration of X. In some of the examples, however, for convenience we will assume that the activity coefficient is unity so that the molar concentration and the activity of a species are identical.

How do the activities of reactants and products affect the potential of an electrochemical cell? We will use as an example the cell illustrated in Figure 22-2 for which the spontaneous cell reaction is

$$2AgCl(s) + H_2(g) \rightleftharpoons 2Ag(s) + 2Cl^- + 2H^+$$

The thermodynamic equilibrium constant K for this reaction is given by

$$K = \frac{a_{H^+}^2 \cdot a_{Cl^-}^2 \cdot a_{Ag}^2}{p_{H_2} \cdot a_{AgCl}^2} \tag{22-4}$$

where a's are the activities of the various species indicated by the subscripts and p_{H_2} is the partial pressure of hydrogen in atmospheres.

In Appendix 2[3] we show that the activity of a pure solid is unity when it is present in excess (that is, $a_{Ag} = a_{AgCl} = 1.00$). Therefore, Equation 22-4 simplifies to

$$K = \frac{a_{H^+}^2 \cdot a_{Cl^-}^2}{p_{H_2}} \tag{22-5}$$

It is convenient to define the activity quotient Q such that

$$Q = \frac{(a_{H^+})_i^2 (a_{Cl^-})_i^2}{(p_{H_2})_i} \tag{22-6}$$

The subscript i indicates that the terms in the parentheses are instantaneous activities and not equilibrium activities. The quantity Q, therefore, is not a constant but changes continuously until equilibrium is reached; at that point, Q becomes equal to K, and the i subscripts are deleted.

The change in free energy ΔG for a cell reaction (that is, the maximum work obtainable at constant temperature and pressure) is given by

$$\Delta G = RT \ln Q - RT \ln K = RT \ln \frac{Q}{K} \tag{22-7}$$

where R is the gas constant (8.316 J mol^{-1} K^{-1}) and T is the temperature in kelvins; ln refers to the base e, or natural, logarithm. This relationship implies that the free energy for the system depends on how far the system is from the equilibrium state as indicated by the quotient Q/K. The cell potential E_{cell} is related to the free energy of the reaction by the relationship

$$\Delta G = -nFE_{cell} \tag{22-8}$$

where F is the faraday (96,485 coulombs per mole of electrons) and n is the number of moles of electrons associated with the oxidation-reduction process (in this example, $n = 2$).

When Equations 22-6 and 22-8 are substituted into 22-7 and the resulting equation is rearranged, we find that

$$E_{cell} = -\frac{RT}{nF} \ln Q + \frac{RT}{nF} \ln K$$
$$= -\frac{RT}{nF} \ln \frac{(a_{H^+})_i^2 (a_{Cl^-})_i^2}{(p_{H_2})_i} + \frac{RT}{nF} \ln K \tag{22-9}$$

The last term in this equation is a constant, which is called the standard potential for the cell E_{cell}^0. That is,

$$E_{cell}^0 = \frac{RT}{nF} \ln K \tag{22-10}$$

Substitution of Equation 22-10 into Equation 22-9 yields

$$E_{cell} = E_{cell}^0 - \frac{RT}{nF} \ln \frac{(a_{H^+})_i^2 (a_{Cl^-})_i^2}{(p_{H_2})_i} \tag{22-11}$$

Note that the standard potential is equal to the *cell potential when the reactants and products are at unit activity and pressure.*

Equation 22-11 is a form of the *Nernst equation,*[4] named in honor of Walther Nernst (1864–1941), German physical chemist and winner of the 1920 Nobel Prize in Chemistry. This equation is used throughout electroanalytical chemistry and forms the basis for many applications.

---

[3] See page 967.

[4] Equations similar in form to Equation 22-11 are described as *nernstian* relationships; electrodes and sensors whose behavior can be described by the Nernst equation are said to exhibit nernstian behavior, and RT/nF (2.303 RT/nF when base 10 logarithms are used) is often called the *nernstian factor.*

22B-2 Liquid-Junction Potentials

When two electrolyte solutions of different composition are in contact with one another, there is a potential difference across the interface. This *junction potential* is the result of an unequal distribution of cations and anions across the boundary due to differences in the rates at which these species diffuse.

Consider the liquid junction in the system

$$HCl(1\ M)|HCl(0.01\ M)$$

Both hydrogen ions and chloride ions tend to diffuse across this boundary from the more concentrated to the more dilute solution, and the driving force for this movement is proportional to the concentration difference. The diffusion rates of various ions under the influence of a fixed force vary considerably (that is, the mobilities are different). In this example, hydrogen ions are about a factor of 5 more mobile than chloride ions. As a result, hydrogen ions tend to diffuse faster than the chloride ions, and this difference in diffusion rate produces a separation of charge (see Figure 22-4). The more dilute side (the right side in the figure) of the boundary becomes positively charged because hydrogen ions diffuse more rapidly. The concentrated side, therefore, acquires a negative charge from the excess of slower-moving chloride ions. The charge separation that occurs tends to counteract the differences in mobilities of the two ions, and as a result, equilibrium is soon achieved. The junction potential difference resulting from this charge separation may amount to 30 mV or more.

In a simple system such as that shown in Figure 22-4, the magnitude of the junction potential can be calculated from the mobilities of the two ions involved. However, it is seldom that a cell of analytical importance has a sufficiently simple composition to permit such a computation.[5]

Experiment shows that the magnitude of the junction potential can be greatly reduced by introduction of a concentrated electrolyte solution (a salt bridge) between the two solutions. The effectiveness of a salt bridge improves not only as the concentration of the salt increases but also as the difference between the mobilities of the positive and negative ions of the salt approaches zero. A saturated potassium chloride solution is particularly effective from both standpoints. Its

<hr />

[5]For methods for approximating junction potentials, see A. J. Bard and L. R. Faulkner, *Electrochemical Methods*, 2nd ed., pp. 69–72, New York: Wiley, 2001.

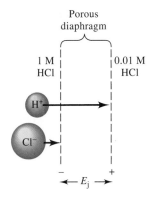

FIGURE 22-4 Schematic representation of a liquid junction showing the source of the junction potential E_j. The length of the arrows corresponds to the relative mobility of the two ions.

concentration is greater than 4 M at room temperature, and equally important, the mobilities of potassium and chloride ions differ by only 4%. When chloride ion interferes with a particular experiment, a concentrated solution of potassium nitrate can be substituted. With such bridges, the net junction potential is typically a few millivolts or less, which is negligible in most analytical measurements.

22C ELECTRODE POTENTIALS

It is useful to think of the cell reaction of an electrochemical cell as being made up of two *half-cell reactions*, each of which has a characteristic *electrode potential* associated with it. As will be shown later, these electrode potentials measure the driving force for the two half-reactions *when, by convention, they are both written as reductions*. Thus, the two half-cell or electrode reactions for the cell

$$Pt,H_2(1\ atm)|H^+(0.01\ M),Cl^-(0.01\ M),AgCl(sat'd)|Ag$$

shown in Figure 22-2, are

$$2AgCl(s) + 2e^- \rightleftharpoons 2Ag(s) + 2Cl^-$$
$$2H^+ + 2e^- \rightleftharpoons H_2(g)$$

To obtain the spontaneous cell reaction, the second half-reaction is subtracted from the first to give

$$2AgCl(s) + H_2 \rightleftharpoons 2Ag(s) + 2H^+ + 2Cl^-$$

If the electrode potentials $E_{AgCl/Ag}$ and E_{H^+/H_2} are known for the two half-reactions, we may then find the cell potential E_{cell} by subtracting the electrode potential for the second reaction from the first. That is,

$$E_{cell} = E_{AgCl/Ag} - E_{H^+/H_2}$$

A more general statement of the last relationship is

$$E_{cell} = E_{right} - E_{left} \qquad (22\text{-}12)$$

where E_{right} is the potential of the half-cell written on the right in the diagram of the cell or in the shorthand representation, and E_{left} is the electrode potential for the half-reaction written on the left. We discuss this convention further in Section 22C-5.

22C-1 Nature of Electrode Potentials

The potential of an electrochemical cell is the difference between the potential of one of the electrodes and the potential of the other, and it is therefore important to have a clear idea of what is meant by the potential of an electrode. This potential is a measure of an electrode's electron energy. For a metallic conductor immersed in a solution of an electrolyte, all excess charge resides on its surface, and it is possible to adjust the charge density on this surface by adjusting the output of an external power supply attached to the conductor. As this external source forces electrons onto the surface of an electrode, the electrons become more crowded, and their energy increases because of coulombic repulsion. The potential of the electrode thus becomes more negative. If the external circuitry withdraws enough electrons from the electrode, the surface will have a positive charge, and the electrode becomes more positive. It is also possible to vary the electron energy, and thus the potential, of a metallic electrode by varying the composition of the solution that surrounds it. For example, there is a potential at a platinum electrode immersed in a solution containing hexacyanoferrate(II) and hexacyanoferrate(III) as result of the equilibrium

$$Fe(CN)_6^{3-} + e^- \rightleftharpoons Fe(CN)_6^{4-}$$

If the concentration of $Fe(CN)_6^{4-}$ is made much larger than that of $Fe(CN)_6^{3-}$, hexacyanoferrate(II) has a tendency to donate electrons to the metal, thus creating a negative charge at the surface of the metal. Under this condition, the potential of the electrode is negative. On the other hand, if $Fe(CN)_6^{3-}$ is present in large excess, there is a tendency for electrons to be removed from the electrode, causing surface layer ions to

form in the solution and leaving a positive charge on the surface of the electrode. The platinum electrode then exhibits a positive potential.

We must emphasize that *no method can determine the absolute value of the potential of a single electrode*, because all voltage-measuring devices determine only *differences in potential*. One conductor from such a device is connected to the electrode under study. To measure a potential difference, however, the second conductor must make contact with the electrolyte solution of the half-cell under study. This second contact inevitably creates a solid-solution interface and hence acts as a second half-cell in which *a chemical reaction must also take place if charge is to flow*. A potential is associated with this second reaction. Thus, we cannot measure the absolute value for the desired half-cell potential. Instead, we can measure only the difference between the potential of interest and the half-cell potential for the contact between the voltage-measuring device and the solution.

Our inability to measure absolute potentials for half-cell processes is not a serious problem because relative half-cell potentials, measured versus a common reference electrode, are just as useful. These relative potentials can be combined to give real cell potentials. In addition, they can be used to calculate equilibrium constants of oxidation-reduction processes.

To develop a useful list of relative half-cell or electrode potentials, we must have a carefully defined reference electrode that is adopted by the entire chemical community. The standard hydrogen electrode, or the normal hydrogen electrode, is such a half-cell.

22C-2 The Standard Hydrogen Electrode

Hydrogen gas electrodes were widely used in early electrochemical studies not only as reference electrodes but also as indicator electrodes for determining pH. The composition of this type of electrode can be represented as

$$Pt, H_2(p \text{ atm}) | H^+(a_{H^+} = x)$$

The terms in parentheses suggest that the potential at the platinum surface depends on the hydrogen ion activity of the solution and on the partial pressure of the hydrogen used to saturate the solution.

The half-cell shown on the left in Figure 22-5 shows the components of a typical hydrogen electrode. The conductor is made of platinum foil that has been *platinized*. Platinum electrodes are platinized by coating their surfaces with a finely divided layer of platinum by

Simulation: Learn more about **electrochemical cell potentials**.

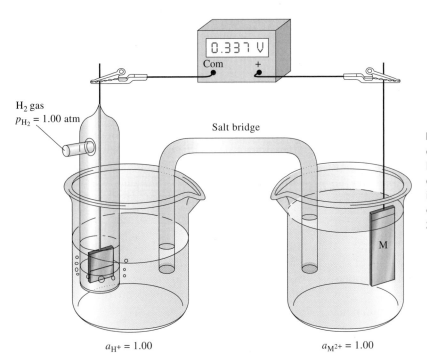

FIGURE 22-5 Measurement of the electrode potential for an M electrode. If the M^{2+} ion activity in the right-hand compartment is 1.00, the cell potential is the standard electrode potential of the half-reaction $M^{2+}(aq) + 2e^- \rightleftharpoons M(s)$.

rapid chemical or electrochemical reduction of H_2PtCl_6. The finely divided platinum on the surface of the electrode does not reflect light as does polished platinum, so the electrode appears black. Because of its appearance, the deposited platinum is called *platinum black*. Platinum black has a very large surface area to ensure that the reaction

$$2H^+ + 2e^- \rightleftharpoons H_2(g)$$

is rapid at the electrode surface. As was pointed out earlier, the stream of hydrogen simply keeps the solution adjacent to the electrode saturated with the gas.

The hydrogen electrode may act as the positive or the negative electrode, depending on the half-cell with which it is coupled with the salt bridge shown in Figure 22-5. When the cell is short circuited, hydrogen is oxidized to hydrogen ions when the electrode is an anode; the reverse reaction takes place when the electrode is a cathode. Under proper conditions, then, the hydrogen electrode is electrochemically reversible. When the cell is connected as shown in Figure 22-5, there is essentially no current in the cell because of the very high impedance of the meter. Under short circuit, the meter is replaced with a wire or low-resistance load as shown in Figure 22-1b, and the reaction proceeds.

The potential of a hydrogen electrode depends on the temperature, the hydrogen ion activity in the solu-

tion, and the pressure of the hydrogen at the surface of the electrode. Values for these experimental variables must be defined carefully for the half-cell to serve as a reference. Specifications for the *standard hydrogen electrode* (SHE) call for a hydrogen ion activity of unity and a partial pressure for hydrogen of exactly one atmosphere. *By convention, the potential of this electrode is assigned the value of exactly zero volt at all temperatures.*

22C-3 Practical Reference Electrodes

Although the SHE has great fundamental importance, the difficulty in preparing the electrode surface and controlling the activities of the reactants make it so impractical that it is rarely used for routine measurements. Instead, reference electrodes simple to prepare, more rugged, and easier to use are nearly always substituted for the hydrogen gas electrode. One of the most common of these is the silver–silver chloride electrode. This electrode can be prepared by applying an oxidizing voltage to a silver wire immersed in a dilute solution of hydrochloric acid. A thin coating of silver chloride forms that adheres tightly to the wire. The wire is then immersed in a saturated solution of potassium chloride. A salt bridge connects the potassium chloride solution to the electrode system being studied. The potential of this electrode is about +0.22 V with respect to the SHE. The electrode half-reaction is

$$AgCl(s) + e^- \rightleftharpoons Cl^- + Ag(s)$$

A second widely used reference electrode is the saturated calomel electrode (SCE), which consists of a pool of mercury in contact with a solution that is saturated with mercury(I) chloride (calomel) as well as potassium chloride. Platinum wire dipping in the mercury provides electrical contact to the other conductor, and a salt bridge to the second electrolyte completes the circuit. The potential of this reference is about 0.24 V positive. The electrode reaction is

$$Hg_2Cl_2(s) + 2e^- \rightleftharpoons 2Cl^- + 2Hg(l)$$

Section 23A gives more detailed descriptions of the silver–silver chloride and the calomel reference electrode systems. Both reference electrodes can be purchased from suppliers of electrochemical equipment.

22C-4 Definition of Electrode Potential

An *electrode potential* is defined as the potential of a cell in which the electrode under investigation is the right-hand electrode and the SHE is the left-hand electrode.[6] We must emphasize that, in spite of its name, an electrode potential is in fact the potential of an electrochemical cell that contains a carefully defined reference electrode on the left. Electrode potentials could more properly be called *relative electrode potentials*, but they seldom are. Note that this cell potential may be positive or negative depending on the electron energy of the electrode under study. Thus, when this energy is greater than that of the SHE, the electrode potential is negative; when the electron energy of the electrode in question is less than that of the SHE, the electrode potential is positive.

The cell in Figure 22-5 illustrates the definition of the electrode potential for the half-reaction

$$M^{2+} + 2e^- \rightleftharpoons M(s)$$

In this figure, the half-cell on the right consists of a strip of the metal M in contact with a solution of M^{2+}. The half-cell on the left is a SHE. By definition, the potential E observed on the voltmeter is the electrode potential for the M^{2+}/M *couple*. In this general example, we assume that the junction potentials across the salt

bridge are zero. If we further assume that the activity of M^{2+} in the solution is exactly 1.00, the potential is called the *standard electrode potential* for the system and is given the symbol E^0. That is, the standard electrode potential for a half-reaction is the electrode potential when the reactants and products are all at unit activity.

If M in the figure is copper, and if the copper ion activity in the solution is 1.00, the compartment on the right is positive, and the observed potential is +0.337 V as shown in the figure. The spontaneous cell reaction, which would occur if the voltmeter were replaced by a wire, is

$$Cu^{2+} + H_2(g) \rightleftharpoons Cu(s) + 2H^+$$

Because the hydrogen electrode is on the left, the measured potential is, by definition, the electrode potential for the Cu-Cu^{2+} half-cell. Note that the copper electrode is positive with respect to the hydrogen electrode; thus, we write

$$Cu^{2+} + 2e^- \rightleftharpoons Cu(s) \qquad E^0 = +0.337 \text{ V}$$

If M in Figure 22-5 is cadmium instead of copper, and the solution has a cadmium ion activity of 1.00, the potential is observed to be −0.403 V. In this case, the cadmium electrode is negative, and the cell potential has a negative sign. The spontaneous cell reaction would be

$$Cd(s) + 2H^+ \rightleftharpoons Cd^{2+} + H_2(g)$$

and we may write

$$Cd^{2+} + 2e^- \rightleftharpoons Cd(s) \qquad E^0 = -0.403 \text{ V}$$

A zinc electrode in a solution of zinc ion at unity activity exhibits a potential of −0.763 V when coupled with the SHE. The zinc electrode is the negative electrode in the galvanic cell, and its electrode potential is also negative.

The standard electrode potentials for the four half-cells just described can be arranged in the order

$$Cu^{2+} + 2e^- \rightleftharpoons Cu(s) \qquad E^0 = +0.337 \text{ V}$$
$$2H^+ + 2e^- \rightleftharpoons H_2(g) \qquad E^0 = 0.000 \text{ V}$$
$$Cd^{2+} + 2e^- \rightleftharpoons Cd(s) \qquad E^0 = -0.403 \text{ V}$$
$$Zn^{2+} + 2e^- \rightleftharpoons Zn(s) \qquad E^0 = -0.763 \text{ V}$$

The magnitudes of these standard electrode potentials show the relative strengths of the four ionic species as electron acceptors (oxidizing agents). In other words, we may arrange the ions in order of their decreasing strengths as oxidizing agents, $Cu^{2+} > H^+ > Cd^{2+} > Zn^{2+}$, or alternatively, the elements in order of

---

[6] A. D. McNaught and A. Wilkinson, *Compendium of Chemical Terminology* (The Gold Book), 2nd ed., p. 61, Malden, MA: Blackwell, 1997.

their increasing strengths as reducing agents, Cu < H_2 < Cd < Zn.

22C-5 Sign Conventions for Electrode Potentials

When we consider a normal chemical reaction, we speak of the reaction occurring from reactants on the left side of the arrow to products on the right side. By the International Union of Pure and Applied Chemistry (IUPAC) sign convention, when we consider an electrochemical cell and its resulting potential, we consider the cell reaction to occur in a certain direction as well. The convention for cells is called the *plus right rule*. This rule implies that we always measure the cell potential by connecting the positive lead of the voltmeter to the right-hand electrode in the schematic or cell drawing (for example, the Ag electrode in the example below) and the common, or ground, lead of the voltmeter to the left-hand electrode (the Cu electrode below).

$$Cu|Cu^{2+}(0.0200 \text{ M})\|Ag^{+}(0.0200 \text{ M})|Ag$$

If we always follow this convention, the value of E_{cell} is a measure of the tendency of the cell reaction to occur spontaneously in the direction written from left to right. That is, the direction of the overall process has Cu metal being oxidized to Cu^{2+} in the left-hand compartment and Ag^+ being reduced to Ag metal in the right-hand compartment. In other words, the reaction for the process occurring in the cell is considered to be

$$Cu(s) + 2Ag^{+} \rightleftharpoons Cu^{2+} + 2Ag(s)$$

Implications of the IUPAC Convention

Several implications of the sign convention may not be obvious. First, if the measured value of E_{cell} is positive, the right-hand electrode is positive with respect to the left-hand electrode, and the free energy change for the reaction as written is negative according to $\Delta G = -nFE_{cell}$ (Equation 22-8). Hence, the reaction in the direction being considered will occur spontaneously if the cell is short-circuited or connected to some device to perform work (e.g., light a lamp, power a radio, start a car). On the other hand, if E_{cell} is negative, the right-hand electrode is negative with respect to the left-hand electrode, the free energy change is positive, and the reaction in the direction considered (oxidation on the left, reduction on the right) is not the spontaneous cell reaction. For our previous copper-silver cell, $E_{cell} = +0.412$ V, and the oxidation of Cu and reduction of

Ag^+ occur spontaneously when the cell is connected to a device and allowed to do so.

The IUPAC convention is consistent with the signs that the electrodes actually exhibit in a galvanic cell. That is, in the Cu-Ag cell, the Cu electrode is electron rich (negative) because of the tendency of Cu to be oxidized to Cu^{2+}, and the Ag electrode is electron deficient (positive) because of the tendency for Ag^+ to be reduced to Ag. As the galvanic cell discharges spontaneously, the silver electrode is the cathode, and the copper electrode is the anode. Note that for the same cell written in the opposite direction

$$Ag|Ag^{+}(0.0200 \text{ M})\|Cu^{2+}(0.0200 \text{ M})|Cu$$

the measured cell potential would be $E_{cell} = -0.412$ V, and the reaction considered is

$$2Ag(s) + Cu^{2+} \rightleftharpoons 2Ag^{+} + Cu(s)$$

This reaction is *not* the spontaneous cell reaction because E_{cell} is negative and ΔG is thus positive. It does not matter to the cell which electrode is written in the schematic on the right and which is written on the left. The spontaneous cell reaction is *always*

$$Cu(s) + 2Ag^{+} \rightleftharpoons Cu^{2+} + 2Ag(s)$$

By convention, we just measure the cell in a standard manner and consider the cell reaction in a standard direction. Finally, we must emphasize that no matter how we write the cell schematic or arrange the cell in the laboratory, if we connect a wire or a low-resistance circuit to the cell, the spontaneous cell reaction will occur. The only way to achieve the reverse reaction is to connect an external voltage source and force the electrolytic reaction $2Ag(s) + Cu^{2+} \rightleftharpoons 2Ag^{+} + Cu(s)$ to occur.

Half-Cell Potentials

The potential of a cell such as that shown in Figure 22-5 is the difference between two half-cell or single-electrode potentials, one associated with the half-reaction at the right-hand electrode (E_{right}), the other associated with the half-reaction at the left-hand electrode (E_{left}). According to the IUPAC sign convention, as long as the liquid-junction potential is negligible or there is no liquid junction, we may write the cell potential E_{cell} as $E_{cell} = E_{right} - E_{left}$ as given by Equation 22-12. Although we cannot determine absolute potentials of electrodes such as these, we can easily determine relative electrode potentials as discussed in Section 22C-4.

Any sign convention must be based on half-cell processes written in a single way — that is, entirely as oxidations or as reductions. According to the IUPAC convention, the term electrode potential (or more exactly, relative electrode potential) is reserved exclusively for half-reactions written as reductions. There is no objection to using the term *oxidation potential* to indicate an electrode process written in the opposite sense, but *an oxidation potential should never be called an electrode potential.*

22C-6 Effect of Activity on Electrode Potential

Let us consider the half-reaction

$$pP + qQ + \cdots + ne^- \rightleftharpoons rR + sS$$

where the capital letters represent formulas of reacting species (whether charged or uncharged), e^- represents the electron, and the lowercase italic letters indicate the number of moles of each species (including electrons) participating in the half-cell reaction. By invoking the same arguments that we used in the case of the silver–silver chloride–SHE cell in Section 22B-1, we obtain

$$E = E^0 - \frac{RT}{nF} \ln \frac{(a_R)_i^r \cdot (a_S)_i^s \cdots}{(a_P)_i^p \cdot (a_Q)_i^q \cdots}$$

At room temperature (298 K), the collection of constants in front of the logarithm has units of joules per coulomb or volt. Therefore,

$$\frac{RT}{nF} = \frac{8.316 \, \text{J mol}^{-1}\text{K}^{-1} \times 298 \, \text{K}}{n \times 96487 \, \text{C mol}^{-1}}$$

$$= \frac{2.568 \times 10^{-2} \, \text{J C}^{-1}}{n} = \frac{2.568 \times 10^{-2}}{n} \, \text{V}$$

When we convert from natural (ln) to base ten logarithms (log) by multiplying by 2.303, the previous equation can be written as

$$E = E^0 - \frac{0.0592}{n} \log \frac{(a_R)^r \cdot (a_S)^s \cdots}{(a_P)^p \cdot (a_Q)^q \cdots} \quad (22\text{-}13)$$

For convenience, we have also deleted the i subscripts, which were inserted earlier as a reminder that the bracketed terms represented nonequilibrium concentrations. Hereafter, *the subscripts will not be used*; you should, however, keep in mind that the quotients that appear in this type of equation are *not equilibrium constants*, despite their similarity in appearance.

Equation 22-13 is a general statement of the *Nernst equation*, which can be applied to both half-cell reactions and cell reactions.

22C-7 The Standard Electrode Potential, E^0

Equation 22-13 reveals that the constant E^0 is equal to the half-cell potential when the logarithmic term is zero. This condition occurs whenever the activity quotient is equal to unity, such as, for example, when the activities of all reactants and products are unity. Thus, the standard potential is often defined as the electrode potential of a half-cell reaction (versus SHE) when all reactants and products have unit activity.

The standard electrode potential is an important physical constant that gives a quantitative description of the relative driving force for a half-cell reaction. Keep in mind the following four facts regarding this constant. (1) The electrode potential is temperature dependent; if it is to have significance, the temperature at which it is determined must be specified. (2) The standard electrode potential is a relative quantity in the sense that it is really the potential of an electrochemical cell in which the left electrode is a carefully specified reference electrode — the SHE — whose potential is *assigned* a value of zero. (3) The sign of a standard potential is identical with that of the conductor in the right-hand electrode in a galvanic cell, with the left electrode being the SHE. (4) The standard potential is a measure of the driving force for a half-reaction. As such, it is independent of the notation used to express the half-cell process. Thus, although the potential for the process

$$Ag^+ + e^- \rightleftharpoons Ag(s) \qquad E^0 = +0.799 \, \text{V}$$

depends on the concentration of silver ions, it is the same regardless of whether we write the half-reaction as the preceding or as

$$100 \, Ag^+ + 100 \, e^- \rightleftharpoons 100 \, Ag(s) \qquad E^0 = +0.799 \, \text{V}$$

The Nernst equation must be consistent with the half-reaction as it is written. For the reaction written for one mole of silver, the Nernst equation is

$$E = 0.799 - \frac{0.0592}{1} \log \frac{1}{a_{Ag^+}}$$

and for 100 moles of silver,

$$E = 0.799 - \frac{0.0592}{100} \log \frac{1}{(a_{Ag^+})^{100}}$$

$$= 0.799 - \frac{0.0592}{\cancel{100}} \left[\cancel{100} \log \frac{1}{a_{Ag^+}} \right]$$

$$= 0.799 - \frac{0.0592}{1} \log \frac{1}{a_{Ag^+}}$$

Standard electrode potentials have been tabulated for many half-reactions. Many have been determined directly from voltage measurements of cells with a hydrogen electrode or other reference electrode as reference. It is possible, however, to calculate E^0 values from equilibrium studies of oxidation-reduction systems and from thermochemical data relating to such reactions. Many of the values found in the literature were calculated in this way.[7]

For illustrative purposes, a few standard electrode potentials are given in Table 22-1; a more comprehensive table is found in Appendix 3. The species in the upper-left-hand part of the equations in Table 22-1 are most easily reduced, as indicated by the large positive E^0 values; they are therefore the most effective oxidizing agents. Proceeding down the left-hand side of the table, each succeeding species is a less effective acceptor of electrons than the one above it, as indicated by its increasingly negative standard potential. The half-cell reactions at the bottom of the table have little tendency to take place as *reductions* as written. On the other hand, they do tend to proceed in the opposite sense, as *oxidations*. The most effective reducing agents, then, are species that appear in the lower-right-hand side of the equations in the table.

Compilations of standard potentials provide information regarding the extent and direction of electron-transfer reactions between the tabulated species. Table 22-1 suggests, for example, that zinc is more easily oxidized than cadmium, and we conclude that when a piece of zinc is immersed in a solution of cadmium ions, metallic cadmium will deposit on the surface of the zinc, which will dissolve and produce zinc ions as long as there is an appreciable concentration of cadmium ions in the solution. On the other hand, cadmium has no tendency to reduce zinc ions, so a piece of cadmium immersed in a solution of zinc ions will remain unaffected. Table 22-1 also shows that iron(III) is a better oxidizing agent than triiodide ion. Therefore, in a solution containing an equilibrium mixture of iron(III), iodide, iron(II), and triiodide ions, we predict that iron(II) and triiodide will predominate.

22C-8 Measuring Electrode Potentials

Although the SHE is the universal reference standard, we must emphasize once again that the electrode, as described, is almost never used in the laboratory; it is a

TABLE 22-1 Standard Electrode Potentials

Reaction	E^0 at 25°C, V
$Cl_2(g) + 2e^- \rightleftharpoons 2Cl^-$	+1.359
$O_2(g) + 4H^+ + 4e^- \rightleftharpoons 2H_2O$	+1.229
$Br_2(aq) + 2e^- \rightleftharpoons 2Br^-$	+1.087
$Br_2(l) + 2e^- \rightleftharpoons 2Br^-$	+1.065
$Ag^+ + e^- \rightleftharpoons Ag(s)$	+0.799
$Fe^{3+} + e^- \rightleftharpoons Fe^{2+}$	+0.771
$I_3^- + 2e^- \rightleftharpoons 3I^-$	+0.536
$Cu^{2+} + 2e^- \rightleftharpoons Cu(s)$	+0.337
$Hg_2Cl_2(s) + 2e^- \rightleftharpoons 2Hg(l) + 2Cl^-$	+0.268
$AgCl(s) + e^- \rightleftharpoons Ag(s) + Cl^-$	+0.222
$Ag(S_2O_3)_2^{3-} + e^- \rightleftharpoons Ag(s) + 2S_2O_3^{2-}$	+0.010
$2H^+ + 2e^- \rightleftharpoons H_2(g)$	0.000
$AgI(s) + e^- \rightleftharpoons Ag(s) + I^-$	−0.151
$PbSO_4(s) + 2e^- \rightleftharpoons Pb(s) + SO_4^{2-}$	−0.350
$Cd^{2+} + 2e^- \rightleftharpoons Cd(s)$	−0.403
$Zn^{2+} + 2e^- \rightleftharpoons Zn(s)$	−0.763

See Appendix 3 for a more extensive list.

hypothetical electrode to which experimentally determined potentials can be referred only by suitable computation. The reason that the electrode, as defined, cannot be prepared is that we are unable to prepare a solution with a hydrogen ion activity of exactly unity. Neither the Debye-Hückel theory (see Appendix 2) nor any other theory of electrolyte solutions permits the determination of the activity coefficient of hydrogen ions in solutions with ionic strength approaching unity, as required by the definition of the SHE. Thus, the *concentration* of HCl or another acid required to give a hydrogen ion activity of unity cannot be calculated accurately. In spite of this limitation, data for more dilute solutions of acid, for which activity coefficients can be determined, can be used to compute *hypothetical* potentials at unit activity. The example that follows illustrates how standard potentials can be obtained in this way.

EXAMPLE 22-1

D.A. MacInnes[8] found that a cell without liquid junction similar to that shown in Figure 22-2 has a potential of 0.52053 V. The cell is described by

$$Pt, H_2(1.00 \text{ atm}) | HCl(3.215 \times 10^{-3} \text{ M}), AgCl(\text{sat'd}) | Ag$$

---

[7]Comprehensive sources for standard electrode potentials include *Standard Electrode Potentials in Aqueous Solutions*, A. J. Bard, R. Parsons, and J. Jordan, eds., New York: Dekker, 1985; G. Milazzo, S. Caroli, and V. K. Sharma, *Tables of Standard Electrode Potentials*, New York: Wiley-Interscience, 1978; M. S. Antelman and F. J. Harris, *Chemical Electrode Potentials*, New York: Plenum Press, 1982. Some compilations are arranged alphabetically by element; others are tabulated according to the value of E^0.

[8]D. A. MacInnes, *The Principles of Electrochemistry*, p. 187, New York: Reinhold, 1939.

Calculate the standard electrode potential for the half-reaction

$$AgCl(s) + e^- \rightleftharpoons Ag(s) + Cl^-$$

▶ *Solution*

The electrode potential for the right-hand electrode is

$$E_{\text{right}} = E^0_{\text{AgCl}} - 0.0592 \log a_{\text{Cl}^-}$$
$$= E^0_{\text{AgCl}} - 0.0592 \log \gamma_{\text{Cl}^-} c_{\text{HCl}}$$

where γ_{Cl^-} is the activity coefficient of Cl^-. The left half-cell reaction is

$$H^+ + e^- \rightleftharpoons \frac{1}{2}H_2(g)$$

and

$$E_{\text{left}} = E^0_{\text{H}_2} - \frac{0.0592}{1} \log \frac{P_{\text{H}_2}^{1/2}}{a_{\text{H}^+}}$$
$$= E^0_{\text{H}_2} - \frac{0.0592}{1} \log \frac{P_{\text{H}_2}^{1/2}}{\gamma_{\text{H}^+} c_{\text{HCl}}}$$

The measured potential is the difference between these half-cell potentials (Equation 22-12):

$$E_{\text{cell}} = (E^0_{\text{AgCl}} - 0.0592 \log \gamma_{\text{Cl}^-} c_{\text{HCl}})$$
$$- \left(0.000 - 0.0592 \log \frac{P_{\text{H}_2}^{1/2}}{\gamma_{\text{H}^+} c_{\text{HCl}}} \right)$$

Combining the two logarithmic terms gives

$$E_{\text{cell}} = E^0_{\text{AgCl}} - 0.0592 \log \frac{\gamma_{\text{H}^+} \gamma_{\text{Cl}^-} c_{\text{HCl}}^2}{P_{\text{H}_2}^{1/2}}$$

The activity coefficients for H^+ and Cl^- can be calculated from Equation a2-3 (Appendix 2) taking 3.215×10^{-3} for the ionic strength μ; these values are 0.945 and 0.939, respectively. If we substitute these activity coefficients and the experimental data into the previous equation and rearrange, we find

$$E^0_{\text{AgCl}} = 0.5203$$
$$+ 0.0592 \log \left[\frac{(3.215 \times 10^{-3})^2 (0.945)(0.939)}{1.00^{1/2}} \right]$$
$$= 0.2223 \approx 0.222 \text{ V}$$

(The mean for this and similar measurements at other concentrations of HCl was 0.222 V.)

22C-9 Calculating Half-Cell Potentials from E^0 Values

We illustrate typical applications of the Nernst equation to calculate half-cell potentials with the following examples.

EXAMPLE 22-2

What is the electrode potential for a half-cell consisting of a cadmium electrode immersed in a solution that is 0.0150 M in Cd^{2+}?

▶ *Solution*

From Table 22-1, we find

$$Cd^{2+} + e^- \rightleftharpoons Cd(s) \qquad E^0 = -0.403 \text{ V}$$

We will assume that $a_{\text{Cd}^{2+}} \approx [Cd^{2+}]$ and write

$$E_{\text{Cd}} = E^0_{\text{Cd}} - \frac{0.0592}{2} \log \frac{1}{[Cd^{2+}]}$$

Substituting the Cd^{2+} concentration into this equation gives

$$E_{\text{Cd}} = -0.403 - \frac{0.0592}{2} \log \frac{1}{0.0150} = -0.457 \text{ V}$$

The sign for the potential calculated in Example 22-2 indicates the direction of the reaction when this half-cell is coupled with the SHE. The fact that it is negative shows that the reaction

$$Cd(s) + 2H^+ \rightleftharpoons H_2(g) + Cd^{2+}$$

occurs spontaneously. Note that the calculated potential is a larger negative number than the standard electrode potential itself. This follows from mass-law considerations because the half-reaction, as written, has less tendency to occur with the lower cadmium ion concentration.

EXAMPLE 22-3

Calculate the potential for a platinum electrode immersed in a solution prepared by saturating a 0.0150 M solution of KBr with Br_2.

▶ *Solution*

The half-reaction is

$$Br_2(l) + 2e^- \rightleftharpoons 2Br^- \qquad E^0 = 1.065 \text{ V}$$

Notice that the *l* following Br_2 indicates that the aqueous solution is saturated with *liquid* Br_2. Thus, the overall process is the sum of the two equilibria

$$Br_2(l) \rightleftharpoons Br_2(\text{sat'd } aq)$$

$$Br_2(\text{sat'd } aq) + 2e^- \rightleftharpoons 2Br^-$$

Assuming that $[Br^-] = a_{Br^-}$, the Nernst equation for the overall process is

$$E = 1.065 - \frac{0.0592}{2} \log \frac{[Br^-]^2}{1.00}$$

The activity of Br_2 in the pure liquid is constant and equal to 1.00 by definition. Thus,

$$E = 1.065 - \frac{0.0592}{2} \log (0.0150)^2$$

$$= 1.173 \text{ V}$$

EXAMPLE 22-4

Calculate the potential for a platinum electrode immersed in a solution that is 0.0150 M in KBr and 1.00×10^{-3} M in Br_2.

Solution

In this example, the half-reaction used in the preceding example *does not apply because the solution is no longer saturated in Br₂*. Table 22-1, however, contains the half-reaction

$$Br_2(aq) + 2e^- \rightleftharpoons 2Br^- \qquad E^0 = 1.087 \text{ V}$$

The *aq* implies that all of the Br_2 present is in solution and that 1.087 V is the electrode potential for the half-reaction when the Br^- and $Br_2(aq)$ activities are 1.00 mol/L. It turns out, however, that the solubility of Br_2 in water at 25°C is only about 0.18 mol/L. Therefore, the recorded potential of 1.087 V is based on a hypothetical system that cannot be achieved experimentally. Nevertheless, this potential is useful because it provides the means by which potentials for undersaturated systems can be calculated. Thus, if we assume that activities of solutes are equal to their molar concentrations, we obtain

$$E = 1.087 - \frac{0.0592}{2} \log \frac{[Br^-]^2}{[Br_2]}$$

$$E = 1.087 - \frac{0.0592}{2} \log \frac{(1.50 \times 10^{-2})^2}{1.00 \times 10^{-3}}$$

$$= 1.106 \text{ V}$$

In this case, the Br_2 activity is 1.00×10^{-3} rather than 1.00, as was the situation when the solution was saturated and excess $Br_2(l)$ was present.

22C-10 Electrode Potentials in the Presence of Precipitation and Complex-Forming Reagents

The following example shows that reagents that react with the participants of an electrode process have a significant effect on the potential for the process.

EXAMPLE 22-5

Calculate the potential of a silver electrode in a solution that is saturated with silver iodide and has an iodide ion activity of exactly 1.00 (K_{sp} for AgI = 8.3×10^{-17}).

$$Ag^+ + e^- \rightleftharpoons Ag(s) \qquad E^0 = +0.799 \text{ V}$$

$$E = +0.799 - 0.0592 \log \frac{1}{a_{Ag^+}}$$

Solution

We may calculate a_{Ag^+} from the solubility-product constant. Thus,

$$a_{Ag^+} = \frac{K_{sp}}{a_{I^-}}$$

Substituting this expression into the Nernst equation gives

$$E = +0.799 - \frac{0.0592}{1} \log \frac{a_{I^-}}{K_{sp}}$$

This equation may be rewritten as

$$E = +0.799 + 0.0592 \log K_{sp} - 0.0592 \log a_{I^-} \quad (22\text{-}14)$$

If we substitute 1.00 for a_{I^-} and use 8.3×10^{-17} for K_{sp}, the solubility product for AgI at 25.0°C, we obtain

$$E = +0.799 + 0.0592 \log 8.3 \times 10^{-17}$$

$$-0.0592 \log 1.00 = -0.151 \text{ V}$$

This example shows that the half-cell potential for the reduction of silver ion becomes smaller in the presence of iodide ions. Qualitatively, this is the expected effect because decreases in the concentration of silver ions diminish the tendency for their reduction.

Equation 22-14 relates the potential of a silver electrode to the iodide ion activity of a solution that is also saturated with silver iodide. *When the iodide ion activity is unity*, the potential is the sum of two constants; it is thus the standard electrode potential for the half-reaction

$$AgI(s) + e^- \rightleftharpoons Ag(s) + I^- \qquad E^0 = -0.151 \text{ V}$$

where

$$E^0_{AgI} = +0.799 + 0.0592 \log K_{sp}$$

The Nernst relationship for the silver electrode in a solution *saturated with silver iodide* can then be written as

$$E = E^0 - 0.0592 \log a_{I^-} = -0.151 - 0.0592 \log a_{I^-}$$

Thus, when in contact with a solution saturated with silver iodide, the potential of a silver electrode can be described *either* in terms of the silver ion activity (with the standard electrode potential for the simple silver half-reaction) *or* in terms of the iodide ion activity (with the standard electrode potential for the silver–silver iodide half-reaction). The silver–silver iodide half-reaction is usually more convenient.

The potential of a silver electrode in a solution containing an ion that forms a soluble complex with silver ion can be handled in a way analogous to the treatment above. For example, in a solution containing thiosulfate and silver ions, complex formation occurs:

$$Ag^+ + 2S_2O_3^{2-} \rightleftharpoons Ag(S_2O_3)_2^{3-}$$

$$K_f = \frac{a_{Ag(S_2O_3)_2^{3-}}}{a_{Ag^+} \cdot (a_{S_2O_3^{2-}})^2}$$

where K_f is the formation constant for the complex. The half-reaction for a silver electrode in such a solution can be written as

$$Ag(S_2O_3)_2^{3-} + e^- \rightleftharpoons Ag(s) + 2S_2O_3^{2-}$$

The standard electrode potential for this half-reaction is the electrode potential when both the complex and the complexing anion are at unit activity. Using the same approach as in the previous example, we find that

$$E^0 = +0.799 + 0.0592 \log \frac{1}{K_f} \qquad (22\text{-}15)$$

Data for the potential of the silver electrode in the presence of selected ions are given in the tables of standard electrode potentials in Appendix 3 and in Table 22-1. Similar information is also provided for other electrode systems. These data often simplify the calculation of half-cell potentials.

22C-11 Some Limitations to the Use of Standard Electrode Potentials

Standard electrode potentials are of great importance in understanding electroanalytical processes. We should emphasize certain inherent limitations to the use of these data, however.

Substitution of Concentrations for Activities

As a matter of convenience, molar concentrations — rather than activities — of reactive species are usually used in making calculations with the Nernst equation. Unfortunately, these two quantities are identical *only in dilute solutions*. With increasing electrolyte concentrations, potentials calculated using molar concentrations are in general quite different from those obtained by experiment.

To illustrate, the standard electrode potential for the half-reaction

$$Fe^{3+} + e^- \rightleftharpoons Fe^{2+}$$

is +0.771 V. Neglecting activities, we predict that a platinum electrode immersed in a solution that contains 1 mol/L each of Fe^{3+} and Fe^{2+} ions in addition to perchloric acid would have a potential numerically equal to this value relative to the SHE. In fact, however, we measure a potential of +0.732 V when the perchloric acid concentration is 1 M. We can see the reason for the discrepancy if we write the Nernst equation in the form

$$E = E^0 - 0.0592 \log \frac{\gamma_{Fe^{2+}}[Fe^{2+}]}{\gamma_{Fe^{3+}}[Fe^{3+}]}$$

where $\gamma_{Fe^{2+}}$ and $\gamma_{Fe^{3+}}$ are the respective activity coefficients. The activity coefficients of the two species are less than one in this system because of the high ionic strength resulting from the perchloric acid and the iron salts. What is more important, however, is that the activity coefficient of the iron(III) ion is smaller than that of the iron(II) ion. As a result, the ratio of the activity coefficients as they appear in the Nernst equation is larger than one, and the potential of the half-cell is smaller than the standard potential.

Activity coefficient data for ions in solutions of the types commonly used in oxidation-reduction titrations and electrochemical work are somewhat limited. As a result, we must use molar concentrations rather than activities in many calculations. Using molar concentrations may cause appreciable errors. Such calculations

 Tutorial: Learn more about **standard cell potentials**.

are still useful, however, because relative changes in values are often close to correct, and the direction and magnitude of changes are sufficiently accurate for many purposes.

Effect of Other Equilibria

The application of standard electrode potentials is further complicated by solvation, dissociation, association, and complex-formation reactions involving the species of interest. An example of this problem is the behavior of the potential of the iron(III)-iron(II) couple. As noted earlier, an equimolar mixture of these two ions in 1 M perchloric acid has an electrode potential of +0.732 V. Substituting hydrochloric acid of the same concentration for perchloric acid alters the observed potential to +0.700 V, and we observe a value of +0.600 V in 1 M phosphoric acid. These changes in potential occur because iron(III) forms more stable complexes with chloride and phosphate ions than does iron(II). As a result, the actual concentration of *uncomplexed* iron(III) in such solutions is less than that of *uncomplexed* iron(II), and the net effect is a shift in the observed potential.

Potentials can be corrected for this effect only if the equilibria involved are known and equilibrium constants for the processes are available. Often, however, such information is unavailable, and we are forced to neglect such effects and hope that serious errors do not appear in the calculated results.

Formal Potentials

To compensate partially for activity effects and errors resulting from side reactions, such as those described in the previous section, Swift proposed substituting a quantity called the *formal potential $E^{0'}$* in place of the standard electrode potential in oxidation-reduction calculations.[9] The formal potential, sometimes referred to as the *conditional potential*, of a system is the potential of the half-cell with respect to the SHE when the concentrations of reactants and products are 1 M and the concentrations of any other constituents of the solution are carefully specified. Thus, for example, the formal potential for the reduction of iron(III) is +0.732 V in 1 M perchloric acid and +0.700 V in 1 M hydrochloric acid. Using these values in place of the standard electrode potential in the Nernst equation will yield better agreement between calculated and experimental potentials, if the electrolyte concentration of the solution approximates that for which the formal potential was measured. Using formal potentials with systems differing significantly in composition and concentration of the electrolyte can, however, lead to errors greater than those produced by using standard potentials. The table in Appendix 3 contains selected formal potentials as well as standard potentials, and in subsequent chapters, we will use whichever is more appropriate.

Reaction Rates

Note that the appearance of a half-reaction in a table of electrode potentials does not necessarily imply that there is a real electrode whose potential corresponds to the half-reaction. Many of the data in such tables were calculated from equilibrium or thermal data, and thus, no actual potential measurements on the electrode system of interest were ever made. For some half-reactions, no suitable electrode is known. For example, the standard electrode potential for the process

$$2CO_2 + 2H^+ + 2e^- \rightleftharpoons H_2C_2O_4 \qquad E^0 = -0.49 \text{ V}$$

was determined indirectly. The electrode reaction is not reversible, and the rate at which carbon dioxide combines to give oxalic acid is negligibly slow. There is no known electrode system with a potential that varies in the expected way with the ratio of the activities of the reactants and products of this reaction. In spite of the absence of direct measurements, the tabulated potential is useful for computational purposes.

22D CALCULATION OF CELL POTENTIALS FROM ELECTRODE POTENTIALS

An important use of standard electrode potentials is the calculation of the potential of a galvanic cell or the potential required to operate an electrolytic cell. These calculated potentials (sometimes called thermodynamic potentials) are theoretical in the sense that they refer to cells in which there is no current.[10] Additional factors must be taken into account when there is a current in the cell. Furthermore, these potentials do not take into account junction potentials within the cell. Normally, junction potentials can be made small enough to be neglected without serious error.

---

[9]E. H. Swift, *A System of Chemical Analysis*, p. 50, San Francisco: Freeman, 1939.

[10]In fact, with modern high-impedance (up to 10^{15} Ω) voltmeters, cell potentials can be measured under essentially zero-current conditions, so the distinction between theoretical and practical cell potentials becomes negligible.

We calculate the voltage of a cell from the difference between two half-cell potentials as shown in Equation 22-12.

$$E_{cell} = E_{right} - E_{left}$$

where E_{right} and E_{left} are the *electrode potentials* for the two half-reactions constituting the cell.

Consider the hypothetical cell

$$Zn|ZnSO_4(a_{Zn^{2+}} = 1.00)\|CuSO_4(a_{Cu^{2+}} = 1.00)|Cu$$

Because the activities of the two ions are both 1.00, the standard potentials are also the electrode potentials. Thus, using E^0 data from Table 22-1,

$$E_{cell} = E_{right} - E_{left} = E_{Cu}^0 - E_{Zn}^0$$
$$= +0.337 - (-0.763) = +1.100\,V$$

The positive sign for the cell potential indicates that the reaction

$$Zn(s) + Cu^{2+} \rightarrow Zn^{2+} + Cu(s)$$

occurs spontaneously under standard conditions, and elemental zinc is oxidized to zinc(II) and copper(II) is reduced to metallic copper.

If the preceding cell is written in the reverse sense, that is, as

$$Cu|Cu^{2+}(a_{Cu^{2+}} = 1.00)\|Zn^{2+}(a_{Zn^{2+}} = 1.00)|Zn$$

we write the cell potential as

$$E_{cell} = E_{right} - E_{left} = E_{Zn}^0 - E_{Cu}^0$$
$$= -0.763 - (+0.337) = -1.100\,V$$

The negative sign indicates that the following reaction is not spontaneous under standard conditions.

$$Cu(s) + Zn^{2+} \not\rightarrow Cu^{2+} + Zn(s)$$

Thus, we must apply an external voltage greater than 1.100 V to cause this reaction to occur.

EXAMPLE 22-6

Calculate potentials for the following cell using (a) concentrations and (b) activities:

$$Zn|ZnSO_4(c_{ZnSO_4}),PbSO_4(sat'd)|Pb$$

where $c_{ZnSO_4} = 5.00 \times 10^{-4}$, 2.00×10^{-3}, 1.00×10^{-2}, 2.00×10^{-2}, and 5.00×10^{-2}.

Solution

(a) In a neutral solution, little HSO_4^- will be formed; thus, we may assume that

$$[SO_4^{2-}] = c_{ZnSO_4} = 5.00 \times 10^{-4}$$

The half-reactions and standard potentials are

$$PbSO_4(s) + 2e^- \rightleftharpoons Pb(s) + SO_4^{2-} \quad E^0 = -0.350\,V$$
$$Zn^{2+} + 2e^- \rightleftharpoons Zn \quad E^0 = -0.763\,V$$

The potential of the lead electrode is given by

$$E_{Pb} = -0.350 - \frac{0.0592}{2} \log 5.00 \times 10^{-4}$$
$$= -0.252\,V$$

The zinc ion concentration is also 5.00×10^{-4} and

$$E_{Zn} = -0.763 - \frac{0.0592}{2} \log \frac{1}{5.00 \times 10^{-4}}$$
$$= -0.860\,V$$

Because the Pb electrode is specified as the right-hand electrode,

$$E_{cell} = -0.252\,V - (-0.860\,V) = 0.608\,V$$

Cell potentials at the other concentrations can be calculated in the same way. Their values are given in column (a) of Table 22-2.

(b) To calculate activity coefficients for Zn^{2+} and SO_4^{2-}, we must first find the ionic strength using Equation a2-3 (Appendix 2). We assume that the concentrations of Pb^{2+}, H^+, and OH^- are negligible compared to the concentrations of Zn^{2+} and SO_4^{2-}. Thus, the ionic strength is

$$\mu = \frac{1}{2}[5.00 \times 10^{-4} \times (2)^2 + 5.00 \times 10^{-4} \times (2)^2]$$
$$= 2.00 \times 10^{-3}$$

In Table a2-1, we find for SO_4^{2-}, $\alpha_A = 0.4$, and for Zn^{2+}, $\alpha_A = 0.6$. Substituting these values into Equation a2-3 gives for sulfate ion

$$-\log \gamma_{SO_4^{2-}} = \frac{0.0509 \times 2^2 \times \sqrt{2.00 \times 10^{-3}}}{1 + 3.28 \times 0.4 \times \sqrt{2.00 \times 10^{-3}}}$$
$$= 0.859 \times 10^{-2}$$
$$\gamma_{SO_4^{2-}} = 0.820$$
$$a_{SO_4^{2-}} = 0.820 \times 5.00 \times 10^{-4}$$
$$= 4.10 \times 10^{-4}$$

Repeating the calculations using $\alpha_A = 0.6$ for Zn^{2+} yields

$$\gamma_{Zn^{2+}} = 0.825$$
$$a_{Zn^{2+}} = 4.13 \times 10^{-4}$$

TABLE 22-2 Calculated Potentials for a Cell Based on (a) Concentrations and (b) Activities (see Example 22-6)

c_{ZnSO_4}, M	μ	(a) E (calc), V	(b) E (calc), V	E (exptl)*, V
5.00×10^{-4}	2.00×10^{-3}	0.608	0.613	0.611
2.00×10^{-3}	8.00×10^{-3}	0.572	0.582	0.583
1.00×10^{-2}	4.00×10^{-2}	0.531	0.549	0.553
2.00×10^{-2}	8.00×10^{-2}	0.513	0.537	0.542
5.00×10^{-2}	2.00×10^{-1}	0.490	0.521	0.529

*Experimental data from I. A. Cowperthwaite and V. K. LaMer, *J. Amer. Chem. Soc.*, **1931**, *53*, 4333.

Note that the potential of this cell drops dramatically when there is a current in the cell.

The Nernst equation for the Pb electrode now becomes

$$E_{Pb} = -0.350 - \frac{0.0592}{2} \times \log 4.10 \times 10^{-4}$$

$$= -0.250$$

For the zinc electrode

$$E_{Zn} = -0.763 - \frac{0.0592}{2} \times \log \frac{1}{4.13 \times 10^{-4}}$$

$$= -0.863$$

and

$$E_{cell} = -0.250 - (-0.863) = 0.613 \text{ V}$$

Note that the value calculated using concentrations (0.608 V) is about 1% different from this value. Values at other concentrations are listed in column (b) of Table 22-2.

It is interesting to compare the calculated cell potentials shown in the columns labeled (a) and (b) in Table 22-2 with the experimental results shown in the last column. Clearly, the use of activities provides a significant improvement at higher ionic strengths.

22E CURRENTS IN ELECTROCHEMICAL CELLS

Only one general type of electroanalytical method is based on measurements made in the absence of appreciable current: potentiometric methods, which are discussed in Chapter 23. The remaining electroanalytical methods, considered in Chapters 24 and 25, all involve electrical currents and current measurements. Thus, we must discuss the behavior of cells with significant currents.

As noted earlier, electricity is carried within a cell by the movement of ions. With small currents, Ohm's law (See Section 2A-1) is usually obeyed and we may write $E = IR$, where E is the potential difference in volts responsible for movement of the ions, I is the current in amperes, and R is the resistance in ohms of the electrolyte to the current. The resistance depends on the nature and concentrations of ions in the solution.

When there is a direct current in an electrochemical cell, the measured cell potential is usually different from the thermodynamic potential calculated as demonstrated in Section 22B. This difference can be attributed to a number of phenomena, including ohmic resistance and several *polarization effects*, such as *charge-transfer overvoltage*, *reaction overvoltage*, *diffusion overvoltage*, and *crystallization overvoltage*. Generally, these phenomena reduce the voltage of a galvanic cell or increase the voltage required to produce current in an electrolytic cell.

22E-1 Ohmic Potential: IR Drop

To develop a current in either a galvanic or an electrolytic cell, a driving force in the form of a voltage is required to overcome the resistance of the ions to movement toward the anode and the cathode. Just as in metallic conduction, this force follows Ohm's law and is equal to the product of the current in amperes and the resistance of the cell in ohms. This voltage is generally referred to as the *ohmic potential*, or the *IR drop*.

The net effect of *IR* drop is to increase the potential required to operate an electrolytic cell and to decrease the measured potential of a galvanic cell. Therefore, the *IR* drop is always subtracted from the theoretical cell potential. That is,[11]

$$E_{cell} = E_{right} - E_{left} - IR \qquad (22\text{-}16)$$

[11] In this and subsequent discussions we assume that the junction potential is negligible compared to other potentials.

EXAMPLE 22-7

The following cell has a resistance of 4.00 Ω. Calculate its potential when it is producing a current of 0.100 A.

$$Cd|Cd^{2+} (0.0100 \text{ M})\|Cu^{2+} (0.0100 \text{ M})|Cu$$

Solution

By substituting the standard potentials and concentrations into the Nernst equation, we find that the electrode potential for the Cu electrode is 0.278 V, and for the Cd electrode it is −0.462 V. Thus, the thermodynamic cell potential is

$$E = E_{Cu} - E_{Cd} = 0.278 - (-0.462) = 0.740 \text{ V}$$

and the potential to yield the desired current is

$$E_{cell} = 0.740 \text{ V} - IR$$
$$= 0.740 \text{ V} - (0.100 \text{ A} \times 4.00 \text{ Ω}) = 0.340 \text{ V}$$

EXAMPLE 22-8

Calculate the potential required to generate a current of 0.100 A in the reverse direction in the cell shown in Example 22-7.

Solution

$$E = E_{Cd} - E_{Cu}$$
$$= -0.462 \text{ V} - 0.278 \text{ V} = -0.740 \text{ V}$$
$$E_{cell} = E - IR$$
$$= -0.740 \text{ V} - (0.100 \text{ A} \times 4.00 \text{ Ω}) = -1.140 \text{ V}$$

In this example, an external potential greater than 1.140 V is needed to cause Cd^{2+} to deposit and Cu to dissolve at a rate required for a current of 0.100 A.

22E-2 Polarization

In several important electroanalytical methods we measure the current in a cell as a function of potential and construct current versus voltage curves from the data. Equation 22-16 predicts a linear relationship between the cell voltage and the current at constant electrode potentials. In fact, current-voltage curves are frequently nonlinear at the extremes; under these circumstances, the cell is *polarized*. Polarization may occur at one or both electrodes.

To begin our discussion, it is useful to consider current-voltage curves for an *ideal polarized* and an *ideal nonpolarized electrode*. Polarization at a single electrode can be studied by coupling it with an electrode that is not easily polarized. Such electrodes have large surface areas and have half-cell reactions that are rapid and reversible. Design details of nonpolarized electrodes are described in subsequent chapters.

Ideal Polarized and Nonpolarized Electrodes and Cells

The ideal polarized electrode is one in which current remains constant and independent of potential over a wide range. Figure 22-6a is a current-voltage curve for an electrode that is ideally polarized in the region between A and B. Figure 22-6b shows the current-voltage relationship for a nonpolarized electrode that behaves ideally in the region between A and B. For this electrode, the potential is independent of the current.

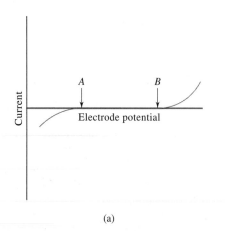

(a)

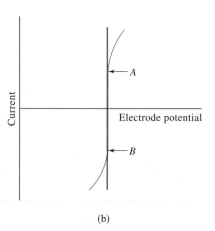

(b)

FIGURE 22-6 Current-voltage curves in blue for an ideal (a) polarized and (b) nonpolarized electrode. Gray lines show departure from ideal behavior by real electrodes.

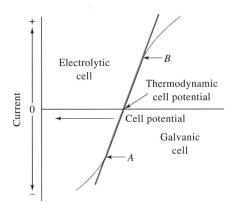

FIGURE 22-7 Current-voltage curve for a cell showing nonpolarized behavior between *A* and *B* (blue line) and polarized behavior (gray line).

Figure 22-7 is a current-voltage curve for a cell having electrodes that exhibit nonpolarized behavior between points *A* and *B*. Because of the internal resistance of the cell, the current-voltage curve has a finite slope equal to *R* (Equation 22-16) rather than the infinite slope for the ideal nonpolarized electrode shown in Figures 22-6b, and it obeys Ohm's law. Beyond points *A* and *B*, polarization occurs at one or both electrodes, resulting in deviations from the ideal straight line. The upper half of the curve gives the current-voltage relationship when the cell is operating as an electrolytic cell; the lower half describes its behavior as a galvanic cell.[12] Note that when polarization occurs in an electrolytic cell, a higher potential than that predicted from the ideal line is required to achieve a given current. Similarly, polarization of a galvanic cell produces a potential that is lower than expected.

Sources of Polarization in Electrolytic Cells

Figure 22-8 shows three regions of a half-cell in an electrolytic cell where polarization can occur. These regions include the electrode itself, a surface film of solution immediately adjacent to the electrode, and the bulk of the solution. For this half-cell, the overall electrode reaction is

$$Ox + ne^- \rightleftharpoons Red$$

Any one of the several intermediate steps shown in the figure may, however, limit the rate of this overall reac-

tion, which thus limits the magnitude of the current. One of these steps in the reaction, called *mass transfer*, involves movement of Ox from the bulk of the solution to the surface film. When this step (or the reverse mass transfer of Red to the bulk) limits the rate of the overall reaction (and thus the current), we have *concentration polarization*. Some half-cell reactions proceed by an intermediate chemical reaction in which species such as Ox′ or Red′ form; this intermediate is then the actual participant in the electron-transfer process. If the rate of formation or decomposition of the intermediate limits the current, *reaction polarization* occurs. In some instances, the rate of a physical process such as adsorption, desorption, or crystallization limits the current. In these cases, *adsorption*, *desorption*, or *crystallization polarization* occurs. Finally, when the current is limited by a slow rate of electron transfer from the electrode to the oxidized species in the surface film or from the reduced species to the electrode, we have *charge-transfer polarization*. It is not unusual to find several types of polarization occurring simultaneously.

Overvoltage

The degree of polarization of an electrode in an electrolytic cell is measured by the *overvoltage*, or *overpotential*, η, which is the difference between the actual electrode potential E and the thermodynamic, or equilibrium, potential E_{eq}. That is,

$$\eta = E - E_{eq} \qquad (22\text{-}17)$$

Here, η is the additional potential beyond the thermodynamic value needed to cause the reaction to occur at an appreciable rate. For cathodic reactions where E_{eq} is negative, *E must be more negative than* E_{eq}, and η is negative. For anodic reactions where E_{eq} is positive, E must be more positive than E_{eq}, and η is positive.

Concentration Polarization

Concentration polarization occurs when the rate of transport of reactive species to the electrode surface is not sufficient to maintain the current required by Equation 22-16. When concentration polarization begins, a diffusion overvoltage appears.

For example, consider a cell made up of an ideal nonpolarized anode and a polarizable cathode consisting of a small cadmium electrode immersed in a solution of cadmium ions. The reduction of cadmium ions is a rapid and reversible process so that when a potential is applied to this electrode, the surface layer of the solution comes to equilibrium with the electrode essentially

---

[12] In this book we follow the convention that cathodic currents are positive and anodic currents negative. This practice came about for historical reasons, because reductions were most often studied. Some electrochemists prefer to consider anodic currents as positive. When looking at current-potential curves, it is always wise to decide which convention is being followed.

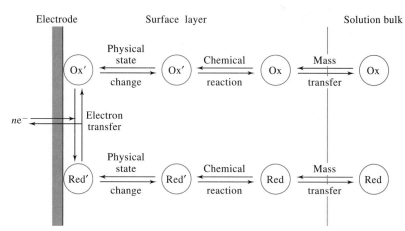

FIGURE 22-8 Steps in the reaction Ox + ne^- ⇌ Red at an electrode. Note that the surface layer is only a few molecules thick. (Adapted from A. J. Bard and L. R. Faulkner, *Electrochemical Methods,* 2nd ed., p. 23, New York: Wiley, 2001. Reprinted by permission of John Wiley & Sons, Inc.)

instantaneously. That is, a brief current is generated that reduces the surface concentration of cadmium ions to the equilibrium concentration, c_0, given by

$$E_{Cd} = E_{Cd}^0 - \frac{0.0592}{2} \log \frac{1}{c_0} \qquad (22\text{-}18)$$

If there were no mechanism to transport cadmium ions from the bulk of the solution to the surface film, the current would rapidly decrease to zero as the concentration of the film approached c_0. As we shall see, however, several mechanisms bring cadmium ions from the bulk of the solution into the surface layer at a constant rate. Therefore, the large initial current decreases rapidly to a constant level determined by the rate of ion transport.

It is important to appreciate that for a rapid and reversible electrode reaction, the concentration of the surface layer may always be considered to be the equilibrium concentration, which is determined by the instantaneous electrode potential (Equation 22-18). It is also important to realize that the surface concentration c_0 is often far different from the concentration in the bulk of the solution. This is true because, even though surface equilibrium is achieved essentially instantaneously, equilibrium between the electrode and the bulk of the solution is established quite slowly and may require minutes or even hours.

For a current of the magnitude required by Equation 22-16 to be maintained, reactant must be brought from the bulk of the solution to the surface layer at a rate dn_A/dt that is given by

$$I = dQ/dt = nF\, dn_A/dt$$

where dQ/dt is the rate of flow of electrons in the electrode (or the current I), n is the number of electrons appearing in the half-reaction, n_A is the number of moles of analyte brought to the surface, and F is the faraday. The rate of reaction (mole/s) can be written as

$$\frac{dn_A}{dt} = AJ$$

where A is the surface area of the electrode (cm^2) and J is the flux ($mol\ s^{-1}\ cm^{-2}$). The two equations can then be combined to give

$$I = nFAJ \qquad (22\text{-}19)$$

When the mass-transport process cannot meet the demand for reactant, the IR drop in Equation 22-16 becomes smaller than the theoretical value, and a diffusion overvoltage appears that just offsets the decrease in IR. Here, we consider an electrolytic cell to which we apply a negative voltage to produce a reduction at the cathode. We assume that the anode is nonpolarized. Thus, with the appearance of concentration polarization, Equation 22-16 becomes

$$E_{cell} = E_{cathode} - E_{anode} - IR + \eta_{cathode}$$

where $\eta_{cathode}$ represents the overpotential associated with the cathode. A more general equation for a cell in which both electrodes are polarized is

$$E_{cell} = (E_{cathode} - E_{anode}) + [(\eta_{cathode} - \eta_{anode}) - IR] \qquad (22\text{-}20)$$

where η_{anode} is the anodic overvoltage. Note that the overvoltage associated with the cathode is negative and the overvoltage at the anode is positive. The overvoltages at each electrode have the effect of reducing the overall potential of the cell. Analogous expressions can be written for a cell in which the reaction under study is an oxidation at the anode.

 Tutorial: Learn more about **cells under non-standard conditions**.

22E-3 Mechanisms of Mass Transport

It is important now to investigate the mechanisms by which ions or molecules are transported from the bulk of the solution to a surface layer (or the reverse) because these mechanisms provide insights into how concentration polarization can be prevented or induced as required. As we noted in Section 22A-7, there are three mechanisms of mass transport: (1) *diffusion*, (2) *migration*, and (3) *convection*.

Whenever there is a concentration difference between two regions of a solution, as happens when a species is reduced at a cathode surface (or oxidized at an anode surface), diffusion causes ions or molecules to move from the more concentrated region to the more dilute. The rate of diffusion dc/dt is given by

$$dc/dt = k(c - c_0) \qquad (22\text{-}21)$$

where c is the reactant concentration in the bulk of the solution, c_0 is its equilibrium concentration at the electrode surface, and k is a proportionality constant. As shown earlier, *the value of c_0 is fixed by the potential of the electrode and can be calculated from the Nernst equation.* As increasingly higher potentials are applied to the electrode, c_0 becomes smaller and smaller, and the diffusion rate becomes greater and greater. Ultimately, however, c_0 becomes negligible with respect to c, and the rate then becomes constant. That is, when $c_0 \to 0$,

$$dc/dt = kc \qquad (22\text{-}22)$$

Under this condition, concentration polarization is said to be complete, and the electrode operates as an ideal polarized electrode.

Migration is the process by which ions move under the influence of an electrostatic field. It is often the primary mass-transfer process in the bulk of the solution in a cell. The electrostatic attraction (or repulsion) between a particular ionic species and the electrode becomes smaller as the total electrolyte concentration of the solution becomes greater. It may approach zero when the reactive species is only a small fraction, say 1/100, of the total concentration of ions with a given charge.

Reactants can also be transferred mechanically to or from an electrode. Thus, forced convection, such as stirring or agitation, tends to decrease concentration polarization. Natural convection resulting from temperature or density differences also contributes to material transport.

 Animation: Learn more about **mass transport mechanisms**.

To summarize, concentration polarization is observed when diffusion, migration, and convection are insufficient to transport the reactant to or from an electrode surface at a rate demanded by the theoretical current. Because of concentration polarization, a larger potential must be applied to an electrolytic cell than the value predicted from the thermodynamic potential and the *IR* drop.

Concentration polarization is important in several electroanalytical methods. In some applications, steps are taken to eliminate it; in others, however, it is essential to the method, and every effort is made to promote it. The degree of concentration polarization is influenced experimentally by (1) the reactant concentration, with polarization becoming more pronounced at low concentrations; (2) the total electrolyte concentration, with polarization becoming more pronounced at high concentrations; (3) mechanical agitation, with polarization decreasing in well-stirred solutions; and (4) electrode size, with polarization effects decreasing as the electrode surface area increases.

22E-4 Charge-Transfer Polarization

Charge-transfer polarization occurs when the rate of the oxidation or reduction reaction at one or both electrodes is not sufficiently rapid to yield currents of the size suggested by theory. Overvoltage caused by charge-transfer polarization has the following characteristics:

1. Overvoltages increase with current density (current density is defined as the current per unit area [A/cm^2] of electrode surface).
2. Overvoltages usually decrease with increases in temperature.
3. Overvoltages vary with the chemical composition of the electrode, often being most pronounced with softer metals such as tin, lead, zinc, and particularly mercury.
4. Overvoltages are especially significant for electrode processes that yield gaseous products such as hydrogen or oxygen, but they are frequently negligible when a metal is being deposited or when an ion is undergoing a change of oxidation state.
5. The magnitude of overvoltage in any given situation cannot be exactly predicted because it is determined by a number of uncontrollable variables.[13]

---

[13] Overvoltage data for various gaseous species at different electrode surfaces are in *Analytical Chemistry Handbook*, J. A. Dean, ed., pp. 14.96–14.97, New York: McGraw-Hill, 1995.

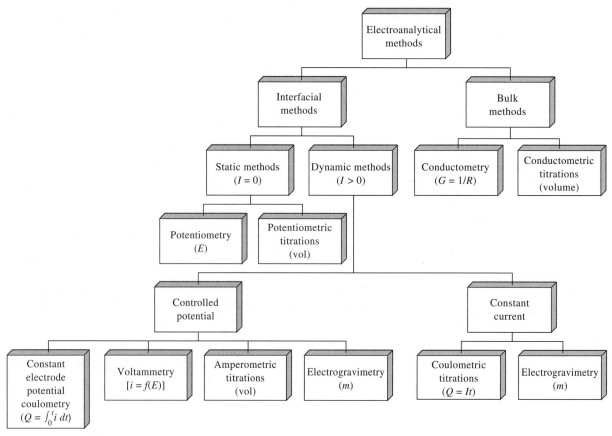

FIGURE 22-9 Summary of common electroanalytical methods. Quantity measured given in parentheses. (I or i = current, E = potential, R = resistance, G = conductance, Q = quantity of charge, t = time, vol = volume of a standard solution, m = mass of an electrodeposited species.)

The overvoltage for the evolution of hydrogen and oxygen is especially significant. The difference between the overvoltage of these gases on smooth and on platinized platinum surfaces is striking. For example, for a smooth Pt electrode immersed in 1 M H_2SO_4 at a current density of 1.0 A/cm^2, the overvoltage is 0.68 V; for a platinized Pt electrode under the same conditions, the overvoltage is only 0.048 V. This difference is primarily due to the much larger surface area associated with platinized electrodes, which produces a real current density significantly smaller than is apparent from the overall dimensions of the electrode. Platinized surfaces are always used in constructing hydrogen reference electrodes to lower the current density to a point where the overvoltage is negligible.

The high overvoltage associated with the formation of hydrogen permits the electrolytic deposition of sev-eral metals that require potentials at which hydrogen would otherwise be expected to interfere. For example, the standard potentials of hydrogen and zinc suggest that rapid evolution of hydrogen should occur well below the potential required for the deposition of zinc from a neutral solution. Nevertheless, zinc can be deposited quantitatively using a mercury or copper electrode. Because of the high overvoltage of hydrogen on these metals (1.07 V and 1.23 V, respectively, under the conditions cited in the previous paragraph), little or no gas is evolved during the electrodeposition.

The magnitude of overvoltage can, at best, be only crudely approximated from empirical information available in the literature. Calculation of cell potentials in which overvoltage plays a part cannot, therefore, be very accurate.

22F TYPES OF ELECTROANALYTICAL METHODS

Many types of electroanalytical methods have been developed. Many of the methods that are generally useful, and are discussed in this book, are shown in Figure 22-9. These methods are divided into interfacial methods and bulk methods. Interfacial methods, which are more widely used than bulk methods, are based on phenomena that occur at the interface between electrode surfaces and the thin layer of solution just adjacent to these surfaces. Bulk methods, in contrast, are based on phenomena that occur in the bulk of the solution; every effort is made to avoid interfacial effects.

Interfacial methods can be divided into two major categories, static methods and dynamic methods, depending on whether there is a current in the electrochemical cells. The static methods, which require potentiometric measurements, are extremely important because of their speed and selectivity. We will discuss potentiometric methods in Chapter 23.

Dynamic interfacial methods, in which currents in electrochemical cells play a vital part, are of several types. In three of the methods shown on the left in Figure 22-9, the potential of the cell is controlled while measurements of other variables are made. Generally, these methods are sensitive and have relatively wide dynamic ranges (typically, 10^{-3} to 10^{-8} M). Furthermore, many of these procedures can be carried out with microliter or even nanoliter volumes of sample. Thus, these methods may achieve detection limits in the picomole range.

In constant-current dynamic methods, the current in the cell is held constant while data are collected. Dynamic methods of both kinds are discussed in Chapters 24 and 25. Most of the electroanalytical techniques shown in Figure 22-9 have been used as detectors in various chromatographic procedures (see Chapters 25, 27, and 28).

QUESTIONS AND PROBLEMS

*Answers are provided at the end of the book for problems marked with an asterisk.

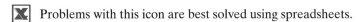 Problems with this icon are best solved using spreadsheets.

*22-1 Calculate the electrode potentials of the following half-cells.
    (a) $Ag^+(0.0261\ M)|Ag$
    (b) $Fe^{3+}(6.72 \times 10^{-4}\ M),Fe^{2+}(0.100\ M)|Pt$
    (c) $AgBr(sat'd),Br^-(0.050\ M)|Ag$

*22-2 Calculate the electrode potentials of the following half-cells.
    (a) $HCl(1.76\ M)|H_2(0.987\ atm),Pt$
    (b) $IO_3^-(0.194\ M),I_2(2.00 \times 10^{-4}\ M),H^+(3.50 \times 10^{-3}\ M)|Pt$
    (c) $Ag_2CrO_4(sat'd),CrO_4^{2-}(0.0520\ M)|Ag$

*22-3 For each of the following half-cells, compare electrode potentials calculated from (1) concentration and (2) activity data.
    (a) $HCl(0.0200\ M),NaCl(0.0300\ M)|H_2(1.00\ atm),Pt$
    (b) $Fe(ClO_4)_2(0.0111\ M),Fe(ClO_4)_3(0.0111\ M)|Pt$

*22-4 For each of the following half-cells, compare electrode potentials calculated from (1) concentration and (2) activity data.
    (a) $Sn(ClO_4)_2(3.00 \times 10^{-5}\ M),Sn(ClO_4)_4(6.00 \times 10^{-5}\ M)|Pt$
    (b) $Sn(ClO_4)_2(3.00 \times 10^{-5}\ M),Sn(ClO_4)_4(6.00 \times 10^{-5}\ M),NaClO_4(0.0800\ M)|Pt$

*22-5 Calculate the potential of a silver electrode in contact with the following:
    (a) a solution that is 0.0150 M in I_2 and saturated with AgI.
    (b) a solution that is 0.0040 M in CN^- and 0.0600 M in $Ag(CN)_2^-$.

(c) the solution that results from mixing 25.0 mL of 0.0500 M KBr with 20.0 mL of 0.100 M Ag^+.

(d) the solution that results from mixing 25.0 mL of 0.0500 M Ag^+ with 20.0 mL of 0.100 M KBr.

*22-6   Calculate the electrode potentials for the following systems:

(a) $Cr_2O_7^{2-}(4.00 \times 10^{-3}$ M$),Cr^{3+}(2.00 \times 10^{-2}$ M$),H^+(0.100$ M$)|Pt$

(b) $UO_2^{2+}(0.200$ M$),U^{4+}(0.100$ M$),H^+(0.500$ M$)|Pt$

*22-7   Calculate the theoretical potential of each of the following cells. Is the cell reaction spontaneous as written or spontaneous in the opposite direction?

(a) $Pt|Cr^{3+}(1.00 \times 10^{-4}$ M$),Cr^{2+}(2.00 \times 10^{-3}$ M$)||Pb^{2+}(5.60 \times 10^{-2}$ M$)|Pb$

(b) $Hg|Hg_2^{2+}(2.00 \times 10^{-2}$ M$)||H^+(1.00 \times 10^{-2}$ M$),V^{3+}(3.00 \times 10^{-2}$ M$),VO^{2+}$ $(2.00 \times 10^{-3}$ M$)|Pt$

(c) $Pt|Fe^{3+}(4.00 \times 10^{-2}$ M$),Fe^{2+}(3.00 \times 10^{-5}$ M$)||Sn^{2+}(5.50 \times 10^{-2}$ M$),$ $Sn^{4+}(3.50 \times 10^{-4}$ M$)|Pt$

*22-8   Calculate the theoretical potential of each of the following cells. Is the cell reaction spontaneous as written or spontaneous in the opposite direction?

(a) $Bi|BiO^+(0.0400$ M$),H^+(0.200$ M$)||I^-(0.100$ M$),AgI(sat'd)|Ag$

(b) $Zn|Zn^{2+}(7.50 \times 10^{-4}$ M$)||Fe(CN)_6^{4-}(4.50 \times 10^{-2}$ M$),Fe(CN)_6^{3-}(7.00 \times 10^{-2}$ M$)|Pt$

(c) $Pt,H_2(0.200$ atm$)|HCl(7.50 \times 10^{-4}$ M$), AgCl(sat'd)|Ag$

*22-9   Compute E^0 for the process

$$Ni(CN)_4^{2-} + 2e^- \rightleftharpoons Ni(s) + 4CN^-$$

given that the formation constant for the complex is 1.0×10^{22}.

*22-10   The solubility product constant for PbI_2 is 7.1×10^{-9} at 25°C. Calculate E^0 for the process

$$PbI_2(s) + 2e^- \rightleftharpoons Pb(s) + 2I_2$$

*22-11   Calculate the standard potential for the half-reaction

$$BiOCl(s) + 2H^+ + 3e^- \rightleftharpoons Bi(s) + Cl^- + H_2O$$

given that K_{sp} for BiOCl has a value of 8.1×10^{-19}.

*22-12   Calculate the standard potential for the half-reaction

$$Al(C_2O_4)_2^- + 3e^- \rightarrow Al(s) + 2C_2O_4^{2-}$$

if the formation constant for the complex is 1.3×10^{13}.

*22-13   From the standard potentials

$$Tl^+ + e^- \rightleftharpoons Tl(s) \qquad E^0 = -0.336 \text{ V}$$
$$TlCl(s) + e^- \rightleftharpoons Tl(s) + Cl^- \qquad E^0 = -0.557 \text{ V}$$

calculate the solubility product constant for TlCl.

*22-14   From the standard potentials

$$Ag_2SeO_4(s) + 2e^- \rightleftharpoons 2Ag(s) + SeO_4^{2-} \qquad E^0 = 0.355 \text{ V}$$
$$Ag^+ + e^- \rightleftharpoons Ag(s) \qquad E^0 = 0.799 \text{ V}$$

calculate the solubility product constant for Ag_2SeO_4.

*22-15 Suppose that we wish to produce a current of 0.0750 A in the cell

$$Pt|V^{3+}(3.7 \times 10^{-5} \text{ M}),V^{2+}(4.48 \times 10^{-1} \text{ M})\|Br^-(0.0850 \text{ M}),AgBr(\text{sat'd})|Ag$$

As a result of its design, the cell has an internal resistance of 4.87 Ω. Calculate the initial potential of the cell.

*22-16 The cell

$$Pt|V(OH)_4^+(2.67 \times 10^{-4} \text{ M}),VO^{2+}(3.42 \times 10^{-2} \text{ M}),$$
$$H^+(4.81 \times 10^{-3} \text{ M})\|Cu^{2+}(2.50 \times 10^{-2} \text{ M})|Cu$$

has an internal resistance of 3.81 Ω. What will be the initial potential if a current of 0.0750 A is drawn from this cell?

*22-17 The resistance of the galvanic cell

$$Pt|Fe(CN)_6^{4-}(4.42 \times 10^{-2}\text{M}),Fe(CN)_6^{3-}(8.93 \times 10^{-3}\text{M})\|Ag^+(5.75 \times 10^{-2}\text{M})|Ag$$

is 3.85 Ω. Calculate the initial potential when 0.0442 A is drawn from this cell.

22-18 The following data are similar to those given in Example 22-1.[14]

c_{HCl}, m	$\gamma_\pm$	E	E^0
0.003215	0.9418	0.52053	0.22255
0.004488	0.9328	0.50384	0.22251
0.005619	0.9259	0.49257	0.22241
0.007311	0.9173	0.47948	0.22236
0.009138	0.9094	0.46860	0.22250
0.011195	0.9031	0.45861	0.22258
0.013407	0.8946	0.44974	0.22248
0.01710	0.8843	0.43783	0.22247
0.02563	0.8660	0.41824	0.22260
0.05391	0.8293	0.38222	0.22256
0.1238	0.7877	0.34199	0.22244

(a) Create a spreadsheet to calculate the standard electrode potential for the Ag-AgCl electrode using the method described in Example 22-1. Make columns for the activity coefficients and the standard potential. Calculate γ values for H^+ and Cl^- for each molality. Then find $\gamma_\pm$ at each molality. Use the measured values of E to find E^0 at each molality.

(b) Compare your values for the activity coefficients and standard potential with those of MacInnes, and if there are any differences between your values and those in the table above, suggest possible reasons for the discrepancies.

(c) Use the Descriptive Statistics function of Data Analysis Toolpak[15] to find the mean, standard deviation, 95% confidence interval, and other useful statistics for the standard potential of the Ag-AgCl electrode.

(d) Comment on the results of your analysis and, in particular, the quality of MacInnes's results.

Challenge Problems

22-19 As a part of a study to measure the dissociation constant of acetic acid, Harned and Ehlers[16] needed to measure E^0 for the following cell:

---

[14] D. A. MacInnes, *The Principles of Electrochemistry*, Table I, p. 187, New York: Reinhold, 1939.

[15] S. R. Crouch and F. J. Holler, *Applications of Microsoft® Excel in Analytical Chemistry*, Belmont, CA: Brooks/Cole, 2004, pp. 32–34.

[16] H. S. Harned and R. W. Ehlers, *J. Am. Chem. Soc.*, **1932**, *54* (4), 1350–57.

$$Pt,H_2(1 \text{ atm})|HCl(m),AgCl(\text{sat'd})|Ag$$

(a) Write an expression for the potential of the cell.
(b) Show that the expression can be arranged to give

$$E = E^0 - \frac{RT}{F} \ln \gamma_{H_3O^+} \gamma_{Cl^-} m_{H_3O^+} m_{Cl^-}$$

where $m_{H_3O^+}$ and m_{Cl^-} are the molal (mole solute per kilogram solvent) concentrations.

(c) Under what circumstances is this expression valid?
(d) Show that the expression in (b) may be written

$$E + 2k \log m = E^0 - 2k \log \gamma_{\pm}$$

where $k = \ln 10 RT/F$.

(e) A considerably simplified version of the Debye-Hückel expression that is valid for very dilute solutions is $\log \gamma = -0.5\sqrt{m} + bm$, where c is a constant. Show that the expression for the cell potential in (d) may be written as

$$E + 2k \log m - k\sqrt{m} = E^0 - 2kcm$$

(f) The previous expression is a "limiting law" that becomes linear as the concentration of the electrolyte approaches zero. The equation assumes the form $y = ax + b$, where $y = E + 2k \log m - k\sqrt{m}$; $x = m$, the slope; $a = -2kc$; and the y-intercept $b = E^0$. Harned and Ehlers very accurately measured the potential of the cell without liquid junction presented at the beginning of the problem as a function of concentration of HCl (molal) and temperature and obtained the data in the following table. For example, they measured the potential of the cell at 25°C with an HCl concentration of 0.01 m and obtained a value of 0.46419 volts.

Potential Measurements of Cell $Pt,H_2(1 \text{ atm})|HCl(m),AgCl(\text{sat'd})|Ag$ without Liquid Junction as a Function of Concentration (molality) and Temperature (°C)

m, molal	E_T, volts							
	E_0	E_5	E_{10}	E_{15}	E_{20}	E_{25}	E_{30}	E_{35}
0.005	0.48916	0.49138	0.49338	0.49521	0.44690	0.49844	0.49983	0.50109
0.006	0.48089	0.48295	0.48480	0.48647	0.48800	0.48940	0.49065	0.49176
0.007	0.4739	0.47584	0.47756	0.47910	0.48050	0.48178	0.48289	0.48389
0.008	0.46785	0.46968	0.47128	0.47270	0.47399	0.47518	0.47617	0.47704
0.009	0.46254	0.46426	0.46576	0.46708	0.46828	0.46937	0.47026	0.47103
0.01	0.4578	0.45943	0.46084	0.46207	0.46319	0.46419	0.46499	0.46565
0.02	0.42669	0.42776	0.42802	0.42925	0.42978	0.43022	0.43049	0.43058
0.03	0.40859	0.40931	0.40993	0.41021	0.41041	0.41056	0.41050	0.41028
0.04	0.39577	0.39624	0.39668	0.39673	0.39673	0.39666	0.39638	0.39595
0.05	0.38586	0.38616	0.38641	0.38631	0.38614	0.38589	0.38543	0.38484
0.06	0.37777	0.37793	0.37802	0.37780	0.37749	0.37709	0.37648	0.37578
0.07	0.37093	0.37098	0.37092	0.37061	0.37017	0.36965	0.36890	0.36808
0.08	0.36497	0.36495	0.36479	0.36438	0.36382	0.36320	0.36285	0.36143
0.09	0.35976	0.35963	0.35937	0.35888	0.35823	0.35751	0.35658	0.35556
0.1	0.35507	0.35487	0.33451	0.35394	0.35321	0.35240	0.35140	0.35031
E^0	0.23627	0.23386	0.23126	0.22847	0.22550	0.22239	0.21918	0.21591

Construct a plot of $E + 2k \log m - k\sqrt{m}$ versus m, and note that the plot is quite linear at low concentration. Extrapolate the line to the y-intercept, and estimate a value for E^0. Compare your value with the value of Harned and Ehlers, and explain any difference. Also compare the value to the one shown in Table 22-1. The simplest way to carry out this exercise is to place the data in a spreadsheet, and use the Excel function INTERCEPT(known_y's, known_x's) to determine the extrapolated value for E^0.[17] Use only the data from 0.005–0.01 m to find the intercept.

(g) Enter the data for all temperatures into the spreadsheet and determine values for E^0 at all temperatures from 5°C to 35°C. Alternatively, you may download an Excel spreadsheet containing the entire data table. Use your web browser to connect to http://www.thomsonedu.com/chemistry/skoog, and select your course, Instrumental Analysis. Finally, navigate to the links for Chapter 22, and click on the spreadsheet link for this problem.

(h) Two typographical errors in the preceding table appeared in the original published paper. Find the errors, and correct them. How can you justify these corrections? What statistical criteria can you apply to justify your action? In your judgment, is it likely that these errors have been detected previously? Explain your answer.

(i) Why do you think that these workers used molality in their studies rather than molarity or weight molarity? Explain whether it matters which of these concentration units are used.

22-20 As we saw in Problem 22-19, as a preliminary experiment in their effort to measure the dissociation constant of acetic acid, Harned and Ehlers[18] measured E^0 for the cell without liquid junction shown. To complete the study and determine the dissociation constant, these workers also measured the potential of the following cell:

$$Pt,H_2(1 \text{ atm})|HOAc(m_1),NaOAc(m_2),NaCl(m_3),AgCl(\text{sat'd})|Ag$$

(a) Show that the potential of this cell is given by

$$E = E^0 - \frac{RT}{F} \ln \gamma_{H_3O^+} \gamma_{Cl^-} m_{H_3O^+} m_{Cl^-}$$

where $\gamma_{H_3O^+}$ and γ_{Cl^-} are the activity coefficients of hydronium ion and chloride ion, respectively, and $m_{H_3O^+}$ and m_{Cl^-} are their respective molal (mole solute per kilogram solvent) concentrations.

(b) The dissociation constant for acetic acid is given by

$$K = \frac{\gamma_{H_3O^+} \gamma_{OAc^-}}{\gamma_{HOAc}} \frac{m_{H_3O^+} m_{OAc^-}}{m_{HOAc}}$$

where γ_{OAc^-} and γ_{HOAc} are the activity coefficients of acetate ion and acetic acid, respectively, and m_{OAc^-} and m_{HOAc} are their respective equilibrium

---

[17] S. R. Crouch and F. J. Holler, *Applications of Microsoft® Excel in Analytical Chemistry*, Belmont, CA: Brooks/Cole, 2004, p. 67.
[18] See note 16.

molal (mole solute per kilogram solvent) concentrations. Show that the potential of the cell in part (a) is given by

$$E - E^0 + \frac{RT}{F} \ln \frac{m_{HOAc} m_{Cl^-}}{m_{OAc^-}} = -\frac{RT}{F} \ln \frac{\gamma_{H_3O^+} \gamma_{Cl^-} \gamma_{HOAc}}{\gamma_{H_3O^+} \gamma_{OAc^-}} - \frac{RT}{F} \ln K$$

(c) As the ionic strength of the solution approaches zero, what happens to the right-hand side of the preceding equation?

(d) As a result of the answer to part (i) in Problem 22-19, we can write the right-hand side of the equation as $-(RT/F)\ln K'$. Show that

$$K' = \exp\left[-\frac{(E - E^0)F}{RT} \ln\left(\frac{m_{HOAc} m_{Cl^-}}{m_{OAc^-}} \right) \right]$$

(e) The ionic strength of the solution in the cell without liquid junction calculated by Harned and Ehlers is

$$\mu = c_{NaCl} + [H^+] + [OAc^-]$$

Show that this expression is correct.

(f) These workers prepared solutions of various molal analytical concentrations of acetic acid, sodium acetate, and sodium chloride and measured the potential of the cell presented at the beginning of this problem. Their results are shown in the following table.

Potential Measurements of Cell Pt,H_2(1 atm)|HOAc(c_{HOAc}),NaOAc(c_{NaOAc}),NaCl(c_{NaCl}), AgCl(sat'd)|Ag without Liquid Junction as a Function of Ionic Strength (molality) and Temperature (°C)

c_{HOAc}, m	c_{NaOAc}, m	c_{NaCl}, m	E_0	E_5	E_{10}	E_{15}	E_{20}	E_{25}	E_{30}	E_{35}
0.004779	0.004599	0.004896	0.61995	0.62392	0.62789	0.63183	0.63580	0.63959	0.64335	0.64722
0.012035	0.011582	0.012326	0.59826	0.60183	0.60538	0.60890	0.61241	0.61583	0.61922	0.62264
0.021006	0.020216	0.021516	0.58528	0.58855	0.59186	0.59508	0.59840	0.60154	0.60470	0.60792
0.04922	0.04737	0.05042	0.56546	0.56833	0.57128	0.57413	0.57699	0.57977	0.58257	0.58529
0.08101	0.07796	0.08297	0.55388	0.55667	0.55928	0.56189	0.56456	0.56712	0.56964	0.57213
0.09056	0.08716	0.09276	0.55128	0.55397	0.55661	0.55912	0.56171	0.56423	0.56672	0.56917

Calculate the ionic strength of each of the solutions using the expression for the K_a of acetic acid to calculate $[H_3O^+]$, $[OAc^-]$, and $[HOAc]$ with the usual suitable approximations and a provisional value of $K_a = 1.8 \times 10^{-5}$. Use the potentials in the table for 25°C to calculate values for K' with the expression in part (j). Construct a plot of K' versus μ, and extrapolate the graph to infinite dilution ($\mu = 0$) to find a value for K_a at 25°C. Compare the extrapolated value to the provisional value used to calculate μ. What effect does the provisional value of K_a have on the extrapolated value of K_a? You can perform these calculations most easily using a spreadsheet.

(g) If you have made these computations using a spreadsheet, determine the dissociation constant for acetic acid at all other temperatures for which data are available. How does K_a vary with temperature? At what temperature does the maximum in K_a occur?

CHAPTER TWENTY-THREE

Potentiometry

*P*otentiometric methods of analysis are based on measuring the potential of electrochemical cells without drawing appreciable current. For nearly a century, potentiometric techniques have been used for the location of end points in titrations. More recently, ion concentrations have been measured directly from the potential of an ion-selective membrane electrode. Such electrodes are relatively free from interference and provide a rapid and convenient means for quantitative estimations of numerous important anions and cations.

![logo] Throughout this chapter, this logo indicates an opportunity for online self-study at **www .thomsonedu.com/chemistry/skoog**, linking you to interactive tutorials, simulations, and exercises.

The equipment required for potentiometric methods is simple and inexpensive and includes an *indicator electrode*, a *reference electrode*, and a *potential measuring device*. The design and properties of each of these components are described in the initial sections of this chapter. Following these discussions, we investigate analytical applications of potentiometric measurements.[1]

23A GENERAL PRINCIPLES

In Chapter 22, we state that absolute values for individual half-cell potentials cannot be determined in the laboratory. That is, only relative cell potentials can be measured experimentally. Figure 23-1 shows a typical cell for potentiometric analysis. This cell can be represented as

$$\text{reference electrode} \mid \text{salt bridge} \mid \text{analyte solution} \mid \text{indicator electrode}$$
$$\underbrace{\qquad}_{E_{\text{ref}}} \quad \underbrace{\qquad}_{E_{\text{j}}} \quad \underbrace{\qquad\qquad}_{E_{\text{ind}}}$$

The *reference electrode* in this diagram is a half-cell with an accurately known electrode potential, E_{ref}, that is independent of the concentration of the analyte or any other ions in the solution under study. It can be a standard hydrogen electrode but seldom is because a standard hydrogen electrode is somewhat troublesome to maintain and use. By convention, the reference electrode is always treated as the left-hand electrode in potentiometric measurements. The *indicator electrode*, which is immersed in a solution of the analyte, develops a potential, E_{ind}, that depends on the activity of the analyte. Most indicator electrodes used in potentiometry are selective in their responses. The third component of a potentiometric cell is a salt bridge that prevents the components of the analyte solution from mixing with those of the reference electrode. As noted in Section 22B-2, a potential develops across the liquid junctions at each end of the salt bridge. These two potentials tend to cancel one another if the mobilities of the cation and the anion in the bridge solution are approximately the same. Potassium chloride is a nearly ideal electrolyte for the salt bridge because the mobilities of the K^+ ion and the Cl^- ion are nearly equal. The net potential difference across the salt bridge E_{j} is thereby reduced to a few millivolts or less. For most electroanalytical methods, the junction potential is small enough to be

---

[1] For more information, see R. S. Hutchins and L. G. Bachas, in *Handbook of Instrumental Techniques for Analytical Chemistry*, F. A. Settle, ed., Chap. 38, pp. 727–48, Upper Saddle River, NJ: Prentice-Hall, 1997.

659

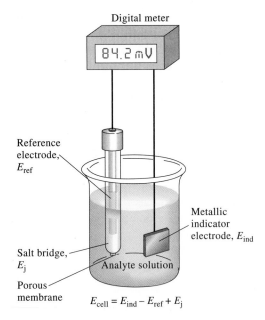

FIGURE 23-1 A cell for potentiometric determinations.

Digital meter

84.2 mV

Reference electrode, E_{ref}

Metallic indicator electrode, E_{ind}

Salt bridge, E_j

Analyte solution

Porous membrane

$$E_{cell} = E_{ind} - E_{ref} + E_j$$

neglected. In the potentiometric methods discussed in this chapter, however, the junction potential and its uncertainty can be factors that limit the measurement accuracy and precision.

The potential of the cell we have just considered is given by the equation

$$E_{cell} = (E_{ind} - E_{ref}) + E_j \qquad (23\text{-}1)$$

The first term in this equation, E_{ind}, contains the information that we are looking for — the concentration of the analyte. To make a potentiometric determination of an analyte, then, we must measure a cell potential, correct this potential for the reference and junction potentials, and compute the analyte concentration from the indicator electrode potential. Strictly, the potential of a galvanic cell is related to the activity of the analyte. Only through proper calibration of the electrode system with solutions of known concentration can we determine the concentration of the analyte. In the sections that follow, we discuss the nature and origin of the three potentials shown on the right side of Equation 23-1.

23B REFERENCE ELECTRODES

The ideal reference electrode has a potential that is known, constant, and completely insensitive to the composition of the solution under study. In addition,

this electrode should be rugged and easy to assemble and should maintain a constant potential even when there is a net current in the cell.

23B-1 Calomel Electrodes

Calomel reference electrodes consist of mercury in contact with a solution that is saturated with mercury(I) chloride (calomel) and that also contains a known concentration of potassium chloride. Calomel half-cells can be represented as follows:

$$Hg|Hg_2Cl_2(\text{sat'd}),KCl(x\text{M})\|$$

where x represents the molar concentration of potassium chloride in the solution.[2] The electrode potential for this half-cell is determined by the reaction

$$Hg_2Cl_2(s) + 2e^- \rightleftharpoons 2Hg(l) + 2Cl^-$$

and depends on the chloride concentration x. Thus, the KCl concentration must be specified in describing the electrode.

Table 23-1 lists the composition and the potentials for three common calomel electrodes. Note that each solution is saturated with mercury(I) chloride (calomel) and that the cells differ only with respect to the potassium chloride concentration.

The saturated calomel electrode (SCE) is widely used because of the ease with which it can be prepared.[3] Compared with the other calomel electrodes, however, its temperature coefficient is significantly larger (see Table 23-1). A further disadvantage is that when the temperature is changed, the potential comes to a new value only slowly because of the time required for solubility equilibrium for the potassium chloride and for the calomel to be reestablished. The potential of the SCE at 25°C is 0.2444 V.

Several convenient calomel electrodes, such as the electrode illustrated in Figure 23-2a, are available commercially. The H-shape body of the electrode is made of glass of dimensions shown in the diagram. The right arm of the electrode contains a platinum electrical contact, a small quantity of mercury–mercury(I) chloride paste in saturated potassium chloride, and a few

---

[2] By convention, a reference electrode is always the left-hand electrode, as shown in Figure 23-1. This practice is consistent with the International Union of Pure and Applied Chemistry (IUPAC) convention for electrode potentials, discussed in Section 22C-4 in which the reference is the standard hydrogen electrode and is the electrode on the left in a cell diagram.
[3] Note that the term "saturated" in the name refers to the concentration of KCl (about 4.6 M) and not to the concentration of Hg_2Cl_2; all calomel electrodes are saturated with Hg_2Cl_2.

TABLE 23-1 Potentials of Reference Electrodes in Aqueous Solutions

Temperature, °C	Electrode Potential vs. SHE, V				
	0.1 M[c] Calomel[a]	3.5 M[c] Calomel[b]	Saturated[c] Calomel[a]	3.5 M[b,c] Ag-AgCl	Saturated[b,c] Ag-AgCl
10	—	0.256	—	0.215	0.214
12	0.3362	—	0.2528	—	—
15	0.3362	0.254	0.2511	0.212	0.209
20	0.3359	0.252	0.2479	0.208	0.204
25	0.3356	0.250	0.2444	0.205	0.199
30	0.3351	0.248	0.2411	0.201	0.194
35	0.3344	0.246	0.2376	0.197	0.189
38	0.3338	—	0.2355	—	0.184
40	—	0.244	—	0.193	—

[a]Data from R. G. Bates, in *Treatise on Analytical Chemistry*, 2nd ed., I. M. Kolthoff and P. J. Elving, eds., Part I, Vol. 1, p. 793, New York: Wiley, 1978.

[b]Data from D. T. Sawyer, A. Soblowski, and J. L. Roberts Jr., *Experimental Electrochemistry for Chemists*, 2nd ed., p. 192, New York: Wiley, 1995.

[c]"M" and "saturated" refer to the concentration of KCl and *not* Hg_2Cl_2.

crystals of KCl. The tube is filled with saturated KCl to act as a salt bridge through a piece of porous Vycor ("thirsty glass") sealed in the end of the left arm. This type of junction has a relatively high resistance (2000 to 3000 Ω) and a limited current-carrying capacity, but

contamination of the analyte solution due to leakage of potassium chloride is minimal. Other configurations of SCEs are available with much lower resistance and better electrical contact to the analyte solution, but they tend to leak small amounts of saturated potassium

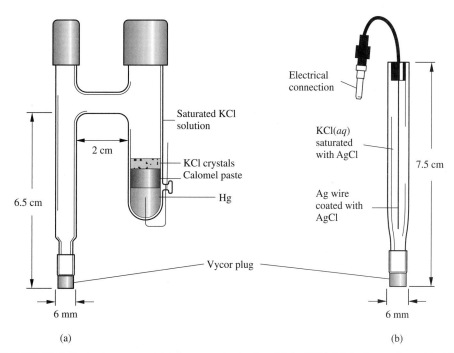

FIGURE 23-2 Typical commercial reference electrodes. (a) An SCE. (b) A silver–silver chloride electrode. (Adapted with permission of Bioanalytical Systems, West Lafayette, IN.)

chloride into the sample. Because of concerns with mercury contamination, SCEs are less common than they once were, but for some applications, they are superior to Ag-AgCl reference electrodes, which are described next.

23B-2 Silver–Silver Chloride Electrodes

The most widely marketed reference electrode system consists of a silver electrode immersed in a solution of potassium chloride that has been saturated with silver chloride

$$Ag|AgCl(sat'd),KCl(xM)\|$$

The electrode potential is determined by the half-reaction

$$AgCl(s) + e^- \rightleftharpoons Ag(s) + Cl^-$$

Normally, this electrode is prepared with either a saturated or a 3.5-M potassium chloride solution; potentials for these electrodes are given in Table 23-1. Figure 23-2b shows a commercial model of this electrode, which is little more than a piece of glass tubing that has a narrow opening at the bottom connected to a Vycor plug for making contact with the analyte solution. The tube contains a silver wire coated with a layer of silver chloride that is immersed in a potassium chloride solution saturated with silver chloride.

Silver–silver chloride electrodes have the advantage that they can be used at temperatures greater than 60°C, whereas calomel electrodes cannot. On the other hand, mercury(II) ions react with fewer sample components than do silver ions (which can react with proteins, for example); such reactions can lead to plugging of the junction between the electrode and the analyte solution.

23B-3 Precautions in the Use of Reference Electrodes

In using reference electrodes, such as those shown in Figure 23-2, the level of the internal liquid should always be kept above the level of the liquid in the sample solution to prevent contamination of the electrode solution and plugging of the junction due to reaction of the analyte solution with silver or mercury(I) ions from the internal solution. Junction plugging is probably the most common source of erratic cell behavior (noise) in potentiometric measurements. Many schemes have

Tutorial: Learn more about **potentiometric electrodes**.

been developed for preventing plugging and maintaining good contact between reference electrodes and analyte solutions. The Vycor plugs shown in Figure 23-2 provide excellent contact, and if they are kept moist, they provide a reproducible, low-noise junction. For some applications using ion-selective electrodes, the reference electrode may need a special low-resistance flowing junction to reduce electrical noise such as the free diffusion junction described in Section 23H-5 and illustrated in Figure 23-18.

With the liquid level above the analyte solution, some contamination of the sample is inevitable. In most instances, the amount of contamination is too slight to be of concern. In determining ions such as chloride, potassium, silver, and mercury, however, precaution must often be taken to avoid this source of error. A common way is to interpose a second salt bridge between the analyte and the reference electrode; this bridge should contain a noninterfering electrolyte, such as potassium nitrate or sodium sulfate. Double-junction electrodes based on this design are offered by several manufacturers.

23C METALLIC INDICATOR ELECTRODES

An ideal indicator electrode responds rapidly and reproducibly to changes in activity of the analyte ion. Although no indicator electrode is absolutely specific in its response, a few are now available that are remarkably selective. There are two types of indicator electrodes: metallic and membrane. This section deals with metallic indicator electrodes.

It is convenient to classify metallic indicator electrodes as electrodes of the first kind, electrodes of the second kind, electrodes of the third kind, and inert redox electrodes.

23C-1 Electrodes of the First Kind

A metallic electrode of the first kind is a pure metal electrode in direct equilibrium with its cation in solution. Here, a single reaction is involved. For example, for a copper indicator electrode, we may write

$$Cu^{2+} + 2e^- \rightleftharpoons Cu(s)$$

The potential E_{ind} of this electrode is given by

$$E_{ind} = E^0_{Cu} - \frac{0.0592}{2} \log \frac{1}{a_{Cu^{2+}}}$$

$$= E^0_{Cu} - \frac{0.0592}{2} pCu \qquad (23\text{-}2)$$

where pCu is the negative logarithm of the copper(II) ion activity $a_{Cu^{2+}}$. Thus, the copper electrode provides a direct measure of the pCu of the solution.[4]

Electrode systems of the first kind are not widely used for potentiometric analyses for several reasons. For one, they are not very selective and respond not only to their own cations but also to other more easily reduced cations. For example, a copper electrode cannot be used for the determination of copper(II) ions in the presence of silver(I) ions, which are also reduced at the copper surface. In addition, many metal electrodes, such as zinc and cadmium, can be used only in neutral or basic solutions because they dissolve in the presence of acids. Third, some metals are so easily oxidized that their use is restricted to solutions that have been deaerated. Finally, certain harder metals — such as iron, chromium, cobalt, and nickel — do not provide reproducible potentials. Moreover, for these electrodes, plots of pX versus activity yield slopes that differ significantly and irregularly from the theoretical ($-0.0592/n$). For these reasons, the only electrode systems of the first kind that have been used are Ag-Ag$^+$ and Hg-Hg$_2^{2+}$ in neutral solutions, and CuCu^{2+}, Zn-Zn^{2+}, Cd-Cd^{2+}, Bi-Bi^{3+}, Tl-Tl$^+$, and Pb-Pb^{2+} in deaerated solutions.

Recall that the nernstian factor 0.0592/2 appearing in Equation 23-2 and throughout this chapter is equal to $2.303RT/2F$. We do not refer to the temperature dependence of this "constant" routinely, but *be aware that all electrode measurements are subject to errors caused by temperature fluctuations that inevitably occur in the laboratory, in the field, and thus in analytical samples.* For routine low-precision work, small temperature fluctuations have negligible effect on results, but when large temperature changes occur or when high-precision measurements are required, sample temperatures must be measured and suitable corrections must be applied to the measurements. Many commercial instruments for potentiometric measurements have built-in software or circuits for monitoring temperature

and correcting the output for temperature change (see Section 23G-2).

23C-2 Electrodes of the Second Kind

A metal electrode can often be made responsive to the activity of an anion with which its ion forms a precipitate or a stable complex ion. For example, silver can serve as an *electrode of the second kind* for halide and halide-like anions. To prepare an electrode to determine chloride ion it is necessary only to saturate the layer of the analyte solution adjacent to a silver electrode with silver chloride. The electrode reaction is then written as

$$AgCl(s) + e^- \rightleftharpoons Ag(s) + Cl^- \qquad E^0 = 0.222 \text{ V}$$

We apply the Nernst equation to this reaction, which yields

$$\begin{aligned} E_{ind} &= 0.222 - 0.0592 \log a_{Cl^-} \\ &= 0.222 + 0.0592 \, pCl \end{aligned} \qquad (23\text{-}3)$$

A convenient way of preparing a chloride-sensitive electrode is to make a pure silver wire the anode in an electrolytic cell containing potassium chloride. The wire becomes coated with an adherent silver halide deposit, which will rapidly equilibrate with the surface layer of a solution in which it is immersed. Because the solubility of the silver chloride is low, an electrode formed in this way can be used for many measurements.

An important electrode of the second kind for measuring the activity of ethylenediaminetetraacetic acid (EDTA) anion Y^{4-} is based on the response of a mercury electrode in the presence of a small concentration of the stable EDTA complex of Hg(II). The half-reaction for the electrode process can be written as

$$HgY^{2-} + 2e^- \rightleftharpoons Hg(l) + Y^{4-} \qquad E^0 = 0.21 \text{ V}$$

for which

$$E_{ind} = 0.21 - \frac{0.0592}{2} \log \frac{a_{Y^{4-}}}{a_{HgY^{2-}}}$$

To use this electrode system, it is necessary to introduce a small concentration of HgY^{2-} into the analyte solution at the outset. The complex is so stable (for HgY^{2-}, $K_f = 6.3 \times 10^{21}$) that its activity remains essentially constant over a wide range of Y^{4-} activities. Therefore, the potential equation can be written in the form

$$E_{ind} = K - \frac{0.0592}{2} \log a_{Y^{4-}} = K + \frac{0.0592}{2} pY \quad (23\text{-}4)$$

---

[4]The results of potentiometric measurements are usually expressed in terms of a variable, the *p-function*, which is proportional to the measured potential. The p-function then provides a measure of activity in terms of a convenient, small, and ordinarily positive number. Thus, for a solution with a calcium ion activity of 2.00×10^{-6} M, we may write pCa = $-\log (2.00 \times 10^{-6}) = 5.699$. Note that as the activity of calcium increases, its p-function decreases. Note also that because the activity was given to three significant figures, we are entitled to keep three figures to the right of the decimal point in the computed pCa because these are the only numbers that carry information about the original 2.00. The 5 in the value for pCa provides only information about the position of the decimal point in the original number.

where the constant K is

$$K = 0.21 - \frac{0.0592}{2} \log \frac{1}{a_{HgY^{2-}}}$$

This electrode is useful for locating end points for EDTA titrations.

23C-3 Electrodes of the Third Kind

Under some circumstances, a metal electrode can be made to respond to a different cation. It then becomes an electrode of the third kind. As an example, a mercury electrode has been used for the determination of the pCa of calcium-containing solutions. As in the previous example, a small concentration of the EDTA complex of Hg(II) is introduced into the solution. As before (Equation 23-4), the potential of a mercury electrode in this solution is given by

$$E_{ind} = K - \frac{0.0592}{2} \log a_{Y^{4-}}$$

If, in addition, a small volume of a solution containing the EDTA complex of calcium is introduced, a new equilibrium is established, namely,

$$CaY^{2-} \rightleftharpoons Ca^{2+} + Y^{4-} \qquad K_f = \frac{a_{Ca^{2+}} a_{Y^{4-}}}{a_{CaY^{2-}}}$$

Combining the formation-constant expression for CaY^{2-} with the potential expression yields

$$E_{ind} = K - \frac{0.0592}{2} \log \frac{K_f a_{CaY^{2-}}}{a_{Ca^{2+}}}$$

which can be written as

$$E_{ind} = K - \frac{0.0592}{2} \log K_f a_{CaY^{2-}} - \frac{0.0592}{2} \log \frac{1}{a_{Ca^{2+}}}$$

If a constant amount of CaY^{2-} is used in the analyte solution and in the solutions for standardization, we may write

$$E_{ind} = K' - \frac{0.0592}{2} pCa$$

where

$$K' = K - \frac{0.0592}{2} \log K_f a_{CaY^{2-}}$$

Thus, the mercury electrode has become an electrode of the third kind for calcium ion.

23C-4 Metallic Redox Indicators

Electrodes fashioned from platinum, gold, palladium, or other inert metals often serve as indicator electrodes for oxidation-reduction systems. In these applications, the inert electrode acts as a source or sink for electrons transferred from a redox system in the solution. For example, the potential of a platinum electrode in a solution containing Ce(III) and Ce(IV) ions is given by

$$E_{ind} = E^0 - 0.0592 \log \frac{a_{Ce^{3+}}}{a_{Ce^{4+}}}$$

Thus, a platinum electrode can serve as the indicator electrode in a titration in which Ce(IV) serves as the standard reagent.

Note, however, that electron-transfer processes at inert electrodes are frequently not reversible.[5] As a result, inert electrodes do not respond in a predictable way to many of the half-reactions found in a table of electrode potentials. For example, a platinum electrode immersed in a solution of thiosulfate and tetrathionate ions does not exhibit reproducible potentials because the electron-transfer process

$$S_4O_6^{2-} + 2e^- \rightleftharpoons 2S_2O_3^{2-}$$

is slow and, therefore, not reversible at the electrode surface.

23D MEMBRANE INDICATOR ELECTRODES

A wide variety of membrane electrodes are available from commercial sources that permit the rapid and selective determination of numerous cations and anions by direct potentiometric measurements.[6] Often, membrane electrodes are called *ion-selective electrodes* because of the high selectivity of most of these devices.

---

[5] Here, we refer to practical thermodynamic reversibility; that is, if there is a small change in the electrode potential, equilibrium is reestablished relatively rapidly. If the electron-transfer reaction in either direction is too slow, equilibrium is not achieved, and the reaction is said to be irreversible, or quasi-reversible. For details on the thermodynamics and reversibility of electron-transfer processes, see A. J. Bard and L. R. Faulkner, *Electrochemical Methods*, 2nd ed., New York: Wiley, 2001, Chap. 2, pp. 44–48.

[6] Some sources for additional information on this topic are R. S. Hutchins and L. G. Bachas, in *Handbook of Instrumental Techniques for Analytical Chemistry*, F. A. Settle, ed., Upper Saddle River, NJ: Prentice-Hall, 1997; J. Koryta, *Ions, Electrodes, and Membranes*, 2nd ed., New York: Wiley, 1991; A. Evans, *Potentiometry and Ion-Selective Electrodes*, New York: Wiley, 1987; D. Ammann, *Ion-Selective Microelectrodes: Principles, Design, and Application*, New York: Springer, 1986.

TABLE 23-2 Types of Ion-Selective Membrane Electrodes

A. Crystalline Membrane Electrodes
   1. Single crystal
      Example: LaF_3 for F^-
   2. Polycrystalline or mixed crystal
      Example: Ag_2S for S^{2-} and Ag^+
B. Noncrystalline Membrane Electrodes
   1. Glass
      Examples: silicate glasses for Na^+ and H^+
   2. Liquid
      Examples: liquid ion exchangers for Ca^{2+} and
      neutral carriers for K^+
   3. Immobilized liquid in a rigid polymer
      Examples: PVC matrix for Ca^{2+} and NO_3^-

They are also referred to as *pIon electrodes* because their output is usually recorded as a p-function, such as pH, pCa, or pNO_3 (see note 4).

23D-1 Classification of Membranes

Table 23-2 lists the various types of ion-selective membrane electrodes that have been developed. These differ in the physical or chemical composition of the membrane. The general mechanism by which an ion-selective potential develops in these devices depends on the nature of the membrane and is entirely different from the source of potential in metallic indicator electrodes. We have seen that the potential of a metallic electrode arises from the tendency of an oxidation-reduction reaction to occur at an electrode surface. In membrane electrodes, in contrast, the observed potential is a kind of junction potential that develops across a membrane that separates the analyte solution from a reference solution.

23D-2 Properties of Ion-Selective Membranes

All of the ion-selective membranes in the electrodes shown in Table 23-2 share common properties, which lead to the sensitivity and selectivity of membrane electrodes toward certain cations or anions. These properties include the following:

1. **Minimal solubility.** A necessary property of an ion-selective medium is that its solubility in analyte solutions (usually aqueous) approaches zero. Thus, many membranes are formed from large molecules or molecular aggregates such as silica glasses or polymeric resins. Ionic inorganic compounds of low solubility, such as the silver halides, can also be converted into membranes.

2. **Electrical conductivity.** A membrane must exhibit some electrical conductivity, albeit small. Generally, this conduction takes the form of migration of singly charged ions within the membrane.

3. **Selective reactivity with the analyte.** A membrane or some species contained within the membrane matrix must be capable of selectively binding the analyte ion. Three types of binding are encountered: ion-exchange, crystallization, and complexation. The former two are the more common, and we will largely focus on these types of bindings.

23D-3 The Glass Electrode for pH Measurements

We begin our discussion of how pIon membrane electrodes are constructed and how they function by examining in some detail the glass electrode for pH measurements. The glass pH electrode predates all other membrane electrodes by several decades, and it is the most widely used electrode in the world.[7]

Since the early 1930s, the most convenient way for determining pH has been by measuring the potential difference across a glass membrane separating the analyte solution from a reference solution of fixed acidity. Cremer[8] first recognized the phenomenon underlying this measurement in 1906, and Haber[9] systematically explored it a few years later. General implementation of the glass electrode for pH measurements did not occur for two decades, however, until the invention of the vacuum tube permitted the convenient measurement of potentials across glass membranes having resistances of 100 MΩ or more. Systematic studies of the pH sensitivity of glass membranes led ultimately in the late 1960s to the development and marketing of membrane electrodes for two dozen or more ions such as K^+, Na^+, Ca^{2+}, F^-, and NO_3^-.

Figure 23-3a shows a typical cell for measuring pH. The cell consists of a glass indicator electrode and a silver–silver chloride or a saturated calomel reference electrode immersed in the solution whose pH is to be determined. The indicator electrode consists of a thin,

---

[7]H. Galster, *pH Measurement: Fundamentals, Methods, Applications, Instrumentation*, New York: Wiley, 1991; R. G. Bates, *Determination of pH*, 2nd ed., New York: Wiley, 1973.
[8]M. Cremer, *Z. Biol.*, **1906**, *47*, 562.
[9]F. Haber and Z. Klemensiewicz, *Z. Phys. Chem.*, **1909**, *67*, 385.

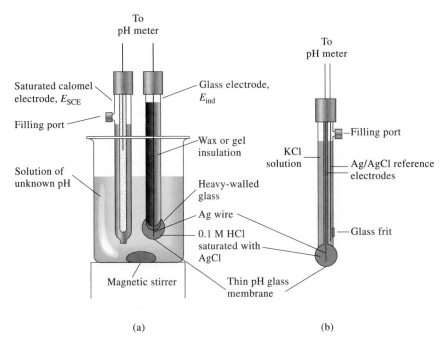

FIGURE 23-3 Typical electrode system for measuring pH. (a) Glass electrode (indicator) and SCE (reference) immersed in a solution of unknown pH. (b) Combination probe consisting of both an indicator glass electrode and a silver–silver chloride reference. A second silver–silver chloride electrode serves as the internal reference for the glass electrode. The two electrodes are arranged concentrically with the internal reference in the center and the external reference outside. The reference makes contact with the analyte solution through the glass frit or other suitable porous medium. Combination probes are the most common configuration of glass electrode and reference for measuring pH.

pH-sensitive glass membrane sealed onto one end of a heavy-walled glass or plastic tube. A small volume of dilute hydrochloric acid saturated with silver chloride is contained in the tube (the inner solution in some electrodes is a buffer containing chloride ion). A silver wire in this solution forms a silver–silver chloride reference electrode, which is connected to one of the terminals of a potential-measuring device. The reference electrode is connected to the other terminal.

Figure 23-4, which is a schematic representation of the cell in Figure 23-3a, shows that this cell contains two reference electrodes: (1) the external silver–silver chloride electrode (ref 1) and (2) the internal silver–silver chloride electrode (ref 2). Although the internal reference electrode is a part of the glass electrode, it is not the pH-sensing element. Instead, it is the thin glass membrane at the tip of the electrode that responds to pH.

In Figure 23-3b, we see the most common configuration for measuring pH with a glass electrode. In this

arrangement, the glass electrode and its Ag-AgCl internal reference electrode are positioned in the center of a cylindrical probe. Surrounding the glass electrode is the external reference electrode, which is most often of the Ag-AgCl type. The presence of the external reference electrode is not as obvious as in the dual-probe arrangement of Figure 23-3a, but the single-probe variety is considerably more convenient and can be made much smaller than the dual system. The pH-sensitive glass membrane is attached to the tip of the probe. These probes are manufactured in many different physical shapes and sizes (5 cm to 5 mm) to suit a broad range of laboratory and industrial applications.

The Composition and Structure of Glass Membranes

There has been a good deal of research devoted to the effects of glass composition on the sensitivity of membranes to protons and other cations, and a number of formulations are now used for the manufacture of elec-

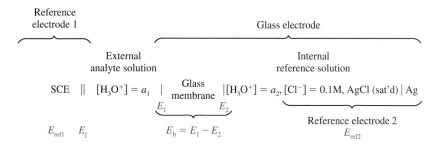

FIGURE 23-4 Diagram of glass-calomel cell for the measurement of pH. E_{SCE} is the potential of the reference electrode, E_j is the junction potential, a_1 is the activity of hydronium ions in the analyte solution, E_1 and E_2 are the potentials on either side of the glass membrane, E_b is the boundary potential, and a_2 is the activity of hydronium ion in the internal reference solution.

trodes. Corning 015 glass, which has been widely used for membranes, consists of approximately 22% Na_2O, 6% CaO, and 72% SiO_2. This membrane is specific in its response toward hydrogen ions up to a pH of about 9. At higher pH values, however, the glass becomes somewhat responsive to sodium, as well as to other singly charged cations. Other glass formulations are now in use in which sodium and calcium ions are replaced to various degrees by barium and lithium ions. These membranes have superior selectivity at high pH.

Figure 23-5 is a two-dimensional view of the structure of a silicate glass membrane. Each silicon atom is shown as being bonded to three oxygen atoms in the plane of the paper. In addition, each is bonded to

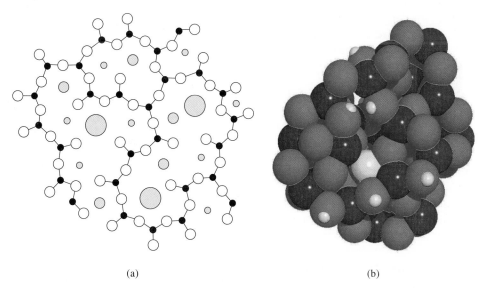

(a)   (b)

FIGURE 23-5 (a) Cross-sectional view of a silicate glass structure. In addition to the three Si—O bonds shown, each silicon is bonded to an additional oxygen atom, either above or below the plane of the paper. (Adapted with permission from G. A. Perley, *Anal. Chem.*, **1949**, *21*, 395. Copyright 1949 American Chemical Society.) (b) Model showing three-dimensional structure of amorphous silica with Na^+ ion (large light gray) and several H^+ ions (small light gray) incorporated. Note that the Na^+ ion is surrounded by a cage of oxygen atoms (light blue) and that each proton in the amorphous lattice is attached to an oxygen. The cavities in the structure, the small size, and the high mobility of the proton ensure that protons can migrate deep into the surface of the silica. Other cations and water molecules may be incorporated into the interstices of the structure as well.

another oxygen above or below the plane. Thus, the glass consists of an infinite three-dimensional network of SiO_4^- groups in which each silicon is bonded to four oxygens and each oxygen is shared by two silicons. Within the interstices of this structure are sufficient cations to balance the negative charge of the silicate groups. Singly charged cations, such as sodium and lithium, are mobile in the lattice and are responsible for electrical conduction within the membrane.

The Hygroscopicity of Glass Membranes

The surface of a glass membrane must be hydrated before it will function as a pH electrode. The amount of water involved is approximately 50 mg per cubic centimeter of glass. Nonhygroscopic glasses show no pH function. Even hygroscopic glasses lose their pH sensitivity after dehydration by storage over a desiccant. The effect is reversible, however, and the response of a glass electrode can be restored by soaking it in water.

The hydration of a pH-sensitive glass membrane involves an ion-exchange reaction between singly charged cations in the glass lattice and protons from the solution. The process involves univalent cations exclusively because di- and trivalent cations are too strongly held within the silicate structure to exchange with ions in the solution. In general, then, the ion-exchange reaction can be written as

$$\underset{\text{soln}}{H^+} + \underset{\text{glass}}{Na^+Gl^-} \rightleftharpoons \underset{\text{soln}}{Na^+} + \underset{\text{glass}}{H^+Gl^-} \quad (23\text{-}5)$$

The equilibrium constant for this process is so large that the surface of a hydrated glass membrane ordinarily consists entirely of silicic acid (H^+Gl^-) groups. An exception to this situation exists in highly alkaline media, where the hydrogen ion concentration is extremely small and the sodium ion concentration is large; here, a significant fraction of the sites are occupied by sodium ions.

Electrical Conduction across Glass Membranes

To serve as an indicator for cations, a glass membrane must conduct electricity. Conduction within the hydrated gel layer involves the movement of hydrogen ions. Sodium ions are the charge carriers in the dry interior of the membrane. Conduction across the solution-gel interfaces occurs by the reactions

$$\underset{\text{soln}_1}{H^+} + \underset{\text{glass}_1}{Gl^-} \rightleftharpoons \underset{\text{glass}_1}{H^+Gl^-} \quad (23\text{-}6)$$

$$\underset{\text{glass}_2}{H^+Gl^-} \rightleftharpoons \underset{\text{soln}_2}{H^+} + \underset{\text{glass}_2}{Gl^-} \quad (23\text{-}7)$$

where subscript 1 refers to the interface between the glass and the analyte solution and subscript 2 refers to the interface between the internal solution and the glass. The positions of these two equilibria are determined by the hydrogen ion activities in the solutions on the two sides of the membrane. The surface at which the greater dissociation occurs becomes negative with respect to the other surface where less dissociation has taken place. A boundary potential E_b thus develops across the membrane. The magnitude of the boundary potential depends on the ratio of the hydrogen ion activities of the two solutions. It is this potential difference that serves as the analytical parameter in potentiometric pH measurements with a membrane electrode.

Membrane Potentials

The lower part of Figure 23-4 shows four potentials that develop in a cell when pH is being determined with a glass electrode. Two of these, E_{ref1} and E_{ref2}, are reference electrode potentials. The third potential is the junction potential E_j across the salt bridge that separates the calomel electrode from the analyte solution. Junction potentials are found in all cells used for the potentiometric measurement of ion concentration. Proper selection of the junction for the reference electrode is necessary to keep E_j low and stable. A poorly chosen salt bridge can cause high resistance to ion flow, leading to noise and unstable readings. Manufacturers offer many choices of junctions for different applications. The fourth, and most important, potential shown in Figure 23-4 is the *boundary potential, E_b, which varies with the pH of the analyte solution.* The two reference electrodes simply provide electrical contacts with the solutions so that changes in the boundary potential can be measured.

Figure 23-4 reveals that the potential of a glass electrode has two components: the fixed potential of a silver–silver chloride electrode E_{ref2} and the pH-dependent boundary potential E_b.

The Boundary Potential

As shown in Figure 23-4, the boundary potential consists of two potentials, E_1 and E_2, each of which is associated with one of the two glass surfaces. The boundary potential is simply the difference between these potentials:

$$E_b = E_1 - E_2 \quad (23\text{-}8)$$

It can be demonstrated from thermodynamic considerations[10] that E_1 and E_2 in Equation 23-8 are related to the hydrogen ion activities at each surface by nernstian relationships:

$$E_1 = j_1 - \frac{0.0592}{n} \log \frac{a_1'}{a_1} \qquad (23\text{-}9)$$

$$E_2 = j_2 - \frac{0.0592}{n} \log \frac{a_2'}{a_2} \qquad (23\text{-}10)$$

where j_1 and j_2 are constants and a_1 and a_2 are activities of H^+ in the *solutions* on the external and internal sides of the membrane, respectively. The terms a_1' and a_2' are the activities of H^+ at the external and internal *surfaces* of the glass making up the membrane.

If the two membrane surfaces have the same number of negatively charged sites (as they normally do) from which H^+ can dissociate, then j_1 and j_2 are identical; so also are a_1' and a_2'. By substituting Equations 23-9 and 23-10 into Equation 23-8, substituting the equalities $j_1 = j_2$ and $a_1' = a_2'$, and rearranging the resulting equation, we find that

$$E_b = E_1 - E_2 = 0.0592 \log \frac{a_1}{a_2} \qquad (23\text{-}11)$$

Thus, the boundary potential E_b depends only on the hydrogen ion activities of the solutions on either side of the membrane. For a glass pH electrode, the hydrogen ion activity of the internal solution a_2 is held constant so that Equation 23-11 simplifies to

$$E_b = L' + 0.0592 \log a_1 = L' - 0.0592 \, \mathrm{pH} \quad (23\text{-}12)$$

where

$$L' = -0.0592 \log a_2$$

The boundary potential is then a measure of the hydrogen ion activity of the external solution (a_1).

The significance of the potentials and the differences shown in Equation 23-11 is illustrated by the potential profiles shown in Figure 23-6. The profiles are plotted across the membrane from the analyte solution on the left through the membrane to the internal solution on the right.

The Asymmetry Potential

When identical solutions are placed on the two sides of a glass membrane, the boundary potential should in principle be zero (see Figure 23-6b). In fact, however,

[10] G. Eisenman, *Biophys. J.*, **1962**, *2* (part 2), 259.

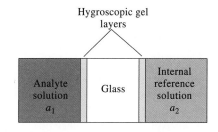

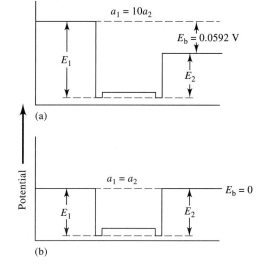

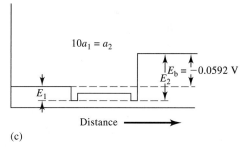

FIGURE 23-6 Potential profiles across a glass membrane from the analyte solution to the internal reference solution. The reference electrode potentials are not shown.

we frequently encounter a small asymmetry potential that changes gradually with time.

The sources of the asymmetry potential are obscure but undoubtedly include such causes as differences in strain on the two surfaces of the membrane created during manufacture, mechanical abrasion on the outer surface during use, and chemical etching of the outer surface. To eliminate the bias caused by the asymmetry potential, all membrane electrodes must be calibrated against one or more standard analyte solutions.

Such calibrations should be carried out at least daily, and more often when the electrode receives heavy use.

The Glass Electrode Potential

As noted earlier, the potential of a glass indicator electrode E_{ind} has three components: (1) the boundary potential, given by Equation 23-12, (2) the potential of the internal Ag-AgCl reference electrode E_{ref2}, and (3) the small asymmetry potential E_{asy}. In equation form,

$$E_{ind} = E_b + E_{ref2} + E_{asy}$$

Substitution of Equation 23-12 for E_b gives

$$E_{ind} = L' + 0.0592 \log a_1 + E_{ref2} + E_{asy}$$

or

$$E_{ind} = L + 0.0592 \log a_1 = L - 0.0592 \, pH \quad (23\text{-}13)$$

where L is a combination of the three constant terms. That is,

$$L = L' + E_{ref2} + E_{asy} \quad (23\text{-}14)$$

Note the similarity between Equation 23-13 and Equations 23-2 and 23-4 for metallic cation indicator electrodes. It is important to emphasize that although Equations 23-2 and 23-4 are similar in form to Equation 23-13, the sources of the potential of the electrodes that they describe are totally different — one is a redox potential, and the other is a boundary potential.

The Alkaline Error

Glass electrodes respond to the concentration of both hydrogen ion and alkali metal ions in basic solution. The magnitude of this alkaline error for four different glass membranes is shown in Figure 23-7 (curves C to F). These curves refer to solutions in which the sodium ion concentration was held constant at 1 M while the pH was varied. Note that the error is negative (that is, measured pH values were lower than the true values), which suggests that the electrode is responding to sodium ions as well as to protons. This observation is confirmed by data obtained for solutions containing different sodium ion concentrations. Thus at pH 12, the electrode with a Corning 015 membrane (curve C in Figure 23-7) registered a pH of 11.3 when immersed in a solution having a sodium ion concentration of 1 M but registered 11.7 in a solution that was 0.1 M in this ion. All singly charged cations induce an alkaline error whose magnitude depends on both the cation in question and the composition of the glass membrane.

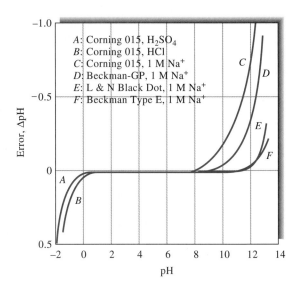

FIGURE 23-7 Acid and alkaline error of selected glass electrodes at 25°C. (From R. G. Bates, *Determination of pH*, 2nd ed., p. 365. New York: Wiley, 1973. With permission.)

The alkaline error can be satisfactorily explained by assuming that there is an exchange equilibrium between the hydrogen ions on the glass surface and the cations in solution. This process is simply the reverse of that shown in Equation 23-5, or

$$\underset{glass}{H^+Gl^-} + \underset{soln}{B^+} \rightleftharpoons \underset{glass}{B^+Gl^-} + \underset{soln}{B^+}$$

where B^+ represents any singly charged cation, such as sodium ion. In this case, the activity of the sodium ions relative to that of the hydrogen ions becomes so large that the electrode responds to both species.

Glass formulations are available from many manufacturers that minimize alkaline error at high pH. Many glass electrodes for routine use are still made of glasses with characteristics similar to Corning 015, however.

Selectivity Coefficients

The effect of an alkali metal ion on the potential across a membrane can be accounted for by inserting an additional term in Equation 23-12 to give

$$E_b = L' + 0.0592 \log(a_1 + k_{H,B}b_1) \quad (23\text{-}15)$$

where $k_{H,B}$ is the selectivity coefficient for the electrode and b_1 is the activity of the alkali metal ion. Equation 23-15 applies not only to glass indicator electrodes for hydrogen ion but also to all other types of membrane electrodes. Selectivity coefficients range from zero (no interference) to values greater than unity. A selectivity coefficient of unity means the electrode responds equally to the analyte ion and the interfering ion. If an electrode for ion A responds 20 times more strongly to ion B than to ion A, then $k_{A,B}$ has a value of 20. If the response of the electrode to ion C is 0.001 of its response to A (a much more desirable situation), $k_{A,C}$ is 0.001.[11]

The product $k_{H,B}b_1$ for a glass pH electrode is usually small relative to a_1 provided that the pH is less than 9; under these conditions, Equation 23-15 simplifies to Equation 23-13. At high pH values and at high concentrations of a singly charged ion, however, the second term in Equation 23-15 assumes a more important role in determining E_b, and an alkaline error is observed. For electrodes specifically designed for work in highly alkaline media (curve E in Figure 23-7), the magnitude of $k_{H,B}b_1$ is appreciably smaller than for ordinary glass electrodes.

The Acid Error

As shown in Figure 23-7, the typical glass electrode exhibits an error, opposite in sign to the alkaline error, in solutions of pH less than about 0.5; pH readings tend to be too high in this region. The magnitude of the error depends on a variety of factors and is generally not very reproducible. Not all of the causes of the acid error are well understood, but one source is a saturation effect that occurs when all the surface sites on the glass are occupied with H^+ ions. Under these conditions, the electrode no longer responds to further increases in the H^+ concentration and the pH readings are too high.

The data in Figure 23-7 are somewhat dated, and the glass electrode models listed are no longer available, but Corning 015 and similar glasses are still used to fabricate many of these devices. The errors shown in the figure draw our attention to regions of the pH scale where we must be cautious. Recent glass formulations by many manufacturers extend the usable range of glass electrodes at either the high or the low end, but there is no doubt that the relationship between pH and glass electrode potential becomes nonlinear at the extremes of the scale. It is important to exercise extraordinary caution both in calibrating the pH meter and in interpreting the results when measurements are made at the extremes of the pH scale.

The departures from linearity at the extremes in Figure 23-7 should be constant reminders that pH values below 0 and above 12 must be viewed with a very critical eye. Theoretical descriptions of even the simplest solutions in these pH regions are complex, and physical interpretation of the results is difficult.[12] Measurements on real samples in these pH regions should be regarded as qualitative or semiquantitative at best.

23D-4 Glass Electrodes for Other Cations

The alkaline error in early glass electrodes led to investigations concerning the effect of glass composition on the magnitude of this error. One consequence has been the development of glasses for which the alkaline error is negligible below about pH 12. Other studies have discovered glass compositions that permit the determination of cations other than hydrogen. This application requires that the hydrogen ion activity a_1 in Equation 23-15 be negligible relative to $k_{H,B}b_1$; under these circumstances, the potential is independent of pH and is a function of pB instead. Incorporation of Al_2O_3 or B_2O_3 in the glass has the desired effect. Glass electrodes for the direct potentiometric measurement of singly charged species such as Na^+, K^+, NH_4^+, Rb^+, Cs^+, Li^+, and Ag^+ have been developed. Some of these glasses are reasonably selective for certain singly charged cations. Glass electrodes for Na^+, Li^+, NH_4^+, and total concentration of univalent cations are available from commercial sources.

23D-5 Crystalline Membrane Electrodes

The most important type of crystalline membranes is manufactured from an ionic compound or a homogeneous mixture of ionic compounds. In some instances the membrane is cut from a single crystal; in others, disks are formed from the finely ground crystalline solid by high pressures or by casting from a melt. Typical membranes have a diameter of about 10 mm and a thickness of 1 or 2 mm. To form an electrode, a membrane is sealed to the end of a tube made from a chemically inert plastic such as Teflon or polyvinyl chloride (PVC).

Conductivity of Crystalline Membranes

Most ionic crystals are insulators and do not have sufficient electrical conductivity at room temperature to be used as membrane electrodes. Those that are con-

---

[11] For a collection of selectivity coefficients for all types of ion-selective electrodes, see Y. Umezawa, *CRC Handbook of Ion Selective Electrodes: Selectivity Coefficients*, Boca Raton, FL: CRC Press, 1990.

[12] S. L. Clegg, J. A. Rard, and K. S. Pitzer, *J. Chem. Soc., Faraday Trans.*, **1994**, *90*, 1875.

ductive are characterized by having a small singly charged ion that is mobile in the solid phase. Examples are fluoride ion in certain rare earth fluorides, silver ion in silver halides and sulfides, and copper(I) ion in copper(I) sulfide.

The Fluoride Electrode

Lanthanum fluoride, LaF_3, is a nearly ideal substance for the preparation of a crystalline membrane electrode for the determination of fluoride ion. Although this compound is a natural conductor, its conductivity can be enhanced by doping with europium fluoride, EuF_2. Membranes are prepared by cutting disks from a single crystal of the doped compound.

The mechanism of the development of a fluoride-sensitive potential across a lanthanum fluoride membrane is quite analogous to that described for glass, pH-sensitive membranes. That is, at the two interfaces, ionization creates a charge on the membrane surface as shown by the equation

$$\underset{\text{solid}}{LaF_3} \rightleftharpoons \underset{\text{solid}}{LaF_2^+} + \underset{\text{soln}}{F^-}$$

The magnitude of the charge depends on the fluoride ion concentration of the solution. Thus, the side of the membrane in contact with the lower fluoride ion concentration becomes positive with respect to the other surface. This charge produces a potential difference that is a measure of the difference in fluoride concentration of the two solutions. The potential of a cell containing a lanthanum fluoride electrode is given by an equation analogous to Equation 23-13. That is,

$$E_{ind} = L - 0.0592 \log a_{F^-} = L + 0.0592 \, pF \quad (23\text{-}16)$$

Note that the signs of the second terms on the right are reversed because an anion is being determined (see also Equation 23-3).

Commercial lanthanum fluoride electrodes come in various shapes and sizes and are available from several sources. Most are rugged and can be used at temperatures between $0°$ and $80°C$. The response of the fluoride electrode is linear down to 10^{-6} M (0.02 ppm), where the solubility of lanthanum fluoride begins to contribute to the concentration of fluoride ion in the analyte solution. The only ion that interferes directly with fluoride measurements is hydroxide ion; this interference becomes serious at pH > 8. At pH < 5, hydrogen ions also interfere in total fluoride determinations. Under these conditions, undissociated hydrogen fluoride forms, and the electrode does not respond to this species. In most respects, the fluoride ion electrode approaches the ideal for selective electrodes.

Electrodes Based on Silver Salts

Membranes prepared from single crystals or pressed disks of the various silver halides are selective toward silver and halide ions. Their behavior is generally far from ideal, however, because of low conductivity, low mechanical strength, and a tendency to develop high photoelectric potentials. These disadvantages are minimized if the silver salts are mixed with crystalline silver sulfide in an approximately 1:1 molar ratio. Homogeneous mixtures are formed from equimolar solutions of sulfide and halide ions by precipitation with silver nitrate. After washing and drying, the product is shaped into disks under a pressure of about 10^5 pounds per square inch. The resulting disk exhibits good electrical conductivity because of the mobility of the silver ion in the sulfide matrix.

Membranes constructed either from silver sulfide or from a mixture of silver sulfide and another silver salt are useful for the determination of both sulfide and silver ions. Toward silver ions, the electrical response is similar to a metal electrode of the first kind (although the mechanism of activity is totally different). The electrical response of a silver sulfide membrane to sulfide ions is similar to that of an electrode of the second kind (Section 23B-2). When the membrane is immersed in the analyte solution, a minuscule amount of silver sulfide dissolves and quickly saturates the film of liquid adjacent to the electrode. The solubility, and thus the silver ion concentration, however, depends on the sulfide concentration of the analyte.

Crystalline membranes are also available that consist of a homogeneous mixture of silver sulfide with sulfides of copper(II), lead, or cadmium. Toward these divalent cations, electrodes from these materials have electrical responses similar to electrodes of the third kind (Section 23C-3). Note that these divalent sulfides, by themselves, are not conductors and thus do not exhibit ion-selective activity.

Table 23-3 is a representative list of solid-state electrodes that are available from commercial sources.

23D-6 Liquid-Membrane Electrodes

Liquid membranes are formed from immiscible liquids that selectively bond certain ions. Membranes of this type are particularly important because they permit the direct potentiometric determination of the activi-

TABLE 23-3 Characteristics of Solid-State Crystalline Electrodes

Analyte Ion	Concentration Range, M	Major Interferences
Br^-	10^0 to 5×10^{-6}	CN^-, I^-, S^{2-}
Cd^{2+}	10^{-1} to 1×10^{-7}	$Fe^{2+}, Pb^{2+}, Hg^{2+}, Ag^+, Cu^{2+}$
Cl^-	10^0 to 5×10^{-5}	$CN^-, I^-, Br^-, S^{2-}, OH^-, NH_3$
Cu^{2+}	10^{-1} to 1×10^{-8}	Hg^{2+}, Ag^+, Cd^{2+}
CN^-	10^{-2} to 1×10^{-6}	S^{2-}, I^-
F^-	Sat'd to 1×10^{-6}	OH^-
I^-	10^0 to 5×10^{-8}	CN^-
Pb^{2+}	10^{-1} to 1×10^{-6}	Hg^{2+}, Ag^+, Cu^{2+}
Ag^+/S^{2-}	Ag^+: 10^0 to 1×10^{-7} S^{2-}: 10^0 to 1×10^{-7}	Hg^{2+}
SCN^-	10^0 to 5×10^{-6}	I^-, Br^-, CN^-, S^{2-}

From *Orion Guide to Ion Analysis*, Boston, MA: Thermo Orion, 1992. With permission of Thermo Electron Corp., Waltham, MA.

ties of several polyvalent cations and of certain singly charged anions and cations as well.

Divalent Cation Electrodes

Liquid membranes are prepared from immiscible, liquid ion exchangers, which are retained in a porous inert solid support. As shown schematically in Figure 23-8, a porous, hydrophobic (that is, water-repelling), plastic disk (typical dimensions: 3×0.15 mm) holds the organic layer between the two aqueous solutions. For divalent cation determinations, the inner tube contains an aqueous standard solution of MCl_2, where M^{2+} is the cation whose activity is to be determined. This solution is also saturated with AgCl to form a Ag-AgCl reference electrode with the silver lead wire.

As an alternative to the use of a porous disk as a rigid supporting medium, it is possible to immobilize liquid exchangers in tough PVC membranes. In this type of electrode, the liquid ion exchanger and PVC are dissolved in a solvent such as tetrahydrofuran. The solvent is evaporated to leave behind a flexible membrane that can be cut, shaped, and bonded to the end of a glass or plastic tube. Membranes formed in this way behave in much the same way as those in which the ion exchanger is encased as a liquid in the pores of a disk. Most liquid-membrane electrodes are of this newer type.

The active substances in liquid membranes are of three kinds: (1) cation exchangers; (2) anion exchangers; and (3) neutral macrocyclic compounds, which selectively complex certain cations.

One of the most important liquid-membrane electrodes is selective toward calcium ion in neutral media. The active ingredient in the membrane is a cation exchanger consisting of an aliphatic diester of phosphoric acid dissolved in a polar solvent. The diester contains a single acidic proton; thus, two molecules react with the

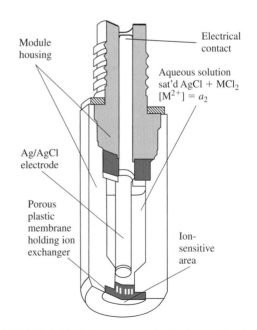

FIGURE 23-8 Liquid-membrane electrode sensitive to M^{2+}. (Courtesy of Thermo Orion Corp., Waltham, MA.)

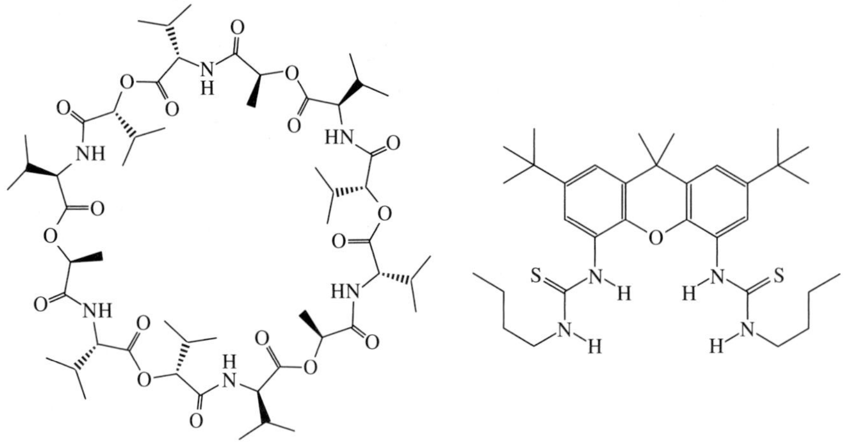

FIGURE 23-9 Typical ionophores (a) valinomycin and (b) a bis-thiourea.

divalent calcium ion to form a dialkyl phosphate with the structure

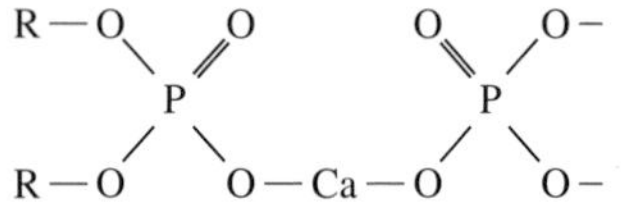

calcium dialkyl phosphate

Here, R is an aliphatic group containing from eight to sixteen carbon atoms. In commercial electrodes of this type, R is usually a C_9 group. The internal aqueous solution in contact with the exchanger (see Figure 23-8) contains a fixed concentration of calcium chloride and a silver–silver chloride reference electrode. The porous disk (or the PVC membrane) containing the ion-exchange liquid separates the analyte solution from the reference calcium chloride solution. The equilibrium established at each interface can be represented as

$$\underset{\text{organic}}{[(RO)_2POO]_2Ca} \rightleftharpoons \underset{\text{organic}}{2(RO)_2POO^-} + \underset{\text{aqueous}}{Ca^{2+}}$$

Note the similarity of this equilibrium to Equation 23-7 for the glass electrode. The relationship between potential and pCa is also analogous to that for the glass electrode (Equation 23-12). Thus,

$$E_{\text{ind}} = L + \frac{0.0592}{2} \log a_1$$

$$= L' - \frac{0.0592}{2} pCa \qquad (23\text{-}17)$$

In this case, however, the second term on the right is divided by two because the cation is divalent.

The calcium membrane electrode is a valuable tool for physiological studies because calcium plays important roles in nerve conduction, bone formation, muscle contraction, cardiac conduction and contraction, and renal tubular function. At least some of these processes are influenced more by calcium ion activity than by calcium ion concentration; activity, of course, is measured by the electrode.

Ionophore-Based Ion-Selective Electrodes

When neutral, lipophilic compounds called *ionophores* that form complexes with target ions are incorporated in liquid or polymer membranes along with a small amount of a lipophilic ion exchanger, the target ions are carried across the solution-membrane boundary by the formation of the complex. The selectivity of the membrane toward a given ion is governed by the stability of the complex formed between the ionophore and the target ion, and the separation of charge across the solution-membrane barrier produces a nernstian response similar in form to Equations 23-16 and 23-17. Ionophores are designed and synthesized to maximize selectivity toward particular ions, and hundreds of examples have appeared in the literature along with their selectivity coefficients and other characteristics.[13] Figure 23-9a shows valinomycin, an uncharged macrocyclic ether and antibiotic that has very high selectivity for potassium. Whereas Figure 23-9b

[13] P. Buhlmann, E. Pretsch, and E. Bakker, *Chem. Rev.*, **1998**, *98*, 1593; Y. Umezawa et al., *Pure Appl. Chem.*, **2000**, *72*, 1851; *Pure Appl. Chem.*, **2002**, *74*, 923; *Pure Appl. Chem.*, **2002**, *74*, 995.

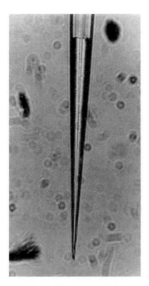

FIGURE 23-10 Photograph of a potassium liquid-ion exchanger (valinomycin) microelectrode with 125 mm of ion exchanger inside the tip. The magnification of the original photo was 400×. (From J. L. Walker, *Anal. Chem.*, **1971**, *43* [3], 91A. Reproduced by permission of the American Chemical Society.)

shows a bis-thiourea that is especially selective for chloride.

Potassium ion-selective electrodes have great value for physiological studies. The selectivity of a liquid membrane for potassium relative to sodium is especially important because both of these ions are present in all living systems and play important roles in neural transmission. Valinomycin is unquestionably the most widely used ionophore for potassium. It is about 10^4 times as responsive to potassium ion as to sodium ion and 10^7 times as responsive to potassium as to calcium and magnesium.[14] Figure 23-10 is a photomicrograph of a valinomycin microelectrode for monitoring the potassium activity in the interior of a single cell. In this case, no physical membrane is needed to separate the internal solution from the analyte because of the small diameter of the opening at the tip (<1 µm) and because the interior of the glass was made hydrophobic by a coating of silicone. Valinomycin-based membrane electrodes are used in many clinical analyzers such as the i-STAT unit discussed in Section 23F-2. It has been estimated that in 1990 alone, more than 64 million valinomycin-based electrodes were used.[15]

Table 23-4 lists some typical commercially available liquid-membrane electrodes. The anion-sensitive electrodes contain a solution of an anion exchanger in an organic solvent. As mentioned earlier, many of the so-called liquid-membrane electrodes are in fact solids in which the liquid is contained in a polymer (plastic) matrix. The first and most widely used polymer for membrane electrodes is PVC,[16] but other materials have been used as well for compatibility with ionophores and fabrication materials. Polymer-based electrodes are somewhat more convenient to use and more rugged than the older porous disk electrodes. All electrodes listed in Table 23-4 are of the plastic-membrane type.

23E ION-SELECTIVE FIELD-EFFECT TRANSISTORS

In Section 2C-3 (Figure 2-19), we described the metal oxide semiconductor field-effect transistor (MOSFET), which is widely used in computers and other electronic circuits as a switch to control current in circuits. One of the problems in using this type of device in electronic circuits has been its pronounced sensitivity to ionic surface impurities, and a great deal of money and effort has been expended by the electronics industry in minimizing or eliminating this sensitivity to produce stable transistors.

Since 1970 much has been accomplished in exploiting the sensitivity of MOSFETs to surface ionic impurities for the selective potentiometric determination of various ions. These studies have led to the development of a number of different *ion-selective field-effect transistors* (ISFETs). The theory of their ion-selective sensitivity is well understood and is described in the section that follows.[17]

23E-1 Mechanism of ISFET Ion-Selective Behavior

An ISFET is very similar in construction and function to an *n*-channel enhancement mode MOSFET (Section 2C-3). The ISFET differs only in that variation in the concentration of the ions of interest provides the variable gate voltage to control the conductivity of the

[14]M. S. Frant and J. W. Ross Jr., *Science*, **1970**, *167*, 987.
[15]See note 13.

[16]G. J. Moody, R. B. Oke, and J. D. R. Thomas, *Analyst*, **1970**, *95*, 910.
[17]For a review of ISFETs, see P. Bergveld, *Sens. Actuators, B*, **2003**, *88*, 1. For an introduction to the theory and operation of ISFETs, see J. Janata, *Principles of Chemical Sensors*, pp. 125–41, New York: Plenum, 1989.

TABLE 23-4 Characteristics of Liquid-Membrane Electrodes

Analyte Ion	Concentration Range, M[†]	Major Interferences[‡]
NH_4^+	10^0 to 5×10^{-7}	<1 H^+, 5×10^{-1} Li^+, 8×10^{-2} Na^+, 6×10^{-4} K^+, 5×10^{-2} Cs^+, >1 Mg^{2+}, >1 Ca^{2+}, >1 Sr^{2+}, >0.5 Sr^{2+}, 1×10^{-2} Zn^{2+}
Cd^{2+}	10^0 to 5×10^{-7}	Hg^{2+} and Ag^+ (poisons electrode at $>10^{-7}$ M), Fe^{3+} (at $>0.1[Cd^{2+}]$), Pb^{2+} (at $>[Cd^{2+}]$), Cu^{2+} (possible)
Ca^{2+}	10^0 to 5×10^{-7}	10^{-5} Pb^{2+}; 4×10^{-3} Hg^{2+}, H^+, 6×10^{-3} Sr^{2+}; 2×10^{-2} Fe^{2+}; 4×10^{-2} Cu^{2+}; 5×10^{-2} Ni^{2+}; 0.2 NH_3; 0.2 Na^+; 0.3 $Tris^+$; 0.3 Li^+; 0.4 K^+; 0.7 Ba^{2+}; 1.0 Zn^{2+}; 1.0 Mg^{2+}
Cl^-	10^0 to 5×10^{-6}	Maximum allowable ratio of interferent to $[Cl^-]$: OH^- 80, Br^- 3×10^{-3}, I^- 5×10^{-7}, S^{2-} 10^{-6}, CN^- 2×10^{-7}, NH_3 0.12, $S_2O_3^{2-}$ 0.01
BF_4^-	10^0 to 7×10^{-6}	5×10^{-7} ClO_4^-; 5×10^{-6} I^-; 5×10^{-5} ClO_3^-; 5×10^{-4} CN^-; 10^{-3} Br^-; 10^{-3} NO_2^-; 5×10^{-3} NO_3^-; 3×10^{-3} HCO_3^-, 5×10^{-2} Cl^-; 8×10^{-2} $H_2PO_4^-$, HPO_4^{2-}, PO_4^{3-}; 0.2 OAc^-; 0.6 F^-; 1.0 SO_4^{2-}
NO_3^-	10^0 to 7×10^{-6}	10^{-7} ClO_4^-; 5×10^{-6} I^-; 5×10^{-5} ClO_3^-; 10^{-4} CN^-; 7×10^{-4} Br^-; 10^{-3} HS^-; 10^{-2} HCO_3^-, 2×10^{-2} CO_3^{2-}; 3×10^{-2} Cl^-; 5×10^{-2} $H_2PO_4^-$, HPO_4^{2-}; PO_4^{3-}; 0.2 OAc^-; 0.6 F^-; 1.0 SO_4^{2-}
NO_2^-	1.4×10^{-6} to 3.6×10^{-6}	7×10^{-1} salicylate, 2×10^{-3} I^-, 10^{-1} Br^-, 3×10^{-1} ClO_3^-, 2×10^{-1} acetate, 2×10^{-1} HCO_3^-, 2×10^{-1} NO_3^-, 2×10^{-1} SO_4^{2-}, 1×10^{-1} Cl^-, 1×10^{-1} ClO_4^-, 1×10^{-1} F^-
ClO_4^-	10^0 to 7×10^{-6}	2×10^{-3} I^-; 2×10^{-2} ClO_3^-; 4×10^{-2} CN^-, Br^-; 5×10^{-2} NO_2^-, NO_3^-; 2 HCO_3^-, CO_3^{2-}; Cl^-, $H_2PO_4^-$, HPO_4^{2-}, PO_4^{3-}, OAc^-, F^-, SO_4^{2-}
K^+	10^0 to 1×10^{-6}	3×10^{-4} Cs^+; 6×10^{-3} NH_4^+, Tl^+; 10^{-2} H^+; 1.0 Ag^+, $Tris^+$; 2.0 Li^+, Na^+
Water hardness $(Ca^{2+} + Mg^{2+})$	10^{-3} to 6×10^{-6}	3×10^{-5} Cu^{2+}, Zn^{2+}; 10^{-4} Ni^{2+}; 4×10^{-4} Sr^{2+}; 6×10^{-5} Fe^{2+}; 6×10^{-4} Ba^{2+}; 3×10^{-2} Na^+; 0.1 K^+

All electrodes are the plastic-membrane type.

[†]From product catalog, Boston, MA: Thermo Orion, 2006. With permission of Thermo Electron Corp., Waltham, MA.

[‡]From product instruction manuals, Boston, MA: Thermo Orion, 2003. With permission of Thermo Electron Corp., Waltham, MA.

channel. As shown in Figure 23-11, instead of the usual metallic contact, the gate of the ISFET is covered with an insulating layer of silicon nitride (Si_3N_4). The analyte solution, containing hydrogen ions in this example, is in contact with this insulating layer and with a reference electrode. The surface of the gate insulator functions very much like the surface of a glass electrode. Protons from the hydrogen ions in the test solution are adsorbed on available microscopic sites on the silicon nitride. Any change in the hydronium ion concentration of the solution results in a change in the concentration of adsorbed protons. The change in concentration of adsorbed protons then gives rise to a changing electrochemical potential between the gate and source, which in turn changes the conductivity of the channel of the ISFET. The conductivity of the channel can be monitored electronically to provide a signal that is proportional to the logarithm of the concentration of H^+ in the solution. Note that the entire ISFET except the gate insulator is coated with a polymeric encapsulant to insulate all electrical connections from the analyte solution.

23E-2 Application of ISFETs

The ion-sensitive surface of the ISFET is naturally sensitive to pH changes, but the device may be rendered sensitive to other species by coating the silicon nitride gate insulator with a polymer containing molecules that tend to form complexes with species other than hydronium ion. Furthermore, several ISFETs may be fabricated on the same substrate so that multiple measurements may be made simultaneously. All of the ISFETs may detect the same species to enhance accuracy and reliability, or each ISFET may be coated with a different polymer so that measurements of several different species may be made.

ISFETs offer a number of significant advantages over membrane electrodes, including ruggedness, small

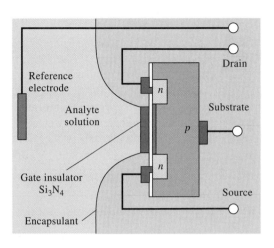

FIGURE 23-11 An ISFET for measuring pH.

size, inertness toward harsh environments, rapid response, and low electrical impedance. In contrast to membrane electrodes, ISFETs do not require hydration before use and can be stored indefinitely in the dry state. Despite these many advantages, no ISFET-specific-ion electrode appeared on the market until the early 1990s, more than 20 years after their invention. The reason for this delay is that manufacturers were unable to develop the technology of encapsulating the devices to create a product that did not exhibit drift and instability. The only significant disadvantage of ISFETs other than drift appears to be that they require a more or less traditional reference electrode. This requirement places a lower limit on the size of the ISFET probe. Work continues on the development of a differential pair of ISFETs, one selective for the analyte ion and the other not. The second reference ISFET is called a REFET, and a differential amplifier is used to measure the voltage difference between the ISFET and the REFET, which is proportional to pX.[18] Many ISFET-based devices have appeared on the market over the past decade for the determination of pH, and research has continued on the development of devices that are selective for other analytes. Well over 150 ISFET patents have been filed over the last three decades, and more than twenty companies manufacture ISFETs in various forms. The promise of a tiny, rugged sensor that can be used in a broad range of harsh and unusual environments is being achieved as the reference electrode problem is being solved.[19]

[18]P. Bergveld, *Sens. Actuators, B*, **2003**, *88*, 9.
[19]Ibid., pp. 18–19.

23F MOLECULAR-SELECTIVE ELECTRODE SYSTEMS

Two types of membrane electrode systems have been developed that act selectively toward certain types of *molecules*. One of these is used for the determination of dissolved gases, such as carbon dioxide and ammonia. The other, which is based on biocatalytic membranes, permits the determination of a variety of organic compounds, such as glucose and urea.

23F-1 Gas-Sensing Probes

During the past three decades, several gas-sensing electrochemical devices have become available from commercial sources. In the manufacturer literature, these devices are often called gas-sensing "electrodes." Figure 23-12 shows that these devices are not, in fact, electrodes but instead are electrochemical cells made up of a specific-ion electrode and a reference electrode immersed in an internal solution retained by a thin gas-permeable membrane. Thus, *gas-sensing probes* is a more suitable name for these gas sensors.

Gas-sensing probes are remarkably selective and sensitive devices for determining dissolved gases or

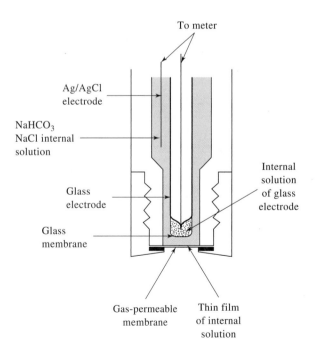

FIGURE 23-12 Schematic of a gas-sensing probe for carbon dioxide.

ions that can be converted to dissolved gases by pH adjustment.

Membrane Probe Design

Figure 23-12 is a schematic showing details of a gas-sensing probe for carbon dioxide. The heart of the probe is a thin, porous membrane, which is easily replaceable. This membrane separates the analyte solution from an internal solution containing sodium bicarbonate and sodium chloride. A pH-sensitive glass electrode having a flat membrane is held in position so that a very thin film of the internal solution is sandwiched between it and the gas-permeable membrane. A silver–silver chloride reference electrode is also located in the internal solution. It is the pH of the film of liquid adjacent to the glass electrode that provides a measure of the carbon dioxide content of the analyte solution on the other side of the membrane.

Gas-Permeable Membranes

There are two common types of membrane material: microporous and homogeneous. Microporous materials are manufactured from hydrophobic polymers such as polytetrafluoroethylene or polypropylene, which have a porosity (void volume) of about 70% and a pore size of less than 1 μm. Because of the nonpolar, water-repellent properties of the film, water molecules and electrolyte ions are excluded from the pores; gaseous molecules, on the other hand, are free to move in and out of the pores by *effusion* and thus across this barrier. Typically, the thickness of microporous membranes is about 0.1 mm.

Homogeneous films, in contrast, are solid polymeric substances through which the analyte gas passes by dissolving in the membrane, diffusing, and then desolvating into the internal solution. Silicone rubber is the most widely used material for construction of these films. Homogeneous films are generally thinner than microporous membranes (0.01 to 0.03 mm) to hasten the transfer of gas and thus the rate of response of the system.

Mechanism of Response

When a solution containing dissolved carbon dioxide is brought into contact with the microporous membrane shown in Figure 23-12, the gas effuses through the membrane, as described by the reactions

$$\underset{\substack{\text{analyte}\\\text{solution}}}{CO_2(aq)} \rightleftharpoons \underset{\substack{\text{membrane}\\\text{pores}}}{CO_2(g)}$$

$$\underset{\substack{\text{membrane}\\\text{pores}}}{CO_2(g)} \rightleftharpoons \underset{\substack{\text{internal}\\\text{solution}}}{CO_2(aq)}$$

$$\underset{\substack{\text{internal}\\\text{solution}}}{CO_2(aq) + 2H_2O} \rightleftharpoons \underset{\substack{\text{internal}\\\text{solution}}}{HCO_3^- + H_3O^+}$$

The last equilibrium causes the pH of the internal surface film to change. This change is then detected by the internal glass–calomel electrode system. A description of the overall process is obtained by adding the equations for the three equilibria to give

$$\underset{\substack{\text{analyte}\\\text{solution}}}{CO_2(aq) + 2H_2O} \rightleftharpoons \underset{\substack{\text{internal}\\\text{solution}}}{HCO_3^- + H_3O^+}$$

The thermodynamic equilibrium constant K for this overall reaction is

$$K = \frac{(a_{H_3O^+})_{int}(a_{HCO_3^-})_{int}}{(a_{CO_2})_{ext}}$$

For a neutral species such as CO_2, $a_{CO_2} = [CO_2(aq)]$, so

$$K = \frac{(a_{H_3O^+})_{int}(a_{HCO_3^-})_{int}}{[CO_2(aq)]_{ext}}$$

where $[CO_2(aq)]_{ext}$ is the molar concentration of the gas in the analyte solution. For the measured cell potential to vary linearly with the logarithm of the carbon dioxide concentration of the external solution, the hydrogen carbonate activity of the internal solution must be sufficiently large that it is not altered significantly by the carbon dioxide entering from the external solution. Assuming then that $(a_{HCO_3^-})_{int}$ is constant, we can rearrange the previous equations to

$$\frac{(a_{H_3O^+})_{int}}{[CO_2(aq)]_{ext}} = \frac{K}{(a_{HCO_3^-})_{int}} = K_g \quad (23\text{-}18)$$

If we allow a_1 to be the hydrogen ion activity of the internal solution, we rearrange this equation to give

$$(a_{H_3O^+})_{int} = a_1 = K_g[CO_2(aq)]_{ext} \quad (23\text{-}19)$$

By substituting Equation 21-19 into Equation 21-13, we find that

$$\begin{aligned} E_{ind} &= L + 0.0592 \log a_1 \\ &= L + 0.0592 \log K_g[CO_2(aq)]_{ext} \\ &= L + 0.0592 \log K_g + 0.0592 \log[CO_2(aq)]_{ext} \end{aligned}$$
$$(23\text{-}20)$$

Combining the two constant terms to give a new constant L' leads to

$$E_{ind} = L' + 0.0592 \log[CO_2(aq)]_{ext} \quad (23\text{-}21)$$

Finally, because

$$E_{cell} = E_{ind} - E_{ref}$$

then

$$E_{cell} = L' + 0.0592 \log [CO_2(aq)]_{ext} - E_{ref} \quad (23\text{-}22)$$

or

$$E_{cell} = L'' + 0.0592 \log [CO_2(aq)]_{ext}$$

where

$$L'' = L + 0.0592 \log K_g - E_{ref}$$

Thus, the potential between the glass electrode and the reference electrode in the internal solution is determined by the CO_2 concentration in the external solution. Note once again that *no electrode comes in direct contact with the analyte solution.*

The only species that interfere with the response of these sensors to CO_2 are other dissolved gases that permeate the membrane and then affect the pH of the internal solution. The specificity of gas probes depends only on the permeability of the gas membrane. Membrane-based sensors for CO_2 have been microfabricated for use in clinical analyzers as described in Section 23F-2.[20]

It is possible to increase the selectivity of the gas-sensing probe by using an internal electrode sensitive to some species other than hydrogen ion; for example, a nitrate-sensing electrode can be used to construct a probe that is sensitive to nitrogen dioxide. For this device, the equilibrium is

$$\underset{\text{external}}{2NO_2(aq)} + 2H_2O \rightleftharpoons \underset{\text{internal solution}}{NO_2^- + NO_3^- + 2H_3O^+}$$

This electrode permits the determination of NO_2 in the presence of gases such as SO_2, CO_2, and NH_3, which would also alter the pH of the internal solution.

Table 23-5 lists representative gas-sensing probes that are commercially available. An oxygen-sensitive probe is also on the market; it, however, is based on a voltammetric measurement and is discussed in Chapter 25.

23F-2 Biosensors

Over the past three decades, considerable effort has been devoted to combining the selectivity of biochemical materials and reactions with electrochemical trans-

TABLE 23-5 Commercial Gas-Sensing Probes

Gas	Equilibrium in Internal Solution	Sensing Electrode
NH_3	$NH_3 + H_2O \rightleftharpoons NH_4^+ + OH^-$	Glass, pH
CO_2	$CO_2 + H_2O \rightleftharpoons HCO_3^- + H^+$	Glass, pH
HCN	$HCN \rightleftharpoons H^+ + CN^-$	Ag_2S, pCN
HF	$HF \rightleftharpoons H^+ + F^-$	LaF_3, pF
H_2S	$H_2S \rightleftharpoons 2H^+ + S^{2-}$	Ag_2S, pS
SO_2	$SO_2 + H_2O \rightleftharpoons HSO_3^- + H^+$	Glass, pH
NO_2	$2NO_2 + H_2O \rightleftharpoons$ $NO_2^- + NO_3^- + 2H^+$	Immobilized ion exchange, pNO_3

ducers (See Figure 1-7) to give highly selective biosensors for the determination of biological and biochemical compounds.[21] These materials include enzymes,[22] DNA, antigens, antibodies, bacteria, cells,[23] and whole samples of animal and plant tissue. When analyte molecules react with these materials, the interaction triggers the production of species that can be monitored directly or indirectly by one of the ion- or molecular-selective electrodes previously discussed. The selectivity advantage of biosensors is counterbalanced by the limited stability of many biochemical substances, which inhibits their use in harsh environments. This disadvantage is being overcome to some extent by the development of synthetic materials that have molecular recognition characteristics similar to naturally occurring molecules but that are more robust.

Enzyme-Based Biosensors

The most widely studied and perhaps the most useful biosensor is based on enzymes. In these devices the sample is brought into contact with an *immobilized enzyme*, which reacts with the analyte to yield a species such as ammonia, carbon dioxide, hydrogen ions, or hydrogen peroxide. The concentration of this product, which is proportional to the analyte concentration, is then determined by the transducer. The most common

---

[20] M. E. Meyerhoff, *Clin. Chem.*, **1990**, *36*, 1567.

[21] For reviews of biosensors, see E. Bakker and M. Telting-Diaz, *Anal. Chem.*, **2002**, *74*, 2781. For extensive treatments of the subject see *Biosensors and Modern Biospecific Analytical Techniques*, Vol. 44 (Comprehensive Analytical Chemistry), L. Gorton, ed., Amsterdam: Elsevier, 2004; *Biosensor Technology: Fundamentals and Applications*, R. P. Buck, W. E. Hatfield, M. Umaña, and E. F. Bowden, eds., New York: Dekker, 1990; *Biosensors: Fundamentals and Applications*, A. P. F. Turner, I. Karube, and G. S. Wilson, eds., New York: Oxford University Press, 1987.

[22] G. G. Guilbault and J. G. Montalvo, *J. Am. Chem. Soc.*, **1969**, *91*, 2164.

[23] G. A. Rechnitz, R. K. Kobos, S. J. Riechel, and C. R. Gebauer, *Anal. Chim. Acta*, **1977**, *94*, 357.

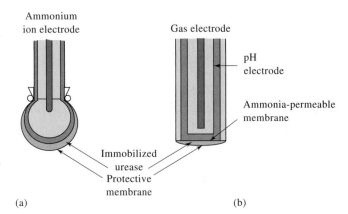

FIGURE 23-13 Enzyme electrodes for measuring urea. (Reprinted with permission from D. N. Gray, M. H. Keyes, and B. Watson, *Anal. Chem.*, **1977**, *49*, 1067A. Copyright 1977 American Chemical Society.)

transducers in these devices are membrane electrodes, gas-sensing probes, and voltammetric devices, which are discussed in Chapter 25.

Biosensors based on membrane electrodes are attractive from several perspectives. First, complex organic molecules can be determined with the convenience, speed, and ease that characterize ion-selective measurements of inorganic species. Second, enzyme-catalyzed reactions occur under mild conditions of temperature and pH and at relatively low substrate concentrations. Third, combining the selectivity of the enzymatic reaction and the electrode response provides results that are free from most interferences.

The main limitation to enzymatic procedures is the high cost of enzymes, particularly when used for routine or continuous measurements. This disadvantage has led to the use of immobilized enzyme media in which a small amount of enzyme can be used for the repetitive analysis of hundreds of samples. Two general techniques are used. In one, the sample is passed through a fixed bed of immobilized enzyme and then to the detector. In the second, a porous layer of the immobilized enzyme is attached directly to the surface of the ion-selective electrode, thus forming an *enzyme electrode*. In such devices, the reaction product reaches the selective membrane surface by diffusion.

Immobilization of enzymes can be accomplished in several ways, including physical entrapment in a polymer gel, physical adsorption on a porous inorganic support such as alumina, covalent bonding of the enzyme to a solid surface such as glass beads or a polymer, or copolymerization of the enzyme with a suitable monomer.

Figure 23-13 shows two types of enzyme electrodes for the determination of blood urea nitrogen (BUN),

an important routine clinical test. In the presence of the enzyme urease, urea is hydrolyzed according to the reaction

$$(NH_2)_2CO + H_3O^+ + H_2O \rightarrow 2NH_4^+ + HCO_3^-$$
$$\upharpoonleft\downharpoonright \; + 2H_2O$$
$$2NH_3 + 2H_3O^+ \quad (23\text{-}23)$$

The electrode in Figure 23-13a is a glass electrode that responds to the ammonium ion formed by the reaction shown in the upper part of Equation 23-23. The electrode in Figure 23-13b is an ammonia gas probe that responds to the molecular ammonia in equilibrium with the ammonium ion. Unfortunately, both electrodes have limitations. The glass electrode responds to all monovalent cations, and its selectivity coefficients for NH_4^+ over Na^+ and K^+ are such that interference occurs in most biological media (such as blood). The ammonia gas probe has a different problem — the pH of the probe is incompatible with the enzyme. The enzyme requires a pH of about 7 for maximum catalytic activity, but the sensor's maximum response occurs at a pH that is greater than 8 to 9 (where essentially all of the NH_4^+ has been converted to NH_3). Thus, the sensitivity of the electrode is limited. Both limitations are overcome by use of a fixed-bed enzyme system where the sample at a pH of about 7 is pumped over the enzyme. The resulting solution is then made alkaline and the liberated ammonia determined with an ammonia gas probe. Automated instruments (see Chapter 33) based on this technique have been on the market for several years.

Factors that affect the detection limits of enzyme-based biosensors include the inherent detection limits of the ion-selective electrode coupled with the enzyme layer, the kinetics of the enzyme reaction, and the mass-

TABLE 23-6 Potentiometric Enzyme-Based Biosensors*

Analyte	Enzyme	Reaction	ISE detectors
Urea	Urease	Urea $\longrightarrow 2NH_3 + CO_2$	NH_4^+–glass or polymer NH_3–gas sensor H^+–glass or polymer
Creatinine	Creatininase	Creatinine $\longrightarrow$ N-methylhydantoin + NH_3 NH_3–gas sensor	NH_4^+–glass or polymer
L- or D-Amino acids	L- or D-Amino acid oxidase	L (D) AA $\longrightarrow$ RCOCOOH + NH_3 + H_2O_2 NH_3–gas sensor	NH_4^+–glass or polymer
L-Glutamine	Glutaminase	L-Glutamine $\longrightarrow$ glutamic acid + NH_3 NH_3–gas sensor	NH_4^+–glass or polymer
Adenosine	Adenosine deaminase	Adenosine $\longrightarrow$ inosine + NH_3 NH_3–gas sensor	NH_4^+–glass or polymer
L-Glutamate	Glutamate decarboxylase	L-Glutamate $\longrightarrow$ GABA + CO_2	CO_2–gas sensor
Amygdalin	β-Glucosidase	amygdalin $\longrightarrow$ HCN + $2C_6H_{12}O_6$ + benzaldehyde	CN-solid-state
Glucose	Glucose oxidase	glucose + O_2 $\longrightarrow$ gluconic acid + H_2O_2	H^+–glass or polymer
Penicillin	Penicillinase	Penicillin $\longrightarrow$ Penicilloic acid	H^+–glass or polymer

*From E. Bakker and M. E. Meyerhoff, in *Encyclopedia of Electrochemistry*, A. J. Bard., Ed., Vol. 9, *Bioelectrochemistry*, G. S. Wilson, Ed., New York: Wiley, 2002, p. 305, with permission.

transfer rate of substrate into the layer.[24] Despite a large body of research on these devices and, as shown in Table 23-6, the several potentially useful enzyme-based membranes, there have been only a few commercial potentiometric enzyme electrodes, due at least in part to limitations such as those cited previously. The enzymatic determination of urea nitrogen in clinical applications discussed in the following paragraphs is an exceptional example of this type of sensor. A number of commercial sources offer enzymatic electrodes based on voltammetric measurements on various enzyme systems. These electrodes are discussed in Chapter 25.

Applications in Clinical Analysis

An example of the application of a potentiometric enzyme-based biosensor in clinical analysis is an automated monitor designed specifically to analyze blood samples at the bedside of patients. The i-STAT Portable Clinical Analyzer, shown in Figure 23-14a, is a handheld device capable of determining a broad range of clinically important analytes such as potassium, sodium, chloride, pH, pCO$_2$, pO$_2$, urea nitrogen, and glucose. In addition, the computer-based analyzer calculates bicarbonate, total carbon dioxide, base excess, O$_2$ saturation, and hemoglobin in whole blood. Relative standard deviations on the order of 0.5% to 4% are obtained with these devices. Studies have shown that results are sufficiently reliable and cost effective to substitute for similar measurements made in a traditional, remote clinical laboratory.[25]

Most of the analytes (pCO$_2$, Na$^+$, K$^+$, Ca^{2+}, urea nitrogen, and pH) are determined by potentiometric measurements using microfabricated membrane-based ion-selective electrode technology. The urea nitrogen sensor consists of a polymer layer containing urease overlying an ammonium ion-selective electrode. The chemistry of the BUN determination is described in the previous section. The hematocrit is measured by electrolytic conductivity detection, and pO$_2$ is determined with a Clark voltammetric sensor (see Section 25C-4). Other results are calculated from these data. The central component of the monitor is the single-use disposable electrochemical i-STAT sensor array depicted in Figure 23-14b. The individual sensor electrodes are located on silicon chips along a narrow flow channel, as shown in the figure. The

[24]E. Bakker and M. E. Meyerhoff, in *Encyclopedia of Electrochemistry*, A. J. Bard, ed., Vol. 9, *Bioelectrochemistry*, G. S. Wilson, ed., New York: Wiley, 2002, p. 303; P. W. Carr and L. D. Bowers, *Immobilized Enzymes in Analytical and Clinical Chemistry: Fundamentals and Applications*, New York: Wiley, 1980.

[25]J. N. Murthy, J. M. Hicks, and S. J. Soldin, *Clin. Biochem.*, **1997**, *30*, 385.

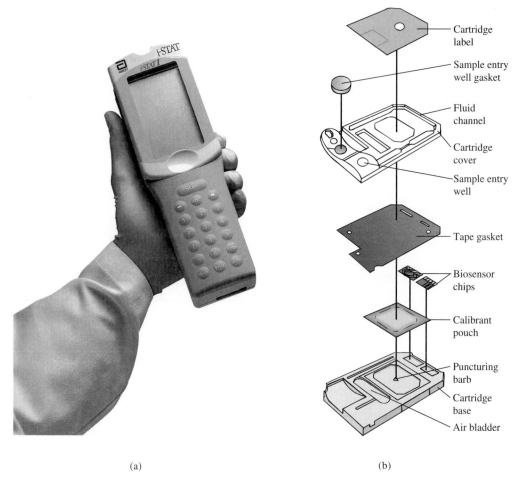

(a)   (b)

FIGURE 23-14 (a) Photograph of i-STAT 1 portable clinical analyzer. (b) Exploded view of i-STAT sensor array cartridge. (Abbott Point of Care Inc.)

integrated biosensors are manufactured by a patented microfabrication process.[26]

Each new sensor array is automatically calibrated prior to the measurement step. A blood sample withdrawn from the patient is deposited into the sample entry well, and the cartridge is inserted into the i-STAT analyzer. The calibrant pouch, which contains a standard buffered solution of the analytes, is punctured by the i-STAT analyzer and is compressed to force the calibrant through the flow channel across the surface of the sensor array. When the calibration step is complete, the analyzer compresses the air bladder, which forces the blood sample through the flow channel to ex-

pel the calibrant solution to waste and to bring the blood (20 μL to 100 μL) into contact with the sensor array. Electrochemical measurements are then made, and the results are calculated and presented on the liquid crystal display of the analyzer. The results are stored in the memory of the analyzer and may be transmitted to the hospital laboratory data management system for permanent storage and retrieval (see Section 4H-2). Cartridges are available that are configured for different combinations of more than twenty different analytes and measurements that can be made with this system.

Instruments similar to the i-STAT bedside monitor have been available for some time for in vitro potentiometric determination of various analytes in biomedical and biological systems. In recent years, there

[26]S. N. Cozzette et al. for i-STAT Corporation, East Windsor, NJ. U.S. Patent 5 466 575, 1995.

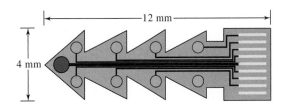

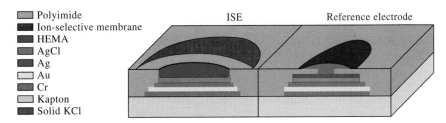

Polyimide
Ion-selective membrane
HEMA
AgCl
Ag
Au
Cr
Kapton
Solid KCl

FIGURE 23-15 A microfabricated potentiometric sensor for in vivo determination of analytes. The sensor was fabricated on a flexible plastic substrate using thick-film techniques. There are nine ion-selective electrodes (ISEs) on one side of the film and corresponding reference electrodes on the opposite side. (From E. Lindner and R. P. Buck, *Anal. Chem.*, **2000**, *72*, 336A, with permission. Copyright 2000 American Chemical Society.)

has been much interest in the development of miniature potentiometric sensors for in vivo studies as well. Figure 23-15 shows a small array (4 × 12 mm) of nine potentiometric electrodes fabricated on a thin Kapton® polyimide film.[27] Nine separate electrodes (along with corresponding reference electrodes on the opposite side of the array) are fabricated by thick-film techniques similar to those used for printed circuit boards. A chromium adhesion layer and a gold layer are vapor deposited on the polymer to provide electrical connection to the sensor areas. The array is next coated with photoresist and exposed to UV light through a mask that defines the electrode areas. The photoresist is developed to expose the electrode pads. Silver is then deposited electrochemically followed by silver chloride to form the internal reference electrode for each sensor. A mixture of hydroxyethyl methacrylate (HEMA) and an electrolyte is added to provide a salt bridge between the internal reference electrode and the ion-selective membrane. Finally, the ion-selective membrane is applied over the internal reference electrode along with a protective layer of polyimide. The sensor array has been used successfully to monitor H^+, K^+, Na^+, Ca^{2+}, and other analytes in heart muscle.

Other disposable electrochemical cells based on ion-selective electrodes, which are designed for the routine determination of various ions in clinical samples, have been available for some time. These systems are described briefly in Section 33D-3.

Light Addressable Potentiometric Sensor

An intriguing and useful application of semiconductor devices and potentiometric measurement principles is the light addressable potentiometric sensor (LAPS).[28] This device is fabricated from a thin, flat plate of *p*- or *n*-type silicon with a 1000-Å-thick coating of silicon oxynitride on one side of the plate. If a bias voltage is applied between a reference electrode immersed in a solution in contact with the insulating oxynitride layer and the silicon substrate, the semiconductor is depleted of majority carriers (see Section 2C-1). When light from a modulated source strikes the plate from either side, a corresponding photocurrent is produced. By measuring the photocurrent as a function of bias voltage, the surface potential of the device can be determined. Because the oxynitride layer is pH sensitive, the surface potential provides a measure of pH with a nernstian response over several decades.

[27] E. Lindner and R. P. Buck, *Anal. Chem.*, **2000**, *72*, 336A.

[28] D. A. Hafeman, J. W. Parce, and H. M. McConnell, *Science*, **1988**, *240*, 1182.

The versatility of the device is enhanced by attaching to the underside of the silicon substrate an array of light-emitting diodes. By modulating the diodes in sequence, different regions of the surface of the device can be interrogated for pH changes. Furthermore, by applying membranes containing various ionophores or enzymes to the oxynitride layer, each LED can monitor a different analyte. When this device is used to monitor the response of living cells to various biochemical stimuli, it is referred to as a *microphysiometer*. For a more complete description of the LAPS and its bioanalytical applications, see the Instrumental Analysis in Action feature following Chapter 25.

23G INSTRUMENTS FOR MEASURING CELL POTENTIALS

An important consideration in the design of an instrument for measuring cell potentials is that its resistance be large with respect to the cell. If it is not, the IR drop in the cell produces significant error (see Section 2A-3). We demonstrate this effect in the example that follows.

EXAMPLE 23-1

The true potential of a glass-calomel electrode system is 0.800 V; its internal resistance is 20 MΩ. What would be the relative error in the measured potential if the measuring device has a resistance of 100 MΩ?

Solution

The following schematic diagram shows that the measurement circuit can be considered as a voltage source E_s and two resistors in series: the source resistance R_s and the internal resistance of the measuring device R_M.

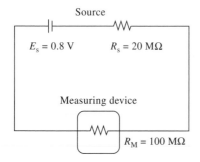

From Ohm's law, we may write

$$E_s = IR_s + IR_M$$

where I is the current in this circuit consisting of the cell and the measuring device. The current is then given by

$$I = \frac{0.800\ \text{V}}{(20 + 100) \times 10^6\ \Omega} = 6.67 \times 10^{-9}\ \text{A}$$

The potential drop across the measuring device (which is the potential indicated by the device, E_M) is IR_M. Thus,

$$E_M = (6.67 \times 10^{-9}\ \text{A})(100 \times 10^6\ \Omega) = 0.667\ \text{V}$$

and

$$\text{rel error} = \frac{0.667\ \text{V} - 0.800\ \text{V}}{0.800\ \text{V}} \times 100\%$$

$$= \frac{-0.133\ \text{V}}{0.800\ \text{V}} \times 100\% = -17\%$$

We can easily show that to reduce the loading error to 1%, the resistance of the voltage measuring device must be about 100 times greater than the cell resistance; for a relative error of 0.1%, the resistance must be 1000 times greater. Because the electrical resistance of cells containing ion-selective electrodes may be 100 MΩ or more, voltage measuring devices to be used with these electrodes generally have internal resistances of at least $10^{12}\ \Omega$.

It is important to appreciate that an error in measured voltage, such as that shown in Example 23-1 (-0.133 V), would have an enormous effect on the accuracy of a concentration measurement based on that potential. Thus, as shown in Section 23H-2, a 0.001 V uncertainty in potential leads to a relative error of about 4% in the determination of the hydrogen ion concentration of a solution by potential measurement with a glass electrode. An error of the size found in Example 23-1 would result in a concentration uncertainty of two orders of magnitude or more.

Direct-reading digital voltmeters with high internal resistances are used almost exclusively for pH and pIon measurements. We describe these devices in the next section.

23G-1 Direct-Reading Instruments

Numerous direct-reading pH meters are available commercially. Generally, these are solid-state devices with a field-effect transistor or a voltage follower as the first amplifier stage to provide the necessary high input resistance. Figure 23-16 is a schematic of a simple, battery-operated pH meter that can be built for about

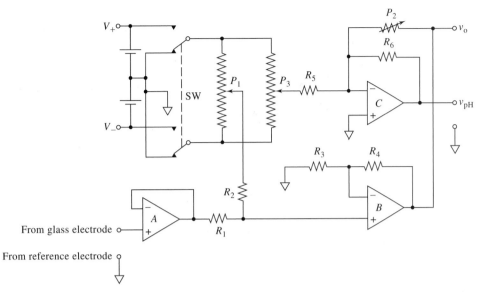

FIGURE 23-16 A pH meter based on a quad junction-FET input operational amplifier integrated circuit. Resistors R_1–R_6 and potentiometers P_1–P_3 are 50 kΩ. The circuit may be powered by batteries as shown or an appropriate power supply. See the original paper for descriptions of the circuit operation and the calibration procedure. (Adapted from D. L. Harris and D. C. Harris, *J. Chem. Educ.*, **1992**, *69*, 563. With permission.)

$10, exclusive of the pH probe and digital multimeter, which is used as the readout.[29] The output of the probe is connected to a high-resistance voltage follower *A* (see Sections 3B-2 and 3C-2), which has an input resistance of 10^{12} Ω. Operational amplifiers *B* and *C* provide gain (slope, or temperature control) and offset (calibration) for the circuit readout. The inverting amplifier *C* (see Section 3B-3) inverts the sense of the signal so that an increase in pH produces a positive increase in the output voltage of the circuit. The circuit is calibrated with buffers to display a range of output voltages extending from 100 to 1400 mV, corresponding to a pH range of 1 to 14. The three junction-FET operational amplifiers are located in a single quad integrated circuit package such as the LF347 (made by National Semiconductor).

23G-2 Commercial Instruments

The range of pIon meters available from instrument manufacturers is simply astonishing.[30] We can classify four groups of meters based on price and readability.

The groups include utility meters, which are portable, usually battery-operated instruments that range in price from less than $50 to more than $500. The low-end meters are intended for the consumer market and are about the size of a digital medical thermometer. Generally, utility meters can resolve about 0.5 pH unit or better. General-purpose meters are line-operated instruments, which can measure pH to 0.05 pH unit or better. Many offer such features as digital readout, automatic temperature compensation (ATC), scale expansions so that full scale covers 1.4 units instead of 0 to 14 units, and a millivolt scale. Currently, prices for general-purpose meters range from $100 to more than $1000. Expanded-scale instruments are generally readable to 0.01 pH unit or better and cost about $800 to $2000. Most offer full-scale ranges of 0.5 to 2 pH units (as well as 0-to-7 and 0-to-14 ranges), four-digit readout, push-button control, millivolt scale, and automatic temperature compensation. Research meters are readable to 0.001 pH unit or better, generally have a five-digit display, and cost $2000 to $3000. Note that the readability of these instruments is usually significantly better than the sensitivity of most ion-selective electrodes.

The ATC feature in many commercial instruments usually consists of a thermistor or thermocouple and associated circuitry (Section 3C-3) that provide a

---

[29] D. L. Harris and D. C. Harris, *J. Chem. Educ.*, **1992**, *69*, 563.
[30] To develop a feeling for the tremendous number of models of pH and pIon meters that are available, perform Google searches on the terms "pH meter" and "ion meter."

signal proportional to temperature. This signal is then used to correct the output by adjusting the slope of the transfer function of the instrument ($2.303RT/F = 0.0592$ at 25°C) using either software calculation or analog circuitry. The thermistor or thermocouple may be incorporated in the instrument console, in which case it merely monitors the ambient temperature, or it may be a probe that is inserted directly into the analyte solution. For routine work, temperature compensation is not essential, but for high-quality potentiometric measurements, it is vital.

23H DIRECT POTENTIOMETRIC MEASUREMENTS

The determination of an ion or molecule by direct potentiometric measurement is rapid and simple, requiring only a comparison of the potential developed by the indicator electrode in the test solution with its potential when immersed in one or more standard solutions of the analyte. Because most indicator electrodes are selective, preliminary separation steps are seldom required. In addition, direct potentiometric measurements are rapid and readily adapted to the continuous and automatic monitoring of ion activities.

23H-1 The Sign Convention and Equations for Direct Potentiometry

The sign convention for potentiometry is consistent with the convention described in Chapter 22 for standard electrode potentials.[31] In this convention, the indicator electrode is the right-hand electrode and the reference electrode is on the left.[32] For direct potentiometric measurements, the potential of a cell is then expressed in terms of the indicator electrode potential, the reference electrode potential, and the junction potential as shown in Equation 23-1:

$$E_{cell} = (E_{ind} - E_{ref}) + E_j \qquad (23\text{-}24)$$

---

[31] The convention described here has been endorsed by the National Institute of Standards and Technology (NIST) and the IUPAC. See R. P. Buck and E. Lindner, *Pure Appl. Chem.*, **1994**, *66*, 2527 (http://www.iupac.org/publications/pac/1994/pdf/6612x2527.pdf).

[32] In effect, the sign convention for electrode potentials described in Section 22C-5 also designates the indicator electrode as the right-hand electrode by stipulating that half-reactions must always be written as reductions; the standard hydrogen electrode, which is the reference electrode in this case, is then the left-hand electrode.

In Sections 23B and 23C, we describe the response of various types of indicator electrodes to analyte activities. For the cation X^{n+} at 25°C, the electrode response takes the general nernstian form

$$E_{ind} = L - \frac{0.0592}{n} \log \frac{1}{a_X}$$
$$= L - \frac{0.0592}{n} pX \qquad (23\text{-}25)$$

where L is a constant and a_X is the activity of the cation. For metallic indicator electrodes, L is ordinarily the standard electrode potential (see Equation 23-2); for membrane electrodes, L is the sum of several constants, including the time-dependent asymmetry potential of uncertain magnitude (see Equation 23-14).

By substituting Equation 23-25 into Equation 23-24 and rearranging, we find

$$pX = -\log a_X = -\frac{E_{cell} - [(E_j - E_{ref}) + L]}{0.0592/n}$$

The constant terms in square brackets can be combined to give a new constant K:

$$pX = -\log a_X = -\frac{n(E_{cell} - K)}{0.0592} \qquad (23\text{-}26)$$

where

$$K = (E_j - E_{ref}) + L \qquad (23\text{-}27)$$

For an anion A^{n-}, the sign of Equation 23-26 is reversed, so

$$pA = \frac{n(E_{cell} - K)}{0.0592} \qquad (23\text{-}28)$$

All direct potentiometric methods are based on Equation 23-26 or 23-28. The difference in sign in the two equations has a subtle but important consequence in the way that ion-selective electrodes are connected to pH meters and pIon meters. When we solve the two equations for E_{cell}, we find that for cations

$$E_{cell} = K - \frac{0.0592}{n} pX \qquad (23\text{-}29)$$

and for anions

$$E_{cell} = K + \frac{0.0592}{n} pA \qquad (23\text{-}30)$$

Equation 23-29 shows that an increase in pX produces a decrease in E_{cell} with a cation-selective electrode. Thus, when a high-resistance voltmeter is connected to the cell in the usual way, with the indicator electrode attached to the positive terminal, the meter reading decreases as pX increases. To eliminate this problem, instrument manufacturers may reverse the leads so that cation-sensitive electrodes are connected to the negative terminal of the voltage-measuring device. Alternatively, the signal may be inverted between the input stage and the meter as shown in Section 23F-1. Meter readings then increase with increases in pX. Anion-selective electrodes, on the other hand, are connected to the positive terminal of the meter so that increases in pA also produce larger positive readings.

23H-2 The Electrode Calibration Method

The constant K in Equations 23-29 and 23-30 contains several other constants, at least one of which, the junction potential, cannot be accurately computed from theory or measured directly. Therefore, before these equations can be used for the determination of pX or pA, K must be evaluated experimentally with one or more standard solutions of the analyte.

In the electrode-calibration method, K is determined by measuring E_{cell} for one or more standard solutions of known pX or pA. We then assume that K is unchanged when the standard is replaced with analyte. The calibration is usually performed at the time pX or pA for the unknown is determined. With membrane electrodes recalibration may be necessary if measurements extend over several hours because of the slowly changing asymmetry potential.

The direct electrode calibration method offers the advantages of simplicity, speed, and applicability to the continuous monitoring of pX or pA. The method has two important disadvantages, however. One of these is that the accuracy of a measurement obtained by this procedure is limited by the inherent uncertainty caused by the junction potential E_j, and unfortunately, this uncertainty can never be totally eliminated. The second disadvantage of this procedure is that results of an analysis are activities rather than concentrations (for some applications this is an advantage rather than a disadvantage).

Inherent Error in the Electrode Calibration Procedure

A serious disadvantage of the electrode calibration method is the inherent uncertainty that results from the assumption that K in Equation 23-26 or 23-28 remains constant between calibration and analyte determina-

tion. This assumption can seldom, if ever, be exactly valid because the electrolyte composition of the unknown is almost always different from the composition of the solution used for calibration. The junction potential E_j contained in K (Equation 23-27) varies slightly as a result, even though a salt bridge is used. This uncertainty is frequently on the order of 1 mV or more. Unfortunately, because of the nature of the potential-activity relationship, this uncertainty has an amplified effect on the inherent accuracy of the analysis. The magnitude of the uncertainty in analyte concentration can be estimated by differentiating Equation 23-26 with respect to K while holding E_{cell} constant:

$$-\log_{10} e \, \frac{da_X}{a_X} = -0.434 \frac{da_X}{a_X} = -\frac{dK}{0.0592/n}$$

$$\frac{da_X}{a_X} = \frac{ndK}{0.0257} = 38.9ndK$$

By replacing da_X and dK with finite increments and multiplying both sides of the equation by 100%, we obtain

$$\% \text{ rel error} = \frac{\Delta a_X}{a_X} \times 100\%$$
$$= 3.89 \times 10^3 \, n\Delta K\% \quad (23\text{-}31)$$

The quantity $\Delta a_X/a_X$ is the relative error in a_X associated with an absolute uncertainty ΔK in K. If, for example, ΔK is ± 0.001 V, a relative error in activity of about $\pm 4n\%$ can be expected. *It is important to appreciate that this uncertainty is characteristic of all measurements involving cells that contain a salt bridge and that it cannot be eliminated by even the most careful measurements of cell potentials or the most sensitive and precise measuring devices.* In addition, it is apparently impossible to devise a method that eliminates the uncertainty in K that is the source of this problem.

Activity versus Concentration

Electrode response is related to activity rather than to analyte concentration. We are usually interested in concentration, and the determination of this quantity from a potentiometric measurement requires activity coefficient data. Often, activity coefficients are unavailable because the ionic strength of the solution is either unknown or so high that the Debye-Hückel equation is not applicable. Unfortunately, the assumption that activity and concentration are identical may lead to serious errors, particularly when the analyte is polyvalent.

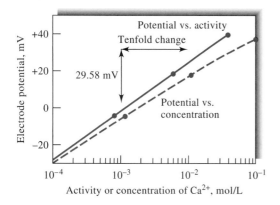

FIGURE 23-17 Response of a liquid-membrane electrode to variations in the concentration and activity of calcium ion. (Courtesy of Thermo Orion, Boston, MA.)

The difference between activity and concentration is illustrated by Figure 23-17, where the lower curve gives the change in potential of a calcium electrode as a function of calcium chloride concentration (note that the activity or concentration scale is logarithmic). The nonlinearity of the curve is due to the increase in ionic strength — and the consequent decrease in the activity coefficient of the calcium — as the electrolyte concentration becomes larger. When these concentrations are converted to activities, the data produce the upper line with the nernstian slope of 0.0296 (0.0592/2).

Activity coefficients for singly charged ions are less affected by changes in ionic strength than are coefficients for species with multiple charges. Thus, the effect shown in Figure 23-17 is less pronounced for electrodes that respond to H^+, Na^+, and other univalent ions.

In potentiometric pH measurements, the pH of the standard buffer used for calibration is generally based on the activity of hydrogen ions. Thus, the resulting hydrogen ion results are also on an activity scale. If the unknown sample has a high ionic strength, the hydrogen ion concentration will differ appreciably from the activity measured.

23H-3 Calibration Curves for Concentration Measurement

Potentiometric measurements can be corrected to give results in terms of concentration using an empirical calibration curve such as the lower curve in Figure 23-17. For this approach to be successful, however, it is essential that the ionic composition of the standards approx-

 Simulation: Learn more about **calibration of potentiometric systems**.

imate that of the analyte — a condition that is difficult to achieve experimentally for complex samples.

Where electrolyte concentrations are not too great, it is often helpful to swamp both the samples and the calibration standards with a measured excess of an inert electrolyte. Under these circumstances, the added effect of the electrolyte in the sample becomes negligible, and the empirical calibration curve yields concentration data. This approach has been used for the potentiometric determination of fluoride in public water supplies with a lanthanum fluoride electrode. In this determination, both samples and standards are diluted on a 1:1 basis with a solution containing sodium chloride, a citrate buffer, and an acetate buffer. This mixture, which fixes both ionic strength and pH, is sold under the name Total Ionic Strength Adjustment Buffer, or TISAB. The diluent is sufficiently concentrated so that the samples and standards do not differ significantly in ionic strength. This procedure permits a rapid measurement of fluoride ion in the 1-ppm range, with a precision of about 5% relative. Calibration curves are also useful for electrodes that do not respond linearly to pA.

23H-4 Standard-Addition Method

The standard-addition method, described in Section 1D-3, is equally applicable to potentiometric determinations. In this method, the potential of the electrode system is measured before and after addition of a small volume (or volumes) of a standard to a known volume of the sample. We assume that this addition does not change the ionic strength and thus the activity coefficient γ_X of the analyte. We further assume that the added standard does not significantly alter the junction potential.

The standard-addition method has been applied to the determination of chloride and fluoride in samples of commercial phosphors.[33] In this application, solid-state indicator electrodes for chloride and fluoride were used in conjunction with a reference electrode; the added standard contained known quantities of the two anions. The relative standard deviation for the measurement of replicate standard samples was 0.7% for fluoride and 0.4% for chloride. In routine use with real samples, the standard-addition method yielded relative standard deviations of 1.1% for fluoride and 0.8% for chloride. When similar samples were analyzed by the usual electrode calibration methods, the

[33] L. G. Bruton, *Anal. Chem.*, **1971**, *43*, 579.

precision was worse by about a factor of 2. Similar methods have been used to determine fluoride and chloride in geological materials with limits of detection of 5–10 µg/g and 40–100 µg/g, respectively.

23H-5 Potentiometric pH Measurements with a Glass Electrode

The glass electrode[34] is unquestionably the most important indicator electrode for hydrogen ion. It is convenient to use and is subject to few of the interferences that affect other pH-sensing electrodes. Glass electrodes are available at relatively low cost and come in many shapes and sizes. A common variety is illustrated in Figure 23-3; the reference electrode is usually a silver–silver chloride electrode.

The glass electrode is a remarkably versatile tool for the measurement of pH under many conditions. The electrode can be used without interference in solutions containing strong oxidants, reductants, gases, and proteins (a calomel reference electrode, rather than a silver–silver chloride reference electrode, is normally used in the presence of proteins because silver ions react with proteins). The pH of viscous or even semisolid fluids can be determined in this way. Electrodes for special applications are available. Examples of these are small electrodes for pH measurements in a drop (or less) of solution or in a cavity of a tooth, microelectrodes that permit the measurement of pH inside a living cell, systems for insertion in a flowing liquid stream to provide a continuous monitoring of pH, a small glass electrode that can be swallowed to indicate the acidity of the stomach contents (the reference electrode is kept in the mouth), and combination electrodes that contain both indicator and reference electrodes in a single probe.[35]

Summary of Errors Affecting pH Measurements with Glass Electrode

The ubiquity of the pH meter and the general applicability of the glass electrode tend to lull us into the attitude that any measurement obtained with such an instrument is surely correct. It is a good idea to guard against this false sense of security because there are distinct limitations to the electrode system. These have been discussed in earlier sections and are summarized here.

1. **The alkaline error.** Modern glass electrodes become somewhat sensitive to alkali-metal ions at pH values greater than 11 to 12.
2. **The acid error.** At a pH less than 0.5, values obtained with a glass electrode tend to be somewhat high.
3. **Dehydration.** Dehydration of the electrode may cause unstable performance and errors.
4. **Errors in low-ionic-strength solutions.** Significant errors (as much as 1 or 2 pH units) may occur when the pH of low-ionic-strength samples, such as lake or stream samples, are measured with a glass electrode system.[36] The prime source of such errors is nonreproducible junction potentials, which result from partial clogging of the fritted plug or porous fiber that is used to restrict the flow of liquid from the salt bridge into the analyte solution. To overcome this problem, *free diffusion junctions* of various types have been designed, and one is produced commercially. In the commercial junction, an electrolyte solution is dispensed from a syringe cartridge through a capillary tube, the tip of which is in contact with the sample solution (see Figure 23-18). Before each measurement, 6 µL of electrolyte is dispensed so that a fresh portion of electrolyte is in contact with the analyte solution.
5. **Variation in junction potential.** We reemphasize that variation in the junction potential between standard and sample leads to a fundamental uncertainty in the measurement of pH that is not correctable. Absolute measurements more reliable than 0.01 pH unit are generally unobtainable. Even reliability to 0.03 pH unit requires tremendous care. On the other hand, it is often possible to detect pH differences between similar solutions or pH changes in a single solution that are as small as 0.001 unit. For this reason, many pH meters are designed to produce readings to less than 0.01 pH unit. The free diffusion junction described in Section 23H-5 and illustrated in Figure 23-18 can minimize the variation in the junction potential.
6. **Error in the pH of the standard buffer.** Any inaccuracies in the preparation of the buffer used for

---

[34]For a detailed discussion of the definitions, standards, procedures, and errors associated with pH measurement, see R. P. Buck et al., *Pure Appl. Chem.*, **2002**, *74*, 2169 (http://www.iupac.org/publications/pac/2002/pdf/7411x2169.pdf) and R. G. Bates, *Determination of pH: Theory and Practice*, 2nd ed. New York: Wiley, 1973.

[35]For details of the construction of a typical pH probe, see http://www.radiometer-analytical.com/pdf/general/Electrode_manufacturing.pdf.

[36]See W. Davison, *Trends Anal. Chem.*, **1990**, *9*, 80; W. Davison and C. Woof, *Anal. Chem.*, **1985**, *57*, 2567; T. R. Harbinson and W. Davison, *Anal. Chem.*, **1987**, *59*, 2450; A. Kopelove, S. Franklin, and G. M. Miller, *Amer. Lab.*, **1989** (6), 40.

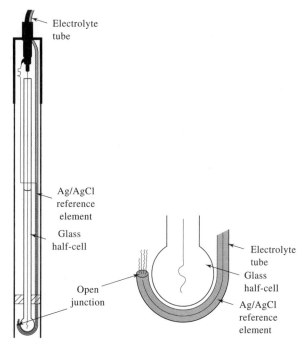

Electrolyte
tube

Ag/AgCl
reference
element

Glass
half-cell

Open
junction

Electrolyte
tube

Glass
half-cell

Ag/AgCl
reference
element

FIGURE 23-18 A combination pH electrode system with a free diffusion junction. (Courtesy of Hach Company, Loveland, CO. With permission.)

calibration, or changes in its composition during storage, are propagated as errors in pH measurements. A common cause of deterioration is the action of bacteria on organic components of buffers.

7. **Errors resulting from temperature changes.** When pH measurements are made at temperatures other than 25°C, pH meters must be adjusted to compensate for the change in the nernstian response of the glass electrode. Compensation is often accomplished with ATC circuitry or software in modern meters, but with older meters, manual adjustment of the slope (or temperature) control may be necessary.

The Operational Definition of pH

The utility of pH as a measure of the acidity or alkalinity of aqueous media, the wide availability of commercial glass electrodes, and the proliferation of inexpensive solid-state pH meters have made the potentiometric measurement of pH one of the most common analytical techniques in all of science. It is thus extremely important that pH be defined in a manner that is easily duplicated at various times and vari-

ous laboratories throughout the world. To meet this requirement, it is necessary to define pH in operational terms — that is, by the way the measurement is made. Only then will the pH measured by one worker be the same as that measured by another.

The operational definition of pH endorsed by the National Institute of Standards and Technology (NIST), similar organizations in other countries, and the International Union of Pure and Applied Chemistry (IUPAC) is based on the direct calibration of the meter with carefully prescribed standard buffers followed by potentiometric determination of the pH of unknown solutions.

Consider, for example, the glass-reference systems in Figure 23-3. When these electrodes are immersed in a standard buffer at 25°C, Equation 23-26 applies and we can write

$$pH_S = -\frac{E_S - K}{0.0592}$$

where E_S is the cell potential when the electrodes are immersed in the standard buffer. Similarly, if the cell potential is E_U when the electrodes are immersed in a solution of unknown pH also at 25°C, we have

$$pH_U = -\frac{(E_U - K)}{0.0592}$$

By subtracting the first equation from the second and solving for pH_U, we find

$$pH_U = pH_S - \frac{(E_U - E_S)}{0.0592} \qquad (23\text{-}32)$$

Equation 23-32 has been adopted throughout the world as the operational definition of pH.[37]

Workers at NIST and elsewhere have used cells without liquid junctions to study primary-standard buffers extensively. Some of the properties of these buffers are presented in Table 23-7 and discussed in detail elsewhere.[38] For general use, the buffers can be prepared from relatively inexpensive laboratory

---

[37] Equation 23-32 applies only to solutions at 25°C. A more general equation is

$$pH_U = pH_S - \frac{(E_U - E_S)F}{2.303RT} = pH_S - \frac{(E_U - E_S)F}{1.984 \times 10^{-4}\,T}$$

where T is the temperature of the sample and the standard buffer and pH_S is the pH of the standard at temperature T.

[38] R. G. Bates, *Determination of pH*, 2nd ed., Chap. 4, New York: Wiley, 1973.

TABLE 23-7 Values of NIST Standard-Reference pH Solutions from 0°C to 60°C[a]

Temperature, °C	0.05 m[a] KH₃(C₂O₄)₂· 2H₂O 189b	0.01 m KH Tartrate 188	0.05 m KH Phthalate 185h	0.025 m KH₂PO₄/ 0.025 m Na₂HPO₄ 186g	0.008695 m KH₂PO₄/ 0.03043 m Na₂HPO₄ 186g	0.01 m Na₂B₄O₇ 187e	0.025 m NaHCO₃/ 0.025 m[a] Na₂CO₃ 191a/192a
0	—	3.711	—	—	—	—	—
5	1.709	3.689	4.004	6.952	7.503	9.393	—
10	1.709	3.671	4.001	6.923	7.475	9.333	—
15	1.711	3.657	4.001	6.899	7.451	9.278	10.117
20	1.714	3.647	4.003	6.880	7.432	9.230	10.063
25	1.719	3.639	4.008	6.864	7.416	9.186	10.012
30	1.724	3.635	4.015	6.853	7.404	9.146	—
35	1.731	3.632	4.023	6.844	7.396	9.110	9.928
37	—	3.631[d]	4.027	6.841	7.394	9.096	—
40	1.738	3.632	4.034	6.837	7.390	9.077	—
45	1.746	3.635	4.046	6.835	7.387	9.047	—
50	1.754	3.639	4.060	6.833	7.385	9.020	—
55	—	3.644	—	—	—	—	—
60	—	3.651	—	—	—	—	—
Buffer capacity,[b] mol/pH unit	0.070	0.027	0.016	0.029	0.016	0.020	0.029
$\Delta pH_{1/2}$ for 1:1 dilution[c]	0.186	0.049	0.052	0.080	0.07	0.01	0.079
Temperature coefficient, pH/K	+0.001	−0.0014	−0.0012	−0.0028	−0.0028	−0.0082	−0.0096

Adapted from Y. C. Wu, W. F. Koch, and R. A. Durst, *Standard Reference Materials: Standardization of pH Measurements*, NIST Special Publication 260-53. Gaithersburg, MD: U.S. Department of Commerce, 1988, http://ts.nist.gov/ts/htdocs/230/232/SP_PUBLICATIONS/documents/SP260-53.pdf. Updated values from most recent NIST SRM certificates.

m = molality (mol solute/kg H₂O).

[a] All uncertainties are ±0.005 pH except for solutions of $KHC_2O_4 \cdot 2H_2C_2O_4$, which have uncertainty of ±0.01 pH.

[b] The buffer capacity is the number of moles of a monoprotic strong acid or base that causes 1.0 L to change pH by 1.0.

[c] Change in pH that occurs when one volume of buffer is diluted with one volume of H₂O.

[d] 38°C.

reagents. For careful work, certified buffers can be purchased from NIST.[39]

23I POTENTIOMETRIC TITRATIONS

The potential of a suitable indicator electrode is convenient for determining the equivalence point for a titration (a potentiometric titration).[40] A potentiometric titration provides different information than does a direct potentiometric measurement. For example, the

direct measurement of 0.100 M acetic and 0.100 M hydrochloric acid solutions with a pH-sensitive electrode yield widely different pH values because acetic acid is only partially dissociated. On the other hand, potentiometric titrations of equal volumes of the two acids require the same amount of standard base for neutralization.

The potentiometric end point is widely applicable and provides inherently more accurate data than the corresponding method with indicators. It is particularly useful for titration of colored or turbid solutions and for detecting the presence of unsuspected species in a solution. Unfortunately, it is more time-consuming than

[39] See https://srmors.nist.gov/detail.cfm, keyword pH.
[40] For a monograph on this method, see E. P. Sergeant, *Potentiometry and Potentiometric Titrations*, New York: Wiley, 1984, reprint: Melbourne, FL: Krieger, 1991.

 Simulation: Learn more about **potentiometric titrations**.

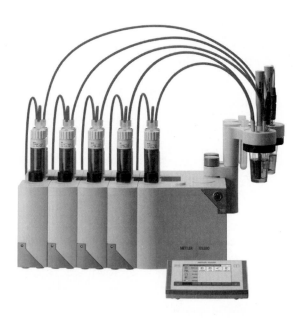

FIGURE 23-19 High-performance automated titrator. Accommodates up to six buret drives, has user-programmable software, permits high sample throughput with an automatic sample turntable (not shown). The unit can be adapted for several different types of titrations. (Copyright 2006 Mettler-Toledo, Inc.)

 Simulation: Learn more about **derivative titrations**.

an indicator titration unless an automatic titrator is used. We will not discuss details of the potentiometric titration techniques here because most elementary analytical texts treat this subject thoroughly.[41]

Automated titrators are available from a number of manufacturers. One such instrument appears in the photo of Figure 23-19. These devices are equipped with a titrant reservoir, a titrant delivery device such as a syringe pump, and an arrangement of tubes and valves to deliver measured volumes of titrant at an appropriate rate. The titrator acquires data from a pH probe or other sensor to monitor the progress of the titration. The end point of the titration is determined either by a built-in program based on first- and second-derivative techniques[42] or by another method specified or programmed by the user. Many of these devices are capable of calculating analyte concentrations from standard-concentration data entered by the user, and experimental results are printed, stored, or transmitted via a serial or other appropriate interface to a laboratory information management system (see Section 4H-2). Automatic titrators are especially useful in industrial or clinical laboratories where large numbers of samples must be routinely analyzed.

---

[41] For example, see D. A. Skoog, D. M. West, F. J. Holler, and S. R. Crouch, *Fundamentals of Analytical Chemistry*, 8th ed., Chap. 21, Belmont, CA: Brooks/Cole, 2004.
[42] S. R. Crouch and F. J. Holler, *Applications of Microsoft® Excel in Analytical Chemistry*, Belmont, CA: Brooks/Cole, 2004, pp. 134–42.

QUESTIONS AND PROBLEMS

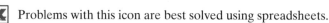

*Answers are provided at the end of the book for problems marked with an asterisk.

X Problems with this icon are best solved using spreadsheets.

23-1 What is meant by nernstian behavior of an indicator electrode?

23-2 What is the source of the alkaline error in pH measurement with a glass electrode?

23-3 Compare the average temperature coefficients in the range of 15°C to 35°C for the five reference electrodes listed in Table 23-1.

23-4 What occurs when the pH-sensitive tip of a newly manufactured glass electrode is immersed in water?

23-5 Differentiate between an electrode of the first kind and an electrode of the second kind.

23-6 How does a gas-sensing probe differ from other membrane electrodes?

23-7 What is the source of
  (a) the asymmetry potential in a membrane electrode?
  (b) the boundary potential in a membrane electrode?
  (c) a junction potential in a glass-reference electrode system?
  (d) the potential of a crystalline membrane electrode used to determine the concentration of F^-?

23-8 List several sources of uncertainty in pH measurements with a glass-reference electrode system.

23-9 What is an ionophore? What is the purpose of ionophores in membrane electrodes?

23-10 List the advantages and disadvantages of a potentiometric titration relative to a titration with visual indicators.

23-11 What is the *operational definition of pH* and how is it used?

23-12 What are the advantages of microfabricated ion-selective electrodes? Describe typical applications of this type of sensor.

*23-13 Calculate the theoretical potential of the following cells. (In each case assume that activities are approximately equal to molar concentrations and that the temperature is 25°C.)
  (a) $SCE \| Fe^{3+}(0.0250\ M), Fe^{2+}(0.0150\ M) | Pt$
  (b) $SCE \| Zn^{2+}(0.00228\ M) | Zn$
  (c) Saturated $Ag\text{-}AgCl$ reference $\| Ti^{3+}(0.0250\ M), Ti^{2+}(0.0450\ M) | Pt$
  (d) Saturated $Ag\text{-}AgCl$ reference $\| I_3^-\ (0.00667\ M), I^-(0.00433\ M) | Pt$

*23-14 (a) Calculate the standard potential for the reaction

$$CuBr(s) + e^- \rightleftharpoons Cu(s) + Br_2$$

  For CuBr, $K_{sp} = 5.2 \times 10^{-9}$.
  (b) Give a schematic representation of a cell with a copper indicator electrode and a reference SCE that could be used for the determination of Br_2.
  (c) Derive an equation that relates the measured potential of the cell in (b) to pBr (assume that the junction potential is zero).
  (d) Calculate the pBr of a bromide-containing solution that is saturated with CuBr and contained in the cell described in (b) if the resulting potential is -0.095 V.

*23-15 (a) Calculate the standard potential for the reaction

$$Ag_3AsO_4(s) + 3e^- \rightleftharpoons 3Ag(s) + AsO_4^{3-}$$

  For Ag_3AsO_4, $K_{sp} = 1.2 \times 10^{-22}$.
  (b) Give a schematic representation of a cell with a silver indicator electrode and an SCE as reference that could be used for determining AsO_4^{3-}.
  (c) Derive an equation that relates the measured potential of the cell in (b) to $pAsO_4$ (assume that the junction potential is zero).
  (d) Calculate the $pAsO_4$ of a solution that is saturated with Ag_3AsO_4 and contained in the cell described in (b) if the resulting potential is 0.247 V.

*23-16 The following cell was used for the determination of $pCrO_4$:

$$SCE \| CrO_4^{2-}(x M), Ag_2CrO_4(sat'd) | Ag$$

Calculate $pCrO_4$ if the cell potential is -0.386 V.

*23-17 The following cell was used to determine the pSO$_4$ of a solution:

$$SCE \| SO_4^{2-}(xM), Hg_2SO_4(sat'd) | Hg$$

Calculate the pSO$_4$ if the potential was -0.537 V.

*23-18 The formation constant for the mercury(II) acetate complex is

$$Hg^{2+} + 2OAc^- \rightleftharpoons Hg(OAc)_2(aq) \qquad K_f = 2.7 \times 10^8$$

Calculate the standard potential for the half-reaction

$$Hg(OAc)_2(aq) + 2e^- \rightleftharpoons Hg(l) + 2OAc^-$$

*23-19 The standard electrode potential for the reduction of the Cu(II) complex of EDTA is given by

$$CuY^{2-} + 2e^- \rightleftharpoons Cu(s) + Y^{4-} \qquad E^0 = 0.13 \text{ V}$$

Calculate the formation constant for the reaction

$$Cu^{2+} + Y^{4-} \rightleftharpoons CuY^{2-}$$

*23-20 The cell

$$Ag | AgCl(sat'd) \| H^+(a = x) | glass\ electrode$$

has a potential of -0.2094 V when the solution in the right-hand compartment is a buffer of pH 4.006. The following potentials are obtained when the buffer is replaced with unknowns: (a) -0.2806 V and (b) -0.2132 V. Calculate the pH and the hydrogen ion activity of each unknown. (c) Assuming an uncertainty of 0.001 V in the junction potential, what is the range of hydrogen ion activities within which the true value might be expected to lie?

*23-21 The following cell was found to have a potential of 0.124 V:

$$Ag | AgCl(sat'd) \| Cu^{2+}(3.25 \times 10^{-3} \text{ M}) | membrane\ electrode\ for\ Cu^{2+}$$

When the solution of known copper activity was replaced with an unknown solution, the potential was found to be 0.086 V. What was the pCu of this unknown solution? Neglect the junction potential.

*23-22 The following cell was found to have a potential of -1.007 V:

$$SCE \| X^-(0.0200 \text{ M}), CdX_2(sat'd) | Cd$$

Calculate the solubility product of CdX$_2$, neglecting the junction potential.

*23-23 The following cell was found to have a potential of -0.492 V:

$$Ag | AgCl(sat'd) \| HA(0.200 \text{ M}), NaA(0.300 \text{ M}) | H_2(1.00 \text{ atm}), Pt$$

Calculate the dissociation constant of HA, neglecting the junction potential.

 23-24 A 40.00-mL aliquot of 0.0500 M HNO$_2$ is diluted to 75.0 mL and titrated with 0.0800 M Ce^{4+}. Assume the hydrogen ion concentration is at 1.00 M throughout the titration. (Use 1.44 V for the formal potential of the cerium system.)
(a) Calculate the potential of the indicator electrode with respect to a Ag-AgCl(sat'd) reference electrode after the addition of 5.00, 10.00, 15.00, 20.00, 25.00, 40.00, 45.00, 49.00, 49.50, 49.60, 49.70, 49.80, 49.90, 49.95, 49.99, 50.00, 50.01, 50.05, 50.10, 50.20, 50.30, 50.40, 50.50, 51.00, 55.00, 60.00, 75.00, and 90.00 mL of cerium(IV).
(b) Construct a titration curve for these data.

(c) Generate a first- and second-derivative curve for these data.[43] Does the volume at which the second-derivative curve crosses zero correspond to the theoretical equivalence point? Why or why not?

*23-25 The following cell was found to have a potential of 0.2897 V:

$$\text{SCE} \| \text{Mg}^{2+}(a = 3.32 \times 10^{-3}\,\text{M}) | \text{membrane electrode for Mg}^{2+}$$

(a) When the solution of known magnesium activity was replaced with an unknown solution, the potential was found to be 0.2041 V. What was the pMg of this unknown solution?
(b) Assuming an uncertainty of ± 0.002 V in the junction potential, what is the range of Mg^{2+} activities within which the true value might be expected?
(c) What is the relative error in $[\text{Mg}^{2+}]$ associated with the uncertainty in E_j?

*23-26 A 0.400-g sample of toothpaste was boiled with a 50-mL solution containing a citrate buffer and NaCl to extract the fluoride ion. After cooling, the solution was diluted to exactly 100 mL. The potential of an ion-selective electrode with a Ag-AgCl(sat'd) reference electrode in a 25.0-mL aliquot of the sample was found to be -0.1823 V. Addition of 5.0 mL of a solution containing 0.00107 mg F^-/mL caused the potential to change to -0.2446 V. Calculate the mass percentage of F^- in the sample.

X *Challenge Problem*

23-27 Ceresa, Pretsch, and Bakker[44] investigated three ion-selective electrodes for determining calcium concentrations. All three electrodes used the same membrane, but differed in the composition of the inner solution. Electrode 1 was a conventional ion-selective electrode with an inner solution of 1.00×10^{-3} M CaCl_2 and 0.10 M NaCl. Electrode 2 (low activity of Ca^{2+}) had an inner solution containing the same analytical concentration of CaCl_2, but with 5.0×10^{-2} M EDTA adjusted to a pH of 9.0 with 6.0×10^{-2} M NaOH. Electrode 3 (high Ca^{2+} activity) had an inner solution of 1.00 M $\text{Ca(NO}_3)_2$.
(a) Determine the Ca^{2+} concentration in the inner solution of Electrode 2.
(b) Determine the ionic strength of the solution in Electrode 2.
(c) Use the Debye-Hückel equation and determine the activity of Ca^{2+} in Electrode 2. Use 0.6 nm for the α_X value for Ca^{2+} (see Appendix 2).
(d) Electrode 1 was used in a cell with a calomel reference electrode to measure standard calcium solutions with activities ranging from 0.001 M to 1.00×10^{-9} M. The following data were obtained.

Activity of Ca^{2+}, M	Cell Potential, mV
1.0×10^{-3}	93
1.0×10^{-4}	73
1.0×10^{-5}	37
1.0×10^{-6}	2
1.0×10^{-7}	-23
1.0×10^{-8}	-51
1.0×10^{-9}	-55

[43]S. R. Crouch and F. J. Holler, *Applications of Microsoft® Excel in Analytical Chemistry*, pp. 134–40. Belmont, CA: Brooks/Cole, 2004.
[44]A. Ceresa, E. Pretsch, and E. Bakker, *Anal. Chem.*, **2000**, *72*, 2054.

Plot the cell potential versus the pCa and determine the pCa value where the plot deviates more than 5% from linearity (the limit of linearity; see Section 1E-2). For the linear portion, determine the slope and intercept of the plot. Does the plot obey Equation 23-29 as expected?

(e) For Electrode 2, the following results were obtained.

Activity of Ca^{2+}, M	Cell Potential, mV
1.0×10^{-3}	228
1.0×10^{-4}	190
1.0×10^{-5}	165
1.0×10^{-6}	139
5.6×10^{-7}	105
3.2×10^{-7}	63
1.8×10^{-7}	36
1.0×10^{-7}	23
1.0×10^{-8}	18
1.0×10^{-9}	17

Again plot cell potential versus pCa and determine the range of linearity for Electrode 2. Determine the slope and intercept for the linear portion. Does this electrode obey Equation 21-24 for the higher Ca^{2+} activities?

(f) Electrode 2 is said to be super-nernstian for concentrations from 10^{-7} M to 10^{-6} M. Why is this term used? If you have access to a library that subscribes to *Analytical Chemistry* or has web access to the journal, read the article. This electrode is said to have Ca^{2+} uptake. What does this mean and how might it explain the response?

(g) Electrode 3 gave the following results.

Activity of Ca^{2+}, M	Cell Potential, mV
1.0×10^{-3}	175
1.0×10^{-4}	150
1.0×10^{-5}	123
1.0×10^{-6}	88
1.0×10^{-7}	75
1.0×10^{-8}	72
1.0×10^{-9}	71

Plot the cell potential versus pCa and determine the range of linearity. Again determine the slope and intercept. Does this electrode obey Equation 23-29?

(h) Electrode 3 is said to have Ca^{2+} release. Explain this term from the article and describe how it might explain the response.

(i) Does the article give any alternative explanations for the experimental results? If so, describe these alternatives.

CHAPTER TWENTY-FOUR

Coulometry

The three methods generally have moderate selectivity, sensitivity, and speed; in many instances, they are among the most accurate and precise methods available, with uncertainties of a few tenths of a percent relative being common. Finally, in contrast to all of the other methods discussed in this text, these three require no calibration against standards; that is, the functional relationship between the quantity measured and the mass of analyte can be calculated from theory. Applications of electrogravimetric methods are found in many elementary textbooks,[1] so we will not discuss them in any detail here. However, before discussing the two coulometric methods, we explore the processes that occur in an electrolytic deposition.

Three electroanalytical methods are based on electrolytic oxidation or reduction of an analyte for a sufficient period to assure its quantitative conversion to a new oxidation state. These methods are **constant-potential coulometry; constant-current coulometry,** or **coulometric titrations; and electrogravimetry.** In electrogravimetric methods, the product of the electrolysis is weighed as a deposit on one of the electrodes. In the two coulometric procedures, on the other hand, the quantity of electricity needed to complete the electrolysis is a measure of the amount of analyte present.

24A CURRENT-VOLTAGE RELATIONSHIPS DURING AN ELECTROLYSIS

An electrolysis can be performed in one of three ways: (1) with constant applied cell voltage, (2) with constant electrolysis current, or (3) with constant working electrode potential. It is useful to consider the consequences of each of these modes of operation. For all three modes, the behavior of the cell is governed by an equation similar to Equation 22-20.

$$E_{appl} = (E_r - E_l) + (\eta_{rc} + \eta_{rk}) - (\eta_{lc} + \eta_{lk}) - IR \tag{24-1}$$

where E_{appl} is the applied voltage from an external source and E_r and E_l are the reversible, or thermodynamic, potentials associated with the right- and left-hand electrodes, respectively. The terms η_{rc} and η_{rk} are overvoltages resulting from concentration polarization and kinetic polarization at the right-hand electrode; η_{lc} and η_{lk} are the corresponding overvoltages at the left-hand electrode (see Section 22E-2). The values for E_r and E_l can be calculated from standard potentials using the Nernst equation. In many cases, only the right-hand electrode is polarizable because the left-hand electrode is a nonpolarizable reference electrode. The overvoltage η is the extra voltage, above the thermodynamic potential, required to drive the electrode reaction at a certain rate and thus produce current in the cell.[2] Of the terms on the right side of Equation 24-1, only E_r and E_l

[1]For example, see D. A. Skoog, D. M. West, F. J. Holler, and S. R. Crouch, *Fundamentals of Analytical Chemistry*, 8th ed., Chap. 22, Belmont, CA: Brooks-Cole, 2004.

[2]Throughout this chapter, junction potentials are neglected because they are usually small enough to be of no significance in the discussion.

can be calculated from theory; the other terms must be determined experimentally.

24A-1 Operation of a Cell at a Fixed Applied Potential

The simplest way of performing an analytical electrolysis is to maintain the applied cell potential at a constant value. In practice, electrolysis at a constant cell potential is limited to the separation of easily reduced cations from those that are more difficult to reduce than hydrogen ion or nitrate ion. The reason for this limitation is illustrated in Figure 24-1, which shows the changes of current, IR drop, and cathode potential E_c during electrolysis in the cell for the determination of copper(II). The cell consists of two platinum electrodes, each with a surface area of 150 cm^2, immersed in 200 mL of a solution that is 0.0220 M in copper(II) ion and 1.00 M in hydrogen ion. The cell resistance is 0.50 Ω. When a suitable potential difference is applied between the two electrodes, copper is deposited on the cathode, and oxygen is evolved at a partial pressure of 1.00 atm at the anode. The overall cell reaction is

$$Cu^{2+} + H_2O \rightarrow Cu(s) + \tfrac{1}{2}O_2(g) + 2H^+$$

The analyte here is copper(II) ions in a solution containing an excess of sulfuric or nitric acid.

Initial Thermodynamic Potential of the Cell

Standard potential data for the two half-reactions in the cell under consideration are

$$Cu^{2+} + 2e^- \rightleftharpoons Cu(s) \qquad E^0 = 0.34\ V$$
$$\tfrac{1}{2}O_2(g) + 2H^+ + 2e^- \rightleftharpoons H_2O \qquad E^0 = 1.23\ V$$

Using the method shown in Example 22-6, the thermodynamic potential for this cell can be shown to be −0.94 V. Thus, we expect no current at less-negative applied potentials; at greater potentials, a linear increase in current should be observed in the absence of kinetic or concentration polarization. When a potential of about −2.5 V is applied to the cell, the initial current is about 1.5 A, as shown in Figure 24-1a. The electrolytic deposition of copper is then completed at this applied potential.

As shown in Figure 24-1b, the IR drop decreases continually as the reaction proceeds. The reason for this decrease is primarily concentration polarization at the cathode, which limits the rate at which copper ions are brought to the electrode surface and thus the current. As shown by Equation 24-1, the decrease in IR

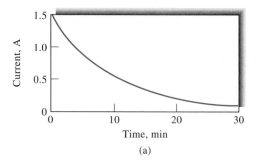

(a)

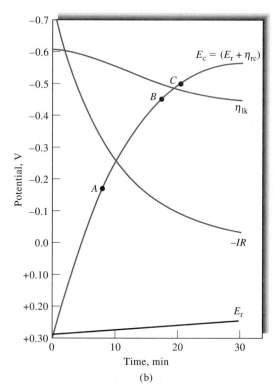

(b)

FIGURE 24-1 Changes in (a) current and (b) potentials during the electrolytic deposition of Cu^{2+}. Points A and B on the cathode potential curve are potentials at which Pb and Cd would begin to codeposit if present. Point C is the potential at which H_2 might begin to form at the cathode. Any of these processes would distort curve ABC.

must be offset by an increase in the cathode potential (more negative) because the applied cell potential is constant.

Ultimately, the decrease in current and the increase in cathode potential are slowed at point B by the reduction of hydrogen ions. Because the solution contains a large excess of acid, the current is now no longer limited by concentration polarization, and codeposition of copper and hydrogen goes on simultaneously until the remainder of the copper ions are deposited. Under these conditions, the cathode is said to be depolarized by hydrogen ions.

Consider now the fate of some metal ion, such as lead(II), that begins to deposit at point A on the cathode potential curve. Lead(II) would codeposit well before copper deposition was complete and would therefore interfere with the determination of copper. In contrast, a metal ion such as cobalt(II) that reacts at a cathode potential corresponding to point C on the curve would not interfere because depolarization by hydrogen gas formation prevents the cathode from reaching this potential. At best, an electrolysis at constant potential can be used only to separate easily reduced cations, such as Pb(II), Cd(II), Ag(I), Tl(I), Cu(II), and Zn(II), from those that are more difficult to reduce than hydrogen ion, such as Al(III). As suggested in the previous paragraph, hydrogen evolution occurs near the end of the electrolysis and prevents interference by cations that are reduced at more negative potentials.

Codeposition of hydrogen during electrolysis often leads to formation of deposits that do not adhere well. These are usually unsatisfactory for analytical purposes. This problem can be resolved by introducing another species that is reduced at a less-negative potential than hydrogen ion and does not adversely affect the physical properties of the deposit. One such cathode depolarizer is nitrate ion. Hydrazine and hydroxylamine are also commonly used.

In practice, with modern operational amplifier-based instrumentation, electrolytic cells are seldom operated at constant applied cell potential. Instead, the potential at the working electrode is controlled using a device called a *potentiostat*, which we describe in Section 24C-1. A discussion of electrolytic deposition follows, but the principles also apply to coulometry and coulometric titrations.

Current Changes during an Electrolysis at Constant Applied Potential

It is useful to consider the changes in current in the cell under discussion when the potential is held constant at -2.5 V throughout the electrolysis. Under these conditions, the current decreases with time as a result of the depletion of copper ions in the solution as well as the increase in cathodic concentration polarization. In

fact, with the onset of concentration polarization, the current decrease becomes exponential in time. That is,

$$I_t = I_0 e^{-kt}$$

where I_t is the current t min after the onset of polarization and I_0 is the initial current. Lingane[3] showed that values for the constant k can be computed from the relationship

$$k = \frac{25.8DA}{V\delta}$$

where D is the diffusion coefficient (cm^2/s), or the rate at which the reactant diffuses under a unit concentration gradient. The quantity A is the electrode surface area (cm^2), V is the volume of the solution (cm^3), and δ is the thickness of the surface layer (cm) in which the concentration gradient exists. Typical values for D and δ are 10^{-5} cm^2/s and 2×10^{-3} cm. (The constant 25.8 includes the factor of 60 (s/min) to make k compatible with the units of t in the equation for I_t.) When the initial applied potential is -2.5 V, we find that concentration polarization, and thus an exponential decrease in current, occurs essentially immediately after applying the potential. Figure 24-1a depicts this behavior; the data for the curve shown were computed for the cell using the preceding two equations. After 30 min, the current decreases from the initial 1.5 A to 0.08 A; by this time, approximately 96% of the copper has been deposited.

24A-2 Constant-Current Electrolysis

The analytical electrodeposition that we are discussing, as well as others, can be carried out by maintaining the current, rather than the applied voltage, at a constant level. To maintain constant current, we must increase the applied voltage periodically as the electrolysis proceeds.

In the preceding section, we showed that concentration polarization at the cathode causes a decrease in current. Initially, this effect can be partially offset by increasing the applied potential. Electrostatic forces would then postpone the onset of concentration polarization by enhancing the rate at which copper ions are brought to the electrode surface. Soon, however, the solution becomes so depleted of copper ions that diffusion, electrostatic attraction, and stirring cannot keep the electrode surface supplied with sufficient copper

 Simulation: Learn more about **electrolysis**.

[3]See J. J. Lingane, *Electroanalytical Chemistry*, 2nd ed., pp. 223–29, New York: Interscience, 1958.

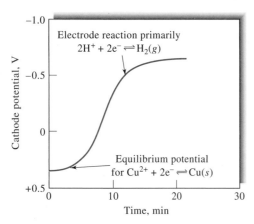

FIGURE 24-2 Changes in cathode potential during the deposition of copper with a constant current of 1.5 A. The cathode potential is equal to $E_r + \eta_{rc}$.

ions to maintain the desired current. When this occurs, further increases in E_{appl} cause rapid changes in η_{rc} and thus the cathode potential; codeposition of hydrogen (or other reducible species) then takes place. The cathode potential ultimately becomes stabilized at a level fixed by the standard potential and the overvoltage for the new electrode reaction; further large increases in the cell potential are no longer necessary to maintain a constant current. Copper continues to deposit as copper(II) ions reach the electrode surface; the contribution of this process to the total current, however, becomes smaller and smaller as the deposition becomes more and more nearly complete. An alternative process, such as reduction of hydrogen or nitrate ions, soon predominates. The changes in cathode potential under constant current conditions are shown in Figure 24-2.

24A-3 Electrolysis at Constant Working Electrode Potentials

From the Nernst equation, we see that a tenfold decrease in the concentration of an ion being deposited requires a negative shift in potential of only $0.0592/n$ V. Electrolytic methods, therefore, are reasonably selective. For example, as the copper concentration of a solution is decreased from 0.10 M to 10^{-6} M, the thermodynamic cathode potential E_r changes from an initial value of +0.31 to +0.16 V. In theory, then, it should be feasible to separate copper from any element that does not deposit within this 0.15-V potential

range. Species that deposit quantitatively at potentials more positive than +0.31 V could be eliminated with a prereduction; ions that require potentials more negative than +0.16 V would not interfere with the copper deposition. Thus, if we are willing to accept a reduction in analyte concentration to 10^{-6} M as a quantitative separation, it follows that divalent ions differing in standard potentials by about 0.15 V or greater can, theoretically, be separated quantitatively by electrodeposition, provided their initial concentrations are about the same. Correspondingly, about 0.30- to 0.10-V differences are required for univalent and trivalent ions, respectively.

An approach to these theoretical separations, within a reasonable electrolysis period, requires a more sophisticated technique than the ones thus far discussed because concentration polarization at the cathode, if unchecked, will prevent all but the crudest of separations. The change in cathode potential is governed by the decrease in IR drop (Figure 24-1b). Thus, for instances in which relatively large currents are applied initially, the change in cathode potential can ultimately be expected to be large. On the other hand, if the cell is operated at low current levels so that the variation in cathode potential is decreased, the time required for completion of the deposition may become prohibitively long. A straightforward solution to this dilemma is to begin the electrolysis with an applied cell potential that is sufficiently high to ensure a reasonable current; the applied potential is then continuously decreased to keep the cathode potential at the level necessary to accomplish the desired separation. Unfortunately, it is not feasible to predict the required changes in applied potential on a theoretical basis because of uncertainties in variables affecting the deposition, such as overvoltage effects and perhaps conductivity changes. Nor, indeed, does it help to measure the potential across the two electrodes, because such a measurement gives only the overall cell potential, E_{appl}. The alternative is to measure the potential of the working electrode against a third electrode whose potential in the solution is known and constant — that is, a reference electrode. The voltage applied across the working electrode and its counter electrode can then be adjusted to the level that will control the cathode (or anode) at the desired potential with respect to the reference electrode. This technique is called *controlled-potential electrolysis*, or sometimes *potentiostatic electrolysis*.

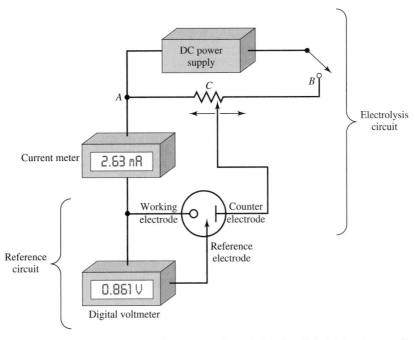

FIGURE 24-3 Apparatus for controlled-potential electrolysis. The digital voltmeter monitors the potential between the working and the reference electrode. The voltage applied between the working and the counter electrode is varied by adjusting contact *C* on the potentiometer to maintain the working electrode (cathode in this example) at a constant potential versus a reference electrode. The current in the reference electrode is essentially zero at all times. Modern potentiostats are fully automatic and often computer controlled. The electrode symbols shown ($-\!\!\!\bigcirc$ Working, $\rightarrow$ Reference, $\vdash$ Counter) are the currently accepted notation.

Experimental details for performing a controlled-cathode-potential electrolysis are presented in Section 24C-1. For the present, it is sufficient to note that the potential difference between the reference electrode and the cathode is measured with a voltmeter. The voltage applied between the working electrode and its counter electrode is controlled with a voltage divider so that the cathode potential is maintained at a level suitable for the separation. Figure 24-3 is a schematic of a manual apparatus that permits deposition at a constant cathode potential.

An apparatus of the type shown in Figure 24-3 can be operated at relatively high initial applied potentials to give high currents. As the electrolysis progresses, however, a decrease in the voltage applied across AC is required. This decrease, in turn, decreases the current. Completion of the electrolysis is indicated by the approach of the current to zero. The changes that occur

 Tutorial: Learn more about **electrolysis**.

in a typical constant-cathode-potential electrolysis are depicted in Figure 24-4. In contrast to the electrolytic methods described earlier, implementing this technique manually would demand constant attention. Fortunately, the constant-potential electrolysis method can be readily automated. A potentiostat, such as that shown in Figure 24-6, is suitable for the control of the working electrode potential with respect to the reference electrode.

24B AN INTRODUCTION TO COULOMETRIC METHODS OF ANALYSIS

Coulometry encompasses a group of analytical methods that involve measuring the quantity of electricity (in coulombs) needed to convert the analyte quantitatively to a different oxidation state. Like gravimetric methods, coulometry has the advantage that the proportionality constant between the measured quantity (charge in coulombs) and the mass of analyte can be computed from known physical constants; thus, calibration or standardization is not usually necessary. Coulometric methods are often as accurate as gravimetric or volumetric procedures, and they are usually faster and more convenient than gravimetric methods. Finally, coulometric procedures are easily automated.[4]

24B-1 Units for Quantity of Electricity

The quantity of electricity or charge is measured in units of coulombs (C). A coulomb is the quantity of charge transported in one second by a constant current of one ampere. Thus, for a constant current of I amperes for t seconds, the charge in coulombs Q is given by the expression

$$Q = It \qquad (24\text{-}3)$$

For a variable current i, the charge is given by the integral

$$Q = \int_0^t i\, dt \qquad (24\text{-}4)$$

The faraday F is the charge in coulombs of one mole of electrons. The charge of the electron is 1.60218×10^{-19} C, so we may therefore write

$$F = 6.02214 \times 10^{23} \frac{e^-}{\text{mol } e^-} \times 1.60218 \times 10^{-19} \frac{C}{e^-}$$

$$= 96{,}485 \frac{C}{\text{mol } e^-}$$

Faraday's law relates the number of moles of the analyte n_A to the charge

$$n_A = \frac{Q}{nF} \qquad (24\text{-}5)$$

where n is the number of moles of electrons in the analyte half-reaction. As shown in Example 24-1, we can use these definitions to calculate the mass of a chemical species that is formed at an electrode by a current of known magnitude.

---

[4]For additional information about coulometric methods, see J. A. Dean, *Analytical Chemistry Handbook*, Section 14, pp. 14.118–14.133, New York: McGraw-Hill, 1995; D. J. Curran, in *Laboratory Techniques in Electroanalytical Chemistry*, 2nd ed., P. T. Kissinger and W. R. Heinemann, eds., pp. 739–68, New York: Marcel Dekker, 1996; J. A. Plambeck, *Electroanalytical Chemistry*, Chap. 12, New York: Wiley, 1982.

EXAMPLE 24-1

A constant current of 0.800 A was used to deposit copper at the cathode and oxygen at the anode of an electrolytic cell. Calculate the mass of each product that was formed in 15.2 min, assuming that no other redox reactions occur.

Solution

The equivalent masses are determined from the two half-reactions

$$Cu^{2+} + 2e^- \rightarrow Cu(s)$$

$$2H_2O \rightarrow 4e^- + O_2(g) + 4H^+$$

Thus, 1 mol of copper is equivalent to 2 mol of electrons and 1 mol of oxygen corresponds to 4 mol of electrons.

From Equation 24-3, we find

$$Q = 0.800 \text{ A} \times 15.2 \text{ min} \times 60 \text{ s/min}$$

$$= 729.6 \text{ A}\cdot\text{s} = 729.6 \text{ C}$$

We can find the number of moles of Cu and O_2 from Equation 24-5:

$$n_{Cu} = \frac{Q}{nF} = \frac{729.6 \ \cancel{C}}{2 \ \cancel{\text{mol } e^-}/\text{mol Cu} \times 96{,}485 \ \cancel{C}/\cancel{\text{mol } e^-}}$$

$$= 3.781 \times 10^{-3} \text{ mol Cu}$$

$$n_{O_2} = \frac{Q}{nF} = \frac{729.6 \ \cancel{C}}{4 \ \cancel{\text{mol } e^-}/\text{mol } O_2 \times 96{,}485 \ \cancel{C}/\cancel{\text{mol } e^-}}$$

$$= 1.890 \times 10^{-3} \text{ mol } O_2$$

The masses of Cu and O_2 are given by

$$m_{Cu} = n_{Cu}\mathcal{M}_{Cu} = 3.781 \times 10^{-3} \ \cancel{\text{mol Cu}} \times \frac{63.55 \text{ g Cu}}{1 \ \cancel{\text{mol Cu}}}$$

$$= 0.240 \text{ g Cu}$$

$$m_{O_2} = n_{O_2}\mathcal{M}_{O_2} = 1.890 \times 10^{-3} \ \cancel{\text{mol } O_2} \times \frac{32.00 \text{ g } O_2}{1 \ \cancel{\text{mol } O_2}}$$

$$= 0.0605 \text{ g } O_2$$

24B-2 Types of Coulometric Methods

Two general techniques are used for coulometric analysis: *controlled-potential (potentiostatic) coulometry* and *controlled-current (amperostatic) coulometry*. In controlled-potential coulometry, the potential of the working electrode (the electrode at which the analytical reaction occurs) is maintained at a constant

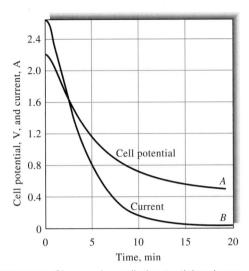

FIGURE 24-4 Changes in applied potential and current during a controlled-cathode-potential electrolysis. Deposition of copper on a cathode maintained at −0.36 versus a saturated calomel electrode. (Experimental data from J. J. Lingane, *Anal. Chem. Acta*, **1948**, *2*, 590, with permission.)

level such that quantitative oxidation or reduction of the analyte occurs without involvement of less-reactive species in the sample or solvent. In this method, the current is initially high but decreases rapidly and approaches zero as the analyte is removed from the solution (see Figure 24-4). The quantity of charge required is usually measured with an electronic integrator.

Controlled-current coulometry uses a constant current, which passes through a cell until an indicator signals completion of the analytical reaction. The quantity of charge required to reach the end point is then calculated from the magnitude of the current and the time that the current passes. This method has enjoyed wider application than potentiostatic coulometry. It is frequently called a *coulometric titration* for reasons that we discuss in Section 24D.

A fundamental requirement of all coulometric methods is that the analyte must react with 100% current efficiency. This requirement means that each faraday of charge must bring about a chemical change in $1/n$ moles of analyte, where n is the number of electrons that is equivalent to one mole of analyte. Current efficiency of 100% does not, however, imply that the analyte must necessarily participate directly in the electron-transfer process at the electrode. Indeed, more often than not, the analyte participates, at least

 Tutorial: Learn more about **coulometry**.

in part, in a reaction that is secondary to the electrode reaction. For example, at the beginning of the oxidation of iron(II) at a platinum anode, all of the current results from the reaction

$$Fe^{2+} \rightleftharpoons Fe^{3+} + e^-$$

As the concentration of iron(II) decreases, however, concentration polarization will cause the anode potential to rise until decomposition of water occurs as a competing process. That is,

$$2H_2O \rightleftharpoons O_2(g) + 4H^+ + 4e^-$$

The charge required to completely oxidize all of the iron(II) in the solution would then exceed that demanded by theory. To avoid the resulting error, an unmeasured excess of cerium(III) can be introduced at the start of the electrolysis. This ion is oxidized at a lower anode potential than is water:

$$Ce^{3+} \rightleftharpoons Ce^{4+} + e^-$$

The cerium(IV) produced diffuses rapidly from the electrode surface, where it then oxidizes an equivalent amount of iron(II):

$$Ce^{4+} + Fe^{2+} \longrightarrow Ce^{3+} + Fe^{3+}$$

The net effect is an electrochemical oxidation of iron(II) with 100% current efficiency even though only a fraction of the iron(II) ions are directly oxidized at the electrode surface.

The coulometric determination of chloride provides another example of an indirect process. Here, a silver electrode is the anode, and silver ions are produced by the current. These cations diffuse into the solution and precipitate the chloride. A current efficiency of 100% with respect to the chloride ion is achieved even though this ion is neither oxidized nor reduced in the cell.

24C CONTROLLED-POTENTIAL COULOMETRY

In *controlled-potential coulometry*, the potential of the working electrode is maintained at a constant level such that the analyte conducts charge across the electrode-solution interface. The charge required to convert the analyte to its reaction product is then determined by integrating the current-versus-time curve during the electrolysis. An analysis of this kind has all the advantages of an electrogravimetric method, but it is not necessary to weigh a product. The technique can therefore be applied to systems that yield deposits with poor physical properties as well as to reactions that

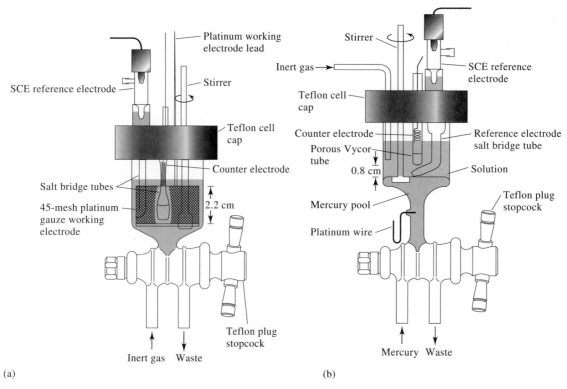

FIGURE 24-5 Electrolysis cells for potentiostatic coulometry. Working electrode: (a) platinum gauze; (b) mercury pool. (Reprinted with permission from J. E. Harrar and C. L. Pomernacki, *Anal. Chem.*, **1973**, *45*, 57. Copyright 1973 American Chemical Society.)

yield no solid product at all. For example, arsenic may be determined coulometrically by the electrolytic oxidation of arsenous acid (H_3AsO_3) to arsenic acid (H_3AsO_4) at a platinum anode. Similarly, the analytical conversion of iron(II) to iron(III) can be accomplished with suitable control of the anode potential.

24C-1 Instrumentation

The instrumentation for potentiostatic coulometry consists of an electrolysis cell, a potentiostat (see Section 3D-1), and an electronic integrator for determining the charge consumed.

Cells

Figure 24-5 illustrates two types of cells that are used for potentiostatic coulometry. The first consists of a platinum-gauze working electrode and a platinum-wire counter electrode, which is separated from the test solution by a porous tube containing the same supporting electrolyte as the test solution (Figure 24-5a). Separating the counter electrode is sometimes neces-

sary to prevent its reaction products from interfering in the analysis. A saturated calomel or a Ag-AgCl reference electrode is connected to the test solution with a salt bridge. Often the bridge contains the same electrolyte as the test solution.

The second type of cell is a mercury pool type. A mercury cathode is particularly useful for separating easily reduced elements as a preliminary step in an analysis. For example, copper, nickel, cobalt, silver, and cadmium are readily separated from ions such as aluminum, titanium, the alkali metals, and phosphates. The precipitated elements dissolve in the mercury; little hydrogen evolution occurs even at high applied potentials because of large overvoltage effects. A coulometric cell such as that shown in Figure 24-5b is also useful for coulometric determination of metal ions and certain types of organic compounds as well.

Potentiostats

A potentiostat is an electronic device that maintains the potential of a working electrode at a constant level relative to a reference electrode. Two such devices are

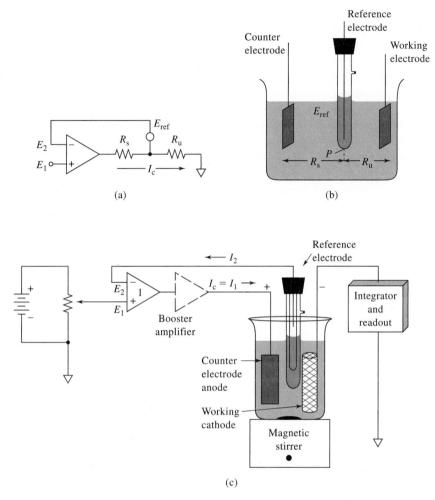

FIGURE 24-6 Schematic of a system for controlled-potential coulometry. (a) Equivalent circuit. (b) Resistances within the cell. (c) Practical circuit. The cell current I_c passes to the integrator-readout module, which provides a number proportional to the total quantity of charge that passes through the cell.

shown in Figure 3-14. Figure 24-6c is a schematic of an apparatus for controlled-potential coulometry, which contains a somewhat different type of potentiostat. To understand how this circuit works, consider the equivalent circuit shown in Figure 24-6a. The two resistances in this diagram correspond to resistances in two parts of the electrochemical cell shown in Figure 24-6b. Here, R_s is the cell resistance between the counter electrode and the tip P of the reference electrode, and R_u is the so-called *uncompensated cell resistance*, which is the cell resistance between P and the working electrode. Because of the extremely high resistance of the inputs to the operational amplifier, there is no current in the feedback loop to the invert-

ing input, but the potential difference between P and the inverting input of the operational amplifier is simply the reference electrode potential E_{ref}.

Recall that in the noninverting configuration, the operational amplifier works to keep E_1 and E_2 equal and that the cell current I_c is supplied by the operational amplifier to maintain this condition. If we consider the path between the inverting input and the circuit common at the output, we see that

$$E_2 = E_1 = E_{ref} + I_c R_u = E_{ref} + E_c$$

where E_c, the cathode potential, is essentially equal to the potential difference between P and the working cathode (see Figure 24-6b). Because E_1 and E_{ref} are constant, I_cR_u must also be constant. If R_u or R_s change in any way during the electrolysis, the operational amplifier output voltage changes in such a way as to keep $E_c = I_cR_u$ at a constant level. If R_u increases as a result of an increase in the cell resistance or concentration polarization, the output voltage of the operational amplifier decreases, which leads to a decrease in I_c. If R_u decreases, the operational amplifier output voltage increases correspondingly to maintain E_c constant.

The practical circuit in Figure 24-6c shows other components necessary in potentiostatic coulometry. This circuit includes a variable voltage source at the noninverting input of the operational amplifier so that the potentiostat control potential can be varied, a booster amplifier to supply the high currents that are often necessary, and an integrator and readout device. The presence of the booster amplifier has no effect on the potential control circuit. In the circuit, $I_1 = I_c - I_2$, but because the input bias current I_2 of operational amplifier 1 is negligibly small, $I_c \approx I_1$, which passes to the integrator and readout.

Integrators

As shown in Section 3E-3, analog integrators can be constructed from operational amplifier circuits. Most modern devices for potentiostatic coulometry, however, have digital integrators to determine the quantity of charge required to complete an electrolysis. One type of digital integrator consists of a voltage-to-frequency converter attached to a counter. The cell current I_c in Figure 24-6c passes through a small standard resistor R in the integrator-readout module, and the voltage I_cR is passed to the voltage-to-frequency converter, which produces a frequency that is proportional to the instantaneous current I_c. The total number of cycles of the voltage-to-frequency converter occurring during an experiment is proportional to the quantity of charge necessary to complete the electrolysis. These pulses are counted to produce a number in the readout that is proportional to the charge, which is in turn proportional to the amount of analyte.

In computerized coulometric instruments, the voltage I_cR is converted to a number with an analog-to-digital converter, and the resulting voltage-versus-time data are integrated using any of the standard algorithms for digital integration to obtain a number proportional to the charge.[5]

24C-2 Applications

Controlled-potential coulometric methods have been applied to more than fifty elements in inorganic compounds.[6] Historically, mercury was widely used as the working electrode material, and methods for the deposition of two dozen or more metals at this electrode have been described. In recent years, many other electrode materials have been used, including platinum and various forms of carbon. Coulometry is used widely in the nuclear energy field for the relatively interference-free determination of uranium and plutonium. For example, the uranium-to-oxygen ratio in spent nuclear fuel has been determined by controlled-potential coulometry with a precision of 0.06% relative standard deviation (RSD).[7]

Potentiostatic coulometry also offers possibilities for the electrolytic determination (and synthesis) of organic compounds. For example, Meites and Meites[8] demonstrated that trichloroacetic acid and picric acid are quantitatively reduced at a mercury cathode whose potential is suitably controlled:

$$Cl_3CCOO^- + H^+ + 2e^- \rightarrow Cl_2HCCOO^- + Cl_2$$

picric acid

Coulometric measurements permit the determination of these compounds with a relative error of a few tenths of a percent.

Coulometry has been used for detection in liquid chromatography (Section 28C-6). A relatively recent innovation is the coulometric array detector that con-

---

[5]S. R. Crouch and F. J. Holler, *Applications of Microsoft® Excel in Analytical Chemistry*, Belmont, CA: Brooks/Cole, 2004, pp. 201–6.

[6]For a summary of applications, see J. A. Dean, *Analytical Chemistry Handbook*, Section 14, pp. 14.119–14.123, New York: McGraw-Hill, 1995; A. J. Bard and L. R. Faulkner, *Electrochemical Methods*, 2nd ed., pp. 427–31, New York: Wiley, 2001.

[7]S. R. Sarkar, K. Une, and Y. Tominaga, *J. Radioanal. Nuc. Chem.*, **1997**, *220*, 155.

[8]T. Meites and L. Meites, *Anal. Chem.*, **1955**, *27*, 1531; **1956**, *28*, 103.

sists of a number (four, eight, twelve, or sixteen) of pairs of special miniature high-surface-area porous graphite electrodes in series at the outlet of a liquid chromatograph.[9] The pairs of electrodes are connected to a multichannel, computer-controlled potentiostat that maintains each pair at a different voltage, thus providing selectivity in the detection of electroactive species. Currents are monitored at each electrode pair with near 100% efficiency and integrated to provide detection of up to sixteen species simultaneously. The data-analysis software provides the capability to numerically manipulate the signal from each channel. For example, if two compounds are electroactive at one channel and only one of the compounds is detected at the other channel, the two channels can be subtracted to give a signal proportional to the second compound. This system has been used to determine all components in mixtures of guanosine, kynurenine, 3-nitro-tyrosine, 5-hydroxyindole-3-acetic acid, homovanillic acid, and serotonin with precisions of 2%– 5% RSD at concentrations of 4 ng/mL to 4 µg/mL.[10] The dynamic range of concentrations accessible by the coulometric array detector is from femtomoles to micromoles in the same sample. These devices are being used in the detection of a broad range of important electroactive compounds in biology and medicine.

24D COULOMETRIC TITRATIONS

In coulometric titrations, a constant current generates the titrant electrolytically. In some analyses, the active electrode process involves only generation of the reagent.[11] An example is the titration of halides by silver ions produced at a silver anode. In other titrations, the analyte may also be directly involved at the generator electrode. An example of this type of titration is the coulometric oxidation of iron(II) — in part by electrolytically generated cerium(IV) and in part by direct electrode reaction (Section 24B-2). Under any circumstance, the net process must approach 100% current efficiency with respect to a single chemical change in the analyte.

The current in a coulometric titration is carefully maintained at a constant and accurately known level by means of an amperostat. The product of this current

[9]CoulArray® Multi-channel Electrode Array Detector for HPLC, ESA Biosciences, Inc., Chelmsford, MA.
[10]J. K. Yaoa, P. Cheng, *J. Chromatography B*, **2004**, *810*, 93.
[11]For further details on this technique, see D. J. Curran, in *Laboratory Techniques in Electroanalytical Chemistry*, 2nd ed., P. T. Kissinger and W. R. Heineman, eds., pp. 750–68, New York: Marcel Dekker, 1996.

in amperes and the time in seconds required to reach an end point is the number of coulombs, which is proportional to the quantity of analyte involved in the electrolysis. The constant-current aspect of this operation precludes the quantitative oxidation or reduction of the unknown species at the generator electrode because concentration polarization of the solution is inevitable before the electrolysis can be complete. The electrode potential must then rise if a constant current is to be maintained (Section 24A-2). Unless this increase in potential produces a reagent that can react with the analyte, the current efficiency will be less than 100%. In a coulometric titration, then, at least part (and frequently all) of the reaction involving the analyte occurs away from the surface of the working electrode.

A coulometric titration, like a more conventional volumetric procedure, requires some means of detecting the point of chemical equivalence. Most of the end-point detection methods applicable to volumetric analysis are equally satisfactory here. Visual observations of color changes of indicators, as well as potentiometric, amperometric, and photometric measurements have all been used successfully.

24D-1 Electrical Apparatus

Coulometric titrators are available from several manufacturers. In addition, they can be assembled easily from components available in most analytical laboratories. Figure 24-7 is a conceptual diagram of a coulometric titration apparatus showing its main com-

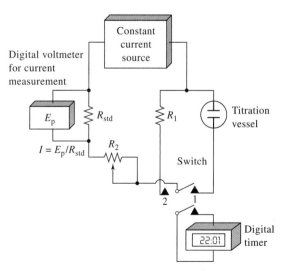

FIGURE 24-7 Schematic of a manual coulometric titration apparatus. Commercial coulometric titrators are totally electronic and usually computer controlled.

ponents. Included are a constant current source and a switch that simultaneously connects the current and starts an electronic timer. Also shown is a meter for accurately measuring the current. In Figure 24-7, the voltage drop across the standard resistor R_{std} is used for this measurement. Many electronic amperostats are described in the literature. These devices can be easily constructed using inexpensive operational amplifiers (see, for example, Figure 3-15).

Figure 24-8 is a photo of one of a number of automated coulometric titrators on the market. Most of these devices use a potentiometric end point. Some of the commercial instruments are multipurpose and can be used for the determination of a variety of species. Others are designed for a single analysis. Examples of the latter include chloride titrators, in which silver ion is generated coulometrically; sulfur dioxide monitors, where anodically generated bromine oxidizes the analyte to sulfate ions; carbon dioxide monitors in which the gas, absorbed in monoethanolamine, is titrated with coulometrically generated base; and water titrators in which the Karl Fischer reagent[12] is generated electrolytically.

The instrument in Figure 24-8 is specifically designed for the Karl Fischer determination of water. These devices are often connected to a printer for data output and for displaying titration curves. They are often equipped with a serial interface for connection to an external computer, but instrument operation is under the control of an internal computer. The device shown has five programmed standard methods and can store up to 50 user-defined methods. The precision of determinations is between 0.5% and 5% RSD for the range of 10 ppm to 100 ppm water. In addition to water determination, the device has programmed methods for the determination of the so-called *bromine index*, which is essentially a measure of the concentration of double bonds, or degree of saturation, in a sample of a mixture of organic compounds, such as gasoline. The titrant is electrolytically generated bromine.

Cells for Coulometric Titrations

A typical coulometric titration cell is shown in Figure 24-9. It consists of a generator electrode at which the reagent is formed and an auxiliary electrode to complete the circuit. The generator electrode, which should have a relatively large surface area, is often a

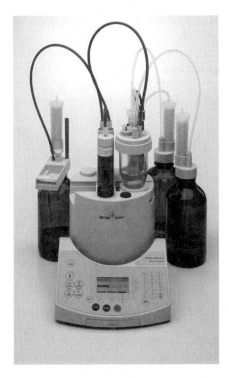

FIGURE 24-8 An automated coulometric titrator. All instrument functions are under control of an internal computer, and results of determinations appear on the LCD display. (Courtesy of Mettler-Toledo, Inc., Columbus, OH.)

rectangular strip or a wire coil of platinum. A gauze electrode such as the cathode shown in Figure 24-5a can also be used. The products formed at the second electrode frequently represent potential sources of interference. For example, the anodic generation of oxidizing agents is often accompanied by the evolution of hydrogen from the cathode. Unless this gas is allowed to escape from the solution, it is likely to react with the oxidizing agent. To eliminate this type of difficulty, the second electrode is isolated by a sintered disk or some other porous medium. An alternative to isolation of the auxiliary electrode is a device such as that shown in Figure 24-10 in which the reagent is generated externally. The apparatus is arranged so that flow of the electrolyte continues briefly after the current is discontinued, thus flushing the residual reagent into the titration vessel. Note that the apparatus shown in Figure 24-10 provides either hydrogen or hydroxide ions depending on which arm is used. The cell has also been used to generate other reagents such as iodine produced by oxidation of iodide at the anode.

[12]D. A. Skoog, D. M. West, F. J. Holler, and S. R. Crouch, *Fundamentals of Analytical Chemistry*, 8th ed., Section 20C-5, pp. 580–82, Belmont, CA: Brooks-Cole, 2004.

TABLE 24-2 Summary of Coulometric Titrations Involving Oxidation-Reduction Reactions

Reagent	Generator Electrode Reaction	Substance Determined
Br_2	$2Br^- \rightleftharpoons Br_2 + 2e^-$	As(III), Sb(III), U(IV), Ti(I), I$^-$, SCN$^-$, NH$_3$, N$_2$H$_4$, NH$_2$OH, phenol, aniline, mustard gas, mercaptans, 8-hydroxyquinoline, oleins
Cl_2	$2Cl^- \rightleftharpoons Cl_2 + 2e^-$	As(III), I$^-$, styrene, fatty acids
I_2	$2I^- \rightleftharpoons I_2 + 2e^-$	As(III), Sb(III), S$_2$O$_3^{2-}$, H$_2$S, ascorbic acid
Ce^{4+}	$Ce^{3+} \rightleftharpoons Ce^{4+} + e^-$	Fe(II), Ti(III), U(IV), As(III), I$^-$, Fe(CN)$_6^{4-}$
Mn^{3+}	$Mn^{2+} \rightleftharpoons Mn^{3+} + e^-$	H$_2$C$_2$O$_4$, Fe(II), As(III)
Ag^{2+}	$Ag^+ \rightleftharpoons Ag^{2+} + e^-$	Ce(III), V(IV), H$_2$C$_2$O$_4$, As(III)
Fe^{2+}	$Fe^{3+} + e^- \rightleftharpoons Fe^{2+}$	Cr(VI), Mn(VII), V(V), Ce(IV)
Ti^{3+}	$TiO^{2+} + 2H^+ + e^- \rightleftharpoons Ti^{3+} + H_2O$	Fe(III), V(V), Ce(IV), U(VI)
$CuCl_3^{2-}$	$Cu^{2+} + 3Cl^- + e^- \rightleftharpoons CuCl_3^{2-}$	V(V), Cr(VI), IO$_3^-$
U^{4+}	$UO_2^{2+} + 4H^+ + 2e^- \rightleftharpoons U^{4+} + 2H_2O$	Cr(VI), Ce(IV)

particularly useful among the oxidizing agents and forms the basis for a host of methods. Of interest also are some of the unusual reagents not ordinarily used in volumetric analysis because of the instability of their solutions. These reagents include dipositive silver ion, tripositive manganese, and the chloride complex of unipositive copper.

An interesting recent application of coulometric titrations is the determination of total antioxidant capacity in human serum.[16] Many antioxidants appear in serum: low-molecular-mass compounds such as tocopherols, ascorbate, β-carotene, glutathione, uric acid, and bilirubin and proteins such as albumin, transferrin, caeruloplasmin, ferritin, superoxide dismutase, catalase, and glutathione peroxidase. By far the greatest contributor to total antioxidant capacity is serum albumin. Coulometric titration of serum samples with electrogenerated bromine gives a value of antioxidant capacity, which may be an important clinical variable in assessing the status of patients undergoing hemodialysis. This method for determining antioxidant capacity is straightforward and quite precise ($\pm 0.04\%$ RSD).

Comparison of Coulometric and Volumetric Titrations

There are several interesting analogies between coulometric and volumetric titration methods. Both require a detectable end point and are subject to a titration

error as a consequence. Furthermore, in both techniques, the amount of analyte is determined through evaluation of its combining capacity — in the one case, for a standard solution and, in the other, for electrons. Also, similar demands are made of the reactions; that is, they must be rapid, essentially complete, and free of side reactions. Finally, there is a close analogy between the various components of the apparatus shown in Figure 24-7 and the apparatus and solutions employed in a conventional volumetric analysis. The constant-current source of known magnitude has the same function as the standard solution in a volumetric method. The timer and switch correspond closely to the buret, the switch performing the same function as a stopcock. During the early phases of a coulometric titration, the switch is kept closed for extended periods. As the end point is approached, however, small additions of "reagent" are achieved by closing the switch for shorter and shorter intervals. This process is analogous to the operation of a buret.

Coulometric titrations have some important advantages in comparison with the classical volumetric process. The most important of these advantages is the elimination of problems associated with the preparation, standardization, and storage of standard solutions. This advantage is particularly important with labile reagents such as chlorine, bromine, or titanium(III) ion. Because of their instability, these species are inconvenient as volumetric reagents. Their use in coulometric analysis is straightforward, however, because they undergo reaction with the analyte immediately after being generated.

[16] G. K. Ziyatdinova, H. C. Budnikov, V. I. Pogorel'tzev, and T. S. Ganeev, *Talanta*, **2006**, *68*, 800.

Where small quantities of reagent are required, a coulometric titration offers a considerable advantage. By proper choice of current, microquantities of a substance can be introduced with ease and accuracy. The equivalent volumetric process requires dispensing small volumes of very dilute solutions, which is always difficult.

A single constant current source can be used to generate precipitation, complex formation, oxidation-reduction, or neutralization reagents. Furthermore, the coulometric method adapts easily to automatic titrations, because current can be controlled quite easily.

Coulometric titrations are subject to five potential sources of error: (1) variation in the current during electrolysis, (2) departure of the process from 100% current efficiency, (3) error in the current measurement, (4) error in the measurement of time, and (5) titration error due to the difference between the equivalence point and the end point. The last of these difficulties is common to volumetric methods as well. For situations in which the indicator error is the limiting factor, the two methods are likely to have comparable reliability.

With simple instrumentation, currents constant to 0.2% relative are easily achieved, and with somewhat more sophisticated apparatus, current may be controlled to 0.01%. In general, then, errors due to current fluctuations are seldom significant.

Although it is difficult to generalize when discussing the magnitude of uncertainty associated with the electrode process, current efficiencies of 99.5% to better than 99.9% are often reported in the literature. Currents can be measured to ±0.1% relative quite easily.

To summarize, then, the current-time measurements required for a coulometric titration are inherently as accurate as or more accurate than the comparable volume-molarity measurements of a classical volumetric analysis, particularly where small quantities of reagent are involved. Often, however, the accuracy of a titration is not limited by these measurements but by the sensitivity of the end point; in this respect, the two procedures are equivalent.

QUESTIONS AND PROBLEMS

*Answers are provided at the end of the book for problems marked with an asterisk.

$\boxed{\mathbf{X}}$ Problems with this icon are best solved using spreadsheets.

*24-1 Lead is to be deposited at a cathode from a solution that is 0.100 M in Pb^{2+} and 0.200 M in $HClO_4$. Oxygen is evolved at a pressure of 0.800 atm at a 30-cm^2 platinum anode. The cell has a resistance of 0.950 Ω. (a) Calculate the thermodynamic potential of the cell. (b) Calculate the IR drop if a current of 0.250 A is to be used.

*24-2 Calculate the minimum difference in standard electrode potentials needed to lower the concentration of the metal M_1 to 2.00×10^{-4} M in a solution that is 1.00×10^{-1} M in the less-reducible metal M_2 where (a) M_2 is univalent and M_1 is divalent, (b) M_2 and M_1 are both divalent, (c) M_2 is trivalent and M_1 is univalent, (d) M_2 is divalent and M_1 is univalent, (e) M_2 is divalent and M_1 is trivalent.

24-3 It is desired to separate and determine bismuth, copper, and silver in a solution that is 0.0550 M in BiO^+, 0.125 M in Cu^{2+}, 0.0962 M in Ag^+, and 0.500 M in $HClO_4$. (a) Using 1.00×10^{-6} M as the criterion for quantitative removal, determine whether separation of the three species is feasible by controlled-potential electrolysis. (b) If any separations are feasible, evaluate the range (versus Ag-AgCl) within which the cathode potential should be controlled for the deposition of each.

24-4 Halide ions can be deposited at a silver anode, the reaction being

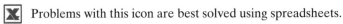

$$Ag(s) + X^- \longrightarrow AgX(s) + e^-$$

Suppose that a cell was formed by immersing a silver anode in an analyte solution that was 0.0250 M Cl^-, Br^-, and I^- ions and connecting the half-cell to a saturated calomel cathode via a salt bridge.

(a) Which halide would form first and at what potential? Is the cell galvanic or electrolytic?

(b) Could I^- and Br^- be separated quantitatively? (Take 1.00×10^{-5} M as the criterion for quantitative removal of an ion.) If a separation is feasible, what range of cell potential could be used?

(c) Repeat part (b) for I^- and Cl^-.

(d) Repeat part (b) for Br^- and Cl^-.

*24-5 What cathode potential (versus SCE) would be required to lower the total Hg(II) concentration of the following solutions to 1.00×10^{-6} M (assume reaction product in each case is elemental Hg):

(a) an aqueous solution of Hg^{2+}?

(b) a solution with an equilibrium SCN^- concentration of 0.150 M?

$$Hg^{2+} + 2SCN^- \rightleftharpoons Hg(SCN)_2(aq) \qquad K_f = 1.8 \times 10^7$$

(c) a solution with an equilibrium Br^- concentration of 0.150 M?

$$HgBr_4^{2-} + 2e^- \rightleftharpoons Hg(l) + 4Br^- \qquad E^0 = 0.223 \text{ V}$$

*24-6 Calculate the time required for a constant current of 0.750 A to deposit 0.270 g of (a) Co(II) as the element on a cathode and (b) as Co_3O_4 on an anode. Assume 100% current efficiency for both cases.

*24-7 Calculate the time required for a constant current of 0.905 A to deposit 0.300 g of (a) Tl(III) as the element on a cathode, (b) Tl(I) as the Tl_2O_3 on an anode, and (c) Tl(I) as the element on a cathode.

⊠ 24-8 At a potential of -1.0 V (versus SCE), carbon tetrachloride in methanol is reduced to chloroform at a Hg cathode:

$$2CCl_4 + 2H^+ + 2e^- + 2Hg(l) \longrightarrow 2CHCl_3 + Hg_2Cl_2(s)$$

At -1.80 V, the chloroform further reacts to give methane:

$$2CHCl_3 + 6H^+ + 6e^- + 6Hg(l) \longrightarrow 2CH_4 + 3Hg_2Cl_2(s)$$

Several 0.750-g samples containing CCl_4, $CHCl_3$, and inert organic species were dissolved in methanol and electrolyzed at -1.0 V until the current approached zero. A coulometer indicated the charge required to complete the reaction, as given in the second column of the following table. The potential of the cathode was then adjusted to -1.80 V. The additional charge required to complete the reaction at this potential is given in the third column of the table. Calculate the percent CCl_4 and $CHCl_3$ in each mixture.

Sample No.	Charge Required at -1.0 V, C	Charge Required at -1.8 V, C
1	11.63	68.60
2	21.52	85.33
3	6.22	45.98
4	12.92	55.31

*24-9 A 6.39-g sample of an ant-control preparation was decomposed by wet ashing with H_2SO_4 and HNO_3. The arsenic in the residue was reduced to the trivalent state with hydrazine. After the excess reducing agent had been removed, the arsenic(III) was oxidized with electrolytically generated I_2 *in a faintly alkaline medium*:

$$HAsO_3^{2-} + I_2 + 2HCO_3^- \longrightarrow HAsO_4^{2-} + 2I^- + 2CO + H_2O$$

The titration was complete after a constant current of 127.6 mA had been passed for 11 min and 54 s. Express the results of this analysis in terms of the percentage As_2O_3 in the original sample.

*24-10 A 0.0809-g sample of a purified organic acid was dissolved in an alcohol-water mixture and titrated with coulometrically generated hydroxide ions. With a current of 0.0441 A, 266 s was required to reach a phenolphthalein end point. Calculate the equivalent mass of the acid.

*24-11 Traces of aniline can be determined by reaction with an excess of electrolytically generated Br_2:

$$C_6H_5NH_2 + 3Br_2 \longrightarrow C_6H_2Br_3NH_2 + 3H^+ + 3Br^-$$

The polarity of the working electrode is then reversed, and the excess bromine is determined by a coulometric titration involving the generation of Cu(I):

$$Br_2 + 2Cu^+ \longrightarrow 2Br^- + 2Cu^{2+}$$

Suitable quantities of KBr and copper(II) sulfate were added to a 25.0-mL sample containing aniline. Calculate the mass in micrograms of $C_6H_5NH_2$ in the sample from the accompanying data:

Working Electrode Functioning As	Generation Time (min) with a Constant Current of 1.00 mA
Anode	3.76
Cathode	0.270

 24-12 Construct a coulometric titration curve of 100.0 mL of a 1 M H_2SO_4 solution containing Fe(II) titrated with Ce(IV) generated from 0.075 M Ce(III). The titration is monitored by potentiometry. The initial amount of Fe(II) present is 0.05182 mmol. A constant current of 20.0 mA is used. Find the time corresponding to the equivalence point. Then, for about ten values of time before the equivalence point, use the stoichiometry of the reaction to calculate the amount of Fe^{3+} produced and the amount of Fe^{2+} remaining. Use the Nernst equation to find the system potential. Find the equivalence point potential in the usual manner for a redox titration. For about ten times after the equivalence point, calculate the amount of Ce^{4+} produced from the electrolysis and the amount of Ce^{3+} remaining. Plot the curve of system potential versus electrolysis time.

X *Challenge Problem*

24-13 Sulfide ion (S^{2-}) is formed in wastewater by the action of anaerobic bacteria on organic matter. Sulfide can be readily protonated to form volatile, toxic H_2S. In addition to the toxicity and noxious odor, sulfide and H_2S cause corrosion problems because they can be easily converted to sulfuric acid when conditions change to aerobic. One common method to determine sulfide is by coulometric titration with generated silver ion. At the generator electrode, the reaction is $Ag \rightarrow Ag^+ + e^-$. The titration reaction is $S^{2-} + 2Ag^+ \rightarrow Ag_2S(s)$.

(a) A digital chloridometer was used to determine the mass of sulfide in a wastewater sample. The chloridometer reads out directly in ng Cl^-. In chloride determinations, the same generator reaction is used, but the titration reaction is $Cl^- + Ag^+ \rightarrow AgCl(s)$. Derive an equation that relates the desired quantity, mass S^{2-} (ng), to the chloridometer readout in mass Cl^- (ng).

(b) A particular wastewater standard gave a reading of 1689.6 ng Cl^-. What total charge in coulombs was required to generate the Ag^+ needed to precipitate the sulfide in this standard?

(c) The following results were obtained on 20.00-mL samples containing known amounts of sulfide.[17] Each standard was analyzed in triplicate and the mass of chloride recorded. Convert each of the chloride results to mass S^{2-}(ng).

Known Mass S^{2-}, ng	Mass Cl^- Determined, ng		
6365	10447.0	10918.1	10654.9
4773	8416.9	8366.0	8416.9
3580	6528.3	6320.4	6638.9
1989	3779.4	3763.9	3936.4
796	1682.9	1713.9	1669.7
699	1127.9	1180.9	1174.3
466	705.5	736.4	707.7
373	506.4	521.9	508.6
233	278.6	278.6	247.7
0	−22.1	−19.9	−17.7

(d) Determine the average mass of S^{2-} (ng), the standard deviation, and the %RSD of each standard.

(e) Prepare a plot of the average mass of S^{2-} determined (ng) versus the actual mass (ng). Determine the slope, the intercept, the standard error, and the R^2 value. Comment on the fit of the data to a linear model.

(f) Determine the detection limit (ng) and in parts per million using a k factor of 2 (see Equation 1-12).

(g) An unknown wastewater sample gave an average reading of 893.2 ng Cl^-. What is the mass of sulfide (ng)? If 20.00 mL of the wastewater sample was introduced into the titration vessel, what is the concentration of S^{2-} in parts per million?

[17] D. T. Pierce, M. S. Applebee, C. Lacher, and J. Bessie, *Environ. Sci. Technol.*, **1998**, *32*, 1734.

CHAPTER TWENTY-FIVE

Voltammetry

V oltammetry comprises a group of electro-analytical methods in which information about the analyte is obtained by measuring current as a function of applied potential under conditions that promote polarization of an indicator, or working, electrode. When current proportional to analyte concentration is monitored at fixed potential, the technique is called **amperometry.** *Generally, to enhance polarization, working electrodes in voltammetry and amperometry have surface areas of a few square millimeters at the most and, in some applications, a few square micrometers or less.*

Throughout this chapter, this logo indicates an opportunity for online self-study at **www .thomsonedu.com/chemistry/skoog**, linking you to interactive tutorials, simulations, and exercises.

Let us begin by pointing out the basic differences between voltammetry and the two types of electrochemical methods that we discussed in earlier chapters. Voltammetry is based on the measurement of the current that develops in an electrochemical cell under conditions where concentration polarization exists. Recall from Section 22E-2 that a polarized electrode is one to which we have applied a voltage in excess of that predicted by the Nernst equation to cause oxidation or reduction to occur. In contrast, potentiometric measurements are made at currents that approach zero and where polarization is absent. Voltammetry differs from coulometry in that, with coulometry, measures are taken to minimize or compensate for the effects of concentration polarization. Furthermore, in voltammetry there is minimal consumption of analyte, whereas in coulometry essentially all of the analyte is converted to another state.

Voltammetry is widely used by inorganic, physical, and biological chemists for nonanalytical purposes, including fundamental studies of oxidation and reduction processes in various media, adsorption processes on surfaces, and electron-transfer mechanisms at chemically modified electrode surfaces.

Historically, the field of voltammetry developed from *polarography*, which is a particular type of voltammetry that was invented by the Czechoslovakian chemist Jaroslav Heyrovsky in the early 1920s.[1] Polarography differs from other types of voltammetry in that the working electrode is the unique *dropping mercury electrode*. At one time, polarography was an important tool used by chemists for the determination of inorganic ions and certain organic species in aqueous solutions. In the late 1950s and the early 1960s, however, many of these analytical applications were replaced by various spectroscopic methods, and polarography became a less important method of analysis except for certain special applications, such as the determination of molecular oxygen in solutions. In the mid-1960s, several major modifications of classical voltammetric techniques were developed that enhanced significantly the sensitivity and selectivity of the method. At about this same time, the advent of low-cost operational amplifiers made possible the commercial development of relatively inexpensive instruments that incorporated many of these modifications and made them available to all chemists. The result was a resurgence of interest in

---

[1] J. Heyrovsky, *Chem. Listy*, **1922**, *16*, 256. Heyrovsky was awarded the 1959 Nobel Prize in Chemistry for his discovery and development of polarography.

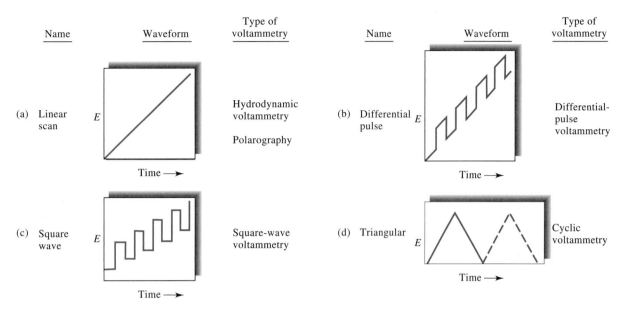

Name	Waveform	Type of voltammetry
(a) Linear scan	E	Hydrodynamic voltammetry Polarography
	Time $\longrightarrow$	
(b) Differential pulse	E	Differential-pulse voltammetry
	Time $\longrightarrow$	
(c) Square wave	E	Square-wave voltammetry
	Time $\longrightarrow$	
(d) Triangular	E	Cyclic voltammetry
	Time $\longrightarrow$	

FIGURE 25-1 Voltage versus time excitation signals used in voltammetry.

applying polarographic methods to the determination of a host of species, particularly those of pharmaceutical, environmental, and biological interest.[2] Since the invention of polarography, at least 60,000 research papers have appeared in the literature on the subject. Research activity in this field, which dominated electroanalytical chemistry for more than five decades, peaked with nearly 2,000 published journal articles in 1973. Since that time, interest in polarographic methods has steadily declined, at a rate nearly twice the rate of growth of the general chemical literature, until in 2005 only about 300 papers on these methods appeared. This decline has been largely a result of concerns about the use of large amounts of mercury in the laboratory as well as in the environment, the somewhat cumbersome nature of the apparatus, and the broad availability of faster and more convenient (mainly spectroscopic) methods. For these reasons, we will discuss polarography only briefly and, instead, refer you to the many sources that are available on the subject.[3]

Although polarography declined in importance, voltammetry and amperometry at working electrodes other than the dropping mercury electrode have grown at an astonishing pace.[4] Furthermore, voltammetry and amperometry coupled with liquid chromatography have become powerful tools for the analysis of complex mixtures. Modern voltammetry also continues to be an excellent tool in diverse areas of chemistry, biochemistry, materials science and engineering, and the environmental sciences for studying oxidation, reduction, and adsorption processes.[5]

25A EXCITATION SIGNALS IN VOLTAMMETRY

In voltammetry, a variable potential excitation signal is impressed on a working electrode in an electrochemical cell. This excitation signal produces a characteristic current response, which is the measurable quantity in this method. The waveforms of four of the most common excitation signals used in voltammetry are shown in Figure 25-1. The classical voltammetric excitation

[2] A. Bond, *Broadening Electrochemical Horizons: Principles and Illustration of Voltammetric and Related Techniques*, New York: Oxford, 2003; C. M. A. Brett and A. M. Oliveira Brett, in *Encyclopedia of Electrochemistry*, A. J. Bard and M. Stratmann, eds., Vol. 3, *Instrumentation and Electroanalytical Chemistry*, P. Unwin, ed., New York: Wiley, 2002, pp. 105–24.[3] A. J. Bard and L. R. Faulkner, *Electrochemical Methods*, 2nd ed., New York: Wiley, 2001, Chap. 7, pp. 261–304; *Laboratory Techniques in Electroanalytical Chemistry*, 2nd ed., P. T. Kissinger and W. R. Heineman, eds., New York: Dekker, 1996, pp. 444–61.

[4] From 1973 to 2005, the annual number of journal articles on voltammetry and amperometry grew at three times and two and one-half times, respectively, the rate of production of articles in all of chemistry.
[5] Some general references on voltammetry include A. J. Bard and L. R. Faulkner, *Electrochemical Methods*, 2nd ed., New York: Wiley, 2001; S. P. Kounaves, in *Handbook of Instrumental Techniques for Analytical Chemistry*, Frank A. Settle, ed., Upper Saddle River, NJ: Prentice-Hall, 1997, pp. 711–28; *Laboratory Techniques in Electroanalytical Chemistry*, 2nd ed., P. T. Kissinger and W. R. Heineman, eds., New York: Dekker, 1996; *Analytical Voltammetry*, M. R. Smyth and F. G. Vos, eds., New York: Elsevier, 1992.

signal is the linear scan shown in Figure 25-1a, in which the voltage applied to the cell increases linearly (usually over a 2- to 3-V range) as a function of time. The current in the cell is then recorded as a function of time, and thus as a function of the applied voltage. In amperometry, current is recorded at fixed applied voltage.

Two pulse excitation signals are shown in Figure 25-1b and c. Currents are measured at various times during the lifetime of these pulses. With the triangular waveform shown in Figure 25-1d, the potential is cycled between two values, first increasing linearly to a maximum and then decreasing linearly with the same slope to its original value. This process may be repeated numerous times as the current is recorded as a function of time. A complete cycle may take 100 or more seconds or be completed in less than 1 second.

To the right of each of the waveforms of Figure 25-1 is listed the types of voltammetry that use the various excitation signals. We discuss these techniques in the sections that follow.

25B VOLTAMMETRIC INSTRUMENTATION

Figure 25-2 is a schematic showing the components of a modern operational amplifier potentiostat (see Section 24C-1) for carrying out linear-scan voltammetric measurements. The cell is made up of three electrodes immersed in a solution containing the analyte and also an excess of a nonreactive electrolyte called a *supporting electrolyte*. One of the three electrodes is the work-

ing electrode, whose potential is varied linearly with time. Its dimensions are kept small to enhance its tendency to become polarized (see Section 22E-2). The second electrode is a reference electrode (commonly a saturated calomel or a silver–silver chloride electrode) whose potential remains constant throughout the experiment. The third electrode is a counter electrode, which is often a coil of platinum wire that simply conducts electricity from the signal source through the solution to the working electrode.

The signal source is a linear-scan voltage generator similar to the integration circuit shown in Figure 3-16c. The output from this type of source is described by Equation 3-22. Thus, for a constant dc input potential of E_i, the output potential E_o is given by

$$E_o = -\frac{E_i}{R_iC_f}\int_0^t dt = -\frac{E_it}{R_iC_f} \qquad (25\text{-}1)$$

The output signal from the source is fed into a potentiostatic circuit similar to that shown in Figure 25-2 (see also Figure 24-6c). The electrical resistance of the control circuit containing the reference electrode is so large ($>10^{11}\ \Omega$) that it draws essentially no current. Thus, the entire current from the source is carried from the counter electrode to the working electrode. Furthermore, the control circuit adjusts this current so that the potential difference between the working electrode and the reference electrode is identical to the output voltage from the linear voltage generator. The resulting current, which is directly proportional to the potential difference between the working electrode–reference

 Tutorial: Learn more about **voltammetric instrumentation and waveforms**.

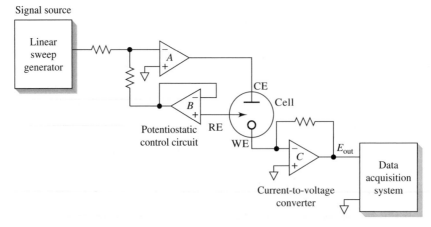

FIGURE 25-2 An operational amplifier potentiostat. The three-electrode cell has a working electrode (WE), reference electrode (RE), and a counter electrode (CE).

electrode pair, is then converted to a voltage and recorded as a function of time by the data-acquisition system.[6] It is important to emphasize that the independent variable in this experiment is the potential of the working electrode versus the reference electrode and not the potential between the working electrode and the counter electrode. The working electrode is at virtual common potential throughout the course of the experiment (see Section 3B-3).

25B-1 Working Electrodes

The working electrodes used in voltammetry take a variety of shapes and forms.[7] Often, they are small flat disks of a conductor that are press fitted into a rod of an inert material, such as Teflon or Kel-F, that has embedded in it a wire contact (see Figure 25-3a). The conductor may be a noble metal, such as platinum or gold; a carbon material, such as carbon paste, carbon fiber, pyrolytic graphite, glassy carbon, diamond, or carbon nanotubes; a semiconductor, such as tin or indium oxide; or a metal coated with a film of mercury. As shown in Figure 25-4, the range of potentials that can be used with these electrodes in aqueous solutions varies and depends not only on electrode material but also on the composition of the solution in which it is immersed. Generally, the positive potential limitations are caused by the large currents that develop because of oxidation of the water to give molecular oxygen. The negative limits arise from the reduction of water to produce hydrogen. Note that relatively large negative potentials can be tolerated with mercury electrodes because of the high overvoltage of hydrogen on this metal.

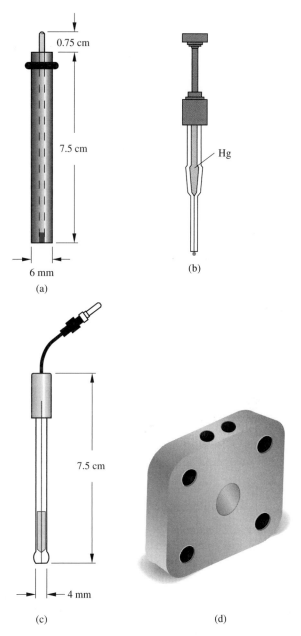

FIGURE 25-3 Some common types of commercial voltammetric electrodes: (a) a disk electrode; (b) a hanging mercury drop electrode (HMDE); (c) a microelectrode; (d) a sandwich-type flow electrode. (Electrodes [a], [c], and [d] courtesy of Bioanalytical Systems, Inc., West Lafayette, IN, with permission.)

Mercury working electrodes have been widely used in voltammetry for several reasons. One is the relatively large negative potential range just described.

---

[6] Early voltammetry was performed with a two-electrode system rather than the three-electrode system shown in Figure 25-2. With a two-electrode system, the second electrode is either a large metal electrode or a reference electrode large enough to prevent its polarization during an experiment. This second electrode combines the functions of the reference electrode and the counter electrode in Figure 25-2. In the two-electrode system, we assume that the potential of this second electrode is constant throughout a scan so that the working electrode potential is simply the difference between the applied potential and the potential of the second electrode. With solutions of high electrical resistance, however, this assumption is not valid because the *IR* drop is significant and increases as the current increases. Distorted voltammograms are the result. Almost all voltammetry is now performed with three-electrode systems.

[7] Many of the working electrodes that we describe in this chapter have dimensions in the millimeter range. There is now intense interest in studies with electrodes having dimensions in the micrometer range and smaller. We will term such electrodes *microelectrodes*. Such electrodes have several advantages over classical working electrodes. We describe some of the unique characteristics of microelectrodes in Section 25I.

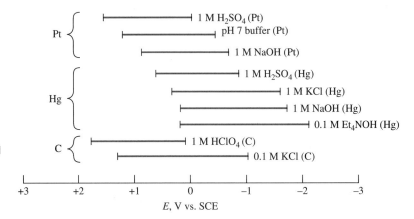

FIGURE 25-4 Potential ranges for three types of electrodes in various supporting electrolytes. (Adapted from A. J. Bard and L. R. Faulkner, *Electrochemical Methods*, 2nd ed., back cover, New York: Wiley, 2001. Reprinted by permission of John Wiley & Sons, Inc.)

Furthermore, a fresh metallic surface is readily formed by simply producing a new drop. The ability to obtain a fresh surface readily is important because the currents measured in voltammetry are quite sensitive to cleanliness and freedom from irregularities. An additional advantage of mercury electrodes is that many metal ions are reversibly reduced to amalgams at the surface of a mercury electrode, which simplifies the chemistry. Mercury electrodes take several forms. The simplest of these is a mercury film electrode formed by electrodeposition of the metal onto a disk electrode, such as that shown in Figure 25-3a. Figure 25-3b illustrates a hanging mercury drop electrode (HMDE). The electrode, which is available from commercial sources, consists of a very fine capillary tube connected to a mercury-containing reservoir. The metal is forced out of the capillary by a piston arrangement driven by a micrometer screw. The micrometer permits formation of drops having surface areas that are reproducible to 5% or better.

Figure 25-3c shows a typical commercial microelectrode. Such electrodes consist of small-diameter metal wires or fibers (5–100 μm) sealed within tempered glass bodies. The flattened end of the microelectrode is polished to a mirror finish, which can be maintained using alumina or diamond polish or both. The electrical connection is a 0.060 in. gold-plated pin. Microelectrodes are available in a variety of materials, including carbon fiber, platinum, gold, and silver. Other materials can be incorporated into microelectrodes if they are available as a wire or a fiber and form a good seal with epoxy. The electrode shown is approximately 7.5 cm long and has a 4-mm outside diameter.

Figure 25-3d shows a commercially available sandwich-type working electrode for voltammetry (or amperometry) in flowing streams. The block is made of polyetheretherketone (PEEK) and is available in several formats with different size electrodes (3 mm and 6 mm; see the blue area in the figure) and various arrays (dual 3 mm and quad 2 mm). See Figure 25-17 for a diagram showing how the electrodes are used in flowing streams. The working electrodes may be made of glassy carbon, carbon paste, gold, copper, nickel, platinum, or other suitable custom materials.

25B-2 Modified Electrodes

An active area of research in electrochemistry is the development of electrodes produced by chemical modification of various conductive substrates.[8] Such electrodes have been tailored to accomplish a broad range of functions. Modifications include applying irreversibly adsorbing substances with desired functionalities, covalent bonding of components to the surface, and coating the electrode with polymer films or films of other substances. The covalent attachment process is shown in Figure 25-5 for a metallic electrode and a carbon electrode. First, the surface of the electrode is oxidized to create functional groups on the surface as shown in Figure 25-5a and b. Then, linking agents such as organosilanes (Figure 25-5c) or amines (Figure 25-5d) are attached to the surface prior to attaching the target group. Polymer films can be prepared from dissolved polymers by dip coating, spin coating, electrodeposition, or covalent attachment. They can also be produced from the monomer by thermal, plasma, pho-

---

[8]For more information, see R. W. Murray, "Molecular Design of Electrode Surfaces," *Techniques in Chemistry*, Vol. 22, W. Weissberger, founding ed., New York: Wiley, 1992; A. J. Bard, *Integrated Chemical Systems*, New York: Wiley, 1994.

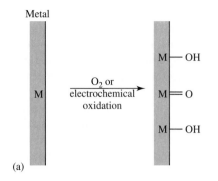

(a)

(b)

(c) Pt/PtO—OSi(CH$_2$)$_3$NH(CH$_2$)$_2$NHC(=O)—CH$_2$—C$_6$H$_4$—FeCp$_2$

(d) graphite—C(=O)—NHCH$_2$—⟨C$_6$H$_4$⟩—NRu(NH$_3$)$_5$$^{2+}$

FIGURE 25-5 Functional groups formed on (a) a metal or (b) a carbon surface by oxidation. (With permission from A. J. Bard, *Integrated Chemical Systems*, New York: Wiley, 1994.) (c) A linking agent such as the organosilane shown is often bonded to the functionalized surface. Reactive components, such as ferrocenes, viologens, and metal bipyridine complexes, are then attached to form the modified surfaces. A Pt electrode is shown with a ferrocene attached. (With permission from J. R. Lenhard and R. W. Murray, *J. Am. Chem. Soc.*, **1978**, *100*, 7870.) In (d), a graphite electrode is shown with attached py-Ru(NH$_3$)$_5$. (With permission from C. A. Koval and F. C. Anson, *Anal. Chem.*, **1978**, *50*, 223.)

tochemical, or electrochemical polymerization methods. Immobilized enzyme biosensors, such as the amperometric sensors described in Section 25C-4, are a type of modified electrode. These can be prepared by covalent attachment, adsorption, or gel entrapment. Another mode of attachment for electrode modification is by *self-assembled monolayers*, or SAMs.[9] In the most common procedure, a long-chain hydrocarbon with a thiol group at one end and an amine or carboxyl group at the other is applied to a pristine gold or mercury film electrode. The hydrocarbon molecules assemble themselves into a highly ordered array with the thiol group attached to the metal surface and the chosen functional group exposed. The arrays may then be further functionalized by covalent attachment or adsorption of the desired molecular species.

Modified electrodes have many applications. A primary interest has been in the area of electrocatalysis. In this application, electrodes capable of reducing oxygen to water have been sought for use in fuel cells and batteries. Another application is in the production of electrochromic devices that change color on oxidation and reduction. Such devices are used in displays or *smart windows* and *mirrors*. Electrochemical devices that could serve as molecular electronic devices, such as diodes and transistors, are also under intense study. Finally, the most important analytical use for such electrodes is as analytical sensors selective for a particular species or functional group (see Figure 1-7).

25B-3 Voltammograms

Figure 25-6 illustrates the appearance of a typical linear-scan voltammogram for an electrolysis involving the reduction of an analyte species A to give a product P at a mercury film electrode. Here, the working electrode is assumed to be connected to the negative terminal of the linear-scan generator so that the applied potentials are given a negative sign as shown. By convention, cathodic currents are taken to be positive, whereas anodic currents are given a negative sign.[10] In this hypothetical experiment, the solution is assumed to be about 10^{-4} M in A, zero M in P, and 0.1 M in

[9]H. O. Finklea, in *Electroanalytical Chemistry: A Series of Advances*, A. J. Bard and I. Rubinstein, eds., Vol. 19, pp. 109–335, New York: Dekker, 1996.

[10]This convention, which originated from early polarographic studies, is not universally accepted. Many workers prefer to assign a positive sign to anodic currents. When reading the literature, it is important to establish which convention is being used. See A. J. Bard and L. R. Faulkner, *Electrochemical Methods*, 2nd ed., p. 6, New York: Wiley, 2001.

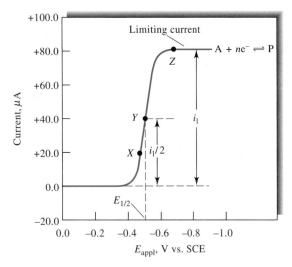

FIGURE 25-6 Linear-sweep voltammogram for the reduction of a hypothetical species A to give a product P. The limiting current i_l is proportional to the analyte concentration and is used for quantitative analysis. The half-wave potential $E_{1/2}$ is related to the standard potential for the half-reaction and is often used for qualitative identification of species. The half-wave potential is the applied potential at which the current i is $i_l/2$.

KCl, which serves as the supporting electrolyte. The half-reaction at the working electrode is the reversible reaction

$$A + ne^- \rightleftharpoons P \qquad E^0 = -0.26 \text{ V} \qquad (25\text{-}2)$$

For convenience, we have neglected the charges on A and P and also have assumed that the standard potential for the half-reaction is -0.26 V.

Linear-scan voltammograms generally have a sigmoid shape and are called *voltammetric waves*. The constant current beyond the steep rise is called the diffusion-limited current, or simply the *limiting current* i_l because the rate at which the reactant can be brought to the surface of the electrode by mass-transport processes limits the current. Limiting currents are usually directly proportional to reactant concentration. Thus, we may write

$$i_l = kc_A$$

where c_A is the analyte concentration and k is a constant. Quantitative linear-scan voltammetry relies on this relationship.

The potential at which the current is equal to one half the limiting current is called the *half-wave potential* and given the symbol $E_{1/2}$. After correction for the

reference electrode potential (0.242 V with a saturated calomel electrode), the half-wave potential is closely related to the standard potential for the half-reaction but is usually not identical to it. Half-wave potentials are sometimes useful for identification of the components of a solution.

Reproducible limiting currents can be achieved rapidly when either the analyte solution or the working electrode is in continuous and reproducible motion. Linear-scan voltammetry in which the solution or the electrode is in constant motion is called *hydrodynamic voltammetry*. In this chapter, we will focus much of our attention on hydrodynamic voltammetry.

25B-4 Circuit Model of a Working Electrode

It is often useful and instructive to represent the electrochemical cell as an electrical circuit that responds to excitation in the same way as the cell. In this discussion, we focus only on the working electrode and assume that the counter electrode is a large inert electrode, that it is nonpolarizable, and that it serves only to make contact with the analyte solution. Figure 25-7 shows a schematic of three of a number of possible circuit models for the electrochemical cell. In Figure 25-7a, we present the *Randles circuit*,[11] which consists of the solution resistance R_Ω, the double-layer capacitance C_d, and the *faradaic impedance* Z_f. The physical diagram of the electrode above the circuit shows the correspondence between the circuit elements and the characteristics of the electrode. Although R_Ω and C_d represent the behavior of real electrodes quite accurately over a broad frequency range, and their values are independent of frequency, the faradaic impedance does not. This is because, in general, Z_f must model any electron- and mass-transfer processes that occur in the cell, and these processes are frequency dependent. The simplest representation of the faradaic impedance contains a series resistance R_s and the *pseudocapacitance* C_s (Figure 25-7b), which is so called because of its frequency dependence.[12]

In the past, clever methods were devised to determine values for R_Ω and C_d. For example, suppose that we apply a small-amplitude sinusoidal excitation signal to a cell containing the working electrode represented by Figure 25-7b. We further assume that the value of the applied voltage is insufficient to initiate faradaic

[11] J. E. B. Randles, *Disc. Faraday Soc.*, **1947**, *1*, 11.
[12] A. J. Bard and L. R. Faulkner, *Electrochemical Methods*, 2nd ed., p. 376, New York: Wiley, 2001.

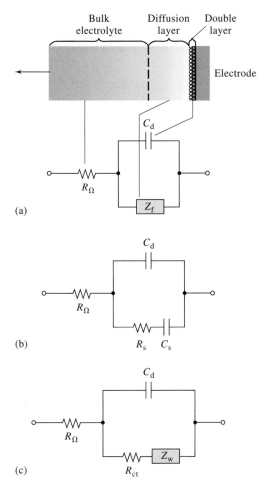

FIGURE 25-7 Circuit model of an electrochemical cell. R_Ω is the cell resistance, C_d is the double-layer capacitance, and Z_f is the faradaic impedance, which may be represented by either of the equivalent circuits shown. R_s is the cell resistance, C_s is the so-called pseudocapacitance, and Z_w is the Warburg impedance. (Adapted from A. J. Bard and L. R. Faulkner, *Electrochemical Methods*, 2nd ed., p. 376, New York: Wiley, 2001. Reprinted by permission of John Wiley & Sons, Inc.)

processes. At relatively high frequencies, the capacitive reactance $X_C = 1/(2\pi f C_d)$ will be quite small (see Section 2B-4), and C_d acts essentially as a short circuit. If we then measure the peak current I_p in the circuit, we find $R_\Omega = V_p/I_p$. This is actually a widely used method for determining solution conductance $G = 1/R_\Omega = I_p/V_p$. In a similar way, each of the circuit elements of the model can be isolated and their values determined by artful application of *ac* circuit measurement techniques.

Figure 25-7c shows a third model for the working electrode in which the faradaic impedance is represented as the series combination of the charge-transfer resistance R_{ct} and the *Warburg impedance* Z_w. The charge-transfer resistance is given by $R_{ct} = -\eta/i$, where η is the overpotential for the faradaic process occurring at the working electrode (see Section 22E-2), and i is the current.[13] The Warburg impedance is a frequency-dependent circuit analog of the resistance of the working electrode to mass transport of analyte molecules across the electrode-solution interface. When experimental conditions are such that the Warburg impedance can be neglected, the charge-transfer resistance can be easily measured, for example, at excitation frequencies near zero.

With modern computerized frequency-analysis instrumentation and software, it is possible to acquire impedance data on cells and extract the values for all components of the circuit models of Figure 25-7. This type of analysis, which is called *electrochemical impedance spectroscopy*, reveals the nature of the faradaic processes and often aids in the investigation of the mechanisms of electron-transfer reactions.[14] In the section that follows, we explore the processes at the electrode-solution interface that give rise to the faradaic impedance.

25C HYDRODYNAMIC VOLTAMMETRY

Hydrodynamic voltammetry is performed in several ways. In one method the solution is stirred vigorously while it is in contact with a fixed working electrode. A typical cell for hydrodynamic voltammetry is pictured in Figure 25-8. In this cell, stirring is accomplished with an ordinary magnetic stirrer. Another approach is to rotate the working electrode at a constant high speed in the solution to provide the stirring action (Figure 25-21a). Still another way of doing hydrodynamic voltammetry is to pass an analyte solution through a tube fitted with a working electrode (Figure 25-17). The last technique is widely used for detecting oxidizable or reducible analytes as they exit from a liquid chromatographic column (Section 28C-6) or a flow-injection manifold.

As described in Section 22E-3, during an electrolysis, reactant is carried to the surface of an electrode by

---

[13] Ibid., p. 102.
[14] Ibid., pp. 383–88; M. E. Orazem and B. Tribollet, *Electrochemical Impedance Spectroscopy*, New York: Wiley, 2006.

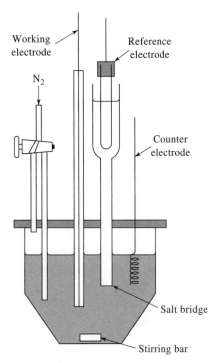

FIGURE 25-8 A three-electrode cell for hydrodynamic voltammetry.

three mechanisms: *migration* under the influence of an electric field, *convection* resulting from stirring or vibration, and *diffusion* due to concentration differences between the film of liquid at the electrode surface and the bulk of the solution. In voltammetry, we attempt to minimize the effect of migration by introducing an excess of an inactive supporting electrolyte. When the concentration of supporting electrolyte exceeds that of the analyte by 50- to 100-fold, the fraction of the total current carried by the analyte approaches zero. As a result, the rate of migration of the analyte toward the electrode of opposite charge becomes essentially independent of applied potential.

25C-1 Concentration Profiles at Electrode Surfaces

Throughout this discussion we will consider that the electrode reaction shown in Equation 25-2 takes place at an electrode in a solution of A that also contains an excess of a supporting electrolyte. We will assume that the initial concentration of A is c_A and that of the product P is zero. We also assume that the reduction

 Simulation: Learn more about **diffusion at electrodes**.

reaction is rapid and reversible so that the concentrations of A and P in the film of solution immediately adjacent to the electrode is given at any instant by the Nernst equation:

$$E_{appl} = E_A^0 - \frac{0.0592}{n} \log \frac{c_P^0}{c_A^0} - E_{ref} \quad (25\text{-}3)$$

where E_{appl} is the potential between the working electrode and the reference electrode and c_P^0 and c_A^0 are the molar concentrations of P and A in a thin layer of solution at the electrode surface only. We also assume that because the electrode is quite small, the electrolysis, over short periods, does not alter the bulk concentration of the solution appreciably. As a result, the concentration of A in the bulk of the solution c_A is unchanged by the electrolysis and the concentration of P in the bulk of the solution c_P continues to be, for all practical purposes, zero ($c_P \approx 0$).

Profiles for Planar Electrodes in Unstirred Solutions

Before describing the behavior of an electrode in this solution under hydrodynamic conditions, it is instructive to consider what occurs when a potential is applied to a planar electrode, such as that shown in Figure 25-3a, in the absence of convection — that is, in an unstirred solution. Under these conditions mass transport of the analyte to the electrode surface occurs by diffusion alone.

Let us assume that a square-wave excitation potential E_{appl} is applied to the working electrode for a period of time t as shown in Figure 25-9a. Let us further assume that E_{appl} is large enough that the ratio c_P^0/c_A^0 in Equation 25-3 is 1000 or greater. Under this condition, the concentration of A at the electrode surface is, for

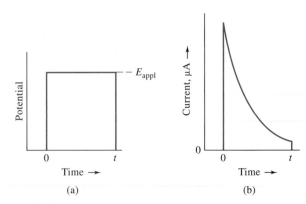

FIGURE 25-9 Current response to a stepped potential for a planar electrode in an unstirred solution. (a) Excitation potential. (b) Current response.

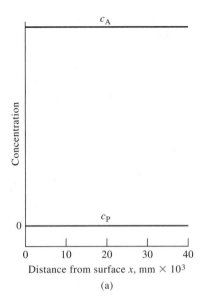
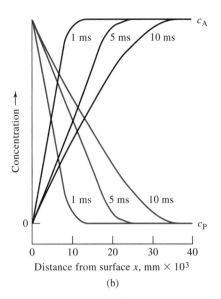

FIGURE 25-10 Concentration distance profiles during the diffusion-controlled reduction of A to give P at a planar electrode. (a) $E_{appl} = 0$ V. (b) $E_{appl} =$ point Z in Figure 25-6; elapsed time: 1, 5, and 10 ms.

all practical purposes, immediately reduced to zero $(c_A^0 \rightarrow 0)$. The current response to this step-excitation signal is shown in Figure 25-9b. Initially, the current rises to a peak value that is required to convert essentially all of A in the surface layer of solution to P. Diffusion from the bulk of the solution then brings more A into this surface layer where further reduction occurs. The current required to keep the concentration of A at the level required by Equation 25-3 decreases rapidly with time, however, because A must travel greater and greater distances to reach the surface layer where it can be reduced. Thus, as seen in Figure 25-9b, the current drops off rapidly after its initial surge.

Figure 25-10 shows concentration profiles for A and P after 0, 1, 5, and 10 ms of electrolysis in the system under discussion. In this example, the concentration of A (solid black lines) and P (solid blue lines) are plotted as a function of distance from the electrode surface. Figure 25-10a shows that the solution is homogeneous before application of the stepped potential with the concentration of A being c_A at the electrode surface and in the bulk of the solution as well; the concentration of P is zero in both of these regions. One millisecond after application of the potential (Figure 25-10b), the profiles have changed dramatically. At the surface of the electrode, the concentration of A

 Simulation: Learn more about **concentration profiles at electrodes**.

has been reduced to essentially zero and the concentration of P has increased and become equal to the original concentration of A; that is, $c_P^0 = c_A$. Moving away from the surface, the concentration of A increases linearly with distance and approaches c_A at about 0.01 mm from the surface. A linear decrease in the concentration of P occurs in this same region. As shown in the figure, with time, these concentration gradients extend farther and farther into the solution. The current i required to produce these gradients is proportional to the slopes of the straight-line portions of the solid lines in Figure 25-10b. That is,

$$i = nFAD_A\left(\frac{\partial c_A}{\partial x}\right) \qquad (25\text{-}4)$$

where i is the current in amperes, n is the number of moles of electrons per mole of analyte, F is the faraday, A is the electrode surface area (cm^2), D_A is the diffusion coefficient for A (cm^2/s), and c_A is the concentration of A (mol/cm^3). As shown in Figure 25-10b, these slopes $(\partial c_A / \partial x)$ become smaller with time, as does the current. The product $D_A(\partial c_A / \partial x)$ is called the *flux*, which is the number of moles of A per unit time per unit area diffusing to the electrode.

It is not practical to obtain limiting currents with planar electrodes in unstirred solutions because the currents continually decrease with time as the slopes of the concentration profiles become smaller.

FIGURE 25-11 Visualization of flow patterns in a flowing stream. Turbulent flow, shown on the right, becomes laminar flow as the average velocity decreases to the left. In turbulent flow, the molecules move in an irregular, zigzag fashion and there are swirls and eddies in the movement. In laminar flow, the streamlines become steady as layers of liquid slide by each other in a regular manner. (From *An Album of Fluid Motion*, assembled by Milton Van Dyke, no. 152, photograph by Thomas Corke and Hassan Nagib, Stanford, CA: Parabolic Press, 1982.)

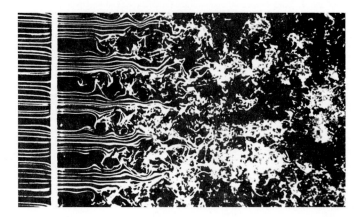

Profiles for Electrodes in Stirred Solutions

Let us now consider concentration-distance profiles when the reduction described in the previous section is performed at an electrode immersed in a solution that is stirred vigorously. To understand the effect of stirring, we must develop a picture of liquid flow patterns in a stirred solution containing a small planar electrode. We can identify two types of flow depending on the average flow velocity, as shown in Figure 25-11. *Laminar flow* occurs at low flow velocities and has smooth and regular motion as depicted on the left in the figure. *Turbulent flow*, on the other hand, happens at high velocities and has irregular, fluctuating motion as shown on the right. In a stirred electrochemical cell, we have a region of turbulent flow in the bulk of solution far from the electrode and a region of laminar flow

close to the electrode. These regions are illustrated in Figure 25-12. In the laminar-flow region, the layers of liquid slide by one another in a direction parallel to the electrode surface. Very near the electrode, at a distance δ centimeters from the surface, frictional forces give rise to a region where the flow velocity is essentially zero. The thin layer of solution in this region is a stagnant layer called the *Nernst diffusion layer*. It is only within the stagnant Nernst diffusion layer that the concentrations of reactant and product vary as a function of distance from the electrode surface and that there are concentration gradients. That is, throughout the laminar-flow and turbulent-flow regions, convection maintains the concentration of A at its original value and the concentration of P at a very low level.

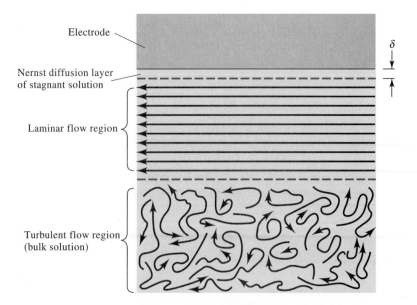

Electrode

Nernst diffusion layer of stagnant solution

Laminar flow region

δ

FIGURE 25-12 Flow patterns and regions of interest near the working electrode in hydrodynamic voltammetry.

Turbulent flow region (bulk solution)

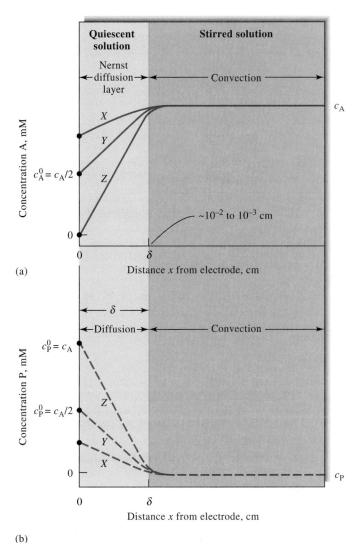

(a)

(b)

FIGURE 25-13 Concentration profiles at an electrode-solution interface during the electrolysis $A + ne^- \rightarrow P$ from a stirred solution of A. See Figure 25-6 for potentials corresponding to curves X, Y, and Z.

Figure 25-13 shows two sets of concentration profiles for A and P at three potentials shown as X, Y, and Z in Figure 25-6. In Figure 25-13a, the solution is divided into two regions. One makes up the bulk of the solution and consists of both the turbulent- and laminar-flow regions shown in Figure 25-12, where mass transport takes place by mechanical convection brought about by the stirrer. The concentration of A throughout this region is c_A, whereas c_P is essentially zero. The second region is the Nernst diffusion layer, which is immediately adjacent to the electrode surface and has a thickness of δ centimeters. Typically, δ ranges from 10^{-2} to 10^{-3} cm, depending on the efficiency of the stirring and the viscosity of the liquid. Within the static diffusion layer, mass transport takes

place by diffusion alone, just as was the case with the unstirred solution. With the stirred solution, however, diffusion is limited to a narrow layer of liquid, which even with time cannot extend indefinitely into the solution. As a result, steady, diffusion-controlled currents appear shortly after applying a voltage.

As is shown in Figure 25-13, at potential X, the equilibrium concentration of A at the electrode surface has been reduced to about 80% of its original value and the equilibrium concentration P has increased by an equivalent amount; that is, $c_P^0 = c_A - c_A^0$. At potential Y, which is the half-wave potential, the equilibrium concentrations of the two species at the surface are approximately the same and equal to $c_A/2$. Finally, at potential Z and beyond, the surface concentration of A approaches zero, and that of P approaches the original concentration of A, c_A. Thus, at potentials more negative than Z, essentially all A ions entering the surface layer are instantaneously reduced to P. As is shown in Figure 25-13b, at potentials greater than Z the concentration of P in the surface layer remains constant at $c_P^0 = c_A$ because of diffusion of P back into the stirred region.

25C-2 Voltammetric Currents

The current at any point in the electrolysis we have just discussed is determined by the rate of transport of A from the outer edge of the diffusion layer to the electrode surface. Because the product of the electrolysis P diffuses from the surface and is ultimately swept away by convection, a continuous current is required to maintain the surface concentrations demanded by the Nernst equation. Convection, however, maintains a constant supply of A at the outer edge of the diffusion layer. Thus, a steady-state current results that is determined by the applied potential. This current is a quantitative measure of how fast A is being brought to the surface of the electrode, and this rate is given by $\partial c_A / \partial x$ where x is the distance in centimeters from the electrode surface. For a planar electrode, the current is given by Equation 25-4.

Note that $\partial c_A / \partial x$ is the slope of the initial part of the concentration profiles shown in Figure 25-13a, and these slopes can be approximated by $(c_A - c_A^0)/\delta$. When this approximation is valid, Equation 25-4 reduces to

$$i = \frac{nFAD_A}{\delta}(c_A - c_A^0) = k_A(c_A - c_A^0) \quad (25\text{-}5)$$

where the constant k_A is equal to $nFAD_A/\delta$.

Equation 25-5 shows that as c_A^0 becomes smaller as a result of a larger negative applied potential the current increases until the surface concentration approaches zero, at which point the current becomes constant and independent of the applied potential. Thus, when $c_A^0 \to 0$, the current becomes the limiting current i_l, and Equation 25-5 reduces to Equation 25-6.[15]

$$i_l = \frac{nFAD_A}{\delta}c_A = k_A c_A \quad (25\text{-}6)$$

---

[15] Careful analysis of the units of the variables in this equation leads to

$$n\left(\frac{\text{mol } e^-}{\text{mol analyte}}\right)F\left(\frac{C}{\text{mol } e^-}\right)A(\text{cm}^2)D_A\left(\frac{\text{cm}^2}{\text{s}}\right)c_A\left(\frac{\text{mol analyte}}{\text{cm}^3}\right)/\delta\,(\text{cm})$$

$$= i_l\left(\frac{C}{s}\right)$$

By definition, one coulomb per second is one ampere.

This derivation is based on an oversimplified picture of the diffusion layer in that the interface between the moving and stationary layers is viewed as a sharply defined edge where transport by convection ceases and transport by diffusion begins. Nevertheless, this simplified model does provide a reasonable approximation of the relationship between current and the variables that affect the current.

Current-Voltage Relationships for Reversible Reactions

To develop an equation for the sigmoid curve shown in Figure 25-6, we substitute Equation 25-6 into Equation 25-5 and rearrange, which gives

$$c_A^0 = \frac{i_l - i}{k_A} \quad (25\text{-}7)$$

The surface concentration of P can also be expressed in terms of the current by using a relationship similar to Equation 25-5. That is,

$$i = -\frac{nFAD_P}{\delta}(c_P - c_P^0) \quad (25\text{-}8)$$

where the minus sign results from the negative slope of the concentration profile for P. Note that D_P is now the diffusion coefficient of P. But we have said earlier that throughout the electrolysis the concentration of P approaches zero in the bulk of the solution and, therefore, when $c_P \approx 0$,

$$i = \frac{-nFAD_P c_P^0}{\delta} = k_P c_P^0 \quad (25\text{-}9)$$

where $k_P = -nFAD_P/\delta$. Rearranging gives

$$c_P^0 = i/k_P \quad (25\text{-}10)$$

Substituting Equations 25-7 and 25-10 into Equation 25-3 yields, after rearrangement,

$$E_{appl} = E_A^0 - \frac{0.0592}{n}\log\frac{k_A}{k_P} - \frac{0.0592}{n}\log\frac{i}{i_l - i} - E_{ref} \quad (25\text{-}11)$$

When $i = i_l/2$, the third term on the right side of this equation becomes equal to zero, and, by definition, E_{appl} is the half-wave potential. That is,

$$E_{appl} = E_{1/2} = E_A^0 - \frac{0.0592}{n}\log\frac{k_A}{k_P} - E_{ref} \quad (25\text{-}12)$$

Substituting this expression into Equation 25-11 gives an expression for the voltammogram in Figure 25-6. That is,

$$E_{appl} = E_{1/2} - \frac{0.0592}{n} \log \frac{i}{i_l - i} \quad (25\text{-}13)$$

Often, the ratio k_A/k_P in Equation 25-11 and in Equation 25-12 is nearly unity, so that we may write for the species A

$$E_{1/2} \approx E_A^0 - E_{ref} \quad (25\text{-}14)$$

Current-Voltage Relationships for Irreversible Reactions

Many voltammetric electrode processes, particularly those associated with organic systems, are irreversible, which leads to drawn-out and less well-defined waves. To describe these waves quantitatively requires an additional term in Equation 25-12 involving the activation energy of the reaction to account for the kinetics of the electrode process. Although half-wave potentials for irreversible reactions ordinarily show some dependence on concentration, diffusion currents remain linearly related to concentration. Some irreversible processes can, therefore, be adapted to quantitative analysis if suitable calibration standards are available.

Voltammograms for Mixtures of Reactants

The reactants of a mixture generally behave independently of one another at a working electrode. Thus, a voltammogram for a mixture is just the sum of the waves for the individual components. Figure 25-14 shows the voltammograms for a pair of two-component mixtures. The half-wave potentials of the two reactants differ by about 0.1 V in curve A and by about 0.2 V in curve B. Note that a single voltammogram may permit the quantitative determination of two or more species provided there is sufficient difference between succeeding half-wave potentials to permit evaluation of individual diffusion currents. Generally, a difference of 0.1 to 0.2 V is required if the more easily reducible species undergoes a two-electron reduction; a minimum of about 0.3 V is needed if the first reduction is a one-electron process.

Anodic and Mixed Anodic-Cathodic Voltammograms

Anodic waves as well as cathodic waves are encountered in voltammetry. An example of an anodic wave is illustrated in curve A of Figure 25-15, where the electrode reaction is the oxidation of iron(II) to iron(III) in the presence of citrate ion. A limiting current is observed at about +0.1 V (versus a saturated calomel electrode [SCE]), which is due to the half-reaction

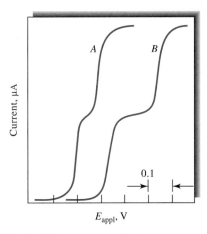

FIGURE 25-14 Voltammograms for two-component mixtures. Half-wave potentials differ by 0.1 V in curve A, and by 0.2 V in curve B.

$$Fe^{2+} \rightleftharpoons Fe^{3+} + e^-$$

As the potential is made more negative, a decrease in the anodic current occurs; at about -0.02 V, the current becomes zero because the oxidation of iron(II) ion has ceased.

Curve C represents the voltammogram for a solution of iron(III) in the same medium. Here, a cathodic wave results from reduction of iron(III) to iron (II). The half-wave potential is identical with that for the anodic wave, indicating that the oxidation and reduc-

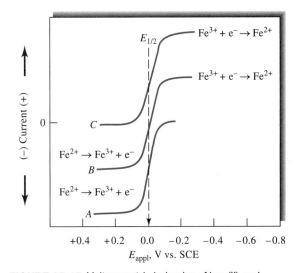

FIGURE 25-15 Voltammetric behavior of iron(II) and iron(III) in a citrate medium. Curve A: anodic wave for a solution in which $c_{Fe^{2+}} = 1 \times 10^{-4}$ M. Curve B: anodic-cathodic wave for a solution in which $c_{Fe^{2+}} = c_{Fe^{3+}} = 0.5 \times 10^{-4}$ M. Curve C: cathodic wave for a solution in which $c_{Fe^{3+}} = 1 \times 10^{-4}$ M.

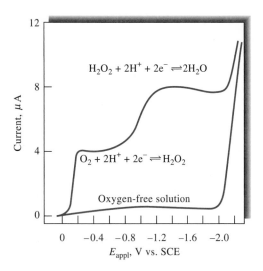

FIGURE 25-16 Voltammogram for the reduction of oxygen in an air-saturated 0.1-M KCl solution. The lower curve is for a 0.1-M KCl solution in which the oxygen is removed by bubbling nitrogen through the solution.

tion of the two iron species are perfectly reversible at the working electrode.

Curve B is the voltammogram of an equimolar mixture of iron(II) and iron(III). The portion of the curve below the zero-current line corresponds to the oxidation of the iron(II); this reaction ceases at an applied potential equal to the half-wave potential. The upper portion of the curve is due to the reduction of iron(III).

25C-3 Oxygen Waves

Dissolved oxygen is easily reduced at many working electrodes. Thus, as shown in Figure 25-16, an aqueous solution saturated with air exhibits two distinct oxygen waves. The first results from the reduction of oxygen to hydrogen peroxide:

$$O_2(g) + 2H^+ + 2e^- \rightleftharpoons H_2O_2$$

The second wave corresponds to the further reduction of the hydrogen peroxide:

$$H_2O_2 + 2H^+ + 2e^- \rightleftharpoons 2H_2O$$

Because both reactions are two-electron reductions, the two waves are of equal height.

Voltammetric measurements offer a convenient and widely used method for determining dissolved oxygen in solutions. However, the presence of oxygen often interferes with the accurate determination of other species. Thus, oxygen removal is usually the first step in amperometric procedures. Oxygen can be re-

moved by passing an inert gas through the analyte solution for several minutes (*sparging*). A stream of the same gas, usually nitrogen, is passed over the surface of the solution during analysis to prevent reabsorption of oxygen. The lower curve in Figure 25-16 is a voltammogram of an oxygen-free solution.

25C-4 Applications of Hydrodynamic Voltammetry

The most important uses of hydrodynamic voltammetry include (1) detection and determination of chemical species as they exit from chromatographic columns or flow-injection apparatus; (2) routine determination of oxygen and certain species of biochemical interest, such as glucose, lactose, and sucrose; (3) detection of end points in coulometric and volumetric titrations; and (4) fundamental studies of electrochemical processes.

Voltammetric Detectors in Chromatography and Flow-Injection Analysis

Hydrodynamic voltammetry is widely used for detection and determination of oxidizable or reducible compounds or ions that have been separated by liquid

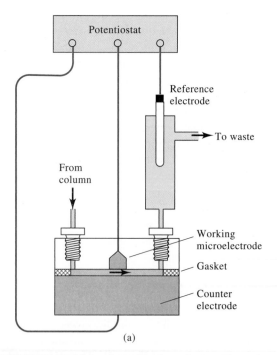

(a)

FIGURE 25-17 (a) A schematic of a voltammetric system for detecting electroactive species as they elute from a column. The cell volume is determined by the thickness of the gasket.

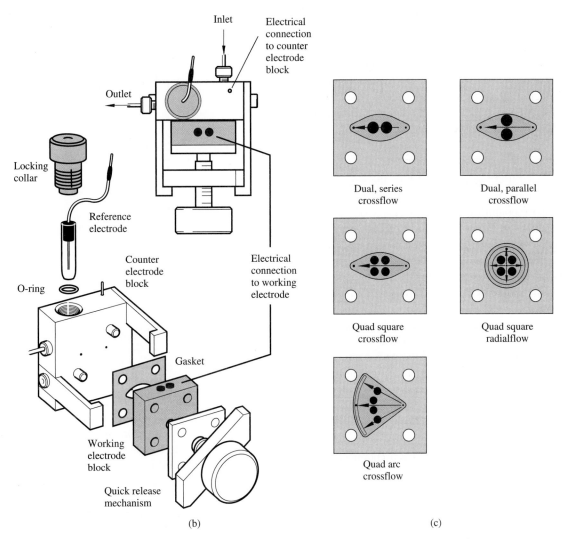

FIGURE 25-17 (*continued*) (b) Detail of a commercial flow cell assembly. (c) Configurations of working electrode blocks. Arrows show the direction of flow in the cell. ([b] and [c] courtesy of Bioanalytical Systems, Inc., West Lafayette, IN.)

chromatography or that are produced by flow-injection methods.[16] A thin-layer cell such as the one shown schematically in Figure 25-17a is used in these applications. The working electrode in these cells is usually embedded in the wall of an insulating block separated

from a counter electrode by a thin spacer as shown. The volume of such a cell is typically 0.1 to 1 μL. A voltage corresponding to the limiting-current region for analytes is applied between the working electrode and a silver–silver chloride reference electrode that is located downstream from the detector. We present an exploded view of a commercial flow cell in Figure 25-17b, which shows clearly how the sandwiched cell is assembled and held in place by the quick-release mechanism. A locking collar in the counter electrode block, which is electrically connected to the potentiostat, retains the reference electrode. Five different configurations of working electrode are shown in

---

[16]Voltammetric detectors are a particular type of transducer called *limiting-current transducers*. In this discussion and subsequent discussions involving voltammetric transducers, we use the more common term *voltammetric detector*. When a voltammetric transducer is inherently selective for a particular species by virtue of control of various experimental variables or when it is covered with a chemically selective layer of polymer or other membranous material, we refer to it as a *voltammetric sensor*. For a discussion of transducers, detectors, sensors, and their definitions, see Section 1C-4.

Figure 25-17c. These configurations permit optimization of detector sensitivity under a variety of experimental conditions. Working electrode blocks and electrode materials are described in Section 25B-1. This type of application of voltammetry (or amperometry) has detection limits as low as 10^{-9} to 10^{-10} M. We discuss voltammetric detection for liquid chromatography in more detail in Section 28C-6.

Voltammetric and Amperometric Sensors

In Section 23F-2, we described how the specificity of potentiometric sensors could be enhanced by applying molecular recognition layers to the electrode surfaces. There has been much research in recent years in applying the same concepts to voltammetric electrodes.[17] A number of voltammetric systems are available commercially for the determination of specific species in industrial, biomedical, environmental, and research applications. These devices are sometimes called electrodes or detectors but are, in fact, complete voltammetric cells and are better referred to as sensors. In the sections that follow, we describe two commercially available sensors and one that is under development in this rapidly expanding field.

Oxygen Sensors. The determination of dissolved oxygen in a variety of aqueous environments, such as seawater, blood, sewage, effluents from chemical plants, and soils, is of tremendous importance to industry, biomedical and environmental research, and clinical medicine. One of the most common and convenient methods for making such measurements is with the Clark oxygen sensor, which was patented by L. C. Clark Jr. in 1956.[18] A schematic of the Clark oxygen sensor is shown in Figure 25-18. The cell consists of a cathodic platinum-disk working electrode embedded in a centrally located cylindrical insulator. Surrounding the lower end of this insulator is a ring-shape silver anode. The tubular insulator and electrodes are mounted inside a second cylinder that contains a buffered solution of potassium chloride. A thin (~20 μm), replaceable, oxygen-permeable membrane of Teflon or polyethylene is held in place at the bottom end of the tube by an O-ring. The thickness of the electrolyte solution between the cathode and the membrane is approximately 10 μm.

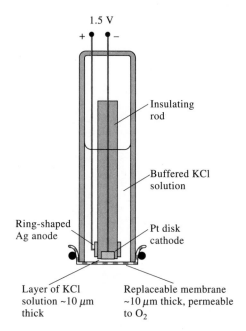

FIGURE 25-18 The Clark voltammetric oxygen sensor. Cathodic reaction: $O_2 + 4H^+ + 4e^- \rightleftharpoons 2H_2O$. Anodic reaction: $Ag + Cl^- \rightleftharpoons AgCl(s) + e^-$.

When the oxygen sensor is immersed in a flowing or stirred solution of the analyte, oxygen diffuses through the membrane into the thin layer of electrolyte immediately adjacent to the disk cathode, where it diffuses to the electrode and is immediately reduced to water. In contrast with a normal hydrodynamic electrode, two diffusion processes are involved — one through the membrane and the other through the solution between the membrane and the electrode surface. For a steady-state condition to be reached in a reasonable period (10 to 20 s), the thickness of the membrane and the electrolyte film must be 20 μm or less. Under these conditions, it is the rate of equilibration of the transfer of oxygen across the membrane that determines the steady-state current that is reached.

Enzyme-Based Sensors. A number of enzyme-based voltammetric sensors are available commercially. An example is a glucose sensor that is widely used in clinical laboratories for the routine determination of glucose in blood serum. This device is similar in construction to the oxygen sensor shown in Figure 25-18. The membrane in this case is more complex and consists of three layers. The outer layer is a polycarbonate film that is permeable to glucose but impermeable to proteins and other constituents of blood. The middle layer

[17]For a recent review of electrochemical sensors, see E. Bakker and Yu Qin, *Anal. Chem.*, **2006**, *78*, 3965.
[18]For a detailed discussion of the Clark oxygen sensor, see M. L. Hitchman, *Measurement of Dissolved Oxygen*, Chaps. 3–5, New York: Wiley, 1978.

is an immobilized enzyme (see Section 23F-2), glucose oxidase in this example. The inner layer is a cellulose acetate membrane, which is permeable to small molecules, such as hydrogen peroxide. When this device is immersed in a glucose-containing solution, glucose diffuses through the outer membrane into the immobilized enzyme, where the following catalytic reaction occurs:

$$\text{glucose} + O_2 \xrightarrow{\text{glucose oxidase}} H_2O_2 + \text{gluconic acid}$$

The hydrogen peroxide then diffuses through the inner layer of membrane and to the electrode surface, where it is oxidized to give oxygen. That is,

$$H_2O_2 + OH^- \longrightarrow O_2 + H_2O + 2e^-$$

The resulting current is directly proportional to the glucose concentration of the analyte solution. A variation on this type of sensor is often found in home glucose monitors widely used by diabetic patients. This device is one of the largest-selling chemical instruments in the world.[19]

Several other sensors are available that are based on the voltammetric measurement of hydrogen peroxide produced by enzymatic oxidations of other species of clinical interest. These analytes include sucrose, lactose, ethanol, and L-lactate. A different enzyme is, of course, required for each species. In some cases, enzyme electrodes can be based on measuring oxygen or on measuring pH as discussed in Section 23F-2.

Immunosensors. Sensor specificity is often achieved by using molecular recognition elements that react exclusively with the analyte. Antibodies are proteins that have exceptional specificity toward analytes and are the recognition elements most commonly used in *immunosensors*.[20] Immunosensors are typically fabricated by immobilizing antibodies on the sensor surface either through adsorption, covalent attachment, polymer entrapment, or other methods.

A variety of assay formats are used with immunosensors. One of the most common methods uses two antibodies, one that is immobilized on the sensor surface and is used to capture the target analyte, and one that is labeled and is used to detect the captured analyte (a *sandwich assay*). Traditionally, radionuclides were

used as labels in immunoassays, but these have largely been replaced by more convenient labels such as fluorescent molecules and enzymes. Thus, immunosensors can use a variety of detection methods, such as optical and electrochemical techniques. More recently, label-free immunosensors have been developed based on piezoelectric (Section 1C-4), surface plasmon resonance (Section 21E-1), and other detection strategies.

Figure 25-19 shows a scheme and an amperometric biosensor for determining specific proteins using a sandwich assay.[21] In this scheme (Figure 25-19a), an antibody appropriate for the desired analyte is immobilized on the surface of an electrode (A). In this example, the antibody is immobilized by physical adsorption.[22] When the electrode is in contact with a solution containing the analyte, which is represented as blue triangles in the figure, it binds preferentially with the antibody (B). The electrode is then rinsed and brought into contact with a second antibody that has been tagged, or labeled, which is indicated by the stars in the figure (C). In this example, the antibody is tagged with the enzyme alkaline phosphatase, which catalyzes the conversion of hydroquinone diphosphate to hydroquinone. When a voltage of 320 mV versus Ag-AgCl is applied to the working electrode, hydroquinone undergoes a two-electron oxidation to quinone (D). The resulting current is directly proportional to the original concentration of the analyte.

Figure 25-19b is a photo of the biosensor array used to carry out the immunoassay described in the previous paragraph. The biosensor consists of an array of electrodes, with each electrode being an independent immunosensor. This arrangement permits multiple proteins to be determined simultaneously. The array was fabricated on glass using photolithographic techniques that are common in the semiconductor industry. In this procedure, the glass substrate was patterned with a polymer (photoresist) that is easily removed by solvent. The pattern of the polymer was the negative of the pattern of the electrodes. Iridium (~100 nm) was then deposited over the entire substrate using sputtering. The polymer was then removed to leave the iridium working and counter electrodes on the substrate. The reference electrode was prepared using a similar procedure and depositing silver. The reference

---

[19]See Figure 33-19 for a photo of a home glucose monitor and Section 33D-3 for a discussion of discrete clinical analyzers.

[20]For a review of the principles and applications of immunosensors, see P. B. Luppa, L. J. Sokoll, and D. W. Chan, *Clinica Chimica Acta*, **2001**, *314*, 1.

[21]M. S. Wilson and W. Nie, *Anal. Chem.*, **2006**, *78*, 2507.

[22]For a discussion of the terms and definitions associated with chemically modified electrodes, see R. A. Durst, A. J. Baumner, R. W. Murray, R. P. Buck, and C. P. Andrieux, *Pure and Applied Chemistry*, **1997**, *69*, 1317. http://www.iupac.org/publications/pac/1997/pdf/6906x1317.pdf.

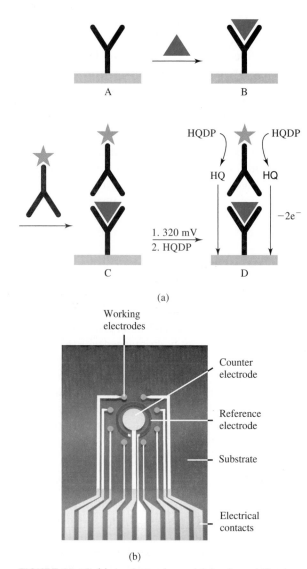

(a)

(b)

FIGURE 25-19 (a) A: electrode containing immobilized antibody (Y); B: binding of target analyte (▼) to electrode-bound antibody; C: binding of alkaline phosphatase–labeled antibody to electrode-bound analyte; D: application of 320 mV to the electrode and addition of hydroquinone diphosphate (HQDP). Electrochemical oxidation of AP-generated hydroquinone (HQ) generates a current at the electrode that is proportional to the amount of analyte bound to the electrode. (b) Photograph of the biosensor showing the arrangement of IrOx 1-mm-diameter working electrodes, 4-mm-diameter counter electrode, 7-mm-outside-diameter Ag-AgCl reference electrode, and electrical contacts on the substrate (28 × 35 × 1 mm). For clarity, the sample well is not shown. (Adapted from M. S. Wilson and W. Nie, *Anal. Chem.*, **2006**, *78*, 2507, with permission of the American Chemical Society.) Image reprinted with permission from *Anal. Chem.* Copyright 2006 American Chemical Society.

electrode was then completed by electrochemical deposition of AgCl from a solution containing KCl and added Ag^+. The array was coated with an insulating polymer film leaving the electrode and contact areas exposed, and finally, a cylindrical polycarbonate well (not shown) was attached over the electrodes to contain the analyte solution.

The 0.785-mm^2 Ir working electrodes were then electrochemically activated to build up a layer of iridium oxide on each. The porous oxide was grown to increase the surface area available for antibody immobilization on the electrode. Drops of solution containing the appropriate antibodies were then applied over the working electrodes, and the array was incubated at 4°C to complete the immobilization.

Analyses were performed on serum samples containing a mixture of goat IgG, mouse IgG, human IgG, and chicken IgY antibodies (0–125 ng/mL). In this assay, the target analytes were also antibodies (IgG and IgY), and so the capture and detection antibodies were antibodies toward antibodies (anti-IgG and anti-IgY). Analyte solutions were added to the sensor well, followed by a mixture of detection antibodies. To perform the electrochemical measurements, the biosensor was connected to a multichannel computer-controlled potentiostat via the edge contacts of the array, and current measurements were made on all eight working electrodes.

The precision of the measurements ranged from 1.9% to 8.2% relative standard deviation, and the detection limit was 3 ng/mL for all analytes. The results compare quite favorably (2.4%–6.5% difference) with results from commercial single-analyte assays using spectroscopic enzyme-linked immunosorbent assays (ELISAs). In this application, four analytes were determined in duplicate, but any combination of up to eight analytes may be determined simultaneously. It is anticipated that when the array biosensor is coupled with a microfluidics system (Section 33C) for solution handling, the combined device will become a powerful tool for the routine determination of this important class of biochemical analytes.

Amperometric Titrations

Hydrodynamic voltammetry can be used to estimate the equivalence point of titrations if at least one of the participants or products of the reaction involved is oxidized or reduced at a working electrode. Here, the current at some fixed potential in the limiting-current region is measured as a function of the reagent volume

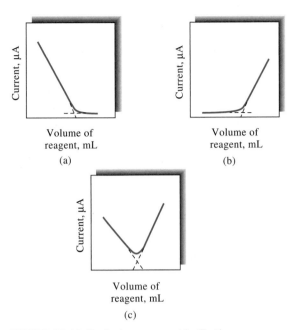

FIGURE 25-20 Typical amperometric titration curves: (a) analyte is reduced, reagent is not; (b) reagent is reduced, analyte is not; (c) both reagent and analyte are reduced.

or of time if the reagent is generated by a constant-current coulometric process. Plots of the data on either side of the equivalence point are straight lines with different slopes; the end point is established by extrapolation to the intersection of the lines.[23]

Amperometric titration curves typically take one of the forms shown in Figure 25-20. Figure 25-20a represents a titration in which the analyte reacts at the electrode but the reagent does not. Figure 25-20b is typical of a titration in which the reagent reacts at the working electrode and the analyte does not. Figure 25-20c corresponds to a titration in which both the analyte and the titrant react at the working electrode.

There are two types of amperometric electrode systems. One uses a single polarizable electrode coupled to a reference electrode; the other uses a pair of identical solid-state electrodes immersed in a stirred solution. For the first, the working electrode is often a rotating platinum electrode constructed by sealing a platinum wire into the side of a glass tube that is connected to a stirring motor.

Amperometric titrations with one indicator electrode have, with one notable exception, been confined

[23]S. R. Crouch and F. J. Holler, *Applications of Microsoft® Excel in Analytical Chemistry*, Belmont, CA: Brooks/Cole, 2004, pp. 214–18.

to titrations in which a precipitate or a stable complex is the product. Precipitating reagents include silver nitrate for halide ions, lead nitrate for sulfate ion, and several organic reagents, such as 8-hydroxyquinoline, dimethylglyoxime, and cupferron, for various metallic ions that are reducible at working electrodes. Several metal ions have also been determined by titration with standard solutions of ethylenediaminetetraacetic acid (EDTA). The exception just noted involves titrations of organic compounds, such as certain phenols, aromatic amines, and olefins; hydrazine; and arsenic(III) and antimony(III) with bromine. The bromine is often generated coulometrically. It has also been formed by adding a standard solution of potassium bromate to an acidic solution of the analyte that also contains an excess of potassium bromide. Bromine is formed in the acidic medium by the reaction

$$BrO_3^- + 5Br^- + 6H^+ \longrightarrow 3Br_2 + 3H_2O$$

This type of titration has been carried out with a rotating platinum electrode or twin platinum electrodes. There is no current prior to the equivalence point; after the equivalence point, there is a rapid increase in current because of the electrochemical reduction of the excess bromine.

There are two advantages in using a pair of identical metallic electrodes to establish the equivalence point in amperometric titrations: simplicity of equipment and not having to purchase or prepare and maintain a reference electrode. This type of system has been incorporated in instruments designed for routine automatic determination of a single species, usually with a coulometrically generated reagent. An instrument of this type is often used for the automatic determination of chloride in samples of serum, sweat, tissue extracts, pesticides, and food products. The reagent in this system is silver ion coulometrically generated from a silver anode. About 0.1 V is applied between a pair of twin silver electrodes that serve as the indicator system. Short of the equivalence point in the titration of chloride ion, there is essentially no current because no electroactive species is present in the solution. Because of this, there is no electron transfer at the cathode, and the electrode is completely polarized. Note that the anode is not polarized because the reaction

$$Ag \rightleftharpoons Ag^+ + e^-$$

occurs in the presence of a suitable cathodic reactant or depolarizer.

Past the equivalence point, the cathode becomes depolarized because silver ions are present; these ions react to give silver. That is,

$$Ag^+ + e^- \rightleftharpoons Ag$$

This half-reaction and the corresponding oxidation of silver at the anode produce a current whose magnitude is, as in other amperometric methods, directly proportional to the concentration of the excess reagent. Thus, the titration curve is similar to that shown in Figure 25-20b. In the automatic titrator just mentioned, an electronic circuit senses the amperometric detection current signal and shuts off the coulometric generator current. The chloride concentration is then computed from the magnitude of the titration current and the generation time. The instrument has a range of 1 to 999.9 mM Cl^- per liter, a precision of 0.1%, and an accuracy of 0.5%. Typical titration times are about 20 s.

The most common end-point detection method for the Karl Fischer titration for determining water (see Section 24D-1) is the amperometric method with dual polarized electrodes. Several manufacturers offer fully automated instruments for use in performing these titrations. A closely related end-point detection method for Karl Fischer titrations measures the potential difference between two identical electrodes through which a small constant current is passed.

Rotating Electrodes

To carry out theoretical studies of oxidation-reduction reactions, it is often of interest to know how k_A in Equation 25-6 is affected by the hydrodynamics of the system. A common method for obtaining a rigorous description of the hydrodynamic flow of stirred solution is based on measurements made with a rotating disk electrode (RDE), such as the one illustrated in Figure 25-21a and b. When the disk electrode is rotated rapidly, the flow pattern shown by the arrows in the figure is set up. At the surface of the disk, the liquid moves out horizontally from the center of the device, which produces an upward axial flow to replenish the displaced liquid. A rigorous treatment of the hydrodynamics is possible in this case[24] and leads to the *Levich equation*[25]

$$i_l = 0.620\,nFAD\omega^{1/2}\nu^{-1/6}c_A \qquad (25\text{-}15)$$

[24] A. J. Bard and L. R. Faulkner, *Electrochemical Methods*, 2nd ed., New York: Wiley, 2001, pp. 335–39.
[25] V. G. Levich, *Acta Physicochimica URSS*, **1942**, *17*, 257.

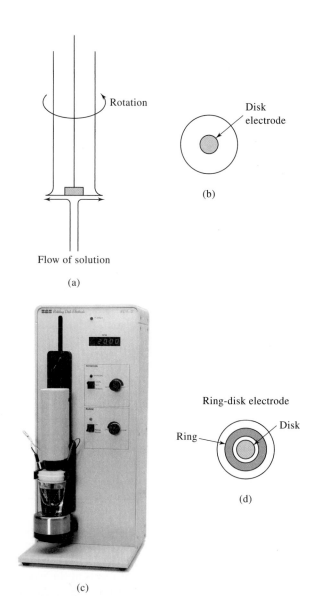

FIGURE 25-21 (a) Side view of an RDE showing solution flow pattern. (b) Bottom view of a disk electrode. (c) Photo of a commercial RDE. (Courtesy of Bioanalytical Systems, Inc., West Lafayette, IN.) (d) Bottom view of a ring-disk electrode.

The terms n, F, A, and D in this equation have the same meaning as in Equation 25-5, ω is the angular velocity of the disk in radians per second, and ν is the *kinematic viscosity* in centimeters squared per second, which is the ratio of the viscosity of the solution to its density. Voltammograms for reversible systems generally have the ideal shape shown in Figure 25-6. Numerous studies of the kinetics and the mechanisms of electrochemical

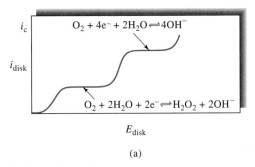

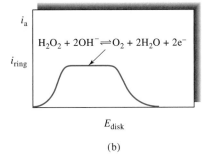

FIGURE 25-22 Disk (a) and ring (b) current for reduction of oxygen at the rotating-ring-disk electrode. (From *Laboratory Techniques in Electroanalytical Chemistry*, 2nd ed., P. T. Kissinger and W. R. Heineman, eds., p. 117, New York: Dekker, 1996. With permission.)

reactions have been performed with RDEs. A common experiment with an RDE is to study the dependence of i_l on $\omega^{1/2}$. A plot of i_l versus $\omega^{1/2}$ is known as a *Levich plot*, and deviations from the linear relationship often indicate kinetic limitations on the electron-transfer process. For example, if i_l becomes independent of ω at large values of $\omega^{1/2}$, the current is not limited by mass transport of the electroactive species to the electrode surface, but instead, the rate of the reaction is the limiting factor. RDEs, such as the versatile commercial model shown in Figure 25-21c, have attracted renewed interest in recent years for both fundamental and quantitative analytical studies as enthusiasm for the dropping mercury electrode (polarography) has faded. RDE detection with a mercury film electrode is sometimes referred to as *pseudopolarography*.

The *rotating-ring-disk electrode* is a modified RDE that is useful for studying electrode reactions; it has little use in analysis. Figure 25-21d shows that a ring-disk electrode contains a second ring-shape electrode that is electrically isolated from the center disk. After an electroactive species is generated at the disk, it is then swept past the ring where it undergoes a second electrochemical reaction. Figure 25-22 shows voltammograms from a typical ring-disk experiment. The curve on the left is the voltammogram for the reduction of oxygen to hydrogen peroxide at the disk electrode. The curve on the right is the *anodic* voltammogram for the oxidation of the hydrogen peroxide as it flows past the ring electrode. Note that when the potential of the disk electrode becomes sufficiently negative so that the reduction product is hydroxide rather than hydrogen peroxide, the current in the ring electrode decreases to zero. Studies of this type provide much useful information about mechanisms and intermediates in electrochemical reactions.

25D CYCLIC VOLTAMMETRY

In *cyclic voltammetry* (CV),[26] the current response of a small stationary electrode in an unstirred solution is excited by a triangular voltage waveform, such as that shown in Figure 25-23. In this example, the potential is first varied linearly from +0.8 V to −0.15 V versus an SCE. When the extreme of −0.15 V is reached, the scan direction is reversed, and the potential is returned to its original value of +0.8 V. The scan rate in either direction is 50 mV/s. This excitation cycle is often repeated several times. The voltage extrema at which

---

[26] For brief reviews, see P. T. Kissinger and W. R. Heineman, *J. Chem. Educ.*, **1983**, *60*, 702; D. H. Evans, K. M. O'Connell, T. A. Petersen, and M. J. Kelly, *J. Chem. Educ.*, **1983**, *60*, 290.

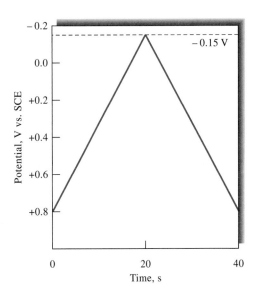

FIGURE 25-23 Cyclic voltammetric excitation signal.

reversal takes place (in this case, -0.15 and $+0.8$ V) are called *switching potentials*. The range of switching potentials chosen for a given experiment is one in which a diffusion-controlled oxidation or reduction of one or more analytes occurs. The direction of the initial scan may be either negative, as shown, or positive, depending on the composition of the sample (a scan in the direction of more negative potentials is termed a *forward scan*, and one in the opposite direction is called a *reverse scan*). Generally, cycle times range from 1 ms or less to 100 s or more. In this example, the cycle time is 40 s.

Figure 25-24b shows the current response when a solution that is 6 mM in $K_3Fe(CN)_6$ and 1 M in KNO_3 is subjected to the cyclic excitation signal shown in Figures 25-23 and 25-24a. The working electrode was a carefully polished stationary platinum electrode and the reference electrode was an SCE. At the initial potential of $+0.8$ V, a tiny anodic current is observed, which immediately decreases to zero as the scan is continued. This initial negative current arises from the oxidation of water to give oxygen (at more positive potentials, this current rapidly increases and becomes quite large at about $+0.9$ V). No current is observed between a potential of $+0.7$ and $+0.4$ V because no reducible or oxidizable species is present in this potential range. When the potential becomes less positive than approximately $+0.4$ V, a cathodic current begins to develop (point B) because of the reduction of the hexacyanoferrate(III) ion to hexacyanoferrate(II) ion. The reaction at the cathode is then

$$Fe(CN)_6^{3-} + e^- \rightleftharpoons Fe(CN)_6^{4-}$$

A rapid increase in the current occurs in the region of B to D as the surface concentration of $Fe(CN)_6^{3-}$ becomes smaller and smaller. The current at the peak is made up of two components. One is the initial current surge required to adjust the surface concentration of the reactant to its equilibrium concentration as given by the Nernst equation. The second is the normal diffusion-controlled current. The first current then decays rapidly (points D to F) as the diffusion layer is extended farther and farther away from the electrode surface (see also Figure 25-10b). At point F (-0.15 V), the scan direction is switched. The current, however, continues to be cathodic even though the scan is toward more positive potentials because the potentials are still negative enough to cause reduction of $Fe(CN)_6^{3-}$. As the potential sweeps in the positive direction, eventu-

 Simulation: Learn more about **cyclic voltammetry**.

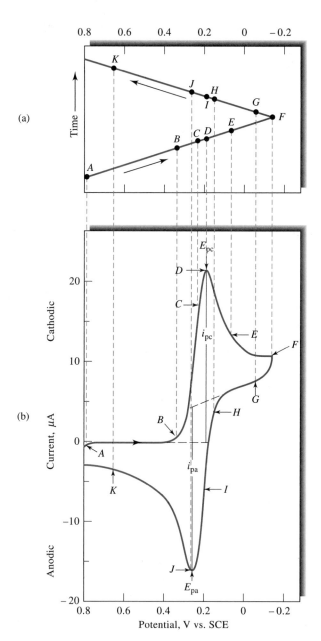

FIGURE 25-24 (a) Potential versus time waveform and (b) cyclic voltammogram for a solution that is 6.0 mM in $K_3Fe(CN)_6$ and 1.0 M in KNO_3. (From P. T. Kissinger and W. H. Heineman, *J. Chem. Educ.*, **1983**, *60*, 702. Copyright 1983; Division of Chemical Education, Inc.)

ally reduction of $Fe(CN)_6^{3-}$ no longer occurs and the current goes to zero and then becomes anodic. The anodic current results from the reoxidation of $Fe(CN)_6^{4-}$ that has accumulated near the surface during the forward scan. This anodic current peaks and then decreases as the accumulated $Fe(CN)_6^{4-}$ is used up by the anodic reaction.

Important variables in a cyclic voltammogram are the cathodic peak potential E_{pc}, the anodic peak potential E_{pa}, the cathodic peak current i_{pc}, and the anodic peak current i_{pa}. The definitions and measurements of these parameters are illustrated in Figure 25-24. For a reversible electrode reaction, anodic and cathodic peak currents are approximately equal in absolute value but opposite in sign. For a reversible electrode reaction at 25°C, the difference in peak potentials, ΔE_p, is expected to be

$$\Delta E_p = |E_{pa} - E_{pc}| = 0.0592/n \quad (25\text{-}16)$$

where n is the number of electrons involved in the half-reaction. Irreversibility because of slow electron-transfer kinetics results in ΔE_p exceeding the expected value. Although an electron-transfer reaction may appear reversible at a slow sweep rate, increasing the sweep rate may lead to increasing values of ΔE_p, a sure sign of irreversibility. Hence, to detect slow electron-transfer kinetics and to obtain rate constants, ΔE_p is measured for different sweep rates.

Quantitative information is obtained from the Randles-Sevcik equation, which at 25°C is

$$i_p = 2.686 \times 10^5 n^{3/2} A c D^{1/2} v^{1/2} \quad (25\text{-}17)$$

where i_p is the peak current (A), A is the electrode area (cm^2), D is the diffusion coefficient (cm^2/s), c is the concentration (mol/cm^3), and v is the scan rate (V/s). CV offers a way of determining diffusion coefficients if the concentration, electrode area, and scan rate are known.

25D-1 Fundamental Studies

The primary use of CV is as a tool for fundamental and diagnostic studies that provides qualitative information about electrochemical processes under various conditions. As an example, consider the cyclic voltammogram for the agricultural insecticide parathion that is shown in Figure 25-25.[27] Here, the switching potentials were about −1.2 V and +0.3 V. The initial forward scan was, however, started at 0.0 V and not +0.3 V. Three peaks are observed. The first cathodic peak (A) results from a four-electron reduction of the parathion to give a hydroxylamine derivative

$$\text{R—C}_6\text{H}_4\text{NO}_2 + 4\text{e}^- + 4\text{H}^+ \rightarrow \text{R—C}_6\text{H}_4\text{NHOH} + \text{H}_2\text{O} \quad (25\text{-}18)$$

[27]This discussion and the voltammogram are from W. R. Heineman and P. T. Kissinger, *Amer. Lab.*, **1982** (11), 29.

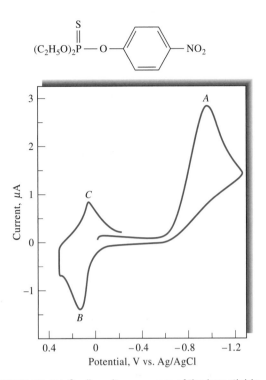

FIGURE 25-25 Cyclic voltammogram of the insecticide parathion in 0.5 M pH 5 sodium acetate buffer in 50% ethanol. Hanging mercury drop electrode. Scan rate: 200 m V/s. (From W. R. Heineman and P. T. Kissinger, *Amer. Lab.*, **1982**, no. 11, 34. Copyright 1982 by International Scientific Communications, Inc.)

The anodic peak at B arises from the oxidation of the hydroxylamine to a nitroso derivative during the reverse scan. The electrode reaction is

$$\text{R—C}_6\text{H}_4\text{NHOH} \rightarrow \text{R—C}_6\text{H}_4\text{NO} + 2\text{H}^+ + 2\text{e}^- \quad (25\text{-}19)$$

The cathodic peak at C results from the reduction of the nitroso compound to the hydroxylamine as shown by the equation

$$\text{R—C}_6\text{H}_4\text{NO} + 2\text{e}^- + 2\text{H}^+ \rightarrow \text{R—C}_6\text{H}_4\text{NHOH} \quad (25\text{-}20)$$

Cyclic voltammograms for authentic samples of the two intermediates confirmed the identities of the compounds responsible for peaks B and C.

CV is widely used in organic and inorganic chemistry. It is often the first technique selected for investigation of a system with electroactive species. For example, CV was used to investigate the behavior of the modified electrodes described in Section 25B-2 and shown in Figure 25-5. The resulting cyclic voltammo-

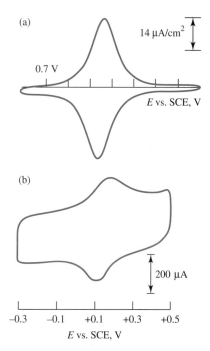

(a)

14 µA/cm^2

0.7 V

E vs. SCE, V

(b)

200 µA

−0.3   −0.1   +0.1   +0.3   +0.5

E vs. SCE, V

FIGURE 25-26 Cyclic voltammograms of the modified electrodes shown in Figure 25-5. (a) Cyclic voltammogram of a Pt electrode with ferrocene attached. (With permission from J. R. Lenhard and R. W. Murray, *J. Am. Chem. Soc.*, **1978**, *100*, 7870.) In (b), cyclic voltammogram of a graphite electrode with attached py-Ru(NH$_3$)$_5$. (With permission from C. A. Koval and F. C. Anson, *Anal. Chem.*, **1978**, *50*, 223.)

grams shown in Figure 25-26 show the characteristic symmetrical peaks of a reversible surface redox couple. Cyclic voltammograms can reveal the presence of intermediates in oxidation-reduction reactions (for example, see Figure 25-25). Platinum electrodes are often used in CV. For negative potentials, mercury film electrodes can be used. Other popular working electrode

materials include glassy carbon, carbon paste, graphite, gold, diamond, and recently, carbon nanotubes. Chemically modified electrodes, which are discussed in several sections of this chapter, have also been used in CV.

25D-2 Determination of Analytes Using CV

Equation 25-17 shows that peak currents in CV are directly proportional to analyte concentration. Although it is not common to use CV peak currents in routine analytical work, occasionally such applications do appear in the literature, and they are appearing with increasing frequency. As we have mentioned, many modified electrodes are being developed as biosensors. An example is an enzyme-amplified, sandwich-type immunosensor for detecting the interaction between an antigen and an antibody using redox mediation to facilitate electron transfer.[28] In this scheme, a self-assembled monolayer is first attached to the surface of a gold electrode to provide a uniform surface with the proper functionality, for attaching subsequent biosensor layers. In the next layer, a ferrocenyl-tethered dendrimer is synthesized as shown in the reaction sequence at the bottom of the page to provide a redox mediator for efficient electron transfer (ferrocene). A *dendrimer* is a spherical polymer molecule that is synthesized in a stepwise fashion to build up onion-like layers of organic functionality. Each successive layer has a larger number of functional sites for binding target molecules.

Approximately 30% of the sixty-four amine groups on the surface of the dendrimer are combined in an imine-formation reaction with the ferrocene carboxaldehyde. Part of the remaining sites are available for binding to the molecular recognition element, which in

---

[28] S. J. Kwon, E. Kim, H. Yang, and J. Kwak, *Analyst*, **2006**, *131*, 402.

(NH$_2$)$_{64}$

Amine-terminated
G4 dendrimer

Ferrocene
carboxaldehyde

NaBH$_4$

Partial ferrocenyl-tethered
dendrimer (Fc-D)

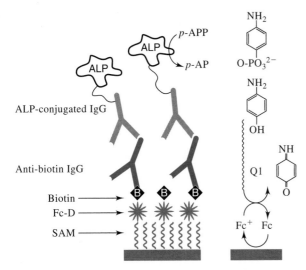

FIGURE 25-27 Schematic illustration of an enzyme-amplified immunosensor using redox mediation of Fc-D. The analyte IgG is shown in blue. (From S. J. Kwon, E. Kim, H. Yang, and J. Kwak, *Analyst*, **2006**, *131*, 402, with permission.)

this example is biotin, on one side of the dendrimer, and part bind to the SAM on the other side.

Figure 25-27 shows how the sandwich arrangement is assembled on the surface of the gold electrode. The SAM layer consists of a 4:1 molar ratio of mercapto-undecanol and mercaptododecanoic acid, so it has on its surface one carboxyl group for every four hydroxyl groups. The ferrocenyl-tethered dendrimer (Fc-D) is then added to the SAM so that a few of the remaining amine groups on the surface of the dendrimer react with carboxyl groups on the SAM to attach the dendrimer to the surface. The final step in the preparation of the working electrode surface is the addition of biotin that is chemically modified with the *N*-hydroxy-succinimide group, with the modified biotin combining with amine groups remaining on the surface of the dendrimer.

In the sandwich immunoassay, the analyte antibody (anti-biotin IgG in the example shown in Figure 25-27) is brought into contact with the working electrode where it binds specifically with the biotin on the surface of the dendrimer. Alkaline phosphatase–conjugated IgG is then added, and it binds specifically with the analyte. The working electrode is then immersed in a buffer containing 1 mM *p*-aminophenylphosphate (*p*-APP) as shown in the figure. This species is converted by alkaline phosphatase to *p*-aminophenol (*p*-AP), which then diffuses to the surface of the dendrimer. Ferrocene (Fc) on the surface of the dendrimer

is oxidized to Fc^+ when an anodic potential is applied to the working electrode, and Fc^+ then oxidizes *p*-aminophenol to the quinoid structure Q1 shown in the figure. The magnitude of the resulting current is a measure of the surface concentration of the analyte IgG.

Cyclic voltammograms of working electrodes in the presence of antibodies other than anti-biotin IgG produced no anodic peaks. For analyte solutions containing anti-biotin IgG, CV peak currents were proportional to the concentration of the analyte, and working dose–response curves of CV peak current versus concentration provided an effective means for completing the immunoassay. The detection limit is 0.1 μg/mL, and the range of the technique is 0.1 to 100 μg/mL of anti-biotin IgG.

25D-3 Digital Simulation of Cyclic Voltammograms

Digital simulation of chemical phenomena is a potent tool in fundamental investigations. Over the past four decades, many user-written software applications have appeared in the literature for the simulation of a broad range of electrochemical processes.[29] In recent years, investigators and students have benefited from the commercial availability of specialized software packages such as DigiSim®.[30] This PC-based software uses the fast implicit finite difference method[31] to simulate cyclic voltammograms for any electrochemical mechanism that can be expressed in terms of single or multiple electron-transfer reactions and to simulate first- and second-order homogeneous chemical reactions. In addition, DigiSim can generate dynamic concentration profiles and display them using a feature called CV—The Movie™. The software can fit simulated data to experimental data that may be imported in a variety of text formats for comparison using least-squares procedures. It can also simulate a range of electrode geometries, finite diffusion, and hydrodynamic mass transport in addition to semi-infinite diffusion.

Figure 25-28 shows an experimental cyclic voltammogram of the $Fe(CN)_6^{3-}$-$Fe(CN)_6^{4-}$ couple at a boron-doped diamond thin-film electrode compared to a cyclic voltammogram simulated by DigiSim. Note the

[29] B. Speiser, in *Electroanalytical Chemistry*, A. J. Bard and I. Rubinstein, eds., Vol. 19, New York: Dekker, 1996, pp. 1–108.

[30] M. Cable and E. T. Smith, *Anal. Chim. Acta*, **2005**, *537*, 299; A. E. Fischer, Y. Show, and G. M. Swain, *Anal. Chem.*, **2004**, *76*, 2553; E. T. Smith, C. A. Davis, and M. J. Barber, *Anal. Biochem.*, **2003**, *323*, 114; for a comprehensive bibliography of principles and applications of DigiSim, see http://www.epsilon-web.net/Ec/digisim/bibdig.html.

[31] M. Rudolph, D. P. Reddy, and S. W. Feldberg, *Anal. Chem.*, **1994**, *66*, 589A.

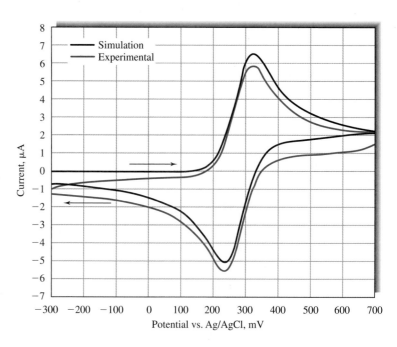

FIGURE 25-28 Experimental and simulated cyclic voltammograms of 0.1 mM $Fe(CN)_6^{3-}$-$Fe(CN)_6^{4-}$ in 1 M KCl at a commercial diamond electrode. (From A. E. Fischer, Y. Show, and G. M. Swain, *Anal. Chem.*, **2004**, *76*, 2553 with permission of the American Chemical Society.)

good agreement in the general shape of the two cyclic voltammograms and the excellent correspondence of E_{pc} and E_{pa}. The peak separation $\Delta E_p = E_{pc} - E_{pa}$ can be used to extract the standard heterogeneous rate constant for the electron-transfer process. The two plots are offset because the experimental data contain both faradaic and nonfaradaic (background) contributions to the current, and the simulation represents only the faradaic current.

25E PULSE VOLTAMMETRY

Many of the limitations of traditional linear-scan voltammetry were overcome by the development of pulse methods. We will discuss the two most important pulse techniques, *differential-pulse voltammetry* and *square-wave voltammetry*. The idea behind all pulse-voltammetric methods is to measure the current at a time when the difference between the desired faradaic curve and the interfering charging current is large. These methods are used with many different types of solid electrodes, the HMDE, and rotating electrodes (Section 25C-4).

25E-1 Differential-Pulse Voltammetry

Figure 25-29 shows the two most common excitation signals used in commercial instruments for differential-pulse voltammetry. The first (Figure 25-29a), which is

usually used in analog instruments, is obtained by superimposing a periodic pulse on a linear scan. The second waveform (Figure 25-29b), which is typically used in digital instruments, is the sum of a pulse and a staircase signal. In either case, a small pulse, typically 50-mV, is applied during the last 50 ms of the period of the excitation signal.

As shown in Figure 25-29, two current measurements are made alternately — one (at S1), which is 16.7 ms prior to the dc pulse and one for 16.7 ms (at S2) at the end of the pulse. The difference in current per pulse (Δi) is recorded as a function of the linearly increasing excitation voltage. A differential curve results, consisting of a peak (see Figure 25-30) whose height is directly proportional to concentration. For a reversible reaction, the peak potential is approximately equal to the standard potential for the half-reaction.

One advantage of the derivative-type voltammogram is that individual peak maxima can be observed for substances with half-wave potentials differing by as little as 0.04 to 0.05 V; in contrast, classical and normal-pulse voltammetry require a potential difference of about 0.2 V for resolving waves. More important, however, differential-pulse voltammetry increases the sensitivity of voltammetry. Typically, differential-pulse voltammetry provides well-defined peaks at a concentration level that is 2×10^{-3} that for the classical voltammetric wave. Note also that the current scale for Δi is in nanoamperes. Generally, detection limits with

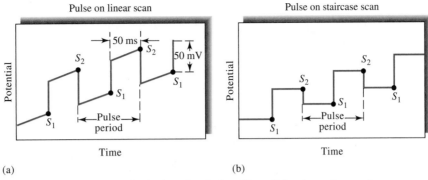

FIGURE 25-29 Excitation signals for differential-pulse voltammetry.

differential-pulse voltammetry are two to three orders of magnitude lower than those for classical voltammetry and lie in the range of 10^{-7} to 10^{-8} M.

The greater sensitivity of differential-pulse voltammetry can be attributed to two sources. The first is an enhancement of the faradaic current, and the second is a decrease in the nonfaradaic charging current. To account for the enhancement, let us consider the events that must occur in the surface layer around an electrode as the potential is suddenly increased by 50 mV. If an electroactive species is present in this layer, there will be a surge of current that lowers the reactant concentration to that demanded by the new potential (see Figure 25-9b). As the equilibrium concentration for that potential is approached, however, the current decays to a level just sufficient to counteract diffusion; that is, to the diffusion-controlled current. In classical voltammetry, the initial surge of current is not observed because the time scale of the measurement is long relative to the lifetime of the momentary current. On the other hand, in pulse voltammetry, the current measurement is made

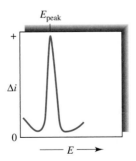

FIGURE 25-30 Voltammogram for a differential-pulse voltammetry experiment. Here, $\Delta i = i_{S2} - i_{S1}$ (see Figure 25-29). The peak potential, E_{peak}, is closely related to the voltammetric half-wave potential.

before the surge has completely decayed. Thus, the current measured contains both a diffusion-controlled component and a component that has to do with reducing the surface layer to the concentration demanded by the Nernst expression; the total current is typically several times larger than the diffusion current. Note that, under hydrodynamic conditions, the solution becomes homogeneous with respect to the analyte by the time the next pulse sequence occurs. Thus, at any given applied voltage, an identical current surge accompanies each voltage pulse.

When the potential pulse is first applied to the electrode, a surge in the nonfaradaic current also occurs as the charge increases. This current, however, decays exponentially with time and approaches zero with time. Thus, by measuring currents at this time only, the nonfaradaic residual current is greatly reduced, and the signal-to-noise ratio is larger. Enhanced sensitivity results.

Reliable instruments for differential-pulse voltammetry are now available commercially at reasonable cost. The method has thus become one of the most widely used analytical voltammetric procedures and is especially useful for determining trace concentrations of heavy metal ions.

25E-2 Square-Wave Voltammetry

Square-wave voltammetry is a type of pulse voltammetry that offers the advantage of great speed and high sensitivity.[32] An entire voltammogram is obtained

---

[32] For further information on square-wave voltammetry, see A. J. Bard and L. R. Faulkner, *Electrochemical Methods*, 2nd ed., New York: Wiley, 2001, Chap. 7, pp. 293–99; J. G. Osteryoung and R. A. Osteryoung, *Anal. Chem.*, **1985**, *57*, 101A.

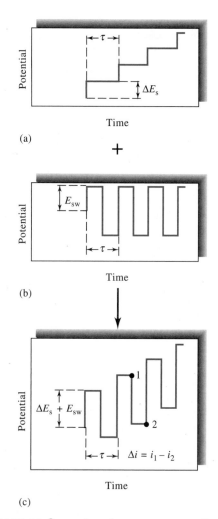

(a)

+

(b)

↓

(c)

FIGURE 25-31 Generation of a square-wave voltammetry excitation signal. The staircase signal in (a) is added to the pulse train in (b) to give the square-wave excitation signal in (c). The current response Δi is equal to the current at potential 1 minus that at potential 2.

in less than 10 ms. Square-wave voltammetry has been used with HMDEs and with other electrodes (see Figure 25-3) and sensors.

Figure 25-31c shows the excitation signal in square-wave voltammetry, which is obtained by superimposing the pulse train shown in 25-31b onto the staircase signal in 25-31a. The length of each step of the staircase and the period τ of the pulses are identical and usually about 5 ms. The potential step of the staircase ΔE_s is typically 10 mV. The magnitude of the pulse $2E_{sw}$ is often 50 mV. Operating under these conditions, which correspond to a pulse frequency of 200 Hz, a

1-V scan requires 0.5 s. For a reversible reduction reaction, the size of a pulse is great enough so that oxidation of the product formed on the forward pulse occurs during the reverse pulse. Thus, as shown in Figure 25-32, the forward pulse produces a cathodic current i_1, and the reverse pulse gives an anodic current i_2. Usually, the difference in these currents Δi is plotted to give voltammograms. This difference is directly proportional to concentration; the potential of the peak corresponds to the voltammetric half-wave potential. Because of the speed of the measurement, it is possible and practical to increase the precision of analyses by signal-averaging data from several voltammetric scans. Detection limits for square-wave voltammetry are reported to be 10^{-7} to 10^{-8} M.

Commercial instruments for square-wave voltammetry are available from several manufacturers, and as a consequence this technique is being used routinely for determining inorganic and organic species. Square-wave voltammetry is also being used in detectors for liquid chromatography.

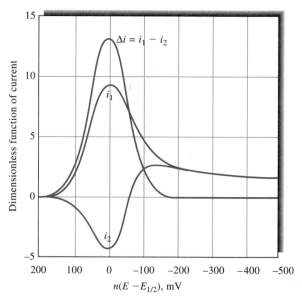

FIGURE 25-32 Current response for a reversible reaction to excitation signal in Figure 25-31c. This theoretical response plots a dimensionless function of current versus a function of potential, $n(E - E_{1/2})$ in millivolts. Here, i_1 = forward current; i_2 = reverse current; $i_1 - i_2$ = current difference. (From J. J. O'Dea, J. Osteryoung, and R. A. Osteryoung, *Anal. Chem.*, **1981**, *53*, 695. With permission. Copyright 1981 American Chemical Society.)

25F HIGH-FREQUENCY AND HIGH-SPEED VOLTAMMETRY

As the science, technology, and art of voltammetric instrumentation and data reduction methods have developed, so too have the time and spatial regimes of voltammetry decreased in scale. Traditionally, voltammetric measurements were made at dc and relatively low frequency. As a result, only relatively slow electron-transfer processes could be explored using these methods.

25F-1 Fourier Transform Voltammetry

Just as Fourier transform (FT) methods revolutionized nuclear magnetic resonance and IR spectroscopies, FT methods may have a similar impact on voltammetry. Bond et al.[33] have shown how off-the-shelf PC-based stereo gear coupled with common data-reduction software (MATLAB and LabVIEW) can be used to perform FT voltammetry at sampling frequencies up to 40 kHz. Excitation waveforms of any shape can be synthesized and applied to potentiostatic circuitry, and the resulting response can be analyzed very rapidly to obtain voltammograms corresponding to each of the frequency components of the excitation waveform. The data output from FT voltammetry provides power spectra for each frequency component of the input wave in a new and visually intriguing format. Patterns in the data can be recognized for various types of electron-transfer mechanisms in the chemical systems under study. These workers give several examples, including FT voltammetric measurements on ferrocene and hexacyanoferrate(III), to illustrate the ways that patterns in the output may be analyzed visually. As more and varied systems are explored, databases of mechanisms are cataloged, and software algorithms are applied to recognition of mechanistic patterns, FT voltammetry may become a mainstay in the electrochemical toolkit.

25F-2 Fast-Scan Cyclic Voltammetry

Fast-scan cyclic voltammetry (FSCV) is being extended into biomedical fields, particularly in the areas of neurophysiology and interdisciplinary *psychoanalytical*

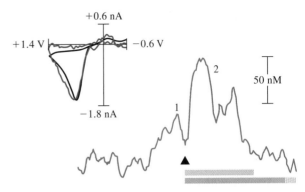

FIGURE 25-33 Dopamine release during cocaine self-administration. Rats were trained to press a lever (arrowhead) to receive a small intravenous injection of cocaine. The lower trace shows changes in dopamine. Peak 1 indicates an increase in dopamine before the rat pressed the lever. The two peaks (2) indicate transients that occurred after the lever press. Underneath the trace, the dark blue bar marks the time the audiovisual cues associated with the lever press were on. The light blue bar indicates when the pump was activated to deliver cocaine. The cyclic voltammogram (black) of behaviorally evoked dopamine matches the electrically evoked voltammogram (blue). (Adapted with permission from P. E. M. Phillips, *Nature*, **2003**, *422*, 614.)

electrochemistry. Venton and Wightman have described how FSCV with carbon fiber electrodes can be used to carry out in vivo measurements of dopamine release in rat brains during behavioral stimulation.[34] In the experiment illustrated by the data in Figure 25-33, a carbon fiber microelectrode (see Section 25I) implanted in the brain of a rat was used as the working electrode as 450-V/s CV scans over +1.4 to −0.6 V were performed at 100-ms intervals. The rat had been previously taught to autoadminister doses of cocaine by pressing a lever.

The two cyclic voltammograms in the upper left of the figure show that dopamine was released both by behavioral stimulation (black trace) and cocaine administration (blue trace). The blue trace in the center shows the concentration of dopamine as a function of time calculated from the peak currents of the voltammograms. The trace was extracted from 200 discrete cyclic voltammograms acquired over the 20-second period shown. The peak at position 1 shows that there

[33]A. M. Bond, M. W. Duffy, S. X. Guo, J. Zhang, and D. Elton, *Anal. Chem.*, **2005**, *77*, 186A.

[34]B. J. Venton and R. M. Wightman, *Anal. Chem.*, **2003**, *75*, 414A.

was a release of dopamine just prior to the rat autoadministering cocaine as indicated by the black triangle. The peaks at position 2 show a dramatic increase in the release of dopamine as the dosage of cocaine continued, as indicated by the light-blue bar. Studies such as this permit researchers to correlate behavior with chemical changes in the brain.[35]

25F-3 Nanosecond Voltammetry

To expand frontier knowledge of the electron-transfer process, it is important to be able to conduct CV experiments on a very fast time scale. Careful consideration of the circuit model of Figure 25-7 suggests that the solution resistance and the double-layer capacitance in a voltammetric cell place limitations on the time scale of CV. The time required to charge the double layer is on the order of $R_\Omega C_d$, which for a CV cell with a 1-mm-radius Pt electrode immersed in a 1-mM solution of a typical electroactive species might be 500 $\Omega \times 0.3$ μF = 0.15 ms. Furthermore, the IR drop across the solution resistance in a CV cell carrying a current of 1 mA would be 0.5 V. Acquiring useful CV data at a scan rate of 1000 V/s would be virtually impossible under these conditions. On the other hand, for a 5-μm-radius-disk microelectrode, the IR drop is 2.5 mV and $R_\Omega C_d = 750$ ns.[36] If, in addition to decreasing the size of the working electrode, iterative positive feedback is applied to the excitation waveform,[37] the IR drop can be compensated completely, and scan rates of up to 2.5 million V/s can be achieved with properly optimized potentiostatic circuitry. Using such an optimized instrumental setup, cyclic voltammograms of pyrylium cation have been acquired, and kinetics of the dimerization of the reduction product have been measured along with the rate constant of the reaction, which is $k_{dim} = 0.9(\pm0.3) \times 10^9$ M/s. This value corresponds to a half-life for the pyrylium radical of about 200 ns. The time scale of these measurements is comparable to that of nanosecond flash photolysis experiments. These innovative techniques have opened an entirely new time window for voltammetric measurements, and it is hoped that the goal of observing single-electron-transfer events may be in sight.[38]

25G APPLICATIONS OF VOLTAMMETRY

In the past, linear-scan voltammetry was used for the quantitative determination of a wide variety of inorganic and organic species, including molecules of biological and biochemical interest. Pulse methods have largely replaced classical voltammetry because of their greater sensitivity, convenience, and selectivity. Generally, quantitative applications are based on calibration curves in which peak heights are plotted as a function of analyte concentration. In some instances the standard-addition method is used in lieu of calibration curves. In either case, it is essential that the composition of standards resemble as closely as possible the composition of the sample, both as to electrolyte concentrations and pH. When this is done, relative precisions and accuracies in the range of 1% to 3% can often be achieved.

25G-1 Inorganic Applications

Voltammetry is applicable to the analysis of many inorganic substances. Most metallic cations, for example, are reduced at common working electrodes. Even the alkali and alkaline-earth metals are reducible, provided the supporting electrolyte does not react at the high potentials required; here, the tetraalkyl ammonium halides are useful electrolytes because of their high reduction potentials.

The successful voltammetric determination of cations frequently depends on the supporting electrolyte that is used. To aid in this selection, tabular compilations of half-wave potential data are available.[39] The judicious choice of anion often enhances the selectivity of the method. For example, with potassium chloride as a supporting electrolyte, the waves for iron(III) and copper(II) interfere with one another; in a fluoride medium, however, the half-wave potential of iron(III) is shifted by about −0.5 V and that for copper(II) is altered by only a few hundredths of a volt. The presence of fluoride thus results in the appearance of well-separated waves for the two ions.

Voltammetry is also applicable to the analysis of such inorganic anions as bromate, iodate, dichromate, vanadate, selenite, and nitrite. In general, voltammograms for these substances are affected by the pH of

[35] P. E. M. Phillips, *Nature*, **2003**, *422*, 614.
[36] C. Amatore and E. Maisonhaute, *Anal. Chem.*, **2005**, 77, 303A.
[37] D. O. Wipf, *Anal. Chem.*, **1996**, *68*, 1871.
[38] See note 36.

[39] For example, see J. A. Dean, *Analytical Chemistry Handbook*, Section 14, pp. 14.66–14.70, New York: McGraw-Hill, 1995; D. T. Sawyer, A. Sobkowiak, and J. L. Roberts, *Experimental Electrochemistry for Chemists*, 2nd ed., pp. 102–30, New York: Wiley, 1995.

the solution because the hydrogen ion is a participant in their reduction. As a consequence, strong buffering to some fixed pH is necessary to obtain reproducible data (see next section).

25G-2 Organic Voltammetric Analysis

Almost from its inception, voltammetry has been used for the study and determination of organic compounds, with many papers being devoted to this subject. Several organic functional groups are reduced at common working electrodes, thus making possible the determination of a wide variety of organic compounds.[40] Oxidizable organic functional groups can be studied voltammetrically with platinum, gold, carbon, or various modified electrodes.

Effect of pH on Voltammograms

Organic electrode processes often involve hydrogen ions, the typical reaction being represented as

$$R + nH^+ + ne^- \rightleftharpoons RH_n$$

where R and RH_n are the oxidized and reduced forms of the organic molecule. Half-wave potentials for organic compounds are therefore pH dependent. Furthermore, changing the pH may produce a change in the reaction product. For example, when benzaldehyde is reduced in a basic solution, a wave is obtained at about -1.4 V, attributable to the formation of benzyl alcohol:

$$C_6H_5CHO + 2H^+ + 2e^- \rightleftharpoons C_6H_5CH_2OH$$

If the pH is less than 2, however, a wave occurs at about -1.0 V that is just half the size of the foregoing one; here, the reaction involves the production of hydrobenzoin:

$$2C_6H_5CHO + 2H^+ + 2e^- \rightleftharpoons C_6H_5CHOHCHOHC_6H_5$$

At intermediate pH values, two waves are observed, indicating the occurrence of both reactions.

We must emphasize that an electrode process that consumes or produces hydrogen ions will alter the pH of the solution at the electrode surface, often drastically, unless the solution is well buffered. These changes affect the reduction potential of the reaction

and cause drawn-out, poorly defined waves. Moreover, where the electrode process is altered by pH, as in the case of benzaldehyde, the diffusion current–concentration relationship may also be nonlinear. Thus, in organic voltammetry good buffering is generally vital for the generation of reproducible half-wave potentials and diffusion currents.

Solvents for Organic Voltammetry

Solubility considerations frequently dictate the use of solvents other than pure water for organic voltammetry; aqueous mixtures containing varying amounts of such miscible solvents as glycols, dioxane, acetonitrile, alcohols, Cellosolve, or acetic acid have been used. Anhydrous media such as acetic acid, formamide, diethylamine, and ethylene glycol have also been investigated. Supporting electrolytes are often lithium or tetraalkyl ammonium salts.

Reactive Functional Groups

Organic compounds containing any of the following functional groups often produce one or more voltammetric waves.

1. The carbonyl group, including aldehydes, ketones, and quinones, produce voltammetric waves. In general, aldehydes are reduced at lower potentials than ketones; conjugation of the carbonyl double bond also results in lower half-wave potentials.
2. Certain carboxylic acids are reduced voltammetrically, although simple aliphatic and aromatic monocarboxylic acids are not. Dicarboxylic acids such as fumaric, maleic, or phthalic acid, in which the carboxyl groups are conjugated with one another, give characteristic voltammograms; the same is true of certain keto and aldehydo acids.
3. Most peroxides and epoxides yield voltammetric waves.
4. Nitro, nitroso, amine oxide, and azo groups are generally reduced at working electrodes.
5. Most organic halogen groups produce a voltammetric wave, which results from replacement of the halogen group with an atom of hydrogen.
6. The carbon-carbon double bond is reduced when it is conjugated with another double bond, an aromatic ring, or an unsaturated group.
7. Hydroquinones and mercaptans produce anodic waves.

In addition, a number of other organic groups cause catalytic hydrogen waves that can be used for analysis.

---

[40]For a detailed discussion of organic electrochemistry, see *Encyclopedia of Electrochemistry*, A. J. Bard and M. Stratmann, eds., Vol. 8, *Organic Electrochemistry*, H. J. Schäfer, ed., New York: Wiley, 2002; *Organic Electrochemistry*, 4th ed., H. Lund and O. Hammerich, eds., New York: Dekker, 2001.

These include amines, mercaptans, acids, and heterocyclic nitrogen compounds. Numerous applications to biological systems have been reported.[41]

25H STRIPPING METHODS

Stripping methods encompass a variety of electrochemical procedures having a common, characteristic initial step.[42] In all of these procedures, the analyte is first deposited on a working electrode, usually from a stirred solution. After an accurately measured period, the electrolysis is discontinued, the stirring is stopped, and the deposited analyte is determined by one of the voltammetric procedures that have been described in the previous section. During this second step in the analysis, the analyte is redissolved or stripped from the working electrode; hence the name attached to these methods. In anodic stripping methods, the working electrode behaves as a cathode during the deposition step and as an anode during the stripping step, with the analyte being oxidized back to its original form. In a cathodic stripping method, the working electrode behaves as an anode during the deposition step and as a cathode during stripping. The deposition step amounts to an electrochemical preconcentration of the analyte; that is, the concentration of the analyte in the surface of the working electrode is far greater than it is in the bulk solution. As a result of the preconcentration step, stripping methods yield the lowest detection limits of all voltammetric procedures. For example, anodic stripping with pulse voltammetry can reach nanomolar detection limits for environmentally important species, such as Pb^{2+}, Ca^{2+}, and Tl^+.

Figure 25-34a illustrates the voltage excitation program that is followed in an anodic stripping method for determining cadmium and copper in an aqueous solution of these ions. A linear scan method is often used to complete the analysis. Initially, a constant cathodic potential of about -1 V is applied to the working electrode, which causes both cadmium and copper ions to be reduced and deposited as metals. The electrode is maintained at this potential for several minutes until a significant amount of the two metals has accumulated

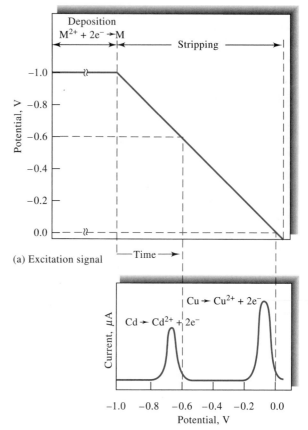

(a) Excitation signal

(b) Voltammogram

FIGURE 25-34 (a) Excitation signal for stripping determination of Cd^{2+} and Cu^{2+}. (b) Stripping voltammogram.

at the electrode. The stirring is then stopped for 30 s or so while the electrode is maintained at -1 V. The potential of the electrode is then decreased linearly to less negative values and the current in the cell is recorded as a function of time, or potential. Figure 25-34b shows the resulting voltammogram. At a potential somewhat more negative than -0.6 V, cadmium starts to be oxidized, causing a sharp increase in the current. As the deposited cadmium is consumed, the current peaks and then decreases to its original level. A second peak for oxidation of the copper is then observed when the potential has decreased to approximately -0.1 V. The heights of the two peaks are proportional to the weights of deposited metal.

Stripping methods are important in trace work because the preconcentration step permits the determination of minute amounts of an analyte with reasonable accuracy. Thus, the analysis of solutions in the range of 10^{-6} to 10^{-9} M becomes feasible by methods that are both simple and rapid.

[41] *Encyclopedia of Electrochemistry*, A. J. Bard and M. Stratmann, eds., Vol. 9, *Bioelectrochemistry*, G. S. Wilson, ed., New York: Wiley, 2002.

[42] For detailed discussions of stripping methods, see H. D. Dewald, in *Modern Techniques in Electroanalysis*, P. Vanysek, ed., Chap. 4, p. 151, New York: Wiley-Interscience, 1996; J. Wang, *Stripping Analysis*, Deerfield Beach, FL: VCH, 1985.

25H-1 Electrodeposition Step

Only a fraction of the analyte is usually deposited during the electrodeposition step; hence, quantitative results depend not only on control of electrode potential but also on such factors as electrode size, time of deposition, and stirring rate for both the sample and standard solutions used for calibration.

Working electrodes for stripping methods have been formed from a variety of materials, including mercury, gold, silver, platinum, and carbon in various forms. The most popular electrode is the HMDE shown in Figure 25-3b. RDEs may also be used in stripping analysis.

To carry out the determination of a metal ion by anodic stripping, a fresh hanging drop is formed, stirring is begun, and a potential is applied that is a few tenths of a volt more negative than the half-wave potential for the ion of interest. Deposition is allowed to occur for a carefully measured period that can vary from a minute or less for 10^{-7} M solutions to 30 min or longer for 10^{-9} M solutions. We should reemphasize that these times seldom result in complete removal of the ion. The electrolysis period is determined by the sensitivity of the method ultimately used for completion of the analysis.

25H-2 Voltammetric Completion of the Analysis

The analyte collected in the working electrode can be determined by any of several voltammetric procedures. For example, in a linear anodic scan procedure, as described at the beginning of this section, stirring is discontinued for 30 s or so after stopping the deposition. The voltage is then decreased at a linear fixed rate from its original cathodic value, and the resulting anodic current is recorded as a function of the applied voltage. This linear scan produces a curve of the type shown in Figure 25-34b. Analyses of this type are generally based on calibration with standard solutions of the cations of interest. With reasonable care, analytical precisions of about 2% relative can be obtained.

Most of the other voltammetric procedures described in the previous section have also been applied in the stripping step. The most widely used of these appears to be an anodic differential-pulse technique. Often, narrower peaks are produced by this procedure, which is desirable when mixtures are analyzed. Another method of obtaining narrower peaks is to use a mercury film electrode. Here, a thin mercury film is electrodeposited on an inert electrode such as glassy carbon. Usually, the mercury deposition is carried out

simultaneously with the analyte deposition. Because the average diffusion path length from the film to the solution interface is much shorter than that in a drop of mercury, escape of the analyte is hastened; the consequence is narrower and larger voltammetric peaks, which leads to greater sensitivity and better resolution of mixtures. On the other hand, the hanging drop electrode appears to give more reproducible results, especially at higher analyte concentrations. Thus, for most applications the hanging drop electrode is used. Figure 25-35 is a differential-pulse anodic stripping voltammogram for five cations in a sample of mineralized honey, which had been spiked with 1×10^{-5} M $GaCl_3$. The voltammogram demonstrates good resolution and adequate sensitivity for many purposes.

Many other variations of the stripping technique have been developed. For example, a number of cations have been determined by electrodeposition on a platinum cathode. The quantity of electricity required to remove the deposit is then measured coulometrically. Here again, the method is particularly advantageous for trace analyses. Cathodic stripping methods for the halides have also been developed. In these methods, the halide ions are first deposited as mercury(I) salts on a mercury anode. Stripping is then performed by a cathodic current.

25H-3 Adsorptive Stripping Methods

Adsorptive stripping methods are quite similar to the anodic and cathodic stripping methods we have just considered. In this technique, a working electrode is immersed in a stirred solution of the analyte for several minutes. Deposition of the analyte then occurs by physical adsorption on the electrode surface rather than by electrolytic deposition. After sufficient analyte has accumulated, the stirring is discontinued and the deposited material determined by linear scan or pulsed voltammetric measurements. Quantitative information is acquired by calibration with standard solutions that are treated in the same way as samples.

Many organic molecules of clinical and pharmaceutical interest have a strong tendency to be adsorbed from aqueous solutions onto a mercury or carbon surface, particularly if the surface is maintained at a voltage where the charge on the electrode is near zero. With good stirring, adsorption is rapid, and only 1 to 5 min is required to accumulate sufficient analyte for analysis from 10^{-7} M solutions and 10 to 20 min for

 Tutorial: Learn more about **stripping methods**.

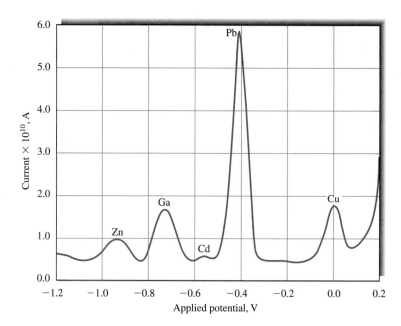

FIGURE 25-35 Differential-pulse anodic stripping voltammogram in the analysis of a mineralized honey sample spiked with GaCl$_3$ (final concentration in the analysis solution: 1×10^{-5} M). Deposition potential: -1.20 V; deposition time: 1200 s in unstirred solution; pulse height: 50 mV; and anodic potential scan rate: 5 mVs^{-1}. (Adapted from G. Sannaa, M. I. Pilo, P. C. Piu, A. Tapparo, and R. Seeber, *Anal. Chim. Acta*, **2000**, *415*, 165, with permission.)

10^{-9} M solutions. Figure 25-36a illustrates the sensitivity of differential-pulse adsorptive stripping voltammetry when it is applied to the determination of calf-thymus DNA in a 0.5 mg/L solution. Figure 25-36b shows the dependence of the signal on deposition time. Many other examples of this type can be found in the recent literature.

Adsorptive stripping voltammetry has also been applied to the determination of a variety of inorganic cations at very low concentrations. In these applications the cations are generally complexed with surface active complexing agents, such as dimethylglyoxime, catechol, and bipyridine. Detection limits in the range of 10^{-10} to 10^{-11} M have been reported.

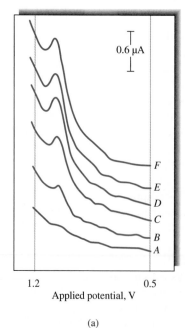

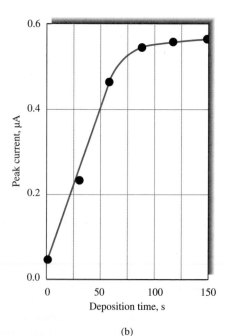

FIGURE 25-36 Effect of preconcentration period on the voltammetric (a) stripping response at the pretreated carbon paste electrode. Preconcentration at $+0.5$ V for (A) 1, (B) 30, (C) 60, (D) 90, (E) 120, and (F) 150 s. Square-wave amplitude, 10 mV; frequency, 40 Hz; 0.5 mg/L calf-thymus DNA. (b) Peak current versus deposition time. (Adapted from J. Wang, X. Cai, C. Jonsson, and M. Balakrishnan, *Electroanalysis*, **1996**, *8*, 20, with permission.)

(a)　　　　　　　　　　　　　(b)

25I VOLTAMMETRY WITH MICROELECTRODES

In recent years, many voltammetric studies have been carried out with electrodes that have dimensions smaller by an order of magnitude or more than normal working electrodes. The electrochemical behavior of these tiny electrodes is significantly different from classical electrodes and appears to offer advantages in certain analytical applications.[43] Such electrodes are often called microscopic electrodes, or *microelectrodes*, to distinguish them from classical electrodes. Figure 25-3c shows one type of commercial microelectrode. The dimensions of these electrodes are typically smaller than about 20 μm and may be as small as 30 nm in diameter and 2 μm long ($A \approx 0.2\ \mu m^2$). Experience has led to an *operational definition of microelectrodes*. A microelectrode is any electrode whose characteristic dimension is, under the given experimental conditions, comparable to or smaller than the diffusion layer thickness, δ. Under these conditions, a steady state or, in the case of cylindrical electrodes, a pseudo–steady state is attained.[44]

25I-1 Voltammetric Currents at Microelectrodes

In section 25C-2, we discussed the nature of the current that is produced at an ordinary planar electrode in voltammetric experiments. Using a more elaborate treatment, it can be shown[45] that the concentration gradient at a spherical electrode following application of a voltage step is

$$\frac{\partial c_A}{\partial x} = c_A^0\left(\frac{1}{\sqrt{\pi Dt}} + \frac{1}{r}\right) = c_A^0\left(\frac{1}{\delta} + \frac{1}{r}\right) \quad (25\text{-}21)$$

where r is the radius of the sphere, $\delta = \sqrt{\pi Dt}$ is the thickness of the Nernst diffusion layer, and t is the time after the voltage is applied. Note here that δ is proportional to $t^{1/2}$. By substituting this relation into Equation 25-4, we obtain the time-dependent faradaic current at the spherical electrode.

$$i = nFADc_A^0\left(\frac{1}{\delta} + \frac{1}{r}\right) \quad (25\text{-}22)$$

Note that if $r \gg \delta$, which occurs at short times, the $1/\delta$ term predominates, and Equation 25-22 reduces to an equation analogous to Equation 25-5. If $r \ll \delta$, which occurs at long times, the $1/r$ term predominates, the electron-transfer process reaches a steady state, and the steady-state current then depends only on the size of the electrode. This means that if the size of the electrode is small compared to the thickness of the Nernst diffusion layer, steady state is achieved very rapidly, and a constant current is produced. Because the current is proportional to the area of the electrode, it also means that microelectrodes produce tiny currents. Expressions similar in form to Equation 25-22 may be formulated for other geometries, and they all have in common the characteristic that the smaller the electrode, the more rapidly steady-state current is achieved.

The advantages of microelectrodes may be summarized[46] as follows:

1. Steady state for faradaic processes is attained very rapidly, often in microseconds to milliseconds. Measurements on this time scale permit the study of intermediates in rapid electrochemical reactions.
2. Because charging current is proportional to the area of the electrode A and faradaic current is proportional to A/r, the relative contribution of charging to the overall current decreases with the size of the microelectrode.
3. Because charging current is minimal with microelectrodes, the potential may be scanned very rapidly.
4. Because currents are so very small (in the picoampere to nanoampere range), the IR drop decreases dramatically as the size of the microelectrode decreases.
5. When microelectrodes operate under steady-state conditions, the signal-to-noise ratio in the current is much higher than is the case under dynamic conditions.
6. The solution at the surface of a microelectrode used in a flow system is replenished constantly, which minimizes δ and thus maximizes faradaic current.
7. Measurements with microelectrodes can be made on incredibly small solution volumes, for example, the volume of a biological cell.

[43] See R. M. Wightman, *Science*, **1988**, *240*, 415; *Anal. Chem.*, **1981**, *53*, 1325A; S. Pons and M. Fleischmann, *Anal. Chem.*, **1987**, *59*, 1391A; J. Heinze, *Agnew. Chem., Int. Ed.*, **1993**, *32*, 1268; R. M. Wightman and D. O. Wipf, in *Electroanalytical Chemistry*, Volume 15, A. J. Bard, ed., New York: Dekker, 1989; A. C. Michael and R. M. Wightman, in *Laboratory Techniques in Electroanalytical Chemistry*, 2nd ed., P. T. Kissinger and W. R. Heinemann, eds., Chap. 12, New York: Dekker, 1996; C. G. Zoski, in *Modern Techniques in Electroanalysis*, P. Vanysek, ed., Chap. 6, New York: Wiley, 1996.

[44] For a discussion of microelectrodes, including terminology, characterization, and applications, see K. Štulík, C. Amatore, K. Holub, V. Mareček, and W. Kutner, *Pure Appl. Chem.*, **2000**, *72*, 1483. http://www.iupac.org/publications/pac/2000/7208/7208pdfs/7208stulik_1483.pdf.

[45] See note 43.

[46] See note 43.

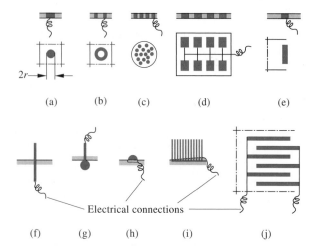

FIGURE 25-37 Most important geometries of micro-electrodes and microelectrode arrays: (a) microdisk, (b) microring; (c) microdisk array (a composite electrode); (d) lithographically produced microband array; (e) micro-band; (f) single fiber (microcylinder); (g) microsphere; (h) microhemisphere; (i) fiber array; (j) interdigitated array. (From K. Štulík, C. Amatore, K. Holub, V. Mareek, and W. Kutner, *Pure Appl. Chem.*, **2000**, *72*, 1483, with permission.)

8. Tiny currents make it possible to make voltammetric measurements in high-resistance, nonaqueous solvents, such as those used in normal-phase liquid chromatography.

As shown in Figure 25-37, microelectrodes take several forms. The most common is a planar electrode formed by sealing a 5-μm-radius carbon fiber or a 0.3- to 20-μm gold or platinum wire into a fine capillary tube; the fiber or wires are then cut flush with the ends of the tubes (see Figures 25-3c and 25-37a and b). Cylindrical electrodes are also used in which a small portion of the wire extends from the end of the tube (Figure 25-37f). This geometry has the advantage of larger currents but the disadvantages of being fragile and difficult to clean and polish. Band electrodes (Figure 25-37d and e) are attractive because they can be fabricated on a nanometer scale in one dimension, and their behavior is determined by this dimension except that the magnitude of their currents increases with length. Electrodes of this type of 20 Å have been constructed by sandwiching metal films between glass or epoxy insulators. Other configurations such as the microdisk array, microsphere, microhemisphere, fiber array, and interdigitated array (Figure 25-37c, g, h, i, and

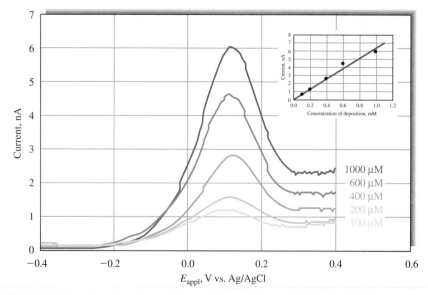

FIGURE 25-38 Detection of dopamine at a multiwall carbon nanotube-based nanoneedle electrode. Differential-pulse voltammograms of dopamine at the nanoneedle electrode in various concentrations from 100 to 1000 μM. Inset is the calibration curve. (From H. Boo et al., *Anal. Chem.*, **2006**, *78*, 617. With permission. Copyright 2006 American Chemical Society.)

j, respectively) have been used successfully. Mercury microelectrodes are formed by electrodeposition of the metal onto carbon or metal electrodes.

25I-2 Applications of Microelectrodes

In Section 25F-2, we described the use of a carbon fiber electrode to monitor concentration of the neurotransmitter dopamine in rat brains in response to behavioral change. In Figure 25-38, we see the results of the differential-pulse voltammetric determination of dopamine at 100–1000 µM levels.[47] The working electrode in this study was a nanoneedle consisting of a multiwall carbon nanotube attached to the end of a tungsten wire tip. This electrode may be the smallest fabricated up to this time. The entire surface of the probe except the nanoneedle (30 nm in diameter and 3 µm long) was coated with a nonconducting UV-hardening polymer. Both CV and differential-pulse voltammetry were performed with the nanoneedle electrode with good results. The inset in the figure shows a working curve of the peak currents from the voltammograms plotted versus concentration. In light of tremendous interest in nanomaterials and

biosensors for determining analytes in minuscule volumes of solution, it is likely that research and development in this fertile area will continue for some time.

25I-3 The Scanning Electrochemical Microscope

Another application of microelectrodes is the *scanning electrochemical microscope* (SECM), introduced by Bard in 1989.[48] The SECM is closely related to the scanning probe microscopes (SPMs) discussed in Section 21G. The SECM works by measuring the current through a microelectrode (the tip) in a solution containing an electroactive species while the tip is scanned over a substrate surface. The presence of the substrate causes a perturbation of the electrochemical response of the tip, which gives information about the characteristics and nature of the surface. Surfaces studied have included solids, such as glasses, polymers, metals, and biological substances, and liquids, such as mercury or oils. The SECM, which is now commercially available, has been used to study conducting polymers, dissolution of crystals, nano materials, and biologically interesting surfaces.

[47]H. Boo et al., *Anal. Chem.*, **2006**, *78*, 617.

[48]A. J. Bard, F.-R. F. Fan, J. Kwak, and O. Lev, *Anal. Chem.* **1989**, *61*, 132.

QUESTIONS AND PROBLEMS

*Answers are provided at the end of the book for problems marked with an asterisk.

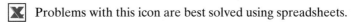 Problems with this icon are best solved using spreadsheets.

25-1 Distinguish between (a) voltammetry and amperometry, (b) linear-scan voltammetry and pulse voltammetry, (c) differential-pulse voltammetry and square-wave voltammetry, (d) an RDE and a ring-disk electrode, (e) faradaic impedance and double-layer capacitance, (f) a limiting current and a diffusion current, (g) laminar flow and turbulent flow, (h) the standard electrode potential and the half-wave potential for a reversible reaction at a working electrode, (i) normal stripping methods and adsorptive stripping methods.

25-2 Define (a) voltammograms, (b) hydrodynamic voltammetry, (c) Nernst diffusion layer, (d) mercury film electrode, (e) half-wave potential, and (f) voltammetric sensor.

25-3 Why is a high supporting electrolyte concentration used in most electroanalytical procedures?

25-4 Why is the reference electrode placed near the working electrode in a three-electrode cell?

25-5 Why is it necessary to buffer solutions in organic voltammetry?

25-6 Why are stripping methods more sensitive than other voltammetric procedures?

25-7 What is the purpose of the electrodeposition step in stripping analysis?

25-8 List the advantages and disadvantages of the mercury film electrode compared with platinum or carbon electrodes.

25-9 Suggest how Equation 25-13 could be used to determine the number of electrons n involved in a reversible reaction at an electrode.

*25-10 Quinone undergoes a reversible reduction at a voltammetric working electrode. The reaction is

$$+ \; 2H^+ + 2e^- \rightleftharpoons \qquad\qquad E^0 = 0.599 \text{ V}$$

(a) Assume that the diffusion coefficient for quinone and hydroquinone are approximately the same and calculate the approximate half-wave potential (versus SCE) for the reduction of hydroquinone at an RDE from a solution buffered to a pH of 7.0.

(b) Repeat the calculation in (a) for a solution buffered to a pH of 5.0.

*25-11 In experiment 1, a cyclic voltammogram at an HMDE was obtained from a 0.167-mM solution of Pb^{2+} at a scan rate of 2.5 V/s. In experiment 2, a second CV is to be obtained from a 4.38-mM solution of Cd^{2+} using the same HMDE. What must the scan rate be in experiment 2 to record the same peak current in both experiments if the diffusion coefficients of Cd^{2+} and Pb^{2+} are $0.72 \times 10^{-5} \text{ cm}^2 \text{ s}^{-1}$ and $0.98 \text{ cm}^2 \text{ s}^{-1}$, respectively. Assume that the reductions of both cations are reversible at the HMDE.

25-12 The working curve for the determination of dopamine at a nanoneedle electrode by differential-pulse voltammetry (Figure 25-38) was constructed from the following table of data.

Concentration Dopamine, mM	Peak Current, nA
0.093	0.66
0.194	1.31
0.400	2.64
0.596	4.51
0.991	5.97

(a) Use Excel to perform a least-squares analysis of the data to determine the slope, intercept, and regression statistics, including the standard deviation about regression.

(b) Use your results to find the concentration of dopamine in a sample solution that produced a peak current of 3.62 nA. This value is the average of duplicate experiments.

(c) Calculate the standard deviation of the unknown concentration and its 95% confidence interval assuming that each of the data in the table was obtained in a single experiment.

*25-13 A solution containing Cd^{2+} was analyzed voltammetrically using the standard addition method. Twenty-five milliliters of the deaerated solution, which was 1 M in HNO_3, produced a net limiting current of 1.78 µA at a rotating mercury film working electrode at a potential of −0.85 V (versus SCE). Following addition of 5.00 mL of a 2.25×10^{-3} M standard Cd^{2+} solution, the resulting solution produced a current of 4.48 µA. Calculate the concentration of Cd^{2+} in the sample.

 25-14 Sulfate ion can be determined by an amperometric titration procedure using Pb^{2+} as the titrant. If the potential of a rotating mercury film electrode is adjusted to −1.00 V versus SCE, the current can be used to monitor the Pb^{2+} concentration during the titration. In a calibration experiment, the limiting current, after correction for background and residual currents, was found to be related to the Pb^{2+} concentration by $i_l = 10c_{Pb^{2+}}$, where i_l is the limiting current in mA and $c_{Pb^{2+}}$ is the Pb^{2+} concentration in mM. The titration reaction is

$$SO_4^{2-} + Pb^{2+} \rightleftharpoons PbSO_4(s) \qquad K_{sp} = 1.6 \times 10^{-8}$$

If 25 mL of 0.025 M Na_2SO_4 is titrated with 0.040 M $Pb(NO_3)_2$, develop the titration curve in spreadsheet format and plot the limiting current versus the volume of titrant.

25-15 Suppose that a spherical electrode can be fabricated from a single nano onion, $C_{60}-C_{240}-C_{540}-C_{960}-C_{1500}-C_{2160}-C_{2940}-C_{3840}-C_{4860}$, shown here. Nano onions comprise concentric fullerenes of increasingly larger size as indicated in the formula. Assume that the nano onion has been synthesized with a carbon nanotube tail of sufficient size and strength that it can be electrically connected to a 0.1 µm tungsten needle and that the needle and nanotube tail can be properly insulated so that only the surface of the nano onion is exposed.

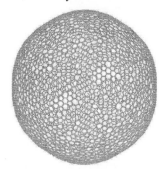

(a) Given that the radius of the nano onion is 3.17 nm, find the surface area in cm^2, neglecting the area of the nanotube attachment.
(b) If the diffusion coefficient of analyte A is 8×10^{-10} m^2/s, calculate the concentration gradient and the current for A at a concentration of 1.00 mM at the following times after the application of a voltage at which A is reduced: 1×10^{-8} s, 1×10^{-7} s, 1×10^{-6} s, 1×10^{-5} s, 1×10^{-4} s, 1×10^{-3} s, 1×10^{-2} s, 1×10^{-1} s, 1 s, and 10 s.
(c) Find the steady-state current.
(d) Find the time required for the electrode to achieve steady-state current following the application of the voltage step.
(e) Repeat these calculations for a 3-µm spherical platinum electrode and for a spherical iridium electrode with a surface area of 0.785 mm^2.
(f) Compare the results for the three electrodes, and discuss any differences that you find.

25-16 (a) What are the advantages of performing voltammetry with microelectrodes?
(b) Is it possible for an electrode to be too small? Explain your answer.

X *Challenge Problem*

25-17 A new method for determining ultrasmall (nL) volumes by anodic stripping voltammetry has been proposed (W. R. Vandaveer and I. Fritsch, *Anal. Chem.*, **2002**, *74*, 3575). In this method, a metal is exhaustively deposited from the small volume to be measured onto an electrode, from which it is later stripped. The solution volume V_s is related to the total charge Q required to strip the metal by

$$V_s = \frac{Q}{nFC}$$

where n is the number of moles of electrons per mole of analyte, F is the faraday, and C is the molar concentration of the metal ion before electrolysis.

(a) Beginning with Faraday's law (see Equation 22-8), derive the above equation for V_s.

(b) In one experiment, the metal deposited was $Ag(s)$ from a solution that was 8.00 mM in $AgNO_3$. The solution was electrolyzed for 30 min at a potential of -0.700 V versus a gold top layer as a pseudoreference. A tubular nanoband electrode was used. The silver was then anodically stripped off the electrode using a linear sweep rate of 0.10 V/s. The following table represents idealized anodic stripping results. By integration, determine the total charge required to strip the silver from the tubular electrode. You can do a manual Simpson's rule integration or do the integration with Excel.[49] From the charge, determine the volume of the solution from which the silver was deposited.

Potential, V	Current, nA	Potential, V	Current, nA
-0.50	0.000	-0.123	-1.10
-0.45	-0.02	-0.10	-0.80
-0.40	-0.001	-0.115	-1.00
-0.30	-0.10	-0.09	-0.65
-0.25	-0.20	-0.08	-0.52
-0.22	-0.30	-0.065	-0.37
-0.20	-0.44	-0.05	-0.22
-0.18	-0.67	-0.025	-0.12
-0.175	-0.80	0.00	-0.05
-0.168	-1.00	0.05	-0.03
-0.16	-1.18	0.10	-0.02
-0.15	-1.34	0.15	-0.005
-0.135	-1.28	—	—

(c) Suggest experiments to show whether all the Ag^+ was reduced to $Ag(s)$ in the deposition step.

(d) Would it matter if the droplet were not a hemisphere? Why or why not?

(e) Describe an alternative method against which you might test the proposed method.

---

[49]S. R. Crouch and F. J. Holler, *Applications of Microsoft® Excel in Analytical Chemistry*, Chap. 11, Belmont, CA: Brooks/Cole, 2004.

Measuring the Parts to Understand the Whole: The Microphysiometer

Traditionally, the biological sciences, biomedical sciences, and indeed science itself have relied on chemical analysis to provide fundamental information on the details of complex systems. Armed with analytical information often obtained at widely different times and many different places, scientists have sought to develop models of complex systems that might lead to understanding the detailed interactions of living things with their environments. In recent years, the emergent field of *systems biology* has approached the problem from the top down rather than the bottom up. By bringing to bear knowledge and tools from diverse fields, such as systems analysis, computation, experimental design, genomics, and analytical chemistry, models of intracellular biochemical and genetic interactions are being developed to aid in drug discovery and medical diagnosis. These models strongly depend on the availability of instrumentation to provide rapid, real-time, high-throughput, accurate, multivariate measurements on living organisms.[1]

Throughout Section 4, we have explored many electrochemical instruments and devices capable of rapidly and accurately determining analytes at low concentrations. One such device is the light addressable potentiometric sensor, or LAPS (see Section 23F-2). There have been a number of implementations of this device, and they have been used for studying a variety of biological systems.[2] These devices have been especially useful for real-time monitoring of extracellular acidification and changes in ionic composition resulting from stimulation of metabolic processes in living cells. To understand the significance of such measurements, it is useful to consider Figure IA4-1, which shows some of the processes that produce extracellular acidification.

For example, when membrane-bound receptors bind to appropriate ligands, metabolic processes such as respiration or glycolysis are stimulated to produce acidic waste products inside the cell. Protons are exchanged with sodium ions, and bicarbonate ion and chloride ion are exchanged across the cell membrane as a result. These exchange processes cause a decrease in the pH of the solution surrounding the cell. The LAPS-based microphysiometer is used to monitor these pH changes and can also measure changes in the concentrations of other important ionic species.

How the System Works

As described in Section 23F-2, the LAPS is a semiconductor device that is pH sensitive. When light impinges on its surface, which is maintained at a certain bias voltage versus a reference electrode, a photocurrent is produced that is proportional to the pH of the solution. Figure IA4-2 shows the basic elements of one implementation of a LAPS chip designed to monitor simultaneously the concentrations of H^+, Ca^{2+}, and K^+.[3]

Without modification, the LAPS chip is selective for H^+. As shown in Figure IA4-2, a potentiostat is used to bias the chip versus a reference electrode (RE). When light from one of the LEDs falls on the surface of the biased chip, a photocurrent appears between the counter electrode (CE) and the working electrode (WE). The current-to-voltage converter of the potentiostat produces a voltage proportional to pH that is then passed to a lock-in amplifier whose output is recorded with a digital oscilloscope. The oscilloscope is interfaced to a computer for data storage and manipulation. The response of the LAPS to changes in pH is shown in Figure IA4-3. In the experiments shown, the LAPS was exposed to a buffer, and the photocurrent was recorded as a function of bias voltage; the experiment was then repeated for two different buffers.

The vertical line indicates a bias voltage that is appropriate for measuring changes in pH from the change in the photocurrent (and resulting voltage change ΔV). The LAPS is made selective for other species by applying ionophores to different areas of the surface of the device (see Section 23D-6). LEDs are then positioned opposite each of the ionophoric areas as shown in Figure IA4-2 so that by alternately switching on each light source, photocurrents corresponding to each of selected species may be recorded in turn. Response curves such as those in Figure IA4-3 are recorded for each of the ionophores to calibrate the instrument for each detected species.

Although the three sensor areas of the LAPS can be addressed sequentially, they can also be addressed simultaneously using the arrangement shown in Figure IA4-2. In this scheme, the three LEDs corresponding to K^+, Ca^{2+}, and H^+ are modulated independently by a signal generator with three frequency outputs: 3 kHz, 3.5 kHz, and 4 kHz. Concentration information for each of the species is encoded in one

[1] H. Kitano, *Science*, **2002**, *295*, 1662.
[2] F. Hafner, *Biosens. Bioelectron.*, **2000**, *15*, 149.

[3] W. Yicong, W. Ping, Y. Xuesong, Z. Qingtao, L. Rong, Y. Weimin, and Z. Xiaoxiang, *Biosens. Bioelectron.*, **2001**, *16*, 277.

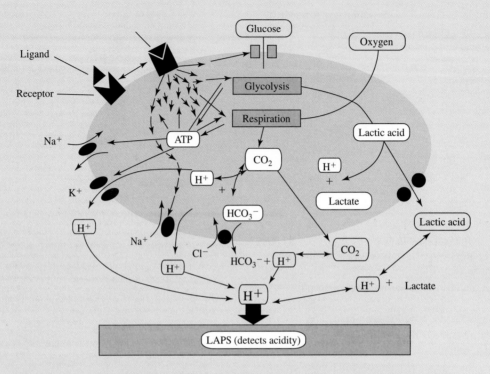

FIGURE IA4-1 The cell biology of extracellular acidification. When a receptor is stimulated, signal transduction pathways are induced. Adenosine triphosphate (ATP) consumption is then compensated by the increased uptake and metabolism of glucose, which results in an increase in the excretion of acid waste products. The extracellular acidification is measured by the LAPS. (From F. Hafner, *Biosens. Bioelectron.*, **2000**, *15*, 149. With permission.)

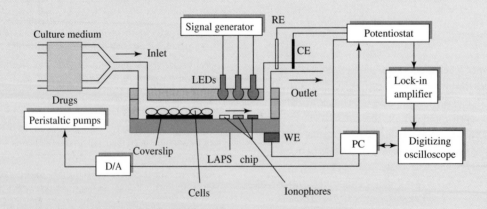

FIGURE IA4-2 Schematic diagram of a LAPS-based system for monitoring changes in extracellular H^+, Ca^{2+}, and K^+ concentrations in studies of the effects of drugs on living cells. RE = reference electrode, CE = counter electrode, WE = working electrode, PC = personal computer, D/A = digital-to-analog converter. (Adapted from W. Yicong et al., *Biosens. Bioelectron.*, **2001**, *16*, 277. With permission.)

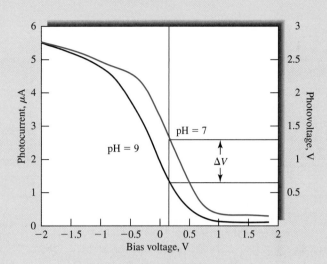

FIGURE IA4-3 The response of the LAPS to changes in bias voltage for two buffer solutions. The change in output voltage ΔV of the system at a constant bias voltage of 0.25 V is proportional to the change in pH on the surface of the LAPS. (Adapted from W. Yicong et al., *Biosens. Bioelectron.*, **2001**, *16*, 277. With permission.)

of these frequency components. For example, K^+ is encoded at 3 kHz, Ca^{2+} is encoded at 3.5 kHz, and H^+ is encoded at 4 kHz. The signal at the output of the potentiostat for a solution containing all three species is shown in Figure IA4-4a. This complex waveform is then subjected to Fourier transformation and digital filtering (see Section 5C-2) to produce the frequency spectrum of Figure IA4-4b. The amplitude of the signal for each species is then extracted from the spectrum, and its inverse transform provides a measure of the concentration of the selected species. At a data acquisition rate of 100 kHz, each 1024-point time-domain signal is acquired in about 10 ms. For experiments in which the concentrations of the three species are to be monitored as a function of time, time-domain signals are acquired and stored on the hard drive of the computer for subsequent analysis and plotting of the results.

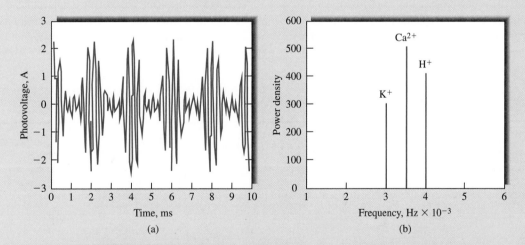

FIGURE IA4-4 (a) Time-domain LAPS output for a solution containing K^+, Ca^{2+}, and H^+. The signal was sampled at 100 kHz, and the time-domain signal contained 1024 points. (b) Fourier transform of the signal in (a). (Adapted from W. Yicong et al., *Biosens. Bioelectron.*, **2001**, *16*, 277. With permission.)

Studying the Effects of Drugs on Living Cells

The LAPS system described in the previous paragraphs can be used to study the effects of drugs on living cells placed directly on the sensor, as shown in Figure IA4-2. Two peristaltic pumps, which are controlled by the system computer, cause either or both of two solutions to flow past the cells to the active area of the LAPS. The first solution is RPMI 1640 culture medium (CM), and the other solution contains the drug being studied, which in this example is phenobarbital (PB). In a typical experiment, rat cardiac muscle cells were placed in the LAPS flow channel, and CM and PB were alternately passed over the cells and across the active area of the LAPS for 3-minute periods. Results from the LAPS data analysis are shown in Figure IA4-5, in which concentrations of K^+, Ca^{2+}, and H^+ are plotted versus time.

First, CM is passed over the cells for 3 minutes. During this time, all three ions generally decrease in concentration. Then, PB is passed over the cells for 3 minutes. About halfway through the PB period, Ca^{2+} and H^+ concentrations begin to increase and K^+ begins to decrease. These changes continue into the second CM period, and the change reverses for all three ions beginning at about the same point during the period. Subsequent periods repeat the same behavior as CM and PB are alternately passed over the cells.

In similar experiments, cells were exposed to other drugs, and the physiological responses of the cells were compared. Drugs that produce similar physiological responses are presumed to operate by similar mechanisms. Dilantin, for example, produces a somewhat different response pattern for the three monitored ions, which suggests that a different mechanism is active.

The LAPS microphysiometer combines several interesting technologies and measurement strategies to produce

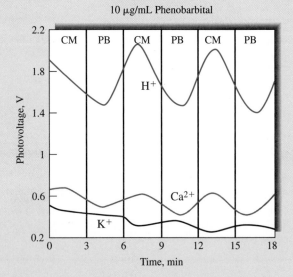

FIGURE IA4-5 Results of LAPS monitoring of K^+, Ca^{2+}, and H^+ while administering phenobarbital to rat cardiac muscle cells. (Adapted from W. Yicong et al., *Biosens. Bioelectron.*, **2001**, *16*, 277. With permission.)

multivariate experimental data on the production of ions on the surface of living cells. This example[4] shows how microminiaturization, advances in solid-state transducers and sensors, and modern data-acquisition and -manipulation methods continue to provide new tools for investigations at the forefront of chemistry, biology, biochemistry, systems biology, and other biosciences.

---

[4]For other related systems, see S. E. Eklund et al., *J. Electroanal. Chem.*, **2006**, *587*, 333; *Anal. Chem.*, **2004**, *76*, 519.

Photo courtesy of Perkin-Elmer Corp.

The photo shows a state-of-the-art LC system. On the right from top to bottom are solvent reservoirs, an automated sampler area, and two pumps that can be used to produce a binary gradient. At the top left is the thermostatted column area. A diode-array UV-visible detector is shown on the bottom left. (Courtesy of Perkin-Elmer, Inc., Shelton, CT.)

The methods described in this section are used to separate various components of analytical samples prior to their determination by instrumental methods. Chapter 26 begins this section with an introduction to the terminology and theory of chromatographic separations with an emphasis on optimization of experimental variables for efficient qualitative and quantitative analysis. This treatment is followed by a thorough discussion of the theory and practice of gas chromatography in Chapter 27. Liquid chromatography, the mainstay of analytical chromatography, is the subject of Chapter 28. Chapter 29 describes the techniques of supercritical fluid chromatography and supercritical fluid extraction. This section concludes with a discussion in Chapter 30 of the emerging fields of capillary electrophoresis, capillary electrochromatography, and field-flow fractionation. Applications of each method are included in the appropriate chapter.

CHAPTER TWENTY-SIX

An Introduction to Chromatographic Separations

T here are very few, if any, methods for chemical analysis that are specific for a single chemical species. At best, analytical methods are selective for a few species or a class of species. Consequently, the separation of the analyte from potential interferences is quite often a vital step in analytical procedures. Until the middle of the twentieth century, analytical separations were largely carried out by such classical methods as precipitation, distillation, and extraction. Today, however, analytical separations are most commonly carried out by chromatography and electrophoresis, particularly with samples that are multicomponent and complex.

Chromatography is a powerful separation method that finds applications in all branches of science. Chromatography was invented and named by the Russian botanist Mikhail Tswett shortly after the turn of the last century. He employed the technique to separate various plant pigments such as chlorophylls and xanthophylls by passing solutions of these compounds through a glass column packed with finely divided calcium carbonate. The separated species appeared as colored bands on the column, which accounts for the name he chose for the method (Greek *chroma* meaning "color" and *graphein* meaning "writing").

The applications of chromatography have grown explosively in the last half century, due not only to the development of several new types of chromatographic techniques but also to the growing need by scientists for better methods for characterizing complex mixtures. The tremendous impact of these methods on science is attested by the 1952 Nobel Prize in Chemistry that was awarded to A. J. P. Martin and R. L. M. Synge for their discoveries in the field. Many of the Nobel Prizes awarded since that time have been based on work in which chromatography played a vital role.

26A GENERAL DESCRIPTION OF CHROMATOGRAPHY

Chromatography encompasses a diverse and important group of methods that allow the separation, identification, and determination of closely related components of complex mixtures; many of these separations are impossible by other means.[1] In all chromatographic separations the sample is dissolved in a *mobile phase*, which may be a gas, a liquid, or a supercritical fluid. This mobile phase is then forced through an immiscible *stationary phase*, which is fixed in place in a column or on a solid surface. The two phases are chosen so that the components of the sample distribute themselves between the mobile and stationary phases to varying degrees. Those components strongly *retained* by the sta-

Throughout this chapter, this logo indicates an opportunity for online self-study at **www.thomsonedu.com/chemistry/skoog**, linking you to interactive tutorials, simulations, and exercises.

[1] Some general references on chromatography include J. M. Miller, *Chromatography: Concepts and Contrasts*, 2nd ed., New York: Wiley, 2005; *Chromatography: Fundamentals of Chromatography and Related Differential Migration Methods*, E. F. Heftman, ed., Amsterdam, Boston: Elsevier, 2004; C. F. Poole, *The Essence of Chromatography*, Amsterdam, Boston: Elsevier, 2003; J. Cazes and R. P. W. Scott, *Chromatography Theory*, New York: Dekker, 2002; A. Braithwaite and F. J. Smith, *Chromatographic Methods*, 5th ed., London: Blackie, 1996; R. P. W. Scott, *Techniques and Practice of Chromatography*, New York: Marcel Dekker, 1995; J. C. Giddings, *Unified Separation Science*, New York: Wiley, 1991.

TABLE 26-1 Classification of Column Chromatographic Methods

General Classification	Specific Method	Stationary Phase	Type of Equilibrium
1. Gas chromatography (GC)	a. Gas-liquid chromatography (GLC)	Liquid adsorbed or bonded to a solid surface	Partition between gas and liquid
	b. Gas-solid	Solid	Adsorption
2. Liquid chromatography (LC)	a. Liquid-liquid, or partition	Liquid adsorbed or bonded to a solid surface	Partition between immiscible liquids
	b. Liquid-solid, or adsorption	Solid	Adsorption
	c. Ion exchange	Ion-exchange resin	Ion exchange
	d. Size exclusion	Liquid in interstices of a polymeric solid	Partition/sieving
	e. Affinity	Group specific liquid bonded to a solid surface	Partition between surface liquid and mobile liquid
3. Supercritical fluid chromatography (SFC; mobile phase: supercritical fluid)		Organic species bonded to a solid surface	Partition between supercritical fluid and bonded surface

tionary phase move only slowly with the flow of mobile phase. In contrast, components that are weakly held by the stationary phase travel rapidly. As a consequence of these differences in migration rates, sample components separate into discrete *bands*, or *zones*, that can be analyzed qualitatively and quantitatively.

26A-1 Classification of Chromatographic Methods

Chromatographic methods can be categorized in two ways. The first classification is based on the physical means by which the stationary and mobile phases are brought into contact. In *column chromatography*, the stationary phase is held in a narrow tube through which the mobile phase is forced under pressure. In *planar chromatography*, the stationary phase is supported on a flat plate or in the interstices of paper; here, the mobile phase moves through the stationary phase by capillary action or under the influence of gravity. The discussion in this and the next three chapters focuses on column chromatography. Section 28I is devoted to planar methods. It is important to point out here, however, that the equilibria on which the two types of chromatography are based are identical and that the theory developed for column chromatography is readily adapted to planar chromatography.

A more fundamental classification of chromatographic methods is one based on the types of mobile and stationary phases and the kinds of equilibria involved in the transfer of solutes between phases. Table 26-1

lists three general categories of chromatography: *gas chromatography* (GC), *liquid chromatography* (LC), and *supercritical fluid chromatography* (SFC). As the names imply, the mobile phases in the three techniques are gases, liquids, and supercritical fluids, respectively. As shown in column 2 of the table, several specific chromatographic methods fall into each of the first two general categories.

Note that only LC can be performed either in columns or on planar surfaces; GC and SFC, on the other hand, are restricted to column procedures so that the column walls contain the mobile phase.

26A-2 Elution in Column Chromatography

Figure 26-1 shows schematically how two substances A and B are separated on a packed column by *elution*. The column consists of a narrow-bore tubing packed with a finely divided inert solid that holds the stationary phase on its surface. The mobile phase occupies the open spaces between the particles of the packing. Initially, a solution of the sample containing a mixture of A and B in the mobile phase is introduced at the head of the column as a narrow plug, as shown in Figure 26-1 at time t_0. Here, the two components distribute themselves between the mobile and the stationary phases. Elution involves washing a species through a column by continuous addition of fresh mobile phase. In modern chromatography, continuous addition is accomplished by means of a pump in LC and SFC or by the application of pressure in GC.

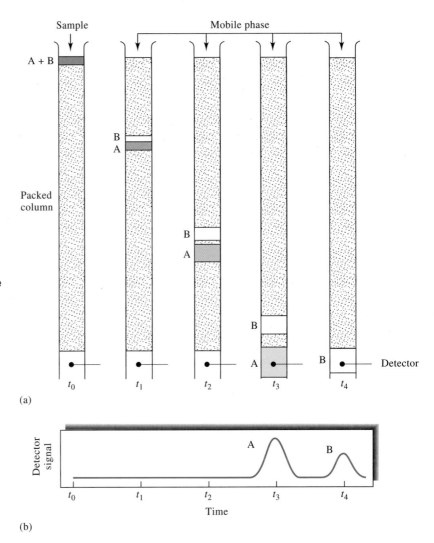

FIGURE 26-1 (a) Diagram showing the separation of a mixture of components A and B by column elution chromatography. (b) The output of the signal detector at the various stages of elution shown in (a).

With the first introduction of fresh mobile phase, the *eluent*— the portion of the sample contained in the mobile phase — moves down the column, where it further *partitions* between the mobile phase and the stationary phase (time t_1). Partitioning between the fresh mobile phase and the stationary phase takes place simultaneously at the site of the original sample.

As fresh mobile phase flows through the column, it carries solute molecules down the column in a continuous series of transfers between the two phases. Because solute movement can occur only in the mobile phase, the average *rate* at which a solute zone migrates down the column *depends on the fraction of time it spends in that phase*. This fraction is small for solutes

Simulation: Learn more about **elution chromatography**.

strongly retained by the stationary phase (component B in Figure 26-1) and large when the solute resides mostly in the mobile phase (component A). Ideally, the resulting differences in rates cause the components in a mixture to separate into *bands*, or *zones*, along the length of the column (see time t_2 in Figure 26-1). Isolation of the separated species is then accomplished by passing a sufficient quantity of mobile phase through the column to cause the individual zones to pass out the end (to be *eluted* from the column), where they can be detected or collected (times t_3 and t_4 in Figure 26-1).

Analyte Dilution

Figure 26-1 illustrates an important general characteristic of the chromatographic process — namely, dilution of analytes always accompanies chromatographic

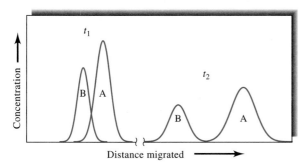

FIGURE 26-2 Concentration profiles of solute bands A and B at two different times in their migration down the column in Figure 26-1. The times t_1 and t_2 are indicated in Figure 26-1.

separation. Thus, the size of the original zone containing the analytes in the figure is noticeably smaller than either of the two zones that reach the detector, meaning that significant dilution of the analytes has occurred while they were being separated. As a result, the detectors employed for separated analytes must often be more sensitive than would be required if the separation process were unnecessary.

Chromatograms

If a detector that responds to solute concentration is placed at the end of the column and its signal is plotted as function of time (or of volume of the added mobile phase), a series of peaks is obtained, as shown in the lower part of Figure 26-1. Such a plot, called a *chromatogram*, is useful for both qualitative and quantitative analysis. The positions of peaks on the time axis can be used to identify the components of the sample; the areas under the peaks provide a quantitative measure of the amount of each component.

Improving Column Performance

Figure 26-2 shows concentration profiles for the bands containing solutes A and B in Figure 26-1 at time t_1 and a later time t_2.[2] Because species B is more strongly retained by the stationary phase than is A, B lags during the migration. Note that movement down the column increases the distance between the two bands. At the

same time, however, broadening of both zones takes place, which lowers the efficiency of the column as a separating device. While band broadening is inevitable, conditions can ordinarily be found where it occurs more slowly than band separation. Thus, as shown in Figures 26-1 and 26-2, a clean separation of species is often possible provided the column is sufficiently long.

Several chemical and physical variables influence the rates of band separation and band broadening. As a consequence, improved separations can often be realized by the control of variables that either (1) increase the rate of band separation or (2) decrease the rate of band spreading. These alternatives are illustrated in Figure 26-3. Variables that influence the relative migration rate of solute zones through a stationary phase are described in the next section. In Section 26C, consideration is given to factors that influence the rate of zone broadening.

26B   MIGRATION RATES OF SOLUTES

The effectiveness of a chromatographic column in separating two solutes depends in part on the relative rates at which the two species are eluted. These rates are

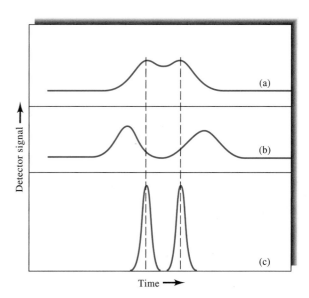

FIGURE 26-3 Two-component chromatogram illustrating two methods for improving separation: (a) original chromatogram with overlapping peaks; (b) improvement brought about by an increase in band separation; (c) improvement brought about by a decrease in the widths.

---

[2] Note that the relative positions of bands for A and B in the concentration profile shown in Figure 26-2 appear to be reversed from their positions in the lower part of Figure 26-1. The difference is that the abscissa represents distance along the column in Figure 26-2 but time in Figure 26-1. Thus, in Figure 26-1 the front of a peak lies to the left and the tail to the right; in Figure 26-2, the reverse is true.

determined by the magnitude of the equilibrium constants for the reactions by which the solutes distribute themselves between the mobile and stationary phases.

26B-1 Distribution Constants

Often, the distribution equilibria involved in chromatography are described by relatively straightforward equations that involve the transfer of an analyte between the mobile and stationary phases. Thus, for the solute species A, we may write

$$A_{mobile} \rightleftharpoons A_{stationary}$$

The equilibrium constant K_c for the distribution of species A between the two phases is called the *distribution constant*, which is defined as

$$K_c = \frac{(a_A)_S}{(a_A)_M} \tag{26-1}$$

where $(a_A)_S$ is the activity of solute A in the stationary phase and $(a_A)_M$ is the activity in the mobile phase. When concentrations are low or when nonionic species are involved, activity coefficients (see Appendix 2) are nearly unity. Under these conditions, we often substitute for activities c_S, the molar analytical concentration of the solute in the stationary phase, and c_M, its molar analytical concentration in the mobile phase. Hence, we often write Equation 26-1 as

$$K_c = \frac{c_S}{c_M} = \frac{n_S/V_S}{n_M/V_M} \tag{26-2}$$

where n_S and n_M are the number of moles of analyte in the two phases and V_S and V_M are the volumes of the two phases. Ideally, the distribution constant, sometimes called a *partition ratio* or *partition coefficient*,[3] is constant over a wide range of solute concentrations; that is, c_S is directly proportional to c_M. Chromatography in which Equation 26-2 applies is termed *linear chromatography* and results in such characteristics as symmetric Gaussian-type peaks and retention times independent of the amount of analyte injected. We limit our theoretical discussions here to linear chromatography.

---

[3] For recommendations of the IUPAC Commission on Analytical Nomenclature for chromatography, see L. S. Ettre, *Pure Appl. Chem.*, **1993**, *65*, 819. The committee recommends the use of the term distribution constant K_c rather than the older term "partition coefficient," or "partition ratio," K. Be aware that both terms are found in the chromatographic literature.

Note that K_c is the fundamental quantity affecting distribution of components between phases and thus separations. By appropriate choice of the mobile phase, the stationary phase, or both, the distribution constant can be manipulated within limits. By adjusting the volume of a phase, we can alter the molar ratio in the two phases.

26B-2 Retention Time

Although the distribution constant is fundamental to chromatographic separations, it is not readily measured. Instead, we can measure a quantity called the *retention time* that is a function of K_c. To see how this is done, let us refer to Figure 26-4, a simple chromatogram made up of two peaks. The small peak on the left is for a species that is *not* retained by the column. Often, the sample or the mobile phase contains an unretained species. When it does not, such a species may be added to aid in peak identification. The time t_M for the unretained species to reach the detector is sometimes called the *dead*, or *void*, *time*. The dead time provides a measure of the average rate of migration of the mobile phase and is an important parameter in identifying analyte peaks. All components spend time t_M in the mobile phase. The larger peak on the right in Figure 26-4 is that of an analyte species. The time required for this zone to reach the detector after sample injection is called the *retention time* and is given the symbol t_R. The analyte has been retained because it spends a time t_S in the stationary phase. The retention time is then

$$t_R = t_S + t_M \tag{26-3}$$

The average linear rate of solute migration through the column $\bar{v}$ (usually cm/s) is

$$\bar{v} = \frac{L}{t_R} \tag{26-4}$$

where L is the length of the column packing. Similarly, the average linear velocity u of the mobile-phase molecules is

$$u = \frac{L}{t_M} \tag{26-5}$$

26B-3 The Relationship between Volumetric Flow Rate and Linear Flow Velocity

Experimentally, in chromatography the mobile-phase flow is usually characterized by the volumetric flow rate F (cm³/min) at the column outlet. For an open tu-

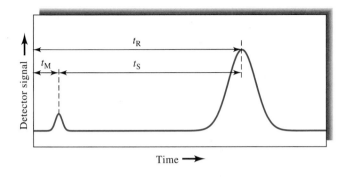

FIGURE 26-4 A typical chromatogram for a two-component mixture. The small peak on the left represents a solute that is not retained on the column and so reaches the detector almost immediately after elution is begun. Thus, its retention time t_M is approximately equal to the time required for a molecule of the mobile phase to pass through the column.

bular column, F is related to the linear velocity at the column outlet u_o

$$F = u_o A = u_o \times \pi r^2 \qquad (26\text{-}6)$$

where A is the cross-sectional area of the tube (πr^2). For a packed column, the entire column volume is not available to the liquid and so Equation 26-6 must be modified to

$$F = \pi r^2 u_o \varepsilon \qquad (26\text{-}7)$$

where ε is the fraction of the total column volume available to the liquid (column porosity).

26B-4 The Relationship between Retention Time and Distribution Constant

To relate the rate of migration of a solute to its distribution constant, we express the rate as a fraction of the velocity of the mobile phase:

$$\bar{v} = u \times (\text{fraction of time solute spends in mobile phase})$$

This fraction, however, equals the average number of moles of solute in the mobile phase at any instant divided by the total number of moles of solute in the column:

$$\bar{v} = u \times \frac{\text{moles of solute in mobile phase}}{\text{total moles of solute}}$$

The total number of moles of solute in the mobile phase is equal to the molar concentration, c_M, of the solute in that phase multiplied by its volume, V_M. Similarly, the number of moles of solute in the stationary phase is given by the product of the concentration, c_S, of the solute in the stationary phase and its volume, V_S. Therefore,

$$\bar{v} = u \times \frac{c_M V_M}{c_M V_M + c_S V_S} = u \times \frac{1}{1 + c_S V_S / c_M V_M}$$

Substitution of Equation 26-2 into this equation gives an expression for the rate of solute migration as a function of its distribution constant as well as a function of the volumes of the stationary and mobile phases:

$$\bar{v} = u \times \frac{1}{1 + K_c V_S / V_M} \qquad (26\text{-}8)$$

The two volumes can be estimated from the method by which the column is prepared.

26B-5 The Rate of Solute Migration: The Retention Factor

The *retention factor k* is an important experimental quantity widely used to compare the migration rates of solutes in columns.[4] The reason that k is so useful is that it does not depend on column geometry or on volumetric flow rate. This means that for a given combination of solute, mobile phase, and stationary phase any column of any geometry operated at any mobile-phase flow rate will give the same retention factor. For solute A, the retention factor k_A is defined as

$$k_A = \frac{K_A V_S}{V_M} \qquad (26\text{-}9)$$

where K_A is the distribution constant for solute A. Substitution of Equation 26-9 into 26-8 yields

$$\bar{v} = u \times \frac{1}{1 + k_A} \qquad (26\text{-}10)$$

---

[4]In the older literature, this constant was called the capacity factor and symbolized by k'. In 1993, however, the IUPAC Commission on Analytical Nomenclature recommended that this constant be termed the *retention factor* and symbolized by k.

To show how k_A can be derived from a chromatogram, we substitute Equations 26-4 and 26-5 into Equation 26-10:

$$\frac{L}{t_R} = \frac{L}{t_M} \times \frac{1}{1 + k_A} \qquad (26\text{-}11)$$

This equation rearranges to

$$k_A = \frac{t_R - t_M}{t_M} \qquad (26\text{-}12)$$

Note that the time spent in the stationary phase, $t_R - t_M$, is sometimes called the *adjusted retention time* and given the symbol t'_R.

As shown in Figure 26-4, t_R and t_M are readily obtained from a chromatogram. The retention factor is then found from these quantities and Equation 26-12. A retention factor much less than unity means that the solute emerges from the column at a time near that of the void time. When the retention factor is larger than perhaps 20 to 30, elution times become inordinately long. Ideally, separations are performed under conditions in which the retention factors for the solutes in a mixture lie in the range between 1 and 10. Optimization of separations (see Section 26D) is then done by optimizing k values for the components of interest.

26B-6 Relative Migration Rates: The Selectivity Factor

The *selectivity factor* α of a column for the two solutes A and B is defined as

$$\alpha = \frac{K_B}{K_A} \qquad (26\text{-}13)$$

where K_B is the distribution constant for the more strongly retained species B and K_A is the constant for the less strongly held or more rapidly eluted species A. According to this definition, α *is always greater than unity*.

Substitution of Equation 26-9 and the analogous equation for solute B into Equation 26-13 provides a relationship between the selectivity factor for two solutes and their retention factors:

$$\alpha = \frac{k_B}{k_A} \qquad (26\text{-}14)$$

where k_B and k_A are the retention factors for B and A, respectively. Substitution of Equation 26-12 for the

two solutes into Equation 26-14 gives an expression that permits the determination of α from an experimental chromatogram:

$$\alpha = \frac{(t_R)_B - t_M}{(t_R)_A - t_M} \qquad (26\text{-}15)$$

In Section 26D-2, we show how we use the selectivity and retention factors to compute the resolving power of a column.

26C BAND BROADENING AND COLUMN EFFICIENCY

The efficiency of a chromatographic column is affected by the amount of band broadening that occurs as a compound passes through the column. Before defining column efficiency in more quantitative terms, let us examine the reasons that bands become broader as they move down a column.

26C-1 The Rate Theory of Chromatography

The *rate theory* of chromatography describes the shapes and breadths of elution bands in quantitative terms based on a random-walk mechanism for the migration of molecules through a column. A detailed discussion of the rate theory is beyond the scope of this text. We can, however, give a qualitative picture of why bands broaden and what variables improve column efficiency.[5]

If you examine the chromatograms shown in this and the next chapter, you will see that the elution peaks look very much like the Gaussian or normal error curves (see Appendix 1). As shown in Appendix 1, Section a1B-1, normal error curves are rationalized by assuming that the uncertainty associated with any single measurement is the summation of a much larger number of small, individually undetectable and random uncertainties, each of which has an equal probability of being positive or negative. In a similar way, the typical Gaussian shape of a chromatographic band can be attributed to the additive combination of the random motions of the various molecules as they move down the column. We assume in the following discussion that a narrow zone has been introduced so that

---

[5]For more information, see J. C. Giddings, *Unified Separation Science*, New York: Wiley, 1991, pp. 94–96.

the injection width is not the limiting factor determining the overall width of the band that elutes. It is important to realize that the widths of eluting bands can never be *narrower* than the width of the injection zone.

It is instructive to consider a single solute molecule as it undergoes many thousands of transfers between the stationary and the mobile phases during elution. Residence time in either phase is highly irregular. Transfer from one phase to the other requires energy, and the molecule must acquire this energy from its surroundings. Thus, the residence time in a given phase may be transitory after some transfers and relatively long after others. Recall that movement down the column can occur *only while the molecule is in the mobile phase*. As a consequence, certain particles travel rapidly by virtue of their accidental inclusion in the mobile phase for most of the time whereas others lag because they happen to be incorporated in the stationary phase for a greater-than-average length of time. The result of these random individual processes is a symmetric spread of velocities around the mean value, which represents the behavior of the average analyte molecule.

The breadth of a zone increases as it moves down the column because more time is allowed for spreading to occur. Thus, the zone breadth is directly related to residence time in the column and inversely related to the velocity of the mobile-phase flow.

As shown in Figure 26-5, some chromatographic peaks are nonideal and exhibit *tailing* or *fronting*. In the former case the tail of the peak, appearing to the right on the chromatogram, is drawn out and the front is steepened. With fronting, the reverse is the case. A common cause of tailing and fronting is a distribution constant that varies with concentration. Fronting also arises when the amount of sample introduced onto a column is too large. Distortions of this kind are undesirable because they lead to poorer separations and less reproducible elution times. In the discussion that follows, tailing and fronting are assumed to be minimal.

26C-2 A Quantitative Description of Column Efficiency

Two related terms are widely used as quantitative measures of chromatographic column efficiency: (1) *plate height H* and (2) *plate count*, or *number of theoretical plates*, *N*. The two are related by the equation

$$N = \frac{L}{H} \qquad (26\text{-}16)$$

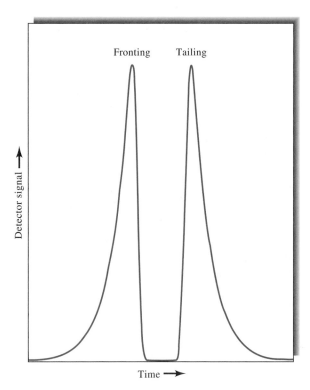

FIGURE 26-5 Illustration of fronting and tailing in chromatographic peaks.

where *L* is the length (usually in centimeters) of the column packing. The efficiency of chromatographic columns increases as the plate count *N* becomes greater and as the plate height *H* becomes smaller. Enormous differences in efficiencies are encountered in columns as a result of differences in column type and in mobile and stationary phases. Efficiencies in terms of plate numbers can vary from a few hundred to several hundred thousand; plate heights ranging from a few tenths to one thousandth of a centimeter or smaller are not uncommon.

The genesis of the terms "plate height" and "number of theoretical plates" is a pioneering theoretical study of Martin and Synge in which they treated a chromatographic column as if it were similar to a distillation column made up of numerous discrete but contiguous narrow layers called *theoretical plates*.[6] At each plate, equilibration of the solute between the mobile and stationary phases was assumed to take place. Movement of the solute down the column was then treated as a stepwise transfer of equilibrated mobile phase from one plate to the next.

[6] A. J. P. Martin and R. L. M. Synge, *Biochem. J.*, **1941**, *35*, 1358.

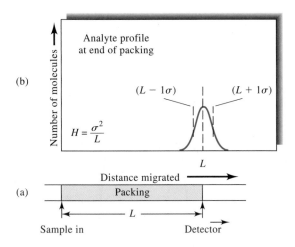

FIGURE 26-6 Definition of plate height $H = \sigma^2/L$. In (a) the column length is shown as the distance from the sample entrance point to the detector. In (b) the Gaussian distribution of sample molecules is shown.

The *plate theory* successfully accounts for the Gaussian shape of chromatographic peaks and their rate of movement down a column. The theory was ultimately abandoned in favor of the *rate theory*, however, because it fails to account for peak broadening in a mechanistic way. Nevertheless, the original terms for efficiency have been carried over to the rate theory. This nomenclature is perhaps unfortunate because it tends to perpetuate the myth that a column contains plates where equilibrium conditions exist. In fact, the equilibrium state can never be realized with the mobile phase in constant motion.

The Definition of Plate Height

As shown in Section a1B-1 of Appendix 1, the breadth of a Gaussian curve is described by its standard deviation σ or its variance σ^2. Because chromatographic bands are often assumed to be Gaussian in shape, it is convenient to define the efficiency of a column in terms of variance per unit length of column. That is, the plate height H is given by

$$H = \frac{\sigma^2}{L} \qquad (26\text{-}17)$$

This definition of column efficiency is illustrated in Figure 26-6a, which shows a column having a packing L cm in length. Above this schematic (Figure 26-6b) is a plot showing the distribution of molecules along the length of the column at the moment the analyte peak reaches the end of the packing (that is, at the retention time). The curve is Gaussian, and the locations of $L + 1\sigma$ and $L - 1\sigma$ are indicated as broken vertical lines. Note that L carries units of centimeters and σ^2 units of centimeters squared; thus H represents a linear distance in centimeters as well (Equation 26-17). In fact, the plate height can be thought of as the length of column that contains a fraction of the analyte that lies between L and $L - \sigma$. Because the area under a normal error curve bounded by plus or minus one standard deviation ($\pm 1\sigma$) is about 68% of the total area, the plate height, as defined, contains approximately 34% of the analyte.

The Experimental Evaluation of H and N

Figure 26-7 is a typical chromatogram with time as the abscissa. The variance of the solute peak, which can be obtained by a straightforward graphical procedure, has units of seconds squared and is usually designated as τ^2 to distinguish it from σ^2, which has units of centimeters squared. The two standard deviations τ and σ are related by

$$\tau = \frac{\sigma}{L/t_R} \qquad (26\text{-}18)$$

FIGURE 26-7 Determination of the number of plates $N = 16\left(\dfrac{t_R}{W}\right)^2$.

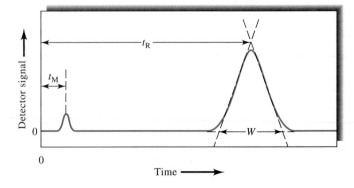

where L/t_R is the average linear velocity $\bar{\nu}$ of the solute in centimeters per second (Equation 26-4).

Figure 26-7 illustrates a method for approximating τ and σ from an experimental chromatogram. Tangents at the inflection points on the two sides of the chromatographic peak are extended to form a triangle with the baseline of the chromatogram. The area of this triangle can be shown to be approximately 96% of the total area under the peak if the peak is Gaussian. In Section a1B-1 of Appendix 1 it is shown that about 96% of the area under a Gaussian peak is included within plus or minus two standard deviations ($\pm 2\sigma$) of its maximum. Thus, the intercepts shown in Figure 26-7 occur at approximately plus or minus two standard deviations of the solute peak ($\pm 2\tau$) from the maximum, and $W = 4\tau$, where W is the magnitude of the base of the triangle. Substituting this relationship into Equation 26-18 and rearranging yields

$$\sigma = \frac{LW}{4t_R} \qquad (26\text{-}19)$$

Substitution of this equation for σ into Equation 26-17 gives

$$H = \frac{LW^2}{16t_R^2} \qquad (26\text{-}20)$$

To obtain N, we substitute into Equation 26-16 and rearrange to get

$$N = 16\left(\frac{t_R}{W}\right)^2 \qquad (26\text{-}21)$$

Thus, N can be calculated from two time measurements, t_R and W; to obtain H, the length of the column packing L must also be known. Note that these calculations are only approximate and assume Gaussian peak shapes.

Another method for approximating N, which some workers believe to be more reliable, is to determine $W_{1/2}$, the width of the peak at half its maximum height. The plate count is then given by

$$N = 5.54\left(\frac{t_R}{W_{1/2}}\right)^2 \qquad (26\text{-}22)$$

Because the experimental determinations of H and N described here are based on Gaussian chromatographic peaks, the calculations are considered estimates. More accurate methods for dealing with skewed Gaussian peaks have been described in the literature; they are based on finding the variance of the peaks using statistical second-moment calculations.[7]

TABLE 26-2 Variables That Influence Column Efficiency

Variable	Symbol	Usual Units
Linear velocity of mobile phase	u	cm s^{-1}
Diffusion coefficient in mobile phase*	D_M**	cm^2 s^{-1}
Diffusion coefficient in stationary phase*	D_S	cm^2 s^{-1}
Retention factor (Equation 26-12)	k	unitless
Diameter of packing particles	d_p	cm
Thickness of liquid coating on stationary phase	d_f	cm

*Increases as temperature increases and viscosity decreases.

**In liquids, D_M values for small and medium-size molecules are on the order of 10^{-5} cm^2 s^{-1}; for gases, D_M values are ~10^5 times larger.

The plate count N and the plate height H are widely used in the literature and by instrument manufacturers. They can be quite useful quantities in comparing separating power and efficiencies among columns. However, for these numbers to be meaningful in comparing columns, it is essential that they be determined with the *same compound*.

26C-3 Kinetic Variables Affecting Column Efficiency

Band broadening reflects a loss of column efficiency. The slower the rate of mass-transfer processes occurring while a solute migrates through a column, the broader the band at the column exit. Some of the variables that affect mass-transfer rates are controllable and can be exploited to improve separations. Table 26-2 lists the most important of these variables. Their effects on column efficiency, as measured by plate height H, are described in the paragraphs that follow.

The Effect of Mobile-Phase Flow Rate

The extent of band broadening depends on the length of time the mobile phase is in contact with the stationary phase, which in turn depends on the flow rate of the mobile phase. For this reason, efficiency studies have generally been carried out by determining H (by means

[7] J. P. Foley and J. G. Dorsey, *Anal. Chem.*, **1983**, *55*, 730; *J. Chromatogr. Sci.*, **1984**, *22*, 40. For a spreadsheet approach to nonideal peak shapes, see S. R. Crouch and F. J. Holler, *Applications of Microsoft® Excel in Analytical Chemistry*, Belmont, CA: Brooks/Cole, 2004.

 Tutorial: Learn more about **column efficiency**.

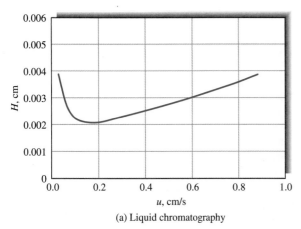

(a) Liquid chromatography

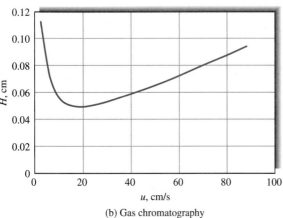

(b) Gas chromatography

FIGURE 26-8 Effect of mobile-phase flow rate on plate height for (a) LC and (b) GC. Note very different flow rate and plate height scales.

of Equations 26-21 or 26-22 and Equation 26-16) as a function of mobile-phase velocity. The plots for LC and for GC shown in Figure 26-8 are typical of the data obtained from such studies. Although both show a minimum in H (or a maximum in efficiency) at low linear flow rates, the minimum for LC usually occurs at flow rates well below those for GC. Often these flow rates are so low that the minimum H is not observed for LC under normal operating conditions. The rate theory of chromatographic zone broadening, which is described in this section, accurately predicts the shape of a plot of H versus u, which is often called a *van Deemter plot* after the originator of the theory.

As suggested by Figure 26-8, flow rates for LC are significantly smaller than those used in GC. This means that gas chromatographic separations are typically completed in shorter times than are liquid chromatographic separations. Furthermore, as shown in

the figure, plate heights for liquid chromatographic columns are an order of magnitude or more smaller than those encountered with gas chromatographic columns. Offsetting this advantage, however, is the fact that LC columns longer than about 25 cm are impractical because of high pressure drops, whereas GC columns may be 50 m long or more. Consequently, the total number of plates is greater, and thus overall column efficiency is usually superior with gas chromatographic columns. Thus, one comparison between GC and LC is that GC is capable of faster separations and higher-efficiency separations, although not necessarily both simultaneously.

Theory of Band Broadening

Over the last 40 years, an enormous amount of theoretical and experimental effort has been devoted to developing quantitative relationships describing the effects of the experimental variables listed in Table 26-2 on plate heights for various types of columns. Perhaps a dozen or more expressions for calculating plate height have been proposed and applied with various degrees of success. None of these is entirely adequate to explain the complex physical interactions and effects that lead to zone broadening and thus lower column efficiencies. Some of the equations, though imperfect, have been quite useful, however, in pointing the way toward improved column performance.

The efficiency of chromatographic columns can be approximated by the expression

$$H = A + \frac{B}{u} + C_S u + C_M u \qquad (26\text{-}23)$$

where H is the plate height in centimeters and u is the linear velocity of the mobile phase in centimeters per second. The quantity A is a coefficient that describes *multiple path effects* (eddy diffusion) as discussed later, B is the *longitudinal diffusion coefficient*, and C_S and C_M are *mass-transfer coefficients* for the stationary and mobile phases, respectively. Equation 26-23 is equivalent to the well-known *van Deemter equation*, developed by Dutch chemical engineers in the 1950s and often used to describe chromatographic efficiency.

More recent studies have led to elaborations of the basic van Deemter expression, but it has been shown experimentally that Equation 26-23 is quite satisfactory in explaining column efficiency.[8] Note that the van Deemter expression contains terms linearly and in-

---

[8] E. Katz, K. L. Ogan, and R. P. W. Scott, *J. Chromatogr.*, **1983**, *270*, 51.

26C Band Broadening and Column Efficiency **773**

TABLE 26-3 Processes That Contribute to Band Broadening

Process	Term in Equation 26-23	Relationship to Column* and Analyte Properties
Multiple flow paths	A	$A = 2\lambda d_p$
Longitudinal diffusion	B/u	$\dfrac{B}{u} = \dfrac{2\gamma D_M}{u}$
Mass transfer to and from stationary phase	$C_S u$	$C_S u = \dfrac{f(k)d_f^2}{D_S}u$
Mass transfer in mobile phase	$C_M u$	$C_M u = \dfrac{f'(k)d_p^2}{D_M}u$

*u, D_S, D_M, d_f, d_p, and k are as defined in Table 26-2.

$f(k)$ and $f'(k)$ are functions of k.

λ and γ are constants that depend on quality of the packing.

B is the coefficient of longitudinal diffusion.

C_S and C_M are coefficients of mass transfer in stationary and mobile phase, respectively.

versely proportional to, as well as independent of, the mobile-phase velocity.

Let us now examine in some detail the variables that affect the four terms in Equation 26-23. These terms are identified in Table 26-3 and described in the paragraphs that follow.[9]

The Multipath Term A. Zone broadening in the mobile phase is due in part to the multitude of pathways by which a molecule (or ion) can find its way through a packed column. As shown in Figure 26-9, the length of these pathways may differ significantly; thus, the residence time in the column for molecules of the same species is also variable. Solute molecules then reach the end of the column over a time interval, which leads to a broadened band. This multiple path effect, which is sometimes called *eddy diffusion*, would be independent of solvent velocity if it were not partially offset by ordinary diffusion, which results in molecules being transferred from a stream following one pathway to a stream following another. If the velocity of flow is very low, a large number of these transfers will occur, and each molecule in its movement down the column will sample numerous flow paths, spending a brief time in each. As a result, the rate at which each molecule

[9]For a spreadsheet approach to calculations using the van Deemter equation, see S. R. Crouch and F. J. Holler, *Applications of Microsoft® Excel in Analytical Chemistry*, Belmont, CA: Brooks/Cole, 2004.

moves down the column tends to approach that of the average. Thus, at low mobile-phase velocities, the molecules are not significantly dispersed by the multiple-path effect. At moderate or high velocities, however, sufficient time is not available for diffusion averaging to occur, and band broadening due to the different path lengths is observed. At sufficiently high velocities, the effect of eddy diffusion becomes independent of flow rate.

Superimposed on the eddy-diffusion effect is one that arises from stagnant pools of the mobile phase retained in the stationary phase. Thus, when a solid serves as the stationary phase, its pores are filled with *static* volumes of mobile phase. Solute molecules must then diffuse through these stagnant pools before transfer can occur between the *moving* mobile phase and the stationary phase. This situation applies not only to solid stationary phases but also to liquid stationary phases immobilized on porous solids because the immobilized liquid does not usually fully fill the pores.

The presence of stagnant pools of mobile phase slows the exchange process and produces a contribution to the plate height directly proportional to the mobile-phase velocity and inversely proportional to the diffusion coefficient for the solute in the mobile phase. An increase in internal volume then accompanies increases in particle size.

The Longitudinal Diffusion Term B/u. Diffusion is a process in which species migrate from a more concentrated part of a medium to a more dilute region. The

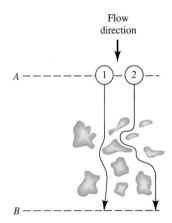

FIGURE 26-9 Typical pathways of two molecules during elution. Note that the distance traveled by molecule 2 is greater than that traveled by molecule 1. Thus, molecule 2 will arrive at B later than molecule 1.

rate of migration is proportional to the concentration difference between the regions and to the *diffusion coefficient D_M* of the species. The latter, which is a measure of the mobility of a substance in a given medium, is a constant for a given species equal to the velocity of migration under a unit concentration gradient.

In chromatography, longitudinal diffusion results in the migration of a solute from the concentrated center of a band to the more dilute regions on either side (that is, toward and opposed to the direction of flow). Longitudinal diffusion is a common source of band broadening in GC because gaseous molecules diffuse at relatively high rates. The phenomenon is of little significance in LC where diffusion rates are much smaller. The magnitude of the *B* term in Equation 26-23 is largely determined by the diffusion coefficient D_M of the analyte in the mobile phase and is directly proportional to this constant.

As shown by Equation 26-23, the contribution of longitudinal diffusion to plate height is inversely proportional to the linear velocity of the eluent. This relationship is not surprising because the analyte is in the column for a shorter period when the flow rate is high. Thus, diffusion from the center of the band to the two edges has less time to occur.

The initial decreases in *H* shown in both curves in Figure 26-8 are a direct result of the longitudinal diffusion. Note that the effect is much less pronounced in LC because of the much lower diffusion rates in a liquid mobile phase. The striking difference in plate heights shown by the two curves in Figure 26-8 can also be explained by considering the relative rates of longitudinal diffusion in the two mobile phases. That is, diffusion coefficients in gaseous media are orders of magnitude larger than in liquids. Thus band broadening occurs to a much greater extent in GC than in LC.

The Stationary-Phase Mass-Transfer Term $C_S u$. When the stationary phase is an immobilized liquid, the mass-transfer coefficient is directly proportional to the square of the thickness of the film on the support particles d_f^2 and inversely proportional to the diffusion coefficient D_S of the solute in the film. These effects can be understood by realizing that both reduce the average frequency at which analyte molecules reach the liquid-liquid interface where transfer to the mobile phase can occur. That is, with thick films, molecules must on the average travel farther to reach the surface, and with smaller diffusion coefficients, they travel

slower. The consequence is a slower rate of mass transfer and an increase in plate height.

When the stationary phase is a solid surface, the mass-transfer coefficient C_S is directly proportional to the time required for a species to be adsorbed or desorbed, which in turn is inversely proportional to the first-order rate constant for the processes.

The Mobile-Phase Mass-Transfer Term $C_M u$. The mass-transfer processes that occur in the mobile phase are sufficiently complex that we do not yet have a complete quantitative description. On the other hand, we have a good qualitative understanding of the variables affecting zone broadening from this cause, and this understanding has led to vast improvements in all types of chromatographic columns.

The mobile-phase mass-transfer coefficient C_M is known to be inversely proportional to the diffusion coefficient of the analyte in the mobile phase D_M. For packed columns, C_M is proportional to the square of the particle diameter of the packing material, d_p^2. For open tubular columns, C_M is proportional to the square of the column diameter, d_c^2.

The contribution of mobile-phase mass transfer to plate height is the product of the mass-transfer coefficient C_M (which is a function of solvent velocity) and the velocity of the solvent itself. Thus, the net contribution of $C_M u$ to plate height is not linear in u (see the curve labeled $C_M u$ in Figure 26-10) but bears a complex dependency on solvent velocity.

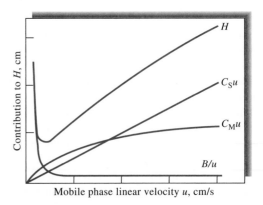

FIGURE 26-10 Contribution of various mass-transfer terms to plate height. $C_S u$ arises from the rate of mass transfer to and from the stationary phase, $C_M u$ comes from a limitation in the rate of mass transfer in the mobile phase, and B/u is associated with longitudinal diffusion.

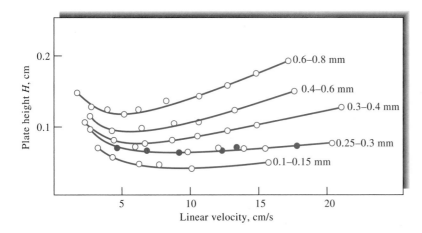

FIGURE 26-11 Effect of particle size on plate height for a packed GC column. The numbers to the right of each curve are particle diameters. (From J. Boheman and J. H. Purnell, in *Gas Chromatography 1958*, D. H. Desty, ed., New York: Academic Press, 1958. With permission.)

Effect of Mobile-Phase Velocity on Terms in Equation 26-23. Figure 26-10 shows the variation of the three terms in Equation 26-23 as a function of mobile-phase velocity. The top curve is the summation of these various effects. Note that there is an optimal flow rate at which the plate height is a minimum and the separation efficiency is a maximum.

Summary of Methods for Reducing Band Broadening. For packed columns, the most important variable that affects column efficiency is the diameter of the particles making up the packing. For open tubular columns, the diameter of the column itself is an important variable. The effect of particle diameter is demonstrated by the data shown in Figure 26-11 for GC. A similar plot for LC is shown in Figure 28-2. To take advantage of the effect of column diameter, narrower and narrower columns have been used in recent years.

With gaseous mobile phases, the rate of longitudinal diffusion can be reduced appreciably by lowering the temperature and thus the diffusion coefficient. When operating at low mobile-phase velocities where the B/u term is important (see Figure 26-10), lower temperatures can give smaller plate heights. This effect is usually not noticeable in LC because diffusion is slow enough so that the longitudinal diffusion term has little effect on overall plate height. Also, most chromatography is done with mobile-phase velocities higher than the optimal velocity so that mass-transfer terms (C terms) control column efficiency.

With liquid stationary phases, the thickness of the layer of adsorbed liquid should be minimized because C_S in Equation 26-23 is proportional to the square of this variable.

26D OPTIMIZATION OF COLUMN PERFORMANCE

A chromatographic separation is optimized by varying experimental conditions until the components of a mixture are separated cleanly in a minimum amount of time. Optimization experiments are aimed at either (1) reducing zone broadening or (2) altering relative migration rates of the components. As we have shown in Section 26C, zone broadening is increased by those kinetic variables that increase the plate height of a column. Migration rates, on the other hand, are varied by changing those variables that affect retention and selectivity factors (Section 26B).

26D-1 Column Resolution

The *resolution* R_s of a column tells us how far apart two bands are relative to their widths. The resolution provides a quantitative measure of the ability of the column to separate two analytes. The significance of this term is illustrated in Figure 26-12, which consists of chromatograms for species A and B on three columns with different resolving powers. The resolution of each column is defined as

$$R_s = \frac{\Delta Z}{\frac{W_A}{2} + \frac{W_B}{2}} = \frac{2\Delta Z}{W_A + W_B}$$

$$= \frac{2[(t_R)_B - (t_R)_A]}{W_A + W_B} \tag{26-24}$$

where all of the terms on the right side are as defined in the figure.

 Tutorial: Learn more about **column resolution**.

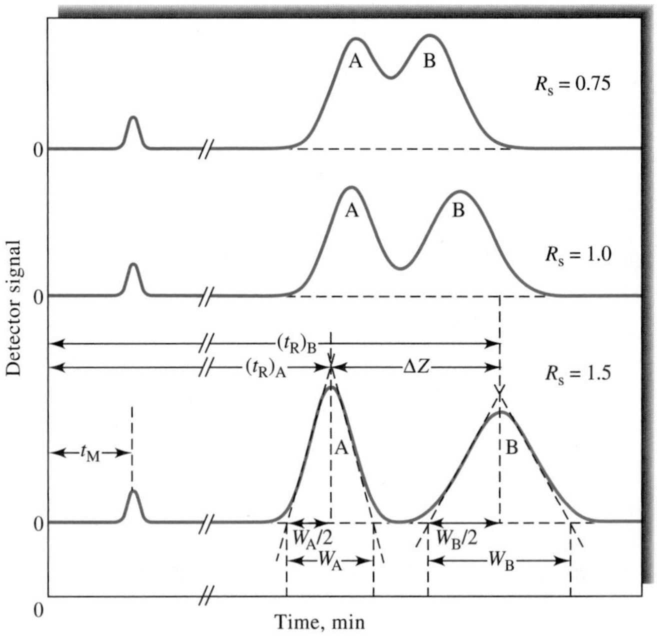

FIGURE 26-12 Separation at three resolution values: $R_s = 2\Delta Z/(W_A + W_B)$. The peak heights have been adjusted to be approximately the same in each case.

Note from Figure 26-12 that a resolution of 1.5 gives an essentially complete separation of A and B, whereas a resolution of 0.75 does not. At a resolution of 1.0, zone A contains about 4% B and zone B contains about 4% A. At a resolution of 1.5, the overlap is about 0.3%. The resolution for a given stationary phase can be improved by lengthening the column, which increases the number of plates. Adding more theoretical plates, however, comes at a cost of increasing the time required for separating the components.

26D-2 The Effect of Retention and Selectivity Factors on Resolution

It is useful to develop a mathematical relationship between the resolution of a column and the retention factors k_A and k_B for two solutes, the selectivity factor α, and the number of plates N making up the column. We will assume that we are dealing with two solutes A and B having retention times close enough to one another that we can assume

$$W_A = W_B \approx W$$

Equation 26-24 then takes the form

$$R_s = \frac{(t_R)_B - (t_R)_A}{W}$$

Tutorial: Learn more about the **selectivity factor**.

We solve Equation 26-21 for W in terms of $(t_R)_B$ and N and then substitute the resulting expression into the foregoing equation to obtain

$$R_s = \frac{(t_R)_B - (t_R)_A}{(t_R)_B} \times \frac{\sqrt{N}}{4}$$

Substituting Equation 26-12 into this equation and rearranging leads to an expression for R_s in terms of the retention factors for A and B. That is,

$$R_s = \frac{k_B - k_A}{1 + k_B} \times \frac{\sqrt{N}}{4}$$

Let us eliminate k_A from this expression by substituting Equation 26-14 and rearranging. Thus,

$$R_s = \frac{\sqrt{N}}{4}\left(\frac{\alpha - 1}{\alpha}\right)\left(\frac{k_B}{1 + k_B}\right) \quad (26\text{-}25)$$

where k_B is the retention factor of the slower-moving species and α is the selectivity factor. Equation 26-25 can be rearranged to give the number of plates needed to realize a given resolution:

$$N = 16R_s^2\left(\frac{\alpha}{\alpha - 1}\right)^2\left(\frac{1 + k_B}{k_B}\right)^2 \quad (26\text{-}26)$$

Simplified forms of Equations 26-25 and 26-26 are sometimes encountered where these equations are applied to a pair of solutes whose distribution constants are similar enough to make their separation difficult. Thus, when $K_A \approx K_B$, it follows from Equation 26-9 that $k_A \approx k_B = k$ and from Equation 26-13, $\alpha \rightarrow 1$. With these approximations, Equations 26-25 and 26-26 reduce to

$$R_s = \frac{\sqrt{N}}{4}(\alpha - 1)\left(\frac{k}{1 + k}\right) \qquad (26\text{-}27)$$

$$N = 16R_s^2\left(\frac{1}{\alpha - 1}\right)^2\left(\frac{1 + k}{k}\right)^2 \qquad (26\text{-}28)$$

where k is the average of k_A and k_B.

26D-3 The Effect of Resolution on Retention Time

Before considering in detail the significance of the four equations just derived, it is worthwhile to develop an equation for a related performance characteristic for a column, the time required to complete the separation of solutes A and B. The goal in chromatography is the highest possible resolution in the shortest possible time. Unfortunately, these goals tend to be incompatible, and a compromise between the two is usually necessary.

The completion time of a separation is determined by the velocity $\bar{\nu}_B$ of the slower-moving solute, as given in Equation 26-4. That is,

$$\bar{\nu}_B = \frac{L}{(t_R)_B}$$

Combining this expression with 26-10 and 26-16 yields, after rearranging

$$(t_R)_B = \frac{NH(1 + k_B)}{u}$$

where $(t_R)_B$ is the time required to bring the peak for B to the end of the column when the velocity of the mobile phase is u. When this equation is combined with Equation 26-26 and rearranged, we find that

$$(t_R)_B = \frac{16R_s^2 H}{u}\left(\frac{\alpha}{\alpha - 1}\right)^2\frac{(1 + k_B)^3}{(k_B)^2} \qquad (26\text{-}29)$$

Because H is a function of u and α is a function of k, such an equation is difficult to solve and use except when doing comparisons where several of the factors may cancel in a ratio.

Several of these relationships are applied in Example 26-1.

EXAMPLE 26-1

Substances A and B have retention times of 16.40 and 17.63 min, respectively, on a 30.0-cm column. An unretained species passes through the column in 1.30 min. The peak widths (at base) for A and B are 1.11 and 1.21 min, respectively. Calculate (a) the column resolution, (b) the average number of plates in the column, (c) the plate height, (d) the length of column required to achieve a resolution of 1.5, (e) the time required to elute substance B on the column that gives an R_s value of 1.5, and (f) the plate height required for a resolution of 1.5 on the original 30-cm column and in the original time.

▸ *Solution*

(a) Using Equation 26-24, we find

$$R_s = \frac{2(17.63 - 16.40)}{1.11 + 1.21} = 1.06$$

(b) Equation 26-21 permits computation of N:

$$N = 16\left(\frac{16.40}{1.11}\right)^2 = 3493 \quad \text{and} \quad N = 16\left(\frac{17.63}{1.21}\right)^2 = 3397$$

$$N_{av} = \frac{3493 + 3397}{2} = 3445$$

(c) $H = \dfrac{L}{N} = \dfrac{30.0}{3445} = 8.7 \times 10^{-3}$ cm

(d) k and α do not change greatly with increasing N and L. Thus, substituting N_1 and N_2 into Equation 26-25 and dividing one of the resulting equations by the other yields

$$\frac{(R_s)_1}{(R_s)_2} = \frac{\sqrt{N_1}}{\sqrt{N_2}}$$

where the subscripts 1 and 2 refer to the original and longer columns, respectively. Substituting the appropriate values for N_1, $(R_s)_1$, and $(R_s)_2$ gives

$$\frac{1.06}{1.5} = \frac{\sqrt{3445}}{\sqrt{N_2}}$$

$$N_2 = 3445\left(\frac{1.5}{1.06}\right)^2 = 6.9 \times 10^3$$

But

$$L = NH = 6.9 \times 10^3 \times 8.7 \times 10^{-3} = 60 \text{ cm}$$

(e) Substituting $(R_s)_1$ and $(R_s)_2$ into Equation 26-29 and dividing yields

$$\frac{(t_R)_1}{(t_R)_2} = \frac{(R_s)_1^2}{(R_s)_2^2} = \frac{17.63}{(t_R)_2} = \frac{(1.06)^2}{(1.5)^2}$$

$$(t_R)_2 = 35 \text{ min}$$

Thus, to obtain the improved resolution, the column length and, consequently, the separation time must be doubled.

(f) Substituting H_1 and H_2 into Equation 26-29 and dividing one of the resulting equations by the second gives

$$\frac{(t_R)_B}{(t_R)_B} = \frac{(R_s)_1^2}{(R_s)_2^2} \times \frac{H_1}{H_2}$$

where the subscripts 1 and 2 refer to the original and the new plate heights, respectively. Rearranging gives

$$H_2 = H_1 \frac{(R_s)_1^2}{(R_s)_2^2} = 8.7 \times 10^{-3} \text{ cm} \times \frac{(1.06)^2}{(1.5)^2}$$

$$= 4.3 \times 10^{-3} \text{ cm}$$

Thus, to achieve a resolution of 1.5 in 17.63 min on a 30-cm column, the plate height would need to be halved.

26D-4 Variables That Affect Column Performance

Equations 26-25 and 26-29 are significant because they serve as guides to the choice of conditions that are likely to allow the user of chromatography to achieve the sometimes elusive goal of a clean separation in a minimum amount of time. An examination of these equations reveals that each is made up of three parts. The first, which is related to the kinetic effects that lead to band broadening, consists of $\sqrt{N}$ or H/u. The second and third terms are related to the thermodynamics of the constituents being separated — that is, to the relative magnitude of their distribution constants and the volumes of the mobile and stationary phases. The second term in Equations 26-25 and 26-29, which is the quotient containing α, is a selectivity term. For a given mobile-phase and stationary-phase combination, the α term depends solely on the properties of the two solutes. The third term, which is the quotient containing k_B, depends on the properties of both the solute and the column.

In seeking optimal conditions for achieving a desired separation, it must be kept in mind that the fundamental parameters α, k, and N (or H) can be adjusted more or less independently. Thus, α and k can be varied most easily by varying temperature or the composition of the mobile phase. Less conveniently, a different type of column packing can be employed. As we have seen, it is possible to change N by changing the length of the column and to change H by altering the flow rate of the mobile phase, the particle size of the packing, the viscosity of the mobile phase (and thus D_M or D_S), and the thickness of the film of adsorbed liquid constituting the stationary phase (see Table 26-3).

Variation in N

An obvious way to improve resolution is to increase the number of plates in the column (Equation 26-25). As shown by Example 26-1, this approach usually requires more time to complete the separation unless the increase in N is accomplished by reducing H rather than by lengthening the column.

Variation in H

In Example 26-1f, we showed that resolution can be significantly improved at no cost in time if the plate height can be reduced. Table 26-3 reveals the variables available for accomplishing this end. Note that decreases in particle size of the packing lead to substantial improvements in H. For liquid mobile phases, where B/u is usually negligible, reduced plate heights can also be achieved by reducing the solvent viscosity, thus increasing the diffusion coefficient in the mobile phase.

Variation in the Retention Factor

Often, a separation can be improved significantly by manipulating the retention factor k_B. Increases in k_B generally enhance resolution (but at the expense of elution time). To determine the optimal range of values for k_B, it is convenient to write Equation 26-25 in the form

$$R_s = \frac{Q k_B}{1 + k_B}$$

where $Q = \frac{\sqrt{N}}{4}\left(\frac{\alpha - 1}{\alpha}\right)$, and Equation 26-29 as

$$(t_R)_B = Q' \frac{(1 + k_B)^3}{(k_B)^2}$$

where

$$Q' = \frac{16 R_s^2 H}{u}\left(\frac{\alpha}{\alpha - 1}\right)^2$$

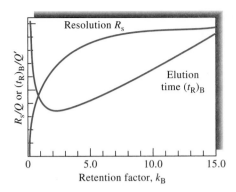

FIGURE 26-13 Effect of retention factor k_B on resolution R_s and elution time $(t_R)_B$. It is assumed that Q and Q' remain constant with variations in k_B.

Figure 26-13 is a plot of R_s/Q and $(t_R)_B/Q'$ as a function of k_B, assuming that Q and Q' remain approximately constant. Note that values of k_B greater than about 10 should be avoided because they provide little increase in resolution but noticeably increase the time required for separations. The minimum in the elution time curve occurs at a value of k_B of about 2. The curves of Figure 26-13 suggest that the optimal value of k_B, taking into account both resolution and time, lies in the range of 1 to 5.

Usually, the easiest way to improve resolution is by optimizing k. For gaseous mobile phases, k can often be improved by temperature changes. For liquid mobile phases, changes in the solvent composition often permit manipulation of k to yield better separations. An example of the dramatic effect that relatively simple solvent changes can bring about is demonstrated in Figure 26-14. Here, modest variations in the methanol-to-water ratio convert unsatisfactory chromatograms (a and b) to ones with well-separated peaks for each component (c and d). For most purposes, the chromatogram shown in (c) is best because it shows adequate resolution in minimum time. The retention factor is also influenced by the stationary-phase film thickness.

Variation in the Selectivity Factor

Optimizing k and increasing N are not sufficient to give a satisfactory separation of two solutes in a reasonable time when α approaches unity. Here, a means must be sought to increase α while maintaining k in the range of 1 to 10. Several options are available; in decreasing order of their desirability as determined by promise

and convenience, the options are (1) changing the composition of the mobile phase, (2) changing the column temperature, (3) changing the composition of the stationary phase, and (4) using special chemical effects.

An example of the use of option 1 has been reported for the separation of anisole ($C_6H_5OCH_3$) and benzene.[10] With a mobile phase that was a 50% mixture of water and methanol, k was 4.5 for anisole and 4.7 for benzene, and α was only 1.04. Substitution of an aqueous mobile phase containing 37% tetrahydrofuran gave k values of 3.9 and 4.7 and an α value of 1.20. Peak overlap was significant with the first solvent system and negligible with the second.

A less convenient but often highly effective method of improving α while maintaining values for k in their optimal range is to alter the chemical composition of the stationary phase. To take advantage of this option, most laboratories that carry out chromatographic separations frequently maintain several columns that can be interchanged conveniently.

Increases in temperature usually cause decreases in k and may affect α values in liquid-liquid and liquid-solid chromatography. In contrast, with ion-exchange chromatography, temperature effects can be large enough to make exploration of this option worthwhile before resorting to a change in column packing material.

A final method for enhancing resolution is to incorporate into the stationary phase a species that complexes or otherwise interacts with one or more components of the sample. A well-known example of the use of this option occurs when an adsorbent impregnated with a silver salt is used to improve the separation of olefins. The improvement results from the formation of complexes between the silver ions and unsaturated organic compounds.

26D-5 The General Elution Problem

Figure 26-15 shows hypothetical chromatograms for a six-component mixture made up of three pairs of components with widely different distribution constants and thus widely different retention factors. In chromatogram (a), conditions have been adjusted so that the retention factors for components 1 and 2 (k_1 and k_2) are in the optimal range of 1 to 5. The factors for

---

[10]L. R. Snyder and J. J. Kirkland, *Introduction to Modern Liquid Chromatography*, 2nd ed., p. 75, New York: Wiley, 1979.

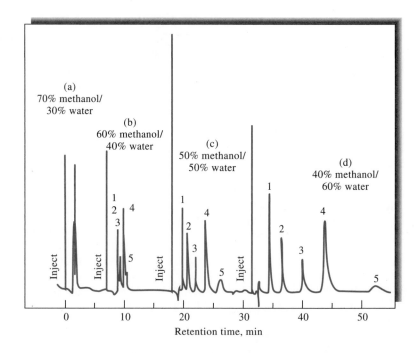

FIGURE 26-14 Effect of solvent variation on chromatograms. Analytes: (1) 9,10-anthraquinone; (2) 2-methyl-9, 10-anthraquinone; (3) 2-ethyl-9,10-anthraquinone; (4) 1,4-dimethyl-9, 10-anthraquinone; (5) 2-*t*-butyl-9, 10-anthraquinone.

the other components are far larger than the optimal, however. Thus, the bands corresponding to components 5 and 6 appear after a significant period of time has passed; furthermore, the bands are so broad that they may be difficult to identify unambiguously.

As shown in chromatogram (b), changing conditions to optimize the separation of components 5 and 6 bunches the peaks for the first four components to the point where their resolution is unsatisfactory. Here, however, the total elution time is ideal.

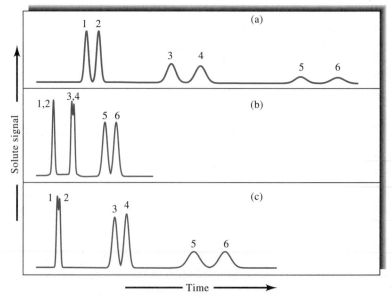

FIGURE 26-15 The general elution problem in chromatography.

A third set of conditions, in which k values for components 3 and 4 are optimal, results in chromatogram (c). Again, separation of the other two pairs is not entirely satisfactory.

The phenomenon illustrated in Figure 26-15 is encountered often enough to be given a name, the *general elution problem*. A common solution to this problem is to change conditions that determine the values of k as the separation proceeds. These changes can be performed in a stepwise manner or continuously. Thus, for the mixture shown in Figure 26-15, conditions at the beginning of the separation could be those producing chromatogram (a). Immediately after the elution of components 1 and 2, conditions could be changed to those that are optimal for separating components 3 and 4, as in chromatogram (c). With the appearance of peaks for these components, the elution could be completed under the conditions used for producing chromatogram (b). Often, such a procedure leads to satisfactory separation of all the components of a mixture in minimal time.

For LC, variations in k can be achieved by varying the composition of the mobile phase during elution. Such a procedure is called *gradient elution* or *solvent programming*. Elution with constant mobile-phase composition is called *isocratic elution*. For GC, the temperature can be changed in a known fashion to bring about changes in k. This *temperature programming* mode can help achieve optimal conditions for many separations. These modes are discussed in more detail in Chapters 27 and 28.

26E SUMMARY OF CHROMATOGRAPHIC RELATIONSHIPS

The number of quantities, terms, and relationships employed in chromatography is large and often confusing. Tables 26-4 and 26-5 summarize the most important definitions and equations that will be used in the next three chapters.

26F APPLICATIONS OF CHROMATOGRAPHY

Chromatography has grown to be the premiere method for separating closely related chemical species. In addition, it can be employed for qualitative identification and quantitative determination of separated species. This section considers some of the general characteristics of chromatography as a tool for completion of an analysis.

26F-1 Qualitative Analysis

A chromatogram provides only a single piece of qualitative information about each species in a sample, its retention time or its position on the stationary phase after a certain elution period. Additional data can, of course, be derived from chromatograms involving different mobile and stationary phases and various elution temperatures. Still, the amount of information revealed by chromatography is small compared with

TABLE 26-4 Important Chromatographic Quantities and Relationships

Name	Symbol of Experimental Quantity	Determined From
Migration time, unretained species	t_M	Chromatogram (Figure 26-7)
Retention time, species A and B	$(t_R)_A, (t_R)_B$	Chromatogram (Figures 26-7 and 26-12)
Adjusted retention time for A	$(t_R')_A$	$(t_R')_A = (t_R)_A - t_M$
Peak widths for A and B	W_A, W_B	Chromatogram (Figures 26-7 and 26-12)
Length of column packing	L	Direct measurement
Volumetric flow rate	F	Direct measurement
Linear flow velocity	u	F and column dimensions (Equations 26-6 and 26-7)
Stationary-phase volume	V_S	Packing preparation data
Concentration of analyte in mobile and stationary phases	c_M, c_S	Analysis and preparation data

TABLE 26-5 Important Derived Quantities and Relationships

Name	Calculation of Derived Quantities	Relationship to Other Quantities
Linear mobile-phase velocity	$u = \dfrac{L}{t_M}$	
Volume of mobile phase	$V_M = t_M F$	
Retention factor	$k = \dfrac{t_R - t_M}{t_M}$	$k = \dfrac{K V_S}{V_M}$
Distribution constant	$K = \dfrac{k V_M}{V_S}$	$K = \dfrac{c_S}{c_M}$
Selectivity factor	$\alpha = \dfrac{(t_R)_B - t_M}{(t_R)_A - t_M}$	$\alpha = \dfrac{k_B}{k_A} = \dfrac{K_B}{K_A}$
Resolution	$R_s = \dfrac{2[(t_R)_B - (t_R)_A]}{W_A + W_B}$	$R_s = \dfrac{\sqrt{N}}{4}\left(\dfrac{\alpha - 1}{\alpha}\right)\left(\dfrac{k_B}{1 + k_B}\right)$
Number of plates	$N = 16\left(\dfrac{t_R}{W}\right)^2$	$N = 16 R_s^2\left(\dfrac{\alpha}{\alpha - 1}\right)^2\left(\dfrac{1 + k_B}{k_B}\right)^2$
Plate height	$H = \dfrac{L}{N}$	
Retention time	$(t_R)_B = \dfrac{16 R_s^2 H}{u}\left(\dfrac{\alpha}{\alpha - 1}\right)^2 \dfrac{(1 + k_B)^3}{(k_B)^2}$	

the amount provided by a single IR, nuclear magnetic resonance, or mass spectrum. Furthermore, spectral wavelength or frequency data can be determined with much higher precision than can their chromatographic counterpart t_R.

This should not be interpreted to mean that chromatography lacks important qualitative applications. Indeed, it is a widely used tool for recognizing the presence or absence of components of mixtures containing a limited number of possible species whose identities are known. For example, thirty or more amino acids in a protein hydrolysate can be identified with a relatively high degree of certainty by using chromatography. Even in such cases, confirmation of identity requires spectral or chemical investigation of the isolated components. Note, however, that positive spectroscopic identification would ordinarily be impossible on as complex a sample as a protein hydrolysate without a preliminary chromatographic separation. Thus, chromatography is often an essential step in qualitative spectroscopic analyses.

It is important to note that, although chromatograms may not lead to positive identification of species present in a sample, they often provide sure evidence of the *absence* of certain compounds. Thus, if the sample does not produce a peak at the same retention time as a standard run under identical conditions, it can be assumed that the compound in question is either absent or is present at a concentration level below the detection limit of the procedure.

26F-2 Quantitative Analysis

Chromatography owes its rapid growth during the past four decades in part to its speed, simplicity, relatively low cost, and wide applicability as a tool for separations. It is doubtful, however, that chromatography would have become as widespread had it not been able to provide useful quantitative information about the separated species. It is important, therefore, to discuss some of the quantitative aspects that apply to all types of chromatography.

Quantitative column chromatography is based on a comparison of either the height or the area of the analyte peak with that of one or more standards. For planar chromatography, the area covered by the separated species serves as the analytical variable. If conditions are properly controlled, these variables vary linearly with concentration.

Analyses Based on Peak Height

The height of a chromatographic peak is obtained by connecting the baselines on either side of the peak by a straight line and measuring the perpendicular dis-

 Tutorial: Learn more about **quantitative chromatographic analysis**.

tance from this line to the peak. This measurement can often be made with reasonably high precision. It is important to note, however, that peak heights are inversely related to peak widths. Thus, accurate results are obtained with peak heights only if variations in column conditions do not alter the peak widths during the period required to obtain chromatograms for sample and standards. The variables that must be controlled closely are column temperature, eluent flow rate, and rate of sample injection. In addition, care must be taken to avoid overloading the column. The effect of sample injection rate is particularly critical for the early peaks of a chromatogram. Relative errors of 5% to 10% due to this cause are not unusual with syringe injection.

Analyses Based on Peak Areas

Peak areas are independent of broadening effects due to the variables mentioned in the previous paragraph. From this standpoint, therefore, areas are a more satisfactory analytical parameter than peak heights. On the other hand, peak heights are more easily measured and, for narrow peaks, more accurately determined. However, peak heights are affected by changes in retention times or column efficiency whereas peak areas are not. Hence, peak areas are usually the preferred method of quantitation.

Most modern chromatographic instruments are equipped with computers or digital electronic integrators that permit precise estimation of peak areas. If such equipment is not available, a manual estimate must be made. A simple method, which works well for symmetric peaks of reasonable widths, is to multiply the height of the peak by its width at one half the peak height. Older methods involved the use of a planimeter or cutting out the peak and determining its mass relative to the mass of a known area of recorder paper. In general, manual integration techniques provide areas that are reproducible to 2% to 5%; digital integrators are at least an order of magnitude more precise.[11]

Calibration and Standards

The most straightforward method for quantitative chromatographic analyses involves the preparation of a series of external-standard solutions that approxi-

mate the composition of the unknown. Chromatograms for the standards are then obtained and peak heights or areas are plotted as a function of concentration. A plot of the data should yield a straight line passing through the origin; determinations are based on this calibration curve. Frequent restandardization is necessary for highest accuracy.

The most important source of error in analyses by the method just described is usually the uncertainty in the volume of sample; occasionally, the rate of injection is also a factor. Samples are usually small ($\sim 1\ \mu L$), and the uncertainties associated with injection of a reproducible volume of this size with a microsyringe may amount to several percent relative. The situation is even worse in GC, where the sample must be injected into a heated sample port; here, evaporation from the needle tip may lead to large variations in the volume injected. Errors in sample volume can be reduced to perhaps 1% to 2% relative by means of autosamplers or a rotary sample valve such as that described in Chapter 27.

The Internal-Standard Method

The highest precision for quantitative chromatography is obtained using internal standards because the uncertainties introduced by sample injection are avoided. In this procedure, a carefully measured quantity of an internal-standard substance is introduced into each standard and sample, and the ratio of analyte to internal standard peak areas (or heights) serves as the analytical variable. For this method to be successful, the internal-standard peak must be well separated from the peaks of all other components of the sample ($R_s >$ 1.25); the internal-standard peak should, on the other hand, appear close to the analyte peak. With a suitable internal standard, precisions of better than 1% relative can usually be achieved.

The Area-Normalization Method

Another approach that avoids the uncertainties associated with sample injection is the area-normalization method. Complete elution of all components of the sample is required. In the normalization method, the areas of all eluted peaks are computed; after correcting these areas for differences in the detector response to different compound types, the concentration of the analyte is found from the ratio of its area to the total area of all peaks. The following example illustrates the procedure.

---

[11]See N. A. Dyson, *Chromatographic Integration Methods*, 2nd ed., London: Royal Society of Chemistry, 1998.

EXAMPLE 26-2

The area normalization method was applied to the determination of normal-, secondary-, iso-, and tertiary-butyl alcohol. To determine the relative response factor for the alcohols, a standard solution of the alcohols was prepared and its gas chromatogram observed. The results were as follows:

Alcohol	Weight Taken, g	Weight % Alcohol	Peak Area A, cm^2	Weight % Area	Relative Response Factor F
n-Butyl	0.1731	24.61	3.023	8.141	1.000
i-Butyl	0.1964	27.92	3.074	9.083	1.116
s-Butyl	0.1514	21.52	3.112	6.915	0.849
t-Butyl	0.1826	25.96	3.004	8.642	1.062
	Σ wt = 0.7035	Σ % = 100.00	ΣA = 12.213		

The relative response factors were obtained by dividing the data in column 5 by 8.141, the first entry in column 5.

A sample containing only the four alcohols yielded the area data in the second column below. Calculate the weight percent of each alcohol present.

Solution

The results are shown in column 4 below.

Alcohol	Peak Area, cm^2	Area $\times F$	Weight % Alcohol
n-Butyl	1.731	1.731	18.18
i-Butyl	3.753	4.188	43.99
s-Butyl	2.845	2.415	25.36
t-Butyl	1.117	1.186	12.46
		Σ = 9.521	99.99

Unfortunately, it is often not practical to arrange conditions so that all of the components of a mixture are eluted from a column in a reasonable period. As a result, the area-normalization method has limited applications.

QUESTIONS AND PROBLEMS

*Answers are provided at the end of the book for problems marked with an asterisk.

 Problems with this icon are best solved using spreadsheets.

26-1 Define
(a) elution
(b) mobile phase
(c) stationary phase
(d) distribution constant
(e) retention time
(f) retention factor
(g) selectivity factor
(h) plate height
(i) longitudinal diffusion
(j) eddy diffusion
(k) column resolution
(l) eluent

26-2 Describe the general elution problem.

26-3 List the variables that lead to zone broadening in chromatography.

26-4 What are the major differences between gas-liquid and liquid-liquid chromatography?

26-5 What are the differences between liquid-liquid and liquid-solid chromatography?

26-6 What variables are likely to affect the selectivity factor α for a pair of analytes?

26-7 Describe how the retention factor for a solute can be manipulated.

26-8 Describe a method for determining the number of plates in a column.

26-9 Name two general methods for improving the resolution of two substances on a chromatographic column.

26-10 Why does the minimum in a plot of plate height versus flow rate occur at lower flow rates with LC than with GC?

26-11 What is gradient elution?

26-12 List the variables in chromatography that lead to zone separation.

26-13 What would be the effect on a chromatographic peak of introducing the sample at too slow a rate?

*26-14** The following data are for a liquid chromatographic column

Length of packing	24.7 cm
Flow rate	0.313 mL/min
V_M	1.37 mL
V_S	0.164 mL

A chromatogram of a mixture of species A, B, C, and D provided the following data:

	Retention Time, min	Width of Peak Base (W), min
Nonretained	3.1	—
A	5.4	0.41
B	13.3	1.07
C	14.1	1.16
D	21.6	1.72

Calculate
(a) the number of plates from each peak.
(b) the mean and the standard deviation for N.
(c) the plate height for the column.

*26-15 From the data in Problem 26-14, calculate for A, B, C, and D
(a) the retention factor.
(b) the distribution constant.

*26-16 From the data in Problem 26-14, calculate for species B and C
(a) the resolution.
(b) the selectivity factor α.
(c) the length of column necessary to separate the two species with a resolution of 1.5.
(d) the time required to separate the two species on the column in part (c).

*26-17 From the data in Problem 26-14, calculate for species C and D
(a) the resolution.
(b) the length of column necessary to separate the two species with a resolution of 1.5.

☒ *26-18 The following data were obtained by gas-liquid chromatography on a 40-cm packed column:

Compound	t_R, min	W, min
Air	1.9	—
Methylcyclohexane	10.0	0.76
Methylcyclohexene	10.9	0.82
Toluene	13.4	1.06

Calculate
(a) an average number of plates from the data.
(b) the standard deviation for the average in (a).
(c) an average plate height for the column.

☒ *26-19 Referring to Problem 26-18, calculate the resolution for
(a) methylcyclohexene and methylcyclohexane.
(b) methylcyclohexene and toluene.
(c) methylcyclohexane and toluene.

☒ *26-20 If a resolution of 1.5 is desired in separating methylcyclohexane and methyl-cyclohexene in Problem 26-18,
(a) how many plates are required?
(b) how long must the column be if the same packing is employed?
(c) what is the retention time for methylcyclohexane on the column of part (b)?

☒ *26-21 If V_S and V_M for the column in Problem 26-18 are 19.6 and 62.6 mL, respectively, and a nonretained air peak appears after 1.9 min, calculate
(a) the retention factor for each compound.
(b) the distribution constant for each compound.
(c) the selectivity factor for methylcyclohexane and methylcyclohexene.

X *26-22 The relative areas for the five gas chromatographic peaks obtained in the separation of five steroids are given below. Also shown are the relative responses of the detector to the five compounds. Calculate the percentage of each component in the mixture.

Compound	Peak Area, Relative	Detector Response, Relative
Dehydroepiandrosterone	27.6	0.70
Estradiol	32.4	0.72
Estrone	47.1	0.75
Testosterone	40.6	0.73
Estriol	27.3	0.78

Challenge Problem

26-23 A chromatogram of a two-component mixture on a 25-cm packed LC column is shown in the figure below. The flow rate was 0.40 mL/min.
   (a) Find the times that components A and B spend in the stationary phase.
   (b) Find the retention times for A and B.
   (c) Determine the retention factors for the two components.
   (d) Find the full widths of each peak and the full width at half-maximum values.
   (e) Find the resolution of the two peaks.

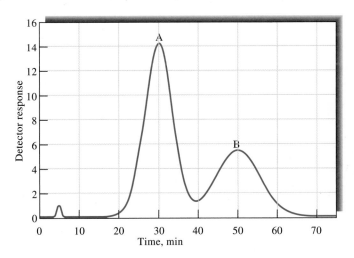

   (f) Find the average number of plates for the column.
   (g) Find the average plate height.
   (h) What column length would be needed to achieve a resolution of 1.75?
   (i) What time would be required to achieve the resolution in part (h)?
   (j) Assume that the column length is fixed at 25 cm and the packing material is fixed. What measures could you take to increase the resolution to achieve baseline separation?
   (k) Are there any measures you could use to achieve a better separation in a shorter time with the same column as in part (j)?

CHAPTER TWENTY-SEVEN

Gas Chromatography

I*n gas chromatography, the components of a vaporized sample are separated as a consequence
of being partitioned between a mobile gaseous
phase and a liquid or a solid stationary phase held
in a column. In performing a gas chromatographic
separation, the sample is vaporized and injected
onto the head of a chromatographic column. Elution
is brought about by the flow of an inert gaseous mobile phase. In contrast to most other types of chromatography, the mobile phase does not interact
with molecules of the analyte; its only function is to
transport the analyte through the column.*

Throughout this chapter, this logo indicates
an opportunity for online self-study at **www
.thomsonedu.com/chemistry/skoog**, linking you to
interactive tutorials, simulations, and exercises.

There are two types of gas chromatography: *gas-liquid
chromatography* (GLC) and *gas-solid chromatography*
(GSC). GLC finds widespread use in all fields of science; its name is usually shortened to *gas chromatography* (GC).[1] GSC is based on a solid stationary phase in
which retention of analytes occurs because of physical
adsorption. The application of GSC is limited because
of semipermanent retention of active or polar molecules and severe tailing of elution peaks. Tailing is a result of the nonlinear nature of the adsorption process.
Thus, this technique is not widely used except for the
separation of certain low-molecular-mass gaseous species; we discuss the method briefly in Section 27F.

In GLC the analyte is partitioned between a gaseous
mobile phase and a liquid phase immobilized on the
surface of an inert solid packing or on the walls of a capillary tubing. The concept of GLC was first suggested in
1941 by Martin and Synge, who were also responsible
for the development of liquid-liquid partition chromatography. More than a decade was to elapse, however, before the value of GLC was demonstrated experimentally[2] and this technique began to be used as a
routine laboratory tool. In 1955 the first commercial
apparatus for GLC appeared on the market. Since that
time, the growth in applications of this technique has
been phenomenal. Currently, nearly a million gas chromatographs are in use throughout the world.

27A PRINCIPLES OF GLC

The general principles of chromatography, which were
developed in Chapter 26, and the mathematical relationships summarized in Section 26E are applicable to
GC with only minor modifications that arise from the
compressibility of gaseous mobile phases.

27A-1 Retention Volumes

To take into account the effects of pressure and temperature in GC, it is often useful to use retention volumes rather than the retention times that were dis-

---

[1] For detailed treatment of GC, see R. L. Grob and E. F. Barry, eds., *Modern
Practice of Gas Chromatography*, 4th ed., New York: Wiley-Interscience,
2004; H. M. McNair and J. M. Miller, *Basic Gas Chromatography*, New
York: Wiley, 1998; R. P. W. Scott, *Introduction to Analytical Gas Chromatography*, 2nd ed., New York: Marcel Dekker, 1997; W. Jennings, E. Mittle-
fehldt, and P. Stremple, *Analytical Gas Chromatography*, 2nd ed., Orlando,
FL: Academic Press, 1997.
[2] A. T. James and A. J. P. Martin, *Analyst*, **1952**, 77, 915–932.

cussed in Section 26B. The relationship between the two is given in Equations 27-1 and 27-2

$$V_R = t_R F \quad (27\text{-}1)$$

$$V_M = t_M F \quad (27\text{-}2)$$

where F is the average volumetric flow rate within the column; V and t are retention volumes and times, respectively; and the subscripts R and M refer to species that are retained and not retained on the column. The flow rate within the column is not directly measurable. Instead, the rate of gas flow as it exits the column is determined experimentally with a flow meter, which is discussed in Section 27B. For popular soap-bubble-type flow meters, where the gas is saturated with water, the average flow rate F is related to the measured flow rate F_m by

$$F = F_m \times \frac{T_c}{T} \times \frac{(P - P_{H_2O})}{P} \quad (27\text{-}3)$$

where T_c is the column temperature in kelvins, T is the temperature at the flow meter, and P is the gas pressure at the end of the column. Usually P and T are the ambient pressure and temperature. The term involving the vapor pressure of water, P_{H_2O}, is a correction for the pressure used when the gas is saturated with water.

Both V_R and V_M depend on the average pressure within the column—a quantity that lies intermediate between the inlet pressure P_i and the outlet pressure P (atmospheric pressure). The *pressure drop correction factor j*, also known as the *compressibility factor*, accounts for the pressure within the column being a nonlinear function of the P_i/P ratio. Corrected retention volumes V_R^0 and V_M^0, which correspond to volumes at the average column pressure, are obtained from the relationships

$$V_R^0 = jt_R F \quad \text{and} \quad V_M^0 = jt_M F \quad (27\text{-}4)$$

where j can be calculated from the relationship

$$j = \frac{3[(P_i/P)^2 - 1]}{2[(P_i/P)^3 - 1]} \quad (27\text{-}5)$$

The *specific retention volume V_g* is then defined as

$$V_g = \frac{V_R^0 - V_M^0}{m_S} \times \frac{273}{T_c} = \frac{jF(t_R - t_M)}{m_S} \times \frac{273}{T_c} \quad (27\text{-}6)$$

where m_S is the mass of the stationary phase, a quantity determined at the time of column preparation.

27A-2 Relationship between V_g and K

The specific retention volume V_g can be related to the distribution constant K_c. To do so, we substitute the expression relating t_R and t_M to k (Equation 26-12) into Equation 27-6, which gives

$$V_g = \frac{jFt_M k}{m_S} \times \frac{273}{T_c}$$

Combining this expression with Equation 27-4 yields

$$V_g = \frac{V_M^0 k}{m_S} \times \frac{273}{T_c}$$

Substituting Equation 26-9 for k gives (here, V_M^0 and V_M are identical)

$$V_g = \frac{KV_S}{m_S} \times \frac{273}{T_c}$$

The density of the liquid on the stationary phase ρ_S is given by

$$\rho_S = \frac{m_S}{V_S}$$

where V_S is the stationary-phase volume. Thus,

$$V_g = \frac{K}{\rho_S} \times \frac{273}{T_c} \quad (27\text{-}7)$$

Note that V_g at a given temperature depends only on the distribution constant of the solute and the density of the liquid making up the stationary phase.

27A-3 Effect of Mobile-Phase Flow Rate

Equation 26-23 and the relationships shown in Table 26-3 are fully applicable to GC. The longitudinal diffusion term (B/u) is more important in GLC, however, than in other chromatographic processes because of the much larger diffusion rates in gases (10^4–10^5 times greater than liquids). As a result, the minima in curves relating plate height H to flow rate (van Deemter plots) are usually considerably broadened in GC (see Figure 26-8).

27B INSTRUMENTS FOR GLC

Many changes and improvements in gas chromatographic instruments have appeared in the marketplace since their commercial introduction. In the 1970s, electronic integrators and computer-based data-processing

FIGURE 27-1 Block diagram of a typical gas chromatograph.

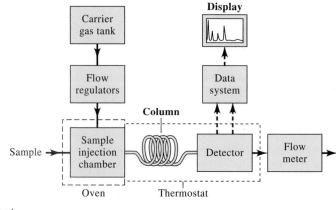

Display

equipment became common. The 1980s saw computers being used for automatic control of most instrument variables, such as column temperature, flow rates, and sample injection; the development of very high-performance instruments at moderate costs; and perhaps most important, the development of open tubular columns capable of separating components of complex mixtures in relatively short times. Today, more than 50 instrument manufacturers offer several hundred different models of gas chromatographic equipment at costs that vary from about $1000 to more than $50,000.

The basic components of a typical instrument for performing GC are shown in Figure 27-1. A description of each component follows.

27B-1 Carrier Gas System

The mobile-phase gas in GC is called the *carrier gas* and must be chemically inert. Helium is the most common mobile-phase gas used, although argon, nitrogen, and hydrogen are also used. These gases are available in pressurized tanks. Pressure regulators, gauges, and flow meters are required to control the flow rate of the gas. In addition, the carrier gas system often contains a molecular sieve to remove impurities and water.

Flow rates are normally controlled by a two-stage pressure regulator at the gas cylinder and some sort of pressure regulator or flow regulator mounted in the chromatograph. Inlet pressures usually range from 10 to 50 psi (lb/in.^{2}) above room pressure, which lead to flow rates of 25 to 150 mL/min with packed columns and 1 to 25 mL/min for open tubular capillary columns. Generally, it is assumed that flow rates will be constant if the inlet pressure remains constant. Flow rates can be established by a rotometer at the column head; this device, however, is not as accurate as the

 Exercise: Learn more about **gas chromatography**.

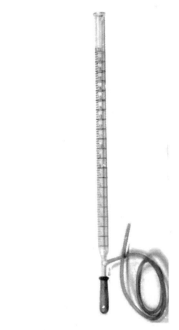

FIGURE 27-2 A soap-bubble flow meter. (Courtesy Agilent Technologies.)

simple soap-bubble meter shown in Figure 27-2. Usually, the flow meter is located at the end of the column as shown in Figure 27-1. A soap film is formed in the path of the gas when a rubber bulb containing an aqueous solution of soap or detergent is squeezed; the time required for this film to move between two graduations on the buret is measured and converted to volumetric flow rate (see Figure 27-2). Note that volumetric flow rates and linear flow velocities are related by Equation 26-6 or 26-7. Many modern computer-controlled gas chromatographs are equipped with electronic flow meters that can be regulated to maintain the flow rate at the desired level.

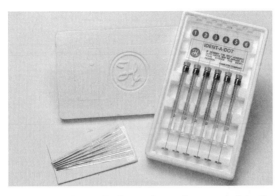

FIGURE 27-3 A set of microsyringes for sample injection. (Courtesy Hamilton Company.)

27B-2 Sample Injection Systems

To achieve high column efficiency, the sample must be of a suitable size and introduced as a "plug" of vapor; slow injection or oversize samples cause band spreading and poor resolution. Calibrated microsyringes, such as those shown in Figure 27-3, are used to inject liquid samples through a rubber or silicone diaphragm, or septum, into a heated sample port located at the head of the column. The sample port (Figure 27-4) is ordinarily about 50°C above the boiling point of the least volatile component of the sample. For ordinary packed analytical columns, sample sizes range from a few tenths of a microliter to 20 μL. Capillary columns require samples that are smaller by a factor of 100 or more. Here, a *sample splitter* is often needed to deliver a small known fraction (1:50 to 1:500) of the injected sample, with the remainder going to waste. Commercial gas chromatographs intended for use with capillary columns incorporate such splitters; they also allow for *splitless injection* to improve sensitivity or for use with packed columns. With splitless inlets, the purge valve closes at injection and stays closed for 30–60 seconds. During this time the sample vapor can go only onto the column. When the purge valve opens, any remaining vapor is rapidly vented. *On-column inlets* are also available for capillary GC. With this type of inlet, the entire sample is injected onto the column as a liquid, which is later vaporized by temperature programming of the column or inlet. With on-column inlets, the analyte is separated from the solvent by thermal and solvent effects.[3]

---

[3]N. H. Snow, in *Modern Practice of Gas Chromatography*, R. L. Grob and E. F. Barry, eds., 4th ed., Chap. 9, New York: Wiley-Interscience, 2004.

For quantitative work, more reproducible sample sizes for both liquids and gases can be obtained with a gas sampling valve such as that shown in Figure 27-5. With such devices, sample sizes can be reproduced to better than 0.5% relative. Autoinjectors with automatic sampling trays are available for most of the higher-end gas chromatographs. These can substantially improve the precision of the injected volume over manual syringe injection.

27B-3 Column Configurations and Column Ovens

Two general types of columns are used in GC, *packed* and *open tubular*, also called *capillary*. In the past, the vast majority of gas chromatographic analyses used packed columns. For most current applications, packed columns have been replaced by the more efficient open tubular columns.

Packed chromatographic columns vary in length from 1 m to 5 m, and capillary columns can range from a few meters to 100 m. Most columns are constructed of fused silica or stainless steel, although glass and Teflon are also used. To fit into an oven for thermostatting, they are usually formed as coils having diameters of 10 to 30 cm (Figure 27-6). A detailed discussion of

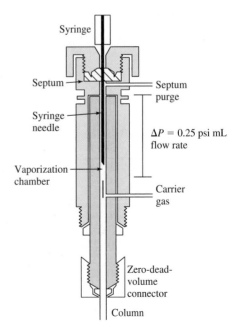

FIGURE 27-4 Cross-sectional view of a microflash vaporizer direct injector.

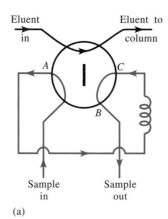
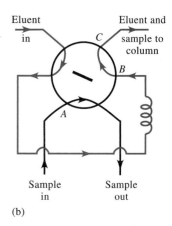

FIGURE 27-5 A rotary sample valve: valve position (a) is for filling the sample loop ACB; position (b) is for introduction of sample into column.

columns, column packings, and stationary phases is found in Section 27C.

Column temperature is an important variable that must be controlled to a few tenths of a degree for precise work. Thus, the column is ordinarily housed in a thermostatted oven. The optimal column temperature depends on the boiling point of the sample and the degree of separation required. Roughly, a temperature

FIGURE 27-6 Fused-silica capillary columns. (Courtesy of Restek Corp., Bellefonte, PA.)

equal to or slightly above the average boiling point of a sample results in a reasonable elution time (2 to 30 min). For samples with a broad boiling range, it is often desirable to employ *temperature programming*, in which the column temperature is increased either continuously or in steps as the separation proceeds. Figure 27-7 shows the improvement in a chromatogram brought about by temperature programming.

In general, optimal resolution is associated with minimal temperature; the cost of lowered temperature, however, is an increase in elution time and therefore the time required to complete an analysis. Figures 27-7a and 27-7b illustrate this principle.

Analytes of limited volatility can sometimes be determined by forming derivatives that are more volatile. Likewise, derivatization is sometimes used to enhance detection or chromatographic performance.

27B-4 Detection Systems

Dozens of detectors have been investigated and used with gas chromatographic separations.[4] We first describe the ideal characteristics of a gas chromatographic detector and then discuss the most widely used detection systems. In some cases, gas chromatographs are coupled to spectroscopic instruments such as mass and infrared spectrometers. With such systems, the spectral device not only detects the appearance of the analytes as they elute from the column but also helps to identify them.

---

[4]See L. A. Colon and L. J. Baird, in *Modern Practice of Gas Chromatography*, R. L. Grob and E. F. Barry, eds., 4th ed., Chap. 6, New York: Wiley-Interscience, 2004.

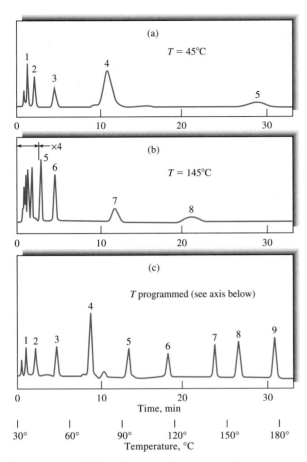

FIGURE 27-7 Effect of temperature on gas chromatograms: (a) isothermal at 45°C; (b) isothermal at 145°C; (c) programmed at 30°C to 180°C. (From W. E. Harris and H. W. Habgood, *Programmed Temperature Gas Chromatography*, New York: Wiley, 1966, p. 10. Reprinted with permission.)

Characteristics of the Ideal Detector

The ideal detector for GC has the following characteristics:

1. Adequate sensitivity. Just what constitutes adequate sensitivity cannot be described in quantitative terms. For example, the sensitivities of the detectors described in this section vary by a factor of 10^7. Yet all are widely used and clearly adequate for certain tasks; the least sensitive are not, however, satisfactory for certain applications. In general, the sensitivities of present-day detectors lie in the range of 10^{-8} to 10^{-15} g solute/s.
2. Good stability and reproducibility.
3. A linear response to solutes that extends over several orders of magnitude.
4. A temperature range from room temperature to at least 400°C.
5. A short response time independent of flow rate.
6. High reliability and ease of use. The detector should be foolproof in the hands of inexperienced operators, if possible.
7. Similarity in response toward all solutes or alternatively a highly predictable and selective response toward one or more classes of solutes.
8. The detector should be nondestructive.

Unfortunately, no detector exhibits all of these characteristics. Some of the more common detectors are listed in Table 27-1. Several of the most widely used detectors are described in the paragraphs that follow.

Flame Ionization Detectors

The flame ionization detector (FID) is the most widely used and generally applicable detector for GC. With an FID such as the one shown in Figure 27-8, effluent from

TABLE 27-1 Typical Gas Chromatographic Detectors

Type	Applicable Samples	Typical Detection Limit
Flame ionization	Hydrocarbons	1 pg/s
Thermal conductivity	Universal detector	500 pg/mL
Electron capture	Halogenated compounds	5 fg/s
Mass spectrometer (MS)	Tunable for any species	0.25 to 100 pg
Thermionic	Nitrogen and phosphorous compounds	0.1 pg/s (P), 1 pg/s (N)
Electrolytic conductivity (Hall)	Compounds containing halogens, sulfur, or nitrogen	0.5 pg Cl/s, 2 pg S/s, 4 pg N/s
Photoionization	Compounds ionized by UV radiation	2 pg C/s
Fourier transform IR (FTIR)	Organic compounds	0.2 to 40 ng

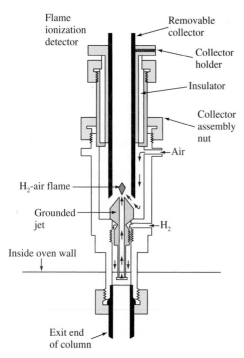

FIGURE 27-8 A typical flame ionization detector. (Courtesy of Agilent Technologies.)

the column is directed into a small air-hydrogen flame. Most organic compounds produce ions and electrons when pyrolyzed at the temperature of an air-hydrogen flame. Detection involves monitoring the current produced by collecting these charge carriers. A few hundred volts applied between the burner tip and a collector electrode located above the flame causes the ions and electrons to move toward the collector. The resulting current ($\sim 10^{-12}$ A) is then measured with a high-impedance picoammeter.

The ionization of carbon compounds in the FID is not fully understood, although the number of ions produced is roughly proportional to the number of *reduced* carbon atoms in the flame. Because the FID responds to the number of carbon atoms entering the detector per unit of time, it is a *mass-sensitive* rather than a *concentration-sensitive* device. As such, this detector has the advantage that changes in flow rate of the mobile phase have little effect on detector response.

Functional groups, such as carbonyl, alcohol, halogen, and amine, yield fewer ions or none at all in a flame. In addition, the detector is insensitive toward noncombustible gases such as H_2O, CO_2, SO_2, CO, noble gases, and NO_x. These properties make the FID a most useful general detector for the analysis of most organic samples, including those contaminated with water and the oxides of nitrogen and sulfur.

The FID exhibits a high sensitivity ($\sim 10^{-13}$ g/s), large linear response range ($\sim 10^7$), and low noise. It is generally rugged and easy to use. Disadvantages of the flame ionization detector are that it destroys the sample during the combustion step and requires additional gases and controllers.

Thermal Conductivity Detectors

The *thermal conductivity detector* (TCD), which was one of the earliest detectors for GC, is still widely used. This device contains an electrically heated source whose temperature at constant electrical power depends on the thermal conductivity of the surrounding gas. The heated element may be a fine platinum, gold, or tungsten wire or, alternatively, a small thermistor.

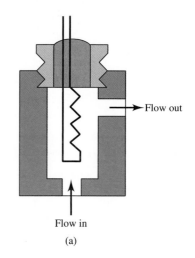

(a)

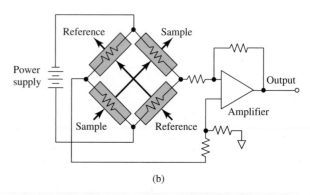

(b)

FIGURE 27-9 Schematic of (a) a TCD cell, and (b) an arrangement of two sample detector cells and two reference detector cells. (From J. Hinshaw, *LC-GC*, **1990**, *8*, 298. With permission.)

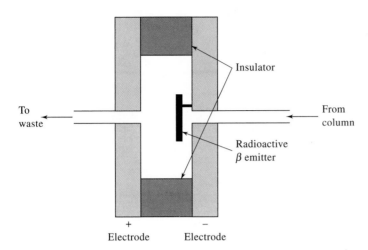

To
waste

From
column

Insulator

Radioactive
β emitter

+
Electrode

−
Electrode

FIGURE 27-10 Schematic diagram of an ECD.

The electrical resistance of this element depends on the thermal conductivity of the gas. Figure 27-9a shows a cross-sectional view of one of the temperature-sensitive elements in a TCD.

Twin detectors are usually used, one being located ahead of the sample-injection chamber and the other immediately beyond the column. The detector elements are labeled *sample* and *reference* in Figure 27-9b. Alternatively, the gas stream can be split. The detectors are incorporated in two arms of a bridge circuit. The bridge circuit is arranged so that the thermal conductivity of the carrier gas is canceled. In addition, the effects of variations in temperature, pressure, and electrical power are minimized.

Modulated single-filament TCDs are also available. Here, the analytical and reference gases are passed alternately over a tiny filament held in a low-volume ($\sim$5-μL) cell. The gases are switched at a frequency of 10 Hz. The output is thus a 10-Hz signal whose amplitude is proportional to the difference in thermal conductivity of the analytical and reference gases. Because the amplifier responds only to a 10-Hz signal, thermal noise in the system is largely eliminated.

The thermal conductivities of helium and hydrogen are roughly six to ten times greater than those of most organic compounds. Thus, even small amounts of organic species cause relatively large decreases in the thermal conductivity of the column effluent, which results in a marked rise in the temperature of the detector. Detection by thermal conductivity is less satisfactory with carrier gases whose conductivities closely resemble those of most sample components.

The advantages of the TCD are its simplicity, its large linear dynamic range ($\sim10^5$), its general response

to both organic and inorganic species, and its nondestructive character, which permits collection of solutes after detection. Its chief limitation is its relatively low sensitivity ($\sim10^{-8}$ g solute/mL carrier gas). Other detectors exceed this sensitivity by factors of 10^4 to 10^7. It should be noted that the low sensitivity of TCDs often precludes their use with capillary columns where sample amounts are very small.

Electron-Capture Detectors

The electron-capture detector (ECD)[5] has become one of the most widely used detectors for environmental samples because it selectively responds to halogen-containing organic compounds, such as pesticides and polychlorinated biphenyls. As shown in Figure 27-10, the sample eluate from a column is passed over a radioactive β emitter, usually nickel-63. An electron from the emitter causes ionization of the carrier gas (often nitrogen) and the production of a burst of electrons. In the absence of organic species, a constant standing current between a pair of electrodes results from this ionization process. The current decreases significantly, however, in the presence of organic molecules containing electronegative functional groups that tend to capture electrons.

The ECD is selective in its response. Compounds such as halogens, peroxides, quinones, and nitro groups are detected with high sensitivity. The detector is insensitive to functional groups such as amines, alcohols, and hydrocarbons. An important application of the ECD is for the detection and quantitative determination of chlorinated insecticides.

[5]For a description of commercial electron-capture detectors, see D. Noble, *Anal. Chem.*, **1995**, *67*, 442A.

ECDs are highly sensitive and have the advantage of not altering the sample significantly (in contrast to the flame ionization detector, which consumes the sample). The linear response of the detector, however, is limited to about two orders of magnitude.

Thermionic Detectors

The thermionic detector is selective toward organic compounds containing phosphorus and nitrogen. Its response to a phosphorus atom is approximately 10 times greater than to a nitrogen atom and 10^4 to 10^6 times larger than to a carbon atom. Compared with the FID, the thermionic detector is approximately 500 times more sensitive to phosphorus-containing compounds and 50 times more sensitive to nitrogen-bearing species. These properties make thermionic detection particularly useful for sensing and determining the many phosphorus-containing pesticides.

A thermionic detector is similar in structure to the FID shown in Figure 27-8. The column effluent is mixed with hydrogen, passes through the flame tip assembly, and is ignited. The hot gas then flows around an electrically heated rubidium silicate bead, which is maintained at about 180 V with respect to the collector. The heated bead forms a plasma having a temperature of 600°C–800°C. Exactly what occurs in the plasma to produce unusually large numbers of ions from phosphorus- or nitrogen-containing molecules is not fully understood; but large ion currents result, which are useful for determining compounds containing these two elements.

Electrolytic Conductivity Detectors

In the Hall electrolytic conductivity detector, compounds containing halogens, sulfur, or nitrogen are mixed with a reaction gas in a small reactor tube, usually made of nickel. The reaction tube is kept at 850°C–1000°C. The products are then dissolved in a liquid, which produces a conductive solution. The change in conductivity as a result of the ionic species in the conductance cell is then measured. A typical detector is illustrated in Figure 27-11.

In the halogen mode, hydrogen is used as the reaction gas. Halogen-containing compounds are converted to HX and dissolved in n-propyl alcohol as the conductivity solvent. In this mode, sulfur-containing compounds are converted to H_2S and nitrogen-containing compounds to NH_3, which do not give significant responses because both are poorly ionized in the solvent. The limit of detection is ~0.5 pg Cl/s, and the linear range is 10^6.

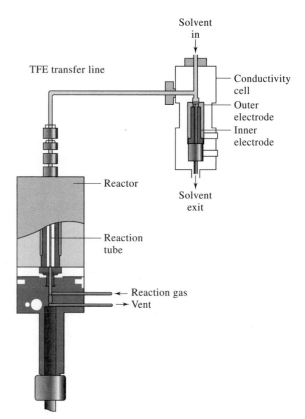

FIGURE 27-11 Diagram of a Hall electrolytic conductivity detector. (Courtesy of ThermoElectron Corp.)

In the sulfur mode, the reaction gas is air, which converts the sulfur-containing compounds to SO_2. The conductivity solvent is methyl alcohol with a small amount of water. The SO_2 in the presence of water is converted to sulfite and sulfate ions. Nitrogen-containing compounds are converted to N_2 and nitrogen oxides and show little or no response. Halogen-containing compounds are converted to HX and must be removed with a postreaction scrubber prior to detection. In the sulfur mode, approximately 2 pg S/s can be detected with a linear range of three orders of magnitude.

In the nitrogen mode, hydrogen is used as the reaction gas, as in the halogen mode. However, here, water containing a small amount of an organic solvent is used as the conductivity solvent. In this solvent, the NH_3 produced is converted to NH_4^+. The HX and H_2S produced from halogen- and sulfur-containing compounds must be removed with a postreaction scrubber. The limit of detection is ~4 pg N/s with a linear range of three orders of magnitude.

Dry electrolytic conductivity detectors are also available. These differ from the conventional detectors

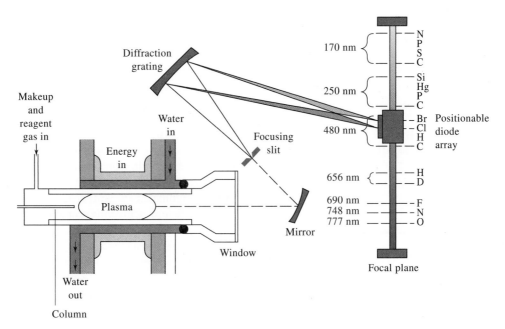

FIGURE 27-12 An AED for GC. (Courtesy of Agilent Technologies.)

in that they do not use a solvent, but instead detect the product ions in the gas phase. The dry detector is responsive to chlorine- and bromine-containing compounds. It can be used in series with a flame ionization detector.

Photoionization Detector

In the photoionization detector, molecules eluting from the GC column are photoionized by ultraviolet radiation from a 10.2 eV hydrogen or a 11.7 eV argon lamp. This source ionizes species with an ionization potential below the lamp energy. Compounds with a higher ionization potential do not absorb the energy and thus are not detected. The ions and electrons produced by photoionization are then collected at a pair of biased electrodes. The detector is most sensitive for aromatic hydrocarbons and organosulfur or organophosphorus compounds that are easily photoionized. The linear range is as high as seven orders of magnitude.

Atomic Emission Detectors

In the atomic emission detector (AED), the effluent from the GC column is introduced into a microwave-induced plasma (MIP), an inductively coupled plasma (ICP), or a direct current plasma (DCP). The MIP has been most widely used and is available commercially. The MIP is used in conjunction with a diode array or charge-coupled-device atomic emission spectrometer as shown in Figure 27-12. The plasma is sufficiently en-ergetic to atomize all of the elements in a sample and to excite their characteristic atomic emission spectra. Hence, the AED is an *element-selective detector*. As shown on the right of the figure, the positionable diode array is capable of monitoring simultaneously several elements at any given setting.

Figure 27-13 illustrates the power of element-selective detection. The sample in this case consisted of a gasoline containing a small concentration of methyl tertiary butyl ether (MTBE), an antiknock agent, as well as several aliphatic alcohols in low concentrations. The upper chromatogram, obtained by monitoring the carbon emission line at 198 nm, consists of a myriad of peaks that would be very difficult to sort out and identify. In contrast, when the oxygen line at 777 nm is used to record the chromatogram (Figure 27-13b), peaks for the alcohols and for MTBE are evident and readily identifiable.

Flame Photometric Detector

The *flame photometric detector* (FPD) has been widely applied to the analysis of air and water pollutants, pesticides, and coal hydrogenation products. It is a selective detector that is primarily responsive to compounds containing sulfur and phosphorus. In this detector, the eluent is passed into a low-temperature hydrogen-air flame, which converts part of the phosphorus to an HPO species that emits bands of radiation centered at about 510 and 526 nm. Sulfur in the

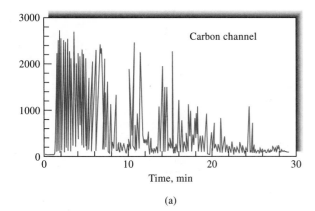

(a)

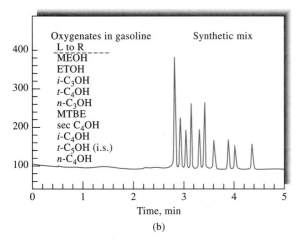

(b)

FIGURE 27-13 Chromatogram for a gasoline sample containing a small amount of MTBE and several aliphatic alcohols: (a) monitoring a carbon emission line; (b) monitoring an oxygen emission line. (Courtesy of Agilent Technologies.)

sample is simultaneously converted to S_2, which emits a band centered at 394 nm. However, the sulfur chemiluminescence detector discussed later in this section provides lower detection limits and wider linear working range than does the FPD. Suitable filters are used to isolate the appropriate bands, and their intensity is recorded photometrically. Other elements that have been detected by flame photometry include the halogens, nitrogen, and several metals, such as tin, chromium, selenium, and germanium.

Mass Spectrometry Detectors

One of the most powerful detectors for GC is the mass spectrometer. Mass spectrometers and applications of mass spectrometry are discussed in Chapters 11 and 20. The combination of GC with mass spectrometry is known as *GC/MS*.[6] Currently, nearly fifty instrument companies offer GC/MS equipment. The flow rate from capillary columns is generally low enough that the column output can be fed directly into the ionization chamber of the mass spectrometer. A schematic of a typical system is shown in Figure 27-14. Prior to the advent of capillary columns in GC, when packed columns were used, it was necessary to minimize the large volumes of carrier gas eluting from the GC. Various jet, membrane, and effusion separators were used for this purpose. However, in many cases such devices also removed a significant amount of the analyte and were thus quite inefficient. Currently, capillary columns are invariably used in GC/MS instruments, and such separators are no longer needed.

Thermal degradation of components can be a difficulty in GC/MS. Not only can the GC injection port and GC column cause degradation but also the heated metal surfaces in the mass spectrometer ion source may cause problems. Lowering the temperature can minimize degradation. Often, however, the mass spectrometer can be used to identify decomposition products, which can lead to chromatographic modifications that solve the degradation problem.

The most common ion sources used in GC/MS are electron-impact ionization and chemical ionization. Ion sources for mass spectrometry are discussed in detail in Section 20B. The most common mass analyzers are quadrupole and ion-trap analyzers. These analyzers are described in Sections 11B-2 and 20C-3. Time-of-flight mass analyzers are also used, but not as frequently as quadrupoles and ion traps.

In GC/MS, the mass spectrometer scans the masses repetitively during a chromatographic experiment. If the chromatographic run is 10 minutes, for example, and a scan is taken each second, 600 mass spectra are recorded. The data can be analyzed by the data system in several different ways. First, the ion abundances in each spectrum can be summed and plotted as a function of time to give a *total-ion chromatogram*. This plot is similar to a conventional chromatogram. One can also display the mass spectrum at a particular time during the chromatogram to identify the species eluting at that time. Finally, a single mass-to-charge (m/z)

────────
[6]For additional information, see J. Masucci and G. W. Caldwell, in *Modern Practice of Gas Chromatography*, 4th ed., R. L. Grob and E. F. Barry, eds., Chap. 7, New York: Wiley-Interscience, 2004; H-J. Hubschmann, *Handbook of GC/MS: Fundamentals and Applications*, Weinheim, Germany: Wiley-VCH, 2001; M. McMaster and C. McMaster, *GC/MS: A Practical User's Guide*, New York, Wiley-VCH, 1998.

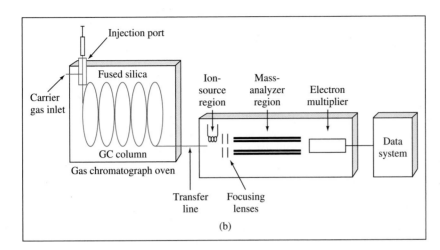

FIGURE 27-14 Schematic of a typical capillary GC/MS system. The effluent from the GC is passed into the inlet of the mass spectrometer, where the molecules in the gas are fragmented, ionized, analyzed, and detected.

value can be selected and monitored throughout the chromatographic experiment, a technique known as *selected-ion monitoring*. Mass spectra of selected ions obtained during a chromatography experiment are known as *mass chromatograms*.

GC/MS instruments have been used for the identification of thousands of components that are present in natural and biological systems. For example, these procedures have permitted characterization of the odor and flavor components of foods, identification of water pollutants, medical diagnosis based on breath components, and studies of drug metabolites.

An example of one application of GC/MS is shown in Figure 27-15. The upper figure is the total-ion

 Animation: Learn more about **GC/MS**.

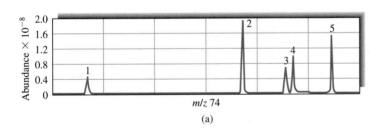

(a)

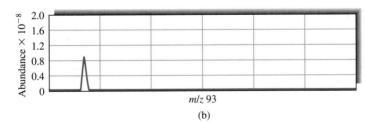

(b)

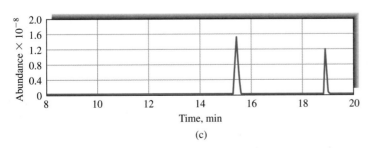

(c)

FIGURE 27-15 Typical outputs for a GC/MS system. In (a), the total ion current chromatogram of a five-component mixture is shown. The components were 1, *N*-nitrosodimethylamine, 2, bis(2-chloroethyl)ether, 3, bis(2-chloroisopropyl)ether, 4, *N*-nitrosodi-*n*-propylamine, and 5, bis(2-chloroethoxy)methane. In (b), the mass chromatogram at $m/z = 74$ is shown. The peak is due to the parent ion of *n*-nitrosodimethylamine ($C_2H_6N_2O$). A selected-ion chromatogram at $m/z = 93$ is shown in (c). Peaks 2 and 5 give a response at this m/z value due to fragmentation products. (With permission from J. A. Masucci and G. W. Caldwell, in *Modern Practice of Gas Chromatography*, 4th ed., R. L. Grob and E. F. Barry, eds., New York: Wiley-Interscience, 2004, p. 356.)

chromatogram of a five-component mixture. Also shown are mass chromatograms at $m/z = 74$ and $m/z = 93$. From these, the identities of components 1, 2, and 5 can be determined.

Mass spectrometry can also be used to obtain information about incompletely separated components. For example, the mass spectrum of the front edge of a GC peak may be different from that of the middle part of the peak or the trailing edge if the peak is due to more than one component. With mass spectrometry, we can not only determine that a peak is due to more than one species but also identify the various unresolved components. GC has also been coupled to tandem mass spectrometers or to Fourier transform mass spectrometers to give GC/MS/MS or GC/MS^n systems. These are extremely powerful tools for identifying components in mixtures.

GC Coupled with Spectroscopic Detection

GC is often coupled with the selective techniques of spectroscopy and electrochemistry to provide powerful tools for separating and identifying the components of complex mixtures. Combinations of GC with mass spectrometry (GC/MS), Fourier transform infrared spectroscopy (GC/FTIR), nuclear magnetic resonance spectroscopy, and electroanalytical methods are sometimes termed *hyphenated methods*.

In early systems, the eluates from the GC column were collected as separate fractions in a cold trap, and a nondestructive, nonselective detector was used to indicate their appearance. The composition of each fraction was then investigated by nuclear magnetic resonance spectrometry, infrared spectroscopy, or by electroanalytical measurements. A serious limitation to this approach was the very small (usually micromolar) quantities of solute contained in a fraction.

Most modern hyphenated methods monitor the effluent from the chromatographic column continuously by spectroscopic methods. The combination of two techniques based on different principles can achieve tremendous selectivity. Today's computer-based GC instruments incorporate large databases for comparing spectra and identifying compounds.

Other Types of Detectors

Several other types of GC detectors are useful for specific applications. The *sulfur chemiluminescence detector* is based on the reaction between certain sulfur compounds and ozone. The resulting luminescence intensity is proportional to the concentration of sulfur. This detector has proven particularly useful for the determination of pollutants such as mercaptans. In the sulfur chemiluminescence detector, the eluent is mixed with hydrogen and air, and combustion takes place as in the flame ionization detector. The resulting gases are then mixed with ozone, and the intensity of the emission produced is measured. The linear range is about five orders of magnitude, and the limit of detection for sulfur is approximately 0.5 pg/s. The sulfur chemiluminescence detector has also been adapted to supercritical fluid chromatography.

The nitrogen-specific chemiluminescence detector is quite similar to the sulfur detector. The nitrous oxide combustion product reacts with ozone to produce chemiluminescence. The detector responds linearly to nitrogen over about four orders of magnitude. The limit of detection for nitrogen is about 5 pg/s. The detector can be used for organic nitrogen compounds and for inorganic compounds such as ammonia, hydrazine, HCN, and nitrogen oxides.

27C GAS CHROMATOGRAPHIC COLUMNS AND STATIONARY PHASES

The pioneering gas-liquid chromatographic studies in the early 1950s were carried out on packed columns in which the stationary phase was a thin film of liquid retained by adsorption on the surface of a finely divided, inert solid support. From theoretical studies made during this early period, it became apparent that unpacked columns having inside diameters of a few tenths of a millimeter could provide separations superior to those on packed columns in both speed and column efficiency. In such *capillary columns*, the stationary phase was a film of liquid a few tenths of a micrometer thick that uniformly coated the interior of a capillary tubing. In the late 1950s such *open tubular columns* were constructed and the predicted performance characteristics were confirmed experimentally in several laboratories, with open tubular columns having 300,000 plates or more being described.[7] Today, open tubular columns pre-

---

[7] In 1987 a world record for length of an open tubular column and number of theoretical plates was set, as attested in the *Guinness Book of Records*, by Chrompack International Corporation of the Netherlands. The column was a fused-silica column drawn in one piece and having an internal diameter of 0.32 mm and a length of 2.1 km, or 1.3 miles. The column was coated with a 0.1 m film of polydimethyl siloxane. A 1.3-km section of this column contained more than 2 million plates.

dominate in GC because, with no packing, columns can be narrower and longer, leading to higher efficiencies than with packed columns.

Despite such spectacular performance characteristics, capillary columns did not gain widespread use until more than two decades after their invention. The reasons for the delay were several, including small sample capacities; fragility of columns; mechanical problems associated with sample introduction and connection of the column to the detector; difficulties in coating the column reproducibly; short lifetimes of poorly prepared columns; tendencies of columns to clog; and patents, which limited commercial development to a single manufacturer (the original patent expired in 1977). By the late 1970s these problems had become manageable and several instrument companies began to offer open tubular columns at a reasonable cost. As a consequence, we have seen a major growth in the use of capillary columns since then.[8]

27C-1 Open Tubular Columns

Open tubular, or capillary, columns are of two basic types: *wall-coated open tubular* (WCOT) and *support-coated open tubular* (SCOT) columns.[9] Wall-coated columns are simply capillary tubes coated with a thin layer of the stationary phase. In support-coated open tubular columns, the inner surface of the capillary is lined with a thin film ($\sim$30 μm) of a support material, such as diatomaceous earth. This type of column holds several times as much stationary phase as does a wall-coated column and thus has a greater sample capacity. Generally, the efficiency of a SCOT column is less than that of a WCOT column but significantly greater than that of a packed column.

Early WCOT columns were constructed of stainless steel, aluminum, copper, or plastic. Later, glass columns began to be used. Often, the glass was etched with gaseous hydrochloric acid, strong aqueous hydrochloric acid, or potassium hydrogen fluoride to give a rough surface, which bonded the stationary phase more tightly. The most widely used capillary columns are *fused-silica wall-coated* (FSWC) *open tubular columns*. Fused-silica capillaries are drawn from specially purified silica that contains minimal amounts of

metal oxides. These capillaries have much thinner walls than glass columns. The tubes are given added strength by an outside protective polyimide coating, which is applied as the capillary tubing is drawn. The resulting columns are quite flexible and can be bent into coils with diameters of a few inches. Figure 27-6 shows fused-silica open tubular columns. Silica open tubular columns are available commercially and offer several important advantages such as physical strength, much lower reactivity toward sample components, and flexibility. For most applications, they have replaced the older-type WCOT glass columns.

The most widely used silica open tubular columns have inside diameters of 0.32 and 0.25 mm. Higher-resolution columns are also available with diameters of 0.20 and 0.15 mm. Such columns are more troublesome to use and are more demanding on the injection and detection systems. Thus, a sample splitter must be used to reduce the size of the sample injected onto the column and a more sensitive detector system with a rapid response time is required.

Recently, 530-μm capillaries, sometimes called *megabore columns*, have appeared on the market. These columns will tolerate sample sizes that are similar to those for packed columns. The performance characteristics of megabore open tubular columns are not as good as those of smaller-diameter columns but are significantly better than those of packed columns.

Table 27-2 compares the performance characteristics of fused silica capillary columns with other types of wall-coated columns as well as with support-coated and packed columns.

27C-2 Packed Columns

Modern packed columns are fabricated from glass or metal tubing; they are typically 2 to 3 m long and have inside diameters of 2 to 4 mm. These tubes are densely packed with a uniform, finely divided packing material, or solid support, coated with a thin layer (0.05 to 1 μm) of the stationary liquid phase. The tubes are usually formed as coils with diameters of roughly 15 cm to permit convenient thermostatting in an oven.

Solid Support Materials

The packing, or solid support in a packed column, holds the liquid stationary phase in place so that as large a surface area as possible is exposed to the mobile phase. The ideal support consists of small, uniform, spherical particles with good mechanical strength and a specific

---

[8]For more information on columns in GC, see E. F. Barry, in *Modern Practice of Gas Chromatography*, R. L. Grob and E. F. Barry, eds., 4th ed., Chap. 3, New York: Wiley-Interscience, 2004.

[9]For a detailed description of open tubular columns, see M. L. Lee, F. J. Yang, and K. D. Bartle, *Open Tubular Column Gas Chromatography: Theory and Practice*, New York: Wiley, 1984.

TABLE 27-2 Properties and Characteristics of Typical GC Columns

	Type of Column			
	FSWC*	WCOT†	SCOT‡	Packed
Length, m	10–100	10–100	10–100	1–6
Inside diameter, mm	0.1–0.3	0.25–0.75	0.5	2–4
Efficiency, plates/m	2000–4000	1000–4000	600–1200	500–1000
Sample size, ng	10–75	10–1000	10–1000	$10–10^6$
Relative pressure	Low	Low	Low	High
Relative speed	Fast	Fast	Fast	Slow
Flexibility?	Yes	No	No	No
Chemical inertness	Best ──────────────────────────────────→ Poorest			

*Fused silica, wall-coated open tubular column.

†Wall-coated open tubular metal, plastic, or glass columns.

‡Support-coated open tubular column (also called porous-layer open tubular, or PLOT).

surface area of at least 1 m²/g. In addition, the material should be inert at elevated temperatures and be uniformly wetted by the liquid phase. No material is yet available that meets all of these criteria perfectly.

The earliest, and still the most widely used, packings for GC were prepared from naturally occurring diatomaceous earth, which consists of the skeletons of thousands of species of single-celled plants that once inhabited ancient lakes and seas. Figure 27-16 is an enlarged photo of a diatom obtained with a scanning electron microscope. Such plants received their nutrients and disposed of their wastes via molecular diffusion through their pores. As a result, their remains are well-suited as support materials because GC is also based on the same kind of molecular diffusion.

Particle Size of Supports

As shown in Figure 26-11, the efficiency of a gas chromatographic column increases rapidly with decreasing particle diameter of the packing. The pressure difference required to maintain an acceptable flow rate of carrier gas, however, varies inversely as the square of the particle diameter; the latter relationship has placed lower limits on the size of particles used in GC because it is not convenient to use pressure differences that are greater than about 50 psi. As a result, the usual support particles are 60 to 80 mesh (250 to 170 μm) or 80 to 100 mesh (170 to 149 μm).

27C-3 Adsorption on Column Packings or Capillary Walls

A problem that has plagued GC from its beginning has been the physical adsorption of polar or polarizable analyte species, such as alcohols or aromatic hydrocarbons, on the silicate surfaces of column packings or capillary walls. Adsorption causes distorted peaks, which are broadened and often exhibit a tail (recall Figure 26-5). Adsorption occurs with silanol groups that form on the surface of silicates by reaction with

FIGURE 27-16 A photomicrograph of a diatom. Magnification 5000×. (© Dr. Anne Smith/Photo Researchers, Inc.)

moisture. Thus, a fully hydrolyzed silicate surface has the structure

The SiOH groups on the support surface have a strong affinity for polar organic molecules and tend to retain them by adsorption.

Support materials can be deactivated by silanization with dimethylchlorosilane (DMCS). The reaction is

When the support is washed with methanol, the second chloride is replaced by a methoxy group. That is,

Silanized surfaces of column packings may still show a residual adsorption, which apparently occurs with metal oxide impurities in the diatomaceous earth. Acid washing prior to silanization removes these impurities. Fused silica used for manufacturing open tubular columns is largely free of this type of impurity. Because of this, fewer problems with adsorption arise with fused-silica columns.

27C-4 The Stationary Phase

Desirable properties for the immobilized liquid phase in a gas-liquid chromatographic column include (1) *low volatility* (ideally, the boiling point of the liquid should be at least 100°C higher than the maximum operating temperature for the column); (2) *thermal stability*; (3) *chemical inertness*; (4) *solvent characteristics* such that k and α (Sections 26B-5 and 26B-6, respectively) values for the solutes to be resolved fall within a suitable range.

Many liquids have been proposed as stationary phases in the development of GLC. Currently, fewer than a dozen are commonly used. The proper choice of stationary phase is often crucial to the success of a separation. Qualitative guidelines for stationary-phase selection can be based on a literature review, an Internet search, prior experience, or advice from a vendor of chromatographic equipment and supplies.

The retention time for an analyte on a column depends on its distribution constant, which in turn is related to the chemical nature of the liquid stationary phase. To separate various sample components, their distribution constants must be sufficiently different to accomplish a clean separation. At the same time, these constants must not be extremely large or extremely small because large distribution constants lead to prohibitively long retention times and small constants produce such short retention times that separations are incomplete.

To have a reasonable residence time in the column, an analyte must show some degree of compatibility (solubility) with the stationary phase. Here, the principle of "like dissolves like" applies, where "like" refers to the polarities of the analyte and the immobilized liquid. The polarity of a molecule, as indicated by its dipole moment, is a measure of the electric field produced by separation of charge within the molecule. Polar stationary phases contain functional groups such as —CN, —CO, and —OH. Hydrocarbon-type stationary phases and dialkyl siloxanes are nonpolar, whereas polyester phases are highly polar. Polar analytes include alcohols, acids, and amines; solutes of medium polarity include ethers, ketones, and aldehydes. Saturated hydrocarbons are nonpolar. Generally, the polarity of the stationary phase should match that of the sample components. When the match is good, the order of elution is determined by the boiling point of the eluents.

Classification of Stationary Phases

Many different schemes have been reported to classify stationary phases and thereby simplify stationary-phase selection. Most of these are based on solute probes that test specific interactions between the solute and the liquid phase by measuring solute retention

TABLE 27-3 Some Common Liquid Stationary Phases for GLC

Stationary Phase	Common Trade Name	Maximum Temperature, °C	Common Applications
Polydimethyl siloxane	OV-1, SE-30	350	General-purpose nonpolar phase, hydrocarbons, polynuclear aromatics, steroids, PCBs
5% Phenyl-polydimethyl siloxane	OV-3, SE-52	350	Fatty acid methyl esters, alkaloids, drugs, halogenated compounds
50% Phenyl-polydimethyl siloxane	OV-17	250	Drugs, steroids, pesticides, glycols
50% Trifluoropropyl-polydimethyl siloxane	OV-210	200	Chlorinated aromatics, nitroaromatics, alkyl substituted benzenes
Polyethylene glycol	Carbowax 20M	250	Free acids, alcohols, ethers, essential oils, glycols
50% Cyanopropyl-polydimethyl siloxane	OV-275	240	Polyunsaturated fatty acids, rosin acids, free acids, alcohols

characteristics. Two of the most important classifications are based on the work of Rohrschneider and McReynolds.[10] The result was the production of lists of stationary phases and the compound classes that can be separated by each phase (see Table 27-3). Likewise, numerical values, known as McReynolds constants, are available that can guide the user in selecting a stationary phase to separate analytes having different functional groups, such as alcohols from aldehydes or ketones.[11]

Some Widely Used Stationary Phases

Table 27-3 lists the most widely used stationary phases for both packed and open tubular column GC in order of increasing polarity. These six liquids can probably provide satisfactory separations for 90% or more of samples encountered.

Five of the liquids listed in Table 27-3 are polydimethyl siloxanes that have the general structure

$$R-\underset{\underset{R}{|}}{\overset{\overset{R}{|}}{Si}}-O-\left[\underset{\underset{R}{|}}{\overset{\overset{R}{|}}{Si}}-O\right]_n\underset{\underset{R}{|}}{\overset{\overset{R}{|}}{Si}}-R$$

In the first of these, polydimethyl siloxane, the —R groups are all —CH_3, giving a liquid that is relatively nonpolar. In the other polysiloxanes shown in the table, a fraction of the methyl groups are replaced by functional groups such as phenyl (—C_6H_5), cyanopropyl (—C_3H_6CN), and trifluoropropyl (—$C_3H_6CF_3$). The percentage description in each case gives the amount of substitution of the named group for methyl groups on the polysiloxane backbone. Thus, for example, 5% phenyl-polydimethyl siloxane has a phenyl ring bonded to 5% (by number) of the silicon atoms in the polymer. These substitutions increase the polarity of the liquids to various degrees.

The fifth entry in Table 27-3 is a polyethylene glycol with the structure

$$HO-CH_2-CH_2-(O-CH_2-CH_2)_n-OH$$

It finds widespread use for separating polar species. Figure 27-17 illustrates applications of the phases listed in Table 27-3 for open tubular columns.

Bonded and Cross-Linked Stationary Phases

Commercial columns are advertised as having bonded or cross-linked stationary phases. The purpose of bonding and cross-linking is to provide a longer-lasting stationary phase that is not disrupted at elevated temperatures or during temperature programming. With use, untreated columns slowly lose their stationary

[10]L. Rohrschneider, *J. Chromatogr.*, **1966**, *22*, 6; W. O. McReynolds, *J. Chromatogr. Sci.*, **1970**, *8*, 685.
[11]J. A. Dean, *Analytical Chemistry Handbook*, pp. 4.34–4.37, New York: McGraw-Hill, 1995.

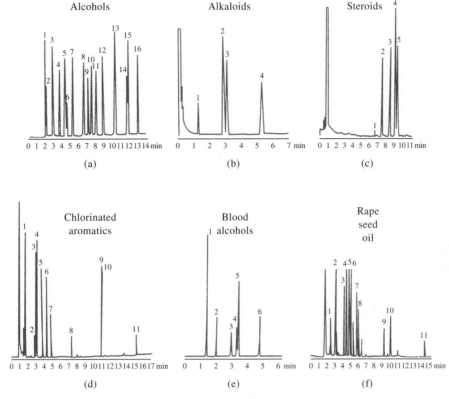

FIGURE 27-17 Typical chromatograms from open tubular columns coated with (a) polydimethyl siloxane; (b) 5% (phenyl methyldimethyl) siloxane; (c) 50% (phenyl methyldimethyl) siloxane; (d) 50% poly(trifluoropropyl-dimethyl) siloxane; (e) polyethylene glycol; (f) 50% poly(cyanopropyl-dimethyl) siloxane. (Courtesy of J&W Scientific.)

phase because of "bleeding," in which a small amount of immobilized liquid is carried out of the column during the elution process. Such columns are also recommended for on-column injection where a large solvent volume is used. Indeed, cross-linked or bonded columns may be backflushed to remove contaminants without significant loss of stationary phase.

Bonding involves attaching a monomolecular layer of the stationary phase to the silica surface of the column by a chemical reaction. For commercial columns, the nature of the reaction is usually proprietary.

Cross-linking is carried out *in situ* after a column is coated with one of the polymers listed in Table 27-3. One way of cross-linking is to incorporate a peroxide into the original liquid. When the film is heated, reaction between the methyl groups in the polymer chains is initiated by a free radical mechanism. The polymer molecules are then cross-linked through carbon-to-carbon bonds. The resulting films are less

extractable and have considerably greater thermal stability than do untreated films. Cross-linking has also been initiated by exposing the coated columns to gamma radiation.

Film Thickness

Commercial columns are available having stationary phases that vary in thickness from 0.1 to 5 μm. Film thickness primarily affects the retentive character and the capacity of a column as discussed in Section 26C-3. Thick films are used with highly volatile analytes because such films retain solutes for a longer time, thus providing a greater time for separation to take place. Thin films are useful for separating species of low volatility in a reasonable length of time. For most applications with 0.25- or 0.32-mm columns, a film thickness of 0.25 μm is recommended. With megabore columns, 1- to 1.5-μm films are often used. Today, columns with 8-μm films are marketed.

Chiral Stationary Phases

In recent years, much effort has been devoted to developing methods for the separation of enantiomers by gas or liquid chromatography.[12] Two approaches have been used. One is based on forming derivatives of the analyte with an optically active reagent that forms a pair of diastereomers that can be separated on an achiral column. The alternative method is to use a chiral liquid as the stationary phase. A number of amino acid–derived chiral phases have been developed for this purpose, and others are becoming available commercially.

27D APPLICATIONS OF GC

To evaluate the importance of GC, we must distinguish between the two roles the method plays. First, GC is a tool for performing separations. In this role, GC methods are unsurpassed when applied to complex organic, metal-organic, and biochemical systems made up of volatile species or species that can be derivatized to yield volatile substances. The second role that GC plays is in the completion of an analysis. In this role, retention times or volumes are used for qualitative identification, and peak heights or peak areas provide quantitative information. For qualitative purposes, GC is much more limited than most of the spectroscopic methods considered in earlier chapters. Thus, an important trend in the field has been in the direction of combining the remarkable separation capabilities of GC with the superior identification properties of such instruments as mass, IR, and nuclear magnetic resonance spectrometers (see Section 27B-4).

27D-1 Qualitative Analysis

Gas chromatograms are widely used to establish the purity of organic compounds. The appearance of additional peaks reveals any contaminants present, and the areas under these peaks provide estimates of the extent of contamination. Such areas are only estimates because different components may have widely differing detector response factors. Gas chromatographic techniques are also useful for evaluating the effectiveness of purification procedures.

In theory, GC retention times should be useful for identifying components in mixtures. In fact, however, the applicability of such data is limited by the number of variables that must be controlled to obtain reproducible results. Nevertheless, GC provides an excellent means of confirming the presence or absence of a suspected compound in a mixture, provided that an authentic sample of the substance is available. No new peaks in the chromatogram of the mixture should appear on addition of the known compound, and enhancement of an existing peak should be observed. The evidence is particularly convincing if the effect can be duplicated on different columns and at different temperatures.

Selectivity Factors

We have seen (Section 26B-6) that the selectivity factor α for compounds A and B is given by the relationship

$$\alpha = \frac{K_B}{K_A} = \frac{(t_R)_B - t_M}{(t_R)_A - t_M} = \frac{(t'_R)_B}{(t'_R)_A}$$

where $(t'_R)_A = [(t_R)_A - t_M]$ is the adjusted retention time for species A. If a standard substance is chosen as compound B, then α can provide an index for identification of compound A, which is largely independent of column variables other than temperature. Numerical tabulations of selectivity factors for pure compounds relative to a common standard can be prepared and then used for the characterization of solutes. Unfortunately, finding a universal standard that yields selectivity factors of reasonable magnitude for all types of analytes is impossible. Thus, the amount of selectivity factor data available in the literature is presently limited.

The Retention Index

The retention index I was first proposed by E. Kovats in 1958 for identifying solutes from chromatograms.[13] The retention index for any given solute can be calculated from a chromatogram of a mixture of that solute with at least two normal alkanes having retention times that bracket that of the solute. The retention index scale is based on normal alkanes. By definition, the retention index for a normal alkane is equal to 100 times the number of carbons in the compound *regardless of the column packing, the temperature, or other chromatographic conditions*. The retention indexes for all compounds other than normal alkanes vary, often

---

[12]For a review of chiral stationary-phase separations by GC, see J. V. Hinshaw, *LC-GC*, **1993**, *11*, 644; E. F. Barry, in *Modern Practice of Gas Chromatography*, R. L. Grob and E. F. Barry, eds., 4th ed., Chap. 3, New York: Wiley-Interscience, 2004.

[13]E. Kovats, *Helv. Chim. Acta*, **1958**, *41*, 1915.

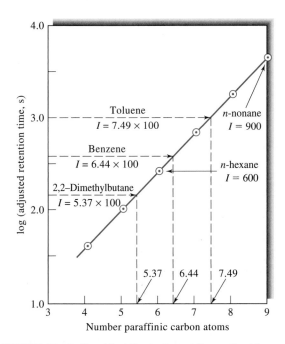

FIGURE 27-18 Graphical illustration of the method for determining retention indexes for three compounds. Stationary phase: squalane. Temperature: 60°C. Retention indexes for normal alkane standards nonane and hexane are indicated.

by several hundred retention index units, with column variables.

It has long been known that within a homologous series, a plot of the logarithm of adjusted retention time ($t'_R = t_R - t_M$) versus the number of carbon atoms is linear, provided the lowest member of the series is excluded. Such a plot for C_4 to C_9 normal alkane standards is shown in Figure 27-18. Also indicated on the ordinate are log adjusted retention times for three compounds on the same column and at the same temperature. Their retention indexes are then obtained by multiplying the corresponding abscissa values by 100. Thus, the retention index for toluene is 749, and for benzene it is 644.

Normally, a graphical procedure is not required to determine retention indexes. Instead, adjusted retention data are calculated by interpolation from a chromatogram of a mixture of the solute of interest and two or more alkane standards.

It is important to reiterate that the retention index *for a normal alkane* is independent of temperature and column packing. Thus, I for heptane, by definition, is always 700. In contrast, retention indexes of all other solutes may, and often do, vary widely from one column to another. For example, the retention index for ace-

naphthene on a cross-linked polydimethyl siloxane stationary phase at 140°C is 1460. With 5% phenylpolydimethyl siloxane as the stationary phase, it is 1500 at the same temperature, and with polyethylene glycol as the stationary phase, the retention index is 2084.

The retention index system has the advantage of being based on readily available reference materials that cover a wide boiling range. In addition, the temperature dependence of retention indexes is relatively small. In 1984 Sadtler Research Laboratories introduced a library of retention indexes measured on four types of fused-silica open tubular columns. The computerized format of the database allows retention index searching and possible identity recall with a desktop computer.[14] Measurement of retention indexes is the basis of the Rohrschneider-McReynolds scheme for classification of stationary phases in GC (see Section 27C-4).

The use of retention data on two or more GC columns can improve the chances of correctly identifying an unknown compound. The columns can be used in separate experiments or sometimes they can be employed in tandem. The use of two or more columns in series is termed *multidimensional chromatography*.[15] Likewise, the responses of two or more GC detectors can greatly aid in qualitative identification.

The combination of GC with various spectroscopic detectors, particularly with mass spectrometry, can greatly aid in identifying components. In fact, GC/MS is now the premier technique for separating and identifying species in mixtures.

27D-2 Quantitative Analysis

The peak height or peak area of an eluate from a GC column has been widely used for quantitative and semiquantitative analyses. An accuracy of 1% relative is attainable under carefully controlled conditions using either the external or the internal standard method. As with most analytical tools, reliability is directly related to the control of variables; the nature of the sample also plays a part in determining the potential accuracy. The general discussion of quantitative chromatographic analysis given in Section 26F-2 applies to GC as well as to other types; therefore, no further consideration of this topic is given here.

---

[14] *The Sadtler Standard Gas Chromatography Retention Index Library*, Vols. 1–4, Philadelphia: Bio-Rad Laboratories, Sadtler Division, 1984–85.
[15] For reviews of two-dimensional GC, see M. Adahchour, J. Beens, R. J. J. Vreuls, U. A. Th. Brinkman, *Trends Anal. Chem. (TRAC)*, **2006**, *25*, 438; **2006**, *25*, 540; **2006**, *25*, 726.

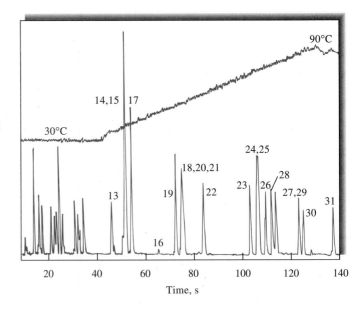

FIGURE 27-19 High-speed chromatogram obtained with isothermal operation (30°C) for 37 s followed by a 35°C/min temperature ramp to 90°C. (From H. Smith and R. D. Sacks, *Anal. Chem.*, **1998**, *70*, 4960. Copyright 1998 American Chemical Society.)

27E ADVANCES IN GC

Although GC is quite a mature technique, there have been many developments in recent years in theory, instrumentation, columns, and practical applications. We discuss here some developments in high-speed GC and in miniaturized GC systems.

27E-1 High-Speed GC

Researchers in GC have often focused on achieving ever higher resolution to separate more and more complex mixtures.[16] In most separations, conditions are varied to separate the most difficult-to-separate pair of components, the so-called *critical pair*. Many of the components of interest, under these conditions, are highly overseparated. The basic idea of high-speed GC is that, for many separations of interest, higher speed can be achieved albeit at the expense of some selectivity and resolution.

The principles of high-speed separations can be demonstrated by substituting Equation 26-5 into Equation 26-11

$$\frac{L}{t_R} = u \times \frac{1}{1 + k_n} \qquad (27\text{-}8)$$

where k_n is the retention factor for the last component of interest in the chromatogram. If we rearrange Equation 27-8 and solve for the retention time of the last component of interest, we obtain

$$t_R = \frac{L}{u} \times (1 + k_n) \qquad (27\text{-}9)$$

Equation 27-9 tells us that we can achieve faster separations by using short columns, higher-than-usual carrier gas velocities and small retention factors. For example, if we reduce the column length L by a factor of 4 and increase the carrier-gas velocity u by a factor of 5, the analysis time t_R is reduced by a factor of 20. The price paid is reduced resolving power caused by increased band broadening and reduced peak capacity (the number of peaks that will fit in the chromatogram).

Research workers in the field have been designing instrumentation and chromatographic conditions to optimize separation speed at the lowest cost in terms of resolution and peak capacity.[17] They have designed systems to achieve tunable columns and high-speed temperature programming. A tunable column is a series combination of a polar and a nonpolar column. Figure 27-19 shows the separation of twelve compounds prior to initiating a programmed temperature ramp and nineteen compounds after the temperature program was begun. The total time required was 140 s.

[16] For more information, see R. D. Sacks, in *Modern Practice of Gas Chromatography*, R. L. Grob and E. F. Barry, eds., 4th ed., New York: Wiley-Interscience, 2004, Chap. 5.

[17] H. Smith and R. D. Sacks, *Anal. Chem.*, **1998**, *70*, 4960.

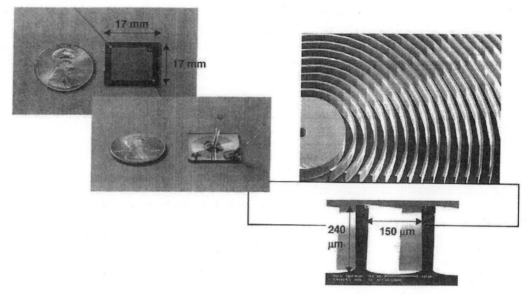

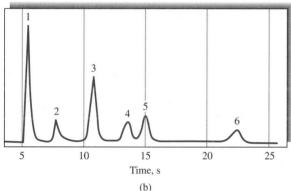

(b)

FIGURE 27-20 Microfabricated columns (a) and chromatogram (b). The columns in (a) were 0.9-m-long spiral and serpentine channels. The mixture (b) was 1, acetone; 2, 2-butanone; 3, benzene; 4, trichloroethylene; 5, 2,5-dimethylfuran; and 6, toluene. Air was used as the carrier gas with an outlet pressure of 0.5 atm. (From R. Sacks, in *Modern Practice of Gas Chromatography*, R. L. Grob and E. F. Barry, eds., 4th ed., New York: Wiley-Interscience, 2004, p. 269. With permission.)

These workers have also been using high-speed GC with mass spectrometry detection including time-of-flight detection.[18]

27E-2 Miniaturized GC Systems

For many years there has been a desire to miniaturize GC systems to the microchip level. Miniature GC systems are useful in space exploration, in portable instruments for field use, and in environmental monitoring. Early work reported on GC columns etched onto a microchip.[19] However, the relatively poor chromatographic performance of such devices led to removal of

the column from the chip in the commercial systems that were produced.

In recent work, microfabricated columns have been designed using substrates of silicon, several metals, and polymers.[20] Relatively deep, narrow channels are etched into the substrate. These channels have low dead volume to reduce band broadening and high surface area to increase stationary-phase volume. Figure 27-20 shows photomicrographs of columns made from etched silicon wafers and a chromatogram recorded using a 0.9 m column. A photoionization detector was used. A complete microfabricated GC system has been described.[21]

[18]C. Leonard and R. Sacks, *Anal. Chem.*, **1999**, *71*, 5177.
[19]See S. C. Terry, J. H. Jerman, and J. B. Angell, *IEEE Trans. Electron Devices*, **1979**, *26*, 1880; J. B. Angell, S. C. Terry, and P. W. Barth, *Sci. Am.*, **1983**, *248* (4), 44.

[20]G. Lambertus, A. Elstro, K. Sensenig, J. Potkay, M. Agah, S. Scheuering, K. Wise, F. Dorman, and R. Sacks, *Anal. Chem.*, **2004**, *76*, 2629.
[21]See notes 16 and 20.

27F   GAS-SOLID CHROMATOGRAPHY

GSC is based on adsorption of gaseous substances on solid surfaces. Distribution constants are generally much larger than those for GLC. As a result, GSC is useful for the separation of species that are not retained by gas-liquid columns, such as the components of air, hydrogen sulfide, carbon disulfide, nitrogen oxides, carbon monoxide, carbon dioxide, and the rare gases.

GSC is performed with both packed and open tubular columns. For the latter, a thin layer of the adsorbent is affixed to the inner walls of the capillary. Such columns are sometimes called *porous-layer open tubular*, or PLOT, columns.

27F-1 Molecular Sieves

Molecular sieves are aluminum silicate ion exchangers, whose pore size depends on the kind of cation present. Commercial preparations of these materials are available in particle sizes of 40 to 60 mesh to 100 to 120 mesh. The sieves are classified according to the maximum diameter of molecules that can enter the pores. Commercial molecular sieves come in pore sizes of 4, 5, 10, and 13 Å. Molecules smaller than these dimensions penetrate into the interior of the particles where adsorption takes place. For such molecules, the surface area is enormous when compared with the area available to larger molecules. Thus, molecular sieves can be used to separate small molecules from large. For example, a 6-ft, 5-Å packing at room temperature will easily separate a mixture of helium, oxygen, nitrogen, methane, and carbon monoxide in the order given.

Figure 27-21a shows a typical molecular sieve chromatogram. In this application two packed columns were used, one an ordinary gas-liquid column and the other a molecular sieve column. The former retains only the carbon dioxide and passes the remaining gases at rates corresponding to the carrier rate. When the carbon dioxide elutes from the first column, a switch briefly directs the flow around the second column to avoid permanent adsorption of the carbon dioxide on the molecular sieve. After the carbon dioxide signal has returned to zero, the flow is switched back through the second column, thereby permitting separation and elution of the remainder of the sample components.

27F-2 Porous Polymers

Porous polymer beads of uniform size are manufactured from styrene cross-linked with divinylbenzene (Section 28F-2). The pore size of these beads is uniform and is controlled by the amount of cross-linking. Porous polymers have found considerable use in the separation of gaseous polar species such as hydrogen sulfide, oxides of nitrogen, water, carbon dioxide, methanol, and vinyl chloride. A typical application of an open tubular column lined with a porous polymer (PLOT column) is shown in Figure 27-21b.

FIGURE 27-21 Typical gas-solid chromatographic separations: (a) a 5 ft. × 1/8 in. molecular sieve column; (b) a 30 m × 0.53 mm PLOT column. C_n = hydrocarbon with n carbons.

Exhaust mixture: A, 35% H_2; B, 25% CO_2; C, 1% O_2; D, 1% N_2; E, 1% C_2; F, 30% CH_4; G, 3% CO; H, 1% C_3; I, 1% C_4; J, 1% i-C_5; K, 1% n-C_5.

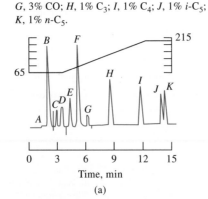

(a)

1, air; 2, methane; 3, carbon dioxide; 4, ethylene; 5, ethane.

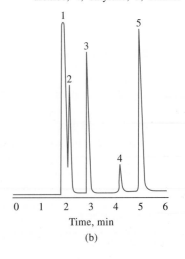

(b)

QUESTIONS AND PROBLEMS

*Answers are provided at the end of the book for problems marked with an asterisk.

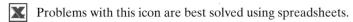 Problems with this icon are best solved using spreadsheets.

27-1 How do gas-liquid and gas-solid chromatography differ?

27-2 How does a soap-bubble flow meter work?

27-3 What is meant by temperature programming in GC? Why is it frequently used?

27-4 Define (a) retention volume, (b) corrected retention volume, (c) specific retention volume.

27-5 What is the difference between a concentration-sensitive and a mass-sensitive detector? Are the following detectors mass or concentration sensitive? (a) thermal conductivity, (b) atomic emission, (c) thermionic, (d) electron captive, (e) flame photometric, (f) flame ionization.

27-6 Describe the principles of operation for the detectors listed in Question 27-5.

27-7 What are the major advantages and the principal limitations of each of the detectors listed in Question 27-5?

27-8 What is the difference between a total-ion chromatogram and a mass chromatogram?

27-9 Discuss why the combination of GC and mass spectrometry is so powerful.

27-10 What are *hyphenated* GC methods? Briefly describe two hyphenated methods.

27-11 What is the packing material used in most packed GC columns?

27-12 How do the following open tubular columns differ?
(a) PLOT columns   (b) WCOT columns   (c) SCOT columns

27-13 What are megabore open tubular columns? Why are they used?

27-14 What are the advantages of fused-silica capillary columns compared with glass or metal columns?

27-15 What properties should the stationary-phase liquid for GC possess?

27-16 What is the effect of stationary-phase film thickness on gas chromatograms?

27-17 Why are gas chromatographic stationary phases often bonded and cross-linked? What do these terms mean?

27-18 List the variables that lead to (a) band broadening and (b) band separation in GLC.

27-19 What are retention indexes? Describe how they are determined.

27-20 The same polar compound is gas chromatographed on an SE-30 (very nonpolar) column and then on a Carbowax 20M (very polar column). How will $K = c_S/c_M$ vary between the two columns?

*27-21 Use the retention data given in the following table to calculate the retention index of 1-hexene.

Sample	Retention Time, min
Air	0.571
n-pentane	2.16
n-hexane	4.23
1-hexene	3.15

 *27-22 A GC column was operated under the following conditions:

column: 1.10 m × 2.0 mm, packed with Chromosorb P; weight of stationary liquid added, 1.40 g; density of liquid, 1.02 g/mL
pressures: inlet, 26.1 psi above room; room, 748 torr
measured outlet flow rate: 25.3 mL/min
temperature: room, 21.2°C; column, 102.0°C
retention times: air, 18.0 s; methyl acetate, 1.98 min; methyl propionate, 4.16 min; methyl *n*-butyrate, 7.93 min
peak widths of esters at base: 0.19, 0.39, and 0.79, respectively

Calculate
(a) the average flow rate in the column.
(b) the corrected retention volumes for air and the three esters.
(c) the specific retention volumes for the three components.
(d) the distribution constants for each of the esters.
(e) a corrected retention volume and retention time for methyl *n*-hexanoate.

 *27-23 From the data in Problem 27-22, calculate
(a) the retention factor k for each component.
(b) selectivity factor α for each adjacent pair of compounds.
(c) the average number of theoretical plates and plate height for the column.
(d) the resolution for each adjacent pair of compounds.

27-24 The stationary-phase liquid in the column described in Problem 27-23 was didecylphthalate, a solvent of intermediate polarity. If a nonpolar solvent such as a silicone oil had been used instead, would the retention times for the three compounds be larger or smaller? Why?

 27-25 One method for quantitative determination of the concentration of constituents in a sample analyzed by GC is the area-normalization method. Here, complete elution of all of the sample constituents is necessary. The area of each peak is then measured and corrected for differences in detector response to the different eluates. This correction involves dividing the area by an empirically determined correction factor. The concentration of the analyte is found from the ratio of its corrected area to the total corrected area of all peaks. For a chromatogram containing three peaks, the relative areas were found to be 16.4, 45.2, and 30.2 in the order of increasing retention time. Calculate the percentage of each compound if the relative detector responses were 0.60, 0.78, and 0.88, respectively.

27-26 Determine the concentration of species in a sample using the peak areas and relative detector responses for the five gas chromatographic peaks given in the following table. Use the area-normalization method described in Problem 27-25.

Also shown are the relative responses of the detector. Calculate the percentage of each component in the mixture.

Compound	Relative Peak Area	Relative Detector Response
A	32.5	0.70
B	20.7	0.72
C	60.1	0.75
D	30.2	0.73
E	18.3	0.78

27-27 What would be the effect of the following on the plate height of a column? Explain.
(a) Increasing the weight of the stationary phase relative to the packing weight.
(b) Decreasing the rate of sample injection.
(c) Increasing the injection port temperature.
(d) Increasing the flow rate.
(e) Reducing the particle size of the packing.
(f) Decreasing the column temperature.

27-28 What kinds of mixtures are separated by GSC?

27-29 Why is GSC not used nearly as extensively as GLC?

 Challenge Problem

27-30 Cinnamaldehyde is the component responsible for cinnamon flavor. It is also a potent antimicrobial compound present in essential oils (see M. Friedman, N. Kozukue, and L. A. Harden, *J. Agric. Food Chem.*, **2000**, *48*, 5702). The GC response of an artificial mixture containing six essential oil components and methyl benzoate as an internal standard is shown in part (a) of the figure.

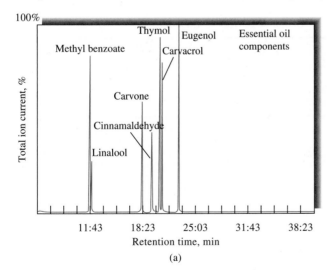

(a)

(a) An idealized enlargement of the region near the cinnamaldehyde peak is given in part (b) of the figure. Determine the retention time for cinnamaldehyde.

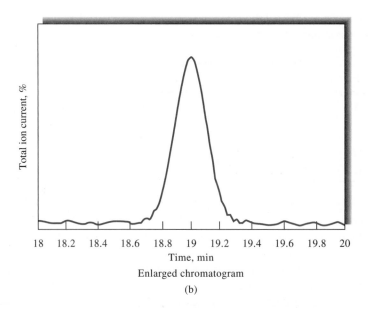

Enlarged chromatogram

(b)

(b) From part (b) of the figure, determine the number of theoretical plates for the column.

(c) The fused-silica column was 0.25 mm × 30 cm with a 0.25-μm film. Determine the height equivalent to a theoretical plate from the data in parts (a) and (b).

(d) Quantitative data were obtained by using methyl benzoate as the internal standard. The following results were found for calibration curves of cinnamaldehyde, eugenol, and thymol. The values under each component represent the peak area of the component divided by the peak area of the internal standard.

Concentration, mg sample/ 200 μL	Cinnamaldehyde	Eugenol	Thymol
0.50	—	0.4	—
0.65	—	—	1.8
0.75	1.0	0.8	—
1.10	—	1.2	—
1.25	2.0	—	—
1.30	—	—	3.0
1.50	—	1.5	—
1.90	3.1	2.0	4.6
2.50	4.0	—	5.8

Determine the calibration curve equations for each component. Include the R^2 values.

(e) From the data in part (d), determine which of the components has the highest calibration curve sensitivity? Which has the lowest?

(f) A sample containing the three essential oils in part (d) gave the peak areas relative to the internal standard area: cinnamaldeyde, 2.6; eugenol, 0.9; thymol, 3.8. Determine the concentrations of each of the oils in the sample and the standard deviations in concentration.

(g) A study was made of the decomposition of cinnamaldehyde in cinnamon oil. The oil was heated for various times at different temperatures. The following data were obtained:

Temp, °C	Time, min	Cinnamaldehyde, %
25, initial		90.9
40	20	87.7
—	40	88.2
—	60	87.9
60	20	72.2
—	40	63.1
—	60	69.1
100	20	66.1
—	40	57.6
—	60	63.1
140	20	64.4
—	40	53.7
—	60	57.1
180	20	62.3
—	40	63.1
—	60	52.2
200	20	63.1
—	40	64.5
—	60	63.3
210	20	74.9
—	40	73.4
—	60	77.4

Determine whether temperature has a statistical effect on the decomposition of cinnamaldehyde using analysis of variance (ANOVA). (For how to perform ANOVA, see S. R. Crouch and F. J. Holler, *Applications of Microsoft® Excel in Analytical Chemistry*, Chap. 3, Belmont, CA: Brooks/ Cole, 2004.) In the same way, determine if time of heating has an effect.

(h) Using the data in part (g), assume that decomposition begins at 60°C and test the hypothesis that there is no effect of temperature or time.

CHAPTER TWENTY-EIGHT

Liquid Chromatography

In several basic types of chromatography the mobile phase is a liquid. These types are often classified by separation mechanism or by the type of stationary phase. The varieties include (1) **partition chromatography**; (2) **adsorption**, or **liquid-solid chromatography**; (3) **ion-exchange**, or **ion chromatography**; (4) **size-exclusion chromatography**; (5) **affinity chromatography**, and (6) **chiral chromatography**. Most of this chapter deals with column applications of these important types of chromatography. The final section, however, presents a brief description of planar liquid chromatography because this technique provides a simple and inexpensive way of determining likely optimal conditions for column separations.

Throughout this chapter, this logo indicates an opportunity for online self-study at **www.thomsonedu.com/chemistry/skoog**, linking you to interactive tutorials, simulations, and exercises.

Early liquid chromatography (LC) was carried out in glass columns with diameters of 10 to 50 mm. The columns were packed with 50- to 500-cm lengths of solid particles coated with an adsorbed liquid that formed the stationary phase. To ensure reasonable flow rates through this type of stationary phase, the particle size of the solid was kept larger than 150 to 200 μm; even then, flow rates were at best a few tenths of a milliliter per minute. Thus, separation times were long — often several hours. Attempts to speed up this classic procedure by application of vacuum or pressure were not effective because increases in flow rates tended to increase plate heights beyond the minimum in the typical plate height versus flow rate curve (see Figure 26-8a); decreased efficiencies were the result.

Early in the development of liquid chromatography, scientists realized that major increases in column efficiency could be achieved by decreasing the particle size of packings. It was not until the late 1960s, however, that the technology for producing and using packings with particle diameters as small as 3 to 10 μm was developed. This technology required sophisticated instruments operating at high pressures, which contrasted markedly with the simple glass columns of classic gravity-flow liquid chromatography. The name *high-performance liquid chromatography* (HPLC) was originally used to distinguish these newer procedures from the original gravity-flow methods. Today, virtually all LC is done using pressurized flow, and we use the abbreviations LC and HPLC interchangeably.[1]

28A SCOPE OF HPLC

LC is the most widely used of all of the analytical separation techniques. The reasons for the popularity of the method are its sensitivity, its ready adaptability to accurate quantitative determinations, its ease of automation, its suitability for separating nonvolatile species or thermally fragile ones, and above all, its widespread applicability to substances that are important to industry, to many fields of science, and to the public. Examples of such materials include amino acids, proteins, nucleic acids, hydrocarbons, carbohydrates, drugs, terpenoids, pesticides, antibiotics, steroids, metal-organic species, and a variety of inorganic substances.

[1] For detailed discussions of HPLC, see L. R. Snyder and J. J. Kirkland, *Introduction to Modern Liquid Chromatography*, 2nd ed., New York: Wiley, 1979; R. P. W. Scott, *Liquid Chromatography for the Analyst*, New York: Dekker, 1995; S. Lindsay, *High Performance Liquid Chromatography*, New York: Wiley, 1992.

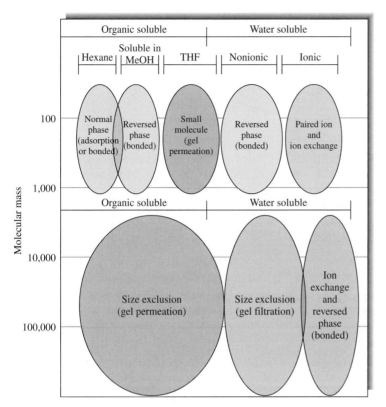

FIGURE 28-1 Selection of LC modes. Methods can be chosen based on solubility and molecular mass. In most cases for non-ionic small molecules ($M < 2000$), reversed-phase methods are suitable. Techniques toward the bottom of the diagram are best suited for species of high molecular mass ($M > 2000$). (Adapted from *High Performance Liquid Chromatography*, 2nd ed., S. Lindsay and J. Barnes, eds., New York: Wiley, 1992. With permission.)

Figure 28-1 reveals that the various liquid chromatographic procedures are complementary in their application. Thus, for solutes having molecular masses greater than 10,000, size-exclusion chromatography is often used, although it is now becoming possible to handle such compounds by reversed-phase chromatography as well. For lower-molecular-mass ionic species, ion-exchange chromatography is widely used. Small polar but nonionic species are best handled by reversed-phase methods. In addition, this procedure is frequently useful for separating members of a homologous series. Adsorption chromatography was once used for separating nonpolar species, structural isomers, and compound classes such as aliphatic hydrocarbons from aliphatic alcohols. Because of problems with retention reproducibility and irreversible adsorption, adsorption chromatography with solid stationary phases has been largely replaced by normal-phase (bonded-phase) chromatography. Among the specialized forms of LC, affinity chromatography is widely used for isolation and preparation of biomolecules, and chiral chromatography is employed for separating enantiomers.

28B COLUMN EFFICIENCY IN LC

The discussion on band broadening in Section 26C-3 is generally applicable to LC. Here, we illustrate the important effect of stationary-phase particle size and describe two additional sources of zone spreading that are sometimes of considerable importance in LC.

28B-1 Effects of Particle Size of Packings

The mobile-phase mass-transfer coefficient (see Table 26-3) reveals that C_M in Equation 26-23 is directly related to the square of the diameter d_p of the particles making up a packing. Because of this, the efficiency of an LC column should improve dramatically as the particle size decreases. Figure 28-2 is an experimental demonstration of this effect, where it is seen that a reduction of particle size from 45 to 6 μm results in a tenfold or more decrease in plate height. Note that none of the plots in this figure exhibits the minimum that is predicted by Equation 26-23. Such minima are, in fact, observable in LC (see Figure 26-8a) but usually at flow rates too low for most practical applications.

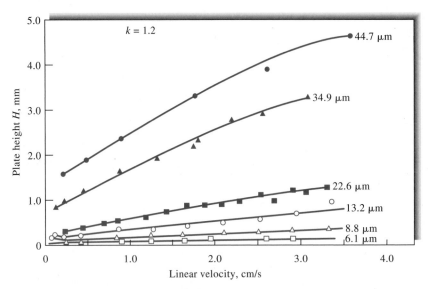

FIGURE 28-2 Effect of particle size of packing and flow rate on plate height H in LC. Column dimensions: 30 cm × 2.4 mm. Solute: *N,N′*-diethyl-*p*-aminoazobenzene. Mobile phase: mixture of hexane, methylene chloride, isopropyl alcohol. (From R. E. Majors, *J. Chromatogr. Sci.*, **1973**, *11*, 88. With permission.)

28B-2 Extracolumn Band Broadening in LC

In LC, significant band broadening sometimes occurs outside the column packing itself. This *extracolumn band broadening* occurs as the solute is carried through open tubes such as those found in the injection system, the detector region, and the piping connecting the various components of the system. Here, broadening arises from differences in flow rates between layers of liquid adjacent to the wall and the center of the tube. As a result, the center part of a solute band moves more rapidly than the peripheral part. In GC, extracolumn spreading is largely offset by diffusion. Diffusion in liquids, however, is significantly slower, and band broadening of this type often becomes noticeable.

It has been shown that the contribution of extracolumn effects H_{ex} to the total plate height is given by[2]

$$H_{ex} = \frac{\pi r^2 u}{24 D_M} \qquad (28\text{-}1)$$

where u is the linear-flow velocity (cm/s), r is the radius of the tube (cm), and D_M is the diffusion coefficient of the solute in the mobile phase (cm²/s).

Extracolumn broadening can become quite serious when small-bore columns are used. Here, the radius of the extracolumn components should be reduced to

---
[2]R. P. W. Scott and P. Kucera, *J. Chromatogr. Sci.*, **1971**, *9*, 641.

0.010 inch or less, and the length of extracolumn tubing made as small as feasible to minimize this source of broadening.

28C LC INSTRUMENTATION

Pumping pressures of several hundred atmospheres are required to achieve reasonable flow rates with packings of 3 to 10 μm, which are common in modern LC. Because of these high pressures, the equipment for HPLC tends to be more elaborate and expensive than equipment for other types of chromatography. Figure 28-3 is a diagram showing the important components of a typical LC instrument.

28C-1 Mobile-Phase Reservoirs and Solvent Treatment Systems

A modern LC apparatus is equipped with one or more glass reservoirs, each of which contains 500 mL or more of a solvent. Provisions are often included to remove dissolved gases and dust from the liquids. Dissolved gases can lead to irreproducible flow rates and band spreading; in addition, both bubbles and dust interfere with the performance of most detectors. Degassers may consist of a vacuum pumping system, a distillation system, a device for heating and stirring, or as

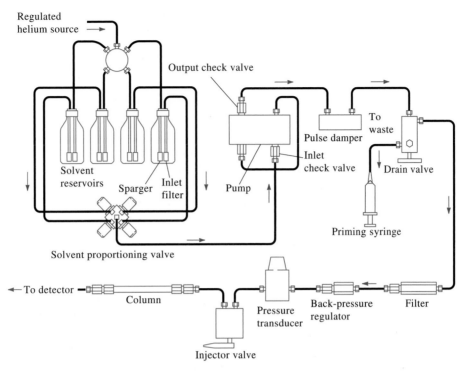

FIGURE 28-3 Block diagram showing components of a typical apparatus for HPLC. (Courtesy of Perkin-Elmer Corp., Norwalk, CT).

shown in Figure 28-3, a system for *sparging*, in which the dissolved gases are swept out of solution by fine bubbles of an inert gas that is not soluble in the mobile phase. Often the systems also contain a means of filtering dust and particulate matter from the solvents to prevent these particles from damaging the pumping or injection systems or clogging the column. It is not necessary that the degassers and filters be integral parts of the HPLC system as shown in Figure 28-3. For example, a convenient way of treating solvents before introduction into the reservoir is to filter them through a millipore filter under vacuum. This treatment removes gases as well as suspended matter.

An elution with a single solvent or solvent mixture of constant composition is termed an *isocratic elution*. In *gradient elution*, two (and sometimes more) solvent systems that differ significantly in polarity are used and varied in composition during the separation. The ratio of the two solvents is varied in a preprogrammed way, sometimes continuously and sometimes in a series of steps. Modern HPLC instruments are often equipped with proportioning valves that introduce liquids from two or more reservoirs at ratios that can be varied continuously (Figure 28-3). The volume ratio of the solvents can be altered linearly or exponentially with time.

Figure 28-4 illustrates the advantage of a gradient eluent in the separation of a mixture of chlorobenzenes. Isocratic elution with a 50:50 (v/v) methanol-water solution yielded the curve in Figure 28-4b. The curve in Figure 28-4a is for gradient elution, which was initiated with a 40:60 mixture of the two solvents; the methanol concentration was then increased at the rate of 8%/min. Note that gradient elution shortened the time of separation significantly without sacrificing the resolution of the early peaks. Note also that gradient elution produces effects similar to those produced by temperature programming in gas chromatography (see Figure 27-7).

28C-2 Pumping Systems

The requirements for liquid chromatographic pumps include (1) the generation of pressures of up to 6000 psi (lb/in.^{2}), or 414 bar, (2) pulse-free output, (3) flow rates ranging from 0.1 to 10 mL/min, (4) flow reproducibilities of 0.5% relative or better, and (5) resistance to corrosion by a variety of solvents. The high pressures

 Simulation: Learn more about **liquid chromatography**.

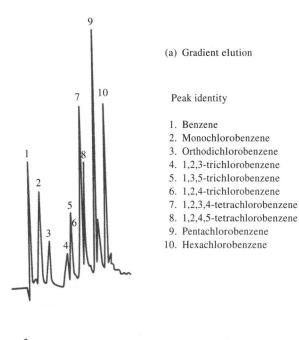

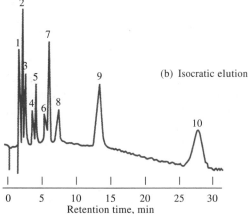

(a) Gradient elution

Peak identity

1. Benzene
2. Monochlorobenzene
3. Orthodichlorobenzene
4. 1,2,3-trichlorobenzene
5. 1,3,5-trichlorobenzene
6. 1,2,4-trichlorobenzene
7. 1,2,3,4-tetrachlorobenzene
8. 1,2,4,5-tetrachlorobenzene
9. Pentachlorobenzene
10. Hexachlorobenzene

(b) Isocratic elution

FIGURE 28-4 Improvement in separation effectiveness by gradient elution. Column: 1 m × 2.1 mm inside-diameter, precision-bore stainless steel; packing: 1% Permaphase® ODS (C_{18}). Sample: 5 μL of chlorinated benzenes in isopropanol. Detector: UV photometer (254 nm). Conditions: temperature, 60°C, pressure, 1200 psi. (From J. J. Kirkland, *Modern Practice of Liquid Chromatography*, p. 88, New York: Interscience, 1971. Reprinted by permission of John Wiley & Sons, Inc.)

generated by liquid chromatographic pumps are not an explosion hazard because liquids are not very compressible. Thus, rupture of a component results only in solvent leakage. However, such leakage may constitute a fire or environmental hazard with some solvents.

Two major types of pumps are used in LC: the screw-driven syringe type and the reciprocating pump. Reciprocating pumps are used in almost all modern commercial chromatographs.

Reciprocating Pumps

Reciprocating pumps usually consist of a small chamber in which the solvent is pumped by the back and forth motion of a motor-driven piston (see Figure 28-5). Two ball check valves, which open and close alternately, control the flow of solvent into and out of a cylinder. The solvent is in direct contact with the piston. As an alternative, pressure may be transmitted to the solvent via a flexible diaphragm, which in turn is hydraulically pumped by a reciprocating piston. Reciprocating pumps have the disadvantage of producing a pulsed flow, which must be damped because the pulses appear as baseline noise on the chromatogram. Modern LC instruments use dual pump heads or elliptical cams to minimize such pulsations. The advantages of reciprocating pumps include their small internal volume (35 to 400 μL), their high output pressures (up to 10,000 psi), their adaptability to gradient elution, their large solvent capacities, and their constant flow rates, which are largely independent of column back pressure and solvent viscosity.

Displacement Pumps

Displacement pumps usually consist of large, syringe-like chambers equipped with a plunger activated by a screw-driven mechanism powered by a stepping motor. Displacement pumps also produce a flow that tends to be independent of viscosity and back pressure. In ad-

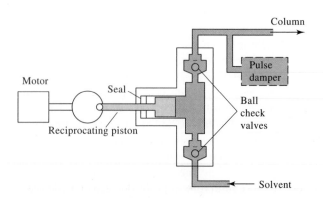

FIGURE 28-5 A reciprocating pump for HPLC.

dition, the output is pulse free. Disadvantages include limited solvent capacity (~250 mL) and considerable inconvenience when solvents must be changed.

Flow Control and Programming Systems

As part of their pumping systems, many commercial instruments are equipped with computer-controlled devices for measuring the flow rate by determining the pressure drop across a restrictor located at the pump outlet. Any difference in signal from a preset value is then used to increase or decrease the speed of the pump motor. Most instruments also have a means for varying the composition of the solvent either continuously or in a stepwise fashion. For example, the instrument shown in Figure 28-3 contains a proportioning valve that permits mixing of up to four solvents in a preprogrammed and continuously variable way.

28C-3 Sample-Injection Systems

Often, the limiting factor in the precision of liquid chromatographic measurements is the reproducibility with which samples can be introduced onto the column packing. The problem is exacerbated by band broadening, which accompanies a lengthy sample injection plug. Thus, sample volumes must be very small — a few tenths of a microliter to perhaps 500 μL. Furthermore, it is convenient to be able to introduce the sample without depressurizing the system.

The most widely used method of sample introduction in LC is based on sampling loops, such as that shown in Figures 28-6 and 27-5. These devices are often an integral part of liquid-chromatographic equipment and have interchangeable loops providing a choice of sample sizes from 1 μL to 100 μL or more. Sampling loops of this type permit the introduction of samples at pressures up to 7000 psi with relative standard deviations of a few tenths of a percent.

Most chromatographs today are sold with autoinjectors. Such units are capable of injecting samples into the LC from vials on a sample carousel or from microtiter plates. They usually contain sampling loops and a syringe pump for injection volumes from less than 1 μL to more than 1 mL. Some have controlled-temperature environments that allow for sample storage and for carrying out derivatization reactions prior to injection. Most are programmable to allow for unattended injections into the LC system.

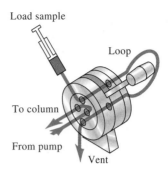

Load sample

Loop

To column

From pump

Vent

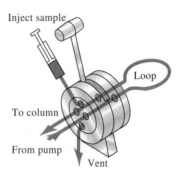

Inject sample

Loop

To column

From pump

Vent

FIGURE 28-6 A sampling loop for LC. With the valve handle as shown on the left, the loop is filled from the syringe, and the mobile phase flows from pump to column. When the valve is placed in the position on the right, the loop is inserted between the pump and the column so that the mobile phase sweeps the sample onto the column. (Courtesy of Beckman-Coulter, Inc.)

28C-4 Columns for HPLC

Liquid-chromatographic columns[3] are usually constructed from smooth-bore stainless steel tubing. HPLC columns are sometimes made from heavy-walled glass tubing and polymer tubing, such as polyetheretherketone (PEEK). In addition, stainless steel columns lined with glass or PEEK are also available. Hundreds of packed columns differing in size and packing are available. The cost of standard-sized, nonspeciality columns ranges from $200 to more than $500. Specialized columns, such as chiral columns, can cost more than $1000.

---

[3]For more information, see U. D. Neue, *HPLC Columns: Theory, Technology, and Practice*, New York: Wiley-VCH, 1997.

Analytical Columns

Most liquid-chromatographic columns range from 5 to 25 cm long. Straight columns are invariably used. Sometimes length is added by coupling two or more columns. The inside diameter of analytical columns is often 3 to 5 mm; the most common particle size of packings is 3 or 5 μm. The most common columns are 10 or 15 cm long, 4.6 mm in inside diameter, and packed with 5-μm particles. Columns of this type generate 40,000 to 70,000 plates/meter (typically about 10,000 plates/column).

In the 1980s microcolumns became available with inside diameters of 1 to 4.6 mm and lengths of 3 to 7.5 cm. These columns, which are packed with 3- or 5-μm particles, achieve as many as 100,000 plates/m and have the advantage of speed and minimal solvent consumption. This latter property is of considerable importance because the high-purity solvents required for LC are expensive to purchase and to dispose of after use. Figure 28-7 illustrates the speed with which a separation can be performed on a microbore column. In this example, eight diverse components are separated in about 15 s. The column was 4 cm long and had an inside diameter of 4 mm; it was packed with 3-μm particles.

Guard Columns

Usually, a short *guard column* is introduced before the analytical column to increase the life of the analytical column by removing not only particulate matter and contaminants from the solvents but also sample components that bind irreversibly to the stationary phase. The composition of the guard-column packing should be similar to that of the analytical column; the particle size is usually larger, however, to minimize pressure drop. When the guard column has become contaminated, it is repacked or discarded and replaced with a new one of the same type. Thus, the guard column is sacrificed to protect the more expensive analytical column.

Column Temperature Control

For some applications, close control of column temperature is not necessary and columns are operated at room temperature. Often, however, better, more reproducible chromatograms are obtained by maintaining constant column temperature. Most modern commercial instruments are now equipped with heaters that control column temperatures to a few tenths of a degree from near ambient to 150°C. Columns may

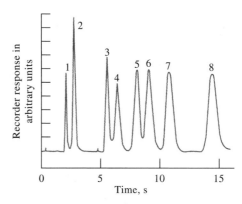

FIGURE 28-7 High-speed isocratic separation. Column dimensions: 4 cm long, 0.4 cm inside diameter; packing: 3-μm spherisorb; mobile phase: 4.1% ethyl acetate in *n*-hexane. Compounds: (1) *p*-xylene, (2) anisole, (3) benzyl acetate, (4) dioctyl phthalate, (5) dipentyl phthalate, (6) dibutyl phthalate, (7) dipropyl phthalate, (8) diethyl phthalate. (From R. P. W. Scott, *Small Bore Liquid Chromatography Columns: Their Properties and Uses*, New York: Wiley, 1984, p. 156. Reprinted with permission of John Wiley & Sons, Inc.)

also be fitted with water jackets fed from a constant-temperature bath to give precise temperature control. Many chromatographers consider temperature control to be essential for reproducible separations.

28C-5 Types of Column Packings

Two basic types of packings have been used in LC, *pellicular* and *porous particle*. The original pellicular particles were spherical, nonporous, glass or polymer beads with typical diameters of 30 to 40 μm. A thin, porous layer of silica, alumina, a polystyrene-divinylbenzene synthetic resin, or an ion-exchange resin was deposited on the surface of these beads. Small porous microparticles have completely replaced these large pellicular particles. In recent years, small (~5 μm) pellicular packings have been reintroduced for separation of proteins and large biomolecules.

The typical porous particle packing for LC consists of porous microparticles having diameters ranging from 3 to 10 μm; for a given size particle, a very narrow particle size distribution is desirable. The particles are composed of silica, alumina, the synthetic resin polystyrene-divinylbenzene, or an ion-exchange resin. Silica is by far the most common packing in LC. Silica particles are prepared by agglomerating submicron silica particles under conditions that lead to larger

TABLE 28-1 Performance of HPLC Detectors

HPLC Detector	Commercially Available	Mass LOD* (typical)	Linear Range[†] (decades)
Absorbance	Yes	10 pg	3–4
Fluorescence	Yes	10 fg	5
Electrochemical	Yes	100 pg	4–5
Refractive index	Yes	1 ng	3
Conductivity	Yes	100 pg–1 ng	5
Mass spectrometry	Yes	<1 pg	5
FTIR	Yes	1 μg	3
Light scattering	Yes	1 μg	5
Optical activity	No	1 ng	4
Element selective	No	1 ng	4–5
Photoionization	No	<1 pg	4

Sources: From manufacturer's literature; *Handbook of Instrumental Techniques for Analytical Chemistry*, F. Settle, ed., Upper Saddle River, NJ: Prentice-Hall, 1997; E. S. Yeung and R. E. Synovec, *Anal. Chem.*, **1986**, *58*, 1237A.

*Mass LODs (limits of detection) depend on compound, instrument, and HPLC conditions, but those given are typical values with commercial systems when available.

[†]Typical values from the preceding sources.

particles having highly uniform diameters. The resulting particles are often coated with thin organic films, which are chemically or physically bonded to the surface. Column packings for specific chromatographic modes are discussed in later sections of this chapter.

28C-6 Detectors

No detectors for LC are as universally applicable as the flame ionization and thermal conductivity detectors for gas chromatography described in Section 27B-4. GC detectors were specifically developed to measure small concentrations of analytes in flowing gas streams. On the other hand, LC detectors have often been traditional analytical instruments adapted with flow cells to measure low concentrations of solutes in liquid streams. A major challenge in the development of LC has been in adapting and improving such devices.[4]

Characteristics of the Ideal Detector

The ideal detector for LC should have all of the properties listed in Section 27B-4 for GC detectors with the exception that an LC detector need not be responsive over as great a temperature range. In addition, an HPLC detector should have minimal internal volume

[4]For more detailed discussions, see R. P. W. Scott, *Chromatographic Detectors*, New York: Dekker, 1996; R. P. W. Scott, *Liquid Chromatography Detectors*, 2nd ed., Amsterdam: Elsevier, 1986.

to reduce zone broadening (Section 28B-2) and should be compatible with liquid flow.

Types of Detectors

Liquid chromatographic detectors are of two basic types. *Bulk-property detectors* respond to a mobile-phase bulk property, such as refractive index, dielectric constant, or density, that is modulated by the presence of solutes. In contrast, *solute-property detectors* respond to some property of solutes, such as UV absorbance, fluorescence, or diffusion current, that is not possessed by the mobile phase.

Table 28-1 lists the most common detectors for HPLC and some of their most important properties. The most widely used detectors for LC are based on absorption of ultraviolet or visible radiation (see Figure 28-8). Fluorescence, refractive-index, and electrochemical detectors are also widely used. Mass spectrometry (MS) detectors are currently quite popular. Such LC/MS systems can greatly aid in identifying the analytes exiting from the HPLC column as discussed later in this section.

UV-Visible Absorption Detectors

Figure 28-8 is a schematic of a typical, Z-shape, flow-through cell for absorption measurements on eluents from a chromatographic column. To minimize extracolumn band broadening, the volume of such a cell should

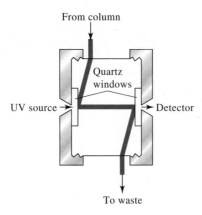

From column

Quartz
windows

UV source

Detector

To waste

FIGURE 28-8 A UV-visible absorption cell for HPLC.

be kept as small as possible. Typical cell volumes are 1 to 10 μL with cell path lengths of 2 to 10 mm.

Many absorption detectors are double-beam devices in which one beam passes through the eluent cell and the other beam is a reference beam. Matched photoelectric detectors are then used to compare the intensities of the two beams. Alternatively, a chopped beam system similar to that shown in Figure 9-13b (see also Figure 13-13b) is used in conjunction with a single phototransducer. In either case, the chromatogram consists of a plot of the absorbance (logarithm of the ratio of the two transduced signals) as a function of time. Single-beam instruments are also encountered. Here, intensity measurements of the solvent system are stored in a computer memory and ultimately recalled for the calculation of absorbance.

UV Absorption Detectors with Filters. The earliest absorption detectors were filter photometers with a mercury lamp as the source. Most commonly, the intense line at 254 nm was isolated by filters; with some instruments, lines at 250, 313, 334, and 365 nm could also be used by substitution of filters. Obviously, this type of detector is restricted to solutes that absorb at one of these wavelengths. As shown in Section 14B, several organic functional groups and a number of inorganic species exhibit broad absorption bands that encompass one or more of these ultraviolet wavelengths.

Deuterium or tungsten filament sources with interference filters were also used to provide a simple means of detecting absorbing species as they eluted from a column. Some instruments were dual wavelength devices, or they were equipped with filter wheels containing several filters that could be rapidly switched to detect various species as they eluted. Today, filter instruments

have been almost entirely replaced by scanning and diode array spectrometers.

Absorption Detectors with Scanning Capabilities. Most HPLC manufacturers offer detectors that consist of a scanning spectrophotometer with grating optics. Some are limited to ultraviolet radiation; others encompass both ultraviolet and visible radiation. Several operational modes can be chosen. For example, the entire chromatogram can be obtained at a single wavelength; alternatively, when eluent peaks are sufficiently separated in time, different wavelengths can be chosen for each peak. Here again, computer control is often used to select the best wavelength for each eluent. Where entire spectra are desired for identification purposes, the flow of eluent can be stopped for a sufficient period to permit scanning the wavelength region of interest.

The most powerful ultraviolet spectrophotometric detectors are array-based instruments as described in Section 7E-3 and Figure 13-25.[5] Several manufacturers offer such instruments, which permit collection of an entire spectrum in approximately 1 second. Thus, spectral data for each chromatographic peak can be collected and stored as it appears at the end of the column. One form of presentation of the spectral data, which is helpful in identification of species and for choosing conditions for quantitative determination, is a three-dimensional plot such as that shown in Figure 28-9. Here, spectra were obtained at successive 5-second intervals. The appearance and disappearance of each of the three steroids in the eluent can be seen.

Infrared Absorption Detectors. Two types of infrared detectors have been offered commercially. The first is a filter instrument similar in design to that shown in Figure 16-13. The second, and more sophisticated, type of infrared detector is based on Fourier transform instruments similar to those discussed in Section 16B-1. Several of the manufacturers of Fourier transform infrared (FTIR) instruments offer accessories that permit their use as HPLC detectors. Infrared detector cells are similar in construction to those used with ultraviolet radiation except that windows are constructed of sodium chloride or calcium fluoride. Cell path lengths range from 0.2 to 1.0 mm and volumes from 1.5 to 10 μL.

[5] See R. P. W. Scott, *Chromatographic Detectors*, New York: Dekker, 1996; *Diode Array Detection in HPLC*, L. Huber and S. A. George, eds., New York: Dekker, 1993.

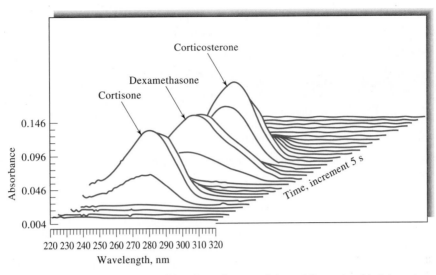

FIGURE 28-9 Absorption spectra of the eluent from a mixture of three steroids taken at 5-second intervals. (Courtesy of Hewlett-Packard Co., Palo Alto, CA.)

A major limitation to the use of IR detectors lies in the low transparency of many useful solvents. For example, the broad infrared absorption bands for water and the alcohols largely preclude the use of this detector for many applications. Also, the use of aqueous mobile phases can lead to rapid deterioration of cell materials unless special windows are used. The poor limits of detection for many solutes has likewise limited the use of IR detectors. The introduction of mass spectrometric detectors for LC (discussed later in this section) has led to a dramatic decline in applications of IR detection.

Fluorescence Detectors

Fluorescence detectors for HPLC are similar in design to the fluorometers and spectrofluorometers described in Section 15B-2. In most, fluorescence is observed by a photoelectric transducer located at 90° to the excitation beam. The simplest detectors use a mercury excitation source and one or more filters to isolate a band of emitted radiation. More sophisticated instruments are based on a xenon source and use a grating monochromator to isolate the fluorescence radiation. Laser-induced fluorescence is also used because of its sensitivity and selectivity.

As was pointed out in Chapter 15, an inherent advantage of fluorescence methods is their high sensitivity, which is typically greater by more than an order of magnitude than most absorption procedures. This advantage has been exploited in LC for the separation and determination of the components of samples that fluoresce. As was discussed in Section 15C-2, fluorescent compounds are frequently encountered in the analysis of such materials as pharmaceuticals, natural products, clinical samples, and petroleum products. Often, the number of fluorescing species can be enlarged by preliminary treatment of samples with reagents that form fluorescent derivatives. For example, dansylchloride (5-dimethylamino-naphthalene-1-sulphonyl chloride), which reacts with primary and secondary amines, amino acids, and phenols to give fluorescent compounds, has been widely used for the detection of amino acids in protein hydrolyzates.

Refractive-Index Detectors

Figure 28-10 shows a schematic diagram of a differential refractive-index detector in which the solvent passes through one half of the cell on its way to the column; the eluate then flows through the other chamber. The two compartments are separated by a glass plate mounted at an angle such that bending of the incident beam occurs if the two solutions differ in refractive index. The resulting displacement of the beam with respect to the photosensitive surface of a detector causes variation in the output signal, which, when amplified and recorded, provides the chromatogram.

Refractive-index detectors have the significant advantage of responding to nearly all solutes. That is, they are general detectors analogous to flame or thermal conductivity detectors in gas chromatography. In

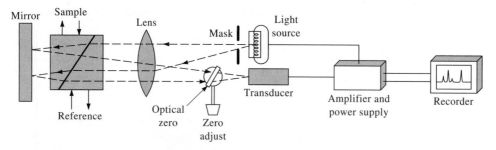

FIGURE 28-10 Schematic diagram of a differential refractive-index detector. (Courtesy of Waters Associates, Inc., Milford, MA.)

addition, they are reliable and unaffected by flow rate. They are, however, highly temperature dependent and must be maintained at a constant temperature to a few thousandths of a degree centigrade. Furthermore, they are not as sensitive as many other types of detectors and are not compatible with gradient elution methods. Refractive-index detectors find application in the determination of some analytes such as sugars.

Evaporative Light Scattering Detector

One of the newer types of detectors for HPLC is the *evaporative light scattering detector* (ELSD). In this detector, the column effluent is passed into a nebulizer where it is converted into a fine mist by a flow of nitrogen or air. The fine droplets are then carried through a controlled-temperature drift tube where evaporation of the mobile phase occurs, leading to formation of fine particles of the analyte. The cloud of analyte particles then passes through a laser beam. The scattered radiation is detected at right angles to the flow by a silicon photodiode.

A major advantage of this type of detector is that its response is approximately the same for all nonvolatile solutes. In addition it is significantly more sensitive than the refractive-index detector, with detection limits of 0.2 ng/μL. A disadvantage is that mobile-phase compositions are limited to only volatile components.

Electrochemical Detectors

Electrochemical detectors of several types are currently available from instrument manufacturers.[6] These devices are based on amperometry, voltammetry, coulometry, and conductometry. The first three of these methods are discussed in Chapters 24 and 25.

Although electroanalytical procedures have as yet not been exploited to the extent of optical detectors, they appear to offer advantages, in many instances, of high sensitivity, simplicity, convenience, and widespread applicability. This last property is illustrated in Figure 28-11, which depicts the potential ranges at which oxidation or reduction of sixteen organic functional groups occurs. In principle, species containing any of these groups could be detected by amperometric, voltammetric, or coulometric procedures. A major limitation of electrochemical detectors, however, is that they are not compatible with gradient elution.

Several electrochemical detector cells for HPLC applications have been described in the literature, and several are available from commercial sources. Figure 28-12 is an example of a thin-layer flow-through cell for amperometric detection (see also Figure 25-17). Here, the electrode surface is part of a channel wall formed by sandwiching a 50-μm Teflon gasket between two machined blocks of Kel-F plastic. The indicator electrode is platinum, gold, glassy carbon, or carbon paste. A reference electrode, and often a counter electrode, is located downstream from the indicator electrode block. The cell volume is 1 to 5 μL. A useful modification of this cell, which is available commercially, includes two working electrodes, which can be operated in series or in parallel.[7] The former configuration, in which the eluent flows first over one electrode and then over the second, requires that the analyte undergo a reversible oxidation (or reduction) at the upstream electrode. The second electrode then operates as a cathode (or an anode) to determine the oxidation (or reduction) product. This arrangement enhances the selectivity of the detection system. An interesting application of this system is for the detection and determination of the

[6]For brief descriptions of commercially available detectors, see B. E. Erickson, *Anal. Chem.*, **2000**, *72*, 353A; M. Warner, *Anal. Chem.*, **1994**, *66*, 601A.

[7]D. A. Roston, R. E. Shoup, and P. T. Kissinger, *Anal. Chem.*, **1982**, *54*, 1417A.

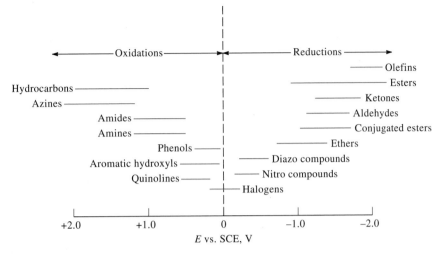

FIGURE 28-11 Potential amperometric detection of organic functional groups. The horizontal lines show the range of oxidation or reduction potentials for electroactive compounds containing the indicated functional groups.

components in mixtures containing both thiols and disulfides. Here, the upstream mercury electrode reduces the disulfides at about -1.0 V. That is,

$$RSSR + 2H^+ + 2e^- \rightarrow 2RSH$$

A downstream mercury electrode is then oxidized in the presence of the thiols from the original sample as well as those formed at the upstream electrode. That is,

$$2RSH + Hg(l) \rightarrow Hg(SR)_2(s) + 2H^+ + 2e^-$$

In the parallel configuration, the two electrodes are rotated so that the axis between them is at 90° to the stream flow. The two can then be operated at different potentials (relative to a downstream reference electrode), which often gives an indication of peak purity. Alternatively, one electrode can be operated as a cathode and the other as an anode, thus making possible simultaneous detection of both oxidants and reductants.

Voltammetric, conductometric and coulometric detectors are also available commercially. Conductometric detectors are discussed further in Section 28F-3.

Mass Spectrometric Detectors

The combination of LC and mass spectrometry would seem to be an ideal merger of separation and detection.[8] Just as in GC, a mass spectrometer can greatly aid in identifying species as they elute from the chromatographic column. There are major problems, however, in the coupling of these two techniques. A gas-phase sample is needed for mass spectrometry, and the output of the LC column is a solute dissolved in a solvent. As a first step, the solvent must be vaporized. When vaporized, however, the LC solvent produces a gas volume that is 10–1000 times greater than the carrier gas in GC. Hence, most of the solvent must also be removed.

There have been several devices developed to solve the problems of solvent removal and LC column interfacing. Today, the most popular approaches use a low-flow-rate atmospheric pressure ionization technique.

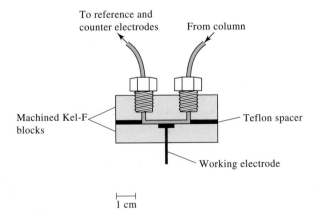

$\vdash\!\!-\!\!\dashv$
1 cm

FIGURE 28-12 Amperometric thin-layer detector cell for HPLC.

[8]M. C. McMaster, *LC/MS: A Practical User's Guide*, Hoboken, NJ: Wiley, 2005; W. M. A. Niessen, *Liquid Chromatography-Mass Spectrometry*, 2nd ed., New York: Dekker, 1999.

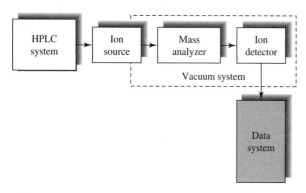

FIGURE 28-13 Block diagram of an LC/MS system. The effluent from the LC column is introduced to an atmospheric pressure ionization source, such as an electrospray or a chemical ionization source. The ions produced are sorted by the mass analyzer and detected by the ion detector.

A block diagram of a typical LC/MS system is shown in Figure 28-13. The HPLC system is often a nanoscale capillary LC system with flow rates in the μL/min range. Alternatively, some interfaces allow flow rates as high as 1 to 2 mL/min, which is typical of conventional HPLC conditions. The most common ionization sources are electrospray ionization and atmospheric pressure chemical ionization (see Section 20B). The combination of HPLC and mass spectrometry gives high selectivity because unresolved peaks can be isolated by monitoring only a selected mass. The LC/MS technique can provide fingerprinting of a particular eluate instead of relying on retention time as in conventional HPLC. The combination also can give molecular mass, structural information, and accurate quantitative analysis.[9]

For some complex mixtures, the combination of LC and MS does not provide enough resolution. In recent years it has become feasible to couple two or more mass analyzers to form tandem mass spectrometers (see Section 20C-5). When combined with LC, the tandem mass spectrometry system is called an LC/MS/MS instrument. Tandem mass spectrometers are usually triple quadrupole systems or quadrupole ion-trap spectrometers. To attain higher resolution than can be achieved with a quadrupole, the final mass analyzer in a tandem MS system can be a time-of-flight mass spectrometer. Sector mass spectrometers can also be combined to give tandem systems. Ion cyclotron resonance and ion-trap

mass spectrometers can be operated in such a way as to provide not only two stages of mass analysis but *n* stages. Such MS[n] systems (see Section 20C-5) provide the analysis steps sequentially within a single mass analyzer. These have been combined with LC systems in LC/MS[n] instruments.

LC/MS systems are invariably computer controlled. With these instruments, both real-time and computer-reconstructed chromatograms and spectra of the eluted peaks can be obtained.

28D PARTITION CHROMATOGRAPHY

The most widely used type of HPLC is *partition chromatography*, in which the stationary phase is a second liquid that is immiscible with the liquid mobile phase. In the past, most of the applications have been to nonionic, polar compounds of low to moderate molecular mass (usually <3000). Recently, however, methods have been developed (derivatization and ion pairing) that have extended partition separations to ionic compounds.

The early forms of partition chromatography used liquid-liquid columns. These have been replaced in modern LC systems by *liquid-bonded-phase* columns. In liquid-liquid chromatography, the liquid was held in place by physical adsorption. In bonded-phase chromatography, on the other hand, it is attached by chemical bonding, resulting in highly stable packings insoluble in the mobile phase. Bonded-phase columns are also compatible with gradient elution techniques. Therefore, our discussion focuses exclusively on bonded-phase partition chromatography.[10]

28D-1 Columns for Bonded-Phase Chromatography

The supports for the majority of bonded-phase packings for partition chromatography are prepared from rigid silica, or silica-based, compositions. These solids are formed as uniform, porous, mechanically sturdy particles commonly having diameters of 1.5–10 μm, with 3- and 5-μm particles being most common. The surface of fully hydrolyzed silica (hydrolyzed by heating with 0.1 M HCl for a day or two) is made up of chemically reactive silanol groups. That is,

[9]For a review of commercial LC/MS systems, see B. E. Erickson, *Anal. Chem.*, **2000**, *72*, 711A.

[10]For a report on retention mechanisms in bonded-phase chromatography, see J. G. Dorsey and W. T. Cooper, *Anal. Chem.*, **1994**, *66*, 857A.

(a)

Normal-phase chromatography

Low-polarity mobile phase

————Time————→

Medium-polarity mobile phase

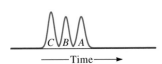

————Time————→

Solute polarities: $A > B > C$

(b)

Reversed-phase chromatography

High-polarity mobile phase

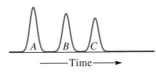

————Time————→

Medium-polarity mobile phase

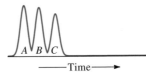

————Time————→

FIGURE 28-14 Relationship between polarity and elution times for normal-phase and reversed-phase chromatography.

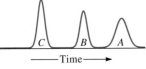

Typical silica surfaces contain about 8 μmol/m² of OH groups and have surface areas of 100 to 300 m²/g.

The most useful bonded-phase coatings are siloxanes formed by reaction of the hydrolyzed surface with an organochlorosilane. For example,

$$—Si—OH + Cl—Si—R → —Si—O—Si—R$$

where R is an alkyl group or a substituted alkyl group.

Surface coverage by silanization is limited to 4 μmol/m² or less because of steric effects. The unreacted SiOH groups, unfortunately, impart an undesirable polarity to the surface, which may lead to tailing of chromatographic peaks, particularly for basic solutes. To lessen this effect, siloxane packings are frequently *capped* by further reaction with chlorotrimethylsilane that, because of its smaller size, can bond some of the unreacted silanol groups.

Normal- and Reversed-Phase Packings

Two types of partition chromatography are distinguishable based on the relative polarities of the mobile and stationary phases. Early work in LC was based on highly polar stationary phases such as triethylene

 Tutorial: Learn more about **partition chromatography**.

glycol or water; a relatively nonpolar solvent such as hexane or *i*-propyl ether then served as the mobile phase. For historic reasons, this type of chromatography is now called *normal-phase chromatography*. In *reversed-phase chromatography*, the stationary phase is nonpolar, often a hydrocarbon, and the mobile phase is a relatively polar solvent (such as water, methanol, acetonitrile, or tetrahydrofuran).[11]

In normal-phase chromatography, the *least* polar component is eluted first; *increasing* the polarity of the mobile phase then *decreases* the elution time. With reversed-phase chromatography, however, the *most* polar component elutes first, and *increasing* the mobile-phase polarity *increases* the elution time. These relationships are illustrated in Figure 28-14. As shown in Figure 28-1, normal-phase partition chromatography and adsorption chromatography overlap considerably. In fact, retention in most types of normal-phase chromatography appears to be governed by adsorption-displacement processes.[12]

Bonded-phase packings are classified as reversed phase when the bonded coating is nonpolar in character and as normal phase when the coating contains polar functional groups. It has been estimated that more than three quarters of all HPLC separations are currently performed in columns with reversed-phase packings. The major advantage of reversed-phase separations is that water can be used as the mobile phase.

[11]For further discussion of reversed-phase HPLC, see L. R. Snyder, J. J. Kirkland, and J. L. Glajch, *Practical HPLC Method Development*, 2nd ed., New York: Wiley, 1997; A. M. Krstulovic and P. R. Brown, *Reversed-Phase High Performance Liquid Chromatography*, New York: Wiley, 1982.
[12]See J. G. Dorsey et al., *Anal. Chem.*, **1994**, *66*, 500R.

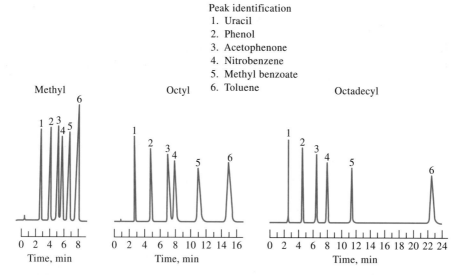

Peak identification
1. Uracil
2. Phenol
3. Acetophenone
4. Nitrobenzene
5. Methyl benzoate
6. Toluene

FIGURE 28-15 Effect of chain length on performance of reversed-phase siloxane columns packed with 5-μm particles. Mobile phase: 50:50 methanol-water. Flow rate: 1.0 ml/min.

Water is an inexpensive, nontoxic, UV-transparent solvent compatible with biological solutes. Also, mass transfer is rapid with nonpolar stationary phases, as is solvent equilibration after gradient elution. Most commonly, the R group of the siloxane in these coatings is a C_8 chain (*n*-octyl) or a C_{18} chain (*n*-octyldecyl).

The mechanism by which these surfaces retain solute molecules is not entirely clear. The retention mechanism appears to be quite complex and very different from bulk phase partitioning because the nonpolar stationary phase is anchored at one end.[13] Regardless of the detailed mechanism of retention, a bonded coating can usually be treated as if it were a conventional, physically retained liquid.

Figure 28-15 illustrates the effect of chain length of the alkyl group on performance. As expected, longer chains produce packings that are more retentive. In addition, longer chain lengths permit the use of larger samples. For example, the maximum sample size for a C_{18} packing is roughly double that for a C_4 preparation under similar conditions.

In most applications of reversed-phase chromatography, elution is carried out with a highly polar mobile phase, for example, an aqueous solution containing various concentrations of such solvents as methanol, acetonitrile, or tetrahydrofuran. In this mode, care

must be taken to avoid pH values greater than about 7.5 because the silica can form soluble silicate species, causing dissolution of the stationary phase. Also, hydrolysis of the siloxane can occur in alkaline solutions, which leads to degradation or destruction of the packing. Acid hydrolysis of the siloxane can limit the pH to about 2.5 in acidic solutions.

In commercial normal-phase bonded packings, the R in the siloxane structure is a polar functional group such as cyano, $-C_2H_4CN$; diol, $-C_3H_6OCH_2CHOHCH_2OH$; amino, $-C_3H_6NH_2$; and dimethylamino, $C_3H_6N(CH_3)_2$. The polarities of these packing materials vary over a considerable range, with the cyano type being the least polar and the amino types the most. Diol packings are intermediate in polarity. With normal-phase packings, elution is carried out with relatively nonpolar solvents, such as ethyl ether, chloroform, and *n*-hexane.

28D-2 Method Development in Partition Chromatography

Method development tends to be more complex in LC than in gas chromatography because in a liquid mobile phase the sample components interact with both the stationary phase and the mobile phase.[14] In contrast, in

[13] L. A. Cole and J. G. Dorsey, *Anal. Chem.*, **1992**, *64*, 1317; L. A. Cole, J. G. Dorsey, and K. A. Dill, *Anal. Chem.*, **1992**, *64*, 1324.

[14] For more information, see L. R. Snyder, J. J. Kirkland, and J. L. Glajch, *Practical HPLC Method Development*, 2nd ed., New York: Wiley, 1997.

gas chromatography, the mobile phase simply carries the sample components through the stationary phase and makes no contribution to the separation process. That is, in gas chromatography separations are not significantly affected by whether the mobile phase is helium, nitrogen, or hydrogen. In marked contrast, the success of a partition chromatographic separation is often critically dependent on whether the mobile phase is, say, acetonitrile, hexane, or dioxane.

Column Selection in Partition Chromatographic Separations

Successful chromatography with interactive mobile phases requires a proper balance of intermolecular forces among the three active participants in the separation process — the solute, the mobile phase, and the stationary phase. These intermolecular forces are described qualitatively in terms of the relative polarity of each of the three reactants. The polarities of the various analyte functional groups increase in the following order: hydrocarbons < ethers < esters < ketones < aldehydes < amides < amines < alcohols. Water is more polar than compounds containing any of the preceding functional groups.

Often, in choosing a column for a partition chromatographic separation, the polarity of the stationary phase is matched roughly with that of the analytes; a mobile phase of considerably different polarity is then used for elution. This procedure is generally more successful than one in which the polarities of the solute and mobile phase are matched but different from that of the stationary phase. Here, the stationary phase often cannot compete successfully for the sample components; retention times then become too short for practical application. At the other extreme, of course, is the situation where the polarity of the solute and stationary phases are too much alike and totally different from that of the mobile phase. Here, retention times become inordinately long.

In summary, then, polarities for solute, mobile phase, and stationary phase must be carefully blended if good partition chromatographic separations are to be realized in a reasonable time. Unfortunately, theories of mobile-phase and stationary-phase interactions with any given set of sample components are imperfect, and at best, the user can only narrow the choice of stationary phase to a general type. Having made this choice, some trial-and-error experiments must be performed in which chromatograms are obtained with various mobile phases until a satisfactory separation is realized. The high degree of correlation between mobile-phase composition and retention factors, however, allows prediction of approximate retention times from only a few trial-and-error experiments. If resolution of all of the components of a mixture proves to be impossible, a different type of column may have to be chosen.

Mobile-Phase Selection in Partition Chromatography

In Section 26D-4, three methods were described for improving the resolution of a chromatographic column; each is based on varying one of the three parameters (N, k, and α) contained in Equation 26-26.[15] In LC, the retention factor k is experimentally the most easily manipulated of the three because of the strong dependence of this constant on the composition of the mobile phase. As noted earlier, for optimal performance, k should be in the ideal range between 2 and 10; for complex mixtures, however, this range must often be expanded to perhaps 0.5 to 20 to provide time for peaks for all of the components to appear.

Sometimes, adjustment of k alone is not sufficient to produce individual peaks with no overlap. In this case we must resort to variations in the selectivity factor α. Here again, the simplest way of bringing about changes in α is by altering the mobile-phase composition, taking care, however, to keep k within a reasonable range. Alternatively, α can be changed by choosing a different column packing.

Effect of Solvent Strength on Retention Factors. Solvents that interact strongly with solutes are often termed "strong" solvents. Strong solvents are often, but not always, polar solvents. Solvent strength depends on the nature of the analyte and stationary phase. Several indexes have been developed for quantitatively describing the polarity of solvents. The most useful of these for partition chromatography is the polarity index P', which was developed by Snyder.[16] This parameter is based on solubility measurements for the substance in question in three solvents: dioxane (a low

---

[15] For more information on selecting stationary and mobile phases, see L. R. Snyder, J. J. Kirkland, and J. L. Glajch, *Practical HPLC Method Development*, 2nd ed., New York: Wiley, 1997.
[16] L. R. Snyder, *J. Chromatogr. Sci.*, **1978**, *16*, 223.

TABLE 28-2 Properties of Common Chromatographic Mobile Phases

Solvent	Refractive Index[a]	Viscosity, cP[b]	Boiling Point, °C	Polarity Index, P'	Eluent Strength,[c] ε^0
Fluoroalkanes[d]	1.27–1.29	0.4–2.6	50–174	<–2	–0.25
Cyclohexane	1.423	0.90	81	0.04	–0.2
n-Hexane	1.372	0.30	69	0.1	0.01
l-Chlorobutane	1.400	0.42	78	1.0	0.26
Carbon tetrachloride	1.457	0.90	77	1.6	0.18
i-Propyl ether	1.365	0.38	68	2.4	0.28
Toluene	1.494	0.55	110	2.4	0.29
Diethyl ether	1.350	0.24	35	2.8	0.38
Tetrahydrofuran	1.405	0.46	66	4.0	0.57
Chloroform	1.443	0.53	61	4.1	0.40
Ethanol	1.359	1.08	78	4.3	0.88
Ethyl acetate	1.370	0.43	77	4.4	0.58
Dioxane	1.420	1.2	101	4.8	0.56
Methanol	1.326	0.54	65	5.1	0.95
Acetonitrile	1.341	0.34	82	5.8	0.65
Nitromethane	1.380	0.61	101	6.0	0.64
Ethylene glycol	1.431	16.5	182	6.9	1.11
Water	1.333	0.89	100	10.2	Large

[a] At 25°C.

[b] The centipoise is a common unit of viscosity; in SI units, 1 cP = 1 mN·s·m^{-2}.

[c] On Al_2O_3. Multiplication by 0.8 gives ε^0 on SiO_2.

[d] Properties depend on molecular mass range of data given.

dipole proton acceptor), nitromethane (a high dipole proton acceptor), and ethyl alcohol (a high dipole proton donor). The polarity index is a numerical measure of the relative polarity of various solvents. Table 28-2 lists polarity indexes (and other properties) for a number of solvents used in partition chromatography. Note that the polarity index varies from 10.2 for highly polar water to −2 for the highly nonpolar fluoroalkanes. Any desired polarity index between these limits can be achieved by mixing two appropriate solvents. Thus, the polarity index P'_{AB} of a mixture of solvents A and B is given by

$$P'_{AB} = \phi_A P'_A + \phi_B P'_B \qquad (28\text{-}2)$$

where P'_A and P'_B are the polarity indexes of the two solvents and ϕ_A and ϕ_B are the volume fractions of solvents A and B.

In Section 26D-4, it was pointed out that the easiest way to improve the chromatographic resolution of two species is by manipulating the retention factor k, which

can in turn be varied by changing the polarity index of the solvent. Here, adjustment of P' is easily accomplished by the use of mobile phases that consist of a mixture of two solvents. Typically, a two-unit change in P' results (very roughly) in a ten-fold change in k. That is, for a normal-phase separation

$$\frac{k_2}{k_1} = 10^{(P'_1 - P'_2)/2} \qquad (28\text{-}3)$$

where k_1 and k_2 are the initial and final values of k for a solute and P'_1 and P'_2 are the corresponding polarity indexes. For a reversed-phase column

$$\frac{k_2}{k_1} = 10^{(P'_2 - P'_1)/2} \qquad (28\text{-}4)$$

It should be emphasized that these equations apply only approximately. Nonetheless, as shown in Example 28-1, they can be useful.

EXAMPLE 28-1

In a reversed-phase column, a solute was found to have a retention time of 31.3 min, and an unretained species required 0.48 min for elution when the mobile phase was 30% (by volume) methanol and 70% water. Calculate (a) k and (b) a water-methanol composition that should bring k to a value of about 5.

Solution

(a) Application of Equation 26-12 yields

$$k = (31.3 - 0.48)/0.48 = 64$$

(b) To obtain P' for the mobile phase we substitute polarity indexes for methanol and water from Table 28-2 into Equation 28-2 to give

$$P' = (0.30 \times 5.1) + (0.70 \times 10.2) = 8.7$$

Substitution of this result into Equation 28-4 gives

$$\frac{5}{64} = 10^{(P'_2 - 8.7)/2}$$

If we take the logarithm of both sides of this equation, we can solve for P'_2

$$-1.11 = \frac{P'_2 - 8.7}{2} = 0.5P'_2 - 4.35$$

$$P'_2 = 6.5$$

Letting x be the volume fraction of methanol in the new solvent mixture and substituting again into Equation 28-2, we find

$$6.5 = (x \times 5.1) + (1 - x) \times 10.2$$

$$x = 0.73, \text{ or } 73\%$$

Often, in reversed-phase separations, we use a solvent mixture containing water and a polar organic solvent. The retention factor can then be manipulated by varying the water concentration as shown by Example 28-1. The effect of such manipulations is shown by the chromatograms in Figure 28-16a and b where the sample was a mixture of six steroids. With a mixture of 41% acetonitrile and 59% water, k had a value of 5 and all of the analytes were eluted in such a short time (~2 min) that the separation was quite incomplete. By increasing the concentration of water to 70%, elution took place over 7 min, which doubled the value of k. Now the total elution time was sufficient to achieve a separation, but the α value for compounds 1 and 3 was not great enough to resolve them.

Effect of Mobile Phase on Selectivities. In many cases, adjusting k to a suitable level is all that is needed to give a satisfactory separation. When two bands still overlap, however, as in Figure 28-16a and b, the selectivity factor α for the two species must be made larger. Such a change can be brought about most conveniently by changing the chemical nature of the mobile phase while holding the predetermined value of k *more or less the same*. The most widely accepted approach for this type of optimization is called the *solvent selectivity triangle*.[17] Here, the mobile-phase effects contributing to selectivity are considered to be the result of proton donor, proton acceptor, and dipolar interactions. Solvents incorporating these interactions are then chosen for statistically designed experiments to find the optimal solvent mixture.[18]

For reversed-phase chromatography, the three solvent modifiers are methanol, acetonitrile, and tetrahydrofuran. Water is then used to adjust the solvent strength of the mixtures and yield a suitable value of k. Three binary solvent compositions (water plus modifier) define the three vertices of the solvent triangle as shown in Figure 28-17. Usually, seven to ten experiments are enough to define a solvent composition that will produce the best selectivity for a suitable k range. Several other statistical design schemes, such as full factorial designs, have also been used, but the solvent-triangle approach requires fewer experiments. Software is also available to aid in automated schemes for solvent optimization. In addition, simulation software, such as the popular DryLab,[19] is widely used for method development and optimization.

Figure 28-16 illustrates the systematic, solvent-triangle approach to the development of a separation of six steroids by reversed-phase chromatography. The first two chromatograms show the results from initial experiments to determine the minimal value of k required. Here, the last eluting component is used to calculate k. With k for component 6 equal to 10, room exists on the time scale for discrete peaks; here, however, the α value for components 1 and 3 (and to a lesser extent 5 and 6) is not great enough for satisfactory resolution. Further experiments were then performed to

[17] For more information, see J. C. Berridge, *Techniques for the Automated Optimization of HPLC Separations*, Chichester, UK: Wiley, 1985.
[18] See J. L. Glajch, J. J. Kirkland, K. M. Squire, and J. M. Minor, *J. Chromatogr.*, **1980**, *199*, 57; L. R. Snyder, *J. Chromatogr. Sci.*, **1978**, *16*, 223–234.
[19] DryLab is now part of Rheodyne, LLC, Rohnert Park, CA. It was developed by LC Resources, Inc., Walnut Creek, CA.

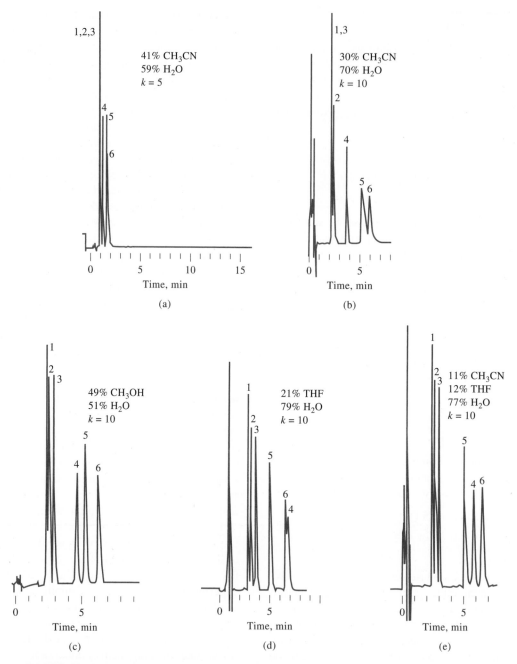

FIGURE 28-16 Systematic approach to the separation of six steroids. The use of water to adjust k is shown in (a) and (b). The effects of varying α at constant k are shown in (b), (c), (d), and (e). Column: 0.4 × 150 mm packed with 5 μm C_8 bonded, reversed-phase particles. Temperature: 50°C. Flow rate: 3.0 cm³/min. Detector: UV 254 nm. THF = tetrahydrofuran. CH_3CN = acetonitrile. Compounds: (1) prednisone, (2) cortisone, (3) hydrocortisone, (4) dexamethasone, (5) corticosterone, (6) cortoexolone.

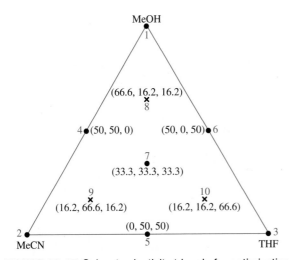

FIGURE 28-17 Solvent selectivity triangle for optimization of reversed-phase separations. Ten mixtures of the three organic solvents (methanol, acetonitrile, and tetrahydrofuran) are shown with the relative proportions indicated in parentheses (MeOH, MeCN, and THF). Water is used to maintain solvent strength and keep the k value within an appropriate range.

find better α values; in each case, the water concentration was adjusted to a level that yielded $k = 10$ for the last eluting component. The results of experiments with methanol-water and tetrahydrofuran-water mixtures are shown in Figure 28-16c and d. Several additional experiments involving the systematic variation of pairs of the three organic solvents were performed (in each case adjusting k to 10 with water). Finally, the mixture shown in Figure 28-16e was chosen as the best mobile phase for the separation of the particular group of compounds.

For normal-phase separations, a similar solvent triangle is used in which the selectivity solvents are ethyl ether, methylene chloride, and chloroform;[20] the solvent strength adjustment is then made with *n*-hexane. With these solvent systems, optimization can be accomplished with a minimal number of experiments.

28D-3 Applications of Partition Chromatography

Reversed-phase bonded packings, when used in conjunction with highly polar solvents (often aqueous), approach the ideal, universal system for LC. Because of their wide range of applicability, their convenience,

and the ease with which k and α can be altered by manipulation of aqueous mobile phases, these packings are frequently applied before all others for exploratory separations with new types of samples.

Table 28-3 lists a few typical examples of the multitude of uses of partition chromatography in various fields. A more complete picture of the many applications of this technique can be obtained by consulting review articles and books.[21] Figure 28-18 illustrates two of many thousands of applications of bonded-phase partition chromatography to the analysis of consumer and industrial materials.

Derivative Formation

In some instances, it is useful to convert the components of a sample to a derivative before, or sometimes after, chromatographic separation. Such treatment may be desirable (1) to reduce the polarity of the species so that partition rather than adsorption or ion-exchange columns can be used; (2) to increase the detector response, and thus sensitivity, for all of the sample components; and (3) to selectively enhance the detector response to certain components of the sample.

Figure 28-19 illustrates the use of derivatives to reduce polarity and enhance sensitivity. The sample was made up of thirty amino acids of physiological

---

[21]See, for example, L. R. Snyder, *Anal. Chem.*, **2000**, *72*, 412A; J. G. Dorsey et al., *Anal. Chem.*, **1998**, *70*, 591R; *HPLC: Practical and Industrial Applications*, J. K. Swadesh, ed., 2nd ed., Boca Raton, FL: CRC Press, 2001; *Handbook of Pharmaceutical Analysis by HPLC*, S. Ahuja and M. W. Dong, eds., San Diego, CA: Elsevier, 2005.

TABLE 28-3 Typical Applications of Partition Chromatography

Field	Typical Mixtures
Pharmaceuticals	Antibiotics, sedatives, steroids, analgesics
Biochemical	Amino acids, proteins, carbohydrates, lipids
Food products	Artificial sweeteners, antioxidants, aflatoxins, additives
Industrial chemicals	Condensed aromatics, surfactants, propellants, dyes
Pollutants	Pesticides, herbicides, phenols, polychlorinated biphenyls
Forensic science	Drugs, poisons, blood alcohol, narcotics
Clinical chemistry	Bile acids, drug metabolites, urine extracts, estrogens

---

[20]Note that chlorinated solvents are now subject to strict environmental regulations.

FIGURE 28-18 Typical applications of bonded-phase chromatography. (a) soft-drink additives. Column: 4.6 × 250 mm packed with polar (nitrile) bonded-phase packing. Isocratic elution with 6% HOAc to 94% H_2O. Flow rate 1.0 mL/min. (Courtesy of BTR Separations, a DuPont ConAgra affiliate.) (b) Organophosphate insecticides. Column 4.5 × 250 mm packed with 5 μm C_8 bonded-phase particles. Gradient elution: 67% CH_3OH to 33% H_2O, to 80% CH_3OH to 20% H_2O. Flow rate 2 mL/min. Both used 254-nm-UV detectors.

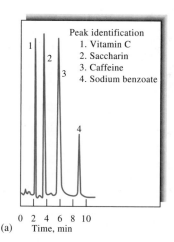

Peak identification
1. Vitamin C
2. Saccharin
3. Caffeine
4. Sodium benzoate

(a)   0 2 4 6 8 10
        Time, min

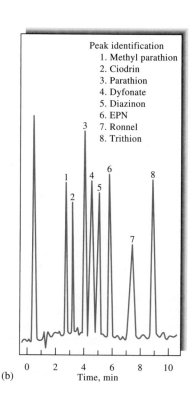

Peak identification
1. Methyl parathion
2. Ciodrin
3. Parathion
4. Dyfonate
5. Diazinon
6. EPN
7. Ronnel
8. Trithion

(b)   0   2   4   6   8   10
            Time, min

importance. Often, such a separation would be performed on an ion-exchange column with photometric detection based on postcolumn reaction of the amino acids with a colorimetric reagent such as ninhydrin. The chromatogram shown in Figure 28-19 was obtained by automatic, precolumn derivative formation with orthophthalaldehyde. The substituted isoindoles[22] formed by the reaction exhibit intense fluorescence at 425 nm, which permits detection down to a few picomoles (10^{-12} mol). Furthermore, the polarity of the derivatives are such that separation on a C_{18} reversed-phase packing becomes feasible. The advantages of this newer procedure are speed and smaller sample size. Derivatives are frequently used with selective detectors such as fluorescence, electrochemical, and mass spectrometric detectors.

Ion-Pair Chromatography

Ion-pair chromatography,[23] sometimes called *paired-ion chromatography*, is a subset of reversed-phase chromatography in which easily ionizable species are separated on reversed-phase columns. In this type of chromatography, an organic salt containing a large organic counterion, such as a quaternary ammonium ion or alkyl sulfonate, is added to the mobile phase as an ion-pairing reagent. Two mechanisms for separation are postulated. In the first, the counterion forms an uncharged ion pair with a solute ion of opposite charge in the mobile phase. This ion pair then partitions into the nonpolar stationary phase, giving differential retention of solutes based on the affinity of the ion pair for the two phases. Alternatively, the counterion is retained strongly by the normally neutral stationary phase and imparts a charge to this phase. Separation of organic solute ions of the opposite charge then occurs by formation of reversible ion-pair complexes with the more strongly retained solutes forming the strongest complexes with the stationary phase. Some unique separations of both ionic and nonionic compounds in the same sample can be accomplished by this form of partition chromatography.

Figure 28-20 illustrates the separation of ionic and nonionic compounds using alkyl sulfonates of various chain lengths as ion-pairing agents. Note that a mixture of C_5- and C_7-alkyl sulfonates gives the best separation results.

Applications of ion-pair chromatography frequently overlap those of ion-exchange chromatography, which are discussed in Section 28F. For the separation of small inorganic and organic ions, ion exchange is usually preferred unless selectivity is a problem.

---

[22] For the structure of these compounds, see S. S. Simons Jr. and D. E. Johnson, *J. Amer. Chem. Soc.*, **1976**, *98*, 7098.

[23] See *Ion Pair Chromatography*, M. T. W. Hearn, ed., New York: Dekker, 1985.

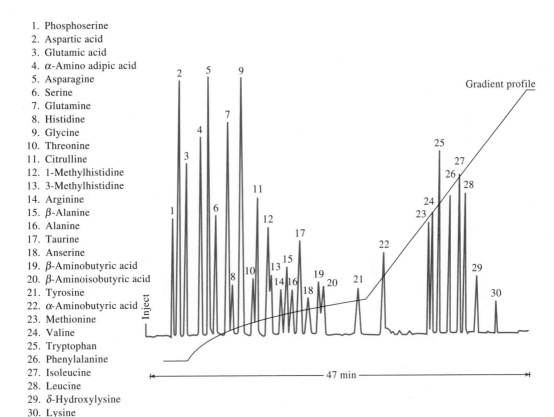

1. Phosphoserine
2. Aspartic acid
3. Glutamic acid
4. α-Amino adipic acid
5. Asparagine
6. Serine
7. Glutamine
8. Histidine
9. Glycine
10. Threonine
11. Citrulline
12. 1-Methylhistidine
13. 3-Methylhistidine
14. Arginine
15. β-Alanine
16. Alanine
17. Taurine
18. Anserine
19. β-Aminobutyric acid
20. β-Aminoisobutyric acid
21. Tyrosine
22. α-Aminobutyric acid
23. Methionine
24. Valine
25. Tryptophan
26. Phenylalanine
27. Isoleucine
28. Leucine
29. δ-Hydroxylysine
30. Lysine

FIGURE 28-19 Chromatogram of orthophthalaldehyde derivatives of 30 amino acids of physiological importance. Column: 5 μm C_{18}, reversed-phase. Solvent A: 0.05 M Na_2HPO_4, pH 7.4, 96:2:2 CH_3OH, THF, H_2O. Fluorescence detector: excitation 334 nm; emission 425 nm. (Reprinted with permission from R. Pfiefer et al., *Amer. Lab.*, **1983**, *15* (3), 86. Copyright 1983 by International Scientific Communications, Inc.)

Chiral Chromatography

Tremendous advances have been made in recent years in separating compounds that are nonsuperimposable mirror images of each other called *chiral compounds*.[24] Such mirror images are called *enantiomers*. Either chiral mobile-phase additives or chiral stationary phases are required for these separations. Preferential complexation between the *chiral resolving agent* (additive or stationary phase) and one of the isomers results in a separation of the enantiomers. The chiral resolving agent must have chiral character itself to recognize the chiral nature of the solute.

Chiral stationary phases have received the most attention.[25] Here, a chiral agent is immobilized on the surface of a solid support. Chiral stationary phases are available from a number of manufacturers. Several different modes of interaction can occur between the chiral resolving agent and the solute.[26] In one type, the interactions are due to attractive forces such as those between π bonds, hydrogen bonds, or dipoles. In another type, the solute can fit into chiral cavities in the stationary phase to form inclusion complexes. No matter what the mode, the ability to separate these very closely related compounds is of extreme importance in many fields. Figure 28-21 shows the separation of a racemic mixture of an ester on a chiral stationary phase. Note the excellent resolution of the *R* and *S* enantiomers.

28E ADSORPTION CHROMATOGRAPHY

Adsorption, or liquid-solid, chromatography is the classic form of LC first introduced by Tswett at the beginning of the last century. Because of the strong

[24]For further information, see S. Ahuja, *Chiral Separations by Chromatography*, New York: Oxford University Press, 2000; *Chiral Separations: Applications and Technology*, S. Ahuja, ed., Washington, DC: American Chemical Society, 1996.

[25]For a recent review on chiral stationary phases, see D. W. Armstrong and B. Zhang, *Anal. Chem.*, **2001**, *73*, 557A.

[26]For a review on chiral interactions, see M. C. Ringo and C. E. Evans, *Anal. Chem.*, **1998**, *70*, 315A.

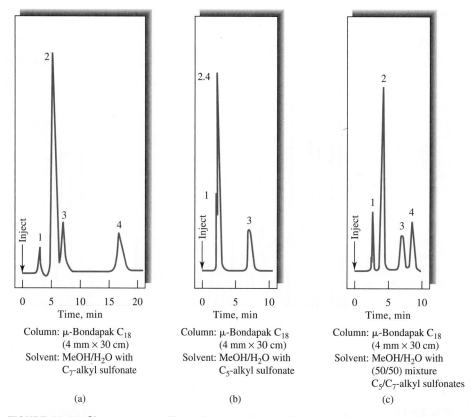

FIGURE 28-20 Chromatograms illustrating separations of mixtures of ionic and nonionic compounds by ion-pair chromatography. Compounds: (1) niacinamide, (2) pyridoxine, (3) riboflavin, (4) thiamine. At pH 3.5, niacinamide is strongly ionized, and riboflavin is nonionic. Pyridoxine and thiamine are weakly ionized. Column: μ-Bondapak, C_{18}, 4 mm × 30 cm. Mobile phase: (a) MeOH and H_2O with C_7-alkyl sulfonate; (b) MeOH and H_2O with C_5-alkyl sulfonate; (c) MeOH and H_2O with 1:1 mixture of C_5- and C_7-alkyl sulfonates. (Courtesy of Waters Associates, Inc.)

FIGURE 28-21 Chromatogram of a racemic mixture of *N*-(1-naphthyl)leucine ester on a dinitrobenzene-leucine chiral stationary phase. The *R* and *S* enantiomers are seen to be well separated. Column: 4.6 × 50 mm; mobile phase, 20% 2-propanol in hexane; flow rate 1.2 mL/min; UV detector at 254 nm. (From L. H. Bluhm, Y. Wang, and T. Li, *Anal. Chem.*, **2000**, *72*, 5201. With permission of American Chemical Society.)

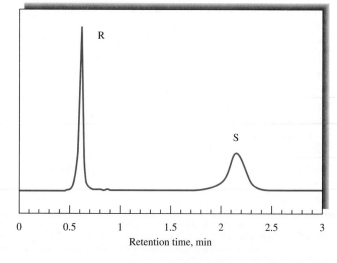

overlap between normal-phase partition chromatography and adsorption chromatography, many of the principles and techniques used for the former apply to adsorption chromatography. In fact, in many normal-phase separations, adsorption-displacement processes govern retention.

Finely divided silica and alumina are the only stationary phases that find use for adsorption chromatography. Silica is preferred for most applications because of its higher sample capacity. The adsorption characteristics of the two substances parallel one another. For both, retention times become longer as the polarity of the analyte increases.

It has been found that the polarity index P', which was described in Section 28D-2, can also serve as a rough guide to the strengths of solvents for adsorption chromatography. A much better index, however, is the *eluent strength* ε^0, which is the adsorption energy per unit area of solvent.[27] The eluent strength depends on the adsorbent, with ε^0 values for silica being about 0.8 of alumina. The values for ε^0 in the last column of Table 28-2 are for alumina. Note that solvent-to-solvent differences in ε^0 roughly parallel those for P'.

Because of the versatility and ready availability of bonded stationary phases, use of traditional adsorption chromatography with solid stationary phases has decreased in recent years in favor of normal-phase chromatography.

28F ION CHROMATOGRAPHY

Ion chromatography refers to modern, efficient methods for separating and determining ions on columns with relatively low ion-exchange capacity. Although ion-exchange separations have been around since ion-exchange resins were developed in the mid-1930s, ion chromatography as it is practiced today was first developed in the mid-1970s when it was shown that anion or cation mixtures can be resolved on HPLC columns packed with anion-exchange or cation-exchange resins. At that time, detection was generally performed with conductivity measurements, which were not ideal because of high electrolyte concentrations in the mobile phase. The development of low-exchange-capacity columns allowed the use of low-ionic-strength mobile phases that could be further deionized (ionization suppressed) to allow high-sensitivity conductivity detection. Currently, several other detector types are

available for ion chromatography, including spectrophotometric and electrochemical.[28]

Ion chromatography was an outgrowth of ion-exchange chromatography, which during the Manhattan project[29] was developed for the separation of closely related rare earth cations with cation-exchange resins. This monumental work, which laid the theoretical groundwork for ion-exchange separations, was extended after World War II to many other types of materials; ultimately, it led to automated methods for the separation and detection of amino acids and other ionic species in complex mixtures. The development of modern HPLC began in the late 1960s, but its application to separation of ionic species was delayed by the lack of a sensitive general method of detecting such eluted ionic species as alkali and alkaline earth cations and halide, acetate, and nitrate anions. This situation was remedied in 1975 by the development by workers at Dow Chemical Company of an eluent suppressor technique, which made possible the conductometric detection of eluted ions.[30] This technique is described in Section 28F-3.

28F-1 Ion-Exchange Equilibria

Ion-exchange processes are based on exchange equilibria between ions in solution and ions of like sign on the surface of an essentially insoluble, high-molecular-mass solid. Natural ion-exchangers, such as clays and zeolites, have been recognized and used for several decades. Synthetic ion-exchange resins were first produced in the mid-1930s for water softening, water deionization, and solution purification. The most common active sites for cation-exchange resins are the sulfonic acid group $—SO_3^- H^+$, a strong acid, and the carboxylic acid group $—COO^- H^+$, a weak acid. Anionic exchangers contain strongly basic tertiary amine groups $—N(CH_3)_3^+ OH^-$ or weakly basic primary amine groups $—NH_3^+ OH^-$.

When a sulfonic acid ion-exchanger is brought in contact with an aqueous solvent containing a cation M^{x+}, an exchange equilibrium is set up that can be described by

$$x\text{RSO}_3^-\text{H}^+ + \text{M}^{x+} \rightleftharpoons (\text{RSO}_3^-)_x\text{M}^{x+} + x\text{H}^+$$
$$\text{solid} \qquad \text{solution} \qquad \text{solid} \qquad \text{solution}$$

---

[27]See L. R. Snyder, *Principles of Adsorption Chromatography*, Chap. 8, New York: Dekker, 1968.

[28]For a brief review of ion chromatography, see J. S. Fritz, *Anal. Chem.*, **1987**, *59*, 335A; P. R. Haddad, *Anal. Chem.*, **2001**, *73*, 266A. For monographs on the subject, see J. S. Fritz and D. T. Gjerde, *Ion Chromatography*, 3rd ed., Weinheim, Germany: Wiley-VCH, 2000; H. Small, *Ion Chromatography*, New York: Plenum Press, 1989.

[29]The Manhattan project was the U.S. effort, with help from Canada and the United Kingdom, during World War II to develop nuclear weapons.

[30]H. Small, T. S. Stevens, and W. C. Bauman, *Anal. Chem.*, **1975**, *47*, 1801.

where $RSO_3^-H^+$ represents *one* of many sulfonic acid groups attached to a large polymer molecule. Similarly, a strong-base exchanger interacts with the anion A^{x-} as shown by the reaction

$$xRN(CH_3)_3OH^- + A^{x-} \rightleftharpoons [RN(CH_3)_3^+]_xA^{x-} + xOH^-$$
solid solution solid solution

As an example of the application of the mass-action law to ion-exchange equilibria, we will consider the reaction between a singly charged ion B^+ with a sulfonic acid resin held in a chromatographic column. From a neutral solution, initial retention of B^+ ions at the head of the column occurs because of the reaction

$$RSO_3^-H^+(s) + B^+(aq) \rightleftharpoons RSO_3^-B^+(s) + H^+(aq)$$
$$(28\text{-}5)$$

Elution with a dilute solution of hydrochloric acid shifts the equilibrium in Equation 28-5 to the left, causing part of the B^+ ions in the stationary phase to be transferred to the mobile phase. These ions then move down the column in a series of transfers between the stationary and mobile phases.

The equilibrium constant K_{ex} for the exchange reaction shown in Equation 28-5 takes the form

$$K_{ex} = \frac{[RSO_3^-B^+]_s[H^+]_{aq}}{[RSO_3^-H^+]_s[B^+]_{aq}} \qquad (28\text{-}6)$$

Here, $[RSO_3^-B^+]_s$ and $[RSO_3^-H^+]_s$ are concentrations (strictly activities) of B^+ and H^+ *in the solid phase*. Rearranging yields

$$\frac{[RSO_3^-B^+]_s}{[B^+]_{aq}} = K_{ex}\frac{[RSO_3^-H^+]_s}{[H^+]_{aq}} \qquad (28\text{-}7)$$

During the elution, the aqueous concentration of hydrogen ions is much larger than the concentration of the singly charged B^+ ions in the mobile phase. Also, the exchanger has an enormous number of exchange sites relative to the number of B^+ ions being retained. Thus, the overall concentrations $[H^+]_{aq}$ and $[RSO_3^-H^+]_s$ are not affected significantly by shifts in the equilibrium 28-5. Therefore, when $[RSO_3^-H^+]_s \gg [RSO_3^-B^+]_s$ and $[H^+]_{aq} \gg [B^+]_{aq}$, the right-hand side of Equation 28-7 is substantially constant, and we can write

$$\frac{[RSO_3^-B^+]_s}{[B^+]_{aq}} = K = \frac{c_S}{c_M} \qquad (28\text{-}8)$$

where K is a constant that corresponds to the distribution constant as defined by Equations 26-1 and 26-2. All of the equations in Table 26-5 (Section 26E) can then be applied to ion-exchange chromatography in

the same way as to the other types, which have already been considered.

Note that K_{ex} in Equation 28-6 represents the affinity of the resin for the ion B^+ relative to another ion (here, H^+). Where K_{ex} is large, a strong tendency exists for the solid phase to retain B^+; where K_{ex} is small, the reverse is true. By selecting a common reference ion such as H^+, distribution ratios for different ions on a given type of resin can be experimentally compared. Such experiments reveal that polyvalent ions are much more strongly held than singly charged species. Within a given charge group, however, differences appear that are related to the size of the hydrated ion as well as to other properties. Thus, for a typical sulfonated cation-exchange resin, values for K_{ex} decrease in the order $Tl^+ > Ag^+ > Cs^+ > Rb^+ > K^+ > NH_4^+ > Na^+ > H^+ > Li^+$. For divalent cations, the order is $Ba^{2+} > Pb^{2+} > Sr^{2+} > Ca^{2+} > Ni^{2+} > Cd^{2+} > Cu^{2+} > Co^{2+} > Zn^{2+} > Mg^{2+} > UO_2^{2+}$.

For anions, K_{ex} for a strong-base resin decreases in the order $SO_4^{2-} > C_2O_4^{2-} > I^- > NO_3^- > Br^- > Cl^- > HCO_2^- > CH_3CO_2^- > OH^- > F^-$. This sequence somewhat depends on the type of resin and reaction conditions and should thus be considered only approximate.

28F-2 Ion-Exchange Packings

Historically, ion-exchange chromatography was performed on small, porous beads formed during emulsion copolymerization of styrene and divinylbenzene. The presence of divinylbenzene (usually ~8%) results in cross-linking, which makes the beads mechanically stable. To make the polymer active toward ions, acidic or basic functional groups are bonded chemically to the structure. The most common groups are sulfonic acid and quaternary amines.

Figure 28-22 shows the structure of a strong acid resin. Note the cross-linking that holds the linear polystyrene molecules together. The other types of resins have similar structures except for the active functional group.

Porous polymeric particles are not entirely satisfactory for chromatographic packings because of the slow rate of diffusion of analyte molecules through the micropores of the polymer matrix and because of the compressibility of the matrix. To overcome this problem, two newer types of packings have been developed and are in more general use than the porous polymer type. One is a polymeric bead packing in which the bead surface is coated with a synthetic ion-exchange

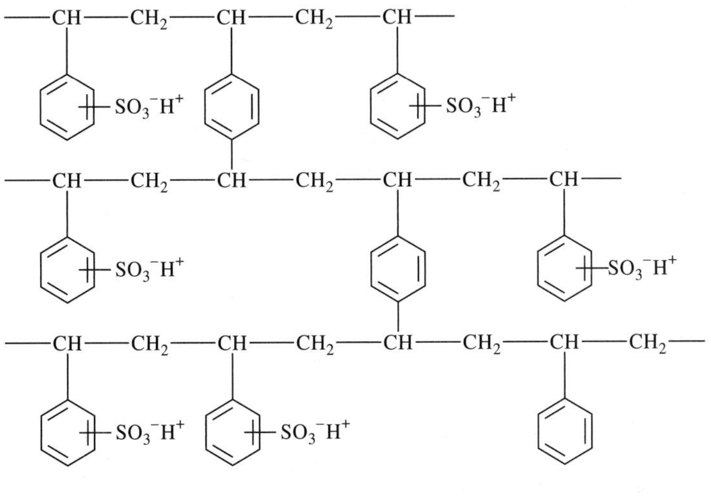

FIGURE 28-22 Structure of a cross-linked polystyrene ion-exchange resin. Similar resins are used in which the $-SO_3^-H^+$ group is replaced by $-COO^-H^+$, $-NH_3^+OH^-$, and $-N(CH_3)_3^+OH^-$ groups.

resin. A second type of packing is prepared by coating porous microparticles of silica, such as those used in adsorption chromatography, with a thin film of the exchanger. Particle diameters are typically 3–10 μm. With either type, faster diffusion in the polymer film leads to enhanced efficiency. Polymer-based packings have higher capacity than silica-based packings and can be used over a broad pH range. Silica-based ion exchangers give higher efficiencies, but suffer from a limited pH range of stability and an incompatibility with suppressor-based detection (see next section).

28F-3 Inorganic-Ion Chromatography

The mobile phase in ion-exchange chromatography must have the same general properties required for other types of chromatography. That is, it must dissolve the sample, have a solvent strength that leads to reasonable retention times (appropriate k values), and interact with solutes in such a way as to lead to selectivity (suitable α values). The mobile phases in ion-exchange chromatography are aqueous solutions that may contain moderate amounts of methanol or other water-miscible organic solvents; these mobile phases also contain ionic species, often in the form of a buffer. Solvent strength and selectivity are determined by the kind and concentration of these added ingredients. In general, the ions of the mobile phase compete with analyte ions for the active sites on the ion-exchange packing.

Two types of ion chromatography are currently in use: *suppressor-based* and *single-column*. They differ in the method used to prevent the conductivity of the eluting electrolyte from interfering with the measurement of analyte conductivities.

Ion Chromatography Based on Suppressors

As noted earlier, the widespread application of ion chromatography for the determination of inorganic species was inhibited by the lack of a good general detector, which would permit quantitative determination of ions on the basis of chromatographic peak areas. Conductivity detectors are an obvious choice for this task. They can be highly sensitive, they are universal for charged species, and as a general rule, they respond in a predictable way to concentration changes. Furthermore, such detectors are simple, inexpensive to construct and maintain, easy to miniaturize, and ordinarily give prolonged, trouble-free service. The only limitation to conductivity detectors proved to be a serious one, which delayed their general use. This limitation arises from the high electrolyte concentration required to elute most analyte ions in a reasonable time. As a result, the conductivity from the mobile-phase components tends to swamp that from analyte ions, which greatly reduces the detector sensitivity.

In 1975 the problem created by the high conductance of eluents was solved by the introduction of an *eluent suppressor column* immediately following the ion-exchange column.[31] The suppressor column is packed with a second ion-exchange resin that effectively converts the ions of the eluting solvent to a molecular species of limited ionization without affecting the conductivity due to analyte ions. For example, when cations are being separated and determined, hydrochloric acid is chosen as the eluting reagent, and the suppressor column is an anion-exchange resin in the hydroxide form.

---

[31] H. Small, T. S. Stevens, and W. C. Bauman, *Anal. Chem.*, **1975**, *47*, 1801.

The product of the reaction in the suppressor is water. That is

$$H^+(aq) + Cl^-(aq) + resin^+OH^-(s) \rightarrow$$
$$resin^+Cl^-(s) + H_2O$$

The analyte cations are not retained by this second column.

For anion separations, the suppressor packing is the acid form of a cation-exchange resin and sodium bicarbonate or carbonate is the eluting agent. The reaction in the suppressor is

$$Na^+(aq) + HCO_3^-(aq) + resin^-H^+(s) \rightarrow$$
$$resin^-Na^+(s) + H_2CO_3(aq)$$

The largely undissociated carbonic acid does not contribute significantly to the conductivity.

An inconvenience associated with the original suppressor columns was the need to regenerate them periodically (typically, every 8 to 10 h) to convert the packing back to the original acid or base form. In the 1980s, however, micromembrane suppressors that operate

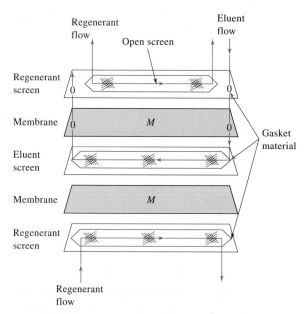

FIGURE 28-23 A micromembrane suppressor. Eluent flows through a narrow channel that contains a plastic screen that reduces the void volume and appears to increase mass-transfer rates. The eluent is separated from the suppressor solution by 50-µm exchange membranes. Regenerant flow is in the direction opposite to eluent flow. (Courtesy of Dionex Corporation, Sunnyvale, CA.)

Concentrations, ppm		Concentrations, ppm	
F^-	3	Ca^{2+}	3
Formate	8	Mg^{2+}	3
BrO_3^-	10	Sr^{2+}	10
Cl^-	4	Ba^{2+}	25
NO_2^-	10		
HPO_4^{2-}	30		
Br^-	30		
NO_3^-	30		
SO_4^{2-}	25		

FIGURE 28-24 Typical applications of ion chromatography. (a) Separation of anions on an anion-exchange column. Eluent: 0.0028 M NaHCO$_3$ to 0.0023 M Na$_2$CO$_3$. Sample size: 50 µL. (b) Separation of alkaline earth ions on a cation-exchange column. Eluent: 0.025 M phenylenediamine dihydrochloride to 0.0025 M HCl. Sample size: 100 µL. (Courtesy of Dionex Corp., Sunnyvale, CA.)

continuously became available.[32] Figure 28-23 illustrates a typical micromembrane suppressor. Here, the eluent and the suppressor solutions flow in opposite directions on either side of the permeable ion-exchange membrane M. For the analysis of anions, the membranes are cation-exchange resins; for cations, they are anion exchangers. When, for example, sodium ions are to be removed from the eluent, acid for regeneration flows continuously in the suppressor stream. Sodium ions from the eluent exchange with hydrogen ions from

[32] For a description of these devices, see G. O. Franklin, *Amer. Lab.*, **1985** (3), 71.

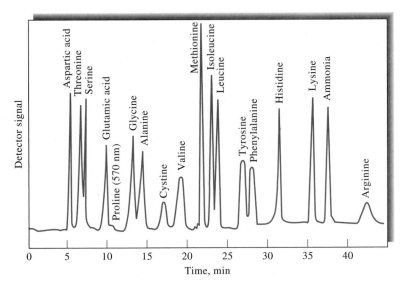

FIGURE 28-25 Separation of amino acids on an ion-exchange column. Packing: cation exchange with particle size of 8 μm. Pressure: 2700 psi. (Reprinted with permission from J. R. Benson, *Amer. Lab.*, **1972**, *4* (10), 60. Copyright 1972 by International Scientific Communications, Inc.)

the membrane and then migrate across the membrane where they exchange with hydrogen ions of the regeneration reagent. The device has a remarkably high exchange rate. For example, it is capable of removing essentially all of the sodium ions from a 0.1 M solution of sodium hydroxide when the eluent flow rate is 2 mL/min.

In recently designed commercial instruments, regeneration of suppressor solutions is performed automatically with electrogenerated hydrogen or hydroxyl ions so that interruptions in the use of the instruments for regeneration are not required. Figure 28-24 shows two applications of ion chromatography based on a suppressor column and conductometric detection. In each case, the ions were present in the parts-per-million range, and the sample sizes were 50 μL in one case and 100 μL in the other. The method is particularly important for anion analysis because no other rapid and convenient method for handling mixtures of this type now exists.

Single-Column Ion Chromatography

Commercial ion chromatography instrumentation that requires no suppressor column is also available. This approach depends on the small differences in conductivity between sample ions and the prevailing eluent ions. To amplify these differences, low-capacity exchangers are used that permit elution with solutions with low electrolyte concentrations. Furthermore, eluents of low conductivity are chosen.[33]

Single-column ion chromatography offers the advantage of not requiring special equipment for suppression. However, it is a somewhat less sensitive method for determining anions than suppressor column methods.

28F-4 Organic and Biochemical Applications of Ion-Exchange Chromatography

Ion-exchange chromatography has been applied to a variety of organic and biochemical systems, including drugs and their metabolites, serums, food preservatives, vitamin mixtures, sugars, and pharmaceutical preparations. An example of one of these applications is shown in Figure 28-25 in which 1×10^{-8} mol each of seventeen amino acids was separated on a cation-exchange column.

28F-5 Ion-Exclusion Chromatography

Ion-exclusion chromatography is not a form of ion chromatography because neutral species rather than ions are being separated. Nevertheless, it is convenient to discuss it here because ion-exclusion chromatography, like ion chromatography, uses ion-exchange columns to achieve separations.

The theory and applications of ion-exclusion chromatography are conveniently illustrated by the separation of simple carboxylic acids shown by the chromatogram in Figure 28-26. Here, the stationary phase was a cation-exchange resin in its acidic form, and elu-

[33] See R. M. Becker, *Anal. Chem.*, **1980**, *52*, 1510; J. R. Benson, *Amer. Lab.*, **1985** (6), 30; T. Jupille, *Amer. Lab.*, **1986** (5), 114.

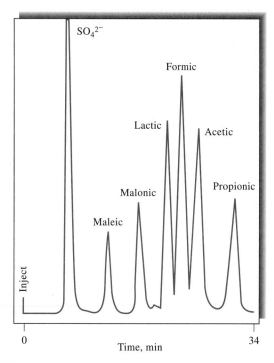

FIGURE 28-26 An ion-exclusion chromatogram for a mixture of six weak acids.

tion was accomplished with a dilute solution of hydrochloric acid. The analytical column was followed by a suppressor column packed with a cation-exchange resin in the silver form. The hydrogen ions of the eluent were exchanged for silver ions, which then precipitated the chloride ions, thus removing the ions contributed by the eluent. The undissociated form of the analyte acids was distributed between the mobile phase in the column and the immobilized liquid held in the pores of the packing. The distribution constants for the various acids are primarily related to the inverse of their dissociation constants, although other factors also play a part in the extent to which various species are distributed between the two phases.

Ion-exclusion chromatography finds numerous applications for identification and determination of acidic species in milk, coffee, wine, and many other products of commerce. Salts of weak acids can be analyzed because they are converted to the corresponding acid by the hydrogen ions in the exchanger. Weak bases and their salts can also be determined by ion-exclusion chromatography. In these applications an anion-exchange column in the hydroxide form is used.

28G SIZE-EXCLUSION CHROMATOGRAPHY

Size-exclusion, or gel, chromatography is a powerful technique that is particularly applicable to high-molecular-mass species.[34] Packings for size-exclusion chromatography consist of small ($\sim$10 µm) silica or polymer particles containing a network of uniform pores into which solute and solvent molecules can diffuse. While in the pores, molecules are effectively trapped and removed from the flow of the mobile phase. The average residence time in the pores depends on the effective size of the analyte molecules. Molecules larger than the average pore size of the packing are excluded and thus suffer essentially no retention; such species are the first to be eluted. Molecules having diameters significantly smaller than the pores can penetrate or permeate throughout the pore maze and are thus entrapped for the greatest time; these are last to be eluted. Between these two extremes are intermediate-size molecules whose average penetration into the pores of the packing depends on their diameters. Within this group, fractionation occurs, which is directly related to molecular size and to some extent molecular shape. Note that size-exclusion separations differ from the other procedures we have been considering in that no chemical or physical interaction between analytes and the stationary phase are involved. Indeed, such interactions are avoided because they lead to lower column efficiencies. Also note that, unlike other forms of chromatography, there is an upper limit to retention time because no analyte species is retained longer than those that totally permeate the stationary phase.

28G-1 Column Packings

Two types of packing for size-exclusion chromatography are encountered: polymer beads and silica-based particles, both of which have diameters of 5 to 10 µm. Silica particles have the advantages of greater rigidity, which leads to easier packing and permits the use of higher pressures; greater stability, which permits the use of a wider range of solvents, including water; more

---

[34]For monographs on this subject, see *Handbook of Size Exclusion Chromatography*, 2nd ed., C. S. Wu, ed., New York: Dekker, 2004; *Column Handbook for Size Exclusion Chromatography*, C. S. Wu, ed., San Diego: Academic Press, 1999; S. Mori and H. G. Barth, *Size Exclusion Chromatography*, New York: Springer, 1999.

rapid equilibration with new solvents; and stability at higher temperatures. The disadvantages of silica-based particles include their tendency to retain solutes by adsorption and their potential for catalyzing the degradation of solute molecules.

Most early size-exclusion chromatography was carried out on cross-linked styrene-divinylbenzene copolymers similar in structure (except that the sulfonic acid groups are absent) to that shown in Figure 28-22. The pore size of these polymers is controlled by the extent of cross-linking and hence the relative amount of divinylbenzene present during manufacture. Polymeric packings having several different average pore sizes are thus marketed. Originally, styrene-divinylbenzene gels were hydrophobic and thus could be used only with nonaqueous mobile phases. Now, however, hydrophilic gels are available, making possible the use of aqueous solvents for the separation of large, water-soluble molecules such as sugars. These hydrophilic gels are sulfonated divinylbenzenes or polyacrylamides. Chromatography based on the hydrophilic packings was once called *gel filtration*, and techniques based on hydrophobic packings were termed *gel permeation*. Today, both techniques are described as size-exclusion methods. With both types of packings many pore diameters are available. Ordinarily, a given packing will accommodate a 2-to-2.5-decade range of molecular mass. The average molecular mass suitable for a given packing may be as small as a few hundred or as large as several million.

Table 28-4 lists the properties of some typical commercial size-exclusion packings.

28G-2 Theory of Size-Exclusion Chromatography

The total volume V_t of a column packed with a porous polymer or silica gel is given by

$$V_t = V_g + V_i + V_o \qquad (28\text{-}9)$$

where V_g is the volume occupied by the solid matrix of the polymer or gel, V_i is the volume of solvent held in its pores, and V_o is the free volume outside the particles. Assuming no mixing or diffusion, V_o also represents the theoretical volume of solvent required to transport through the column those components too large to enter the pores. In fact, however, some mixing and diffusion will occur, and as a result the unretained components will appear in a Gaussian-shape band with a concentration maximum at V_o. For components small enough to enter freely into the pores of the polymer or gel, band maxima will appear at the end of the column at an eluent volume corresponding to (V_i + V_o). Generally, V_i, V_o, and V_g are of the same order of magnitude; thus, a size-exclusion column permits separation of the large molecules of a sample from the small molecules with a minimal volume of eluate.

Molecules of intermediate size are able to transfer to some fraction K of the solvent held in the pores; the elution volume V_e for these retained molecules is

$$V_e = V_o + KV_i \qquad (28\text{-}10)$$

Equation 28-10 applies to all of the solutes on the column. For molecules too large to enter the pores, $K = 0$ and $V_e = V_o$; for molecules that can enter the pores unhindered, $K = 1$ and $V_e = (V_o + V_i)$. In deriving Equation 28-10, the assumption is made that no interaction,

TABLE 28-4 Properties of Typical Packings for Size-Exclusion Chromatography

Type	Particle Size, μm	Average Pore Size, Å	Molecular Mass Exclusion Limit*
Polystyrene-divinylbenzene	10	100	700
		1000	$(0.1 \text{ to } 20) \times 10^4$
		10^4	$(1 \text{ to } 20) \times 10^4$
		10^5	$(1 \text{ to } 20) \times 10^5$
		10^6	$(5 \text{ to } >10) \times 10^6$
Silica	10	125	$(0.2 \text{ to } 5) \times 10^4$
		300	$(0.03 \text{ to } 1) \times 10^5$
		500	$(0.05 \text{ to } 5) \times 10^5$
		1000	$(5 \text{ to } 20) \times 10^5$

*Molecular mass above which no retention occurs.

such as adsorption, occurs between the solute molecules and the polymer or gel surfaces. With adsorption, the amount of interstitially held solute will increase; with small molecules, K will then be greater than unity.

Equation 28-10 rearranges to

$$K = (V_e - V_o)/V_i = c_S/c_M \qquad (28\text{-}11)$$

where K is the distribution constant for the solute (see Equations 26-1 and 26-2). Values of K range from zero for totally excluded large molecules to unity for small molecules. The distribution constant is a valuable parameter for comparing data from different packings. In addition, it makes possible the application of all of the equations in Table 26-5 to exclusion chromatography.

The useful molecular mass range for a size-exclusion packing is conveniently illustrated by means of a calibration curve such as that shown in Figure 28-27a. Here, molecular mass, which is directly related to the size of

 Simulation: Learn more about **size-exclusion chromatography**.

solute molecules, is plotted against retention volume V_R, where V_R is the product of the retention time and the volumetric flow rate. Note that the ordinate scale is logarithmic. The *exclusion limit* is the maximum molecular mass of a species that will penetrate the pores. All species having greater molecular mass than the exclusion limit are so large that they are not retained and elute together to give peak A in the chromatogram shown in Figure 28-27b. Below the *permeation limit*, the solute molecules can penetrate into the pores completely. All molecules below this molecular mass are so small that they elute as the single band labeled D. As molecular masses decrease from the exclusion limit, solute molecules spend more and more time, on the average, in the particle pores and thus move progressively more slowly. It is in the selective permeation region that fractionation occurs, yielding individual solute peaks such as B and C in the chromatogram.

Experimental calibration curves, similar in appearance to the hypothetical one in Figure 28-27a, are usu-

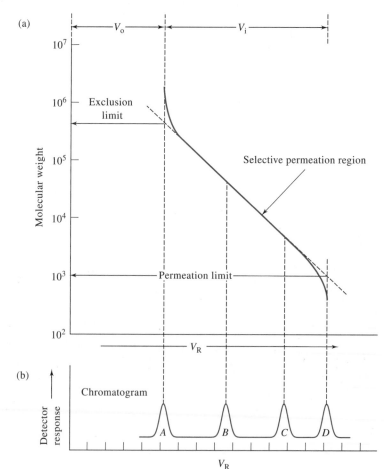

FIGURE 28-27 (a) Calibration curve for a size-exclusion column. (b) Chromatogram showing peak A containing all compounds with molecular masses greater than the exclusion limit, peaks B and C consisting of compounds within the selective permeation region, and peak D containing all compounds smaller than the permeation limit.

ally obtained by means of standards. Often, such curves are supplied by manufacturers of size-exclusion columns. Molecular-mass calibration standards should be as similar as possible in chemical functionality to the sample components.

28G-3 Applications of Size-Exclusion Chromatography

Gel filtration and gel permeation methods are complementary in that gel filtration is applied to water-soluble samples and gel permeation is used for substances in less-polar organic solvents. One useful application of the size-exclusion procedure is to the separation of high-molecular-mass, natural-product molecules from low-molecular-mass species and from salts. For example, a gel with an exclusion limit of several thousand can clearly separate proteins from amino acids and low-molecular-mass peptides.

Another useful application of gel permeation chromatography is to the separation of homologs and oligomers. These applications are illustrated by the two examples shown in Figure 28-28. The first shows the separation of a series of fatty acids ranging in molecular mass M from 116 to 344 on a polystyrene-based packing with an exclusion limit of 1000. The second is a chromatogram of a commercial epoxy resin, again on a polystyrene packing. Here, n refers to the number of monomeric units ($M = 264$) in the molecules.

Another important application of size-exclusion chromatography is to the rapid determination of the molecular mass or the molecular mass distribution of large polymers or natural products. The key to such determinations is an accurate molecular mass calibration. Calibrations can be accomplished by means of standards of known molecular mass (see Figure 28-27) or by the universal calibration method. The latter method relies on the principle that the product of the intrinsic molecular viscosity η and molecular mass M is proportional to hydrodynamic volume (effective volume including solvation sheath). Ideally, molecules are separated in size-exclusion chromatography according to hydrodynamic volume. Hence, a universal calibration curve can be obtained by plotting $\log(\eta M)$ versus the retention volume V_R. Alternatively, absolute calibration can be achieved by using a molar mass-sensitive detector such as a low-angle, light-scattering detector.

The most important advantages of size-exclusion procedures include (1) short and well-defined separation times (all solutes leave the column between V_o and $V_o + V_i$ in Equation 28-9 and Figure 28-27); (2) narrow bands, which lead to good sensitivity; (3) freedom from sample loss because solutes do not interact with the stationary phase; and (4) absence of column deactivation brought about by interaction of solute with the packing.

The disadvantages are (1) only a limited number of bands can be accommodated because the time scale of the chromatogram is short and (2) inapplicability to

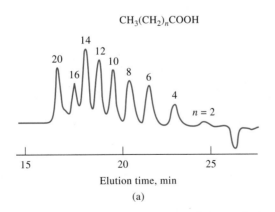

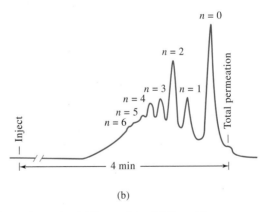

FIGURE 28-28 Applications of size-exclusion chromatography. (a) Separation of fatty acids. Column: polystyrene based, 7.5 × 600 nm, with exclusion limit of 1×10^3. Mobile phase: tetrahydrofuran. Flow rate: 1.2 mL/min. Detector: refractive index. (b) Analysis of a commercial epoxy resin (n = number of monomeric units in the polymer). Column: porous silica 6.2 × 250 mm. Mobile phase: tetrahydrofuran. Flow rate: 1.3 mL/min. Detector: UV absorption. (Courtesy of BTR Separations.)

samples of similar size, such as isomers. Generally, a 10% difference in molecular mass is required for reasonable resolution.

28H AFFINITY CHROMATOGRAPHY

Affinity chromatography involves covalently bonding a reagent, called an *affinity ligand*, to a solid support.[35] Typical affinity ligands are antibodies, enzyme inhibitors, or other molecules that reversibly and selectively bind to analyte molecules in the sample. When the sample passes through the column, only the molecules that selectively bind to the affinity ligand are retained. Molecules that do not bind pass through the column with the mobile phase. After the undesired molecules are removed, the retained analytes can be eluted by changing the mobile-phase conditions.

The stationary phase for affinity chromatography is a solid such as agarose or a porous glass bead to which the affinity ligand is immobilized. The mobile phase in affinity chromatography has two distinct roles to play. First, it must support the strong binding of the analyte molecules to the ligand. Second, once the undesired species are removed, the mobile phase must weaken or eliminate the analyte-ligand interaction so that the analyte can be eluted. Often, changes in pH or ionic strength are used to change the elution conditions during the two stages of the process.

Affinity chromatography has the major advantage of extraordinary specificity. The primary use is in the rapid isolation of biomolecules during preparative work.

28I THIN-LAYER CHROMATOGRAPHY

Planar chromatographic methods include *thin-layer chromatography* (TLC) and *paper chromatography* (PC). Each makes use of a flat, relatively thin layer of material that is either self-supporting or is coated on a glass, plastic, or metal surface. The mobile phase moves through the stationary phase by capillary action, sometimes assisted by gravity or an electrical potential.

Currently, most planar chromatography is based on the thin-layer technique, which is faster, has better resolution, and is more sensitive than its paper chro-

matography equivalent. This section is devoted to thin-layer methods.

28I-1 The Scope of Thin-Layer Chromatography

In terms of theory, types of stationary and mobile phases, and applications, thin-layer and LC are remarkably similar. TLC techniques in fact have been used to develop conditions for HPLC separations. At one time TLC methods were widely used in the pharmaceutical industry. Today, such techniques have largely been replaced by LC methods, which are readily automated and faster. Thin-layer chromatography has found widespread use in clinical laboratories and is the backbone of many biochemical and biological studies. It also finds extensive use in industrial laboratories.[36] Because of these many areas of application, TLC remains a very important technique.

28I-2 Principles of Thin-Layer Chromatography

Typical thin-layer separations are performed on a glass plate coated with a thin and adherent layer of finely divided particles; this layer constitutes the stationary phase. The particles are similar to those described in the discussion of adsorption, normal- and reversed-phase partition, ion-exchange, and size-exclusion-column chromatography. Mobile phases are also similar to those found in HPLC.

Preparation of Thin-Layer Plates

A thin-layer plate is prepared by spreading an aqueous slurry of the finely ground solid on the clean surface of a glass or plastic plate or microscope slide. Often a binder is incorporated into the slurry to enhance adhesion of the solid particles to the glass and to one another. The plate is then allowed to stand until the layer has set and adheres tightly to the surface; for some purposes, it may be heated in an oven for several hours. Several chemical supply houses offer precoated plates of various kinds. Costs are a few dollars per plate. The

---

[35] For details on affinity chromatography see *Handbook of Affinity Chromatography*, T. Kline, ed., New York: Dekker, 1993; *Analytical Affinity Chromatography*, I. M. Chaiken, ed., Boca Raton, FL: CRC Press, 1987; R. R. Walters, *Anal. Chem.*, **1985**, *57*, 1097A.

[36] Books devoted to the principles and applications of thin-layer chromatography include *Handbook of Thin-Layer Chromatography*, 3rd ed., J. Sherma and B. Fried, eds., New York: Dekker, 2003; B. Fried and J. Sherma, *Thin Layer Chromatography*, 4th ed., New York: Dekker, 1999; *Practical Thin-Layer Chromatography*, B. Fried and J. Sherma, eds., Boca Raton, FL: CRC Press, 1996. For recent reviews, see J. Sherma, *Anal. Chem.*, **2006**, *78*, 3841; J. Sherma, *Anal. Chem.*, **2004**, *76*, 3251; J. Sherma, **2002**, *74*, 2653.

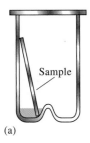

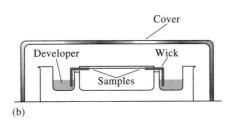

FIGURE 28-29 (a) Ascending-flow developing chamber. (b) Horizontal-flow developing chamber, in which samples are placed on both ends of the plate and developed toward the middle, thus doubling the number of samples that can be accommodated.

common plate sizes in centimeters are 5×20, 10×20, and 20×20.

Commercial plates come in two categories, conventional and high-performance. Conventional plates have thicker layers (200 to 250 μm) of particles having particle sizes of 20 μm or greater. High-performance plates usually have film thicknesses of 100 μm and particle diameters of 5 μm or less. High-performance plates provide sharper separations in shorter times. They suffer, however, from having a significantly smaller sample capacity than conventional plates.

Sample Application

Sample application is perhaps the most critical aspect of thin-layer chromatography, particularly for quantitative measurements. Usually, the sample, as a 0.01% to 0.1% solution, is applied as a spot 1 to 2 cm from the edge of the plate. For best separation efficiency, the spot should have a minimal diameter — about 5 mm for qualitative work and smaller for quantitative analysis. For dilute solutions, three or four repetitive applications are used, with drying between.

Manual application of samples is performed by touching a capillary tube containing the sample to the plate or by use of a hypodermic syringe. A number of mechanical dispensers, which increase the precision and accuracy of sample application, are now offered commercially.

Plate Development

Plate development is the process in which a sample is carried through the stationary phase by a mobile phase; it is analogous to elution in LC. The most common way of developing a plate is to place a drop of the sample near one edge of the plate and mark its position with a pencil. After the sample solvent has evaporated, the plate is placed in a closed container saturated with vapors of the developing solvent. One end of the plate is immersed in the developing solvent, with

care being taken to avoid direct contact between the sample and the developer (Figure 28-29). After the developer has traversed one half or two thirds of the length of the plate, the plate is removed from the container and dried. The positions of the components are then determined in any of several ways.

Locating Analytes on the Plate

Several methods can be used to locate sample components after separation. Two common methods, which can be applied to most organic mixtures, involve spraying with a solution of sulfuric acid or placing the plate in a chamber containing a few crystals of iodine. Both of these reagents react with organic compounds on the plate to yield dark products. Several specific reagents (such as ninhydrin) are also useful for locating separated species.

Another method of detection is based on incorporating a fluorescent material into the stationary phase. After development, the plate is examined under ultraviolet light. The sample components quench the fluorescence of the material so that all of the plate fluoresces except where the nonfluorescing sample components are located.

Figure 28-30 is an idealized drawing showing the appearance of a plate after development and the corresponding chromatogram. Sample 1 contained two components, whereas sample 2 contained one. Frequently, the spots on a real plate exhibit tailing, giving signals that are not symmetric as are those in the figure.

28I-3 Performance Characteristics of Thin-Layer Plates

Most of the terms and relationships developed for column chromatography in Section 26B can, with slight modification, be applied to thin-layer chromatography. One new term, the *retardation factor* or R_F, is required.

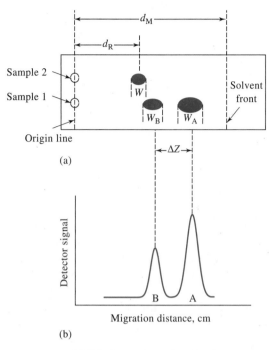

FIGURE 28-30 (a) Thin-layer plate after development. (b) Thin-layer chromatogram for sample 1.

The Retardation Factor

The developed thin-layer plate for a single solute is shown in Figure 28-30a for sample 2. The retardation factor for this solute is given by

$$R_F = \frac{d_R}{d_M} \qquad (28\text{-}12)$$

where d_R and d_M are linear distances measured from the origin line. Values for R_F can vary from 1 for solutes that are not retarded to a value that approaches 0. Note that if the spots are not symmetric, as they are in Figure 28-30a, the measurement of d_R is based on the position of maximum intensity.

The Retention Factor

All of the equations in Table 26-5 can be adapted to thin-layer chromatography. To apply these equations, it is only necessary to relate d_R and d_M as defined in Figure 28-30a to t_R and t_M, which are defined in Figure 26-4. To arrive at these relationships, consider the single solute that appears in sample 2 in Figure 28-30a. Here, t_M and t_R correspond to times required for the mobile phase and the solute to travel a fixed distance — in this case, d_R. For the mobile phase, this time

is equal to the distance divided by its linear velocity u, or

$$t_M = d_R/u \qquad (28\text{-}13)$$

The solute does not reach this same point, however, until the mobile phase has traveled the distance d_M. Therefore,

$$t_R = d_M/u \qquad (28\text{-}14)$$

Substitution of Equations 28-13 and 28-14 into Equation 26-12 yields

$$k = \frac{d_M - d_R}{d_R} \qquad (28\text{-}15)$$

The retention factor k can also be expressed in terms of the retardation factor by rewriting Equation 28-15 in the form

$$k = \frac{1 - d_R/d_M}{d_R/d_M} = \frac{1 - R_F}{R_F} \qquad (28\text{-}16)$$

Retention factors derived in this way can be used for method development in column chromatography as described in 28D-2. Obtaining retention factors by thin-layer chromatography is, however, usually simpler and more rapid than obtaining the data from experiments on a column.

Plate Heights

Approximate plate heights can also be determined for a given type of packing by thin-layer chromatographic measurements. Thus, for sample 2 in Figure 28-30a, the plate count is given by the equation

$$N = 16\left(\frac{d_R}{W}\right)^2 \qquad (28\text{-}17)$$

where d_R and W are defined in the figure. The plate height is then given by

$$H = d_R/N \qquad (28\text{-}18)$$

28I-4 Applications of Thin-Layer Chromatography

The thin-layer method can be applied both to qualitative identification of compounds and to quantitative analysis.

Qualitative TLC

The data from a single chromatogram usually do not provide sufficient information to permit identification of the various species present in a mixture because of the variability of R_F values with sample size, the thin-layer plate, and the development conditions. In addition, there is always the possibility that two quite different solutes may exhibit identical or nearly identical R_F values under a given set of conditions.

Identification Methods. Sometimes, tentative identification can be made by applying to the plate solutions of purified species suspected to be present in the unknown. A match in R_F values provides evidence as to the identity of one of the components. In other cases, the identity of separated species can be confirmed by a scraping-and-dissolution technique. Here, the area containing the analyte is scraped from the plate with a razor blade or spatula and the contents collected on a piece of glazed paper. After transfer to a suitable container, the analyte is dissolved in an appropriate solvent and separated from the stationary phase by centrifugation or filtration. Spectroscopic techniques, such as fluorescence, UV-visible absorption, nuclear magnetic resonance, or FTIR, are then used to identify the species. Thin-layer chromatography has also been on-line coupled to a number of spectrometric detection techniques. The coupling to TLC to mass spectrometers is an active area of research.[37]

Two-Dimensional Planar Chromatography. Figure 28-31 illustrates the separation of amino acids in a mixture by development in two dimensions. The sample was placed in one corner of a square plate and development was performed in the ascending direction with solvent A. This solvent was then removed by evaporation, and the plate was rotated 90°, following which an ascending development with solvent B was performed. After solvent removal, the positions of the amino acids were determined by spraying with ninhydrin, a reagent that forms a pink to purple product with amino acids. The spots were identified by comparison

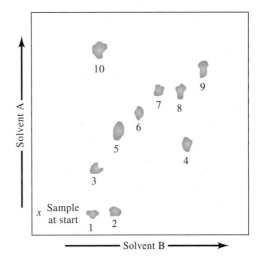

FIGURE 28-31 Two-dimensional thin-layer chromatogram (silica gel) of some amino acids. Solvent A: toluene, 2-chloroethanol, pyridine. Solvent B: chloroform, benzyl alcohol, acetic acid. Amino acids: (1) aspartic acid, (2) glutamic acid, (3) serine, (4) β-alanine, (5) glycine, (6) alanine, (7) methionine, (8) valine, (9) isoleucine, (10) cysteine.

of their positions with those of standards. Imaging techniques have been used for quantitative analysis of two-dimensional separations.[38]

Quantitative Analysis

A semiquantitative estimate of the amount of a component present can be obtained by comparing the area of a spot with that of a standard. More accurate results can be obtained by scraping the spot from the plate, extracting the analyte from the stationary-phase solid, and measuring the analyte by a suitable physical or chemical method. In a third method, a scanning densitometer can be used to measure fluorescence or absorption of the spot.

[37] J. Sharma, *Anal. Chem.*, **2006**, *78*, 3841.

[38] M. Medic-Saric et al., *J. Planar. Chromatogr.-Mod. TLC*, **2004**, *17*, 459.

QUESTIONS AND PROBLEMS

*Answers are provided at the end of the book for problems marked with an asterisk.

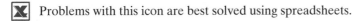 Problems with this icon are best solved using spreadsheets.

28-1 List the types of substances to which each of the following chromatographic methods is most applicable:
(a) gas-liquid
(b) liquid adsorption

(c) liquid-liquid partition
(d) reversed-phase partition
(e) ion exchange
(f) gel permeation
(g) gas-solid
(h) gel filtration
(i) ion-pair

28-2 Describe three general methods for improving resolution in partition chromatography.

28-3 Describe a way to manipulate the retention factor of a solute in partition chromatography.

28-4 How can the selectivity factor be manipulated in (a) gas chromatography and (b) LC?

28-5 In preparing a hexane-acetone gradient for an alumina HPLC column, is it desirable to increase or decrease the proportion of hexane as the column is eluted?

28-6 What is meant by the linear-response range of a detector?

28-7 Define
(a) isocratic elution.
(b) gradient elution.
(c) stop-flow injection.
(d) reversed-phase packing.
(e) normal-phase packing.
(f) ion-pairing chromatography.
(g) ion chromatography.
(h) bulk property detector.
(i) solute property detector.
(j) sparging.

28-8 What is a guard column in partition chromatography?

28-9 In what way are normal-phase partition chromatography and adsorption chromatography similar?

28-10 What is the order in which the following compounds would be eluted from an HPLC column containing a reversed-phase packing?
(a) benzene, diethyl ether, *n*-hexane
(b) acetone, dichloroethane, acetamide

*28-11 What is the order of elution of the following compounds from a normal-phase packed HPLC column?
(a) ethyl acetate, acetic acid, dimethylamine
(b) propylene, hexane, benzene, dichlorobenzene

28-12 Describe the fundamental difference between adsorption and partition chromatography.

28-13 Describe the fundamental difference between ion-exchange and size-exclusion chromatography.

28-14 What types of species can be separated by HPLC but not by GC?

28-15 Describe the various kinds of pumps used in HPLC. What are the advantages and disadvantages of each?

28-16 Describe the differences between single-column and suppressor-column ion chromatography.

28-17 Mass spectrometry is an extremely versatile detection system for gas chromatography. However, interfacing an HPLC system to a mass spectrometer is a much more difficult task. Describe the major reasons why it is more difficult to combine HPLC with mass spectrometry than it is to combine GC with mass spectrometry.

28-18 Which of the GC detectors in Table 27-1 are suitable for HPLC? Why are some of the detectors listed unsuitable for HPLC?

28-19 Although temperature does not have nearly the effect on HPLC separations that it has on GC separations, it nonetheless can play an important role. Discuss how and why temperature might or might not influence the following separations:
(a) a reversed-phase chromatographic separation of a steroid mixture.
(b) an adsorption chromatographic separation of a mixture of closely related isomers.

✖ *28-20** Two components in an HPLC separation have retention times that differ by 15 s. The first peak elutes in 9.0 min and the peak widths are approximately equal. The dead time t_M was 65 s. Use a spreadsheet to find the minimal number of theoretical plates needed to achieve the following resolution R_s values: 0.50, 0.75, 0.90, 1.0, 1.10, 1.25, 1.50, 1.75, 2.0, 2.5. How would the results change if peak 2 were twice as broad as peak 1?

✖ *28-21** An HPLC method was developed for the separation and determination of ibuprofen in rat plasma samples as part of a study of the time course of the drug in laboratory animals. Several standards were chromatographed and the following results obtained:

Ibuprofen Concentration, μg/mL	Relative Peak Area
0.5	5.0
1.0	10.1
2.0	17.2
3.0	19.8
6.0	39.7
8.0	57.3
10.0	66.9
15.0	95.3

Next, a 10 mg/kg sample of ibuprofen was administered orally to a laboratory rat. Blood samples were drawn at various times after administration of the drug and subjected to HPLC analysis. The following results were obtained:

Time, h	Peak Area
0	0
0.5	91.3
1.0	80.2
1.5	52.1
2.0	38.5
3.0	24.2
4.0	21.2
6.0	18.5
8.0	15.2

Find the concentration of ibuprofen in the blood plasma for each of the times given above and plot the concentration versus time. On a percentage basis, during what half-hour period (first, second, third, etc.) is most of the ibuprofen lost?

*28-22 In a normal-phase partition column, a solute was found to have a retention time of 29.1 min, and an unretained sample had a retention time of 1.05 min when the mobile phase was 50% by volume chloroform and 50% n-hexane. Calculate (a) k for the solute and (b) a solvent composition that would bring k down to a value of about 10.

Challenge Problem

28-23 Assume for simplicity that the HPLC plate height, H, can be given by Equation 26-23 as

$$H = \frac{B}{u} + C_S u + C_M u = \frac{B}{u} + Cu$$

where $C = C_S + C_M$.

(a) By using calculus to find the minimum H, show that the optimal velocity u_{opt} can be expressed as

$$u_{opt} = \sqrt{\frac{B}{C}}$$

(b) Show that this leads to a minimum plate height H_{min} given by

$$H_{min} = 2\sqrt{BC}$$

(c) Under some conditions for chromatography, C_S is negligible compared to C_M. For packed HPLC columns, C_M is given by

$$C_M = \frac{\omega d_p^2}{D_M}$$

where ω is a dimensionless constant, d_p is the particle size of the column packing, and D_M is the diffusion coefficient in the mobile phase. The B coefficient can be expressed as

$$B = 2\gamma D_M$$

where γ is also a dimensionless constant. Express u_{opt} and H_{min} in terms of D_M, d_p, and the dimensionless constants γ and ω.

(d) If the dimensionless constants are on the order of unity, show that u_{opt} and H_{min} can be expressed as

$$u_{opt} \approx \frac{D_M}{d_p} \quad \text{and} \quad H_{min} \approx d_p$$

(e) Under the preceding conditions, how could the plate height be reduced by one third? What would happen to the optimal velocity under these conditions? What would happen to the number of theoretical plates N for the same length column?

(f) For the conditions in part (e), how could you maintain the same number of theoretical plates while reducing the plate height by one third?

(g) The preceding discussion assumes that all band broadening occurs within the column. Name two sources of extracolumn band broadening that might also contribute to the overall width of HPLC peaks.

CHAPTER TWENTY-NINE

Supercritical Fluid Chromatography and Extraction

Two techniques based on the use of super-critical fluids were developed during the 1970s and early 1980s and today play an important role in the analysis of environmental, biomedical, and food samples. These new methods are **supercritical fluid chromatography** (SFC) and **supercritical fluid extraction** (SFE). Commercial instrumentation for both of these techniques became available in the mid-1980s, and their use has grown rapidly in the analytical community. In this chapter we describe the theory, instrumentation, and applications of both SFC and SFE.

Throughout this chapter, this logo indicates an opportunity for online self-study at **www.thomsonedu.com/chemistry/skoog**, linking you to interactive tutorials, simulations, and exercises.

29A PROPERTIES OF SUPERCRITICAL FLUIDS

A *supercritical fluid* is formed whenever a substance is heated above its critical temperature. The *critical temperature* of a substance is the temperature above which a distinct liquid phase cannot exist, regardless of pressure. The vapor pressure of a substance at its critical temperature is its *critical pressure*. At temperatures and pressures above its critical temperature and pressure (its critical point), a substance is called a *supercritical fluid*. Supercritical fluids have densities, viscosities, and other properties that are intermediate between those of the substance in its gaseous and liquid states. Table 29-1 compares some properties of supercritical fluids to those of typical gases and liquids. These properties are important in gas, liquid, and supercritical fluid chromatography and extractions.

Table 29-2 lists properties of four out of perhaps two dozen compounds that have been used as mobile phases in SFC. Note that their critical temperatures, and the pressures at these temperatures, are well within the operating conditions of ordinary high-performance liquid chromatography (HPLC).

An important property of supercritical fluids, and one that is related to their high densities (0.2 to 0.5 g/cm^3), is their ability to dissolve large nonvolatile molecules. For example, supercritical carbon dioxide readily dissolves *n*-alkanes containing from five to twenty-two carbon atoms, di-*n*-alkylphthalates in which the alkyl groups contain four to sixteen carbon atoms, and various polycyclic aromatic hydrocarbons consisting of several rings. Certain important industrial processes are based on the high solubility of organic species in supercritical carbon dioxide. For example, this medium has been employed for extracting caffeine from coffee beans to give decaffeinated coffee and for extracting nicotine from cigarette tobacco.

A second important property of many supercritical fluids is that analytes dissolved in them can be easily recovered by simply allowing the solutions to equilibrate with the atmosphere at relatively low temperatures. Thus, an analyte dissolved in supercritical carbon dioxide, the most commonly used solvent, can be recovered by simply reducing the pressure and allowing the fluid to evaporate under ambient laboratory conditions. This property is particularly important with thermally unstable analytes. Another advantage of many supercritical fluids is that they are inexpensive, innocuous, and generally nontoxic substances. Supercritical fluid car-

TABLE 29-1 Comparison of Properties of Supercritical Fluids with Liquids and Gases

Property	Gas (STP)	Supercritical Fluid	Liquid
Density, g/cm^3	$(0.6-2) \times 10^{-3}$	$0.2-0.5$	$0.6-2$
Diffusion coefficient, cm^2/s	$(1-4) \times 10^{-1}$	$10^{-3}-10^{-4}$	$(0.2-2) \times 10^{-5}$
Viscosity, $g\ cm^{-1}\ s^{-1}$	$(1-3) \times 10^{-4}$	$(1-3) \times 10^{-4}$	$(0.2-3) \times 10^{-2}$

Note: All data are to an order of magnitude only.

bon dioxide is especially attractive for extractions and chromatography. Finally, supercritical fluids have the advantage that solute diffusivities are an order of magnitude higher than in liquid solvents and viscosities are an order of magnitude lower. These last two advantages are important in both chromatography and extractions with supercritical fluids.

29B SUPERCRITICAL FLUID CHROMATOGRAPHY

SFC, in which the mobile phase is a supercritical fluid, is a hybrid of gas chromatography (GC) and liquid chromatography (LC) that combines some of the best features of each. For certain applications, it is superior to both GC and HPLC.[1] In 1985 several instrument manufacturers began to offer equipment specifically designed for SFC, and the use of such equipment is expanding at a rapid pace. The SFC technique has become an important tool in many industrial, regulatory, and academic laboratories.

---

[1] For additional information on SFC, see *Practical Supercritical Fluid Chromatography and Extraction*, M. Caude and D. Thiebaut, eds., Amsterdam: Harwood, 2000; *Supercritical Fluid Chromatography with Packed Columns, Techniques and Applications*, K. Anton and C. Berger, eds., New York: Dekker, 1998; L. Taylor, in *Handbook of Instrumental Techniques for Analytical Chemistry*, F. Settle, ed., Chap. 11, Upper Saddle River, NJ: Prentice Hall, 1997.

SFC is important because it permits the separation and determination of a group of compounds not conveniently handled by either GC or LC. These compounds (1) are either nonvolatile or thermally unstable so that GC procedures are inapplicable and (2) contain no functional groups that make possible detection by the spectroscopic or electrochemical techniques used in HPLC. Chester[2] has estimated that up to 25% of all separation problems faced by scientists today involve mixtures containing such intractable species.

29B-1 Instrumentation and Operating Variables

As mentioned earlier, the pressures and temperatures required for creating supercritical fluids derived from several common gases and liquids lie well within the operating limits of ordinary HPLC equipment. Thus, as shown in Figure 29-1, instruments for SFC are similar in most aspects to the instruments for HPLC described in Section 28C.[3] There are two important differences between the two techniques, however. First, a thermostatted column oven, similar to that used in GC (Section 27B-3), is required to provide precise temperature control of the mobile phase; second, a restrictor,

---

[2] T. L. Chester, *J. Chromatogr. Sci.*, **1986**, *24*, 226.

[3] For descriptions of commercial instrumentation for SFC, see C. M. Harris, *Anal. Chem.*, **2002**, *74*, 87A; B. Erikson, *Anal. Chem.*, **1997**, *69*, 683A.

TABLE 29-2 Properties of Some Supercritical Fluids

Fluid	Critical Temperature, °C	Critical Pressure, atm	Critical Point Density, g/mL	Density at 400 atm, g/mL
CO_2	31.3	72.9	0.47	0.96
N_2O	36.5	71.7	0.45	0.94
NH_3	132.5	112.5	0.24	0.40
n-Butane	152.0	37.5	0.23	0.50

Source: From M. L. Lee and K. E. Markides, *Science*, **1987**, *235*, 1345, with permission. Data taken from *Matheson Gas Data Book* and *CRC Handbook of Chemistry and Physics*.

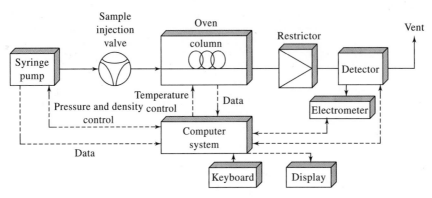

FIGURE 29-1 Block diagram of an instrument for SFC.

or back-pressure device, is used to maintain the desired pressure in the column and to convert the eluent from a supercritical fluid to a gas for transfer to the detector. A typical restrictor for a 50- or 100-μm open tubular column consists of a 2- to 10-cm length of 5- to 10-μm capillary tubing attached directly to the end of the column. This permits the use of interchangeable restrictors having different inside diameters, thus providing a range of flow rates at any given pumping pressure. Alternatively, the restrictor may be an integral part of the column formed by drawing down the end of the column in a flame.

As shown in Figure 29-1, in a commercial instrument for SFC, such instrument variables as pumping pressure, oven temperature, and detector performance are computer controlled.

Effects of Pressure

Pressure changes in SFC have a pronounced effect on the retention factor k and thus the retention time t_R. The density of a supercritical fluid increases rapidly and nonlinearly with increases in pressure. Such density increases cause a rise in the solvent power of the mobile phase, which in turn shortens elution time. For example, the elution time for hexadecane is reported to decrease from 25 to 5 min as the pressure of carbon dioxide is raised from 70 to 90 atm. An effect similar to that of temperature programming in GC and gradient elution in HPLC can be achieved by linearly increasing the column pressure or by regulating the pressure to create linear density increases. Figure 29-2 illustrates the improvement in chromatograms realized by pres-

Simulation: Learn more about **supercritical fluid chromatography**.

sure programming. The decompression of fluids as they travel through the column can cause temperature changes that can affect separations and thermodynamic measurements. The most common pressure profiles used in SFC are often constant (isobaric) for a given length of time followed by a linear or asymptotic increase to a final pressure. In addition to pressure programming, temperature programming and mobile-phase gradients can be used.

Sample:	1. cholesteryl octanoate
	2. cholesteryl decylate
	3. cholesteryl laurate
	4. cholesteryl myristate
	5. cholesteryl palmitate
	6. cholesteryl stearate
Column:	DB-1
Mobile phase:	CO_2
Temperature:	90°C
Detector:	FID Flame ionization

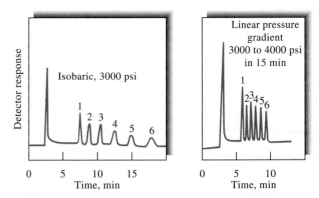

FIGURE 29-2 Effect of pressure programming in SFC. Note the shorter time for the pressure-gradient chromatogram on the right compared with the constant pressure (isobaric) chromatogram on the left. (Courtesy of Brownlee Labs, Santa Clara, CA.)

Stationary Phases

Both packed columns and open tubular columns are used in SFC. Packed columns can provide more theoretical plates and handle larger sample volumes than open tubular columns. Because of the low viscosity of supercritical media, columns can be much longer than those used in LC, and column lengths of 10 to 20 m and inside diameters of 50 or 100 μm are common with open tubular columns. For difficult separations, columns 60 m or longer have been used. Open tubular columns are similar to the fused-silica wall-coated (FSWC) open tubular columns described in Section 27C-1. Packed columns are usually made of stainless steel, 10–25 cm long. More than 100,000 plates have been achieved with packed columns.

Many of the column coatings used in LC have been applied to SFC as well. Typically, these are polysiloxanes (see Section 27C-3) chemically bonded to the surface of silica particles or to the inner silica wall of capillary tubing. Film thicknesses are 0.05 to 0.4 μm.

Mobile Phases

The most widely used mobile phase for SFC is carbon dioxide. It is an excellent solvent for a variety of nonpolar organic molecules. In addition, it transmits in the ultraviolet region and is odorless, nontoxic, readily available, and remarkably inexpensive compared to other chromatographic solvents. Its critical temperature of 31°C and its critical pressure of 72.9 atm permit a wide selection of temperatures and pressures without exceeding the operating limits of modern HPLC equipment. In some applications, polar organic modifiers, such as methanol, are introduced in small concentrations (~1%) to modify α values for analytes.

Ethane, pentane, dichlorodifluoromethane, diethyl ether, ammonia, and tetrahydrofuran have also served as mobile phases in SFC.

Detectors

A major advantage of SFC over HPLC is that the flame ionization detector (FID) of GC can be used. As discussed in Section 27B-4, the FID exhibits a *general response* to organic compounds, is highly sensitive, and is largely trouble free. Mass spectrometers are also more easily adapted as detectors for SFC than HPLC. Several of the detectors used in LC find use in SFC as well, including UV and IR absorption, fluorescence emission, thermionic, and flame photometric detectors.

29B-2 Comparison to Other Types of Chromatography

The data in Tables 29-1 and 29-2 reveal that several physical properties of supercritical fluids are intermediate between gases and liquids. Hence, this type of chromatography combines some of the characteristics of both GC and LC. For example, like GC, SFC is inherently faster than LC because the lower viscosity makes possible the use of higher flow rates. Diffusion rates in supercritical fluids are intermediate between those in gases and in liquids. As a result, band broadening is greater in supercritical fluids than in liquids but less than in gases. Thus, the intermediate diffusivities and viscosities of supercritical fluids should in theory produce faster separations than are achieved with LC and be accompanied by lower zone spreading than found in GC.

Figures 29-3 and 29-4 compare the performance characteristics of a packed column when elution is performed with supercritical carbon dioxide and a conventional liquid mobile phase. Figure 29-3 shows that at a linear mobile-phase velocity of 0.6 cm/s, the supercritical column yields a plate height of 0.013 mm whereas

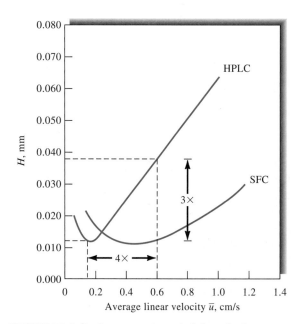

FIGURE 29-3 Performance characteristics of a 5-μm octadecyl (C_{18}) bonded silica column when elution is carried out with a conventional mobile phase (HPLC) and supercritical carbon dioxide SFC. (From D. R. Gere, *Application Note 800-3*, Hewlett-Packard Corp., Palo Alto, CA, 1983. With permission.)

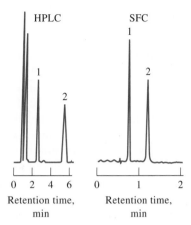

FIGURE 29-4 Comparison of chromatograms obtained by conventional partition chromatography (HPLC) and SFC. Column: 20 cm × 4.6 mm with 10-μm reversed-phase bonded packing. Analytes: (1) biphenyl; (2) terphenyl. For HPLC: mobile phase, 65% CH_3OH to 35% H_2O; flow rate, 4 mL/min; linear velocity, 0.55 cm/s; sample size, 10 μL. For SFC: mobile phase, CO_2; flow rate, 5.4 mL/min; linear velocity, 0.76/s; sample size, 3 μL. (From D. R. Gere, T. J. Stark, and T. N. Tweeten, *Application Note 800-4*, Hewlett-Packard Corp., Palo Alto, CA, 1983. With permission.)

the plate height with a liquid eluent is three times as large, or 0.039 mm. Thus, a reduction in peak width by a factor of $\sqrt{3}$ should be realized with SFC. Alternatively, the linear velocity can be increased by a factor of 4 at the plate height corresponding to the minimum in the HPLC curve; this increase would result in a reduction of analysis time by a factor of 4. These advantages are reflected in the two chromatograms shown in Figure 29-4.

The mobile phase plays different roles in GC, LC, and SFC. Ordinarily, in GC the mobile phase serves but one purpose — zone movement. As we have seen in Chapter 28, in LC the mobile phase provides not only transport of solute molecules but also interactions with solutes that influence selectivity factors (α values). When a molecule dissolves in a supercritical medium, the process resembles volatilization but at a much lower temperature than would normally be used in GC. Thus, at a given temperature, the vapor pressure for a large molecule in a supercritical fluid may be 10^{10} times greater than in the absence of the fluid. Because of this, high-molecular-mass compounds, thermally unstable species, polymers, and large biological molecules can be eluted from a column at relatively low temperatures. Interactions between solute molecules and the molecules of a supercritical fluid must occur to account for their solubility in these media. The

solvent power is thus a function of the chemical composition and the density of the fluid. Therefore, in contrast to GC, it is possible to vary α values by changing the mobile phase.

Figure 29-5 compares the application range of SFC with GC, LC, and size-exclusion chromatography (SEC). Note that LC and SFC are applicable over molecular mass ranges that are several orders of magnitude greater than GC. As we have noted earlier, SEC can be applied to even larger molecules.

29B-3 Applications

The SFC technique has been applied to a wide variety of materials, including natural products, drugs, foods, pesticides and herbicides, surfactants, polymers and polymer additives, fossil fuels, and explosives and propellants. One of the most important applications of SFC is to chiral separations, particularly in the pharmaceutical industry.[4] The increased diffusion rates of SFC over LC eluents leads to higher resolution and sharper peaks to enable accurate measurements of enantiomeric purity. Figure 29-6 shows the separation of the enantiomers of metoprolol, a beta-blocking drug, by both HPLC and SFC. Although both methods achieve complete resolution of the enantiomers, the separation requires only 6 min in SFC compared to 22 min in HPLC. Note also the higher resolution and more symmetric peaks obtained by SFC. In addition to higher resolution, method development time can also

---

[4]K. W. Phinney, *Anal. Chem.*, **2000**, *72*, 204A.

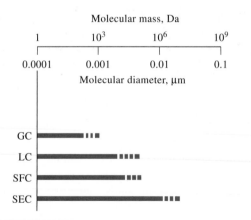

FIGURE 29-5 Range of molecular masses and sizes over which column chromatographic techniques can be applied. (From M. L. Lee and K. E. Markides, *Science*, **1987**, 235. With permission.)

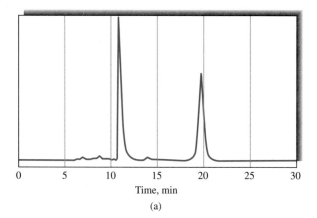

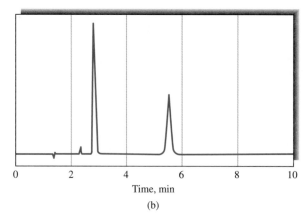

FIGURE 29-6 Separation of enantiomers of metoprolol by HPLC (a) and SFC (b) on a Chiralcel OD stationary phase. In (a) the mobile phase was 20% 2-propanol in hexane with 1% diethylamine; selectivity $\alpha = 2.67$ and resolution $R_s = 4.8$. In (b), CO_2 was used containing 20% methanol with 0.5% isopropylamine; $\alpha = 2.77$ and $R_s = 12.7$. (Adapted from M. S. Villeneuve and R. J. Anderegg, *J. Chromatogr. A*, **1998**, *826*, 217.)

be decreased with SFC compared to HPLC methods because the desired resolution for enantiomers can often be achieved without investigating as many stationary phases and mobile phases as in HPLC. In cases where a single column is insufficient to produce the desired separation, the lower pressure drop in SFC makes it easier to couple columns with different stationary phases. Preparative chiral separations by SFC are very attractive because the mobile phase can be readily removed by simply venting the CO_2.

SFC is also very useful for achiral separations. Figures 29-7, 29-8, and 29-9 illustrate three typical and diverse applications of SFC. Figure 29-7 shows the separation of a series of dimethylpolysiloxane oligomers

ranging in molecular mass from 400 to 700 Da. This chromatogram was obtained by using a 10 m $\times$ 100 μm inside-diameter fused-silica capillary coated with a 0.25-μm film of 5% phenylpolysiloxane. The mobile phase was CO_2 at 140°C, and the following pressure program was used: 80 atm for 20 min, then a linear gradient from 80 to 280 atm at 5 atm/min. A flame ionization detector was used.

Figure 29-8 illustrates the separation of polycyclic aromatic hydrocarbons extracted from a carbon black. Detection was by fluorescence excited at two different wavelengths. Note the selectivity provided by this technique. The chromatogram was obtained by using a 40 m $\times$ 50 μm inside-diameter capillary coated with a 0.25-μm film of 50% phenylpolysiloxane. The mobile phase was pentane at 210°C, and the following program was used: initial mobile-phase density held at

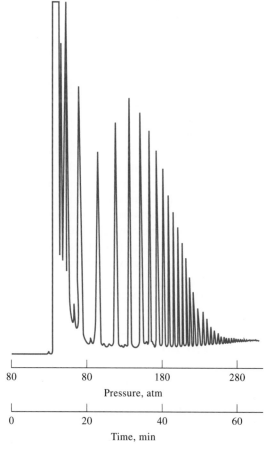

FIGURE 29-7 Separation of oligomers of dimethylpolysiloxane by SFC. (From C. M. White and R. K. Houck, *HRC&CC*, **1986**, *9*, 4. With permission.)

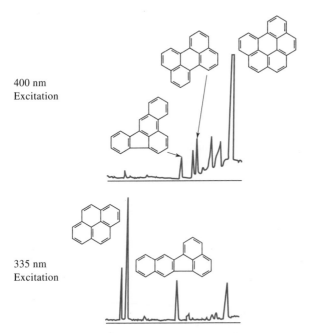

400 nm
Excitation

335 nm
Excitation

FIGURE 29-8 Portions of the supercritical fluid chromatograms of polycyclic aromatics in a carbon-black extract, illustrating the selectivity achieved by fluorescence excitation at two wavelengths. (From C. M. White and R. K. Houck, *HRC&CC*, **1986**, *9*, 4. With permission.)

0.07 g/mL for 24 min, then an asymptotic density program to 0.197 g/mL.

Figure 29-9 illustrates a separation of the oligomers in a sample of the nonionic surfactant Triton X-100. Detection involved measuring the total ion current produced by chemical ionization mass spectrometry. The mobile phase was carbon dioxide containing 1% by volume of methanol. The column was a 30-m capillary column coated with a 1-μm film of 5% phenylpolysiloxane. The column pressure was increased linearly at a rate of 2.5 bar/min.

Several reviews on applications of SFC are available in the recent literature.[5]

29C   SUPERCRITICAL FLUID EXTRACTION

Often, the analysis of complex materials requires as a preliminary step separation of the analyte or analytes from a sample matrix. Ideally, an analytical separation

method should be rapid, simple, and inexpensive; should give quantitative recovery of analytes without loss or degradation; should yield a solution of the analyte that is sufficiently concentrated to permit the final measurement to be made without the need for concentration; and should generate little or no laboratory wastes.

For many years, one of the most common methods for performing analytical separations on complex environmental, pharmaceutical, food, and petroleum samples was based on extraction of bulk samples with hydrocarbon or chlorinated organic solvents using a Soxhlet extractor. Unfortunately, liquid extractions frequently fail to meet several of the ideal criteria listed in the previous paragraph. They usually require several hours or more to achieve satisfactory recoveries of analytes, and sometimes never do. The solvent costs are often high. The solution of the recovered analyte is often so dilute that a concentration step must follow the extraction. Analyte degradation or loss as well as atmospheric pollution may accompany this concentration step.

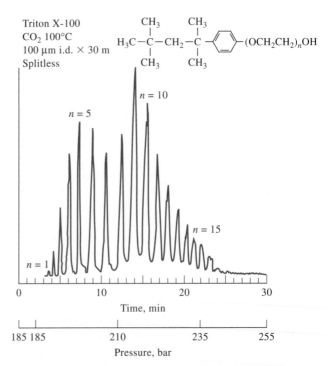

FIGURE 29-9 Chromatograms for the nonionic surfactant Triton X-100 with total current mass spectrometric detection. i.d. = inside diameter. (Reprinted with permission from R. D. Smith and H. R. Udseth, *Anal. Chem.*, **1987**, *59*, 17. Copyright 1981 American Chemical Society.)

[5]M. C. Henry and C. R. Yonker, *Anal. Chem.*, **2006**, *78*, 3909; T. L. Chester and J. D. Pinkston, *Anal. Chem.*, **2004**, *76*, 4606; T. L. Chester and J. D. Pinkston, *Anal. Chem.*, **2002**, *74*, 2901; T. L. Chester and J. D. Pinkston **2000**, *72*, 129R; T. L. Chester, J. D. Pinkston, and D. B. Raynie, *Anal. Chem.*, **1998**, *70*, 301R.

Solid-phase extraction, or liquid-solid extraction, can overcome several of these problems.[6] Solid-phase extraction techniques use membranes or small, disposable, syringe-barrel columns or cartridges. A hydrophobic organic compound is coated or chemically bonded to powdered silica to form the solid extracting phase. The compounds can be nonpolar, moderately polar, or polar. For example, an octadecyl (C_{18}) bonded silica is a common packing. The functional groups bonded to the packing attract hydrophobic compounds in the sample by van der Waals interactions and extract them from the aqueous solution.

SFE is another way to avoid many of the problems that plague organic liquid extractants. Beginning in the mid-1980s, chemists began to investigate the use of supercritical fluids for separating analytes. Such extractions were shown to have many advantages as discussed in the following sections.[7]

29C-1 Advantages of SFE

Some of the advantages of SFE include the following:

1. SFE is generally fast. The rate of mass transfer between a sample matrix and an extraction fluid is determined by the rate of diffusion of a species in the fluid and the viscosity of the fluid — the greater the diffusion rate and the lower the viscosity, the greater the rate of mass transfer. As we have noted earlier, both of these variables are more favorable for supercritical fluids than for typical liquid solvents. Because of this, SFE can generally be completed in 10 to 60 minutes, whereas extractions with an organic liquid may require several hours or even days.

2. The solvent strength of a supercritical fluid can be varied by changes in the pressure and to a lesser extent in the temperature. In contrast, the solvent strength of an organic liquid is essentially constant regardless of conditions. This property allows the

conditions for extraction with a supercritical fluid to be optimized for a given class of analytes.

3. Many supercritical fluids are gases at ambient conditions. Thus, recovery of analytes is simple compared to organic liquids, which must be vaporized by heating, leading to possible decomposition of thermally unstable analytes or loss of volatile analytes. In contrast, a supercritical fluid can be separated from the analyte by simply releasing pressure. Alternatively, the analyte stream can be bubbled through a small vial containing a good solvent for the analyte, which will dissolve in the small volume of solvent.

4. Some supercritical fluids are cheap, inert, and nontoxic. Thus, they are readily disposed of after an extraction is completed by allowing them to evaporate into the atmosphere.

29C-2 Instrumentation

The instrumentation for SFE can be relatively simple as shown in Figure 29-10. Instrument components include a fluid source, most commonly a tank of carbon dioxide; a syringe pump having a pressure rating of at least 400 atm and a flow rate for the pressurized fluid of at least 2 mL/min; a valve to control the flow of the critical fluid into a heated extraction cell having a capacity of a few milliliters; and an exit valve leading to a flow restrictor that depressurizes the fluid and transfers it into a collection device. In the simplest instruments, the flow restrictor is 10 to 50 cm of capillary tubing. In modern sophisticated commercial instruments, the restrictors are variable and controlled manually or automatically. Several instrument manufacturers offer various types of SFE apparatus.[8]

An SFE system can be operated in one of two ways. In the dynamic extraction mode, the valve between the extraction cell and the restrictor remains open so that the sample is continually supplied with fresh supercritical fluid and the extracted material flows into the collection vessel where depressurization occurs. In the static extraction mode, the valve between the extraction cell and the restrictor is closed and the extraction cell is pressurized under static conditions. After a suitable period, the exit valve is opened and the cell contents are transferred through the restrictor by a dynamic flow of fluid from the pump. The dynamic mode is more widely used than the static mode.

---

[6]For more information see D. A. Skoog, D. M. West, F. J. Holler, and S. R. Crouch, *Fundamentals of Analytical Chemistry*, 8th ed., Belmont, CA: Brooks/Cole, 2004, Chap. 30.

[7]See *Supercritical Fluids as Solvents and Reaction Media*, G. Brunner, ed., Amsterdam: Elsevier, 2004, Chap. 4; *Practical Supercritical Fluid Chromatography and Extraction*, M. Caude and D. Thiebaut, eds., Amsterdam: Harwood, 2000; *Supercritical Fluids in Chromatography and Extraction*, R. M. Smith and S. B. Hawthorne, eds., Amsterdam: Elsevier, 1997; L. Taylor, in *Handbook of Instrumental Techniques for Analytical Chemistry*, F. Settle, ed., Chap. 11, Upper Saddle River, NJ: Prentice-Hall, 1997; L. T. Taylor, *Supercritical Fluid Extraction*, New York: Wiley, 1996.

[8]See B. E. Erickson, *Anal. Chem.*, **1998**, *70*, 333A; F. Wach, *Anal. Chem.*, **1994**, *66*, 369A.

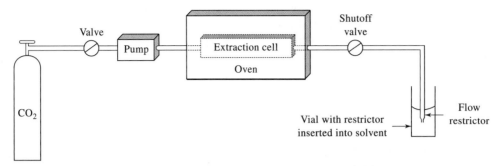

FIGURE 29-10 A typical arrangement for off-line SFE. The shutoff valve is required for static SFE but not dynamic SFE.

29C-3 Supercritical Fluid Choice

Two dozen or more supercritical fluids have been described as extraction media, but as in SFC, by far the most widely used substance is carbon dioxide alone or containing an organic modifier. The best choice of fluid is determined by a number of variables, including polarity and solubility of the analytes and the matrix components, physical nature of the matrix, concentration of the analytes, moisture content of the sample, and kinetic considerations.[9] Unfortunately, the current theory of SFE is not well developed, and final conditions in most cases must be determined empirically. Carbon dioxide has been the fluid of choice in most studies. It is an excellent solvent for nonpolar species, such as alkanes and terpenes, and a decent extraction medium for moderately polar species, such as polycyclic aromatic hydrocarbons, polychlorinated biphenyls, aldehydes, esters, alcohols, organic chloro pesticides, and fats. It is generally not a good extraction medium for highly polar compounds unless strongly polar modifiers such as methanol are added. Modifiers can be introduced into the extraction system either by a second pump or by injection into the sample prior to extraction.

A variety of modifiers have been used to enhance the polarity of supercritical fluid carbon dioxide, including several of the lower-molecular-mass alcohols, propylene carbonate, 2-methoxyethanol, methylene chloride, and certain organic acids. The most common modifier is methanol. Figure 29-11 demonstrates the improved efficiency achieved by the presence of a small amount of methanol in the extraction of various materials from soil samples.

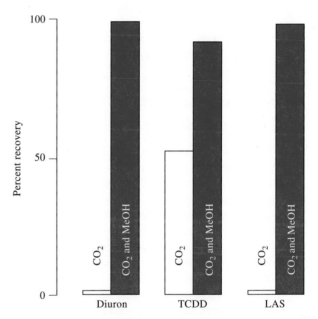

FIGURE 29-11 Comparison of extraction efficiencies obtained by using CO_2 and CO_2 modified with methanol. A soil sample was used. All extractions were for 30 min. Diuron is a common herbicide that is an aromatic substituted derivative of urea. TCDD is 2,3,7,8-tetrachloro-dibenzo-*p*-dioxin. LAS is linear alkylbenzenesulfonate detergent.

29C-4 Off-Line and On-Line Extractions

Two types of methods have been used to collect analytes after extraction: *off-line* and *on-line*. In off-line collection, which is the simpler of the two, the analytes are collected by immersing the restrictor in a few milliliters of solvent and allowing the gaseous supercritical fluid to escape into the atmosphere (see Fig-

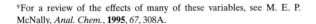

[9]For a review of the effects of many of these variables, see M. E. P. McNally, *Anal. Chem.*, **1995**, *67*, 308A.

TABLE 29-3 Some Typical Analytical Applications of SFE

Material	Analyte*	Supercritical Fluid	Extraction Time, min	Off-line (1), On-line (2)
Soils	Pesticides	CO_2	20	1
River sediments	PAHs	CO_2/5% MeOH	120	1
Smoke, urban dust	PAHs	CO_2	15	2
Railroad bed soil	PCBs, PAHs	CO_2/MeOH	45	1
Foods	Fats	CO_2/MeOH	12	1
Spices, bubble gum	Aromas and fragrances	CO_2	10	2
Serum	Cholesterol	CO_2	30	1
Coal, fly ash	PCBs, dioxins	CO_2	15	2
Polymers	Additives and oligomers	CO_2	15	2
Animal tissue	Drug residues	CO_2	9	1

*PAHs = polycyclic aromatic hydrocarbons; PCBs = polychlorinated biphenyls.

ure 29-10). Analytes have also been collected on adsorbents, such as silica. The adsorbed analytes are then eluted with a small volume of a liquid solvent. In either case the separated analytes are then identified by any of several optical, electrochemical, or chromatographic methods.

In the on-line method, the effluent from the restrictor, after depressurization, is transferred directly to a chromatograph system. In most cases the latter is a GC or an SFC instrument, although occasionally an HPLC system has been used. The principal advantages of an on-line system are the elimination of sample handling between the extraction and the measurement and the potential for enhanced sensitivity because no dilution of the analyte occurs.

29C-5 Typical Applications of SFE

Hundreds of applications of SFE have appeared in the literature. Most applications are for the analysis of environmental samples. Others have been for the analysis of foods, biomedical samples, and industrial samples. Table 29-3 provides a few typical applications of off-line and on-line SFE. In addition to the analytical uses of SFE, there are many preparative and isolation uses in the pharmaceutical and polymer industries.

QUESTIONS AND PROBLEMS

*Answers are provided at the end of the book for problems marked with an asterisk.

☒ Problems with this icon are best solved using spreadsheets.

29-1 Define
   (a) critical temperature and critical pressure of a gas.
   (b) supercritical fluid.

29-2 What properties of a supercritical fluid are important in chromatography?

29-3 How do instruments for SFC differ from those for (a) HPLC and (b) GC?

29-4 Describe the effect of pressure on supercritical fluid chromatograms.

29-5 List some of the advantageous properties of supercritical CO_2 as a mobile phase for chromatographic separations.

29-6 Compare SFC with other column chromatographic methods.

29-7 For supercritical carbon dioxide, predict the effect that the following changes will have on the elution time in an SFC experiment:
(a) increasing the flow rate (at constant temperature and pressure).
(b) increasing the pressure (at constant temperature and flow rate).
(c) increasing the temperature (at constant pressure and flow rate).

29-8 For SFE, differentiate between
(a) on-line and off-line processes.
(b) static and dynamic extractions.

29-9 List the advantages and any disadvantages of SFE compared to liquid-liquid extractions.

29-10 How are analytes usually recovered after an SFE?

Challenge Problem

29-11 In a recent paper, Zheng and coworkers (J. Zheng, L. T. Taylor, J. D. Pinkston, and M. L. Mangels, *J. Chromatogr. A*, **2005**, *1082*, 220) discuss the elution of polar and ionic compounds in SFC.
(a) Why are highly polar or ionic compounds usually not eluted in SFC?
(b) What types of mobile-phase additives have been used to improve the elution of highly polar or ionic compounds?
(c) Why is ion-pairing SFC not often used?
(d) Why are ammonium salts sometimes added as mobile-phase modifiers in SFC?
(e) The authors describe an SFC system that uses mass spectrometry (MS) as a detector. Discuss the interfacing of an SFC unit to a mass spectrometer. Compare the compatibility of SFC with MS to that of HPLC and GC with MS.
(f) The authors studied the effect of column outlet pressure on the elution of sodium 4-dodecylbenzene sulfonate on three different stationary phases with five mobile-phase additives. What effect was observed, and what was the explanation for the effect?
(g) What elution mechanisms were considered by the authors?
(h) Which mobile-phase additive gave the fastest elution of the sulfonate salts? Which provided the longest retention times?
(i) Did a silica column give results similar to or different from a cyano bonded-phase column?

CHAPTER THIRTY

Capillary Electrophoresis, Capillary Electrochromatography, and Field-Flow Fractionation

*T*his chapter deals with three relatively new separation methods. We first consider the principles of electrophoretic separations, with particular emphasis on **capillary electrophoresis**, and the applications of this versatile technique to various types of analytical problems. Capillary zone electrophoresis, capillary gel electrophoresis, capillary isotachophoresis, capillary isoelectric focusing, and micellar electrokinetic chromatography are described. A brief discussion of electrochromatography is then given. The chapter concludes with a discussion of the principles and applications of **field-flow fractionation** techniques, which are used in the separation of polymers, colloids, and other macromolecules.

Throughout this chapter, this logo indicates an opportunity for online self-study at **www .thomsonedu.com/chemistry/skoog**, linking you to interactive tutorials, simulations, and exercises.

In capillary electrophoresis and electrochromatography, separations occur in a buffer-filled capillary tube under the influence of an electric field as seen in the schematic of Figure 30-1. Separations in field-flow fractionation, on the other hand, occur in a thin ribbon-like flow channel under the influence of a sedimentation, electrical, or thermal field applied perpendicular to the flow direction.

30A AN OVERVIEW OF ELECTROPHORESIS

Electrophoresis is a separation method based on the differential rate of migration of charged species in an applied dc electric field. This separation technique was first developed by the Swedish chemist Arne Tiselius in the 1930s for the study of serum proteins; he was awarded the 1948 Nobel Prize in Chemistry for this work.

Electrophoresis on a macro scale has been applied to a variety of difficult analytical separation problems: inorganic anions and cations, amino acids, catecholamines, drugs, vitamins, carbohydrates, peptides, proteins, nucleic acids, nucleotides, polynucleotides, and numerous other species.

A particular strength of electrophoresis is its unique ability to separate charged macromolecules of interest in biochemical, biological, and biomedical research and the biotechnology industry. For many years, electrophoresis has been the powerhouse method for separating proteins (enzymes, hormones, antibodies) and nucleic acids (DNA, RNA) with unparalleled resolution. For example, to sequence DNA it is necessary to distinguish between long-chain polynucleotides that have as many as perhaps 200 to 500 bases and that differ by only a single nucleotide. Only electrophoresis has sufficient resolving power to handle this problem. Without electrophoresis, for example, the Human Genome Project would have been nearly impossible because human DNA contains some three billion nucleotides.

An electrophoretic separation is performed by injecting a small band of the sample into an aqueous buffer solution contained in a narrow tube or on a flat porous support medium such as paper or a semisolid gel. A high voltage is applied across the length of the buffer by means of a pair of electrodes located at each

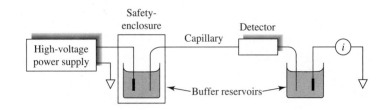

FIGURE 30-1 Schematic of a capillary electrophoresis system.

end of the buffer. This field causes ions of the sample to migrate toward one or the other of the electrodes. The rate of migration of a given species depends on its charge and its size. Separations are then based on differences in charge-to-size ratios for the various analytes in a sample. The larger this ratio, the faster an ion migrates in the electric field.

30A-1 Types of Electrophoresis

Electrophoretic separations are currently performed in two quite different formats: one is called *slab electrophoresis* and the other *capillary electrophoresis*. The first is the classical method that has been used for many years to separate complex, high-molecular-mass species of biological and biochemical interest. Slab separations are carried out on a thin flat layer or slab of a porous semisolid gel containing an aqueous buffer solution within its pores. This slab has dimensions of a few centimeters on a side and, like a chromatographic thin-layer plate, is capable of separating several samples simultaneously. Samples are introduced as spots or bands on the slab, and a dc electric field is applied across the slab for a fixed period. When the separations are complete, the field is discontinued and the separated species are visualized by staining in much the same way as was described for thin-layer chromatography in Section 28I-2.

Slab electrophoresis is now the most widely used separation tool in biochemistry and biology. Monographs, textbooks, and journals in the life sciences contain hundreds of photographs of developed electrophoretic slabs. Capillary electrophoresis, which is an instrumental version of electrophoresis, was developed in the mid-to-late 1980s. It has become an important tool for a wide variety of analytical separation problems. In many cases, this new method of performing electrophoretic separations is a satisfactory substitute for slab electrophoresis with several important advantages that are described later in this chapter.

30A-2 The Basis for Electrophoretic Separations

The migration rate v of an ion (cm/s) in an electric field is equal to the product of the field strength E (V cm^{-1}) and the electrophoretic mobility μ_e (cm^2 V^{-1} s^{-1}). That is,

$$v = \mu_e E \qquad (30\text{-}1)$$

The electrophoretic mobility is in turn proportional to the ionic charge on the analyte and inversely proportional to frictional retarding factors. The electric field acts on only ions. If two species differ either in charge or in the frictional forces they experience while moving through the buffer, they will be separated from each other. Neutral species are not separated. The frictional retarding force on an analyte ion is determined by the size and shape of the ion and the viscosity of the migration medium. For ions of the same size, the greater the charge, the greater the driving force and the faster the rate of migration. For ions of the same charge, the smaller the ion, the smaller the frictional forces and the faster the rate of migration. The ion's *charge-to-size ratio* combines these two effects. Note that in contrast to chromatography, only one phase is involved in an electrophoretic separation.

30B CAPILLARY ELECTROPHORESIS

As useful as conventional slab electrophoresis is, this type of electrophoretic separation is typically slow, labor intensive, and difficult to automate. Slab electrophoresis does not yield very precise quantitative information. During the mid-to-late 1980s, there was explosive growth in research and application of electrophoresis performed in capillary tubes, and several commercial instruments appeared. *Capillary electrophoresis* (CE) yields high-speed, high-resolution separations on exceptionally small sample volumes (0.1 to 10 nL in contrast to slab electrophoresis, which requires samples in the µL range). Additionally, the separated species are eluted from one end of the capillary, so quantitative detectors, similar to those found in high-performance liquid chromatography (HPLC), can be used instead of the cumbersome staining techniques of slab electrophoresis.[1]

30B-1 Migration Rates in CE

As Equation 30-1 shows, the migration rate of an ion v depends on the electric field strength. The electric field in turn is proportional to the magnitude of the applied voltage V and inversely proportional to the length L over which it is applied. Thus

$$v = \mu_e \times \frac{V}{L} \tag{30-2}$$

This relationship indicates that high applied voltages are desirable to achieve rapid ionic migration and a fast separation. It is desirable to have rapid separations, but it is even more important to achieve high-resolution separations. So we must examine the factors that determine resolution in electrophoresis.

30B-2 Plate Heights in CE

In chromatography, both longitudinal diffusion and mass-transfer resistance contribute to band broadening. However, because only a single phase is used in electrophoresis, in theory only longitudinal diffusion needs to be considered. In practice, however, Joule heating can add variance as well as the injection process. Although CE is not a chromatographic process, separations are often described in a manner similar to chromatography. For example, in electrophoresis, we calculate the plate count N by

$$N = \frac{\mu_e V}{2D} \tag{30-3}$$

where D is the diffusion coefficient of the solute (cm^2 s^{-1}). Because resolution increases as the plate count increases, it is desirable to use high applied voltages to

achieve high-resolution separations. Note that for electrophoresis, contrary to the situation in chromatography, the plate count does not increase with the length of the column.

With gel slab electrophoresis, joule heating limits the magnitude of the applied voltage to about 500 V. Here, one of the strengths of the capillary format compared with the slab format is realized. Because the capillary is quite long and has a small cross-sectional area, the solution resistance through the capillary is exceptionally high. Because power dissipation is inversely proportional to resistance ($P = I^2/R$), much higher voltages can be applied to capillaries than to slabs for the same amount of heating. Additionally, the high surface-to-volume ratio of the capillary provides efficient cooling. As a result of these two factors, band broadening due to thermally driven convective mixing does not occur to a significant extent in capillaries. Electric fields of $100-400$ V/cm are typically used. High-voltage power supplies of $10-25$ kV are normal. The high fields lead to corresponding improvements in speed and resolution over those seen in the slab format. CE peak widths often approach the theoretical limit set by longitudinal diffusion. CE normally yields plate counts in the range of 100,000 to 200,000, compared to the 5,000 to 20,000 plates typical for HPLC. Plate counts of 3,000,000 have been reported for capillary zone electrophoresis of dansylated amino acids,[2] and plate counts of 10,000,000 have been reported for capillary gel electrophoresis of polynucleotides.[3]

30B-3 Electroosmotic Flow

A unique feature of CE is *electroosmotic flow*. When a high voltage is applied across a fused-silica capillary tube containing a buffer solution, electroosmotic flow usually occurs, in which the bulk liquid migrates toward the cathode. The rate of migration can be substantial. For example, a 50 mM pH 8 buffer flows through a 50-cm capillary toward the cathode at approximately 5 cm/min with an applied voltage of 25 kV.[4]

As shown in Figure 30-2, the cause of electroosmotic flow is the electric double layer that develops at the silica-solution interface. At pH values higher than 3, the

[1]For additional discussion of CE, see *Analysis and Detection by Capillary Electrophoresis*, M. L. Marina, A. Rios, and M. Valcarcel, eds., Vol. 45 of *Comprehensive Analytical Chemistry*, D. Barcelo, ed., Amsterdam: Elsevier, 2005; *Capillary Electrophoresis of Proteins and Peptides*, M. A. Strege and A. L. Lagu, eds., Totowa, NJ: Humana Press, 2004; *Clinical and Forensic Applications of Capillary Electrophoresis*, J. R. Petersen and A. A. Mohamad, eds., Totowa, NJ: Humana Press, 2001; R. Weinberger, *Practical Capillary Electrophoresis*, 2nd ed., New York: Academic Press, 2000; *High Performance Capillary Electrophoresis*, M. G. Khaledi, ed., New York: Wiley, 1998.

[2]R. D. Smith, J. A. Olivares, N. T. Nguyen, and H. R. Udseth, *Anal. Chem.*, **1988**, *60*, 436.

[3]A. Guttman, A. S. Cohen, D. N. Heiger, and B. L. Karger, *Anal. Chem.*, **1990**, *62*, 137.

[4]J. D. Olechno, J. M. Y. Tso, J. Thayer, and A. Wainright, *Amer. Lab.*, **1990**, *22* (17), 51.

Electroosmotic flow

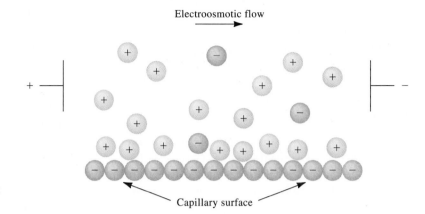

FIGURE 30-2 Charge distribution at a silica-capillary interface and resulting electroosmotic flow. (From A. G. Ewing, R. A. Wallingford, and T. M Olefirowicz, *Anal. Chem.*, **1989**, *61*, 298A. Copyright 1989 American Chemical Society.)

Capillary surface

inside wall of a silica capillary is negatively charged because of ionization of the surface silanol groups (Si—OH). Buffer cations congregate in the electrical double layer adjacent to the negative surface of the silica capillary. The cations in the diffuse outer layer of the double layer are attracted toward the cathode, or negative electrode, and because they are solvated, they drag the bulk solvent along with them. As shown in Figure 30-3, electroosmosis leads to bulk solution flow that has a flat profile across the tube because flow originates at the walls of the tubing. This profile is in contrast to the laminar (parabolic) profile observed with the pressure-driven flow encountered in HPLC. Because the profile is essentially flat, electroosmotic flow does not contribute significantly to band broadening the way pressure-driven flow does in liquid chromatography.

The rate of electroosmotic flow is generally greater than the electrophoretic migration velocities of the individual ions and effectively becomes the mobile-phase pump of CE. Even though analytes migrate according to their charges within the capillary, the electroosmotic flow rate is usually sufficient to sweep all positive, neutral, and even negative species toward the same end of the capillary, so all can be detected as they pass by a common point (see Figure 30-4). The resulting *electropherogram* looks like a chromatogram but with narrower peaks.

The electroosmotic flow velocity v is given by an equation similar to Equation 30-1. That is,

$$v = \mu_{eo}E \tag{30-4}$$

In the presence of electroosmosis, the velocity of an ion is the sum of its migration velocity and the electroosmotic flow velocity. Thus,

$$v = (\mu_e + \mu_{eo})E \tag{30-5}$$

As a result of electroosmosis, order of elution in a typical electrophoretic separation is, first, the fastest cation followed by successively slower cations, then all the neutrals in a single zone, and finally the slowest anion followed by successively faster anions (see Figure 30-4). In some instances, the rate of electroosmotic flow may not be great enough to surpass the rate at which some of the anions move toward the anode, in which case these species move in that direction instead of toward the cathode.

The migration time t_m in CE is the time it takes for a solute to migrate from the point of introduction to the detector. If a capillary of total length L is used and the length to the detector is l, the migration time is

$$t_m = \frac{l}{(\mu_e + \mu_{eo})E} = \frac{lL}{(\mu_e + \mu_{eo})V} \tag{30-6}$$

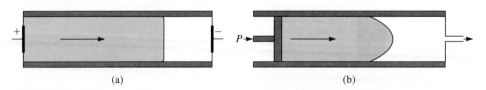

(a)   (b)

FIGURE 30-3 Flow profiles for liquids under (a) electroosmotic flow and (b) pressure-induced flow.

The number of theoretical plates in the presence of electroosmotic flow can be found from an expression analogous to Equation 26-21:

$$N = 16\left(\frac{t_m}{W}\right)^2 \qquad (30\text{-}7)$$

where W, as in chromatography, is the peak width measured at the base of the peak.

It is possible to reverse the direction of the normal electroosmotic flow by adding a cationic surfactant to the buffer. The surfactant adsorbs on the capillary wall and makes the wall positively charged. Now buffer anions congregate near the wall and are swept toward the cathode, or positive electrode. This ploy is often used to speed up the separation of anions.

Electroosmosis is often desirable in certain types of CE, but in other types it is not. Electroosmotic flow

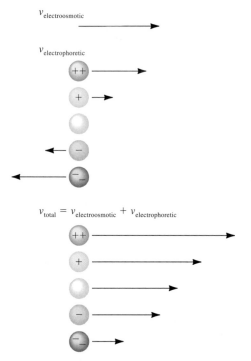

FIGURE 30-4 Velocities in the presence of electroosmotic flow. The length of the arrow next to an ion indicates the magnitude of its velocity; the direction of the arrow indicates the direction of motion. The negative electrode is to the right and the positive electrode to the left of this section of solution.

 Simulation: Learn more about **capillary electrophoresis**.

can be minimized by modifying the inside capillary walls with a reagent like trimethylchlorosilane that bonds to the surface and reduces the number of surface silanol groups (see Section 28D-1).

30B-4 Instrumentation for CE

As shown in Figure 30-1, the instrumentation for CE is relatively simple.[5] A buffer-filled fused-silica capillary, typically 10 to 100 μm in internal diameter and 30 to 100 cm long, extends between two buffer reservoirs that also hold platinum electrodes. Like the capillary tubes used in gas chromatography (GC), the outside walls of the fused-silica capillary are typically coated with polyimide for durability, flexibility, and stability. The sample is introduced at one end and detection occurs at the other. A voltage of 5 to 30 kV dc is applied across the two electrodes. The polarity of this high voltage can be as indicated in Figure 30-1 or can be reversed to allow rapid separation of anions. High-voltage electrophoresis compartments are usually safety interlocked to protect the user.

Although the instrumentation is conceptually simple, significant experimental difficulties in sample introduction and detection arise due to the very small volumes involved. Because the volume of a normal capillary is 4 to 5 μL, injection and detection volumes must be on the order of a few nanoliters or less.

Sample Introduction

The most common sample-introduction methods are *electrokinetic injection* and *pressure injection*. With electrokinetic injection, one end of the capillary and its electrode are removed from their buffer compartment and placed in a small cup containing the sample. A voltage is then applied for a measured time, causing the sample to enter the capillary by a combination of ionic migration and electroosmotic flow. The capillary end and electrode are then returned to the regular buffer solution for the duration of the separation. This injection technique discriminates by injecting larger amounts of the more mobile ions relative to the slower-moving ions.

In pressure injection, the sample-introduction end of the capillary is also placed in a small cup containing the sample, but here a pressure difference drives the sample solution into the capillary. The pressure difference can be produced by applying a vacuum at the de-

[5]For a review of commercially available capillary electrophoresis instruments, see L. DeFrancesco, *Anal. Chem.*, **2001**, *73*, 497A.

tector end, by pressurizing the sample, or by elevating the sample end (hydrodynamic injection). Pressure injection does not discriminate because of ion mobility, but it cannot be used in gel-filled capillaries.

For both electrokinetic injection and pressure injection, the volume injected is controlled by the duration of the injection. Injections of 5 to 50 nL are common, and volumes below 100 pL have been reported. For a buffer with density and viscosity near the values for water, a height differential of 5 cm for 10 s injects about 6 nL with a 75-μm inside-diameter capillary.

Microinjection tips constructed from capillaries drawn to very small diameters allow sampling from picoliter environments such as single cells or substructures within single cells. This technique has been used to study amino acids and neurotransmitters from single cells. Other novel injection techniques have been described in the literature.[6] Commercial CE systems are available with thermostatted multiposition carousels for automated sampling.

Detection

Because the separated analytes move past a common point in most types of CE, detectors are similar in design and function to those described for HPLC. Table 30-1 lists several of the detection methods that have been reported for CE. The second column of the table shows representative detection limits for these detectors.

Absorption Methods. Both fluorescence and absorption detectors are widely used in CE, although absorption methods are more common because they are more generally applicable. To keep the detection volume on the nanoliter scale or smaller, detection is performed on-column. In this case a small section of the protective polyimide coating is removed from the exterior of the capillary by burning or etching. That section of the capillary then serves as the detector cell. Unfortunately, the path length for such measurements is no more than 50 to 100 μm, which restricts detection limits in concentration terms; because such small volumes are involved, however, mass detection limits are equal to or better than those for HPLC.

Several cell designs have been used for increasing the measurement path length to improve the sensitivity of absorption methods. Three of these are shown in

TABLE 30-1 Detectors for CE

Type of Detector	Representative Detection Limit* (attomoles detected)
Spectrometry	
Absorption[†]	1–1000
Fluorescence	1–0.01
Thermal lens[†]	10
Raman[†]	1000
Chemiluminescence[†]	1–0.0001
Mass spectrometry	1–0.01
Electrochemical	
Conductivity[†]	100
Potentiometry[†]	1
Amperometry	0.1

Sources: B. Huang, J. J. Li, L. Zhang, J. K. Cheng, *Anal. Chem.*, **1996**, *68*, 2366; S. C. Beale, *Anal. Chem.*, **1998**, *70*, 279R. S. N. Krylov and N. J. Dovichi, *Anal. Chem.*, **2000**, *72*, 111R; S. Hu and N. J. Dovichi, *Anal. Chem.*, **2002**, *74*, 2833.

*Detection limits quoted have been determined with injection volumes ranging from 18 pL to 10 nL.

[†]Mass detection limit converted from concentration detection limit using a 1-nL injection volume.

Figure 30-5. In the commercial detector shown in Figure 30-5a, the end of the capillary is bent into a Z shape, which produces a path length as long as ten times the capillary diameter. Increases in path length can lead to decreases in peak efficiency and thus resolution. In some cases, special lenses, such as spherical ball lenses, are inserted between the source and the z cell and between the cell and the detector.[7] Such lenses improve sensitivity by focusing the light into the cell and onto the detector.

Figure 30-5b shows a second way to increase the absorption path length. In this example, a bubble is formed near the end of the capillary. In the commercial version of this technique, the bubble for a 50-μm capillary has an inside diameter of 150 μm, which gives a threefold increase in path length.

A third method for increasing the path length of radiation by reflection is shown in Figure 30-5c. In this technique, a reflective coating of silver is deposited on the end of the capillary. The source radiation then undergoes multiple reflections during its transit through the capillary, which significantly increase the path length.

[6]L. M. Ponton and C. E. Evans, *Anal. Chem.*, **2001**, *73*, 1974; R. Kuldvee and M. Kaljurand, *Crit. Rev. Anal. Chem.*, **1999**, *29*, 29.

[7]D. M. Spence, A. M. Sekelsky, and S. R. Crouch, *Instrum. Sci. Technol.*, **1996**, *24*, 103.

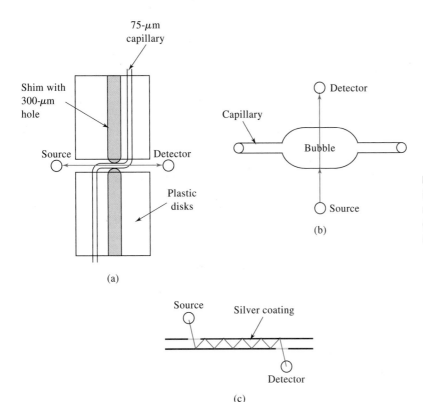

(a)

(b)

(c)

FIGURE 30-5 Three types of cells for improving the sensitivity of absorption measurements in CE: (a) a 3-mm z cell, (b) a 150-μm bubble cell, (c) a multi-reflection cell.

Commercial CE systems are available with diode array detectors that allow spectra to be collected over the UV-visible range in less than 1 s.

Indirect Detection. Indirect absorbance detection has been used for species of low molar absorptivity that are difficult to detect without derivatization. An ionic chromophore is placed in the electrophoresis buffer. The detector then receives a constant signal due to the presence of this substance. The analyte displaces some of these ions, just as in ion-exchange chromatography, so that the detector signal decreases during the passage of an analyte band through the detector. The analyte is then determined from the decrease in absorbance. The electropherogram in Figure 30-6 was generated by using indirect absorbance detection with 4-mM chromate ion as the chromophore; this ion absorbs radiation strongly at 254 nm in the buffer. Although the peaks obtained are negative (decreasing A) peaks, they appear as positive peaks in Figure 30-6 because the detector polarity was reversed.

Fluorescence Detection. Just as in HPLC, fluorescence detection yields increased sensitivity and selectivity for fluorescent analytes or fluorescent derivatives.

Laser-based instrumentation is preferred to focus the excitation radiation on the small capillary and to achieve the low detection limits available from intense sources. Laser-induced-fluorescence attachments are

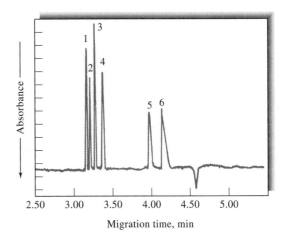

FIGURE 30-6 Electropherogram of a six-anion mixture by indirect detection with 4-mM chromate ion at 254 nm. Peak: (1) bromide (4 ppm), (2) chloride (2 ppm), (3) sulfate (4 ppm), (4) nitrate (4 ppm), (5) fluoride (1 ppm), (6) phosphate (6 ppm).

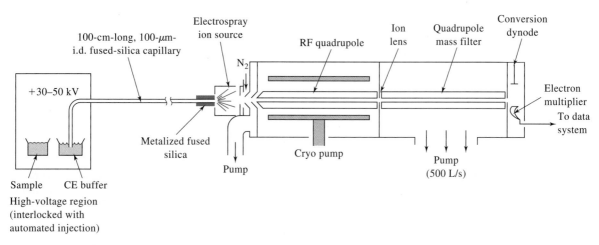

FIGURE 30-7 An instrument for CE/MS. The high-voltage (anode) end was maintained at 30–50 kV in an electrically isolated, interlocked box. Electrical contact at the low-voltage (cathode) end was made by silver deposited on the capillary and a stainless steel sheath. This electrical contact was at 3 to 5 kV with respect to common, which also charged the electrospray. The flow of nitrogen at ~70°C for desolvation was 3 to 6 L/min. (From R. D. Smith, J. A. Olivares, N. T. Nguyen, and H. R. Udseth, *Anal. Chem.*, **1988**, *60*, 436. With permission.)

available that couple with commercial CE instruments. Laser-induced fluorescence has allowed detection of as little as 10 zeptomoles, or 6000 molecules.[8]

Electrochemical Detection. Two types of electrochemical detection have been used with CE: conductivity and amperometry. One of the problems with electrochemical detection has been that of isolating the detector electrodes from the high voltage required for the separation. One method for isolation involves inserting a porous glass or graphite joint between the end of the capillary and a second capillary containing the detector electrodes.

Mass Spectrometric Detection. The very small volumetric flow rates of less than 1 μL/min from electrophoresis capillaries make it feasible to couple the effluent directly to the ionization source of a mass spectrometer. The most common sample-introduction and ionization interface for this purpose is currently electrospray (Section 20B-4), although fast atom bombardment, matrix-assisted laser desorption-ionization (MALDI) spectrometry, and inductively coupled plasma mass spectrometry (ICPMS) have also been used. Because the liquid sample must be vaporized before entering the mass spectrometry (MS) system,

it is important that volatile buffers be used. Capillary electrophoresis–mass spectrometry (CE/MS) systems have become quite important in the life sciences for determining large biomolecules that occur in nature, such as proteins, DNA fragments, and peptides.[9]

Figure 30-7 is a schematic of a typical electrospray interface coupled to a quadrupole mass spectrometer. Note that the capillary is positioned between the isolated high-voltage region and the electrospray source. The high-voltage end of the capillary was at 30 to 50 kV with respect to common. The low-voltage end was maintained at 3–5 kV and charged the droplets. Similar electrospray instruments are available commercially coupled with either quadrupole or ion-trap mass spectrometers.[10] Ion-trap mass spectrometers can allow CE/MS/MS or CE/MS^n operation.

Figure 30-8 shows the electrospray mass spectrum obtained for vasotocin, a polypeptide having a molecular mass of 1050. Note the presence of doubly and triply charged species. With higher-molecular-mass species, ions are often observed with charges of +12 or more. Ions with such large charges make it possible to detect high-molecular-mass analytes with a quadru-

[8] S. Wu and N. Dovichi, *J. Chromatogr.*, **1989**, *480*, 141.

[9] For more information on mass spectrometric detection, see J. C. Severs and R. D. Smith, in *Handbook of Capillary Electrophoresis*, 2nd ed., J. P. Landers, ed., Chap. 28, Boca Raton, FL: CRC Press, 1997.

[10] Agilent Technologies, Wilmington, DE; Beckman Coulter, Inc., Fullerton, CA.

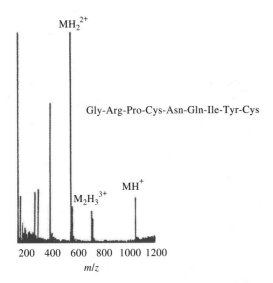

FIGURE 30-8 Electrospray ionization mass spectrum for vasotocin. (From R. D. Smith, J. A. Olivares, N. T. Nguyen, and H. R. Udseth, *Anal. Chem.*, **1988**, *60*, 436. With permission.)

pole instrument with a relatively modest mass range. Typical detection limits for CE/MS are of a few tens of femtomoles for molecules with molecular masses of 100,000 or more.

Commercial CE Systems

Currently, fewer than ten companies worldwide manufacture CE instruments. Some two dozen companies offer supplies and accessories for CE. The initial cost of equipment and the expense of maintenance for CE are generally significantly lower than those for ion chromatographic and atomic spectroscopic instruments. Thus, commercial CE instruments with standard absorption or fluorescence detectors cost $10,000 to $65,000.[11] Addition of mass spectrometric detection can raise the cost significantly.

30C APPLICATIONS OF CE

Capillary electrophoretic separations can be performed in several different modes. These include capillary zone electrophoresis, capillary gel electrophoresis, capillary isoelectric focusing, capillary isotachophoresis, and micellar electrokinetic chromatography. The

[11] See L. DeFrancesco, *Anal. Chem.*, **2001**, *73*, 497A.

sections that follow illustrate typical applications of each of these techniques.

30C-1 Capillary Zone Electrophoresis

In capillary zone electrophoresis (CZE), the buffer composition is constant throughout the region of the separation. The applied field causes each of the different ionic components of the mixture to migrate according to its own mobility and to separate into zones that may be completely resolved or that may be partially overlapped. Completely resolved zones have regions of buffer between them as illustrated in Figure 30-9a. The situation is analogous to elution column chromatography, where regions of mobile phase are located between zones containing separated analytes.

Separation of Small Ions

For most electrophoretic separations of small ions, the smallest analysis time occurs when the analyte ions move in the same direction as the electroosmotic flow. Thus, for cation separations, the walls of the capillary are untreated and the electroosmotic flow and the cation movement are toward the cathode. For the separation of anions, on the other hand, the electroosmotic flow is usually reversed by treating the walls of the capillary with an alkyl ammonium salt, such as cetyl trimethylammonium bromide. The positively charged ammonium ions become attached to the silica surface, yielding a positively charged, immobile surface layer. This, in turn, creates a negatively charged, mobile solution layer, which is attracted toward the anode, reversing the electroosmotic flow.

In the past, the most common method for analysis of small anions has been ion chromatography. For cations, the preferred techniques have been atomic absorption spectroscopy and ICPMS. Figure 30-10 illustrates the speed and resolution of electrophoretic separations of small anions. Here, thirty anions were separated cleanly in just more than 3 minutes. Typically, an ion-exchange separation of only three or four anions can be accomplished in this brief period. Figure 30-11 further illustrates the speed at which separations can be carried out. In this example, nineteen cations were separated in less than 2 minutes. CE methods were once predicted to replace the more established methods because of lower equipment costs, smaller-sample-size requirements, and shorter analysis times. However, because variations in electroosmotic flow rates make reproducing CE separations difficult, LC methods and

(a) Zone electrophoresis

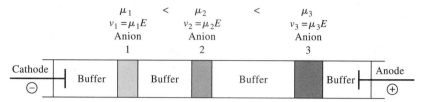

FIGURE 30-9 Three modes of separation by electrophoresis. In zone electrophoresis (a), the ions separate into zones 1, 2, and 3. The zones shown are completely resolved with buffer between each zone. In isotachophoresis (b), the sample is injected between a leading, high-mobility buffer and a trailing, low-mobility buffer. In isoelectric focusing (c), a continuous pH gradient exists along the length of the capillary. Analyte ions migrate to the pH corresponding to their isoelectric points.

(b) Isotachophoresis

(c) Isoelectric focusing

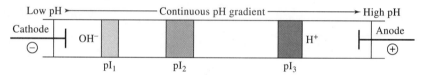

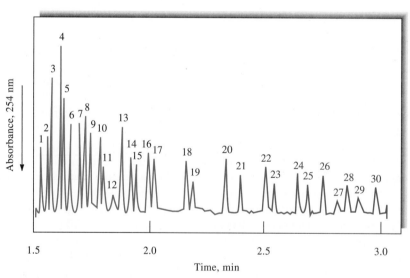

FIGURE 30-10 Electropherogram showing the separation of thirty anions. Capillary internal diameter 50 μm (fused silica). Detection: indirect UV, 254 nm. Peaks 1 = thiosulfate (4 ppm), 2 = bromide (4 ppm), 3 = chloride (2 ppm), 4 = sulfate (4 ppm), 5 = nitrite (4 ppm), 6 = nitrate (4 ppm), 7 = molybdate (10 ppm), 8 = azide (4 ppm), 9 = tungstate (10 ppm), 10 = monofluorophosphate (4 ppm), 11 = chlorate (4 ppm), 12 = citrate (2 ppm), 13 = fluoride (1 ppm), 14 = formate (2 ppm), 15 = phosphate (4 ppm), 16 = phosphite (4 ppm), 17 = chlorite (4 ppm), 18 = galactarate (5 ppm), 19 = carbonate (4 ppm), 20 = acetate (4 ppm), 21 = ethanesulfonate (4 ppm), 22 = propionate (5 ppm), 23 = propanesulfonate (4 ppm), 24 = butyrate (5 ppm), 25 = butanesulfonate (4 ppm), 26 = valerate (5 ppm), 27 = benzoate (4 ppm), 28 = l-glutamate (5 ppm), 29 = pentanesulfonate (4 ppm), 30 = d-gluconate (5 ppm). (From W. A. Jones and P. Jandik, *J. Chromatogr.*, **1991**, *546*, 445. With permission.)

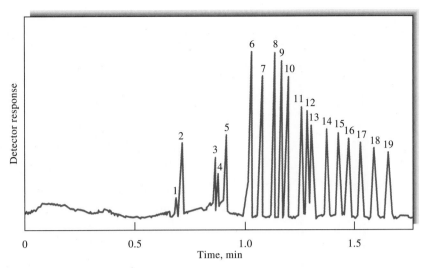

FIGURE 30-11 Separation of alkali, alkaline earths, and lanthanides. Capillary: 36.5 cm ×
75 μm fused silica, +30 kV. Injection: hydrostatic, 20 s at 10 cm. Detection: indirect UV, 214 nm.
Peaks 1 = rubidium (2 ppm), 2 = potassium (5 ppm), 3 = calcium (2 ppm), 4 = sodium (1 ppm),
5 = magnesium (1 ppm), 6 = lithium (1 ppm), 7 = lanthanum (5 ppm), 8 = cerium (5 ppm), 9 =
praseodymium (5 ppm), 10 = neodymium (5 ppm), 11 = samarium (5 ppm), 12 = europium
(5 ppm), 13 = gadolinium (5 ppm), 14 = terbium (5 ppm), 15 = dysprosium (5 ppm), 16 =
holmium (5 ppm), 17 = erbium (5 ppm), 18 = thulium (5 ppm), 19 = ytterbium (5 ppm). (From
P. Jandik, W. R. Jones, O. Weston, and P. R. Brown, *LC-GC*, **1991**, *9*, 634. With permission.)

atomic spectrometric methods are still widely used for
small inorganic ions.

Separation of Molecular Species

A variety of synthetic herbicides, pesticides, and pharmaceuticals that are ions or can be derivatized to yield
ions have been separated and analyzed by CZE. Figure 30-12 illustrates this type of application, in which
three anti-inflammatory drugs, carboxylic acids and
carboxylate salts, are separated in less than 15 min.

Proteins, amino acids, and carbohydrates have all
been separated in minimum times by CZE. In the case
of neutral carbohydrates, the separations are preceded
by formation of negatively charged borate complexes.
Protein mixtures can be separated, as illustrated in Figure 30-13. Capillary gel electrophoresis is widely used
in DNA sequencing as discussed in the next section.

30C-2 Capillary Gel Electrophoresis

Capillary gel electrophoresis (CGE) is generally performed in a porous gel polymer matrix with a buffer
mixture that fills the pores of the gel. In early slab
electrophoresis studies, the primary purpose of the

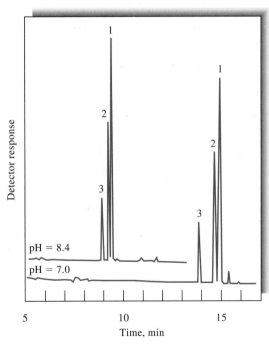

FIGURE 30-12 Separation of anti-inflammatory drugs
by CZE. Detection: UV at 215 nm. Analytes: (1) naproxen,
(2) ibuprofen, (3) tolmetin. (Reprinted with permission from
A. Wainright, *J. Microcolumn. Sep.*, **1990**, *2*, 166.)

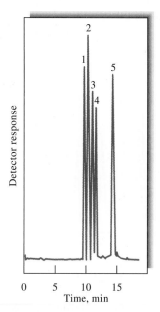

FIGURE 30-13 CZE separation of a model protein mixture. Conditions: pH 2.7 buffer; absorbance detection at 214 nm; 22 kV, 10 A. Peaks are identified in the following table:

Model Proteins Separated at pH 2.7

Peak No.	Protein	Molecular Weight	Isoelectric Point, pH
1	Cytochrome *c*	12,400	10.7
2	Lysozyme	14,100	11.1
3	Trypsin	24,000	10.1
4	Trypsinogen	23,700	8.7
5	Trypsin inhibitor	20,100	4.5

polymeric medium was to reduce analyte dispersion by convection and diffusion and to provide a convenient medium for detection and scanning. This type of medium provided a molecular sieving action that retarded the migration of analyte species to various extents depending on the pore size of the polymer and the size of the analyte ions. This sieving action is particularly helpful in separating macromolecules such as proteins, DNA fragments, and oligonucleotides that have similar charges but differ in size. Most macroscale electrophoresis separations are now carried out on a gel slab, but some capillary electrophoretic separations of species that differ in size are also performed in gels contained in capillary tubes.

Types of Gels

The most common type of gel used in electrophoresis is a polyacrylamide polymer formed by polymerizing acrylamide (CH_2=CH—CO—NH$_2$) in the presence of a cross-linking agent. The pore size of the polymer depends on the ratio of monomer to cross-linking agent. Increases in the amount of the cross-linking agent produce smaller pore size. Other gels that have been used for CGE include agarose (a polysaccharide extracted from a marine alga), methyl cellulose, various derivatives of cellulose, polyvinyl alcohol, polyethyleneoxide, dextran, polydimethylacrylamide, and polyethylene glycol. Because of the difficulty of filling capillaries with rigid gels, entangled polymer gel networks are often used.[12] These allow easy filling and flushing of the capillary. A typical separation of a standard protein mixture in a polyethylene glycol gel is illustrated in Figure 30-14.

CE in DNA Sequencing

A major goal of the Human Genome Project was to determine the order of occurrence of the four bases, adenine (A), cytosine (C), guanine (G), and thymine (T), in DNA molecules. The sequence defines an individual's genetic code. The need for sequencing DNA has spawned the development of several new analytical instruments. Among the most attractive of these approaches is capillary array electrophoresis.[13] In this technique, as many as ninety-six capillaries are operated in parallel. The capillaries are filled with a separation matrix, normally a linear polyacrylamide gel. The capillaries have inner diameters of 35–75 μm and are 30–60 cm long.

In sequencing, DNA extracted from cells is fragmented by various approaches. Depending on the terminal base in the fragment, one of four fluorescent dyes is attached to the various fragments. The sample contains many different-size fragments, each with a fluorescent label. Under the influence of the electrophoretic field, lower-molecular-mass fragments move faster and arrive at the detector sooner than higher-molecular-mass fragments. The DNA sequence is determined by the dye color sequence of the eluting fragments. Lasers are used to excite the dye fluorescence. Several different techniques have been described for detecting the

---

[12]M. Zhu, D. L. Hansen, S. Burd, and F. Gannon, *J. Chromatogr.*, **1989**, *480*, 311.
[13]For a review, see I. Kheterpal and R. A. Mathies, *Anal. Chem.*, **1999**, *71*, 31A.

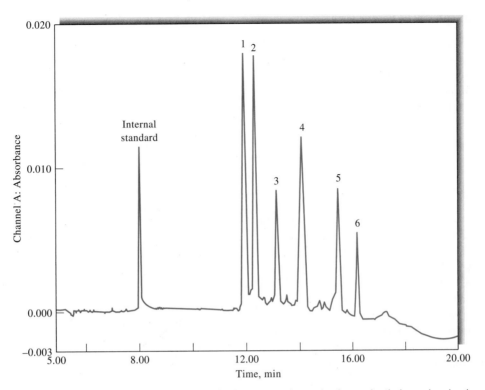

FIGURE 30-14 Capillary gel separation of SDS denatured proteins in a polyethylene glycol column. UV absorption detection at 214 nm. Proteins: (1) α-lactalbumin, (2) soybean trypsin inhibitor, (3) carbonic anhydrase, (4) ovalbumin, (5) bovine serum albumin, (6) phosphorylase B. (Reprinted from K. Ganzler et al., *Anal. Chem.*, **1992**, *64*, 2665. Copyright 1992 American Chemical Society.)

fluorescence. One method uses a scanning system in which the capillary bundle is moved relative to the excitation laser and a four-wavelength detection system. In the detection system illustrated in Figure 30-15, a laser beam is focused onto the capillary array by a lens. The region illuminated by the laser is imaged onto a charge-coupled-device (CCD) detector (see Section 7E-3). Filters allow wavelength selection to detect the four colors. Simultaneous separation of eleven DNA fragments in 100 capillaries has been reported.[14] Other designs include sheath-flow detector systems and a detector that uses two diode lasers for excitation. Commercial instrumentation is available with prices ranging from $85,000 to more than $300,000.[15] DNA sequencers and other genetic analyzers have been miniaturized using lab-on-a-chip technology.[16] Such miniature systems will eventually become portable so that they can

be used in the field. CE played a major role in identifying human remains in the debris of the World Trade Center disaster.

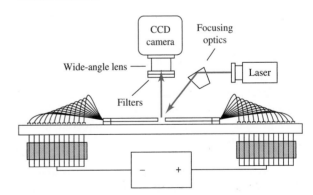

FIGURE 30-15 On-column laser fluorescence detection system for capillary array electrophoresis. A laser is focused as a line onto the array of capillaries at a 45° angle. The fluorescence is filtered and detected by a CCD camera through a wide-angle lens. (From K. Ueno and E. S. Yeung, *Anal. Chem.*, **1994**, *66*, 1424. Copyright 1994 American Chemical Society.)

[14] K. Ueno and E. S. Yeung, *Anal. Chem.*, **1994**, *66*, 1424.

[15] For a review of commercial sequencers, see J. P. Smith and V. Hinson-Smith, *Anal. Chem.*, **2001**, *73*, 327A.

[16] C. A. Emrich, H. Tian, I. L. Medintz, and R. A. Mathies, *Anal. Chem.*, **2002**, *74*, 5076; Agilent Technologies, Palo Alto, CA.

30C-3 Capillary Isotachophoresis

In capillary isotachophoresis all analyte bands ultimately migrate at the same velocity; hence, the name from *iso* for same and *tach* for speed. In any particular application, either cations or anions can be separated, but not both at the same time. In an isotachophoresis separation, the sample is injected between two buffers: a leading one containing ions of a higher mobility than any of the analyte ions and a terminating one with ions of a lower mobility than the sample ions (Figure 30-9b). For example, in separating anions, chloride ions might be contained in the leading buffer, and slow-moving heptanoate ions in the terminating buffer. For a separation of anions, the leading electrolyte solution is connected to the anode, and the terminating one to the cathode.

When the electric field is first applied in an isotachophoretic separation, analyte ions migrate as in zone electrophoresis, each ion with its unique velocity given by the product $\mu_e E$. This difference in migration rates results in the separation of the various analyte species into adjacent bands, with fastest species located in a band directly adjacent to the leading buffer and the slowest just ahead of the terminal buffer. After the bands have formed, they then move at the same velocity. The reason that the bands have the same velocity is that the electric field becomes smaller for the more mobile bands and greater for the slower bands, so that the current is the same, as it must be, in all parts of the buffer. The ionic current that results from the flow of ions in the buffer is analogous to the dc current in a circuit consisting of several resistors connected in series to a battery. Here, the current must be identical in all of the resistors. Hence, the potential across each of the resistors must vary in such a way that Ohm's law is obeyed.

When equilibrium is reached in an isotachophoretic experiment, each sample component is migrating in a band sandwiched between a band that contains the next-slower-moving ions and the next-faster-moving band, as shown in Figure 30-9b. The boundary between bands is sharp. If a solute species starts to move into the next-faster band, it encounters a lower field, which reduces its velocity until it drops back into its original band. Note in Figure 30-9b that, in contrast to zone electrophoresis or elution chromatography, the analyte bands are immediately adjacent to one another and are not separated by bands of the buffer.

30C-4 Capillary Isoelectric Focusing

Capillary isoelectric focusing is used to separate amphiprotic species, such as amino acids and proteins that contain a weak carboxylic acid group and a weak-base amine group.

Properties of Amphiprotic Compounds

An *amphiprotic* compound is a species that in solution is capable of both donating and accepting a proton. A typical amino acid, such as glycine, is an amphiprotic compound. When glycine is dissolved in water, three important equilibria operate:

$$NH_2CH_2COOH \rightleftharpoons NH_3^+CH_2COO^-$$

$$NH_3^+CH_2COO^- + H_2O \rightleftharpoons NH_2CH_2COO^- + H_3O^+$$

$$K_a = \frac{[H_3O^+][NH_2CH_2COO^-]}{[NH_3^+CH_2COO^-]} = 2 \times 10^{-10} \quad (30\text{-}8)$$

$$NH_3^+CH_2COO^- + H_2O \rightleftharpoons NH_3^+CH_2COOH + OH^-$$

$$K_b = \frac{[OH^-][NH_3^+CH_2COOH]}{[NH_3^+CH_2COO^-]} = 2 \times 10^{-12} \quad (30\text{-}9)$$

The first reaction constitutes a kind of internal acid-base reaction and is analogous to the reaction one would observe between a simple carboxylic acid and an amine. The typical aliphatic amine has a base dissociation constant of 10^{-4} to 10^{-5}, and many carboxylic acids have acid dissociation constants of about the same magnitude. The result is that the first reaction proceeds far to the right, with the product or products being the predominant species in the solution.

The amino acid product in the first reaction, bearing both a positive and a negative charge, is called a *zwitterion*. As shown by the equilibrium constants for the second and third reactions (Equations 30-8 and 30-9), the zwitterion of glycine is stronger as an acid than as a base. Thus, an aqueous solution of glycine is somewhat acidic.

The zwitterion of an amino acid, containing as it does a positive and a negative charge, has no tendency to migrate to an electric field, but the singly charged anionic and cationic species are attracted to oppositely charged electrodes. No net migration of the amino acid occurs in an electric field when the pH of the solvent is such that the concentrations of the anionic and cationic forms are identical. The pH at which no net migration occurs is called the *isoelectric point* (pI) and is an important physical constant for characterizing amino acids. The isoelectric point is readily related to the ionization constants for the species. Thus, for glycine, at the isoelectric point,

$$[\mathrm{NH_2CH_2COO^-}] = [\mathrm{NH_3^+CH_2COOH}]$$

If we divide K_a by K_b and substitute this equality, we obtain at the isoelectric point

$$\frac{K_a}{K_b} = \frac{[\mathrm{H_3O^+}]_{iso}\, \overline{[\mathrm{NH_2CH_2COO^-}]}}{[\mathrm{OH^-}]_{iso}\, \overline{[\mathrm{NH_3^+CH_2COOH}]}} = \frac{[\mathrm{H_3O^+}]_{iso}}{[\mathrm{OH^-}]_{iso}} \tag{30-10}$$

Substituting $K_w/[\mathrm{H_3O^+}]_{iso}$ for $[\mathrm{OH^-}]_{iso}$ in Equation 30-10 and rearranging yields

$$[\mathrm{H_3O^+}]_{iso} = \sqrt{\frac{K_a K_w}{K_b}} \tag{30-11}$$

We can convert Equation 30-11 into the isoelectric point pH (pI) by taking the negative logarithms of both sides. Thus, pI can be expressed as

$$\mathrm{pI} = -\log[\mathrm{H_3O^+}]_{iso} = \frac{(-\log K_a - \log K_w + \log K_b)}{2}$$

$$\mathrm{pI} = \frac{(\mathrm{p}K_a + \mathrm{p}K_w - \mathrm{p}K_b)}{2} \tag{30-12}$$

For glycine, $\mathrm{p}K_a = -\log(2 \times 10^{-10}) = 9.7$, $\mathrm{p}K_b = 11.7$ and $\mathrm{p}K_w = 14.0$. Thus,

$$\mathrm{pI} = (9.7 + 14.0 - 11.7)/2 = 6.0$$

Hence, the isoelectric point pI for glycine occurs at a pH of 6.0.

Separation of Amphiprotic Species

In isoelectric separation of amphiprotic species, the separation is performed in a buffer mixture that continuously varies in pH along its length. This pH gradient is prepared from a mixture of several different *ampholytes* in an aqueous solution. Ampholytes are amphoteric compounds usually containing carboxylic and amino groups. Ampholyte mixtures having different pH ranges can be prepared or are available from several commercial sources.

To perform an isoelectric focusing experiment in a capillary tube, the analyte mixture is dissolved in a dilute solution of the ampholytes, which is then transferred to the tube. One end of the capillary is then inserted in a solution of strong base, such as sodium hydroxide, that also holds the cathode. The other end of the tube is immersed in a solution of a strong acid, such as phosphoric, that also holds the anode. When the electric field is applied, hydrogen ions begin to migrate from the anode compartment toward the cathode. Hydroxide ions from the cathode begin to move in the opposite direction. If a component of the ampholyte or the analyte has a net negative charge, it migrates toward the positive anode. As it migrates it passes into continuously lower pH regions, where progressive protonation of the species occurs, which lowers its negative charge. Ultimately, it reaches the pH where its net charge is zero (its isoelectric point). Migration of the species then ceases. This process goes on for each ampholyte species and ultimately provides a continuous pH gradient throughout the tube. Analyte ions also migrate until they reach their isoelectric points. These processes then result in the separation of each analyte into a narrow band that is located at the pH of its isoelectric point (see Figure 30-9c). Very sharp focusing is realized in such systems. Note that isoelectric focusing separations are based on differences in equilibrium properties of the analytes (K_a, K_b) rather than on differences in rates of migration. Once each analyte has migrated to a region where it is neutral, the positions of bands become constant and no longer change with time.

Mobilization of Focused Bands

To detect the focused bands in a capillary isoelectric focusing separation, it is necessary to move, or mobilize, the contents of the capillary so that the bands pass the detector located at one end. This mobilization can be accomplished by applying a pressure difference, just as for sample loading, or by simply changing the solution in the electrode compartment. During the focusing step, equal numbers of $\mathrm{H^+}$ and $\mathrm{OH^-}$ ions enter opposite ends of the capillary, so the pH gradient remains stable. Suppose that sodium chloride is added to the sodium hydroxide solution after focusing is finished. Now both $\mathrm{Cl^-}$ and $\mathrm{OH^-}$ migrate into the cathode end of the column, and the sum of these two concentrations is balanced by $\mathrm{H^+}$ entering the opposite end. That means that there is now less $\mathrm{OH^-}$ than $\mathrm{H^+}$ flowing into the capillary. The pH decreases at the cathode end. The pH gradient is no longer stable. It moves toward the cathode end, and along with it go the focused bands. The bands that pass the detector first are the ones corresponding to proteins with the most alkaline isoelectric points. Figure 30-16 shows an electropherogram for the separation of several proteins by capillary isoelectric focusing. The pI for each protein is shown above the peak. Mobilization was accomplished by adding sodium chloride to the anode compartment.

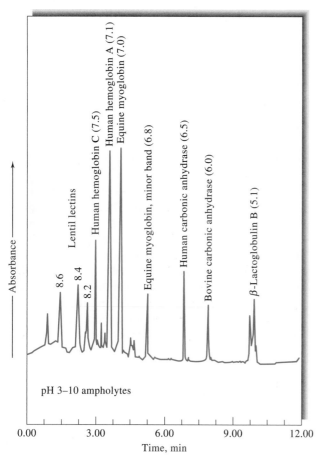

FIGURE 30-16 Capillary isoelectric focusing of proteins. Isoelectric pI listed above the peaks. Detection was by UV absorption. (From T. Wehr, M. Zhu, R. Rodriguez, D. Burke, and K. Duncan, *Amer. Biotech. Lab.*, **1990**, *8*, 22. With permission.)

30C-5 Micellar Electrokinetic Chromatography

In 1984, Terabe and collaborators[17] described a modification of CE that permitted the separation of low-molecular-mass aromatic phenols and nitro compounds with equipment such as shown in Figure 30-1. *Micellar electrokinetic chromatography* (MEKC) is a type of CE that offers several unique features.[18] Like CE, MEKC provides highly efficient separations on microvolumes of sample solution without the need for

the high-pressure pumping system required for HPLC. Unlike normal CE, however, MEKC allows uncharged species to be separated as well as charged species. In MEKC a mobile phase is transported across a stationary phase by electroosmotic flow. As shown in Figure 30-3, electroosmotic pumping leads to a flat-plug profile rather than the parabolic profile of pressure-induced flow. The flat profile of osmotic pumping leads to narrow bands and thus high separation efficiencies.

Micelles form in aqueous solutions when the concentration of an ionic species having a long-chain hydrocarbon tail is increased above a certain level called the *critical micelle concentration* (CMC). At this point the surfactant begins to form spherical aggregates made up of 40 to 100 ions with their hydrocarbon tails in the interior of the aggregate and their charged ends exposed to water on the outside. Micelles constitute a stable second phase that can incorporate nonpolar compounds in the hydrocarbon interior of the particles, thus *solubilizing* the nonpolar species. Solubilization is commonly encountered when a greasy material or surface is washed with a detergent solution.

In MEKC, surfactants are added to the operating buffer in amounts that exceed the CMC. For most applications to date, the surfactant has been sodium dodecyl sulfate (SDS). The surface of an ionic micelle of this type has a large negative charge, which gives it a large electrophoretic mobility. Most buffers, however, exhibit such a high electroosmotic flow rate toward the negative electrode that the anionic micelles are carried toward that electrode also, but at a much reduced rate. Thus, during an experiment, the buffer mixture consists of a faster-moving aqueous phase and a slower-moving micellar phase. When a sample is introduced into this system, the components distribute themselves between the aqueous phase and the hydrocarbon phase in the interior of the micelles. The positions of the resulting equilibria depend on the polarity of the solutes. With polar solutes the aqueous solution is favored; with nonpolar compounds, the hydrocarbon environment is preferred.

The phenomena just described are quite similar to what occurs in an LC column except that the "stationary phase" moves along the length of the column but at a much slower rate than the mobile phase. The mechanism of separation is identical in the two cases and depends on differences in distribution constants for analytes between the mobile aqueous phase and the hydrocarbon *pseudostationary phase*. The process is thus true chromatography; hence, the name "micellar

[17] S. Terabe, K. Otsuka, K. Ichikawa, A. Tsuchiya, and T. Ando, *Anal. Chem.*, **1984**, *56*, 111; S. Terabe, K. Otsuka, and T. Ando, *Anal. Chem.*, **1985**, *57*, 841. See also K. R. Nielsen and J. P. Foley, in *Capillary Electrophoresis*, P. Camilleri, ed., Boca Raton, FL: CRC Press, 1993, Chap. 4.
[18] For a brief review, see S. Terabe, *Anal. Chem.*, **2004**, *76*, 240A.

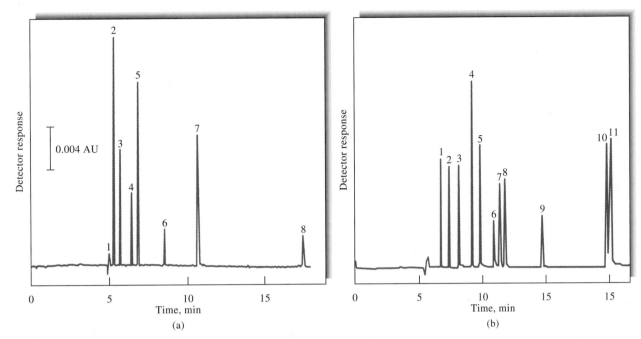

FIGURE 30-17 Typical separations by MEKC. (a) Some test compounds: 1 = methanol, 2 = resorcinol, 3 = phenol, 4 = *p*-nitroaniline, 5 = nitrobenzene, 6 = toluene, 7 = 2-naphthol, 8 = Sudan III; capillary, 50-μm inside diameter, 500 mm to the detector; applied voltage, ~15 kV; detection UV absorption at 210 nm. (b) Analysis of a cold medicine: 1 = acetaminophen, 2 = caffeine, 3 = sulpyrine, 4 = naproxen, 5 = guaiphenesin, 10 = noscapine, 11 = chlorpheniramine and tipepidine; applied voltage, 20 kV; capillary, as in (a); detection UV absorption at 220 nm. (From S. Terabe, *Trends Anal. Chem.*, **1989**, *8*, 129. With permission.)

electrokinetic *chromatography*." Figure 30-17 illustrates two typical separations by MEKC.

MEKC has become important for chiral separations.[19] Here, chiral resolving agents are used as in HPLC to preferentially complex one of the isomers. Either a chiral resolving agent with detergent properties, such as a bile acid, is used to form the micelles or a resolving agent, such as a cyclodextrin, is added to a detergent that is itself achiral. In many aspects, chiral separations are easier to develop by MEKC than by LC, although LC is still the dominant technique in industry because of its familiarity.

MEKC appears to have a promising future. One advantage that this hybrid technique has over HPLC is much higher column efficiencies (100,000 plates or more). In addition, changing the second phase in MEKC is simple, involving only changing the micellar composition of the buffer. In contrast, in HPLC, the second phase can be altered only by changing the type of column packing or column. The MEKC technique appears particularly useful for separating small molecules that are impossible to separate by traditional gel electrophoresis. Recent advances include on-line preconcentration to enhance sensitivity and mass spectrometric detection.[20]

30D PACKED COLUMN ELECTROCHROMATOGRAPHY

Electrochromatography is a hybrid of HPLC and CE that offers some of the best features of the two methods. Like HPLC and MEKC, it is applicable to the separation of neutral species or charged species. Electrochromatography with packed columns is, however, the least mature of the various electroseparation techniques. In this method, a polar solvent is usually driven by electroosmotic flow through a capillary packed with

[19] R. Vaspalec and P. Boček, *Chem. Rev.*, **2000**, *100*, 3715.

[20] S. Terabe, *Anal. Chem.*, **2004**, *76*, 240A.

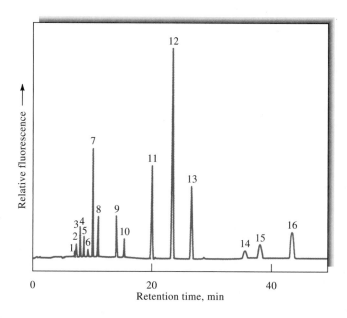

FIGURE 30-18 Electrochromatogram showing the electrochromatographic separation of 16 PAHs (~10^{-6} to 10^{-8} M of each compound). The peaks are identified as follows: (1) naphthalene, (2) acenaphthylene, (3) acenaphthene, (4) fluorene, (5) phenanthrene, (6) anthracene, (7) fluoranthene, (8) pyrene, (9) benz[*a*]anthracene, (10) chrysene, (11) benzo[*b*]fluoranthene, (12) benzo[*k*]fluoranthene, (13) benzo[*a*]pyrene, (14) dibenz[*a,h*]anthracene, (15) benzo[*ghi*]perylene, and (16) indeo[1,2,3-*cd*]pyrene. (From C. Yan, R. Dadoo, H. Zhao, D. J. Rakestraw, and R. N. Zare, *Anal. Chem.*, **1995**, *67*, 2026. With permission.)

a reversed-phase HPLC packing. Separations depend on the distribution of the analyte species between the mobile phase and the liquid stationary phase held on the packing. Figure 30-18 shows a typical electrochromatogram for the separation of sixteen polyaromatic hydrocarbons (PAHs) in a 33-cm-long capillary having an inside diameter of 75 μm. The mobile phase consisted of acetonitrile in a 4-mM sodium borate solution. The stationary phase consisted of 3-μm octadecylsilica particles.

30E FIELD-FLOW FRACTIONATION

Field-flow fractionation (FFF) describes a group of analytical techniques that are becoming quite useful in the separation and characterization of dissolved or suspended materials such as polymers, large particles, and colloids. Although the FFF concept was first described by Giddings in 1966,[21] only recently have practical applications and advantages over other methods been shown.[22]

30E-1 Separation Mechanisms

Separations in FFF occur in a thin ribbon-like flow channel such as that shown in Figure 30-19. The channel is typically 25–100 cm long and 1–3 cm wide. The thick-

[21] J. C. Gidding, *Sep. Sci.*, **1966**, *1*, 123.
[22] For a review of FFF methods, see J. C. Giddings, *Anal. Chem.*, **1995**, *67*, 592A.

ness of the ribbon-like structure is usually 50–500 μm. The channel is usually cut from a thin spacer and sandwiched between two walls. An electrical, thermal, or centrifugal field is applied perpendicular to the flow direction. Alternatively, a cross flow perpendicular to the main flow direction can be used.

In practice, the sample is injected at the inlet to the channel. The external field is next applied across the face of the channel as illustrated in Figure 30-19. In the presence of the field, sample components migrate toward the *accumulation wall* at a velocity determined by the strength of the interaction of the component with the field. Sample components rapidly reach a steady-state concentration distribution near the accumulation wall as shown in Figure 30-20. The mean thickness of the component layer *l* is related to the diffusion coefficient of the molecule *D* and to the field-induced velocity *U* toward the wall. The faster the component moves in the field, the thinner the layer near the wall. The larger the diffusion coefficient, the thicker the layer. Because the sample components have different values of *D* and *U*, the mean layer thickness will vary among components.

Once components have reached their steady-state profiles near the accumulation wall, the channel flow is begun. The flow is laminar, resulting in the parabolic profile shown on the left in Figure 30-20. The main carrier flow has its highest velocity in the center of the channel and its lowest velocity near the walls. Components that interact strongly with the field are com-

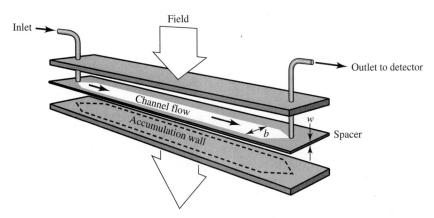

FIGURE 30-19 Schematic diagram of FFF flow channel sandwiched between two walls. An external field (electrical, thermal, centrifugal) is applied perpendicular to the flow direction.

pressed very near the wall as shown by component A in Figure 30-21. Here, they are eluted by slow-moving solvent. Components B and C protrude more into the channel and experience a higher solvent velocity. The elution order is thus C, then B, then A. Components separated by FFF flow through a UV-visible absorption, refractive index, or fluorescence detector located at the end of the flow channel and similar to those used in HPLC separations. The separation results are revealed by a plot of detector response versus time, called a *fractogram*, similar to a chromatogram in chromatography.

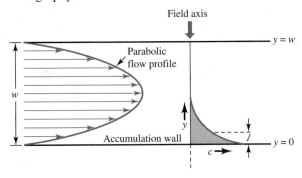

FIGURE 30-20 When the field is applied in FFF, components migrate to the accumulation wall where an exponential concentration profile exists as seen on the right. Components extend a distance y into the channel. The average thickness of the layer is l, which differs for each component. The main channel flow is then turned on and the parabolic flow profile of the eluting solvent is shown on the left.

 Simulation: Learn more about **field flow fractionation**.

30E-2 FFF Methods

Different FFF subtechniques result from the application of different types of fields or gradients.[23] To date, the methods that have been developed are *sedimentation FFF*, *electrical FFF*, *thermal FFF*, and *flow FFF*.

Sedimentation FFF

Sedimentation FFF is by far the most widely used form. In this technique, the channel is coiled and made to fit inside a centrifuge basket as illustrated in Figure 30-22. Components with the highest mass and density are driven to the wall by the sedimentation (centrifugation) force and elute last. Low-mass species are eluted first. There is relatively high selectivity between particles of different size in sedimentation FFF. A separation of polystyrene beads of various diameters by sedimentation FFF is shown in Figure 30-23.

Because the centrifugation forces are relatively weak for small molecules, sedimentation FFF is most applicable for molecules with molecular masses exceeding 10^6. Such systems as polymers, biological macromolecules, natural and industrial colloids, emulsions, and subcellular particles appear to be amenable to separation by sedimentation FFF.

Electrical FFF

In electrical FFF, an electric field is applied perpendicular to the flow direction. Retention and separation occur based on electrical charge. Species with the

[23] For a discussion of the various FFF methods, see J. C. Giddings, *Unified Separation Science*, New York: Wiley, 1991, Chap. 9; M. E. Schimpf, K. Caldwell, and J. C. Giddings, eds., *Field-Flow Fractionation Handbook*, New York: Wiley, 2000.

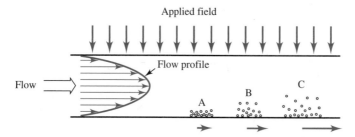

Applied field

Flow profile

Flow

A   B   C

FIGURE 30-21 Three components A, B, and C are shown compressed against the accumulation wall in FFF to different amounts because of different interactions with the external field. When the flow is begun, component A experiences the lowest solvent velocity because it is the closest to the wall. Component B protrudes more into the channel where it experiences a higher flow velocity. Component C, which interacts the least with the field, experiences the highest solvent-flow velocity and thus is displaced the most rapidly by the flow.

FIGURE 30-22 Sedimentation FFF apparatus. (Courtesy of Postnova Analytics.)

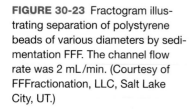

FIGURE 30-23 Fractogram illustrating separation of polystyrene beads of various diameters by sedimentation FFF. The channel flow rate was 2 mL/min. (Courtesy of FFFractionation, LLC, Salt Lake City, UT.)

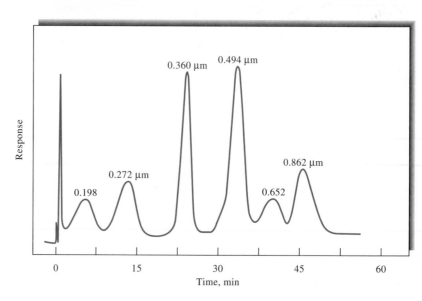

Response

$0.360\ \mu m$ $0.494\ \mu m$

$0.272\ \mu m$

$0.862\ \mu m$

0.198

0.652

0   15   30   45   60

Time, min

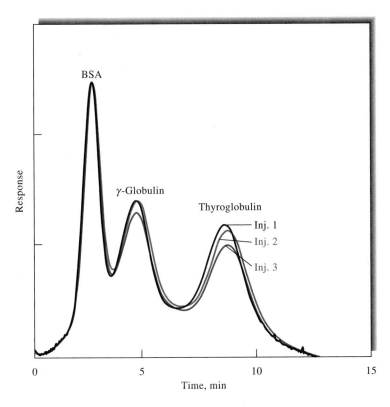

FIGURE 30-24 Separation of three proteins by flow FFF. Three separate injections are shown. In the experiment shown, the sample was concentrated at the head of the channel by means of an opposing flow. BSA = bovine serum albumin. (From H. Lee, S. K. R. Williams, and J. C. Giddings, *Anal. Chem.*, **1998**, *70*, 2495. Copyright 1998 American Chemical Society.)

highest charge are driven most effectively toward the accumulation wall. Species of lower charge are not as compacted and protrude more into the higher-flow region. Hence, species of the lowest charge are eluted first, with highly charged species retained the most.

Because electric fields are quite powerful, even small ions should be amenable to separation by electrical FFF. However, electrolysis effects have limited the application of this method to the separation of mixtures of proteins and other large molecules.

Thermal FFF

In thermal FFF, a thermal field is applied perpendicular to the flow direction by forming a temperature gradient across the FFF channel. The temperature difference induces thermal diffusion in which the velocity of movement is related to the thermal diffusion coefficient of the species.

Thermal FFF is particularly well suited for the separation of synthetic polymers with molecular masses in the range of 10^3 to 10^7. The technique has significant advantages over size-exclusion chromatography for high-molecular-mass polymers. On the other hand, low-molecular-mass polymers appear to be better separated by size-exclusion methods. In addition to polymers, particles and colloids have been separated by thermal FFF.[24]

Flow FFF

Perhaps the most versatile of all the FFF subtechniques is flow FFF, in which the external field is replaced by a slow cross flow of the carrier liquid. The perpendicular flow transports material to the accumulation wall in a nonselective manner. However, steady-state layer thicknesses are different for various components because they depend not only on the transport rate but also on molecular diffusion. Exponential distributions of differing thicknesses are formed as in normal FFF.

Flow FFF has been applied to the separation of proteins, synthetic polymers, and a variety of colloidal particles. Figure 30-24 illustrates the separation of three proteins by flow FFF. The reproducibility is illustrated by the fractograms for the three injections.

---

[24]P. M. Shiundu, G. Liu, and J. C. Giddings, *Anal. Chem.*, **1995**, *67*, 2705.

30E-3 Advantages of FFF over Chromatographic Methods

FFF appears to have several advantages over ordinary chromatographic methods for some applications. First, no packing material or stationary phase is needed for separation to occur. In some chromatographic systems, there may be undesirable interactions between the packing material or stationary phase and the sample constituents. Some solvents or sample materials adsorb or react with the stationary phase or its support. Macromolecules and particles are particularly prone to such adverse interactions.

The geometry and flow profiles involved in FFF are well characterized. Likewise, the effects of most external fields can be readily modeled. As a result, fairly exact theoretical predictions of retention and plate height can be made in FFF. Chromatographic predictions are still rather inexact in comparison.

Finally, the external field governs FFF retention. With electrical, sedimentation, and flow FFF, the perpendicular forces can be varied rapidly and in a time-programmed fashion. This gives FFF a certain versatility in adapting to different types of samples. Methods can also be readily optimized for resolution and separation speed.

Although FFF is a fairly recent addition to the repertoire of analytical separation methods, it has been shown to be highly complementary to chromatography. The FFF methods are best suited for macromolecules and particles that are for the most part beyond the molecular mass range of chromatographic methods. On the other hand, chromatographic methods are clearly superior for low-molecular-mass substances.

QUESTIONS AND PROBLEMS

*Answers are provided at the end of the book for problems marked with an asterisk.

[X] Problems with this icon are best solved using spreadsheets.

30-1 What is electroosmotic flow? Why does it occur?

30-2 Suggest how electroosmotic flow might be suppressed.

30-3 Why does pH affect separation of amino acids by electrophoresis?

30-4 What is the principle of separation by CZE?

***30-5** A certain inorganic cation has an electrophoretic mobility of $4.31 \times 10^{-4} \text{ cm}^2 \text{ s}^{-1} \text{ V}^{-1}$. This same ion has a diffusion coefficient of $9.8 \times 10^{-6} \text{ cm}^2 \text{ s}^{-1}$. If this ion is separated from other cations by CZE with a 50.0-cm capillary, what is the expected plate count N at applied voltages of
(a) 5.0 kV?
(b) 10.0 kV?
(c) 30.0 kV?

***30-6** The cationic analyte of Problem 30-5 was separated by CZE in a 50.0-cm capillary at 10.0 kV. Under the separation conditions, the electroosmotic flow rate was 0.85 mm s^{-1} toward the cathode. If the detector were placed 40.0 cm from the injection end of the capillary, how long would it take in minutes for the analyte cation to reach the detector after the field is applied?

30-7 What is the principle of micellar electrokinetic capillary chromatography? How does it differ from CZE?

30-8 Describe a major advantage of micellar electrokinetic capillary chromatography over conventional liquid chromatography.

30-9 Three large proteins are ionized at the pH at which an electrical FFF separation is carried out. If the ions are designated A^{2+}, B^+, and C^{3+}, predict the order of elution.

30-10 What determines the elution order in sedimentation FFF?

30-11 List the major advantages and limitations of FFF compared to chromatographic methods.

Challenge Problem

30-12 Doxorubicin (DOX) is a widely used anthracycline that has been effective in treatments of leukemia and breast cancer in humans (A. B. Anderson, C. M. Ciriaks, K. M. Fuller, and E. A. Ariaga, *Anal. Chem.*, **2003**, *75*, 8). Unfortunately, side effects, such as liver toxicity and drug resistance, have been reported. In a recent study, Anderson et al. used laser-induced fluorescence (LIF) as a detection mode for CE to investigate metabolites of DOX in single cells and subcellular fractions. The following are results similar to those obtained by Anderson et al. for quantifying doxorubicin by LIF. The CE peak areas were measured as a function of the DOX concentration to construct a calibration curve.

DOX Concentration, nM	Peak Area
0.10	0.10
1.00	0.80
5.00	4.52
10.00	8.32
20.00	15.7
30.00	26.2
50.00	41.5

(a) Find the equation for the calibration curve and the standard deviations of the slope and intercept. Find the R^2 value.

(b) Rearrange the equation found in part (a) to express concentration in terms of the measured area.

(c) The limit of detection (LOD) for DOX was found to be 3×10^{-11} M. If the injection volume was 100 pL, what was the LOD in moles?

(d) Two samples of unknown DOX concentration were injected and peak areas of 11.3 and 6.97 obtained. What were the concentrations and their standard deviations?

(e) Under certain conditions, the DOX peak required 300 s to reach the LIF detector. What time would be required if the applied voltage were doubled? What time would be required if the capillary length were doubled at the same applied voltage?

(f) The capillary used in part (e) under normal conditions had a plate count of 100,000. What would N be if the capillary length were doubled at the same applied voltage? What would N be if the applied voltage were doubled at the original capillary length?

(g) For a 40.6-cm-long capillary of inside diameter 50 μm, what would the plate height be for a capillary with $N = 100,000$?

(h) For the same capillary as in part (g), what is the variance σ^2 of a typical peak?

Discovering Acrylamide

Introduction

In 1997 in southwestern Sweden, residents near the Hallandsas ridge began finding dead rainbow trout in nearby streams and paralyzed cows in grasslands used for grazing. In addition, workers building a tunnel through the ridge for the Swedish railway experienced tingling and numbness in their extremities. The problem was soon traced to chemicals that had leached from a grouting agent injected to firm up the soft rock layers being bored near the tunnel entrance. The discovery of high levels of acrylamide ($CH_2{=}CH{-}CO{-}NH_2$), a *neurotoxin* and possible *carcinogen*, in the streams, the groundwater, and the grazing grasses prompted a shutdown of the tunnel construction. Acrylamide is the monomer of polyacrylamide, a widely used industrial polymer. In 2002 acrylamide was identified in human foods by workers at the Swedish National Food Administration and the University of Stockholm.[1] Since that time, relatively high concentrations of acrylamide have been discovered in a variety of processed foods, such as potato chips, coffee, breakfast cereals, peanut butter, and pastries. In high doses, acrylamide has been found to cause cancer in laboratory animals. The chemical has been classified as a group 2A carcinogen ("possibly carcinogenic to humans").[2] Analytical methods to determine acrylamide are clearly quite important in establishing its origins, in studying its health risks, and in finding means to reduce human exposure to its effects.

Analytical Methods for Acrylamide

Several different analytical approaches have been proposed for the determination of acrylamide.[3] The analysis of food samples presents a special challenge because of the high levels of matrix interferences, particularly in starchy foods. Early methods used GC or HPLC in the determination. Although these methods were successful in water, agricultural samples, and environmental samples, the selectivity was not sufficient for samples of cooked foods where the matrices are quite complex. The most successful methods for food samples have involved MS coupled with GC or LC separation methods.

GC/MS Methods

In these methods, the analyte has either been derivatized by bromination or determined by a direct method after liquid-liquid extraction. In the direct method, the extract is separated by capillary GC, and the eluent is monitored by chemical-ionization mass spectrometry (CI/MS). A typical GC/CI/MS chromatogram of standards is shown in Figure IA5-1. In these experiments, a Carbowax capillary column was used with positive-ion CI/MS detection. Bromination methods, although more tedious and time-consuming than other methods, have several advantages. They are more selective, the analyte is more volatile, and they are more sensitive than direct methods. Usually, the ions $(CH_2{-}CHBr{-}CONH_2)^+$ and $(CH_2{=}CHBr)^+$ are measured in the selective-ion monitoring mode using a basic quadrupole MS instrument. Improved results have been obtained for acrylamide by GC/MS/MS or GC coupled with high-resolution MS. Detection limits are less than 10 µg/kg.

LC/MS Methods

Most LC/MS methods use solid-phase extraction (SPE) for sample cleanup and analyte preconcentration. A major advantage of SPE is the retention of the many interfering species. Often, tandem MS is used following the extraction to provide additional resolution and freedom from interferences due to coextractants. Ethyl acetate extractions from an aqueous phase have also been used to remove salts, sugars, starches, and amino acids. Such cleanup procedures are often necessary with difficult matrices such as coffee, chocolates, and cocoa powders. With extensive cleanup of the sample, single-stage mass spectrometers have been more successful than MS/MS systems. Detection limits by both single-stage and tandem MS approach the 10-µg/kg level. An important advantage of LC methods is that derivatization is not necessary, which can simplify the procedure and decrease the analysis time. Typical LC/MS chromatograms of a control standard and an acrylamide-spiked potato chip extract are shown in Figure IA5-2. Electrospray ionization and selected-ion monitoring were used. A completely aqueous mobile phase was employed with a fluorinated reversed-phase LC column.

On-Line Monitoring

Proton-transfer-reaction mass spectrometry (PTR/MS) has been used for on-line monitoring of acrylamide without any sample pretreatment.[4] The formation of acrylamide

[1] E. Tareke, P. Rydberg, P. Karlsson, S. Eriksson, and M. Törnqvist, *J. Agric. Food Chem.*, **2002**, *50*, 4998.
[2] International Agency for Research on Cancer, *Monographs on Evaluation of Carcinogenic Risk of Chemicals to Humans*, **1994**, *60*, 435.
[3] D. Taeymans, J. Wood, P. Ashby, I. Blank, et al. *Crit. Rev. Food Sci. Nutr.*, **2004**, *44*, 323.
[4] P. Pollien, C. Lindinger, C. Yeretzian, and I. Blank, *Anal. Chem.*, **2003**, *75*, 5488.

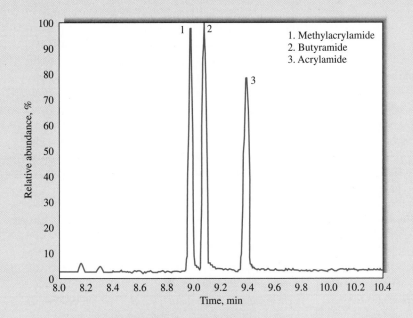

FIGURE IA5-1 A typical GC/CI/MS chromatogram of standards.

1. Methylacrylamide
2. Butyramide
3. Acrylamide

released into the headspace above a potato sample was monitored in real-time during thermal treatment. The amount of acrylamide released correlated positively with the treatment temperature.

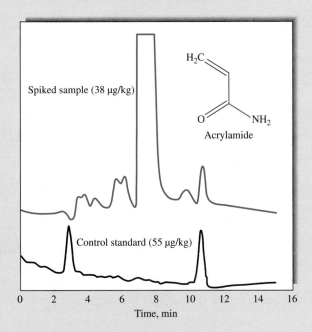

FIGURE IA5-2 A chromatogram of control standards and an acrylamide-spiked potato chip.

Interlaboratory Studies

Although substantial progress has been made in determining acrylamide, a recent interlaboratory comparison showed that many current methods are not satisfactory for samples in complex matrices.[5] Even with relatively simple samples, the range of results was too wide to be acceptable. Thus, more improvements in analytical methodology are certainly in order. A fully validated improved LC/MS/MS method was recently reported that uses isotope dilution, LC, and electrospray ionization MS/MS.[6] Excellent results were obtained on chocolate powders, cocoa, and coffee.

The Origin of Acrylamide in Foods and the Environment

With analytical methods based on GC/MS or LC/MS, it has been possible to discover how acrylamide originates in foods and the environment. In the environment, acrylamide arises from agricultural, water-treatment, and industrial sources. Among its many applications, polyacrylamide is used in water purification to aggregate suspended organic compounds, in irrigation water to improve soil texture, in

[5]T. Wenzl, B. de la Calle, R. Gatermann, K. Hoenicke, F. Ulberth, and E. Anklam, *Anal. Bioanal. Chem.*, **2004**, *379*, 449.
[6]T. Delatour, A. Perisset, T. Goldmann, S. Riediker, and R. H. Stadler, *J. Agric. Food Chem.*, **2004**, *52*, 4625.

pesticide formulations to limit spray drifting, and in biology and chemistry to prepare gels used in electrophoresis (see Section 30C-2). Although polyacrylamide is generally classified as a nontoxic additive, it can be contaminated with the toxic monomer. An arbitrary limit of 500 ppm acrylamide in polyacrylamide preparations has been set for agricultural and water-treatment uses. The polyacrylamide preparation used in irrigation and water treatment is often an acrylic acid copolymer, which can also release acrylates known to cause birth defects (a *teratogen*).

In foods, carbohydrate-rich foods cooked at high temperatures and under low-moisture conditions have been found to contain acrylamide.[7] The compound has not been found in raw and boiled foods. Researchers have established that acrylamide in foods arises from reactions related to the Maillard reaction, a nonenzymatic reaction responsible for the brown color of cooked foods. The Maillard reaction involves the initial combination of an amino acid and a reducing sugar. In the case of foods, the amino acid has been found to be asparagine. A complex series of steps leading to acrylamide has been proposed. Heating time and temperature are crucial. The role of water and the influence of side reactions are less well understood.

Health Risks Studies

Acrylamide is known to cause damage to the nervous system in humans and animals. A joint study of the UN Food and Agriculture Organization (FAO) and the World Health Organization (WHO) reported in 2002 on the health implications of acrylamide in foods.[8] The average daily food intake of acrylamide in the general population was estimated to be below the level that could be expected to cause neurotoxic effects. Regarding genetic damage that may lead to cancer, the study concluded that the evidence from human studies did not show a link between acrylamide levels and incidences of cancer, although animal studies have suggested such a link. The study also called for additional research by the international community and for better sharing of scientific information on acrylamide. It was also suggested that the public try to minimize exposure by not cooking foods excessively, by lowering consumption of fried and fatty foods, and by eating a balanced diet.

Future Prospects

Current research is aimed at further improvements in the analytical methodology for acrylamide, modeling methods to help elucidate the kinetics of its formation, understanding its health risks, and methods for reducing its levels in foods. Once the formation kinetics and mechanism are better understood, reduction of acrylamide levels can be pursued on a sound basis. For example, on the basis of current knowledge, methods have been suggested for reducing levels of acrylamide in potato products by storing raw potatoes under controlled temperatures and by modifying the conditions used in potato processing. Enzymatic treatments of potato products with asparaginase have also been suggested. Clearly, more scientific research must be done in this area, and analytical chemistry will play a leading role.

---

[7] I. Blank, *Ann. N.Y. Acad. Sci.*, **2005**, *1043*, 30.

[8] *Health Implications of Acrylamide in Foods*, Report of a joint FAO/WHO Consultation, World Health Organization, Geneva, 2002.

A modern thermogravimetric analyzer. The instrument has a temperature-controlled thermobalance and operates under conditions of controlled humidity. It has a multiposition auto-sampler capable of automated analysis of up to twenty-five samples. One version of this system features an infrared furnace for heating from ambient temperature to 1200°C. The entire system is under computer control with software designed for specific techniques and analytical methods. (Courtesy of TA Instruments, New Castle, DE.)

Section 6 consists of four chapters devoted to miscellaneous instrumental methods. Thermogravimetric analysis, differential thermal analysis, differential scanning calorimetry, and microthermal analysis are discussed in Chapter 31. The theory and practice of radiochemical methods, including neutron activation analysis and isotope dilution techniques, are discussed in Chapter 32. In Chapter 33 the principles, instrumentation, and applications of automated analyzers are described. Flow injection analyzers, microfluidic systems, and discrete analyzers are included. Chapter 34, the final chapter in this section and in the book, examines analytical methods to determine particle size. The methods include low-angle laser light scattering, dynamic light scattering, and photosedimentation. The Instrumental Analysis in Action feature presents the John Vollman case, the first murder investigation in which neutron activation analysis was used.

CHAPTER THIRTY-ONE

Thermal Methods

T hermal analysis techniques are those in which a physical property of a substance or its reaction products is measured as a function of temperature. Usually, the substance is subjected to a controlled temperature program during the analysis. Although there are more than a dozen thermal analysis techniques, we confine our discussion in this chapter to four methods that provide primarily chemical rather than physical information about samples of matter. These methods include **thermogravimetric analysis**, **differential thermal analysis**, **differential scanning calorimetry**, and **microthermal analysis**.

Throughout this chapter, this logo indicates an opportunity for online self-study at **www.thomsonedu.com/chemistry/skoog**, linking you to interactive tutorials, simulations, and exercises.

Thermal methods differ in the properties measured and the temperature programs applied.[1] The four methods discussed here find widespread use for both quality control and research applications on polymers, pharmaceutical preparations, clays, minerals, metals, and alloys.

31A THERMOGRAVIMETRIC ANALYSIS

In a thermogravimetric analysis (TGA) the mass of a sample in a controlled atmosphere is recorded continuously as a function of temperature or time as the temperature of the sample is increased (usually linearly with time). A plot of mass or mass percentage as a function of time is called a *thermogram* or a *thermal decomposition curve*.

31A-1 Instrumentation

Commercial instruments for TGA consist of (1) a sensitive microbalance, called a thermobalance; (2) a furnace; (3) a purge-gas system for providing an inert, or sometimes reactive, atmosphere; and (4) a computer system for instrument control, data acquisition, and data processing. A purge-gas switching system is a common option for applications in which the purge gas must be changed during an experiment.

The Thermobalance

A number of different thermobalance designs available commercially are capable of providing quantitative information about samples ranging in mass from less than 1 mg to 100 g. The usual range of thermobalances, however, is from 1 to 100 mg. Many of the balances can detect changes in mass as small as 0.1 μg. Although the sample holder must be housed in the furnace, the rest of the balance must be thermally isolated from the furnace. Figure 31-1 is a schematic diagram of one thermobalance design. A change in sample mass causes a deflection of the beam, which interposes a light shutter between a lamp and one of two photodiodes. The resulting imbalance in the photodiode

---

[1] For a detailed description of thermal methods, see *Principles of Thermal Analysis and Calorimetry*, P. J. Haines, ed., Cambridge, UK: Royal Society of Chemistry, 2002; P. J. Haines, *Thermal Methods of Analysis*, London: Blackie, 1995; B. Wunderlich, *Thermal Analysis*, Boston: Academic Press, 1990; W. W. Wendlandt, *Thermal Analysis*, 3rd ed., New York: Wiley, 1985. For recent reviews, see S. Vyazovkin, *Anal. Chem.*, **2006**, *78*, 3875; *Anal. Chem.*, **2004**, *76*, 3299; *Anal. Chem.*, **2002**, *74*, 2749. For a description of thermal analysis instruments, see B. E. Erickson, *Anal. Chem.*, **1999**, *71*, 689A.

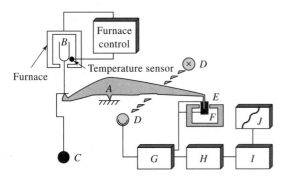

FIGURE 31-1 Thermobalance components. The balance beam is shown as A. The sample cup and holder are B; C is a counterweight. D is a lamp and photodiodes, E is a magnetic coil, and F is a permanent magnet. The computer data-acquisition, data-processing, and control systems are components G, H, and I. Component J is the printer and display unit. (Courtesy of Mettler-Toledo.)

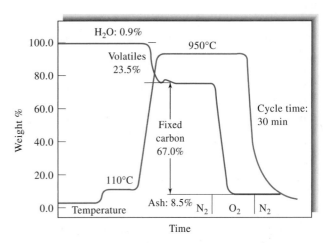

FIGURE 31-2 Controlled atmospheric thermogram for a bituminous coal sample. A nitrogen atmosphere was used for about 18 min followed by an oxygen atmosphere for 4 to 5 min. The experiment was then completed in nitrogen. (Reprinted with permission from C. M. Earnest, *Anal. Chem.*, **1984**, *56*, 1471A.)

current is amplified and fed into coil E, which is situated between the poles of a permanent magnet F. The magnetic field generated by the current in the coil restores the beam to its original position. The amplified photodiode current is monitored and transformed into mass or mass-loss information by the data-processing system. In most cases mass versus temperature data can either be plotted in real time or stored for further manipulation or display at a later time.

The Furnace

Furnaces for TGA typically cover the range from ambient temperature to 1000°C, although some can be used for temperatures up to 1600°C. Heating rates can often be selected from 0.1°C/min to 100°C/min. Some units can heat as rapidly as 200°C/min. Insulation and cooling of the exterior of the furnace is required to avoid heat transfer to the balance. Nitrogen or argon is usually used to purge the furnace and prevent oxidation of the sample. For some analyses, it is desirable to switch purge gases as the analysis proceeds. Figure 31-2 provides an example in which the purge gas was automatically switched from nitrogen to oxygen and then back to nitrogen. The sample in this case was a bituminous coal. The purge gas was nitrogen during the first 18 min, while the moisture content and the percentage of volatiles were recorded. The gas was then switched to oxygen for 4 to 5 min, which caused oxidation of carbon to carbon dioxide. Finally, the experiment was concluded with a nitrogen purge to give a measure of the ash content.

The temperature axis is usually calibrated by using a material of known Curie point or sometimes a melting point standard. The *Curie point* is the temperature at which a ferromagnetic material becomes paramagnetic. The ferromagnetic material is suspended in a magnetic field oriented such that a vertical component of magnetic force acts on the sample. The magnetic force acts as an equivalent magnetic mass on the TGA balance beam to indicate an apparent sample mass. When the sample is heated through its Curie point, the magnetic mass is lost, and the balance indicates an apparent loss in mass. Temperature accuracy is better than 1°C with most furnaces, and temperature precision is typically ±0.1°C. Furnaces are usually cooled by forced air. Most furnaces can be cooled from 1000°C to 50°C in less than 20 minutes.

Sample Holders

Samples are typically contained in sample pans made of platinum, aluminum, or alumina. Platinum is most often used because of its inertness and ease of cleaning. The volumes of sample pans range from 40 μL to more than 500 μL. Autosamplers are available as attachments for most TGA systems. With the majority of these units, all aspects are automated under software control. The sample-pan taring, loading, and weighing; the furnace heating and cooling; and the pan unloading are totally automatic.

Temperature Control and Data Processing

The temperature recorded in a thermogram is ideally the actual temperature of the sample. This temperature can, in principle, be obtained by immersing a small thermocouple directly in the sample. Such a procedure is seldom followed, however, because of possible catalytic decomposition of samples, potential contamination of samples, and weighing errors resulting from the thermocouple leads. Because of these problems, recorded temperatures are generally measured with a small thermocouple located as close as possible to the sample container. The recorded temperatures then generally lag or lead the actual sample temperature.

Modern TGA systems use a computerized temperature control routine that automatically compares the voltage output of the thermocouple with a voltage-versus-temperature table stored in computer memory. The computer uses the difference between the temperature of the thermocouple and the temperature specified to adjust the voltage to the heater. In some systems the same thermocouple behaves as the heating element and the temperature sensor. With modern control systems, it is possible to achieve excellent agreement between the specified temperature program and the temperature of the sample. Typical run-to-run reproducibility for a particular program falls within 2°C throughout an instrument's entire operating range.

Combined Thermal Instruments

Several manufacturers offer systems that provide simultaneous measurement of heat flow (see Section 31C for differential scanning calorimetry) and mass change (that is, TGA) or of energy change (see Section 31B for differential thermal analysis) and mass change. Such instruments can not only track the loss of material or a vaporization phenomenon with temperature but also reveal transitions associated with these processes. These combination units can eliminate the effects of changes in sample size, homogeneity, and geometry. Many TGA systems produce the derivative of the thermogram as well as the thermogram itself. Such derivative plots are not true differential thermograms as produced in differential thermal analysis, but they provide similar qualitative information. They are often called single differential thermal analysis (SDTA) plots.

 Tutorial: Learn more about **thermogravimetric analysis**.

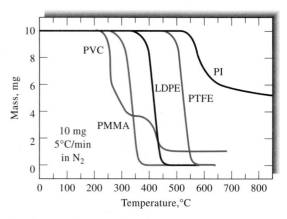

FIGURE 31-3 Thermograms for some common polymeric materials. PVC = polyvinyl chloride; PMMA = polymethylmethacrylate; LDPE = low-density polyethylene; PTFE = polytetrafluoroethylene; PI = aromatic polypyromelitimide. (From J. Chiu, in *Thermoanalysis of Fiber-Forming Polymers*, R. F. Schwenker, ed., New York: Wiley, 1966.)

TGA/MS and TGA/FTIR

TGA is used to determine the loss in mass at particular temperatures, but TGA cannot identify the species responsible. To obtain this type of information, the output of a thermogravimetric analyzer is often connected to a Fourier transform infrared (FTIR) or a mass spectrometer (MS). Several instrument companies offer devices to interface the TGA unit to a spectrometer. Some even claim true integration of the software and hardware of the TGA/MS or TGA/FTIR systems.

High-Resolution TGA

In high-resolution TGA, the sample heating rate fluctuates so that the sample is heated more rapidly during periods of constant mass than during periods when mass changes occur. This allows higher resolution to be obtained during the interesting periods and reduces the time of inactivity.

31A-2 Applications

Because TGA monitors the mass of the analyte with temperature, the information provided is quantitative, but limited to decomposition and oxidation reactions and to such physical processes as vaporization, sublimation, and desorption. Among the most important applications of TGA[2] are compositional analysis and decomposition profiles of multicomponent systems.

---

[2]For a discussion of applications of thermal methods, see A. J. Pasztor, in *Handbook of Instrumental Techniques for Analytical Chemistry*, F. Settle, ed., Upper Saddle River, NJ: Prentice-Hall, 1997, Chap. 50.

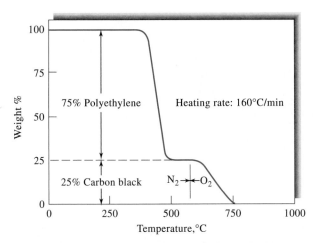

FIGURE 31-4 Thermogravimetric determination of carbon black in polyethylene. (From J. Gibbons, *Amer. Lab.*, **1981**, *13* (1), 33. Copyright 1981 by International Scientific Communications, Inc.)

In polymer studies, thermograms provide information about decomposition mechanisms for various polymeric preparations. In addition, the decomposition patterns are characteristic for each kind of polymer and can sometimes be used for identification purposes. Figure 31-3 shows decomposition patterns for five polymers obtained by thermogravimetry.

Figure 31-4 illustrates how a thermogram is used for compositional analysis of a polymeric material. The sample is a polyethylene that has been formulated with fine carbon-black particles to inhibit degradation from exposure to sunlight. This analysis would be difficult by most other analytical methods.

Figure 31-5 is a recorded thermogram obtained by increasing the temperature of pure $CaC_2O_4 \cdot H_2O$ at a rate of 5°C/min. The clearly defined horizontal regions correspond to temperature ranges in which the indi-

cated calcium compounds are stable. This figure illustrates the use of TGA in defining the thermal conditions needed to produce a pure species.

Figure 31-6a illustrates an application of TGA to the quantitative analysis of a mixture of calcium, strontium, and barium ions. The three are first precipitated as the monohydrated oxalates. The mass in the temperature range between 320°C and 400°C is that of the three anhydrous compounds, CaC_2O_4, SrC_2O_4, and BaC_2O_4, and the mass between about 580°C and 620°C corresponds to the three carbonates. The mass change in the next two steps results from the loss of carbon dioxide, as first CaO and then SrO are formed. Sufficient data are available in the thermogram to calculate the mass of each of the three elements present in the sample.

Figure 31-6b is the derivative of the thermogram shown in (a). The derivative curve can sometimes reveal information that is not detectable in the ordinary thermogram. For example, the three peaks at 140°C, 180°C, and 205°C suggest that the three hydrates lose moisture at different temperatures. However, all appear to lose carbon monoxide simultaneously and thus yield a single sharp peak at 450°C.

Because TGA can provide quantitative information, determination of moisture levels is another important application. Levels of 0.5% and sometimes less can be determined.

31B DIFFERENTIAL THERMAL ANALYSIS

Differential thermal analysis (DTA) is a technique in which the difference in temperature between a substance and a reference material is measured as a function of temperature while the substance and reference material are subjected to a controlled temperature program. Usually, the temperature program involves

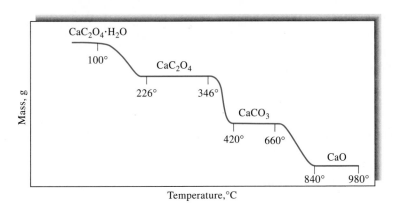

FIGURE 31-5 Thermogram for decomposition of $CaC_2O_4 \cdot H_2O$ in an inert atmosphere. (From S. Peltier and C. Duval, *Anal. Chim. Acta*, **1947**, *1*, 345. With permission.)

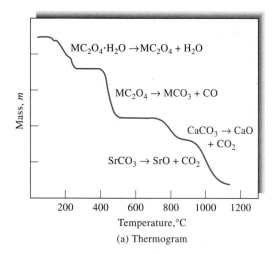

$$MC_2O_4 \cdot H_2O \rightarrow MC_2O_4 + H_2O$$

$$MC_2O_4 \rightarrow MCO_3 + CO$$

$$CaCO_3 \rightarrow CaO + CO_2$$

$$SrCO_3 \rightarrow SrO + CO_2$$

Mass, m

Temperature, °C

(a) Thermogram

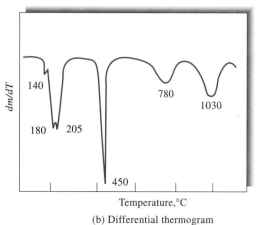

dm/dT

140

780

1030

180   205

450

Temperature, °C

(b) Differential thermogram

FIGURE 31-6 Decomposition of $CaC_2O_4 \cdot H_2O$, $SrC_2O_4 \cdot H_2O$, and $BaC_2O_4 \cdot H_2O$. From L. Erdey, G. Liptay, G. Svehla, and F. Paulike, *Talanta*, **1962**, *9*, 490. With permission.)

heating the sample and reference material in such a way that the temperature of the sample T_s increases linearly with time. The difference in temperature ΔT between the sample temperature and the reference temperature T_r ($\Delta T = T_s - T_r$) is then monitored and plotted against sample temperature to give a differential thermogram such as that shown in Figure 31-7. (The significance of the various parts of this curve is given in Section 31B-2.)

31B-1 Instrumentation

Figure 31-8 is a schematic of the furnace compartment of a differential thermal analyzer. A few milligrams of the sample (S) and an inert reference substance (R) are contained in small aluminum dishes located above sample and reference thermocouples in an electrically

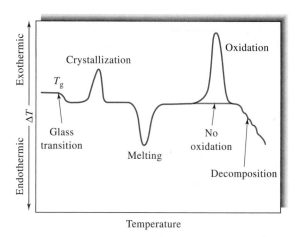

FIGURE 31-7 Differential thermogram showing types of changes encountered with polymeric materials. (From R. M. Schulken Jr., R. E. Roy Jr., and R. H. Cox, *J. Polymer Sci., Part C*, **1964**, *6*, 18. Reprinted by permission of John Wiley & Sons, Inc.)

heated furnace. The reference material is an inert substance such as alumina, silicon carbide, or glass beads.

The digitized output voltage E_s from the sample thermocouple is the input to a computer. The computer controls the current input to the furnace in such a way that the sample temperature increases linearly and at a predetermined rate. The sample thermocouple signal is also converted to temperature T_s, which

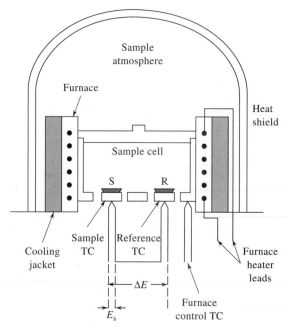

FIGURE 31-8 Schematic diagram of a typical instrument for DTA. TC = thermocouple.

is used as the x-axis of the differential thermogram. The output across the sample and reference thermocouples ΔE is amplified and converted to a temperature difference ΔT, which serves as the y-axis of the thermogram.

Generally, the sample and reference chamber in DTA are designed to permit the circulation of an inert gas, such as nitrogen, or a reactive gas, such as oxygen or air. Some systems also have the capability of operating at high and low pressures.

31B-2 General Principles

Figure 31-7 is an idealized differential thermogram obtained by heating a polymer over a sufficient temperature range to cause its ultimate decomposition. The initial decrease in ΔT is due to the glass transition, a phenomenon observed in the initial segments of many differential thermograms of polymers. The glass transition temperature T_g is the characteristic temperature at which glassy amorphous polymers become flexible or rubberlike because of the onset of the concerted motion of large segments of the polymer molecules. When heated to the glass transition temperature T_g, the polymer changes from a glass to a rubber. Such a transition involves no absorption or evolution of heat so that no change in enthalpy results — that is, $\Delta H = 0$. The heat capacity of the rubber is, however, different from that of the glass, which results in the lowering of the baseline, as shown in the figure. No peak appears during this transition, however, because of the zero enthalpy change.

Two maxima and a minimum are observed in the thermogram in Figure 31-7. The two maxima are the result of exothermic processes in which heat is evolved from the sample, thus causing its temperature to rise. The minimum labeled "melting" is the result of an endothermic process in which heat is absorbed by the analyte. When heated to a characteristic temperature, many amorphous polymers begin to crystallize as microcrystals, giving off heat in the process. Crystal formation is responsible for the first exothermic peak shown in Figure 31-7.

The second peak in the figure is endothermic and involves melting of the microcrystals formed in the initial exothermic process. The third peak is exothermic and is encountered only if the heating is performed in the presence of air or oxygen. This peak is the result of the exothermic oxidation of the polymer. The final negative change in ΔT results from the endothermic decomposition of the polymer to produce a variety of products.

As suggested in Figure 31-7, DTA peaks result from both physical changes and chemical reactions induced by temperature changes in the sample. Physical processes that are endothermic include fusion, vaporization, sublimation, absorption, and desorption. Adsorption and crystallization are generally exothermic. Chemical reactions may also be exothermic or endothermic. Endothermic reactions include dehydration, reduction in a gaseous atmosphere, and decomposition. Exothermic reactions include oxidation in air or oxygen, polymerization, and catalytic reactions.

31B-3 Applications

In general, DTA is considered a qualitative technique. Although able to measure the temperatures at which various changes occur, DTA is unable to measure the energy associated with each event.

DTA is a widely used tool for studying and characterizing polymers. Figure 31-7 illustrates the types of physical and chemical changes in polymeric materials that can be studied by differential thermal methods. Note that thermal transitions for a polymer often take place over an extended temperature range because even a pure polymer is a mixture of homologs and not a single chemical species.

DTA is also widely used in the ceramics and metals industry. The technique is capable of studying high-temperature processes (up to 2400°C for some units) and relatively large sample sizes (hundreds of milligrams). For such materials, DTA is used to study decomposition temperatures, phase transitions, melting and crystallization points, and thermal stability.

An important use of DTA is for the generation of phase diagrams and the study of phase transitions. An example is shown in Figure 31-9, which is a differential thermogram of sulfur, in which the peak at 113°C corresponds to the solid-phase change from the rhombic to the monoclinic form. The peak at 124°C corresponds to the melting point of the element. Liquid sulfur is known to exist in at least three forms, and the peak at 179°C apparently involves these transitions, whereas the peak at 446°C corresponds to the boiling point of sulfur.

The DTA method also provides a simple and accurate way of determining the melting, boiling, and decomposition points of organic compounds. Generally, the data appear to be more consistent and reproducible than those obtained with a hot stage or a capillary tube.

Tutorial: Learn more about **differential thermal analysis**.

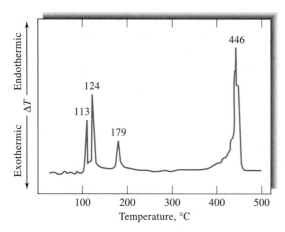

FIGURE 31-9 Differential thermogram for sulfur. (Reprinted with permission from J. Chiu, *Anal. Chem.*, **1963**, *35*, 933. Copyright 1963 American Chemical Society.)

Figure 31-10 shows thermograms for benzoic acid at atmospheric pressure (*A*) and at 13.79 bar (*B*). The first peak corresponds to the melting point and the second to the boiling point of the acid.

31C DIFFERENTIAL SCANNING CALORIMETRY

Differential scanning calorimetry (DSC) is the most often used thermal analysis method, primarily because of its speed, simplicity, and availability. In DSC a

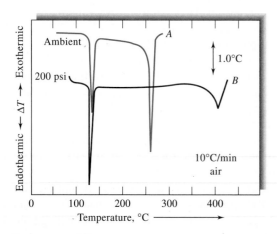

FIGURE 31-10 Differential thermogram for benzoic acid. Curve *A* taken at atmospheric pressure. Curve *B* taken at a pressure of 13.79 bar (200 lbs/in.²). (From P. Levy, G. Nieuweboer, and L. C. Semanski, *Thermochim. Acta*, **1970**, *1*, 433. With permission.)

sample and a reference are placed in holders in the instrument. Heaters either ramp the temperature at a specified rate (e.g., 5°C/min) or hold the DSC at a given temperature. The instrument measures the difference in the heat flow between the sample and the reference. The basic difference between DSC and DTA is that DSC is a calorimetric method in which differences in *energy* are measured. In contrast, in DTA, differences in temperature are recorded. The temperature programs for the two methods are similar. DSC is considered to be a quantitative technique, in contrast to DTA.

31C-1 Instrumentation

There are three different types of DSC instruments: *power-compensated DSC*, *heat-flux DSC*, and *modulated DSC*. Each produces a plot of power or heat flow versus temperature, called a *thermogram*.[3]

Power-Compensated DSC Instruments

In power-compensated DSC, the temperatures of the sample and reference are kept equal to each other while both temperatures are increased or decreased linearly. The power needed to maintain the sample temperature equal to the reference temperature is measured.

A diagram of a power-compensated DSC sample holder and heating unit is shown in Figure 31-11. Two independent heating units are employed. These heating units are quite small, allowing for rapid rates of heating, cooling, and equilibration. The heating units are embedded in a large temperature-controlled heat sink. The sample and reference holders have platinum resistance thermometers to continuously monitor the temperature of the materials. Both sample and reference are maintained at the programmed temperature by applying power to the sample and reference heaters. The instrument records the power difference needed to maintain the sample and reference at the same temperature as a function of the programmed temperature.

Power-compensated DSC has lower sensitivity than heat-flux DSC, but its response time is more rapid. This makes power-compensated DSC well suited for kinetics studies in which fast equilibrations to new temperature settings are needed. Power-compensated DSC is also capable of higher resolution than heat-flux DSC.

---

[3]For descriptions of commercially available DSC instruments, see B. E. Erickson, *Anal. Chem.*, **1999**, *71*, 689A; D. Noble, *Anal. Chem.*, **1995**, *67*, 323A.

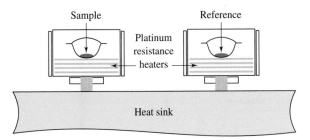

FIGURE 31-11 Power-compensated DSC sample and reference holders and heaters. A temperature program is generated by the computer system. Platinum resistance thermometers, in contact with the sample and reference holders, sense any difference between the programmed temperature and the temperatures of the sample and reference. The error signal is used to adjust the power applied to the sample and the power applied to the reference platinum resistance heaters. The DSC output signal is the difference in the power required between the sample and the reference so that both equal the programmed temperature.

Heat-Flux DSC Instruments

In heat-flux DSC, the difference in heat flow into the sample and reference is measured while the sample temperature is changed at a constant rate. Both sample and reference are heated by a single heating unit. Heat flows into both the sample and reference material via an electrically heated constantan thermoelectric disk, as shown in Figure 31-12. Small aluminum sample and reference pans sit on raised platforms on the constantan disk.[4] Heat is transferred through the disks and up into the material via the two pans. The differential heat flow to the sample and reference is monitored by Chromel-constantan area thermocouples formed by the junction between the constantan platform and Chromel disks attached to the underside of the platforms. The differential heat flow into the two pans is directly proportional to the difference in the outputs of the two thermocouple junctions. The sample temperature is estimated by the Chromel-alumel junction under the sample disk.

In heat-flux DSC, we can write the total heat flow dH/dt as

$$\frac{dH}{dt} = C_p \frac{dT}{dt} + f(T, t) \qquad (31\text{-}1)$$

---

[4]Constantan is an alloy of 60% copper and 40% nickel. Chromel is a trademark for a series of alloys containing chromium, nickel, and at times, iron. Alumel is an alloy of nickel with 2% aluminum, 2% manganese, and 1% silicon.

where H is the enthalpy in J mol^{-1}, C_p is the specific heat capacity (heat capacity per mole) in J K^{-1} mol^{-1}, and $f(T, t)$ is the kinetic response of the sample in J mol^{-1}. We can thus see that the total heat flow is the sum of two terms, one related to the heat capacity, and one related to the kinetic response. A typical DSC thermogram is shown in Figure 31-13. We can see in the plot several processes that occur as the temperature is changed. An increase in heat flow signifies an exothermic process, and a decrease indicates an endothermic process.

Modulated DSC Instruments

Modulated DSC (MDSC) uses the same heating and cell arrangement as the heat-flux DSC method. In MDSC, a sinusoidal function is superimposed on the overall temperature program to produce a micro heating and cooling cycle as the overall temperature is steadily increased or decreased. Using Fourier transform methods, the overall signal is mathematically deconvoluted into two parts, a reversing heat flow signal and a nonreversing heat flow signal. The reversing heat flow signal is associated with the heat capacity component of the thermogram, and the nonreversing heat flow is related to kinetic processes. Usually, step transitions, such as the glass transition, appear only in the reversing heat flow signal, and exothermic or endothermic events may appear in either or in both signals.

A plot showing the decomposition of the total heat flow signal into reversing and nonreversing components is shown in Figure 31-14. Note that the transition near 60°C appears in the reversing heat flow component associated with the heat capacity part of the total heat flow. The endothermic process near 250°C also appears in the reversing component, and the exothermic event near 150°C appears in the nonreversing component.

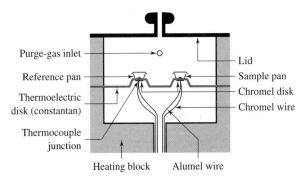

FIGURE 31-12 Heat-flux DSC. (Courtesy of TA Instruments, New Castle, DE.)

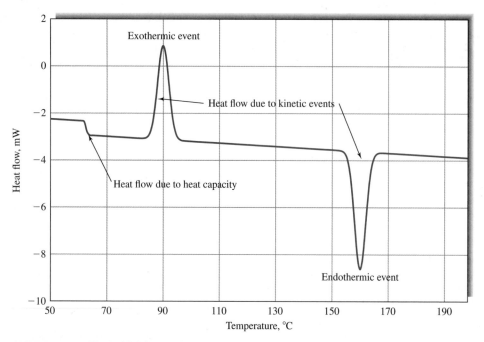

FIGURE 31-13 Typical DSC scan for a polymeric material. Note the step transition at about 63°C. There is an exothermic event at approximately 90°C and an endothermic event at 160°C. Note that the thermogram represents the sum of the heat flow due to heat capacity and kinetic processes. (Courtesy of TA Instruments, New Castle, DE.)

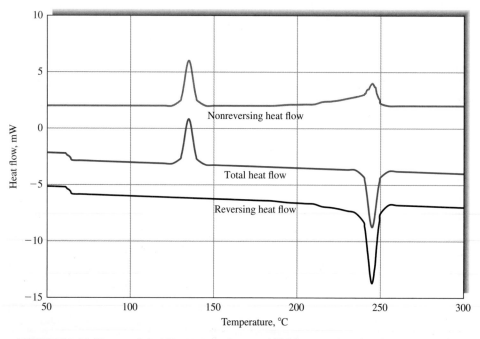

FIGURE 31-14 Deconvoluted thermogram from an MDSC apparatus showing nonreversing and reversing heat flow components. (Courtesy of TA Instruments, New Castle, DE.)

The DSC Experiment

A typical DSC experiment begins by weighing an empty sample pan. The pans are most often made of aluminum, although gold, stainless steel, and glass ampoules can be used. Hermetic pans are often used. A small portion of the sample is then transferred to the pan so as to provide a uniform layer at the bottom of the pan. The pan is hermetically sealed and then weighed again. When the sample and reference pans are in position, the purge gas is applied. Some DSC units have autosamplers that allow multiple sample pans placed in a holder to be run sequentially. Helium or nitrogen is the usual purge gas, although occasionally air or oxygen is employed to study oxidation processes. The purge gas helps to rid the sample of moisture and oxygen and aids in transferring heat to the sample pan.

Modern DSC instruments are computer controlled so that the experiment is performed automatically after the user enters parameters such as the temperature program and various calibration parameters.

DSC Calibration

The DSC system is usually calibrated in several ways. Baseline calibration is performed with no pans in place. The calibration measures the baseline slope and offset over the temperature range of interest. The computer system controlling the DSC stores these values and subtracts baseline slope and offset from subsequent sample runs to minimize their effects.

After baseline calibration is performed, heat flow calibration is done by melting a known quantity of a material with a well-known heat of fusion. Indium is the most often used standard. Indium is placed in the sample pan and scanned against an empty reference pan. The area of the melting peak is related to the known enthalpy of fusion by a calibration factor known as the *cell constant*. This procedure also calibrates the temperature axis from the known melting temperature of indium. Temperature calibration should also be performed over a wider temperature range, by measuring the melting points of several well-known standards.

For heat capacity determinations with normal DSC and MDSC systems, heat capacity calibration is performed by scanning a heat capacity standard, such as sapphire. This calibrates the system for C_p values and is used in separating the heat capacity component from the total heat flow.

DSC Data Analysis

With modern DSC instruments, software is available to aid the user in determining melting points, glass transition temperatures, and heat capacity values. The temperatures of step transitions and kinetic events are usually determined as onset temperatures. The onset temperature is defined as the temperature at which a line tangent to the baseline intersects another line tangent to the slope of the transition, as illustrated in Figure 31-15. In some cases temperatures of transitions,

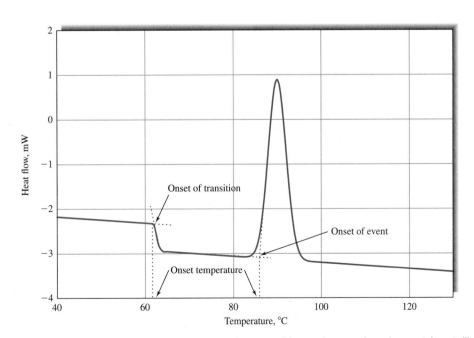

FIGURE 31-15 Determination of onset temperatures for a transition and an exothermic event (crystallization).

such as the glass transition, are taken as the midpoint of the transition rather than the onset.

The step in a transition, such as the glass transition, is related to the change in heat capacity as the material goes through the transition from one state to another. The change in heat capacity is usually determined as the difference between the heat capacity at the onset and at the end of the transition. The enthalpy of melting or crystallization is determined by finding the area of the representative endotherm or exotherm through integration.

31C-2 Applications

DSC finds many applications in characterizing materials. Quantitative applications include the determination of heats of fusion and the extent of crystallization for crystalline materials. Glass transition temperatures and melting points are useful for qualitative classification of materials, although thermal methods cannot be used alone for identification. Melting points are also very useful in establishing the purity of various preparations. Hence, thermal methods are often used in quality control applications.

Glass Transition Temperatures

Determination of the glass transition temperature T_g is one of the most important applications of DSC. The physical properties of a polymer undergo dramatic changes at T_g, where the material goes from a glassy to a rubbery state. At the glass transition, the polymer undergoes changes in volume and expansion, heat flow and heat capacity. The change in heat capacity is readily measured by DSC.

Crystallinity and Crystallization Rate

With crystalline materials, the level of crystallinity is an important factor for determining polymer properties. Degrees of crystallinity can be determined by IR spectroscopy, X-ray diffraction, density measurements, and thermal methods. In most cases DSC is one of the easiest methods for determining levels of crystallinity. The crystallinity level is obtained by measuring the enthalpy of fusion for a sample $(\Delta H_f)_{sample}$ and comparing it to the enthalpy of fusion for the fully crystalline material $(\Delta H_f)_{crystal}$. The fractional crystallinity is then given by

$$\text{Fractional crystallinity} = (\Delta H_f)_{sample} / (\Delta H_f)_{crystal} \quad (31\text{-}2)$$

 Tutorial: Learn more about **differential scanning calorimetry**.

Calorimetric methods are also used to study crystallization rates. Crystallization is an exothermic event, as shown in Figure 31-15. The rate of heat release and thus the crystallization kinetics can be followed by DSC.

Reaction Kinetics

Many chemical reactions, such as polymer formation reactions, are exothermic and readily monitored by DSC methods. Here, the determination of the rate of heat release, dH/dt, is used to determine the extent of reaction as a function of time. Polymerization kinetics can be studied in both a temperature scanning and an isothermal mode. With some polymer systems, factors such as monomer volatility and viscosity can affect the measured kinetics.

31D MICROTHERMAL ANALYSIS

Microthermal analysis[5] combines thermal analysis with atomic force microscopy. It is actually a family of scanning thermal microscopy techniques in which thermal properties of a surface are measured as a function of temperature and used to produce a thermal image. In microthermal analysis the tip of an atomic force microscope is replaced by a thermally sensitive probe such as a thermistor or thermocouple. The surface temperature can be changed externally or by the probe acting both as a heater and as a temperature-measuring device.

A photomicrograph of a thermal probe is shown in Figure 31-16. The most common type of thermal probe is the resistive probe based on a Wollaston wire. This wire has a thick coating of silver on top of a thin core of platinum or a platinum-rhodium alloy. At the tip of the probe, the silver is etched away to expose the bare wire. Micromachined probes have also been developed. With these probes, almost all of the electrical resistance is located at the tip. As a result, when an electric current is applied, only the tip becomes hot. The electrical resistance of the tip is also a measure of the temperature.

A microthermal analysis apparatus, pictured in Figure 31-17, can be operated in either a constant-temperature mode or a constant-current mode. The constant-temperature mode is simplest and most often

[5]For reviews on microthermal analysis and its applications, see H. M. Pollock and A. Hammiche, *J. Phys. D: Appl. Phys.*, **2001**, *34*, R23; C. Q. M. Craig, V. L. Kett, C. S. Andrews, and P. G. Royall, *J. Pharm. Sci.*, **2002**, *91*, 1201.

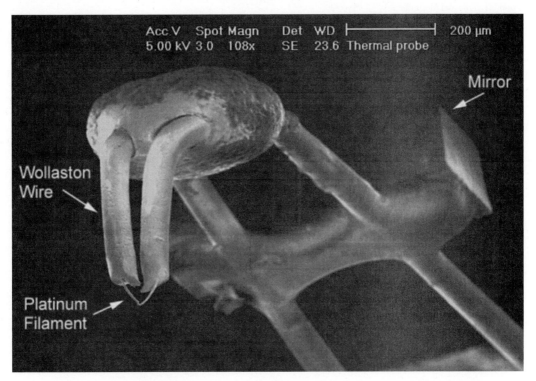

FIGURE 31-16 A microthermal analysis probe. (Reprinted from P. G. Royall, D. Q. M. Craig, and D. B. Grandy, *Thermochim. Acta*, **2001**, *380* (2), 165–173 with permission from Elsevier.)

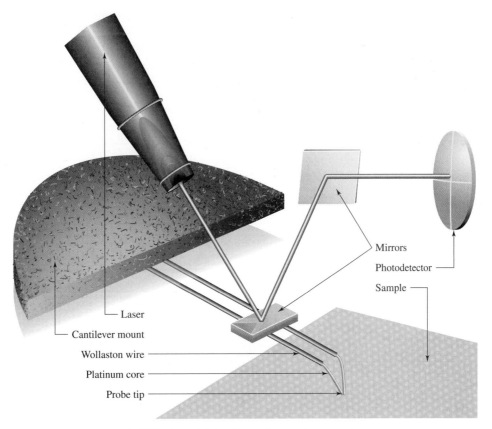

FIGURE 31-17 Microthermal analysis apparatus.

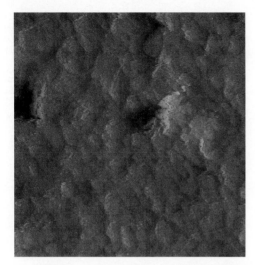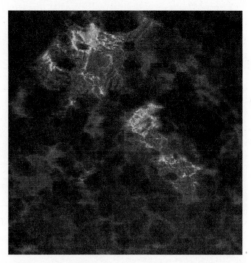

FIGURE 31-18 Comparison of conventional atomic force microscopy topographic image (a) with thermal image (b) of a paracetamol pharmaceutical tablet. (Reprinted from H. M. Pollock and A. Hammiche, *J. Phys. D: Appl. Phys.*, **2001**, *34*, R23. With permission.)

used. In constant-temperature mode, the electrical power needed to keep the probe temperature constant is obtained as the probe is rastered over the sample surface in contact mode (see Section 21G-2). As the probe encounters parts of the surface that differ in thermal properties, varying amounts of heat flow from the probe to the sample. When the probe touches a region of high thermal conductivity, it cools off, and more power is required to keep it at a constant temperature. Conversely, when it touches a region of low thermal conductivity, less power is required. Hence, the thermal conductivity of a sample surface can be im-

aged. Figure 31-18 compares the thermal image of a pharmaceutical preparation to that of the conventional topographic image. Various modulation techniques are also employed to allow the thermal conductivity at different depths of the sample to be monitored by varying the modulation frequency.

Although microthermal analysis is a very new technique, commercial instruments are available. Applications to pharmaceuticals, polymers, and foods have been reported. The technique also has applications in the ceramics industry and in imaging biomedical samples.

QUESTIONS AND PROBLEMS

*Answers are provided at the end of the book for problems marked with an asterisk.

☒ Problems with this icon are best solved using spreadsheets.

31-1 Describe what quantity is measured and how the measurement is performed for each of the following techniques: (a) TGA; (b) DTA; (c) DSC; (d) microthermal analysis.

*31-2 A 0.6025-g sample was dissolved, and the Ca^{2+} and Ba^{2+} ions present were precipitated as $BaC_2O_4 \cdot H_2O$ and $CaC_2O_4 \cdot H_2O$. The oxalates were then heated in a thermogravimetric apparatus leaving a residue that weighed 0.5713 g in the range of 320°C to 400°C and 0.4673 g in the range of 580°C to 620°C. Calculate the percentage Ca and percentage Ba in the sample.

31-3 The following table summarizes some data about three iron(III) chlorides.

Compound	Molecular Mass	Melting Point, °C
$FeCl_3 \cdot 6H_2O$	270	37
$FeCl_3 \cdot \frac{5}{2}H_2O$	207	56
$FeCl_3$	162	306

Sketch the thermogravimetric curve anticipated when a 25.0-mg sample of $FeCl_3 \cdot 6H_2O$ is heated from 0°C to 400°C.

31-4 Why is the low-temperature endotherm for ambient pressure in Figure 31-10 found at the same temperature as that at 13.79 bar, whereas the high-temperature peaks for the two pressures are found at different temperatures?

31-5 It should be possible to at least partially characterize an oil shale sample using techniques discussed in this chapter. Briefly discuss two techniques appropriate to use for this purpose. Sketch typical thermal curves and discuss the information that might be obtained and problems that might be anticipated.

31-6 In thermal analysis methods, why is the thermocouple for measuring sample temperature seldom immersed directly into the sample?

31-7 List the types of physical changes that can yield exothermic and endothermic peaks in DTA and DSC.

31-8 List the types of chemical changes that can yield exothermic and endothermic peaks in DTA and DSC.

31-9 Why are the applications of TGA more limited than those for DSC?

31-10 Why does the glass transition for a polymer yield no exothermic or endothermic peak?

31-11 Describe the difference between power-compensated and heat-flux DSC instruments.

Challenge Problem

31-12 In the pharmaceutical industry, close attention must be paid to drug purity, quality, stability, and safety. Pharmaceutical compounds often have several different structural forms with different molecular shapes. These compounds are also susceptible to thermal degradation, to the pickup and retention of water, and to photodecomposition. One of the best methods to characterize pharmaceuticals from raw product to finished product stage is thermal analysis. The following questions deal with applications of thermal analysis methods to pharmaceuticals.

(a) Determining the purity of drugs is one of the most important tests done by the pharmaceutical industry. One such test is based on a determination of the melting point of the drug of interest by DSC and the following relationship from thermodynamics:

$$\frac{d \ln X_1}{dT} = \frac{\Delta H_f}{RT^2}$$

where X_1 is the mole fraction of the drug whose purity is being determined, ΔH_f is the enthalpy of fusion, R is the gas constant, and T is temperature. From this equation, derive the following modified van't Hoff equation:

$$\frac{1}{f} = \frac{\Delta H_f}{R} \frac{(T_0 - T)}{T_0^2} \frac{1}{X_2^0}$$

where f is the fraction of sample melting at temperature T, T_0 is the melting point of the pure drug, and X_2^0 is the mole fraction of the impurity in the original compound.

Hint: Assume the drug is close to pure so that $\ln X_1$ can be approximated by

$$\ln X_1 = \ln(1 - X_2) \approx -X_2$$

As the temperature is increased toward the true melting point T_0, the mole fraction in the liquid state X_2 is constantly reduced according to

$$X_2 = X_2^0(1/f)$$

(b) The fraction f can be found from A/A_T, where A is the area of the melting endotherm up to temperature T and A_T is the total area of the melting endotherm. By dividing the endotherm into partial melting areas, the fraction f can be determined at various temperatures. Rearrange the modified van't Hoff equation to give the melting temperature T as a function of $1/f$. Show that a plot of T versus $1/f$ should be linear with an intercept of T_0 and a slope of $-RT_0X_2/\Delta H_f$. If the heat of fusion is known, X_2 can be determined.

(c) Look up the article by H. Staub and W. Perron, *Anal. Chem.*, **1974**, *46*, 128. Give some of the limitations of the DSC approach to impurity determinations. Describe the "step heating" method used by the authors. How does this differ from normal DSC?

(d) Which type of DSC, power compensated or heat flux, would be most useful in determining purity? Why?

(e) What thermal analysis technique would be suitable for characterizing and quantifying the moisture content in a pharmaceutical sample? Why?

(f) Describe how thermal analysis could be used to determine the degree of hydration of a pharmaceutical compound.

(g) When heated, many pharmaceutical materials show a variety of thermal events, including melting, transitions to glassy states, moisture loss, thermal relaxation, and in some cases, decomposition. How could irreversible kinetic events be distinguished from reversible heat flow events?

(h) Describe how microthermal analysis might be useful in studies of pharmaceutical materials. Consider problems such as polymorphic forms, impurities, and identification of glassy states in your answer.

CHAPTER THIRTY-TWO

Radiochemical Methods

The availability of natural and artificial radioactive isotopes has made possible the development of sensitive and specific radiochemical methods. These procedures are usually accurate and of widespread applicability. Some radiochemical methods also minimize or eliminate the separations required in other analytical methods. This chapter introduces the principles, instrumentation, and applications of radiochemical methods.

Throughout this chapter, this logo indicates an opportunity for online self-study at **www.thomsonedu.com/chemistry/skoog**, linking you to interactive tutorials, simulations, and exercises.

Radiochemical methods are of three types, based on the origin of the radioactivity.[1] In *activation analysis*, radioactivity is induced in one or more elements of the sample by irradiation with suitable radiation or particles (most commonly, neutrons from a nuclear reactor). The resulting radioactive emissions are then measured. In the second category are methods in which the radioactivity is physically introduced into the sample by adding a measured amount of a radioactive species called a *tracer*. The most important quantitative methods based on introducing radiotracers are *isotope dilution* methods, in which a weighed quantity of radioactively tagged analyte having a known activity is added to a measured amount of the sample. After thorough mixing to assure homogeneity, a fraction of the component of interest is isolated and purified. The activity of the isolated fraction is then measured and used to calculate the mass of the component of interest in the original sample. Tracers are also used by organic and inorganic chemists to elucidate reaction mechanisms. The third class of methods involves measurement of naturally occurring radioactivity in a sample. Examples of this type of method are the measurement of radon in household air and uranium in pottery and ceramic materials.

32A RADIOACTIVE NUCLIDES

Except for hydrogen, which consists of a proton only, all atomic nuclei are made up of a collection of protons and neutrons. The chemical properties of an atom are determined by its atomic number Z, which is the number of protons contained in its nucleus. A *nuclide* is characterized by the number of protons and neutrons in the nucleus. The sum of the number of neutrons and protons in a nucleus is the mass number A.[2] Isotopes of an element are atoms having the same atomic number but different mass numbers. That is, the nuclei of isotopes of an element contain the same number of protons but different numbers of neutrons.

[1] For a detailed treatment of radiochemical methods, see K. H. Lieser, *Nuclear and Radiochemistry*, 2nd ed., Weinheim, Germany: Wiley-VCH, 2001; *Handbook of Radioactivity Analysis*, M. F. L'Annunziata, ed., San Diego: Academic Press, 1998; *Chemical Analysis by Nuclear Methods*, Z. B. Alfassi, ed., Chichester, UK: Wiley, 1994; W. D. Ehmann and D. E. Vance, *Radiochemistry and Nuclear Methods of Analysis*, New York: Wiley, 1991.

[2] The nuclear composition of the isotopes of the element X is indicated by the symbol $^A_Z X$ or sometimes simply as $^A X$, where A is the number of protons plus the number of neutrons (the mass number), and Z is the atomic number, or the number of protons.

TABLE 32-1 Characteristics of Common
Radioactive Decay Products

Product	Symbol	Charge	Mass Number
Alpha particle	α	+2	4
Beta particles			
Negatron	β^-	−1	1/1840 (~0)
Positron	β^+	+1	1/1840 (~0)
Gamma ray	γ	0	0
X-Ray	X	0	0
Neutron	n	0	1
Neutrino/ antineutrino	$\nu/\bar{\nu}$	0	0

Stable nuclides are those that have never been observed to decay spontaneously. *Radionuclides*, in contrast, undergo spontaneous disintegration, which ultimately leads to stable nuclides. The disintegration, or *radioactive decay*, occurs with the emission of electromagnetic radiation in the form of X-rays or gamma rays (γ rays); with the formation of electrons, positrons, and the helium nucleus; or by *fission*, in which a nucleus breaks up into smaller nuclei.

32A-1 Radioactive Decay Products

Table 32-1 lists the most important (from a chemist's viewpoint) types of radiation from radioactive decay. Four of these types — alpha particles, beta particles, gamma-ray photons, and X-ray photons — can be detected and recorded by the detector systems described in Section 12B-4. Most radiochemical methods are based on counting the electronic signals produced when these decay particles or photons strike a radiation detector.

32A-2 Decay Processes

Several types of radioactive decay processes yield the products listed in Table 32-1.

Alpha Decay

Alpha decay is a common radioactive process encountered with heavier radionuclides. Those with mass numbers less than about 150 ($Z \approx 60$) seldom yield alpha particles. The alpha particle is a helium nucleus having a mass of 4 and a charge of +2. An example of alpha decay is shown by the equation

$$^{238}_{92}U \rightarrow {}^{234}_{90}Th + {}^4_2He \quad \text{or} \quad {}^{238}U \rightarrow {}^{234}Th + \alpha \quad (32\text{-}1)$$

Here, uranium-238 (^{238}U) is converted to thorium-234 (^{234}Th), a *daughter* nuclide having an atomic number that is two less than the *parent*.

Alpha particles from a particular decay process are either monoenergetic or are distributed among relatively few discrete energies. For example, the decay process shown as Equation 32-1 proceeds by two distinct pathways. The first, which accounts for 77% of the decays, produces an alpha particle with an energy of 4.196 MeV.[3] The second pathway (23% of decays) produces an alpha particle having an energy of 4.149 MeV and is accompanied by the release of a 0.047-MeV gamma ray.

Alpha particles progressively lose their energy as a result of collisions as they pass through matter, and they are ultimately converted into helium atoms through capture of two electrons from their surroundings. Their relatively large mass and charge render alpha particles highly effective in producing ion pairs within the matter through which they pass; this property facilitates their detection and measurement. Because of their high mass and charge, alpha particles have a low penetrating power in matter. The identity of a radionuclide that is an alpha emitter can often be established by measuring the length (or range) over which the emitted alpha particles produce ion pairs within a particular medium (often air). Alpha particles from radioactive decay are relatively ineffective for producing artificial nuclides except in the lightest elements. Alpha particles have, however, been used as sources in the alpha proton X-ray spectrometers used on the Mars Pathfinder mission and on Mars rovers.

Beta Decay

Any radioactive decay process in which the atomic number Z changes but the mass number A does not is classified as β decay. Three types of β decay are encountered: *negatron emission*, *positron emission*, and *electron capture*. Examples of the three processes are the following:

$$^{14}_6C \rightarrow {}^{14}_7N + \beta^- + \bar{\nu}$$
$$^{65}_{30}Zn \rightarrow {}^{65}_{29}Cu + \beta^+ + \nu$$
$$^{48}_{24}Cr + {}^0_1e^- \rightarrow {}^{48}_{23}V + \text{X-rays}$$

Here, $\bar{\nu}$ and ν represent an antineutrino and a neutrino, respectively. The third equation illustrates the

---

[3]Energies associated with nuclear reactions are usually reported in millions of electron volts (MeV) or thousands of electron volts (keV).

beta-decay process called *electron capture*. In this re-action, the capture of an atomic electron by the $^{48}_{24}Cr$ nucleus produces $^{48}_{23}V$, but this process leaves one of the atomic orbitals (usually the 1*s*, or K, orbital, in which case the process is called *K capture*) of the vana-dium deficient by one electron. X-ray emission results when an electron from one of the outer orbitals fills the void left by the capture process. Note that the emission of an X-ray photon is not a nuclear process, but the capture of an electron by the nucleus is.

Two types of β particles can be created by radioac-tive decay. Negatrons (β^-) are electrons that form when one of the neutrons in the nucleus is converted to a proton. In contrast, the positron (β^+), also with the mass of the electron, forms when the number of pro-tons in the nucleus is decreased by one. The positron has a transitory existence, its ultimate fate being anni-hilation in a reaction with an electron to yield two 0.511-MeV photons.

In contrast to alpha emission, beta emission is char-acterized by production of particles with a continuous spectrum of energies ranging from nearly zero to some maximum that is characteristic of each decay process. The β particle is not nearly as effective as the alpha par-ticle in producing ion pairs in matter because of its small mass (about 1/7000 that of an alpha particle). At the same time, its penetrating power is substantially greater than that of the alpha particle. Beta-particle energies are frequently related to the thickness of an absorber, ordinarily aluminum, required to stop the particle.

Gamma-ray Emission

Many alpha- and beta-emission processes leave the nucleus in an excited state, which then decays to the ground state in one or more quantized steps with the release of monoenergetic gamma-ray photons. It is important to note that gamma rays, except for their source, are indistinguishable from X-rays of the same energy. Thus, gamma rays are produced by nuclear relaxations, and X-rays derive from electronic relax-ations. The gamma-ray emission spectrum is charac-teristic for each nuclide and is thus useful for iden-tifying radionuclides.

Not surprisingly, gamma radiation is highly pene-trating. When gamma rays interact with matter, they lose energy by three mechanisms. The predominant mode depends on the energy of the gamma-ray pho-ton. With low-energy gamma radiation, the *photoelec-tric effect* predominates. Here, the gamma-ray photon disappears after ejecting an electron from an atomic

orbital (usually a K level electron) of the target atom. The photon energy is totally consumed in overcoming the binding energy of the electron and in imparting kinetic energy to the ejected electron. With relatively energetic gamma rays, the *Compton effect* is encoun-tered. In this instance, an electron is also ejected from an atom but acquires only a part of the photon energy. The photon, now with diminished energy, recoils from the electron and then goes on to further Compton or photoelectric interactions. If the gamma-ray photon possesses sufficiently high energy (at least 1.022 MeV), *pair production* can occur. Here, the photon is totally absorbed in creating a positron and an electron in the field of a nucleus.

X-ray Emission

Many processes result in loss of inner-shell electrons from an atom. X-rays are then formed from electronic transitions in which outer electrons fill the vacancies created by the nuclear process. One of the processes is *electron capture*, which was discussed earlier. A second process, which can lead to X-rays, is *internal conver-sion*, a type of nuclear process that is an alternative to gamma-ray emission. In this instance, the electromag-netic interaction between the excited nucleus and an extranuclear electron results in the ejection of an or-bital electron with a kinetic energy equal to the differ-ence between the energy of the nuclear transition and the binding energy of the electron (see Section 12A-3). The emission of this so-called *internal conversion elec-tron* leaves a vacancy in the K, L, or a higher level. The result is X-ray emission as the vacancy is filled by an electronic transition.

32A-3 Radioactive Decay Rates

Radioactive decay is a completely random process. Thus, although no prediction can be made concerning the lifetime of an individual nucleus, the behavior of a large ensemble of identical nuclei can be described by the first-order rate expression

$$-\frac{dN}{dt} = \lambda N \qquad (32\text{-}2)$$

where *N* represents the number of radioactive nuclei of a particular kind in the sample at time *t* and λ is the char-acteristic *decay constant* for the radionuclide. After

 Simulation: Learn more about **counting and decay rates**.

rearranging this equation and integrating over the interval between $t = 0$ and $t = t$ (during which the number of radioactive nuclei in the sample decreases from N_0 to N), we obtain

$$\ln \frac{N}{N_0} = -\lambda t \tag{32-3}$$

or

$$N = N_0 e^{-\lambda t} \tag{32-4}$$

The *half-life* $t_{1/2}$ of a radioactive nuclide is defined as the time required for one half the number of radioactive atoms in a sample to undergo decay. The half-life is thus the time required for N to decrease to $N_0/2$. Substitution of $N_0/2$ for N in Equation 32-3 leads to

$$t_{1/2} = \frac{\ln 2}{\lambda} = \frac{0.693}{\lambda} \tag{32-5}$$

Half-lives of radioactive species range from small fractions of a second to many billions of years.

The activity A of a radionuclide is defined as its disintegration rate. Thus, from Equation 32-2, we may write

$$A = -\frac{dN}{dt} = \lambda N \tag{32-6}$$

Activity is given in units of reciprocal seconds. The *becquerel* (Bq) corresponds to one decay per second. That is, $1 \text{ Bq} = 1 \text{ s}^{-1}$. An older, but still widely used, unit of activity is the *curie* (Ci), which was originally defined as the activity of 1 g of radium-226. One curie is exactly equal to 3.7×10^{10} Bq. In analytical radiochemistry, activities of analytes typically range from a nanocurie or less to a few microcuries.

In the laboratory, absolute activities are seldom measured because detection efficiencies are seldom 100%. Instead, the *counting rate* R is used, where $R = cA$. Substituting this relationship into Equation 32-6 yields

$$R = cA = c\lambda N \tag{32-7}$$

Here, c is a constant called the *absolute detection efficiency*, which depends on the nature of the detector, the geometric arrangement of sample and detector, and other factors. The decay law given by Equation 32-4 can then be written in the form

$$R = R_0 e^{-\lambda t} \tag{32-8}$$

Example 32-1 illustrates the use of the decay rate equations.

EXAMPLE 32-1

In an initial counting period, the counting rate for a particular sample was found to be 453 cpm (counts per minute). In a second experiment performed 420 min later, the same sample exhibited a counting rate of 285 cpm. If it can be assumed that all counts are the result of the decay of a single radionuclide, what is the half-life of that radionuclide?

Solution

Equation 32-8 can be used to calculate the decay constant, λ,

$$285 \text{ cpm} = 453 \text{ cpm } e^{-\lambda(420 \text{ min})}$$

$$\ln\left(\frac{285 \text{ cpm}}{453 \text{ cpm}}\right) = -\lambda(420 \text{ min})$$

$$\lambda = 1.10 \times 10^{-3} \text{ min}^{-1}$$

Equation 32-5 can be used to calculate the half-life

$$t_{1/2} = \frac{\ln 2}{\lambda} = \frac{0.693}{1.10 \times 10^{-3} \text{ min}^{-1}} = 630 \text{ min}$$

32A-4 Counting Statistics

As will be shown in Section 32B, radioactivity is measured by means of a detector that produces an electronic pulse, or *count*, each time radiation strikes the detector. Quantitative information about decay rates is obtained by counting these pulses for a specified

TABLE 32-2 Variations in 1-Minute Measurements from a Radioactive Source

Minute	Counts	Minute	Counts
1	180	7	168
2	187	8	170
3	166	9	173
4	173	10	132
5	170	11	154
6	164	12	167

Total counts = 2004.

Average counts/min = $\bar{x} = 167$.

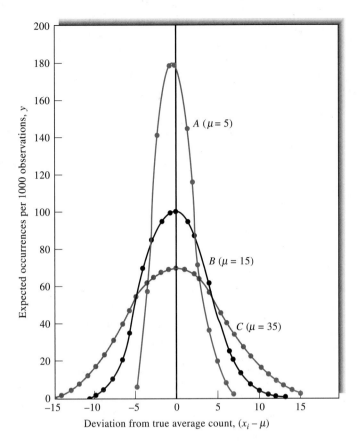

FIGURE 32-1 Expected frequency of deviations from the true average count $(x_i - \mu)$ versus the deviations. Although smooth curves are shown, the Poisson function is defined only for integer values shown as the circled points.

period.[4] Table 32-2 shows typical decay data obtained by successive 1-minute measurements of a radioactive source. Because the decay process is random, considerable variation among the data is observed. Thus, in Table 32-2, the rates range from a low of 132 to a high of 187 counts per minute (cpm).

Although radioactive decay is random, the data, particularly for low number of counts, are not distributed according to Equation a1-14 (Appendix 1), because the decay process does not follow Gaussian behavior. The reason that decay data are not normally distributed is that radioactivity consists of a series of discrete events that cannot vary continuously as can the indeterminate errors for which the Gaussian distribution applies. Furthermore, negative numbers of counts are not possible. Therefore, the data cannot be distributed symmetrically about the mean.

To describe accurately the behavior of a radioactive source, it is necessary to assume a *Poisson distribution*, which is given by the equation

$$y = \frac{\mu^{x_i}}{x_i!} e^{-\mu} \tag{32-9}$$

where y is the frequency of occurrence of a given count x_i and μ is the mean for a large set of counting data.[5]

The data plotted in Figure 32-1 were obtained with the aid of Equation 32-9. These curves show the deviation $(x_i - \mu)$ from the true average value that would be expected if 1000 replicate observations were made on the same sample. Curve A gives the distribution for a substance for which the true average count μ for a selected period is 5; curves B and C correspond to samples having true means of 15 and 35. Note that the *absolute* deviations become greater with increases in μ, but the *relative* deviations become smaller. Note also the lack of symmetry about the mean for the two smaller count numbers.

---

[4]For a more complete discussion, see G. Friedlander, J. W. Kennedy, E. S. Macias, and J. M. Miller, *Nuclear and Radiochemistry*, 3rd ed., Chap. 9, New York: Wiley, 1981.

[5]In the derivation of Equation 32-9, it is assumed that the counting period is short with respect to the half-life so that no significant change in the number of radioactive atoms occurs. Further restrictions include a detector that responds to the decay of a single radionuclide only and an invariant counting geometry so that the detector responds to a constant fraction of the decay events that occur.

Standard Deviation of Counting Data

In contrast to Equation a1-14 (Appendix 1) for a Gaussian distribution, Equation 32-9 for a Poisson distribution contains no corresponding standard deviation term. It can be shown that the breadth of curves such as those in Figure 32-1 depend only on the total number of counts for any given period.[6] That is,

$$\sigma_M = \sqrt{M} \qquad (32\text{-}10)$$

where M is the number of counts for any given period and σ_M is the standard deviation for a Poisson distribution.

The relative standard deviation σ_M/M is given by

$$\frac{\sigma_M}{M} = \frac{\sqrt{M}}{M} = \frac{1}{\sqrt{M}} \qquad (32\text{-}11)$$

Thus, although the standard deviation increases with the number of counts, the relative standard deviation decreases.

The counting rate R is equal to M/t. To obtain the standard deviation in R, we apply Equation a1-29 (Appendix 1), which gives

$$\sigma_R^2 = \left(\frac{\partial R}{\partial M}\right)^2 \sigma_M^2 + \left(\frac{\partial R}{\partial t}\right)^2 \sigma_t^2$$

Generally, time can be measured with such high precision that $\sigma_t^2 \approx 0$. The partial derivative of R with respect to M is $1/t$. Thus,

$$\sigma_R^2 = \frac{\sigma_M^2}{t^2}$$

Taking the square root of this equation and substituting Equation 32-10 gives

$$\sigma_R = \frac{\sqrt{M}}{t} = \frac{\sqrt{Rt}}{t} = \sqrt{\frac{R}{t}} \qquad (32\text{-}12)$$

$$\frac{\sigma_R}{R} = \frac{\sqrt{R/t}}{R} = \sqrt{\frac{1}{Rt}} \qquad (32\text{-}13)$$

Example 32-2 illustrates the use of these equations for counting statistics.

EXAMPLE 32-2

Calculate the absolute and relative standard deviations in the counting rate for (a) the first entry in Table 32-2 and (b) the mean of all of the data in the table.

---
[6]See note 4.

Solution

(a) Applying Equation 32-12 gives

$$\sigma_R = \frac{\sqrt{M}}{t} = \frac{\sqrt{180}}{1\ \text{min}} = 13.4\ \text{cpm}$$

$$\frac{\sigma_R}{R} = \frac{13.4\ \text{cpm}}{180\ \text{cpm}} \times 100\% = 7.4\%$$

(b) For the entire set,

$$\sigma_R = \frac{\sqrt{2004}}{12} = 3.73\ \text{cpm} = 3.7\ \text{cpm}$$

$$\frac{\sigma_R}{R} = \frac{3.73}{167} \times 100\% = 2.2\%$$

Confidence Interval for Counts

In Section a1B-2 (Appendix 1), the confidence interval for a measurement is defined as the limits around a measured quantity within which the true mean can be expected to fall with a stated probability. When the measured standard deviation is believed to be a good approximation of the true standard deviation ($s \rightarrow \sigma$), the confidence interval CI is given by Equation a1-20:

$$\text{CI for } \mu = \bar{x} \pm z\sigma$$

For counting rates, this equation takes the form

$$\text{CI for } R = R \pm z\sigma_R \qquad (32\text{-}14)$$

where z depends on the desired level of confidence. Some values for z are given in Table a1-3. Example 32-3 illustrates the calculation of confidence intervals.

EXAMPLE 32-3

Calculate the 95% confidence interval for (a) the first entry in Table 32-2 and (b) the mean of all the data in the table.

Solution

(a) In Example 32-2, we found that $\sigma_R = 13.4$ cpm. Table a1-3 (Appendix 3) reveals that $z = 1.96$ at the 95% confidence level. Thus, for R

$$95\% \text{ CI} = 180\ \text{cpm} \pm (1.96 \times 13.4\ \text{cpm})$$
$$= 180 \pm 26\ \text{cpm}$$

(b) In this instance σ_R was found to be 3.73 cpm and

$$95\% \text{ CI for } R = 167\ \text{cpm} \pm (1.96 \times 3.73\ \text{cpm})$$
$$= 167 \pm 7\ \text{cpm}$$

Thus, there are 95 chances in 100 that the true rate for R (for the average of 12 min of counting) lies between 160 and 174 counts/min. For the single count in part (a), 95 out of 100 times, the true rate will lie between 154 and 206 counts/min.

Figure 32-2 illustrates the relationship between total counts and tolerable levels of uncertainty as calculated from Equation 32-14. Note that the horizontal axis is logarithmic. Thus, a decrease in the relative uncertainty by a factor of 10 requires that the number of counts be increased by a factor of 100.

Background Corrections

The number of counts recorded in a radiochemical analysis includes a contribution from sources other than the sample. Background activity can be traced to the presence of minute quantities of radon radionuclides in the atmosphere, to the materials used in construction of the laboratory, to accidental contamination within the laboratory, to cosmic radiation, and to the release of radioactive materials into the Earth's atmosphere. To obtain an accurate determination, then, it is necessary to correct the measured counting rate for background contributions. The counting period required to establish the background correction frequently differs from the counting period for the sample. As a result, it is more convenient to employ counting rates as shown in Equation 32-15.

$$R_c = R_x - R_b \qquad (32\text{-}15)$$

where R_c is the corrected counting rate and R_x and R_b are the rates for the sample and the background, respectively. The standard deviation of the corrected counting rate can be obtained by application of Equation (1) in Table a1-6 (Appendix 1). Thus,

$$\sigma_{R_c} = \sqrt{\sigma_{R_x}^2 + \sigma_{R_b}^2}$$

Substituting Equation 32-12 into this equation leads to

$$\sigma_{R_c} = \sqrt{\frac{R_x}{t_x} + \frac{R_b}{t_b}} \qquad (32\text{-}16)$$

EXAMPLE 32-4

A sample yielded 1800 counts in a 10-min period. Background was found to be 80 counts in 4 min. Calculate the absolute uncertainty in the corrected counting rate at the 95% confidence level.

▶ *Solution*

$$R_x = \frac{1800}{10} = 180 \text{ cpm}$$

$$R_b = \frac{80}{4} = 20 \text{ cpm}$$

Substituting into Equation 32-16 yields

$$\sigma_{R_c} = \sqrt{\frac{180}{10} + \frac{80}{4}} = 6.2 \text{ cpm}$$

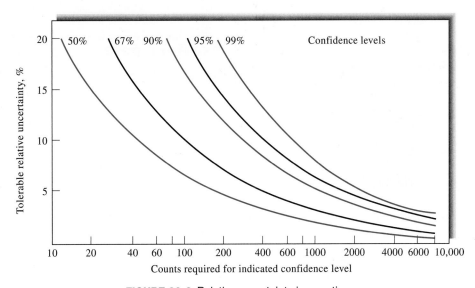

FIGURE 32-2 Relative uncertainty in counting.

At the 95% confidence level,

$$\text{CL for } R_c = (180 - 20) \pm 1.96 \times 6.2$$
$$= 160 \pm 12 \text{ cpm}$$

Here, the chances are 95 in 100 that the true count lies between 148 and 172 cpm.

Note that the inclusion of background contributions invariably leads to an increase in the reported uncertainty of the determination.

32B INSTRUMENTATION

Radiation from radioactive sources can be detected and measured in essentially the same way as X-radiation (Sections 12B-4 and 12B-5). Gas-filled chambers, scintillation counters, and semiconductor detectors are all sensitive to alpha and beta particles and to gamma rays because absorption of these particles produces ionization or photoelectrons, which can in turn produce thousands of ion pairs. A detectable electrical pulse is thus produced for each particle reaching the transducer.

32B-1 Measurement of Alpha Particles

To minimize self-absorption, alpha-emitting samples are generally counted as thin deposits prepared by electrodeposition or by vaporization. Often, these deposits are then sealed with thin layers of material and counted in windowless gas-flow proportional counters or ionization chambers. Alternatively, they are placed immediately adjacent to a solid-state detector, often in a vacuum for counting. Liquid scintillation counting (see next section) is becoming increasingly important for counting alpha emitters because of the ease of sample preparation and the higher sensitivity for alpha-particle detection.

Because alpha-particle spectra consist of characteristic, discrete energy peaks, they are quite useful for identification. Pulse-height analyzers (Section 12B-5) permit the recording of alpha-particle spectra.

32B-2 Measurement of Beta Particles

For beta sources having energies greater than about 0.2 MeV, a uniform layer of the sample is ordinarily counted with a thin-windowed Geiger or proportional

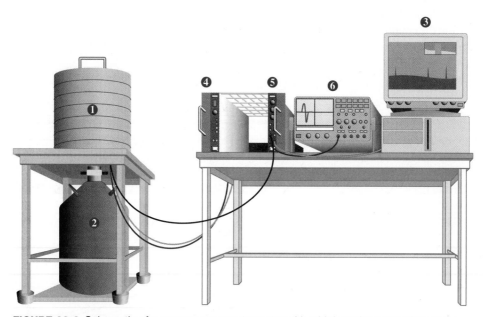

FIGURE 32-3 Schematic of a gamma-ray spectrometer with a high-purity germanium detector: (1) high-purity germanium detector with lead shield, (2) Dewar flask with cryostat and preamplifier, (3) computer with plug-in analog-to-digital converter and multichannel analyzer software, (4) detector bias supply, (5) amplifier, (6) oscilloscope. (Adapted from Nuclear Physics Laboratory, University of Cyprus, with permission.)

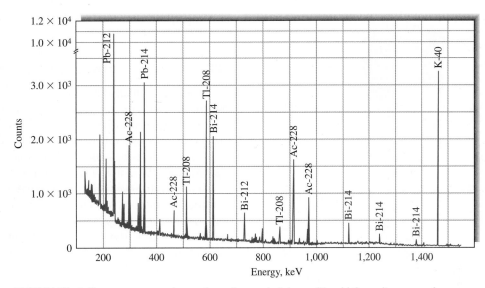

FIGURE 32-4 Gamma-ray spectrum of a soil sample taken with a high-purity germanium detector. (Adapted from Nuclear Physics Laboratory, University of Cyprus, with permission.)

tube counter. For low-energy beta emitters, such as carbon-14, sulfur-35, and tritium, liquid scintillation counters (page 315) are preferable. Here, the sample is dissolved in a solution of the scintillating compound. A vial containing the solution is then placed between two photomultiplier tubes housed in a light-tight compartment. The output from the two tubes is fed into a co-incidence counter, an electronic device that records a count only when pulses from the two transducers arrive at the same time. The coincidence counter reduces background noise from the detectors and amplifiers because of the low probability that such noise will affect both systems simultaneously. Liquid scintillation counting accounts for the majority of β-particle determinations because of the widespread use of this technique in clinical laboratories. Because beta spectra are ordinarily continuous, pulse-height analysis is not as useful as with alpha particles.

32B-3 Measurement of Gamma Radiation

Gamma radiation is detected and measured by the methods described in Sections 12B-4 and 12B-5 for X-radiation. Interference from alpha and beta radiation is easily avoided by filtering the radiation with a thin piece of aluminum or plastic.

Gamma-ray spectrometers[7] are similar to the pulse-height analyzers described in Section 12B-5. A typical arrangement is shown in Figure 32-3. Figure 32-4 shows a gamma-ray spectrum of a soil sample obtained with a high-resolution spectrometer capable of measuring gamma rays in the range of 50 to 3000 keV. Here, the characteristic peaks for the various elements are superimposed on a continuum that arises primarily from the Compton effect.

Figure 32-5 is a schematic of a well-type scintillation counter that is used for gamma-ray counting. Here, the sample is contained in a small vial and placed in a cylindrical hole, or well, in the scintillating crystal of the counter.

Gamma-ray spectrometers have been successfully used in the exploration of the lunar and Martian surfaces, and a gamma-ray spectrometer was a significant component of the Mars Surveyor mission. Its goals were to quantitatively determine elemental abundances on the Martian surface, to map the distribution of water, to study the seasonal polar caps, and to investigate the nature of cosmic gamma-ray bursts.

---

[7]See G. Gilmore and J. D. Hemingway, *Practical Gamma-Ray Spectrometry*, New York: Wiley, 1995.

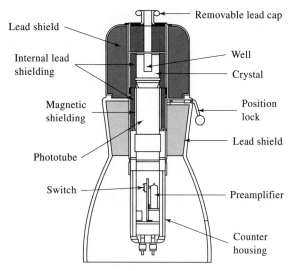

Lead shield

Internal lead shielding

Magnetic shielding

Phototube

Switch

Removable lead cap

Well

Crystal

Position lock

Lead shield

Preamplifier

Counter housing

FIGURE 32-5 A well-type scintillation counter with a NaI(Tl) crystal.

32C   NEUTRON ACTIVATION METHODS

Activation methods are based on the measurement of the radioactivity or radiation produced in samples when they are irradiated with neutrons or charged particles, such as hydrogen, deuterium, or helium ions.[8] An overview of the most common type of neutron activation is shown in Figure 32-6. Here, a neutron is captured by the target nucleus to form an excited compound nucleus. The compound nucleus de-excites almost instantaneously by emission of one or more characteristic *prompt gamma rays*. In many cases a new radioactive nucleus is formed, which can undergo β^- decay to an exited product nucleus with the emission of another characteristic *delayed gamma ray*.

Neutron activation analysis (NAA) can be accomplished by measurement of the prompt gamma rays during irradiation or by measurement of the delayed

gamma rays. Measurement of the delayed gamma-ray emission is more common.

32C-1 Neutrons and Neutron Sources

Three sources of neutrons are employed in neutron activation methods, *reactors*, *radionuclides*, and *accelerators*. All three produce highly energetic neutrons (in the mega-electron-volts range), which are usually passed through a moderating material that reduces their energies to a few hundredths of an electron volt. Energy loss to the moderator occurs *by elastic scattering*, in which neutrons bounce off nuclei in the moderator material, transferring part of their kinetic energy to each nucleus they strike. Ultimately, the nuclei come to thermal equilibrium with their surroundings. Neutrons having these temperatures (energies of about 0.04 eV) are called *slow* or *thermal neutrons*, and the process of moderating high-energy neutrons to thermal conditions is called *thermalization*. The most efficient moderators are hydrogen-containing substances, such as water, deuterium oxide, and paraffin.

Most activation methods are based on thermal neutrons, which react efficiently with most elements of analytical interest. For lighter elements, such as nitrogen, oxygen, fluorine, and silicon, *fast neutrons* having energies of about 14 MeV are more efficient for inducing radioactivity. Such high-energy neutrons are commonly produced at particle accelerators.

Reactors

Nuclear reactors can produce significant fluxes of neutrons and are widely used for activation analyses. A typical research reactor will have a thermal neutron flux of 10^{11} to 10^{14} n cm^{-2}s^{-1} (n is the number of neutrons). These high neutron flux densities lead to detection limits that for many elements range from 10^{-3} to 10 µg.

Radioactive Neutron Sources

Radionuclides are convenient and relatively inexpensive sources of neutrons for activation analyses. Their neutron flux densities range from perhaps 10^5 to 10^{10} n cm^{-2}s^{-1}. Because of the lower fluxes, detection limits are generally not as good as those in which a reactor serves as a source.

One common radioactive source of neutrons is a transuranium element that undergoes spontaneous fission to yield neutrons. The most common example of this type of source consists of ^{252}Cf (californium-

[8] Additional information on activation methods may be found in *Handbook of Prompt Gamma Activation Analysis with Neutron Beams*, G. Molnar, ed., Dordrecht, Netherlands: Kluwer Academic Publishers, 2004; Z. B. Alfassi and C. Chung, *Prompt Gamma Neutron Activation Analysis*, Boca Raton, FL: CRC Press, 1995; S. J. Parry, *Activation Spectrometry in Chemical Analysis*, New York: Wiley, 1991; *Activation Analysis*, Vols. 1 and 2, Z. B. Alfassi, ed., Boca Raton, FL: CRC Press, 1989.

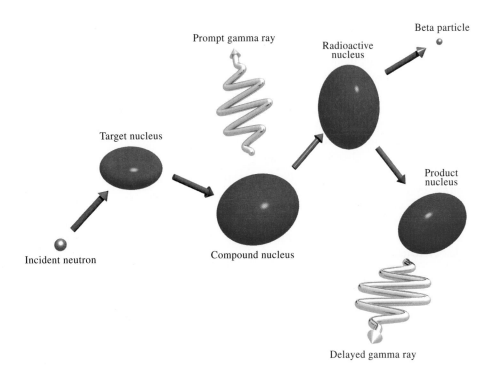

FIGURE 32-6 Overview of the neutron activation process. The incident neutron is captured by the target nucleus to produce an excited compound nucleus, which de-excites with emission of a prompt gamma ray. The radioactive nucleus formed decays by emitting a beta particle. If an excited product nucleus is formed, a delayed gamma ray can be emitted. If decay is directly to the ground state of the product nucleus, no gamma ray is emitted.

252), which has a half-life of 2.6 years. About 3% of its decay occurs by spontaneous fission, which yields 3.8 neutrons per fission. Thermal neutron flux densities of about $10^7 \, n \, \text{cm}^{-2} \, \text{s}^{-1}$ or higher are obtainable from this type of source.

Neutrons can also be produced by preparing an intimate mixture of an alpha emitter such as plutonium, americium, or curium with a light element such as beryllium. A commonly used source of this kind is based on the reaction

$$_4^9\text{Be} + _2^4\text{He} \rightarrow _6^{12}\text{C} + _0^1 n + 5.7 \, \text{MeV}$$

To produce thermal neutrons, paraffin is typically employed as a moderator.

Accelerators

Benchtop charged particle accelerators are commercially available for the generation of neutrons. A typical generator consists of an ion source that delivers deuterium ions to an area where they are accelerated by a potential difference of a few hundred kV to a target containing tritium adsorbed on titanium or zirconium. The reaction is

$$_1^2\text{H} + _1^3\text{H} \rightarrow _2^4\text{He} + _0^1 n$$

The higher-energy neutrons produced in this way are particularly useful for activating the lighter elements.

32C-2 Interactions of Neutrons with Matter

The basic characteristics of neutrons are given in Table 32-1. Free neutrons are not stable, and they decay with a half-life of about 10.2 min to give protons and electrons. Free neutrons do not, however, generally exist long enough to disintegrate in this way because of their great tendency to react with ambient material. The high reactivity of neutrons arises from their lack of charge, which permits them to approach charged nuclei without interference from coulombic forces.

Neutron capture is the most important reaction for activation methods. Here, a neutron is captured by the analyte nucleus to give an isotope with the same atomic number, but with a mass number that is greater by 1. The new nuclide is in a highly excited state because it has acquired several MeV of energy by binding the neutron. This excess energy is released by *prompt gamma-ray emission* or emission of one or more nuclear particles, such as protons or alpha particles. An example of a reaction that produces prompt gamma rays is

$$^{23}_{11}\text{Na} + {}^{1}_{0}n \rightarrow {}^{24}_{11}\text{Na} + \gamma$$

Usually, equations of this type are written in the abbreviated form

$$^{23}_{11}\text{Na}(n, \gamma)^{24}_{11}\text{Na}$$

The prompt gamma rays formed by capture reactions are of analytical interest in some cases, but the radionuclide produced (^{24}Na) and its decay radiations are more often used in NAA.

32C-3 Theory of Activation Methods

When exposed to neutrons, the rate of formation of radioactive nuclei from a single isotope is given by

$$\frac{dN^*}{dt} = N\phi\sigma$$

where dN^*/dt is the reaction rate (s^{-1}), N is the number of stable target atoms, ϕ is the average flux density ($n\ cm^{-1}\ s^{-1}$), and σ is the capture cross section (cm^2).[9] The capture cross section is a measure of the probability of the nuclei reacting with a neutron at the neutron energy employed. Tables of reaction cross sections for thermal neutrons list values for σ in *barns* b, where $1\ b = 10^{-24}\ cm^2$.

Once formed, the radioactive nuclei decay at a rate $-dN^*/dt$ given by Equation 32-2. That is,

$$-\frac{dN^*}{dt} = \lambda N^*$$

Thus, during irradiation with a uniform flux of neutrons, the net rate of formation of active particles is

$$\frac{dN^*}{dt} = N\phi\sigma - \lambda N^*$$

Integrating this equation from time 0 to t, gives

$$N^* = \frac{N\phi\sigma}{\lambda}[1 - \exp(-\lambda t)]$$

If we substitute Equation 32-5 into the exponential term, we obtain

$$N^* = \frac{N\phi\sigma}{\lambda}\left[1 - \exp\left(-\frac{0.693t}{t_{1/2}}\right)\right]$$

The last equation can be rearranged to give the product λN^*, which is the activity A (see Equation 32-6). Thus,

$$A = \lambda N^* = N\phi\sigma\left[1 - \exp\left(-\frac{0.693t}{t_{1/2}}\right)\right] = N\phi\sigma S \tag{32-17}$$

where S is the saturation factor, which is equal to 1 minus the exponential term.

Equation 32-17 can be written in terms of experimental rate measurements by substituting Equation 32-7 to give

$$R = cN\phi\sigma\left[1 - \exp\left(-\frac{0.693t}{t_{1/2}}\right)\right] = N\phi\sigma cS \tag{32-18}$$

Figure 32-7 is a plot of this relationship at three levels of neutron flux density. The abscissa is the ratio of the irradiation time to the half-life of the isotope ($t/t_{1/2}$). In each case, the counting rate approaches a constant value where the rates of formation and disintegration of the radionuclide approach one another. Clearly, irradiation for periods beyond four or five half-lives for an isotope will result in little improvement in sensitivity.

In many analyses, irradiation of the samples and standards is carried out for a long enough period to reach saturation so that S approaches unity. Under this circumstance, all of the terms except N on the right side of Equation 32-18 are constant, and the number of analyte radionuclides is directly proportional to the counting rate. If the parent, or target, nuclide is naturally occurring, the mass of the analyte m can be obtained from N by multiplying it by Avogadro's number, the natural abundance of the analyte nuclide, and the atomic mass. Because all of these are constants, the

---

[9]Here, we use the symbol N^* to distinguish the number of radioactive nuclei from the number of stable nuclei N.

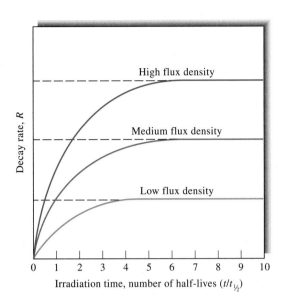

FIGURE 32-7 The effect of neutron flux density and time on the activity induced in a sample.

32C-4 Experimental Considerations in Activation Methods

Figure 32-8 is a block diagram showing the flow of sample and standards in the two most common types of activation methods, *destructive* and *nondestructive*. In both procedures the sample and one or more standards are irradiated simultaneously with neutrons (or other types of radiation). The samples may be solids, liquids, or gases, although the first two are more common. The standards should physically and chemically approximate the sample as closely as possible. Generally, the samples and standards are contained in small polyethylene vials; heat-sealed quartz vials are also used on occasion. Care must be taken to ensure that the samples and standards are exposed to the same neutron flux. The time of irradiation depends on a variety of factors and often is determined empirically. Frequently, an exposure time of roughly three to five times the half-life of the analyte product is used (see Figure 32-7). Irradiation times generally vary from a few minutes to several hours.

After irradiation is terminated, the samples and standards are often allowed to decay (or "cool") for a period that again varies from a few minutes to several hours or more. During cooling, short-lived interferences decay so that they do not affect the outcome of the analysis. Another reason for allowing an irradiated sample to cool is to reduce the health hazard associated with higher levels of radioactivity.

Nondestructive Methods

As shown in Figure 32-8, in the nondestructive method the sample and standards are counted directly after cooling. Here, the ability of a gamma-ray spec-

mass of analyte is directly proportional to the counting rate. If we use the subscripts x and s to represent sample and standard, respectively, we can write

$$R_x = km_x \tag{32-19}$$

$$R_s = km_s \tag{32-20}$$

where k is a proportionality constant. Dividing one equation by the other and rearranging leads to the basic equation for computing the mass of analyte in an unknown:

$$m_x = \frac{R_x}{R_s} m_s \tag{32-21}$$

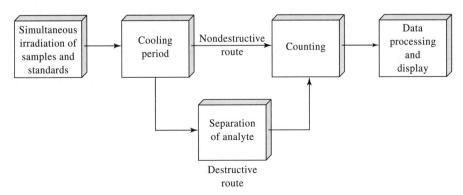

FIGURE 32-8 Flow diagram for two types of neutron activation methods.

trometer to discriminate among the radiations of different energies provides selectivity. Equation 32-21 is then used to calculate the amount of analyte in the unknown as shown in Example 32-5.

EXAMPLE 32-5

Two 5.00-mL aliquots of river water were taken for NAA. Exactly 1.00 mL of a standard solution containing 1.00 µg of Al^{3+} was added to one aliquot, and 1.00 mL of deionized water was introduced into the other. The two samples were then irradiated simultaneously in a homogeneous neutron flux. After a brief cooling period, the gamma radiation from the decay of ^{28}Al was counted for each sample. The solution diluted with water gave a counting rate of 2315 cpm, whereas the solution containing the added Al^{3+} gave 4197 cpm. Calculate the mass of Al in the 5.00-mL sample.

▸ *Solution*

Here, we are dealing with a simple standard-addition problem that can be solved by substituting into Equations 32-19 and 32-20. Thus,

$$2315 = km_x$$
$$4197 = k(m_x + m_s) = k(m_x + 1.00)$$

Solving these two equations leads to

$$m_x = 1.23 \text{ µg}$$

Success of the nondestructive method requires that the spectrometer be able to isolate the gamma-ray signal produced by the analyte from signals arising from the other components. Whether adequate resolution is possible depends on the complexity of the sample, the presence or absence of elements that produce gamma rays of about the same energy as that of the element of interest, and the resolving power of the spectrometer. Improvements in resolving power, which have been made in the past several decades because of the development of high-purity germanium detectors (Section 12B-4), have greatly broadened the scope of the nondestructive method. The great advantage of the nondestructive approach is its simplicity in terms of sample handling and the minimal time required to complete an analysis. In fact, modern activation analysis equipment is largely automated. The nondestructive purely instrumental technique is often termed *instrumental neutron activation analysis*, or INAA.

Destructive Methods

As shown in the lower pathway in Figure 32-8, a destructive method requires that the analyte be separated from the other components of the sample prior to counting. If a chemical separation method is used, this technique is called *radiochemical neutron activation*. In this case a known amount of the irradiated sample is dissolved and the analyte separated by precipitation, extraction, ion exchange, or chromatography. The isolated material or a known fraction thereof is then counted for its gamma — or beta — activity. As in the nondestructive method, standards may be irradiated simultaneously and treated in an identical way. Equation 32-21 is then used to calculate the results of the analysis.

32C-5 Applications of Neutron Activation

Neutron activation methods offer several advantages, including high sensitivity, minimal sample preparation, and ease of calibration. Often, these procedures are nondestructive and for this reason are applied to the analysis of art objects, coins, forensic samples, and archaeological specimens. The major disadvantages of activation methods are their need for large and expensive equipment and special facilities for handling and disposing of radioactive materials. Another shortcoming is the long time required to complete analyses when long-lived radionuclides are used.

Scope

Figure 32-9 illustrates that neutron activation is potentially applicable to the determination of sixty-nine elements. In addition, four of the inert gases form radionuclides with thermal neutrons and thus can also be determined. Finally, three additional elements — oxygen, nitrogen, and yttrium — can be activated with fast neutrons. A list of types of materials to which the method has been applied is impressive and includes metals, alloys, archaeological objects, semiconductors, biological specimens, rocks, minerals, and water. Acceptance of evidence developed from activation analysis by courts of law has led to its widespread use in forensic chemistry. Here, the high sensitivities and nondestructive aspect of the method are particularly useful. Most applications have involved the determination of traces of various elements.[10]

---

[10]For the first forensic application of NAA, see the Instrumental Analysis in Action feature at the end of Section 6.

FIGURE 32-9 Estimated sensitivities of neutron activation methods. Numbers in blue correspond to β sensitivities in micrograms; numbers in black to γ sensitivities in micrograms. In each case samples were irradiated for 1 h or less in a thermal neutron flux of 1.8×10^{12} n/cm^2/s. (From V. P. Guinn and H. R. Lukens Jr., in *Trace Analysis: Physical Methods*, G. H. Morrison, ed., p. 345, New York: Wiley, 1965.)

Element	β	γ
Na	5×10^{-3}	5×10^{-3}
Mg	5×10^{-1}	5×10^{-1}
K	5×10^{-2}	5×10^{-2}
Ca	1.0	5
Sc	1×10^{-2}	5×10^{-2}
Ti	5×10^{-1}	5×10^{-2}
V	5×10^{-3}	1×10^{-2}
Cr	$-$	1
Mn	5×10^{-5}	5×10^{-5}
Fe	50	200
Co	5×10^{-3}	1×10^{-1}
Ni	5×10^{-2}	1×10^{-1}
Cu	1×10^{-3}	1×10^{-3}
Zn	1×10^{-1}	1×10^{-1}
Al	1×10^{-1}	1×10^{-2}
Si	5×10^{-2}	500
P	5×10^{-1}	$-$
S	5	200
Cl	1×10^{-2}	1×10^{-1}
F	$-$	1
Rb	5×10^{-3}	5
Sr	5×10^{-3}	5×10^{-3}
Zr	1	1
Nb	5×10^{-3}	1
Mo	5×10^{-1}	1×10^{-1}
Ru	1×10^{-2}	5×10^{-2}
Rh	1×10^{-3}	5×10^{-4}
Pd	5×10^{-4}	5
Ag	5×10^{-3}	5×10^{-3}
Cd	5×10^{-2}	5×10^{-1}
In	5×10^{-5}	1×10^{-4}
Ga	5×10^{-3}	5×10^{-3}
Ge	5×10^{-3}	5×10^{-3}
As	1×10^{-3}	5×10^{-3}
Se	$-$	5
Br	5×10^{-3}	5×10^{-3}
Sn	5×10^{-3}	5×10^{-1}
Sb	5×10^{-3}	1×10^{-2}
Te	5×10^{-2}	5×10^{-2}
I	5×10^{-2}	5×10^{-2}
Cs	5×10^{-1}	5×10^{-1}
Ba	5×10^{-2}	1×10^{-1}
La	1×10^{-3}	5×10^{-3}
Hf	$-$	1
Ta	5×10^{-2}	5×10^{-1}
W	1×10^{-3}	5×10^{-3}
Re	5×10^{-4}	1×10^{-3}
Os	5×10^{-2}	$-$
Ir	1×10^{-4}	1×10^{-3}
Pt	5×10^{-2}	5×10^{-1}
Au	5×10^{-4}	5×10^{-4}
Hg	$-$	1×10^{-2}
Pb	10	$-$
Bi	5×10^{-1}	$-$
Ce	1×10^{-1}	1×10^{-1}
Pr	5×10^{-4}	5×10^{-2}
Nd	1×10^{-1}	1×10^{-1}
Sm	5×10^{-4}	5×10^{-3}
Eu	5×10^{-6}	5×10^{-4}
Gd	1×10^{-2}	5×10^{-2}
Tb	5×10^{-2}	1×10^{-1}
Dy	1×10^{-6}	5×10^{-6}
Ho	1×10^{-4}	1×10^{-4}
Er	1×10^{-3}	1×10^{-3}
Tm	1×10^{-2}	1×10^{-1}
Yb	1×10^{-3}	1×10^{-3}
Lu	5×10^{-5}	5×10^{-5}
Th	5×10^{-2}	5×10^{-2}
U	5×10^{-3}	5×10^{-3}

Accuracy

The principal errors in activation analyses arise from self-shielding, unequal neutron flux at the sample and standard, counting uncertainties, and errors in counting due to scattering, absorption, and differences in geometry between sample and standard. The errors from these causes can usually be reduced to less than 10% relative. Frequently, uncertainties in the range of 1% to 3% are obtainable.

Sensitivity

The most important characteristic of the neutron activation method is its remarkable sensitivity for many elements. Note in Figure 32-9, for example, that as little as 10^{-5} µg of several elements can be detected. Note also the wide variations in sensitivities among the elements; thus, detection requires about 50 µg of iron, but only 10^{-6} µg for europium.

The efficiency of chemical recovery, if required, may limit the sensitivity of an activation analysis. Other factors include the sensitivity of the detection equipment for the emitted radiation, the extent to which activity in the sample decays between irradiation and assay, the time available for counting, and the magnitude of the background counting rate with respect to that for the sample. A high rate of decay is desirable from the standpoint of minimizing the duration of the counting period. With high decay rates, however, comes the need to establish accurately the time between stopping the irradiation and starting the counting process. A further potential complication is associated with the *dead time* of the counting system, during which the counter is impervious to events. If counting rates are too rapid, events may be missed. Here, a correction must be introduced to account for the difference between elapsed (clock) and live (real) counting times.

32D ISOTOPE DILUTION METHODS

Isotope dilution methods, which predate activation procedures, have been and still are extensively applied to problems in all branches of chemistry. These methods are among the most selective available to chemists. Both stable and radioactive nuclides are employed in the isotope dilution technique. Radioactive nuclides are the more convenient, however, because of the ease with which the concentration of the isotope can be de-

Tutorial: Learn more about **radiochemical methods**.

termined. We limit the discussion here to methods employing radioactive species.

32D-1 Principles of the Isotope Dilution Procedure

Isotope dilution methods require the preparation of a quantity of the analyte in a radioactive form. A known mass of this isotopically labeled species is then mixed with the sample. After treatment to assure homogeneity between the active and nonactive species, a part of the analyte is isolated chemically in the form of a purified compound. By counting a known mass of this product, the extent of dilution of the active material can be calculated and related to the amount of nonactive substance in the original sample. Note that quantitative recovery of the species is not required. Thus, in contrast to the typical analytical separation, steps can be employed to assure a highly pure product on which to base the analysis. It is this independence from the need for quantitative isolation that leads to the high selectivity of the isotope dilution method.

In developing an equation that relates the activity of the isolated and purified mixture of analyte and tracer to the original amount of the analyte, let us assume that a mass m_t of the tracer having a counting rate of R_t is added to a sample containing a mass m_x of inactive analyte. The counting rate for the resulting $m_x + m_t$ grams of the mixture will be the same as that of the m_t grams of the tracer, or R_t. If now m_m grams of the isolated and purified mixture of the active and inactive species are counted, the counting rate R_m will be $m_m/(m_x + m_t)$ of R_t because of dilution. Therefore we can write

$$R_m = R_t\left(\frac{m_m}{m_x + m_t}\right) \qquad (32\text{-}22)$$

which rearranges to

$$m_x = \frac{R_t}{R_m}m_m - m_t \qquad (32\text{-}23)$$

Thus, the mass of the species originally present is obtained from the four measured quantities on the right-hand side of Equation 32-23 as illustrated in Example 32-6.

EXAMPLE 32-6

To a sample of a protein hydrolysate, an analyst added 1.00 mg of tryptophan, which was labeled with ^{14}C and exhibited a counting rate of 584 cpm above background. After this labeled compound was thoroughly mixed with the sample, the mixture was passed through an ion-exchange column. The fraction of effluent containing only tryptophan was collected, and from it an 18.0 mg sample of pure tryptophan was isolated. The isolated sample had a counting rate of 204 cpm in the same counter. What was the weight of tryptophan in the original sample?

▸ *Solution*

Substituting into Equation 32-23, we find

$$m_x = \left(\frac{584\,\text{cpm}}{204\,\text{cpm}} \times 18.0\,\text{mg} \right) - 1.00\,\text{mg} = 50.5\,\text{mg}$$

32D-2 Application of the Isotope Dilution Method

The isotope dilution technique has been employed for the determination of about thirty elements in a variety of matrix materials. Isotopic dilution procedures have also been most widely used for the determination of compounds of interest in organic chemistry and biochemistry. Thus, methods have been developed for the determination of such diverse substances as vitamin D, vitamin B_{12}, sucrose, insulin, penicillin, various amino acids, corticosterone, various alcohols, and thyroxine. Isotope dilution analysis has experienced less widespread application since the advent of activation methods. Continued use of the procedure can be expected, however, because of the relative simplicity of the equipment required. In addition, isotope dilution is often applicable where activation analysis is not.

An example of a technique that incorporates the isotope dilution principle with antibody-antigen reactions is radioimmunoassay (RIA); the development of this technique brought Rosalyn Yalow the Nobel Prize in Physiology or Medicine in 1977. Here, a mixture is prepared of a radioactive antigen (usually containing ^{125}I or ^{131}I) and antibodies specific for that antigen. Known amounts of unlabeled antigen are then added, which compete for antibody binding sites. As the concentration of unlabeled antigen increases, more radioactive antigen is displaced from the antibodies, and the ratio of free to bound antigen increases. To determine this ratio, the free antigen and antibody-bound antigen are separated and the gamma emission from ^{125}I or ^{131}I is measured. A standard binding curve is prepared from the known amounts of antigen added. The unknown samples are then put through the same procedure, and the ratio of free to antibody-bound antigen is determined. The amount of unlabeled antigen in the samples is then found from the binding curve.

It is of interest that the dilution principle has also had other, nonradioactive applications. One example is its use in the estimation of the size of salmon spawning runs in Alaskan coastal streams. Here, a small fraction of the salmon are trapped, mechanically tagged, and returned to the river. A second trapping then takes place perhaps 10 miles upstream, and the fraction of tagged salmon is determined. The total salmon population is readily calculated from this information and from the number originally tagged. The assumption must, of course, be made that the fish population becomes homogenized during its travel between trapping stations.

QUESTIONS AND PROBLEMS

*Answers are provided at the end of the book for problems marked with an asterisk.

[X] Problems with this icon are best solved using spreadsheets.

32-1 Identify X in each of the following nuclear reactions:
(a) $^{68}_{30}\text{Zn} + ^{1}_{0}n \rightarrow ^{65}_{28}\text{Ni} + X$
(b) $^{30}_{15}\text{P} \rightarrow ^{30}_{14}\text{Si} + X$
(c) $^{214}_{82}\text{Pb} \rightarrow ^{214}_{83}\text{Bi} + X$
(d) $^{235}_{92}\text{U} + ^{1}_{0}n \rightarrow 4(^{1}_{0}n) + ^{72}_{30}\text{Zn} + X$
(e) $^{130}_{52}\text{Te} + ^{2}_{1}\text{H} \rightarrow ^{131}_{53}\text{I} + X$
(f) $^{64}_{29}\text{Cu} + X \rightarrow ^{64}_{28}\text{Ni}$

[X] **32-2** Potassium-42 is a β emitter with a half-life of 12.36 h. Calculate the fraction of this radionuclide remaining in a sample after 1, 10, 20, 30, 40, 50, 60, 70, and 100 h. Plot the fraction remaining versus time. Determine the time required for the fraction to fall to 1%.

 32-3 Calculate the fraction of each of the following radionuclides that remains after 1 day, 2 days, 3 days, and 4 days (half-lives are given in parentheses): iron-59 (44.51 days), titanium-45 (3.078 h), calcium-47 (4.536 days), and phosphorus-33 (25.3 days).

*32-4 A $PbSO_4$ sample contains 1 microcurie of Pb-200 ($t_{1/2} = 21.5$ h). What storage period is needed to assure that its activity is less than 0.01 microcurie?

*32-5 Estimate the standard deviation and the relative standard deviation associated with counts of (a) 100, (b) 750, (c) 7.00×10^3, (d) 2.00×10^4.

*32-6 Estimate the absolute and relative uncertainty associated with a measurement involving 800 counts at the
(a) 90% confidence level.
(b) 95% confidence level.
(c) 99% confidence level.

*32-7 For a particular radioactive sample, the total counting rate (sample plus background) was 450 cpm, and this value was obtained over a 15.0-min counting period. The background was counted for 2.0 min and gave 7 cpm. Estimate
(a) the corrected counting rate R_c.
(b) the standard deviation associated with the corrected counting rate σ_{R_c}.
(c) the 95% confidence interval associated with the corrected counting rate.

*32-8 The background counting rate of a laboratory was found to be approximately 9 cpm when measured over a 3-min period. The goal is to keep the relative standard deviation of the corrected counting rate at less than 5% ($\sigma_{R_c}/R_c < 0.05$). What total number of counts should be collected if the total counting rate is
(a) 70 cpm and (b) 400 cpm?

*32-9 A sample of ^{64}Cu exhibits 3250 cpm. After 10.0 h, the same sample gave 2230 cpm. Calculate the half-life of ^{64}Cu.

*32-10 One half of the total activity in a particular sample is due to ^{38}Cl ($t_{1/2} = 87.2$ min). The other half of the activity is due to ^{35}S ($t_{1/2} = 37.5$ days). The beta emission of ^{35}S must be measured because this nuclide emits no gamma photons. Therefore, it is desirable to wait until the activity of the ^{38}Cl has decreased to a negligible level. How much time must elapse before the activity of the ^{38}Cl has decreased to only 0.1% of the remaining activity because of ^{35}S?

32-11 Prove that the relative standard deviation of the counting rate σ_R/R is simply $M^{-1/2}$, where M is the number of counts.

32-12 Under what conditions can the second term in Equation 32-23 be ignored?

*32-13 A 2.00-mL solution containing 0.120 microcurie per milliliter of tritium was injected into the bloodstream of a dog. After allowing time for homogenization, a 1.00-mL sample of the blood was found to have a counting rate of 15.8 counts per second (cps). Calculate the blood volume of the animal.

*32-14 The penicillin in a mixture was determined by adding 0.981 mg of the ^{14}C-labeled compound having a specific activity of 5.42×10^3 cpm/mg. After equilibration, 0.406 mg of pure crystalline penicillin was isolated. This material had a net activity of 343 cpm. Calculate the mass in milligrams of penicillin in the sample.

*32-15 In an isotope dilution experiment, chloride was determined by adding 5.0 mg of sodium chloride containing ^{38}Cl ($t_{1/2} = 37.3$ min) to a sample. The specific activity of the added NaCl was 4.0×10^4 cps/mg. What was the total amount of chloride present in the original sample if 400 mg of pure AgCl was isolated and if this material had a counting rate of 35 cps above background 148 min after the addition of the radiotracer?

32-16 A 10.0-g sample of protein was hydrolyzed. A 3.0-mg portion of ^{14}C-labeled threonine, with specific activity of 1000 cpm/mg, was added to the hydrolysate. After mixing, 11.5 mg of pure threonine was isolated and found to have a specific activity of 20 cpm/mg.
  (a) What percentage of this protein is threonine?
  (b) Had ^{11}C ($t_{1/2} = 20.3$ min) been used instead of ^{14}C, how would the answer change?
  Assume all numerical values are the same as in part (a) and that the elapsed time between the two specific activity measurements was 32 min.

*32-17 The streptomycin in 500 g of a broth was determined by addition of 1.34 mg of the pure antibiotic containing ^{14}C. The specific activity of this preparation was found to be 223 cpm/mg for a 30-min count. From the mixture, 0.112 mg of purified streptomycin was isolated, which produced 654 counts in 60.0 min. Calculate the concentration in parts per million streptomycin in the sample.

32-18 Show, via a calculation, that the average kinetic energy of a population of thermal neutrons is approximately 0.04 eV.

32-19 Naturally occurring manganese is 100% ^{55}Mn. This nuclide has a thermal neutron capture cross section of 13.3×10^{-24} cm^2. The product of neutron capture is ^{56}Mn, which is a beta and gamma emitter with a 2.50-h half-life. When Figure 32-9 was produced, it was assumed that the sample was irradiated for 1 h at a neutron flux of 1.8×10^{12} neutrons/(cm^2 s) and that 10 cpm above background could be detected.
  (a) Calculate the minimum mass of manganese that could be detected.
  (b) Suggest why the calculated value is lower than the tabulated value.

Challenge Problem

32-20 (a) By referring to a reference on NAA or by using an Internet search engine to find the information, describe in some detail the instrumental and radiochemical differences between prompt gamma-ray neutron activation and delayed gamma-ray neutron activation.
  (b) What are the types of elements for which prompt gamma-ray activation analysis is most applicable?
  (c) Why is delayed gamma-ray emission most often used in NAA?
  (d) Why is NAA considered to be a very selective and sensitive method?
  (e) A crystal of potassium fluoride is to be studied via NAA. The following table summarizes the behavior of all naturally occurring isotopes in the crystal.

Natural Abundance

100%	$^{19}_{9}\text{F}\ (n, \gamma)^{20}_{9}\text{F}$	$\xrightarrow{\ t_{1/2} = 11\ \text{s}\ }$	$^{20}_{10}\text{Ne} + \beta^- + \bar{\nu}$
93%	$^{39}_{19}\text{K}\ (n, \gamma)^{40}_{19}\text{K}$		
0.01%	$^{40}_{19}\text{K}\ (n, \gamma)^{41}_{19}\text{K}$		
7%	$^{41}_{19}\text{K}\ (n, \gamma)^{42}_{19}\text{K}$	$\xrightarrow{\ t_{1/2} = 12.4\ \text{h}\ }$	$^{42}_{20}\text{Ca} + \beta^- + \bar{\nu}$

^{19}F, ^{20}Ne, ^{39}K, ^{41}K, and ^{42}Ca are stable, and we will assume that ^{40}K is also stable because it has a half-life of 1.3×10^9 years. What sort of irradiation and detection sequence would you use if you wanted to base your analysis on fluorine?

(f) What sort of irradiation and detection sequence would you use if you wanted to base the analysis on potassium?

(g) Refer to part (e) and calculate the activity due to ^{20}F and ^{42}K in a 58-mg (1.0-millimole) sample of pure potassium fluoride that has been irradiated for 60 s. The thermal neutron cross sections for ^{19}F and ^{41}K are $0.0090 \times 10^{-24}\ \text{cm}^2$ and $1.1 \times 10^{-24}\ \text{cm}^2$, respectively. Assume a flux density of 1.0×10^{13} neutrons $\text{cm}^{-2}\ \text{s}^{-1}$.

(h) Find a method in the literature that describes the use of NAA for determining selenium in freshwater ecosystems. Describe the method in detail. Include the neutron source, the irradiation time, and the calculations used. Give the advantages and disadvantages of the method over other analytical techniques.

CHAPTER THIRTY-THREE

Automated Methods of Analysis

O*ne of the major developments in analytical chemistry during the last few decades has been the appearance of commercial automated analytical systems, which provide analytical and control information with a minimum of operator intervention. Automated systems first appeared in clinical laboratories, where thirty or more species are routinely determined for diagnostic and screening purposes. Laboratory automation soon spread to industrial process control and later to pharmaceutical, environmental, forensic, governmental, and university research laboratories. Today, many routine determinations as well as many of the most demanding analyses are made with totally or partially automated systems.*

Throughout this chapter, this logo indicates an opportunity for online self-study at **www .thomsonedu.com/chemistry/skoog**, linking you to interactive tutorials, simulations, and exercises.

33A OVERVIEW

There are many definitions of automation, but the practical meaning is the performance of operations without human intervention. In the context of analytical chemistry, automation may involve operations like the preparation of samples, the measurement of responses, and the calculation of results.[1] Figure 33-1 illustrates the common steps in a typical chemical analysis and the automation methods used. In some cases, one or more of the steps shown on the left of the figure can be omitted. For example, if the sample is already a liquid, the dissolution step can be omitted. Likewise, if the method is highly selective, the separation step may be unnecessary. The steps listed in Figure 33-1 are sometimes called *unit operations*.

We discuss in this chapter analyzers that are highly automated, such as flow injection and discrete analyzers. In addition, laboratory robotic systems that are becoming more and more commonplace for sample handling and preparation are also described. The latest advances in automation involve the development of microfluidic systems, which are sometimes called *lab-on-a-chip* or *micro total analysis systems*. These recent developments are also described here. It is important to note that the same principles of automatic analysis discussed here also apply to process control systems, which analyze the current state of a process and then use feedback to alter experimental variables in such a way that the current state and the desired state are nearly identical.

33A-1 Advantages and Disadvantages of Automated Analyses

Automated instruments can offer a major economic advantage because of their savings in labor costs. This advantage is realized when the volume of work is large enough to offset the original capital investment. For clinical and testing laboratories, in which large numbers of routine analyses are performed daily, the savings achieved by automation can be very large.

A second major advantage of automated instruments is that the number of determinations per day (throughput) can be much higher than with manual

---

[1] For monographs on laboratory automation, see *Laboratory Automation in the Chemical Industries*, D. G. Cork and T. Sugawara, eds., New York: Dekker, 2002; *Automation in the Laboratory*, W. J. Hurst, ed., New York: Wiley, 1995; V. Cerda and G. Ramis, *An Introduction to Laboratory Automation*, New York: Wiley, 1990.

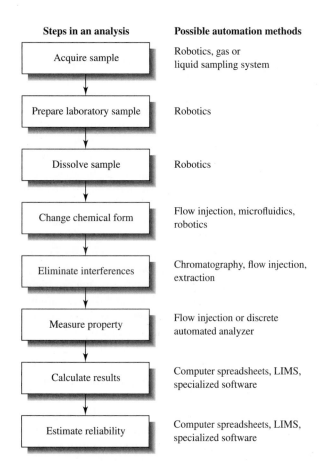

Steps in an analysis	Possible automation methods
Acquire sample	Robotics, gas or liquid sampling system
Prepare laboratory sample	Robotics
Dissolve sample	Robotics
Change chemical form	Flow injection, microfluidics, robotics
Eliminate interferences	Chromatography, flow injection, extraction
Measure property	Flow injection or discrete automated analyzer
Calculate results	Computer spreadsheets, LIMS, specialized software
Estimate reliability	Computer spreadsheets, LIMS, specialized software

FIGURE 33-1 Steps in the analytical process showing possible automation methods. LIMS = laboratory information management system.

methods. In some cases, this is due to the speed of an automated analyzer being significantly greater than that of the equivalent manual device. In other cases, the increased throughput arises because the automated instrument can be used many more hours per day than can a human employee. Indeed, the higher throughput often makes possible the continuous monitoring of the composition of products as they are being manufactured. This information in turn permits alteration of conditions to improve quality or yield. Continuous monitoring is also useful in medicine, where analytical results can be used to determine patients' current conditions and their response to therapy, and in environmental monitoring, where the results can be used to measure the effectiveness of clean-up procedures.

A third advantage of automation is that a well-designed analyzer often provides more reproducible results over a long period than can an operator using a manual instrument. Two reasons can be cited for the higher precision of an automated device. First, machines do not suffer from fatigue or boredom, which have been demonstrated to affect adversely results obtained manually. A more important contributing factor to precision is the high reproducibility of the timing sequences of automated instruments, which can seldom be matched in manual operations. For example, automatic analyzers permit the use of colorimetric reactions that are incomplete or that yield products whose stabilities are inadequate for manual measurement. Similarly, separation techniques, such as solvent extraction or dialysis, where analyte recoveries are incomplete, are still applicable when automated systems are used. In both instances, the high reproducibility of the timing of the operational sequences assures that samples and standards are processed in the same way and for the same length of time.

Other advantages of automation include the ability to process samples in situations that would be dangerous for humans, the minimization of calculation errors, and the direct recording of results in databases and archival storage systems. In some cases procedures that are more lengthy and more complicated than those performed manually can be used with automated systems.

In the early days of automation in laboratories, it was thought that automated instruments would replace human analysts or downgrade their role. It has been found, however, that the role of the analytical chemist has not been downgraded in automated laboratories but merely revised. Analytical chemists are ultimately responsible for the status and quality of the laboratory results. This means the analyst must be fully aware of the strengths and weaknesses of the automated methods employed and be intimately familiar with calibration and validation methods. In fact, modern analytical chemists are now responsible not only for data quality but also for specifying new instrumentation, for adapting or modifying procedures, and for investigating new methods for handling samples, separating interferences, and assessing data reliability — a different, but still vitally important, role.

33A-2 Types of Automatic Systems

Automatic analytical systems are of two general types: *discrete analyzers* and *continuous flow analyzers*; occasionally, the two are combined. In a discrete instrument, individual samples are maintained as separate entities and kept in separate vessels throughout each unit operation listed in Figure 33-1. In continuous flow systems, in contrast, the sample becomes a part of a flowing stream where several of the steps take place as the sample is carried from the injection or sample introduction point to a flow-through measuring unit and thence to waste. Both discrete and continuous flow instruments are generally computer controlled.

Because discrete instruments use individual containers, cross-contamination among samples is totally eliminated. On the other hand, interactions among samples are always a concern in continuous flow systems, particularly as sample throughput increases. Here, special precautions are required to minimize sample contamination.

Modern continuous flow analyzers are generally mechanically simpler and less expensive than their discrete counterparts. Indeed, in many continuous flow systems, the only moving parts are pumps and switching valves. Both of these components are inexpensive and reliable. In contrast, discrete systems often have a number of moving parts such as syringes, valves, and mechanical devices for transporting samples or packets of reagents from one part of the system to another. In the most sophisticated discrete systems, unit operations are performed by versatile computerized robots, in much the same way human operators would.

Some unit operations are not possible with continuous flow systems because such systems are capable of handling only fluid samples. Thus, when solid materials are to be analyzed or when grinding, weighing, ignition, fusion, or filtration is required in an analysis, automation is possible only by using discrete systems or by combining robotics with a continuous flow unit.

In this chapter, we first discuss flow injection analysis (FIA), a recent and important type of continuous flow method. We next consider microfluidic systems, which are miniaturized types of continuous flow units. We then describe several types of discrete automatic systems, several of which are based on laboratory robotics.

33B FLOW INJECTION ANALYSIS

Flow injection methods were first described by Ruzicka and Hansen in Denmark and Stewart and coworkers in the United States in the mid-1970s.[2] Flow injection

methods are an outgrowth of segmented-flow procedures, which were widely used in clinical laboratories in the 1960s and 1970s for automatic routine determination of a variety of species in blood and urine samples for medical diagnostic purposes. In segmented-flow systems, samples were carried through the system to a detector by a flowing aqueous solution that contained closely spaced air bubbles. The purpose of the air bubbles was to minimize sample dispersion, to promote mixing of samples and reagents, and to prevent cross-contamination between successive samples. The air bubbles had to be removed prior to detection by using a *debubbler* or the effects of the bubbles had to be removed electronically. The discoverers of flow injection analysis found, however, that continuous flow systems could be greatly simplified by leaving out the air bubbles and by employing reproducible timing through injection of the sample into a flowing stream.[3] Such systems could be designed with minimal cross-contamination of successive samples.

The absence of air bubbles imparts several important advantages to flow injection measurements, including (1) higher analysis rates (typically 100 to 300 samples/h), (2) enhanced response times (often less than 1 min between sample injection and detector response), (3) much more rapid start-up and shutdown times (less than 5 min for each), and (4) except for the injection system, simpler and more flexible equipment. The last two advantages are particularly important because they make it feasible and economic to apply automated measurements to a relatively few samples of a nonroutine kind. That is, continuous flow methods are no longer restricted to situations where the number of samples is large and the analytical method highly routine. In addition, the precise timing in flow injection leads to *controlled dispersion*, which can be advantageous in promoting reproducible mixing, producing sample or reagent concentration gradients, and minimizing reagent consumption. Because of these advantages, many segmented-flow systems have been replaced by flow injection methods (and also by discrete systems based on robotics). Complete FIA instruments are now available from several different companies.

[2]K. K. Stewart, G. R. Beecher, and P. E. Hare, *Anal. Biochem.*, **1976**, *70*, 167; J. Ruzicka and E. H. Hansen, *Anal. Chim. Acta*, **1975**, *78*, 145.

[3]For monographs on flow injection analysis, see M. Trojanowicz, *Flow Injection Analysis*, River Edge, NJ: World Scientific Publishing Co., 2000; B. Karlberg and G. E. Pacey, *Flow Injection Analysis. A Practical Guide*, New York: Elsevier, 1989; J. Ruzicka and E. H. Hansen, *Flow Injection Analysis*, 2nd ed., New York: Wiley, 1988; M. Valcarcel and M. D. Luque de Castro, *Flow Injection Analysis: Principles and Applications*, Chichester, UK: Ellis Horwood, 1987.

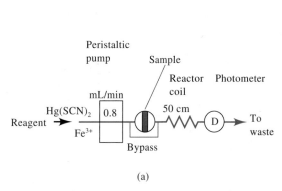
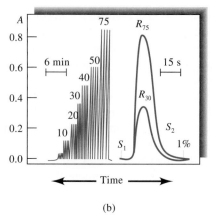

FIGURE 33-2 Flow injection determination of chloride: (a) flow diagram; (b) recorder traces for standards containing 5 to 75 ppm chloride ion (left) and fast scan of two of the standards to demonstrate the low analyte carryover (less than 1%) from run to run (right). Note that the point marked 1% corresponds to where the response would just begin for a sample injected at time S_2. (From J. Ruzicka and E. H. Hansen, *Flow Injection Methods*, 2nd ed., p. 16, New York: Wiley, 1988. Reprinted by permission of John Wiley & Sons, Inc.)

33B-1 Instrumentation

Figure 33-2a is a flow diagram of a basic flow injection system. Here, a peristaltic pump moves colorimetric reagent for chloride ion directly into a valve that permits injection of samples into the flowing stream. The sample and reagent then pass through a 50-cm reactor coil where the reagent diffuses into the sample plug and produces a colored product by the sequence of reactions

$$Hg(SCN)_2(aq) + 2Cl^- \rightleftharpoons HgCl_2(aq) + 2SCN^-$$
$$Fe^{3+} + SCN^- \rightleftharpoons Fe(SCN)^{2+}$$

From the reactor coil, the solution passes into a flow-through photometer equipped with a 480-nm interference filter.

The signal output from this system for a series of standards containing from 5 to 75 ppm chloride is shown on the left of Figure 33-2b. Note that four injections of each standard were made to demonstrate the reproducibility of the system. The two curves to the right are high-speed scans of one of the samples containing 30 ppm (R_{30}) and another containing 75 ppm (R_{75}) chloride. These curves demonstrate that cross-contamination is minimal in an unsegmented stream. Thus, less than 1% of the first analyte is present in the flow cell after 28 s, the time of the next injection (S_2). This system has been successfully used for the routine determination of chloride ion in brackish and waste waters as well as in serum samples.

Sample and Reagent Transport System

The solution in a flow injection analysis is usually pumped through flexible tubing in the system by a peristaltic pump, a device in which a fluid (liquid or gas) is squeezed through plastic tubing by rollers. Figure 33-3 illustrates the operating principle of the peristaltic pump. Here, the spring-loaded cam, or band, pinches the tubing against two or more of the rollers at all times, thus forcing a continuous flow of fluid through the tubing. Modern pumps generally have eight to ten rollers, arranged in a circular configuration so that half are squeezing the tube at any instant. This design leads to a flow that is relatively pulse free. The flow rate is controlled by the speed of the motor, which should be greater than 30 rpm, and the inside diameter (i.d.) of the tube. A wide variety of tube sizes (i.d. = 0.25 to 4 mm) are available commercially that permit flow rates as small as 0.0005 mL/min and as great as 40 mL/min. Flow injection manifolds have been miniaturized through the use of fused silica capillaries (i.d. = 25–100 µm) or through *lab-on-a-chip* technology (see Section 33C). The rollers of typical commercial peristaltic pumps are long enough so that several reagent and sample streams can be pumped simultaneously. Syringe pumps and electroosmosis are also used to induce flow in flow injection systems.

As shown in Figure 33-2a, flow injection systems often contain a coiled section of tubing (typical coil diameters are about 1 cm or less) whose purpose is to enhance axial dispersion and to increase radial mixing

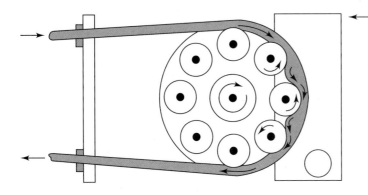

FIGURE 33-3 Diagram showing one channel of a peristaltic pump. Several additional tubes may be located under the one shown (below the plane of the diagram) to carry multiple channels of reagent or sample. (From B. Karlberg and G. E. Pacey, *Flow Injection Analysis: A Practical Guide*, p. 34, New York: Elsevier, 1989. With permission of Elsevier Science Publishers.)

of the sample and reagent, both of which lead to more symmetric peaks.

Sample Injectors and Detectors

The injectors and detectors employed in flow injection analysis are similar in kind and performance requirements to those used in high-performance liquid chromatography (HPLC). Sample sizes for flow injection procedures range from less than 1 µL to 200 µL, with 10 to 30 µL being typical for most applications. For a successful analysis, it is vital that the sample solution be injected rapidly as a pulse or plug of liquid; in addition, the injections must not disturb the flow of the carrier stream. The most useful and convenient injector systems are based on sampling loops similar to those used in chromatography (see, for example, Figures 27-5 and 28-6). The method of operation of the sampling loop is illustrated in Figure 33-2a. With the sampling valve in the position shown, reagents flow through the bypass. When a sample has been injected into the loop and the valve turned 90°, the sample enters the flow as a single, well-defined zone. For all practical purposes, flow through the bypass ceases with the valve in this position because the diameter of the sample loop is significantly greater than that of the bypass tubing.

The most common detectors in flow injection are spectrophotometers, photometers, and fluorometers. Electrochemical, chemiluminescence, atomic emission, and atomic absorption detectors have also been used.

Separations in FIA

Separations by dialysis, by liquid-liquid extraction, and by gaseous diffusion are easily accomplished automatically with flow injection systems.[4]

[4]For a review of applications of FIA for sample preparation and separations, see G. D. Clark, D. A. Whitman, G. D. Christian, and J. Ruzicka, *Crit. Rev. Anal. Chem.*, **1990**, *21* (5), 357.

Dialysis and Gas Diffusion. Dialysis is often used in continuous flow methods to separate inorganic ions, such as chloride or sodium, or small organic molecules, such as glucose, from high-molecular-mass species, such as proteins. Small ions and molecules diffuse relatively rapidly through hydrophilic membranes of cellulose acetate or nitrate but large molecules do not. Dialysis usually precedes the determination of ions and small molecules in whole blood or serum.

Figure 33-4 is a diagram of a dialysis module in which analyte ions or small molecules diffuse from the sample solution through a membrane into a reagent stream, which often contains a species that reacts with the analyte to form a colored species, which can then be determined photometrically. Large molecules that interfere with the determination remain in the original stream and are carried to waste. The membrane is supported between two Teflon plates in which com-

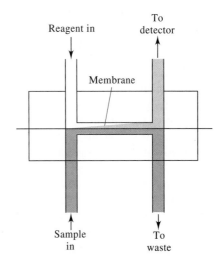

FIGURE 33-4 A dialysis flow module. The membrane is supported between two grooved Teflon blocks.

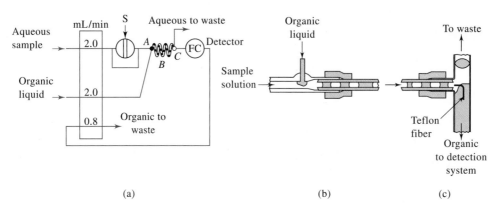

FIGURE 33-5 (a) Flow diagram of a flow injection system containing an extraction module (ABC). (b) Details of A, the organic injector system. (c) Details of C, the separator. (Adapted from J. Ruzicka and E. H. Hansen, *Flow Injection Analysis*, 2nd ed., New York: Wiley, 1988. With permission of John Wiley & Sons.)

plementary channels have been cut to accommodate the two streams on opposite sides of the membrane. The transfer of smaller species through the membrane is usually incomplete (often less than 50%). Thus, successful quantitative analysis requires close control of temperature and flow rates for both samples and standards. Such control is easily accomplished in automated flow injection systems.

Gas diffusion from a donor stream containing a gaseous analyte to an acceptor stream containing a reagent that permits its determination is a highly selective technique that is often used in flow injection analysis. The separations are carried out in a module similar to that shown in Figure 33-4. In this application, however, the membrane is usually a hydrophobic microporous material, such as Teflon or isotactic polypropylene. The determination of total carbonate in an aqueous solution is an example of this type of separation technique. Here, the sample is injected into a carrier stream of dilute sulfuric acid, which is then directed into a gas-diffusion module, where the liberated carbon dioxide diffuses into an acceptor stream containing an acid-base indicator. This stream then passes through a photometric detector that yields a signal proportional to the carbonate content of the sample.

Solvent Extraction. Solvent extraction is another common separation technique that can be easily adapted to continuous flow methods. Figure 33-5a shows a flow diagram for a system for the colorimetric determination of an inorganic cation by extracting an aqueous solution of the sample with chloroform containing a complexing agent, such as 8-hydroxyquinoline. At point A, the organic solution is injected into the sample-containing carrier stream. Figure 33-5b shows that the stream becomes segmented at this point and is made up of successive bubbles of the aqueous solution and the organic solvent. Extraction of the metal complex occurs in the reactor coil. Separation of the immiscible liquids takes place in the T-shape separator shown in Figure 33-5c. The separator contains a Teflon strip or fiber that guides the heavier organic layer out of the lower arm of the T, where it then flows through the detector labeled FC in Figure 33-5a. This type of separator can be used for low-density liquids by inverting the separator.

It is important to reiterate that none of the separation procedures in FIA methods are complete. Because unknowns and standards are treated identically, incomplete separation is unimportant. Reproducible timing in FIA ensures that, even though separations are incomplete, there is no loss of precision and accuracy as would occur with manual operations.

33B-2 Principles of FIA

Immediately after injection with a sampling valve, the sample zone in a flow-injection apparatus has the rectangular concentration profile shown in Figure 33-6a. As it moves through the tubing, band broadening, or *dispersion*, takes place. The shape of the resulting zone is determined by two phenomena. The first is convection arising from laminar flow in which the center of

 Animation: Learn more about **flow injection analysis**.

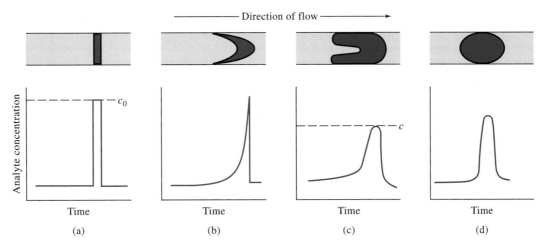

FIGURE 33-6 Effect of convection and diffusion on concentration profiles of analytes at the detector: (a) no dispersion, (b) dispersion by convection, (c) dispersion by convection and radial diffusion, and (d) dispersion by diffusion. (Reprinted with permission from D. Betteridge, *Anal. Chem.*, **1978**, *50*, 836A. Copyright 1978 American Chemical Society.)

the fluid moves more rapidly than the liquid adjacent to the walls, creating the parabolic front and the skewed zone profile shown in Figure 33-6b. Diffusion also causes broadening. Two types of diffusion can, in principle, occur: radial, which is perpendicular to the flow direction, and longitudinal, which is parallel to the flow. It has been shown that longitudinal diffusion is insignificant in narrow tubing, and radial diffusion is much more important. In fact, at low flow rates, radial diffusion is the major source of dispersion. Under such conditions, the symmetrical distribution shown in Figure 33-6d is approached. In fact, flow injection analyses are usually performed under conditions in which dispersion by both convection and radial diffusion occurs; peaks like that in Figure 33-6c are then obtained. Here, the radial dispersion from the walls toward the center essentially frees the walls of analyte and nearly eliminates cross-contamination between samples.

Dispersion

Dispersion D is defined by the equation

$$D = c_0/c$$

where c_0 is the analyte concentration of the injected sample and c is the peak concentration at the detector (see Figure 33-6a and c). Dispersion is measured by injecting a dye solution of known concentration c_0 and then recording the absorbance in a flow-through cell. After calibration, c is calculated from Beer's law.

Dispersion is influenced by three interrelated and controllable variables: sample volume, tubing length, and flow rate. The effect of sample volume on dispersion is shown in Figure 33-7a, where tubing length and flow rate are constant. Note that at large sample volumes, the dispersion becomes unity. Under these circumstances, no appreciable mixing of sample and carrier takes place, and thus no sample dilution has occurred. Most flow injection analyses, however, involve interaction of the sample with the carrier or an injected reagent. Here, a dispersion value greater than unity is necessary. For example, a dispersion value of 2 is required to mix sample and carrier in a 1:1 ratio.

The dramatic effect of sample volume on peak height shown in Figure 33-7a emphasizes the need for highly reproducible injection volumes when D values of 2 and greater are used. Other conditions also must be closely controlled for good precision.

Figure 33-7b demonstrates the effect of tubing length on dispersion when sample size and flow rate are constant. Here, the number above each peak gives the length of sample travel in centimeters.

33B-3 Applications of FIA

Flow injection applications tend to fall into three categories: *low dispersion*, *medium dispersion*, and *large dispersion*.

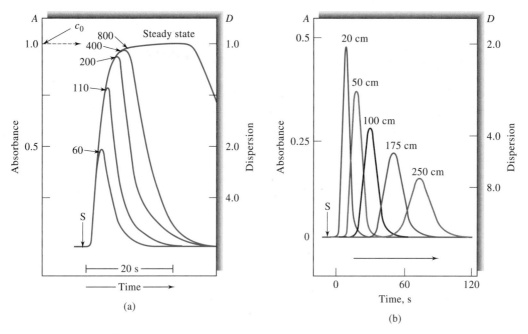

FIGURE 33-7 Effect of sample volume and length of tubing on dispersion. (a) Tube length: 20 cm; flow rate: 1.5 mL/min; indicated volumes are in μL. (b) Sample volume: 60 μL; flow rate: 1.5 mL/min. (From J. Ruzicka and E. H. Hansen, *Anal. Chim. Acta*, **1980**, *114*, 21. With permission.)

Low- and Medium-Dispersion Applications

Low-dispersion flow injection techniques (dispersion values of 1–3) have been used for high-speed sample introduction to such detector systems as inductively coupled plasma atomic emission, flame atomic absorption, and specific-ion electrodes. The justification for using flow injection methods for electrodes such as pH and pCa is the small sample size required (~25 μL) and the short measurement time (~10 s). That is, measurements are made well before steady-state equilibria are established, which for many specific-ion electrodes may require a minute or more. With flow injection measurements, transient signals for sample and standards provide excellent accuracy and precision. For example, it has been reported that pH measurements on blood serum can be accomplished at a rate of 240/h with a precision of ±0.002 pH.

In general, limited-dispersion conditions are realized by reducing as much as possible the distance between injector and detector, slowing the flow rate, and increasing the sample volume. Thus, for the pH measurements just described, the length of 0.5-mm tubing was only 10 cm and the sample size was 30 μL.

Medium dispersion corresponds to D values of 3–10. Figure 33-8a illustrates a medium-dispersion system for the colorimetric determination of calcium in serum, milk, and drinking water. A borax buffer and a color reagent are combined in a 50-cm mixing coil A prior to sample injection. The output for three samples in triplicate and four standards in duplicate is shown in Figure 33-8b.

Figure 33-9 illustrates a more complicated flow injection system designed for the automatic spectrophotometric determination of caffeine in acetylsalicylic acid drug preparations after extraction of the caffeine into chloroform. The chloroform solvent, after cooling in an ice bath to minimize evaporation, is mixed with the alkaline sample stream in a T-tube (see lower insert). After passing through the 2-m extraction coil L, the mixture enters a T-tube separator, which is differentially pumped so that about 35% of the organic phase containing the caffeine passes into the flow cell, and the other 65% accompanying the aqueous solution containing the rest of the sample flows to waste. To avoid contaminating the flow cell with water, Teflon fibers, which are not wetted by water, are twisted into

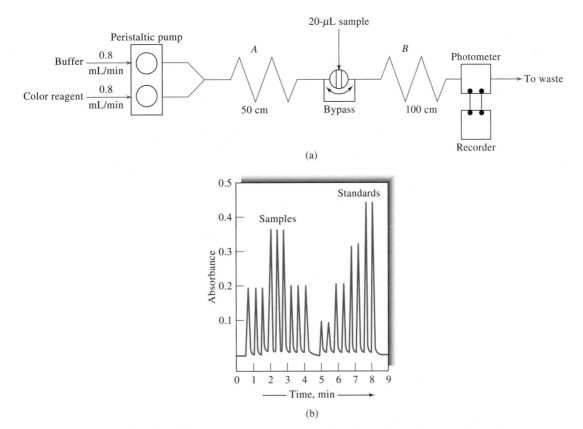

FIGURE 33-8 (a) Flow injection apparatus for determining calcium in water by formation of a colored complex with *o*-cresolphthalein complexone at pH 10. All tubing had an inside diameter of 0.5 mm. *A* and *B* are reaction coils having the indicated lengths. (b) Recorded output. Three sets of curves at left are for triplicate injections of three samples. Four sets of peaks on the right are for duplicate injections of standards containing 5, 10, 15, and 20 ppm calcium. (From E. H. Hansen, J. Ruzicka, and A. K. Ghose, *Anal. Chim. Acta*, **1978**, *100*, 151. With permission.)

a thread and inserted in the inlet to the T-tube in such a way as to form a smooth downward bend. The chloroform flow then follows this bend to the photometer cell where the caffeine concentration is determined on the basis of its absorption at 275 nm.

Stopped-Flow Methods

As discussed earlier, dispersion in small-diameter tubing decreases with decreasing flow rate. In fact, it has been found that dispersion ceases almost entirely when the flow is stopped. This phenomenon has been exploited to increase the sensitivity of measurements by allowing time for reactions to go further toward completion without dilution of the sample zone by dispersion. In this type of application, a timing device is required to turn the pump off at precisely regular intervals.

A second application of the stopped-flow technique is for kinetic measurements. In this application, the flow is stopped with the reaction mixture in the flow cell where the changes in the concentration of reactants or products can be monitored as a function of time. The stopped-flow technique has been used for the enzymatic determination of glucose, urea, galactose, and many other substances of interest in clinical chemistry.

Flow Injection Titrations

Titrations can also be performed continuously in a flow injection apparatus. In these methods, the injected sample is combined with a carrier in a mixing chamber that promotes large dispersion. The mixture is then transported to a confluence fitting, where it is mixed with the reagent containing an indicator. If the

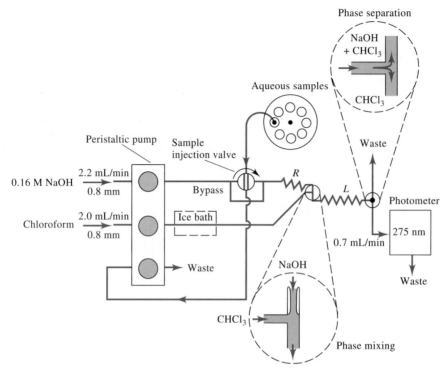

FIGURE 33-9 Flow injection apparatus for the determination of caffeine in acetylsalicylic acid preparations. With the valve rotated 90° the flow in the bypass is essentially zero because of its small diameter. R and L are Teflon coils with 0.8-mm inside diameters; L is 2 m, and the distance from the injection point through R to the mixing point is 0.15 m. (Adapted from B. Karlberg and S. Thelander, *Anal. Chim. Acta*, **1978**, *98*, 2. With permission.)

detection is set to respond to the color of the indicator in the presence of excess analyte, peaks such as those shown in Figure 33-10 are obtained. In this example, an acid is being titrated with a standard solution of sodium hydroxide, which contains bromothymol blue

indicator. With injection of samples, the solution changes from blue to yellow and remains yellow until the acid is consumed and the solution again becomes blue. As shown in the figure, the concentration of analyte is determined from the widths of the peaks at half

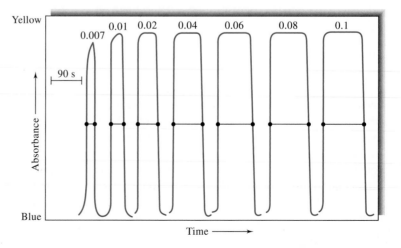

FIGURE 33-10 Flow injection titration of HCl with 0.001 M NaOH. The molarities of the HCl solutions are shown at the top of the figure. The indicator was bromothymol blue. The time interval between the points is a measure of the acid concentration. (From J. Ruzicka, E. H. Hansen, and M. Mosback, *Anal. Chim. Acta*, **1980**, *114*, 29. With permission.)

height. Titrations of this kind can be performed at a rate of sixty samples/h in a conventional FIA system.

Merging Zones FIA

The merging zones principle was first introduced to economize on the use of expensive reagents. When a sample is injected into a carrier stream of reagent, the reagent is usually pumped continuously even when the sample zone is not present. In merging zones FIA, the sample and the reagent are injected into two carrier streams, which are then allowed to merge downstream. Because the carrier streams contain only water or an inert buffer, the expensive reagent is conserved and only a limited quantity injected. By choosing different lengths of the reagent zone and letting it overlap in different ways with the sample zone, differing concentrations of the sample and reagent can be brought together to produce data to construct calibration curves or to study concentration effects.

33B-4 Variants of FIA

Since the introduction of FIA in the mid-1970s, several variations have appeared on normal FIA, which employs continuous unidirectional pumping.

Flow Reversal FIA

One variation on normal FIA, termed *flow reversal FIA*, was introduced by Betteridge and coworkers.[5] Reversing the direction of flow allows the effective reaction coil length to be varied without a physical change in the FIA apparatus. Multiple pumps can be used or the direction of flow can be reversed by means of valves as shown in Figure 33-11a. Flow recycling can also be accomplished by switching two valves as shown in Figure 33-11b. Reversing the flow or recycling the sample plug can allow automated optimization of FIA methods using a variety of different software approaches. Alternatively, kinetics data can be obtained without stopping the flow.

Sequential Injection Analysis

In 1990 Ruzicka and Marshall[6] introduced a variation of flow injection called *sequential injection analysis* (SIA), which can overcome some of the reagent waste of FIA. Instead of the continuous, unidirectional flow typical of FIA, SIA uses discontinuous, bidirectional

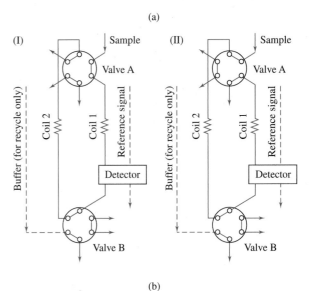

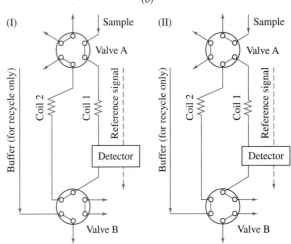

FIGURE 33-11 Configurations for flow reversals (a) and flow recycles (b). In both cases, six-port valves are used. The initial valve positions are shown in diagram I, and diagram II shows the valve configuration during the reversal or recycle. In the recycle configuration (b), both valves turn, but in the reversal configuration, valve B is stationary. All arrows exiting valves go to waste. (From E. B. Townsend and S. R. Crouch, *Trends Anal. Chem.*, **1992**, *11*, 90. With permission.)

flow. A typical apparatus is illustrated in Figure 33-12. The SIA system offers precise, low-volume delivery of reagents and reproducible flow reversals because of the syringe pump. The volumes introduced (usually microliters or less) are controlled by the time the port

[5] D. Betteridge, P. B. Oates, and A. P. Wade, *Anal. Chem.*, **1987**, *59*, 1236.
[6] J. Ruzicka and G. Marshall, *Anal. Chim. Acta*, **1990**, *237*, 329.

 Tutorial: Learn more about **sequential injection analysis**.

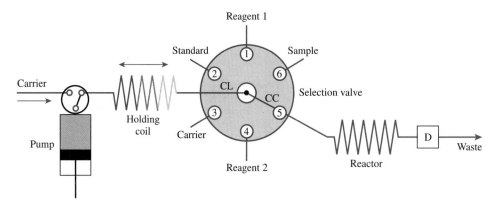

FIGURE 33-12 Sequential injection apparatus based on a bidirectional syringe pump and a six-port selection valve. The valve is equipped with a central communication channel (CC), which can address any of the ports, and a communication line (CL) connected to a holding coil (HC) and the syringe pump. When the communication channel is directed to the various ports, sample and reagent zones are sequentially drawn into the holding coil and stacked one after another. Switching the communication channel to position 5 causes the segments to flow into the reactor toward the detector (D). During this flow, the segments undergo dispersion, partial mixing, and chemical reaction.

is connected to the communication channel and the pump flow rate. Dispersion begins to take place when the sample and reagents are drawn into the holding coil. The flow reversal aids the mixing as the zones travel into the reaction coil and the detector.

Sequential injection has several advantages when compared with traditional flow injection. Not only is reagent usage lower with SIA but the manifolds used are simpler. It is also easy to change from one analytical procedure to another by altering the flow program. With FIA, the entire manifold usually has to be changed when a new procedure is called for. With SIA, the injection valve of FIA is replaced with the selection valve, which allows various reagents or calibration standards to be introduced. Hence, automation of entire procedures, including calibration, can be easily accommodated with SIA. Like FIA, complete SIA instruments, including selection valve, pump, reactor, and detector, are available commercially.

Sequential injection has proven itself especially useful for a variety of separation and preconcentration schemes. It has been applied to analytical procedures involving such methods as membrane separations, pH adjustments, solid-phase extraction, precipitation, and titration. In addition to colorimetry, ion-selective electrodes, amperometry, fluorescence, IR absorption, chemiluminescence, and conductimetry have been used as detection methods with SIA.

Lab-on-a-Valve Technology

The most recent variation of FIA is the so-called *lab-on-a-valve* (LOV) technology introduced by Ruzicka.[7] The LOV idea incorporates an integrated microconduit on top of the selection valve used in SIA. The microconduit is designed to handle all the unit operations needed for a given analytical procedure. Mixing points for analyte and reagents, column reactors, bead reactors, separation columns, and membranes can all be accommodated within the LOV system. In some cases, microfluidic flow systems are used in ways similar to those described in the next section.

33C MICROFLUIDICS

The development of microfluidic systems, in which operations are miniaturized to the scale of an integrated circuit, has enabled the fabrication of a complete *laboratory-on-a-chip* or *micro total analysis system* (μTAS).[8] Miniaturization of laboratory operations to a chip scale can reduce analysis costs by lowering reagent

[7] J. Ruzicka, *Analyst*, **2000**, *125*, 1053.
[8] For reviews of these systems, see P. S. Dittrich, K. Tachikawa, and A. Manz, *Anal. Chem.*, **2006**, *78*, 3887; T. Vilkner, D. Janasek, and A. Manz, *Anal. Chem.*, **2004**, *76*, 3373; D. R. Reyes, D. Iossifidis, P. A. Auroux, and A. Manz, *Anal. Chem.*, **2002**, *74*, 2623; P. A. Auroux, D. Iossifidis, D. R. Reyes, and A. Manz, *Anal. Chem.*, **2002**, *74*, 2637.

consumption and waste production, by automating parallel procedures, and by increasing the numbers of analyses per day. There have been several approaches to implementing the lab-on-a-chip concept. The most successful use the same photolithography technology as is used for preparing electronic integrated circuits. This technology is used to produce the valves, propulsion systems, and reaction chambers needed for performing chemical analyses. The development of microfluidic devices is an active research area involving scientists and engineers from academic and industrial laboratories.[9]

At first, microfluidic flow channels and mixers were coupled with traditional macroscale fluid propulsion systems and valves. The downsizing of the fluid flow channels showed great promise, but the advantages of low reagent consumption and complete automation were not realized. However, in more recent developments, monolithic systems have been used in which the propulsion systems, mixers, flow channels, and valves are integrated into a single structure.[10]

Several different fluid propulsion systems have been investigated for microfluidic systems, including electroosmosis (see Chapter 30), microfabricated mechanical pumps, and hydrogels that emulate human muscles. Flow injection techniques as well as such separation methods as liquid chromatography (Chapter 28), capillary electrophoresis, and micellar electrokinetic chromatography (Chapter 30) have been implemented. Figure 33-13 shows the layout of a microstructure used for FIA. The monolithic unit is made of two permanently bonded polydimethyl siloxane (PDMS) layers. The fluidic channels are 100 μm wide and 10 μm high. The entire device is only 2.0 cm by 2.0 cm. A glass cover allows for optical imaging of the channels by fluorescence excited with an Ar ion laser.

Mixing in most microfluidic systems can be problematic. Slow mixing and significant dispersion because of laminar flow has hampered the use of microfluidics for many applications, including those involving the measurement of kinetics. In an interesting approach to solving these problems, aqueous plugs of solution are segmented by an immiscible organic fluid (perfluorodecaline) to eliminate dispersion.[11] Mixing then occurs by a process known as *chaotic advection*. Mixing times on a millisecond time scale can be achieved with such a

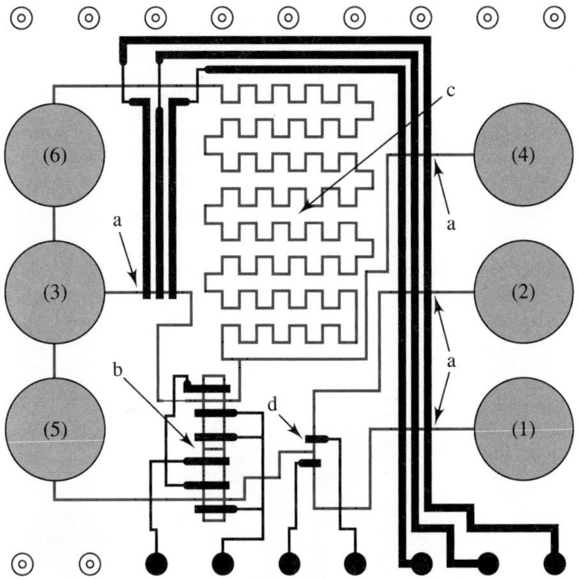

FIGURE 33-13 Layout of a microfabricated structure for FIA. Microfluidic channels are shown in blue, and control channels (pumps and valves) are shown in black. (a) Peristaltic pump, (b) injection valve, (c) mixing and reaction chamber, (d) sample selector. Blue circles indicate fluid reservoirs (1 and 2), samples (3), carrier (4), reactant (5), and waste (6). The entire structure is 2.0 cm by 2.0 cm. (From A. M. Leach, A. R. Wheeler, and R. N. Zare, *Anal. Chem.*, **2003**, *75*, 967.)

droplet-based system. Figure 33-14 gives a comparison between the usual laminar-flow system and the droplet-based system for a reaction between two reactants. This type of segmented system can be used for enzyme determinations, for substrate determinations, and for studies of protein crystallization[12] and adsorption.[13]

Lab-on-a-chip analyzers are now available from several instrument companies. One commercial analyzer allows the analysis of DNA, RNA, proteins, and cells. Another commercial microfluidics device is used for nanoflow liquid chromatography and provides an interface to an electrospray mass spectrometry detector. Lab-on-a-chip analyzers are envisioned for drug screening, for DNA sequencing, and for detecting life forms on Earth, Mars, and other planets. These devices should become more important as the technology matures.

[9]See N. A. Polson and M. A. Hayes, *Anal. Chem.*, **2001**, *73*, 313A.
[10]A. M. Leach, A. R. Wheeler, and R. N. Zare, *Anal. Chem.*, **2003**, *75*, 967.
[11]H. Song and R. F. Ismagilov, *J. Am. Chem. Soc.*, **2003**, *125*, 14613.

[12]B. Zheng, J. D. Tice, and R. F. Ismagilov, *Anal. Chem.*, **2004**, *76*, 4977.
[13]L. S. Roach, H. Song, and R. F. Ismagilov, *Anal. Chem.*, **2005**, *77*, 785.

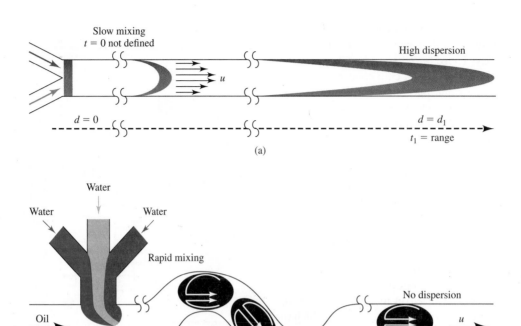

FIGURE 33-14 Comparison of a reaction between two components in a standard pressure-driven microfluidic system (a) and in a droplet-based system (b). In (a), the reaction time is not well defined. In (b), the components and an immiscible fluid form droplet plugs containing the two reagents. Flow-through regions of different geometry cause internal recirculation within the plugs and rapid mixing. The reaction time is equal to the distance d divided by the flow velocity u. (From H. Song, J. D. Tice, and R. F. Ismagilov, *Angew. Chem. Int. Ed.*, **2003**, *42*, 768.)

33D DISCRETE AUTOMATIC SYSTEMS

A wide variety of discrete automatic systems are offered by numerous instrument manufacturers. Some of these devices are designed to perform one or more of the unit operations given in Figure 33-1; others are capable of carrying out an entire analysis automatically. Some discrete systems are intended for a specific analysis only — for example, the determination of nitrogen in organic compounds or the determination of glucose in blood. Others can perform a variety of analyses of a given general type. For example, several automatic titrators can perform neutralization, precipitation, complex formation, and oxidation-reduction titrations as directed by a user-programmed computer. In this section, we describe a few typical discrete systems.

33D-1 Automatic Sampling and Sample Definition of Liquids and Gases

Several dozen automatic devices for sampling liquids and gases are currently available from instrument manufacturers. Figure 33-15a illustrates the principle of reversible pump samplers. This device consists of a movable probe, which is a syringe needle or a piece of fine plastic tubing supported by an arm that periodically lifts the tip of the needle or tube from the sample container and positions it over a second container in which the analysis is performed. This motion is synchronized with the action of a reversible peristaltic pump. As shown in Figure 33-15a, with the probe in the sample container, the pump moves the liquid from left to right for a brief period. The probe is then lifted and positioned over the container on the right, and the direction of pumping is reversed. Pumping is contin-

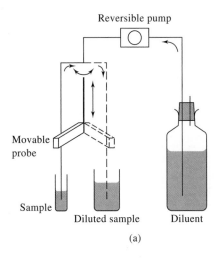

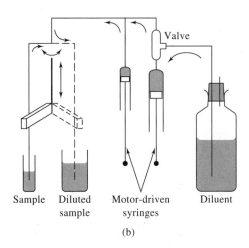

FIGURE 33-15 Automatic samplers: (a) reversible pump type; (b) syringe type.

ued in this direction until the sample and the desired volume of diluent have been delivered. The probe then returns to its original position to sample the next container. The sample volume is always kept small enough so that none of the sample ever reaches the pump or diluent container. Reversible pump samplers are often used in conjunction with a circular rotating sample table. Tables of this kind generally accommodate forty or more samples in plastic or glass cups or tubes. The rotation of the table is synchronized with the moving arm of the sampler so that samples are withdrawn sequentially.

Figure 33-15b illustrates a typical syringe-based sampler and diluter. Here again, a movable sample probe is used. With the probe in the sample cup, the motor-driven syringe on the left withdraws a fixed volume of sample. Simultaneously, the syringe on the right withdraws a fixed volume of diluent (usually larger). The valve shown in the figure permits these two processes to go on independently. When the probe moves over the container on the right, both syringes empty, dispensing the two liquids into the analytical vessel.

Generally, syringe-type injectors are driven by computer-controlled stepping motors, which force the liquid out of the syringe body in a series of identically sized pulses. Thus, for example, a 1-mL syringe powered by a motor that requires 1000 steps to empty the syringe is controllable to 1×10^{-3} mL, or 1 μL. For a 5000-step motor, the precision would be 0.2 μL.

 Tutorial: Learn more about **discrete analysis**.

33D-2 Robotics

With solids, sample preparation, definition, and dissolution involve such unit operations as grinding, homogenizing, drying, weighing, igniting, fusing, and treating with solvents. Each of these individual procedures has been automated. Only recently, however, have instruments appeared that can be programmed to perform several of these operations sequentially and without operator intervention. Generally, such instruments are based on small laboratory robots, which first appeared on the market in the mid-1980s.[14] Figure 33-16 is a schematic of an automated laboratory system based on one of these robots. Central to the system is a horizontal arm, which is mounted on two vertical pillars and has four degrees of freedom in its movement. Its 360° rotational motion covers a circumference of about 150 in.; its maximum reach is then 24 in. The device is equipped with a tonged hand, which can move with a 360° wrist-like motion that permits manipulating vials or tubes, pouring liquids or solids, and shaking and swirling liquids in tubes. Consequently, the arm and hand are capable of carrying out many of the manual operations that are performed by a laboratory scientist. One important feature of the

[14]For a description of laboratory robots, see M. D. Luque de Castro and P. Torres, *Trends Anal. Chem.*, **1995**, *14*, 492; V. Berry, *Anal. Chem.*, **1990**, *62*, 337A; W. J. Hurst and J. W. Mortimer, *Laboratory Robotics: A Guide to Planning, Programming and Applications*, New York: VCH, 1987.

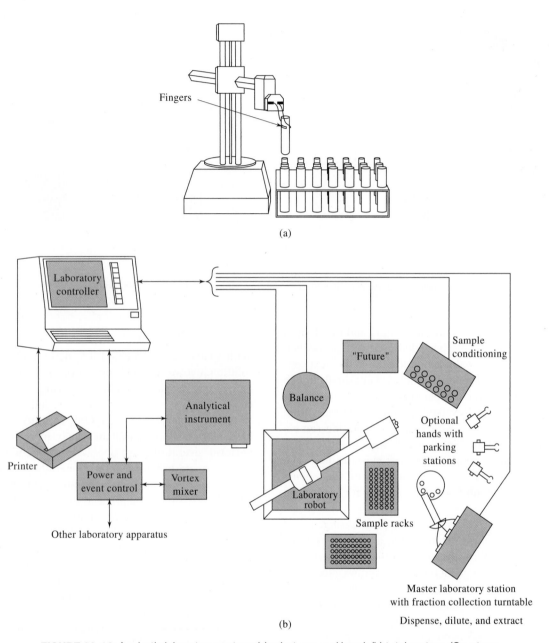

(a)

(b)

FIGURE 33-16 A robotic laboratory system: (a) robot arm and hand; (b) total system. (Courtesy of Caliper Life Sciences, Hopkinton, MA.)

device is its ability to change hands. Thus, the robot can leave its tonged hand on the table and attach a syringe in its place for pipetting liquids.

The robotic system is controlled by a computer that can be user programmed. Thus, the instrument can be instructed to bring samples to the master laboratory station where they can be diluted, filtered, parti-

tioned, ground, centrifuged, homogenized, extracted, and treated with reagents. The device can also be instructed to heat and shake samples, dispense measured volumes of liquids, inject samples into a chromatographic column, and collect fractions from a column. In addition, the robot can be interfaced with an automatic electronic balance for weighing samples.

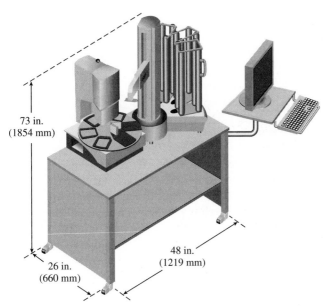

FIGURE 33-17 Sketch of Caliper Staccato Replication System. The unit is a ninety-six-channel device designed to handle loading and unloading of microtiter plates and deep well plates. It can accommodate various reagent reservoirs, pipette tips, and waste disposal units. (Courtesy of Caliper Life Sciences, Hopkinton, MA.)

In addition to general-purpose laboratory robots, robotic units are designed for specific tasks such as loading and unloading of microtiter and deep well plates. The system shown in Figure 33-17 can dispense ninety-six channels simultaneously at volumes of 2 to 200 μL. The robot dispenser is a cylindrical design that can access a wide variety of storage racks arranged in a circular pattern. The dispensing head remains stationary while the well holder rotates to selectable positions.

33D-3 Discrete Clinical Analyzers

Clinical instruments have long been the focus for laboratory automation. Indeed, the Technicon AutoAnalyzer, an air-segmented continuous flow system,[15] was the first truly automated instrument for a clinical laboratory. In 1968 DuPont introduced the ACA (Automated Clinical Analyzer), the first totally automated discrete analyzer. The analyzer was based on prepack-

[15]L. T. Skeggs, *Am. J. Clin. Pathol.*, **1957**, *28*, 311.

ged reagents and a conveyor-belt operation in which n aliquot of the sample was obtained, mixed with ppropriate reagents, reacted, and moved to a photometer for measurement of absorbance. Later discrete analyzers were based on multilayer films (Kodak Ektachem) or on the use of a centrifuge to mix the amples and reagents (centrifugal fast analyzers).

Today, a wide variety of discrete analyzers are available for clinical laboratories. Some of these are general purpose and capable of performing many different determinations, often on a random access basis. Others are of more limited use intended for one or a few specific determinations. Many of the general-purpose analyzers use a combination of traditional chemical tests and immunoassays in which antibody-antigen interactions are used in the determination. Several general-purpose analyzers are capable of performing more than 100 tests, including the determinations shown in Table 33-1.

With some of the analyzers, routine tests can be interrupted to do stat testing (*stat* is an abbreviation of the Latin word *statim*, meaning "immediately"). In other cases, instruments dedicated to a few high-frequency tests are used for stat measurements.

Many modern analyzers use a closed-tube technology to minimize exposure to biohazards and to reduce manual manipulations. Samples and reagents are dispensed automatically, the measurements made by photometry or ion-selective electrodes, and the results computed. Most have bar-coding capabilities to reduce errors from incorrect patient identification. A typical chemistry-immunochemistry automated analyzer is shown in Figure 33-18.

Many dedicated immunochemical analyzers use microtiter plates or wells in which the reagents are immobilized. Adding the sample begins the immunochemical reaction. A well or microplate reader is then used to measure absorbance or in some cases fluorescence. Automatic loading and unloading via robotic dispensing units (see Section 33D-2) is available with most systems.

Several companies now make automated DNA sequencers for clinical and forensic applications. Sequencing is accomplished by incorporating labeled nucleotides into a copy of the DNA piece. The sequence can then be obtained from the positions of the labeled nucleotides. An enzyme is used to make complementary copies of the strands containing the labeled nucleotides. The strands are then separated by gel electrophoresis.

TABLE 33-1 Typical Analyzer Determinations

Blood	Drug of Abuse	Special Chemistry	Therapeutic Drug
Albumin	Alcohol	Ammonia	Acetaminophen
Amylase	Amphetamine	C-reactive protein	Caffeine
Bicarbonate	Barbiturate	Glucose-6-phosphate dehydrogenase	Digoxin
Bilirubin	Cannabinoids	Immunoglobulin A	Lidocaine
Calcium	Cocaine metabolite	Immunoglobulin G	Phenobarbital
Chloride	LSD	Immunoglobulin M	Primidone
Cholesterol	Methadone	Lactate	Quinidine
Cholinesterase	Methaqualone	Microalbumin	Salicylate
Creatinine	Opiates	Rheumatoid factor	Theophylline
Glucose	PCP	Thyroid (T4)	Vancomycin
HDL cholesterol	Urine bleach		
Phosphate	Urine chromate		
Iron	Urine nitrates		
LDL cholesterol	Urine pH		
Lithium	Urine specific gravity		
Magnesium			
Potassium			
Sodium			
Total protein			
Triglyceride			
Urea nitrogen			
Uric acid			

FIGURE 33-18 Picture of integrated chemistry-immunochemistry testing system. (Courtesy of Beckman Coulter, Inc., Fullerton, CA.)

Radioactive nucleotides were traditionally used for this DNA sequencing, but these have been largely replaced by alternative labeling procedures, such as fluorescence labeling. Laser-excited fluorescence is then used to read the sequence. Commercial instruments can be partially or fully automated. With fully automated systems, a sample tray is loaded with template DNA. All the labeling and analysis steps are performed automatically. Many new techniques, such as tandem mass spectrometry, are under development for simplifying and further automating DNA sequencing.

Special-purpose clinical instruments are available for the most-often-used diagnostic tests (glucose, blood gases, urea nitrogen, etc.). Many clinical analyzers have now been miniaturized to the point where they can be used in the field, at a patient's bedside, or at home. Figure 33-19 shows a glucose monitoring system for in-home use. Immobilized glucose oxidase reacts with glucose to produce products that can be detected electrochemically.[16] Electrodes are screen-printed using ink-jet printing technology. The meters have memory so that results can be averaged and stored for later recall. The glucose monitor shown in Figure 33-19 is available for about $50.

33D-4 Automatic Organic Elemental Analyzers

Several manufacturers produce automatic instruments for analyzing organic compounds for one or more of the common elements, including carbon, hydrogen, oxygen, sulfur, and nitrogen. All of these instruments are based on high-temperature oxidation of the organic compounds, which converts the elements of interest to gaseous molecules. In some instruments, the gases are separated on a chromatographic column, and in others separations are based on specific absorbents. In most instruments, thermal conductivity detection completes the determinations. Often, these instruments are equipped with devices that automatically load the weighed samples into the combustion area.

Figure 33-20 is a schematic of a commercial automatic instrument for the determination of carbon, hydrogen, and nitrogen. In this instrument, samples are oxidized at 900°C under static conditions in a pure oxygen environment that produces a gaseous mixture of carbon dioxide, carbon monoxide, water, elemental

---

[16] For a review of glucose biosensors, see J. Wang, *Electroanalysis*, **2001**, *13*, 983.

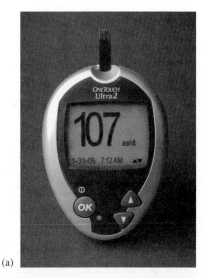

(a)

(b)

FIGURE 33-19 Photograph of a blood glucose monitor for in-home use. The meter (a) contains the electronics and display. A droplet of blood is placed on a test strip (b), which is placed at the top of the meter. The test strip contains immobilized glucose oxidase and electrodes for the amperometric detection system. (Courtesy of Life-Scan, Inc., Milpitas, CA.)

nitrogen, and oxides of nitrogen. After 2 to 6 min in the oxygen environment, the products are swept with a stream of helium through a 750°C tube furnace where hot copper reduces the oxides of nitrogen to the element and also removes the oxygen as copper oxide. Additional copper oxide is also present to convert carbon monoxide to the dioxide. Halogens are removed by a silver wool packing.

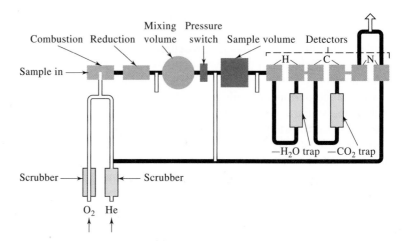

FIGURE 33-20 An automatic C, H, and N analyzer. (Courtesy of Perkin-Elmer Corp., Norwalk, CT.)

The products from the reaction furnace pass into a mixing chamber where they are brought to a constant temperature. The resulting homogeneous mixture is then analyzed by passing it through a series of three precision thermal conductivity detectors, each detector consisting of a pair of sensing cells.

Between the first pair of cells is a magnesium perchlorate absorption trap that removes water. The differential signal then serves as a measure of the hydrogen in the sample. Carbon dioxide is removed in a second absorption trap. Again, the differential signal between the second pair of cells is a measure of carbon in the sample. The remaining gas, consisting of helium and nitrogen, passes through the third detector cell. The output of this cell is compared to that of a reference cell through which pure helium flows. The voltage difference across this pair of cells is related to the amount of nitrogen in the sample.

For oxygen determinations, the reaction tube is replaced by a quartz tube filled with platinized carbon. When the sample is pyrolyzed in helium and swept through this tube, all of the oxygen is converted to carbon monoxide, which is then converted to carbon dioxide by passage over hot copper oxide. The remainder of the procedure is the same as just described, with the oxygen concentration being related to the differential signal before and after absorption of the carbon dioxide.

For a sulfur determination, the sample is combusted in an oxygen atmosphere in a tube packed with tungsten(VI) oxide or copper oxide. Water is removed by a dehydrating reagent located in the cool zone of the same tube. The dry sulfur dioxide is then separated and determined by the differential signal at what is normally the hydrogen detection cell. In this instance, however, the sulfur dioxide is absorbed by a silver oxide reagent.

The instrument shown in Figure 33-20 can be fully automated. In this version up to sixty weighed samples contained in small capsules are loaded into a carousel sample tray for automatic sampling.

QUESTIONS AND PROBLEMS

*Answers are provided at the end of the book for problems marked with an asterisk.

[X] Problems with this icon are best solved using spreadsheets.

33-1 List sequentially a set of laboratory unit operations that might be used to
  (a) ascertain the presence or absence of lead in flakes of dry paint.
  (b) determine the iron content of multiple-vitamin and mineral tablets.
  (c) determine the glucose concentration in a patient's blood.

33-2 Sketch a flow injection system that could be used for the determination of phosphate in a river water sample by means of the molybdenum blue reaction (phosphate + molybdate + ascorbic acid in acid solution give phosphomolybdenum blue, which absorbs at 660 nm).

33-3 Design a flow injection system for determining lead in the aqueous effluent from an industrial plant based on the extraction of lead ions with a carbon tetrachloride solution of dithizone, which reacts with Pb^{2+} to form an intensely colored product.

33-4 Compare and contrast continuous flow and discrete automated systems. Give the advantages and disadvantages of each type of automated system.

33-5 Give the advantages and disadvantages of sequential injection analyzers compared to traditional flow injection analyzers.

33-6 Lab-on-a-chip technology is used in several analyzers made by Agilent Technologies. Use a search engine such as Google to find such an analyzer. Describe in detail the use of microfluidics in one Agilent analyzer. Discuss the types of determinations possible and the advantages and limitations of lab-on-a-chip technology as applied to the analyzer chosen.

33-7 Sketch a flow injection apparatus for the determination of sodium sulfite in aqueous samples.

33-8 Sketch a flow diagram (indicating columns, detectors, and switching valves) designed to satisfy the following requirements. First, the solvent peak must be rapidly separated from two analytes of much lower volatility than the solvent. Second, the solvent peak should not pass into the analytical column. Third, the two analytes, which differ considerably in polarity, must be separated and ultimately determined quantitatively.

Challenge Problem

33-9 Lawrence, Deo, and Wang (*Anal. Chem.*, **2004**, *76*, 3735) have described an electrochemical biosensor for determining glucose based on a carbon paste electrode.
 (a) What electrochemical technique is used with the enzyme electrodes described?
 (b) What are the advantages of incorporating an enzyme in a carbon paste electrode matrix?
 (c) What is a mediator and how is it used with these biosensors?
 (d) What is the advantage of using a mediating paste liquid instead of a conventional carbon paste electrode?
 (e) Give the enzyme and mediator reactions.
 (f) What was cyclic voltammetry used for in the Lawrence, Deo, and Wang study?
 (g) What are the approximate response times for these electrodes?
 (h) Which of the new mediators gave the highest sensitivity? Which gave the wider linear dynamic range?
 (i) Discuss in detail how the current measured with these biosensors is related to the glucose reaction rate. How was the Michaelis constant K_m determined?

CHAPTER THIRTY-FOUR

Particle Size Determination

Particle size information is extremely important for many research projects and industrial processes. Several analytical techniques give particle size information. Some give only a mean size, whereas others give a complete particle size distribution. Hence, the first step in performing a particle size analysis is to choose the appropriate analytical technique. We examine in this chapter some of the considerations that influence this choice and then focus on three techniques: *low-angle laser light scattering*, *dynamic light scattering*, and *photosedimentation*, techniques that we have not yet discussed in this book.

The specific particle sizing method chosen depends on the type of size information needed and the chemical and physical properties of the sample. In addition to the three techniques discussed here, molecular sieving, electrical conductance, microscopy, capillary hydrodynamic chromatography, light obscuration counting, field-flow fractionation, Doppler anemometry, and ultrasonic spectrometry are commonly applied. Each of the particle sizing methods has its advantages and drawbacks for particular samples and analyses.

Throughout this chapter, this logo indicates an opportunity for online self-study at **www .thomsonedu.com/chemistry/skoog**, linking you to interactive tutorials, simulations, and exercises.

34A INTRODUCTION TO PARTICLE SIZE ANALYSIS

Particle size analysis presents a particular dilemma.[1] In general, we try to describe the *particle size* by a single quantity, such as a diameter, volume, or surface area. The *particle size distribution* is a plot of the number of particles having a particular value of the chosen quantity versus that quantity or a cumulative distribution representing the fraction of particles bigger or smaller than a particular size. The dilemma arises because the particles being described are three dimensional. The only three-dimensional object that can be described by a single quantity is a sphere. Given the diameter of a sphere, we can calculate its surface area and volume exactly. If we know the density of a spherical particle, we can also calculate its mass.

How then do we deal with nonspherical particles? The usual approaches are to assume the particles are spherical and then to proceed as we would with spherical particles, or we can convert the measured quantity into that of an equivalent sphere. For example, if we obtain the mass m of a particle, we can convert this into the mass of a sphere because $m = (4/3)\pi r^3 \rho$, where r is the particle radius and ρ is its density. This allows the particle size to be described by only its diameter ($d = 2r$). This diameter then represents the diameter of a sphere of the same mass as the particle of interest. Similar conversions can be made with other measured quantities, such as surface area or volume.

Another basic problem with particle size measurements is that different techniques often give different results for three-dimensional particles. Even if the particle diameter were to be measured directly with a microscope, what dimension is measured? If the particle were cubic, three different dimensions (lengths) could be taken as the diameter. For example, the maximum length could be taken as the diameter. This is equivalent to saying that the particle is a sphere of this diameter. Or the minimum length could be taken as the diameter. Alternatively, the volume or surface area could be used to calculate the diameter of the equivalent sphere. All the results obtained by this single technique, microscopy, will be different but correct for the quantity being measured. It is thus not surprising that

[1]For additional information, see T. Allen, *Particle Size Measurement*, 5th ed., Vol. 1, London: Chapman & Hall, 1997; S. P. Wood and G. A. Von Wald, in *Handbook of Instrumental Techniques for Analytical Chemistry*, F. Settle, ed., Upper Saddle River, NJ: Prentice-Hall, 1997; H. G. Barth, *Modern Methods of Particle Size Analysis*, New York: Wiley, 1984.

different techniques can give different results for particle sizes or particle size distributions.

The only reasonable solution to the size quandary is to use the same measuring technique and measured quantity for comparisons of particle sizes and particle size distributions. Comparisons between instruments that use different measuring procedures are sometimes made but are generally considered to be qualitative.

Some measurement techniques also require physical characteristics, such as density and refractive index. For example, refractive-index information is usually needed for methods based on light scattering, but densities are generally required for those based on sedimentation. Often, assumptions are made about these quantities unless measured values are available. These assumptions can again lead to discrepancies in measured particle sizes.

Before choosing a particle sizing technique, examination of the samples under a microscope is usually wise because the range of sizes and shapes present can then be estimated. Most particle size methods are sensitive to particle shape and all are limited with respect to the particle size range. Particles with approximately spherical shapes are measured most accurately. Needles and other shapes that differ significantly from spherical are often analyzed by microscopy.

For many techniques, particle sizes are best obtained by suspending the particles in a fluid in which the particles are insoluble. The fluid can produce a suspension that is homogeneous in concentration and fairly uniform in size for introduction to the particle sizing instrument. Fluid suspensions also disrupt any cohesive forces that could lead to coagulation or agglomeration of the particles. The fluids chosen must be chemically inert toward the materials contacted in the instrument.

34B LOW-ANGLE LASER LIGHT SCATTERING

The laser *low-angle laser light scattering* (LALLS)[2] technique, also called *laser diffraction*, is one of the most commonly used methods for measuring particle sizes and size distributions from 0.1 μm to 2000 μm. The

technique is popular because of its wide dynamic range, its precision, its ease of use, and its adaptability to samples in various forms. The measurements are made by exposing the sample to a beam of light and sensing the angular patterns of light scattered by particles of different sizes. Because the patterns produced are highly characteristic of the particle size, a mathematical analysis of the light scattering patterns can produce an accurate, reproducible measurement of the size distribution.

34B-1 Instrumentation

A typical laser diffraction apparatus is shown in Figure 34-1. The beam from a continuous-wave (CW) laser, usually a He-Ne laser, is collimated and passed through the sample, where scattering from particles occurs. The beam is then focused on a detector array where the scattering pattern, shown in Figure 34-1 as a diffraction pattern, is measured. The scattering pattern is then analyzed according to theoretical models to give the particle size distribution.

Some instruments monitor the concentration of particles in the beam by means of an *obscuration detector* located at the focal point of the lens. If no particles are present in the beam, all the light falls on the obscuration detector. As soon as particles enter the beam, they block some of the light and scatter light onto the elements of the detector array. The fraction of the light attenuated by scattering, absorption, or both is related to the concentration of particles in the beam (see Section 34D).

Instrument manufacturers provide several different accessories for sample introduction. For example, some introduce the sample directly into the laser beam as an aerosol. Samples can also be passed through a sample cell with transparent windows or suspended in a cuvette while being agitated. Dry powders can be blown through the beam or allowed to fall through the beam by gravity. Particles in a suspension can be recirculated through the beam by means of a pump.

34B-2 Theoretical Models

Most commercial instruments use optical models in their analysis software that are based on Mie scattering theory or Fraunhofer diffraction. *Mie theory* provides a complete solution to the problem of light scattering by a sphere, including the effects of transmitted and absorbed light. On the other hand, the *Fraunhofer dif-*

---

[2]See J. D. Ingle Jr. and S. R. Crouch, *Spectrochemical Analysis*, Upper Saddle River, NJ: Prentice-Hall, 1988, pp. 518–19; P. E. Plantz, in *Modern Methods of Particle Size Analysis*, H. G. Barth, ed., New York: Wiley, 1984, Chap. 6.

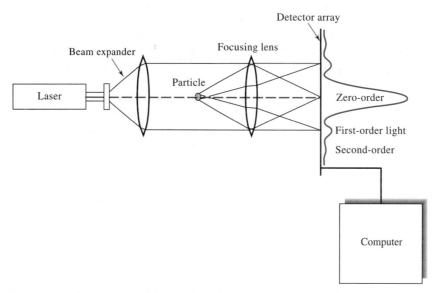

FIGURE 34-1 Laser diffraction apparatus. When a particle enters a laser beam, light is scattered at an angle related to the particle size. The scattered light is collected by a detector and the resulting scattering pattern is analyzed. The scattering pattern of a group of different-size particles is, within limits, the summation of the patterns of the individual particles.

fraction model is much simpler to implement. This model treats particles as opaque, circular apertures obstructing the beam of light.

Mie Theory

In the Mie theory, particles are considered to be finite objects instead of point scatterers. Scattering centers are found in various regions of the particles. When such particles scatter light, the scattering centers are far enough apart that some interference is likely between the rays emitted from one area of the particle and those from another. This condition leads to an intensity distribution that is quite different from the distribution observed from small particles. The scattering that occurs with particles much smaller than the wavelength of the light ($d < 0.05\lambda$) is termed *Rayleigh scattering*.

Figure 34-2a illustrates the interference that can occur in large-particle scattering. As a result, the intensity distribution pattern shifts to one showing predominantly forward scattering as depicted in Figure 34-2b. Scattering from particles with diameters near the wavelength of light ($0.05\lambda < d < \lambda$) is sometimes termed *Debye scattering*. *Mie scattering* occurs for particles with diameters larger than the wavelength of the incident light ($d > \lambda$).

Note in Figure 34-2 that the envelopes for Debye and Mie scattering are similar to Rayleigh scattering in the forward direction but quite different in the reverse direction. Often, Debye scattering is omitted as a separate class and Mie scattering is considered to occur for particles with diameters close to the wavelength of the light and larger. When the particle size is small compared to the incident wavelength, the scattered light shifts toward the side and rear and finally spreads in all directions (Rayleigh scattering). In the Rayleigh scattering limit, the intensity distribution of the front scattering is nearly constant and independent of particle size.

Before the advent of powerful desktop computers, the rigorous Mie theory was difficult to implement for the determination of particle size distributions in laser diffraction systems. The theory assumes that the particles are isotropic and spherical with a smooth surface. Even if these conditions hold, a complex material-dependent Mie parameter must be known. Finally, Mie theory is not applicable for mixtures of different components.

Fraunhofer Diffraction Theory

Fraunhofer theory is a simplification that considers the particles to be transparent, spherical, and much larger than the wavelength of the incident beam. Absorption and interference effects are not considered as they are in Mie theory. Thus, the particle behaves like a circular

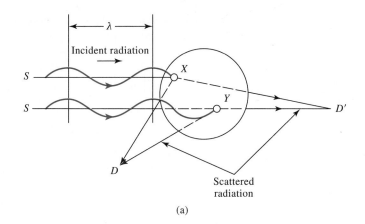

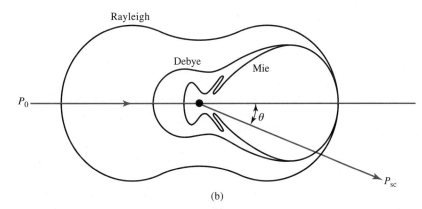

(b)

FIGURE 34-2 Scattering from particles of different sizes. (a) Large-particle scattering showing interference. Rays emitted in a backward direction observed at point D can destructively interfere because of the large path difference between the route SXD and the route SYD. Rays emitted in the forward direction observed at position D' and traveling routes SXD' and SYD' are more likely to constructively interfere because of the much smaller path difference. (b) The distributions of scattered light are shown for three particle sizes, corresponding to Rayleigh scattering, Debye scattering, and Mie scattering. Scattering angle θ is the angle between the incident and scattered rays with radiant powers P_0 and P_{sc} respectively. Thus, θ is 0° for forward scattering and 180° for scattering in the backward direction.

aperture and scattering from it results in a diffraction pattern, known as an *Airy pattern*. The Airy pattern can be expressed as a function of $x = 2\pi r s/\lambda f$, where r is the radius of the particle, s is the radial distance measured from the optical axis, λ is the wavelength of the incident radiation, and f the focal length of the lens (see Figure 34-1). The Airy function can be written as

$$I = I_0 \left(\frac{2J_1(x)}{x} \right)^2 \qquad (34\text{-}1)$$

where I is the scattered intensity, I_0 is the intensity at the center of the pattern, and J_1 is the first-order spherical primary Bessel function. A plot of the Airy function is shown in Figure 34-3. The inset shows an expansion of the y-axis to allow the extrema to be seen more clearly. Note the positions of the maxima and minima.

The Airy patterns are different for different particle radii. Figure 34-4 shows the patterns for particles of radii r, $2r$, and $0.5r$. Note that the pattern is broader for particles of smaller size and narrower for larger particles. Note also that the extrema shift to higher values of x as the particle size decreases. When particles of

different sizes are present, the intensity pattern can be considered to be the summation of the Airy patterns for individual particles.

Particle Size Distribution Analysis

In actuality, particles of various sizes are present in any real sample, and the intensity distribution of the scattered light is used to calculate the particle size distribution. The relationship between the intensity distribution of the scattered light and the particle size distribution for the case of Fraunhofer diffraction is

$$I(x) = I_0 \int_0^\infty \left(\frac{2J_1(x)}{x} \right)^2 f(d)\,dd \qquad (34\text{-}2)$$

where $I(x)$ is the intensity distribution of the scattered light, $f(d)$ the particle size distribution coefficient, and d the particle diameter. We can write Equation 34-2 in terms of the scattering angle θ and a size parameter $\alpha = 2\pi r/\lambda = \pi d/\lambda$

$$I(\theta) = I_0 \int_0^\infty \left(\frac{2J_1(\alpha\theta)}{\alpha\theta} \right)^2 f(d)\,dd \qquad (34\text{-}3)$$

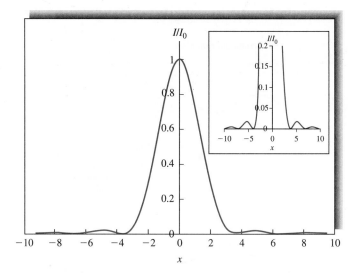

FIGURE 34-3 The Airy function showing a plot of intensity I relative to the central intensity I_0 as a function of the variable x (see text). The inset shows an expansion of the y-axis to show the maxima and minima more clearly.

This equation can be solved in matrix form or by iterative methods.

In most modern instruments, the measurement is performed by an array of N detectors. Also, Mie scattering theory is used in many instruments instead of Fraunhofer diffraction theory. In one popular approach the particles are divided into size intervals and each interval is assumed to generate an intensity distribution according to the average size. In this case, the preceding equation becomes

$$g(N) = \sum_{i=1}^{N} K(N, d_i) f(d_i) \Delta d \qquad (34\text{-}4)$$

where $g(N)$ is the output of the Nth detector, $K(N, d_i)$ is the response coefficient of the Nth detector, d_i is the ith diameter, and Δd is the particle size interval number. The particle size distribution $f(d_i)$ is calculated from the relationship between the output of the detector and its response function. The distribution is usually calculated on the basis of volume. Figure 34-5 shows a plot of the *cumulative undersize distribution*. The value at each particle diameter represents the percentage of particles having diameters less than or equal to the expressed value. A frequency distribution showing the percentage of particles having a particular particle diameter or range of diameters is also com-

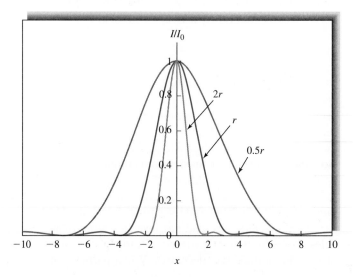

FIGURE 34-4 Airy patterns for particles of three different sizes.

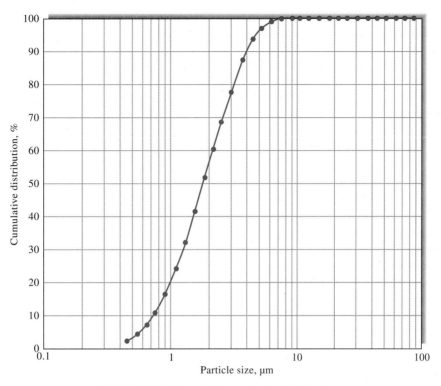

FIGURE 34-5 Cumulative undersize distribution.

monly given. These can be plotted either as histograms or as continuous distributions.

34B-3 Applications

Since the initial introduction of laser diffraction instrumentation in the 1970s, many different applications to particle size analysis have been reported.[3] These have included measurements of size distributions of radioactive tracer particles, ink particles used in photocopy machines, zirconia fibers, alumina particles, droplets from electronic fuel injectors, crystal growth particles, coal powders, cosmetics, soils, resins, pharmaceuticals, metal catalysts, electronic materials, photographic emulsions, organic pigments, and ceramics. About a dozen instrument companies now produce LALLS instruments. Some LALLS instruments have become popular as detectors for size-exclusion chromatography.

34C DYNAMIC LIGHT SCATTERING

Dynamic light scattering (DLS), also known as *photon correlation spectroscopy* (PCS) and *quasi-elastic light scattering* (QELS), is a powerful technique for probing solution dynamics and for measuring particle sizes.[4] The DLS technique can obtain size information in a few minutes for particles with diameters ranging from a few nanometers to about 5 μm.

The DLS technique involves measurement of the Doppler broadening of the Rayleigh-scattered light as a result of Brownian motion (translational diffusion) of the particles. This thermal motion causes time fluctuations in the scattering intensity and a broadening of the Rayleigh line. The Rayleigh line has a Lorentzian line shape. In macromolecular solutions, concentration

---

[3]See, for example, B. B. Weiner, in *Modern Methods of Particle Size Analysis*, H. G. Barth, ed., New York: Wiley, 1984, Chap. 5; P. E. Plantz, ibid., Chap. 6.

---

[4]For additional information, see J. D. Ingle Jr. and S. R. Crouch, *Spectrochemical Analysis*, Upper Saddle River, NJ: Prentice-Hall, 1988, Chap. 16; N. C. Ford, in *Measurements of Suspended Particles by Quasi-elastic Light Scattering*, B. E. Dahenke, ed., New York: Wiley, 1983; M. L. McConnell, *Anal. Chem.*, **1981**, *53*, 1007A; B. J. Berne and R. Pecora, *Dynamic Light Scattering with Applications to Chemistry, Biology and Physics*, New York: Wiley, 1976, reprinted by Dover Publications, Inc., New York.

fluctuations are usually dominant. Under these conditions, the width of the Rayleigh line is directly proportional to the translational diffusion coefficient D_T. The DLS method uses optical mixing techniques and correlation analysis to obtain these diffusion coefficients. The line widths (1 Hz to 1 MHz) are too small to be measured by conventional spectrometers and even interferometers.

34C-1 Principles

The sample in a DLS experiment is well-dispersed in a suspending medium. In a typical DLS experiment, the sample is illuminated by a single-wavelength laser beam. To measure the Doppler widths, the DLS instrument uses optical mixing or light-beating techniques to translate the optical frequencies ($\sim 6 \times 10^{14}$ Hz with the 488-nm line of an Ar^+ laser) to frequencies near 0 Hz (dc) that can be easily measured. In most DLS instruments, a photomultiplier tube (PMT) is used as a nonlinear mixer because its output is proportional to the square of the electric field falling on its photosensitive surface.

To see how the PMT acts as a mixer, let us assume that the scattered radiation contains sine waves of two frequencies ω_1 and ω_2. The electric field vector E can be written as

$$E = E_1 \sin \omega_1 t + E_2 \sin \omega_2 t \qquad (34\text{-}5)$$

where E_1 and E_2 are the amplitudes of the two waves. The PMT output signal $S(\omega)$ is proportional to the square of the electric field and can be expressed as

$$\begin{aligned} S(\omega) = A\{&E_1^2 \sin^2 \omega_1 t + E_2^2 \sin^2 \omega_2 t \\ &+ E_1 E_2 [\cos(\omega_2 - \omega_1)t - \cos(\omega_2 + \omega_1)t]\} \end{aligned} \quad (34\text{-}6)$$

where A is a proportionality constant. The PMT cannot respond directly to frequencies ω_1 and ω_2 or the sum term because these are greater than 10^{14} Hz for visible radiation. The PMT can respond, however, to the difference frequency term ($\omega_2 - \omega_1$), which can be as small as a few Hz. When multiple frequencies are present, a difference spectrum is generated that is centered at 0 Hz. The time dependence of the intensity fluctuations is then used to obtain the particle size information. Optical mixing is accomplished by beating the scattered light against a small portion of the source beam (hetero-

 Tutorial: Learn more about **particle size analysis**.

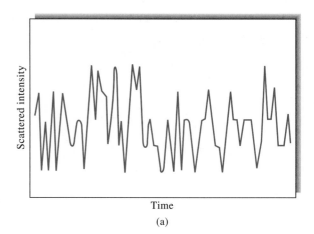

(a)

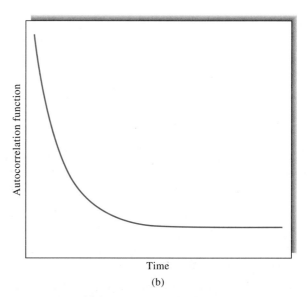

(b)

FIGURE 34-6 (a) Scattered radiation intensity fluctuations from an aqueous solution of 2.02-μm (diameter) polystyrene spheres; (b) autocorrelation function of intensity fluctuations.

dyne detection) or by beating the scattered light against itself (self-beating).

The PMT output signal is proportional to the scattered radiation intensity. Because the dispersed particles are in continuous thermal motion, the observed scattered intensity $I(t)$ fluctuates with time. The intensity-versus-time trace resembles a noise pattern as shown in Figure 34-6a. Small particles cause the intensity to fluctuate more rapidly than large particles.

The next step in the process is to determine the autocorrelation function of the signal. With autocorrelation, the signal is multiplied by a delayed version of itself, and the product is time-averaged. The time-averaged product is obtained at various delay times and plotted against the delay time. The autocorrelation function is the Fourier transform of the power spectrum. Because the scattered radiation has a Lorentzian line shape, its Fourier transform should be an exponential decay, as illustrated in Figure 34-6b. According to the theory of DLS, the time constant of the exponential decay τ is directly related to the translational diffusion coefficient of the isotropic, spherical particles in Brownian motion.

$$\tau = D_T q^2 \tag{34-7}$$

Here, q is called the *modulus of the scattering vector* and is given by

$$q = \frac{4\pi n}{\lambda} \sin\left(\frac{\theta}{2}\right)$$

where n is the refractive index of the suspension liquid, θ is the scattering angle, and λ is the wavelength of the laser radiation.

The particle size is obtained from the translational diffusion coefficient D_T and particle shape information. For a spherical particle, the *Stokes-Einstein relationship* is used to calculate the hydrodynamic particle diameter d_h

$$d_h = \frac{kT}{3\pi\eta D_T} \tag{34-8}$$

where k is Boltzmann's constant, T is the absolute temperature, and η is the viscosity of the medium. Equation 34-8 holds only for noninteracting, spherically shaped particles. For nonspherical particles, the hydrodynamic diameter is the diameter of a hypothetical sphere that would have the same translational diffusion coefficient as the nonspherical particle.

DLS can also be used to give particle size distributions, although these require careful sample preparation and longer measurement times. The DLS data cannot provide a totally accurate particle size distribution because there are many distributions corresponding to similar correlation patterns.

34C-2 Instrumentation

A DLS instrument consists of a laser source, a sample cell, a photodetector, and a computer with an autocorrelator. A typical instrumental setup for DLS is shown in Figure 34-7. A CW laser is used as the source in DLS. He-Ne lasers (632.8 nm) and Ar$^+$ lasers (488.0 nm

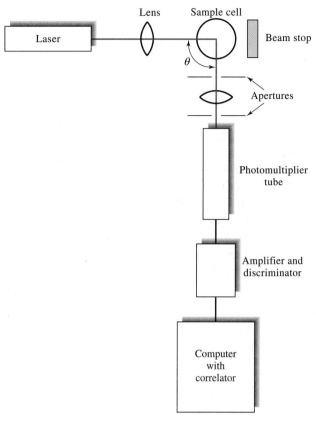

FIGURE 34-7 Typical instrumental arrangement for DLS. A laser source is incident on a sample. The sample is usually a well-dispersed suspension. The scattered radiation containing the Doppler broadening information is incident on a PMT. Photon counting signal processing is used. The autocorrelation function of the scattering signal is calculated and used to obtain the translational diffusion coefficient D_T, which is then related to particle size.

and 514.5 nm) are the most common sources. Diode lasers at 650 nm are also used in some instruments. The laser beam is focused in the middle of the sample cell containing the particles of interest suspended in a liquid.

Sample Cell and Sample Handling

A cuvette-type sample cell is often used. The cell is often surrounded by a liquid maintained at a constant temperature. The refractive index of this liquid is matched to that of the suspending medium.

The sample must be well dispersed in the suspending medium. This is usually accomplished by gentle shaking and sometimes by ultrasonic agitation. Too much agitation can cause aggregation of some materials and particle fracture in others. Solvents must be carefully filtered to avoid dust particles and other contaminants that can cause scattering. There must be a refractive-index difference between the suspending medium and the dispersed phase, and the refractive index of the solvent must be known. The viscosity at the measurement temperature must also be known so that the Stokes-Einstein relationship (Equation 34-8) can be applied.

The sample concentration to be used depends on particle size, laser power, and particle refractive index. The upper concentration limit is determined by multiple scattering phenomena, where light scattered from one particle is rescattered by another. Maximum concentrations are size dependent, but for a strong scatterer a maximum concentration of 0.01% is typical for 100-nm particles. The lower concentration limit is determined by the number of particles in the scattering volume. Typically, at least 1000 particles are needed. Some trial and error is usually needed to obtain the appropriate concentration.

Photodetector

The scattered light is measured at an angle θ, usually 90°, from the incident beam. The scattered light striking the surface of the photodetector acts as a nonlinear mixer. The PMT is the most common photodetector. The PMT output can be processed by photon-counting techniques or as an analog photocurrent (see Section 7F-1). Particle size information is then obtained from a correlation analysis of the processed signal as discussed previously. Photodiodes have also been used in some commercial DLS instruments. In addition, dual photodetectors have been used along with cross-correlation processing to eliminate contributions from multiple scattering.

34C-3 Applications

There are many different applications of the DLS technique. The DLS method has been used to determine the size of polymer lattices and resins and to monitor the growth of particles during processes such as emulsification and polymerization. Micelles and microemulsions have been studied by DLS methods. DLS is also widely applicable to the investigation of biopolymers and biocolloids. It has been used to study natural and synthetic polypeptides, nucleic acids, ribosomes, vesicles, viruses, and muscle fibers.

34D PHOTOSEDIMENTATION

One of the most important types of particle size analyzers is the *photosedimentation* type.[5] A photosedimentation analyzer determines the particle size distribution by measuring photometrically the rate at which particles settle (sediment) through a liquid. Consider various sizes of particles of the same material stirred vigorously in a liquid so that they are distributed homogeneously. When the stirring is stopped, the particles begin to settle, with the largest particles falling most rapidly, intermediate sizes less rapidly, and so on. At any instant after the stirring is stopped, there will be a level below the upper liquid surface with none of the largest particles, a higher level above which there are none of the intermediate-size particles, and still another, higher level above which there are no particles at all. In the photosedimentation analyzer, this changing condition is analyzed to reveal the size and relative population of particle sizes as described below. The force causing a particle to sediment or settle can be a gravitational force or a centrifugal force. Photosedimentation analyzers can measure particle sizes ranging from 0.01 µm to 300 µm.

34D-1 Settling Velocity and Particle Size

When a force F, such as a gravitational force, is applied to a particle in solution, the particle is accelerated. As its velocity increases, the particle experiences more and more retardation due to friction. For low velocities, the frictional force is given by uf, where u is the velocity and f is the *frictional coefficient*. When the velocity gets high enough, the frictional force becomes equal to the applied force and the particle moves with a constant velocity

$$uf = F \tag{34-9}$$

The frictional coefficient f contains information about the size and shape of the particle. For spherical particles, Stokes's law, given in Equation 34-10, holds for laminar-flow conditions

$$f = 6\pi\eta r \tag{34-10}$$

where η is the viscosity and r is the radius of the particle. Here, it is assumed that the drag on the falling particle is due primarily to frictional forces.

[5]For additional information, see T. Allen, *Particle Size Measurement*, Vol. 1, 5th ed., London: Chapman & Hall, 1997; C. Bernhardt, *Particle Size Analysis*, London: Chapman & Hall, 1994.

The Stokes Equations

If a spherical particle settles in a *gravitational field*, the force causing it to settle in the fluid is equal to the effective mass of the particles times the acceleration of gravity. The effective mass is the mass of the particle minus the mass of the fluid displaced by the particle. If the density of the particle is ρ, the force causing the settling is $\frac{4}{3}\pi r^3(\rho - \rho_F)g$, where ρ_F is the fluid density and g is the acceleration due to gravity. When the particle has reached a constant rate of settling, the frictional retarding force is equal to the gravitational force, and

$$\frac{4}{3}\pi r^3(\rho - \rho_F)g = 6\pi\eta ru \qquad (34\text{-}11)$$

By rearranging Equation 34-11, we can solve for the terminal velocity u in terms of the particle radius r or the diameter $d = 2r$

$$u = \frac{(\rho - \rho_F)g}{18\eta}(2r)^2 = \frac{(\rho - \rho_F)g}{18\eta}d^2 \qquad (34\text{-}12)$$

At a sedimentation distance h and time t, the velocity is $u = h/t$, and sedimenting particles with diameters less than or equal to a diameter called the *Stokes diameter* d_{Stokes} will be in the measuring window. The Stokes diameter is given by

$$d_{\text{Stokes}} = \sqrt{\frac{18\eta}{(\rho - \rho_F)g} \times \frac{h}{t}} \qquad (34\text{-}13)$$

In a *centrifugal field*, a particle sedimenting through a viscous medium also reaches a terminal velocity u. The centrifugal acceleration is $\omega^2 r_a$, where ω is the rotational velocity in rad/s, and r_a is the distance from the center of rotation to where the measurement is made, called the *analytical radius*. In this case, the Stokes equation has the form:

$$u = \frac{(\rho - \rho_F)\omega^2 r_a d^2}{18\eta} = \frac{\ln(r_a/r_0)r_a}{t} \qquad (34\text{-}14)$$

Here, t is the time required for a particle of diameter d to move from its starting point radius r_0 to the analytical radius r_a. If the particles have densities greater than that of the fluid, the starting radius r_0 is that of the inner liquid meniscus. From Equation 34-14, we can solve for the Stokes diameter d_{Stokes}.

$$d_{\text{Stokes}} = \sqrt{\frac{18\eta \ln(r_a/r_0)}{\omega^2 t(\rho - \rho_F)}} \qquad (34\text{-}15)$$

The quantity $u/\omega^2 r_a$ is sometimes called the sedimentation coefficient S, with units of seconds. From Equation 34-14, we can see that

$$S = \frac{u}{\omega^2 r_a} = \frac{1}{\omega^2 t}\ln(r_a/r_0) \qquad (34\text{-}16)$$

Assumptions of the Stokes Equations

To apply the Stokes equations to measurements of the size of sedimenting particles, several critical assumptions are made:

1. The particle must be spherical, smooth, and rigid. This assumption is not always valid as discussed previously. For nonspherical particles, the diameter calculated is an *equivalent Stokes diameter*, the diameter of a sphere of the same material with the same sedimentation velocity.
2. The final velocity of the particle is reached instantly. In actuality a finite, but small, time is required for this condition to be met.
3. All particles have the same density ρ and any influences from the wall of the sedimentation vessel are negligible.
4. All particles are assumed to move independently without interference or interaction with other particles in the system. This assumption is good only at low concentrations where there is separation between the particles. Concentrations substantially less than 1% by volume are thus preferred.
5. The fluid behaves as a continuum with a constant viscosity, independent of any velocity and concentration gradients (Newtonian fluid approximation). This assumption is valid when water is the dilution medium.

34D-2 Instrumentation

There are several commercial photosedimentation analyzers. Figure 34-8 shows a schematic diagram of one instrument (the Horiba CAPA-700) that can achieve sedimentation by gravity or by centrifugal force. The components consist of an optical system, a centrifuge, control circuitry, and a computation system.

The rotation speeds can be selected by the user or automatically set by the computer. These speeds correspond to centrifugal forces of 8 G at 300 rpm to 9000 G at 10,000 rpm. Here, G is the gravitational constant equal to 6.6742×10^{-11} N m^2/kg^2. Two cells are mounted on the rotor; the sample cell is filled with the

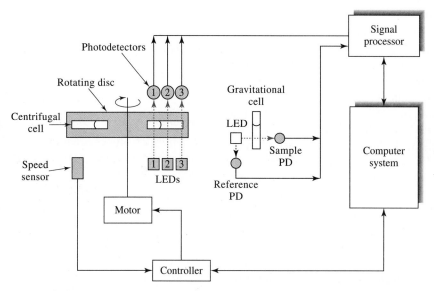

FIGURE 34-8 Diagram of photosedimentation system. The unit can be operated in either the gravitational mode or the centrifugal mode. In the gravitational mode, sedimentation is monitored by the LED source and sample photodetector (PD) shown on the right side of the diagram. The reference photodetector measures the LED intensity. In the centrifugal mode, a motor spins the rotating disc at speeds from 300 to 10,000 rpm. Sedimentation is monitored by the LED-photodetector pair 2. Photodetectors 1 and 3 provide synchronization signals for the reference and the sample, respectively. A speed sensor monitors the rotation speed and feeds the information to the motor controller for maintaining a constant speed. (Courtesy of Horiba, Ltd.)

sample suspension and the opposite reference cell is filled with a clear dispersing fluid. The reference cell is used to balance the centrifuge and provide a constant optical reference. Green LEDs (560 nm) provide the source radiation. In the centrifugal mode the source is continuously illuminated, and in the gravitational mode the source is pulsed at 15 Hz. The computer monitors and sends out signals to control the motor speed. The computer system also carries out the appropriate calculations, as discussed later.

Sedimentation Modes

Sedimentation under a gravitational field can be used for large particles. However, gravitational sedimentation is not feasible for particles much below 5 μm in diameter. For such particles, the sedimentation time is so long that Brownian motion can be a significant contributor to particle movement. With gravitational sedimentation, the only means to vary the particle velocity is by selecting a medium with a different viscosity and density. Centrifugal sedimentation is used for smaller particles. Some instruments even use a *gradi-*

ent mode, in which the centrifuge is accelerated during the analysis.

Distribution Determination

In gravitational sedimentation the particle concentration in the measurement zone remains equal to the initial concentration until the largest particle present in the suspension has settled through the zone. The concentration of particles in the measurement zone at time t represents particles whose size is less than or equal to the Stokes diameter d_{Stokes}. The Stokes diameter is inversely proportional to the square root of time as shown by Equation 34-13. Therefore, a plot of concentration versus d_{Stokes} represents the undersize distribution.

In centrifugal sedimentation, a complication arises because the particle velocity depends on not only particle size, as in gravitational sedimentation, but also the radial position of the particle. As Equation 34-14 shows, the particle velocity increases with increasing radial distance r_a from the axis of rotation. Because of this, the particle concentration below the measurement zone decreases exponentially with the sedimentation

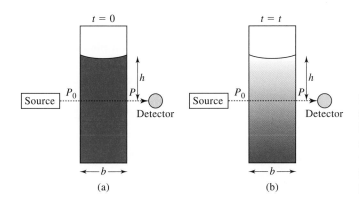

FIGURE 34-9 Photometric determination of particle size distribution. Measurements of the concentrations of the settling particles are made as a function of time by measuring the solution absorbance, $A_t = -\log (P/P_0)_t$. Here, the suspension is shown in (a) as a homogeneous mixture at $t = 0$ and later in (b) as particles sediment ($t = t$).

time. This is often called a *radial dilution effect*. In centrifugal sedimentation with homogeneous suspensions, the radial dilution effect complicates the mathematics required to obtain a particle size distribution. With some units a correction for radial dilution can be applied in the centrifugal sedimentation mode.[6]

Photometric Measurement

In photosedimentation, measurements of the solution absorbance are made as a function of time at a constant distance (gravity field) or radial position (centrifugal field).[7] Figure 34-9 shows the measurement principle for a gravitational field. For suspended particles of concentration c in a cell of path length b, the solution absorbance at time t, A_t, is given by

$$A_t = kcb \sum_{i=0}^{i=\text{Stokes}} K_i n_i d_i^2 \qquad (34\text{-}17)$$

where k is a shape factor, K_i is the absorption coefficient for particles of size d_i, n_i is the number of particles of size d_i in a unit mass of powder, and d_{Stokes} is the Stokes diameter at time t.

Let us now consider a small change in absorbance ΔA_i as the sedimentation time changes from t to $t + \Delta$. Here, the average Stokes diameter in the beam is d_i, and we can write

$$\frac{\Delta A_i}{K_i} = kcbn_i d_i^2 \qquad (34\text{-}18)$$

By measuring absorbance changes as a function of time, the cumulative undersize particle distribution can be obtained by mathematical manipulation. However, in converting Equation 34-18 to the size distribution, it is necessary to know how the absorption coefficient K varies with d_i. Alternatively, an external standard can be used for calibration. If correction is not made for variations in K, results are valid only for comparison purposes.

For incremental, homogeneous, centrifugal sedimentation, the situation is considerably more complex. In general, the solution absorbance is monitored as a function of time. The Stokes diameter is again inversely proportional to the square root of the measurement time. A radial dilution correction should be applied. The measured absorbance changes with time are then related mathematically to the particle size distribution.

34D-3 Applications

Photosedimentation has been used to determine particle sizes of polymers, dye materials, pharmaceutical preparations, and biological materials. It is used in quality control during the production of paints, foods, and ceramics. In the pharmaceutical industry, for example, one of the current areas of interest is the delivery of medicine via inhalation powders. For these, fine particles are prepared and used in an inhaler as dry powders to deliver medicine directly to the lungs (pulmonary delivery). Some preparation processes use supercritical fluids to control powder formation from a wide range of chemicals, such as inorganic and organic materials, polymers, peptides, and proteins. For dry powders to be used as inhalants, the particle size must be within a quite limited range. Hence, techniques such

---

[6] Horiba, Ltd: Notice for CAPA-500, U.S. Version, Code 04169400, December 1993; see also sections 7.13 and 8.6.1 of T. Allen, *Particle Size Measurement*, Vol. 1, 5th ed., London: Chapman & Hall, 1997; H. J. Kamack, *Anal. Chem.*, **1951**, 23, 844.
[7] See T. Allen, *Particle Size Measurement*, Vol. 1, 5th ed., London: Chapman & Hall, 1997, pp. 269–327.

as photosedimentation are used to monitor particle sizes during the preparation process or for quality control purposes.

With dry powder inhalation, protein-based drugs can be delivered without being decomposed by the digestive system before entering the blood stream. Delivering medication in dry powder form thus has advantages over intravenous injection for several types of pharmaceutical preparations. Pulmonary delivery of drugs to treat diabetes (insulin), multiple sclerosis, cystic fibrosis, anemia, asthma, and several other diseases have been implemented or are currently under development. Particle size analysis is playing a significant role in these advances.

QUESTIONS AND PROBLEMS

*Answers are provided at the end of the book for problems marked with an asterisk.

$\boxed{\mathbf{X}}$ Problems with this icon are best solved using spreadsheets.

34-1 What is a normal particle size distribution? What is a cumulative particle size distribution?

34-2 What is the major problem presented by particle size analysis?

34-3 What quantities are used to describe particle size?

34-4 How are nonspherical particles dealt with in particle size analysis?

34-5 Why are comparisons of particle sizes between different instrumental techniques only qualitative?

34-6 Define Mie scattering. For what particle sizes does Mie theory apply?

34-7 What is an Airy pattern? How does it arise in diffraction?

34-8 What is a cumulative undersize particle distribution?

34-9 Discuss the major differences between DLS and LALLS.

*34-10 The translational diffusion coefficient of the enzyme aspartate transcarbamylase in dilute aqueous solution was found to be $3.75 \times 10^{-7} \ cm^2 \ s^{-1}$ at 20°C ($\eta = 1.002$ cP). What is the hydrodynamic diameter of the particles? What does the hydrodynamic diameter mean if the particles are nonspherical?

*34-11 A protein molecule is approximately spherical and has a hydrodynamic diameter of 35 μm. What is the translational diffusion coefficient of the protein at 20°C in dilute aqueous solution?

*34-12 In a particular batch of polystyrene microspheres, their diameter is 10.0 μm and density 1.05 g cm^{-3}. In 20°C water ($\rho_F = 0.998$ g cm^{-3}), what will be the settling velocity of these particles in a 1 G gravitational field? What time will it take for the particles to sediment 10 mm?

*34-13 For polystyrene spheres of 10.0 μm diameter in water at 20°C, how long will it take the particles to move from an initial radius of 70 mm to an analytical radius of 80 mm in a centrifugal field of 10,000 rpm? What is the centrifugal acceleration in G at 10,000 rpm?

*34-14 A polystyrene particle settles, moving from a starting radius of 70 mm to an analytical radius of 80 mm in 2.9 s in a centrifugal field of 9000 G. If the particle is in an aqueous solution at 20°C, what is the Stokes diameter?

Challenge Problem

34-15 (a) Use an Internet search engine to find laser diffraction instruments made by a commercial company (try Malvern, Sympatec, Shimadzu, Beckman Coulter, or Horiba). Choose a specific instrument and describe its operation. What laser is used? What is the detection system? Give typical values of accuracy and precision.

(b) What ranges of particle sizes can the chosen instrument determine?

(c) What types of sampling accessories are available for the instrument? What sample cells are available?

(d) What is the size of the instrument?

(e) What models does the software use to determine particle sizes?

(f) Are any options available to automate particle size analysis?

(g) Use a search engine to find a paper in the literature that uses laser diffraction to determine particle sizes. Describe the application in detail.

The John Vollman Case

Introduction

The John Vollman case in New Brunswick, Canada, in 1958 was the first murder case in which neutron activation analysis (NAA), then labeled in the press as "atomic evidence," played a major role in the outcome. Since then, NAA has been widely used and accepted in forensic science for determining trace elements.

The Crime

In May 1958, a 16-year-old girl, Gaetane Bouchard, left her home in late afternoon to go shopping in the small New Brunswick town of Edmundston. When she failed to return home by 8 p.m., her father telephoned several of her friends, asking if they knew where she was. Some friends mentioned that she might have been with a former boyfriend, John Vollman, who lived across the border in Madawaska, Maine. Mr. Bouchard drove across the border and met with the 20-year-old Vollman who worked a night shift at a printing plant. Vollman claimed that he and Gaetane were no longer seeing each other and that he was now engaged to another woman.

When Bouchard returned to Edmundston, he called the police and then began searching for his daughter. Some friends advised him to search an abandoned gravel pit outside of town. The pit was often a popular "make-out" spot for young couples. Bouchard's flashlight soon illuminated a suede shoe that he recognized as Gaetane's. A few minutes later, he found the body of his daughter. She had been repeatedly stabbed in the chest and back and then dragged off to die. The police were notified, and the pit was secured as a crime scene.

The Evidence

Police searching the pit soon found a dark pool of blood and tire tracks where Gaetane's assault had begun. While making plaster casts of the tire tracks, one of the police officers noted two tiny slivers of green paint possibly chipped off the car by rocks as the car sped off. One of the paint chips was no larger than the head of a pin, while the second was a larger heart-shaped piece.

The next day Gaetane's movements in the hours leading up to her murder were retraced. At 4 p.m., she purchased some chocolate from a local restaurant. A short time later, she was seen talking to the driver of a light-green Pontiac with Maine license plates. About an hour later, two friends saw her seated in a green Pontiac. Another witness saw a green car parked by the gravel pit sometime between 5 and 6 p.m. The medical examiner attributed death to the multiple stabbings and put the time of death no later than 7 p.m.

Detectives then interviewed Vollman at his workplace and examined his recently purchased green 1952 Pontiac. They found a heart-shaped piece of paint missing from just below the passenger door. The paint chip found at the crime scene matched perfectly — both visually and later microscopically — the missing paint particle. The car revealed even more evidence. Inside the glove compartment was a half-eaten bar of chocolate, identical to the type that Gaetane had purchased at the restaurant.

The most important piece of evidence, however, was found clutched in the girl's hand when she was autopsied. Entwined in her fingers was a strand of human hair, a little over 2 inches long. Detectives immediately surmised that Gaetane had pulled the hair from the murderer as she struggled to free herself. The challenge now was to determine whether the hair found had come from Vollman.

Neutron Activation Analysis Results

To establish whether the hair found in Gaetane's hand came from the suspect, investigators turned to the recently introduced and somewhat controversial technique of neutron activation analysis. Neutron activation analysis can identify and determine trace amounts of as many as 14 different elements in a single hair. Theoretical calculations have shown that the probability of two different individuals having the same concentrations of nine of these elements is about one in a million. In many cases, the ratios of two element concentrations are strongly indicative of hair from a particular person. Neutron activation analysis has also been used to determine the presence of poisons, such as arsenic, in the hair of victims.

In the Vollman case, hair samples from the suspect and the victim were compared with the hair found in Gaetane's

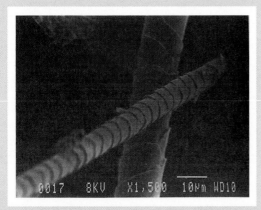

FIGURE IA6-1 SEM photomicrograph of a human hair (left thinner strand) compared with a cat hair. (From D. Owen, *Hidden Evidence: 40 True Crimes and How Forensic Science Helped Solve Them*. Buffalo, NY: Firefly Books, 2000.)

hand. To identify the hair, the ratio of the gamma ray intensity of sulfur to phosphorus (S/P) was measured. In the case of the victim, the ratio was found to be 2.02. The sample found in Gaetane's hand had an S/P intensity ratio of 1.02, while that from Vollman registered 1.07. From these measurements, it was concluded that the hair did not come from Gaetane, but was a close match to that from the suspect.[1] Scanning electron microscopic evidence, similar to that shown in Figure IA6-1, was also consistent with the hair being that of the suspect.

The Trial

John Vollman stood trial for murder in Edmundston in November 1958. He pleaded not guilty and seemed very confident that the "atomic evidence" would not be allowed. Despite the vigorous objections of the defense, however, the NAA evidence was admitted. Scientists eager to explain the NAA procedure took the witness stand. The mood of

the court shifted so dramatically against Vollman that he changed his plea to guilty of manslaughter. He admitted to killing Gaetane, but not intentionally. He claimed that she had initially encouraged his advances, but later changed her mind. As a struggle broke out, Vollman said that he had suffered a blackout. The jury did not believe his story and found him guilty as charged. He was sentenced to death, a sentence that was later commuted to life imprisonment. John Vollman became the first criminal to be convicted by the new technique of neutron activation analysis.

Neutron Activation Analysis in Forensic Science

Although capable of incredible sensitivity and selectivity, neutron activation analysis has never gained great popularity in forensic science. The major reason, of course, is that NAA is expensive and requires access to a nuclear reactor. Today, NAA has been superseded in hair analysis by DNA fingerprinting.[2] The amount of DNA found at the root of one human hair is sufficient for DNA fingerprinting. Such techniques are much less expensive and more readily available than neutron activation analysis.

Neutron activation analysis has been quite useful, however, in special forensic situations. For example, NAA has been used to determine trace amounts of metals in gunshot residues from the hands of suspects and metal fragments taken from the wounds of victims. During the investigation of the assassination of President John F. Kennedy, for example, forensic scientists used NAA to determine silver and antimony in the recovered bullets and in metal fragments from the wounds of President Kennedy and Governor John Connally. The concentrations of these trace elements vary significantly from one bullet to the next, but are quite similar in pieces of metal from a single bullet. The NAA results indicated that a single bullet had wounded both men. The second bullet had struck President Kennedy only. It was also concluded that the two bullets were almost certainly fired by the same gun.[3]

[1] C. Evans, *The Casebook of Forensic Detection: How Science Solved 100 of the World's Most Baffling Crimes*. New York: Wiley, 1996.

[2] D. Owen, *Hidden Evidence: 40 True Crimes and How Forensic Science Helped Solve Them*. Buffalo, NY: Firefly Books, 2000.
[3] P. Moore, *The Forensics Handbook: The Secrets of Crime Scene Investigation*. New York: Barnes & Noble Books, 2004.

APPENDIX ONE EVALUATION OF ANALYTICAL DATA

This appendix describes the types of errors that are encountered in analytical chemistry and how their magnitudes are estimated and reported.[1] Estimation of the probable accuracy of results is a vital part of any analysis because data of unknown reliability are essentially worthless.

a1A PRECISION AND ACCURACY

Two terms are widely used in discussions of the reliability of data: precision and accuracy.

a1A-1 Precision

Precision describes the reproducibility of results, that is, the agreement between numerical values for two or more replicate measurements, or measurements that have been made in *exactly the same way*. Generally, the precision of an analytical method is readily obtained by simply repeating the measurement.

Three terms are widely used to describe the precision of a set of replicate data: standard deviation, variance, and coefficient of variation. These terms have statistical significance and are defined in Section a1B-1.

a1A-2 Accuracy

Accuracy describes the correctness of an experimental result expressed as the closeness of the measurement to the true or accepted value. Accuracy is a relative term in the sense that what is an accurate or inaccurate method very much depends on the needs of the scientist and the difficulty of the analytical problem. For example, an analytical method that yields results that are within ±10%, or 1 part per billion, of the correct amount of mercury in a sample of fish tissue that contains 10 parts per billion of the metal would usually be considered to be reasonably accurate. In contrast, a procedure that yields results that are within ±10% of the correct amount of mercury in an ore that contains 20% of the metal would usually be deemed unacceptably inaccurate.

Accuracy is expressed in terms of either absolute error or relative error. The *absolute error E* of the mean (or average) $\bar{x}$ of a small set of replicate analyses is given by the relationship

$$E = \bar{x} - x_t \qquad (a1\text{-}1)$$

where x_t is the true or accepted value of the quantity being measured. Often, it is useful to express the accuracy in terms of the *relative error* E_r, where

$$E_r = \frac{\bar{x} - x_t}{x_t} \times 100\% \qquad (a1\text{-}2)$$

Frequently, the relative error is expressed as a percentage as shown; in other cases, the quotient is multiplied by 1000 instead of by 100% to give the error in parts per thousand (ppt). Note that both absolute and relative errors bear a sign, a positive sign indicating that the measured result is greater than its true value and a negative sign indicating the reverse.

We will be concerned with two types of errors, *random errors*, often called *indeterminate errors*, and *systematic errors*, often called *determinant errors*.[2] We will give random error the symbol E_d and systematic error the symbol E_s. The error in the mean of a set of replicate measurements is then the sum of these two types of errors:

$$E = E_d + E_s \qquad (a1\text{-}3)$$

Random Errors

Whenever analytical measurements are repeated on the same sample, a scatter of data such as that shown in Table a1-1 are obtained because of the presence

---

[1] For more details, see D. A. Skoog, D. M. West, F. J. Holler, and S. R. Crouch, *Fundamentals of Analytical Chemistry*, 8th ed., Belmont, CA: Brooks/Cole, 2004.

[2] A third type of error that is encountered occasionally is *gross error*, which arises in most instances from a human error. Typical sources include transposition of numbers in recording data, spilling of a sample, using the wrong scale on a meter, accidental introduction of contaminants, and reversing the sign on a meter reading. A gross error in a set of replicate measurements appears as an *outlier*—a data point that is noticeably different from the other data in the set. We will not consider gross errors in this discussion.

TABLE a1-1 Replicate Absorbance Measurements

	A	B	C	D	E	F	G	H
1	Replicate Absorbance Measurements*							
2	Trial	Absorbance		Trial	Absorbance		Trial	Absorbance
3	1	0.488		18	0.475		35	0.476
4	2	0.480		19	0.480		36	0.490
5	3	0.486		20	0.494		37	0.488
6	4	0.473		21	0.492		38	0.471
7	5	0.475		22	0.484		39	0.486
8	6	0.482		23	0.481		40	0.478
9	7	0.486		24	0.487		41	0.486
10	8	0.482		25	0.478		42	0.482
11	9	0.481		26	0.483		43	0.477
12	10	0.490		27	0.482		44	0.477
13	11	0.480		28	0.491		45	0.486
14	12	0.489		29	0.481		46	0.478
15	13	0.478		30	0.469		47	0.483
16	14	0.471		31	0.485		48	0.480
17	15	0.482		32	0.477		49	0.483
18	16	0.483		33	0.476		50	0.479
19	17	0.488		34	0.483			
20	*Data listed in the order obtained							
21	Mean	0.482		Maximum	0.494			
22	Median	0.482		Minimum	0.469			
23	Std. Dev.	0.0056		Spread	0.025			

of random, or indeterminate errors—that is the presence of *random errors* is reflected in the imprecision of the data.[3] The data in columns B, E, and H of the table are absorbances (Section 13A) obtained with a spectrophotometer on 50 replicate red solutions produced by treating identical aqueous samples containing 10 ppm of Fe(III) with an excess of thiocyanate ion. The measured absorbances are directly proportional to the iron concentration.

The distribution of random errors in these data is more easily understood if they are organized into equal-sized, adjacent data groups, or cells, as shown in Table a1-2. The relative frequency of occurrence of results in each cell is then plotted as in Figure a1-1A to give a bar graph called a *histogram*.

We can imagine that, as the number of measurements increases, the histogram approaches the shape of the continuous curve shown as plot B in Figure a1-1.

TABLE a1-2 Frequency Distribution of Data from Table a1-1

Absorbance Range, A	Number in Range, y	Relative Frequency, y/N[a]
0.469–0.471	3	0.06
0.472–0.474	1	0.02
0.475–0.477	7	0.14
0.478–0.480	9	0.18
0.481–0.483	13	0.26
0.484–0.486	7	0.14
0.487–0.489	5	0.10
0.490–0.492	4	0.08
0.493–0.495	1	0.02

[a]N = total number of measurements = 50.

This plot shows a *Gaussian curve*, or *normal error curve*, which applies to an infinitely large set of data. It is found empirically that the results of replicate chemical analyses are frequently distributed in an approximately Gaussian, or normal, form.

---

[3]For more information on the application of spreadsheets in evaluating analytical data, see S. R. Crouch and F. J. Holler, *Applications of Microsoft® Excel in Analytical Chemistry*, Belmont, CA: Brooks/Cole, 2004.

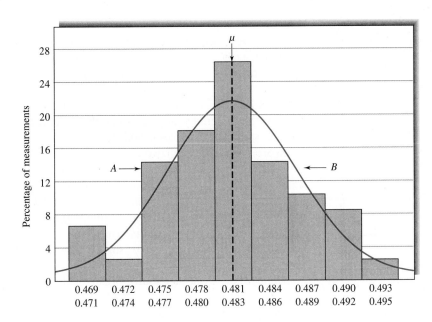

FIGURE a1-1 *A,* Histogram showing distribution of the 50 results in Table a1-1. *B,* Gaussian curve for data with the same mean and standard deviation as those in *A.*

The frequency distribution exhibited by a Gaussian curve has the following characteristics:

1. The most frequently observed result is the mean μ of the set of data.
2. The results cluster symmetrically around this mean value.
3. Small divergences from the central mean value are found more frequently than are large divergences.
4. In the absence of systematic errors, the mean of a large set of data approaches the true value.

Characteristic 4 means that, in principle, it is always possible to reduce the random error of an analysis to something that approaches 0. Unfortunately, it is seldom practical to achieve this goal, however, because to do so requires performing 20 or more replicate analyses. Ordinarily, we can only afford the time for two or three replicated measurements, and a significant random error is to be expected for the mean of such a small number of replicates.

Statisticians usually use μ to represent the mean of an infinite collection of data (see Figure a1-1, curve *B*) and $\bar{x}$ for the mean of a small set of replicate data. In the absence of systematic error, the total error E is just the random error E_d (Equation a1-3). Hence, the error for the mean of a small set is then given by

$$E = \bar{x} - \mu \qquad \text{(a1-4)}$$

The mean of a finite set of data rapidly approaches the true mean when the number of measurements N increases beyond 20 or 30. Thus, as shown in the following example, we can sometimes determine the random error in an individual result or in the mean of a small set of data.

EXAMPLE a1-1

Calculate the random error for (a) the second result in Table a1-1 and (b) the mean for the first three entries in the table.

Solution

For 50 results we can assume the mean $\bar{x} = 0.482 \approx \mu$.

(a) The random error for data point 2 with value x_2 is

$$E = x_2 - \mu = 0.480 - 0.482 = -0.002$$

(b) The mean $\bar{x}$ for the first three results in the table is

$$\bar{x} = \frac{0.488 + 0.480 + 0.486}{3} = 0.485$$

Substituting into Equation a1-4 gives

$$E = \bar{x} - \mu = 0.485 - 0.482 = +0.003$$

The random nature of indeterminate errors makes it possible to treat these effects by statistical methods. Statistical techniques are considered in Section a1B.

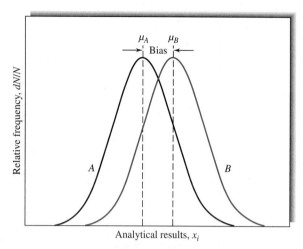

FIGURE a1-2 Illustration of systematic error in analytical results. Curve *A* is the frequency distribution for the accepted value by Method *A*, which has no bias. Curve *B* illustrates the distribution of results by Method *B*, which has a significant bias = $\mu_B - \mu_A$.

Systematic Errors

Systematic errors have a definite value and an assignable cause and are of the same magnitude for replicate measurements made in the same way. Systematic errors lead to *bias* in measurement results. Bias is illustrated by the two curves in Figure a1-2, which show the frequency distribution of replicate results in the analysis of identical samples by two methods that have random errors of identical size. Method *A* has no bias so that the mean μ_A is the true value. Method *B* has a bias that is given by

$$\text{bias} = \mu_B - \mu_A \qquad \text{(a1-5)}$$

Note that bias affects all of the data in a set and that it bears a sign.

Systematic errors are of three types: instrumental, personal, and method.

Instrumental Errors. *Instrumental errors* are caused by nonideal instrument behavior, by faulty calibrations, or by use under inappropriate conditions. Typical sources of instrumental errors include drift in electronic circuits; leakage in vacuum systems; temperature effects on detectors; currents induced in circuits from 110-V power lines; decreases in voltages of batteries with use; and calibration errors in meters, weights, and volumetric equipment.

Systematic instrument errors are commonly detected and corrected by calibration with suitable standards. Periodic calibration of instruments is always desirable because the response of most instruments changes with time due to wear, corrosion, or mistreatment.

Personal Errors. *Personal errors* are those introduced into a measurement by judgments that the experimentalist must make. Examples include estimating the position of a pointer between two scale divisions, the color of a solution at the end point in a titration, the level of a liquid with respect to a graduation in a pipet, or the relative intensity of two light beams. Judgments of this type are often subject to systematic, unidirectional uncertainties. For example, one person may read a pointer consistently high, another may be slightly slow in activating a timer, and a third may be less sensitive to color.

Number bias is another source of personal systematic error that is widely encountered and varies considerably from person to person. The most common bias encountered in estimating the position of a needle on a scale is a preference for the digits 0 and 5. Also prevalent is a preference for small digits over large and even ones over odd. Digital displays and computer-based instruments eliminate number bias.

A near-universal source of personal error is *prejudice*. Most of us, no matter how honest, have a natural tendency to estimate scale readings in a direction that improves the precision in a set of results or causes the results to fall closer to a preconceived notion of the true value for the measurement.

Most personal errors can be minimized by care and self-discipline. Thus, most scientists develop the habit of systematically double-checking instrument readings, notebook entries, and calculations. Robots, automated systems, computerized data collection, and computerized instrument control have the potential of minimizing or eliminating most personal-type systematic errors.

Method Errors. *Method-based errors* are often introduced from nonideal chemical and physical behavior of the reagents and reactions on which an analysis is based. Possible sources include slowness or incompleteness of chemical reactions, losses by volatility, adsorption of the analyte on solids, instability of reagents, contaminants, and chemical interferences.

Systematic method errors are usually more difficult to detect and correct than are instrument and personal errors. The best and surest way involves *validation* of the method by using it for the analysis of standard materials that resemble the samples to be analyzed both in physical state and in chemical composition. The analyte concentrations of these standards must, of course, be known with a high degree of certainty. For simple materials, standards can sometimes be prepared by blending carefully measured amounts of pure compounds. Unfortunately, more often than not, materials to be analyzed are sufficiently complex to preclude this simple approach.

The National Institute of Technology and Standards (NIST) offers for sale a variety of *standard reference materials* (SRMs) that have been specifically prepared for the validation of analytical methods.[4] The concentration of one or more constituents in these materials has been determined by (1) a previously validated reference method; (2) two or more independent, reliable measurement methods; or (3) analyses from a network of cooperating laboratories, technically competent and thoroughly familiar with the material being tested. Most standard reference materials are substances that are commonly encountered in commerce or in environmental, pollution, clinical, biological, and forensic studies. A few examples include trace elements in coal, fuel oil, urban particulate matter, sediments from estuaries, and water; lead in blood samples; cholesterol in human serum; drugs of abuse in urine; and a wide variety of elements in rocks, minerals, and glasses. In addition, several commercial supply houses now offer a variety of analyzed materials for method testing.[5]

a1B STATISTICAL TREATMENT OF RANDOM ERRORS

Randomly distributed data of the kind described previously are conveniently analyzed by statistical techniques, which are considered in the next several sections.[6]

---

[4] See U.S. Department of Commerce, *NIST Standard Reference Materials Catalog*, 1998–99 ed., NIST Special Publication 260-98-99. Washington, DC: U.S. Government Printing Office, 1998. For a description of the reference material programs of NIST, see R. A. Alvarez, S. D. Rasberry, and G. A. Uriano, *Anal. Chem.*, **1982**, *54*, 1226A; see also http://www.nist.gov.
[5] For example, in the clinical and biological sciences area, see Sigma Chemical Co., 3050 Spruce St., St. Louis, MO 63103 and Bio-Rad Laboratories, 1000 Alfred Nobel Dr., Hercules, CA 94547.
[6] For a more detailed treatment of statistics, see J. L. Devore, *Probability and Statistics for Engineering and the Sciences*, 6th ed., Pacific Grove, CA: Duxbury Press at Brooks/Cole, 2004; D. A. Skoog, D. M. West, F. J. Holler, and S. R. Crouch, *Fundamentals of Analytical Chemistry*, 8th ed., Belmont, CA: Brooks/Cole, 2004, Chaps. 5–7; S. R. Crouch and F. J. Holler, *Applications of Microsoft® Excel in Analytical Chemistry*, Belmont, CA: Brooks/Cole, 2004.

a1B-1 Populations and Samples

In the statistical treatment of data, it is assumed that the handful of replicate experimental results obtained in the laboratory is a minute fraction of the infinite number of results that could, in principle, be obtained given infinite time and an infinite amount of sample. Statisticians call the handful of data a *sample* and view it as a subset of an infinite *population*, or *universe*, of data that exists in principle. The laws of statistics apply strictly to populations only; when applying these laws to a sample of laboratory data, we must assume that the sample is truly representative of the population. Because there is no assurance that this assumption is valid, statements about random errors are necessarily uncertain and must be couched in terms of probabilities.

Definition of Some Terms Used in Statistics

Population Mean (μ). The *population mean*, sometimes called the *limiting mean*, of a set of replicate data is defined by

$$\mu = \lim_{N \to \infty} \frac{\sum_{i=1}^{N} x_i}{N} \qquad \text{(a1-6)}$$

where x_i represents the value of the ith measurement. As indicated by this equation, the mean of a set of measurements approaches the population mean as N, the number of measurements, approaches infinity. It is important to add that, in the absence of systematic error, μ *is the true value for the quantity being measured*.

Sample Mean ($\bar{x}$). The sample mean is the mean, or average, of a finite set of data as given by

$$\bar{x} = \frac{\sum_{i=1}^{N} x_i}{N} \qquad \text{(a1-7)}$$

Because N in this case is a finite number, $\bar{x}$ often differs somewhat from the population mean μ, and thus the true value, of the quantity being measured. The use of a different symbol for the sample mean emphasizes this important distinction.

Population Standard Deviation (σ) and Population Variance (σ^2). The population standard deviation and the population variance provide statistically significant measures of the precision of a population of data. The population standard deviation is given by the equation

$$\sigma = \sqrt{\lim_{N \to \infty} \frac{\sum_{i=1}^{N} (x_i - \mu)^2}{N}} \qquad \text{(a1-8)}$$

where x_i is again the value of the ith measurement. Note that the population standard deviation is the root mean square of the individual *deviations from the mean* for the population.

The precision of data is often expressed in terms of the *variance* (σ^2), which is the square of the standard deviation. For independent sources of random error in a system, variances are often additive. That is, if there are n independent sources, the total variance σ_t^2 is given by

$$\sigma_t^2 = \sigma_1^2 + \sigma_2^2 + \cdots + \sigma_n^2 \qquad \text{(a1-9)}$$

where $\sigma_1^2, \sigma_2^2, \ldots, \sigma_n^2$ are the individual variances of the error sources.

Note that the standard deviation has the same units as the data, whereas the variance has the units of the data squared. Scientists tend to use standard deviation rather than variance as a measure of precision. It is easier to relate a measurement and its precision if they both have the same units.

Sample Standard Deviation (s) and Sample Variance (s^2). The *standard deviation* of a sample of data that is of limited size is given by the equation

$$s = \sqrt{\frac{\sum_{i=1}^{N} (x_i - \bar{x})^2}{N - 1}} \qquad \text{(a1-10)}$$

Note that the sample standard deviation differs in three ways from the population standard deviation as defined by Equation a1-8. First, σ is replaced by s in order to emphasize the difference between the two terms. Second, the true mean μ is replaced by $\bar{x}$, the sample mean. Finally, $(N - 1)$, which is defined as the *number of degrees of freedom*, appears in the denominator rather than N.[7]

Relative Standard Deviation (RSD) and Coefficient of Variation (CV). Relative standard deviations are often more informative than are absolute standard

---

[7] By definition, the number of degrees of freedom is the number of values that remain independent when s is calculated. When the sample mean $\bar{x}$ is used in the calculation, only $N - 1$ values are independent, because one value can be obtained from the mean and the other values.

deviations. The relative standard deviation of a data sample is given by

$$\text{RSD} = \frac{s}{x} \times 10^z \qquad \text{(a1-11)}$$

When $z = 2$, the relative standard deviation is given as a percent; when it is 3, the deviation is reported in parts per thousand. The relative standard deviation expressed as a percent is also known as the *coefficient of variation* (CV) for the data. That is,

$$\text{CV} = \frac{s}{x} \times 100\% \qquad \text{(a1-12)}$$

In dealing with a population of data, σ and μ are used in place of s and $\bar{x}$ in Equations a1-11 and a1-12.

Other Ways to Calculate Standard Deviations. Scientific calculators usually have the standard deviation function built in. Many can find the population standard deviation σ as well as the sample standard deviation s. For any small set of data, the sample standard deviation should be used.

To find s with a calculator that does not have a standard deviation key, the following rearrangement of Equation a1-10 is easier to use than Equation a1-10 itself:

$$s = \sqrt{\frac{\sum_{i=1}^{N} x_i^2 - \frac{\left(\sum_{i=1}^{N} x_i\right)^2}{N}}{N - 1}} \qquad \text{(a1-13)}$$

Example a1-2 illustrates the use of Equation a1-13 to find s.

EXAMPLE a1-2

The following results were obtained in the replicate determination of the lead content of a blood sample: 0.752, 0.756, 0.752, 0.751, and 0.760 ppm Pb. Calculate the mean, the standard deviation, and the coefficient of variation for the data.

Solution

To apply Equation a1-13, we calculate $\sum x_i^2$ and $(\sum x_i)^2/N$.

Sample	x_i	x_i^2
1	0.752	0.565504
2	0.756	0.571536
3	0.752	0.565504
4	0.751	0.564001
5	0.760	0.577600
	$\sum x_i = 3.771$	$\sum x_i^2 = 2.844145$

$$\bar{x} = \frac{\sum x_i}{N} = \frac{3.771}{5} = 0.7542 \approx 0.754 \text{ ppm Pb}$$

$$\frac{(\sum x_i)^2}{N} = \frac{(3.771)^2}{5} = \frac{14.220441}{5} = 2.8440882$$

Substituting into Equation a1-13 leads to

$$s = \sqrt{\frac{2.844145 - 2.8440882}{5 - 1}} = \sqrt{\frac{0.0000568}{4}}$$
$$= 0.00377 \approx 0.004 \text{ ppm Pb}$$

$$\text{CV} = \frac{s}{\bar{x}} \times 100\% = \frac{0.00377}{0.7542} \times 100\% = 0.5\%$$

Note in Example a1-2 that the difference between $\sum x_i^2$ and $(\sum x_i)^2/N$ is very small. If we had rounded these numbers before subtracting them, a serious error would have appeared in the computed value of s. To avoid this source of error, *never round a standard deviation calculation until the very end.* Furthermore, and for the same reason, never use Equation a1-13 to calculate the standard deviation of numbers containing five or more digits. Use Equation a1-10 instead.[8] Many calculators and computers with a standard deviation function use a version of Equation a1-13 internally in the calculation. You should always be alert for roundoff errors when calculating the standard deviation of values that have five or more significant figures.

In addition to calculators, computer software is widely used for statistical calculations. Spreadsheet software, such as Microsoft® Excel, can readily obtain a variety of statistical quantities.[9] Some popular dedicated statistics programs include MINITAB, SAS, SYSTAT, Origin, STATISTICA, SigmaStat, SPSS, and STATGRAPHICS Plus. In addition to normal statistics calculations, these programs can be used to carry out least-squares analysis, nonlinear regression, and many advanced functions.

The Normal Error Law

In Gaussian statistics, the results of replicate measurements arising from indeterminate errors are assumed to be distributed according to the normal error law, which states that the fraction of a population of observations, dN/N, whose values lie in the region x to $(x + dx)$ is given by

$$\frac{dN}{N} = \frac{1}{\sigma\sqrt{2\pi}} e^{-(x-\mu)^2/2\sigma^2} dx \qquad \text{(a1-14)}$$

Here, μ and σ are the population mean and the standard deviation, and N is the number of observations. The two curves shown in Figure a1-3a are plots of Equation a1-14. The standard deviation for the data in curve B is twice that for the data in curve A.

Note that $(x - \mu)$ in Equation a1-14 is the *absolute deviation of the individual values of x from the mean* in whatever units are used in the measurement. Often, however, it is more convenient to express the deviations from the mean in units of the variable z, where

$$z = \frac{x - \mu}{\sigma} \qquad \text{(a1-15)}$$

Note that z is the deviation of a data point from the mean relative to one standard deviation. That is, when $x - \mu = \sigma$, z is equal to 1; when $x - \mu = 2\sigma$, z is equal to 2; and so forth. Because z is the deviation from the mean relative to the standard deviation, a plot of relative frequency versus z yields a single Gaussian curve that describes all populations of data regardless of standard deviation. Thus, Figure a1-3b is the normal error curve for both sets of data used to plot curves A and B in Figure a1-3a.

Taking the derivative of Equation a1-15 with respect to x gives

$$dz = \frac{dx}{\sigma} \qquad \text{(a1-16)}$$

Substitution of these two relationships into Equation a1-14 leads to an equation that expresses the distribution in terms of the single variable z. That is,

$$\frac{dN}{N} = \frac{1}{\sqrt{2\pi}} e^{-z^2/2} dz \qquad \text{(a1-17)}$$

[8] In most cases, the first two or three digits in a set of data are identical to each other. As an alternative, then, to using Equation a1-10, these identical digits can be dropped and the remaining digits used with Equation a1-13. For example, the standard deviation for the data in Example a1-2 could be based on 0.052, 0.056, 0.052, and so forth (or even 52, 56, 52, etc.).

[9] S. R. Crouch and F. J. Holler, *Applications of Microsoft® Excel in Analytical Chemistry*, Belmont, CA: Brooks/Cole, 2004.

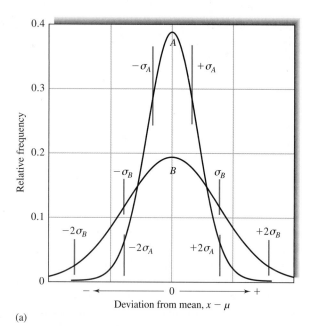

(a)

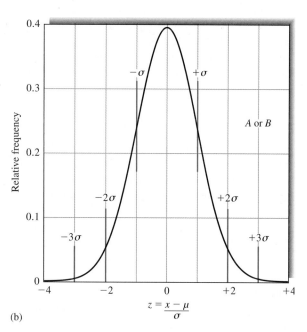

(b)

FIGURE a1-3 Normal error curves. The standard deviation of B is twice that of A; that is, $\sigma_B = 2\sigma_A$. (a) The abscissa is the deviation from the mean in the units of the measurement. (b) The abscissa is the deviation from the mean relative to σ. Thus, A and B produce identical curves when the abscissa is $z = (x - \mu)/\sigma$.

The Normal Error Curve

The normal error curve shown in Figure a1-3b has the following general properties: (1) zero deviation from the mean occurring with maximum frequency, (2) symmetrical distribution of positive and negative deviations about this maximum, and (3) exponential decrease in frequency as the magnitude of the deviations increases. Thus, small random errors are much more common than large errors.

Areas under Regions of the Normal Error Curve. The area under the curve in Figure a1-3b is the integral of Equation a1-17 and is determined as follows:

$$\frac{\Delta N}{N} = \int_{-a}^{a} \frac{1}{\sqrt{2\pi}} e^{-z^2/2} \, dz = \text{erf}\left(\frac{a\sqrt{2}}{2}\right)$$

where $\text{erf}(b)$ is the *error function* given by

$$\text{erf}(b) = \int_0^b \frac{2}{\sqrt{\pi}} e^{-x^2} \, dx$$

The fraction of the population between any specified limits is given by the area under the curve between these limits. For example, the area under the curve between $z = -1$ and $z = +1$ is given by the definite integral

$$\frac{\Delta N}{N} = \int_{-1}^{1} \frac{1}{\sqrt{2\pi}} e^{-z^2/2} \, dz = \text{erf}\left(\frac{\sqrt{2}}{2}\right) = 0.683$$

Thus, $\Delta N/N = 0.683$, which means that 68.3% of a population of data lie within $\pm 1\sigma$ of the mean value. For similar calculations with $z = 2$ and $z = 3$, we find that 95.4% lie within $\pm 2\sigma$ and 99.7% within $\pm 3\sigma$. Values for $(x - \mu)$ corresponding to $\pm 1\sigma$, $\pm 2\sigma$, and $\pm 3\sigma$ are indicated by blue vertical lines in Figure a1-3.

The properties of the normal error curve are useful because they permit statements to be made about the probable magnitude of the net random error in a given measurement or set of measurements *provided the standard deviation is known.* Thus, one can say that it is 68.3% probable that the random error associated with any single measurement is within $\pm 1\sigma$, that it is 95.4% probable that the error is within $\pm 2\sigma$, and so forth. The standard deviation is clearly a useful quantity for estimating and reporting the probable net random error of an analytical method.

Standard Error of the Mean. The probability figures for the Gaussian distribution just cited refer to the probable error of a *single* measurement. If a set of samples, each containing N results, is taken randomly from a population of data, the means of samples will show less and less scatter as N increases. The standard deviation of the means of the samples is known as the *standard error of the mean* and is denoted by σ_m. It can be shown that the standard error is inversely proportional to the square root of the number of data points N used to calculate the means. That is,

$$\sigma_m = \frac{\sigma}{\sqrt{N}} \qquad \text{(a1-18)}$$

For data in which the sample standard deviation s is calculated, Equation a1-18 can be written as

$$s_m = \frac{s}{\sqrt{N}}$$

where s_m is the sample standard deviation of the mean.

The mean and the standard deviation of a set of data are statistics of primary importance in all types of science and engineering. The mean is important because it usually provides the best estimate of the variable of interest. The standard deviation of the mean is equally important because it provides information about the precision and thus the random error associated with the measurement.

Methods to Obtain a Good Estimate of σ

To apply a statistical relationship directly to finite samples of data, it is necessary to know that the sample standard deviation s for the data is a good approximation of the population standard deviation σ. Otherwise, statistical inferences must be modified to take into account the uncertainty in s. In this section, we consider methods for obtaining reliable estimates of σ from small samples of data.

Performing Preliminary Experiments. Uncertainty in the calculated value for s decreases as the number of measurements N in Equation a1-10 increases. Figure a1-4 shows the relative error in s as a function of N. Note that when N is greater than about 20, s and σ can be assumed, for most purposes, to be identical. Thus, when a method of measurement is not excessively time-consuming and when an adequate supply of sample is available, it is sometimes feasible and economical to carry out preliminary experiments whose sole

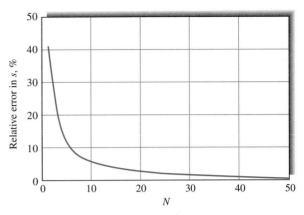

FIGURE a1-4 Relative error in the sample standard deviation s as a function of the number of measurements N.

purpose is to obtain a reliable standard deviation for the method. For example, if the pH of numerous solutions is to be measured in the course of an investigation, it is useful to evaluate s in a series of preliminary experiments. This measurement is straightforward, requiring only that a pair of rinsed and dried electrodes be immersed in the test solution and the pH read from a scale or a display. To determine s, 20 to 30 portions of a buffer solution of fixed pH can be measured with all steps of the procedure being followed exactly. Normally, it is safe to assume that the random error in this test is the same as that in subsequent measurements. The value of s calculated from Equation a1-10 is then a good estimator of the population value, σ.

Pooling Data. If we have several subsets of data, we can get a better estimate of the population standard deviation by pooling (combining) the data than by using only a single data set. Again, we must assume the same sources of random error in all the measurements. This assumption is usually valid if the samples have similar compositions and have been analyzed in exactly the same way. We must also assume that the samples are randomly drawn from the same population and thus have a common value of σ.

The pooled estimate of σ, which we call s_{pooled}, is a weighted average of the individual estimates. To calculate s_{pooled}, deviations from the mean for each subset are squared; the squares of the deviations of all subsets are then summed and divided by the appropriate number of degrees of freedom. The pooled s is obtained by taking the square root of the resulting number. One degree of freedom is lost for each subset. Thus, the number of degrees of freedom for the pooled s is equal to the total number of measurements minus the number of subsets n_t:

$$s_{pooled}$$
$$= \sqrt{\frac{\sum_{i=1}^{N_1}(x_i - \bar{x}_1)^2 + \sum_{j=1}^{N_2}(x_j - \bar{x}_2)^2 + \sum_{k=1}^{N_3}(x_k - \bar{x}_3)^2 + \cdots}{N_1 + N_2 + N_3 + \cdots - n_t}}$$

(a1-19)

Here, the indices i, j, and k refer to the data in each subset, $N_1, N_2, N_3, \ldots, N_{n_t}$ are the numbers of results in each subset. Example a1-3 illustrates the calculation and application of the pooled standard deviation.

EXAMPLE a1-3

Glucose levels are routinely monitored in patients suffering from diabetes. The glucose concentrations in a patient with mildly elevated glucose levels were determined in different months by a spectrophotometric analytical method. The patient was placed on a low-sugar diet to reduce the glucose levels. The following results were obtained during a study to determine the effectiveness of the diet. Calculate a pooled estimate of the standard deviation for the method.

Time	Glucose Concentration, mg/L	Mean Glucose, mg/L	Sum of Squares of Deviations from Mean	Standard Deviation
Month 1	1108, 1122, 1075, 1099, 1115, 1083, 1100	1100.3	1687.43	16.8
Month 2	992, 975, 1022, 1001, 991	996.2	1182.80	17.2
Month 3	788, 805, 779, 822, 800	798.8	1086.80	16.5
Month 4	799, 745, 750, 774, 777, 800, 758	771.9	2950.86	22.2

Note: Total number of measurements = 24; total sum of squares = 6907.89.

Solution

For the first month, the sum of the squares in the next to last column was calculated as follows:

$$\begin{aligned}
\text{Sum of squares} &= (1108 - 1100.3)^2 + (1122 - 1100.3)^2 \\
&+ (1075 - 1100.3)^2 + (1099 - 1100.3)^2 \\
&+ (1115 - 1100.3)^2 + (1083 - 1100.3)^2 \\
&+ (1100 - 1100.3)^2 = 1687.43
\end{aligned}$$

The other sums of squares were obtained similarly. The pooled standard deviation is then

$$s_{pooled} = \sqrt{\frac{6907.89}{24 - 4}} = 18.58 \approx 19 \text{ mg/L}$$

Note this pooled value is a better estimate of σ than any of the individual s values in the last column. Note also that one degree of freedom is lost for each of the four data sets. Because 20 degrees of freedom remain, however, the calculated value of s can be considered a good estimate of σ.

---

a1B-2 Confidence Intervals

In most of the situations encountered in chemical analysis, the true value of the mean μ cannot be determined because a huge number of measurements (approaching infinity) would be required. With statistics, however, we can establish an interval surrounding an experimentally determined mean $\bar{x}$ within which the population mean μ is expected to lie with a certain degree of probability. This interval is known as the *confidence interval*. For example, we might say that it is 99% probable that the true population mean for a set of potassium measurements lies in the interval 7.25 ± 0.15% K. Thus, the mean should lie in the interval from 7.10 to 7.40% K with 99% probability.

The size of the confidence interval, which is computed from the sample standard deviation, depends on how well the sample standard deviation s estimates the population standard deviation σ. If s is a good approximation of σ, the confidence interval can be significantly narrower than if the estimate of σ is based on only a few measurement values.

Confidence Interval When σ Is Known or s Is a Good Estimator of σ

Figure a1-5 shows a series of five normal error curves. In each, the relative frequency is plotted as a function of the quantity z (Equation a1-15), which is the deviation

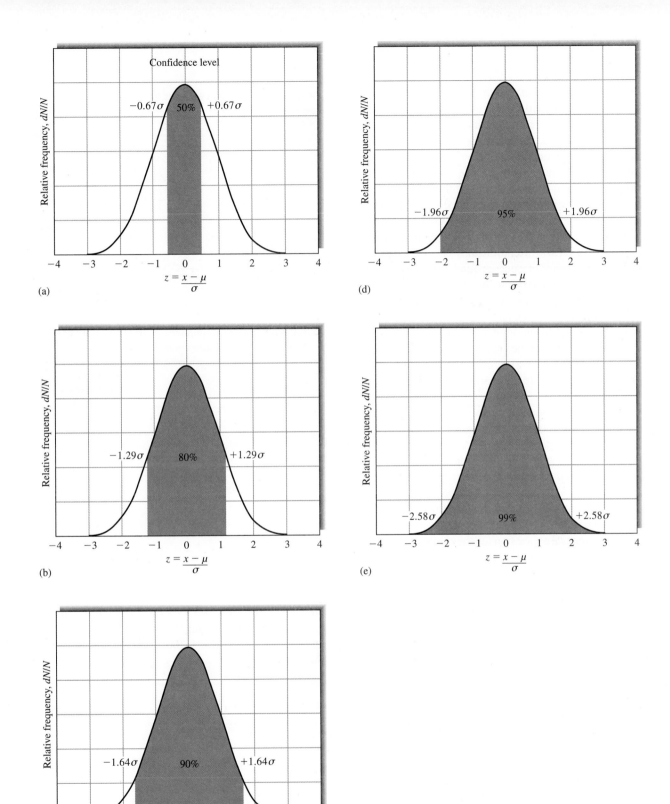

FIGURE a1-5 Areas under a Gaussian curve for various values of $\pm z$. (a) $z = \pm 0.67$; (b) $z = \pm 1.29$; (c) $z = \pm 1.64$; (d) $z = \pm 1.96$; (e) $z = \pm 2.58$.

from the mean *normalized to the population standard deviation*. The shaded areas in each plot lie between the values of $-z$ and $+z$ that are indicated to the left and right of the curves. The numbers within the shaded areas are the percentage of the total area under each curve that is included within these values of z. For example, as shown in curve a, 50% of the area under any Gaussian curve is located between -0.67σ and $+0.67\sigma$. In curves b and c, we see that 80% of the total area lies between -1.28σ and $+1.28\sigma$, and 90% lies between -1.64σ and $+1.64\sigma$. Relationships such as these allow us to define a range of values around a measurement result within which the true mean is likely to lie with a certain probability *provided we have a reasonable estimate of σ*. For example, if we have a result x from a data set with a standard deviation of σ, we may assume that 90 times out of 100, the true mean μ will fall in the interval $x \pm 1.64\sigma$ (see Figure a1-5c). The probability is called the *confidence level*. In the example of Figure a1-5c, the confidence level (CL) is 90% and the *confidence interval* (CI) ranges from -1.64σ to $+1.64\sigma$. The probability that a result is outside the confidence interval is often called the *significance level*.

If we make a single measurement x from a distribution of known σ, we can say that the true mean should lie in the interval $x \pm z\sigma$ with a probability dependent on z. This probability is 90% for $z = 1.64$, 95% for $z = 1.96$, and 99% for $z = 2.58$ as shown in Figure a1-5c–e. We find a general expression for the confidence interval of the true mean based on measuring a single value x by rearranging Equation a1-15 (remember that z can take positive or negative values). Thus,

$$\text{CI for } \mu = x \pm z\sigma \qquad \text{(a1-20)}$$

Rarely do we estimate the true mean from a single measurement, however. Instead, we use the experimental mean $\bar{x}$ of N measurements as a better estimate of μ. In this case, we replace x in Equation a1-20 with $\bar{x}$ and σ with the standard error of the mean, $\sigma/\sqrt{N}$ (Equation a1-18). That is,

$$\text{CI for } \mu = \bar{x} \pm \frac{z\sigma}{\sqrt{N}} \qquad \text{(a1-21)}$$

Values of z at various confidence levels are given in Table a1-3, and the relative size of the confidence interval as a function of N is shown in Table a1-4. Sample calculations of confidence limits and confidence intervals are given in Examples a1-4 and a1-5.

TABLE a1-3 Confidence Levels for Various Values of z

Confidence Level, %	z
50	0.67
68	1.00
80	1.28
90	1.64
95	1.96
95.4	2.00
99	2.58
99.7	3.00
99.9	3.29

TABLE a1-4 Size of Confidence Interval as a Function of the Number of Measurements Averaged

Number of Measurements Averaged	Relative Size of Confidence Interval
1	1.00
2	0.71
3	0.58
4	0.50
5	0.45
6	0.41
10	0.32

EXAMPLE a1-4

Determine the 80% and 95% confidence intervals for (a) the first entry (1108 mg/L glucose) in Example a1-3 and (b) the mean value (1100.3 mg/L) for month 1 in the example.

Solution

We will use the pooled standard deviation, $s_{\text{pooled}} = 19$ as a good estimate of σ.

(a) From Table a1-3, we see that $z = 1.28$ and 1.96 for the 80% and 95% confidence levels. Substituting into Equation a1-20 gives

80% CI = 1108 ± 1.28 × 19 = 1108 ± 24.3 mg/L

95% CI = 1108 ± 1.96 × 19 = 1108 ± 37.2 mg/L

From these calculations, we conclude that it is 80% probable that μ, the population mean (and, *in the absence of determinate error*, the true value), lies in the interval between 1083.7 and 1132.3 mg/L glucose. Furthermore, the probability is 95% that μ lies in the interval between 1070.8 and 1145.2 mg/L glucose.

(b) For the seven measurements,

$$80\% \text{ CL} = 1100.3 \pm \frac{1.28 \times 19}{\sqrt{7}}$$

$$= 1100.3 \pm 9.2 \text{ mg/L}$$

$$95\% \text{ CL} = 1100.3 \pm \frac{1.96 \times 19}{\sqrt{7}}$$

$$= 1100.3 \pm 14.1 \text{ mg/L}$$

Thus, there is an 80% chance that μ is located in the interval between 1091.1 and 1109.5 mg/L glucose and a 95% chance that it lies between 1086.2 and 1114.4 mg/L glucose.

EXAMPLE a1-5

How many replicate measurements in month 1 in Example a1-3 are needed to decrease the 95% confidence interval to 1100.3 ± 10.0 mg/L glucose?

▸ *Solution*

Here, we want the term $\pm z\sigma/\sqrt{N}$ to equal ±10.0 mg/L glucose:

$$\frac{z\sigma}{\sqrt{N}} = \frac{1.96 \times 19}{\sqrt{N}} = 10.0$$

$$\sqrt{N} = \frac{1.96 \times 19}{10.0} = 3.724$$

$$N = (3.724)^2 = 13.9$$

We thus conclude that 14 measurements are needed to provide a slightly better than 95% chance that the population mean will lie within ±10 mg/L glucose of the experimental mean.

Equation a1-21 and Table a1-4 tell us that the confidence interval for an analysis can be halved by carrying out four measurements. Sixteen measurements will narrow the interval by a factor of 4, and so on. We rapidly reach a point of diminishing returns in acquiring additional data. Ordinarily, we take advantage of the relatively large gain attained by averaging two to four measurements, but can seldom afford the time or amount of sample required to obtain narrower confidence intervals through additional replicate measurements.

It is essential to keep in mind at all times that confidence intervals based on Equation a1-21 apply only *in the absence of bias and only if we can assume that s is a good approximation of σ*. We will indicate that s is a good estimate of σ by using the symbol $s \rightarrow \sigma$ (s approaches σ).

Confidence Intervals When σ Is Unknown

Often, limitations in time or in the amount of available sample prevent us from making enough measurements to assume s is a good estimate of σ. In such a case, a single set of replicate measurements must provide not only a mean but also an estimate of precision. As indicated earlier, s calculated from a small set of data may be quite uncertain. Thus, confidence intervals are necessarily broader when we must use a small sample value of s as our estimate of σ.

To account for the variability of s, we use the important statistical quantity t, which is defined in exactly the same way as z (Equation a1-15) except that s is substituted for σ: For a single measurement with result x, we can define t as

$$t = \frac{x - \mu}{s} \qquad (a1-22)$$

For the mean of N measurements,

$$t = \frac{\bar{x} - \mu}{s/\sqrt{N}} \qquad (a1-23)$$

Like z in Equation a1-20, t depends on the desired confidence level. But t also depends on the number of degrees of freedom in the calculation of s. Table a1-5 provides values of t for a few degrees of freedom. More extensive tables are found in various mathematical and statistical handbooks. Note that t approaches z as the number of degrees of freedom becomes large.

The confidence interval for the mean $\bar{x}$ of N replicate measurements can be calculated from t by an equation similar to Equation a1-21:

$$\text{CI for } \mu = \bar{x} \pm \frac{ts}{\sqrt{N}} \qquad (a1-24)$$

Use of the t statistic for confidence intervals is illustrated in Example a1-6.

TABLE a1-5 Values of t for Various Levels of Probability

Degrees of Freedom	80%	90%	95%	99%	99.9%
1	3.08	6.31	12.7	63.7	637
2	1.89	2.92	4.30	9.92	31.6
3	1.64	2.35	3.18	5.84	12.9
4	1.53	2.13	2.78	4.60	8.61
5	1.48	2.02	2.57	4.03	6.87
6	1.44	1.94	2.45	3.71	5.96
7	1.42	1.90	2.36	3.50	5.41
8	1.40	1.86	2.31	3.36	5.04
9	1.38	1.83	2.26	3.25	4.78
10	1.37	1.81	2.23	3.17	4.59
15	1.34	1.75	2.13	2.95	4.07
20	1.32	1.73	2.09	2.84	3.85
40	1.30	1.68	2.02	2.70	3.55
60	1.30	1.67	2.00	2.62	3.46
∞	1.28	1.64	1.96	2.58	3.29

EXAMPLE a1-6

A chemist obtained the following data for the alcohol content of a sample of blood: % C_2H_5OH: 0.084, 0.089, and 0.079. Calculate the 95% confidence interval for the mean assuming (a) the three results obtained are the only indication of the precision of the method and (b) from previous experience on hundreds of samples, we know that the standard deviation of the method $s = 0.005\%$ C_2H_5OH and is a good estimate of σ.

Solution

(a) $\sum x_i = 0.084 + 0.089 + 0.079 = 0.252$

$\sum x_i^2 = 0.007056 + 0.007921 + 0.006241$

$= 0.021218$

$s = \sqrt{\dfrac{0.021218 - (0.252)^2/3}{3 - 1}} = 0.0050\%$ C_2H_5OH

Here, $\bar{x} = 0.252/4 = 0.084$. Table a1-5 indicates that $t = 4.30$ for 2 degrees of freedom and the 95% confidence level. Thus,

$$95\% \text{ CI} = \bar{x} \pm \frac{ts}{\sqrt{N}} = 0.084 \pm \frac{4.30 \times 0.0050}{\sqrt{3}}$$
$$= 0.084 \pm 0.012\% \text{ C}_2\text{H}_5\text{OH}$$

(b) Because $s = 0.0050\%$ is a good estimate of σ,

$$95\% \text{ CI} = \bar{x} \pm \frac{z\sigma}{\sqrt{N}} = 0.094 \pm \frac{1.96 \times 0.0050}{\sqrt{3}}$$
$$= 0.084 \pm 0.006\% \text{ C}_2\text{H}_5\text{OH}$$

Note that a sure knowledge of σ decreases the confidence interval by a significant amount.

a1B-3 Propagation of Measurement Uncertainties

A typical instrumental method of analysis involves several experimental measurements, each of which is subject to an indeterminate uncertainty and each of which contributes to the net indeterminate error of the final result.

Principles

For the purpose of showing how such indeterminate uncertainties affect the outcome of an analysis, let us assume that a result x is dependent on the experimental variables, $p, q, r, \ldots$, each of which fluctuates in a random and independent way. That is, x is a function of $p, q, r, \ldots$, so that we may write

$$x = f(p, q, r, \ldots) \qquad \text{(a1-25)}$$

The uncertainty dx_i (that is, the deviation from the mean) in the ith measurement of x depends on the size and sign of the corresponding uncertainties dp_i, dq_i, $dr_i, \ldots$. That is,

$$dx_i = f(dp_i, dq_i, dr_i, \ldots)$$

The total variation in x as a function of the uncertainties in $p, q, r, \ldots$ can be derived by taking the partial derivative of x (Equation a1-25) with respect to each of the variables $p, q, r, \ldots$. That is,

$$dx = \left(\frac{\partial x}{\partial p}\right)_v dp + \left(\frac{\partial x}{\partial q}\right)_v dq + \left(\frac{\partial x}{\partial r}\right)_v dr + \cdots$$
$$\text{(a1-26)}$$

where the subscript v means the other variables $p, q, r, \ldots$ are held constant.

To develop a relationship between the standard deviation of x and the standard deviations of $p, q,$ and r, we must square Equation a1-26 and sum between the limits of $i = 1$ to $i = N$, where N is the total number of measurements. In squaring Equation a1-26, two types of terms appear on the right-hand side of the equation: square terms and cross terms. Square terms take the form

$$\left(\frac{\partial x}{\partial p}\right)^2 (dp)^2, \left(\frac{\partial x}{\partial q}\right)^2 (dq)^2, \left(\frac{\partial x}{\partial r}\right)^2 (dr)^2, \cdots$$

Such square terms are always positive and *never cancel each other*. In contrast, cross terms may be either positive or negative in sign and take the form

$$\left(\frac{\partial x}{dp}\right)\left(\frac{\partial x}{dq}\right) dp\, dq, \left(\frac{\partial x}{dp}\right)\left(\frac{\partial x}{dr}\right) dp\, dr, \cdots$$

If dp, dq, and dr represent *independent* and *random uncertainties*, some of the cross terms will be negative and others positive. Thus, the sum of all such terms should approach 0, particularly when N is large. Note, however, that if the variables are not independent, but instead are correlated, the cross terms must be kept regardless of the size of N.[10]

As a consequence of the tendency of cross terms to cancel, the sum from $i = 1$ to $i = N$ of the square of Equation a1-26 can be assumed to be made up exclusively of squared terms. This sum then takes the form

$$\sum (dx_i)^2 = \left(\frac{\partial x}{\partial p}\right)^2 \sum (dp_i)^2 + \left(\frac{\partial x}{\partial q}\right)^2 \sum (dq_i)^2$$
$$+ \left(\frac{\partial x}{\partial r}\right)^2 \sum (dr_i)^2 + \cdots \qquad \text{(a1-27)}$$

Dividing through by $N - 1$ gives

$$\frac{\sum (dx_i)^2}{N - 1} = \left(\frac{\partial x}{\partial p}\right)^2 \frac{\sum (dp_i)^2}{N - 1} + \left(\frac{\partial x}{\partial q}\right)^2 \frac{\sum (dq_i)^2}{N - 1}$$
$$+ \left(\frac{\partial x}{\partial r}\right)^2 \frac{\sum (dr_i)^2}{N - 1} + \cdots \qquad \text{(a1-28)}$$

From Equation a1-8, we note that

$$\frac{\sum (dx_i)^2}{N - 1} = \frac{\sum (x_i - \bar{x})^2}{N - 1} = s_x^2$$

where s_x^2 is the variance of x. Similarly,

$$\frac{\sum (dp_i)^2}{N - 1} = s_p^2$$

and so forth. Thus, Equation a1-28 can be written in terms of the variances of the quantities; that is

$$s_x^2 = \left(\frac{\partial x}{\partial p}\right)^2 s_p^2 + \left(\frac{\partial x}{\partial q}\right)^2 s_q^2 + \left(\frac{\partial x}{\partial r}\right)^2 s_r^2 + \cdots \quad \text{(a1-29)}$$

---

[10]See P. R. Bevington and D. K. Robinson, *Data Reduction and Error Analysis for the Physical Sciences*, 2nd ed., pp. 41–50, New York: McGraw-Hill, 1992.

Note that if N is large the sample variances shown in Equation a1-29 can be replaced by population variances. The following example illustrates how Equation a1-29 can be used to give the variance of a quantity calculated from several experimental results.

EXAMPLE a1-7

The number of plates N in a chromatographic column can be computed with Equation 26-21:

$$N = 16\left(\frac{t_R}{W}\right)^2$$

where t_R is the retention time and W is the width of the chromatographic peak in the same units as t_R. The significance of these terms is explained in Figure 26-7.

Hexachlorobenzene exhibited an HPLC peak at a retention time of 13.36 min. The width of the peak at its base was 2.18 min. The standard deviations s for the two time measurements were 0.043 and 0.061 min, respectively. Calculate (a) the number of plates in the column and (b) the standard deviation for the computed result.

Solution

(a) $N = 16\left(\dfrac{13.36}{2.18}\right)^2 = 601$ plates

(b) From Equation a1-29,

$$s_N^2 = \left(\frac{\partial N}{\partial t_R}\right)_W^2 s_{t_R}^2 + \left(\frac{\partial N}{\partial W}\right)_{t_R}^2 s_W^2$$

From the equation for N,

$$\left(\frac{\partial N}{\partial t_R}\right)_W^2 = \frac{32 t_R}{W^2} \quad \text{and} \quad \left(\frac{\partial N}{\partial W}\right)_{t_R}^2 = \frac{-32 t_R^2}{W^3}$$

Substituting these relationships into the previous equation gives

$$s_N^2 = \left(\frac{32 t_R}{W^2}\right) s_{t_R}^2 + \left(\frac{-32 t_R^2}{W^3}\right)^2 s_W^2$$
$$= \left(\frac{32 \times 13.36 \text{ min}}{(2.18 \text{ min})^2}\right)^2 (0.061 \text{ min})^2$$
$$+ \left(\frac{-32(13.36 \text{ min})^2}{(2.18 \text{ min})^3}\right)^2 (0.043 \text{ min})^2 = 592.1$$
$$= \sqrt{592.1} = 24.3 \text{ or } 24 \text{ plates}$$

Thus, $N = 601 \pm 24$ plates.

TABLE a1-6 Error Propagation in Arithmetic Calculations

Type of Calculation	Example[a]	Standard Deviation of y	
Addition or subtraction	$x = p + q - r$	$s_x = \sqrt{s_p^2 + s_q^2 + s_r^2}$	(1)
Multiplication or division	$x = p \times q/r$	$\dfrac{s_x}{x} = \sqrt{\left(\dfrac{s_p}{p}\right)^2 + \left(\dfrac{s_q}{q}\right)^2 + \left(\dfrac{s_r}{r}\right)^2}$	(2)
Exponentiation	$x = p^y$	$\dfrac{s_x}{x} = y\left(\dfrac{s_p}{p}\right)$	(3)
Logarithm	$x = \log_{10} p$	$s_x = 0.434\dfrac{s_p}{p}$	(4)
Antilogarithm	$x = \text{antilog}_{10}\, p$	$\dfrac{s_x}{x} = 2.303\, s_p$	(5)

[a]p, q, and r are experimental variables with standard deviations of s_p, s_q, and s_r, respectively.

Standard Deviation of Calculated Results

Equation a1-29 can be used to derive relationships that permit calculation of standard deviations for the results produced by arithmetic operations. As an example, let us consider the case where a result x is computed by the relationship

$$x = p + q - r$$

where p, q, and r are experimental quantities having sample standard deviations of s_p, s_q, and s_r, respectively. In Equation a1-29, the partial derivatives

$$\left(\frac{\partial x}{dp}\right)_{q,r} = \left(\frac{\partial x}{dq}\right)_{p,r} = 1 \quad \text{and} \quad \left(\frac{\partial x}{\partial r}\right)_{p,q} = -1$$

Therefore, the variance of x is given by

$$s_x^2 = s_p^2 + s_q^2 + s_r^2$$

and the standard deviation of x is

$$s_x = \sqrt{s_p^2 + s_q^2 + s_r^2}$$

Thus, the *absolute* standard deviation of a sum or difference is equal to the square root of the sum of the squares of the absolute standard deviation of the numbers making up the sum or difference.

Proceeding in this same way yields the relationships shown in Table a1-6 for other types of arithmetic operations. Note that in several calculations relative variances such as $(s_x/x)^2$ and $(s_p/p)^2$ are combined rather than absolute standard deviations.

Rounding Data

We often indicate the probable uncertainty associated with an experimental measurement by rounding the result so that it contains only *significant figures*. By definition, the significant figures in a number are all of the digits known with certainty *plus the first uncertain digit*. If we know or have a good estimate of the standard deviation of results, we can tell which digits are uncertain. For example, if the mean of five results was determined to be 2.634 and the standard deviation of the mean was 0.02, we know that the second digit to the right of the decimal point is the uncertain digit. Hence, the result should be expressed as 2.63 ± 0.02. We should note that *it is seldom justifiable to keep more than one significant figure in the standard deviation* because the standard deviation contains error as well. For certain specialized purposes, such as reporting uncertainties in physical constants in research articles, it may be useful to keep two significant figures, and there is certainly nothing wrong with including a second digit in the standard deviation. However, it is important to recognize that the uncertainty usually lies in the first digit.[11]

When the result is computed from experimental quantities with known standard deviations, we can find the standard deviation in the computed result from Equation a1-29 or Table a1-6. This can again guide us in determining which figures are significant and how to round results appropriately as shown in Examples a1-8 and 1-9.

---

[11]For more details on this topic, go to http://www.chem.uky.edu/courses/che226/download/CI_for_sigma.html.

EXAMPLE a1-8

Calculate the standard deviation of the following and determine how to round the result:

$$x = \frac{[14.3(\pm 0.2) - 11.6(\pm 0.2)] \times 0.050(\pm 0.001)}{[820(\pm 10) + 1030(\pm 5)] \times 42.3(\pm 0.4)}$$
$$= 1.725129 \times 10^{-6}$$

Solution

First, we must find the standard deviations of the difference in the numerator and the sum in the denominator. Calling s_d the standard deviation of the difference in the numerator, we find

$$s_d = \sqrt{(\pm 0.2)^2 + (\pm 0.2)^2} = \pm 0.2828$$

For the sum in the denominator, the standard deviation s_s is

$$s_s = \sqrt{(\pm 10)^2 + (\pm 5)^2} = \pm 11.180$$

We can then rewrite the equation as

$$\frac{2.7(\pm 0.2828) \times 0.050(\pm 0.001)}{1850(\pm 11.180) \times 42.3(\pm 0.4)} = 1.725129 \times 10^{-6}$$

This equation now contains only products and quotients, and Equation (2) of Table a1-6 applies:

$$\frac{s_x}{x} =$$

$$\sqrt{\left(\frac{\pm 0.2828}{2.7}\right)^2 + \left(\frac{\pm 0.001}{0.050}\right)^2 + \left(\frac{\pm 11.180}{1850}\right)^2 + \left(\frac{\pm 0.4}{42.3}\right)^2}$$
$$= \pm 0.10722$$

To obtain s_x, we must multiply the preceding relative standard deviation by $x = 1.725129 \times 10^{-6}$

$$s_x = \pm 0.10722 \times 1.725129 \times 10^{-6} = 0.18497 \times 10^{-6}$$

The answer is then $1.7(\pm 0.2) \times 10^{-6}$.

Note in this example that we postponed rounding until the calculation was completed. At least one extra digit beyond the significant digits should be carried through all the calculations in order to avoid a *rounding error*.

EXAMPLE a1-9

Calculate the standard deviations of the results of the following computations and round the results to the appropriate number of significant figures. The absolute standard deviation for each quantity is given in parentheses.

(a) $x = \log[2.00(\pm 0.02) \times 10^{-4}] = -3.69897$
(b) $x = \text{antilog}[1.200(\pm 0.003)] = 15.8493$
(c) $x = [4.73(\pm 0.03) \times 10^{-4}]^3 = 1.0583 \times 10^{-10}$

Solution

(a) Referring to Equation (4) in Table a1-6, we find

$$s_x = \pm 0.434 \times \frac{0.002 \times 10^{-4}}{2.00 \times 10^{-4}} = \pm 0.00434$$

Thus, $\log[2.00(\pm 0.02) \times 10^{-4}] = -3.699 \pm 0.004$.

(b) Using Equation 5 in Table a1-6, we obtain

$$\frac{s_x}{x} = 2.303 \times (\pm 0.003) = \pm 0.006909$$
$$s_x = \pm 0.006909 \times 15.8493 = \pm 0.1095$$

Therefore, $\text{antilog}[1.200(\pm 0.003)] = 15.8 \pm 0.1$.

(c) From Equation (3) in Table a1-6, we find

$$\frac{s_x}{x} = 3\left(\frac{\pm 0.03 \times 10^{-4}}{4.73 \times 10^{-4}}\right) = \pm 0.01903$$
$$s_x = \pm 0.01903 \times 1.0583 \times 10^{-10} = \pm 2.014 \times 10^{-12}$$

Thus, $[4.73(\pm 0.03) \times 10^{-4}]^3 = 1.06(\pm 0.02) \times 10^{-10}$.

The preceding examples show that rounding data when the standard deviations of the quantities are known is relatively straightforward using propagation of error mathematics. Often, however, calculations must be performed with data whose precision is indicated only by the significant-figure convention. In this case, commonsense assumptions must be made as to the uncertainty in each number. With these assumptions, the uncertainty of the final result is then estimated using Equation a1-29. Finally, the result is rounded so that it contains only significant digits. Note that calculators and computer programs generally retain several extra digits that are not significant, and the user must be careful to round final results properly so that only significant figures are included.

a1C HYPOTHESIS TESTING

Hypothesis testing is the basis for many decisions made in science and engineering. Thus, in order to explain an observation, a hypothetical model is advanced and tested experimentally. If the results from these experiments do not support the model, we reject it and seek a new hypothesis. If agreement is found, the hypothetical model serves as the basis for further experiments.

Experimental results seldom agree exactly with those predicted from a theoretical model. Consequently, scientists frequently must judge whether a numerical difference is a result of the random errors inevitable in all measurements or of systematic errors. Certain statistical tests are useful in sharpening these judgments.

Tests of this kind make use of a *null hypothesis*, which assumes that the numerical quantities being compared are, in fact, the same. We then use a probability distribution to calculate the probability that the observed differences are a result of random error. Usually, if the observed difference is greater than or equal to the difference that would occur 5 times in 100 by random chance (a significance level of 0.05), the null hypothesis is considered questionable, and the difference is judged to be significant. Other significance levels, such as 0.01 (1%) or 0.001 (0.1%), may also be adopted, depending on the certainty desired in the judgment. When expressed as a fraction, the significance level is often denoted by the symbol α. The confidence level, CL, as a percentage is related to α by CL = $(1 - \alpha) \times 100\%$.

Specific examples of hypothesis tests that chemists often use include the comparison of (1) the mean of an experimental data set with what is believed to be the true value, (2) the mean to a predicted or cutoff (threshold) value, and (3) the means or the standard deviations from two or more sets of data.

There are two contradictory outcomes that we consider in any hypothesis test. The first, the null hypothesis H_0, states that $\mu = \mu_0$, where μ is our experimental mean and μ_0 is the true value, threshold value, or another mean being compared. The second, the alternative hypothesis H_a, can be stated in several ways. We might reject the null hypothesis in favor of H_a if μ is different from μ_0 ($\mu \neq \mu_0$). In this situation, the test is called a *two-tailed test*. Other alternative hypotheses are $\mu > \mu_0$ and $\mu < \mu_0$. These are called *one-tailed tests*. For example, suppose we are interested in determining whether the concentration of lead in an industrial wastewater discharge exceeds the maximum permissible amount of 0.05 ppm ($\mu_0 = 0.05$ ppm). Our hypothesis test would be written as follows:

$$H_0: \mu = 0.05 \text{ ppm}$$
$$H_a: \mu > 0.05 \text{ ppm}$$

and the test would be one tailed.

Now suppose instead that experiments over a period of several years have determined that the mean lead level is $\mu_0 = 0.02$ ppm. Recently, changes in the industrial process have been made and we suspect that the mean lead level is now different from 0.02 ppm. Here we do not care whether it is higher or lower than 0.02 ppm. Our hypothesis test would be written as

$$H_0: \mu = 0.02 \text{ ppm}$$
$$H_a: \mu \neq 0.02 \text{ ppm}$$

and the test is a two-tailed test.

In order to carry out the statistical test, a test procedure must be implemented. The crucial elements of a test procedure are the formation of an appropriate test statistic and the identification of a rejection region. The test statistic is formulated from the data on which we will base the decision to accept or reject H_0. The rejection region consists of all the values of the test statistic for which H_0 will be rejected. The null hypothesis should be rejected if the test statistic lies within the rejection region. For tests concerning one or two means, the test statistic can be the z statistic, if we have a large number of measurements or if we know σ. Alternatively, we must use the t statistic for small numbers with unknown σ. When in doubt, the t statistic should be used.

For the t statistic, the procedure is as follows:

1. State the null hypothesis: $H_0: \mu = \mu_0$.
2. Form the test statistic: $t = \dfrac{\overline{x} - \mu_0}{s/\sqrt{N}}$.
3. State the alternative hypothesis H_a and determine the rejection region. Here the critical value of t, t_{crit} is the value found from the tables of the t statistic. For $H_a: \mu \neq \mu_0$, reject H_0 if $t \geq t_{\text{crit}}$ or if $t \leq -t_{\text{crit}}$ (two-tailed test).

   For $H_a: \mu > \mu_0$, reject H_0 if $t \geq t_{\text{crit}}$ (one-tailed test).
   For $H_a: \mu < \mu_0$, reject H_0 if $t \leq -t_{\text{crit}}$ (one-tailed test).

We can illustrate the procedure by considering a test for systematic error in an analytical method as illustrated in Figure a1-2. Determination of the analyte in this case gives an experimental mean that is an estimate of the population mean. If the analytical method had no systematic error, or bias, random errors would give the frequency distribution shown by curve A in Figure a1-2 and $\mu_A = \mu_0$, the true value. Method B has some systematic error so that $\overline{x}_B$, which estimates μ_B differs from the accepted value μ_0. The bias is given by

$$\text{Bias} = \mu_B - \mu_0 \qquad (a1\text{-}30)$$

In testing for bias, we do not know initially whether the difference between the experimental mean and the accepted value is due to random error or to an actual systematic error. The t test is used to determine the significance of the difference as illustrated in Example a1-10.

EXAMPLE a1-10

A new procedure for the rapid determination of sulfur in kerosenes was tested on a sample known from its method of preparation to contain 0.123% ($\mu_0 = 0.123\%$) S. The results were % S = 0.112, 0.118, 0.115, and 0.119. Do the data indicate that there is a bias in the method at the 95% confidence level?

Solution

The null hypothesis is H_0: $\mu = 0.123\%$ S, and the alternative hypothesis is H_a: $\mu \neq 0.123\%$ S.

$$\sum x_i = 0.112 + 0.118 + 0.115 + 0.119 = 0.464$$
$$\overline{x} = 0.464/4 = 0.116\% \text{ S}$$
$$\sum x_i^2 = 0.012544 + 0.013924 + 0.013225 + 0.014161$$
$$= 0.53854$$

$$s = \sqrt{\frac{0.053854 - (0.464)^2/4}{4 - 1}}$$
$$= \sqrt{\frac{0.000030}{3}} = 0.0032\% \text{ S}$$

The test statistic can now be calculated as

$$t = \frac{\overline{x} - \mu_0}{s/\sqrt{N}} = \frac{0.116 - 0.123}{0.032/\sqrt{4}} = -4.375$$

From Table a1-5, we find that the critical value of t for 3 degrees of freedom and the 95% confidence level is 3.18. Because $t \leq -3.18$, we conclude that there is a significant difference at the 95% confidence level and, thus, bias in the method. Note that if we were to carry out this test at the 99% confidence level, $t_{crit} = 5.84$ (Table a1-5). Because $-5.84 < -4.375$, we would accept the null hypothesis at the 99% confidence level and conclude there is no difference between the experimental and the accepted values.

Note that in this example the outcome depends on the confidence level that is used. The choice of confidence level depends on our willingness to accept an error in the outcome.[12] The significance level (0.05, 0.01, 0.001, etc.) is the probability of making an error by rejecting the null hypothesis.

Hypothesis testing is widely used in science and engineering.[13] Comparison of one or two samples is carried out as described here. The principles can, however, be extended to comparisons among more than two population means. Multiple comparisons fall under the general category of *analysis of variance* (ANOVA).[14] These methods use a single test to determine whether there is a difference among the population means rather than pairwise comparisons as is done with the t test. After ANOVA indicates a potential difference, *multiple comparison* procedures can be used to identify which specific population means differ from the others. *Experimental design methods* take advantage of ANOVA in planning and performing experiments.

a1D METHOD OF LEAST SQUARES

Most analytical methods are based on a calibration curve in which a measured quantity y is plotted as a function of the known concentration x of a series of standards. Figure a1-6 shows a typical calibration curve, which was computed for the chromatographic determination of isooctane in hydrocarbon samples. The ordinate (the dependent variable) is the area under the chromatographic peak for isooctane, and the abscissa (the independent variable) is the mole percent of isooctane. As is typical, the plot approximates a straight line. Note, however, that not all the data fall exactly on the line because of the random errors in the measurement process. Thus, we must try to find the "best" straight line through the points. *Regression analysis* provides the means for objectively obtaining such a line and also for specifying the uncertainties associated with its subsequent use. We consider here only the basic *method of least squares* for two-dimensional data.

---

[12] For a discussion of errors in hypothesis testing, see J. L. Devore, *Probability and Statistics for Engineering and the Sciences*, 6th ed., Chap. 8, Pacific Grove, CA: Duxbury Press at Brooks/Cole, 2004.

[13] See D. A. Skoog, D. M. West, F. J. Holler, and S. R. Crouch, *Fundamentals of Analytical Chemistry*, 8th ed., Chap. 7, Belmont, CA: Brooks/Cole, 2004; S. R. Crouch and F. J. Holler, *Applications of Microsoft® Excel in Analytical Chemistry*, Belmont, CA: Brooks/Cole, 2004.

[14] For a discussion of ANOVA methods, see D. A. Skoog, D. M. West, F. J. Holler, and S. R. Crouch, *Fundamentals of Analytical Chemistry*, 8th ed., Sec. 7C, Belmont, CA: Brooks/Cole, 2004.

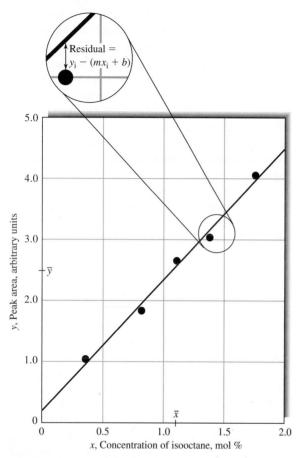

FIGURE a1-6 Calibration curve for determining isooctane in hydrocarbon mixtures.

a1D-1 Assumptions of the Least-Squares Method

Two assumptions are made in using the method of least squares. The first is that there is actually a linear relationship between the measured response y and the standard analyte concentration x. The mathematical relationship that describes this assumption is called the *regression model*, which may be represented as

$$y = mx + b$$

where b is the y intercept (the value of y when x is 0) and m is the slope of the line. We also assume that any deviation of the individual points from the straight line arises from error in the *measurement*. That is, we assume there is no error in x values of the points (concentrations). Both of these assumptions are appropriate for many analytical methods, but bear in mind that whenever there is significant uncertainty in the x data,

basic linear least-squares analysis may not give the best straight line. In such a case, a more complex *correlation analysis* may be necessary. In addition, simple least-squares analysis may not be appropriate when the uncertainties in the y values vary significantly with x. In this case, it may be necessary to apply different weighting factors to the points and perform a *weighted least-squares analysis*.

a1D-2 Finding the Least-Squares Line

As illustrated in Figure a1-6, the vertical deviation of each point from the straight line is called a *residual*. The line generated by the least-squares method is the one that minimizes the sum of the squares of the residuals for all the points. In addition to providing the best fit between the experimental points and the straight line, the method gives the standard deviations for m and b.

The least-squares method finds the sum of the squares of the residuals SS_{resid} and minimizes these according to the minimization technique of calculus.[15] The value of SS_{resid} is found from

$$SS_{resid} = \sum_{i=1}^{N} [y_i - (b + mx_i)]^2$$

where N is the number of points used. The calculation of the slope and intercept is simplified by defining three quantities S_{xx}, S_{yy}, and S_{xy} as follows:

$$S_{xx} = \sum (x_i - \bar{x})^2 = \sum x_i^2 - \frac{(\sum x_i)^2}{N} \quad \text{(a1-31)}$$

$$S_{yy} = \sum (y_i - \bar{y})^2 = \sum y_i^2 - \frac{(\sum y_i)^2}{N} \quad \text{(a1-32)}$$

$$S_{xy} = \sum (x_i - \bar{x})(y_i - \bar{y}) = \sum x_i y_i - \frac{\sum x_i \sum y_i}{N}$$
$$\text{(a1-33)}$$

where x_i and y_i are individual pairs of data for x and y, N is the number of pairs and $\bar{x}$ and $\bar{y}$ are the average values for x and y; that is, $\bar{x} = \sum x_i / N$ and $\bar{y} = \sum y_i / N$.

Note that S_{xx} and S_{yy} are the sums of the squares of the deviations from the mean for individual values of x and y. The expressions shown on the far right in Equations a1-31 through a1-33 are more convenient when a

---

[15]The procedure involves differentiating SS_{resid} with respect to first m and then b and setting the derivatives equal to 0. This yields two equations, called normal equations, in the two unknowns m and b. These are then solved to give the least-squares best estimates of these parameters.

calculator without a built-in regression function is being used.

Six useful quantities can be derived from S_{xx}, S_{yy}, and S_{xy}, as follows:

1. The slope of the line, m:

$$m = \frac{S_{xy}}{S_{xx}} \qquad (a1\text{-}34)$$

2. The intercept, b:

$$b = \bar{y} - m\bar{x} \qquad (a1\text{-}35)$$

3. The standard deviation about regression, s_r:

$$s_r = \sqrt{\frac{S_{yy} - m^2 S_{xx}}{N - 2}} \qquad (a1\text{-}36)$$

4. The standard deviation of the slope, s_m:

$$s_m = \sqrt{\frac{s_r^2}{S_{xx}}} \qquad (a1\text{-}37)$$

5. The standard deviation of the intercept, s_b:

$$s_b = s_r \sqrt{\frac{\sum x_i^2}{N \sum x_i^2 - (\sum x_i)^2}}$$

$$= s_r \sqrt{\frac{1}{N - (\sum x_i)^2 / \sum x_i^2}} \qquad (a1\text{-}38)$$

6. The standard deviation for results obtained from the calibration curve, s_c:

$$s_c = \frac{s_r}{m} \sqrt{\frac{1}{M} + \frac{1}{N} + \frac{(\bar{y}_c - \bar{y})^2}{m^2 S_{xx}}} \qquad (a1\text{-}39)$$

Equation a1-39 allows us to calculate the standard deviation from the mean $\bar{y}_c$ of a set of M replicate analyses of unknowns when a calibration curve that contains N points is used; recall that $\bar{y}$ is the mean value of y for the N calibration points. This equation is only approximate and assumes that the slope and intercept are independent parameters, which is not strictly true.

The standard deviation about regression s_r (Equation a1-36) is the standard deviation for y when the deviations are measured not from the mean of y (as is the usual case), but from the straight line that results from the least-squares prediction. The value of s_r is related to SS_{resid} by

$$s_r = \sqrt{\frac{\sum_{i=1}^{N} [y_i - (b + mx_i)]^2}{N - 2}} = \sqrt{\frac{SS_{\text{resid}}}{N - 2}}$$

In this equation, the number of degrees of freedom is $N - 2$ because one degree of freedom is lost in calculating m and one in determining b. The standard deviation about regression is often called the *standard error of the estimate*. It roughly corresponds to the size of a typical deviation from the estimated regression line. Examples a1-11 and a1-12 illustrate how these quantities are calculated and used.

EXAMPLE a1-11

Carry out a least-squares analysis of the experimental data provided in the first two columns of Table a1-7 and plotted in Figure a1-6.

Solution

Columns 3, 4, and 5 of the table contain computed values for x_i^2, y_i^2, and $x_i y_i$, with their sums appearing as the last entry in each column. Note that the number of digits carried in the computed values should be the *maximum allowed by the calculator or computer*; that is, *rounding should not be performed until the calculation is complete.*

We now substitute into Equations a1-31, a1-32, and a1-33 and obtain

$$S_{yy} = \sum y_i^2 - \frac{(\sum y_i)^2}{N}$$

$$= 36.3775 - \frac{(12.51)^2}{5} = 5.07748$$

$$S_{xy} = \sum x_i y_i - \frac{\sum x_i \sum y_i}{N}$$

$$= 15.81992 - \frac{5.365 \times 12.51}{5} = 2.39669$$

TABLE a1-7 Calibration Data for the Chromatographic Determination of Isooctane in a Hydrocarbon Mixture

Mole Percent Isooctane, x_i	Peak Area, y_i	x_i^2	y_i^2	$x_i y_i$
0.352	1.09	0.12390	1.1881	0.38368
0.803	1.78	0.64481	3.1684	1.42934
1.08	2.60	1.16640	6.7600	2.80800
1.38	3.03	1.90440	9.1809	4.18140
1.75	4.01	3.06250	16.0801	7.01750
5.365	12.51	6.90201	36.3775	15.81992

Substitution of these quantities into Equations a1-34 and a1-35 yields

$$m = \frac{2.39669}{1.14537} = 2.0925 \approx 2.09$$

$$b = \frac{12.51}{5} - 2.0925 \times \frac{5.365}{5} = 0.2567 \approx 0.26$$

Thus, the equation for the least-squares line is

$$y = 2.09x + 0.26$$

Substitution into Equation a1-36 yields the standard deviation about regression:

$$s_r = \sqrt{\frac{S_{yy} - m^2 S_{xx}}{N - 2}} = \sqrt{\frac{5.07748 - (2.0925)^2 \times 1.14537}{5 - 2}}$$

$$= 0.1442 \approx 0.14$$

and substitution into Equation a1-37 gives the standard deviation of the slope:

$$s_m = \sqrt{\frac{s_r^2}{S_{xx}}} = \sqrt{\frac{(0.1442)^2}{1.14537}} = 0.13$$

Finally, we find the standard deviation of the intercept from Equation a1-38:

$$s_b = 0.1442 \sqrt{\frac{1}{5 - (5.365)^2/6.9021}} = 0.16$$

EXAMPLE a1-12

The calibration curve found in Example a1-11 was used for the chromatographic determination of isooctane in a hydrocarbon mixture. A peak area of 2.65 was obtained. Calculate the mole percent of isooctane in the mixture and the standard deviation if the area was (a) the result of a single measurement and (b) the mean of four measurements.

▸ *Solution*

In either case, the unknown concentration is found from rearranging the least-squares equation for the line, which gives

$$x = \frac{y - b}{m} = \frac{y - 0.2567}{2.0925}$$

$$= \frac{2.65 - 0.2567}{2.0925} = 1.144 \text{ mol \%}$$

(a) Substituting into Equation a1-39, we obtain

$$s_c = \frac{0.1442}{2.0925} \sqrt{\frac{1}{1} + \frac{1}{5} + \frac{(2.65 - 12.51/5)^2}{(2.0925)^2 \times 1.145}}$$

$$= 0.076 \text{ mol \%}$$

(b) For the mean of four measurements,

$$s_c = \frac{0.1442}{2.0925} \sqrt{\frac{1}{4} + \frac{1}{5} + \frac{(2.65 - 12.51/5)^2}{(2.0925)^2 \times 1.145}}$$

$$= 0.046 \text{ mol \%}$$

Note that the advent of powerful statistics and spreadsheet software has greatly eased the burden of performing least-squares analysis of data.[16]

---

[16] See S. R. Crouch and F. J. Holler, *Applications of Microsoft® Excel in Analytical Chemistry*, Belmont, CA: Brooks/Cole, 2004.

QUESTIONS AND PROBLEMS

*a1-1 Consider the following sets of replicate measurements:

A	B	C	D
61.45	3.27	12.06	2.7
61.53	3.26	12.14	2.4
61.32	3.24		2.6
	3.24		2.9
	3.28		
	3.23		

For each set, calculate (a) the mean and decide how many degrees of freedom are associated with the calculation of $\bar{x}$; (b) the standard deviation of each set and the number of degrees of freedom associated with calculating s; (c) the coefficient of variation for each data set; and (d) the standard error of the mean of each set.

*a1-2 The accepted values for the sets of data in Problem a1-1 are as follows: A, 61.71; B, 3.28; C, 12.23; and D, 2.75. Find (a) the absolute error for the mean of each set and (b) the percent relative error for each mean.

*a1-3 A particular method for the determination of copper yields results that are low by 0.5 mg. What will be the percent relative error due to this source if the weight of copper in a sample is
(a) 25 mg?   (b) 100 mg?   (c) 250 mg?   (d) 500 mg?

*a1-4 The method described in Problem a1-3 is to be used to analyze an ore that contains about 4.8% copper. What minimum sample weight should be taken if the relative error due to a 0.5-mg loss is to be smaller than
(a) 0.1%?   (b) 0.5%?   (c) 0.8%?   (d) 1.2%?

*a1-5 A certain instrumental technique has a standard deviation of 1.0%. How many replicate measurements are necessary if the standard error of the mean is to be 0.01%?

*a1-6 A certain technique is known to have a mean of 0.500 and standard deviation of 1.84×10^{-3}. It is also known that Gaussian statistics apply. How many replicate determinations are necessary if the standard error of the mean is not to exceed 0.100%?

*a1-7 A constant solubility loss of approximately 1.8 mg is associated with a particular method for the determination of chromium in geological samples. A sample containing approximately 18% Cr was analyzed by this method. Predict the relative error (in parts per thousand) in the results due to this systematic error if the sample taken for analysis weighed 0.400 g.

*a1-8 Analysis of several plant-food preparations for potassium ion yielded the following data:

Sample	Percent K^+
1	5.15, 5.03, 5.04, 5.18, 5.20
2	7.18, 7.17, 6.97
3	4.00, 3.93, 4.15, 3.86
4	4.68, 4.85, 4.79, 4.62
5	6.04, 6.02, 5.82, 6.06, 5.88

The preparations were randomly drawn from the same population.
(a) Find the mean and standard deviation s for each sample.
(b) Obtain the pooled value s_{pooled}.
(c) Why is s_{pooled} a better estimate of σ than the standard deviation from any one sample?

*a1-9 Six bottles of wine of the same variety were analyzed for residual sugar content with the following results:

Bottle	Percent (w/v) Residual Sugar
1	0.99, 0.84, 1.02
2	1.02, 1.13, 1.17, 1.02
3	1.25, 1.32, 1.13, 1.20, 1.12
4	0.72, 0.77, 0.61, 0.58
5	0.90, 0.92, 0.73
6	0.70, 0.88, 0.72, 0.73

(a) Evaluate the standard deviation s for each set of data.

(b) Pool the data to obtain an absolute standard deviation for the method.

*a1-10 Estimate the absolute standard deviation and the coefficient of variation for the results of the following calculations. Round each result so that it contains only significant digits. The numbers in parentheses are absolute standard deviations.

(a) $y = 5.75(\pm0.03) + 0.833(\pm0.001) - 8.021(\pm0.001) = -1.4381$

(b) $y = 18.97(\pm0.04) + 0.0025(\pm0.0001) + 2.29(\pm0.08) = 21.2625$

(c) $y = 66.2(\pm0.3) \times 1.13(\pm0.02) \times 10^{-17} = 7.4806 \times 10^{-16}$

(d) $y = 251(\pm1) \times \dfrac{860(\pm2)}{1.673(\pm0.006)} = 129{,}025.70$

(e) $y = \dfrac{157(\pm6) - 59(\pm3)}{1220(\pm1) + 77(\pm8)} = 7.5559 \times 10^{-2}$

(f) $y = \dfrac{1.97(\pm0.01)}{243(\pm3)} = 8.106996 \times 10^{-3}$

*a1-11 Estimate the absolute standard deviation and the coefficient of variation for the results of the following calculations. Round each result to include only significant figures. The numbers in parentheses are absolute standard deviations.

(a) $y = 1.02(\pm0.02) \times 10^{-8} - 3.54(\pm0.2) \times 10^{-9}$

(b) $y = 90.31(\pm0.08) - 89.32(\pm0.06) + 0.200(\pm0.004)$

(c) $y = 0.0020(\pm0.0005) \times 20.20(\pm0.02) \times 300(\pm1)$

(d) $y = \dfrac{163(\pm0.03) \times 10^{-14}}{1.03(\pm0.04) \times 10^{-16}}$

(e) $y = \dfrac{100(\pm1)}{2(\pm1)}$

(f) $y = \dfrac{2.45(\pm0.02) \times 10^{-2} - 5.06(\pm0.06) \times 10^{-3}}{23.2(\pm0.7) + 9.11(\pm0.08)}$

*a1-12 Consider the following sets of replicate measurements:

A	B	C	D	E	F
3.5	70.24	0.812	2.7	70.65	0.514
3.1	70.22	0.792	3.0	70.63	0.503
3.1	70.10	0.794	2.6	70.64	0.486
3.3		0.900	2.8	70.21	0.497
2.5			3.2		0.472

Calculate the mean and the standard deviation for each of these six data sets. Calculate the 95% confidence interval for each set of data. What does this interval mean?

*a1-13 Calculate the 95% confidence interval for each set of data in Problem a1-12 if s is a good estimate of σ and has the following values: set A, 0.20; set B, 0.070; set C, 0.0090; set D, 0.30; set E, 0.15; and set F, 0.015.

*a1-14 An established method for chlorinated hydrocarbons in air samples has a standard deviation of 0.030 ppm.

(a) Calculate the 95% confidence interval for the mean of four measurements obtained by this method.

(b) How many measurements should be made if the 95% confidence interval is to be ±0.017?

***a1-15** The standard deviation in a method for the determination of carbon monoxide in automotive exhaust gases has been found, on the basis of extensive past experience, to be 0.80 ppm.

(a) Estimate the 90% confidence interval for a triplicate analysis.

(b) How many measurements would be needed for the 90% confidence interval for the set to be 0.50 ppm?

***a1-16** The certified percentage of nickel in a particular NIST reference steel sample is 1.12%, and the standard deviation associated with this number is 0.03%. A new spectrometric method for the determination of nickel produced the following percentages: 1.10, 1.08, 1.09, 1.12, 1.09. Is there an indication of bias in the method at the 95% level?

***a1-17** A titrimetric method for the determination of calcium in limestone was tested by analysis of a NIST limestone containing 30.15% CaO. The mean result of four analyses was 30.26% CaO, with a standard deviation of 0.085%. By pooling data from several analyses, it was found that $s \rightarrow \sigma = 0.094\%$ CaO.

(a) Do the data indicate the presence of a systematic error at the 95% confidence level?

(b) Would the data indicate the presence of a systematic error at the 95% confidence level if no pooled value for s had been available?

***a1-18** To test the quality of the work of a commercial laboratory, duplicate analyses of a purified benzoic acid (68.8% C, 4.953% H) sample was requested. It is assumed that the relative standard deviation of the method is $s_r \rightarrow \sigma = 4$ ppt for carbon and 6 ppt for hydrogen. The means of the reported results are 68.5% C and 4.882% H. At the 95% confidence level, is there any indication of systematic error in either analysis?

***a1-19** The diameter of a sphere has been found to be 2.15 cm, and the standard deviation associated with the mean is 0.02 cm. What is the best estimate of the volume of the sphere, and what is the standard deviation associated with the volume?

***a1-20** A given pH meter can be read with a standard deviation of ±0.01 pH units throughout the range 2 to 12. Calculate the standard deviation of $[H_3O^+]$ at each end of this range.

***a1-21** A solution is prepared by weighing 5.0000 g of compound X into a 100-mL volumetric flask. The balance used had a precision of 0.2 mg reported as a standard deviation, and the volumetric flask could be filled with a precision of 0.15 mL also reported as a standard deviation. What is the estimated standard deviation of concentration (g/mL)?

a1-22 The sulfate ion concentration in natural water can be determined by measuring the turbidity that results when an excess of $BaCl_2$ is added to a measured quantity of the sample. A turbidimeter, the instrument used for this analysis, was calibrated with a series of standard Na_2SO_4 solutions. The following data were obtained in the calibration:

mg SO_4^{2-}/L, C_x	Turbidimeter Reading, R
0.00	0.06
5.00	1.48
10.00	2.28
15.0	3.98
20.0	4.61

Assume that there is a linear relationship between the instrument reading and the concentration.
(a) Plot the data and draw a straight line through the points by eye.
(b) Compute the least-squares slope and intercept for the best straight line among the points.
(c) Compare the straight line from the relationship determined in (b) with that in (a).
(d) Calculate the standard deviation for the slope and intercept of the least-squares line.
(e) Obtain the concentration of sulfate in a sample yielding a turbidimeter reading of 3.67. Find the absolute standard deviation and the coefficient of variation.
(f) Repeat the calculations in (e) assuming that the 3.67 was the mean of six turbidimeter readings.

a1-23 The following data were obtained in calibrating a calcium ion selective electrode for the determination of pCa. A linear relationship between the potential and pCa is known to exist.

pCa = $-\log[Ca^{2+}]$	E, mV
5.00	−53.8
4.00	−27.7
3.00	+2.7
2.00	+31.9
1.00	+65.1

(a) Plot the data and draw a line through the points by eye.
(b) Find the least-squares expression for the best straight line through the points. Plot this line.
(c) Find the standard deviation for the slope and intercept of the least-squares line.
(d) Calculate the pCa of a serum solution in which the electrode potential was 10.7 mV. Find the absolute and relative standard deviations for pCa if the result was from a single voltage measurement.
(e) Find the absolute and relative standard deviations for pCa if the millivolt reading in (d) was the mean of two replicate measurements. Repeat the calculation based on the mean of eight measurements.

*a1-24 The following are relative peak areas for chromatograms of standard solutions of methyl vinyl ketone (MVK):

Concentration MVK, mmol/L	Relative Peak Area
0.500	3.76
1.50	9.16
2.50	15.03
3.50	20.42
4.50	25.33
5.50	31.97

(a) Determine the slope and intercept of the least-squares line.

(b) Plot the least-squares line as well as the experimental points.

(c) A sample containing MVK yielded a relative peak area of 10.3. Calculate the concentration of MVK in the solution.

(d) Assume that the result in (d) represents a single measurement as well as the mean of four measurements. Calculate the respective absolute and relative standard deviations.

(e) Repeat the calculations in (c) and (d) for a sample that gave a peak area of 22.8.

APPENDIX TWO ACTIVITY COEFFICIENTS

The relationship between the *activity* a_X of a species and its molar concentration [X] is given by the expression

$$a_X = \gamma_X[X] \qquad \text{(a2-1)}$$

where γ_X is a dimensionless quantity called the *activity coefficient*. The activity coefficient, and thus the activity of X, varies with the *ionic strength* of a solution such that the use of a_X instead of [X] in an electrode potential calculation, or in equilibrium calculations, renders the numerical value obtained independent of the ionic strength. Here, the ionic strength μ is defined by the equation

$$\mu = \frac{1}{2}\left(c_1 Z_1^2 + c_2 Z_2^2 + c_3 Z_3^2 + \cdots\right) \qquad \text{(a2-2)}$$

where $c_1, c_2, c_3, \ldots$ represent the molar concentration of the various ions in the solution and $Z_1, Z_2, Z_3, \ldots$ are their respective charges. Note that an ionic strength calculation requires taking account of *all* ionic species in a solution, not just the reactive ones.

EXAMPLE a2-1

Calculate the ionic strength of a solution that is 0.0100 M in $NaNO_3$ and 0.0200 M in $Mg(NO_3)_2$.

Solution

Here, we will neglect the contribution of H^+ and OH^- to the ionic strength because their concentrations are so low compared with those of the two salts. The molarities of Na^+, NO_3^-, and Mg^{2+} are 0.0100, 0.0500, and 0.0200, respectively. Then

$$c_{Na^+} \times (1)^2 = 0.0100 \times 1 = 0.0100$$
$$c_{NO_3^-} \times (1)^2 = 0.0500 \times 1 = 0.0500$$
$$c_{Mg^{2+}} \times (2)^2 = 0.0200 \times 2^2 = \underline{0.0800}$$
$$\text{Sum} = 0.1400$$

$$\mu = \frac{1}{2} \times 0.1400 = 0.0700$$

a2A PROPERTIES OF ACTIVITY COEFFICIENTS

Activity coefficients have the following properties:

1. The activity coefficient of a species can be thought of as a measure of the effectiveness with which that species influences an equilibrium in which it is a participant. In very dilute solutions, where the ionic strength is minimal, ions are sufficiently far apart that they do not influence one another's behavior. Here, the effectiveness of a common ion on the position of equilibrium becomes dependent only on its molar concentration and independent of other ions. Under these circumstances, the activity coefficient becomes equal to unity and [X] and a in Equation a2-1 are identical. As the ionic strength becomes greater, the behavior of an individual ion is influenced by its nearby neighbors. The result is a decrease in effectiveness of the ion in altering the position of chemical equilibria. Its activity coefficient then becomes less than unity. We may summarize this behavior in terms of Equation a2-1. At moderate ionic strengths, $\gamma_X < 1$; as the solution approaches infinite dilution $(\mu \to 0), \gamma_X \to 1$ and thus $a_X \to [X]$.

   At high ionic strengths, the activity coefficients for some species increase and may even become greater than 1. The behavior of such solutions is difficult to interpret; we shall confine our discussion to regions of low to moderate ionic strengths (i.e., where $\mu < 0.1$). The variation of typical activity coefficients as a function of ionic strength is shown in Figure a2-1.

2. In dilute solutions, the activity coefficient for a given species is independent of the specific nature of the electrolyte and depends only on the ionic strength.

3. For a given ionic strength, the activity coefficient of an ion departs further from unity as the charge carried by the species increases. This effect is shown in Figure a2-1. The activity coefficient of an uncharged molecule is approximately 1, regardless of ionic strength.

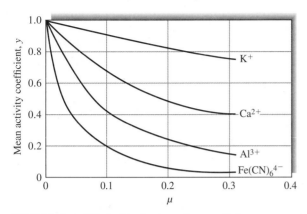

FIGURE a2-1 Effect of ionic strength on activity coefficients.

4. Activity coefficients for ions of the same charge are approximately the same at any given ionic strength. The small variations that do exist can be correlated with the effective diameter of the hydrated ions.
5. The product of the activity coefficient and molar concentration of a given ion describes its effective behavior in all equilibria in which it participates.

a2B EXPERIMENTAL EVALUATION OF ACTIVITY COEFFICIENTS

Although activity coefficients for individual ions can be calculated from theoretical considerations, their experimental measurement is, unfortunately, impossible. Instead, only a mean activity coefficient for the positively and negatively charged species in a solution can be derived.

For the electrolyte $A_m B_n$ the mean activity coefficient $\gamma_\pm$ is defined by the equation

$$\gamma_\pm = (\gamma_A^m \gamma_B^n)^{1/(m+n)}$$

The mean activity coefficient can be measured in any of several ways, but it is impossible experimentally to resolve this term into the individual activity coefficients γ_A and γ_B. For example, if $A_m B_n$ is a precipitate, we can write

$$K_{sp} = [A]^m[B]^n \gamma_A^m \gamma_B^n$$
$$= [A]^m[B]^n \gamma_\pm^{(m+n)}$$

By measuring the solubility of $A_m B_n$ in a solution in which the electrolyte concentration approaches 0 (i.e.,

where γ_A and $\gamma_B \rightarrow 1$), we could obtain K_{sp}. A second solubility measurement at some ionic strength, μ_1, would give values for [A] and [B]. These data would then permit the calculation of $\gamma_A^m \gamma_B^n = \gamma_\pm^{(m+n)}$ for ionic strength μ_1. It is important to understand that there are insufficient experimental data to permit the calculation of the *individual* quantities γ_A and γ_B, however, and that there appears to be no additional experimental information that would permit evaluation of these quantities. This situation is general; the *experimental* determination of individual activity coefficients appears to be impossible.

a2C THE DEBYE-HÜCKEL EQUATION

In 1923, P. Debye and E. Hückel derived the following theoretical expression, which permits the calculation of activity coefficients of ions:[1]

$$-\log \gamma_A = \frac{0.509 Z_A^2 \sqrt{\mu}}{1 + 3.28 \alpha_A \sqrt{\mu}} \qquad (a2\text{-}3)$$

where

γ_A = activity coefficient of the species A
Z_A = charge on the species A
μ = ionic strength of the solution
α_A = the effective diameter of the hydrated ion in nanometers

The constants 0.509 and 0.328 are applicable to solutions at 25°C; other values must be used at different temperatures.

Unfortunately, there is considerable uncertainty in the magnitude of α_A in Equation a2-3. Its value appears to be approximately 3 Å (0.3 nm) for most singly charged ions so that for these species, the denominator of the Debye-Hückel equation reduces to approximately $(1 + \sqrt{\mu})$. For ions with higher charge, α_A may be larger than 10 Å. It should be noted that the second term of the denominator becomes small with respect to the first when the ionic strength is less than 0.01; under these circumstances, uncertainties in α_A are of little significance in calculating activity coefficients.

Kielland[2] has calculated values of α_A for numerous ions from a variety of experimental data. His "best

[1] P. Debye and E. Hückel, *Physik. Z.*, 1923, *24*, 185.
[2] J. Kielland, *J. Amer. Chem. Soc.*, 1937, *59*, 1675.

TABLE a2-1 Activity Coefficients for Ions at 25°C[a]

Ion	α_X, nm	Activity Coefficient at Indicated Ionic Strength				
		0.001	0.005	0.01	0.05	0.1
H_3O^+	0.9	0.967	0.934	0.913	0.85	0.83
Li^+, $C_6H_5COO^-$	0.6	0.966	0.930	0.907	0.83	0.80
Na^+, IO_3^-, HSO_3^-, HCO_3^-, $H_2PO_4^-$, $H_2AsO_4^-$, OAc^-	0.4–0.45	0.965	0.927	0.902	0.82	0.77
OH^-, F^-, SCN^-, HS^-, ClO_3^-, ClO_4^-, BrO_3^-, IO_3^-, MnO_4^-	0.35	0.965	0.926	0.900	0.81	0.76
K^+, Cl^-, Br^-, I^-, CN^-, NO_2^-, NO_3^-, $HCOO^-$	0.3	0.965	0.925	0.899	0.81	0.75
Rb^+, Cs^+, Tl^+, Ag^+, NH_4^+	0.25	0.965	0.925	0.897	0.80	0.75
Mg^{2+}, Be^{2+}	0.8	0.872	0.756	0.690	0.52	0.44
Ca^{2+}, Cu^{2+}, Zn^{2+}, Sn^{2+}, Mn^{2+}, Fe^{2+}, Ni^{2+}, Co^{2+}, Phthalate^{2-}	0.6	0.870	0.748	0.676	0.48	0.40
Sr^{2+}, Ba^{2+}, Cd^{2+}, Hg^{2+}, S^{2-}	0.5	0.869	0.743	0.668	0.46	0.38
Pb^{2+}, CO_3^{2-}, SO_3^{2-}, $C_2O_4^{2-}$	0.45	0.868	0.741	0.665	0.45	0.36
Hg_2^{2+}, SO_4^{2-}, $S_2O_3^{2-}$, Cr_4^{2-}, HPO_4^{2-}	0.40	0.867	0.738	0.661	0.44	0.35
Al^{3+}, Fe^{3+}, Cr^{3+}, La^{3+}, Ce^{3+}	0.9	0.737	0.540	0.443	0.24	0.18
PO_4^{3-}, $Fe(CN)_6^{3-}$	0.4	0.726	0.505	0.394	0.16	0.095
Th^{4+}, Zr^{4+}, Ce^{4+}, Sn^{4+}	1.1	0.587	0.348	0.252	0.10	0.063
$Fe(CN)_6^{4-}$	0.5	0.569	0.305	0.200	0.047	0.020

[a] Calculated from Equation a2-3.

values" for effective diameters are given in Table a2-1. Also presented are activity coefficients calculated from Equation a2-3 using these values for the size parameter.

For ionic strengths up to about 0.01, activity coefficients from the Debye-Hückel equation lead to results from equilibrium calculations that agree closely with experiment; even at ionic strengths of 0.1, major discrepancies are generally not encountered. At higher ionic strengths, however, the equation fails, and experimentally determined mean activity coefficients must be used. Unfortunately, many electrochemical calculations involve solutions of high ionic strength for which no experimental activity coefficients are available. Concentrations must thus be used instead of activities; uncertainties that vary from a few percent relative to an order of magnitude may be expected.

APPENDIX THREE SOME STANDARD AND FORMAL ELECTRODE POTENTIALS

Half-Reaction	E^0, V[a]	Formal Potential, V[b]
Aluminum		
$Al^{3+} + 3e^- \rightleftharpoons Al(s)$	-1.662	
Antimony		
$Sb_2O_5(s) + 6H^+ + 4e^- \rightleftharpoons 2SbO^+ + 3H_2O$	$+0.581$	
Arsenic		
$H_3AsO_4 + 2H^+ + 2e^- \rightleftharpoons H_3AsO_3 + H_2O$	$+0.559$	0.577 in 1 M HCl, HClO$_4$
Barium		
$Ba^{2+} + 2e^- \rightleftharpoons Ba(s)$	-2.906	
Bismuth		
$BiO^+ + 2H^+ + 3e^- \rightleftharpoons Bi(s) + H_2O$	$+0.320$	
$BiCl_4^- + 3e^- \rightleftharpoons Bi(s) + 4Cl^-$	$+0.16$	
Bromine		
$Br_2(l) + 2e^- \rightleftharpoons 2Br^-$	$+1.065$	1.05 in 4 M HCl
$Br_2(aq) + 2e^- \rightleftharpoons 2Br^-$	$+1.087^c$	
$BrO_3^- + 6H^+ + 5e^- \rightleftharpoons \frac{1}{2}Br_2(l) + 3H_2O$	$+1.52$	
$BrO_3^- + 6H^+ + 6e^- \rightleftharpoons Br^- + 3H_2O$	$+1.44$	
Cadmium		
$Cd^{2+} + 2e^- \rightleftharpoons Cd(s)$	-0.403	
Calcium		
$Ca^{2+} + 2e^- \rightleftharpoons Ca(s)$	-2.866	
Carbon		
$C_6H_4O_2 \text{ (quinone)} + 2H^+ + 2e^- \rightleftharpoons C_6H_4(OH)_2$	$+0.699$	0.696 in 1 M HCl, HClO$_4$, H$_2$SO$_4$
$2CO_2(g) + 2H^+ + 2e^- \rightleftharpoons H_2C_2O_4$	-0.49	
Cerium		
$Ce^{4+} + e^- \rightleftharpoons Ce^{3+}$		$+1.70$ in 1 M HClO$_4$; $+1.61$ in 1 M HNO$_3$; $+1.44$ in 1 M H$_2$SO$_4$
Chlorine		
$Cl_2(g) + 2e^- \rightleftharpoons 2Cl^-$	$+1.359$	
$HClO + H^+ + e^- \rightleftharpoons \frac{1}{2}Cl_2(g) + H_2O$	$+1.63$	
$ClO_3^- + 6H^+ + 5e^- \rightleftharpoons \frac{1}{2}Cl_2(g) + 3H_2O$	$+1.47$	
Chromium		
$Cr^{3+} + e^- \rightleftharpoons Cr^{2+}$	-0.408	
$Cr^{3+} + 3e^- \rightleftharpoons Cr(s)$	-0.744	
$Cr_2O_7^{2-} + 14H^+ + 6e^- \rightleftharpoons 2Cr^{3+} + 7H_2O$	$+1.33$	
Cobalt		
$Co^{2+} + 2e^- \rightleftharpoons Co(s)$	-0.277	
$Co^{3+} + e^- \rightleftharpoons Co^{2+}$	$+1.808$	

Half-Reaction	E^0, V[a]	Formal Potential, V[b]
Copper		
$Cu^{2+} + 2e^- \rightleftharpoons Cu(s)$	+0.337	
$Cu^{2+} + e^- \rightleftharpoons Cu^+$	+0.153	
$Cu^+ + e^- \rightleftharpoons Cu(s)$	+0.521	
$Cu^{2+} + I^- + e^- \rightleftharpoons CuI(s)$	+0.86	
$CuI(s) + e^- \rightleftharpoons Cu(s) + I^-$	−0.185	
Fluorine		
$F_2(g) + 2H^+ + 2e^- \rightleftharpoons 2HF(aq)$	+3.06	
Hydrogen		
$2H^+ + 2e^- \rightleftharpoons H_2(g)$	0.000	−0.005 in 1 M HCl, $HClO_4$
Iodine		
$I_2(s) + 2e^- \rightleftharpoons 2I^-$	+0.5355	
$I_2(aq) + 2e^- \rightleftharpoons 2I^-$	+0.615[c]	
$I_3^- + 2e^- \rightleftharpoons 3I^-$	+0.536	
$ICl_2^- + e^- \rightleftharpoons \frac{1}{2}I_2(s) + 2Cl^-$	+1.056	
$IO_3^- + 6H^+ + 5e^- \rightleftharpoons \frac{1}{2}I_2(s) + 3H_2O$	+1.196	
$IO_3^- + 6H^+ + 5e^- \rightleftharpoons \frac{1}{2}I_2(aq) + 3H_2O$	+1.178[c]	
$IO_3^- + 2Cl^- + 6H^+ + 4e^- \rightleftharpoons ICl_2^- + 3H_2O$	+1.24	
$H_5IO_6 + H^+ + 2e^- \rightleftharpoons IO_3^- + 3H_2O$	+1.601	
Iron		
$Fe^{2+} + 2e^- \rightleftharpoons Fe(s)$	−0.440	
$Fe^{3+} + e^- \rightleftharpoons Fe^{2+}$	+0.771	0.700 in 1 M HCl; 0.732 in 1 M $HClO_4$; 0.68 in 1 M H_2SO_4
$Fe(CN)_6^{3-} + e^- \rightleftharpoons Fe(CN)_6^{4-}$	+0.36	0.71 in 1 M HCl; 0.72 in 1 M $HClO_4$, H_2SO_4
Lead		
$Pb^{2+} + 2e^- \rightleftharpoons Pb(s)$	−0.126	−0.14 in 1 M $HClO_4$; −0.29 in 1 M H_2SO_4
$PbO_2(s) + 4H^+ + 2e^- \rightleftharpoons Pb^{2+} + 2H_2O$	+1.455	
$PbSO_4(s) + 2e^- \rightleftharpoons Pb(s) + SO_4^{2-}$	−0.350	
Lithium		
$Li^+ + e^- \rightleftharpoons Li(s)$	−3.045	
Magnesium		
$Mg^{2+} + 2e^- \rightleftharpoons Mg(s)$	−2.363	
Manganese		
$Mn^{2+} + 2e^- \rightleftharpoons Mn(s)$	−1.180	
$Mn^{3+} + e^- \rightleftharpoons Mn^{2+}$		1.51 in 7.5 M H_2SO_4
$MnO_2(s) + 4H^+ + 2e^- \rightleftharpoons Mn^{2+} + 2H_2O$	+1.23	
$MnO_4^- + 8H^+ + 5e^- \rightleftharpoons Mn^{2+} + 4H_2O$	+1.51	
$MnO_4^- + 4H^+ + 3e^- \rightleftharpoons MnO_2(s) + 2H_2O$	+1.695	
$MnO_4^- + e^- \rightleftharpoons MnO_4^{2-}$	+0.564	
Mercury		
$Hg_2^{2+} + 2e^- \rightleftharpoons 2Hg(l)$	+0.788	0.274 in 1 M HCl; 0.776 in 1 M $HClO_4$; 0.674 in 1 M H_2SO_4
$2Hg^{2+} + 2e^- \rightleftharpoons Hg_2^{2+}$	+0.920	0.907 in 1 M $HClO_4$
$Hg^{2+} + 2e^- \rightleftharpoons Hg(l)$	+0.854	
$Hg_2Cl_2(s) + 2e^- \rightleftharpoons 2Hg(l) + 2Cl^-$	+0.268	0.244 in sat'd KCl; 0.282 in 1 M KCl; 0.334 in 0.1 M KCl
$Hg_2SO_4(s) + 2e^- \rightleftharpoons 2Hg(l) + SO_4^{2-}$	+0.615	

Half-Reaction	E^0, V[a]	Formal Potential, V[b]
Nickel		
$Ni^{2+} + 2e^- \rightleftharpoons Ni(s)$	-0.250	
Nitrogen		
$N_2(g) + 5H^+ + 4e^- \rightleftharpoons N_2H_5^+$	-0.23	
$HNO_2 + H^+ + e^- \rightleftharpoons NO(g) + H_2O$	$+1.00$	
$NO_3^- + 3H^+ + 2e^- \rightleftharpoons HNO_2 + H_2O$	$+0.94$	0.92 in 1 M HNO_3
Oxygen		
$H_2O_2 + 2H^+ + 2e^- \rightleftharpoons 2H_2O$	$+1.776$	
$HO_2^- + H_2O + 2e^- \rightleftharpoons 3OH^-$	$+0.88$	
$O_2(g) + 4H^+ + 4e^- \rightleftharpoons 2H_2O$	$+1.229$	
$O_2(g) + 2H^+ + 2e^- \rightleftharpoons H_2O_2$	$+0.682$	
$O_3(g) + 2H^+ + 2e^- \rightleftharpoons O_2(g) + H_2O$	$+2.07$	
Palladium		
$Pd^{2+} + 2e^- \rightleftharpoons Pd(s)$	$+0.987$	
Platinum		
$PtCl_4^{2-} + 2e^- \rightleftharpoons Pt(s) + 4Cl^-$	$+0.73$	
$PtCl_6^{2-} + 2e^- \rightleftharpoons PtCl_4^{2-} + 2Cl^-$	$+0.68$	
Potassium		
$K^+ + e^- \rightleftharpoons K(s)$	-2.925	
Selenium		
$H_2SeO_3 + 4H^+ + 4e^- \rightleftharpoons Se(s) + 3H_2O$	$+0.740$	
$SeO_4^{2-} + 4H^+ + 2e^- \rightleftharpoons H_2SeO_3 + H_2O$	$+1.15$	
Silver		
$Ag^+ + e^- \rightleftharpoons Ag(s)$	$+0.799$	0.228 in 1 M HCl; 0.792 in 1 M $HClO_4$; 0.77 in 1 M H_2SO_4
$AgBr(s) + e^- \rightleftharpoons Ag(s) + Br^-$	$+0.073$	
$AgCl(s) + e^- \rightleftharpoons Ag(s) + Cl^-$	$+0.222$	0.228 in 1 M KCl
$Ag(CN)_2^- + e^- \rightleftharpoons Ag(s) + 2CN^-$	-0.31	
$Ag_2CrO_4(s) + 2e^- \rightleftharpoons 2Ag(s) + CrO_4^{2-}$	$+0.446$	
$AgI(s) + e^- \rightleftharpoons Ag(s) + I^-$	-0.151	
$Ag(S_2O_3)_2^{3-} + e^- \rightleftharpoons Ag(s) + 2S_2O_3^{2-}$	$+0.017$	
Sodium		
$Na^+ + e^- \rightleftharpoons Na(s)$	-2.714	
Sulfur		
$S(s) + 2H^+ + 2e^- \rightleftharpoons H_2S(g)$	$+0.141$	
$H_2SO_3 + 4H^+ + 4e^- \rightleftharpoons S(s) + 3H_2O$	$+0.450$	
$SO_4^{2-} + 4H^+ + 2e^- \rightleftharpoons H_2SO_3 + H_2O$	$+0.172$	
$S_4O_6^{2-} + 2e^- \rightleftharpoons 2S_2O_3^{2-}$	$+0.08$	
$S_2O_8^{2-} + 2e^- \rightleftharpoons 2SO_4^{2-}$	$+2.01$	
Thallium		
$Tl^+ + e^- \rightleftharpoons Tl(s)$	-0.336	-0.551 in 1 M HCl; -0.33 in 1 M $HClO_4$, H_2SO_4
$Tl^{3+} + 2e^- \rightleftharpoons Tl^+$	$+1.25$	0.77 in 1 M HCl
Tin		
$Sn^{2+} + 2e^- \rightleftharpoons Sn(s)$	-0.136	-0.16 in 1 M $HClO_4$
$Sn^{4+} + 2e^- \rightleftharpoons Sn^{2+}$	$+0.154$	0.14 in 1 M HCl

Half-Reaction	E^0, V[a]	Formal Potential, V[b]
Titanium		
$Ti^{3+} + e^- \rightleftharpoons Ti^{2+}$	-0.369	
$TiO^{2+} + 2H^+ + e^- \rightleftharpoons Ti^{3+} + H_2O$	$+0.099$	0.04 in 1 M H_2SO_4
Uranium		
$UO_2^{2+} + 4H^+ + 2e^- \rightleftharpoons U^{4+} + 2H_2O$	$+0.334$	
Vanadium		
$V^{3+} + e^- \rightleftharpoons V^{2+}$	-0.256	0.21 in 1 M $HClO_4$
$VO^{2+} + 2H^+ + e^- \rightleftharpoons V^{3+} + H_2O$	$+0.359$	
$V(OH)_4^+ + 2H^+ + e^- \rightleftharpoons VO^{2+} + 3H_2O$	$+1.00$	1.02 in 1 M HCl, $HClO_4$
Zinc		
$Zn^{2+} + 2e^- \rightleftharpoons Zn(s)$	-0.763	

[a] G. Milazzo, S. Caroli, and V. K. Sharma, *Tables of Standard Electrode Potentials*, London: Wiley, 1978.

[b] E. H. Swift and E. A. Butler, *Quantitative Measurements and Chemical Equilibria*, New York: Freeman, 1972.

[c] These potentials are hypothetical because they correspond to solutions that are 1.00 M in Br_2 or I_2. The solubilities of these two compounds at 25°C are 0.18 M and 0.0020 M, respectively. In saturated solutions containing an excess of $Br_2(l)$ or $I_2(s)$, the standard potentials for the half-reaction $Br_2(l) + 2e^- \rightleftharpoons 2Br^-$ or $I_2(s) + 2e^- \rightleftharpoons 2I^-$ should be used. In contrast, at Br_2 and I_2 concentrations less than saturation, these hypothetical electrode potentials should be employed.

APPENDIX FOUR COMPOUNDS RECOMMENDED FOR THE PREPARATION OF STANDARD SOLUTIONS OF SOME COMMON ELEMENTS[a]

Element	Compound	FW	Solvent[b]	Notes
Aluminum	Al metal	26.98	Hot dil HCl	a
Antimony	$KSbOC_4H_4O_6 \cdot \frac{1}{2}H_2O$	333.93	H_2O	c
Arsenic	As_2O_3	197.84	dil HCl	i,b,d
Barium	$BaCO_3$	197.35	dil HCl	
Bismuth	Bi_2O_3	465.96	HNO_3	
Boron	H_3BO_3	61.83	H_2O	d,e
Bromine	KBr	119.01	H_2O	a
Cadmium	CdO	128.40	HNO_3	
Calcium	$CaCO_3$	100.09	dil HCl	i
Cerium	$(NH_4)_2Ce(NO_3)_6$	548.23	H_2SO_4	
Chromium	$K_2Cr_2O_7$	294.19	H_2O	i,d
Cobalt	Co metal	58.93	HNO_3	a
Copper	Cu metal	63.55	dil HNO_3	a
Fluorine	NaF	41.99	H_2O	b
Iodine	KIO_3	214.00	H_2O	i
Iron	Fe metal	55.85	HCl, hot	a
Lanthanum	La_2O_3	325.82	HCl, hot	f
Lead	$Pb(NO_3)_2$	331.20	H_2O	a
Lithium	Li_2CO_3	73.89	HCl	a
Magnesium	MgO	40.31	HCl	
Manganese	$MnSO_4 \cdot H_2O$	169.01	H_2O	g
Mercury	$HgCl_2$	271.50	H_2O	b
Molybdenum	MoO_3	143.94	1 M NaOH	
Nickel	Ni metal	58.70	HNO_3, hot	a
Phosphorus	KH_2PO_4	136.09	H_2O	
Potassium	KCl	74.56	H_2O	a
	$KHC_8H_4O_4$	204.23	H_2O	i,d
	$K_2Cr_2O_7$	294.19	H_2O	i,d
Silicon	Si metal	28.09	NaOH, concd	
	SiO_2	60.08	HF	j
Silver	$AgNO_3$	169.87	H_2O	a
Sodium	NaCl	58.44	H_2O	i
	$Na_2C_2O_4$	134.00	H_2O	i,d
Strontium	$SrCO_3$	147.63	HCl	a
Sulfur	K_2SO_4	174.27	H_2O	

Tin	Sn metal	118.69	HCl	
Titanium	Ti metal	47.90	H_2SO_4, 1:1	a
Tungsten	$Na_2WO_4 \cdot 2H_2O$	329.86	H_2O	h
Uranium	U_3O_8	842.09	HNO_3	d
Vanadium	V_2O_5	181.88	HCl, hot	
Zinc	ZnO	81.37	HCl	a

[a] The data in this table were taken from a more complete list assembled by B. W. Smith and M. L. Parsons, *J. Chem. Educ.*, **1973**, *50*, 679. Unless otherwise specified, compounds should be dried to constant weight at 110°C.

[b] Unless otherwise specified, acids are concentrated analytical grade.

a Approaches primary standard quality.

b Highly toxic.

c Loses $\frac{1}{2} H_2O$ at 110°C. After drying, molar mass = 324.92. The dried compound should be weighed quickly after removal from the desiccator.

d Available as a primary standard from the National Institute of Standards and Technology.

e H_3BO_3 should be weighed directly from the bottle. It loses $1 H_2O$ at 100°C and is difficult to dry to constant weight.

f Absorbs CO_2 and H_2O. Should be ignited just before use.

g May be dried at 110°C without loss of water.

h Loses both waters at 110°C, molar mass = 293.82. Keep in desiccator after drying.

i Primary standard.

j HF highly toxic and dissolves glass.

ANSWERS TO SELECTED PROBLEMS

Chapter 1

1-9 9.00×10^{-4} M

1-10   **(a)** $m = 0.0701$; $b = 0.0083$
     (b) $s_m = 0.0007$; $s_b = 0.0040$
     (c) 95% CI for $m = 0.0701 \pm 2.78 \times 0.0007 = 0.070 \pm 0.002$
         95% CI for $b = 0.0083 \pm 2.78 \times 0.004 = 0.08 \pm 0.01$
     (d) 4.87 ± 0.09 mM

1-11   **(b)** 0.410 µg/mL   **(c)** $S = 3.16V_s + 3.25$
     (d) 0.410 µg/mL

Chapter 2

2-1   **(a)** $R_1 = 500\ \Omega$; $R_2 = 2.0$ kΩ; $R_3 = 2.5$ kΩ
     (b) 5.0 V   **(c)** 0.002 A (2.0 mA)
     (d) 0.02 W

2-2   **(a)** 4.4 V   **(b)** 0.039 W
     (c) 30%

2-3   **(a)** -15% **(b)** -1.7%
     (c) -0.17%

2-4   **(a)** 74 kΩ   **(b)** 740 kΩ

2-5   **(a)** $V_1 = 1.21$ V; $V_2 = V_3 = 1.73$ V; $V_4 = 12.1$ V
     (b) $I_1 = I_5 = 1.21 \times 10^{-2}$ A; $I_2 = 3.5 \times 10^{-3}$ A; $I_3 = I_4 = 8.6 \times 10^{-3}$ A
     (c) $P = 1.5 \times 10^{-2}$ W   **(d)** 13.8 V

2-6   **(a)** 0.085 W   **(b)** 8.0×10^{-3} A (8.0 mA)
     (c) 8.0 V   **(d)** 5.3 V
     (e) 16 V

2-7   0.535 V

2-10   **(a)** 20 kΩ   **(b)** 2 MΩ

2-13

t, s	i, µA	t, s	i, µA
0.00	2.40	1.0	1.46
0.010	2.39	10	0.0162
0.10	2.28		

2-14   **(a)** 0.69 s   **(b)** 0.069 s
     (c) 6.9×10^{-5} s

2-15   **(a)** 0.15 s   **(b)** 0.015 s
     (c) 1.5×10^{-5} s

Chapter 3

3-1   **(a)** $v_+ > v_-$ by 65 µV for + limit and $v_- > v_+$ by 75 µV for − limit

(b) $v_+ > v_-$ by 26 µV for + limit and $v_- > v_+$ by 28 µV for − limit
(c) $v_+ > v_-$ 8.7 µV for + limit and $v_- > v_+$ by 9.3 µV for − limit

3-2 1.0×10^4 or 80 dB

3-3 2.6×10^6

3-4   **(a)** 0.001%   **(b)** $1.0 \times 10^{-6}\%$

3-6 $R_1 = 2.9$ kΩ; $R_2 = 7.1$ kΩ

3-11 $t_r = 6.7$ ns; $\Delta v/\Delta t = 1500$ V/µs

3-16   **(a)** $v_o = \dfrac{v_1 R_{f1} R_{f2}}{R_1 R_4} + \dfrac{v_2 R_{f1} R_{f2}}{R_2 R_4} - \dfrac{v_3 R_{f2}}{R_3}$
     (b) $v_o = v_1 + 4v_2 - 40v_3$

3-21   34.0 cm from common end

Chapter 4

4-1   **(a)** 11000   **(b)** 1011011
     (c) 10000111   **(d)** 110001100

4-2   **(a)** 0010 1000   **(b)** 1001 0001
     (c) 0001 0011 0101   **(d)** 0011 1001 0110

4-4   **(a)** 5   **(b)** 21
     (c) 117   **(d)** 859

4-5   **(a)** 4   **(b)** 89
     (c) 347   **(d)** 968

4-7   **(a)** $1111_2 = 15_{10}$ **(b)** $101110010_2 = 370_{10}$
     (c) $111111_2 = 63_{10}$ **(d)** $11000_2 = 24_{10}$

4-8   **(a)** 0.039 V   **(b)** 0.0024 V
     (c) 0.00015 V

4-11   **(a)** 1 Hz   **(b)** 20 Hz

4-12   62.5 kHz

Chapter 5

5-7   **(a)** $S/N = 358$ **(b)** $n = 18$

5-8   **(a)** $S/N = 5.3$ **(b)** $n = 29$

5-9 1.28×10^{-4} V; noise reduced by a factor of 100

5-10 $n = 100$

5-11 $(S/N)_{50} = 7.1(S/N)_1$; $(S/N)_{200} = 14.1(S/N)_1$

5-12 $(S/N)_D = 3.9(S/N)_A$

Chapter 6

6-2 $v = 4.80 \times 10^{17}$ Hz; $E = 3.18 \times 10^{-16}$ J; $E = 1.99 \times 10^3$ eV

6-3 $v = 8.524 \times 10^{13}$ s^{-1}; $\bar{v} = 2843$ cm^{-1}; $E = 5.648 \times 10^{-20}$ J

6-4 $\lambda = 81.5$ cm; $E = 5.40 \times 10^{-26}$ J

6-5 $v_{species} = 2.75 \times 10^{10}$ cm s^{-1}; $v = 5.09 \times 10^{14}$ Hz; $\lambda_{species} = 540$ nm

6-6   2.42

6-7   260 nm

6-8   469 nm

6-9   **(a)** 6.02×10^{-20} J   **(b)** 3.64×10^5 m s^{-1}

6-10   1.8 μm

6-11   **(a)** 393 nm   **(b)** 448 nm

6-12   17.3%

6-14   **(a)** 52.7%;   **(b)** 3.17%;   **(c)** 91.4%

6-15   **(a)** 0.524;   **(b)** 0.065;   **(c)** 1.53

6-16   **(a)** 72.6%;   **(b)** 17.8%;   **(c)** 95.6%

6-17   **(a)** 0.825;   **(b)** 0.366;   **(c)** 1.83

6-18 4.98×10^{-3} M

6-19 1.35×10^{-4} M

Chapter 7

7-3   **(a)** 580 nm   **(b)** 967 nm
(c) 1.93 μm

7-4   **(a)** 3.56×10^7 W m^{-2} **(b)** 4.61×10^6 W m^{-2}
(c) 2.88×10^5 W m^{-2}

7-5   **(a)** 1010 nm and 829 nm
(b) 3.86×10^2 W cm^{-2} and 8.54×10^2 W cm^{-2}

7-9   **(a)** 2.43 μm   **(b)** 3.48 μm, 2.17 μm, etc.

7-12   1985 lines/mm

7-13 $\lambda/\Delta\lambda = 1260$; $\Delta\bar{v} = 0.95$ cm^{-1}

7-16   3.26

7-18   **(a)** 4.50×10^4
(b) 0.42 nm/mm and 0.21 nm/mm

7-22   **(a)** 1.6×10^5 Hz   **(b)** 9.6×10^4 Hz
(c) 1.0×10^4 Hz   **(d)** 2.2×10^3 Hz

7-23   **(a)** 2.5 cm   **(b)** 0.075 cm

Chapter 8

8-8   **(a)** 5.7×10^{-3} nm (0.057 Å)
(b) 6.9×10^{-3} nm (0.069 Å)

8-9   for Na and Mg$^+$, respectively, N_j/N_0
(a) 3.8×10^{-6} and 1.2×10^{-12}
(b) 7.6×10^{-4} and 8.0×10^{-8}
(c) 0.10 and 2.5×10^{-3}

Chapter 9

9-11   104 mm

9-12

	Height	T	$N_j/N_0 \times 10^4$	I_x/I_y
(a)	2.0	1973	2.22	1.00
(b)	3.0	2136	4.58	2.06
(c)	4.0	2092	3.81	1.72
(d)	5.0	1998	2.50	1.13

9-14

	Na	Mg$^+$
(a)	2.7×10^{-5}	6.9×10^{-11}
(b)	6.6×10^{-4}	5.9×10^{-8}
(c)	0.051	5.7×10^{-4}

9-15   **(a)** 1.2×10^{-5} **(b)** 1.8×10^{-2}

9-16 0.046 ± 0.009 μg Pb/mL

9-20   0.297 ppm Pb

Chapter 10

10-4   **(a)** 2.9 Å/mm   **(b)** 0.97 Å/mm

Chapter 12

12-1   0.138 Å

12-2

	V_{min} for Kβ_1, kV	V_{min} for Lβ_1, kV
(a)	112	17.2
(b)	3.59	No line
(c)	15.0	1.75
(d)	67.4	9.67

12-5 1.69×10^{-3} cm

12-6   4.92 cm^2/g

12-7 4.82×10^{-3} cm

12-8   **(a)** 1.39%   **(b)** 0.262%

12-9   for Fe, Se, and Ag, respectively,
(a) 80.9°, 42.9°, and 21.1°
(b) 51.8°, 28.5°, and 14.2°
(c) 36.4°, 20.3°, and 10.1°

12-11   **(a)** 4.05 kV   **(b)** 1.32 kV
(c) 20.9 kV   **(d)** 25.0 kV

Chapter 13

13-1   **(a)** 91.6%   **(b)** 11.0%
(c) 39.9%   **(d)** 57.4%
(e) 36.7%   **(f)** 20.3%

13-2   **(a)** 0.801   **(b)** 0.308
(c) 0.405   **(d)** 0.623
(e) 1.07   **(f)** 1.27

13-3 (a) 95.7% (b) 33.2%
(c) 63.2% (d) 75.8%
(e) 60.6% (f) 45.1%

13-4 (a) 0.500 (b) 0.007
(c) 0.103 (d) 0.322
(e) 0.770 (f) 0.968

13-5 1.80×10^4 L mol^{-1} cm^{-1}

13-7 (a) 0.353 (b) 44.4%
(c) 1.52×10^{-5} M

13-8 (a) 0.244 (b) 0.305
(c) 57.0% and 49.5% (d) 0.545

13-9 0.124

13-15 (a) 51.4% T and $A = 0.289$
(b) 0.717 (c) 0.264

Chapter 14

14-1 21.1 ppm

14-2 0.0684%

14-6 3.89×10^{-5} M

14-7 0.0149%

14-8 (a) $c_{Ni} = 6.75 \times 10^{-5}$ M; $c_{Co} = 5.90 \times 10^{-5}$ M
(b) $c_{Co} = 1.88 \times 10^{-4}$ M; $c_{Ni} = 3.99 \times 10^{-5}$ M

14-9 (a) $c_A = 3.95 \times 10^{-5}$ M; $c_B = 5.69 \times 10^{-6}$ M
(b) $c_A = 2.98 \times 10^{-5}$ M; $c_B = 1.23 \times 10^{-6}$ M

14-10 (a) At 485 nm, $\varepsilon_{In} = 150$; $\varepsilon_{HIn} = 974$.
At 625 nm, $\varepsilon_{In} = 1808$; $\varepsilon_{HIn} = 362$.
(b) $K_a = 1.86 \times 10^{-6}$ (c) pH = 4.58
(d) $K_a = 1.81 \times 10^{-6}$
(e) $A_{485} = 0.131$; $A_{625} = 0.391$

14-14 $c_{Al} = 0.57$ μM

14-15 [tryptamine] = 4.5×10^{-2} M

14-21 $K_f = 7.16 \times 10^9$

14-22 $K_f = 3.28 \times 10^5$

Chapter 15

15-7 (b) $F = 22.35c + 3.57 \times 10^{-4}$
(d) 0.544 μmol/L (e) 0.015%
(f) 0.010%

15-9 490 mg

15-10 3.43%

15-11 1.35×10^{-5} M

Chapter 16

16-1 (a) 1.90×10^3 N/m (b) 2083 cm^{-1}

16-2 (a) 4.81×10^2 N/m (b) 2075 cm^{-1}

16-4 1.4×10^4 cm^{-1} or 0.70 μm

16-5 1.3×10^4 cm^{-1} or 0.75 μm

16-6 three vibrational modes and three absorption bands

16-7 (a) inactive (b) active
(c) active (d) active
(e) inactive (f) active
(g) inactive

16-10 (a) $N_1/N_0 = 8.9 \times 10^{-7}$ (b) $N_2/N_0 = 7.9 \times 10^{-13}$

16-13 256 interferograms

16-14 (a) 3.20 kHz (b) 3.27 kHz
(c) 3.33 kHz

Chapter 17

17-2 vinyl alcohol, $CH_2\!=\!CH\!-\!CH_2\!-\!OH$

17-4 acrolein, $CH_2\!=\!CH\!-\!CHO$ with H_2O contaminant

17-5 propanenitrile, $CH_3CH_2C\!\equiv\!N$

17-9 8.9×10^{-3} cm

17-10 0.022 cm

17-11 7.9×10^{-3} cm

Chapter 18

18-3

$\Delta\bar{v}$, cm^{-1}	(a) $\lambda_{ex} = 632.8$ nm		(b) $\lambda_{ex} = 488.0$ nm	
	λ_{st}	λ_{as}	λ_{st}	λ_{as}
218	641.7	624.2	493.2	482.9
314	645.6	620.5	495.6	480.6
459	651.7	614.9	499.2	477.3
762	664.9	603.7	506.8	470.5
790	666.1	620.7	507.6	469.9

18-4 (a) 2.83
(b) Detector efficiency is wavelength dependent.

18-5 (a) 0.342 (20°C) and 0.367 (40°C)
(b) 0.105 and 0.121 (c) 0.0206 and 0.0264

18-7 (a) $\dfrac{I_\perp}{I_\parallel} = 0.77$ (b) $\dfrac{I_\perp}{I_\parallel} = 0.012$ polarized
(c) $\dfrac{I_\perp}{I_\parallel} = 0.076$ polarized (d) $\dfrac{I_\perp}{I_\parallel} = 0.76$

Chapter 19

19-7 (a) 301 MHz (b) 75.5 MHz
(c) 283 MHz (d) 121 MHz

19-8 (a) 60 MHz (b) 200 MHz
(c) 301 MHz (d) 499 MHz

19-9 (a) 299 Hz (b) 450 Hz
(c) 1200 Hz
(d) $\delta = 1.5$ independent of field strength

19-11 $N_j/N_0 = 0.9999878$

19-17 $N_j/N_0 = 0.99998$

Chapter 20

20-5 (a) 0.126–0.498 T (b) 3000–192 V

20-6 5.97 V

20-7   44.3 μs

20-11   **(a)** 2.22×10^3 **(b)** 770
  (c) 7.09×10^4 **(d)** 4.15×10^3

20-13   **(a)** $(M + 2)^+/M^+ = 1.96$
    $(M + 4)^+/M^+ = 0.96$
  (b) $(M + 2)^+/M^+ = 1.30$
    $(M + 4)^+/M^+ = 0.32$
  (c) $(M + 2)^+/M^+ = 0.65$
    $(M + 4)^+/M^+ = 0.106$

Chapter 21

21-4   **(a)** 165.4 eV   **(b)** SO_3^{2+}
  (c) 1306.6 eV   **(d)** 1073.5 eV

21-5   **(a)** 406.3 eV   **(b)** 819.5 eV
  (c) By obtaining the peak with sources of differ-
    ing energy, such as the Al and Mg X-ray tubes.
    Auger peaks do not change with the two sources,
    whereas XPS peaks do.
  (d) 403.4 eV

Chapter 22

22-1   **(a)** 0.705 V   **(b)** 0.642 V
  (c) 0.141 V

22-2   **(a)** 0.015 V   **(b)** 0.826 V
  (c) 0.484 V

22-3   **(a1)** −0.101 V   **(a2)** −0.104 V
  (b1) 0.771 V   **(b2)** 0.750 V

22-4   **(a1)** 0.163 V   **(a2)** 0.159 V
  (b1) 0.163 V   **(b2)** 0.140 V

22-5   **(a)** −0.043 V   **(b)** −0.10 V
  (c) 0.694 V   **(d)** 0.178 V

22-6   **(a)** 1.20 V   **(b)** 0.307 V

22-7   **(a)** 0.326 V   spontaneous as written
  (b) −0.685 V   not spontaneous as written
  (c) −0.867 V   not spontaneous as written

22-8   **(a)** −0.357 V   not spontaneous as written
  (b) 1.23 V   spontaneous as written
  (c) 0.571 V   spontaneous as written

22-9   −0.90 V

22-10   −0.367 V

22-11   −0.037 V

22-12   −1.92 V

22-13 1.85×10^{-4}

22-14 1.0×10^{-15}

22-15   0.268 V

22-16   −0.644 V

22-17   0.24 V

Chapter 23

23-13   **(a)** 0.540 V   **(b)** −1.085 V
  (c) −0.583 V   **(d)** 0.482 V

23-14   **(a)** 0.031 V
  (b) SCE||CuBr(*sat'd*), Br⁻(*x*M)|Cu
  (c) pBr = $(E_{cell} + 0.213)/0.0592$
  (d) 1.99

23-15   **(a)** 0.366 V
  (b) SCE||Ag₃AsO₄(*sat'd*), AsO₄³⁻(*x*M)|Ag
  (c) $pAsO_4 = \dfrac{(E_{cell} - 0.122) \times 3}{0.0592}$
  (d) 6.33

23-16   6.22

23-17   5.61

23-18   0.604 V

23-19 9.8×10^6

23-20   **(a)** 5.21 and 6.18×10^{-6}
  (b) 4.07 and 8.51×10^{-5}
  (c) for (a), 5.19–5.23; for (b), 4.05–4.09

23-21   3.77

23-22 2.75×10^{-16}

23-23 1.69×10^{-5}

23-25   **(a)** 5.4
  (b) $3.64 \times 10^{-6} - 4.97 \times 10^{-6}$
  (c) −23.3% and 16.7%

23-26 4.28×10^{-4}%

Chapter 24

24-1   **(a)** −1.342 V   **(b)** 0.238 V

24-2   **(a)** −0.050 V   **(b)** −0.079 V
  (c) −0.199 V   **(d)** −0.189 V
  (e) −0.043 V

24-5   **(a)** 0.432 V   **(b)** 0.266 V
  (c) −0.101 V

24-6   **(a)** 19.6 min   **(b)** 6.55 min

24-7   **(a)** 7.83 min   **(b)** 5.22 min
  (c) 2.61 min

24-9   0.731%

24-10   665 g/equiv

24-11   33.7 μg

Chapter 25

25-10   **(a)** −0.059 V   **(b)** +0.059 V

25-11   6.7 mV/s

25-13 2.23×10^{-4} M

Chapter 26

26-14   **(a)** $N_A = 2775$ **(b)** $\overline{N} = 2533$ or 2500
    $N_B = 2472$ $s = 200$
    $N_C = 2364$
    $N_D = 2523$
  (c) $H = 0.0097$ cm

26-15 **(a)** $k_A = 0.74$
 $k_B = 3.3$
 $k_C = 3.5$
 $k_D = 6.0$
 (b) $K_A = 6.2$
 $K_B = 27$
 $K_C = 30$
 $K_D = 50$

26-16 **(a)** $R_s = 0.72$ **(b)** $\alpha_{C,B} = 1.08$
 (c) $L = 108$ cm   **(d)** $(t_R)_2 = 61.7$ or 62 min

26-17 **(a)** $R_s = 5.2$ **(b)** $L = 2.0$ cm

26-18 **(a)** $\overline{N} = 2.7 \times 10^3$ **(b)** $s = 100$
 (c) $H = 0.015$

26-19 **(a)** $R_s = 1.1$ **(b)** $R_s = 2.7$
 (c) $R_s = 3.7$

26-20 **(a)** $N = 4.7 \times 10^3$ **(b)** $L = 69$ cm
 (c) $t_R = 19$ min

26-21 **(a)** $k_1 = 4.3$; $k_2 = 4.7$; $k_3 = 6.1$
 (b) $K_1 = 14$; $K_2 = 15$; $K_3 = 19$
 (c) $\alpha_{2,1} = 1.11$; $\alpha_{3,2} = 1.28$

26-22 dehydroepiandrosterone = 14.99%; estradiol = 18.10%; estrone = 27.40%; testosterone = 22.99%; estriol = 16.52%

Chapter 27

27-21 558

27-22 **(a)** 30.2 mL/min
 (b) 8.3 mL; 54.7 mL; 115.0 mL; 219.1 mL
 (c) 24.1 mL/g; 55.5 mL/g; 109.6 mL/g
 (d) $K_1 = 33.8$; $K_2 = 77.8$; $K_3 = 154$
 (e) $t'_R = 35.83$ min; $V_R^0 = 990$ mL

27-23 **(a)** $k_1 = 5.6$; $k_2 = 13$; $k_3 = 25$
 (b) $\alpha_{2,1} = 2.3$; $\alpha_{3,2} = 2.0$
 (c) $\overline{N} = 1.7 \times 10^3$ plates; $H = 0.064$ cm
 (d) $(R_s)_{2,1} = 7.5$; $(R_s)_{3,2} = 6.4$

Chapter 28

28-11 **(a)** ethyl acetate, dimethylamine, acetic acid
 (b) hexane, propylene, benzene, dichlorobenzene

28-20

R_s	N
0.50	5,476
0.75	12,321
0.90	17,742
1.0	21,904
1.10	26,504
1.25	34,225
1.50	49,284
1.75	67,081
2.0	87,616
2.5	136,900

If the second peak were twice as broad as the first, R_s and N would be smaller.

28-21

Time, h	Concentration of Ibuprofen, mg/mL	% Change/0.5 h
0		
0.5	14.05	
1.0	12.28	12.61
1.5	7.79	36.53
2.0	5.62	27.85
3.0	3.34	20.29
4.0	2.86	7.17
6.0	2.43	3.76
8.0	1.90	5.42

From the table, the largest percentage loss occurs between 1.0 and 1.5 h.

28-22 **(a)** $k_1 = 26.7$
 (b) 71% $CHCl_3$ and 29% n-hexane

Chapter 30

30-5 **(a)** 1.1×10^5 **(b)** 2.2×10^5 **(c)** 6.6×10^5

30-6 3.9 min

Chapter 31

31-2 %Ca = 18.1
 %Ba = 22.4

Chapter 32

32-4 5.95 days

32-5 **(a)** 10 counts or 10.0%
 (b) 27.4 counts or 3.65%
 (c) 83.7 counts or 1.20%
 (d) 141 counts or 0.71%

32-6 **(a)** 64 counts or 5.8%   **(b)** 55 counts or 6.9%
 (c) 73 counts or 9.1%

32-7 **(a)** 443 cpm   **(b)** 5.79 cpm
 (c) 95% CI = 443 ± 11 cpm

32-8 **(a)** 708 counts   **(b)** 379 counts

32-9 18.4 h

32-10 14.5 h

32-13 560 mL

32-14 5.31 mg

32-15 36 g

32-17 3.46 ppm

Chapter 34

34-10 11.4 nm; the diameter of a hypothetical sphere with the same translational diffusion coefficient

34-11 1.22×10^{-10} cm²/s

34-12 1.92×10^{-10} cm/s; 5.2×10^9 s

34-13 70.4 min; 8952 G

34-14 145 nm

Appendix 1

a1-1

		A	B	C	D
(a)	$\bar{x}$	61.43	3.25	12.10	2.65
	df	3	6	2	4
(b)	s	0.11	0.02	0.06	0.21
	df	2	5	1	3
(c)	CV	0.17%	0.60%	0.47%	7.9%
(d)	s_m	0.061	0.008	0.040	0.10

a1-2

		A	B	C	D
(a)	Absolute error	−0.28	−0.03	−0.13	−0.10
(b)	Relative error	−0.45%	−0.82%	−1.1%	−3.8%

a1-3 (a) −2.0% (b) −0.5%
(c) −0.2% (d) −0.1%

a1-4 (a) 10 g (b) 2.1 g
(c) 1.3 g (d) 0.87 g

a1-5 $N = 1 \times 10^4$

a1-6 $N = 13.5$ (14 replicates)

a1-7 −25 ppt

a1-8 (a)

Sample	Mean	Standard Deviation
1	5.12	0.08
2	7.11	0.12
3	3.99	0.12
4	4.74	0.10
5	5.96	0.11

(b) $s_{pooled} = 0.11\%$
(c) s_{pooled} is a weighted average of the individual estimates of σ. It uses all the data from the five samples. The reliability of s improves with the number of results.

a1-9 (a)

Bottle	Standard Deviation
1	0.096
2	0.077
3	0.084
4	0.090
5	0.104
6	0.083

(b) $s_{pooled} = 0.088$

a1-10

	s_y	CV	y
(a)	0.030	−2%	$1.44(\pm0.03)$
(b)	0.089	0.42%	$21.26(\pm0.09)$
(c)	0.14×10^{-16}	1.8%	$7.5(\pm0.1) \times 10^{-16}$
(d)	750	0.58%	$1.290(\pm0.008) \times 10^5$
(e)	0.005	6.9%	$7.6(\pm0.5) \times 10^{-2}$
(f)	1.1×10^{-4}	1.34%	$8.1(\pm0.1) \times 10^{-3}$

a1-11

	s_y	CV	y
(a)	3×10^{-10}	−4%	$6.7(\pm0.3) \times 10^{-9}$
(b)	0.1	8.4%	$1.2(\pm0.1)$
(c)	3	25%	$12(\pm3)$
(d)	6.8	4.3%	$158(\pm7)$
(e)	25	50%	$50(\pm25)$
(f)	1.5×10^{-5}	2.4%	$6.0(\pm0.2) \times 10^{-4}$

a1-12

	A	B	C	D	E	F
$\bar{x}$	3.1	70.19	0.82	2.86	70.53	0.494
s	0.37	0.08	0.05	0.24	0.22	0.016
CI	±0.46	±0.20	±0.08	±0.30	±0.34	±0.020

The 95% confidence interval establishes the range about the mean that the true value should lie 95% of the time if the errors are random.

a1-13 Set A, 95% CI = ±0.18; set B, 95% CI = ±0.079; set C, 95% CI = ±0.009; set D, 95% CI = ±0.26; set E, 95% CI = ±0.15; set F, 95% CI = ±0.013

a1-14 (a) $\bar{x} \pm 0.029$
(b) $N = 11.9$ or 12 measurements

a1-15 (a) $\bar{x} \pm 0.76$
(b) $N = 6.9$ or 7 measurements

a1-16 Bias is suggested at the 95% confidence level.

a1-17 (a) Systematic error is suggested.
(b) No systematic error is demonstrated.

a1-18 For C, no systematic error is suggested, but for H, a systematic error is indicated.

a1-19 $V = 5.20 \text{ cm}^3$; $s_V = 0.15 \text{ cm}^3$

a1-20 At pH = 2.0, s for $[H_3O^+] = 2.3 \times 10^{-4}$; at pH 12.0, s for $[H_3O^+] = 2.3 \times 10^{-14}$.

a1-21 $s = 7.5 \times 10^{-5} \text{ g/mL}$

a1-24 **(a)** slope = 5.57, intercept = 0.90

(c) 1.69 mmol/L

(d) For one measurement, s_{MVK} = 0.080; RSD = 0.0482 (CV = 4.8%).

For four replicate measurements, s_{MVK} = 0.052; RSD = 0.308 (CV = 3.1%).

(e) c_{MVK} = 3.93 mmol/L

For one measurement, s_{MVK} = 0.08; RSD = 0.0203 (CV = 2.03%).

For four measurements, s_{MVK} = 0.05; RSD = 0.0126 (CV = 1.26%).

INDEX

USEFUL CONVERSION FACTORS AND RELATIONSHIPS

Length	Mass
SI unit: Meter (m)	**SI unit: Kilogram (kg)**

1 kilometer = 1000 meters
         = 0.62137 mile
     1 meter = 100 centimeters
1 centimeter = 10 millimeters
1 nanometer = 1×10^{-9} meter
1 picometer = 1×10^{-12} meter
     1 inch = 2.54 centimeters (exactly)
1 angstrom = 1×10^{-10} meter
	1 kilogram = 1000 grams
	   1 gram = 1000 milligrams
	1 pound = 453.59237 grams = 16 ounces
	   1 ton = 2000 pounds

Volume	Pressure
SI unit: Cubic meter (m^3)	**SI unit: Pascal (Pa)**

1 liter (L) = 1×10^{-3} m^3
       = 1000 cm^3
       = 1.056688 quarts
1 gallon = 4 quarts
	1 pascal = 1 N/m^2
	      = 1 kg/m · s^2
	1 atmosphere = 101.325 kilopascals
	      = 760 mmHg = 760 torr
	      = 14.70 lb/in^2

Energy	Temperature
SI unit: Joule (J)	**SI unit: kelvin (K)**

1 joule = 1 kg m^2/s^2
     = 0.23901 calorie
     = 1 C × 1 V
1 calorie = 4.184 joules
	$T\,(\text{K}) = T(^{\circ}\text{C})(1\,\text{K}/^{\circ}\text{C}) + 273.15\,\text{K}$
	$T\,(^{\circ}\text{C}) = (5\,^{\circ}\text{C}/9\,^{\circ}\text{F})[T(^{\circ}\text{F}) - 32\,^{\circ}\text{F}]$
	$T\,(^{\circ}\text{F}) = (9\,^{\circ}\text{F}/5\,^{\circ}\text{C})T(^{\circ}\text{C}) + 32\,^{\circ}\text{F}$

SYMBOLS FOR COMMON PHYSICAL AND CHEMICAL QUANTITIES

Symbol	Quantity	Symbol	Quantity
A	absorbance, area	S/N	signal-to-noise ratio
a	absorptivity, activity	T	transmittance, temperature
B	magnetic field strength	t	time
C	capacitance	V	dc voltage, volume
D	diffusion coefficient	v	ac voltage, velocity
d	diameter, spacing	X	reactance
E	electrical potential, energy	$\overline{x}$	sample mean
e^-	electron	Z	impedance
F	faraday		
f	frequency		
G	conductance, free energy		
H	enthalpy	η	(eta) viscosity
I	dc current	γ	(gamma) activity coefficient
i	ac current	ε	(epsilon) molar absorptivity
K	equilibrium constant	λ	(lambda) wavelength
L	inductance	μ	(mu) population mean
n	number of moles, refractive index	ν	(nu) frequency
$\mathbf{n}$	spectral order	$\overline{\nu}$	wavenumber
P	radiant or electrical power	ρ	(rho) density
Q	quantity of dc charge	σ	(sigma) population standard deviation
q	quantity of ac charge	τ	(tau) period, time constant
R	electrical resistance, gas constant	ϕ	(phi) phase angle
s	sample standard deviation	ω	(omega) angular velocity